The Living World

GEORGE B. JOHNSON

Washington University
St. Louis, Missouri

WCB Wm. C. Brown Publishers

Dubuque, IA Bogota Buenos Aires Caracas Chicago Guilford, CT London
Madrid Mexico City Seoul Singapore Sydney Taipei Tokyo Toronto

Project Team

Developmental Editor *Kennie Harris*
Production Editor *Linda M. Jegerlehner*
Marketing Manager *Julie Joyce Keck*
Designer *K. Wayne Harms*
Art Editor *Kathleen M. Timp*
Photo Editor *John C. Leland*
Permissions Coordinator *Patricia Barth*

WCB **Wm. C. Brown Publishers**

President and Chief Executive Officer *Beverly Kolz*
Vice President, Director of Editorial *Kevin Kane*
Vice President, Sales and Market Expansion *Virginia S. Moffat*
Vice President, Director of Production *Colleen A. Yonda*
Director of Marketing *Craig S. Marty*
National Sales Manager *Douglas J. DiNardo*
Executive Editor *Michael D. Lange*
Advertising Manager *Janelle Keeffer*
Production Editorial Manager *Renée Menne*
Publishing Services Manager *Karen J. Slaght*
Royalty/Permissions Manager *Connie Allendorf*

A Times Mirror Company

Copyedited by Laura Beaudoin

Cover: © Charles Lynn Bragg 1995/Licensed by Seiffer & Associates, Inc.

The credits section for this book begins on page 645 and
is considered an extension of the copyright page.

Library of Congress Catalog Card Number: 96–85698

ISBN 0–697–22225–X

Printed in the United States of America by Times Mirror Higher Education Group, Inc.,
2460 Kerper Boulevard, Dubuque, IA 52001

10 9 8 7 6 5 4 3 2 1

Brief Contents

Contents

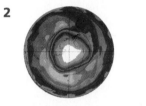

PART TWO • THE CONTINUITY OF LIFE

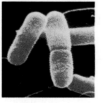

PART THREE • THE DIVERSITY OF LIFE

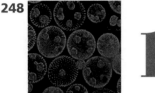

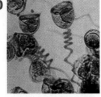

PART FIVE • HUMAN LIFE

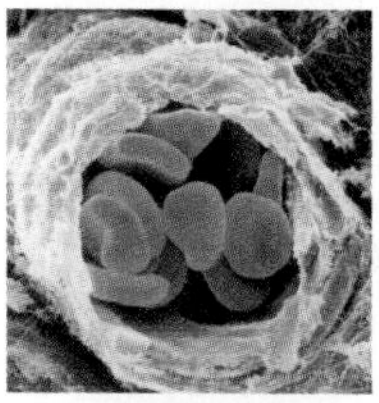

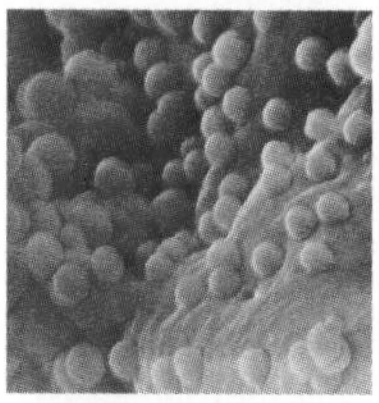

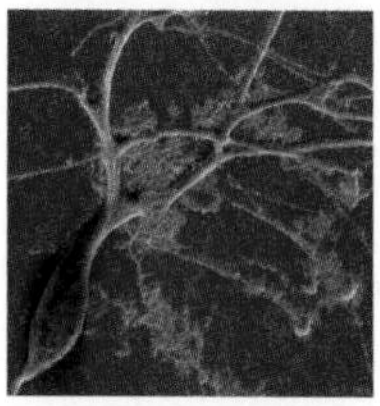

PART SIX • OUR LIVING ENVIRONMENT

556

26

Ecosystems

Ecology is the study of the way organisms interact with one another and with their physical environment. A community of organisms, and the habitat in which they live, is called an ecological system, or ecosystem.

26.1 What Is an Ecosystem? 558
26.2 Energy Flows Through Ecosystems 559
26.3 Materials Cycle Within Ecosystems 562
26.4 Major Kinds of Ecosystems 567

582

27

Living in Ecosystems

The species living in an ecosystem have evolved many accommodations to living together. This coevolution has made the members of ecosystems interdependent in many ways.

27.1 Population Dynamics 584
27.2 How Coevolution Shapes Ecosystems 588
27.3 How Ecosystems Develop Over Time 591
27.4 What Makes an Ecosystem Stable? 598
27.5 How Humans Disrupt Ecosystems 600

606

28

Planet Under Stress

Our planet faces numerous problems that threaten its future existence, including chemical pollution on a global scale and nonsustainable use of the earth's nonrenewable resources.

28.1 Global Change 608
28.2 Saving Our Environment 613
28.3 Solving Environmental Problems 620

Appendix A: Classification of Organisms 624
Appendix B: Answers to Concept Reviews 629
Glossary: Key Terms and Concepts 631
Credits 645
Index 648

Preface

No one who teaches biology today can fail to appreciate how important a subject it has become for our modern world. As a teacher, I have stood in front of classrooms for over 25 years and attempted to explain biology to puzzled and sometimes uninterested students, an experience that has been both fun and frustrating: fun because biology is a joy to teach, rich in ideas and interesting concepts, and increasingly key to many important public issues; frustrating because in every biology class there are always some students who will not pay attention, who not only miss out on the fun but also fail to acquire a tool that will be essential to their futures.

This book, *The Living World,* is my attempt to address that problem. It is short enough to use in one semester, without a lot of added material to intimidate wary students. I have tried to write it in an informal, friendly way, to engage as well as to teach. The focus of the book is on the biology each student ought to know to live as an informed citizen in the twenty-first century. I have at every stage addressed ideas and concepts, rather than detailed information, trying to teach *how* things work and *why* things happen the way they do rather than merely naming parts or giving definitions.

My first step in creating this text was to attack the traditional table of contents (usually a formidable list of chapters covering every subject imaginable). The number of chapters in biology textbooks has grown over the years, until today the most widely used short texts have as many as 38 chapters. I have cut back ruthlessly on this overwhelming amount of information, reducing the number of chapters in *The Living World* to 28. I think it matches closely what is actually being taught in classrooms and, as you will see, all that is really important is preserved.

I have deliberately combined photosynthesis and cellular respiration into a single general chapter in *The Living World,* not because metabolism is unimportant but because the basic principles a student needs to understand are simple and easy to explain, while the details of chemical reactions only get in the way of many students' ability to see concepts. Details are sacrificed, but the basic concepts are presented clearly enough for nonmajors students to grasp.

Another key step in designing *The Living World* was to combine the teaching of evolution and diversity into a single story line. Traditionally, students are exposed to weeks of evolution before tackling animal diversity, struggling past the Hardy-Weinberg equilibrium and population growth equations (microevolution) and on through Darwin's discoveries (macroevolution). Then, when that is all done, they are dragged through a detailed tour of the phyla, followed by a long excursion into botany. In large measure, the three areas are presented as if unrelated to each other. In *The Living World* I have chosen instead to combine all three of these areas into one treatment, presenting biological diversity as an evolutionary journey. It is a lot more fun to teach this way, and I find that my students learn a great deal more, too.

I have also chosen to present ecology in a novel way. Rather than taking a levels-of-organization approach (that is, population ecology, community ecology, ecosystems, and biomes), I focus on the basic unit of ecology, the ecosystem—also the most relevant to the world's problems today. After describing what ecosystems are and how they function, I discuss how the architecture of ecosystems has been shaped by coevolution, how ecosystems respond to disruption, and what factors make some ecosystems more stable to disruption than others. Finally, I present a picture of the kinds of disruptions going on today. The approach throughout is positive and solution-oriented.

Forgive me if you find some favorite topic omitted or treated more lightly than you would wish—keeping this book brief and approachable has limited the amount of material that could be included. But I feel the benefit far outweighs the cost, for this is a book students can learn from.

As you page through the book, you will notice that major topics begin at the start of a page. This is no accident but rather the result of careful planning. Unlike most texts, the pages of *The Living World* were laid out by the author, so that I was able to create a very user-friendly organization of material.

In an effort to make the book more fun for students and at the same time increase its power as a learning tool, I have incorporated interactive CD-ROM *Explorations* into 16 of the chapters. The publisher has also opened a web site on the Internet for each chapter, which students can use to investigate further topics. Although students can use the text perfectly well without these new technological features, they add to the richness of the experience. I hope you will be tempted to try them with your class.

While it is not the first book I have written and probably will not be the last, *The Living World* has been one of the most challenging and rewarding. I hope that you like it—and that you find it an effective teaching tool.

Learning Aids

In the Text

Each chapter of *The Living World* provides the student with several features designed to facilitate study and learning.

Chapter Opener

A "Far Side" cartoon at the beginning of each chapter captures student interest and shows that science can be fun.

Chapter Outline

At the beginning of each chapter, an outline of the main chapter headings introduces the student to the topics about to be discussed.

End-of Chapter Highlights

This concise, visual summary highlights the main headings and key illustrations within the chapter. It recaps the concepts that relate to each main topic and links them to the most important key terms with page references.

Concept Review

Objective questions (multiple-choice and fill-in-the-blank) provide a built-in study guide, enabling students to quiz themselves on the material just read.

Challenge Yourself

Critical thinking questions challenge students to apply the concepts presented in the chapter.

For Further Reading

A list of sources points students and instructors toward additional related material and provides a brief overview of each book or article.

Technology Links

Icons indicate multimedia sources, including *The Living World's* own web site, where students can obtain supplementary material or practice applying the chapter concepts.

In addition, at the end of the book, students will find:

Appendix A: The Classification of Life

Illustrations and narrative explain the six-kingdom classification system used in *The Living World.*

Appendix B: Answers to Concept Reviews

Answers to the objective questions at the end of each chapter are provided so that students can get immediate reinforcement as they quiz themselves on the material.

Glossary of Key Terms and Concepts

This glossary contains definitions and derivations of the most important key terms used in biology and in *The Living World.*

Index

A comprehensive index is provided.

Ancillaries

Instructor's Manual/Test Item File

Prepared by Jennifer Carr Burtwistle of Northeast Community College, the *Instructor's Manual* provides the following instructor aids for each chapter of *The Living World:* chapter overview, learning objectives, student activities, discussion ideas, and a list of audiovisual materials. The test item file offers approximately 35 questions per chapter (multiple-choice, fill-in-the-blank, and essay) as well as answers for all.

Computerized Testing Software

The test item file is also available on disk with a test generator program in either Macintosh or Windows.

Transparencies

Two hundred full-color illustrations from the book are available as transparency acetates and provided free to adopters.

Student Study Guide

The *Student Study Guide,* by Lisa Shimeld of Crafton Hills College, contains chapter concepts, objective questions (multiple-choice, completion, and matching), drawings for students to label, key word searches, crossword puzzles, and a list of related web sites. An appendix provides the answers to all questions and activities.

How to Study Science

How to Study Science, second edition, by Fred Drewes of Suffolk County Community College, contains helpful tips and practical exercises to help students achieve success in biology and other science courses.

Basic Chemistry for Biology

Basic Chemistry for Biology, by Carolyn Chapman of Suffolk County Community College, is a self-paced book that leads students through basic concepts of inorganic chemistry.

How Scientists Think

How Scientists Think, by *The Living World* author George Johnson, is a paperbound text describing 21 experiments that have shaped our understanding of genetics and molecular biology. It fosters critical thinking and reinforces the scientific method.

Biology Study Cards

This boxed set of 300 two-sided study cards provides a quick yet thorough visual synopsis of all key biological terms and concepts in the general biology curriculum. Each card features a masterful illustration pronunciation guide, definition, and description in context.

The Internet Primer: Getting Started on Internet

This primer, by Fritz J. Erickson and John A. Vonk, gives you just enough knowledge for getting started on the Internet, including the basics of how to access and use some of the Internet's most common features.

Multimedia Learning Aids

It is an exciting time for learning, as technology presents us with new tools that are both powerful and fun. The following multimedia products can be used with *The Living World:*

Explorations in Cell Biology & Genetics CD-ROM

*Keyed to the text

Explorations in Human Biology CD-ROM

*Keyed to the text

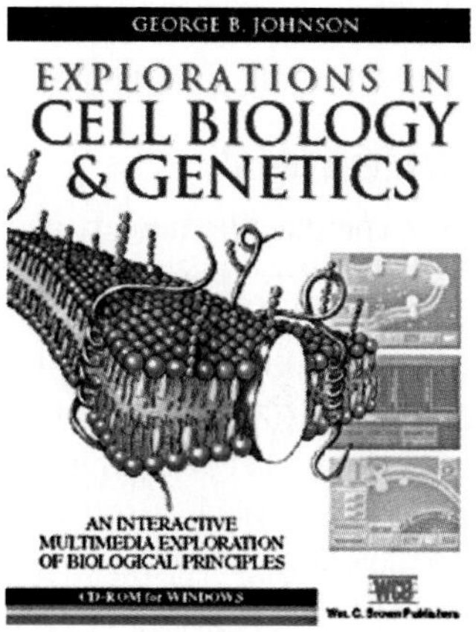

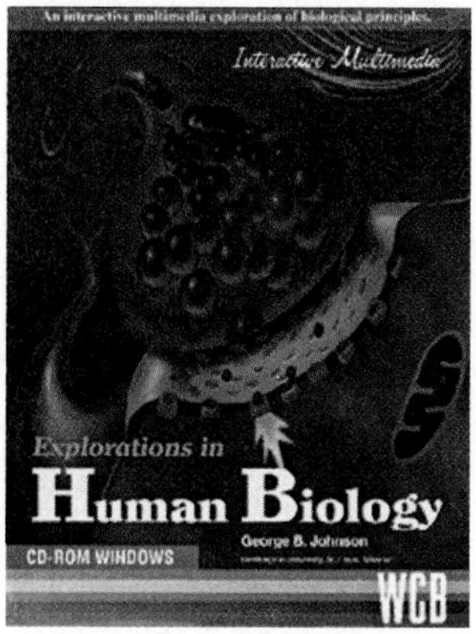

Developed by *The Living World* author George Johnson, these two CDs contain a total of 33 full-color, interactive exercises covering key biological concepts. Screen captures from appropriate modules are used as illustrations in selected chapters of *The Living World,* and the modules are also referenced at the ends of the chapters. *Explorations* can be used by students to enhance learning or by an instructor in lecture and/or a lab or resource center. They operate on either Macintosh or Windows platforms.

The Dynamic Human CD-ROM and Videodisc

*Keyed to the text

This guide to anatomy and physiology interactively illustrates the complex relationships between anatomical structures and their functions in the human body. Realistic, three-dimensional visuals are the premier feature of this learning tool. The program covers every body system, demonstrating to the viewer the anatomy, physiology, histology, and clinical applications of each.

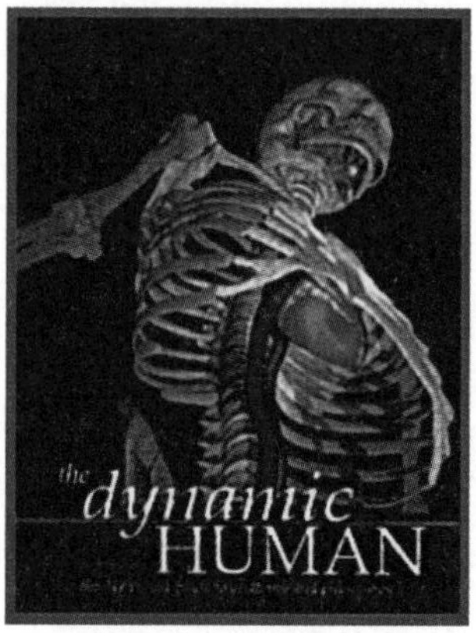

Microbes in Motion CD-ROM

*Keyed to the text

Microbiological topics come to life in this easy-to-use Windows tutorial by Gloria Delisle and Lewis Tomalty. The CD features interactive screens, animations, video, audio, and hyperlinked questions.

Life Science Animations Videotapes

*Keyed to the text

Sixty-six animations of key physiological processes are available on this set of six videotapes:

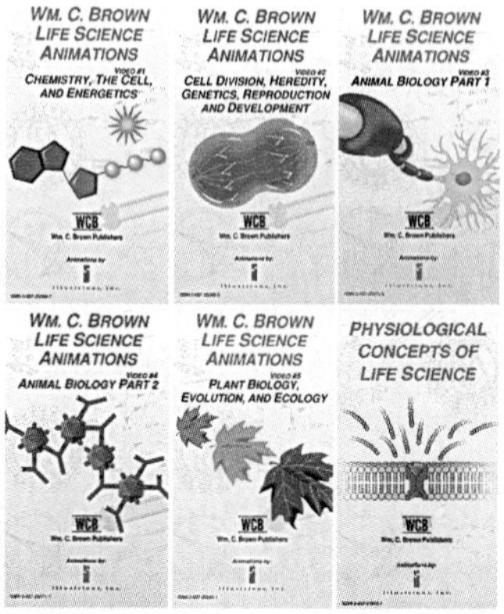

Video 1 *Chemistry, the Cell, and Energetics*

Video 2 *Cell Division/Heredity/Genetics/Reproduction and Development*

Video 3 *Animal Biology 1*

Video 4 *Animal Biology 2*

Video 5 *Plant Biology/Evolution/Ecology*

Video 6 *Physiological Concepts of Life Science*

A Life Science Living Lexicon CD-ROM

A Life Science Living Lexicon CD-ROM, by William Marchuk of Red Deer College, contains a comprehensive collection of life science terms, including definitions of their roots, prefixes, and suffixes as well as audio pronunciations and illustrations. The *Lexicon* is student-interactive, providing quizzing and notetaking capabilities. It contains 4,500 terms, which can be broken down for study into the following categories: anatomy and physiology, botany, cell and molecular biology, genetics, ecology and evolution, and zoology.

BioSource Videodisc

BioSource Videodisc, by Wm. C. Brown Publishers and Sandpiper Multimedia, Inc., features 20 minutes of animations and nearly 10,000 full-color illustrations and photos, many from leading WCB biology textbooks.

Biology Startup

Biology Startup, by Myles Robinson and Kathleen Pace of Grays Harbor College, is a five-disk Macintosh tutorial that helps nonmajors students master challenging biological concepts such as basic chemistry, photosynthesis, and cellular respiration.

Technology and *The Living World*

Two new technology tools play an integral part in your exploration of biology with this text: the Internet and interactive CD-ROMs.

What Is the Internet?

Few innovations have had as dramatic an impact on our lives as the Internet. Starting in the late 1960s, a significant number of scientists began to link their computers for the purposes of sharing data. In the late 1980s, universities and government agencies started joining this network of networks. Soon, average people of all ages, backgrounds, and occupations began to connect as well until by 1995, 5 million different computers were linked to the Internet worldwide. By means of telephone lines, direct wires, fiber optics, and satellite links, the Internet allows instantaneous communication between individuals and gives them unrestricted access to vast amounts of information.

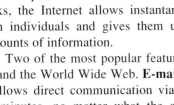

Two of the most popular features of the Internet are e-mail and the World Wide Web. **E-mail** (short for electronic mail) allows direct communication via the Internet, usually within minutes, no matter what the actual distance. Businesses, educational institutions, and private individuals are finding e-mail an inexpensive and convenient way to send memos, letters, and documents. The **World Wide Web** (often abbreviated WWW) is a mega-network created in 1992 that is capable of transferring text, graphics, audio, and video. Its information is presented in the form of a **home page,** also known as a **web site** or **web page.** Through WWW, you can access information as diverse as scientific data and other educational material, your daily newspaper, your favorite mail order catalog, interactive games, and the personal home pages of other Internet users.

Why is the Internet useful to instructors and students using *The Living World?* By visiting *The Living World's* home page, students and their instructors may access study questions specifically correlated to each chapter, the latest developments on topics covered in the text, other biologically relevant web sites, and even a message board where issues may be explored in depth with other students, instructors, and *The Living World* author.

How Does the Internet Work?

Communication within the Internet is possible because all the computers on the network use a common language, known as a **protocol.** The addressing system is very simple—an Internet address, called a **URL** for "uniform resource locator," identifies a specific computer linked to the Internet just as a street address identifies a specific house. For example, the Internet address for *The Living World* home page is:

http://www.wcbp.com/biology/tlw

The *http* stands for Hyper Text Transfer Protocol, the common language used to communicate between World Wide Web servers; *www* stands for World Wide Web; *wcbp.com* stands for Wm. C. Brown Publishers, a commercial organization; and *biology/tlw* specifically identifies *The Living World*'s directory structure and home page apart from those for other textbooks.

How Can I Access the Internet?

In order to use the Internet, you need a computer, a modem, and two kinds of software. The first type of software is an **interface,** a program that determines how the Internet looks to you. Provided by your host site (for example, a school or business) or by a commercial server such as America Online or CompuServe, these programs allow you to access the Internet with either a Windows or Macintosh computer. The second kind of software is called a **browser,** and it enables you to search for and view information. Many browsers are available, including Netscape, Mosaic, and those provided by commercial on-line services.

Once your system is installed, it is easy to access the home page for *The Living World.* Simply double click on the icon for Netscape (or whatever browser you use). When you arrive at a screen like the one shown at the left, type the home page address in the window provided after the words "Go to" or "Location." Then press the "enter" or "return" key on your keyboard. From that point, additional buttons will take you to chapter learning aids, related web sites, and other useful information. (Occasionally, you may not be able to get into Netscape because the lines are busy. If this happens, you will be instructed to try again.) To type in a different address, use your mouse to highlight the old address and then type the new one on top of it.

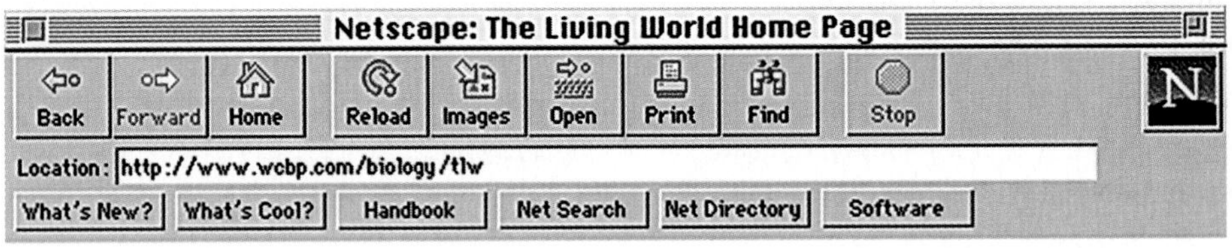

If you don't know a specific address but want to find out about a certain topic, you may use a **search engine.** One way to do this is by clicking on the button labeled "Net Search." A window will appear in which you may type a key word or phrase. Another way to find information on the Web is by using **hyperlinks.** These are words or phrases printed in color on the screen. Clicking on them can lead you deeper into the topic you are currently investigating or on to other related web sites. Navigating the Internet in this manner is known as **surfing.** Other buttons on the tool bar also help you get around on the Internet. For example, "back" and "forward" enable you to move back and forth among pages recently viewed, while "home" will take you back to your home page. With a little experimentation, you will quickly learn how to navigate the Internet.

Interactive CD-ROMs

As every teacher knows, hands-on experience is far and away the most effective way to learn anything. To teach art, let a student paint; to teach driving, let a student drive a car; to teach science, let a student do an experiment. Unfortunately, there is a limit to how much science can be taught "hands on" in a classroom. This is particularly true of biology, where students typically encounter a broad variety of concepts and organisms and can spend only a limited time in a laboratory—if any. Thus, it is with genuine excitement that we now present interactive CD-ROM *Explorations* in this text.

What do CD-ROM *Explorations* provide a classroom that slide collections and laser disc "multimedia" presentation managers don't? In a word, *interactivity*. Just as you cannot make a better painting simply by giving the artist more paint, so you cannot make a more hands-on experience simply by providing the student with more images. What is needed is a back-and-forth interaction, where true inquiry and analysis can take place. This is what a CD-ROM *Exploration* can provide and why CD-ROM technology plays such an important part in the learning experience provided by this text.

As an example, imagine introductory-level students learning in biology class about muscle contraction. Until now, their learning experience consisted of a text diagram illustrating actin and myosin filaments and the teacher's description of how the filaments interact. Using the CD-ROM *Exploration* described in figure 19.20 of *The Living World* (see the screen capture on this page), the class can watch the diagram "come alive," the filaments contracting in a dynamic animation. Then—and this is the key to the hands-on learning experience—the students can manipulate variables and see the effect on the animation. Altering ATP and Ca^{++} levels, the students can see the results of the changes they impose, collecting and analyzing data just as if they were in a laboratory. By exploring the effects of these variables, students learn for themselves how muscle contraction works, instead of simply being told. They gain a grasp of the key concepts far more effectively, and they have fun doing it—the whole interactive exercise need take up only a few minutes of class time.

Sixteen CD-ROM interactive investigations are integrated into *The Living World,* each in a chapter where the subject of the investigation plays a key role. The sixteen investigations have been selected from two CD-ROMs (*Explorations in Cell Biology & Genetics* and *Explorations in Human Biology*) that are part of an inexpensive student ancillary package. A list of all the topics covered appears at right; those printed in bold type are specially featured in certain chapters of *The Living World.* With the wide range of CD-ROM interactive *Explorations* employed in *The Living World,* hands-on learning in the classroom has finally become a practical goal.

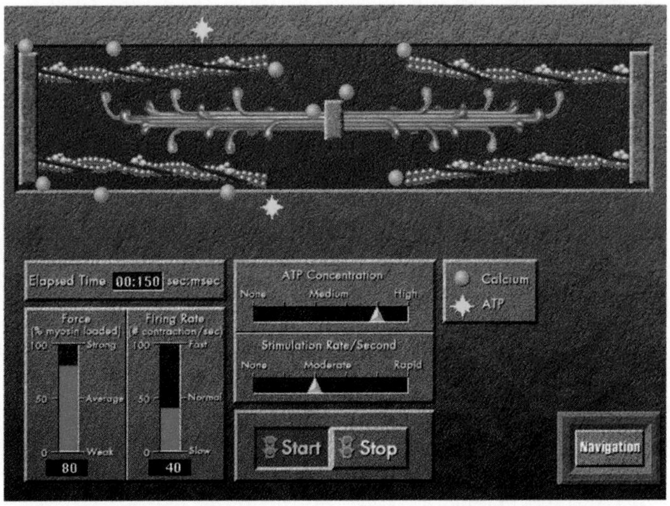

Figure 19.20 from *The Living World.* This is a screen capture from module 4, "Muscle Contraction," in *Explorations in Human Biology.*

Explorations in Cell Biology & Genetics CD-ROM

1. **Hemoglobin, chapter 2**
2. **Cell Size, chapter 3**
3. **Active Transport, chapter 4**
4. Cell-Cell Interactions
5. Mitosis
6. **Thermodynamics, chapter 5**
7. Kinetics
8. Oxidative Respiration
9. **Photosynthesis, chapter 15**
10. Meiosis: Down syndrome
11. Constructing a Genetic Map
12. **Heredity in Families, chapter 6**
13. Gene Segregation
14. **DNA Fingerprinting, chapter 8**
15. Reading DNA
16. **Gene Regulation, chapter 7**
17. Making a Restriction Map

Explorations in Human Biology CD-ROM

1. Cystic Fibrosis
2. **Active Transport, chapter 4**
3. Life Span and Life Style
4. **Muscle Contraction, chapter 19**
5. **Evolution of the Heart, chapter 20**
6. Smoking and Cancer
7. **Diet and Weight Loss, chapter 21**
8. Nerve Conduction
9. Synaptic Transmission
10. **Drug Addiction, chapter 23**
11. **Hormone Action, chapter 24**
12. **Immune Response, chapter 22**
13. **AIDS, chapter 25**
14. Constructing a Genetic Map
15. **Heredity in Families, chapter 6**
16. **Pollution of a Freshwater Lake, chapter 28**

Acknowledgments

Every book has its own particular history. This one began with the appearance of "cut-down" versions of nonmajors biology texts several years ago. I went to my editors and said, "Wouldn't it be nice to write a short book from scratch, at a very approachable level, rather than just cutting the length of a longer, more difficult book?" *The Living World* is the result. For seeing its potential and turning me loose, I am in the debt of Kevin Kane, head of editorial at Wm. C. Brown Publishers. He solidly backed my efforts to develop the interactive CD-ROM software that I have integrated into this text and in many other ways lent support to the project. My editor, Michael Lange, was no less supportive, marshaling considerable resources as well as putting out the fires that I in my enthusiasm seem to forever ignite among editorial and production staff.

Dr. Thomas Emmel of the University of Florida offered significant encouragement to me at an early stage of this project. Reviewing the first draft of the manuscript, he made many valuable suggestions, particularly concerning pedagogy.

Much of the pagemaking task was carried out day-to-day by Megan Jackman, an on-site developmental editor who worked in an office next to mine for a year, hacking and hewing at the manuscript until it "fit." Her intelligence, perseverance, and ability to tolerate my crusty nature played a major role in the high quality of this book. The terrific appearance of the book reflects the efforts of Mark Christianson, who led Megan and me into PageMaker, held our hands as we learned, and also designed the clean, snappy look that helps make this book so approachable. Mark does not know how to say, "It can't be done."

I want to explicitly thank Gary Larson and Toni Carmichael for the use of "The Far Side" cartoons that grace each chapter opening. It has been their policy never to allow more than two of Gary's cartoons in any non-Larson publication, and I am very grateful that they made an exception for *The Living World* and let me use 28! In over 15 years of writing biology texts, few tasks have been more enjoyable than picking those 28 from the wealth of wildly funny cartoons Gary has produced.

As always, I have had the support of a wife and family who have had to live with a distracted daddy spending more hours at the computer than they would like. My girls—seven, ten, and twelve years old—are becoming old enough to appreciate not only the long hours, but also the joy of finally holding the new book in your hands, proud as if it were a new baby. They are quite a fan club.

Acknowledgments would not be complete without thanking the generations of students who have used the many editions of my college biology texts. They have taught me at least as much as I have them, and this book could not have been written without the temper of that experience.

Finally, I need to thank my reviewers. Every text owes a great deal to those faculty across the country who review it. Serving as sensitive antennae for errors and sounding boards for new approaches, reviewers are among the most valuable tools at an author's disposal. *The Living World* has been particularly fortunate in its panel of reviewers. Representing a very diverse array of institutions and interests, they have provided me with invaluable feedback. Many improvements are the direct result of their suggestions. Every one of them has my sincere thanks.

George Kieffer
University of Illinois at Urbana

Bert Atsma
Union County College

Carl Frailey
Johnson County Community College

Richard E. McKeeby
Union College

J. Philip Fawley
Westminster College

James A. Brenneman
University of Evansville

Cran Lucas
Louisiana State University–Shreveport

Vernon L. Wranosky
Colby Community College

Donald L. Terpening
Ulster County Community College

Lisa A. Shimeld
Crafton Hills College

H. Warrington Williams
Westminster College of Missouri

Allan J. Landwer
Hardin-Simmons University

Michael Gipson
Oklahoma Christian University

Quentin Reuer
University of Alaska–Anchorage

Rudolph Prins
Western Kentucky University

Arthur L. Vorhies
Ohio University–Chillicothe Campus

Joseph F. Hawkins
College of Southern Idaho

Edward R. Fliss
Missouri Baptist College

Dennis Anderson
Oklahoma City Community College

John B. Stahl
Southern Illinois University

Kenneth Nickell
Allen County Community College

Teresa Snyder-Leiby
Marist College

Gail F. Baker
LaGuardia Community College

Richard K. Clements
Chattanooga State Technical Community College

Russell Johnson
Ricks College

Michael L. Gleason
Central Washington University

Danny G. Herman
Montcalm Community College

Donald Robbins
Missouri Western State College

Kenneth J. Curry
University of Southern Mississippi

Gene A. Kalland
California State University

David M. Brumagen
Morehead State University

Lester Eddington
Azusa Pacific University

Phillip Eichman
University of Rio Grande

Edward Samuels
Los Angeles Valley College

Glendon R. Miller
Wichita State University

Paul Keith Small
Eureka College

Marirose J. Ethington
Genesee Community College

Terrence Davin
Penn Valley Community College

Brent M. Graves
University of South Dakota

Stan Ivey
Delaware State University

Richard H. Shippee
Vincennes University

Paul Small
Eureka College

1

The Science of Biology

THE FAR SIDE By GARY LARSON

"No doubt about it, Ellington—we've mathematically expressed the purpose of the universe. Gad, how I love the thrill of scientific discovery!"

CHAPTER OUTLINE

1.1 The Scientific Process
- How Scientists Work: Inductive Reasoning
- Science in Action: A Case Study
- Stages of a Scientific Investigation

1.2 Using Science to Make Decisions
- Theory and Certainty
- The Scientific "Method"
- The Limitations of Science

1.3 Biological Principles
- Properties
- Themes

Highlights

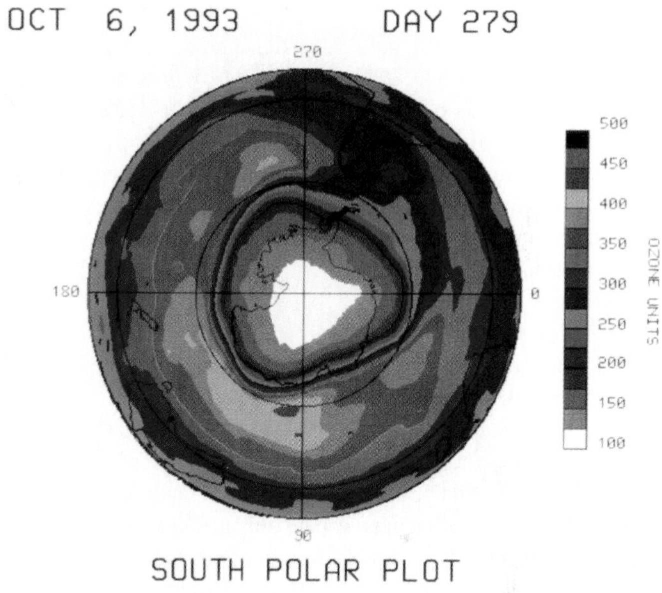

Figure 1.1 The ozone hole.
This satellite picture of the Antarctic ozone hole was taken on October 6, 1993. The ozone hole is an abnormally thin layer of ozone, indicated in white and red. Ozone absorbs harmful radiation from the sun, and its destruction is a serious injury to the environment. There was 15% less ozone over Antarctica in 1993 than a year earlier, leaving the protective shield at less than one-third its normal thickness.

The ozone hole you see in figure 1.1 represents serious damage to the global environment. The shield that protects us from ultraviolet (UV) radiation is being destroyed. The ozone hole bears a clear warning to all of us: Modern society is harming the world in which we all must live, and we must stop before irreparable damage is done. We cannot hope to preserve what we do not understand. In this effort, the study of biology plays a key role. In this chapter we examine how the science of biology is done, and we briefly consider some of the major principles you will encounter in this text.

1.1 The Scientific Process

The green turtle in figure 1.2 laid eggs during the night on the upper beach. As dawn breaks, she returns to the sea, laboriously dragging her 400-pound body across the sand. She is on Ascension Island, in the middle of the Atlantic Ocean, far from her breeding grounds on the coast of Brazil, 1,400 miles away. Every year, great numbers of turtles swim the distance to the island. The journey lasts about two months: two months of churning through strong equatorial currents with no food. The turtles seek the seven small, sandy beaches on this tiny, rocky island as if drawn by a beacon. By tagging the turtles, scientists have learned that individual females usually return faithfully to the same beach, or even the same stretch of beach, for nesting in successive years.

At first glance, the sea turtle in figure 1.2 appears to be just a turtle on a beach. Only observation and study reveal the mystery that is so much a part of this turtle's life. Similar mysteries are common in human life. Thus, the more we know about the biology of other animals, the better we can understand ourselves. The science of biology is devoted to this larger vision. It provides knowledge about the living world of which we are a part, while ceaselessly bombarding us with questions.

Science is a particular way of investigating the world. Not all investigations are scientific. For example, when you want to know how to get to Chicago from St. Louis, you do not conduct a scientific investigation—instead, you look at a map to determine a route. Making individual decisions by applying a "map" of accepted general principles is called **deductive reasoning.** Deductive reasoning is the reasoning of mathematics, of philosophy, of politics, and of ethics; it is the way a computer thinks. All of us rely on deductive reasoning to make everyday decisions. We use general principles as the basis for examining and evaluating these decisions.

How Scientists Work: Inductive Reasoning

Where do general principles come from? Religious and ethical principles often have a religious foundation; political principles reflect social systems. Some general principles, however, are not derived from religion or politics but from observation of the physical world around us. If you drop an apple, it will fall, whether or not you wish it to and despite any laws you may pass forbidding it to do so. Science is devoted to discovering the general principles that govern the operation of the physical world.

How do scientists discover such general principles? Scientists are, above all, observers: they look at the world to understand how it works. It is from their observations that scientists determine the general principles that govern our physical world.

Figure 1.2 End of a long migration.
A female green turtle leaves Ascension Island in the South Atlantic Ocean after laying her eggs in the sand.

This way of discovering general principles by careful examination of specific cases is called **inductive reasoning.** Inductive reasoning first became popular about 400 years ago, when Isaac Newton, Francis Bacon, and others began to conduct experiments and from the results infer general principles about how the world operates. The experiments were sometimes quite simple. Newton's consisted simply of releasing an apple from his hand and watching it fall to the ground. This simple observation is the stuff of science. From a host of particular observations, each no more complicated than the falling of an apple, Newton inferred a general principle—that all objects fall toward the center of the earth. This principle was a possible explanation, or **hypothesis,** about how the world works. Like Newton, scientists today formulate hypotheses, and observations are the materials on which they build them.

Science in Action: A Case Study

In 1985 Joseph Farman, a British earth scientist working in Antarctica, made an alarming discovery. Scanning the Antarctic sky, he found less ozone (O_3, a form of oxygen gas) than should be there—not a slight depletion but a 30% drop from the previous Antarctic reading recorded five years earlier! The swirling colors of figure 1.1 represent different concentrations of ozone over the South Pole as viewed from a satellite. As you can easily see, there is an "ozone hole" over Antarctica, an area about the size of the United States. The ozone concentration there is very

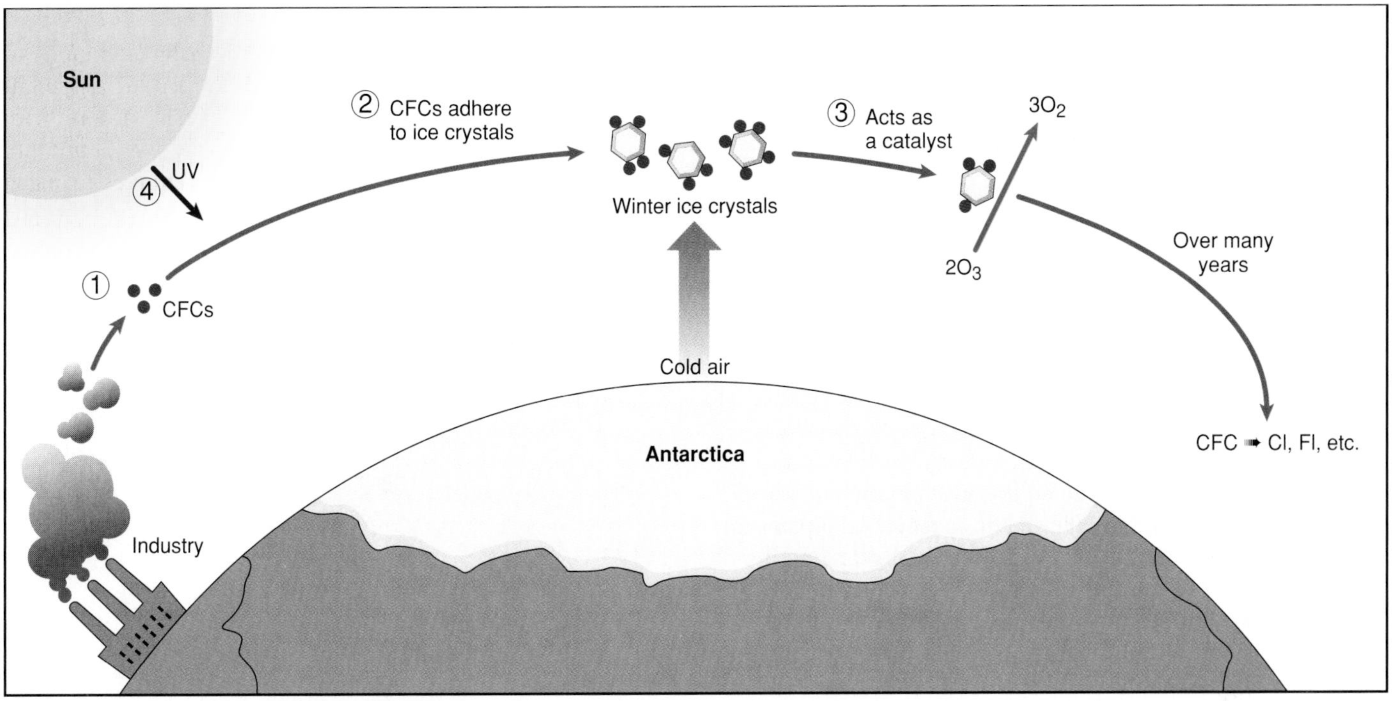

Figure 1.3 How CFCs destroy ozone.
CFCs attack and destroy ozone. CFCs are stable chemicals that accumulate in the atmosphere as a by-product of industrial society (*1*). In the intense cold of the Antarctic, these CFCs adhere to tiny ice crystals in the upper atmosphere (*2*), where they catalytically destroy ozone (*3*). As a result, more harmful UV radiation reaches the earth's surface (*4*).

much less than it is elsewhere. Looking back at satellite data, the hole is first apparent in 1975 and has become larger every year since.

At first it was argued that the ozone hole was an as-yet-unexplained weather phenomenon. Evidence soon mounted, however, implicating synthetic chemicals as the culprit. Spectral analysis of the atmosphere in the zone of the hole revealed a surprising concentration of chlorine, a chemical known to destroy ozone. The source of the chlorine was a class of chemicals called **chlorofluorocarbons (CFCs).** CFCs have been manufactured in large amounts since they were invented in the 1920s, largely for use as coolants in air conditioners, propellants in aerosols, and foaming agents in making Styrofoam. CFCs were widely regarded as harmless, because they were chemically unreactive under normal conditions. But in the intense cold high over Antarctica, CFCs condense onto tiny ice crystals; warmed by the sun in the spring, they attack and destroy ozone without being used up (figure 1.3). Because CFCs are very stable, they can survive in the upper atmosphere 120 years or more, eating away at the earth's ozone.

The thinning of the ozone layer in the upper atmosphere 25 to 40 kilometers above the surface of the earth is a serious matter. This layer of ozone protects life from the harmful UV rays from the sun that bombard the earth continuously in high amounts. Like invisible sunglasses, the ozone layer filters out these dangerous rays. When UV rays strike your skin, they cause it to tan and, if too intense, to sunburn. The damage can also be more serious. When UV rays damage the DNA in skin cells, it can lead to skin cancer, one of the more lethal diseases affecting humans. Every 1% drop in the atmospheric ozone concentration is estimated to lead to a 6% increase in skin cancers. The drop of approximately 3% that has already occurred worldwide, therefore, is estimated to have led to as much as a 20% increase in skin cancers.

The world currently produces about 1 million tons of CFCs annually, three-fourths of it in the United States and Europe. As scientific observations have become widely known, governments have rushed to correct the situation. By 1990, worldwide agreements to phase out production of CFCs by the end of the century had been signed. Nonetheless, most of the CFCs manufactured since they were invented are still in use in air conditioners and aerosols and have not yet reached the atmosphere. As these CFCs, as well as CFCs still being manufactured, move slowly upward through the atmosphere, the problem can be expected to grow worse. Ozone depletion has now been reported over the North Pole as well, and there is serious concern that the Arctic ozone hole will soon extend over densely populated northern Europe and the northeastern United States. Elevated levels of chlorine were reported over northern Europe in 1992, a warning of ozone destruction to come.

Stages of a Scientific Investigation

Joseph Farman, who first reported the ozone hole, is an environmental scientist, and what he was doing in Antarctica was science. Science is a particular way of investigating the world, of forming general rules about why things happen by observing particular situations. A scientist like Farman is an observer, someone who looks at the world in order to understand how it works.

Scientific investigations can be said to have six stages: (1) observing what is going on, (2) forming a set of hypotheses, (3) making predictions, (4) testing them, (5) carrying out controls, and (6) eliminating one or more of the hypotheses (figure 1.4).

1. Observation

The key to any successful scientific investigation is careful **observation.** Farman and other environmental scientists had studied the skies over the Antarctic for many years, noting a thousand details about temperature, light, and levels of chemicals. Had these scientists not kept careful records of what they observed, Farman might not have noticed that ozone levels were dropping.

2. Hypothesis

When the alarming drop in ozone was reported, environmental scientists made a guess about what was destroying the ozone—that perhaps the culprit was CFCs. We call such a guess a hypothesis. A hypothesis is a guess that might be true. What the scientists guessed was that chlorine from CFCs released into the atmosphere was reacting chemically with ozone over the Antarctic, converting ozone (O_3) into oxygen gas (O_2) and in the process removing the ozone shield from our earth's atmosphere. Often, scientists will form **alternative hypotheses** if they have more than one guess about what they observe. In this case, there were several other hypotheses advanced to explain the ozone hole. One suggestion explained it as the result of convection. A hypothesis was proposed that the seeming depletion of ozone was in fact a normal consequence of the spinning of the earth; the ozone spun away from the polar regions much as water spins away from the center as a clothes washer moves through its spin cycle. Another hypothesis was that the ozone hole was a transient phenomenon, due perhaps to sunspots, and would soon disappear.

3. Predictions

If the CFC hypothesis is correct, then several consequences can reasonably be expected. We call these expected consequences **predictions.** A prediction is what you expect to happen if a hypothesis is true. The CFC hypothesis predicts that if CFCs are responsible for producing the ozone hole, then it should be possible to detect CFCs in the upper Antarctic atmosphere as well as the chlorine released from CFCs that attacks the ozone.

4. Testing

Scientists set out to test the CFC hypothesis by attempting to verify some of its predictions. We call the test of a hypothesis an **experiment.** To test the hypothesis, atmospheric samples were collected from the stratosphere over 6 miles up by a high-altitude balloon. Analysis of the samples revealed CFCs, as predicted. Were the CFCs interacting with the ozone? The samples contained free chlorine and fluorine, confirming the breakdown of CFC molecules. The results of the experiment thus support the hypothesis.

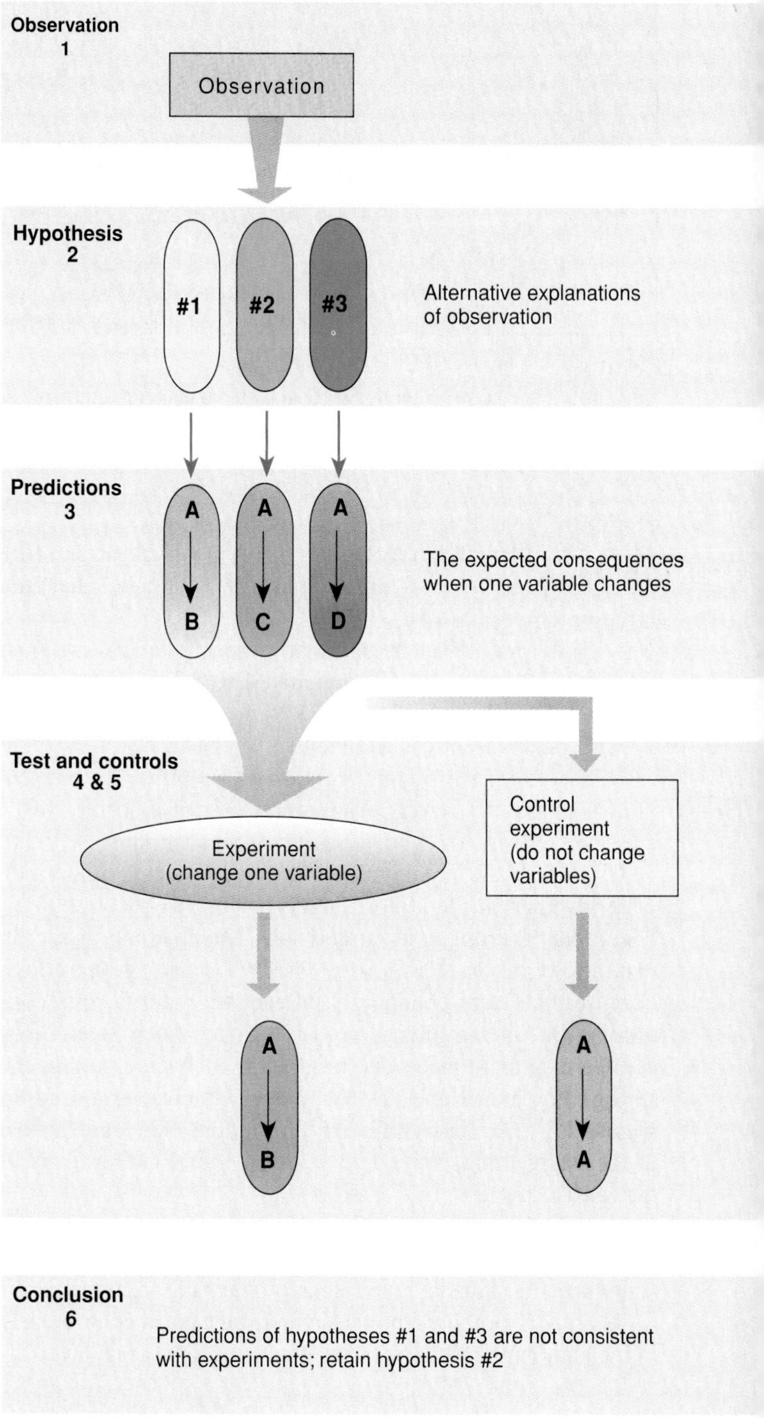

5. Controls

Events in the upper atmosphere can be influenced by many factors. We call each factor that might influence a process a **variable.** To evaluate alternative hypotheses about one variable, all the other variables must be kept constant so that we do not get misled or confused by these other influences. This is done by carrying out two experiments in parallel: in the first experimental test, we alter one variable in a known way to test a particular hypothesis; in the second, called a **control experiment,** we do *not* alter that variable. In all other respects, the two experiments are the same. To further test the CFC hypothesis, scientists carried out control experiments in which the key variable was the amount of CFC in the atmosphere. Working in laboratories, scientists reconstructed the atmospheric conditions, solar bombardment, and extreme temperatures found in the sky far above the Antarctic. If the ozone levels fell without addition of CFCs to the chamber, then CFCs could not be what was attacking the ozone, and the CFC hypothesis must be wrong. Carefully monitoring the chamber, however, scientists detected no drop in ozone levels in the absence of CFCs. Only after CFCs were added did ozone levels begin to fall. The result of the control was thus consistent with the predictions of the hypothesis.

6. Conclusion

A hypothesis that has been tested and not rejected is tentatively accepted. The hypothesis that CFCs released into the atmosphere are destroying the earth's protective ozone shield is now supported by a great deal of experimental evidence and is widely accepted. While other factors have also been implicated in ozone depletion, destruction by CFCs is clearly the dominant phenomenon. A collection of related hypotheses that have been tested many times is called a **theory.** The theory of the ozone shield, that ozone in the upper atmosphere shields the earth's surface from harmful UV rays by absorbing them, is supported by a wealth of observation and experimentation, and is widely accepted.

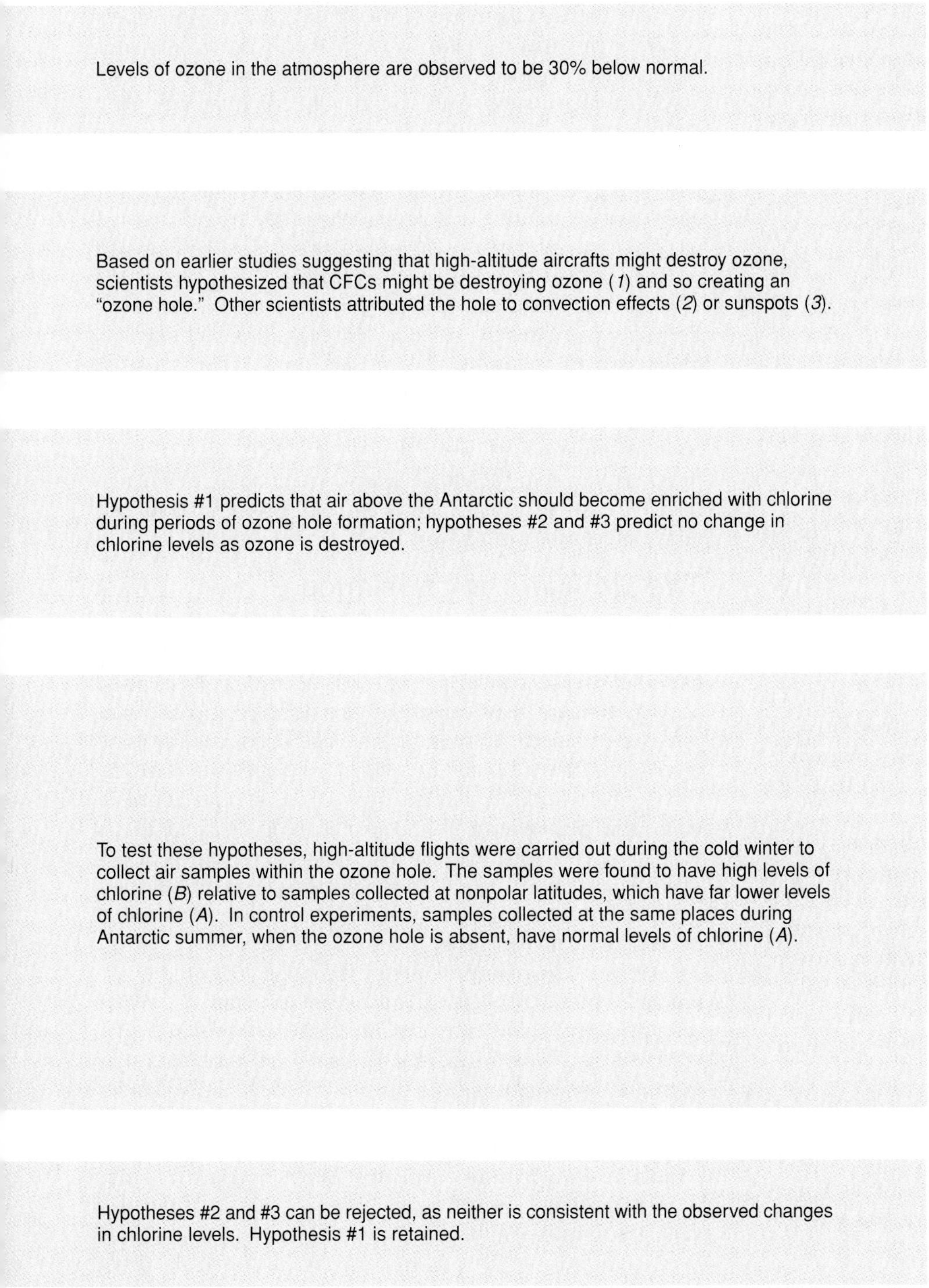

Levels of ozone in the atmosphere are observed to be 30% below normal.
Based on earlier studies suggesting that high-altitude aircrafts might destroy ozone, scientists hypothesized that CFCs might be destroying ozone (*1*) and so creating an "ozone hole." Other scientists attributed the hole to convection effects (*2*) or sunspots (*3*).
Hypothesis #1 predicts that air above the Antarctic should become enriched with chlorine during periods of ozone hole formation; hypotheses #2 and #3 predict no change in chlorine levels as ozone is destroyed.
To test these hypotheses, high-altitude flights were carried out during the cold winter to collect air samples within the ozone hole. The samples were found to have high levels of chlorine (*B*) relative to samples collected at nonpolar latitudes, which have far lower levels of chlorine (*A*). In control experiments, samples collected at the same places during Antarctic summer, when the ozone hole is absent, have normal levels of chlorine (*A*).
Hypotheses #2 and #3 can be rejected, as neither is consistent with the observed changes in chlorine levels. Hypothesis #1 is retained.

Figure 1.4 The scientific process.
This diagram illustrates the six stages of the scientific investigation of the ozone hole. First, observations are made. Then a number of potential explanations (hypotheses) are suggested in response to an observation. Next, predictions are made based on these hypotheses, and experiments (including control experiments) are carried out in an attempt to eliminate one or more of the hypotheses. Finally, any hypothesis that is not eliminated is retained.

1.2 Using Science to Make Decisions

Science does not discover "truth" but rather falsehood. The essence of doing science is rejecting hypotheses that do not match up to observation. When a hypothesis agrees with observation, it is not rejected—yet.

A scientist attempts, in an organized way, to prove that certain hypotheses are not valid by demonstrating that their predictions are not consistent with the results of experiments. An experiment is conducted to demonstrate that one or more alternative hypotheses are inconsistent with observation and thus should be rejected. Hypotheses that are not rejected are retained for further testing and sometimes modification. Environmental scientists, for example, were able in a control experiment to eliminate the hypothesis that ozone destruction occurs in the absence of CFCs. Thus, they retained the alternative hypothesis that ozone destruction occurs because of the presence of CFCs. Scientific progress is made as a wooden statue is made, by chipping away the unwanted bits.

Theory and Certainty

Hypotheses that stand the test of time—their predictions often tested and never rejected—are sometimes combined into general statements called theories. A theory is a unifying explanation for a broad range of observations. Thus we speak of the theory of gravity, the theory of evolution, and the theory of the atom. Theories are the solid ground of science, that of which we are the most certain. There is no absolute truth in science, however, only varying degrees of uncertainty. The possibility always remains that future evidence will cause a theory to be revised. A scientist's acceptance of a theory is always provisional. For example, in another scientist's experiment, evidence that is inconsistent with a theory may be revealed. As information is shared throughout the scientific community, previous hypotheses and theories may be modified, and scientists may formulate new ideas.

Very active areas of science are often alive with controversy, as scientists grope with new and challenging ideas. This uncertainty is not a sign of poor science but rather of the push and pull that is the heart of the scientific process. The hypothesis that the world's climate is growing warmer due to humanity's excessive production of carbon dioxide (CO_2), for example, has been quite controversial, although the weight of evidence has increasingly supported the hypothesis.

The word "theory" is thus used very differently by scientists than by the general public. To a scientist, a theory represents that of which he or she is most certain; to the general public, the word theory implies a *lack* of knowledge or a guess. How often have you heard someone say "Its only a theory!"? As you can imagine, confusion often results. In this text the word "theory" will always be used in its scientific sense, in reference to a generally accepted scientific principle.

The Scientific "Method"

It was once fashionable to claim that scientific progress is the result of applying a series of steps called the **scientific method;** that is, a series of logical "either/or" predictions tested by experiments to reject one alternative. The assumption was that trial-and-error testing would inevitably lead one through the maze of uncertainty that always slows scientific progress. If this were indeed true, a computer would make a good scientist—but science is not done this way! If you ask successful scientists like Farman how they do their work, you will discover that without exception they design their experiments with a pretty fair idea of how they will come out. Environmental scientists understood the chemistry of chlorine and ozone when they formulated the CFC hypothesis, and they could imagine how the chlorine in CFCs would attack ozone molecules. A hypothesis that a successful scientist tests is not just any hypothesis. Rather, it is a "hunch" or educated guess in which the scientist integrates all that he or she knows. The scientist also allows his or her imagination full play, in an attempt to get a sense of what *might* be true. It is because insight and imagination play such a large role in scientific progress that some scientists are so much better at science than others—just as Beethoven and Mozart stand out above most other composers.

The Limitations of Science

Scientific study is limited to organisms and processes that we are able to observe and measure. Supernatural, religious, and unexplained phenomena are beyond the realm of scientific analysis because they cannot be scientifically studied, analyzed, or explained. To some individuals, a nonscientific point of view may have a moral or aesthetic value. However, scientists in their work are limited to objective interpretations of observable phenomena. This does not mean that individuals who are scientists and base their work on the principles of scientific study are less moral. Depending on the society, the culture, and the country, most individuals incorporate many philosophies into their lives.

It is also important to recognize that there are practical limits to what science can accomplish. While scientific study has and continues to revolutionize our world, it cannot be relied upon to solve all problems. For example, we cannot pollute the environment and squander its resources today, in the blind hope that somehow science will make it all right sometime in the future. Nor can science restore an extinct species. Science identifies solutions to problems when solutions exist, but it cannot invent solutions when they don't.

1.3 Biological Principles

The science of biology presents in an organized way what scientists have learned about the living world—what it is like and theories about how and why it works the way it does. In this text we discuss an array of biological **properties** shared by all organisms. As we do this, certain **themes** will emerge about how the living world works.

A house provides a useful analogy of the levels of information you will encounter:

Just as every house has a roof, walls, and a door, so all living organisms share certain fundamental *properties,* such as the ability to grow and reproduce.

Just as every house is organized into thematic areas such as bedrooms, kitchens, and bathrooms, so the living world is organized by major *themes,* such as how energy flows within the living world from one part to another.

Figure 1.5 Metabolism.
These cedar waxwing chicks obtain the energy they need to grow and develop by eating plants. They metabolize this food using chemical processes that occur within cells.

Properties

Biology is the study of life—but what does it mean to be alive? What are the properties that define a living organism? This is not as simple a question as it seems. Some of the most obvious properties of living organisms are also properties of many nonliving things. Three of the most important of these are *complexity* (a computer is complex), *movement* (clouds move in the sky), and *response to stimulation* (a soap bubble pops if you touch it). To appreciate why these three properties, so common among living things, do not help us to define life, imagine a mushroom standing next to a television: the television seems more complex than the mushroom, the picture on the television screen is moving while the mushroom just stands there, and the television responds to a remote control device while the mushroom continues to just stand there—yet it is the mushroom that is alive.

All living things share five more basic properties, passed down over billions of years from the first organisms to evolve on earth: *cellular organization, metabolism, homeostasis, reproduction,* and *heredity*.

Cellular organization. All living things are composed of one or more cells. A cell is a tiny compartment with a thin covering called a *membrane*. Some cells have simple interiors, while others are complexly organized, but all are able to grow and reproduce. Many organisms possess only a single cell; your body contains about 100 trillion—that's how many centimeters long a string would be wrapped around the world 1,600 times!

Metabolism. All living things use energy. Moving, growing, thinking—everything you do requires energy. Where does all this energy come from? It is captured from sunlight by plants and algae. To get the energy that powers our lives, we extract it from plants or from plant-eating animals in a process called *metabolism* (figure 1.5). All organisms use energy to grow, and all organisms transport this energy from one place to another within cells using special energy-carrying molecules called ATP molecules.

Homeostasis. All living things maintain stable internal conditions. While the environment often varies a lot, organisms act to keep their interior conditions relatively constant, a process called *homeostasis*. Your body acts to maintain an internal temperature of 37°C (98.5°F), however hot or cold the weather might be.

Reproduction. All living things reproduce. Bacteria simply split in two, as often as every 15 minutes, while many more complex organisms reproduce sexually (some trees as rarely as once every thousand years).

Heredity. All organisms possess a genetic system that is based on the replication and duplication of a long molecule called *DNA* (*deoxyribonucleic acid*). The information that determines what an individual organism will be like is contained in a code that is dictated by the order of the subunits making up the DNA molecule, just as the order of letters on this page determines the sense of what you are reading. Each set of instructions within the DNA is called a *gene*. Together, the genes determine what the organism will be like. Because DNA is faithfully copied from one generation to the next, any change in a gene is also preserved and passed on to future generations. The transmission of characteristics from parent to offspring is a process called *heredity*.

Themes

The living world teems with a breathtaking variety of creatures—whales, butterflies, mushrooms, and mosquitoes—all of which can be categorized into six groups, or **kingdoms,** of organisms (figure 1.6). In the midst of all this diversity it is easy to lose sight of the key lesson of biology, which is that all living things have much in common. As you study biology in this text, six general themes will emerge repeatedly, themes that serve to both unify and explain biology as a science. The six general themes of biology are

1. levels of organization;
2. the flow of energy;
3. evolution;
4. cooperation;
5. structure determines function;
6. homeostasis.

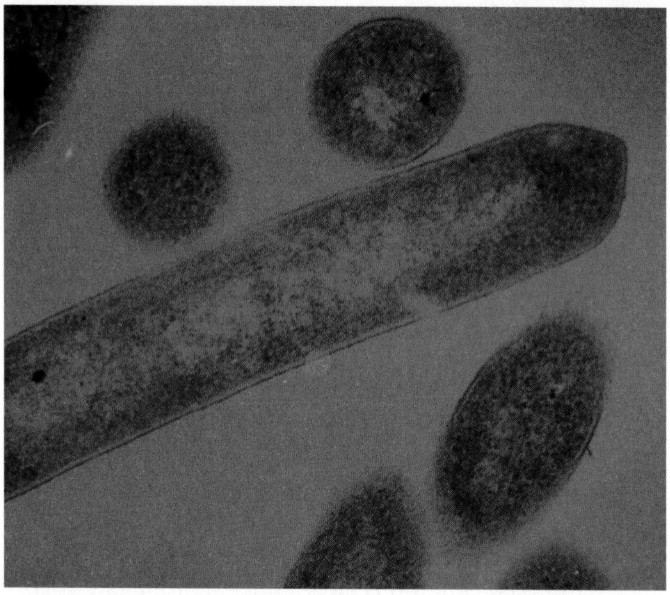

Archaebacteria. This kingdom includes bacteria such as this methanogenic bacterium, which manufactures methane as a result of its metabolic activity.

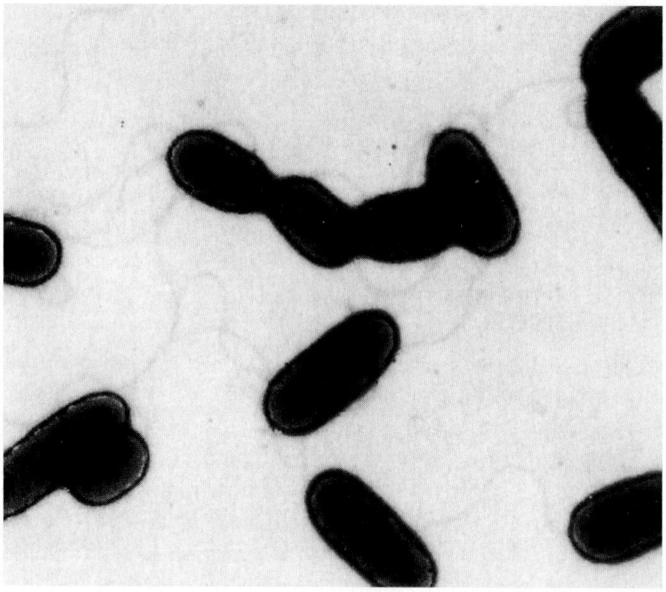

Eubacteria. This group is the second of the two bacterial kingdoms. Shown here is a soil bacterium that is responsible for many plant diseases.

Figure 1.6 The six kingdoms of life. Biologists categorize all living things into six major groups.

Fungi. This kingdom contains nonphotosynthetic multicellular organisms that digest their food externally, such as this cup fungus.

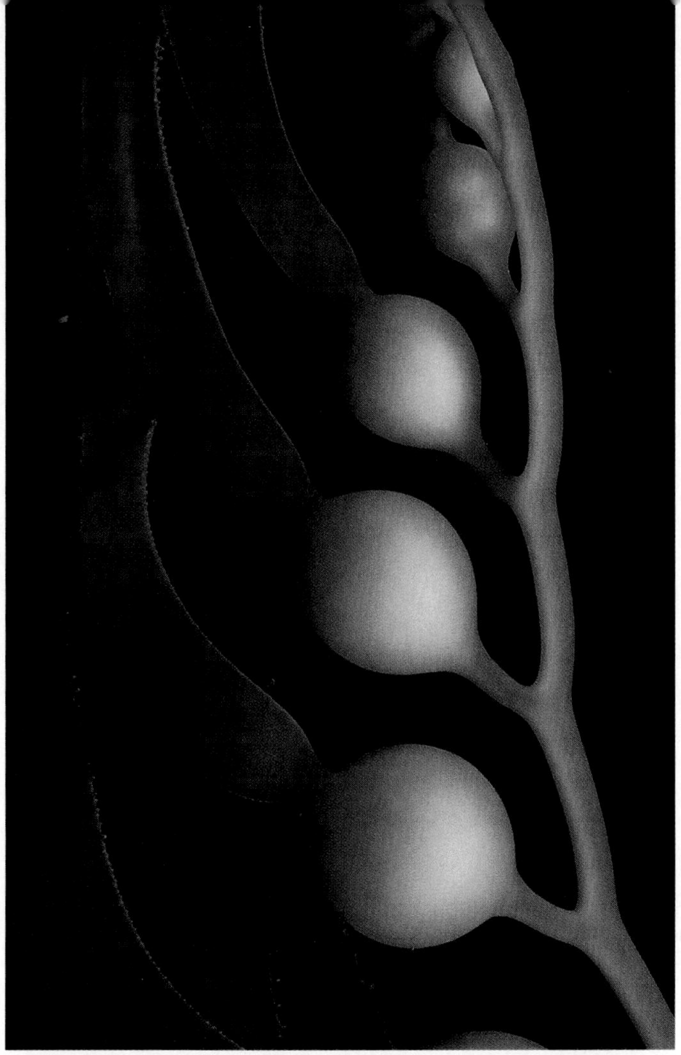

Protista. The unicellular eukaryotes (those whose cells contain a nucleus) are grouped into this kingdom, and so are the algae pictured here.

Plantae. This kingdom contains photosynthetic multicellular organisms that are terrestrial, such as the flowering plant pictured here.

Animalia. Organisms in this kingdom are nonphotosynthetic multicellular organisms that digest their food internally, such as this lizard.

1. Levels of Organization

The organisms of the living world function and interact with each other at many levels (figure 1.7). A key factor in organizing these interactions is the degree of complexity. There is a hierarchy of increasing complexity within cells, from the **molecular level** of DNA, at which the chemistry of life occurs, to the **organelle level,** at which cellular activities are organized, to the **cell,** the smallest level of organization that can be considered alive.

There is a further hierarchy of increasing complexity within multicellular organisms. At the cell level, different cells within the body are specialized to do different things (neurons to conduct signals, for example, and muscle cells to contract). Cells with a similar structure and function are grouped together into **tissues** (for example, muscle is a tissue composed of many muscle cells working together). Different tissues are combined into **organs,** which are biological machines that carry out particular jobs (the heart is an organ composed of muscle, nerve, and other tissues that works as a pump). The various organs that carry out major body functions make up **organ systems** (your heart, blood vessels, and the blood within them, for example, together make up your circulatory system).

There is yet another hierarchy of increasing complexity among different organisms. Individuals living together are called a **population,** and all the populations of a particular kind of organism are members of the same **species**. All the different species that live in a place are called a **community** (a forest community, for example, contains trees and deer and woodpeckers and fungi and many other creatures). A community and the physical environment in which it lives is called an ecological system, or **ecosystem.** Ecosystems that occur again and again worldwide are called **biomes** (desert is a biome, and so is tropical rain forest). Taken together, all of the earth's biomes and other ecosystems compose the **biosphere,** the highest level of complexity in the living world.

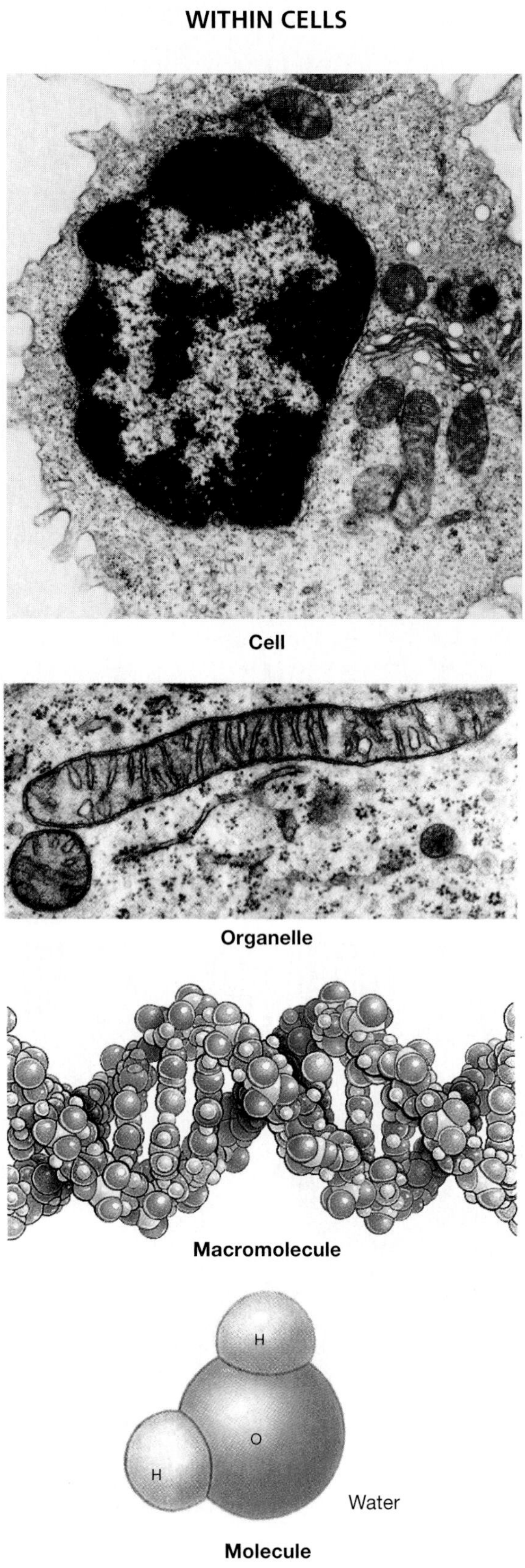

Figure 1.7 Levels of organization.
A traditional and very useful way to sort through the many ways in which the organisms of the living world interact is to organize them in terms of levels of organization, proceeding from the very small and simple to the very large and complex. Here we examine levels of organization within cells, within multicellular organisms, and among organisms.

WITHIN MULTICELLULAR ORGANISMS

Organism

Organ system

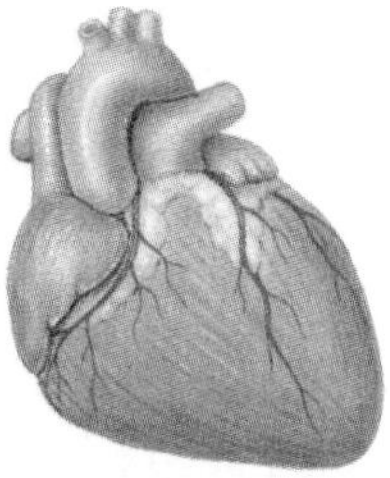

Organ

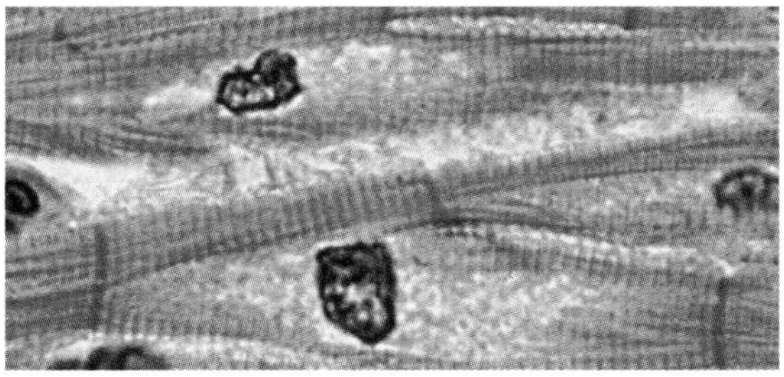

Tissue

AMONG ORGANISMS

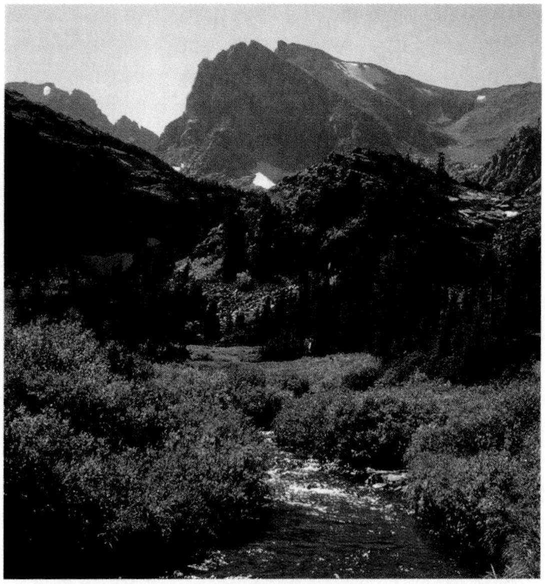

Ecosystem

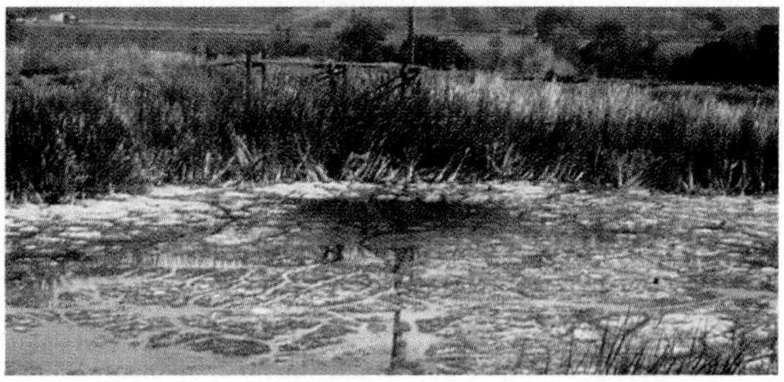

Community

Species

Population

2. The Flow of Energy

All organisms require energy to carry out the activities of living—to build bodies and do work and think thoughts. All of the energy used by organisms comes from the sun and is gradually used up as it passes in one direction through ecosystems. The simplest way to understand the flow of energy through the living world is to look at who uses it. The first stage of energy's journey is its capture by green plants and algae in photosynthesis. Plants then serve as a source of life-driving energy for animals that eat them. Other animals may then eat the plant-eaters. At each stage, some energy is used, some is transferred, and much is lost. The flow of energy is a key factor in shaping ecosystems, affecting how many and what kinds of animals live in a community.

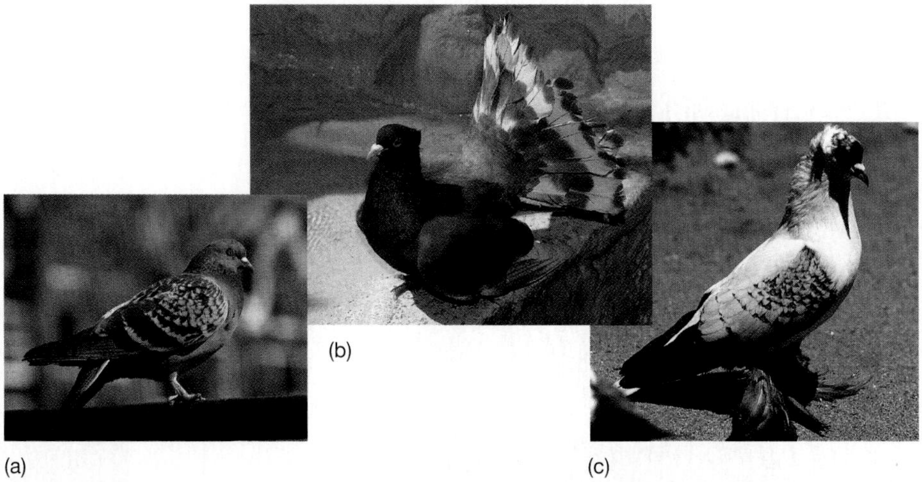

Figure 1.8 Darwin's artificial selection in pigeons.
Darwin's studies of artificial selection in pigeons provided key evidence that selection could produce the sorts of changes predicted by his theory of evolution. In *On the Origin of Species*, Charles Darwin wrote about his attempts to produce differences in domestic pigeons using artificial selection. He wrote that the kinds of pigeons he had produced were so different that "if shown to an ornithologist, and he were told that they were wild birds, would certainly, I think, be ranked by him as well-defined species." The differences that have been obtained by artificial selection of the wild European rock pigeon (*a*) and such domestic races as the red fantail (*b*) and the fairy swallow (*c*) with its fantastic tufts of feathers around its feet, are indeed so great that the birds probably would, if wild, be classified in entirely different major groups.

3. Evolution

Evolution is the change in species over time. Charles Darwin was an English naturalist who, in 1859, proposed the idea that this change is a result of a process called **natural selection.** Simply stated, those organisms better able to successfully respond to the challenges of their environment become more common. Darwin was thoroughly familiar with variation in domesticated animals (in addition to many nondomesticated organisms), and he knew that varieties of pigeons could be selected by breeders to exhibit exaggerated characteristics, a process called **artificial selection** (figure 1.8). We now know that the characteristics selected are passed on through generations because DNA is transmitted from parent to offspring. Darwin then visualized how selection in nature could be similar to that which had produced the different varieties of pigeons. Thus, the many forms of life we see about us on earth today, and the way we ourselves are constructed and function, reflect a long history of natural selection.

4. Cooperation

Cooperation between different kinds of organisms has played a critical role in the evolution of life on earth. For example, animal cells possess organelles that are the descendants of symbiotic bacteria, and symbiotic fungi helped plants first invade land from the sea. The coevolution of flowering plants and insects has been responsible for much of life's great diversity.

5. Structure Determines Function

One of the most obvious lessons of biology is that biological structures are very well suited to their functions. You will see this at every level of organization: Within cells, the shape of the proteins called enzymes that cells use to carry out chemical reactions are precisely suited to match the chemicals the enzymes must manipulate. Within the many kinds of organisms in the living world, body structures seem carefully designed to carry out their functions—the long tongue with which a hummingbird sucks nectar from a deep flower is one example, and the eyes with which you see the hummingbird are another. The superb fit of structure to function in the living world is no accident. Life has existed on earth for over 3 billion years, a long time for evolution to favor changes that better suit organisms to meet the challenges of living. It should come as no surprise to you that after all this honing and adjustment, biological structures carry out their functions well.

6. Homeostasis

The high degree of specialization we see among complex organisms is only possible because these organisms act to maintain a relatively stable internal environment, a process called homeostasis. Without this constancy, many of the complex interactions that need to take place within organisms would be impossible, just as a city cannot function without rules and order. Maintaining homeostasis in a body as complex as yours requires a great deal of signaling back-and-forth between cells.

As already stated, you will encounter these basic properties and themes repeatedly in this text. But just as a budding architect must learn more than the parts of buildings, so your study of biology should teach you more than a list of themes, concepts, and parts of organisms. Biology is a dynamic science that will affect your life in many ways, and that lesson is one of the most important you will learn. It is also an awful lot of fun.

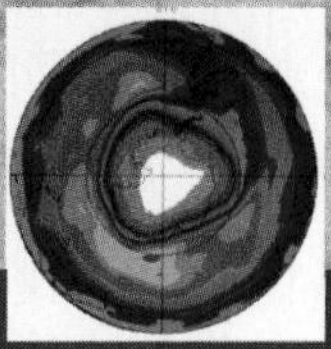

CHAPTER 1

HIGHLIGHTS

1.1 The Scientific Process

Key Terms

- inductive reasoning 4
- hypothesis 4
- control experiment 7
- theory 7

Key Concepts

- The discovery of how CFCs are reducing levels of ozone in the atmosphere is a good example of science in action.
- The scientific process is founded on careful observation.
- In a control experiment, only one variable is allowed to change.
- Scientific progress is made by rejecting hypotheses that are inconsistent with observation.

1.2 Using Science to Make Decisions

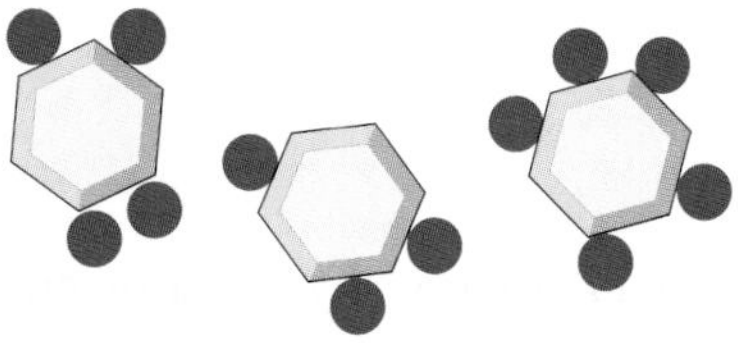

Key Terms

- scientific method 8

Key Concepts

- The acceptance of a hypothesis is always provisional.
- Well-tested hypotheses are often combined into general statements called theories.
- There is no surefire way to do science and no foolproof "method."
- One of the most creative aspects of scientific investigation is the formulation of novel hypotheses.

1.3 Biological Principles

Key Terms

- properties 9
- themes 9

Key Concepts

- All living things share eight fundamental properties:
  - complexity
  - movement
  - response to stimulation
  - cellular organization
  - metabolism
  - homeostasis
  - reproduction
  - heredity
- There are many ways to study biology. Six general themes often used to organize the study of biology are
  - levels of organization
  - the flow of energy
  - evolution
  - cooperation
  - structure determines function
  - homeostasis

1

CONCEPT REVIEW

1. Chlorofluorocarbons (CFCs) are used in
   a. foaming agents.
   b. air conditioners.
   c. aerosols.
   d. all of the above.

2. One of the main functions of the earth's ozone layer is to
   a. prevent global warming.
   b. filter out ultraviolet rays.
   c. absorb pollution.
   d. all of the above.

3. The 3% drop in ozone concentration that has already occurred worldwide has led to an estimated increase in skin cancers of
   a. 1%.
   b. 10%.
   c. 20%.

4. A guess in a scientific process is called a(n)
   a. hypothesis.
   b. observation.
   c. prediction.
   d. test.

5. A collection of hypotheses that have been repeatedly tested without rejection is called a(n)
   a. control.
   b. observation.
   c. test.
   d. theory.

6. Factors that influence a process in a scientific study are
   a. controls.
   b. tests.
   c. theories.
   d. variables.

7. Metabolism refers to an organism's ability to
   a. reproduce.
   b. use energy.
   c. pass on genes.
   d. move.

8. Key terms for homeostasis are
   a. external environment, stable.
   b. internal environment, unstable.
   c. internal environment, stable.
   d. external environment, unstable.

9. Select the smallest level of organization among the following.
   a. cell
   b. organ
   c. organ system
   d. tissue

10. The change in a species through time is
    a. cooperation.
    b. evolution.
    c. homeostasis.
    d. metabolism.

11. In the _________ air over Antarctica, CFCs adhere to crystals and catalyze a reaction that destroys _________.

12. Any good scientific investigation begins with careful _________.

13. A _________ is an experiment in which only a single variable is allowed to change.

14. List five fundamental properties shared by all living organisms on earth.

15. _________ is the complex linear molecule responsible for heredity.

16. A _________ is a tiny living compartment covered with a membrane.

17. List the six kingdoms of life.

18. At each level of organization, _________ determines function.

Answers to the Concept Review questions appear in Appendix B.

CHALLENGE YOURSELF

1. What is the difference between theory and certainty to a scientist? How does the word "hypothesis" fit in with a theory?

2. Imagine that you are a scientist asked to test the following hypothesis: The disappearance of a particular species of fish from a lake in the northeastern United States is due to acid rain resulting from industrial air pollution. What alternative hypotheses could you formulate? What experiments would you conduct to test these hypotheses? How would you use control experiments to isolate the influence of acid rain from that of other variables?

3. How does the human heart show all of the general themes of life: levels of organization, homeostasis, etc.?

4. How do you think that the connection between structure and function is the result of evolution?

5. Why is it correct to state that the process of science does not work to discover truth?

FOR FURTHER READING

Attenborough, D. *Life on Earth.* Boston: Little, Brown, 1979. The companion volume to a television series. A history of nature from the emergence of the first tiny, one-celled organisms to the emergence of upright human beings. Full of interesting evolutionary stories and exceptional photographs.

Blaustein, A., and D. B. Wake. "The Puzzle of Declining Amphibian Populations." *Scientific American,* April 1995, 52–57. The number of frogs, toads, and salamanders is dropping in many areas because of acid rain and depletion of the ozone layer.

Gould, S. "Darwinism Defined: The Difference Between Fact and Theory," *Discover,* January 1987, 64–70. A clear account of what biologists do and do not mean when they refer to the theory of evolution.

Gould, S. *Wonderful Life: The Burgess Shale and the Nature of History.* New York: W. W. Norton, 1989. A marvelous book about the early evolution of animals and about evolution in general.

Moore, J. A. *Science as a Way of Knowing—The Foundations of Modern Biology.* Cambridge, Mass.: Harvard University Press, 1993. An outstanding exposition of evolution and the whole field of biology.

National Research Council. *Ozone Depletion.* Washington, D.C., 1996. A short information-dissemination paper presenting up-to-date information on the ozone hole and what is being done about it.

Trefil, J. "Ah, But There May Have *Been* Life on Mars." *Smithsonian,* August 1995, 70–77. A new space flight will look for signs of past life in the rocks of Mars.

Weuthrich, B. "Trials on Trial." *New Scientist,* May 28, 1994, 14–15. An interesting discussion of the repercussions created by a single incidence of scientific fraud.

TECHNOLOGY LINKS

The Living World Home Page
http://www.wcbp.com/biology/tlw

2

The Chemistry of Life

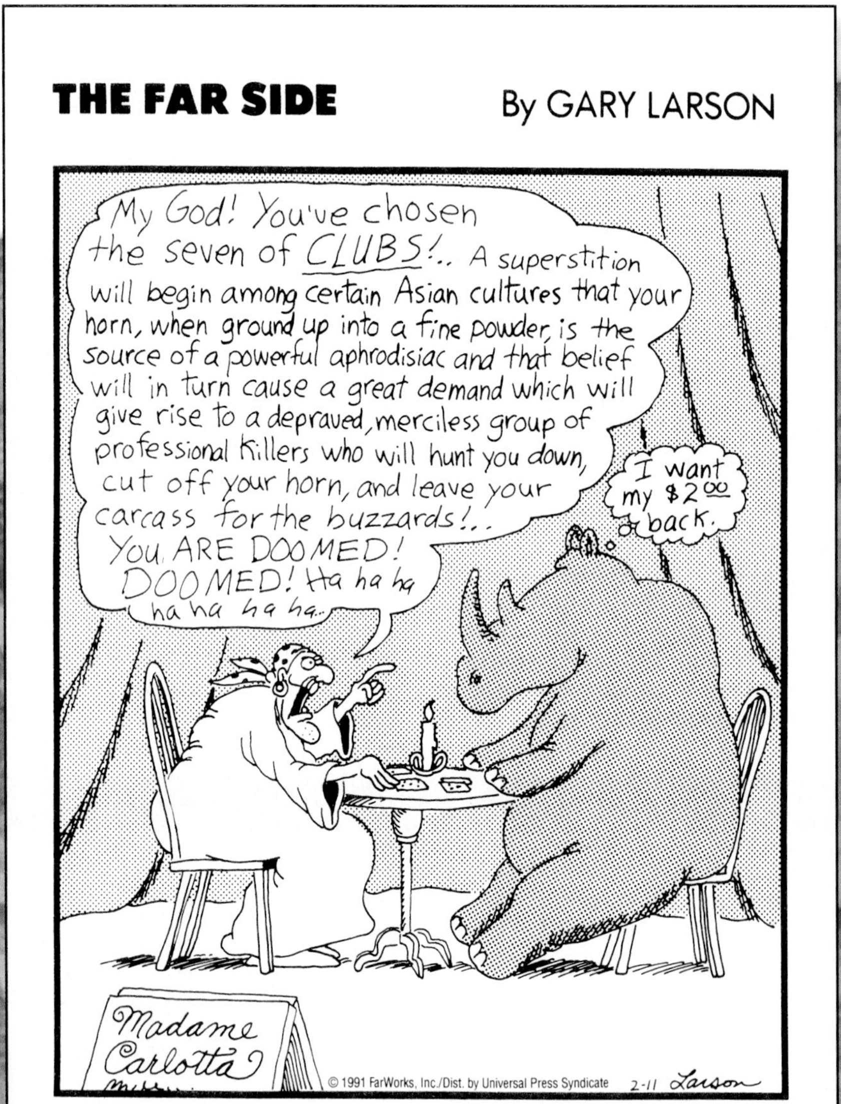

CHAPTER OUTLINE

Figure 2.1 The earth as seen from outer space. Water in oceans, lakes, and rivers is a prominent feature of our planet and comprises the bulk of all living creatures.

The earth was formed about 4.5 billion years ago. We know nothing directly of these very early times, because no rocks older than 3.9 billion years have been found. We do know that life first appeared on earth within the first billion years of our planet's history, because there are fossils of simple living things (bacteria) in rocks 3.5 billion years old, some of the oldest rocks that have persisted on earth. Much of the earth is covered with water (figure 2.1), and life is thought to have evolved spontaneously in the oceans of the young earth.

At the end of this chapter we examine the evidence that life evolved spontaneously from chemicals. Before we do that, however, we first consider the principles of chemistry briefly. Organisms are chemical machines, and to understand them we must learn a little chemistry. We will use what we learn to look at water in some detail, asking how its chemical structure leads to its many important properties. Life on earth has been shaped in large degree by the properties of water. We then go on a short tour of the kinds of chemicals—called macromolecules—that are the chemical building blocks of organisms.

2.1 Some Simple Chemistry

All living organisms are chemical machines, immense collections of molecules that interact in highly coordinated ways. In this chapter we explore very briefly the chemical nature of organisms, starting with the most basic of questions: Where did the first organisms come from?

Did Life Evolve?

No one knows for sure where the first organisms (thought to be like today's bacteria) came from. It is not possible to go back in time and watch how life originated, nor are there any witnesses. Nevertheless, it is difficult to avoid being curious about the origin of life, about what, or who, is responsible for the appearance of the first living organisms on earth. There are, in principle, at least three possibilities:

1. **Extraterrestrial origin.** Life may not have originated on earth at all but instead may have been carried to it, perhaps as an extraterrestrial infection of spores originating on a planet of a distant star. How life came to exist on that planet is a question we cannot hope to answer soon.
2. **Special creation.** Life-forms may have been put on earth by supernatural or divine forces. This viewpoint, called creationism, is common to most Western religions and is the oldest hypothesis. However, almost all scientists reject creationism, preferring evolution as a scientific explanation of life's diversity.
3. **Evolution.** Life may have evolved from inanimate matter, with associations among molecules becoming more and more complex. In this view, the force leading to life was selection; changes in molecules that increased their stability caused the molecules to persist longer.

Figure 2.2 Charles Darwin proposed the theory of evolution.
This newly rediscovered photograph appears to be the last ever taken of the great biologist who proposed the theory of evolution. It was taken in 1881, the year before Darwin died.

In this text we focus on the third possibility and attempt to understand whether the forces of evolution could have led to the origin of life and, if so, how the process might have occurred. The theory of evolution put forth over a century ago by Charles Darwin (figure 2.2) is founded on this same idea of selection. Darwin argued that selection was responsible for the great diversity of life-forms we see today.

This is not to say that the third possibility, evolution, is definitely the correct one. Any one of the three possibilities might be true. Nor does the third possibility preclude religion: a divine agency might have acted via evolution. Rather, we are limiting the scope of our inquiry to scientific matters. Of the three possibilities, only the third permits testable hypotheses to be constructed and so provides the only scientific explanation—that is, one that could potentially be disproved by experiment.

Atoms

If you could shrink yourself down until you were very small, the floor on which you are standing would appear quite different, with dust specks as big as boulders. If you kept on shrinking and shrinking, you would eventually become the size of the littlest thing in the universe, an atom. An **atom** is the smallest particle into which a substance or element can be divided and still retain its chemical properties. You and all other organisms are composed of atoms, linked together in complex assemblies called molecules.

All matter in the universe is composed of atoms. Atoms are extremely small and hard to study, but in the last century scientists have learned a great deal about them. At the center of every atom is a small, very dense **nucleus** formed of two subatomic particles, **protons** and **neutrons** (each of which is made up of even smaller bits, but we won't go into that!). Whizzing around the core is an orbiting cloud of a third kind of subatomic particle, the **electron.** Neutrons have no electrical charge, while protons have a positive charge and electrons a negative one. Every atom has this same basic structure: a core nucleus of protons and neutrons surrounded by a cloud of electrons (figure 2.3).

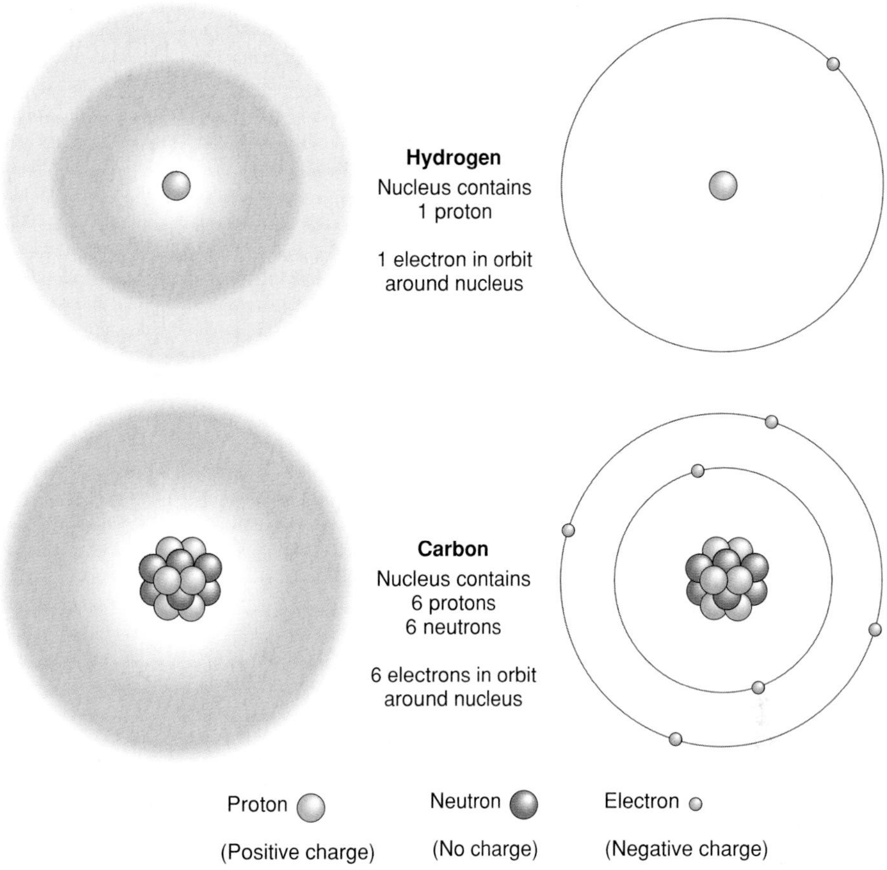

Figure 2.3 Basic structure of atoms.
All atoms have a nucleus consisting of protons and neutrons, except hydrogen, the smallest atom, which has only one proton and no neutrons in its nucleus. Carbon, for example, has six protons and six neutrons in its nucleus. Electrons spin around the nucleus in orbitals a far distance away from the nucleus. The electrons determine how atoms react with each other.

Ions

In a neutral atom (that is, one that carries no electrical charge), there is an orbiting electron for every proton in the nucleus. The electron's negative charge balances the proton's positive charge. Atoms in which the number of electrons does not equal the number of protons are called **ions.** All ions are electrically charged. An atom of sodium, for example, becomes a positively charged sodium ion when it loses an electron, because one proton in the nucleus is left with an unbalanced charge (figure 2.4).

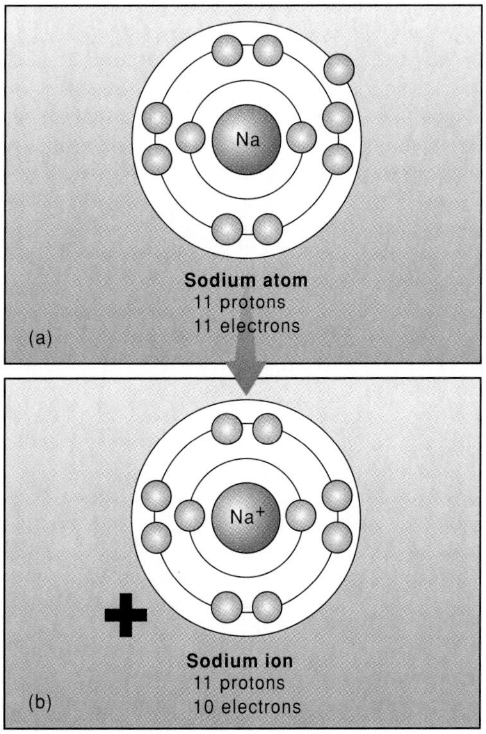

Figure 2.4 Making a sodium ion.
(*a*) An electrically neutral sodium atom with 11 protons and 11 electrons. (*b*) A sodium ion. Sodium ions bear one positive charge when they ionize and lose one electron. This sodium ion has 11 protons and only 10 electrons.

Isotopes

The number of protons in the nucleus of an atom is called the **atomic number.** For example, the atomic number of carbon is 6 because is has six protons. Neutrons are similar to protons in mass, and the number of protons and neutrons in the nucleus of an atom is called the **atomic mass.** A carbon atom that has six protons and six neutrons has an atomic mass of 12. The atomic numbers and mass numbers of the most common elements on earth are shown in table 2.1.

The number of neutrons that atoms of a particular element have can vary without changing the chemical properties of the element. Atoms that have the same number of protons but different numbers of neutrons are called **isotopes.** Isotopes of an atom have the same atomic number but differ in their atomic mass. Most elements in nature exist as mixtures of different isotopes. For example, there are three isotopes of the element carbon, all of which possess six protons (figure 2.5). The most common isotope of carbon has six neutrons. Because its total mass is 12 (six protons plus six neutrons), it is referred to as carbon-12. The isotope carbon-14 is unstable, and its nucleus tends to break up into particles with lower atomic numbers, a process called **radioactive decay.** The use of radioactive isotopes in dating fossils is discussed in chapter 9.

Table 2.1 The Most Common Elements on Earth

Element	Symbol	Atomic Number	Atomic Mass
Oxygen	O	8	15.9994
Silicon	Si	14	28.086
Aluminum	Al	13	26.9815
Iron	Fe	26	55.847
Calcium	Ca	20	40.08
Sodium	Na	11	22.989
Potassium	K	19	39.098
Magnesium	Mg	12	24.305
Hydrogen	H	1	1.0079
Manganese	Mn	25	54.938
Fluorine	F	9	18.9984
Phosphorus	P	15	30.9738
Carbon	C	6	12.0112
Sulfur	S	16	32.064
Chlorine	Cl	17	35.453
Vanadium	V	23	50.942
Chromium	Cr	24	51.996
Copper	Cu	29	63.546
Nitrogen	N	7	14.0067
Boron	B	5	10.811
Cobalt	Co	27	58.933
Zinc	Zn	30	65.38
Selenium	Se	34	78.96
Molybdenum	Mo	42	95.94
Tin	Sn	50	118.69
Iodine	I	53	126.904

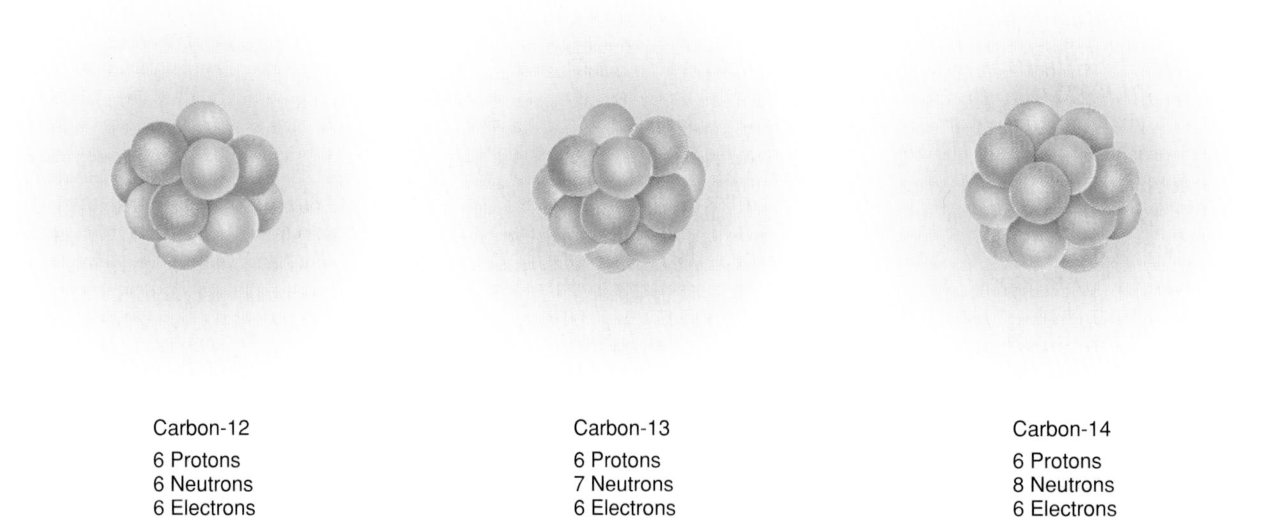

Figure 2.5 Isotopes of the element carbon.
The three most abundant isotopes of carbon are carbon-12, carbon-13, and carbon-14. The yellow "clouds" in the diagrams represent the orbiting electrons, whose numbers are the same for all three isotopes. Protons are shown in pink, and neutrons are shown in green.

Electrons Determine What Atoms Are Like

Electrons have very little mass (only 1/1,840 the mass of a proton). Of all the mass contributing to your weight, the portion that is contributed by electrons is less than the mass of your eyelashes. And yet electrons determine the chemical behavior of atoms. To understand why, it is only necessary to realize that orbiting electrons are very far from the nucleus. Almost all the volume of an atom is empty space. If the nucleus of an atom were the size of an apple, the orbit of the nearest electron would be more than a mile out! Electrons determine chemical behavior because the nuclei of two atoms never come close enough to each other in nature to interact.

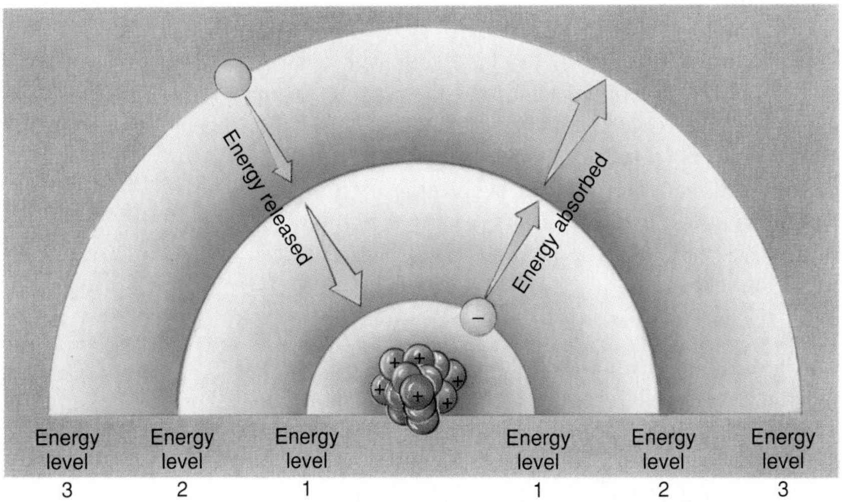

Figure 2.6 The electrons of atoms possess potential energy.
Electrons circulate rapidly around the nucleus in paths called orbitals. Energy level 1 is the lowest potential energy level because it is closest to the nucleus. When an electron absorbs energy, it moves from level 1 to the next higher energy level (level 2). When an electron loses energy, it falls to a lower energy level closer to the nucleus.

Electrons Carry Energy

Because electrons are negatively charged, they are attracted to the positively charged nucleus. It takes work to keep them in orbit, just as it takes work to hold an apple in your hand when gravity is pulling the apple down toward the ground. The apple in your hand is said to possess **energy,** the ability to do work, because of its position—if you were to release it, the apple would fall. Similarly, electrons have energy of position, called potential energy (figure 2.6). It takes work to oppose the attraction of the nucleus, so moving the electron farther out to a more distant orbit requires an input of energy and results in an electron with greater potential energy. For the same reason, it takes energy to lug a bowling ball up to the top of a tall building, and when released from the roof's edge the ball impacts the ground with greater force than one dropped from a distance of a meter. Moving an electron in toward the nucleus has the opposite effect; energy is released, and the electron ends up with less potential energy.

Sometimes an electron is transferred from one atom to another. When an electron is transferred in this way, it keeps its energy of position, like a bowling ball being passed from one person to another. In living organisms, chemical energy is stored by using it to move electrons to more distant "high-energy" orbits, and these energetic electrons are frequently transferred from one atom to another, passing along the energy like so much money. The loss of an electron in such a transfer is called **oxidation;** the gain of an electron is called **reduction** (figure 2.7).

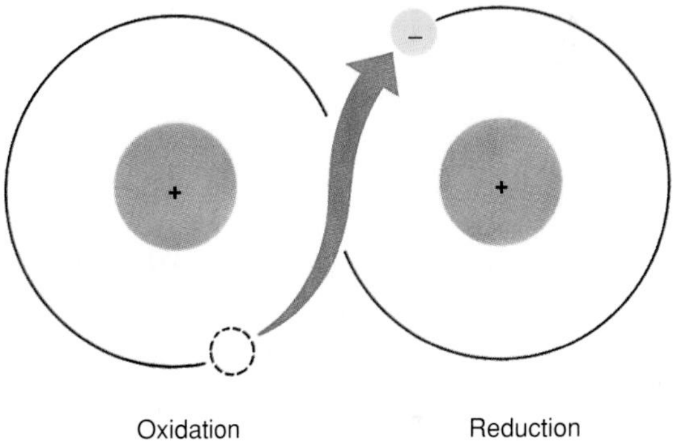

Figure 2.7 Oxidation-reduction.
Oxidation is the loss of an electron; reduction is the gain of one.

While the energy levels of an atom are often visualized as well-defined circular orbits around a central nucleus, such a simple picture is not realistic. These energy levels often consist of complex three-dimensional shapes, and the exact location of an individual electron at any given time is impossible to specify. However, some locations are more probable than others, and it is often possible to say where an electron is *most likely* to be located. The volume of space around a nucleus where an electron is most likely to be found is called the **orbital** of that electron.

Orbitals are located at each energy level, and each energy level has a specific number of orbitals. Each orbital can hold up to two electrons. The first energy level in any atom contains one orbital. Helium, for example, has one energy level with one orbital that contains two electrons. In atoms with more than one energy level, the second energy level contains four orbitals and can contain up to eight electrons. Nitrogen has two energy levels; the first one is completely filled with two electrons, but the four orbitals in the second energy level are not filled because nitrogen's second energy level only contains five electrons. In atoms with more than two energy levels, subsequent energy levels also contain up to four orbitals and a maximum of eight electrons. Atoms with incomplete electron orbitals will tend to be more reactive, and in this way, the electrons of an atom influence the atom's behavior.

Atoms Form Molecules

A **molecule** is a group of atoms held together by energy. The energy acts as "glue," ensuring that the various atoms stick to one another. The force holding two atoms together is called a **chemical bond.** There are three principal kinds of chemical bonds: covalent bonds, where the force results from the sharing of electrons, and ionic and hydrogen bonds, where the force is generated by the attraction of opposite electrical charges.

Ionic Bonds

Chemical bonds called **ionic bonds** form when atoms are attracted to each other by opposite electrical charges. Just as the positive pole of a magnet is attracted to the negative pole of another, so an atom can form a strong link with another atom if they have opposite electrical charges. Because an atom with an electrical charge is an ion, these bonds are called ionic bonds.

Everyday table salt is built of ionic bonds. The sodium and chloride atoms of table salt are ions, sodium having given up the sole electron in its outermost sphere (the sphere underneath has eight) and chloride having gained an electron to complete its outermost sphere (figure 2.8). As a result of this electron hopping, sodium atoms in table salt are positive ions and chloride atoms are negative ions. Because each ion is attracted electrically to surrounding ions of opposite charge, this causes the formation of an elaborate matrix of sodium and chloride ionic bonds—a crystal. That is why table salt is composed of tiny crystals and is not a powder.

The two key properties of ionic bonds that make them form crystals are that they are strong (although not as strong as covalent bonds) and that they are *not* directional. A charged atom is attracted to the electrical field contributed by all nearby atoms of opposite charge. Ionic bonds do not play an important part in most biological molecules because of this lack of directionality. Complex, stable shapes require the more specific associations made possible by directional bonds.

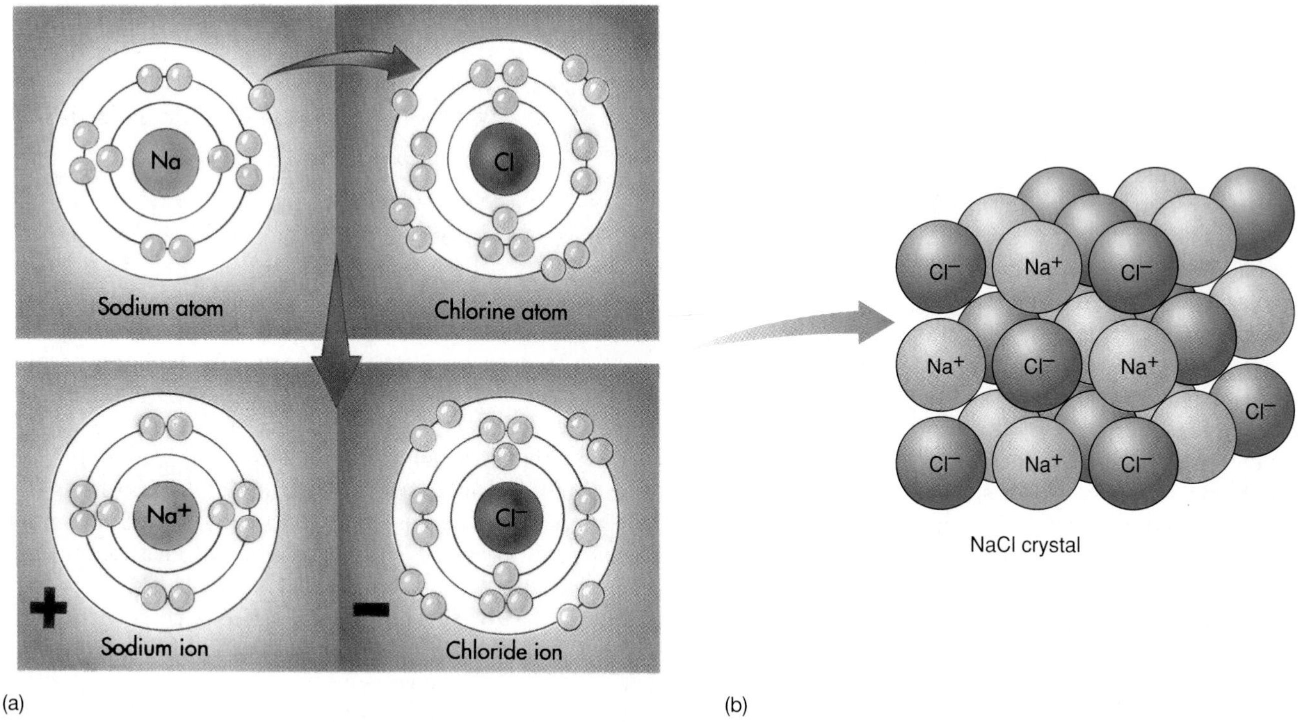

Figure 2.8 The formation of the ionic bond in table salt.
(*a*) When a sodium atom donates an electron to a chlorine atom, the sodium atom, lacking that electron, becomes a positively charged sodium ion. The chlorine atom, having gained an extra electron, becomes a negatively charged chloride ion. (*b*) Sodium chloride forms a highly regular lattice of alternating sodium ions and chloride ions. You are familiar with these crystals as everyday table salt.

Covalent Bonds

Strong chemical bonds called **covalent bonds** form between two atoms when they share electrons. Most of the atoms in your body are linked to other atoms by covalent bonds. Why do atoms in molecules share electrons? All atoms seek to fill up their outermost "sphere" of orbiting electrons, which in all atoms common in living things (except tiny hydrogen) takes eight electrons. If an atom has only seven electrons in its outer orbit, it seeks to share them with an atom that has a single electron in its outer orbit, so that the outer sphere of each atom now has eight electrons at least some of the time.

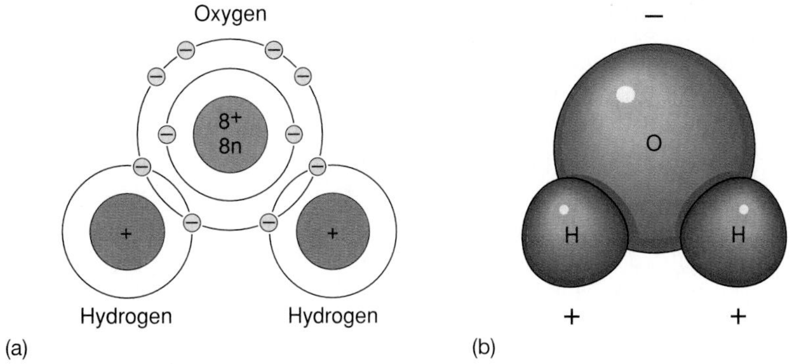

Figure 2.9 Water molecules contain two covalent bonds.
Each water molecule is composed of one oxygen atom and two hydrogen atoms. (*a*) The oxygen atom shares a pair of electrons with each participating hydrogen atom, contributing one electron to each. (*b*) A three-dimensional representation of a water molecule.

In the same way, an atom with six outer-orbit electrons seeks to share them with an atom that has two outer electrons or with two atoms that have single outer electrons. Water (H_2O), for example, is a molecule in which oxygen (six outer electrons) forms covalent bonds with two hydrogens (one outer electron each) (figure 2.9). Because the atom carbon has four electrons in its outermost sphere, carbon can form as many as four covalent bonds in its attempt to fully populate its outermost sphere of electrons. Because there are so many ways that four covalent bonds can form, carbon atoms participate in many different kinds of molecules.

The two key properties of covalent bonds that make them ideal for their molecule-building role in living systems are (1) that they are strong, involving the sharing of lots of energy, and (2) that they are very directional—bonds form between two specific atoms, rather than a generalized attraction of one atom for its neighbors.

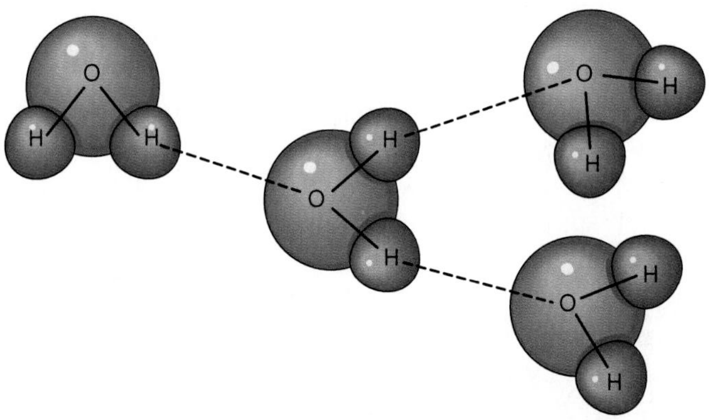

Figure 2.10 Hydrogen bonding in water molecules.
Hydrogen bonds play a key role in determining what shape a protein assumes in the cell. The functional groups of a protein's polar amino acids form many hydrogen bonds with surrounding water molecules, while the nonpolar portions of the polypeptide aggregate in the protein interior.

Hydrogen Bonds

Weak chemical bonds of a very special sort called **hydrogen bonds** play a key role in biology. To understand them we need to look at covalent bonds again briefly. When a covalent bond forms between two atoms, one nucleus may be much better at attracting the shared electrons than the other—in water, for example, the shared electrons are much more strongly attracted to the oxygen atom than to the hydrogen atoms. When this happens, shared electrons spend more time in the vicinity of the more strongly attracting atom, which as a result becomes somewhat negative in charge; they spend less time in the vicinity of the less strongly attracting atom or atoms, and these become somewhat positive in charge (see figure 2.9). The charges are not full electrical charges like ions possess but rather tiny partial charges. What you end up with is a sort of molecular magnet, with positive and negative ends, or "poles." Molecules like this are said to be **polar.** A hydrogen bond occurs when the positive end of one polar molecule is attracted to the negative end of another, like two magnets drawn to each other (figure 2.10).

Two key properties of hydrogen bonds cause them to play an important role in biological molecules. First, they are weak and so are not effective over long distances like more powerful ionic bonds. Second, as a result of their weakness, hydrogen bonds are highly directional—polar molecules must be very close for the weak attraction to be effective. Hydrogen bonds are too weak to actually form stable molecules by themselves. Instead, they act like Velcro, forming a tight bond by the additive effects of *many* weak interactions. Hydrogen bonds stabilize the shapes of many important biological molecules by causing certain parts of a molecule to be attracted to other parts. Proteins, which are huge molecules that play many important roles in organisms, each fold into a particular complex shape because of the formation of hydrogen bonds between parts of the protein molecule.

2.2 Water: Cradle of Life

Of all the molecules that are common on earth, only water exists as a liquid at everyday temperatures. Three-fourths of the earth's surface is covered by liquid water. About two-thirds of your body is water, and you cannot exist long without it. All other organisms also require water. It is no accident that tropical rain forests are bursting with life, whereas dry deserts are almost lifeless except after rain. Farming is only possible in areas where rain is plentiful. The chemistry of life, then, is water chemistry.

Water has a simple atomic structure, an oxygen atom linked to two hydrogen atoms by single covalent bonds. The chemical formula for water is thus H_2O. It is because the oxygen atom attracts the shared electrons more strongly than the hydrogen atoms that water is a *polar molecule* and so can form *hydrogen bonds*.

The single most outstanding chemical property of water is its ability to form hydrogen bonds. This one property of water, which derives directly from its structure, is responsible for much of the organization of living chemistry, from what membranes are like to how proteins fold. Some of these properties of water are illustrated in figure 2.11.

Water Acts Like a Magnet

The weak hydrogen bonds that form between a hydrogen atom of one water molecule and the oxygen atom of another produce a lattice of hydrogen bonds within liquid water. Each of these bonds is individually very weak and short-lived—a single bond lasts only 1/100,000,000,000 of a second. However, like the grains of sand on a beach, the cumulative effect of large numbers of these bonds is enormous. This cumulative effect is responsible for many of the important physical properties of water listed in table 2.2.

Water Clings to Polar Molecules

Because water molecules are very polar, they are attracted to other polar molecules—like glue, hydrogen bonds bind polar molecules to each other. When the other polar molecule is another water molecule, the attraction is called **cohesion.** The surface tension of water is created by cohesion. When the other polar molecule is a different substance, the attraction is called **adhesion.** Capillary action—water rising up a narrow glass tube—is created by adhesion. Water clings to any substance, such as glass, with which it can form hydrogen bonds. Adhesion is why things get "wet" when they are dipped in water and why waxy substances do not—they are composed of nonpolar molecules.

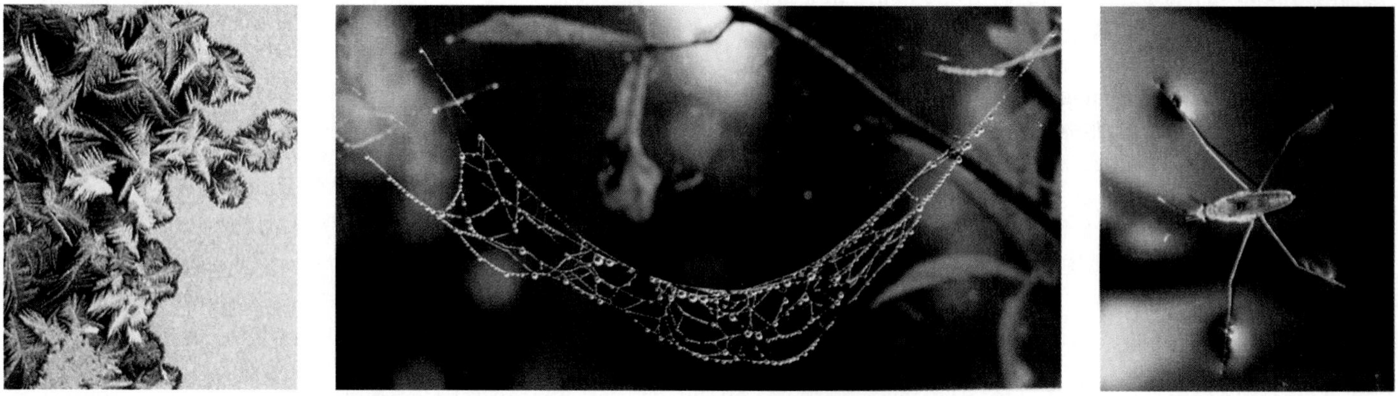

(a) Ice formation (b) Cohesion (c) Surface tension

Figure 2.11 Some physical properties of water that depend on hydrogen bonding.
(*a*) *Ice formation*. When water cools below 0°C, it forms a regular crystal structure in which the partial charges of each atom in the water molecule interact with opposite charges of atoms in other water molecules to form hydrogen bonds. (*b*) *Cohesion*. The ability of water to "stick together" is a result of the collective force of its hydrogen bonds. Water can form droplets that stick together, a property called cohesion. (*c*) *Surface tension*. Some insects, such as this water strider, literally walk on water. You can see the dimpling the strider's feet make on the water as its weight bears down on the surface. Surface tension is a property derived from cohesion—that is, water has a "strong" surface due to the force of its hydrogen bonds. Because the surface tension of the water is greater than the force of one "foot" of the water strider, the water strider does not sink but rather slides along.

Table 2.2 The Properties of Water

Property	Explanation	Example of Benefit to Life
High polarity	Polar water molecules are attracted to ions and polar compounds, making them soluble in water	Many kinds of molecules can move freely in cells, permitting a diverse array of chemical reactions
Heat storage	Hydrogen bonds absorb heat when they break and release heat when they form, minimizing temperature changes	Water stabilizes body temperature, as well as that of the environment
High heat of vaporization	Many hydrogen bonds must be broken for water to evaporate	Evaporation of water cools body surfaces
Lower density of ice	Water molecules in an ice crystal are spaced relatively far apart because of hydrogen bonding	Because ice is less dense than water and floats, lakes do not freeze solid, and they overturn in spring
Cohesion	Hydrogen bonds hold molecules of water together	Water forms thin layers that coat membranes to keep them moist

Water Stores Heat

The temperature of any substance is a measure of how rapidly its individual molecules are moving. Because of the many hydrogen bonds that water molecules form with one another, a large input of thermal energy is required to disrupt the organization of liquid water and raise its temperature. Because of this, water heats up more slowly than almost any other compound and holds its temperature longer. That is why your body is able to maintain a relatively constant internal temperature—the heat generated by your metabolism would cook your cells if water were not able to absorb the bulk of this energy.

If the temperature is low enough, very few hydrogen bonds break in water so that the lattice of these bonds assumes a crystal-like structure, forming a solid we call ice (figure 2.12). Interestingly, ice is less dense than water—that is why icebergs float. Why is ice less dense? Because the hydrogen bonds space the water molecules apart, preventing them from approaching each other more closely.

If the temperature is high enough, many hydrogen bonds break in water so that the liquid is changed into vapor. A considerable amount of heat energy is required to do this—every gram of water that evaporates from your skin removes 586 calories of heat from your body, which is equal to the energy released by lowering the temperature of 586 grams of water 1°C. That is why sweating cools you off.

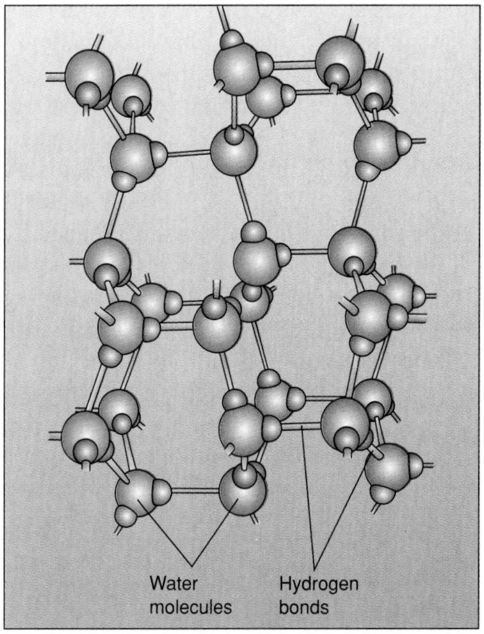

Figure 2.12 The lattice of hydrogen bonds in an ice crystal.
When water cools below 0°C, it forms a regular crystal structure. The individual water molecules are spaced apart and held into position by hydrogen bonds. Because water forms a crystal latticework, ice is less dense than water and floats.

Water Is a Powerful Solvent

Water molecules gather closely around any molecule that exhibits an electrical charge, whether a full charge (ion) or partial charge (polar molecule). When a sugar crystal dissolves in water, what really happens is that individual sugar molecules break off from the crystal and become surrounded by water molecules attracted to its slightly polar hydroxyl (OH^-) groups. Water molecules orient around each sugar molecule like a cloud of flies attracted to honey, and this shell of water molecules prevents the sugar molecule from reassociating with the crystal. Similar shells of water form around all polar molecules, and polar molecules that dissolve in water in this way are said to be **soluble** in water (figure 2.13). Nonpolar molecules like oil are not water-soluble.

Water Organizes Nonpolar Molecules

Water molecules in solution always tend to form the maximum number of hydrogen bonds possible. When nonpolar molecules, which do not form hydrogen bonds, are placed in water, the water molecules shy away, instead forming hydrogen bonds with other water molecules. The nonpolar molecules are forced into association with one another, crowded together to minimize their disruption of the hydrogen bonding of water. It seems almost as if the nonpolar compounds shrink from contact with water, and for this reason they are called **hydrophobic** (from the Greek *hydros*, water + *phobos*, fearing). Many biological structures are shaped by such hydrophobic forces. For example, some of the exterior portions of many proteins are hydrophobic, and by forcing these portions into proximity to one another, water causes these proteins to assume particular shapes in solution.

Polar molecules, on the other hand, form hydrogen bonds and are welcomed by water molecules. For example, ammonia spontaneously forms hydrogen bonds with water molecules. Polar molecules such as ammonia are called hydrophilic (*hydros*, water + *philic*, loving), or water-loving, molecules.

Water Ionizes

The covalent bonds with a water molecule sometimes break spontaneously. When this happens, the nucleus of one of the hydrogen atoms dissociates from the molecule, leaving the shared electron behind. This produces a positively charged H^+ fragment, called a **hydrogen ion,** and a negatively charged OH^- fragment, called a **hydroxide ion.** This process of spontaneous ion formation is called ionization. It can be represented by a simple chemical equation, in which the chemical formulae for water and the two ions are written down, with an arrow showing the direction of the dissociation:

H_2O	$\rightarrow$	OH^-	+	H^+
Water		Hydroxide ion		Hydrogen ion

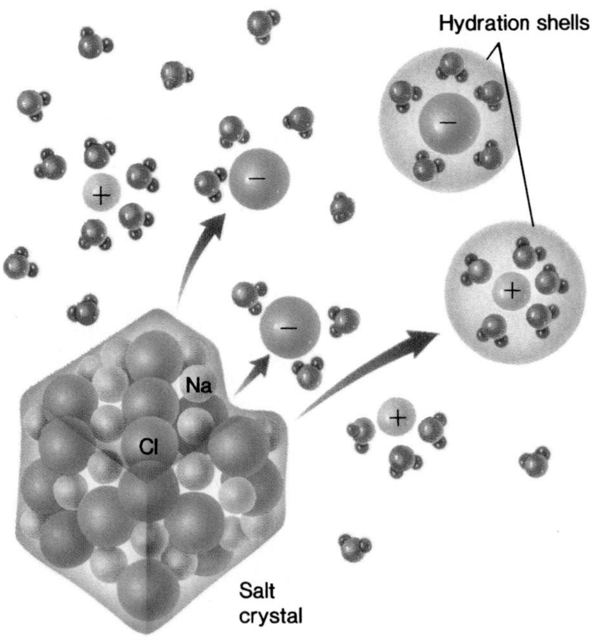

Figure 2.13 How salt dissolves in water.
Salt is soluble in water because the partial charges on water molecules are attracted to the charged sodium and chloride ions. The water molecules surround the ions, forming what are called hydration shells. When all of the ions have been separated from the crystal, the salt is said to be dissolved.

Because covalent bonds are strong, spontaneous ionization is not common. In a liter of water, only roughly 1 molecule out of each 550 million is ionized at any instant in time, corresponding to 1/10,000,000 (that is, 10^{-7}) of a mole of H^+ ions (a mole is defined as the weight in grams that corresponds to the molecular mass of a molecule, which in this case equals 1 gram). The concentration of hydrogen ions in water can be written more easily by simply counting the number of decimal places after the digit "1" in the denominator:

$$[H^+] = \frac{1}{10,000,000}$$

This way of measuring the concentration of hydrogen ions in water is called the **pH scale** (figure 2.14). Since there are seven decimal places, pure water is said to have a pH of 7 (more formally, **pH** is the negative logarithm of the molar concentration). Any substance that has more hydrogen ions than would be present in pure water is called an **acid.** A cup of coffee has 100 times as many hydrogen ions as a glass of water, 1 per 100,000 water molecules, so its pH is 5. The more acid a substance, the lower its pH. A substance with more OH^- ions (and thus fewer H^+ ions) than pure water is called a **base** and has a pH greater than 7.

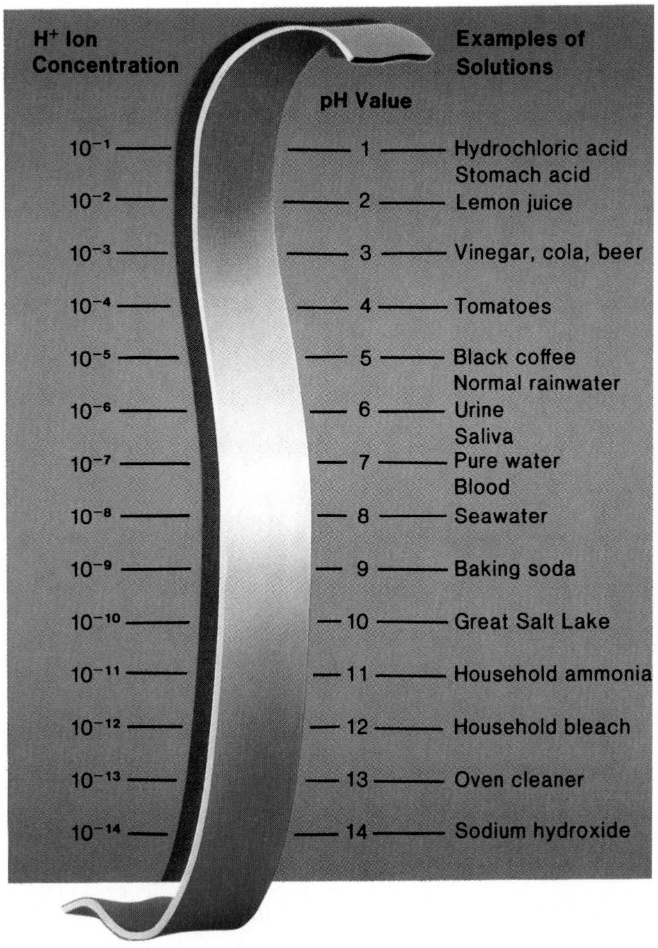

Figure 2.14 The pH scale.

A fluid is assigned a value according to the number of hydrogen ions present in a liter of that fluid. The scale is logarithmic, so that a change of only 1 means a tenfold change in the concentration of hydrogen ions; thus lemon juice is 100 times more acidic than tomatoes, and seawater is 10 times more basic than pure water.

Buffers

The pH inside almost all living cells is about 7, and it is very important to your health that this value not change much. The many proteins that govern your metabolism are all extremely sensitive to pH, and slight alterations in pH change their shape and so disrupt their activities. The pH of your blood, for example, is 7.4, and you would survive only a few minutes if it were to fall to 7.0 or rise to 7.8. And yet we all eat substances that are acidic and basic; Coca-Cola, for example, is a strong (although dilute) acid. In addition, metabolic reactions within cells produce acids and bases all the time. What keeps your body's pH constant? Cells contain chemical substances called **buffers** that minimize changes in concentrations of H^+ and OH^-.

A buffer is a substance that acts as a reservoir for hydrogen ions, donating them to the solution when their concentration falls and taking them from the solution when their concentration rises. For example, the key buffer in human blood is the acid-base pair **carbonic acid** and **bicarbonate.** In water, carbonic acid (H_2CO_3) dissociates to yield bicarbonate ion and a hydrogen ion (figure 2.15). Now if some other source adds H^+ ions to your blood (pH drops), the bicarbonate ion acts as a base and removes the excess H^+ ions from solution by forming H_2CO_3. Similarly, if some process removes H^+ ions from your blood (pH rises), the carbonic acid dissociates, releasing more hydrogen ions into the solution. pH is thus stabilized by the equilibrium between the forward and back reactions converting acid to base or base to acid.

The interaction of carbon dioxide and water has the important consequence that significant amounts of carbon enter into water solution from air in the form of carbonic acid. As we will discuss in section 2.4, biologists believe that life first evolved in the early oceans, which were rich in carbon because of this reaction.

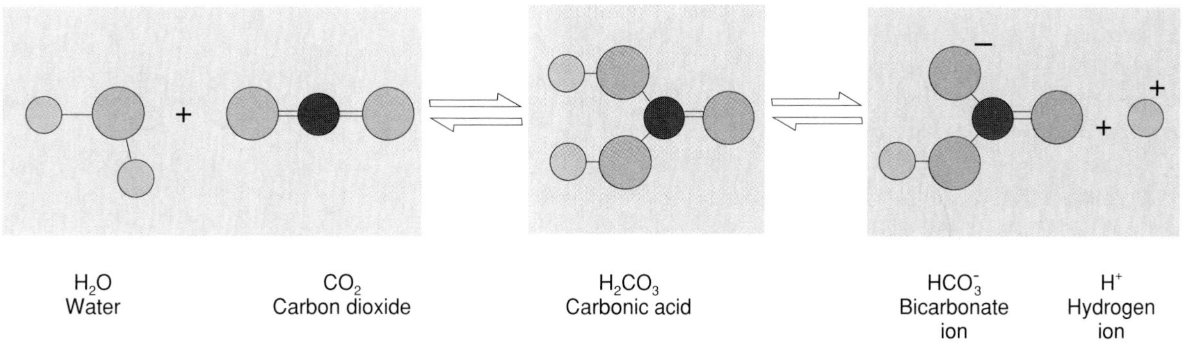

Figure 2.15 The buffering action of carbonic acid and bicarbonate.

Carbon dioxide and water combine chemically to form carbonic acid (H_2CO_3), which dissociates in water, freeing H^+ ions. Now if some other source adds H^+ ions to your blood, the bicarbonate ion acts as a base and removes the excess H^+ ions from the solution, forming H_2CO_3. Similarly, if some process removes H^+ ions from your blood, the carbonic acid dissociates, releasing more hydrogen ions into the solution.

2.3 Macromolecules

The bodies of organisms contain thousands of different kinds of molecules, but much of the body is made of just four kinds: carbohydrates, lipids, proteins, and nucleic acids (table 2.3). Called **macromolecules** because they can be very large, these four are the building materials of cells, the "bricks and steel" that make up the bodies of cells and the machinery that runs within them.

The body's macromolecules are assembled by sticking smaller bits together, much as a train is built by linking railcars together. A molecule built up of long chains of similar subunits is called a **polymer.** An **organic molecule** is a molecule formed by living organisms that consists of a carbon-based core with special groups attached. These groups of atoms have special chemical properties and are referred to as **functional groups.** Functional groups tend to act as units during chemical reactions and to confer specific chemical properties on the molecules that possess them. For example, a hydrogen atom bonded to an oxygen atom (–OH) is a hydroxyl functional group and makes a molecule more basic. The six principal functional groups are listed in figure 2.16.

Group	Structural Formula	Ball-and-Stick Model	Found In:
Hydroxyl	—OH	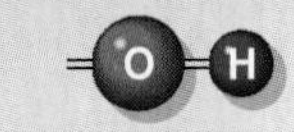	Alcohols
Carbonyl	—C— with =O below (C=O)	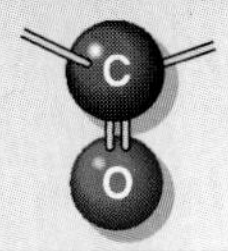	Formaldehyde
Carboxyl	—C(=O)OH	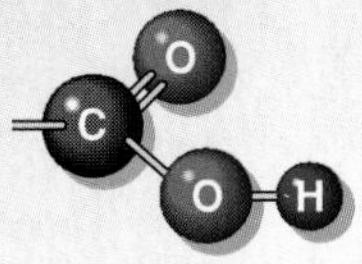	Vinegar
Amino	—N(H)H	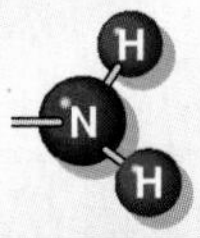	Ammonia
Sulfhydryl	—S—H		Rubber
Phosphate	—O—P(O⁻)(=O)—O⁻	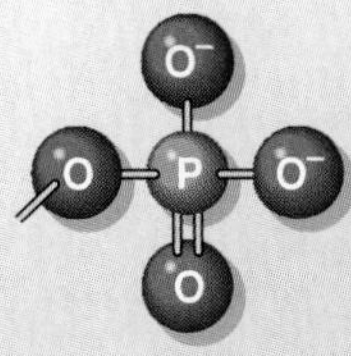	ATP

Figure 2.16 The six principal functional groups.
Most chemical reactions that occur within organisms involve transferring a functional group from one molecule to another or breaking a carbon–carbon bond.

Making (and Breaking) Macromolecules

The four different kinds of macromolecules put their subunits together in the same way: a covalent bond is formed between two subunits in which a hydroxyl group (OH) is removed from one subunit and a hydrogen (H) is removed from the other. This process is called **dehydration synthesis** because, in effect, the removal of the OH and H groups constitutes removal of a molecule of water—the word "dehydration" means taking away water (figure 2.17*a*). This process requires the help of a special class of proteins called **enzymes** to facilitate the positioning of the molecules so that the correct chemical bonds are stressed and broken. The process of tearing down a molecule, such as the protein or fat contained in food that is consumed, is essentially the reverse of dehydration synthesis: instead of removing a water molecule, one is added. When a water molecule comes in, a hydrogen becomes attached to one subunit and a hydroxyl to another, and the covalent bond is broken. The breaking up of a polymer in this way is called **hydrolysis** (figure 2.17*b*).

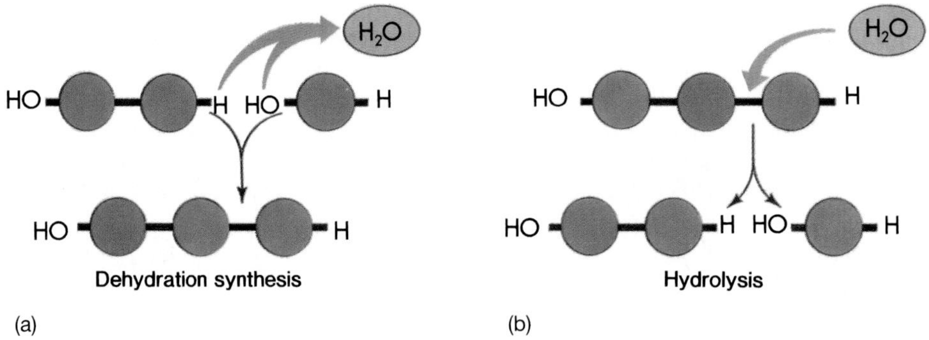

Figure 2.17 Dehydration and hydrolysis.
(*a*) Biological molecules are formed by linking subunits. The covalent bond between subunits is formed in dehydration synthesis, a process during which a water molecule is eliminated. (*b*) Breaking such a bond requires the addition of a water molecule, a reaction called hydrolysis.

Table 2.3 Macromolecules

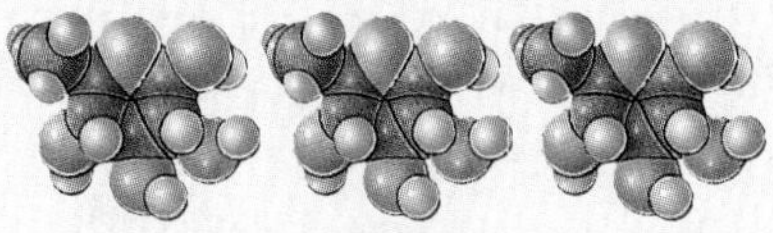

Starch

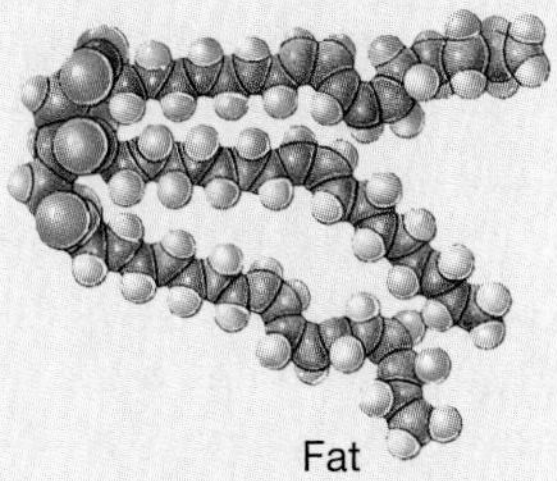

Fat

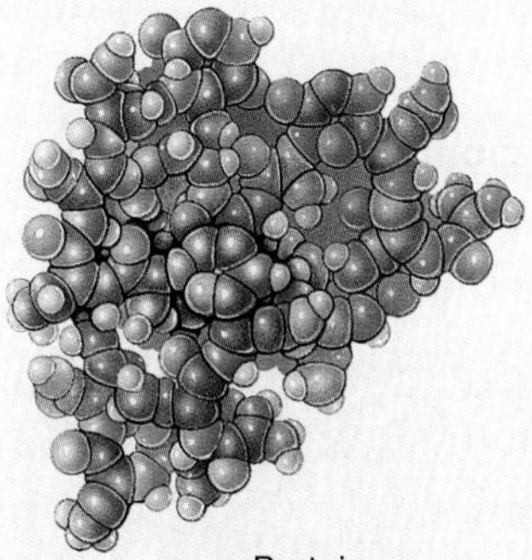

Protein

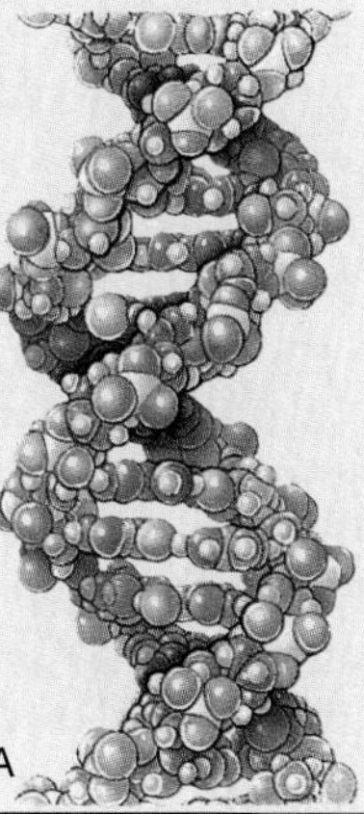

DNA

Macromolecule	Subunit	Function
Carbohydrates		
Starch, glycogen	Glucose	Energy storage
Cellulose	Glucose	Component of plant cell walls
Chitin	Modified glucose	Cell walls of fungi; outer skeleton of insects and related groups
Lipids		
Fats	Glycerol + three fatty acids	Energy storage
Phospholipids	Glycerol + two fatty acids + phosphate	Component of cell membranes
Steroids	Four carbon rings	Message transmission (hormones)
Terpenes	Long carbon chains	Pigments in photosynthesis
Proteins		
Globular	Amino acids	Catalysis
Structural	Amino acids	Support and structure
Nucleic Acids		
DNA	Nucleotides	Encoding of hereditary information
RNA	Nucleotides	Blueprint of hereditary information

Carbohydrates

Polymers called **carbohydrates** make up the structural framework of cells and play a critical role in energy storage. A carbohydrate is any molecule that contains carbon, hydrogen, and oxygen in the ratio 1:2:1. Some carbohydrates are simple, small monomers or dimers and are called **simple carbohydrates.** Others are long polymers and are called **complex carbohydrates.** Because they contain many carbon–hydrogen (C–H) bonds, carbohydrates are well-suited for energy storage. Such C–H bonds are the ones most often broken by organisms to obtain energy.

Simple Carbohydrates

Among the simplest of the carbohydrates are the simple sugars or **monosaccharides** (from the Greek *monos*, meaning single, and *saccharon*, meaning sweet). These molecules consist of one subunit. For example, glucose, the sugar that carries energy to the cells of your body, is made of six carbons and has the chemical formula $C_6H_{12}O_6$ (figure 2.18). Another type of simple carbohydrate is a **disaccharide,** which forms when two monosaccharides link. Sucrose, table sugar, is a disaccharide made of two six-carbon sugars linked together (figure 2.19).

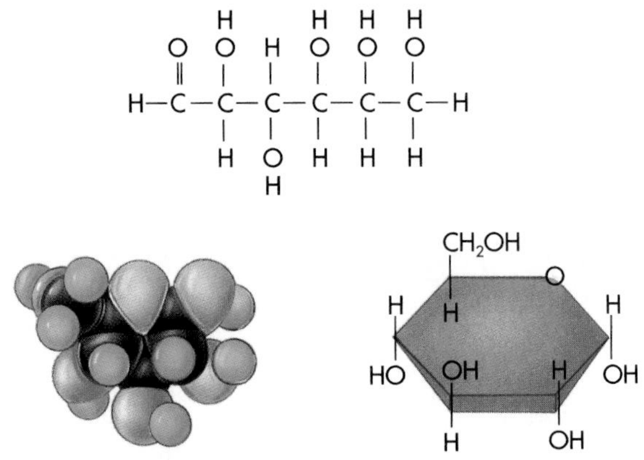

Figure 2.18 The structure of glucose.
Glucose is a monosaccharide and consists of a linear six-carbon molecule that forms a ring when added to water. This illustration shows three ways glucose can be represented diagrammatically.

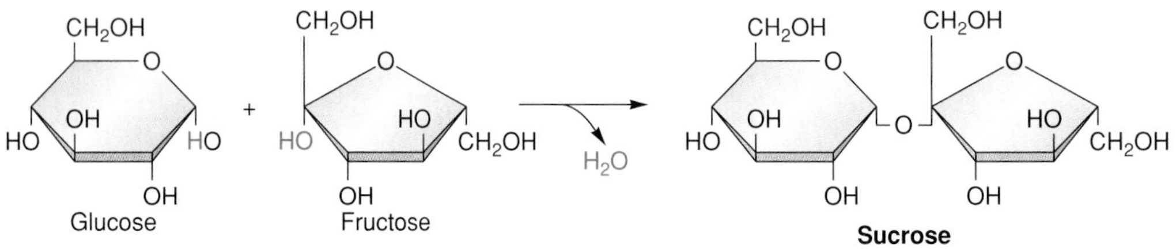

Figure 2.19 Formation of sucrose.
The disaccharide sucrose is formed from glucose and fructose in a dehydration reaction.

Complex Carbohydrates

Organisms store their metabolic energy by converting sugars, which are soluble, into insoluble forms that can be deposited in specific storage areas in the body. This trick is achieved by linking the sugars together into long polymer chains called **polysaccharides.** Polysaccharides formed from glucose are called starches. Many plants store energy in starches—that is why potatoes are "starchy" food. Animals also store energy in glucose chains, but in animals, the chains are very long and highly branched, forming a highly insoluble thicket of glucose chains called **glycogen.**

Plants and animals also use glucose chains as building materials, linking the subunits together in different orientations not recognized by most enzymes. The cellulose that makes up plant cell walls and the chitin of lobster shells (figure 2.20) are both polymers composed of long-chain sugar subunits.

Figure 2.20 This lobster's shell is made of chitin.
Chitin is the principal structural element in the external skeletons of many invertebrates, including insects, and in the cell walls of fungi.

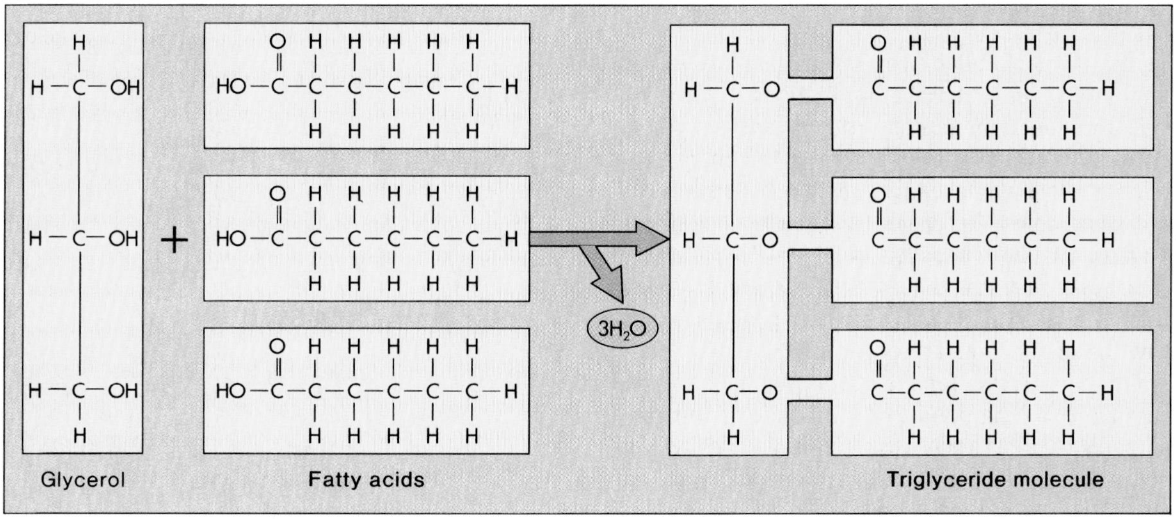

Figure 2.21 Formation of a fat molecule.
This fat molecule, a triglyceride, is formed by dehydration synthesis, in which the glycerol is attached to three fatty acids.

Lipids

For long-term storage, organisms usually convert glucose into fats, another kind of storage molecule that contains more C–H bonds. Fats and all other biological molecules that are not soluble in water but soluble in oil are called **lipids.** Lipids are insoluble in water not because they are long chains like starches but rather because they are nonpolar. In water, fat molecules cluster together because they cannot form hydrogen bonds with water molecules. Olive oil is a lipid, and so is earwax, turpentine, and cholesterol.

Fats

Fat molecules are composed of two subunits: fatty acids and glycerol. A **fatty acid** is a long hydrocarbon chain ending in a carboxyl (–COOH) group. The three carbons of glycerol form the backbone to which three fatty acids are attached in the dehydration reaction that forms the fat molecule (figure 2.21). Because there are three fatty acids, the resulting fat molecule is called a **triglyceride** (or, more formally, a "triacylglycerol").

Fatty acids, with all internal carbon atoms having two hydrogen side groups, contain the maximum number of hydrogen atoms possible. Fats composed of these fatty acids are said to be **saturated** (figure 2.22*a*). On the other hand, fats composed of fatty acids that have double bonds between one or more pairs of successive carbon atoms, and thus contain fewer than the maximum number of hydrogen atoms, are called **unsaturated** (figure 2.22*b*). Many plant fatty acids, such as oleic acid (a vegetable oil) and linolenic acid (a linseed oil), are unsaturated. Animal fats, in contrast, are often saturated and occur as hard fats.

Other Types of Lipids

Fats are just one example of the oily or waxy class of macromolecules called lipids. Your body also contains other types of lipids that play many roles in cells in addition to energy storage. The membranes of cells are made of a modified fat called a **phospholipid.** Phospholipids have a polar group at one end and a long nonpolar tail. In water, the nonpolar ends of phospholipids aggregate, forming two layers of molecules with the nonpolar tails pointed inside—a lipid bilayer. Lipid bilayers are the basic framework of biological membranes.

Membranes also contain a quite different kind of lipid called a **steroid,** composed of four carbon rings. Most animal cell membranes contain the steroid cholesterol. Excess saturated fat intake can cause plugs of cholesterol to form in the blood vessels, which may lead to blockage, high blood pressure, stroke, or heart attack. Experts advise limiting intake of saturated fats to prevent this condition. Male and female sex hormones are also steroids. Other important biological lipids include rubber, waxes, and pigments, such as the chlorophyll that makes plants green and the retinal that your eyes use to detect light.

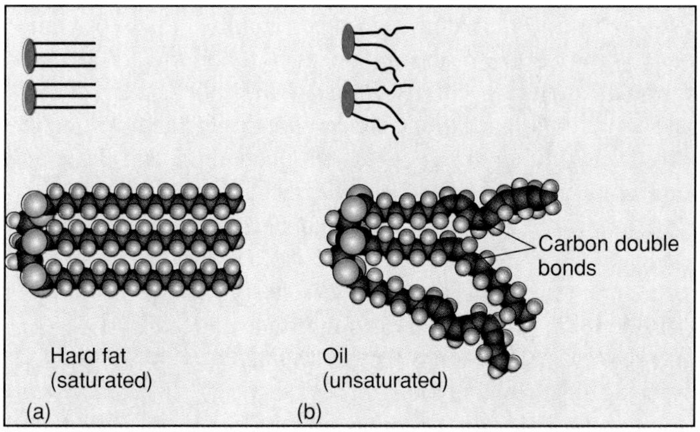

Figure 2.22 Saturated and unsaturated fats.
(*a*) Most animal fats are "saturated" (every carbon atom carries two hydrogens, the maximum load). Because their fatty acid chains can fit closely together, these triglycerides form immobile arrays called hard fats. (*b*) Most plant fats are not saturated, which prevents close association between triglycerides and produces oils.

Proteins

Complex macromolecules called **proteins** are the third major group of macromolecules that make up the bodies of organisms. Perhaps the most important proteins are **enzymes,** which have the key role in cells of lowering the energy required to initiate particular chemical reactions. Other proteins play structural roles. Cartilage, bones, and tendons all contain a **structural protein** called collagen. Keratin, another structural protein, forms the horns of a rhinoceros and the feathers of a bird. Still other proteins act as chemical messengers within the brain and throughout the body.

Despite their diverse functions, all proteins have the same basic structure: a long polymer chain made of subunits called amino acids. **Amino acids** are small molecules with a simple basic structure: a central carbon atom to which an amino group ($-NH_2$), a carboxyl group (–COOH), a hydrogen atom (H), and a functional group, designated *R*, are bonded. There are 20 common kinds of amino acids.

Each amino acid has the same chemical backbone but can be differentiated from other amino acids by its functional group. Six of the amino acid functional groups are nonpolar, differing chiefly in size—the most bulky contain ring structures, and amino acids containing them are called aromatic. Another six are polar but uncharged, and these differ from one another in the strength of their polarity. Five more are polar and are capable of ionizing to a charged form. The remaining three possess special chemical groups that are important in forming links between protein chains or in forming kinks in their shapes.

An individual protein is made by linking specific amino acids together in a particular order, just as a sentence is made by linking a specific sequence of letters of the alphabet together in a particular order. The covalent bond linking two amino acids together is called a **peptide bond** (figure 2.23), and long chains of amino acids linked together by peptide bonds are called **polypeptides.** The hemoglobin proteins in your red blood cells are each composed of four polypeptide chains. The hemoglobin functions as a carrier of oxygen from your lungs to the cells of your body and as a carrier of carbon dioxide wastes from your cells back to your lungs to be expelled from your body. In figure 2.24 the influence of particular amino acids on the functioning of the hemoglobin protein is explored.

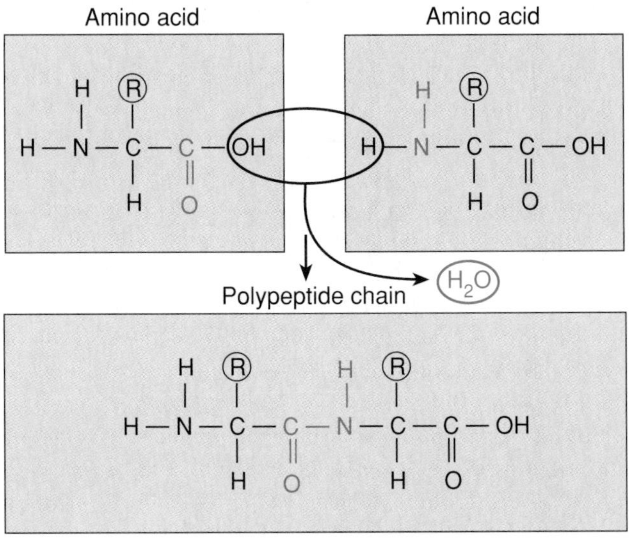

Figure 2.23 The formation of a peptide bond.
Every amino acid has the same basic structure, with an amino group at one end and a carboxyl group at the other. The only variable is the functional, or "R," group. Amino acids are linked together by dehydration synthesis to form peptide bonds. Chains of amino acids linked in this way are called polypeptides and are the basic structural components of proteins.

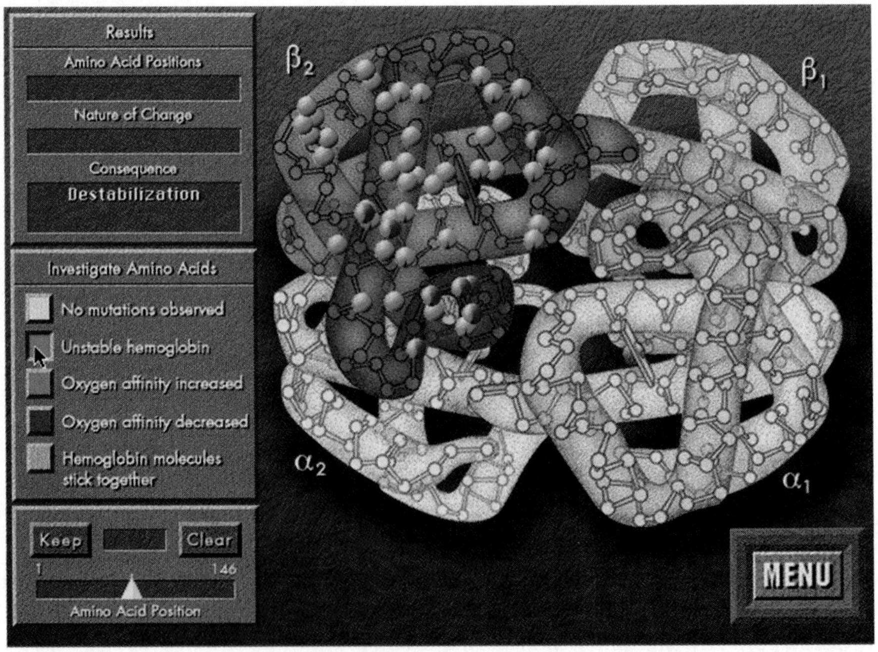

Figure 2.24 Changing a hemoglobin molecule.
The four polypeptide chains in this hemoglobin molecule are labeled alpha-1 (α_1), alpha-2 (α_2), beta-1 (β_1), and beta-2 (β_2). This illustration is a "screen capture" from an interactive CD-ROM exercise that enables you to explore the way in which the hemoglobin molecule works by changing it. By examining the functional consequences of changing which amino acids occur at different positions on the protein, you are able to build a detailed picture of the functional importance of different portions of the hemoglobin molecule. (*Explorations in Cell Biology & Genetics*, Module 1, "Hemoglobin")

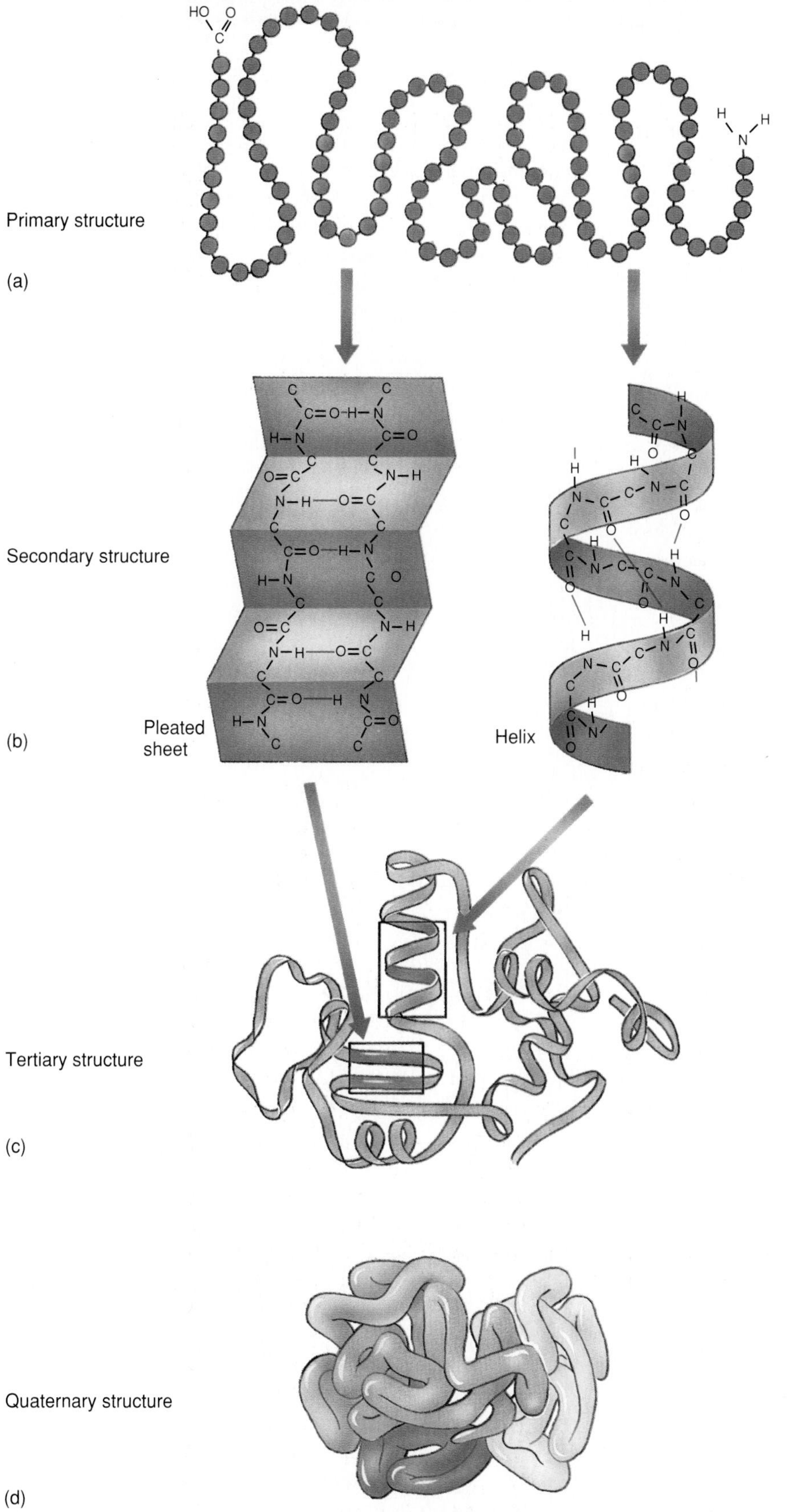

Figure 2.25 How primary structure determines the shape of a protein. The protein illustrated is lysozyme, an enzyme. The *primary structure* of a protein is its sequence of amino acids. Twisting or pleating of the chain of amino acids, called *secondary structure*, is due to the formation of localized hydrogen bonds (red) within the chain. More complex folding of the chain is referred to as *tertiary structure*. Two or more protein chains associated together form a *quaternary structure*.

Protein Structure

The sequence of amino acids of a polypeptide chain is termed the polypeptide's *primary structure* (figure 2.25). Because some of the amino acids are nonpolar and others are not, a protein chain folds up in solution as the nonpolar regions are forced together. This initial folding is called the *secondary structure* of a protein. The final three-dimensional shape, or *tertiary structure,* of the protein, usually folded and twisted into a globular molecule, is determined by exactly where in a protein chain the nonpolar amino acids occur. When a protein is composed of more than one polypeptide chain, the spatial arrangement of the several component chains is called the *quaternary structure* of the protein.

The shape of the protein is largely the result of the interaction of the amino acid functional groups with water, which tends to shove nonpolar portions of the polypeptide into the protein's interior. If the polar nature of the protein's environment changes, the protein may unfold in a process called **denaturation.** When the polar nature of the solvent is reestablished, smaller proteins may spontaneously refold, but larger proteins can rarely refold because of the complex nature of their final shape.

Many structural proteins form long cables that have architectural roles in cells, providing strength and determining shape—the collagen that makes the strings of a tennis racket, the keratin of a bird feather, the silk of a spider's web, and the hair on your head all are structural proteins. The proteins called enzymes carry out the day-to-day chemistry of the cell. These proteins have three-dimensional shapes with grooves or depressions that precisely fit a particular sugar or other chemical; once in the groove, the chemical is encouraged to undergo a reaction—often, one of its chemical bonds is stressed as the chemical is bent by the enzyme, like a foot in a flexing shoe. This process of enhancing chemical reactions is called **catalysis,** and proteins are the catalytic agents of cells, determining what chemical processes take place and where and when.

Nucleic Acids

Very long polymers called **nucleic acids** serve as the information storage devices of cells, just as disks or hard drives store the information that computers use. Nucleic acids are long polymers of repeating subunits called **nucleotides.** Each nucleotide is composed of three parts: a five-carbon sugar, a phosphate group (PO_4), and an organic nitrogen-containing (nitrogenous) base (figure 2.26). In the formation of a nucleic acid, the individual sugars are linked together in a line by the phosphate groups:—[SUGAR]—phosphate—[SUGAR]—phosphate—, in very long **polynucleotide chains.**

How does the long, chainlike structure of a nucleic acid permit it to store the information necessary to specify what a human being is like? If DNA were simply a monotonous repeating polymer, it could not encode the message of life. Imagine trying to write a story using only the letter *E* and no spaces or punctuation. All you could ever say is "EEEEEEE. . . ." You need more than one letter to write—the English alphabet uses 26 letters. Nucleic acids can encode information because they contain more than one kind of nucleotide. There are four different kinds of nucleotides: two big ones called adenine and guanine, and two small ones called cytosine and thymine. Nucleic acids encode information by varying the identity of the nucleotide at each position in the polymer.

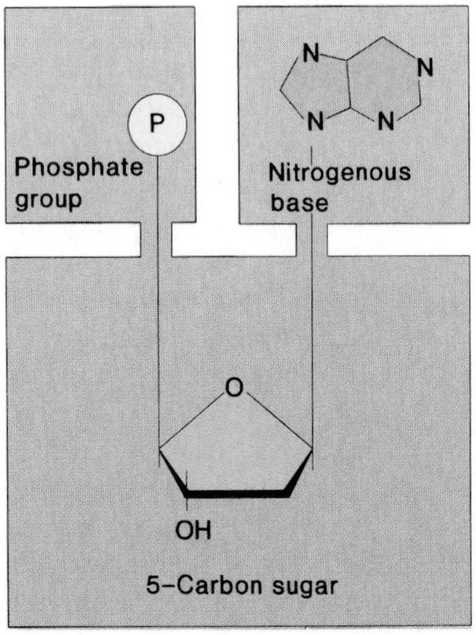

Figure 2.26 The structure of a nucleotide.
Nucleotides are composed of three parts: a five-carbon sugar, a phosphate group, and an organic nitrogenous base.

DNA and RNA

Nucleic acids come in two varieties, **deoxyribonucleic acid (DNA)** and **ribonucleic acid (RNA),** both polymers of nucleotides (figure 2.27). RNA is a long, single strand of nucleotides and is used by cells in making proteins using genetic instructions encoded within DNA. DNA consists of *two* nucleotide strands wound around each other in a **double helix,** like strands of a pearl necklace twisted together (figure 2.28).

The Double Helix

Why is DNA a *double* helix? When scientists looked carefully at the structure of the DNA double helix, they found that the nitrogenous bases of the two chains projected inward from the sugar–phosphate backbone, the bases of each chain pointed toward the other. The bases of the two chains are linked in the middle of the molecule by hydrogen bonds, like two lines of people holding hands. The key to understanding why DNA is a double helix is revealed by looking at the bases: *only two base pairs are possible.* Two big bases cannot pair together—the combination is simply too bulky to fit; similarly, two little ones cannot, as they pinch the helix inward too much. To form a double helix, it is necessary to pair a big base with a little one. *In every DNA double helix, adenine (A) pairs with thymine (T) and guanine (G) pairs with cytosine (C)* (figure 2.29). In case

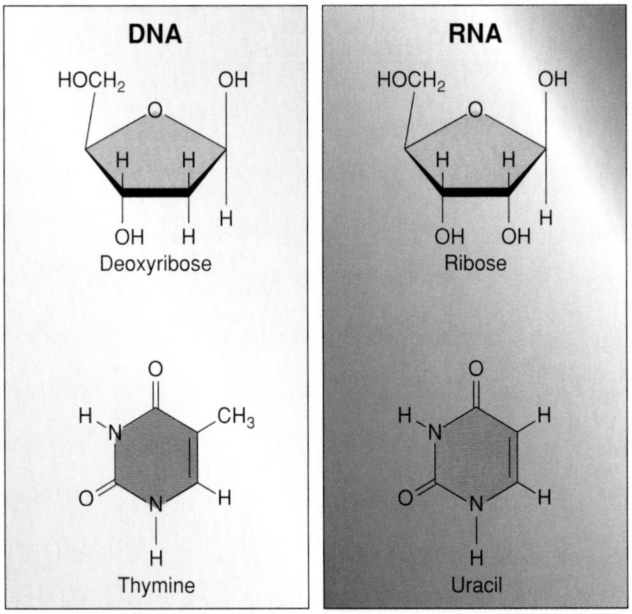

Figure 2.27 How DNA differs from RNA.
DNA is similar in structure to RNA but with two major chemical differences: (1) Both contain ribose (five-carbon) sugars, but in DNA one of the sugar's hydroxyl (–OH) groups is replaced by a hydrogen. (That is why DNA is called *deoxyribo*.) (2) One of the four organic bases of DNA, thymine, is changed slightly in RNA by the removal of a $-CH_3$ group, and is called uracil.

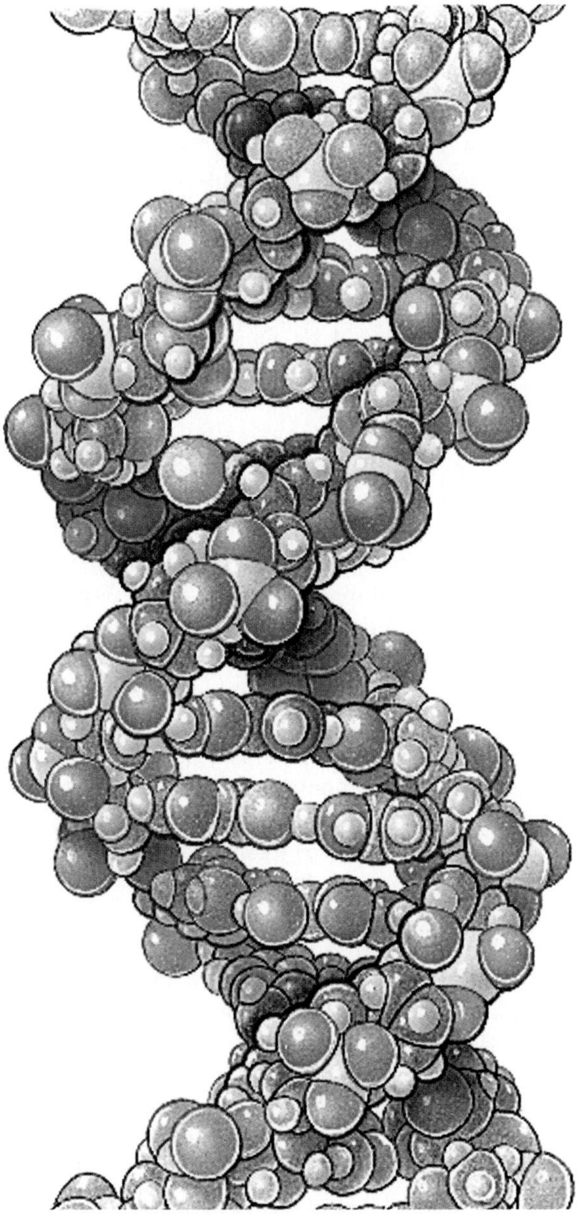

Figure 2.28 A molecular model of the DNA double helix.
The DNA molecule is composed of two nucleotide chains twisted together to form a double helix.

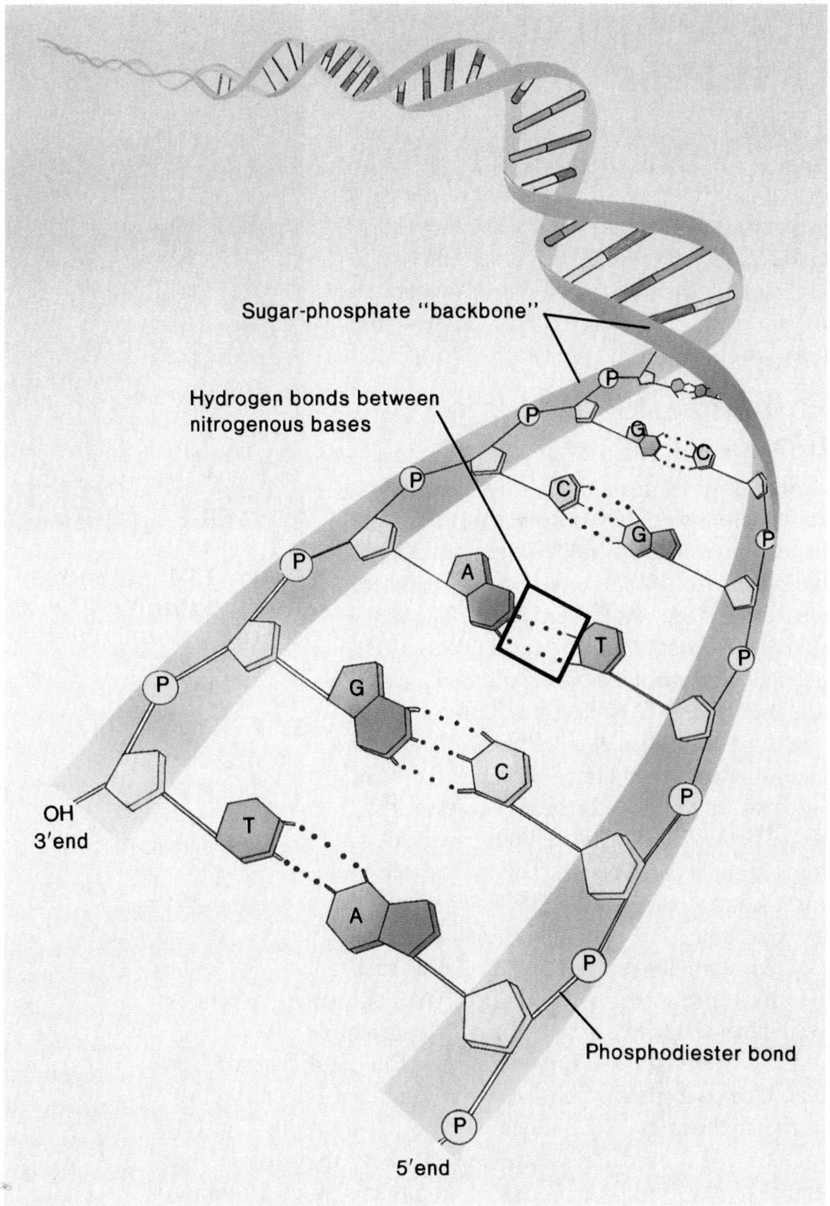

Figure 2.29 The structure of DNA.
The two chains of the double helix are joined by hydrogen bonds between A–T and G–C base pairs.

you're wondering, the reason A doesn't pair with C and G doesn't pair with T is that these base pairs cannot form proper hydrogen bonds—the electron-sharing atoms are not pointed at each other.

The simple A–T, G–C pairs within the DNA double helix allow the cell to copy the information in a very simple way. It just unzips the helix and adds the matching bases to each strand! That is the great advantage of a double helix—it actually contains two copies of the information, one the mirror image of the other. If the sequence of one chain is ATTGCAT, the sequence of its partner in the double helix *must* be TAACGTA. The fidelity with which hereditary information is passed from one generation to the next is a direct result of this simple double-entry bookkeeping, which makes accurate copying of the genetic message possible.

2.4 Origin of the First Cells

All living organisms are constructed of the same four kinds of macromolecules, the bricks and mortar of cells. Where the first macromolecules came from and how they came to be assembled together into cells are among the least understood questions in biology—questions that address the very origin of life itself.

Forming Life's Building Blocks

How can we learn about the origin of the first cells? One way is to try to reconstruct what the earth was like when life originated 3.5 billion years ago. We know from rocks that there was little or no oxygen in the earth's atmosphere then and more of the hydrogen-rich gases hydrogen sulfide (SH_2), ammonia (NH_3), and methane (CH_4). Electrons in these gases would have been frequently pushed to higher energy levels by photons crashing into them from the sun or by electrical energy in lightning (figure 2.30). Today, high-energy electrons are quickly soaked up by the oxygen in earth's atmosphere (air is 21% oxygen, all of it contributed by photosynthesis) because oxygen atoms have a great "thirst" for such electrons. But in the absence of oxygen, high-energy electrons would have been free to help form biological molecules.

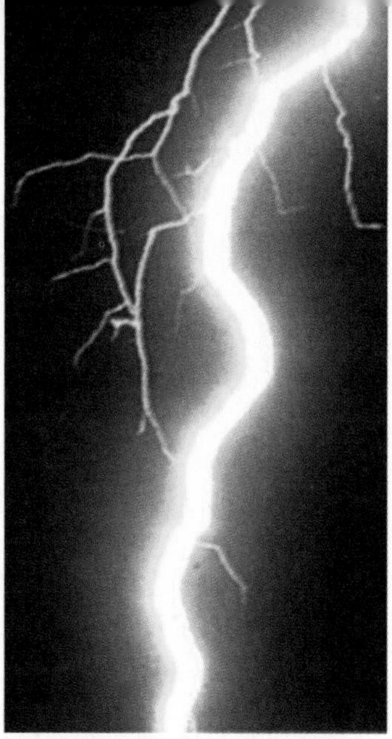

Figure 2.30 In ancient times, lightning provided energy to form molecules.
Before life evolved, the simple molecules in the earth's atmosphere combined to form more complex molecules. The energy that drove some of these chemical reactions is thought to have come from UV radiation, lightning, and other forms of geothermal energy.

When the scientists Stanley Miller and Harold Urey reconstructed the oxygen-free atmosphere of the early earth in their laboratory and subjected it to the lightning and UV radiation it would have experienced then, they found that many of the building blocks of organisms, such as amino acids and nucleotides, formed spontaneously. They concluded that life may have evolved in a "primordial soup" of biological molecules formed in the ancient earth's oceans.

Recent discoveries of 3.5-billion-year-old fossils have caused scientists to reevaluate the primordial soup hypothesis. This allows less than .5 billion years for life to evolve after the molten earth cooled enough to possess oceans. Also, if the earth's atmosphere had no oxygen 4 billion years ago, as Miller and Urey assumed (and most evidence supports this assumption), then there would have been no protective layer of ozone to shield the earth's surface from the sun's damaging ultraviolet radiation. Without an ozone layer, scientists think the ultraviolet radiation would have destroyed any ammonia and methane present in the atmosphere. When these gases are missing, the **Miller-Urey experiment** does not produce key biological molecules such as amino acids. If the necessary ammonia and methane were not in the atmosphere, where were they?

In the last decade, support has grown among scientists for what has been called the **bubble model.** This model suggests that the problems with the primordial soup hypothesis disappear if the model is "stirred up" a bit. The bubble model proposes that the key chemical processes generating the building blocks of life took place not in a primordial soup but rather within bubbles on the ocean's surface. Bubbles produced by wind, wave action, the impact of raindrops, and the eruption of volcanoes cover about 5% of the ocean's surface at any given time. Because water molecules are polar, water bubbles tend to attract other polar molecules, in effect concentrating them within the bubbles. This solves two key problems with the primordial soup hypothesis. First, chemical reactions would proceed much faster in bubbles, where polar reactants would be concentrated, and so life could have originated in a much shorter period of time. Second, inside the bubbles, the methane and ammonia required to produce amino acids would have been protected from destruction by UV radiation.

The Protein Puzzle

How did the first proteins form? Recall from figure 2.23 that making a peptide bond involves producing a molecule of water as one of the products of the reaction. Because this chemical reaction is freely reversible, it should not occur spontaneously in water (an excess of water would push it in the opposite direction). Scientists now suspect that the first macromolecules to form were not proteins but RNA molecules. When "primed" with high-energy phosphate groups (available in many minerals), RNA nucleotides spontaneously form polynucleotide chains that might, folded up, have been capable of catalyzing the formation of the first proteins.

Not everyone accepts the hypothesis that life evolved spontaneously. Those who object say that proteins and RNA could never have assembled spontaneously, for the same reason that shaking a bunch of soft drink cans doesn't spontaneously cause them to form a neat stack—disorder, not order, tends to increase in the universe. Called the **second law of thermodynamics,** this general rule is a basic principle of chemistry and physics. Does the theory of spontaneous origin violate the second law of thermodynamics? Not at all. The second law of thermodynamics applies only to closed systems (ones in which no energy enters or leaves), while earth and its organisms are open systems, with radiant energy from the sun continually entering earth's living systems through photosynthesis.

To better understand why the second law of thermodynamics does not apply to processes involving organisms, consider the changes that occurred as your body developed. Like every human being, you began life as a single fertilized egg. You are now a highly organized creature, far more complex than you were when you started life. You have not violated the second law of thermodynamics because you are an open system. Any open system is capable of increasing in complexity as it absorbs energy from its surroundings. The energy sources present on the early earth were more than ample to enable life's first chemicals to have organized spontaneously.

The First Cells

We don't know how the first cells formed, but most scientists suspect they aggregated spontaneously. When complex carbon-containing macromolecules are present in water, they tend to gather together, much as the people from the same foreign country tend to aggregate within a large city. Sometimes the aggregations form a cluster big enough to see. Try vigorously shaking a bottle of oil-and-vinegar salad dressing—tiny microspheres form spontaneously, suspended in the vinegar. Similar microspheres might have represented the first step in the evolution of cellular organization. Such microspheres have many cell-like properties—their outer boundary resembles the skin of a cell in that it has two layers, and the microspheres can grow and divide. Over millions of years, those **microdrops** better able to incorporate molecules and energy would have tended to persist longer than others, and when a means occurred to transfer these improvements from parent microdrop to offspring, heredity—and life—began.

When we speak of it having taken millions of years for a cell to develop, it is hard to believe there would be enough time for an organism as complicated as a human to develop. But in the scheme of things, human beings are recent additions. If we look at the development of living organisms as a 24-hour clock of biological time (figure 2.31), with the formation of the earth 4.5 billion years ago being midnight, humans do not appear until the day is almost all over, only minutes before its end.

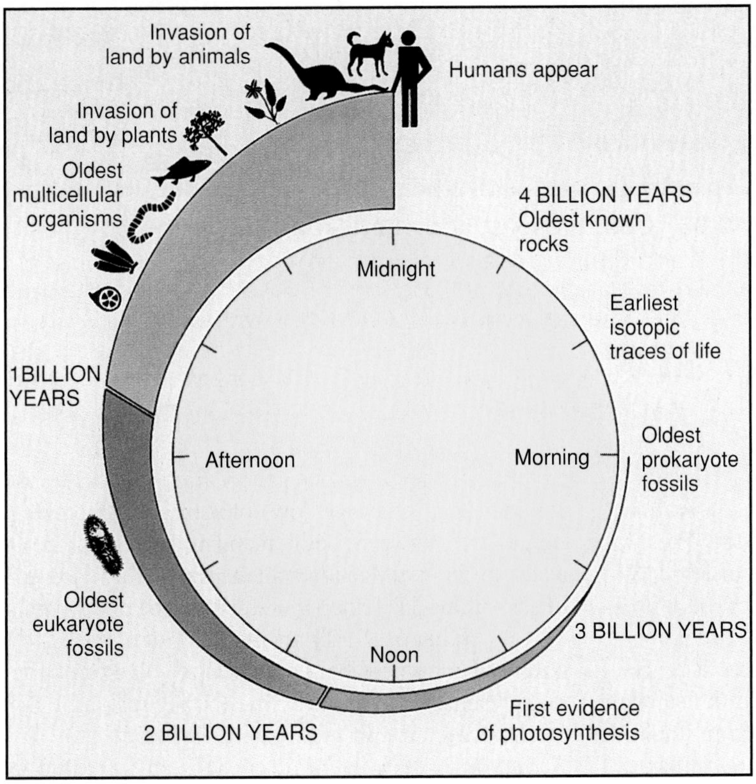

Figure 2.31 A clock of biological time.
A billion seconds ago, it was 1953, and most students using this text had not yet been born. A billion minutes ago, Jesus was alive and walking in Galilee. A billion hours ago, the first human had not been born. A billion days ago, no biped walked on earth. A billion months ago, the first dinosaurs had not yet been hatched. A billion years ago, no creature had ever walked on the surface of the earth.

As you can see, the scientific vision of life's origin is at best a hazy outline. While scientists cannot disprove the hypothesis that life originated naturally and spontaneously, little is known about what actually happened. Many different scenarios seem possible, and some have solid support from experiments. But because we know so little about how DNA, RNA, and hereditary mechanisms first developed, science is currently unable to resolve disputes concerning the origin of life. How life might have originated naturally and spontaneously remains a subject of intense interest, research, and discussion among scientists.

In August 1996, NASA scientists reported that organic molecules called "polycyclic hydrocarbons" have been detected within meteorites from Mars. They speculate that these molecules may have been formed by bacteria! If life did form independently on two separate planets in our own solar system, life might be found elsewhere in the universe as well.

Has Life Evolved Elsewhere?

In a heavily forested mountain valley outside the small town of Arecibo in Puerto Rico rests a huge white dish. One thousand feet across, it is the world's largest radio telescope, and it listens to the stars. In 1992 scientists began to aim it, and other smaller radio telescopes around the world, at 1,000 of the stars nearest earth to listen for signals from other worlds. The **Search for Extra Terrestrial Intelligence (SETI)** project was to be the most comprehensive search ever for signs that intelligent life is out there somewhere in the stars, trying to get in touch with us.

Our galaxy contains millions of stars, many of them resembling our own sun. It is difficult to believe intelligent life has not evolved on some of them (figure 2.32)—and the universe contains over a billion other galaxies like our Milky Way! In photographs of space, individual galaxies resemble grains of sand, every speck a galaxy. Each speck contains countless thousands of stars. On how many planets orbiting these stars do intelligent creatures look to the sky and speculate on our existence?

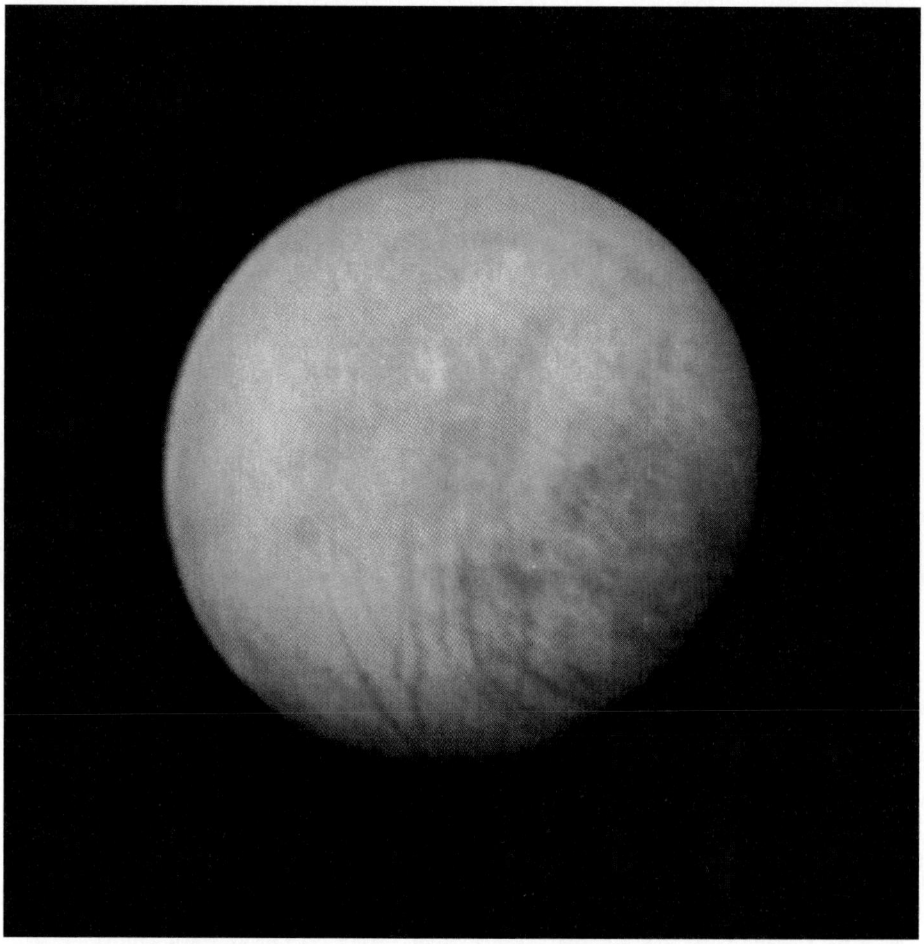

Figure 2.32 Does life exist elsewhere in the universe?
Currently the most likely candidate for life within the solar system is Europa, one of the many moons of the large planet Jupiter. Although most of Jupiter's moons resemble Earth's moon, pockmarked by meteors and devoid of life, Europa is covered with ice. Perhaps below its frozen surface the pressures of gravity create enough heat to maintain water in a liquid form. Life may have evolved in such an environment. The conditions on Europa now are far less hostile to life than the conditions that existed in the oceans of the primitive Earth.

We humans have made earlier attempts to send messages telling of our presence. Every radio and television program ever broadcast travels outward, although the signals are very weak. Elvis Presley's appearance on the *Ed Sullivan* show in 1957 arrived at the star Zeta Herculis six years ago, having traveled 31 light-years (1 light-year = 5,878,000,000,000 miles). In 1995, Zeta Herculis saw President Kennedy assassinated. In 1974, a group of Cornell scientists led by Carl Sagan beamed a signal (much of it in image form) toward the Great Cluster, a massive aggregation of stars 25,000 light-years away. We cannot expect a reply soon. If some intelligent life-form were to respond the same year they receive our message, we will hear from them 50,000 years from now (three times longer than there have been humans in North America!).

Rather than transmitting messages, the SETI project was designed to listen. Using new computer technology called multichannel spectrum analysis, it can simultaneously monitor millions of channels over a frequency range of 9 billion hertz (Hz), the entire range of clear radio frequencies that reach the earth. Static noise in our galaxy limits useful transmission by anybody to a "quiet window" of some 60 billion Hz, and all but frequencies between 1 billion and 10 billion Hz are obscured by water in the earth's atmosphere. Any future look at the rest of the 60-billion-Hz window will have to be done from space.

What is SETI trying to listen *for?* How will we know when we are being spoken to? The only approach that makes sense is to assume that the signals are designed to be easily detected and understood. What do you do when you want your dog's attention? You shout or clap your hands loudly—the sound you make is loud enough and different enough that it doesn't blend in with the rest of the background noise and so it gets the dog's attention. An alien signal ought to be something similar, a patterned pulse, perhaps. Or maybe, like Carl Sagan, they will send a picture of themselves.

CHAPTER 2

HIGHLIGHTS

2.1 Some Simple Chemistry

Key Terms

- atom 21
- electron 21
- oxidation 23
- reduction 23
- chemical bond 24

Key Concepts

- All matter is composed of atoms, made up of protons, neutrons, and electrons.
- Electrons determine the chemical behavior of atoms.
- Molecules are collections of atoms held together by covalent bonds.
- Covalent bonds form when two atoms share electrons.
- Energy can pass from one molecule to another by the transfer of electrons.

2.2 Water: Cradle of Life

Key Terms

- hydrogen bond 25
- polar molecule 25
- hydrophobic 28
- pH 28
- buffer 29

Key Concepts

- Water is a highly polar molecule, a characteristic responsible for many of its properties.
- Polar molecules form hydrogen bonds with water, making it an excellent solvent.
- Nonpolar molecules, called hydrophobic ("water-hating"), tend to aggregate together in water.
- A few molecules of water spontaneously ionize, producing hydrogen ions whose concentration is measured by pH.

2.3 Macromolecules

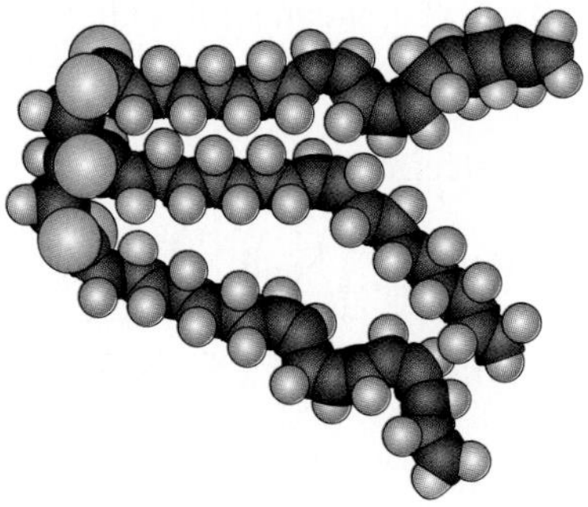

Key Terms

- macromolecule 30
- carbohydrate 32
- lipid 33
- protein 34
- nucleic acid 36
- DNA 36

Key Concepts

- Cells are built of very large macromolecules, principally carbohydrates, lipids, proteins, and nucleic acids.
- Most macromolecules are assembled by forming chains of subunits. Proteins, for example, are chains of amino acids.
- The function of a particular macromolecule in the cell is critically dependent upon its three-dimensional shape.

2.4 Origin of the First Cells

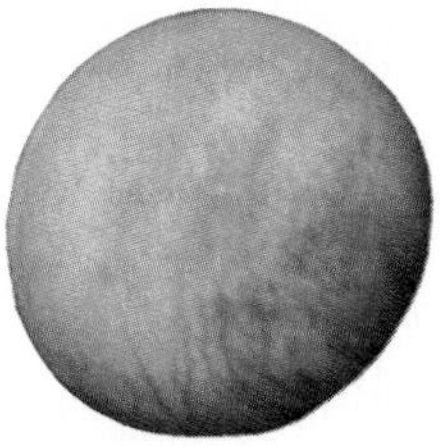

Key Terms

- Miller-Urey experiment 38
- microdrop 39
- SETI 40

Key Concepts

- When conditions resembling those of the early earth are recreated in the laboratory, biological molecules such as amino acids are formed.

2

CONCEPT REVIEW

1. Select the largest chemical structure.
   a. atom
   b. electron
   c. nucleus
   d. proton

2. Carbon-14 is an example of a(n)
   a. radioactive isotope.
   b. ion.
   c. neutron.
   d. oxidation reaction.

3. Select the correct association.
   a. oxidation—gain of an electron
   b. oxidation—loss of an electron
   c. reduction—gain of a neutron
   d. reduction—loss of a neutron

4. An atom has five electrons in its outer orbit. To complete its outer orbit, it needs _________ electrons.
   a. two
   b. three
   c. four
   d. six

5. Which statement about the hydrogen bond is not true?
   a. It occurs with polar molecules.
   b. It is a weak bond.
   c. It is absent in water.
   d. It is found in proteins.

6. Capillary action is due to
   a. adhesion.
   b. cohesion.

7. Hydrophobic means
   a. attracted by water.
   b. water-fearing.
   c. gain of electrons.
   d. loss of electrons.

8. Adding an acid to water _________ its pH.
   a. lowers
   b. raises

9. Select the smallest molecule.
   a. glycogen
   b. sucrose
   c. starch
   d. glucose

10. Amino acids are the subunits of
    a. carbohydrates.
    b. lipids.
    c. nucleic acids.
    d. proteins.

11. Each of the following was a molecule of the earth's early atmosphere except
    a. ammonia.
    b. hydrogen sulfide.
    c. methane.
    d. oxygen.

12. Because _________ is not supported by any scientific observations and does not infer its principles from observation, as does all science, it cannot be called science.

13. _________ is the gain of electrons.

14. Table salt is built from _________ bonds.

15. About _________ of the earth's surface is covered by water.

16. About _________ of the human body consists of water.

17. Polar molecules attract by _________.

18. A _________ is a substance that has a pH greater than 7.

19. There are about _________ kinds of amino acids.

20. Oxygen is supplied to the earth by the process of_______.

Answers to the Concept Review questions appear in Appendix B.

CHALLENGE YOURSELF

1. Carbon (atomic number 6) and silicon (atomic number 14) both have vacancies for four electrons in their outer energy levels. Ammonia (NH_3) is even more polar than water. Why do you suppose life evolved into organisms composed of carbon chains in water solution rather than ones composed of silicon in ammonia?

2. Champagne, a carbonic acid buffer, has a pH of about 2. How can we drink such a strong acid?

3. Carbon atoms can share four electron pairs when forming molecules. Why do you suppose that carbon does not form a bimolecular gas, as hydrogen (one pair of shared electrons), oxygen (two pairs of shared electrons), and nitrogen (three pairs of shared electrons) do?

4. Why do long-distance runners eat complex carbohydrates (that is, starches) in preparation for athletic events?

5. Why do you suppose humans circulate the monosaccharide glucose in their blood, rather than employing a disaccharide such as sucrose as a transport sugar, as do plants?

FOR FURTHER READING

Atkins, P. W. *Molecules.* New York: Scientific American Library, 1987. A delightful journey among the molecules most familiar to us.

Henbest, N. "Organic Molecules from Space Rained Down on Early Earth." *New Scientist,* January 25, 1992, 27. Did life have an extraterrestrial origin? Henbest atttempts to support the argument.

Holzman, D. "Protein Folding." *American Scientist,* May–June 1994, pages 267-74. An insightful discussion of protein-folding mechanisms and the impact they have on biology in general.

Hoyle, F. *The Ten Faces of the Universe.* San Francisco: Freeman, Cooper, and Company, 1980. A readable account of how the universe is thought to have formed, written for the nonscientist.

McKay, D., and others. "Search for Past Life on Mars: Possible Relic Biogenic Activity in Martian Meteorite ALH84001," *Science,* 273, August 1996: 924–30. A meteorite has yielded evidence—but not proof—of ancient life on Mars. The claim has excited scientists, although many remain skeptical.

Mertz, W. "The Essential Trace Elements." *Science*, 213 (1981): 1,332–38. An account of the roles of rare elements in human metabolism and the effects of their absences.

Politzer, B. "Sweet Talk: Sugar and Artificial Sweeteners," *American Health,* September 1991, 40–42. Although principally a discussion of sugars and artifical sweeteners as nutritional evils, this article presents a nice comparison of the major kinds of artificial sweeteners now available.

Tyson, J. L. " Scientists to Turn High Beam on Molecules." *Christian Science Monitor,* March 2, 1994, 14. A very readable discussion of the Advanced Photon Source project underway at the Argonne National Laboratory.

TECHNOLOGY LINKS

The Living World Home Page
http://www.wcbp.com/biology/tlw

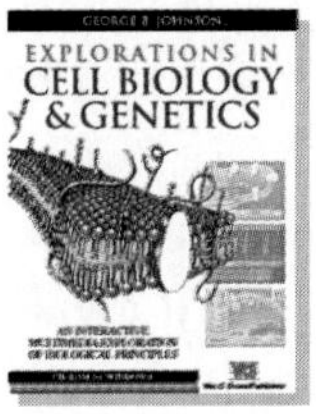

Explorations in Cell Biology & Genetics CD
#1 Hemoglobin

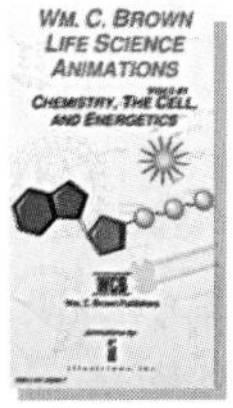

Life Science Animations Videotape 1
#1 Formation of an Ionic Bond

3

Cells

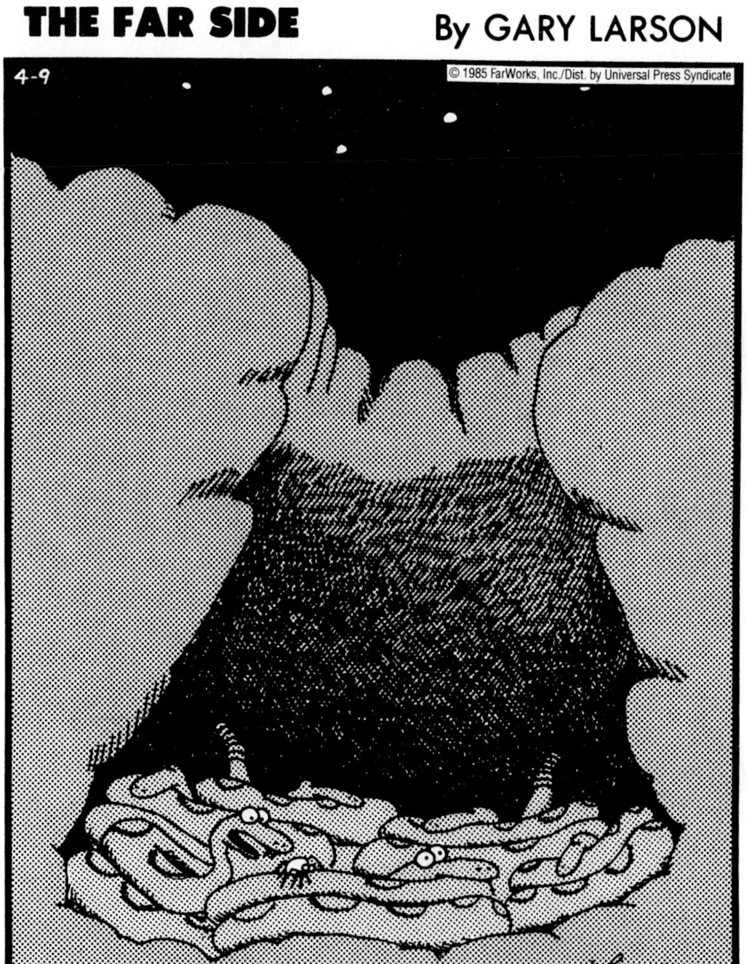

"Doreen! There's a spider on you! One of those big, hairy, brown ones with the long legs that can move like the wind itself!"

CHAPTER OUTLINE

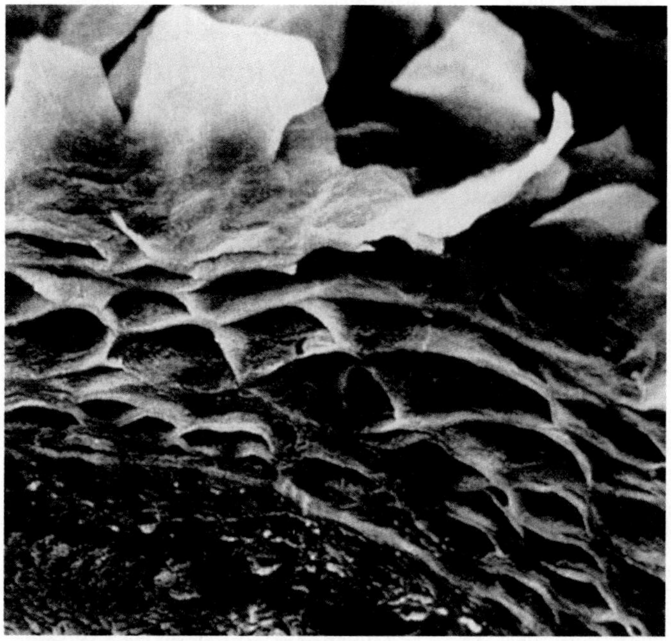

Figure 3.1 Human skin cells.
The skin is the largest organ of the human body but is composed of individual cells each too small to be seen with the naked eye. Each of the flattened structures you see here on the surface of the skin is an epithelial cell.

Every tissue of your body is made of cells, as are the bodies of all organisms. Some organisms are composed of a single cell, and some, such as the snakes and spider in the cartoon at left, are composed of many cells. The tough skin covering your body is actually a sheet of cells (figure 3.1), and so is the glistening layer covering your eyes. The hamburger you eat is composed of cells, whose contents will soon become part of your cells. Your eyelashes and fingernails, your morning orange juice, and the wood in your pencil all were produced by or consist of cells. We cannot imagine an organism that is not cellular in nature.

3.1 The World of Cells

Hold your hand up and look at it closely. What do you see? Skin. It looks solid and smooth, creased with lines and flexible to the touch. But if you were able to remove a bit and examine it under a microscope, it would look very different—a sheet of tiny, irregularly shaped bodies crammed together like shingles on a roof. What you would be seeing are epithelial cells like those shown in figure 3.1. In this chapter we look more closely at cells and learn something of their internal structure. In the following chapter we focus on cells in action—how they communicate with their environment, grow, and reproduce.

Sometimes important things seem so obvious that they are overlooked. In studying cells, for example, it is important that we do not overlook one of their most striking traits—their very small size. Most of the cells of your body are so small that you cannot see them with the naked eye. Your body contains about 100 trillion cells. Not all cells in your body are the same size or shape (figure 3.2).

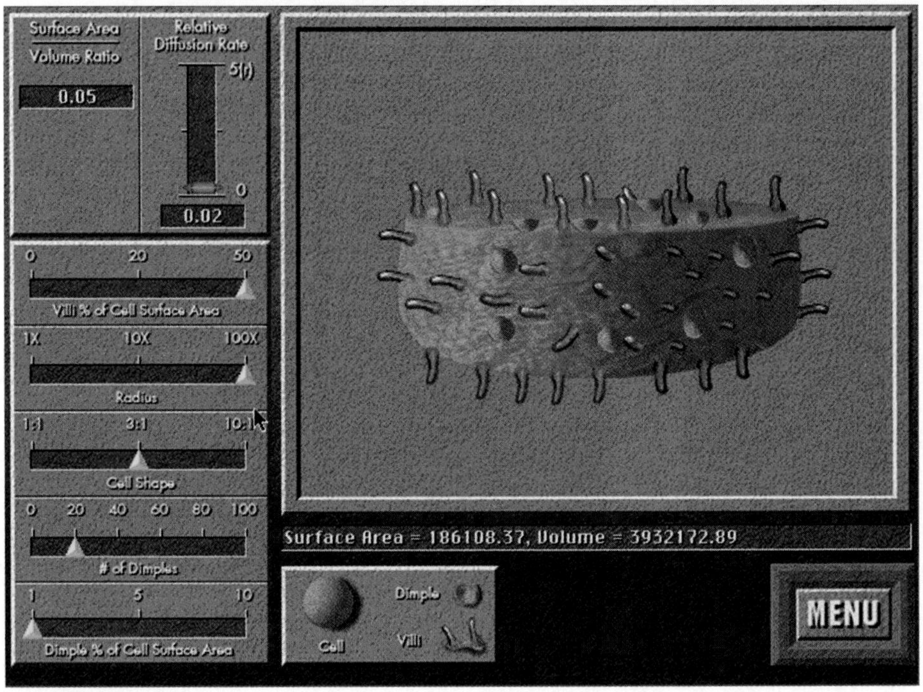

Figure 3.2 Exploring cell size and shape.

The cells of the human body vary widely in size, shape, number of surface dimples, and number of projections. These variables affect how quickly and easily water and nutrients can diffuse into and out of the cell. This illustration is a "screen capture" from an interactive CD-ROM exercise that allows the user to vary cell size and shape and then measure the relative time it takes a molecule to travel from the center of the cell to the nearest surface. You will soon discover a strong relationship between cell surface area and ease of diffusion into and out of the cell. (*Explorations in Cell Biology & Genetics*, Module 2, "Cell Size")

The Cell Theory

Because cells are so small, they were not observed until microscopes were invented in the mid-seventeenth century. Cells were first described by Robert Hooke in 1665. Using a microscope he had built, Hooke observed a honeycomb of tiny empty compartments in a thin slice of cork. He called the compartments in the cork "cellulae," using the Latin word for a small room. His term has come down to us as **cells.** The idea that all organisms are composed of cells is called the **cell theory.**

In its modern form, the cell theory includes four principles:

1. All organisms are composed of one or more cells, within which the processes of life occur.
2. Cells are the smallest living things. Nothing smaller than a cell is considered alive.
3. Life evolved only once, 3.5 billion years ago. All organisms living today represent a continuous line of descent from those early cells.
4. Cells arise only by division of a previously existing cell.

Most Cells Are Very Small

Cells are not all the same size. Individual cells of the marine alga *Acetabularia*, for example, are up to 5 centimeters long—as long as your little finger. In contrast, the cells of your body are typically from 5 to 20 micrometers in diameter, too small to see. If a typical cell in your body were the size of a shoe box, an *Acetabularia* cell to the same scale would be about 2 kilometers high! The cells of bacteria are even smaller than yours, only a few micrometers thick (figure 3.3).

Why Aren't Cells Larger?

Why are most cells so tiny? Most cells are small because larger cells do not function as efficiently. In the center of every cell is a command center that must issue orders to all parts of the cell, directing the synthesis of certain enzymes, the entry of ions from the exterior, and the assembly of new cell parts. These orders must pass by drifting out from the core to all parts of the cell, and it takes them a very long time to reach the periphery of a large cell. For this reason, an organism made up of relatively small cells is at an advantage over one composed of larger cells.

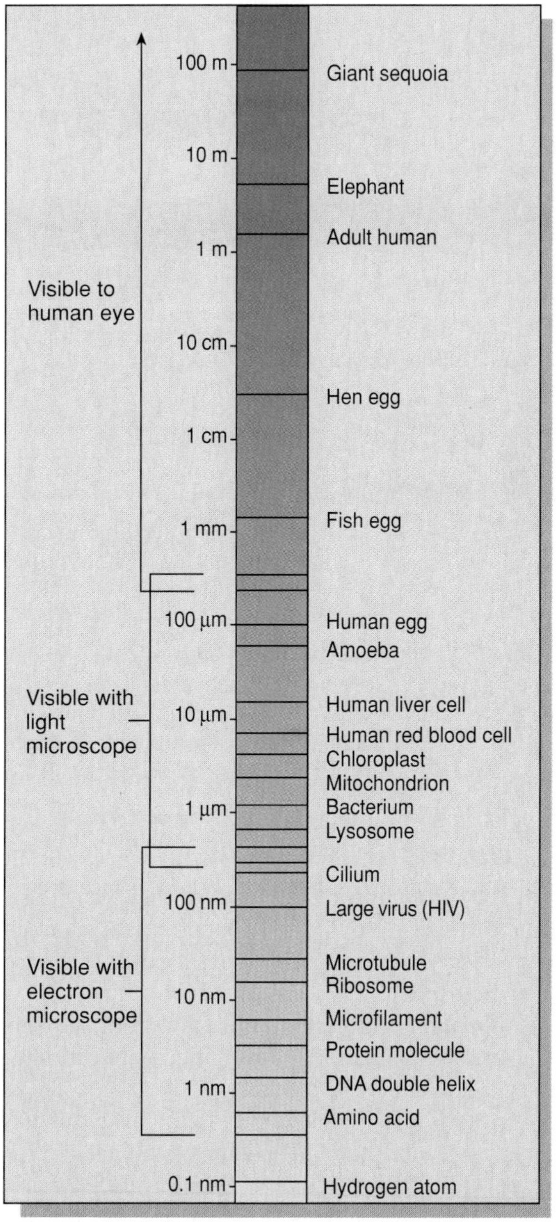

Figure 3.3 The size of cells.
The cells of bacteria are generally 1 to 2 micrometers (µm) across, while human cells are typically orders of magnitude larger. Most cells are microscopic in size, although vertebrate egg cells are typically large enough to be seen with the unaided eye.

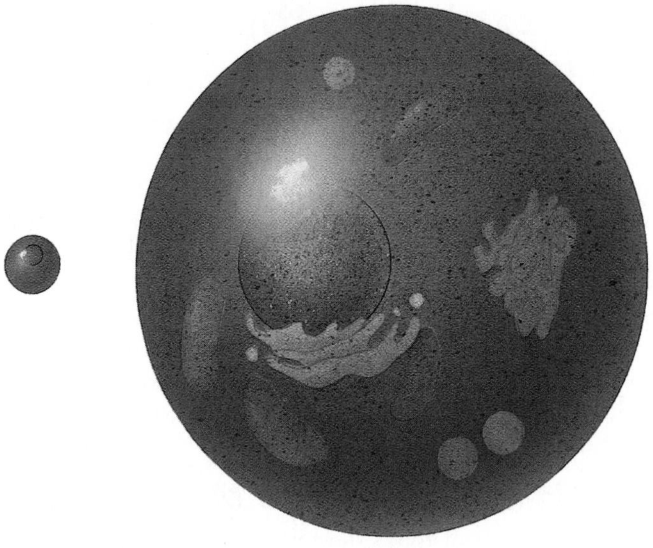

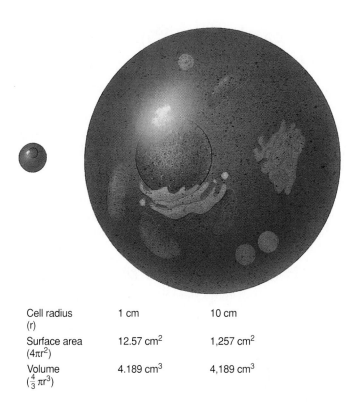

Cell radius (r)	1 cm	10 cm
Surface area ($4\pi r^2$)	12.57 cm^2	1,257 cm^2
Volume ($\frac{4}{3}\pi r^3$)	4.189 cm^3	4,189 cm^3

Figure 3.4 Surface-to-volume ratio.
As a cell gets larger, its volume increases at a faster rate than its surface area. If the cell radius increases by 10 times, the surface area increases by 100 times, but the volume increases by 1,000 times. A cell's surface area must be large enough to meet the needs of its volume.

Another reason cells aren't larger is the advantage of having a greater **surface-to-volume ratio.** As cell size increases, volume grows much more rapidly than surface area (figure 3.4). For a round cell, surface area increases as the square of diameter, whereas volume increases as the cube. Thus a cell with 10 times greater diameter would have 100 (10^2) times the surface area, but 1,000 (10^3) times the volume. A cell's surface provides the interior's only opportunity to interact with the environment, and large cells have far less surface for each unit of volume than do small ones.

An Overview of Cell Structure

All cells are surrounded by a delicate membrane that controls the permeability of the cell to water and dissolved substances. A semifluid matrix called **cytoplasm** fills the interior of the cell. It used to be thought that the cytoplasm was uniform, like jello, but we now know that it is highly organized. Your cells, for example, have an internal framework that both gives the cell its shape and positions materials within its interior. In the following sections, we first explore the surface of cells and then examine in detail their interior.

3.2 Exploring the Cell Surface

Encasing all living cells is a delicate sheet only a few molecules thick called the **plasma membrane.** It would take more than 10,000 of these sheets, which are about 7 nanometers thick, piled on top of one another to equal the thickness of this sheet of paper. But the sheets are not simple in structure, like a soap bubble's skin. Rather, they are made up of a diverse collection of proteins floating within a lipid framework like small boats bobbing on the surface of a pond. Regardless of the kind of cell they enclose, all plasma membranes have the same basic structure, a collection of proteins embedded in a sheet of lipid.

The Lipid Foundation of Membranes

The lipid layer that forms the foundation of a plasma membrane is composed of modified fat molecules called **phospholipids.** One end of a phospholipid molecule has a phosphate chemical group attached to it, making it extremely polar (and thus water-soluble), whereas the other end is composed of two long fatty acid chains that are strongly nonpolar (and thus water-insoluble) (figure 3.5*a*). Because of this structure, with the polar group pointing in one direction and the two nonpolar fatty acids extending in the other direction roughly parallel to each other, phospholipid molecules are usually diagrammed as a (polar) ball with two dangling (nonpolar) tails (figure 3.5*b*).

Imagine what happens when a collection of phospholipid molecules is placed in water. The long nonpolar tails of the phospholipid molecules are pushed away by the water molecules that surround them, shouldered aside as the water molecules seek partners that can form hydrogen bonds. After much shoving and jostling, every phospholipid molecule ends up with its polar head facing water and its nonpolar tail facing away from water. How can this happen? The phospholipid molecules form a *double* layer. Because there are two layers with the tails facing each other, no tails are ever in contact with water. This structure, which forms spontaneously, is called a **lipid bilayer** (figure 3.6).

Because the interior of a lipid bilayer is completely nonpolar, it repels any water-soluble molecules that attempt to pass through it, just as a layer of oil stops the passage of a drop of water (that's why ducks don't get wet). As we will see, it is only because of protein-lined passageways through the bilayer that chemicals are able to enter and leave cells.

Proteins Within the Membrane

The second major component of every biological membrane is a collection of **membrane proteins** that float within the lipid bilayer (figure 3.7). In plasma membranes, these proteins provide channels through which molecules and information pass. Membrane proteins are not fixed into position; instead, they move about freely. Some membranes are crowded with proteins, packed tightly side by side. In other membranes the proteins are more sparsely distributed.

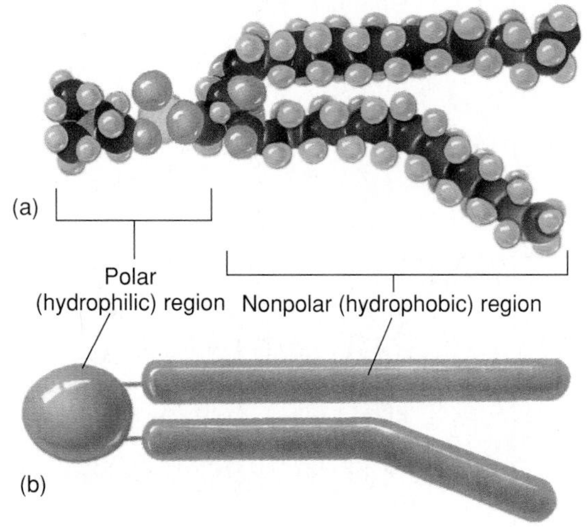

Figure 3.5 Phospholipid structure.
One end of a phospholipid molecule is polar, the other nonpolar. (*a*) The molecular structure is shown by colored spheres representing individual atoms (black for carbon, blue for hydrogen, red for oxygen, and yellow for phosphorus). (*b*) The phospholipid is often depicted diagrammatically as a ball with two tails.

Some membrane proteins project up from the surface of the plasma membrane like buoys, often with carbohydrate chains or lipids attached to their tips like flags. These **cell surface proteins** act as markers to identify particular types of cells or as beacons to bind specific hormones or proteins to the cell. The CD4 protein by which AIDS viruses dock onto human white blood cells is such a beacon.

Other proteins extend all the way across the bilayer, providing channels across which polar ions and molecules can pass into and out of the cell. How do these **transmembrane proteins** manage to span the membrane, rather than just floating on the surface in the way that a drop of water floats on oil? The part of the protein that actually traverses the lipid bilayer is specially constructed, a spiral helix of nonpolar amino acids (figure 3.8). Because water responds to nonpolar amino acids much as it does to nonpolar lipid chains, the helical spiral is held within the lipid interior of the bilayer, anchored there by the strong tendency of water to avoid contact with these nonpolar amino acids. Many transmembrane proteins lock this arrangement in place by positioning amino acids with electrical charges (which are very polar) at the two ends of the helical region.

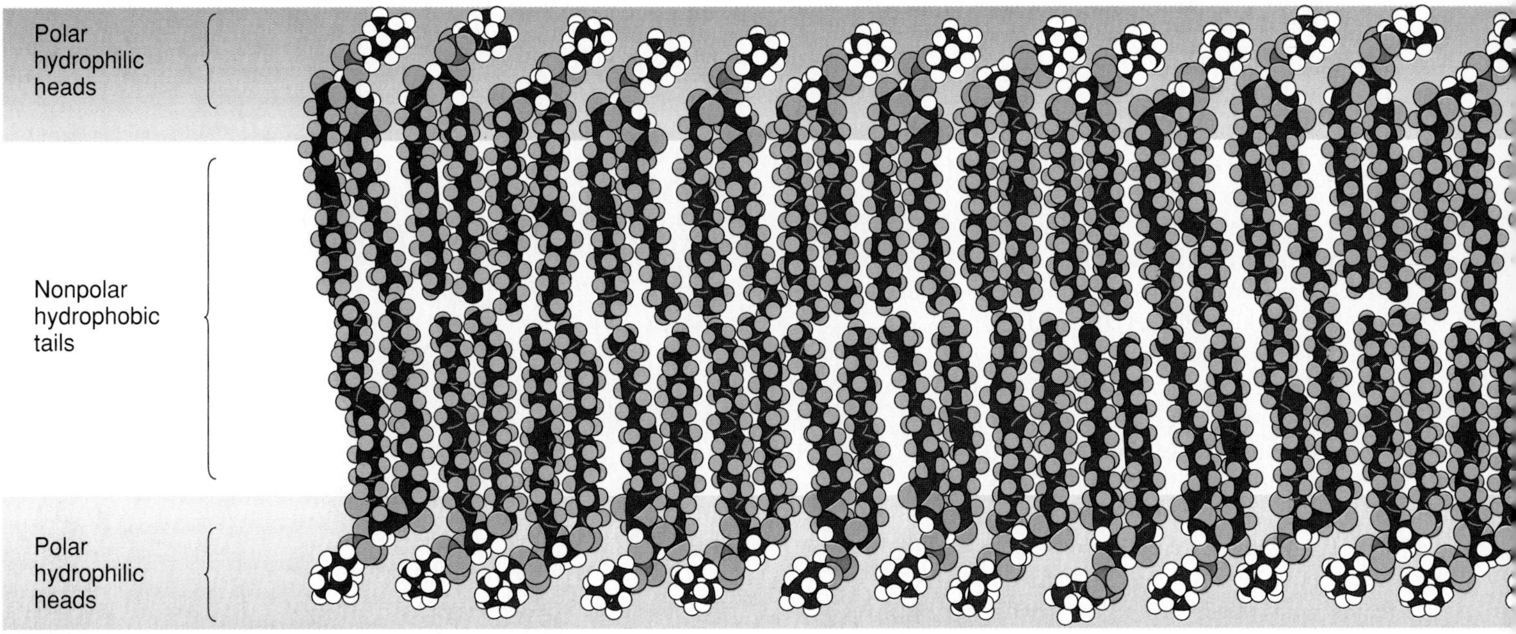

Figure 3.6 The lipid bilayer.
The basic structure of every plasma membrane is a double layer of lipid. This diagram illustrates how phospholipids aggregate to form a bilayer with a nonpolar interior.

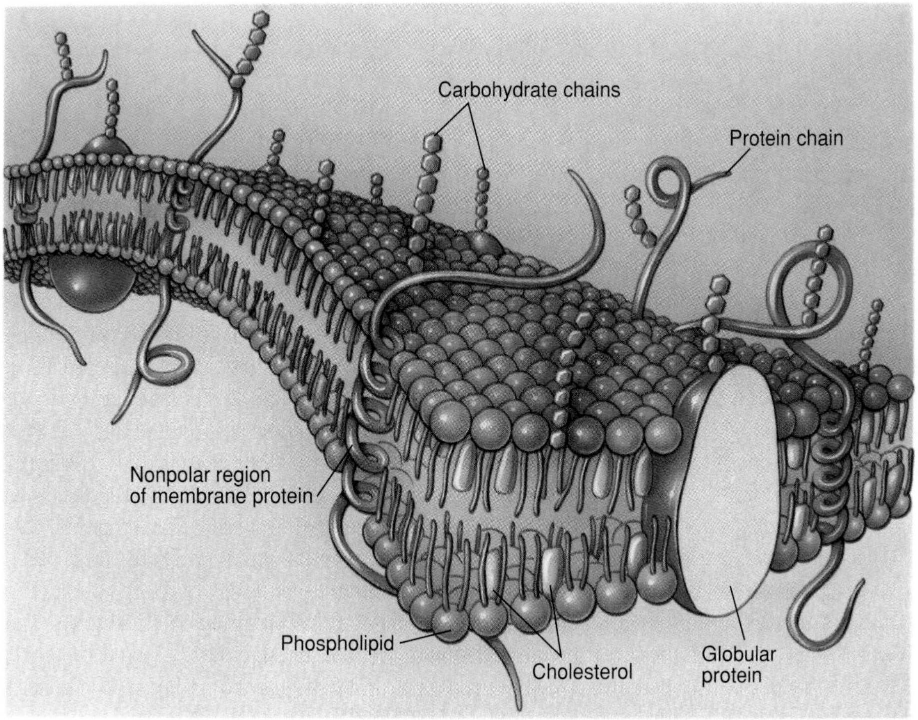

Figure 3.7 Proteins are embedded within the lipid bilayer.
A variety of proteins protrude through the lipid bilayer of animal cells. Membrane proteins function as channels, receptors, and cell surface markers. Carbohydrate chains are often bound to these proteins and to phospholipids in the membrane itself as well. These chains serve as distinctive identification tags, unique to particular types of cells.

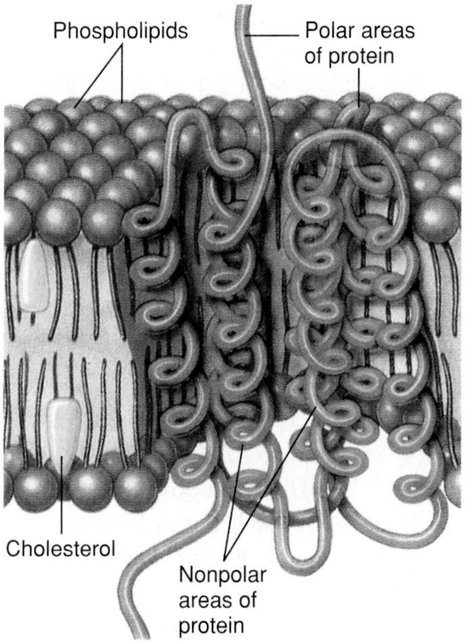

Figure 3.8 Nonpolar regions lock proteins into membranes.
A spiral helix of nonpolar amino acids (red) extends across the nonpolar lipid interior, while polar (purple) portions of the protein protrude out from the bilayer. The protein cannot move in or out because such a movement would drag polar segments of the protein into the nonpolar interior of the membrane.

Membrane Defects Can Cause Disease

The year 1993 marked an important milestone in the treatment of human disease. That year the first attempt was made to cure **cystic fibrosis (CF),** a deadly genetic disorder, by transferring healthy genes into sick individuals. Cystic fibrosis is a fatal disease in which the body cells of affected individuals secrete a thick mucus that clogs the airways of the lungs. These same secretions block the ducts of the pancreas and liver so that the few patients who do not die of lung disease die of liver failure. Cystic fibrosis is usually thought of as a children's disease because until recently few affected individuals lived long enough to become adults. Even today half die before their mid-twenties. There is no known cure.

Figure 3.9 Cystic fibrosis.
Cystic fibrosis is a genetic disorder that results from a defective chloride channel. It leads to a buildup of chloride ions within lung cells, causing water to move into the cells by osmosis. Removing water from the surrounding mucus causes it to thicken, leading to the symptoms of the disorder. This cystic fibrosis patient is breathing into a vitalograph, a device that measures lung function.

Cystic fibrosis results from a defect in a single gene that is passed down from parent to child. It is the most common fatal genetic disease of Caucasians. One in 20 individuals possesses at least one copy of the defective gene. Most carriers are not afflicted with the disease; only those children who inherit two copies of the defective gene, one from each parent, succumb to cystic fibrosis—about 1 in 2,500 U.S. infants.

Cystic fibrosis has proven difficult to study. Many organs are affected, and until recently it was impossible to identify the nature of the defective gene responsible for the disease. In 1985 the first clear clue was obtained. An investigator, Paul Quinton, seized on a commonly observed characteristic of cystic fibrosis patients, that their sweat is abnormally salty, and performed the following experiment. He isolated a sweat duct from a small piece of skin and placed it in a solution of salt (NaCl) that was three times as concentrated as the NaCl inside the duct. He then monitored the movement of ions. Diffusion tends to drive both the sodium (Na^+) and the chloride (Cl^-) ions into the duct because of the higher outer ion concentrations. In skin isolated from normal individuals, Na^+ and Cl^- ions both entered the duct, as expected. In skin isolated from cystic fibrosis individuals, however, only Na^+ ions entered the duct—no Cl^- ions entered. For the first time, the molecular nature of cystic fibrosis became clear. Cystic fibrosis was a defect in a plasma membrane protein that normally admits Cl^- ions into the body's cells. Called the chloride channel, this protein does not function properly in cystic fibrosis patients (figure 3.9).

The defective *cf* gene was isolated in 1987, and its position on a particular human chromosome (chromosome 7) was pinpointed in 1989. In 1990 a working *cf* gene was successfully transferred via adenovirus into human lung cells growing in tissue culture. The defective cells were "cured," becoming able to transport chloride ions across their plasma membranes. Then in 1991 a team of researchers successfully transferred a normal human *cf* gene into the lung cells of a living animal—a rat. The *cf* gene was first inserted into a cold virus that easily infects lung cells, and the virus was inhaled by the rat. Carried piggyback, the *cf* gene entered the rat lung cells and began producing the normal human protein within these cells! Tests of gene transfer into CF patients were begun in 1993, and while a great deal of work remains to be done (the initial experiments were not successful), the future for cystic fibrosis patients for the first time seems bright.

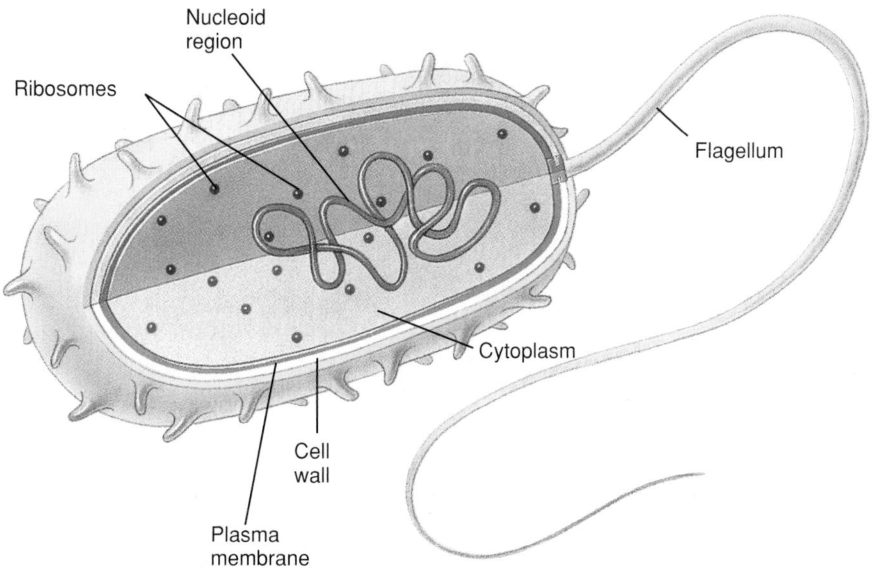

Figure 3.10 Organization of a bacterial cell.
Bacterial cells lack internal compartments. Not all bacterial cells have a flagellum like the one illustrated here, but all do have a nucleoid region, ribosomes, a plasma membrane, cytoplasm, and a cell wall.

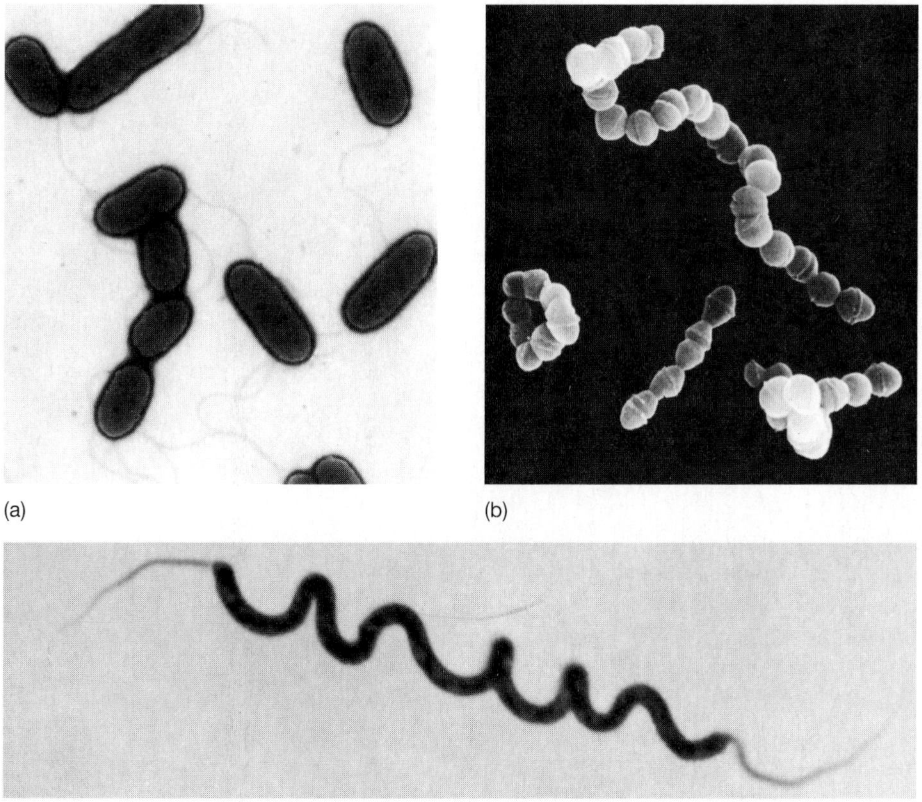

(a) (b) (c)

Figure 3.11 Bacterial cells have different shapes.
(*a*) A rod-shaped bacterium, *Pseudomonas*, is a type associated with many plant diseases. (*b*) *Streptococcus* is a more or less spherical bacterium in which the individuals adhere in chains. (*c*) *Spirillum* is a spiral bacterium; this large bacterium has a tuft of flagella at each end.

3.3 Bacteria Are the Simplest Cells

There are two major kinds of cells: simple prokaryotes and more complex eukaryotes. **Prokaryotes** have a relatively uniform cytoplasm that is not subdivided by interior membranes into separate compartments. They do not, for example, have a *nucleus* (a membrane-bounded compartment that holds the hereditary information). **Eukaryotes** are far more complex. They contain numerous special membrane-bounded compartments called *organelles,* including a nucleus. All bacteria are prokaryotes; all nonbacterial organisms are eukaryotes.

Bacteria are the simplest cellular organisms. Over 2,500 species are recognized, but doubtless many times that number actually exist and have not yet been described properly. Although these species are diverse in form, their organization is fundamentally similar (figure 3.10): small cells about 1 to 10 micrometers thick; enclosed, like all living cells, by a plasma membrane; encased within a rigid cell wall; and with no distinct interior compartments. The *cell wall* consists of a framework of carbohydrates cross-linked into a rigid structure. Bacterial cells assume many shapes (figure 3.11). Sometimes the cells of bacteria adhere in chains or masses, but fundamentally the individual cells are separate from one another.

If you were able to magnify your vision and peer into a bacterial cell, you would be struck by its simple organization. The entire interior of the cell, the cytoplasm, is one unit, with no internal support structure (the rigid wall supports the cell's shape) and no internal compartments bounded by membranes. Scattered throughout the cytoplasm of prokaryotic cells are small structures called *ribosomes,* the site where proteins are made. Ribosomes are not considered organelles because they lack a membrane boundary.

3.4 The Structure of Eukaryotic Cells

For the first 2 billion years of life on earth, all organisms were bacteria, cells with very simple interiors. Then, about 1.5 billion years ago, a new kind of cell appeared for the first time, much larger and with a complex interior organization. All cells alive today except bacteria are of this new kind. Unlike bacteria, these big cells have many membrane-bounded interior compartments and a variety of **organelles** (specialized structures within

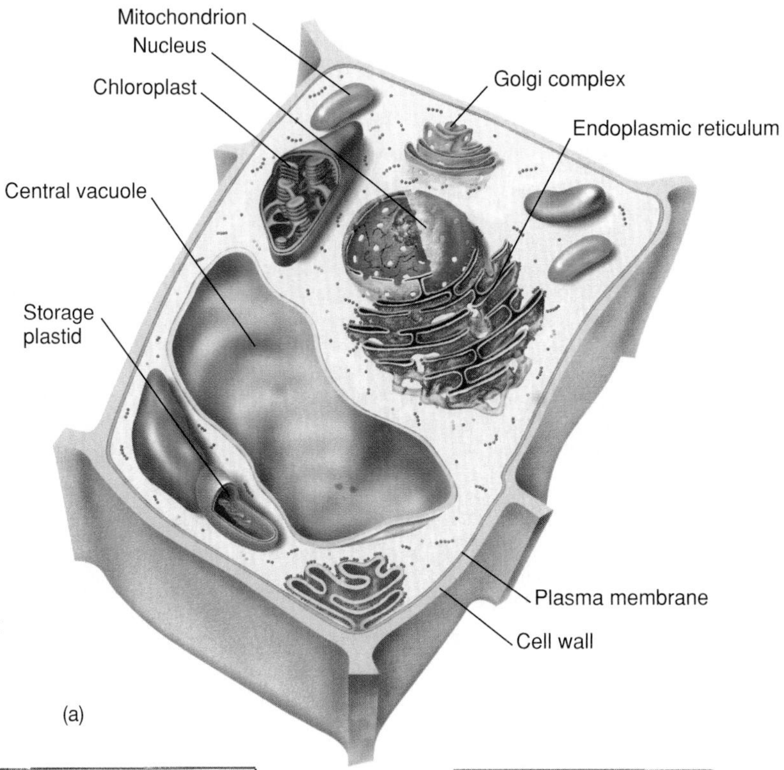

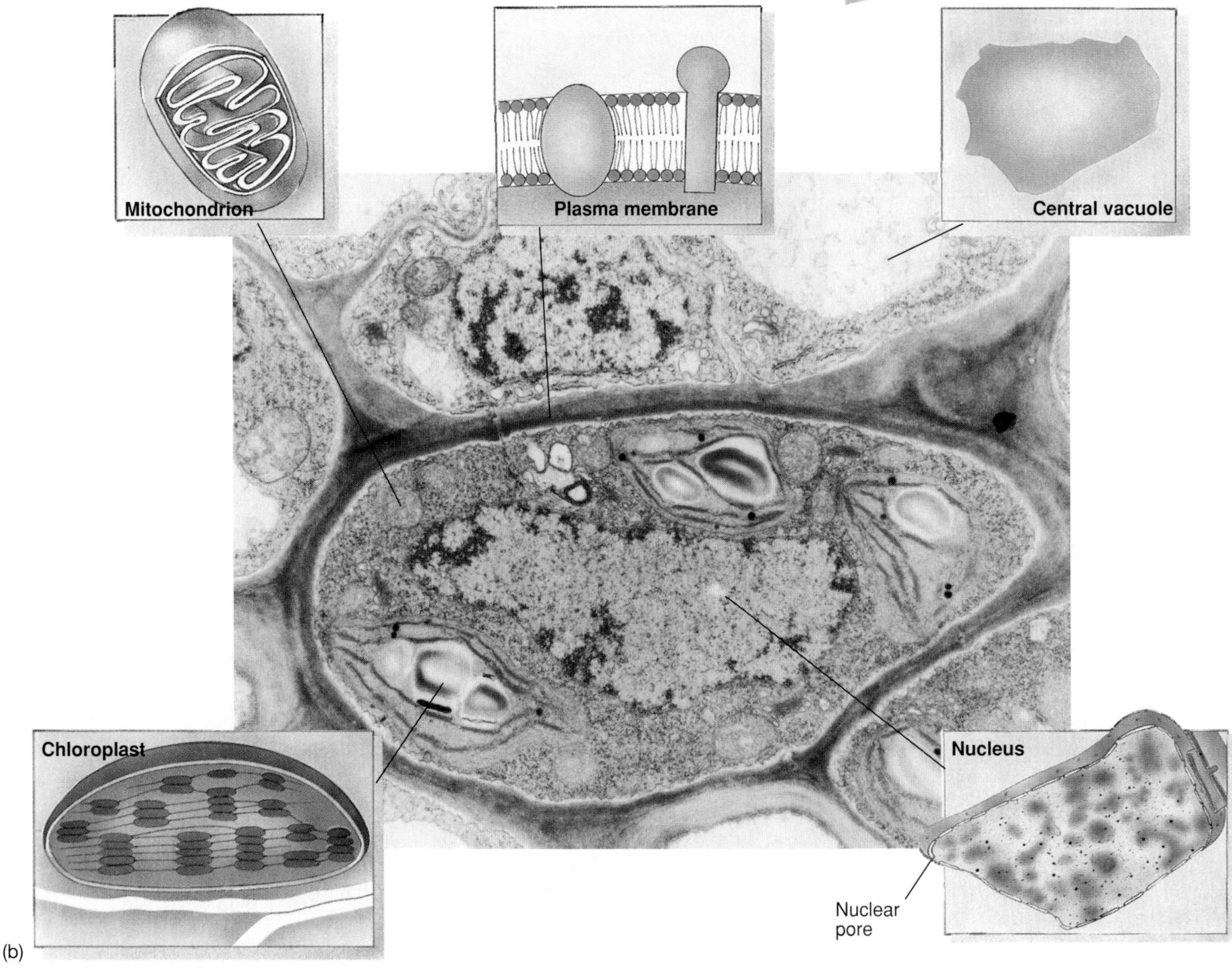

Figure 3.12 Structure of a plant cell.
(*a*) An idealized plant cell. The large central vacuole of a plant cell occupies most of the cell's space, providing an expanded surface area. (*b*) Micrograph of a plant cell with drawings detailing organelles.

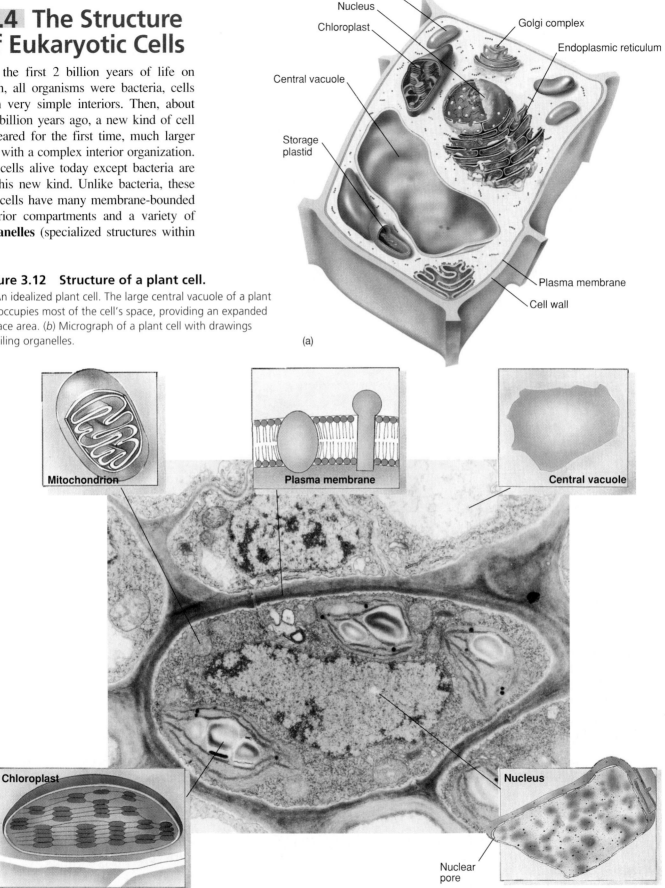

which particular cell processes occur). One of the organelles is very visible when these cells are examined with a microscope, filling the center of the cell like the core of a peach. Seeing it, the English botanist Robert Brown in 1831 called it the *nucleus,* from the Latin word for kernel. All cells with nuclei are called *eukaryotes* (from the Greek words *eu,* true, and *karyon,* nut) while bacteria are called *pro*karyotes ("before the nut"). Both plant cells (figure 3.12) and animal cells (figure 3.13) are eukaryotic, although the organelles found in each differ. We will now journey into the interior of a typical eukaryotic cell and explore its complex organization.

Figure 3.13 Structure of an animal cell.
(*a*) An idealized animal cell. (*b*) Micrograph of an animal cell with drawings detailing organelles.

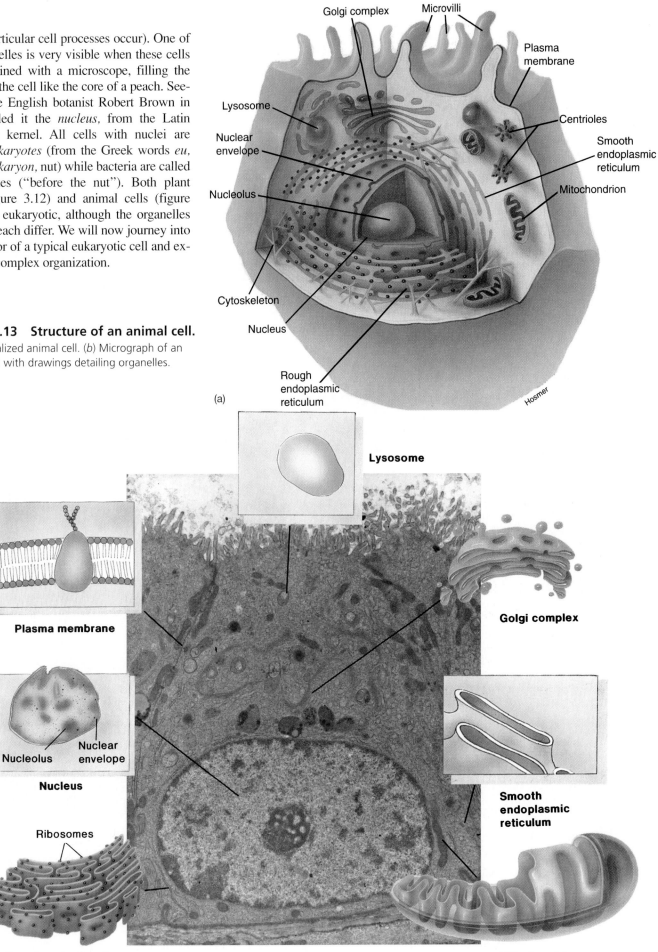

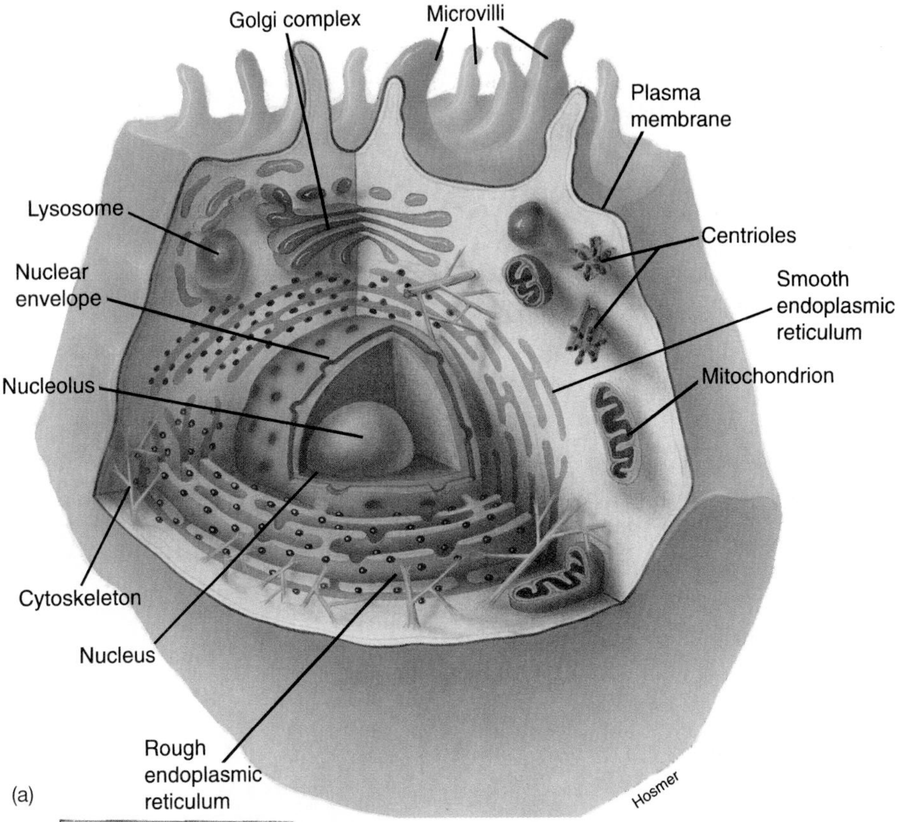

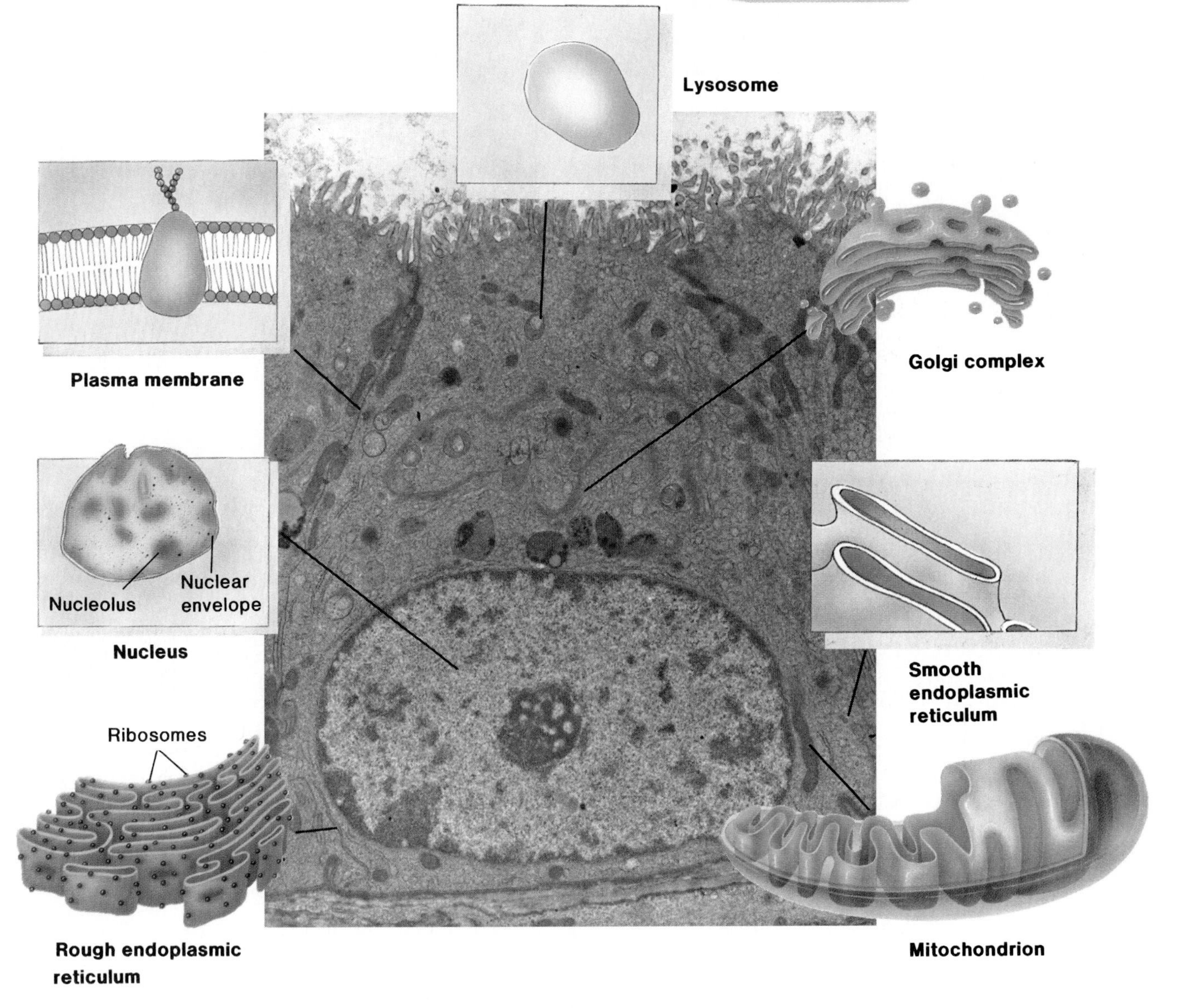

The Cytoskeleton: Interior Framework of the Cell

If you were to shrink down and enter into the interior of a eukaryotic cell, the first thing you would encounter is a dense network of protein fibers called the **cytoskeleton,** which supports the shape of the cell and anchors organelles such as the nucleus to fixed locations (figure 3.14). This network cannot be seen with a light microscope because the individual fibers are single chains of protein, much too fine for microscopes to resolve. To "see" the cytoskeleton, scientists attach fluorescent antibodies to the protein fibers and then photograph them under fluorescent light (figure 3.15).

The protein fibers of the cytoskeleton are a dynamic system, constantly being formed and disassembled. There are three different kinds of protein fibers (figure 3.16): long slender **microfilaments** made of the protein actin, hollow tubes called **microtubules** made of the protein tubulin, and thick ropes called **intermediate fibers.** The filaments, microtubules, and fibers are anchored to membrane proteins embedded within the plasma membrane.

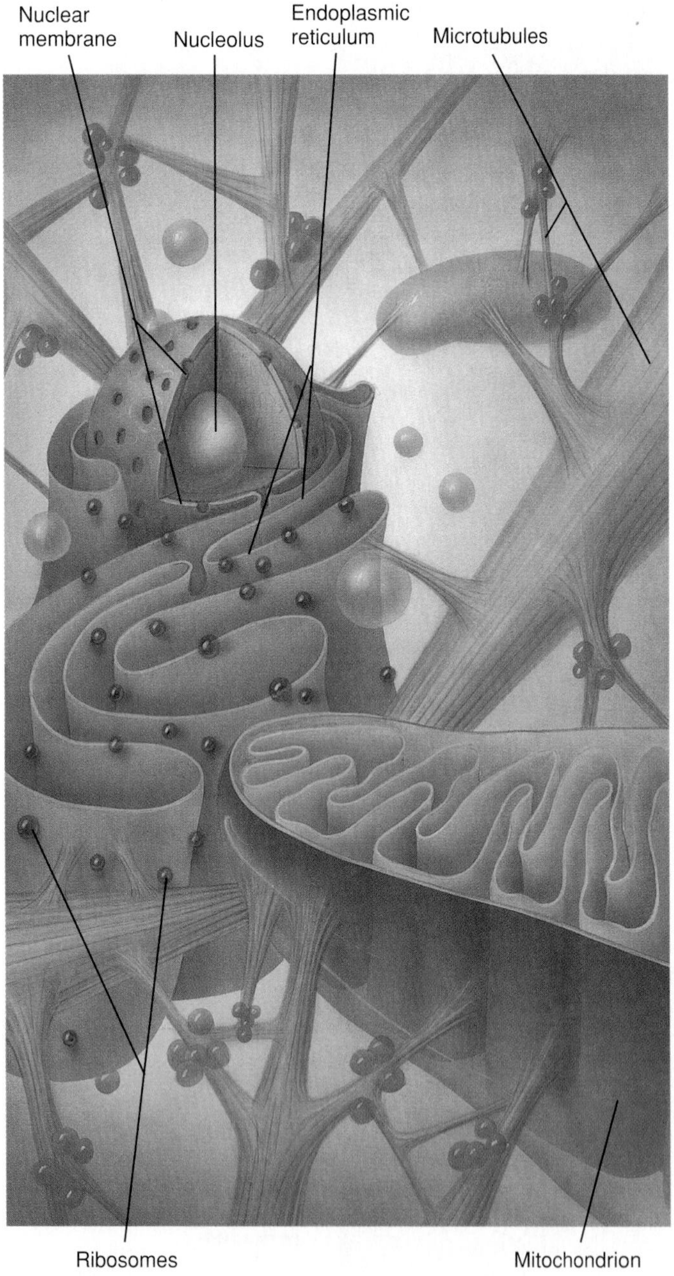

Figure 3.14 The cytoskeleton.
Organelles such as mitochondria are anchored to fixed locations in the cytoplasm by cables called microtubules, which are hollow tubes of protein.

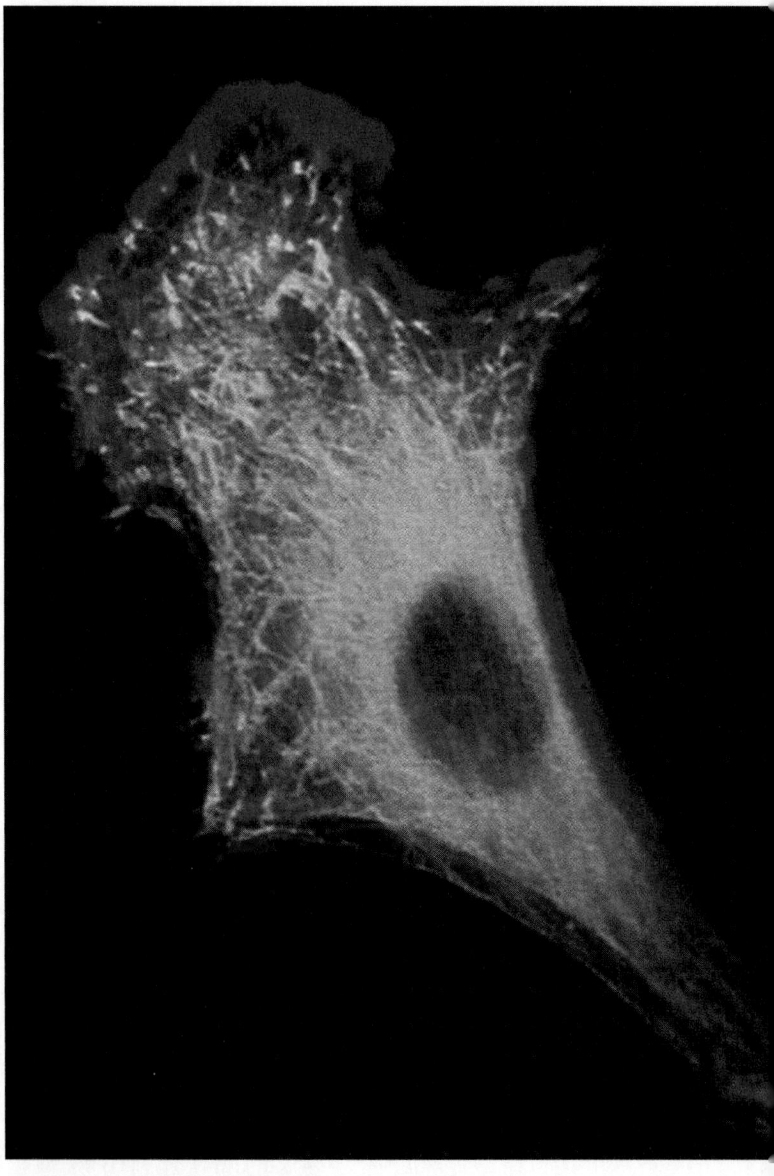

Figure 3.15 Visualizing the cytoskeleton.
A complex web of protein fibers runs through the cytoplasm of the eukaryotic cell. The different proteins are made visible by fluorescence microscopy. Actin appears blue here, microtubules appear green, and intermediate fibers appear red.

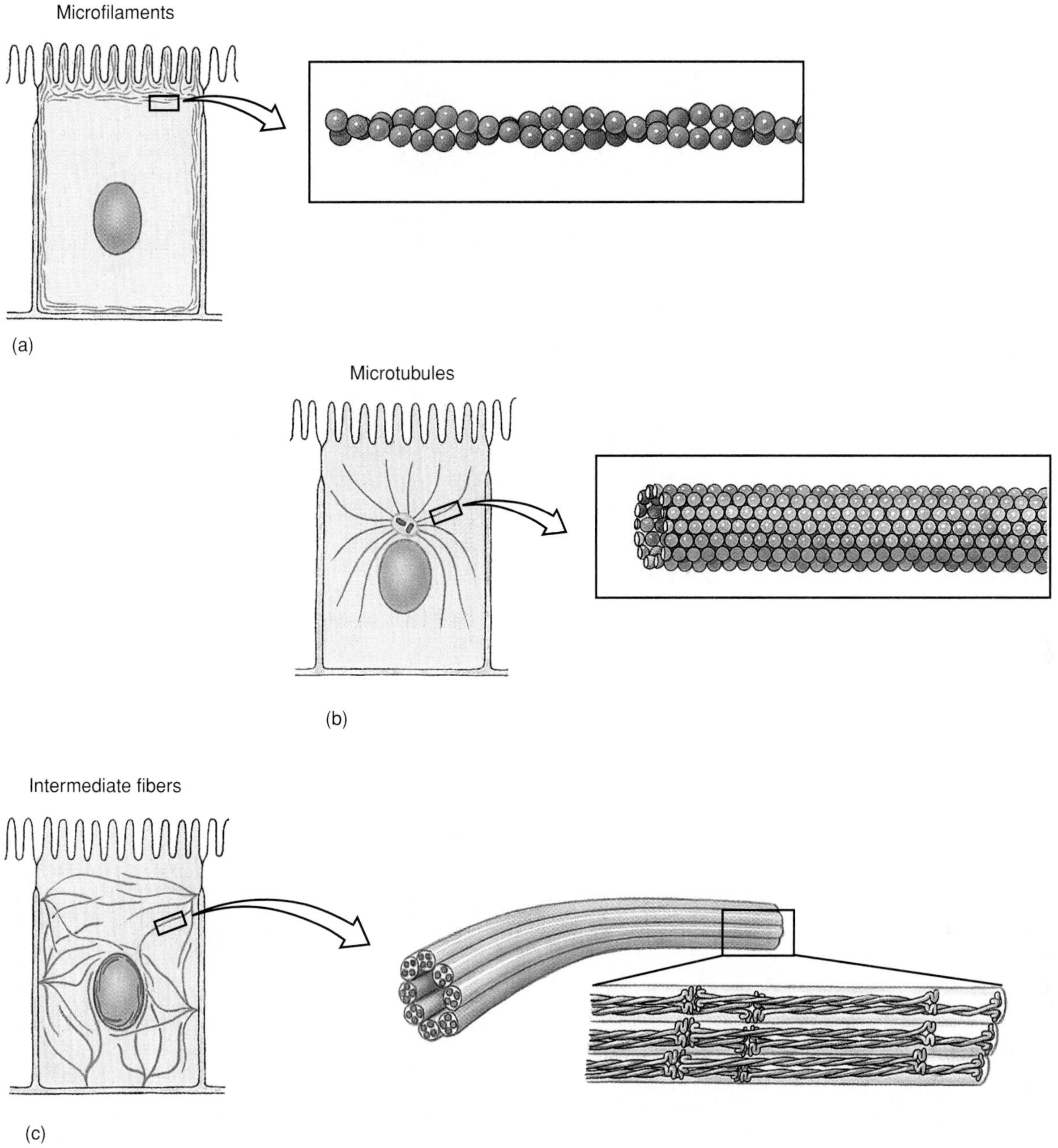

Figure 3.16 The three protein fibers of the cytoskeleton.
(*a*) Microfilaments. Microfilaments parallel the cell surface membrane in bundles of actin known as stress fibers, which may have a contractile function. (*b*) Microtubules. Each microtubule is a hollow tube composed of stacks of a protein called tubulin. There are 13 stacks of tubulin, arrayed side by side in a circle. (*c*) Intermediate fibers. It is not known how the individual subunits are arranged in an intermediate fiber, but the best evidence suggests that three subunits are wound together in a coil, interrupted by uncoiled regions. Intermediate fibers, like microtubules, extend throughout the cytoplasm and probably provide structural reinforcement.

The cytoskeleton plays a major role in determining the shape of animal cells, which lack rigid cell walls. Because filaments can form and dissolve readily, the shape of an animal cell can change rapidly. If you examine the surface of an animal cell with a microscope, you will often find it alive with motion, projections shooting out from the surface and then retracting, only to shoot out elsewhere moments later.

The cytoskeleton is not only responsible for the cell's shape, but it also provides a scaffold both for ribosomes to carry out protein synthesis and for enzymes to be localized within defined areas of the cytoplasm. By anchoring particular enzymes near one another, the cytoplasm participates with organelles in organizing the cell's activities.

Centrioles

Microtubules also play important roles in moving chromosomes during cell division and, as you will see, in the waving of flagella. Microtubules are assembled from tubulin subunits in the cells of animals and most protists by organelles called **centrioles** (figure 3.17). Centrioles occur in pairs within the cytoplasm, usually located at right angles to one another near the nuclear envelope. They are among the most structurally complex organelles of the cell. In cells that contain flagella or cilia, each flagellum is anchored by a form of centriole called a basal body. Most animal and protist cells have both centrioles and basal bodies; plants and fungi lack them, instead organizing microtubules without such structures. Centrioles resemble spirochete bacteria in many respects, and they contain a circular DNA molecule that is involved in the production of their structural proteins. Many biologists believe that centrioles, like mitochondria and chloroplasts, originated as symbiotic bacteria.

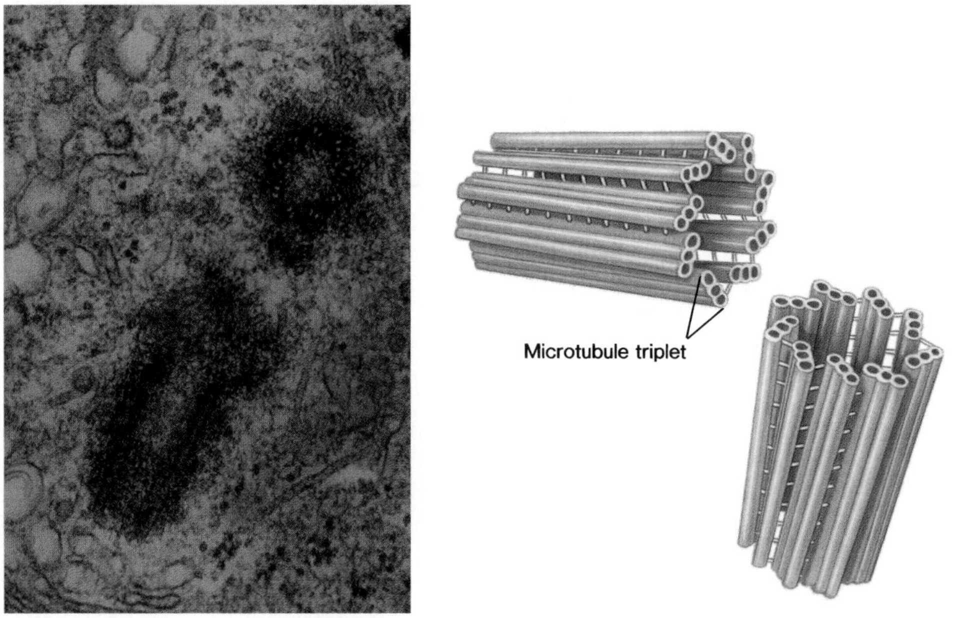

(a) (b)

Figure 3.17 Centrioles.

Centrioles anchor and assemble microtubules. In their anchoring capacity, they can be seen as the basal bodies of eukaryotic flagella. In their organizing capacity, they function during cell division to indicate the plane along which the cell separates. (*a*) A pair of centrioles in a cell are about to divide. Centrioles usually occur in pairs and are located in the cell in characteristic planes. One centriole typically lies parallel to the cell surface, while another lies perpendicular to the surface. (*b*) Centrioles are composed of nine triplets of microtubules.

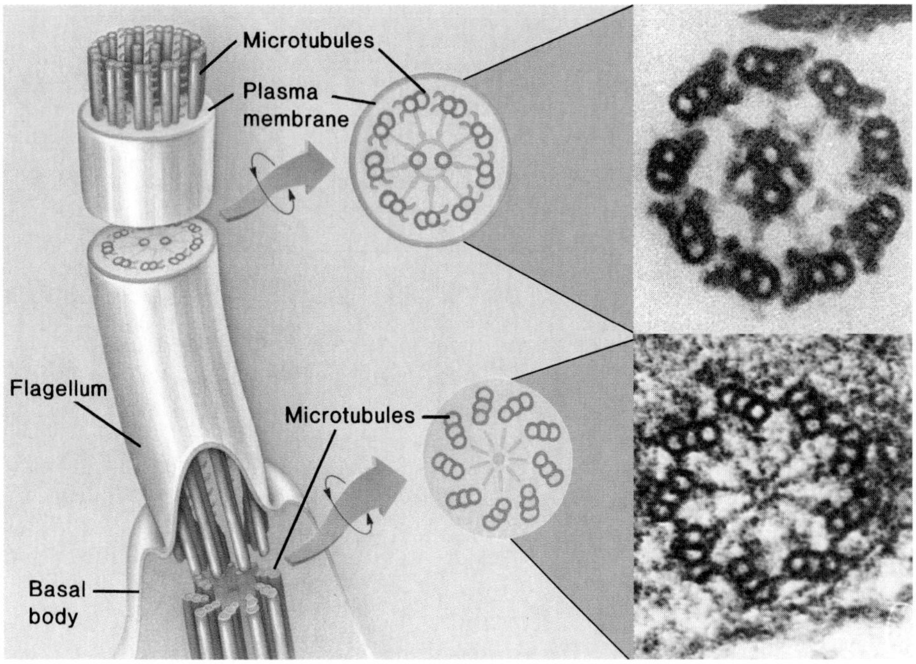

(a)

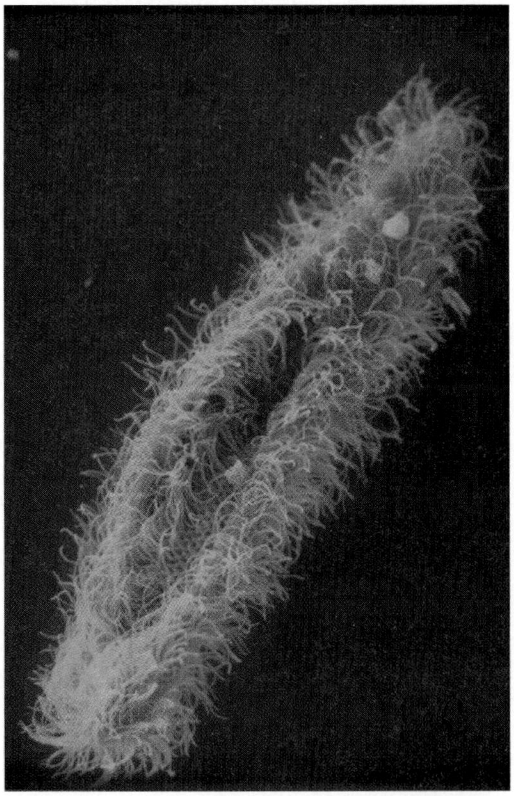

(b)

Figure 3.18 Structure of a eukaryotic flagellum.

(*a*) A eukaryotic flagellum springs directly from a basal body and is composed of a ring of nine pairs of microtubules with two microtubules in its core. (*b*) The surface of this *Paramecium* is covered with a dense forest of cilia.

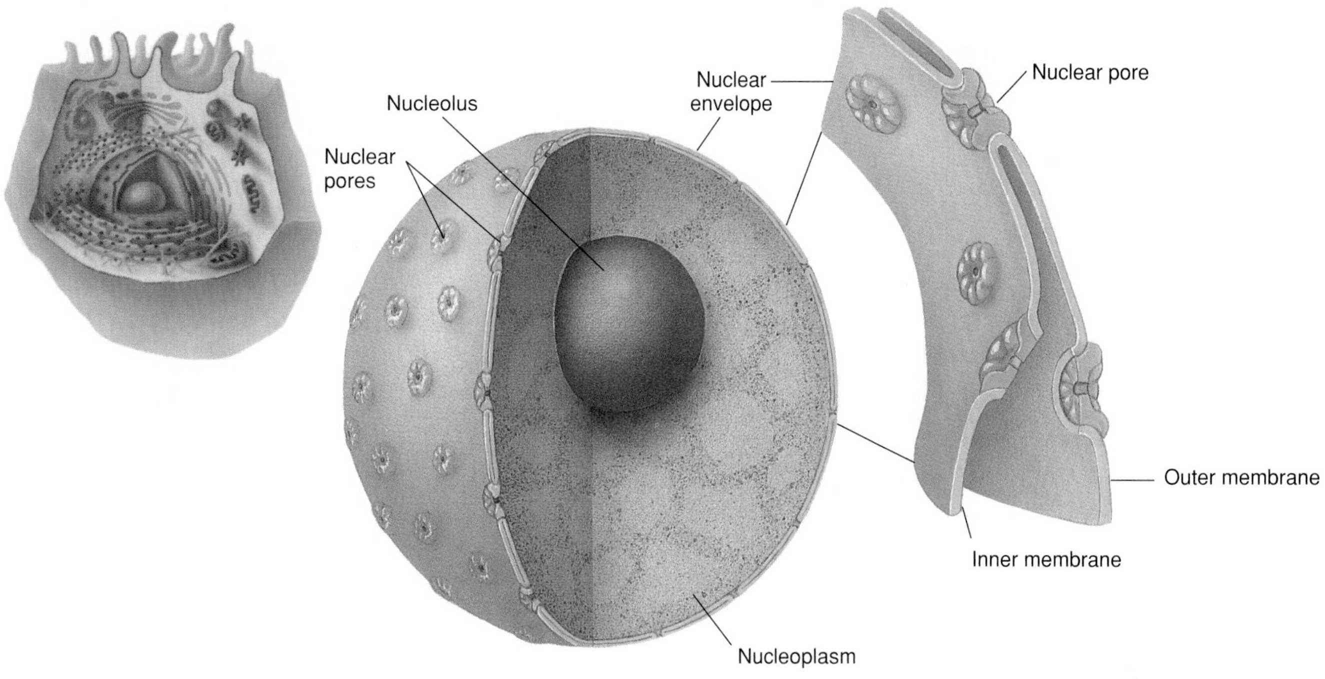

Figure 3.19 The nucleus.

The nucleus is composed of a double membrane, called a nuclear envelope, enclosing a fluid-filled interior containing the chromosomes. In cross section, the individual nuclear pores are seen to extend through the two membrane layers of the envelope; the dark material within the pore is protein, which acts to control access through the pore.

Flagella and Cilia

Flagella (singular, **flagellum**) are fine, long, threadlike organelles protruding out from the cell surface. Each flagellum arises from a structure called a **basal body** and consists of a circle of nine microtubule pairs surrounding two central ones (figure 3.18). This **9 + 2 arrangement** is a fundamental feature of eukaryotes and apparently evolved early in their history. Even in cells that lack flagella, derived structures with the same 9 + 2 arrangement often occur, like in the sensory hairs of the human ear. If flagella are numerous and organized in dense rows, they are called **cilia.** Cilia do not differ from flagella in their structure, but cilia are usually short. In humans, we find a single long flagellum on each sperm cell, and dense mats of cilia line our breathing tube, the trachea, to move mucus and dust particles out of the respiratory tract into the throat (where we can expel these unneeded contaminants by spitting or swallowing).

The Nucleus: The Cell's Control Center

If you were to continue your journey into the cell's interior, you would eventually reach the center of the cell. There you would find, cradled within a network of fine filaments like a ball in a basket, the **nucleus** (figure 3.19). The nucleus is the largest of the organelles within a eukaryotic cell. It is the command and control center of the cell, directing all of its activities. It is also the genetic library where the hereditary information is stored.

The surface of the nucleus is bounded by a special kind of membrane called the **nuclear envelope.** The nuclear envelope is actually *two* membranes, one outside the other, like a sweater over a shirt. Scattered over the surface of this envelope, like the craters of the moon, are shallow depressions called **nuclear pores.** Nuclear pores form when the two membrane layers of the nuclear envelope pinch together. A nuclear pore is not an empty opening like the hole in a doughnut; however, it has many proteins embedded within it that permit proteins and RNA to pass into and out of the nucleus.

In both bacteria and eukaryotes, all hereditary information specifying cell structure and function is encoded in DNA. However, unlike bacterial DNA, the DNA of eukaryotes is divided into several segments and associated with protein, forming **chromosomes.** The proteins in the chromosome permit the DNA to wind up tightly during cell division. Under a light microscope, these condensed chromosomes are readily seen in dividing cells as densely staining rods. After cell division, eukaryotic chromosomes uncoil and can no longer be distinguished individually with a light microscope.

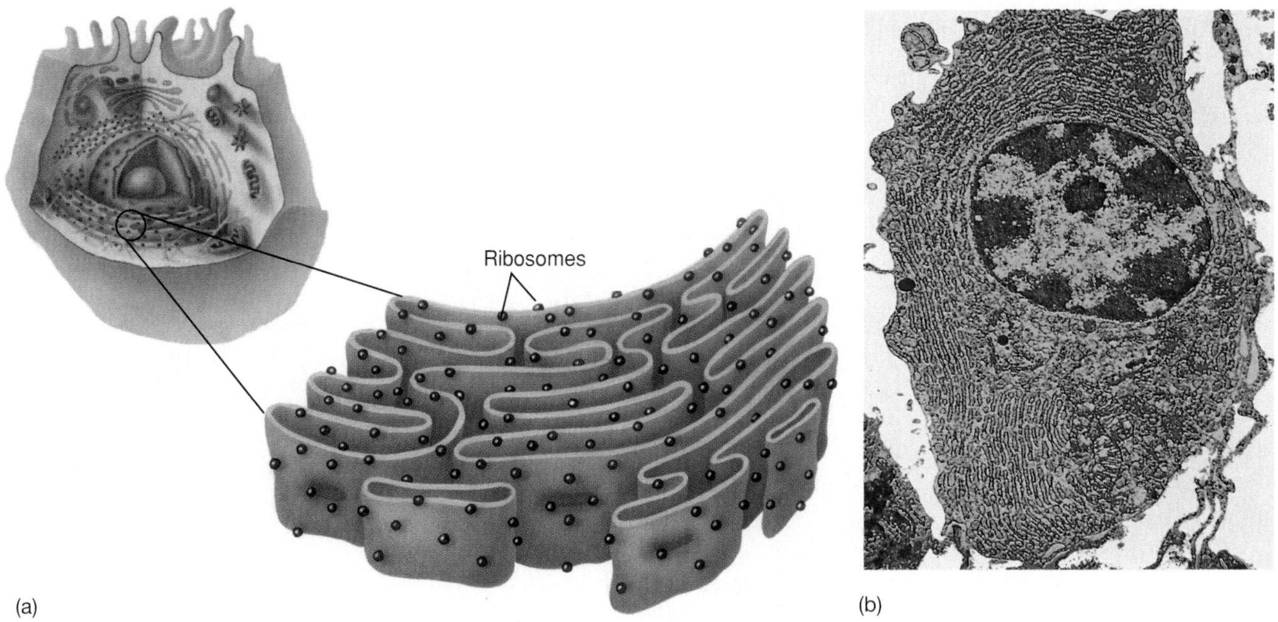

Figure 3.20 The endoplasmic reticulum (ER).
The endoplasmic reticulum provides the cell with an extensive system of internal membranes for the synthesis and transport of materials. (*a*) Diagram of rough ER. (*b*) Micrograph of a cell interior rich in rough ER.

Ribosomes: The Cell's Protein Manufacturing Centers

To make its many proteins, the cell employs a special organelle called a **ribosome,** which reads the RNA copy of a gene and uses that information to direct the construction of a protein. Ribosomes are made up of several special forms of RNA called ribosomal RNA, or rRNA, bound up within a complex of several dozen different proteins.

One region of the nucleus appears darker than the rest; this darker region is called the **nucleolus.** There a cluster of several hundred genes encode rRNA where the ribosome subunits assemble. These subunits leave the nucleus through the nuclear pores and enter the cytoplasm, where final assembly of ribosomes takes place.

Endoplasmic Reticulum: The Transportation System

Surrounding the nucleus within the interior of the eukaryotic cell is a tightly packed mass of membranes. They fill the cell, dividing it into compartments, channeling the transport of molecules through the interior of the cell and providing the surfaces on which enzymes act. The system of internal compartments created by these membranes in eukaryotic cells constitutes the most fundamental distinction between the cells of eukaryotes and prokaryotes. The extensive system of internal membranes is called the **endoplasmic reticulum,** often abbreviated **ER** (figure 3.20). The term *endoplasmic* means "within the cytoplasm," and the term *reticulum* is a Latin word meaning "little net." The ER, weaving in sheets through the interior of the cell, creates a series of channels and interconnections, and it also isolates some spaces as membrane-enclosed sacs called **vesicles.**

The surface of the ER is the place where the cell manufactures many carbohydrates and lipids. It is also where the cell makes proteins intended for export (such as enzymes secreted from the cell surface). The surface of those regions of the ER devoted to the synthesis of such transported proteins is heavily studded with ribosomes and appears pebbly, like the surface of sandpaper, when seen through an electron microscope. For this reason, these regions are called **rough ER.** Regions in which ER-bounded ribosomes are relatively scarce are correspondingly called **smooth ER.**

The Golgi Complex: The Delivery System

As new molecules are made on the surface of the ER, they are passed out across the ER membrane into the vesicle-forming system called the **Golgi complex** (figure 3.21). Animal cells contain 10 to 20 flattened stacks of membranes called Golgi bodies, which are scattered through the cytoplasm and collectively referred to as the Golgi complex. Golgi bodies function in the collection, packaging, and distribution of molecules manufactured in the cell.

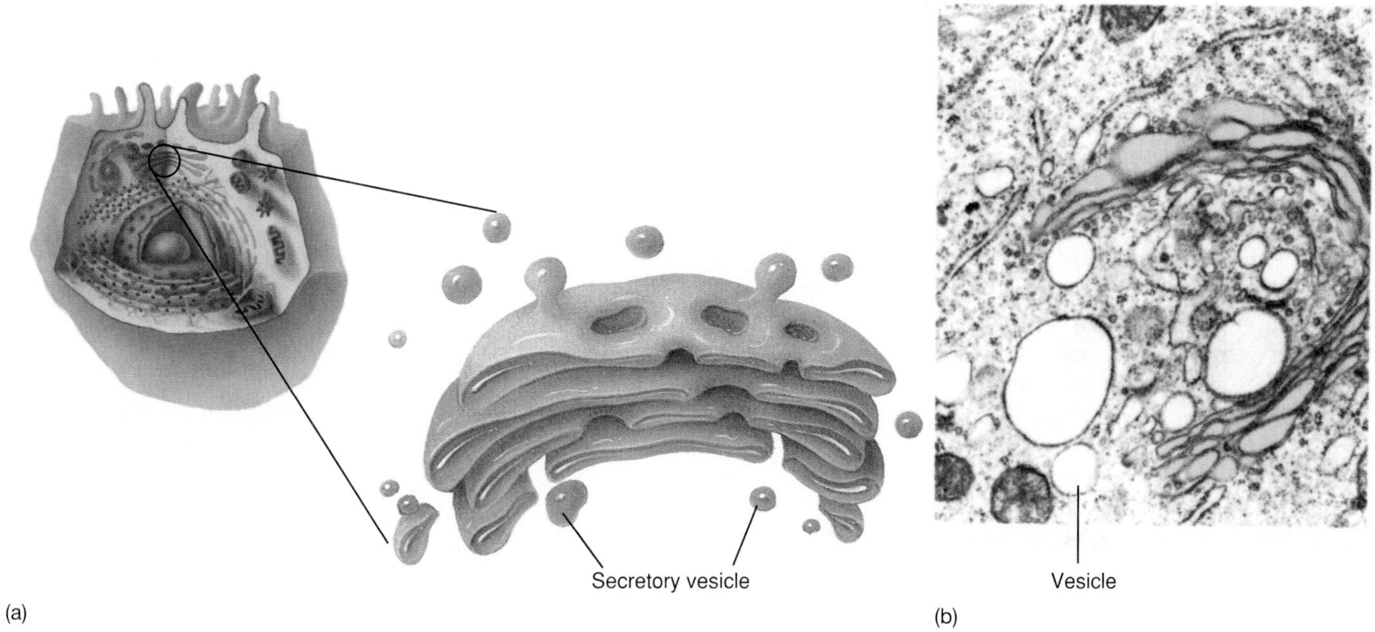

Figure 3.21 Golgi complex.
This vesicle-forming system, called the Golgi complex after its discoverer, is an integral part of the cell's internal membrane system. The Golgi complex processes and packages materials for transport and possible export from the cell. (*a*) Diagram of Golgi complex. (*b*) Micrograph of Golgi complex showing vesicles.

The proteins and lipids that are manufactured on the ER membranes are transported through the channels of the ER, or as vesicles budded off from it, into the Golgi bodies. Within the Golgi bodies, many of these molecules become tagged with carbohydrates. The molecules collect at the ends of the membranous folds of the Golgi bodies; these folds are given the special name cisternae (Latin, "collecting vessels"). Vesicles that pinch off from the cisternae carry the molecules to the different compartments of the cell and to the inner surface of the plasma membrane, where molecules to be secreted are released to the outside.

Peroxisomes: Chemical Specialty Shops

The interior of the eukaryotic cell contains a variety of membrane-bounded spherical organelles derived from the ER that carry out particular chemical functions. Almost all eukaryotic cells, for example, contain **peroxisomes,** vesicles that contain two sets of enzymes. One set converts fats to carbohydrates, and the other set detoxifies various potentially harmful molecules—strong oxidants—that form in cells. They do this by using molecular oxygen to remove hydrogen atoms from specific molecules. These chemical reactions would be very destructive to the cell if not confined to the peroxisomes.

Lysosomes: Recycling Centers

Other organelles called **lysosomes,** about the same size as peroxisomes, contain a concentrated mix of the powerful enzymes that break down macromolecules. Lysosomes are the recycling centers of the cell, digesting worn-out cell components to make way for newly formed ones while recycling the proteins and other materials of the old parts. The large organelles called mitochondria are replaced in some human tissues every 10 days, with lysosomes digesting the old ones as the new ones are produced. Cells can persist for a long time only if their components are constantly renewed. Otherwise, the ravages of use and accident chip away at their metabolic capabilities and slowly degrade the cell's ability to survive. Cells age for the same reason that people do, because of a failure to renew themselves.

It is not known what prevents lysosomes from digesting *themselves,* but the prevention process requires energy, which is the reason that metabolically inactive eukaryotic cells die. Without a constant input of energy, the powerful enzymes of lysosomes digest their membranes from within. When these membranes disintegrate, the digestive enzymes pour out into the cytoplasm and destroy the cell. In contrast, bacteria do not possess lysosomes and do not die when deprived of energy. Instead, they are able to remain quiet but alive until altered conditions provide a source of energy. For us, the very process that repairs the ravages of time within our cells eventually also leads to their destruction. Death is the price we pay for efficiency.

Mitochondria: Powerhouses of the Cell

Most of the membrane-bounded structures within eukaryotic cells are derived from the ER, but there are important exceptions. The eukaryotic cell contains several kinds of complex, cell-like organelles that appear to have been derived from ancient bacteria assimilated by ancestral eukaryotes in the distant past. The three principal kinds of cell-like organelles are mitochondria (which occur in the cells of all but a very few eukaryotes), chloroplasts (which do not occur in animal cells—they occur only in algae and plants), and centrioles (which occur in the cells of all animals and most protists, but not in plants and fungi).

You and all other animals extract energy from organic molecules ("food") in a complex series of chemical reactions called **oxidative metabolism,** which takes place only in mitochondria. **Mitochondria** (singular, mitochondrion) are sausage-shaped organelles about the size of a bacterial cell (figure 3.22). Mitochondria are bounded by two membranes. The outer membrane is smooth and apparently derives from the ER of the host cell that first took up the bacterium long ago. The inner membrane, apparently the plasma membrane of the bacterium that gave rise to the mitochondrion, is bent into numerous folds called **cristae** (singular, **crista**) that resemble the folded plasma membranes in various groups of bacteria. The cristae partition the mitochondrion into two compartments, an inner **matrix** and an outer compartment. As you will learn in chapter 5, this architecture is critical to successfully carrying out oxidative metabolism.

During the 1.5 billion years in which mitochondria have existed in eukaryotic cells, most of their genes have been transferred to the chromosomes of the host cells. But mitochondria still have some of their original genes, contained in a circular, closed, naked molecule of DNA (called mitochondrial DNA, or mtDNA) that closely resembles the circular DNA molecule of a bacterium. On this mtDNA are located several genes that produce some of the proteins essential for oxidative metabolism. In both mitochondria and bacteria, the circular DNA molecule is replicated during the process of division. When a eukaryotic cell divides, its mitochondria split into two by simple fission, dividing much as bacteria do.

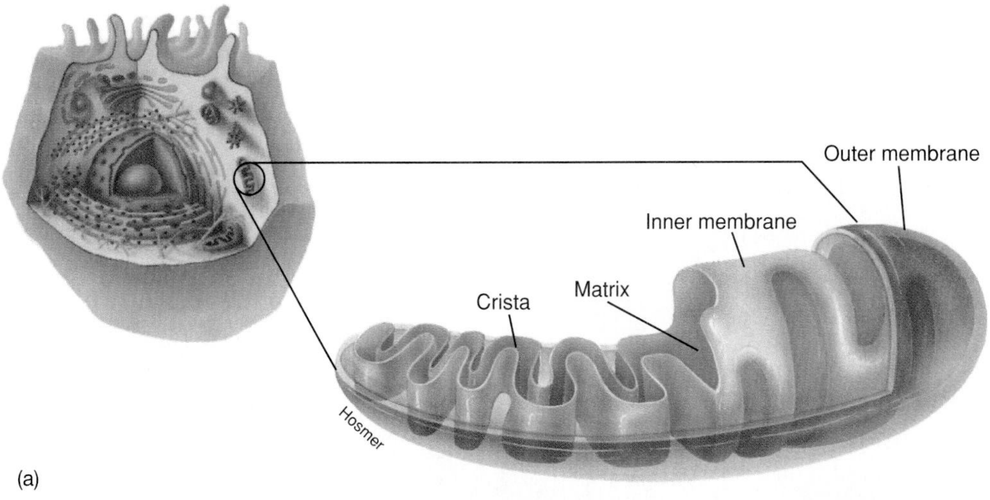

(a)

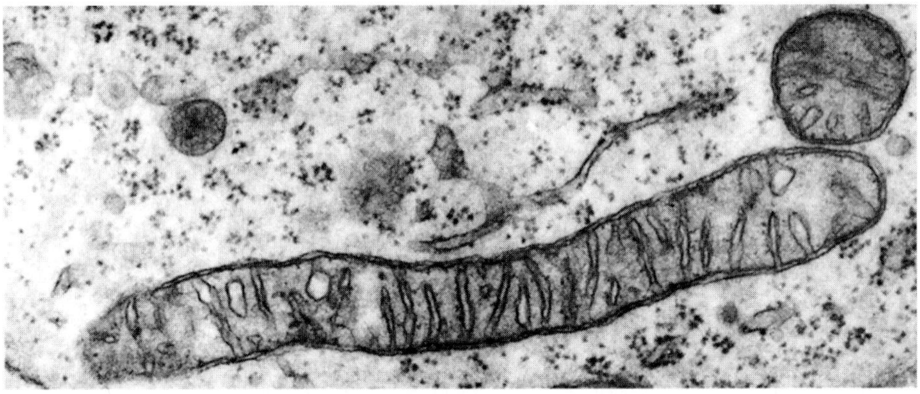

(b)

Figure 3.22 Mitochondria.
The mitochondria of a cell are sausage-shaped organelles within which oxidative metabolism takes place, extracting energy from food using oxygen. (*a*) A mitochondrion has a double membrane. The inner membrane is shaped into folds called cristae. The space within the cristae is called the matrix. The cristae greatly increase the surface area for oxidative metabolism. (*b*) Micrograph of two mitochondria, one in cross-section, the other cut lengthwise.

Chloroplasts: Energy-Capturing Centers

All photosynthesis in plants and algae takes place within another bacteria-like organelle, the **chloroplast** (figure 3.23). There is strong evidence that chloroplasts, like mitochondria, were derived by symbiosis from bacteria. A chloroplast is bounded, like a mitochondrion, by two membranes, the inner derived from the original bacterium and the outer from the host cell's ER. Chloroplasts are larger than mitochondria, and their inner membranes have a more complex organization. The inner membranes are fused to form stacks of closed vesicles called **thylakoids.** Photosynthesis takes place within the thylakoids. The thylakoids are stacked on top of one another to form a column called **granum** (plural, **grana).** The interior of a chloroplast is bathed with a semiliquid substance called the **stroma.**

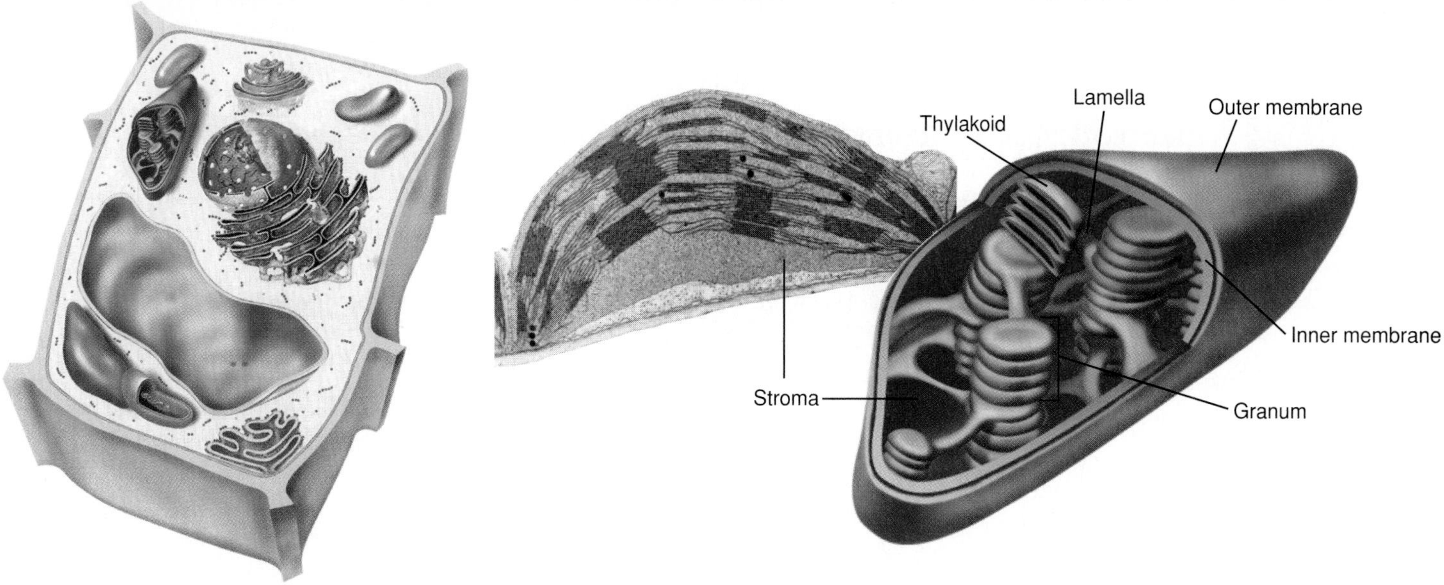

Figure 3.23 A chloroplast.
Bacteria-like organelles called chloroplasts are the sites of photosynthesis in photosynthetic eukaryotes. Like mitochondria, they have a complex system of internal membranes on which chemical reactions take place. The inner membrane of a chloroplast is fused to form stacks of closed vesicles called thylakoids. Photosynthesis occurs within these thylakoids. Thylakoids are stacked one on top of the other in columns called grana. The interior of the chloroplasts is bathed in a semiliquid substance called the stroma.

Also like mitochondria, chloroplasts have a large circular DNA molecule. On this DNA are located the genes coding for the proteins necessary to carry out photosynthesis. Plant cells can contain from one to several hundred chloroplasts, depending on the species. Neither mitochondria nor chloroplasts can be grown in a cell-free culture; they are totally dependent on the cells within which they occur for their survival.

Vacuoles: A Central Storage Compartment

The center of a plant cell usually contains a large, apparently empty space, called the **central vacuole** (see figure 3.12). This vacuole is not really empty; it contains large amounts of water and other materials, such as sugars, ions, and pigments. The central vacuole functions as a storage center for these important substances and also helps to increase the surface-to-volume ratio of the plant cell by applying pressure to the cell membrane. The cell membrane expands outward under this pressure, thereby increasing its surface area.

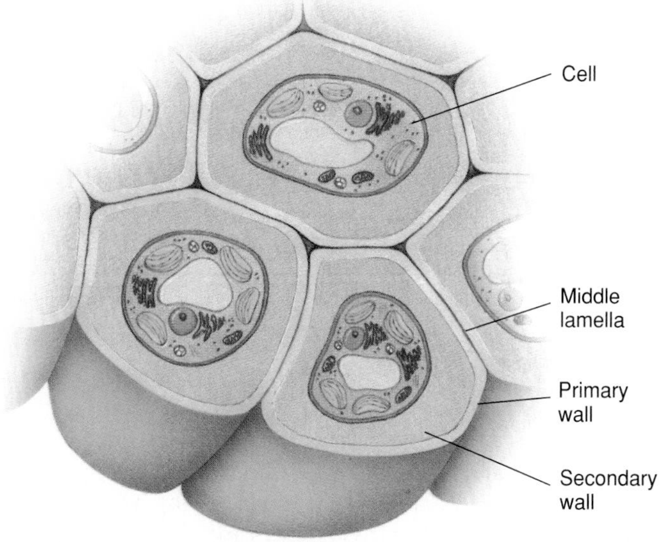

Figure 3.24 Cell walls in plants.
Plant cell walls are fundamentally different from those of bacteria—thicker, stronger, and more rigid.

Cell Walls: Protection and Support

Plant cells share a characteristic with bacteria that is not shared with animal cells—that is, plants have **cell walls,** which protect and support the plant cell. Although bacteria also have cell walls, plant cell walls are chemically and structurally different from bacterial cell walls (figure 3.24). Cell walls are also present in fungi and some protists. In plants, cell walls are composed of fibers of the polysaccharide cellulose. **Primary walls** are laid down when the cell is still growing, and between the walls of adjacent cells is a sticky substance called the **middle lamella,** which glues the cells together. Some plant cells produce strong **secondary walls,** which are deposited inside the primary walls of fully expanded cells.

Table 3.1 Eukaryotic Cell Structures and Their Functions

Structure	Description	Function
Structural Elements		
Cytoskeleton	Network of protein filaments	Structural support; cell movement
Flagella and cilia	Cellular extensions with 9 + 2 arrangement of pairs of microtubules	Motility or moving fluids over surfaces
Interior Membrane System		
Plasma membrane	Lipid bilayer in which proteins are embedded	Regulates what passes into and out of cell; cell-to-cell recognition
Endoplasmic reticulum	Network of internal membranes	Forms compartments and vesicles
Nucleus	Spherical structure bounded by double membrane; contains chromosomes	Control center of cell; directs protein synthesis and cell reproduction
Golgi complex	Stacks of flattened vesicles	Modifies and packages proteins for export from cell; forms secretory vesicles
Lysosomes	Vesicles derived from Golgi complex that contain hydrolytic digestive enzymes	Digest worn-out mitochondria and cell debris; play role in cell death
Energy-Producing Organelles		
Mitochondria	Bacteria-like elements with inner membrane	Power plant of the cell; site of oxidative metabolism
Chloroplast	Bacteria-like organelle found in plants and algae; complex inner membrane consists of stacked vesicles	Site of photosynthesis
Organelles of Gene Expression		
Chromosomes	Long threads of DNA that form a complex with protein	Contain hereditary information
Nucleolus	Site of chromosomes of rRNA synthesis	Assembles ribosomes
Ribosomes	Small, complex assemblies of protein and RNA, often bound to ER	Sites of protein synthesis

An Overall Perspective

The many parts of the cell (table 3.1) are remarkably similar in organisms as diverse as bacteria, trees, and humans. We do not see a great variety of cell organelles from plants to animals, or from paramecia to primates. Instead, mitochondria look similar and carry out the same function of energy production in all organisms. Chloroplasts carry out photosynthesis in algae and giant redwoods. The nucleus and its membrane envelope look very similar in all organisms. These shared properties must have been derived from common ancestral cells that successfully developed the initial organelles several billion years ago.

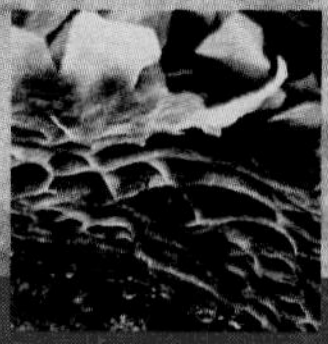

CHAPTER 3

HIGHLIGHTS

3.1 The World of Cells

Key Terms

cell 46
surface-to-volume ratio 47
cytoplasm 47

Key Concepts

- All life is composed of cells—about 100 trillion in the case of a human body.
- Cells tend to be quite small, limited by the distance to which substances must diffuse through the cytoplasm and also by the surface-to-volume ratio as cell size increases.
- Small cells have relatively more surface with which to interact with the environment.

3.2 Exploring the Cell Surface

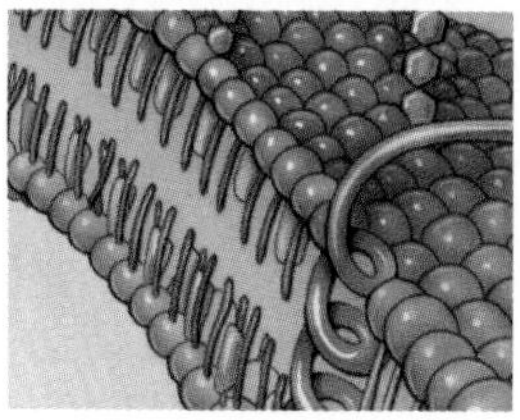

Key Terms

plasma membrane 48
lipid bilayer 48
transmembrane protein 48

Key Concepts

- The cell cytoplasm is bounded by the plasma membrane.
- The plasma membrane is a double layer of modified fat molecules called a lipid bilayer.
- Protein-lined passageways through the bilayer allow chemicals to enter and leave the cell.

3.3 Bacteria Are the Simplest Cells

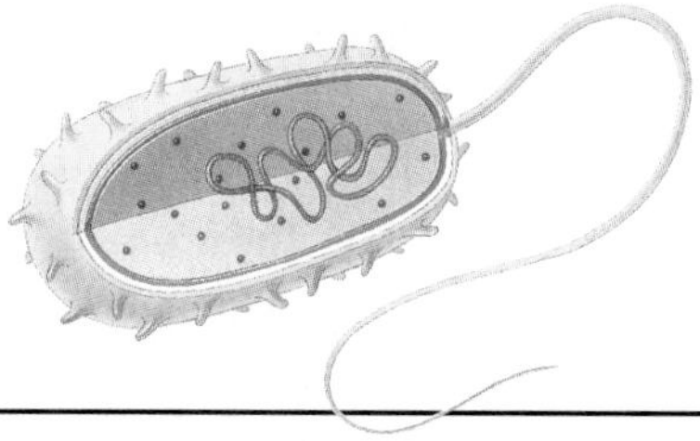

Key Terms

prokaryotes 51

Key Concepts

- The simplest cells are bacteria, called prokaryotes because they lack a nucleus.
- The key characteristic shared by all bacterial cells is that they lack internal compartments.

3.4 The Structure of Eukaryotic Cells

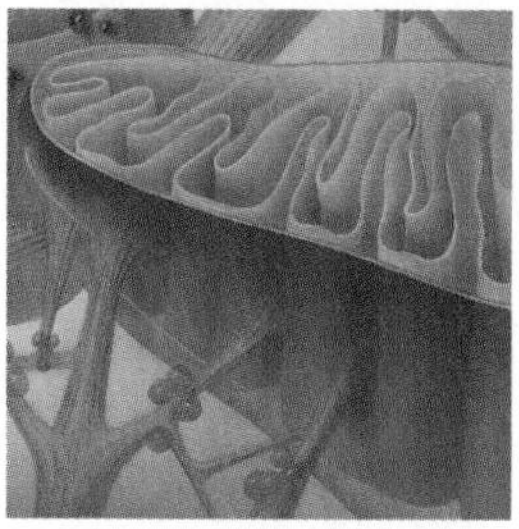

Key Terms

organelles 52
cytoskeleton 54
nucleus 57
ribosome 58
mitochondria 60

Key Concepts

- All nonbacterial cells are eukaryotes.
- The interior of eukaryotic cells is occupied by an extensive membrane system that creates many subcompartments within the cytoplasm.
- The cell's hereditary information, contained within chromosomes, is isolated within a membrane-bounded organelle called the nucleus.
- Proteins are assembled in the cytoplasm on complexes called ribosomes.
- Cells carry out oxidative metabolism within bacteria-like organelles called mitochondria.

3

CONCEPT REVIEW

1. Select the incorrect statement.
a. All organisms consist of one or more cells.
b. Cells are the smallest living things.
c. Life evolved 2 billion years ago.
d. Cells arise only by the division of other cells.

2. Select the molecule not normally found in the plasma membrane.
a. phospholipid
b. freely moving proteins
c. transmembrane proteins
d. DNA

3. A cell with a nucleus is
a. eukaryotic.
b. prokaryotic.

4. Centrioles are involved in
a. storage.
b. energy transfer.
c. organized movement.
d. protein synthesis.

5. The DNA of eukaryotic cells is fragmented into several pieces called _________, unlike the DNA of bacteria, which is mostly contained in a single circular molecule.
a. ribosomes
b. peroxisomes
c. chromosomes
d. chloroplasts

6. Proteins are made at the
a. ribosome.
b. mitochondrion.
c. lysosome.
d. Golgi complex.

7. One word to describe the ER is
a. storage.
b. transport.
c. division.
d. digestion.

8. Each of the following is a function of the Golgi complex with molecules except
a. collection.
b. distribution.
c. packaging.
d. synthesis.

9. Arrange in order the sequence of membrane-bounded structures through which a protein that is going to be secreted by a cell must pass.
a. Golgi body
b. secretory vesicle
c. rough ER
d. smooth ER

10. Cristae are found in the
a. lysosome.
b. mitochondria.
c. centrioles.
d. Golgi complex.

11. Which three of the following properties are exclusive to microtubules, which help make up the cyotskeleton of eukaryotic cells?
a. form spontaneously if the subunits are present
b. made of protein subunits called tubulin
c. anchored to proteins in the plasma membrane
d. hollow tubes about 25 nanometers in diameter
e. play an important role in cell movement

12. The idea that all organisms consist of cells is the cell _________.

13. _________ are organisms with the simplest cell structure.

14. The _________ is the internal framework of the cell.

15. _________ are long, threadlike organelles projecting from the cell that function in cell movement.

16. _________ are one type of human cell organelle known to have their own DNA.

Answers to the Concept Review questions appear in Appendix B.

CHALLENGE YOURSELF

1. Mitochondria and chloroplasts are thought to be the evolutionary descendants of living cells that were engulfed by other cells. Are mitochondria and chloroplasts alive?
2. Some cells are very much larger than others. What would you expect the relationship to be between cell size and level of cell activity?
3. How does a Golgi body "know" where to send what vesicle?
4. Trace a synthesizing protein through the endomembrane system, culminating in release of the protein outside the cell.
5. What mechanism could account for the appearance of the same 9 + 2 structure in three different cellular components (centrioles, flagella, and cilia)?

FOR FURTHER READING

Alberts, B., D. Bray, J. Lewis, M. Raff, K. Roberts, and J. Watson. *Molecular Biology of the Cell*, 3d ed. New York: Garland Press, 1994. A very comprehensive text on modern cell biology.

DeDuve, C. "The Birth of Complex Cells." *Scientific American*, April 1996, 50–57. All animals, plants, and fungi owe their existence to the remnants of tiny primitive bacteria.

Edelson, E. "Conduits for Cell/Cell Communication." *Mosaic* 21 (1990) 48–56. A discussion of the proteinaceous channels between cells that facilitate intercellular communication and transport.

Glover, D., C. Gonzalez, and J. Raff. "The Centrosome," *Scientific American*, June 1993, 62–68. Recent information concerning the structure and function of centrioles.

Greider, C., and E. Blackburn. "Telomeres, Telomerase, and Cancer." *Scientific American*, February 1996, 92–97. Telomerase enzymes act on the ends of chromosomes, shortening them with each cell division.

Rothman, J., and L. Orci. "Budding Vesicles in Living Cells." *Scientific American*, March 1996, 70–75. Complex machinery forms the tiny containers that store proteins and shuttle them to and fro in cells.

Schliwa, M. *The Cytoskeleton: An Introductory Survey*. New York: Springer-Verlag, 1986. A broad-ranging review of the properties of filaments in higher animal cells.

Stossel, T. "How Cells Crawl," *American Scientist*, 1990, 408–23. With the discovery that the cellular "motor" contains muscle proteins, biologists are beginning to understand cell motility in molecular terms.

Stossel, T. "The Machinery of Cell Crawling." *Scientific American*, September 1994, 54–63. The role of protein scaffolding is discussed in the mechanism of cell crawling.

Weber, K., and M. Osborn. "The Molecules of the Cell Matrix." *Scientific American*, October 1985, 110–20. Descriptions of the many different molecules that make up the cell cytoskeleton.

TECHNOLOGY LINKS

The Living World Home Page
http://www.wcbp.com/biology/tlw

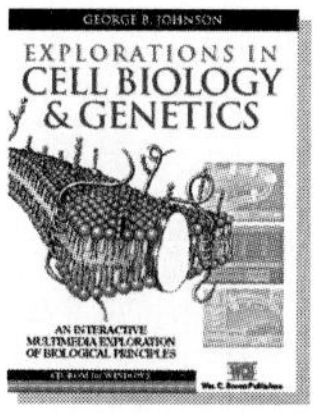

Explorations in Cell Biology & Genetics CD
#2 Cell Size
#4 Cell-Cell Interactions

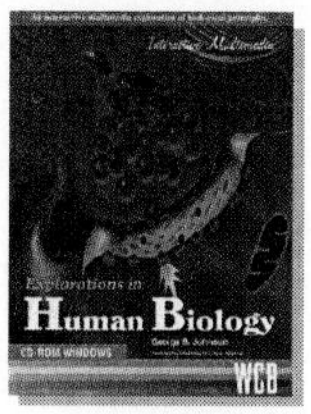

Explorations in Human Biology CD
#1 Cystic Fibrosis

Life Science Animations Videotape 1
#2 Journey into a Cell

4

The Living Cell

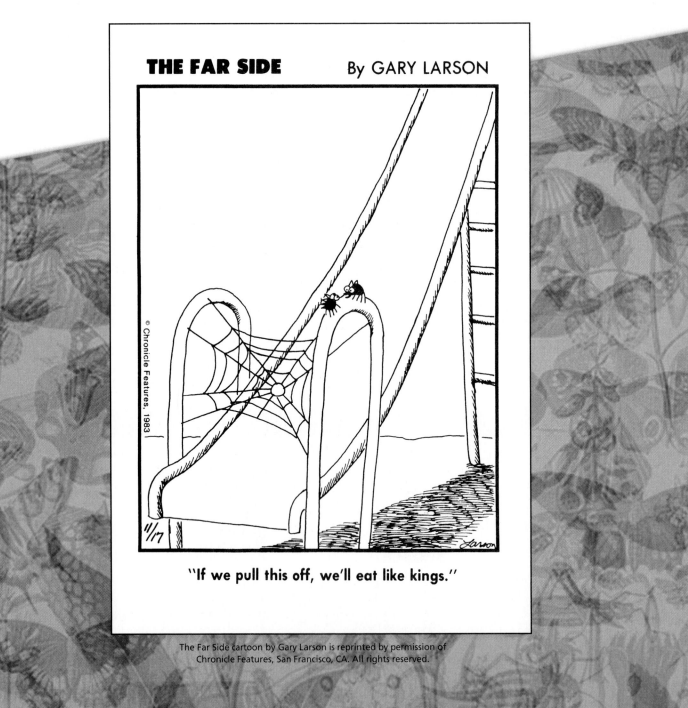

CHAPTER OUTLINE

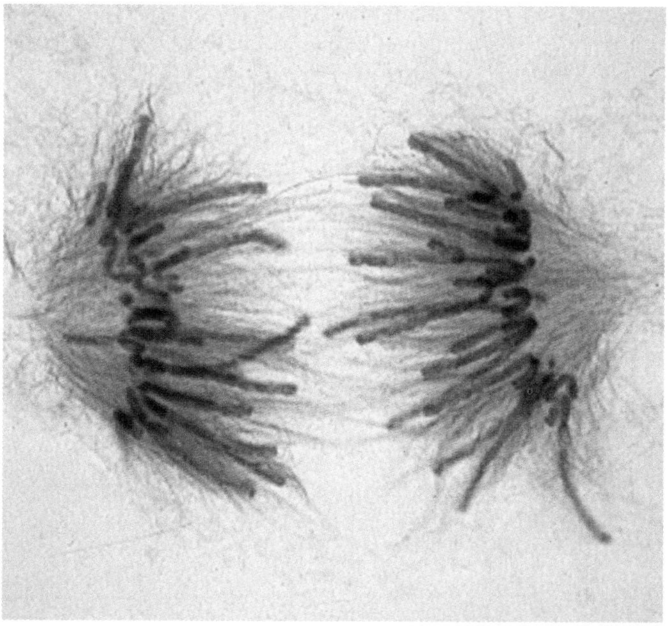

Figure 4.1 Chromosomes in a dividing cell.
The ability to reproduce, as this dividing lily cell is in the process of doing, is one of the hallmarks of living creatures. Note how the chromosomes are pulled to opposite ends of the cell.

All living creatures—elephants and butterflies and tiny bacteria too small to see—share the same tasks of living. To survive, all of them must eat and drink, learn about their environment, and reproduce, as the lily cell in figure 4.1 is doing. While a lily may not seem to have much in common with humans such as the unseen child about to use the slide on the opposite page, all humans are made up of cells too, and every one of the billions of cells in a human must, like the lily cell, carry out the tasks of living. One of the great unifying principles of biology is that all cells face these fundamental tasks in much the same way.

4.1 How Cells Eat and Drink

For cells to survive, food particles, water, and other materials must pass into the cell, and waste materials must be eliminated. All of this moving back and forth across the cell's plasma membrane occurs in one of three ways: (1) water leaks through imperfections in the membrane; (2) food particles and sometimes liquids are engulfed by the membrane folding around them; (3) proteins in the membrane act as doors that admit certain molecules only.

Diffusion and Osmosis

Many of the most important characteristics of living cells are not obvious to us simply because we are so much larger than a single cell. To understand how cells eat and drink—that is, how materials pass into and out of a cell—it is helpful to visualize events from a cell's perspective. Imagine, for example, that you were standing on the surface of a cell, so small that a water molecule is the size of a football. What would you see? You would see swarms of water molecules zipping about in constant motion, racing past and bumping into one another, jostling and shoving as they careen from one place to another. Molecules move like this because of the energy of motion (called kinetic energy) they contain, and at any temperature on earth every molecule contains some kinetic energy and so moves about.

Diffusion

How a molecule moves—just where it goes—is totally random, like shaking marbles in a cup, so if two kinds of molecules are added together, they soon mix. If you place a sugar cube in a glass of water, for example, the molecules of sugar dissolve from the cube into the water. Their random motion soon carries them away from the cube and out into the surrounding water. At first, there are far more sugar molecules in the area immediately surrounding the cube than in the water farther from it—a scientist would say that there was a **concentration gradient** of sugar, that is, progressively less sugar at greater distances from the cube. Eventually, however, all the water in the glass becomes sweet. The random motion of molecules always tends to produce uniform mixtures, causing a net movement of molecules toward zones where that kind of molecule is scarce (that is, *down* the concentration gradient). How does a molecule "know" in what direction to move? It doesn't—molecules don't "know" anything. A molecule is equally likely to move in any direction and is constantly changing course in random ways. There are simply more molecules able to move from where they are common than from where they are scarce. This mixing process is called **diffusion** (figure 4.2). Diffusion is the net movement of molecules down a concentration gradient toward regions of lower concentration (that is, where there are relatively fewer of them) as a result of random motion.

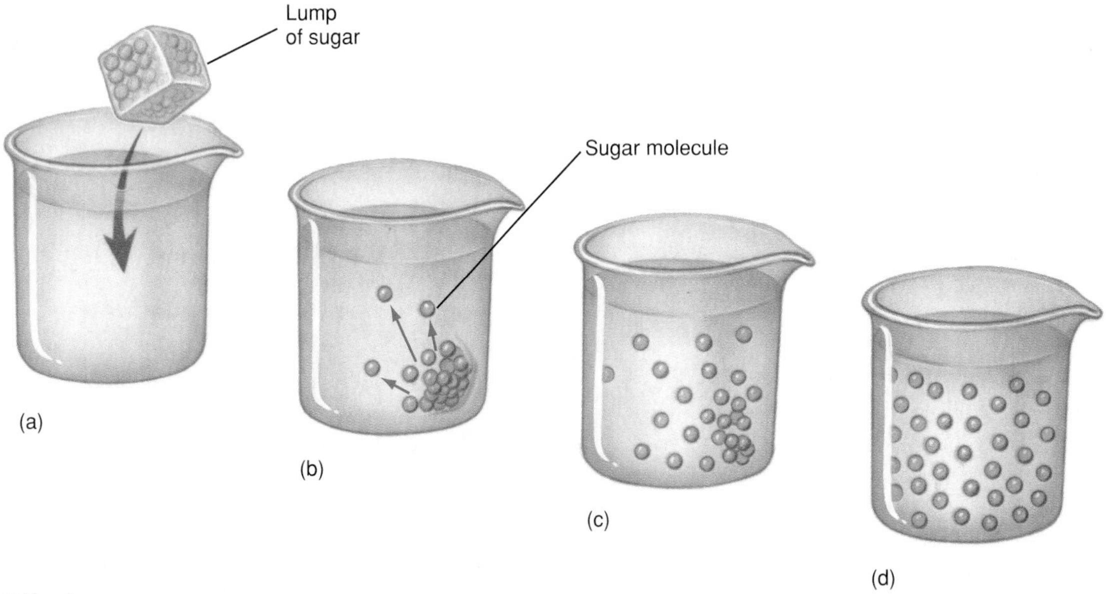

Figure 4.2 Diffusion.
Diffusion is the mixing process that spreads molecules through the cell interior. To see how diffusion works, visualize a simple experiment: drop a lump of sugar into a beaker of water (*a*). Its molecules dissolve and diffuse (*b*, *c*). Eventually, diffusion results in an even distribution of sugar molecules throughout the water (*d*).

Osmosis

Membranes block diffusion, because many molecules cannot pass across them. This is particularly true of food molecules like sugars and proteins, which are polar and thus unable to freely cross the lipid bilayer of the plasma membrane. Because the plasma membrane holds them in, a cell is able to accumulate and store such food molecules within its cytoplasm. The movement of water molecules is not blocked, however, because they are small enough to pass freely through the membrane. This free movement of water across plasma membranes has a very important consequence—it causes cells to absorb water!

To understand this puzzling fact, focus on the water molecules already present inside a cell. What are they doing? Many of them are busily engaged in interacting with the sugars, proteins, and other polar molecules inside. Remember, water is very polar itself and loves to "rub noses" with other polar molecules. These social water molecules are not randomly moving about as they were outside; instead, they remain in place, clustered around the polar molecules they are interacting with. As a result, while water molecules keep coming into the cell by random motion, they don't randomly come out again. Because more water molecules come in than go out, there is a net movement of water into the cell. Diffusion of water across a membrane toward the side with polar molecules that cannot traverse the membrane is called **osmosis** (figure 4.3).

Osmotic Pressure

Movement of water into a cell by osmosis creates pressure, called **osmotic pressure,** which can cause a cell to swell and burst. Most cells cannot withstand osmotic pressure unless their plasma membranes are braced to resist the swelling. If placed in pure water, they soon burst like overinflated balloons (figure 4.4). That is why the cells of so many kinds of organisms have cell walls to stiffen their exteriors. In animals, the fluids bathing the cells have as many polar molecules dissolved in them as the cells do, so the problem doesn't arise.

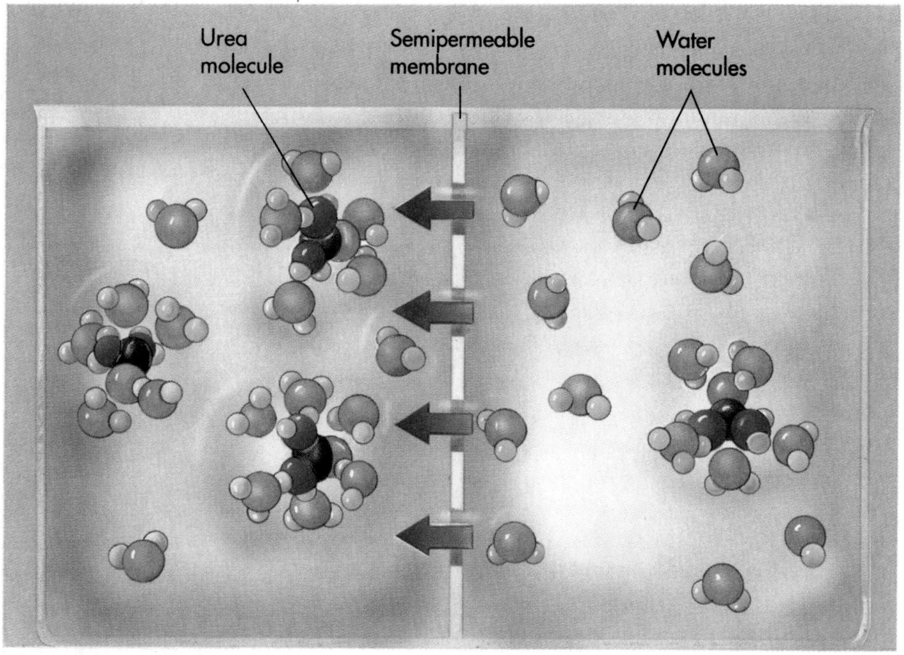

Figure 4.3 Osmosis.
Osmosis is the net movement of water across a membrane toward the side with less "free" water. To visualize this, imagine adding polar urea molecules to a vessel of water divided in the middle by a membrane. When such a polar solute is added, the water molecules that gather around each urea molecule are no longer free to diffuse across the membrane—in effect, the polar solute has reduced the number of free water molecules. Now imagine you added more urea molecules to one side of the membrane than the other. Because the side of the membrane with less solute (on the right) has more unbound water molecules than the side on the left with more solute, water moves by diffusion from the right to the left.

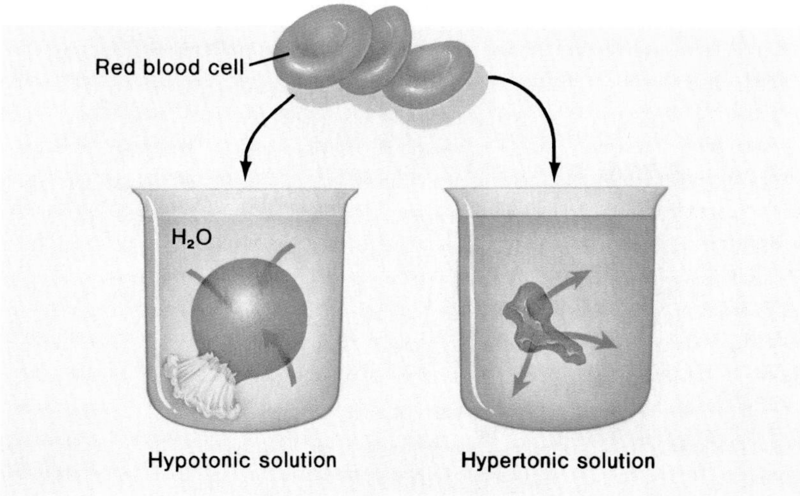

Figure 4.4 The effects of osmotic pressure on an animal cell.
If placed in pure water, an animal cell such as one of these red blood cells will swell and burst. In this case, the surrounding solution is said to be **hypotonic** (from the Greek *hypo*, meaning "lower") with respect to the cell, because it is lower on the concentration gradient of solute molecules (that is, it contains fewer of them). The cell is said to be **hypertonic** (from the Greek *hyper*, meaning "above") with respect to the solution, because it is higher on the solute concentration gradient. There is thus more "free" water on the outside, and water moves down its gradient into the cell. The reverse is true in the solution on the right, in which the cell is hypotonic with respect to the solution.

Engulfing Food Particles

The cells of many eukaryotes take in food and liquids by extending their plasma membranes outward toward food particles. The membrane encircles and engulfs the particle, and when its edges meet on the other side, they fuse together, forming a vesicle—a membrane-bordered sac—around the particle. This process is called **endocytosis** (figure 4.5). The reverse process, ridding a cell of material by discharging it from vesicles at the cell surface, is called **exocytosis** (figure 4.6).

When the vesicle contains a food particle (or sometimes an entire cell such as a bacterium), that particular version of endocytosis is called **phagocytosis** (from the Greek word *phagein,* to eat). If the material brought in is a liquid containing dissolved molecules, the endocytosis is referred to as **pinocytosis** (from the Greek word *pinein,* to drink). Pinocytosis is common among human cells. Human egg cells, for example, are "nursed" by surrounding cells, which secrete nutrients the maturing egg takes up by pinocytosis. Some types of white blood cell ingest 25% of their volume by pinocytosis each hour!

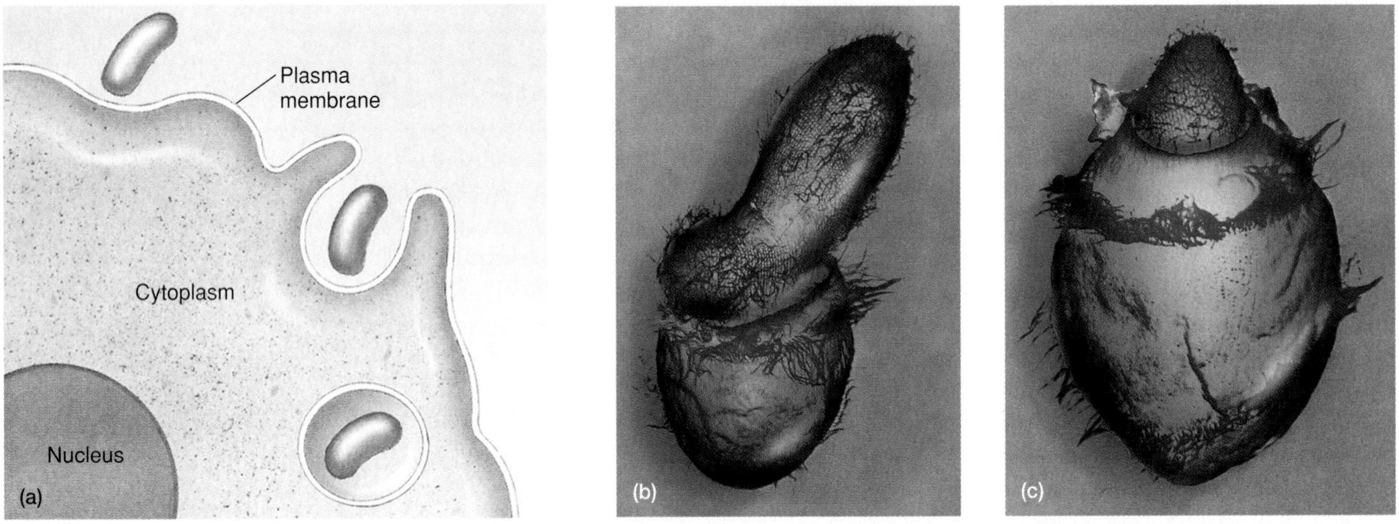

Figure 4.5 Endocytosis.
Endocytosis is the process of engulfing material by folding the plasma membrane around it, forming a vesicle. (*a*) When the material is an organism or some other relatively large fragment of organic matter, the process is called phagocytosis. This form of endocytosis can be quite dramatic. In (*b*), the egg-shaped protist *Didinium nasutum* is ingesting the smaller protist *Paramecium*. A little later (*c*), the *Didinium*'s meal is almost over.

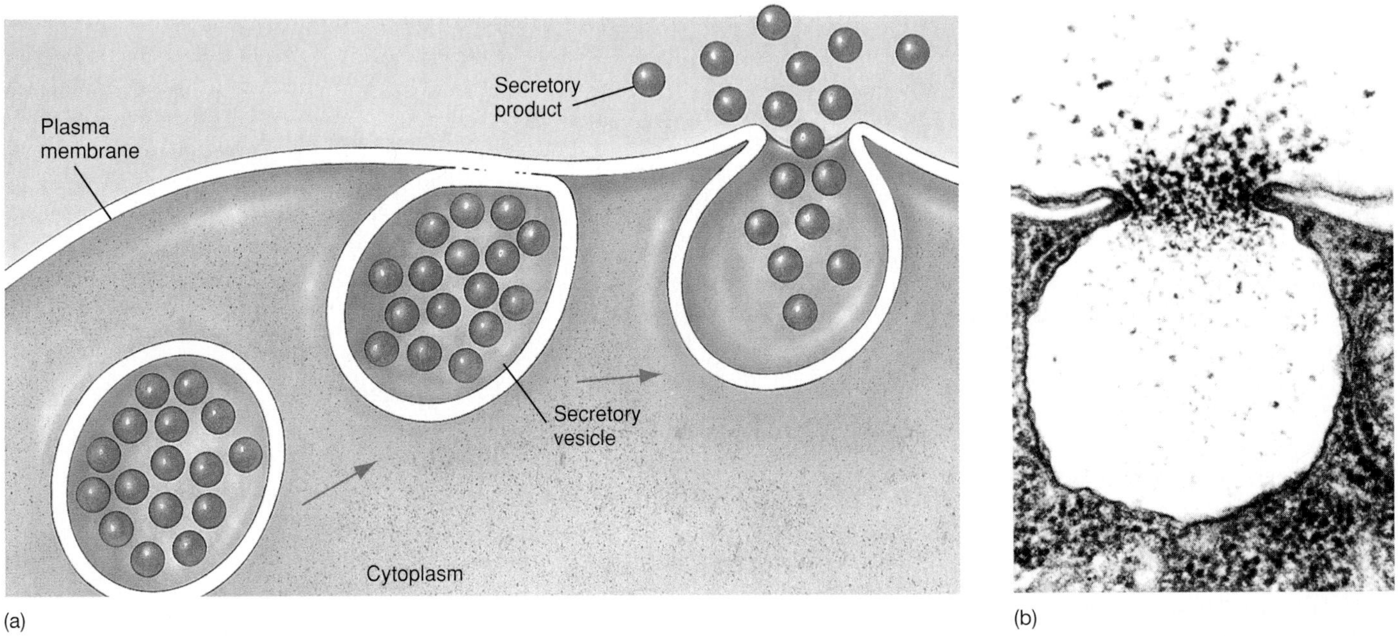

Figure 4.6 Exocytosis.
Exocytosis is the discharge of material from vesicles at the cell surface. (*a*) Proteins and other molecules are secreted from cells in small pockets called secretory vesicles, whose membranes fuse with the cell membrane, thereby allowing the secretory vesicles to release their contents to the cell surface. (*b*) In the photomicrograph, you can see exocytosis taking place explosively.

Selective Transport of Molecules

From the point of view of efficiency, the problem with endocytosis is that it is expensive to carry out—the cell must make and move a lot of membrane. Also, endocytosis is not picky—in pinocytosis particularly, engulfing liquid does not allow the cell to choose which molecules come in. Cells solve this problem by using some of the many proteins that protrude through their plasma membrane as channels to pass molecules into and out of the cell. Because each kind of channel will pass only a certain kind of molecule, the cell can control what enters and leaves.

Facilitated Diffusion

Some channels are open doors. As long as a molecule fits the channel, it is free to pass through in either direction. Diffusion tends to equalize the concentration of such molecules on both sides of the membrane, with more of the molecules moving toward the side where it is scarcest. Because this diffusion could not occur without the channel in the membrane for the molecules to pass across, this form of diffusion is given a special name, **facilitated diffusion** (figure 4.7).

Active Transport

Other channels through the plasma membrane are closed doors. These channels open only when energy is provided. They are designed to enable the cell to maintain high concentrations of certain molecules, much more than exists outside the cell. If the doors were open, the molecules would simply flood out by facilitated diffusion. Instead, like motor-driven turnstiles, the channels only operate when energy is provided, and they only operate in one direction. The operation of these one-way, energy-requiring channels results in **active transport,** the movement of molecules across a membrane to a region of higher concentration by the expenditure of energy (figure 4.8).

Active transport is one of the most important activities of any cell, permitting the cell to accumulate food molecules from outside even when the cell has more of them inside its cytoplasm than are outside. Extracting scarce molecules from the environment is only possible because of active transport.

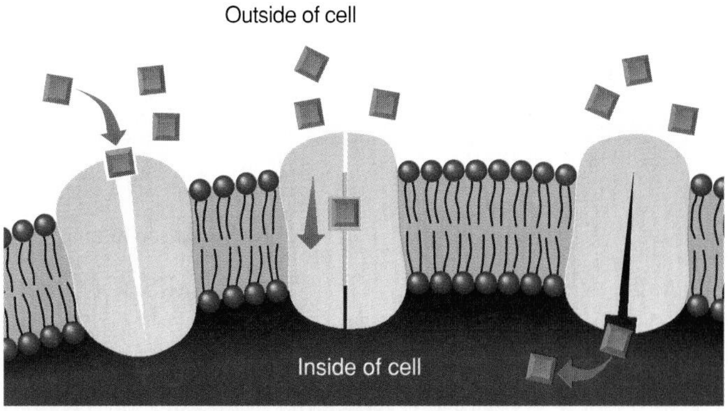

Figure 4.7 Facilitated diffusion.
Facilitated diffusion is diffusion that occurs through protein channels in the membrane. The channels help (facilitate) the diffusion process and do not require energy. The protein channel transports only certain molecules across the membrane but will take them in either direction.

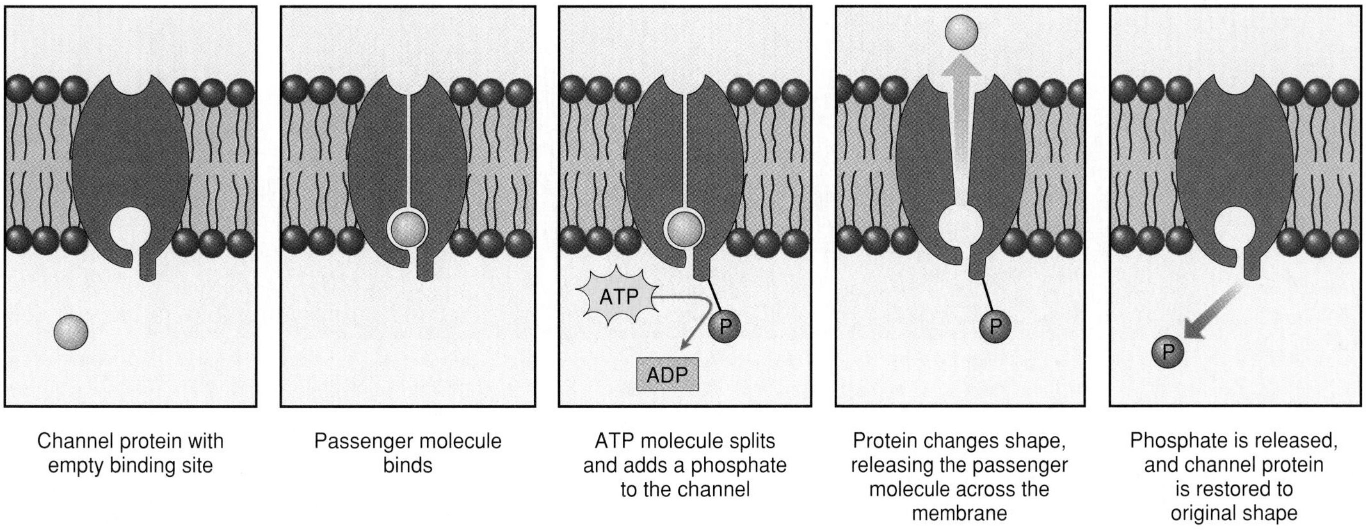

ATP
P
ADP
Channel protein with empty binding site
Passenger molecule binds
ATP molecule splits and adds a phosphate to the channel
Protein changes shape, releasing the passenger molecule across the membrane
Phosphate is released, and channel protein is restored to original shape

Figure 4.8 Active transport.
Active transport is the movement of molecules across a membrane toward a region of higher concentration using energy to drive the process. The active transport channel protein fits only certain molecules or ions (here illustrated by an orange ball). The splitting of an ATP molecule (see page 90) adds a phosphate to the channel, causing it to change shape in such a way that the passenger molecule is released on the opposite side of the membrane.

Active Transport Channels

Because of its overwhelming importance in acquiring all sorts of sugars, amino acids, and other molecules, you might think that the plasma membrane possesses all sorts of active transport channels. You would be wrong. In fact, almost all of the active transport in cells is carried out by only two kinds of channels, the sodium-potassium pump and the proton pump.

The first of these, the **sodium-potassium pump,** expends metabolic energy to actively pump sodium ions (Na^+) out of cells and potassium ions (K^+) in (figure 4.9). More than one-third of all the energy expended by your body's cells is spent driving sodium-potassium pump channels. Each channel can move over 300 sodium ions per second when working full tilt. As a result of all this pumping, there are far fewer sodium ions in the cell. This concentration gradient, paid for by the expenditure of considerable metabolic energy in the form of ATP molecules, is exploited by your cells in many ways. Two of the most important are (1) the conduction of signals along nerve cells (discussed in detail in chapter 23) and (2) the pulling into the cell of valuable molecules such as sugars and amino acids *against* their concentration gradient!

We will focus for a moment on this second process. The plasma membranes of many cells are studded with facilitated diffusion channels, which offer a path for sodium ions that have been pumped out by the Na^+-K^+ pump to diffuse back in. There is a catch, however: these channels require that the sodium ions have a partner in order to pass

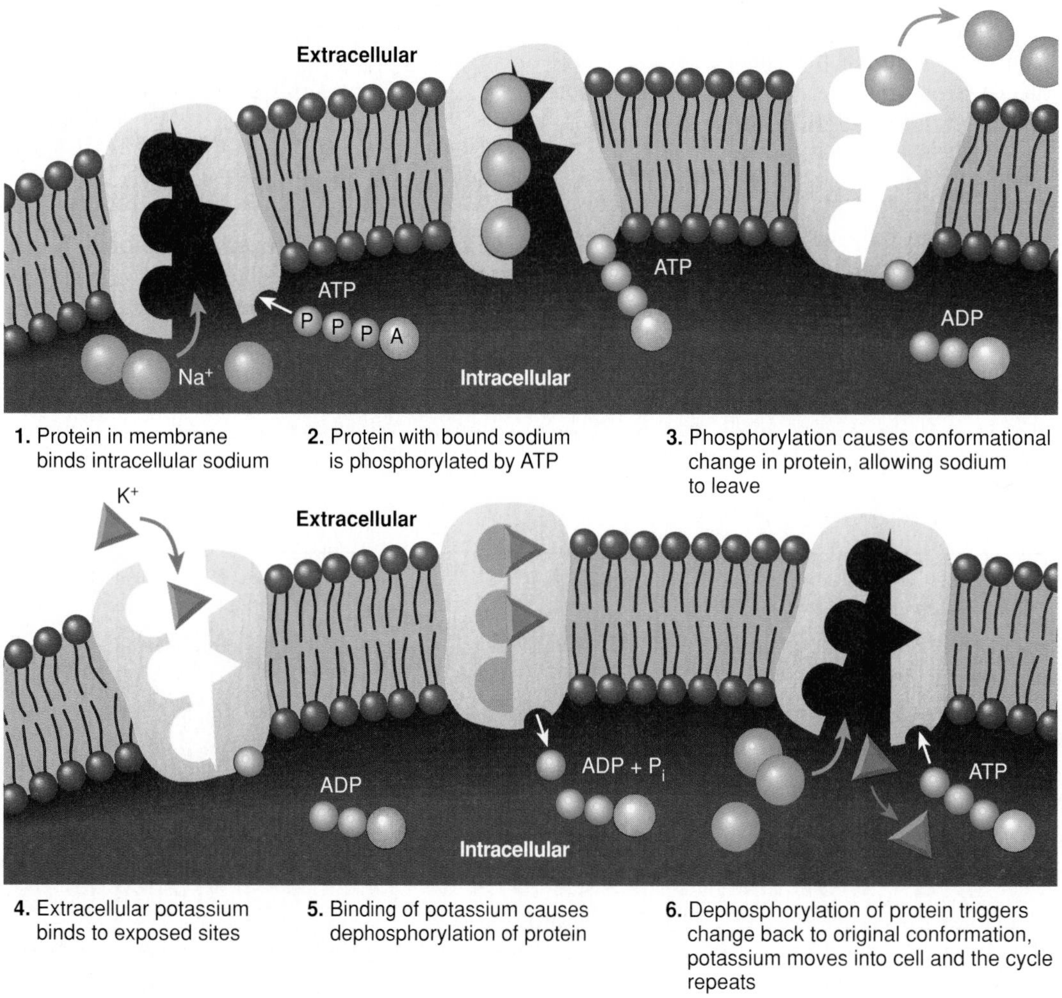
Extracellular
ATP
P P P A
Na+
Intracellular
ATP
ADP
1. Protein in membrane binds intracellular sodium
2. Protein with bound sodium is phosphorylated by ATP
3. Phosphorylation causes conformational change in protein, allowing sodium to leave
K+
Extracellular
ADP
ADP + Pi
ATP
Intracellular
4. Extracellular potassium binds to exposed sites
5. Binding of potassium causes dephosphorylation of protein
6. Dephosphorylation of protein triggers change back to original conformation, potassium moves into cell and the cycle repeats

Figure 4.9 The sodium-potassium pump.

The sodium-potassium pump transports sodium ions (Na^+) out of cells and potassium ions (K^+) in. Na^+ ions first bind to the inner surface of the sodium channel. The splitting of ATP provides energy to change the shape of both channels, causing Na^+ ions to be expelled outward and K^+ ions to bind to the channel. K^+ ions are then released into the interior.

through—like a dancing party where only couples are admitted through the door. These special channels won't let sodium ions across unless another molecule tags along, crossing hand in hand with the sodium ion. In some cases the partner molecule is a sugar, in others an amino acid or other molecule. Because so many sodium ions are trying to get back in, this diffusion pressure drags in the partner molecules as well, even if they are already in high concentration within the cell. In this way, sugars and other actively transported molecules enter the cell—via special **coupled channels** (figure 4.10). Their movement is in fact a form of facilitated diffusion driven by the active transport of sodium ions (figure 4.11).

The second major active transport channel is the **proton pump,** a complex channel that expends metabolic energy to pump protons across membranes. Just as in the sodium-potassium pump, this creates a diffusion pressure that tends to drive protons back across again, but in this case the only channels open to them are not coupled channels but rather channels that make ATP, the energy currency of the cell. This pump is the key to cell metabolism, which is the way cells convert photosynthetic energy or chemical energy from food into ATP. Its activity is referred to as **chemiosmosis.** We discuss chemiosmosis at greater length in the next chapter.

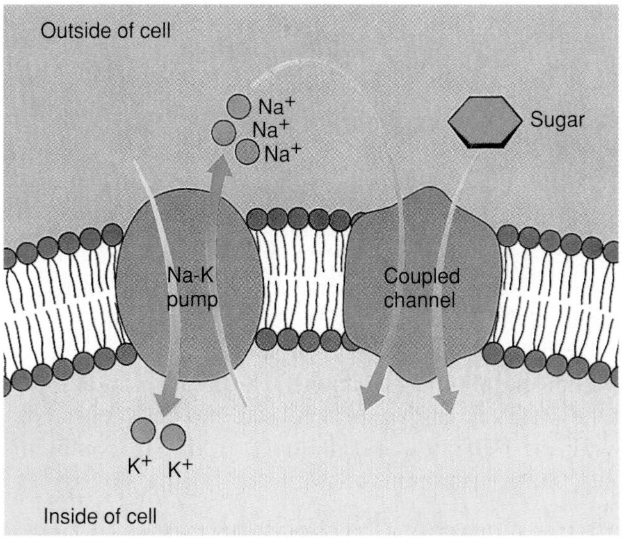

Figure 4.10 A coupled channel.
The active transport of a sugar molecule into a cell typically takes place in two stages—facilitated diffusion of the sugar coupled to active transport of sodium ions. The sodium-potassium pump keeps the Na^+ ion concentration higher outside the cell than inside. There is thus a strong tendency for Na^+ ions to diffuse back in through the coupled channel, but passage through this channel requires the simultaneous transport of a sugar molecule as well. Because the concentration gradient for Na^+ is steeper than the opposing gradient for sugar, Na^+ and sugar move into the cell.

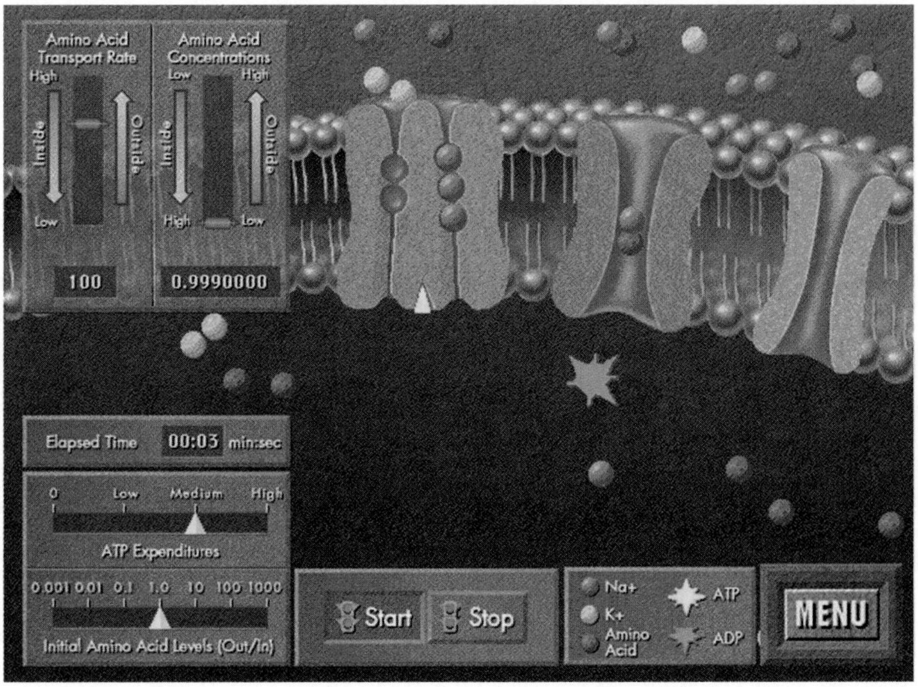

Figure 4.11 Active transport involves coupled channels.
The three channels in this diagram, from the left, are a Na^+-K^+ pump, a coupled channel that transports amino acids in concert with Na^+ ions, and an open K^+ channel. This illustration is a "screen capture" from an interactive CD-ROM exercise that enables you to alter ATP concentrations and the relative inside/outside concentrations of the amino acids. Because the amino acid transport channel is coupled to the ATP-driven sodium-potassium pump, you will discover that both ATP and amino acid levels have important influences. (*Explorations in Human Biology*, Module 2; *Explorations in Cell Biology & Genetics*, Module 3, "Active Transport")

4.2 How Cells Get Information

A cell's ability to respond appropriately to changes in its environment is a key element in its ability to survive. Cells have evolved a variety of ways of sensing things about them. A very few specialized cells have "ears" or "eyes" sensitive to pressure or light. Almost all cells, however, sense their environment primarily by detecting chemical or electrical signals. To do this, cells employ a battery of special proteins called **cell surface proteins** embedded within the plasma membrane. The many proteins that protrude from the surface of a cell are the cell's only contact with the outside world, its only avenue of communication with its environment.

Sensing Chemical Information

Cells sense chemical information by means of cell surface proteins called **receptor proteins** projecting from their plasma membranes. These proteins bind a particular kind of molecule, but they do not provide a channel for the molecule to enter the cell. What the receptors do transmit into the cell is information. Often the information is about the presence of other cells in the vicinity—sensing cellular identity is the basis of the immune system, which defends your body from infection. In other instances the information may be a chemical signal sent from other cells.

Your body uses chemical signals called **hormones,** which provide a good example of how receptor proteins work. The end of a receptor protein exposed to the cell surface has a shape that fits to a specific hormone molecule, like insulin. Most of your cells have only a few insulin receptors, but your liver cells possess as many as 100,000 each! When an insulin molecule encounters an insulin receptor on the surface of a liver cell, the insulin molecule binds to the receptor. This binding produces a change in the shape of the other end of the receptor protein (the end protruding into the interior of the cell), just as stamping hard on your foot causes your mouth to open. This change in receptor shape at the interior end initiates a change in cell activity—in this case, the end protruding into the cytoplasm begins to add phosphate groups to proteins, and so it activates a variety of cell processes involved with regulating glucose levels in the blood.

Sensing Voltage

Many cells can sense electrical as well as chemical information. Embedded within their plasma membranes are special channels for sodium or other ions, channels that are usually closed. Unlike the sodium-potassium pump, these closed channels do not open in response to chemical energy. What does open these channels is voltage. Like little magnets, they flip open or shut in response to electrical signals.

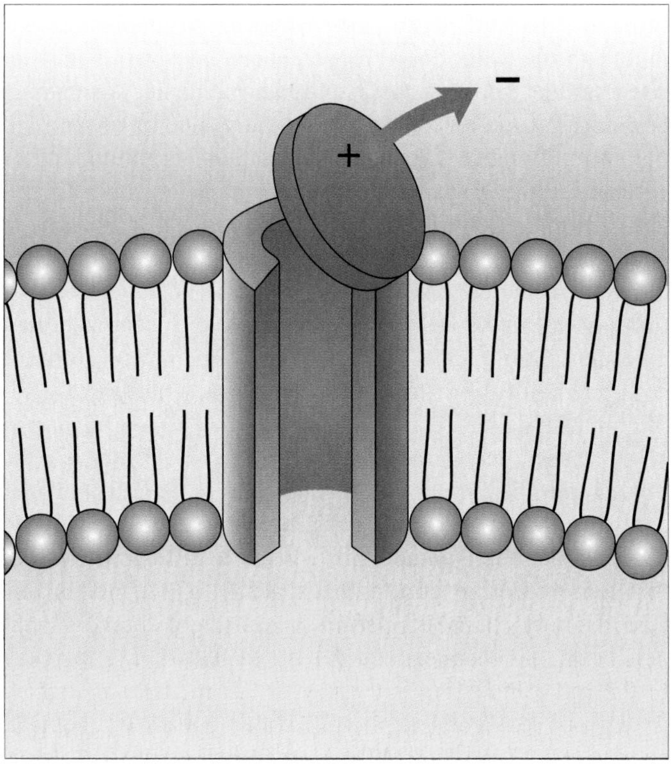

Figure 4.12 Voltage-sensitive channels.
In this diagram, the door of the channel (a positively charged portion of a membrane protein) opens when a negative charge is encountered, and this allows ions to pass through.

It is not difficult to understand how voltage-sensitive channels work. The center of the protein that provides the channel through the membrane is occupied by a voltage-sensitive "door"—a portion of the protein containing charged amino acids. When a current of the opposite charge is encountered, the door flips up out of the way and the channel is open to the passage of sodium or other ions. **Voltage-sensitive channels** play many important roles within the nervous system (figure 4.12).

Sensing Information Within the Cell

In eukaryotic cells, which have many compartments, it is very important that the different parts of the cell be able to sense what is going on elsewhere in the cytoplasm. This sort of communication is provided by the diffusion of molecules within the cytoplasm. Because the endoplasmic reticulum extends throughout the cell's interior, it is the highway over which this information passes. Some of the chemical signals are molecules used in metabolism; others are within-cell hormones; and still others are ions, particularly those that indicate cell pH.

4.3 How Cells Divide

Every living cell is capable of division. Your body contains about 100 trillion cells, all derived from a single cell at the start of your life as a fertilized egg. Many millions of successful cell divisions occurred while your body was reaching its present form. For a cell, reproducing by cell division is one of the essential tasks of living.

Bacteria Simply Split

For bacteria, the process of cell division is simple. The hereditary information—that is, the genes that specify the bacteria—is encoded in a single circle of DNA, attached at one point to the inner surface of the plasma membrane like a rope attached to the inner wall of a tent. Cell division takes place in two stages. First the DNA is copied, and then the cell splits.

Copying the DNA

Before the cell itself divides, the DNA circle makes a copy of itself. Starting at one point, the double helix of DNA begins to unzip, exposing the two strands. A new double helix is then formed on each naked strand by placing on each exposed nucleotide its complementary nucleotide (that is, A with T, G with C). When the unzipping has gone all the way around the circle, the cell possesses two copies of its hereditary information, attached side by side to the interior plasma membrane (figure 4.13).

Splitting the Cell

When the DNA has been copied and the cell reaches an appropriate size, the bacterial cell splits into two equal halves, a process called **binary fission** (figure 4.14). First, new plasma membrane and cell wall are added at a point between where the two DNA copies are attached to the membrane. As new material is added in this zone, the growing plasma membrane pushes inward and the cell is gradually constricted in two, like tightening a belt around a long balloon. Eventually the dividing bacterial cell is pinched into two **daughter cells.** Each contains one of the circles of DNA, and is a complete living cell in its own right.

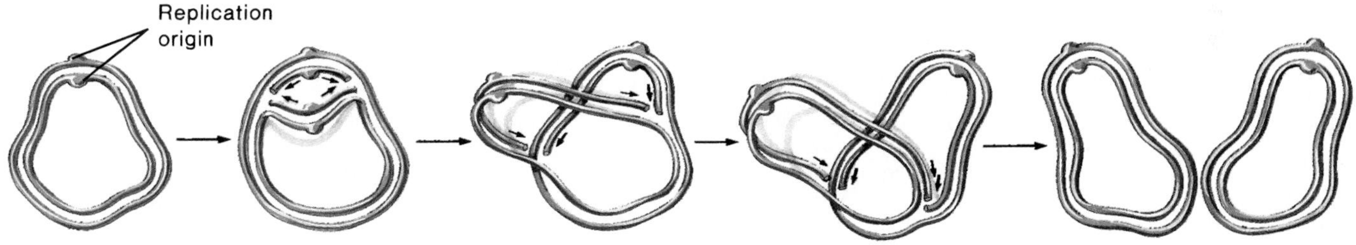

Figure 4.13 How bacterial DNA replicates.
The circular DNA molecule of a bacterium initiates replication at a single site, moving out in both directions. When the two moving replication points meet on the far side of the molecule, its replication has been completed.

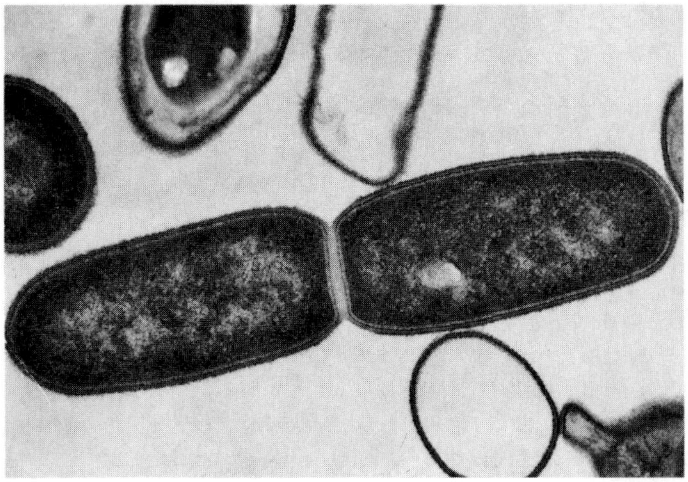

Figure 4.14 Binary fission in bacteria.
Bacteria divide by a process of simple cell fission. Here the daughter cells are about to be pinched apart by the growing plasma membrane.

Eukaryotes Have a Complex Cell Cycle

Cell division in eukaryotes is more complex than in bacteria, both because eukaryotes contain far more DNA and because it is packaged differently. The cells of eukaryotic organisms either undergo mitosis or meiosis to divide up the DNA. **Mitosis** is the mechanism of cell division that occurs in an organism's nonreproductive cells, or **somatic cells.** (A second process, called **meiosis,** operates to divide the DNA in cells that participate in sexual reproduction, or **germ cells.** Meiosis results in the production of gametes, such as sperm and eggs, and is discussed in chapter 6.)

Chromosomes

A eukaryotic cell has far more DNA in it than a bacterium does. The DNA of eukaryotes is divided into several packages, compact rod-shaped masses called **chromosomes** (figure 4.15). Because the phosphate groups of DNA molecules have negative charges, it is impossible to just wind DNA up tightly because all the negative charges would simply repel one another. Instead, the DNA helix is wrapped around proteins with positive charges called **histones,** the positive (histone) and negative (DNA) charges counteracting each other so that the complex has no net charge. This complex of DNA and histone proteins is then coiled tightly to form a compact chromosome. For example, a human

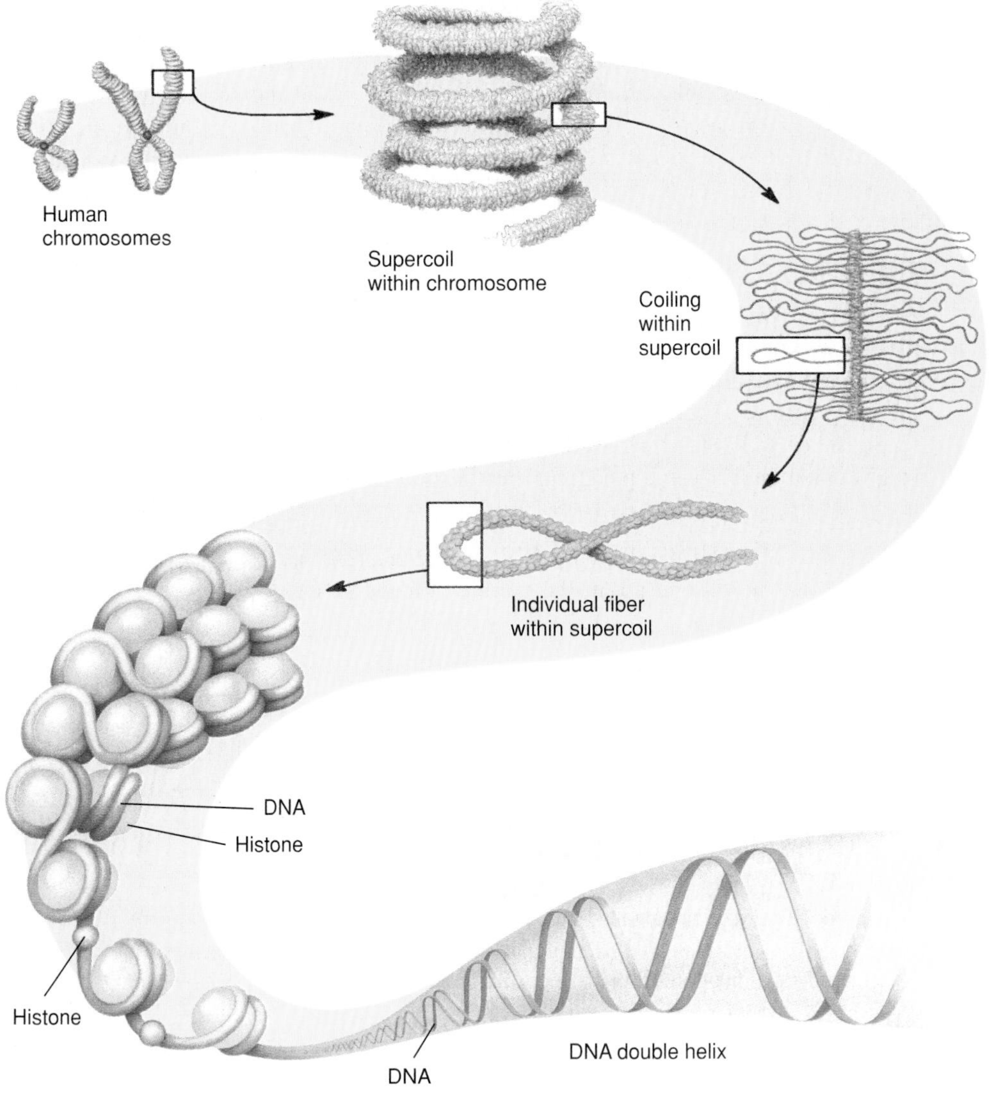

Figure 4.15 Levels of chromosomal organization.
Compact, rod-shaped chromosomes are in fact highly wound-up molecules of DNA.

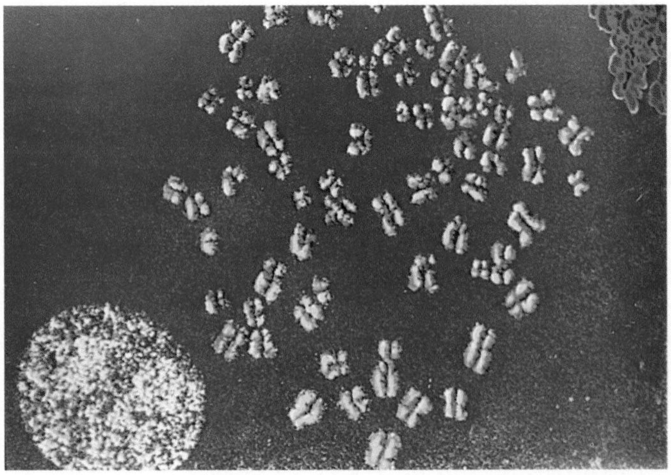

(a)

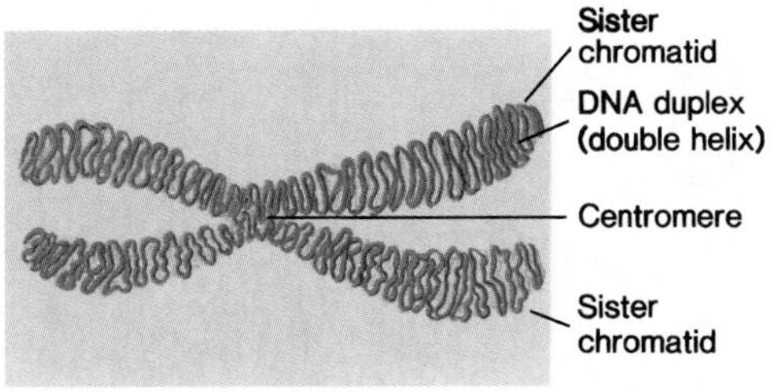

(b)

Figure 4.16 Sister chromatids.
A duplicated chromosome looks somewhat like an X and is composed of two sister chromatids held together by a centromere. (*a*) A collection of sister chromatids appears in the micrograph. (*b*) An illustration of one sister chromatid.

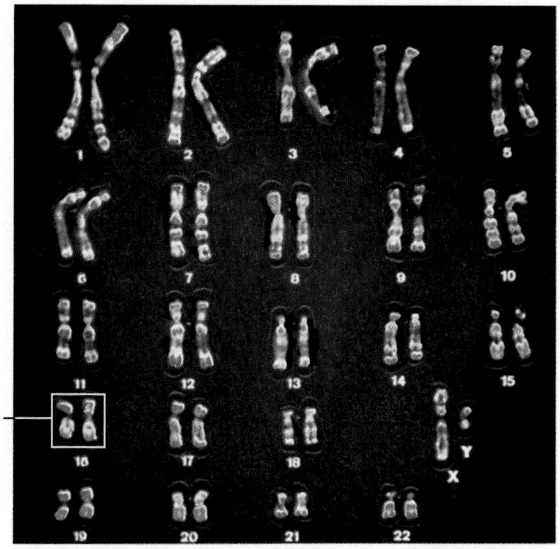

Figure 4.17 The 46 chromosomes of a human.
Each human cell has 23 pairs of homologous chromosomes, for a total of 46. In this presentation, photographs of the individual chromosomes of a human male have been cut out and paired with their homologues, creating an organized display of the chromosomes in their duplicated state called a **karyotype.**

chromosome is about 40% DNA and 60% protein and typically contains about half a billion nucleotides in one long, unbroken DNA helix that would be about 5 centimeters (2 inches) long if laid out in a straight line. The amount of information in one human chromosome would fill about 2,000 printed books of 1,000 pages each!

Chromosomes exist in somatic cells as two nearly identical copies of each other, called **homologous chromosomes,** or **homologues.** Cells that have two of each type of chromosome are called **diploid cells.** Before cell division, each of the two homologues replicates, resulting in two identical copies, called **sister chromatids,** that remain joined together at a special linkage site called the **centromere** (figure 4.16). For example, human body cells have 46 chromosomes. In their unduplicated state, that means you have 23 pairs of homologous chromosomes, one set from your mother and one set from your father (figure 4.17). In their duplicated state, there are 23 pairs of duplicated chromosomes, which consist of two sister chromatids each, for a total of 92 chromatids.

The Cell Cycle

Eukaryotic cell division is more complex than bacterial cell division and requires several predivision growth and preparation phases. The events that prepare the cell for division and the division process itself constitute a **cell cycle.** The five main phases of the eukaryotic cell cycle are shown in figure 4.18. (1) The G_1, or first growth, phase is the cell's primary growth phase. For most organisms, this phase occupies the major portion of the cell's life span. (2) During the S, or synthesis, phase, each chromosome replicates to produce two sister chromatids with identical DNA. (3) Cell division preparation continues in the G_2, or second growth, phase with the replication of mitochondria and other organelles, chromosome condensation, and the synthesis of microtubules. These three phases are collectively referred to as **interphase,** or the period between cell divisions. (4) In the M, or mitosis, phase, a microtubular apparatus is assembled that binds to the chromosomes and moves them apart. (5) Last, in the C, or cytokinesis, phase, the cytoplasm divides, creating two daughter cells.

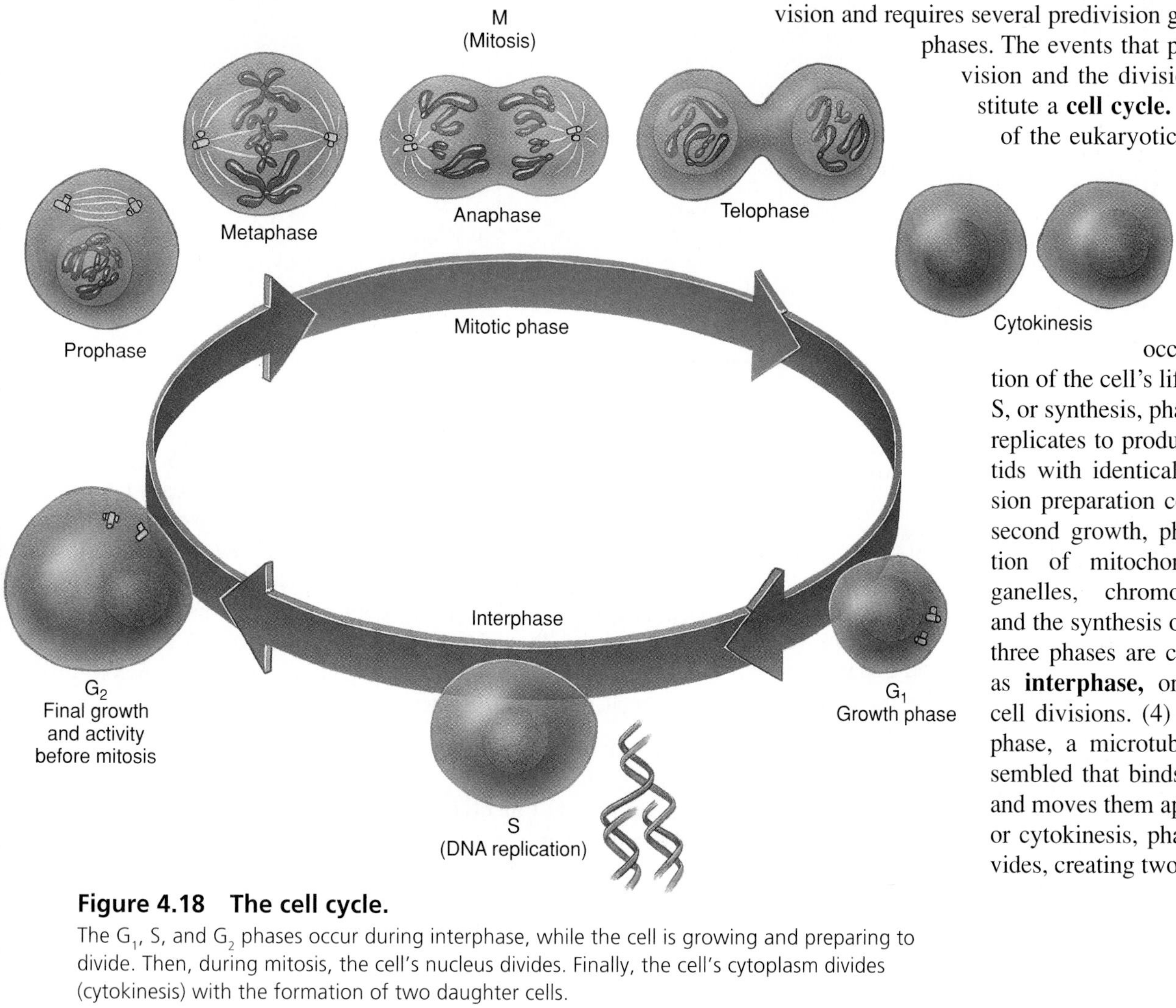

Figure 4.18 The cell cycle.
The G_1, S, and G_2 phases occur during interphase, while the cell is growing and preparing to divide. Then, during mitosis, the cell's nucleus divides. Finally, the cell's cytoplasm divides (cytokinesis) with the formation of two daughter cells.

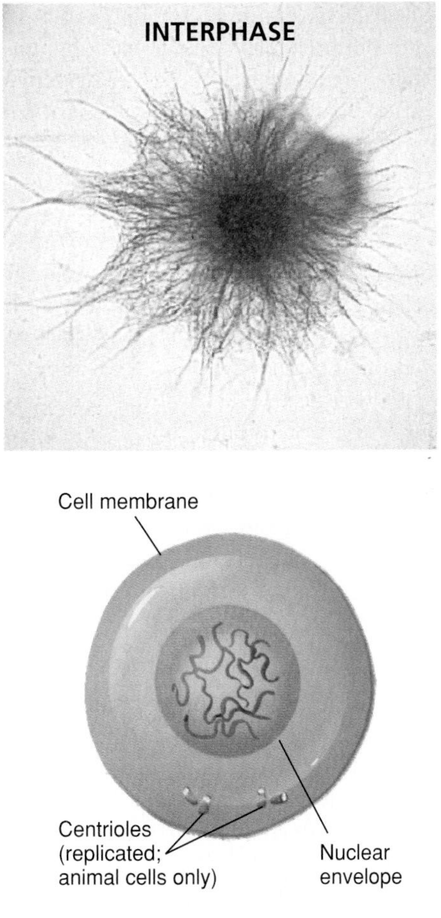

DNA replicates and begins to condense. Centrioles, if present, also replicate, and the cell prepares for division.

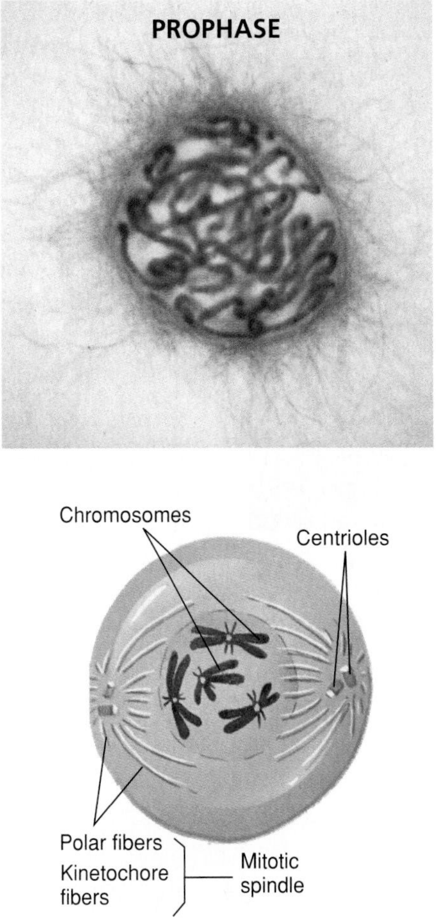

The nuclear envelope begins to break down. DNA further condenses into chromosomes. The mitotic spindle begins to form; it is complete at the end of prophase.

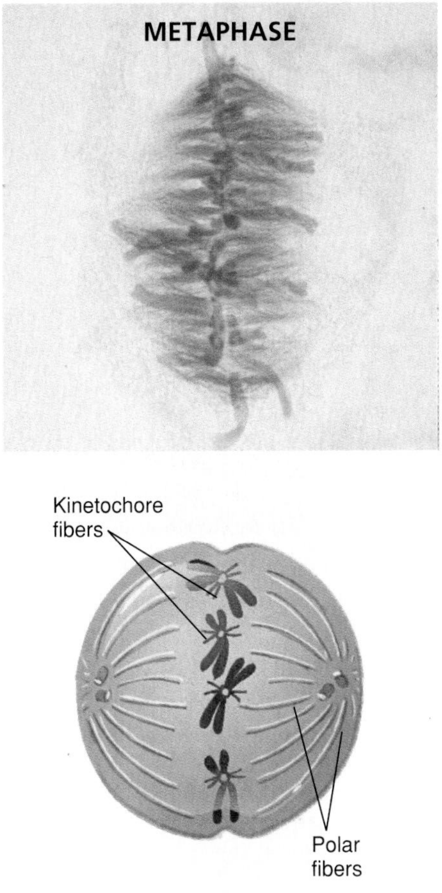

The chromosomes align on a plane in the center of the cell. The kinetochore fibers attach at opposite sides. The centromeres replicate.

How Cells Move Chromosomes

In eukaryotes the mechanical separation of chromosomes into two daughter cells involves a complex series of events. When cell division begins, chromosomes begin to wind up tightly, a process called **condensation.**

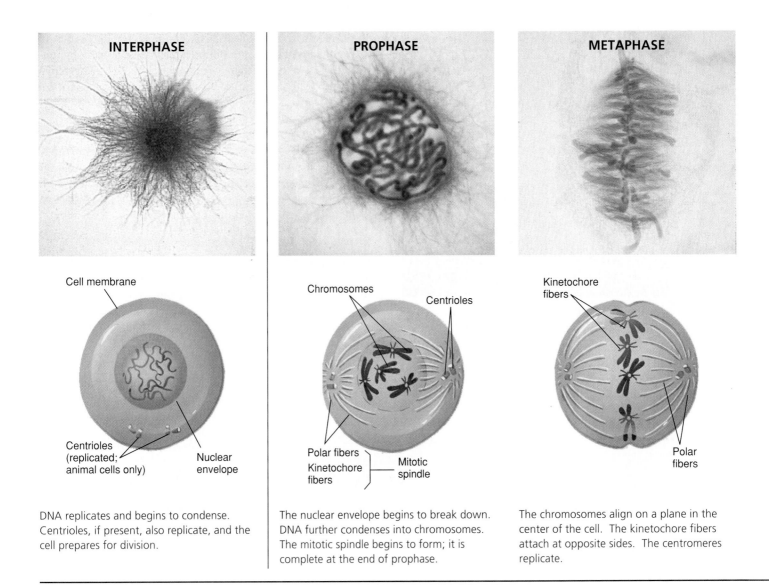

Forming Microtubules

As the chromosomes condense, the cell dismantles the nuclear membrane and begins to assemble the apparatus it will use to pull the replicated daughter chromosomes to opposite ends ("poles") of the cell. In the center of an animal cell, the pair of centrioles starts to separate, the two centrioles moving apart toward opposite poles of the cell, forming between them as they move apart a network of protein cables called the **spindle.** Each cable is called a spindle fiber and is made of microtubules, long hollow tubes of protein. Plant cells lack centrioles and instead brace the ends of the spindle with a support structure called an aster.

Attaching Microtubules to Centromeres

As condensation of the chromosomes continues, a second group of microtubules extends out from the centromere of each chromosome from a disk of protein called a **kinetochore.** The two sets of microtubules extend out from opposite sides of the kinetochore toward opposite poles of the cell. Each set of microtubules continues to grow longer until it makes contact with the pole toward which it is growing. When the process is complete, one daughter chromosome of each pair is attached by microtubules to one pole and the other daughter to the other pole.

Separating Daughter Chromosomes

The kinetochore then splits, freeing the daughter chromosomes from each other. Cell division is now simply a matter of reeling in the microtubules, dragging to the poles the daughter chromosomes like fish on the end of a line—at the poles, the ends of the microtubules are dismantled, one bit after another, making the tubes shorter and shorter and so drawing the chromosome attached to the far end closer and closer to the pole. When they finally arrive, each pole has one complete set of chromosomes.

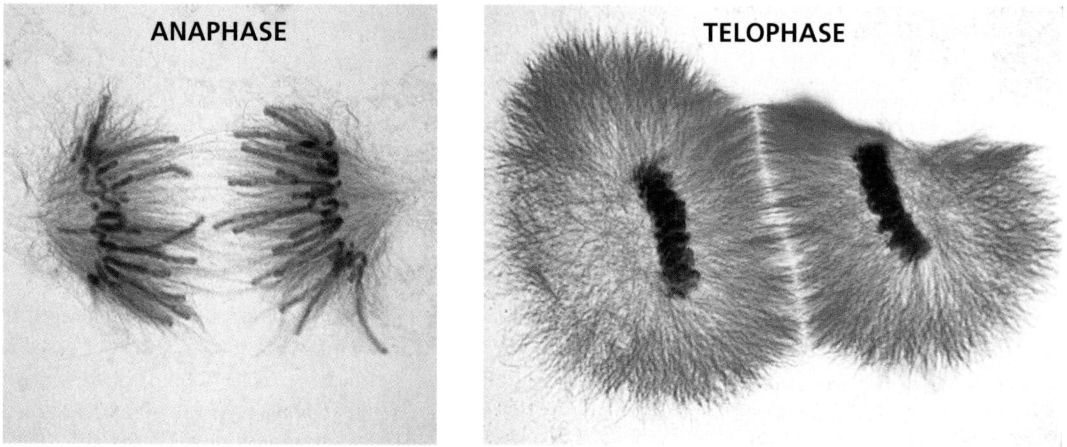

Figure 4.19 Interphase and mitosis.
Interphase and the four stages of mitosis that follow it can be seen clearly in these dividing cells of the African blood lily, *Haemanthus katharinae*. Microtubules are stained red and chromosomes are stained blue. In the drawing, centrioles (present in animals) are indicated, although they are absent in plants.

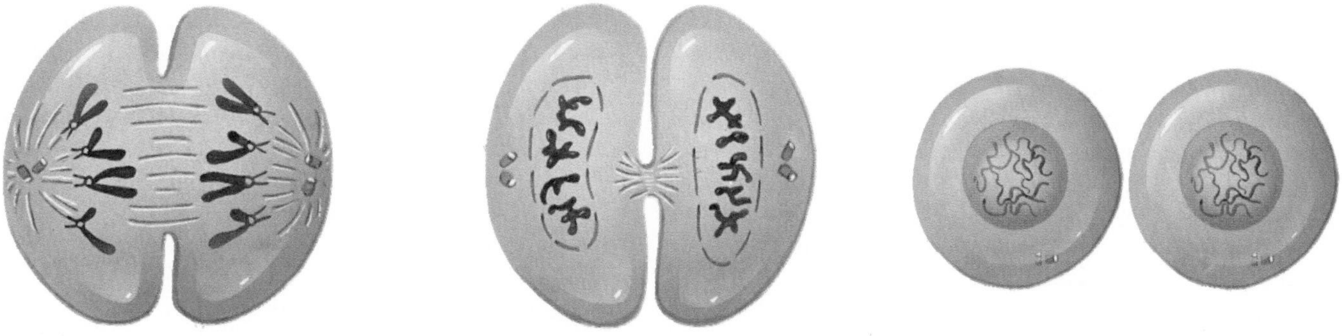

The sister chromatids separate and move to opposite poles.

The nuclear envelope reappears. The chromosomes decondense. As telophase progresses, cytokinesis also occurs.

Two daughter cells have formed. Each cell is a replicate of the parent cell and is diploid.

The Stages of Mitosis

Although the process of mitosis is continuous, with the stages flowing smoothly one into another, for ease of study, mitosis is traditionally subdivided into four stages: prophase, metaphase, anaphase, and telophase (figure 4.19).

Prophase: Mitosis Begins

In **prophase,** the individual condensed chromosomes first become visible with a light microscope. Also in this stage, the nuclear envelope begins to break down and the mitotic spindle that will be used to separate the sister chromatids is assembled. In animal cells and those of most protists, the spindle fibers are associated with the centrioles. After the centrioles are replicated, the pairs separate and move to opposite poles of the cell. In late prophase, a second group of fibers attach to the kinetochores of the individual chromosomes. Two kinetochore fibers extend in opposite directions from each chromosome.

Metaphase: Alignment of the Chromosomes

The second phase of mitosis, **metaphase,** begins when the chromosomes, each consisting of a pair of chromatids, align in the center of the cell. At the end of metaphase, each centromere replicates, freeing the two sister chromatids to be drawn to two opposite poles of the cell in the next phase.

Anaphase: Separation of the Chromatids

In **anaphase,** the sister chromatids move rapidly toward opposite poles of the cell. This is the shortest stage of mitosis. The two sister chromatids separate and become the chromosomes of the daughter cells.

Telophase: Re-formation of the Nuclei

With the movement of the sister chromatids to the opposite poles during anaphase, the only tasks that remain in **telophase** are the dismantling of the stage and the removal of the props. The mitotic spindle is disassembled, and a nuclear envelope forms around each set of chromosomes while they begin to uncoil.

Cytokinesis

At the end of telophase, mitosis is complete. The cell has divided its replicated chromosomes into two nuclei, which are positioned at opposite ends of the cell. Following mitosis, **cytokinesis,** the division of the cytoplasm, occurs, and the cell is cleaved into roughly equal halves. Cytoplasmic organelles have already been replicated and reassorted to the areas that will separate and become the daughter cells.

In animal cells, which lack cell walls, cytokinesis is achieved by pinching the cell in two with a contracting belt of microtubules. As contraction proceeds, a **cleavage furrow** becomes evident around the cell's circumference, where the cytoplasm is being progressively pinched inward by the increasing diameter of the microtubule belt (figure 4.20*a*). The cleavage furrow deepens until the cell is literally pinched in two.

Plant cells have rigid walls that are far too strong to be deformed by microtubule contraction. A different approach to cytokinesis has therefore evolved in plants. Plant cells assemble membrane components in their interior, at right angles to the mitotic spindle. This expanding partition, called a **cell plate,** grows outward until it reaches the interior surface of the cell membrane and fuses with it, at which point it has effectively divided the cell in two (figure 4.20*b*). Cellulose, the major component of cell walls is then laid down on the new membranes, creating two new cells.

Cell Death

Despite the ability to divide, no cell lives forever. The ravages of living slowly tear away at a cell's machinery. To some degree damaged parts can be replaced, but no replacement process is perfect. And sometimes the environment intervenes. If food supplies are cut off, for example, animal cells cannot obtain the energy necessary to maintain their lysosome membranes. The cells die, digested from within by their own enzymes.

During fetal development, many cells are programmed to die. In human embryos, hands and feet appear first as "paddles," but the skin cells between bones die on schedule to form the separated toes and fingers (figure 4.21). In ducks, this cell death is not part of the developmental program, which is why ducks have webbed feet and you don't.

Human cells appear to be programmed to undergo only so many cell divisions and then die, following a plan written into the genes. In tissue culture, cell lines divide about 50 times, and then the entire population of cells dies off. Even if some of the cells are frozen for years, when they are thawed they simply resume where they left off and die on schedule. Only cancer cells appear to thwart these instructions, dividing endlessly. All other cells in your body contain a hidden clock that keeps time by counting cell divisions, and when the alarm goes off the cells die.

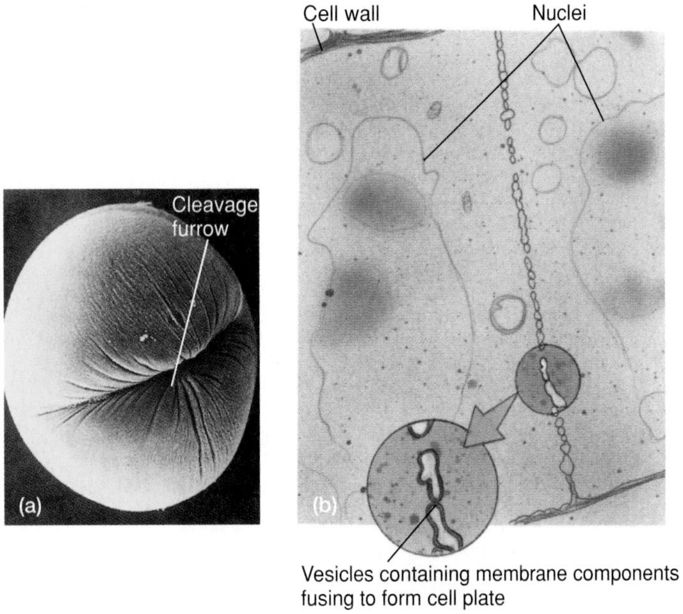

Figure 4.20 Cytokinesis.
The division of cytoplasm that occurs after mitosis is called cytokinesis and cleaves the cell into roughly equal halves. (*a*) In an animal cell, such as this sea urchin egg, a cleavage furrow forms around the dividing cell. (*b*) In this dividing plant cell, a cell plate is forming between the two newly forming daughter cells.

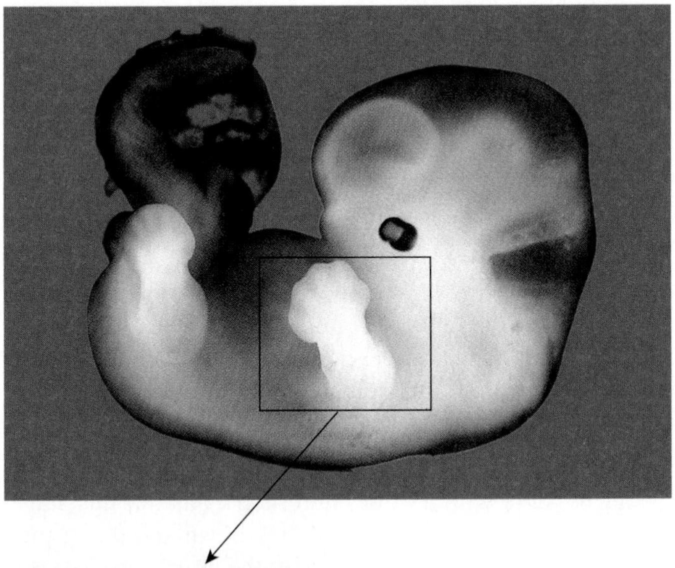

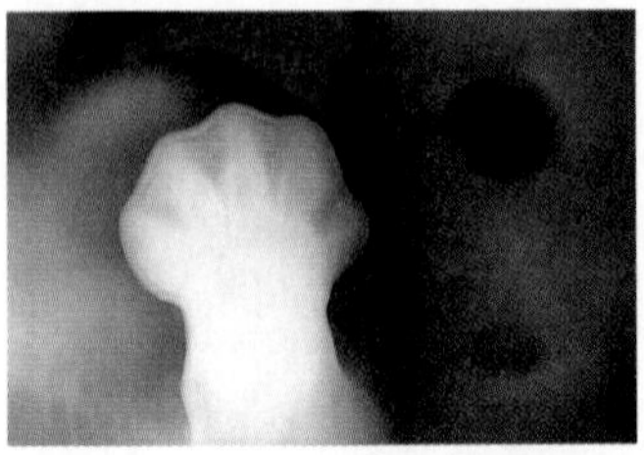

Figure 4.21 Programmed cell death.
In the human embryo, programmed cell death results in the formation of fingers and toes from paddlelike hands and feet.

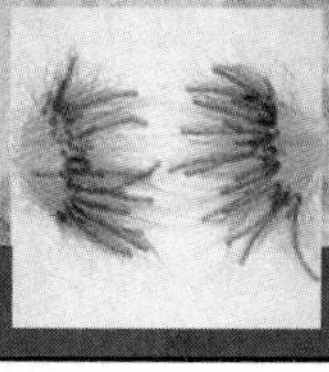

CHAPTER 4

HIGHLIGHTS

Key Terms | Key Concepts

4.1 How Cells Eat and Drink

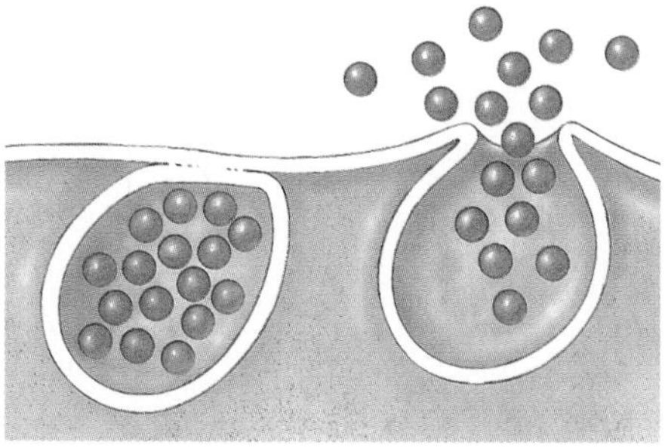

Key Terms

diffusion 68
osmosis 69
endocytosis 70
active transport 71

Key Concepts

- All cells transport water and other molecules across their plasma membranes.
- Molecules move randomly through the cell by diffusion.
- Molecules diffuse down concentration gradients, thus tending to equalize concentrations.
- Osmosis occurs when water molecules are able to move through a membrane and other polar molecules are blocked.
- Selective transport of molecules can occur via facilitated diffusion or via active transport with the expenditure of metabolic energy.

4.2 How Cells Get Information

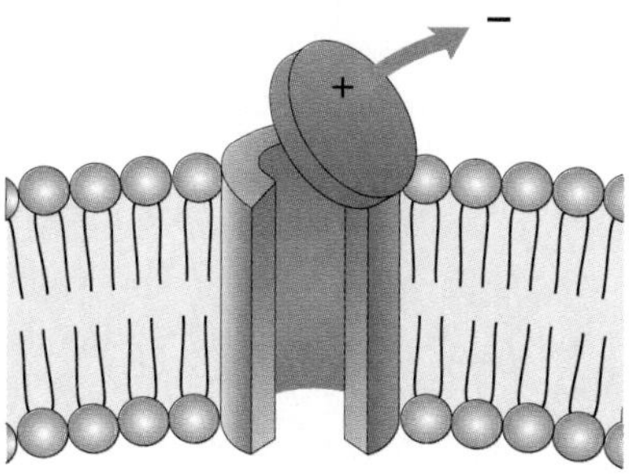

Key Terms

receptor protein 74
hormone 74
voltage-sensitive channel 74

Key Concepts

- Receptor proteins protruding from the surface of cells receive chemical and electrical stimuli and transfer this information to the cell's interior.
- Chemical signals called hormones pass information from one cell to another.
- Voltage-sensitive channels allow a cell to respond to changes in its electrical surroundings.

4.3 How Cells Divide

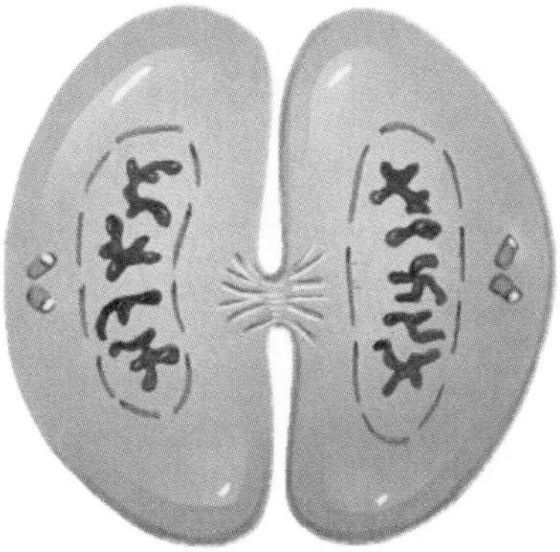

Key Terms

binary fission 75
mitosis 76
chromosome 76
cell cycle 77

Key Concepts

- Bacterial cells divide by simply splitting into two halves.
- Eukaryotic cells divide by mitosis, a complex process that delivers one replica of each chromosome to each of the two daughter cells.
- In humans, cells are programmed to undergo a limited number of cell divisions before dying.

4

CONCEPT REVIEW

1. Which of the following refers specifically to the transport of water?
   a. energy transfer
   b. active transport
   c. diffusion
   d. osmosis

2. How many of the following means of moving things into or out of cells require an expenditure of energy?
   a. diffusion
   b. osmosis
   c. facilitated diffusion
   d. active transport
   e. sodium-potassium pump

3. Each of the following statements describes osmosis except
   a. water molecules diffuse.
   b. water molecules pass through a membrane.
   c. water enters a hypotonic solution.
   d. movement of water solution creates osmotic pressure.

4. If you put an animal cell into a hypotonic solution, it will pop. However, if you put a plant cell into the same hypotonic solution, it will almost never pop. Why?
   a. The plant cell wall can withstand considerable hydrostatic pressure.
   b. Photosynthesis disrupts osmosis.
   c. Plant cells react oppositely from animal cells with regard to hypotonic solutions.
   d. Plant cells contain more water to begin with.

5. Which of the following processes is distinctly different from the other three?
   a. endocytosis
   b. exocytosis
   c. phagocytosis
   d. pinocytosis

6. Each of the following describes active transport except
   a. active transport requires energy.
   b. molecules move from a lower concentration to a higher concentration.
   c. the sodium-potassium pump is one example of active transport.
   d. active transport involves water diffusing through a membrane.

7. Cells can sense chemical information by means of surface receptor proteins.
   a. true
   b. false

8. Which structure is a network of protein cables?
   a. spindle
   b. kinetochore
   c. chromosome
   d. cell membrane

9. During the S phase of the cell cycle,
   a. the cell splits.
   b. DNA is replicated.
   c. the microtubular apparatus is assembled.
   d. cytokinesis occurs.

10. Cells take in liquid with dissolved molecules by _________.

11. Bacteria split by a process called _________ _________.

12. Chromosomes consist of DNA and _________.

13. The _________ phase is the growth phase of the cell cycle.

14. The _________ phase of the cell cycle is when the microtubular apparatus is assembled.

15. ________ is the cytoplasmic division of the cell.

Answers to the Concept Review questions appear in Appendix B.

CHALLENGE YOURSELF

1. When a hypertonic cell is placed in water solution, water molecules move rapidly into the cell. If the introduced cell is hypotonic, however, water molecules leave the cell and enter the surrounding solution. How does an individual water molecule *know* what solutes are on the other side of the membrane?

2. If you could construct an artificial chromosome, what elements would you introduce into it so that it could function normally?

3. What problems could develop in organisms whose cells divided too rapidly by mitosis?

FOR FURTHER READING

Bretscher, M. "How Animal Cells Move." *Scientific American,* December 1987, 72–90. An exciting suggestion that receptor-mediated endocytosis may prove to be the basic mechanism underlying animal cell movement.

Greider, C., and E. Blackburn. "Telomeres, Telomerase, and Cancer." *Scientific American*, February 1996, 92–97. Telomerase enzymes act on the ends of chromosomes, shortening them with each cell division.

Hartewell, L., and M. Kasten. "Cell Cycle Control and Cancer." *Science* 266 (December 1994): 1821–28. Why do cells suddenly divide out of control to produce cancer? The answer lies in regulation of the cell cycle.

Koshland, D. "Mitosis: Back to the Basics." *Cell* 77 (July 1, 1994): 951–54. A detailed overview of the entire process.

Mannuelidis, L. "A View of Interphase Chromosomes." *Science* 250 (December 1990): 1533–40. A review of how chromosomes are organized to express their genes.

Marx, J. "Medicine: A Signal Award for Discovering G Proteins." *Science* 266 (October 1994): 368–69. An account of how Rodbell and Gilman received the 1994 Nobel Prize for Medicine or Physiology for their work on G proteins.

Murray, A., and T. Hunt. *The Cell Cycle: An Introduction*. New York: Freeman, 1993. A nice, updated review of the entire mitotic process by one of the authorities in the field.

Slayman, C. L. "Proton Chemistry and the Ubiquity of Proton Pumps." *Bioscience* (January 1985): 16–47. This entire issue is devoted to a collection of seven articles on proton pumps; together they illustrate the central role these transmembrane channels play in the biology of both bacteria and eukaryotes.

Steahlin, L. A., and B. E. Hull. "Junctions Between Living Cells." *Scientific American,* (May 1978): 140–52. Many passages may exist between adjacent cells that play important roles in the biology of organisms.

Wagner, R. P. *Chromosomes: A Synthesis*. New York: Wilet-Liss, 1993. A discussion of the chromosome, including everything from how it's built to how it functions as part of the genome.

TECHNOLOGY LINKS

The Living World Home Page
http://www.wcbp.com/biology/tlw

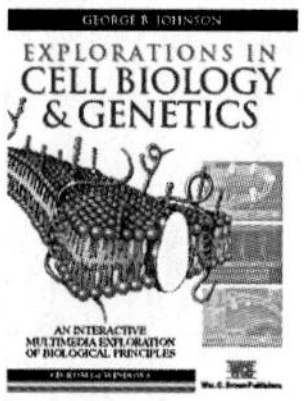

Explorations in Cell Biology & Genetics CD
#3 Active Transport
#5 Mitosis

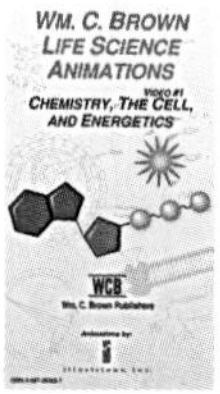

Life Science Animations Videotape 1
#3 Endocytosis

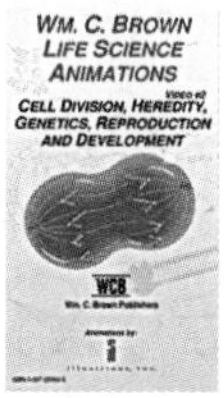

Life Science Animations Videotape 2
#12 Mitosis

Physiological Concepts of Life Science Videotape
#2 Osmosis

5

Energy and Life

CHAPTER OUTLINE

Figure 5.1 The sun's energy drives life.
Less than 1% of all the energy that reaches the earth from the sun is captured in the process of photosynthesis by plants like this sunflower. Yet, this 1% drives all the activities of life on earth.

The life processes of every cell are driven by energy. Cell growth, movement, and the transport of molecules across membranes all require energy. Just as a car is powered by burning gasoline, so the bodies of organisms burn chemical fuel (like the cat the dog on the opposite page dreams of eating). All the food that fuels the lives of animals is first captured from sunlight by photosynthesis, like that being carried out by the sunflowers in figure 5.1. Animals then eat the molecules made by the plants (or other animals that have done so). The energy in these "food" molecules exists in electrons, which spin in energetic orbits around their atomic nuclei. Your cells strip these electrons away and use them to power their lives. Your every thought is fueled by the energy of electrons.

5.1 Cells and Chemistry

Studying the chemistry of the cell may seem to you an uninteresting way to study biology. Indeed, few subjects put biology students off more quickly than chemistry. The prospect of encountering chemical equations seems at the same time both difficult and boring. Here we will try to keep the pain to a minimum, to avoid equations where possible, and to stress ideas rather than formulas. However, there is no avoiding some study of cell chemistry, for the same reason that a successful race car driver must learn how the engine of the car works. We are chemical machines, powered by chemical energy, and if we are to understand ourselves, we must "look under the hood" at the chemical machinery of our cells and see how it operates.

Chemical Reactions

All the chemical activities within cells can be looked on as a series of chemical reactions between molecules. A **chemical reaction** is the making or breaking of chemical bonds—gluing atoms together to form new molecules or tearing molecules apart and sometimes sticking the pieces onto other molecules. For example, making a protein is a matter of sticking amino acid molecules together in long chains, while extracting energy from a sugar like glucose is a matter of ripping apart the carbon backbone of the sugar molecule and sticking the hydrogen atoms stolen from the carbons onto other molecules that carry them piggyback to ATP-making proton pumps. If we are going to understand cell chemistry, we must first look more closely at chemical reactions.

A chemical reaction is one in which some molecules are changed; chemical bonds are formed or broken. The molecules that you start with are called **reactants** or sometimes **substrates,** while the molecules that you end up with, after the reaction is over, are called the **products** of the reaction. Not all chemical reactions are equally likely to occur. Just as a boulder is more likely to roll downhill than uphill, so a reaction is more likely to occur if it releases energy than if it needs to have energy supplied. Reactions that release energy, ones in which the products (what you end up with) contain less energy than the reactants (what you start with) tend to occur by themselves and are called **exergonic.** Harvesting of chemical energy from food, as the squirrel is doing in figure 5.2, is an exergonic process. By contrast, reactions in which the products contain more energy than the reactants, called **endergonic,** do not occur unless energy is supplied from an outside source.

Activation Energy

If all chemical reactions that release energy tend to occur spontaneously, it is fair to ask, "Why haven't all exergonic reactions occurred already?" Clearly they have not. If you ignite gasoline, it still burns with a release of energy. So why doesn't all the gasoline in all the automobiles in the world just burn up right now? It doesn't because the burning of gasoline, and almost all other chemical reactions, requires an input of energy to get it started—a kick in the pants such as a match or spark plug. Even if the new bonds of the product would require less energy than the existing bonds of the reactants, it is first necessary to break those existing bonds, and this takes energy. The extra energy required to destabilize existing chemical bonds and so initiate a chemical reaction is called **activation energy** (figure 5.3*a*). Thus, to roll a boulder downhill, you must first nudge it out of the hole it sits in. Activation energy is simply a chemical nudge.

Figure 5.2 Fueling the body.
The energy this arctic ground squirrel obtains to support the many functions of its body cells is stored in chemical bonds between atoms in the molecules of the food and water it consumes.

Catalysis

Just as all the gasoline in the world doesn't burn up right now because the combustion reaction has a sizable activation energy, so all the exergonic reactions in your cells don't spontaneously happen either. Each reaction waits for something to nudge it along, to supply it with sufficient activation energy to get going. You can think of a cell as crammed with chemical reactions waiting to happen.

One way to make an exergonic reaction more likely to happen is to lower the necessary activation energy. Like digging away the ground below your boulder, lowering activation energy reduces the nudge needed to get things started. The process of lowering the activation energy of a reaction is called **catalysis** (figure 5.4). Catalysis cannot make an endergonic reaction occur spontaneously—you cannot avoid the need to supply energy—but it can make an exergonic reaction proceed much faster, because almost all the time required for chemical reactions is tied up in overcoming the activation energy barrier (see figure 5.3*b*).

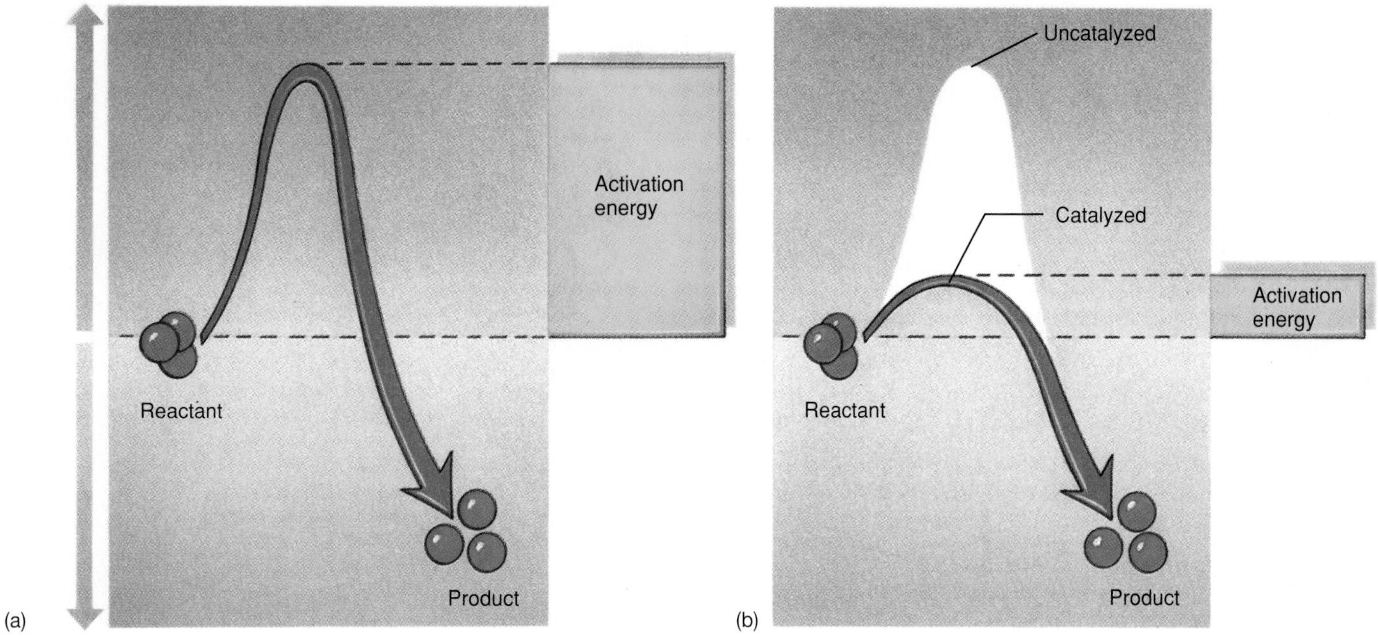

Figure 5.3 Activation energy.
Exergonic reactions do not necessarily proceed rapidly because it takes energy to get them going. (*a*) The "hill" in this energy diagram represents energy that must be supplied to destabilize existing chemical bonds. (*b*) Catalyzed reactions occur faster because the amount of activation energy required to initiate the reaction—the height of the energy hill that must be overcome—is lowered.

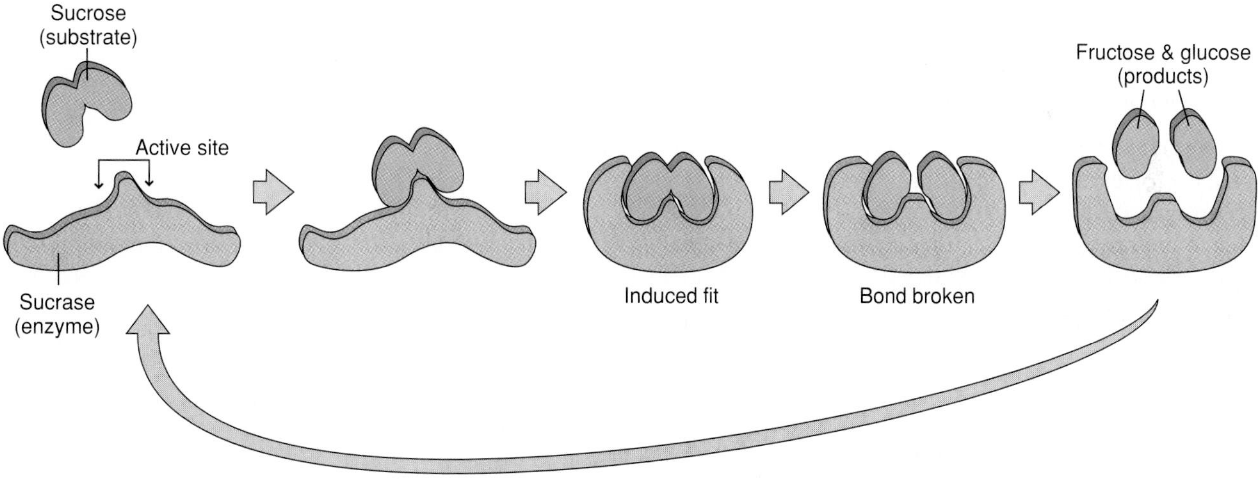

Figure 5.4 How catalysis works.
Catalysis works by lowering activation energy. In the example illustrated here, the substrate (sucrose, the sugar present in most candy) binds to the active site of the enzyme, called sucrase, and induces the enzyme to alter its shape to provide for a tighter fit. Amino acids in the active site, now in close proximity to the bond between the glucose and fructose parts of sucrose, stress the bond and cause it to break. When the products of the reaction are released, the enzyme molecule returns to its original shape.

How Enzymes Work

Proteins called **enzymes** are the catalysts used by cells to touch off particular chemical reactions. By controlling which enzymes are present, and when they are active, cells are able to control what happens within themselves, just as a conductor controls the music an orchestra produces by dictating which instruments play when.

An enzyme works by binding to a specific molecule and stressing the bonds of that molecule in such a way as to make a particular reaction more likely. The key to this activity is the shape of the enzyme (figure 5.5).

Specificity. An enzyme is specific for a particular reactant because the enzyme surface provides a mold that fits the desired reactant very exactly. Other molecules that fit less perfectly simply don't adhere to the enzyme's surface. The site where a reactant binds to an enzyme is called the **binding site.**

Catalysis. An enzyme lowers the activation energy of a particular reaction. It may encourage the breaking of a particular chemical bond in the reactant molecule by weakening it, often by drawing away some of its electrons. Alternatively, an enzyme may encourage the formation of a link between two reactants by holding them near each other. The site on the enzyme surface where the reactant is stressed is called the **active site.**

Enzymes can be very effective catalysts. As an example, consider a common reaction that takes place within cells:

carbon dioxide + water → carbonic acid

The reaction is very slow by itself. In a cell, perhaps 200 molecules of carbonic acid form in an hour under normal circumstances. Things usually happen very fast in cells, and this reaction rate is like a snail racing in the Indianapolis 500. However, in the presence of the right enzyme (in this case, one called carbonic anhydrase), 600,000 molecules of carbonic acid form every second! The enzyme has increased the reaction rate by about 10 million times.

Control. Because an enzyme must have a precise shape in order to work correctly, it is possible for the cell to control when an enzyme is active by altering its shape. Many enzymes have shapes that can be altered by the binding to their surfaces of "signal" molecules. Such enzymes are called *allosteric* (Latin, meaning "other shape"). If the new shape produced by binding the signal molecule is no longer able to fit the reactant, the signal acts as a repressor of the enzyme's activity. If the enzyme is unable to bind its reactant unless the signal molecule is bound to it, the signal acts as an activator of the enzyme's activity. The site where the signal molecule binds to the enzyme surface is called the **allosteric site.**

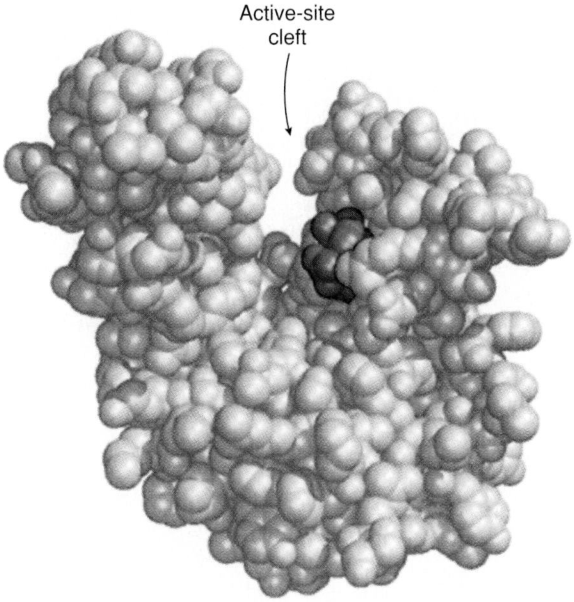

Figure 5.5 Enzyme shape determines its activity. The adenylate kinase enzyme (blue in this diagram) has a deep groove running across it, and at the base of the groove is a pair of substrate binding sites, facing each other. One binds the substrate ATP, the other a second substrate, AMP. By positioning these two molecules so closely together, the enzyme aids the transfer of the terminal phosphate from ATP (three phosphates) to AMP (one phosphate), forming two molecules of ADP (two phosphates). It is the close positioning of these two substrate binding sites that makes the catalysis possible.

A cell contains thousands of different kinds of enzymes, each promoting a different chemical reaction. The enzymes that are active at any one time in a cell determine what happens in that cell, much as traffic lights determine the flow of traffic in a city. Not all cells contain the same enzymes. For example, the chemical reactions going on in a nerve cell are very different from those in a red blood cell because the two kinds of cells possess a different array of enzymes.

Aiding in Catalysis

Enzymes often use additional chemical components called **cofactors** as tools to aid catalysis. Metal ions and coenzymes can both act as cofactors. A **coenzyme** is a nonprotein organic molecule that acts as a cofactor. One of the most important coenzymes is the electron acceptor NAD^+. In many enzyme reactions, energy-bearing electrons are passed from the active site of the enzyme to the coenzyme NAD^+. Electrons are usually transferred along with protons as hydrogen atoms. NAD^+ thus becomes NADH and can then carry the electrons to a different enzyme, release the electrons to that reaction, and return as NAD^+ to the original enzyme for another load of electrons.

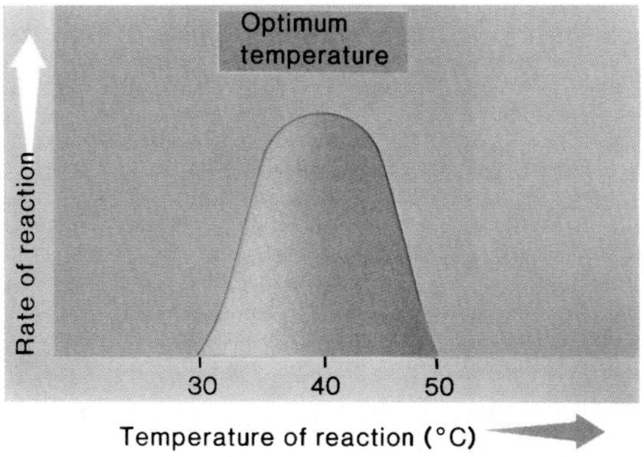

(a)

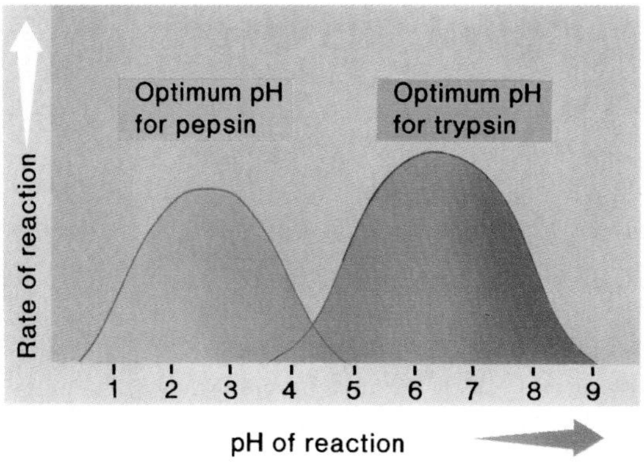

(b)

Figure 5.6 Temperature and pH influence the activity of enzymes.
(*a*) Temperature. The optimum temperatures at which most human enzymes function is between 30° and 50°C.
(*b*) pH. The range of pH values over which these two human digestive enzymes—pepsin and trypsin—work best is quite different. As you can see, pepsin operates at a much lower (acidic) pH than trypsin.

Factors Affecting Enzyme Activity

In addition to coenzymes, temperature and pH can also influence the action of enzymes (figure 5.6). Enzyme activity is affected by any change in condition that alters the enzyme's three-dimensional shape (figure 5.7). When the temperature becomes too low, the bonds that determine enzyme shape are not flexible enough to permit the induced-fit change sometimes necessary for catalysis; at higher temperatures, the bonds are too weak to hold the enzyme's peptide chains in the proper position. As a result, enzymes function best within an optimum temperature range. This range is relatively narrow for most human enzymes. In addition, most enzymes also function within an optimal pH range, because the structural bonds of enzymes are also sensitive to hydrogen ion (H^+) concentration. Most human enzymes, such as the protein-degrading enzyme trypsin, work best within the range of pH 6 to 8. However, some enzymes, such as the digestive enzyme pepsin, are able to function in very acidic environments.

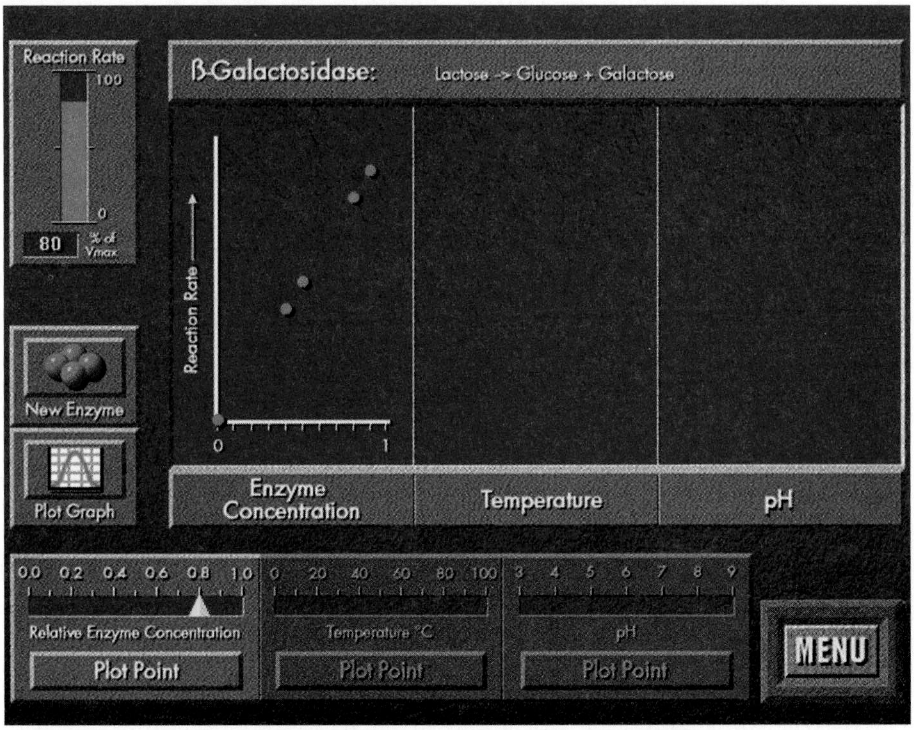

Figure 5.7 How reaction conditions affect enzyme chemistry.
These graphs illustrate the effects of enzyme concentration, temperature, and pH on the rate of an enzyme-catalyzed reaction. This illustration is a "screen capture" from an interactive CD-ROM exercise that allows you to investigate how varying these parameters affects a variety of real enzymes. You will quickly gain an appreciation of how well particular enzymes are suited to the physiological conditions under which they operate. (*Explorations in Cell Biology & Genetics,* Module 6, "Thermodynamics")

5.2 How Cells Use Energy

Every one of us would die if we stopped eating. Why? Each of the significant properties by which we define life—growth, movement, reproduction, heredity—use energy. To keep life going, energy must be continually supplied, like putting more logs on a fire. Life can be viewed as a constant flow of energy, channeled by organisms to do the work of living.

Why We Need Energy

Cells use energy to do all those things that require work. One of the most obvious of these is *movement*. Some bacteria swim about, propelling themselves through the water by rapidly spinning a long, tail-like flagellum, much as a ship moves by spinning a propeller. During your development as an embryo, many of your cells moved about, crawling over one another to reach new positions. When one of your white blood cells engulfs an invading bacterium, it accomplishes this change in shape by extending its cytoskeleton (figure 5.8). Much movement occurs within cells. Mitochondria and cellular materials are passed a meter or more along the narrow nerve cells that connect your feet with your spine. Chromosomes are pulled by microtubules during cell division. All of these movements by cells require the expenditure of energy. The molecule in the body that supplies that energy is **adenosine triphosphate (ATP).**

A second major way cells use energy is to *drive endergonic reactions*. Many of the synthetic activities of the cell are endergonic because building molecules takes energy. Recall that endergonic reactions do not take place spontaneously. The chemical bonds of the products contain more energy than what you started with, so nothing can happen until that extra energy is supplied to the reaction from somewhere. The somewhere is, as you might guess, the cell's supply of ATP.

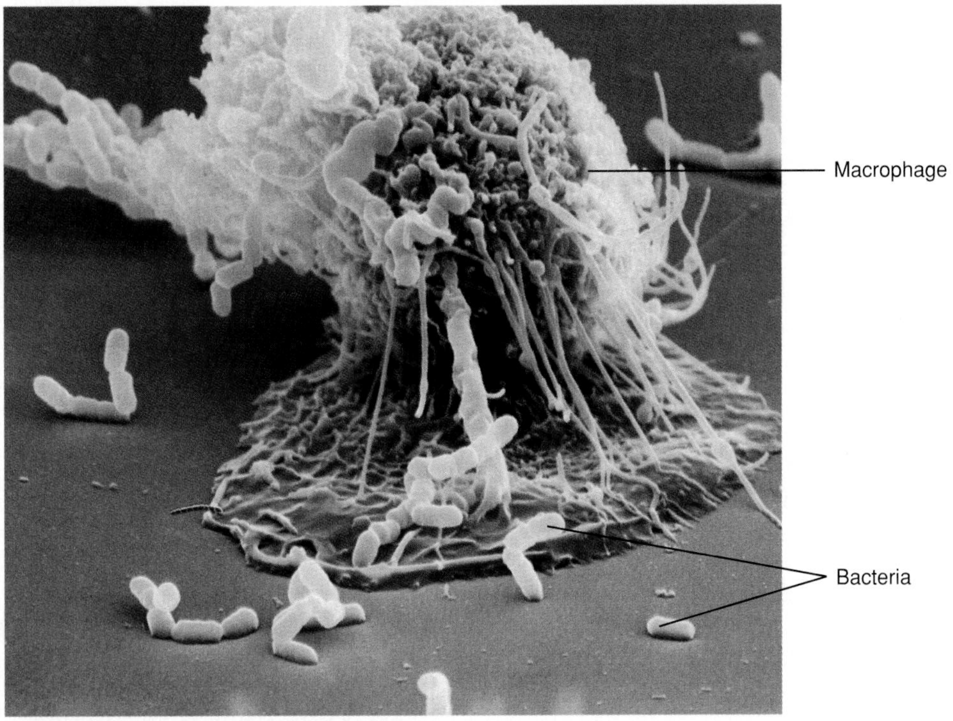

Figure 5.8 A white blood cell engulfing bacteria.
This macrophage (a kind of white blood cell that destroys invading bacteria by phagocytosis) is extending projections that contact and adhere to bacteria. The macrophage then "reels them in," engulfs them by phagocytosis, and destroys them.

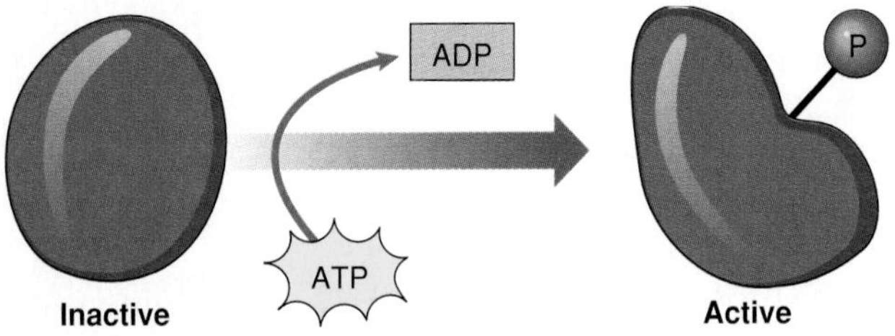

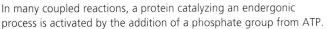

Figure 5.9 A coupled reaction.
In many coupled reactions, a protein catalyzing an endergonic process is activated by the addition of a phosphate group from ATP.

Making Molecules with ATP

Here is how ATP drives an endergonic reaction. The enzyme that catalyzes the endergonic reaction has *two* binding sites on its surface, one for the reactant and another for ATP. The ATP site splits the ATP molecule, liberating a great deal of chemical energy. This energy pushes the reactant at the second site "uphill." In a similar way, you can make water in a swimming pool leap straight up in the air, despite the fact that gravity prevents water from rising spontaneously—just jump in the pool! The energy you add going in more than compensates for the force of gravity holding the water back.

When an energy-requiring reaction in a cell is driven by the splitting of ATP molecules, the reaction is called a **coupled reaction.** The two parts of the reaction—endergonic and ATP-splitting—take place in concert. In some cases the two parts both occur on the surface of the same enzyme; they are physically linked, or "coupled," like two legs walking. In other cases, a high-energy phosphate from ATP is attached to the protein catalyzing the endergonic process, activating it (figure 5.9). The ability to couple energy-requiring reactions to the splitting of ATP in this way is one of the key tricks cells employ to manage energy.

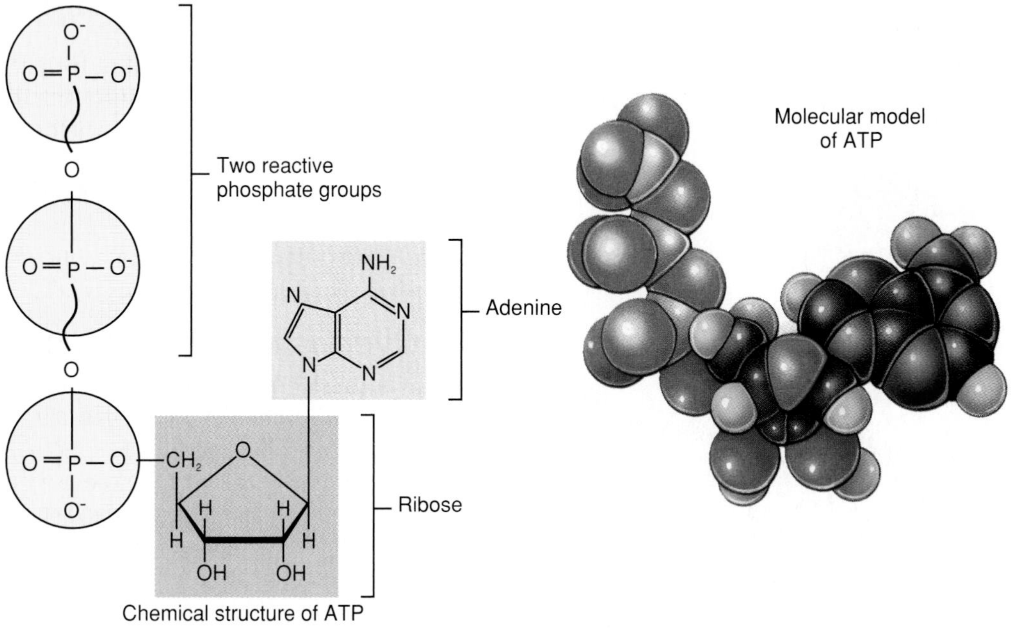

Figure 5.10 The parts of an ATP molecule.
ATP is the primary energy currency of all cells. Structurally, ATP consists of three phosphate groups attached to a ribose (five-carbon sugar) molecule. The ribose molecule is also attached to an adenine molecule (also in one of the nucleotides of DNA). When the outermost phosphate group is split off from the ATP molecule, a great deal of energy is released. The resulting molecule is ADP, or adenosine diphosphate. The second phosphate group can also be removed, yielding additional energy and leaving AMP (adenosine monophosphate). Most energy exchanges in cells involve cleavage of only the outermost bond, converting ATP into ADP and P_i (inorganic phosphate).

The ATP Molecule

As you have seen, ATP is the energy currency of the cell, the source of the force used to power the cell's activities. Each ATP molecule is composed of three parts: (1) a sugar that serves as the backbone to which the other two parts are attached; (2) adenine, one of the four nucleotide bases in DNA; and (3) a chain of three phosphates (figure 5.10).

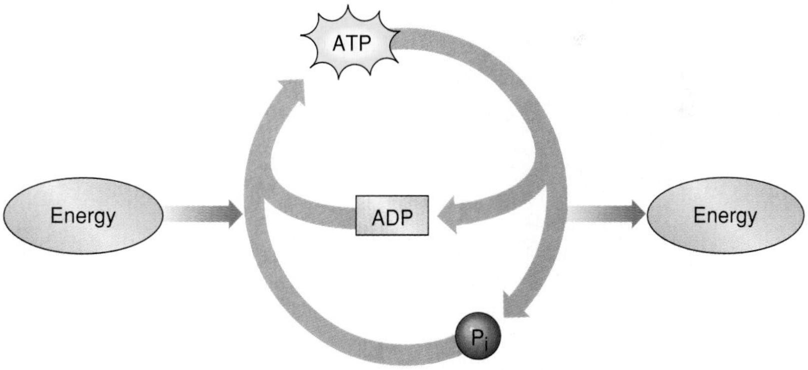

Figure 5.11 The ATP-ADP cycle.
In mitochondria and chloroplasts, chemical or photosynthetic energy is harnessed to form ATP from ADP and inorganic phosphate. When ATP is used to drive the living activities of cells, the molecule is cleaved back to ADP and inorganic phosphate, which are then available to form new ATP molecules.

Because phosphates have negative electrical charges, it takes considerable chemical energy to hold the line of three phosphates next to one another at the end of ATP. Like a coiled spring, the phosphates are poised to push apart. It is for this reason that the chemical bonds linking the phosphates are said to be "high-energy" bonds.

When the endmost phosphate is broken off an ATP molecule, a sizable packet of energy is released. The reaction converts ATP to adenosine diphosphate, ADP. The second phosphate group can also be removed, yielding additional energy and leaving adenosine monophosphate (AMP). Most energy exchanges in cells involve cleavage of only the outermost bond, converting ATP into ADP and P_i, inorganic phosphate (figure 5.11).

$$ATP \rightarrow ADP + P_i + energy$$

Because almost all endergonic reactions in cells require less energy than is released by this reaction, ATP is able to power many of the cell's activities.

5.3 Photosynthesis

Life is powered by sunshine. All of the energy used by living cells comes ultimately from the sun, captured by plants and algae through the process of photosynthesis. Thus, life is only possible because our earth is awash in energy streaming inward from the sun. Each day, the radiant energy that reaches the earth is equal to that of about 1 million Hiroshima-sized atomic bombs. About 1% of it is captured by photosynthesis and provides the energy that drives us all.

An Overview of Photosynthesis

Photosynthesis occurs in plants within leaves. Recall from chapter 3 that the cells of plant leaves contain organelles called chloroplasts that actually carry out the photosynthesis (figure 5.12). No other structure in a plant other than chloroplasts is able to carry out photosynthesis. Photosynthesis takes place in three stages: (1) capturing energy from sunlight; (2) using the energy to make ATP; and (3) using the ATP to power the synthesis of plant molecules from CO_2 in the air.

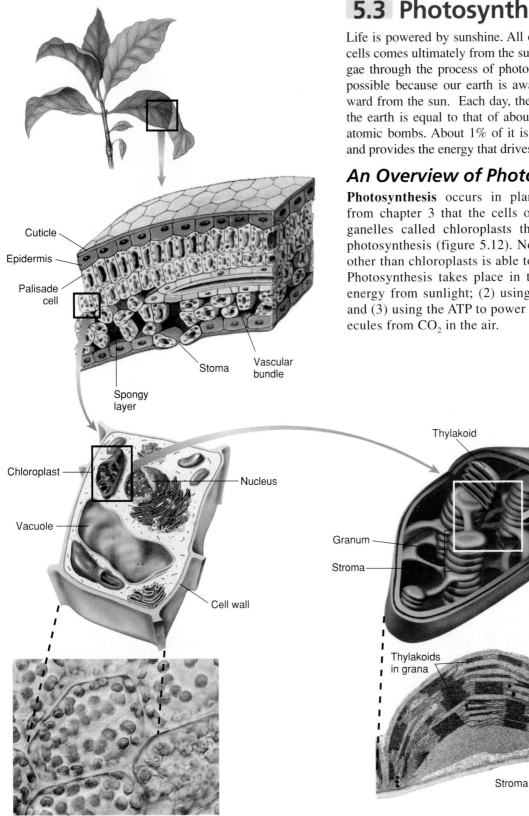

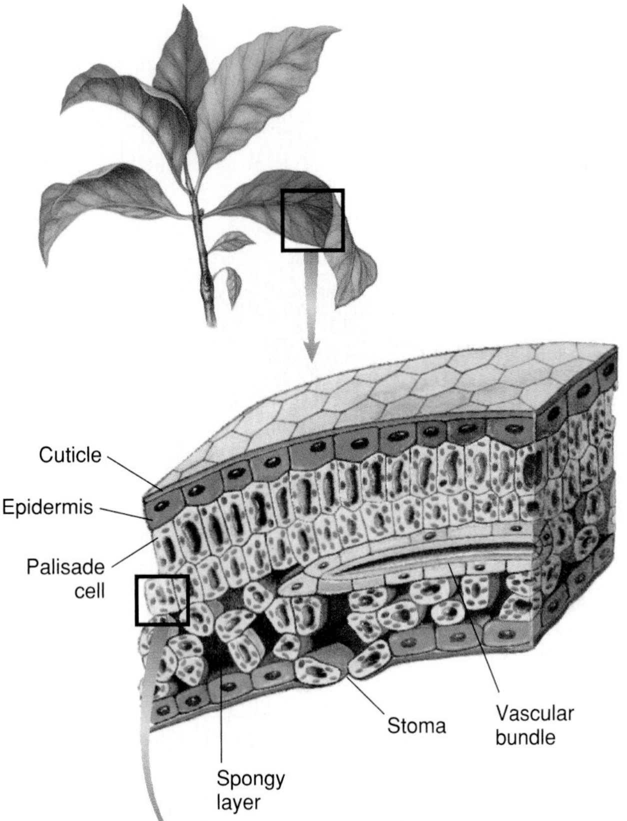

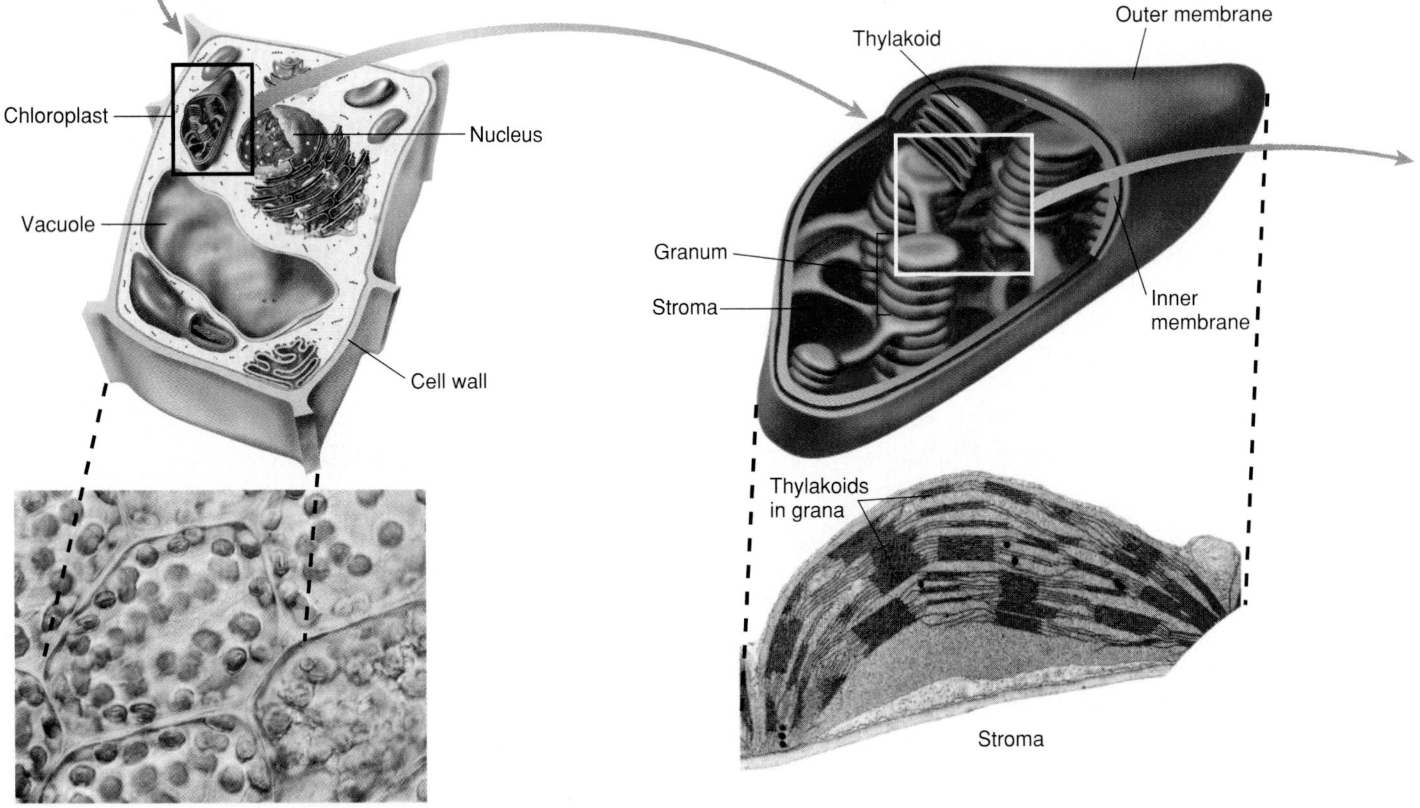

Figure 5.12 Journey into a leaf.
Plant cells within leaves contain chloroplasts, in which thylakoid membranes are stacked. Within the thylakoid, chlorophyll pigments grouped in photosystems drive the reactions of photosynthesis.

The first two processes take place only in the presence of light and are commonly called the **light-dependent reactions.** The third process (formation of organic molecules from atmospheric CO_2) can take place in the light or dark and involves what are called the **light-independent reactions** of photosynthesis. As long as ATP is available, they occur as readily in the absence of light as in its presence.

The overall process of photosynthesis may be summarized by the following simple equation:

$$6CO_2 + 12H_2O + \text{light energy} \rightarrow C_6H_{12}O_6 + 6H_2O + 6O_2$$

carbon dioxide | water | glucose | water | oxygen

Inside the Chloroplast

The internal membranes of chloroplasts are organized into flattened sacs called *thylakoids,* and often numerous thylakoids are stacked on top of one another in columns called *grana*. Surrounding the thylakoid membrane system is a semiliquid substance called *stroma*. In the membranes of thylakoids, chlorophyll pigments are grouped together in a network called a **photosystem.**

Each chlorophyll molecule within the photosystem network is capable of capturing photons. A lattice of proteins holds the pigments of the photosystem in close contact with one another. When light of the proper wavelength strikes any chlorophyll molecule in the photosystem, the resulting excitation passes from one chlorophyll molecule to another. The excited electron does not transfer physically—it is the *energy* that is passed from one chlorophyll to another. A crude analogy to this form of energy transfer is the initial "break" in a game of pool. If the cue ball squarely hits the point of the triangular array of 15 pool balls, the 2 balls at the far corners of the triangle fly off, and none of the central balls move at all. The energy is transferred through the central balls to the most distant ones.

Eventually the energy arrives at a key chlorophyll molecule touching a membrane-bounded protein. It is transferred to that protein, which passes it in turn to a series of proteins that put the energy to work making ATP and building organic molecules, in a way we will discuss. The photosystem thus acts as a large antenna, amplifying the light-gathering powers of individual chlorophyll molecules.

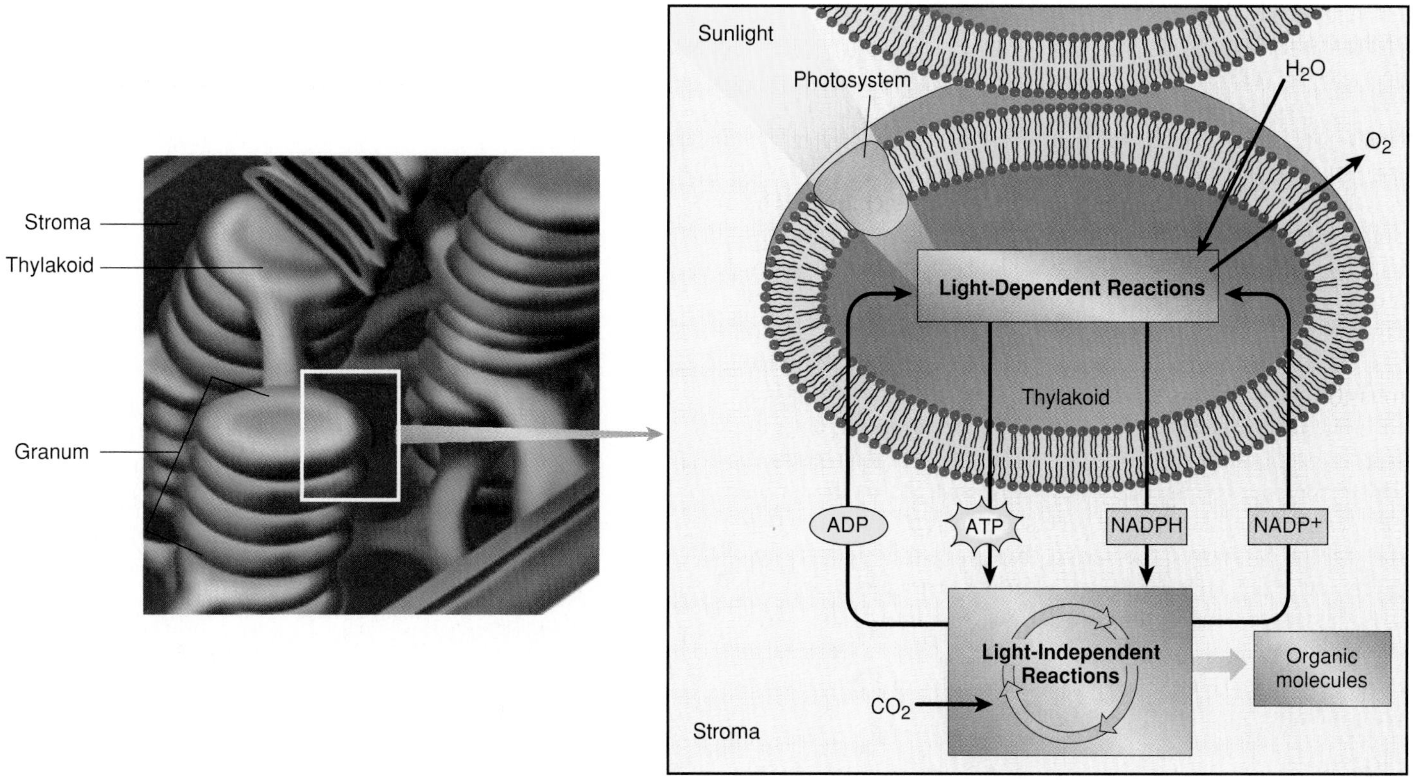

Capturing Energy from Sunlight

Where is the energy in light? What is there about sunlight that a plant can use to create chemical bonds? The revolution in physics in the twentieth century has taught us that light actually consists of tiny packets of energy called **photons.** When light shines on your hand, your skin is being bombarded by a stream of these photons smashing onto its surface.

Sunlight contains photons of many energy levels, only some of which we "see." Some of the photons in sunlight carry a great deal of energy—for example, X rays and "ultraviolet light." Others such as radio waves carry very little. In general, high-energy photons have shorter wavelengths than low-energy photons. Our eyes perceive photons carrying intermediate amounts of energy as **visible light** (figure 5.13), because our eyes only absorb that kind of photon. Plants are even more picky, absorbing mainly blue and red light and reflecting back what is left of the visible light, which is why they appear green (figure 5.14).

Visible light represents only a small portion of the range of photon energies in sunlight. We call the full range of these photons the **electromagnetic spectrum.** The highest-energy photons, which have the shortest wavelengths, are gamma rays with wavelengths of less than 1 nanometer; the lowest-energy photons, with wavelengths of thousands of meters, are radio waves.

How can a leaf or a human eye choose which photons to absorb? The answer to this important question has to do with the nature of atoms. Remember that electrons spin in particular orbits around the atomic nucleus, each at a different energy level. Atoms absorb light by boosting electrons to higher energy levels, using up the energy in the photon to power the move. Boosting the electron requires just the right amount of energy, no more and no less, just as when climbing a ladder you must raise your foot just so far to climb a rung. That is why a particular kind of atom absorbs only certain photons of light—those with the appropriate amount of energy.

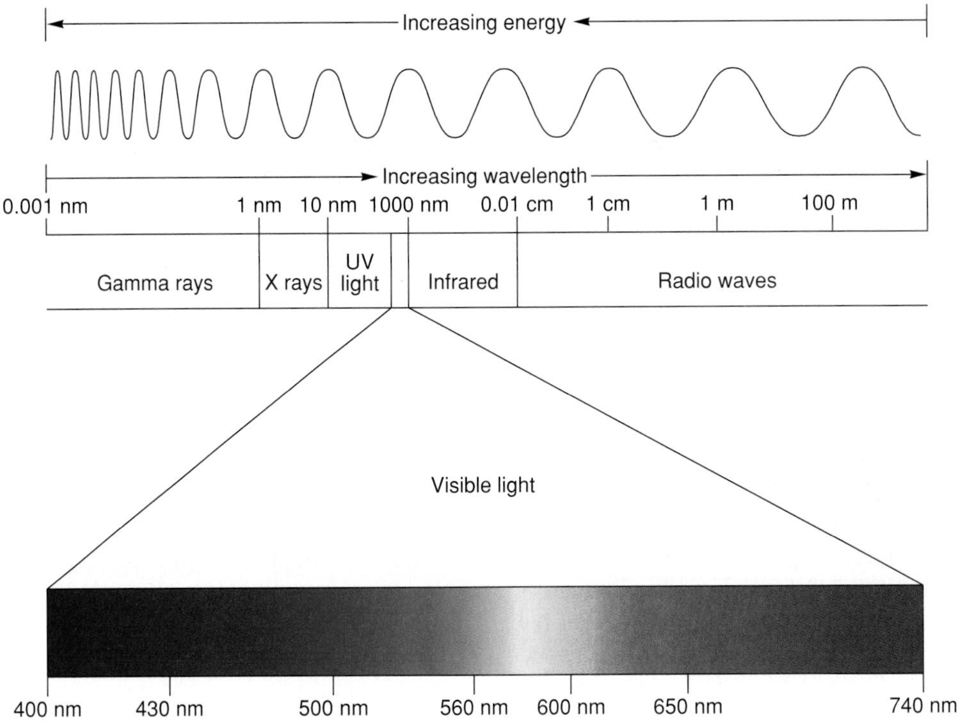

Figure 5.13 Photons of different energy: the electromagnetic spectrum.
Light is composed of packets of energy called photons. Some of the photons in light carry more energy than others. Light, a form of electromagnetic energy, is conveniently thought of as a wave. The shorter the wavelength of light, the greater the energy of its photons. Visible light represents only a small part of the electromagnetic spectrum, that between 380 and 750 nanometers.

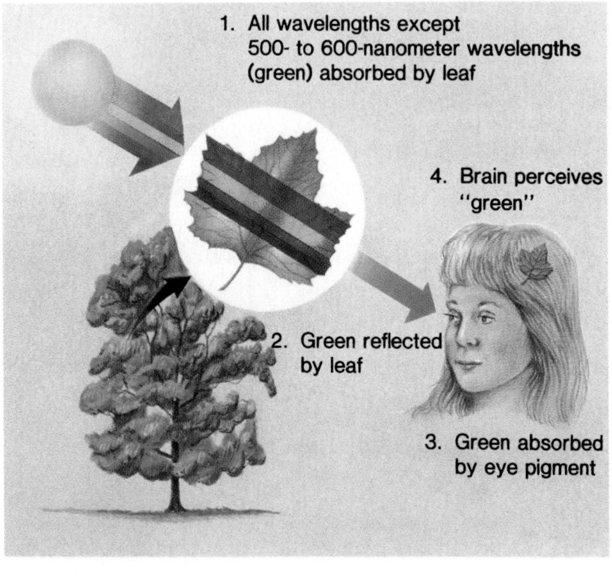

Figure 5.14 Why are plants green?
A leaf containing chlorophyll absorbs a broad range of photons—all the colors in the spectrum except for the photons around 500 to 600 nm. The leaf reflects these colors. These reflected wavelengths are absorbed by the visual pigments in our eyes, and our brains perceive the reflected wavelengths as "green."

Pigments

Molecules that absorb lots of light are called **pigments.** When we speak of visible light, we refer to those wavelengths that the pigment within human eyes, called **retinal,** can absorb—roughly from 380 nanometers (violet) to 750 nanometers (red). Other animals use different pigments for vision and thus "see" a different portion of the electromagnetic spectrum. For example, the pigment in insect eyes absorbs at shorter wavelengths than retinal. That is why bees can see ultraviolet light, which we cannot (it is produced by photons shorter than violet), but are blind to red light, which we can see (produced by photons with relatively long wavelengths).

The main pigment in plants that absorbs light is **chlorophyll,** which occurs in two forms, chlorophyll *a* and chlorophyll *b* (figure 5.15). While it absorbs fewer kinds of photons than retinal, it is very much more efficient at capturing them. The chlorophylls capture photons with a metal ion (magnesium) that lies at the center of a complex ring of carbon molecules. Photons excite electrons of the magnesium ion, which are then channeled away by the carbon atoms. The two kinds of chlorophyll differ in small chemical "side groups" attached to the outside of the ring, which alter slightly the wavelength of the photons they absorb.

Another group of pigments, the **carotenoids,** assists in photosynthesis by capturing energy from light of wavelengths that are not efficiently absorbed by chlorophylls (figure 5.16).

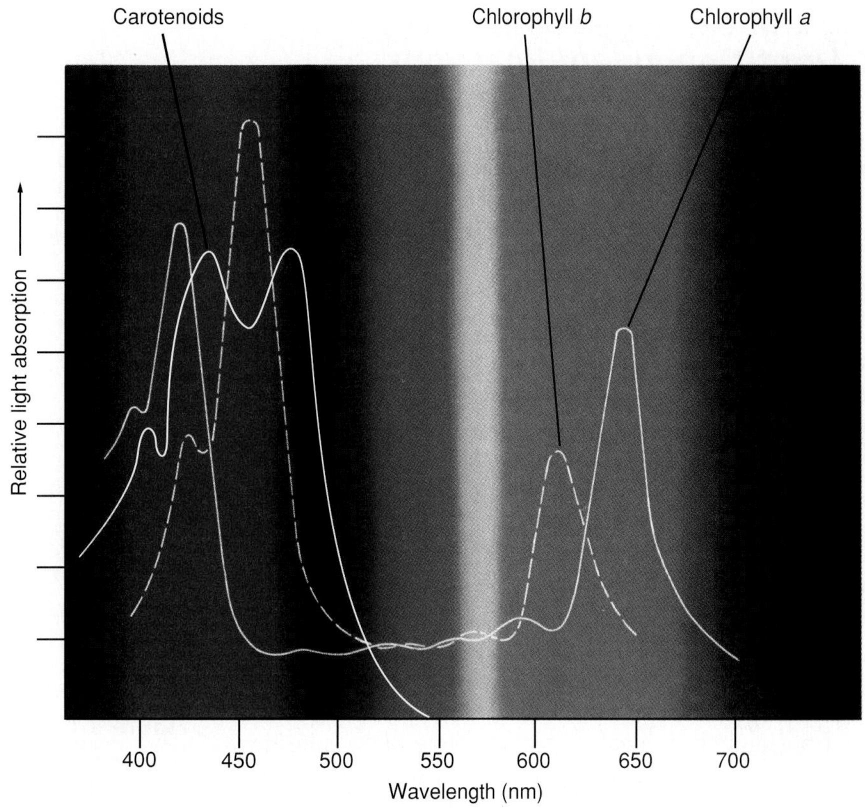

Figure 5.15 The absorption spectrum of chlorophyll.
The peaks you see on the *left* and *right* represent wavelengths of sunlight strongly absorbed by the two common forms of photosynthetic pigment, chlorophyll *a* and chlorophyll *b*. These chlorophylls absorb predominantly violet-blue and red light, in two narrow bands of the spectrum, while they reflect the green light in the middle of the spectrum. Carotenoids (not shown here) absorb mostly blue and green light and reflect orange and yellow light.

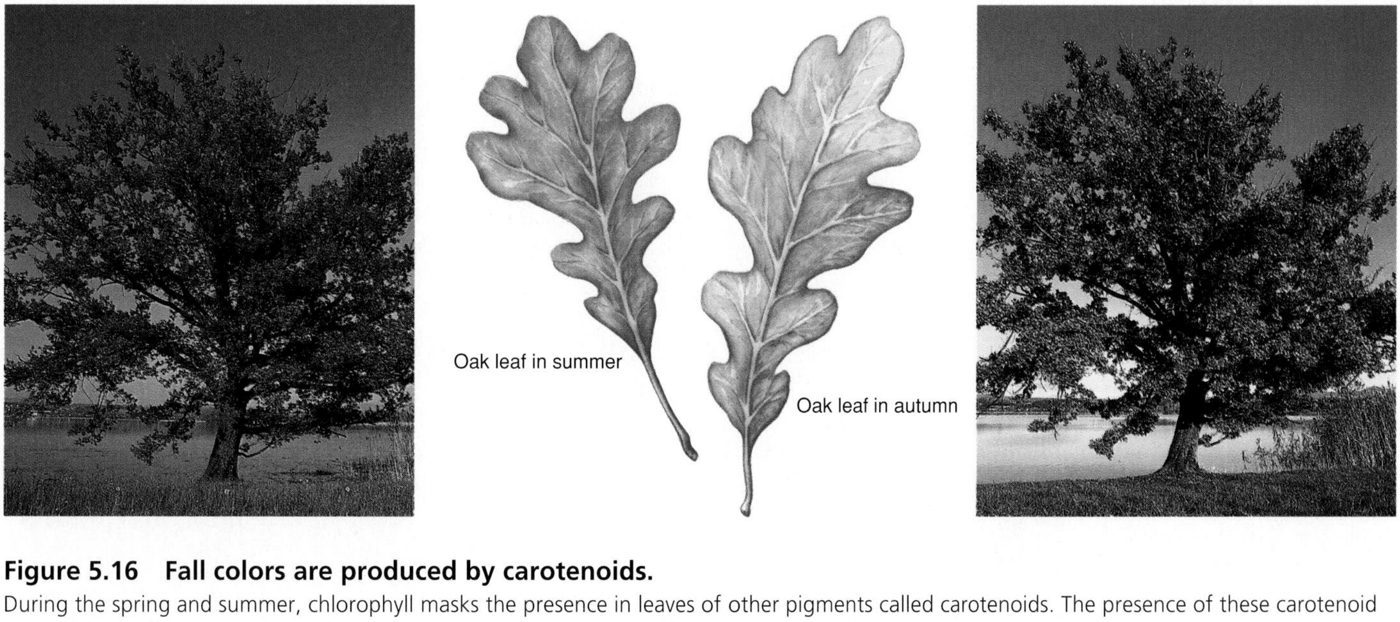

Figure 5.16 Fall colors are produced by carotenoids.
During the spring and summer, chlorophyll masks the presence in leaves of other pigments called carotenoids. The presence of these carotenoid pigments becomes obvious in the fall, when cool temperatures cause leaves to cease manufacturing chlorophyll. With the chlorophyll no longer present to reflect green light, the carotenoids reflect orange and yellow light instead and so give bright colors to the autumn leaves.

Using Light Energy to Make ATP: The Light-Dependent Reactions

The formation of ATP can occur through two different pathways, the cyclic pathway (figure 5.17) and the noncyclic pathway (figure 5.18). When photons of light strike the chloroplasts of plant cells, they bash into chlorophyll molecules in the photosystems and—if they have the right amount of energy—are absorbed, boosting chlorophyll electrons to higher energy levels. The excited electron, traveling as part of a hydrogen atom, quickly jumps to a nearby protein in the chloroplast membrane, which passes it to a neighbor, and so on, like relay racers passing a baton. This series of membrane-embedded electron carriers is collectively called the **electron transport chain.**

Soon the hydrogen carrying the light-excited electron arrives at its destination, a proton pump. This protein responds to its arrival by dropping the excited electron back down to its original orbit and using the energy released to power the pumping of a proton across a membrane into the interior of the thylakoid.

What happens next is called **chemiosmosis** (figure 5.19). As we learned in chapter 4, the proton pump uses this energy from sunlight to pump more and more protons inside the thylakoid vesicle, until the interior is packed with protons. Banging about because of diffusion, the protons escape through the only exit available, a protein channel that uses the force of their exit to add a phosphate group onto an ADP molecule, making ATP.

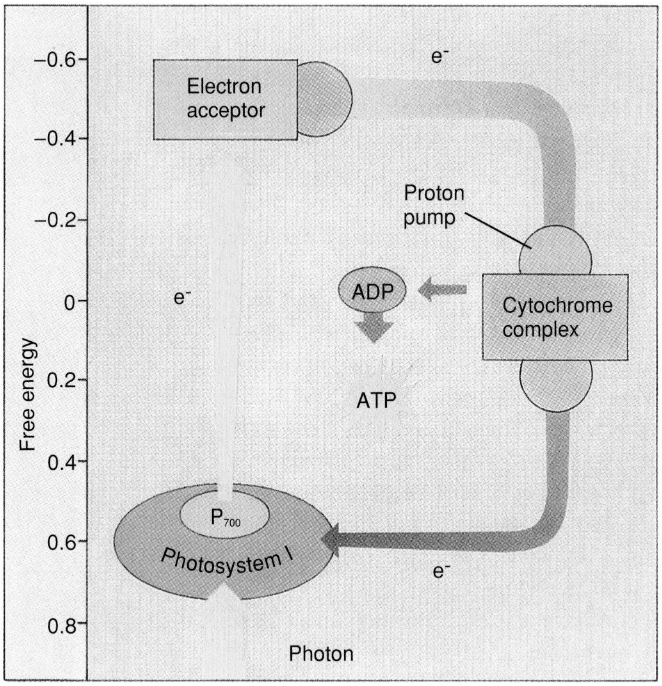

Figure 5.17 The cyclic pathway of ATP production. This bacterial pathway is ancestral to the noncyclic pathways performed by plants, algae, and cyanobacteria. Light is absorbed in **photosystem I** by a chlorophyll molecule labeled P_{700} in the diagram. An electron acceptor molecule takes the excited electron, leaving P_{700} short one electron. Before photosystem I can function again, the electron must be returned. These bacteria channel the electron back to the pigment through the electron transport chain, during which the electron's passage drives the proton pump and the chemiosmotic synthesis of ATP. Thus, the path of the electron originally extracted from P_{700} is circular.

Figure 5.18 The noncyclic pathway of ATP production. Plants, algae, and cyanobacteria use a pathway in which a second photosystem is involved, **photosystem II** (called P_{680} in the diagram). The excited electrons that it captures are passed through an electron transport chain to photosystem I, driving a proton pump and so generating an ATP molecule chemiosmotically. When another photon of light is absorbed, this time by **photosystem I,** the electrons become excited and are channeled down an electron transport chain and ultimately donated to $NADP^+$, thus generating reducing power in the form of NADPH.

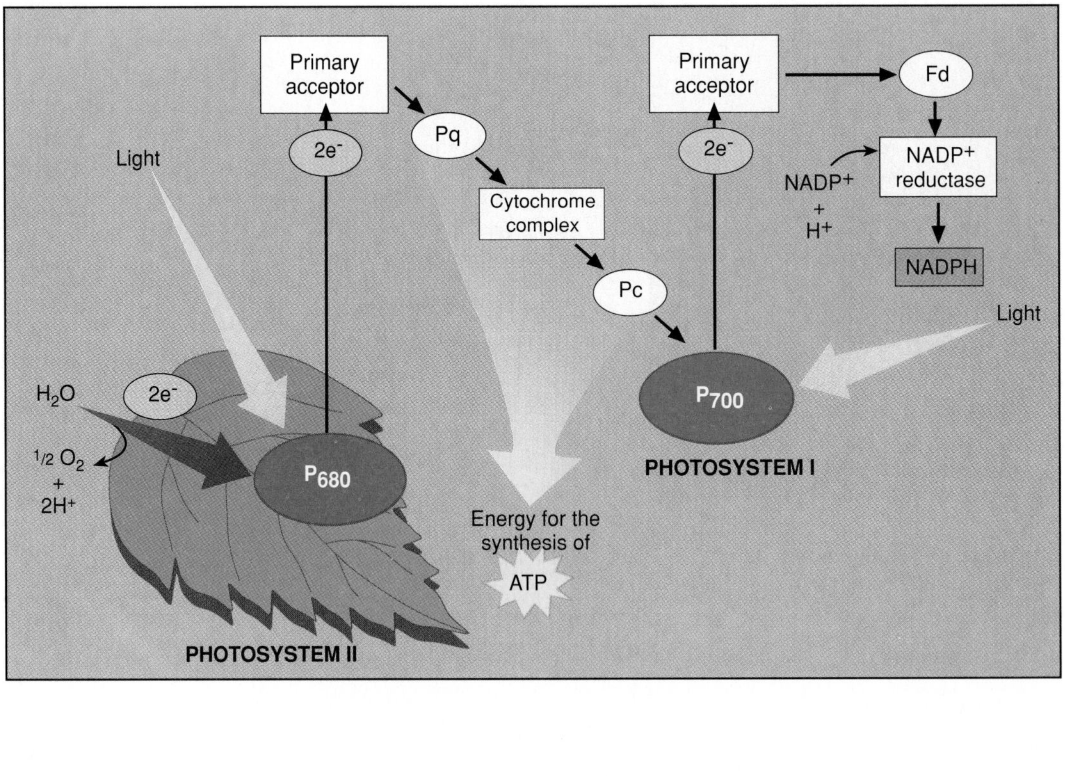

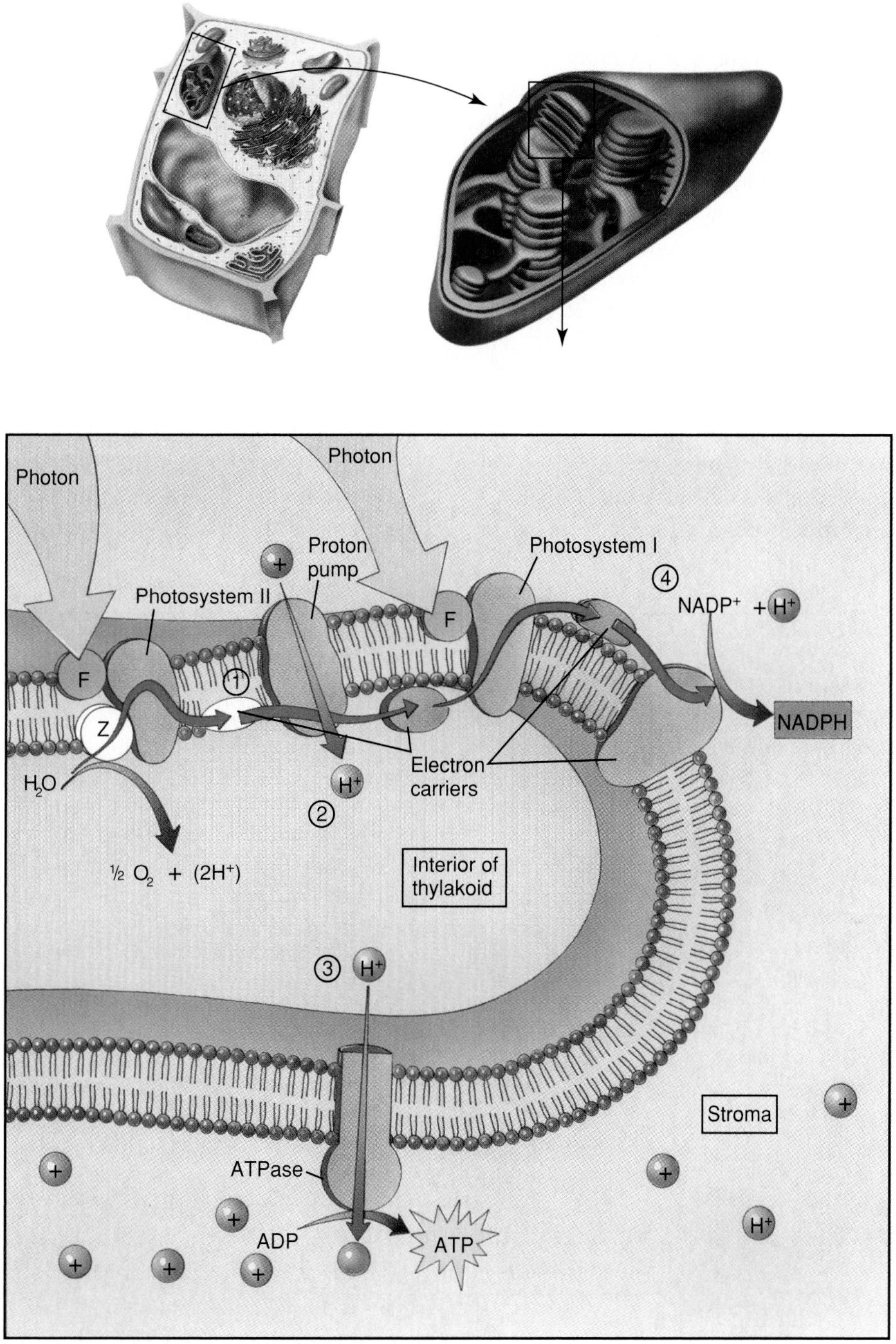

Figure 5.19 How chemiosmosis produces ATP.

When light strikes the chloroplast of a plant and excites electrons, the electrons move through a series of carriers embedded within the membrane of the thylakoid. (*1*) First, a photon of light strikes a pigment molecule in photosystem II, exciting electrons. These electrons are coupled to a proton stripped from water and pass along a chain of membrane-bound cytochrome electron carriers. When water is split, oxygen is released from the cell, and some hydrogen ions remain in the thylakoid space. (*2*) At the proton pump, the energy supplied by the photon is used to transport a proton across the membrane into the thylakoid. The concentration of hydrogen ions within the thylakoid thus increases further. (*3*) Protons (that is, hydrogen ions) increase in concentration, creating a steep gradient of concentration and charge across the thylakoid membrane. The result is a strong tendency for protons to escape through the only protein channels available, ones that use the force of their exit to drive the synthesis of ATP from ADP and an unbound phosphate. The electrons then pass to photosystem I. (*4*) When photosystem I absorbs another photon of light, its pigment passes more high-energy electrons to a reduction complex, which drives NADPH generation.

Building New Molecules: The Light-Independent Reactions

The point of photosynthesis, of course, is not simply to make ATP. Rather, it is to grab carbon atoms from carbon dioxide molecules in the air and stick them together to form new plant molecules, carbon-chain molecules like sugars, proteins, and DNA. This is an endergonic process that takes lots of energy, energy that ATP produced by the pumping of protons provides. But something else is needed too: because sugars, proteins, and other cell molecules have more hydrogen atoms (are more "reduced") than carbon dioxide (carbon dioxide has *no* hydrogen atoms!), a source of hydrogens must be provided. Biologists call such a supply of attachable hydrogens **reducing power.**

Plants obtain reducing power by using a second kind of chlorophyll, one that absorbs photons of higher energy than the ATP-making one. This second chlorophyll channels its light-excited electrons to a hydrogen fuel station, where the hydrogen atom carrying the excited electron is attached to a mobile carrier called **NADPH,** which takes it to where carbon-chain molecules are being built. The energy of the excited electron can then be used to stick the hydrogen onto the growing carbon chain.

The actual assembly of new molecules employs a complex battery of enzymes in what is called the **Calvin cycle** (figure 5.20), or **C_3 photosynthesis.** A carbon atom from a carbon dioxide molecule is first added to a five-carbon sugar, producing two three-carbon sugars. This process is called "fixing carbon" because it attaches a carbon atom that was in a gas to an organic molecule. Then, in a long series of reactions, the carbons are shuffled about. Eventually some of the resulting molecules are channeled off to make sugars, while others are used to re-form the original five-carbon sugar, which is then available to restart the cycle. The cycle has to "turn" six times in order to form a new glucose molecule, because each turn of the cycle adds only one carbon atom from CO_2, and glucose is a six-carbon sugar.

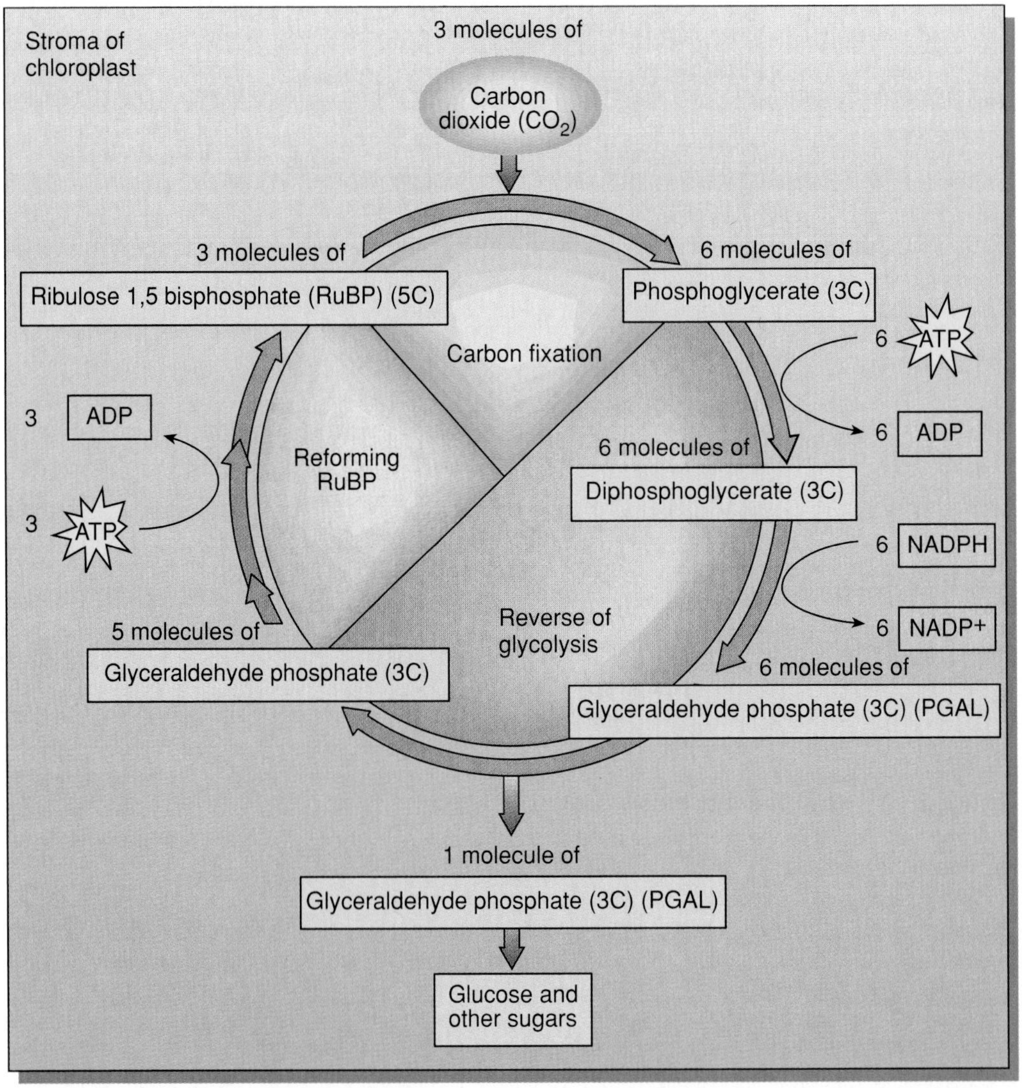

Figure 5.20
The Calvin cycle.
The Calvin cycle takes place in the stroma of the chloroplasts. The NADPH and the ATP that were generated by the light-dependent reactions are used in the Calvin cycle, or light-independent reactions, to build carbon molecules. The number of carbon atoms at each stage is indicated in parentheses. It takes six turns of the cycle to make one molecule of glucose.

C_4 Photosynthesis

Many plants have trouble carrying out C_3 photosynthesis when the weather is hot. As temperature increases, plants partially close their leaf openings, called **stomata,** to conserve water. As a result, CO_2 and O_2 are not able to enter and exit the leaves through these openings. The concentration of CO_2 in the leaves falls, while the concentration of O_2 in the leaves rises. These conditions cause RuBP carboxylase, an enzyme that carries out the first step of the Calvin cycle, to engage in **photorespiration,** removing CO_2 from the product of the reaction rather than adding it to the substrate. Photorespiration thus short-circuits the successful performance of the Calvin cycle.

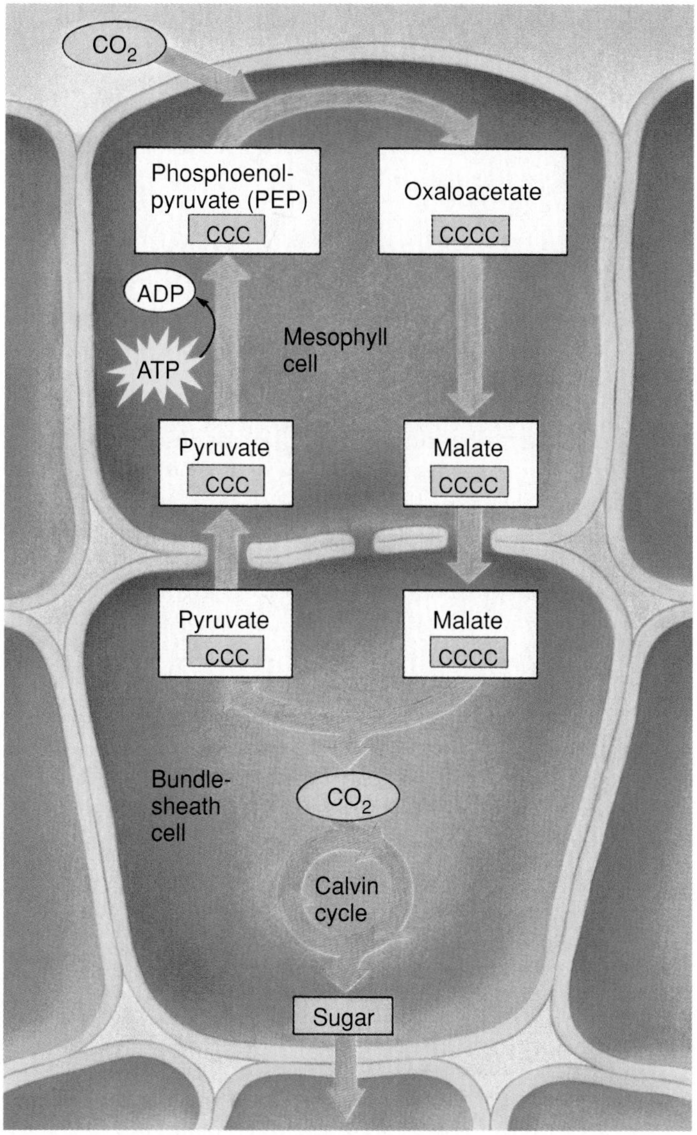

Figure 5.21 The path of carbon fixation in C_4 plants. C_4 plants shuttle a four-carbon molecule (malate) to bundle-sheath cells, where carbon dioxide can be concentrated. In this way, C_4 cells conserve carbon dioxide.

Some plants are able to adapt to climates with higher temperatures by performing **C_4 photosynthesis.** In this process, plants such as sugarcane, corn, and many grasses are able to fix carbon using different types of cells within their leaves and thereby avoiding a reduced yield in photosynthesis due to higher temperatures.

C_4 plants fix CO_2 first as the four-carbon molecule oxaloacetate (hence the name, C_4 photosynthesis), rather than as the three-carbon molecule phosphoglycerate of normal everyday C_3 photosynthesis, such as described previously. C_4 plants carry out this process in the mesophyll cells of their leaves. The oxaloacetate that this process produces is then transferred out to the bundle-sheath cells of the leaf and there broken down to regenerate CO_2, which enters the Calvin cycle (figure 5.21). The bundle-sheath cells are impermeable to carbon dioxide and therefore hold carbon dioxide within them. The concentration of CO_2 increases and thus decreases the occurrence of photorespiration.

In C_4 photosynthesis, the energetic cost is almost doubled compared to that of C_3 photosynthesis. However, in a hot climate in which photorespiration decreases the growth of C_3 plants, C_4 photosynthesis is the best alternative. For this reason, C_4 plants are more abundant in warmer regions than in cooler ones (figure 5.22).

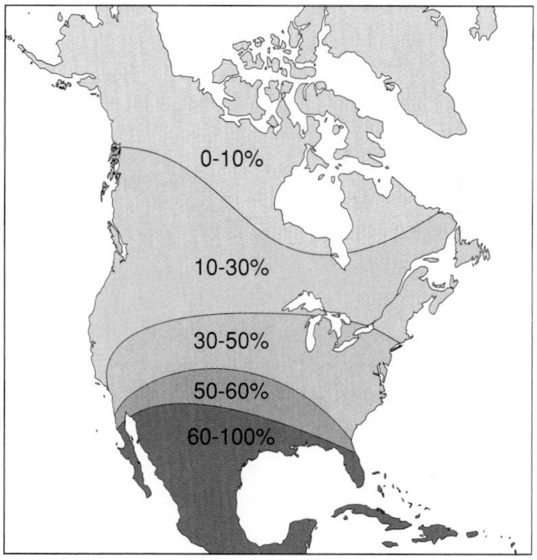

Figure 5.22 Prevalence of C_4 plants in North America. Many more C_4 grasses occur in the South, where the average temperatures during the growing season are higher. At higher temperatures, photorespiration wastes more of the products of photosynthesis, and the ability of C_4 plants to counteract photorespiration is more of an advantage than it is in cooler regions, where C_3 grasses predominate. This map shows the percentage of species of C_4 grasses among all grass species.

5.4 Cellular Respiration

The cells of plants fuel their activities with sugars and other fuel molecules, just as yours do; only the chloroplasts carry out photosynthesis. No light shines on roots below the ground, and yet root cells are just as alive as the cells in the stem and leaves. In both plants and animals, and in fact in almost all organisms, the energy for living is obtained by recycling the organic molecules assembled by chloroplasts. The ATP energy and reducing power invested in building the organic molecules are retrieved by stripping the energetic electrons away and using them to make ATP. This is possible because the electrons are still far from the nucleus, still carrying the energy contributed by their original encounter with the photon of light.

When electrons are stripped away from chemical bonds, the food molecules are being oxidized (remember, oxidation is the loss of electrons). The oxidation of foodstuffs to obtain energy is called **cellular respiration.** Do not confuse this with the breathing of oxygen gas that your lungs carry out, which is called simply respiration.

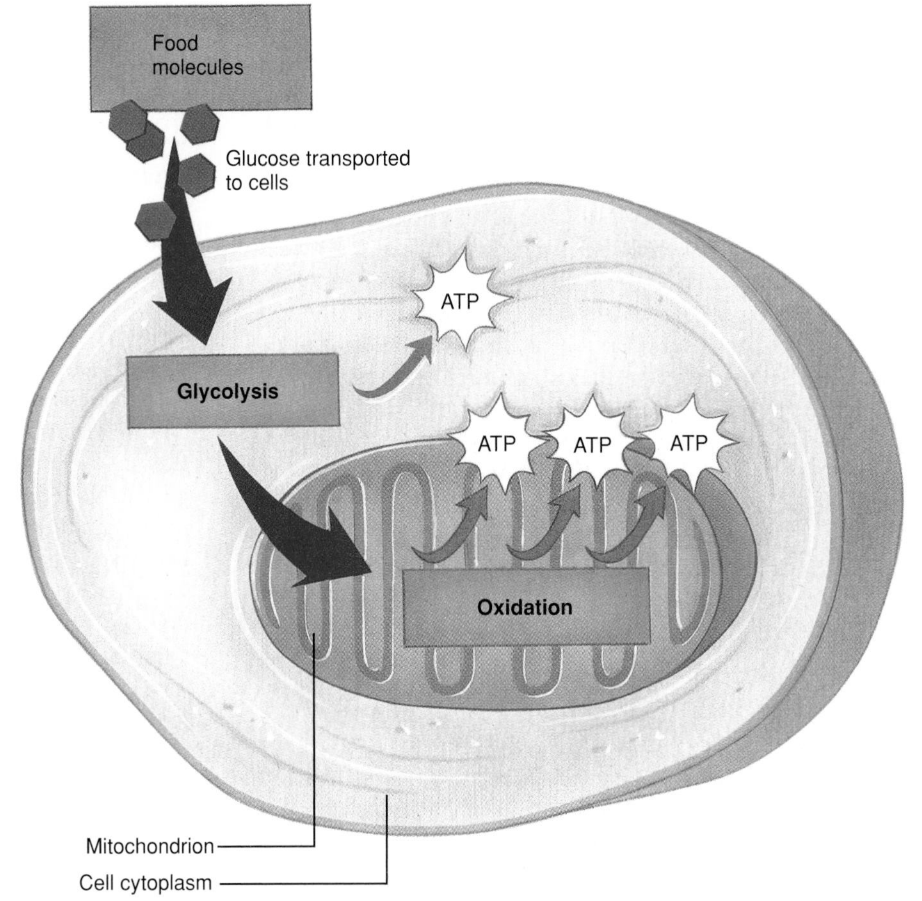

Figure 5.23 An overview of the two stages of cellular respiration. When you take in food, the food is broken down into its various components by your digestive system. Fuel for your cells, in the form of glucose, is delivered to them by your circulatory system. Once inside your cells, the energy stored in glucose is liberated and transferred to ATP molecules by a complex process called *glycolysis*, which takes place in the cytoplasm of the cell, does not require oxygen, and produces only a few ATP molecules. The second phase, *oxidation*, occurs in the cell's mitochondrion, requires oxygen, and produces much larger quantities of ATP.

Cellular respiration is carried out in two stages (figure 5.23). The first, **glycolysis,** takes place in the cell's cytoplasm and does not require oxygen. This ancient energy-extracting process is thought to have evolved over 3 billion years ago, when there was no oxygen in the earth's atmosphere. The second stage, **oxidation,** takes place only within mitochondria. It is far more powerful than glycolysis at recovering energy from food molecules and is where the bulk of the energy used by animal cells is extracted.

Glycolysis

The first stage in cellular respiration, glycolysis, is a series of sequential biochemical reactions, a **biochemical pathway.** In 10 enzyme-catalyzed reactions (figure 5.24), the six-carbon sugar glucose is cleaved into two three-carbon molecules called pyruvate. The biochemical pathway does not employ oxidation to extract energy; it simply shuffles chemical bonds so that two coupled reactions can occur. In each reaction the breaking of a chemical bond contributes enough energy to force the formation of an ATP molecule from ADP. This transfer of a phosphate group from a substrate to ADP is called **substrate-level phosphorylation.** In the process, two energetic electrons are extracted and donated to a carrier molecule called NAD^+. The NAD^+ carries the electrons as NADH to join the other electrons extracted during oxidative respiration, discussed in the following section. Only a small number of ATP molecules are made in glycolysis, two for each molecule of glucose attacked, but in the absence of oxygen it is the only way organisms can get energy from food.

Glycolysis is thought to have been one of the earliest of all biochemical processes to evolve. Every living creature is capable of carrying out glycolysis. Glycolysis, although inefficient, was not discarded during the course of evolution but rather was used as the starting point for the further extraction of energy by oxidation. Nature did not, so to speak, go back to the drawing board and design metabolism from scratch. Rather, the new reactions of the Krebs cycle were added onto the old, just as successive layers of paint can be found in an old apartment.

1. Phosphorylation of glucose by ATP.

2–3. Rearrangement, followed by a second ATP phosphorylation.

4–5. The six carbon molecule is split into two three-carbon molecules of G3P.

6. Oxidation followed by phosphorylation produces two NADH molecules and gives two molecules of BPG, each with one high-energy phosphate bond.

7. Removal of high-energy phosphate by two ADP molecules produces two ATP molecules and gives two 3PG molecules.

8–9. Removal of water gives two PEP molecules, each with a high-energy phosphate bond.

10. Removal of high-energy phosphate by two ADP molecules produces two ATP molecules and gives two pyruvate molecules.

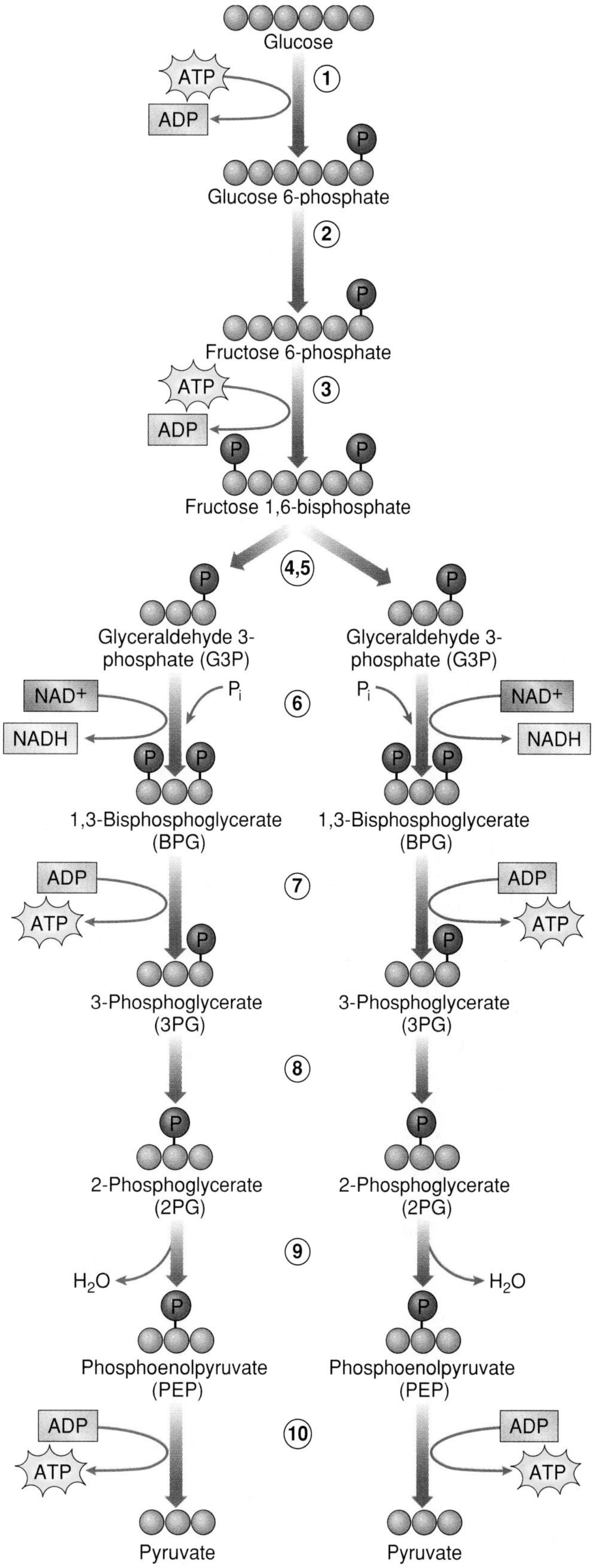

Figure 5.24 Glycolysis.

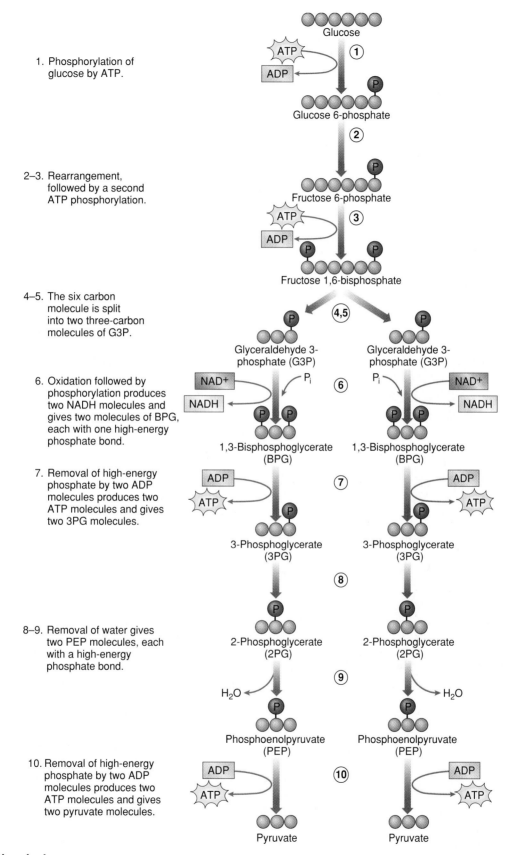

The ten main reactions of glycolysis are described and illustrated above.

Oxidation and the Krebs Cycle

When oxygen is available, the cell need not stop extracting energy after glycolysis. Instead, a second oxidative stage of cellular respiration takes place.

The Oxidation of Pyruvate

The first step of oxidative respiration is to oxidize the three-carbon molecule called pyruvate, which is the end product of glycolysis. Pyruvate molecules produced in the cytoplasm by glycolysis enter the mitochondria, where they are oxidized. A high-energy electron is extracted (we will discuss what happens to it in a moment), and a carbon is expelled as CO_2. If the cell already has plentiful supplies of ATP, the two-carbon fragment that is left, called acetyl coenzyme A (or, more simply, **acetyl-CoA**), is funneled into fat synthesis, its energetic electrons preserved for later needs. If the cell needs ATP now, the fragment is directed to continue the oxidative stage of cellular respiration (figure 5.25).

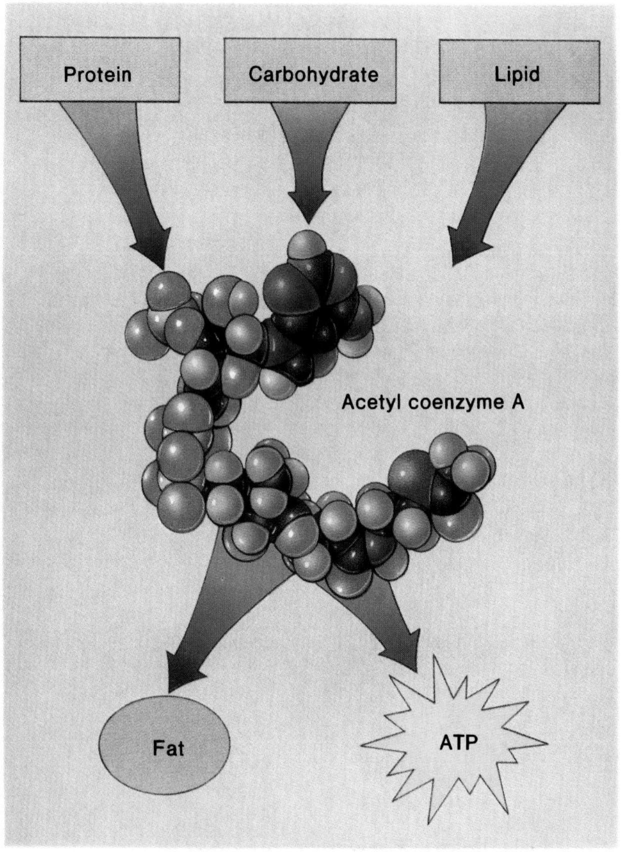

Figure 5.25 The various fates of acetyl-CoA.
Acetyl-CoA (acetyl coenzyme A) is the central molecule of energy metabolism. Almost all the molecules you use as foodstuffs are converted to acetyl-CoA when your body metabolizes them; the acetyl-CoA is then channeled into fat synthesis or into ATP production, depending on your body's energy requirements.

The Krebs Cycle

The rest of oxidative respiration is called the **Krebs cycle,** named after the man who discovered it. The Krebs cycle takes place within the mitochondrion. The cycle starts when the two-carbon acetyl-CoA fragment produced in the preparatory step is stuck onto a four-carbon sugar. Then, in rapid-fire order, a series of eight additional reactions occur. When it is all over, two carbon atoms have been expelled as CO_2, one ATP molecule has been made in a coupled reaction, eight more energetic electrons have been harvested and taken away as NADH or on other carriers, and we are left with the same four-carbon sugar we started with. The process of nine reactions is a cycle, a circle of reactions (figure 5.26).

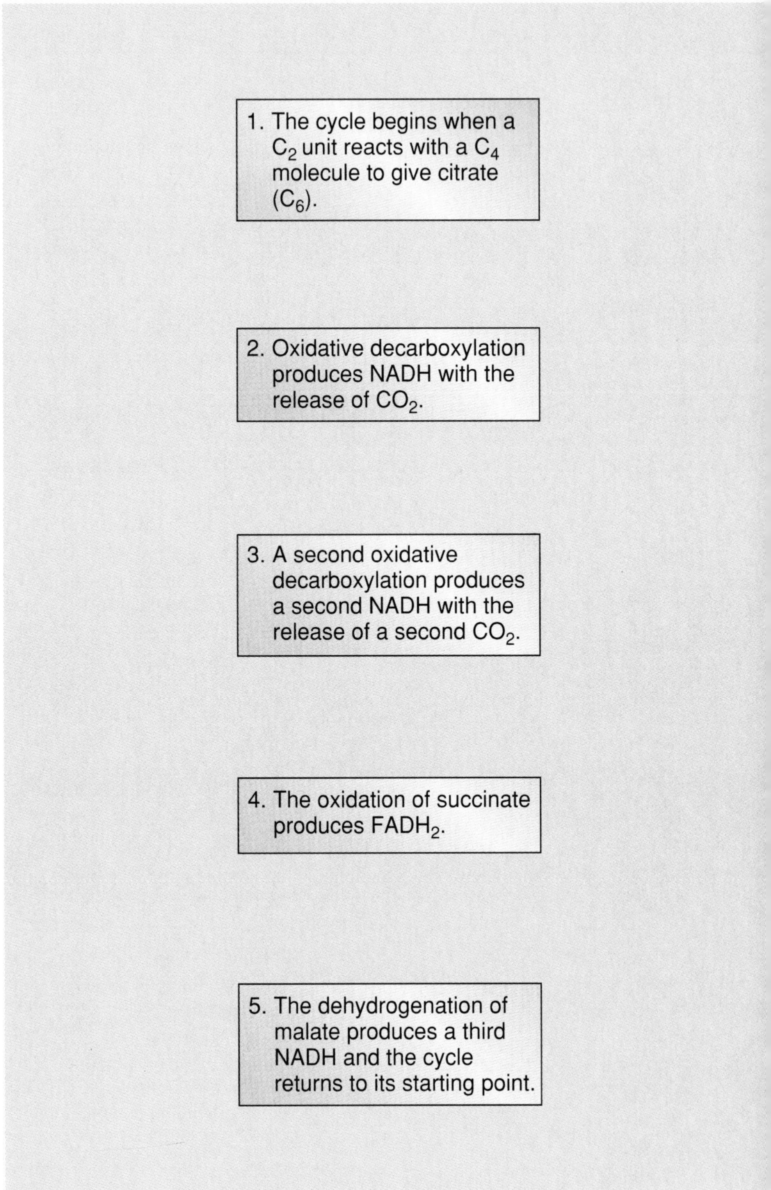

Figure 5.26 The Krebs cycle.
This series of reactions takes place within the mitochondrion. Notice the changes that occur in the carbon skeleton.

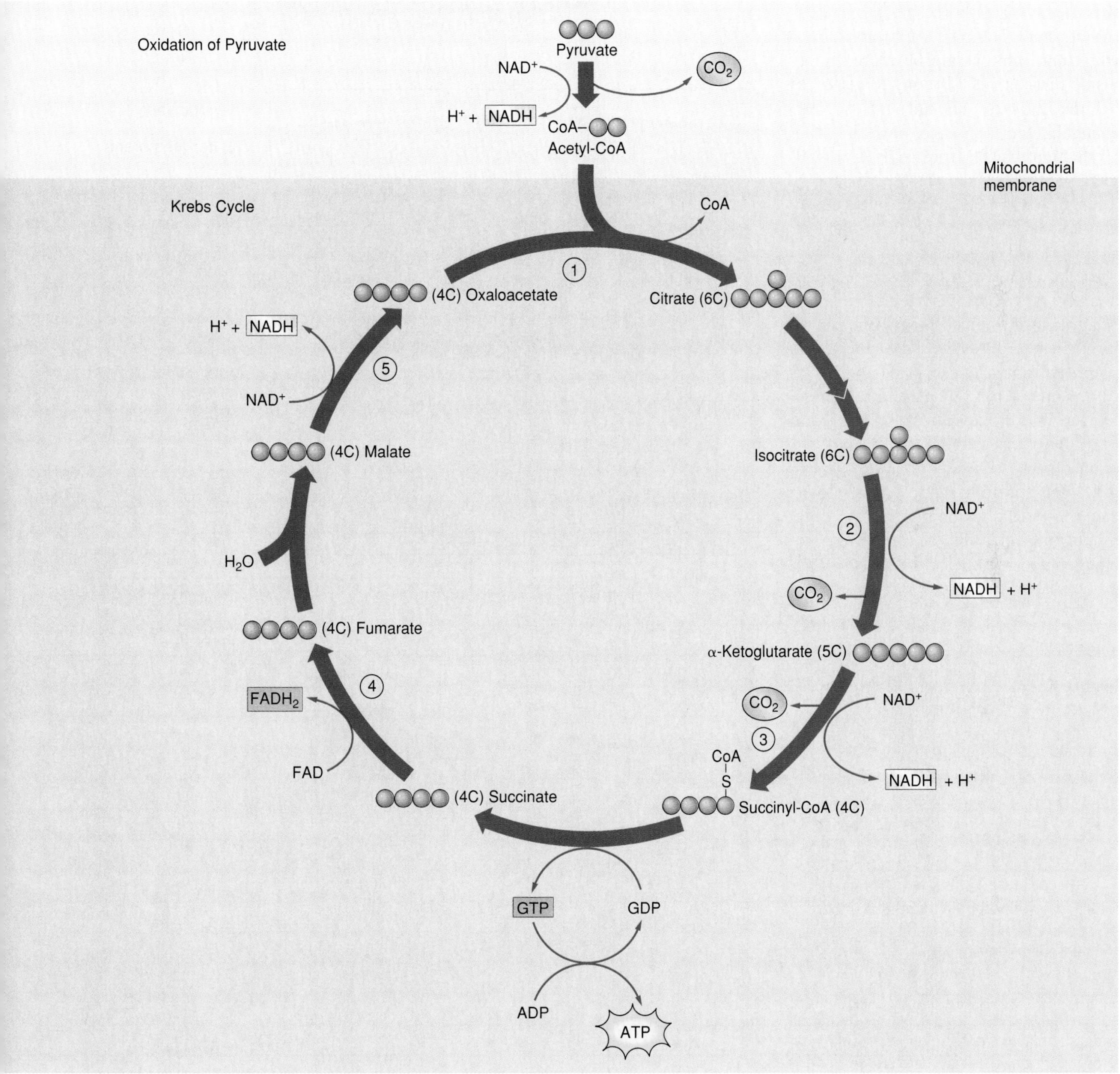

Oxidation of Pyruvate
Pyruvate
NAD⁺
CO_2
H⁺ + NADH
CoA–
Acetyl-CoA
Mitochondrial membrane
Krebs Cycle
CoA
1
(4C) Oxaloacetate
Citrate (6C)
H⁺ + NADH
5
NAD⁺
(4C) Malate
Isocitrate (6C)
NAD⁺
2
H_2O
CO_2
NADH + H⁺
(4C) Fumarate
α-Ketoglutarate (5C)
4
$FADH_2$
CO_2
NAD⁺
3
CoA
S
FAD
NADH + H⁺
(4C) Succinate
Succinyl-CoA (4C)
GTP
GDP
ADP
ATP

Making ATP

Mitochondria use chemiosmosis to make ATP in much the same way that chloroplasts do, although the proton pumps are oriented the opposite way. Mitochondria use energetic electrons extracted from food molecules to power proton pumps that shove protons outside the mitochondrial membrane. A stream of NADH-borne electrons fuels the constant pumping of protons outward, until protons are far more scarce inside than outside. Driving to get back in, the protons jam through special channels, their passage powering the production of ATP from ADP. The ATP then passes outside of the mitochondrion through ATP-passing open channels (figure 5.27).

What happens to the electrons after the proton pumps have expended their energy? The hydrogen atoms carrying them are joined to oxygen gas to form water. Because the electrons stripped from pyruvate need to find a final home, cellular respiration requires oxygen. The energy cannot be extracted from pyruvate without oxygen to siphon off the spent electrons, for otherwise the proton pumps and other electron-ferrying components of the mitochondrion would soon become clogged with spent electrons.

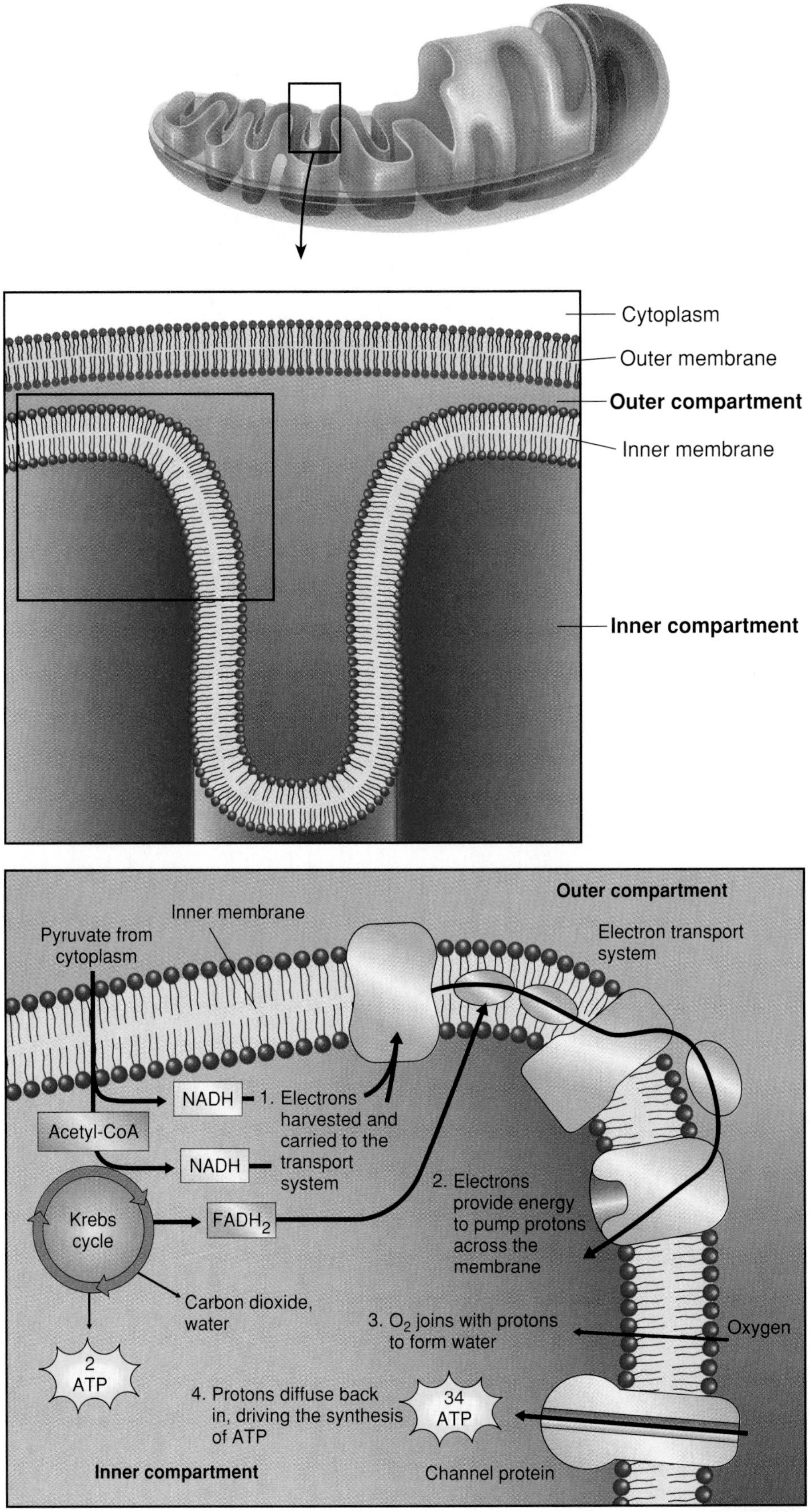

Figure 5.27 How chemiosmosis occurs in a mitochondrion.

The interior of the mitochondrion is separated by double membranes into two compartments. Pyruvate generated from glycolysis in the cell's cytoplasm enters the mitochondrion, is oxidized to acetyl-CoA, and enters the Krebs cycle. The electrons that are harvested by these reactions are carried to the membrane inner surface by NADH and $FADH_2$, which donate them to the electron transport system. This electron transport system channels their energy to drive the pumping of electrons out of the inner compartment. The result is an excess of protons in the outer compartment. Protons reenter the inner compartment, driven by diffusion, through special channels that produce ATP. These ATP molecules can then diffuse back out of the mitochondrion.

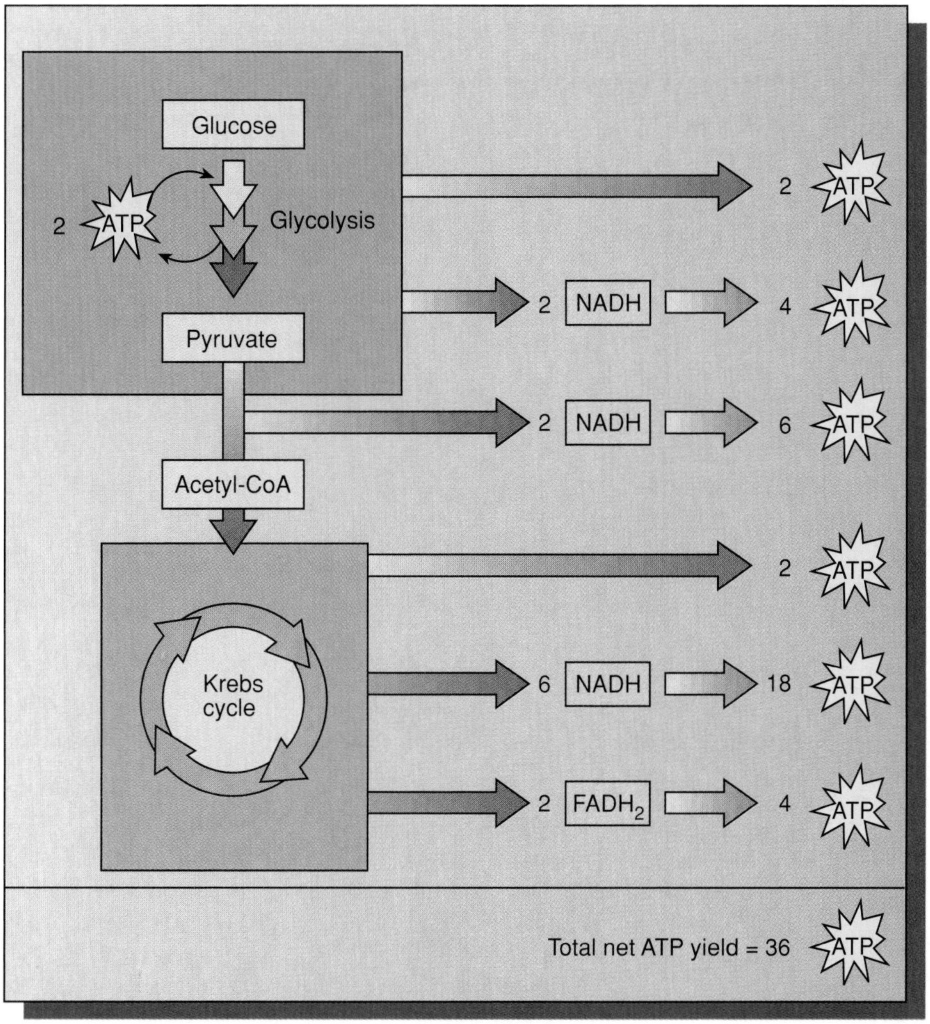

Figure 5.28 Energy extracted during the breakdown of glucose. Most of the energy is extracted during oxidation (oxidation is the removal of electrons). The energetic electrons are carried by NADH and $FADH_2$ and used to produce ATP via chemiosmosis.

An Overview of Cellular Respiration

As an example of metabolism in action, consider the fate of a chocolate bar. The bar is composed of sugar, chocolate, other lipids and fats, protein, and many other molecules. This diverse collection of complex molecules is broken down by the process of digestion into simpler molecules such as glucose, amino acids, and fatty acids. These breakdowns produce little or no energy but prepare the way for three major energy-producing processes. The first process is glycolysis, in which glucose is converted into two molecules of pyruvate, producing 2 ATP and 2 NADH. The 2 NADH enter the electron transport chain to produce another 4 ATP. The second process is the oxidation of the pyruvate molecules to two molecules of acetyl-CoA, producing 2 NADH. The 2 NADH enter the electron transport chain to produce 6 ATP. The third process is the oxidation of acetyl-CoA molecules in the Krebs cycle, producing 2 ATP, 6 NADH, and 2 $FADH_2$. The 6 NADH enter the electron transport chain, producing 18 ATP, and the 2 $FADH_2$ enter the electron transport chain, producing 4 ATP for a grand total of 36 ATP.

Oxidation is very efficient. The breakdown of glucose to pyruvate in glycolysis yields a net of only 2 ATP, while the oxidation of glucose yields an additional 34 ATP molecules (figure 5.28).

Regulating Cellular Respiration

The rate of cellular respiration slows down when your body's cells already have ample supplies of ATP. This is very sensible, but how does each mitochondrion arrive at the appropriate decision? The control works through a system of feedback inhibition in which excess product shuts off the reaction (figure 5.28). Key reactions early in glycolysis and the Krebs cycle are catalyzed by enzymes that have a second, allosteric site. This site is the same shape as ATP, and when ATP levels in the cell are high it is very likely that ATP molecules will become stuck to it. The binding of ATP to the allosteric site goads the protein into changing shape to better accommodate the fit—and the new shape is not active as an enzyme! High levels of ATP thus shut down the processes the cell uses to make ATP. Like a well-designed automobile, your energy-producing machinery only operates when you step on the gas.

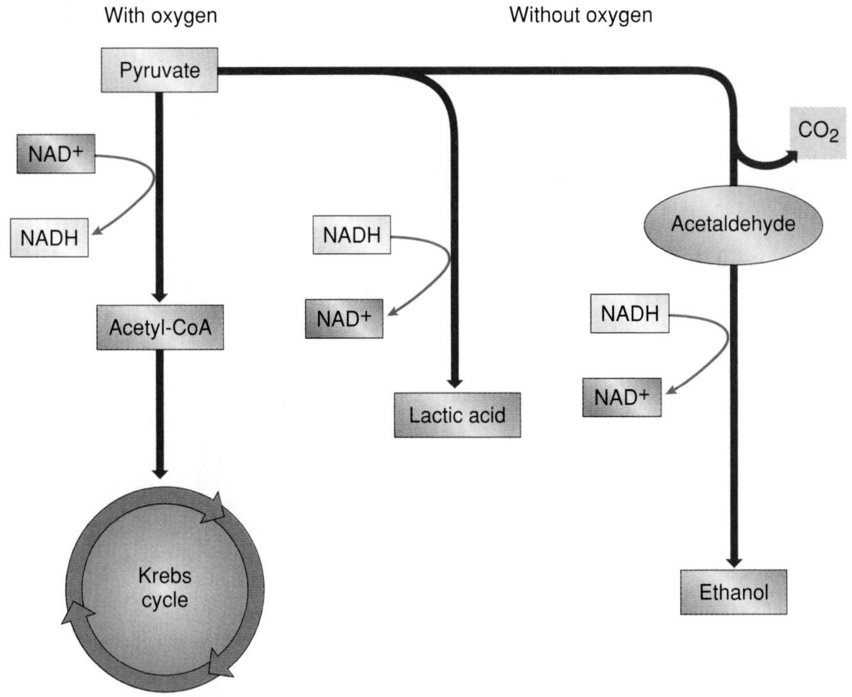

Figure 5.29 Two types of fermentation.
In the presence of oxygen, pyruvate is oxidized to acetyl-CoA and enters the Krebs cycle. In the absence of oxygen, pyruvate instead is reduced, accepting the electrons extracted during glycolysis and carried by NADH. This process is called fermentation. When pyruvate is reduced directly, as it is in your muscles, the product is lactic acid; when CO_2 is first removed from pyruvate and the remainder is reduced, as it is in yeasts, the product is ethanol.

Fermentation

What happens if there is no oxygen to carry out oxidative respiration? Does the pyruvate that is the product of glycolysis and the starting material for oxidative respiration just accumulate in the cytoplasm? No. It has a different fate. Recall that during glycolysis a single energetic electron is extracted at one step, carried away by a carrier molecule called NAD^+. In the absence of oxygen, these electrons are not used in chemiosmosis, and so soon all the cell's NAD^+ becomes saturated with electrons. With no more NAD^+ available to carry away electrons, glycolysis cannot proceed. Clearly, to obtain energy from food in the absence of oxygen, a solution to this problem is needed. A home must be found for these electrons. Adding the extracted electron to an organic molecule, as animals and plants do when they have no oxygen to take it, is called **fermentation.**

Two types of fermentation are common among eukaryotes (figure 5.29). Animals such as ourselves simply add the extracted electrons to pyruvate, forming lactic acid. Later, when oxygen becomes available, the process can be reversed and the electrons used for energy production. This is why your arm muscles would feel tired if you were to lift this book up and down 100 times rapidly. The muscle cells use up all the oxygen, and so they start running on ATP made by glycolysis, storing the pyruvate and electrons as lactic acid. This so-called "oxygen debt" produces the tired, burning feeling in the muscle.

Single-celled fungi called yeasts adopt a different approach to fermentation. First they convert the pyruvate into another molecule, and then they add the electron extracted during glycolysis, producing ethyl alcohol, or ethanol. For centuries, humans have consumed such ethyl alcohol in wine and beer. Yeasts only conduct this fermentation in the absence of oxygen. That is why wine is made in closed containers—to keep oxygen in the air away from the crushed grapes.

CHAPTER 5

HIGHLIGHTS

5.1 Cells and Chemistry

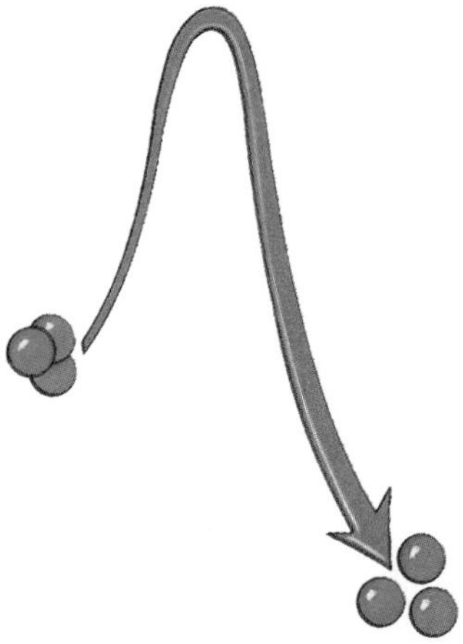

Key Terms

chemical reaction 86
activation energy 86
catalysis 86
enzyme 88

Key Concepts

- The chemistry of cells involves a series of chemical reactions between molecules in which a chemical bond is made or broken.
- Reactions that release energy are called exergonic, and those where the products contain more energy than the reactants are called endergonic.
- Almost all chemical reactions require an input of energy, called activation energy, to start them off.
- Activation energies of cellular reactions are lowered during catalysis by an enzyme.

5.2 How Cells Use Energy

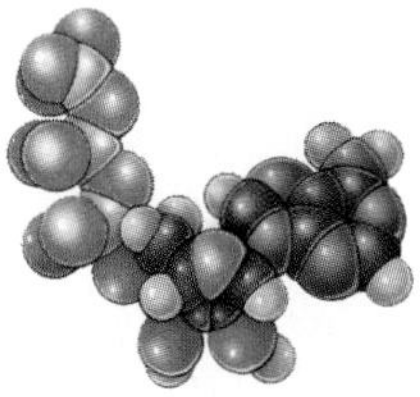

Key Terms

ATP 90
coupled reaction 90

Key Concepts

- Cells need a constant supply of energy to carry out their many activities.
- ATP serves as the molecular energy currency for virtually all of the cell's activities.
- Energy is released when ATP is cleaved into ADP + P.

5.3 Photosynthesis

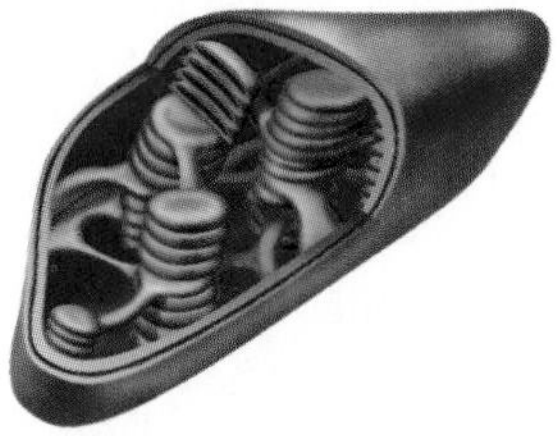

Key Terms

photon 94
pigment 94
chemiosmosis 96
Calvin cycle 98

Key Concepts

- Energy ultimately reaches organisms via photosynthesis, which captures energy from sunlight and uses it to make ATP and NADPH.
- In the Calvin cycle, chloroplasts use the ATP and NADPH to convert CO_2 in the air into organic molecules.
- Ultimately, photosynthesis consumes CO_2 and releases O_2.

5.4 Cellular Respiration

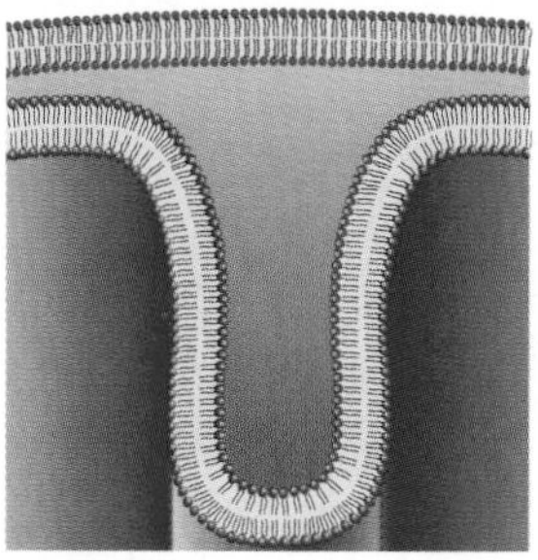

Key Terms

glycolysis 100
oxidation 100
Krebs cycle 102
fermentation 106

Key Concepts

- The first stage of cellular respiration is glycolysis, which does not require oxygen.
- The next stages involve oxidation of the product of glycolysis, first to acetyl-CoA and then, via the Krebs cycle, to CO_2.
- Ultimately, oxidative cellular respiration consumes O_2 and releases CO_2.

5

CONCEPT REVIEW

1. Catalysts _________ the activation energy.
   a. raise
   b. lower
2. The site on the enzyme surface where the reactant is stressed is called the _________ site.
   a. active
   b. allosteric
   c. binding
   d. substrate
3. Select the largest molecule.
   a. ADP
   b. ATP
   c. P
   d. K
4. Carbon fixation requires the expenditure of ATP molecules. This ATP is generated by
   a. formation of glucose during the light-independent reactions.
   b. replenishment of the photosynthetic pigment.
   c. chemiosmotic synthesis during the light-dependent reactions.
   d. none of the above.
5. When P_{700} receives photon-induced excitation energy, what is the form of this energy?
   a. light energy
   b. protons
   c. neutrons
   d. electrons
6. When plants and algae employ both photosystem I and II, _________ and _________ are generated, both of which are required to form organic molecules from atmospheric carbon dioxide.
   a. water
   b. ATP
   c. NADH
   d. NADPH
7. Select the first stage of cellular respiration.
   a. chemiosmosis
   b. glycolysis
   c. Krebs cycle
   d. electron transport
8. The end product of glycolysis is
   a. ADP.
   b. pyruvate.
   c. glucose.
   d. oxygen.
9. Entrance of pyruvate into the second stage of cellular respiration requires the presence of
   a. carbon dioxide.
   b. ATP.
   c. ADP.
   d. oxygen.
10. By the end of oxidative metabolism, all of the carbons from glucose are gone. Where did they go?
    a. carbon dioxide
    b. ATP
    c. to make pyruvate
    d. water
11. An _________ reaction requires energy from an outside source.
12. The _________ site is where a reactant attaches to an enzyme.
13. NAD^+ carries an electron as _________.
14. Light energy is "captured" and used to boost an electron to a higher energy level in molecules called _________.
15. All of the oxygen that you breathe has been produced by the splitting of water during ________.
16. The basic photosynthetic unit of a chloroplast is the _________, a flattened membranous sac that occurs in stacked columns called grana.
17. The second stage of cellular respiration is called the _________ cycle.
18. During fermentation, pyruvate is converted to _________ in yeast cells.

Answers to the Concept Review questions appear in Appendix B.

CHALLENGE YOURSELF

1. Almost no sunlight penetrates into the deep ocean. However, many fish that live there attract prey and potential mates by producing their own light. Where does that light come from? Does its generation require energy?

2. Why could glycolysis be considered an inefficient energy process? Why could the Krebs cycle be considered a somewhat more efficient energy process than glycolysis?

3. If you poke a hole in a mitochondrion, can it still perform oxidative respiration? Can fragments of mitochondria perform oxidative respiration?

4. In theory, a plant kept in total darkness could still manufacture glucose—if it were supplied with *which* molecules?

5. Why are plants that consume 30 ATP molecules to produce 1 molecule of glucose (rather than the usual 18 molecules of ATP per glucose molecule) favored in hot climates but not in cold climates?

FOR FURTHER READING

Barber, J. "A Quantum Step Forward in Understanding Photosynthesis." *Plants Today,* September 1989, 165–69. An engaging account of the discovery of photosystem II, for which the Nobel Prize was given in 1988.

Hall, D., and K. Rao. *Photosynthesis,* 4th ed. Baltimore: Arnold, 1987. A great overview of the photosynthetic process and of how our understanding of it has developed. Highly recommended.

Hinkle, P., and R. McCarty. "How Cells Make ATP." *Scientific American,* March 1978, 104–25. A classic summary of oxidative respiration, with a clear account of the events that happen at the mitochondrial membrane.

Kraut, J. "How Do Enzymes Work?" *Science* 242 (October 1988): 533–40. A clear presentation of the idea that enzymes work by stabilizing transition states.

McCarty, R. "ATPases in Oxidative and Photosynthetic Phosphorylation." *Bioscience* (January 1985): 27–33. An excellent overview of the key protein channels that carry out photosynthesis.

Raven, P., R. Evert, and S. Eichhorn. *Biology of Plants,* 5th ed. New York: Worth, 1992. The definitive botany text, comprehensive and easy to read.

Riddihough, G. "Picture an Enzyme at Work." *Nature,* April 1993, 793. A discussion of how conformational changes in enzymes enable them to function.

Wachtershauser, G. "Evolution of the First Metabolic Cycles." *Proceedings of the National Academy of Science USA* 87 (1990): 200–204. A suggestion that metabolic cycles played a key role in the evolution of life.

TECHNOLOGY LINKS

The Living World Home Page
http://www.wcbp.com/biology/tlw

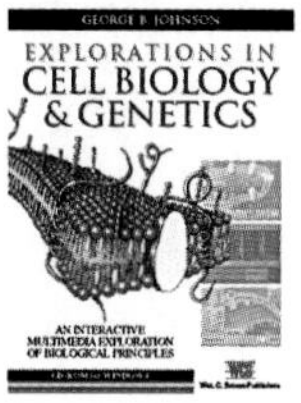

Explorations in Cell Biology & Genetics CD

#6 Thermodynamics

#7 Kinetics

#8 Oxidative Respiration

#9 Photosynthesis

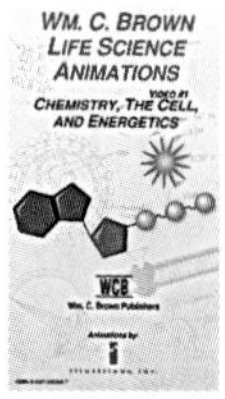

Life Science Animations Videotape 1

#6 Oxidative Respiration

#7 The Electron Transport Chain and the Production of ATP

#9 C_3 Photosynthesis (Calvin Cycle)

6

Foundations of Genetics

"This is your side of the family, you realize."

CHAPTER OUTLINE

Figure 6.1 An inherited trait.
Many human traits are inherited within families. This little redhead is exhibiting one such trait.

When you were born, many things about you—your sex, the color of your hair and eyes, the structure of your face—resembled your mother or father. This tendency for traits to be passed from parent to offspring is called **heredity.** How does heredity happen? Where are the instructions that say that a person will have red hair, and how is a trait like red hair passed on to a person's children, like the child in figure 6.1? Before DNA and chromosomes were discovered, this puzzle was one of the greatest mysteries of science.

6.1 Mendel

The key to understanding the puzzle of heredity was found in the garden of an Austrian monastery over a century ago by a monk named Gregor Mendel (figure 6.2). Crossing pea plants with one another, Mendel developed a series of simple rules that accurately predicted patterns of heredity—that is, how many offspring would be like one parent and how many like the other. When Mendel's rules became widely known, investigators all over the world set out to discover the physical mechanism responsible for them. They learned that hereditary traits are written in *genes*, the instructions carefully laid out in the DNA of the chromosomes the child receives from each parent. Mendel's solution to the puzzle of heredity was the first step on this journey of understanding and one of the greatest intellectual accomplishments in the history of science.

Figure 6.2 Gregor Johann Mendel. The key to understanding the puzzle of heredity was solved by Mendel by cultivating pea plants in the garden of his monastery in Brunn, Austria.

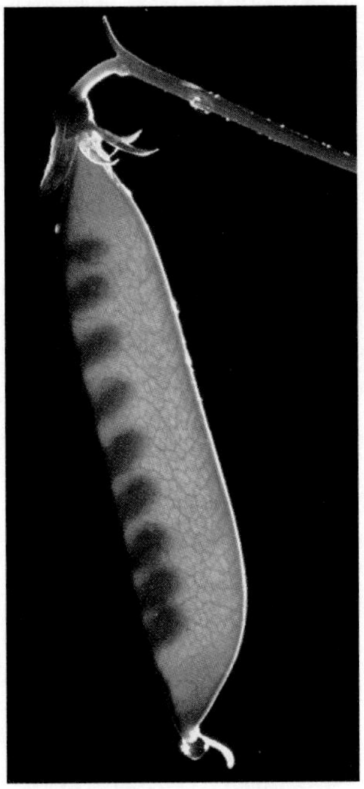

Figure 6.3 The garden pea, *Pisum sativum*. Because it is easy to cultivate and because there are many distinctive varieties, the garden pea had been a popular choice as an experimental subject in investigations of heredity for as long as a century before Gregor Mendel's studies.

Early Ideas About Heredity

Mendel was not the first person to try to understand heredity by crossing pea plants. Over 200 years earlier British farmers had performed similar crosses and obtained results similar to Mendel's. They observed that in crosses between two types—tall and short plants, say—the less frequent kind (in this case, short plants) would disappear in one generation, only to reappear in the next. In the 1790s, for example, the British farmer T. A. Knight crossed a variety of the garden pea that had purple flowers with one that had white flowers. All the offspring of the cross had purple flowers. If two of these offspring were crossed, however, some of *their* offspring were purple and some were white. Knight noted that the purple had a "stronger tendency" to appear than white, but he did not count the numbers of each kind of offspring.

Mendel's Experiments

Gregor Mendel was born in 1822 to peasant parents and was educated in a monastery. He became a monk himself and was sent by the monastery to the University of Vienna to study science and mathematics. Unfortunately for his aspirations to become a scientist and teacher, he failed his university exams for a teaching certificate and returned to the monastery, where he spent the rest of his life, eventually becoming abbot. Upon his return, Mendel joined an informal neighborhood science club, a group of local farmers and others interested in science. Under the patronage of a local nobleman, each member set out to undertake scientific investigations, which were then discussed at meetings and published in the club's own journal. Mendel undertook to repeat the classic series of crosses with pea plants done by Knight and others, but this time he intended to count the numbers of each kind of offspring in the hope that the numbers would give some hint of what was going on. Quantitative approaches to science—measuring and counting—were just becoming fashionable in Europe, so what Mendel proposed to do was on the cutting edge of research at the time. No one could know whether it was worth doing or not.

The Garden Pea

Mendel chose to study the garden pea because, as earlier researchers had discovered, several of its characteristics made it easy to work with (figure 6.3):

1. Many varieties were available. Mendel initially examined 32 varieties and from them selected seven pairs of lines that differed in easily distinguished traits (including the white versus purple flowers that Knight had studied 60 years earlier).

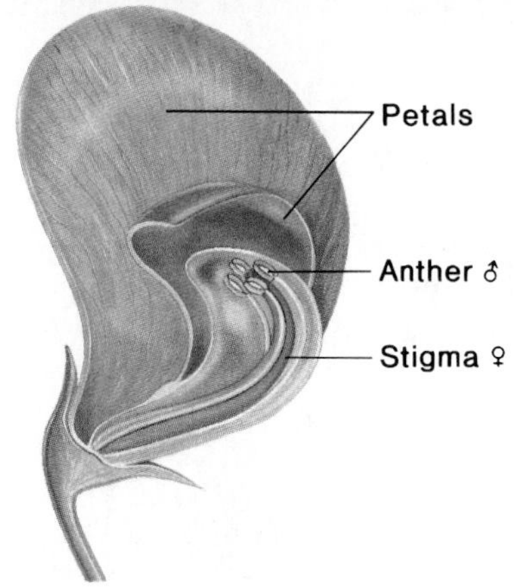

Figure 6.4 The reproductive organs of a pea flower.
In a pea flower, the petals enclose the male (anther) and female (stigma) parts, ensuring that self-fertilization will take place unless the flower is disturbed.

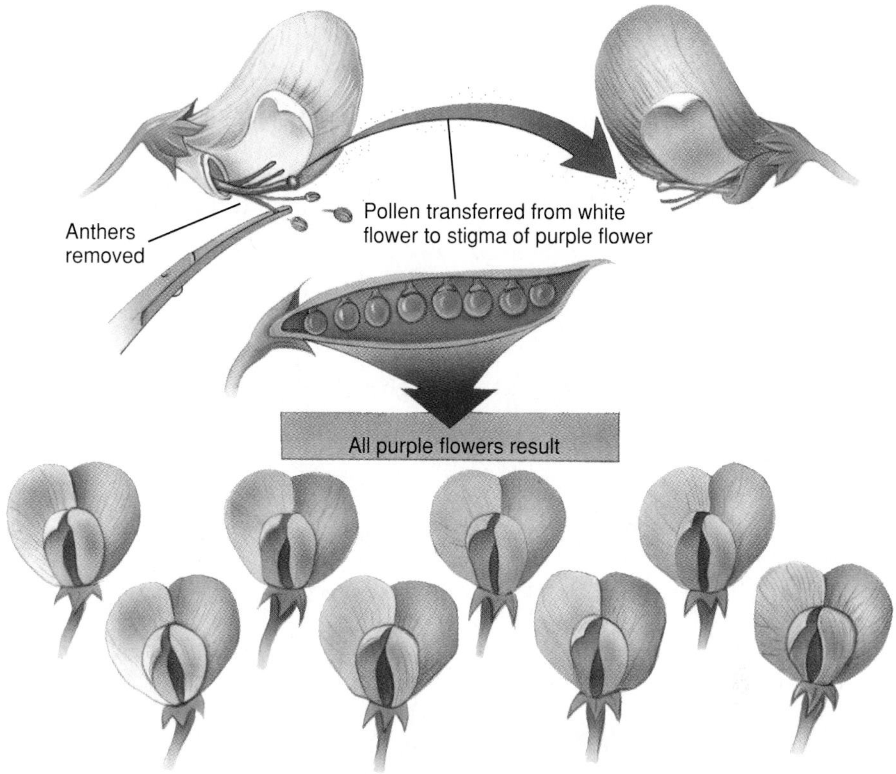

Figure 6.5 How Mendel conducted his experiments.
Mendel pushed aside the petals of a white flower and cut off the anthers, where the male gametes are produced and enclosed in pollen grains. He then placed that pollen onto the female parts of a similarly castrated purple flower, where cross-fertilization took place. All of the seeds in the pod that resulted from this pollination were hybrids with a purple-flowered female parent and a white-flowered male parent. After planting these seeds, Mendel observed what kinds of plants they produced. All of them had purple flowers.

2. Mendel knew from the work of Knight and others who had carried out these same pea plant crosses that he could expect the infrequent version of a trait to disappear in one generation and reappear in the next. He knew, in other words, that he would have something to count.
3. Pea plants are small, easy to grow, produce large numbers of offspring, and mature quickly.
4. The reproductive organs of peas are enclosed within their flowers (figure 6.4), and, left alone, the flowers do not open, simply fertilizing themselves with their own pollen (male gametes). To carry out a cross, Mendel had only to pry the petals apart, reach in with a scissors, and snip off the male organs (anthers); he could then dust the female organs with pollen from another plant to make the cross.

Mendel's Experimental Design

Mendel's experimental design was the same as Knight's, only Mendel counted his plants. The crosses were carried out in three steps:

1. Mendel began by letting each variety self-fertilize for several generations. This ensured that each variety was **true-breeding** and contained no other varieties. The white flower variety, for example, produced only white flowers and no purple ones in each generation. Mendel called these lines the **P generation** (P for parental).
2. Mendel then conducted his experiment: He crossed two pea varieties exhibiting alternative traits, such as white versus purple flowers (figure 6.5). The offspring that resulted he called the **F_1 generation** (F_1 for "first filial" generation, from the Latin word for son or daughter).
3. Finally, Mendel allowed the plants produced in the crosses of step 2 to self-fertilize, and he counted the numbers of each kind of offspring that resulted in this **F_2** ("second filial") **generation.**

Trait	Dominant vs recessive	F_2 generation results: Dominant form	F_2 generation results: Recessive form	Ratio
Flower color	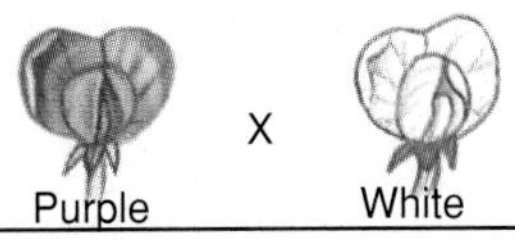	705	224	3.15:1
Seed color		6022	2001	3.01:1
Seed shape	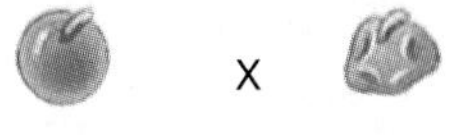	5474	1850	2.96:1
Pod color		428	152	2.82:1
Pod shape	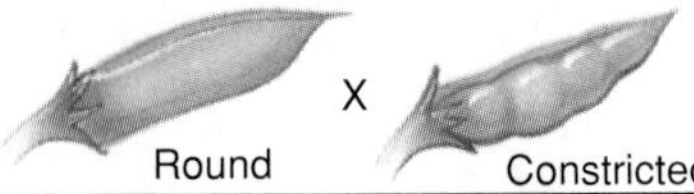	882	299	2.95:1
Flower position	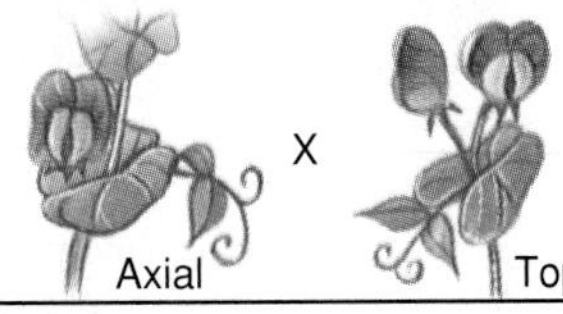	651	207	3.14:1
Plant height	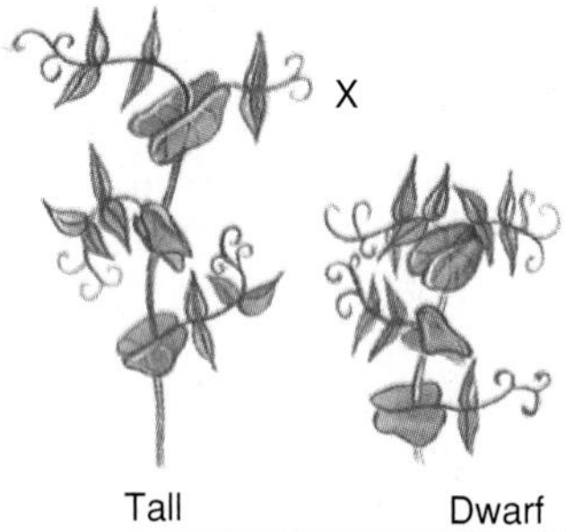	787	277	2.84:1

Figure 6.6 Mendel's experimental results.
The seven pairs of contrasting traits studied by Mendel in the garden pea are illustrated, along with the data he obtained in crosses of these traits. Every one of the seven traits that Mendel studied yielded results very close to a theoretical 3:1 ratio.

What Mendel Found

Mendel examined seven pairs of contrasting traits (figure 6.6). For each pair of contrasting varieties that Mendel crossed, he obtained the same result: In the case of flower color, when he crossed purple and white flowers, all the F_1 generation plants were purple, and the contrasting trait, white flowers, was not seen. Mendel called the trait expressed in the F_1 plants **dominant** and the trait not expressed **recessive.** In this case, purple flower color was dominant and white flower color recessive. Mendel studied several other traits in addition to flower color, and for every pair of contrasting traits Mendel examined, one proved to be dominant and the other recessive.

Allowing F_1 purple plants to self-fertilize, Mendel found (as Knight had earlier) that some F_2 plants exhibited white flowers, the recessive trait. The recessive trait had disappeared in the F_1 generation, only to reappear in the F_2 generation. It must somehow have been present in the F_1 individuals but unexpressed!

At this stage Mendel instituted his radical change in experimental design. He *counted* the number of each type among the F_2 offspring. In the cross of purple-flowered and white-flowered varieties, for example, he counted 929 F_2 individuals—705 of these had purple flowers, 224 had white flowers. He saw immediately that 224 is almost exactly one-fourth of 929—that is, the ratio of purple to white is almost exactly 3:1. Mendel obtained the same result in each of the crosses he carried out. In every case, the ratio of dominant to recessive among F_2 individuals was 3:1. Surely this ratio must mean something important.

Mendel let the F_2 plants self-fertilize for another generation and found that the one-fourth that were recessive were true-breeding—future generations showed nothing but the recessive trait. Thus the white F_2 individuals described previously showed only white flowers in the F_3 generation. Among the three-fourths of the F_2 plants that had shown the dominant trait, only one-third of the individuals were true-breeding. The others showed both traits in the F_3 generation—and when Mendel counted their numbers, he found the ratio of dominant to recessive to again be 3:1! From these results Mendel concluded that the 3:1 ratio he had observed in the F_2 generation was in fact a disguised 1:2:1 ratio (figure 6.7):

1		2		1
true-breeding	:	not-true-breeding	:	true-breeding
dominant		dominant		recessive

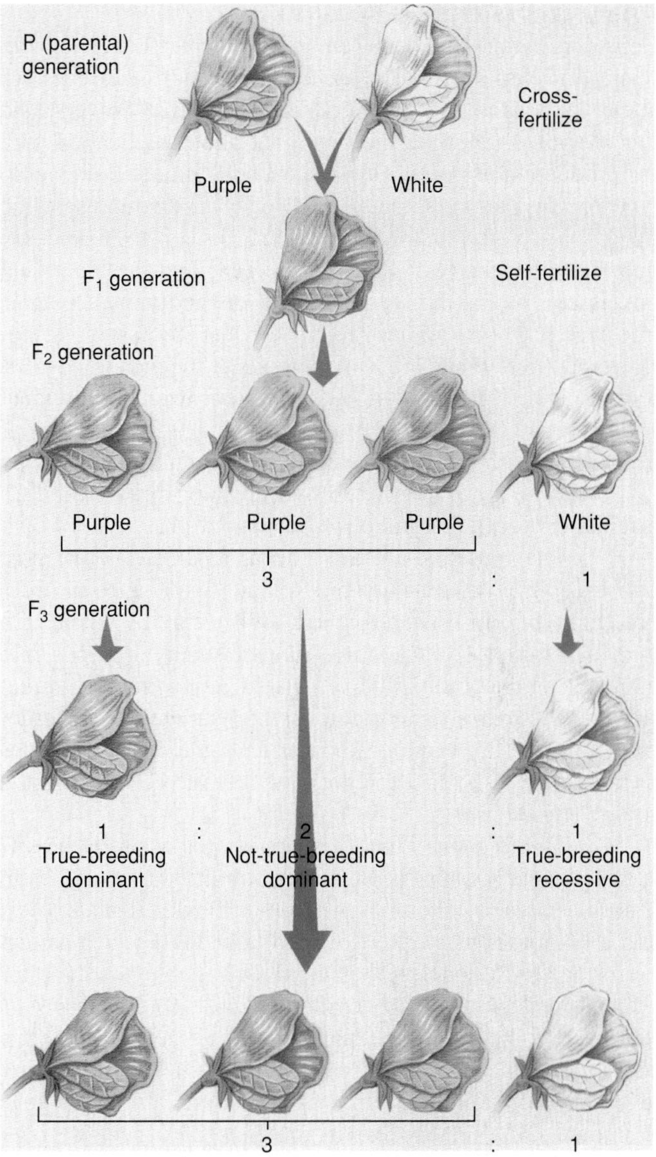

Figure 6.7 The F_2 generation is a disguised 1:2:1 ratio. By allowing the F_2 generation to self-fertilize, Mendel found from the offspring (F_3) that the ratio of F_2 plants was one true-breeding dominant, two not-true-breeding dominant, and one true-breeding recessive.

Mendel's Theory

To explain his results, Mendel proposed a simple set of rules that would faithfully predict the results he had found. It has become one of the most famous theories in the history of science. The model that Mendel proposed has five elements:

1. Parents do not transmit traits directly to their offspring. Rather, they transmit information about the traits, what Mendel called "*merkmal*" (the German word for factor or character). These factors act later, in the offspring, to produce the trait. In modern terminology we call Mendel's factors **genes.**
2. Each parent contains *two* copies of the factor governing each trait, copies that may or may not be the same. If the two copies of the factor are the same (both encoding purple or both white, for example) the individual is said to be **homozygous.** If the two copies of the factor are different (one encoding purple, the other white, for example), the individual is said to be **heterozygous.**
3. The alternative forms of a factor, leading to alternative traits such as white or purple flowers, are called **alleles.** Mendel used lowercase letters to represent recessive alleles and uppercase letters to represent dominant ones. Thus, in the case of purple flowers, the dominant purple flower allele is represented as *W* and the recessive white flower allele is represented as *w*. In modern terms, we call the appearance of an individual, such as possessing white flowers, its **phenotype.** Appearance is determined by which alleles of the flower-color gene the plant receives from its parents, and we call those particular alleles its **genotype.** Thus a pea plant might have the phenotype "white flower" and the genotype *ww*.
4. The two alleles that an individual possesses, one contributed by the male parent and the other by the female parent, do not affect each other, any more than two letters in a mailbox alter each other's contents. Each allele is passed on unchanged when the individual matures and produces its own gametes.
5. The presence of an allele does not ensure that a trait will be expressed in the individual that carries it. In heterozygous individuals, only the dominant allele achieves expression; the recessive allele is present but unexpressed.

Analysis of Mendel's Results

Consider again Mendel's cross of purple-flowered with white-flowered plants. The symbol *w* refers to the recessive allele, associated with the production of white flowers, and the symbol *W* refers to the dominant allele, associated with the production of purple flowers. Mendel indicated the genotype of any particular plant by two letters to represent the two copies of the factor it possessed. Today we would say they represent the two alleles of the gene that the individual possesses. Thus, the genotype of an individual that is true-breeding (or homozygous) for the recessive white-flowered trait would be *ww*. Similarly, the genotype of a true-breeding (or homozygous) purple-flowered individual would be *WW*. A heterozygous individual would be designated *Ww*. With these conventions and a multiplication sign (×) to denote a cross between two breeding lines, Mendel's original cross can be symbolized as *ww* × *WW*.

The possible results from a cross between a true-breeding, white-flowered plant (*ww*) and a true-breeding, purple-flowered plant (*WW*) can be visualized with a **Punnett square.** In a Punnett square, the possible gametes of one individual are listed along the horizontal side of the square, while the possible gametes of the other individual are listed along the vertical side. The genotypes of potential offspring are represented by the cells within the square (figure 6.8).

The frequency that these genotypes occur in the offspring is usually expressed by a **probability.** For example, in a cross between a homozygous white-flowered plant (*ww*) and a homozygous purple-flowered plant (*WW*) such as that performed by Mendel, *Ww* is the only possible genotype for all individuals in the F_1 generation (figure 6.9). Because *W* is dominant to *w*, all individuals in the F_1 generation have purple flowers. When individuals from the F_1 generation are crossed, the probability of obtaining a homozygous dominant (*WW*) individual in the F_2 is 25% because one-fourth of the possible genotypes are *WW*. Similarly, the probability of an individual in the F_2 generation being homozygous recessive (*ww*) is 25%. Because the heterozygous genotype (*Ww*) occurs in half of the cells within the square, the probability of obtaining a heterozygous (*Ww*) individual in the F_2 is 50%.

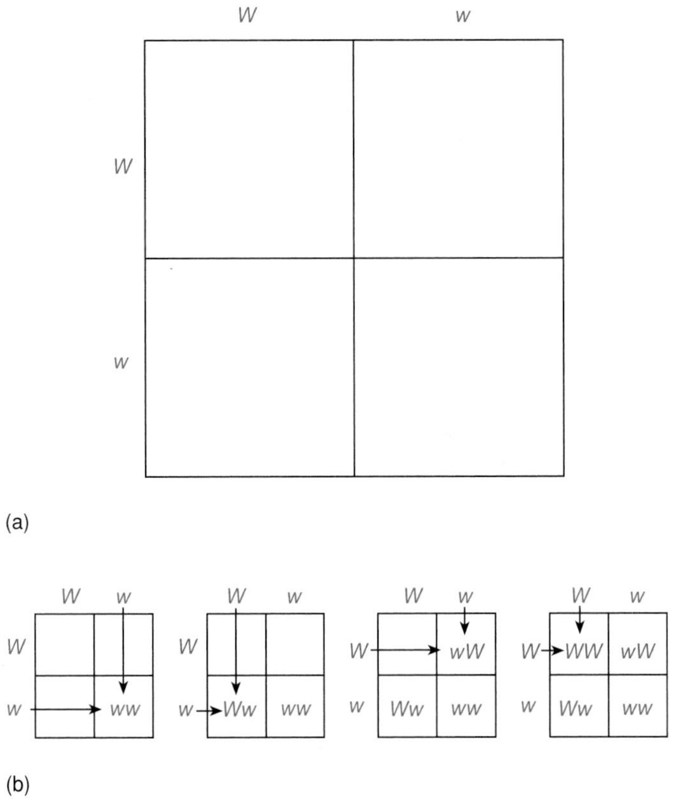

Figure 6.8 A Punnett square.
A Punnett square analysis is an easy way to determine all the possible genotypes of a particular cross. (*a*) The possible gametes for one parent are placed along one side of the square, and the possible gametes for the other parent are placed along the other side of the square. (*b*) The Punnett square can be used like a mathematical table to visualize the genotypes of all potential offspring.

Figure 6.9 How Mendel analyzed flower color.
The only possible offspring of the first cross are *Ww* heterozygotes, purple in color. These individuals are known as the F_1 generation. When two heterozygous F_1 individuals cross, three kinds of offspring are possible: *WW* homozygotes (purple flowers); *Ww* heterozygotes (also purple flowers), which may form two ways; and *ww* homozygotes (white flowers). Among these individuals, known as the F_2 generation, the ratio of dominant phenotype to recessive phenotype is 3:1.

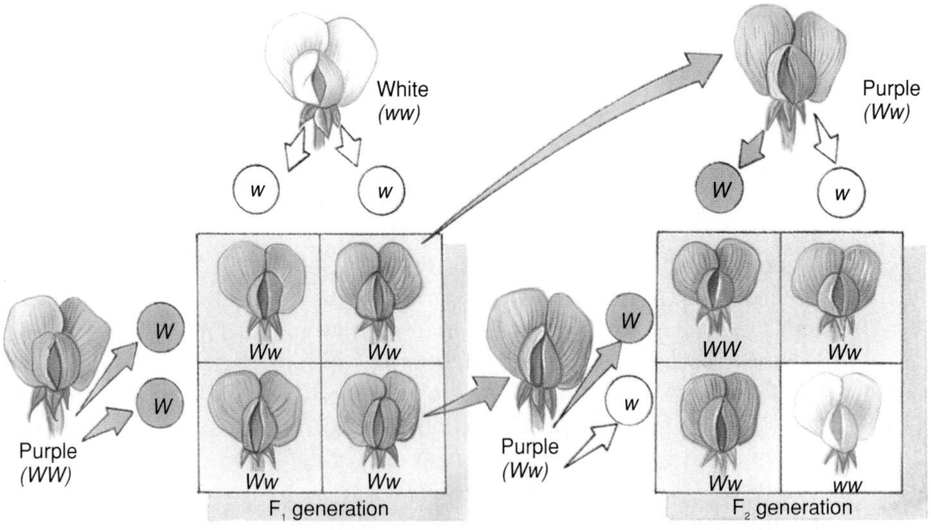

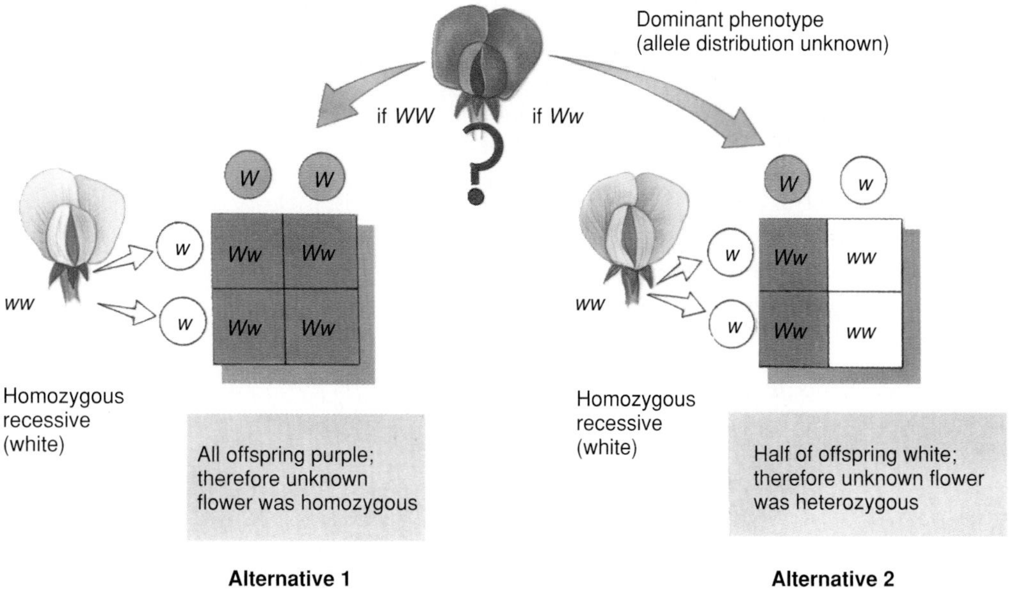

Figure 6.10 How Mendel used the test cross to detect heterozygotes.
To determine whether an individual exhibiting a dominant phenotype, such as purple flowers, is homozygous (*WW*) or heterozygous (*Ww*) for the dominant allele, Mendel devised the test cross. He crossed the individual in question with a known homozygous recessive (*ww*)—in this case, a plant with white flowers.

The Test Cross

How did Mendel know which of the purple-flowered individuals in the F_2 generation (or the P generation) were homozygous (*WW*) and which were heterozygous (*ww*)? It is not possible to tell simply by looking at them. For this reason, Mendel devised a simple and powerful procedure called the **test cross** to determine an individual's actual genetic composition. He crossed a purple-flowered individual (with a genotype of *WW* or *Ww*) with a homozygous recessive (*ww*) individual. By looking at what color flowers the offspring had, Mendel could deduce the genotype of the original purple-flowered plant (figure 6.10).

Mendel's First Law of Heredity: The Law of Segregation

Mendel's model brilliantly predicts the results of his crosses, accounting in a neat and satisfying way for the ratios he observed. Similar patterns of heredity have since been observed in countless other organisms. Traits exhibiting this pattern of heredity are called "Mendelian traits" (figure 6.11). Because of its overwhelming importance, Mendel's theory is often referred to as Mendel's first law, or the **law of segregation.** In modern terms, Mendel's first law states that *only one allele specifying an alternative trait can be carried in a particular gamete, and gametes combine randomly in forming offspring.*

Figure 6.11 A Mendelian trait.
One of the differences among varieties of pea plants that Mendel studied was the shape of the seed. In some varieties the seeds were round, whereas in others they were wrinkled. As you can see, wrinkled seeds look like dried-out, shrunken versions of the round ones.

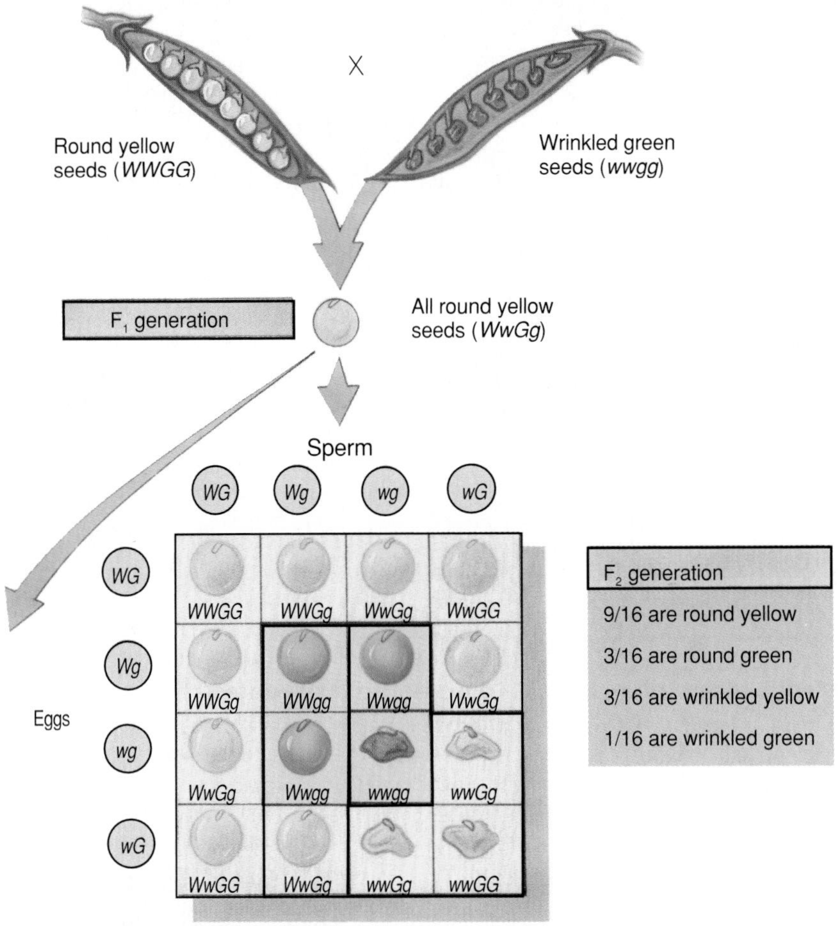

Figure 6.12 Analysis of a dihybrid cross.
This dihybrid cross shows round (*W*) versus wrinkled (*w*) seeds and yellow (*G*) versus green (*g*) seeds. The ratio of the four possible combinations of phenotypes is predicted to be 9:3:3:1, the ratio that Mendel found.

Mendel's Second Law of Heredity: The Law of Independent Assortment

Mendel went on to study how pairs of genes are inherited, such as flower color and plant height. He first established a series of true-breeding lines of peas that differed from one another with respect to two of the seven pairs of characteristics he had studied. Second, he crossed contrasting pairs of true-breeding lines. For example, homozygous individuals with round, yellow seeds crossed with individuals that are homozygous for wrinkled, green seeds produce dihybrid offspring that have round, yellow seeds and are heterozygous for both of these traits. Such F_1 individuals are said to be **dihybrid** (figure 6.12).

Mendel then allowed the dihybrid individuals to self-fertilize. If the segregation of alleles affecting seed shape and seed color were independent, the probability that a particular pair of seed-shape alleles would occur together with a particular pair of seed-color alleles would simply be a product of the two individual probabilities that each pair would occur separately. For example, the probability of an individual with wrinkled, green seed appearing in the F_2 generation would be equal to the probability of an individual with wrinkled seeds (1 in 4) multiplied by the probability of an individual with green seeds (1 in 4), or one in 16.

In his dihybrid crosses, Mendel found that the frequency of phenotypes in the F_2 offspring closely matched the 9:3:3:1 ratio predicted by a Punnett square analysis. He concluded that for the pairs of traits he studied, the inheritance of one trait does not influence the inheritance of any other trait, a result often referred to as Mendel's second law, or the **law of independent assortment.** We now know that this result is only valid for genes not located near one another on the same chromosome. Thus in modern terms Mendel's second law is often stated as follows: *Genes located on different chromosomes are inherited independently of one another.*

Mendel's paper describing his results was published in the journal of his local scientific society in 1866. Unfortunately, his paper failed to arouse much interest, and his work was forgotten. Sixteen years after his death, in 1900, several investigators independently rediscovered Mendel's pioneering paper. They came across it while they were searching the literature in preparation for publishing their own findings, which were similar to those Mendel had quietly presented more than three decades earlier.

6.2 From Genotype to Phenotype

Not all traits have the simple correlation between genotype and phenotype that Mendel saw. Often, the expression of the genotype is not so straightforward. Among the many situations that can complicate matters are *multiple alleles* (the presence of more than two alternative alleles), *epistasis* (one gene modifies the phenotypic expression of another), *continuous variation* (more than one gene contributes to the degree of expression of a trait), *pleiotropy* (an allele affects more than one trait), and *incomplete dominance* (heterozygotes exhibit intermediate phenotypes). We will examine each of these briefly.

Multiple Alleles

Although a diploid individual (one having two sets of genes) may possess no more than two alleles at one time, this does not mean that only two allele alternatives are possible for a given gene in the entire population. On the contrary, almost all genes that have been studied exhibit **multiple alleles.** The gene that determines the human ABO blood group, for example, has three common alleles (figure 6.13).

Epistasis

Few phenotypes are the result of the action of one gene. Most traits reflect the action of many genes that act sequentially or jointly. **Epistasis** is an interaction between the products of two genes in which one of the genes modifies the phenotypic expression produced by the other. For example, some commercial varieties of corn contain a purple pigment in the coats of their grain. Other varieties do not contain the pigment, and their grains are white. In some crosses between true-breeding white varieties, all of the F_1 corn plants have *purple* grains! How can this happen, when both parents are from true-breeding white lines?

It turns out that there are *two* genes that contribute to grain color (figure 6.14). Epistasis occurs because each gene blocks the expression of the other. One of the genes (B) produces an enzyme that permits colored pigment to be produced only if a dominant allele (*BB* or *Bb*) is present. The other gene (A) produces an enzyme that in its dominant form (*AA* or *Aa*) allows the pigment to be deposited on the grain coat color. Thus, an individual with two recessive alleles for gene A (no pigment deposition) will have white grain coats even though it is able to manufacture the pigment because it possesses dominant alleles for gene B (purple pigment production). Similarly, an individual with dominant alleles for gene A (pigment can be deposited) will also have white grain coats if it has only recessive alleles for gene B (pigment production) and cannot manufacture the pigment. For an individual to have purple grain coats, each gene must possess a dominant allele.

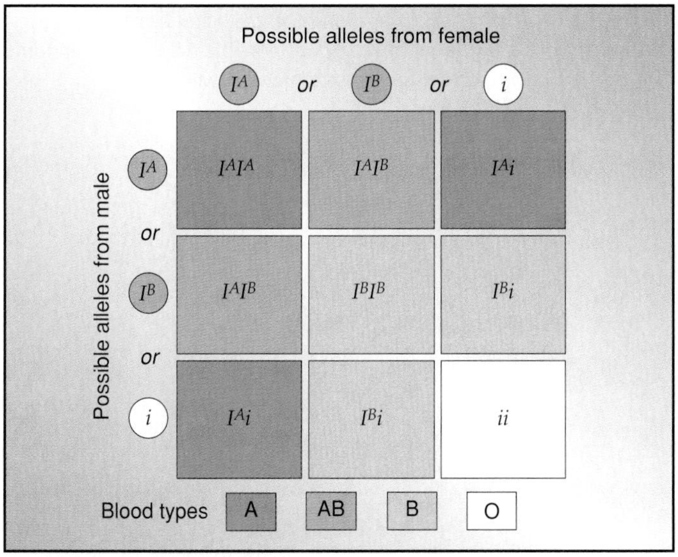

Figure 6.13 Multiple alleles controlling the ABO blood groups.
Three common alleles control the ABO blood group. Different combinations of the three so-called I gene alleles occur in different individuals. An individual may be homozygous for any allele or heterozygous for any two. The different combinations of the three alleles result in four different blood type phenotypes: type A (either I^AI^A homozygotes or I^Ai heterozygotes), type B (either I^BI^B homozygotes or I^Bi heterozygotes), type AB (I^AI^B heterozygotes), and type O (ii homozygotes).

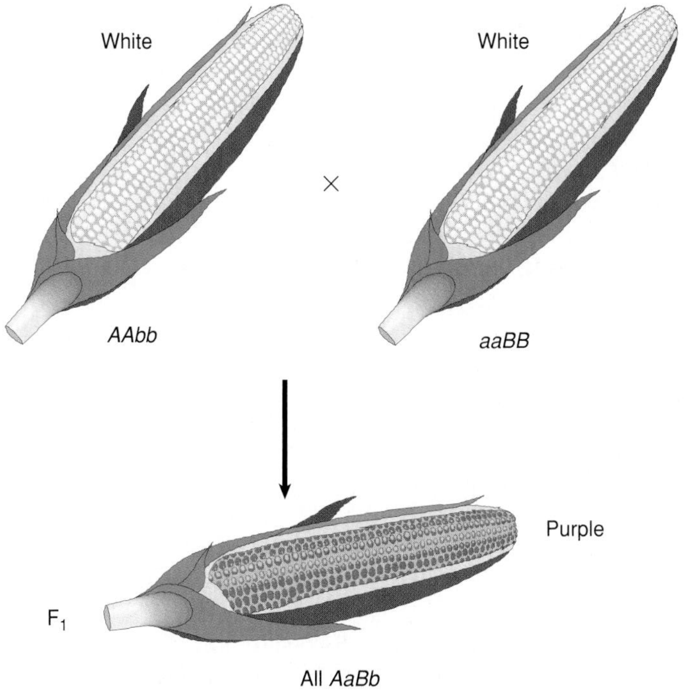

Figure 6.14 How epistasis affects grain color.
The purple pigment found in some varieties of corn is the product of a two-step biochemical pathway. Unless both enzymes are active (a dominant allele at each of the two genes), no pigment is expressed.

Continuous Variation

When multiple genes act jointly to influence a trait, such as height or weight, the resulting phenotypes may consist of a range of small differences. Because all of the genes that play a role in determining phenotypes such as height or weight segregate independently of one another, we see a gradation or a **continuous variation** in degree of difference when many individuals are examined (figure 6.15).

Pleiotropy

Often, an individual allele affects more than one trait. Such an allele is said to be **pleiotropic.** For example, it was not possible for scientists to obtain a true-breeding strain of mice with yellow fur by crossing one yellow mouse with another—individuals that were homozygous for the yellow allele died. The yellow allele was pleiotropic: one effect was yellow color, and the other effect was a lethal developmental defect. A pleiotropic alteration may be dominant with respect to one phenotypic consequence (yellow fur) and recessive with respect to another (lethal developmental defect). Pleiotropic relationships often occur unexpectedly, when a gene also performs other functions about which scientists may be ignorant.

Incomplete Dominance

Not all alternative alleles are fully dominant or recessive in heterozygotes. Sometimes, heterozygous individuals do not resemble one parent precisely. Some pairs of alleles produce instead a heterozygous phenotype that is intermediate between the parents (intermediate, or **incomplete, dominance,** figure 6.16). In many inherited human disorders, the heterozygote resembles one allele closely but can be distinguished from it **(partial dominance).** When it is possible to study the genotype directly by analyzing DNA or proteins, it is often possible to distinguish the contribution of each parental allele to the heterozygote **(codominance).**

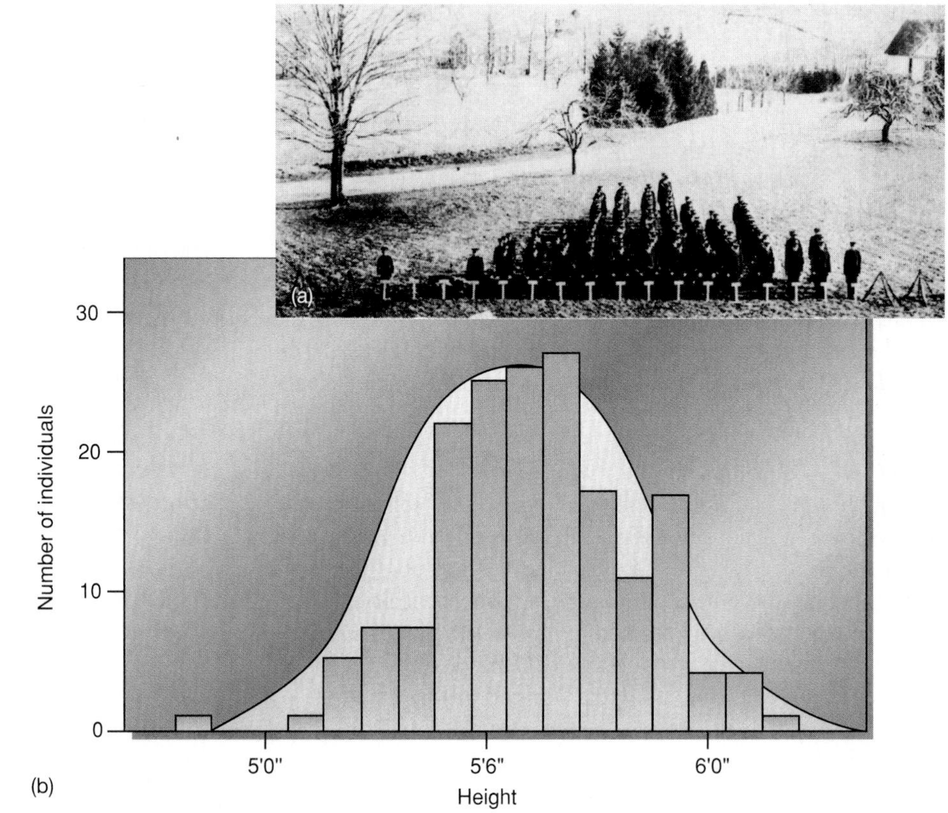

Figure 6.15 Height is a continuously varying trait.
(*a*) Multiple genes acted jointly to influence the variation in height among students of the 1914 class at the Connecticut Agricultural College. Because many genes contribute to height and tend to segregate independently of one another, there are many possible combinations. (*b*) These cumulative contributions form a continuous spectrum of possible heights—a random distribution, in which the extremes are much rarer than the intermediate values.

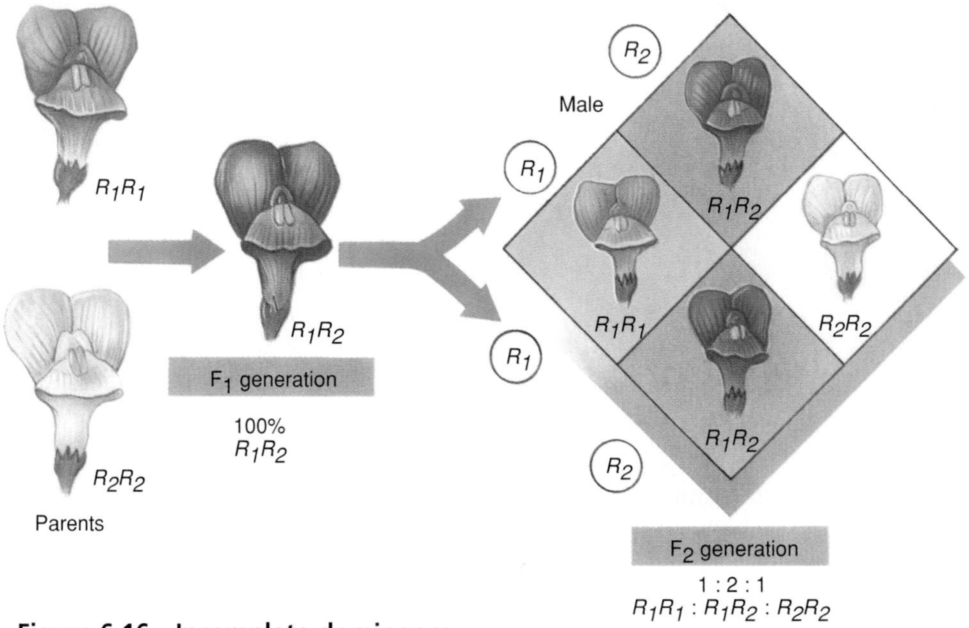

Figure 6.16 Incomplete dominance.
In snapdragons, heterozygous individuals do not resemble either parent. In a cross between a pink-flowered snapdragon, which has the genotype R_1R_1, and a white-flowered one (R_2R_2), neither allele is dominant, and the heterozygotes (genotype R_1R_2) have purple flowers.

6.3 Meiosis

Mendel's theory is one of the most important milestones in the history of human thought. It dispelled forever the mystery of heredity—why traits appear and disappear from one generation to another. Stripped of ratios and symbols, Mendel's work teaches us a very simple and important lesson—that patterns of heredity reflect the donation by parents to offspring of discrete units of encoded information.

Chromosomes: The Vehicles of Inheritance

In place of mystery, Mendel leaves us with a question. What is the physical nature of the factors (genes) that his theory postulates? This question dominated the science of biology for more than half a century after Mendel's work was rediscovered in 1900. We now know that genes are composed of sequences of nucleotides that are part of much longer DNA molecules. These very long DNA molecules contain thousands of other genes, lined up in a long string like railroad cars. While its genes are being used, the DNA molecule is stretched out so that other cell molecules can approach the DNA and "read" the genes. How this happens is the subject of the next chapter. When the cell prepares to divide, the DNA coils up so it can more easily be separated into the two new daughter cells. In doing this, proteins serve to glue the DNA strands together into compact cables called chromosomes. A **chromosome** is a single long DNA molecule packaged with proteins into a compact shape. Biologists soon came to understand that chromosomes were the collections of heredity units (genes) whose presence was demanded by Mendel's theory.

Chromosome Number

Each cell in an organism has a set of chromosomes that carries hereditary information. This hereditary information was used to construct the organism during development and will in turn be passed on to that organism's offspring. **Diploid** individuals, like Mendel's pea plants and like humans, contain two copies of each chromosome. Human gametes—eggs or sperm—are **haploid,** containing only one copy of each chromosome (figure 6.17). Humans have 23 different chromosomes. Because human body cells have two copies of each, they possess a total of 46 chromosomes each.

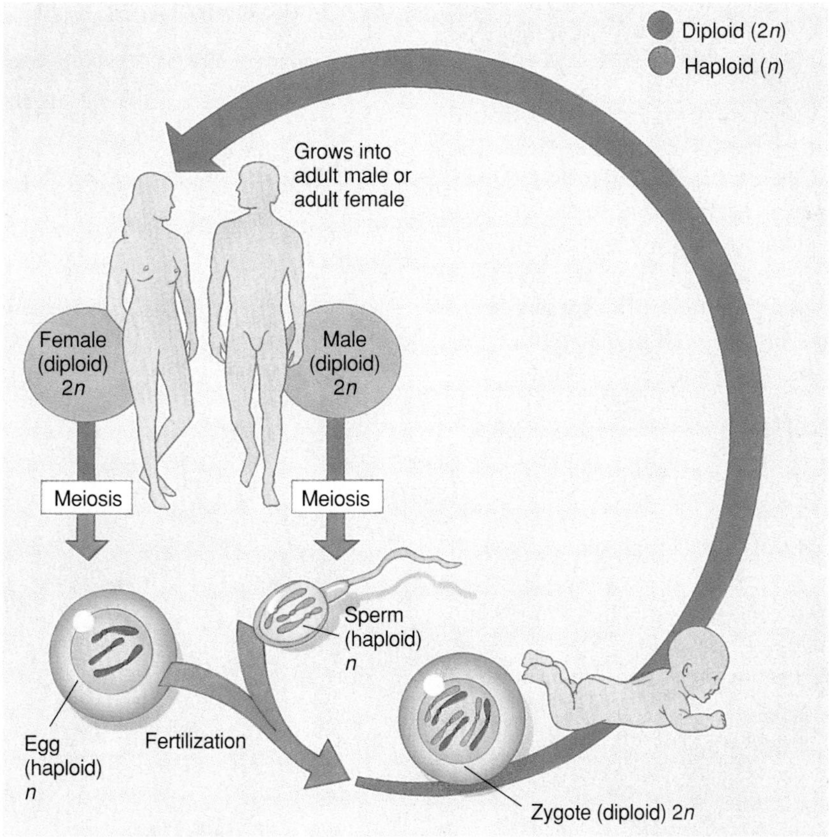

Figure 6.17 The sexual life cycle in animals.
Humans and most other animals are, like Mendel's pea plants, diploid ($2n$) organisms but produce special haploid (n) gamete cells that function in reproduction. During fertilization, two haploid gametes meet and combine their chromosomes. The diploid cell that results contains two versions of each chromosome, one contributed by the haploid egg of the mother, the other by the haploid sperm of the father. In this drawing, n stands for haploid, and $2n$ stands for diploid.

An Overview of Meiosis

Most animals and plants reproduce sexually. In **sexual reproduction,** gametes of opposite sexes unite in the process of fertilization (see figure 6.17). The fusion of gametes forms a new cell, called a **zygote.** Dividing by mitosis, a human zygote eventually gives rise to an adult body with some 100 trillion cells. Except for chance mutations, every one of your 100 trillion cells is genetically identical to the zygote from which you arose. Because the fusion of gametes to form a zygote takes place in every generation, it is clear that something else must happen to reduce the number of chromosomes in gametes—otherwise the number of chromosomes would soon become impossibly large. The mechanism that reduces the number of chromosomes in gametes is a special form of cell division called **meiosis** ("my-o-sis"). Meiosis is a form of cell division in which the number of chromosomes in cells is halved during gamete formation.

Meiosis consists of two rounds of nuclear division similar to that in mitosis (see chapter 4), during the course of which two unique events occur:

1. Early in the first of the two divisions, the two copies of each chromosome pair up. While together, they exchange portions of DNA strands along their length in a process called **crossing over** (figure 6.18).
2. The chromosomes do not replicate between the two divisions!

Just as in mitosis, the chromosomes have replicated before meiosis begins, during a period called interphase. The first of the two divisions of meiosis, called **meiosis I,** serves to separate the two versions of each chromosome; the second, **meiosis II,** serves to separate the two replicas of each version. Thus when meiosis is complete, what started out as one diploid cell ends up as four haploid cells. Because there was one replication of DNA but *two* cell divisions, the process halves the number of chromosomes.

The Stages of Meiosis

Meiosis I, the first meiotic division, is traditionally divided into five stages:

1. **Interphase.** The DNA of each chromosome replicates.
2. **Prophase I.** The two versions of each chromosome pair up and exchange segments.
3. **Metaphase I.** The chromosomes align on a central plane.
4. **Anaphase I.** One version of each chromosome moves to a pole of the cell, and the other version moves to the opposite pole.
5. **Telophase I.** Individual chromosomes gather together at each of the two poles.

We will now examine these stages in more detail (figure 6.19).

Meiosis I

In *prophase I,* individual chromosomes first become visible, as viewed with a light microscope, as their DNA coils more and more tightly. Because the chromosomes (DNA) have replicated before the onset of meiosis, each of the threadlike chromosomes actually consists of two sister chromatids joined at their centromeres. The two homologous chromosomes then line up side by side, and crossing over is initiated, in which DNA is exchanged between the two nonsister chromatids of homologous chromosomes. These crossovers tend to hold the homologous chromosomes together. Late in prophase, the nuclear envelope disperses.

In *metaphase I,* the spindle apparatus forms, but because homologues are held close together by crossovers, spindle fibers can attach to only the outward-facing kinetochore of each centromere. For each pair of homologues, the orientation on the spindle axis is random; which homologue is oriented toward which pole is a matter of chance. Like shuffling a deck of cards, many combinations are possible—in fact, two raised to a power equal to the number of chromosome pairs. In a hypothetical cell that has three chromosome pairs, there are eight possible orientations (2^3). Each orientation results in gametes with different combinations of parental chromosomes. This process is called **independent assortment** (figure 6.20).

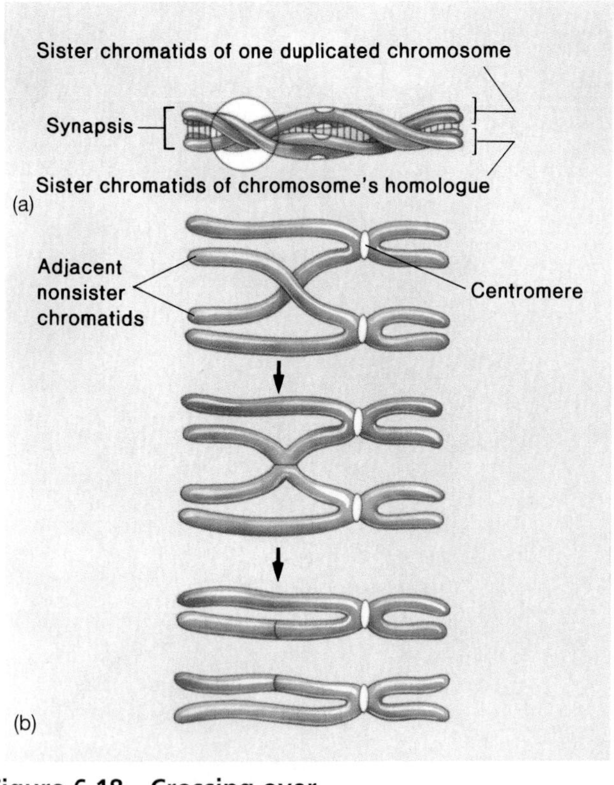

Figure 6.18 Crossing over.
In crossing over, the two copies of each chromosome exchange portions. (*a*) The open circle highlights the complex series of events that occur during close pairing of the two copies, a process called synapsis. (*b*) During the crossing-over process, nonsister chromatids that are next to each other exchange genetic information.

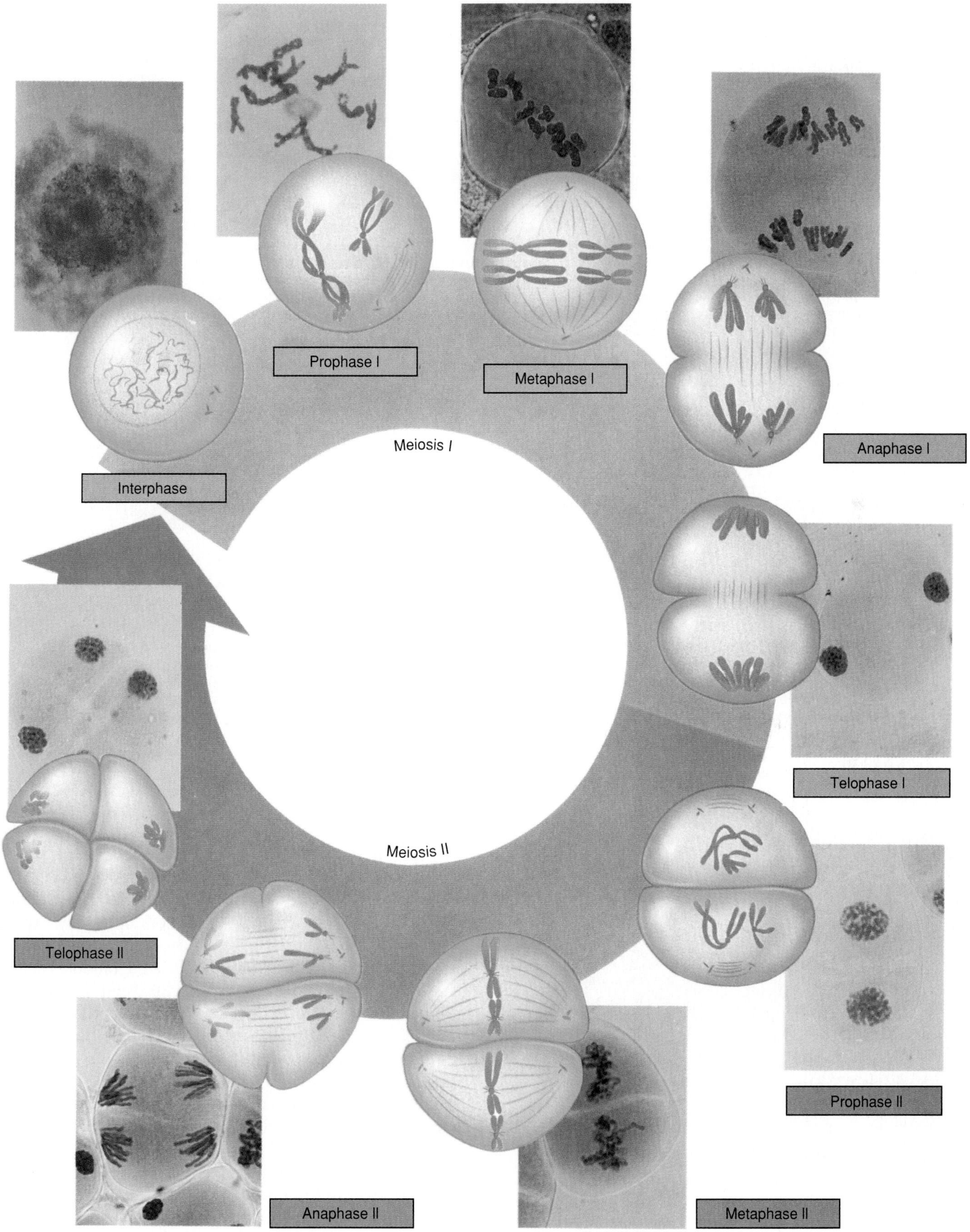

Figure 6.19 The stages of meiosis.
Meiosis consists of two rounds of cell division similar to mitosis and produces four haploid cells.

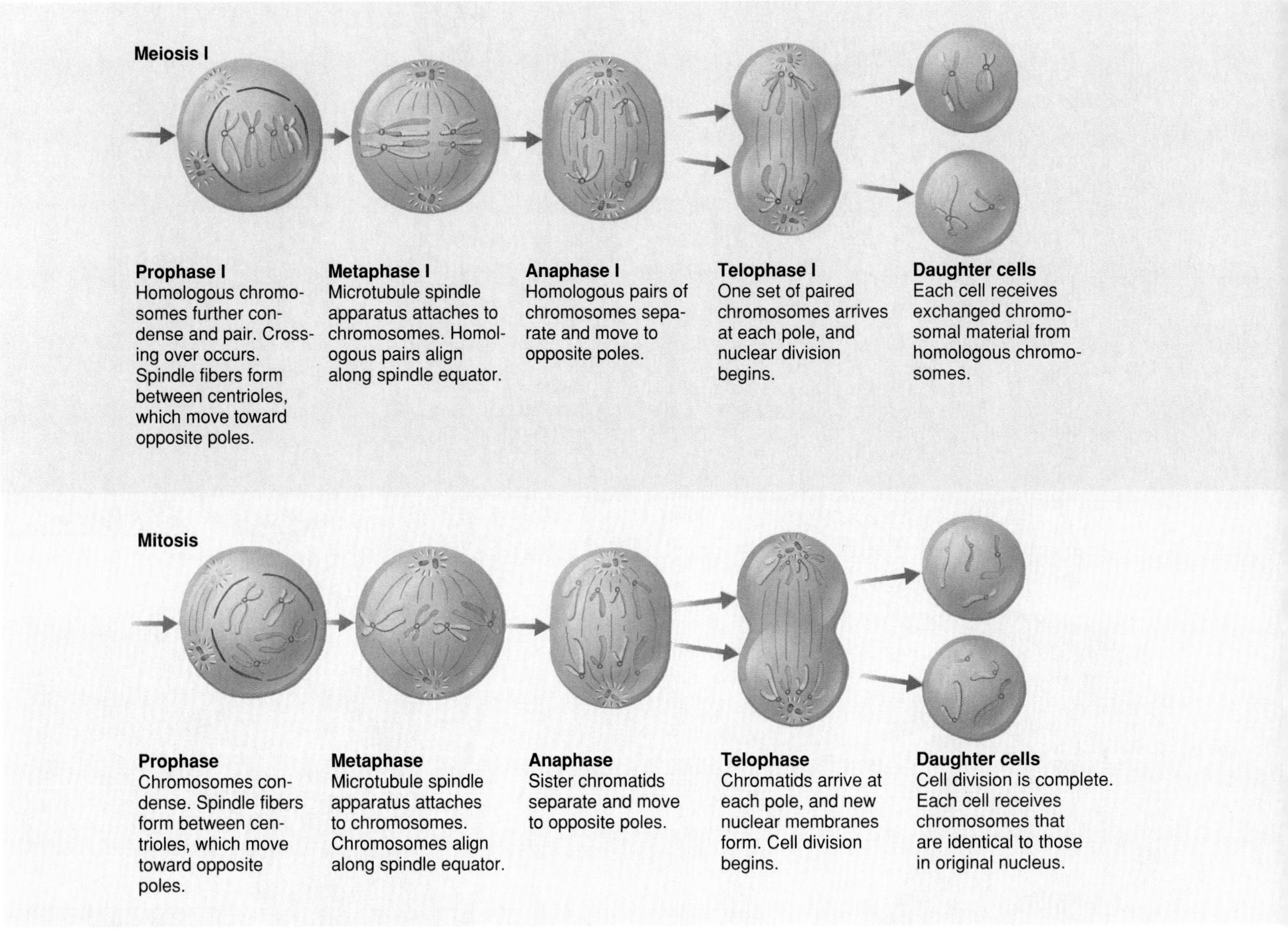

Meiosis I
Prophase I
Homologous chromosomes further condense and pair. Crossing over occurs. Spindle fibers form between centrioles, which move toward opposite poles.
Metaphase I
Microtubule spindle apparatus attaches to chromosomes. Homologous pairs align along spindle equator.
Anaphase I
Homologous pairs of chromosomes separate and move to opposite poles.
Telophase I
One set of paired chromosomes arrives at each pole, and nuclear division begins.
Daughter cells
Each cell receives exchanged chromosomal material from homologous chromosomes.
Mitosis
Prophase
Chromosomes condense. Spindle fibers form between centrioles, which move toward opposite poles.
Metaphase
Microtubule spindle apparatus attaches to chromosomes. Chromosomes align along spindle equator.
Anaphase
Sister chromatids separate and move to opposite poles.
Telophase
Chromatids arrive at each pole, and new nuclear membranes form. Cell division begins.
Daughter cells
Cell division is complete. Each cell receives chromosomes that are identical to those in original nucleus.

Figure 6.20 Independent assortment.
Independent assortment occurs because the orientation of chromosomes (blue in this micrograph) on the metaphase plate is random. Many combinations are possible. Each orientation results in gametes with different combinations of parental chromosomes.

In *anaphase I,* the spindle attachment is complete, and homologues are pulled apart and move toward opposite poles. Sister chromatids are not separated at this stage. Because the orientation along the spindle equator is random, the chromosome that a pole receives from each pair of homologues is also random with respect to all chromosome pairs. At the end of anaphase I, each pole has half as many chromosomes as were present in the cell in which meiosis began. Remember that the chromosomes replicated and thus contained two sister chromatids before the start of meiosis. The purpose of the meiotic stages up to this point has not been to reduce the number of chromosomes but to allow for the exchange of genetic material in crossing over.

In *telophase I,* the chromosomes gather at their respective poles to form two chromosome clusters. After an interval of variable length, meiosis II occurs, in which the number of chromosomes is reduced.

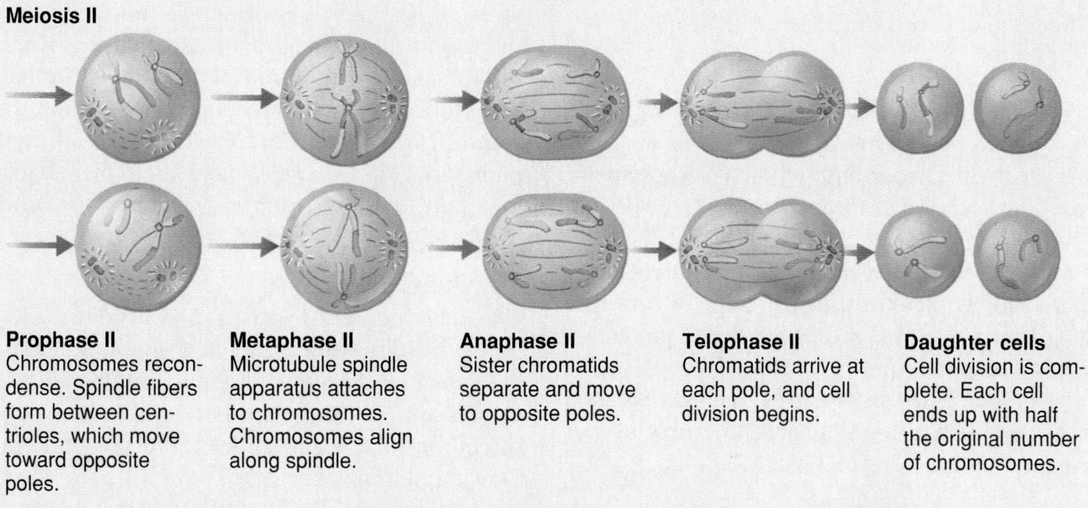
Meiosis II
Prophase II
Chromosomes recondense. Spindle fibers form between centrioles, which move toward opposite poles.
Metaphase II
Microtubule spindle apparatus attaches to chromosomes. Chromosomes align along spindle.
Anaphase II
Sister chromatids separate and move to opposite poles.
Telophase II
Chromatids arrive at each pole, and cell division begins.
Daughter cells
Cell division is complete. Each cell ends up with half the original number of chromosomes.

Figure 6.21 A comparison of mitosis and meiosis.
The four primary differences between these two processes are as follows: (1) Mitosis results in two diploid daughter cells, while meiosis results in four haploid daughter cells. (2) Because the chromosomes line up differently during mitosis than during meiosis I, sister chromatids are pulled apart in mitosis, while homologous pairs are pulled apart in meiosis I. (3) During mitosis, chromosomes do not exchange genetic material; during meiosis, they do. (4) Mitosis occurs in body cells, while meiosis occurs in reproductive cells, such as eggs and sperm.

Meiosis II

Meiosis II is simply a mitotic division involving the products of meiosis I, except that the sister chromatids are not genetically identical, as they are in mitosis, because of crossing over. At the end of anaphase I, each pole has a haploid complement of chromosomes, each of which is still composed of two sister chromatids attached at the centromere. The main purpose of the four stages of meiosis II—*prophase II, metaphase II, anaphase II,* and *telophase II*—is to separate these sister chromatids. At both poles of the original cell, the chromosomes divide mitotically. The result of this division is four haploid complements of chromosomes. The cells that contain these haploid complements may function directly as gametes, as they do in animals, or they may divide again by mitosis, as they do in plants, fungi, and many protists, eventually producing gametes after further mitotic divisions. Figure 6.21 compares and contrasts the various stages of mitosis and meiosis.

The Important Role of Crossing Over

If you think about it, the key to meiosis is that the replicas of each chromosome are not separated from each other in the first division. Why not? What prevents microtubules from attaching to them and pulling them to opposite poles of the cell, just as eventually happens later in the second meiotic division? The answer is the crossing over that occurred early in the first division. By exchanging segments, the two versions of each chromosome are tied together by strands of DNA, like two people sharing a belt. It is because microtubules can gain access to only one side of each replicated chromosome that they cannot pull the two replicas apart! Imagine two people dancing closely—you can tie a rope to the back of each person's belt, but you cannot tie a second rope to their belt buckles because the two dancers are facing each other and are very close. In just the same way, microtubules cannot attach to the inside replica of chromosomes because crossing over holds chromosome replicas together like dancing partners.

6.4 Patterns of Heredity in Humans

Your face may look very much like the face of one of your parents. The degree to which you resemble your father or mother was largely established before your birth by the chromosomes that you received from each of them, just as flower color in Mendel's peas was determined by which chromosomes the plants received. Many of the alleles present in human populations demand more serious concern than the color of a pea's flower. Devastating human disorders result with important functions in our bodies when alleles specify defective forms of proteins. These disorders are passed from parent to child. Recent advances in genetic engineering are beginning to enable us to limit the misery these disorders bring to so many families.

Chromosomes and Development

Each of the 23 different kinds of human chromosome has many hundreds of genes that play important roles in determining what you will be like as your body develops. Because all of these genes need to be present, everyone must have all 23 chromosomes in order to develop correctly—for the same reason that all cars have engines, transmissions, and wheels: all are needed for the car to function properly.

Aneuploidy

Sometimes during meiosis, sister chromatids or homologous chromosomes that paired up during metaphase remain stuck together instead of separating. The failure of chromosomes to separate correctly during either meiosis I or II is called **nondisjunction.** Nondisjunction leads to **aneuploidy,** the condition of having an abnormal number of chromosomes. The offspring that result from the union of a gamete with a normal number of chromosomes and a gamete with an abnormal number of chromosomes often display severe physical and mental disabilities. Individuals who are missing one chromosome, called **monosomics,** do not survive embryonic development.

And just as a car will not function well with two engines stuffed under the hood, or an extra transmission hooked to the axle, so a human embryo does not develop properly with more than two copies of any of its chromosomes. In all but a few cases, individuals who have received an extra chromosome, called **trisomics,** also do not survive. Only the five smallest chromosomes can be present in humans in three copies and still allow the individual to survive for a time; with duplicates of three of these five chromosomes, the infants die within a few months. Only individuals with an extra copy of chromosome 21 (and more rarely chromosome 22) survive to become adults. Such individuals have poorly developed bones, and their mental development is affected—children with an extra copy of chromosome 21 or 22 are always mentally disabled.

Down Syndrome

The developmental defect produced by having an extra copy of chromosome 21 (an example of trisomy) was first described in 1866 by J. Langdon Down and is called **Down syndrome.** A **karyotype** is an individual's particular array of chromosomes, and in the karyotype of an individual with Down syndrome, the extra copy of chromosome 21 can be seen (figure 6.22). Down syndrome occurs in all racial groups at the same frequency, about 1 in 1,000 children. It is much more common in children of older mothers—in mothers under 30 years old, the incidence is only about 1 per 1,500 births, while in mothers 30 to 35 years old, the incidence doubles, to 1 in 750 births. In mothers over 45, the risk is as high as 1 in 16 births. The reason that older mothers are more prone to Down syndrome babies is that all the eggs that a woman will ever produce are present in her ovaries by the time she is born, and as she gets older they accumulate an ever-greater amount of damage. Males by contrast develop new sperm throughout adult life.

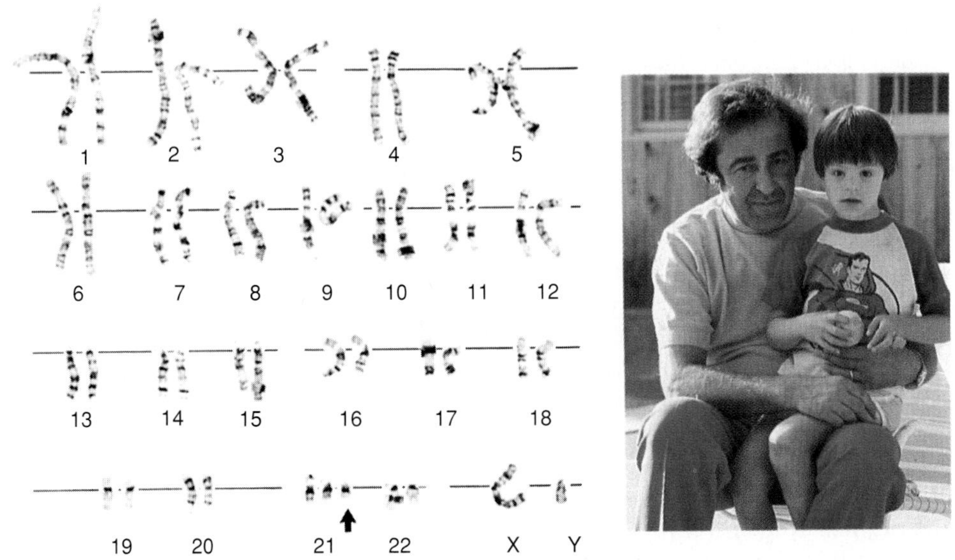

(a) (b)

Figure 6.22 Down syndrome.
(*a*) In this karyotype of an individual with Down syndrome, the trisomy at position 21 can be clearly seen (*arrow*). (*b*) This child shows the physical effects of Down syndrome. Many individuals with Down syndrome can lead productive lives if they receive adequate help and support.

Because of the risk of Down syndrome, pregnant women over the age of 35 are usually advised to have the chromosomal constitution of their fetus checked for an extra copy of chromosome 21. This can be easily done by removing a small volume of fluid from the zone around the fetus and examining fetal cells suspended in it, a process called **amniocentesis** (see figure 6.28).

Sex Chromosomes

Of the 23 pairs of human chromosomes, 22 consist of members that are similar in size and morphology in both males and females; these chromosomes are called **autosomes.** One pair of your chromosomes is unique in that male and female versions can be different. This chromosome, which carries the genes determining sex, exists in two states, a normal state, called an **X chromosome,** and a permanently compressed state in which most of its genes are never expressed, called a **Y chromosome.** In humans, the Y chromosome is not completely inert. The genes that cause the zygote to develop into a male are located there. In 1990 scientists succeeded in isolating one of these genes, called *Sry,* and it is under active investigation.

Thus, any individual containing at least one Y chromosome is a male, and any individual containing no Y chromosome is a female. You can see that the father's gamete determines the sex of the child. Any individual who receives an X chromosome from the father must become a female (she must receive an X chromosome from her mother, as that is the only version of the sex chromosome her mother has), while any individual who receives a Y chromosome from the father becomes a male.

X Chromosome Inactivation

Despite the fact that female mammals have two X chromosomes (XX), only one X chromosome remains active in all but ovarian cells. The other X chromosome condenses into a compact structure called a **Barr body** and becomes inactive. The genes on the Barr body are usually not expressed, although small regions of the chromosome may remain active.

Inactivation of an X chromosome occurs randomly within each cell during the course of early development. As a result, the body of an adult female consists of a mosaic of two types of cells. In one type the active X chromosome was inherited from the father, and in the other type the active X chromosome was inherited from the mother. Thus, a female who is heterozygous for a sex-linked trait expresses one allele in half of her cells and the alternate allele in the other half of her cells. In humans, this mosaic effect can be seen in a recessive X-linked condition in which sweat gland development is prevented. Females who are heterozygous for this condition have patches of skin without sweat glands and patches of normal skin.

Abnormal Numbers of Sex Chromosomes

Individuals who lose a copy of the X chromosome, or gain an extra copy, are not subject to the severe developmental problems associated with similar changes to autosomes. However, sex chromosome aneuploidy often produces sterility (figure 6.23). Females with an extra X chromosome (that is, XXX individuals) are usually sterile although normal in other respects. Males with an extra X chromosome (that is, XXY males) are also sterile and have many female body characteristics. This XXY condition, called **Klinefelter syndrome,** occurs in about 1 in every 1,000 male births. No human can survive without at least one copy of the X chromosome, but it is possible to survive with only one copy even without a Y chromosome. Such individuals, designated XO, are sterile females, typically of short stature and webbed neck. This XO condition, called Turner syndrome, occurs roughly once in every 2,000 female births.

Individuals with an extra copy of the Y chromosome develop into fertile males of normal appearance. It has been reported that the frequency of XYY males in penal and mental institutions is 20 per 1,000, some 20 times more frequent than in the population at large. This observation has led to the suggestion that XYY males are inherently antisocial, a suggestion that has proven highly controversial. This idea has been confirmed by some studies but not by others.

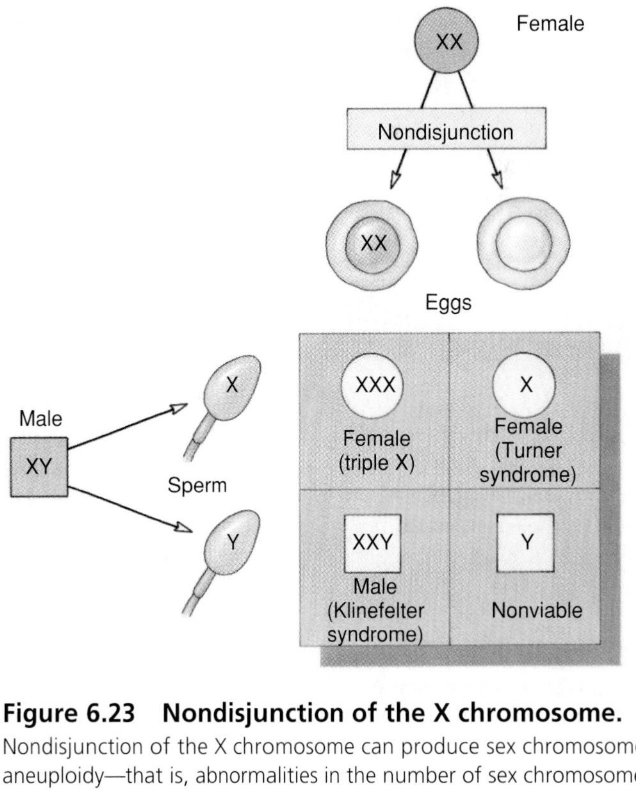

Figure 6.23 Nondisjunction of the X chromosome.
Nondisjunction of the X chromosome can produce sex chromosome aneuploidy—that is, abnormalities in the number of sex chromosomes.

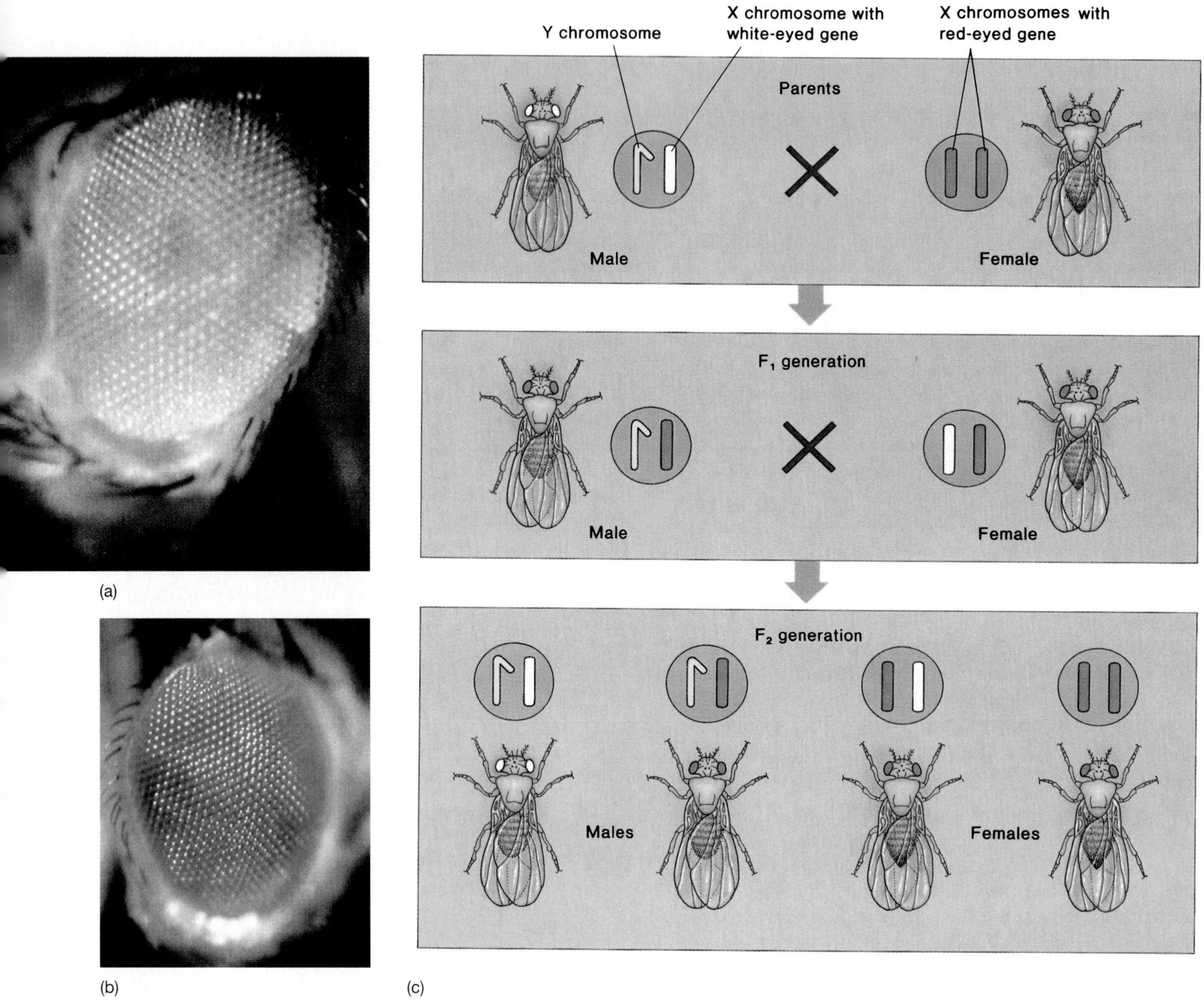

Figure 6.24 Morgan's experiment demonstrating sex-linkage in fruit flies.
In this experiment with fruit flies, *Drosophila*, Morgan demonstrated that the gene for eye color is located on the X chromosome. In *Drosophila*, white eyes (*a*) are mutant, while red eyes (*b*) are normal. The white-eye defect in eye color is hereditary, the result of a mutation in a gene located on the sex-determining X chromosome. (*c*) The white-eyed mutant male fly was crossed with a normal female. The F_1 generation flies all exhibited red eyes, as expected for flies heterozygous for a recessive white-eye allele. In the F_2 generation, all white-eyed flies were male.

Sex-Linkage

Because the Y chromosome contains almost no genes (at least, almost none that are used), genes that function on the X chromosome have no partner on the Y chromosome. In experiments with fruit flies, Thomas Hunt Morgan demonstrated that the gene for eye color is located on the X chromosome. Females that are homozygous recessive for the white-eye allele would have white eyes, but a cross involving a male with red eyes could not produce progeny that included white-eyed females. Such a characteristic, determined by genes on the sex chromosomes, is said to be **sex-linked.** Figure 6.24 illustrates how Morgan used the Y chromosome to explain the sex-linkage he observed in fruit flies. A case of sex-linkage in humans is discussed later in this chapter.

Human Hereditary Disorders

The proteins encoded by most of your genes must function in a very precise fashion in order for you to develop properly and for the many complex processes of your body to function correctly. Unfortunately, genes sometimes sustain damage or are copied incorrectly. We call these accidental changes in genes **mutations.** Mutations occur only rarely, because your cells police your genes and attempt to correct any damage they encounter. Still, some mutations get through. Many of them are bad for you in one way or another. It is easy to see why. Mutations hit genes at random—imagine that you randomly changed the number of a part on a design of a jet fighter. Sometimes it won't matter critically—a seatbelt becomes a radio, say. But what if a key rivet in the wing becomes a roll of toilet paper? The chance of a random mutation in a gene improving the performance of its protein is about the same as that of a bullet shot through a television set improving the picture you see.

Most mutations are rare in human populations. Almost all result in recessive alleles, and so they are not eliminated from the population by evolutionary forces—because they are not expressed in most individuals (heterozygotes) in which they occur. Do you see why they occur mostly in heterozygotes? Because mutant alleles are rare, it is unlikely that a person carrying a copy of the mutant allele will marry someone who also carries it. Instead, he or she will marry someone homozygous normal, so their children cannot be homozygous for the mutant allele.

In some cases, particular mutant alleles have become more common in human populations. In these cases the harmful effects that they produce are called genetic disorders. Some of the most common genetic disorders are listed in table 6.1. To study human heredity, scientists look at the results of crosses that have already been made. They study family trees, or **pedigrees,** to identify which relatives exhibit a trait. Then they can often determine whether the gene producing the trait is sex-linked or autosomal and whether the trait's phenotype is dominant or recessive. Frequently, they can infer which individuals are homozygous and which are heterozygous for the allele specifying the trait.

Table 6.1 Some Important Genetic Disorders

Disorder	Symptom	Defect	Dominant/Recessive	Frequency Among Human Births
Cystic fibrosis	Mucus clogging lungs, liver, and pancreas	Failure of chloride ion transport mechanism	Recessive	1/2,500 (Caucasian)
Sickle-cell anemia	Poor blood circulation	Abnormal hemoglobin molecules	Recessive	1/625 (African Americans)
Tay-Sachs disease	Deterioration of central nervous system in infancy	Defective form of enzyme hexosaminidase A	Recessive	1/3,500 (Ashkenazim Jews)
Phenylketonuria	Failure of brain to develop in infancy	Defective form of enzyme phenylalanine hydroxylase	Recessive	1/12,000
Hemophilia	Failure of blood to clot	Defective form of blood clotting factor VIII	Sex-linked recessive	1/10,000 (Caucasian males)
Huntington's disease	Gradual deterioration of brain tissue in middle age	Production of an inhibitor of brain cell metabolism	Dominant	1/24,000
Muscular dystrophy (Duchenne)	Wasting away of muscles	Degradation of myelin coating of nerves stimulating muscles	Sex-linked recessive	1/3,700 (males)
Hypercholesterolemia	Excessive cholesterol levels in blood, leading to heart disease	Abnormal form of cholesterol cell surface receptor	Dominant	1/500

Sickle-Cell Anemia

An example of a genetic disorder is **sickle-cell anemia.** Sickle-cell anemia is a recessive hereditary disorder in which affected individuals (those homozygous for the defective allele) cannot transport oxygen to their tissues properly because the molecules within their red blood cells that carry oxygen, hemoglobin proteins, are defective (figure 6.25). A mutation altering the identity of one nucleotide in the gene encoding the protein results in the wrong amino acid being at a key position, and the altered hemoglobin protein simply doesn't work. The mutant allele of the hemoglobin gene apparently arose in Africa. In central Africa, up to 45% of people are heterozygous for the sickle-cell hemoglobin allele. Apparently evolution has favored the allele there because people who are heterozygous are not as susceptible to malaria, a leading cause of illness and death in Africa as well as throughout the tropics. (Over a million people worldwide die of malaria each year.)

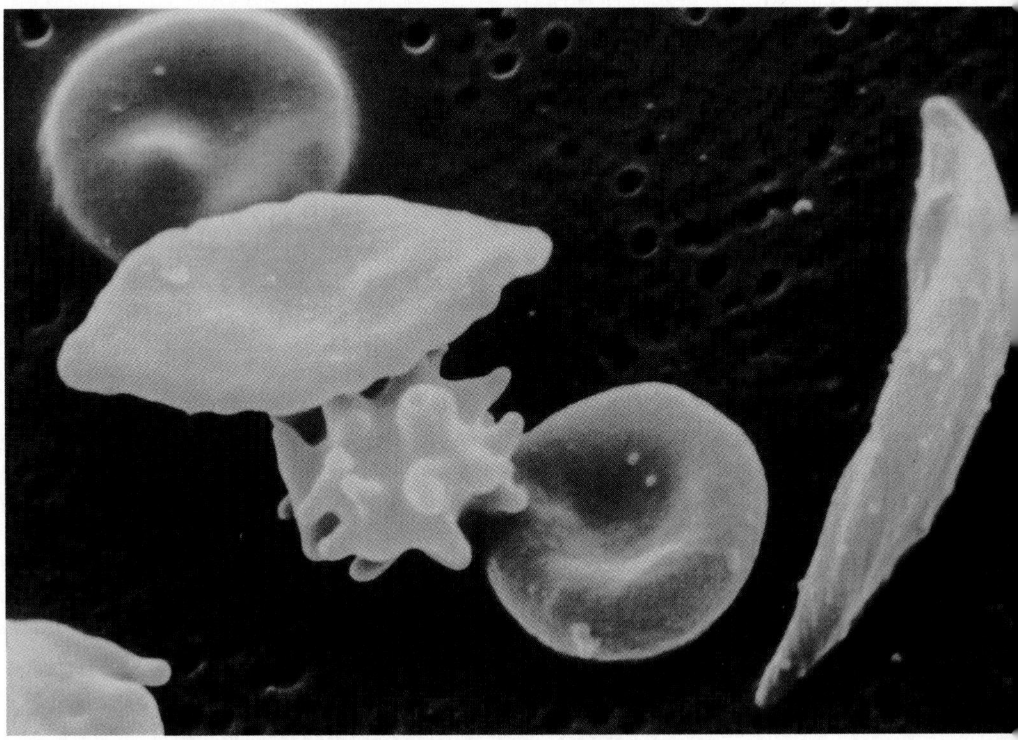

Figure 6.25 Sickle-cell anemia.
The elongated cell is a sickled cell. A normal red blood cell is shaped like a flattened sphere. Its smooth edges pass easily through the body's tiny capillaries. Sickled cells, with their pointed edges, get stuck while passing through capillaries, leading to internal bleeding and anemia.

Hemophilia

A group of recessive genetic disorders in which the blood fails to clot are called hemophilia. In affected individuals, small cuts can be fatal. A dozen genes encode proteins involved in blood clotting, and mutations causing hemophilia can occur in any of them. Analysis of family pedigrees can reveal a great deal about the particular mutation (figure 6.26). Two of the clotting factor genes are located on the X chromosome (that is, they are sex-linked)—in these cases, males invariably show the trait if they have it (there is no normal allele on their Y chromosome to mask the expression of the allele on their X). Thus females with one copy of the allele do not show the trait, while males with one copy do. A famous instance of sex-linked hemophilia arose in the royal family of England in one of the parents of Queen Victoria (figure 6.27). Three of Victoria's nine children had the defective allele and carried it by marriage into many of the royal families of Europe.

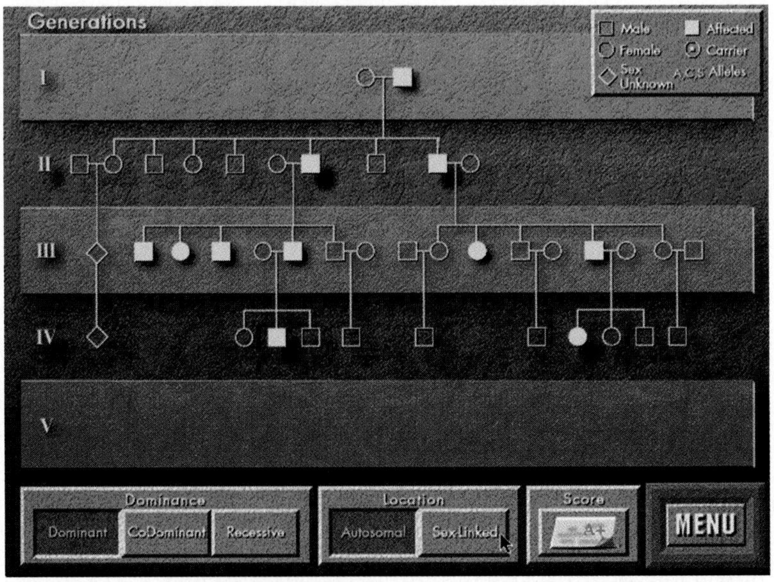

Figure 6.26 A pedigree illustrating heredity in four generations of a family.
Males are represented as *squares*, females as *circles*, and affected individuals as *solid symbols*. This illustration is a "screen capture" from an interactive CD-ROM exercise that allows you to examine real family pedigrees for a variety of inherited human traits. By examining the pedigrees, you are to determine in each case if the trait is dominant or recessive and if it is sex-linked or autosomal. Such pedigrees are often all a human geneticist has to assess the dominance and chromosomal location of a trait. (*Explorations in Human Biology,* Module 15; *Explorations in Cell Biology & Genetics,* Module 12, "Heredity in Families")

(a)

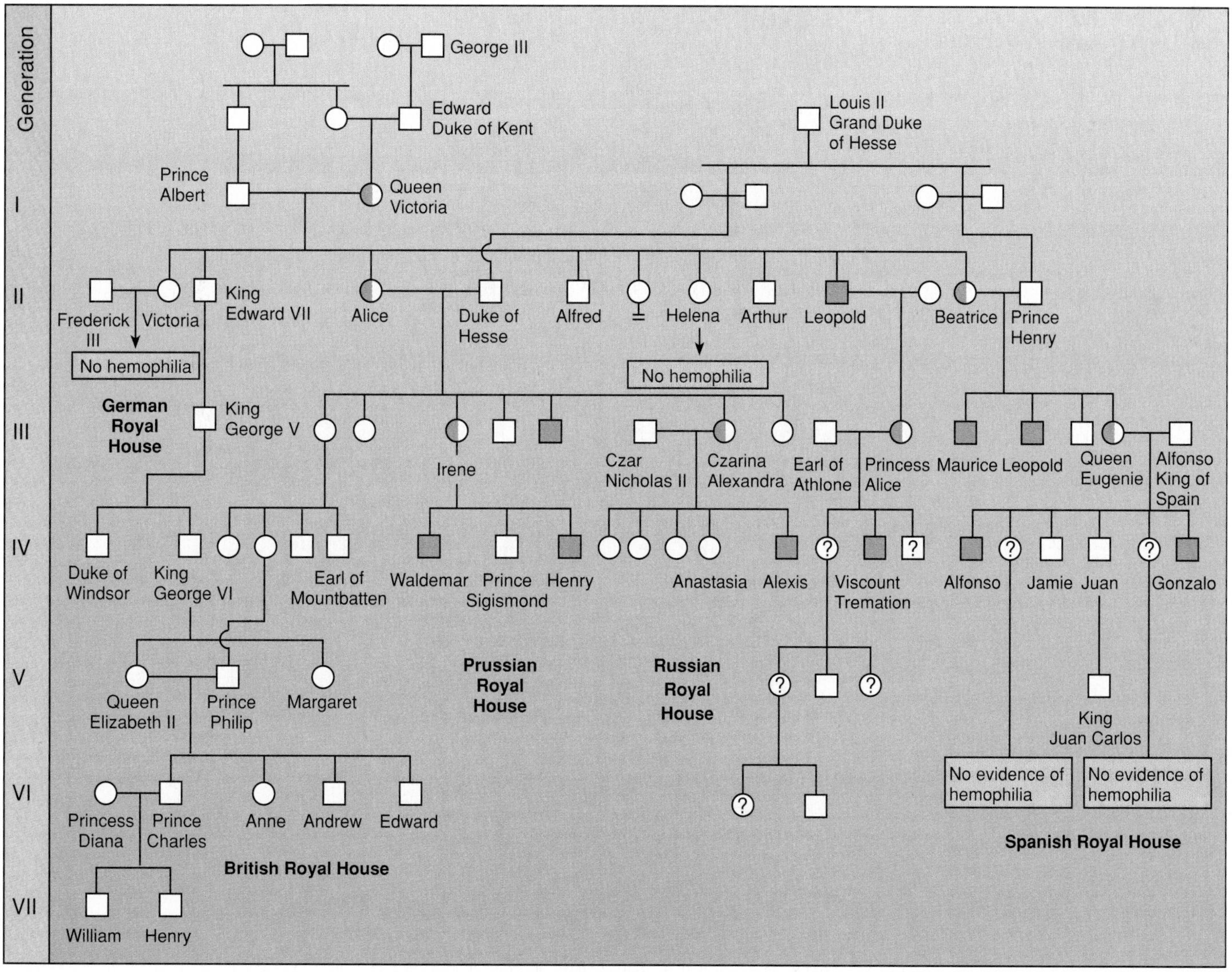

(b)

Figure 6.27 The royal hemophilia pedigree.

(*a*) In this photograph taken in 1894, Queen Victoria of England is surrounded by some of her descendants. Of Victoria's four daughters who lived to bear children, two—Alice and Beatrice—were carriers of hemophilia. Alice's daughters are standing behind Victoria (both are wearing feathered boas). To the *right* is Princess Irene of Prussia, and to the *left* is Alexandra, who would soon become Czarina of Russia. Both Irene and Alexandra were also carriers of hemophilia. Irene carried it into the Prussian royal house, where it affected two of her three sons. Alexandra carried it into the Russian royal house, where it affected her son, Alexis. (*b*) A pedigree of the royal hemophilia.

Genetic Counseling and Therapy

Most hereditary disorders cannot be cured, although progress is being made in many cases. The process of identifying parents at risk for producing children with genetic defects is called **genetic counseling.** If most hereditary disorders are caused by mutations that are recessive alleles, how do potential parents know if they are heterozygous for such an allele? It is usually difficult to be sure, but if one of your relatives has been affected by a disorder like cystic fibrosis, there is a possibility that you are a carrier of the trait. Genetic counselors analyze your pedigree—that is, a diagram of your family tree on which affected individuals are indicated—in order to assess the likelihood that you are a carrier.

Therapy is available for some hereditary disorders if they are diagnosed early enough. A small sample of fetal cells obtained by amniocentesis (figure 6.28) can permit a large battery of tests to be performed. An example of a disorder where such testing can lead to early diagnosis and successful therapy is phenylketonuria, a disorder in which a child's body converts the amino acid phenylalanine into harmful chemicals and produces severe mental retardation. If phenylketonuria is diagnosed in a fetus, the mother can be placed on a low-phenylalanine diet until the child is born. The child is maintained on that diet until age six, when mental development is complete, and so escapes the bad effects of the disorder.

Advances in gene technology, discussed in chapter 8, are rapidly making it possible to correct genetic disorders directly. This is done by transferring healthy copies of the defective gene to the individual. In 1990 this approach was tried for the first time on humans. A healthy copy of a gene encoding a key enzyme was transferred, and the individuals quickly recovered from the effects of their enzyme deficiency. These individuals have remained healthy for five years. Attempts to cure cystic fibrosis and muscular dystrophy in this way are underway but not yet successful. These approaches appear very promising, and research continues at a fever pitch.

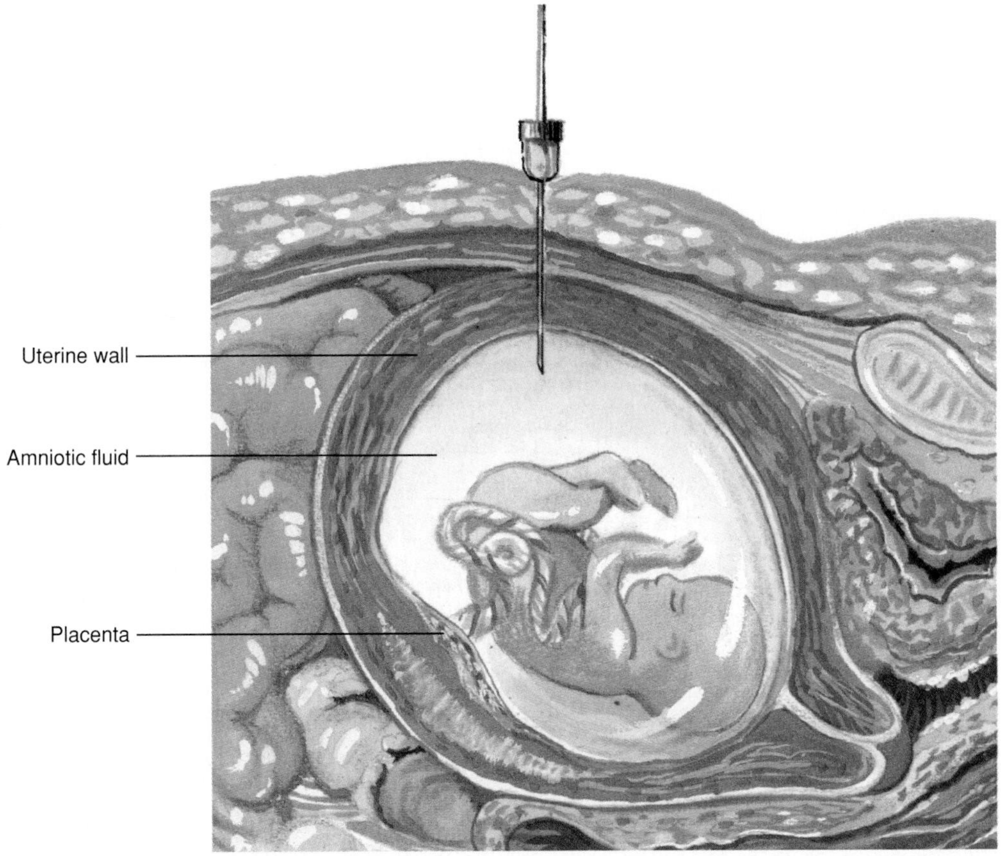

Figure 6.28 Amniocentesis.

To obtain a small sample of fetal cells for diagnosis of hereditary disorders, a needle is inserted into the amniotic cavity, and a sample of amniotic fluid, containing some free cells derived from the embryo, is withdrawn into a syringe. The fetal cells are then grown in tissue culture so that their karyotype and many of their metabolic functions can be examined.

CHAPTER 6

HIGHLIGHTS

Key Terms | Key Concepts

6.1 Mendel

Key Terms

heredity 111
recessive 114
gene 115
heterozygous 115
allele 115

Key Concepts

- In the 1860s, Gregor Mendel conducted genetic crosses of varieties of pea plants.
- Carefully counting the numbers of each kind of offspring, Mendel observed that in crosses of heterozygous parents, one-fourth of the offspring always appear recessive.
- From this simple result, Mendel formulated his theory of heredity.

6.2 From Genotype to Phenotype

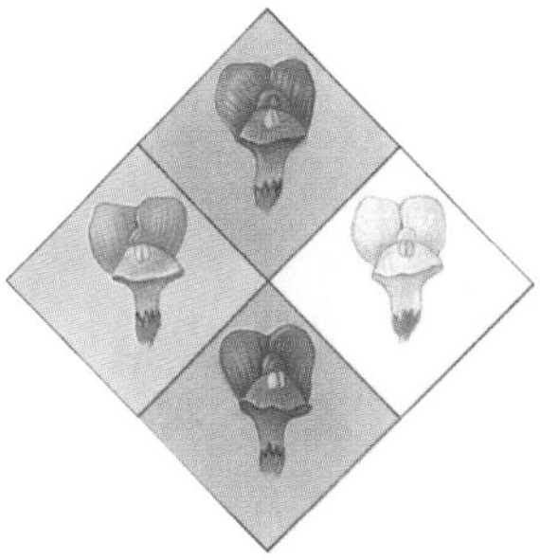

Key Terms

epistasis 119
pleiotropy 120
incomplete dominance 120

Key Concepts

- The essence of Mendel's theory is that traits are determined by *information,* which was soon shown to be stored in genes on chromosomes.
- When the pattern of a trait's heredity reflects chromosome segregation, that trait is said to be "Mendelian."
- Many factors can obscure the underlying pattern of chromosomal segregation, such as when one gene modifies the phenotypic expression of another.

6.3 Meiosis

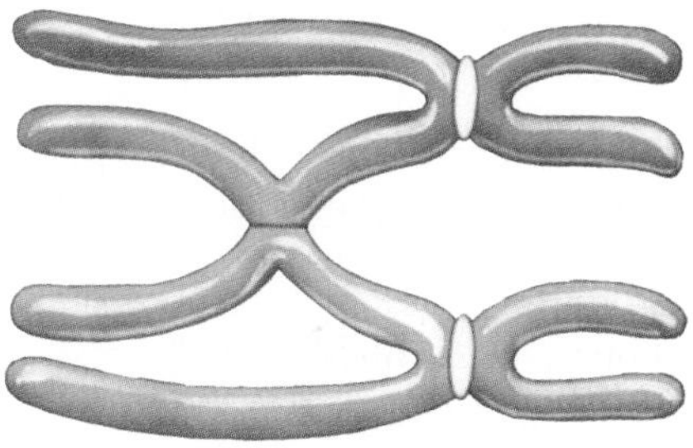

Key Terms

chromosome 121
diploid 121
meiosis 122
crossing over 122
independent assortment 122

Key Concepts

- Haploid gametes are produced from diploid cells by a special form of cell division called meiosis.
- The reduction in chromosome number from diploid to haploid occurs because the chromosomes do not replicate between the two meiotic divisions.
- Early in meiosis, homologous chromosomes pair up closely. At this time, they often exchange segments, a process called crossing over.

6.4 Patterns of Heredity in Humans

Key Terms

nondisjunction 126
aneuploidy 126
karyotype 126
X chromosome 127
sex-linked 128
pedigree 129

Key Concepts

- Extra copies of chromosomes produce developmental abnormalities or sterility.
- Harmful mutations that are inherited are called heredity disorders.
- Genetic counseling through pedigree analysis and techniques like amniocentesis are important aids in predicting the likelihood of producing children expressing hereditary disorders.

6

CONCEPT REVIEW

1. Several people had carried out plant crosses before Mendel, but they are not credited with being the founder of the science of genetics. What did Mendel do that was different?
   a. He used garden peas.
   b. He did specific crosses.
   c. He counted the numbers of different types of offspring.
   d. He removed the male parts from flowers.

2. The phenotypic ratio observed by Mendel in his F_1 plants was 3:1.
   a. true
   b. false

3. Of Mendel's F_2 plants that showed the dominant character, what proportion were true-breeding?
   a. one-fourth
   b. one-third
   c. one-half
   d. all

4. The white color of a homozygous recessive plant is its
   a. phenotype.
   b. genotype.

5. Of the following crosses, which is a test cross?
   a. $WW \times WW$
   b. $WW \times Ww$
   c. $Ww \times ww$
   d. $Ww \times W$

6. Characteristics that exhibit continuous variation are generally controlled by
   a. a single dominant gene.
   b. pleiotropy.
   c. epistatic interactions.
   d. multiple genes.

7. Sexual reproduction in plants and animals involves the production of gametes through a special type of reduction division called _________, followed by the union of haploid gametes to form a _________.
   a. mitosis
   b. cytokinesis
   c. zygote
   d. chromatin
   e. meiosis

8. Which of the following is not true of crossing over?
   a. It occurs during prophase I.
   b. It occurs during prophase II.
   c. It occurs between homologues.
   d. It is seen as X-shaped structures called chiasmata.

9. Down syndrome results from having three copies of chromosome
   a. 13.
   b. 18.
   c. 21.
   d. 22.

10. Which of the following genetic diseases is sex-linked?
    a. hemophilia
    b. Tay-Sachs disease
    c. cystic fibrosis
    d. hypercholesterolemia

11. If two alleles for a trait are the same, the individual is _________.

12. When two heterozygous individuals are crossed, the percentage of the progeny that exhibits the recessive trait is _________.

13. _________ is the observation that different genes on different chromosomes segregate independently of one another in genetic crosses.

14. The ability of one gene to modify the phenotypic expression of another is called _________.

15. Meiotic recombination greatly enhances the ability of organisms to generate genetic _________.

16. The inactive X chromosome seen in the cells of human females is called a _________.

17. The XXY condition produced by nondisjunction of an X chromosome during meiosis is called _________ and occurs in about 1 in every 1,000 male births.

18. _________ is a procedure that permits the prenatal diagnosis of many genetic disorders.

Answers to the Concept Review questions appear in Appendix B.

CHALLENGE YOURSELF

1. Why did Mendel observe only two alleles of any given trait in the crosses that he carried out?
2. How can crossing over be advantageous for natural selection?
3. Humans have 23 pairs of chromosomes. Ignoring the effects of crossing over, what proportion of a woman's eggs contain only chromosomes she received from her mother?
4. Many sexually reproducing lizard species are able to generate local populations that reproduce asexually by parthenogenesis. What do you imagine the sex of these local parthenogenetic populations would be: male, female, or neuter? Why?

FOR FURTHER READING

Carpenter, A. "Chiasma Function." *Cell* 77 (July 1, 1994): 959–62. A discussion of the chiasma and the role it plays in meiosis, crossing over, and mutation.

Cavenee, W., and R. White. "The Genetic Basis of Cancer." *Scientific American*, March 1995, 72–79. Several mutations are required to produce cancer, and some of them are inherited in families.

Corcos, A., and F. Monaghan. "Mendel's Work and Its Rediscovery: A New Perspective." *Critical Reviews in Plant Sciences* 9 (May 1990):197–212. An evaluation of the many myths surrounding Mendel's work.

Diamond, J. "Blood, Genes, and Malaria." *Natural History* 98 (February 1989): 8ff. A lucid account of the evolutionary history of sickle-cell anemia.

Gould, S. J. "Dr. Down's Syndrome." *Natural History* 89 (April 1980): 142–48. An account of the history of Down syndrome, a relatively common chromosomal abnormality that results in severe mental retardation.

John, B. *Meiosis*. New York: Cambridge University Press, 1990. A broad survey of meiosis in the plant and animal kingdoms, with emphasis on the surprising diversity in how this key form of cell division is carried out.

Mulligan, R. "The Basic Science of Gene Therapy." *Science* 260 (May 14, 1993): 926-32. An overview of gene therapy and how it may be used to fight cystic fibrosis.

Thompson, L. *Correcting the Code: Inventing the Genetic Cure for the Human Body*. New York: Simon and Schuster, 1994. This interesting book discusses the potential of gene therapy to cure *everything* that ails you.

Thurman, E. *Human Chromosomes: Structure, Behavior, and Effects*. New York: Springer-Verlag, 1993. A thorough compendium on the human chromosome, discussing meiosis as well as mitosis and other phenomena (such as chromosomal abnormalities) associated with human chromosomes.

TECHNOLOGY LINKS

The Living World Home Page
http://www.wcbp.com/biology/tlw

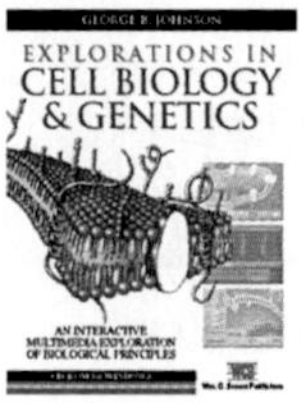

Explorations in Cell Biology & Genetics CD
#10 Meiosis: Down Syndrome
#12 Heredity in Families
#13 Gene Segregation

Life Science Animations Videotape 2
#13 Meiosis
#14 Crossing Over

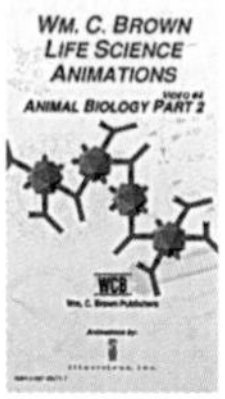

Life Science Animations Videotape 4
#40 A,B,O Blood Types

7

How Genes Work

"Bummer of a birthmark, Hal."

CHAPTER OUTLINE

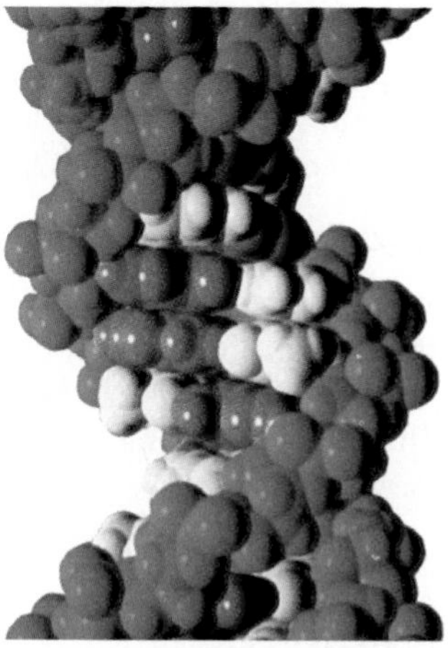

Figure 7.1 Genes are encoded within DNA.
Study of this long spiral molecule has provided the long-sought solution to the puzzle of heredity.

When Mendel solved the puzzle of heredity, he with one stroke removed the mystery from the question of why we resemble our parents. As we learned in chapter 6, we resemble our parents because we are built from copies of their chromosomes. These chromosomes contain sets of instructions called genes that determine what they, and we, are like. This realization that genes on chromosomes determine heredity was one of the most important advances in human thought. Not only did it lead to great progress in agriculture and medicine, but it has profoundly influenced the way we think about ourselves.

However, Mendel's work leaves a key question unanswered: What *is* a gene? After almost a century of research, biologists are now able to answer this question fully. We have learned what genes are made of and how they work. We understand in considerable detail how the information in genes is converted into organisms with eyes, arms, and inquiring minds. The first step in this new understanding came with the discovery that the hereditary information of genes is stored in a long spiral molecule called DNA that forms the core of chromosomes (figure 7.1).

7.1 Genes Are Made of DNA

When biologists began to examine chromosomes in their search for genes, they soon learned that chromosomes are made of two kinds of macromolecules, both of which you encountered in chapter 2: **proteins** (long chains of short subunit molecules called *amino acids* linked together in a string) and **DNA** (deoxyribonucleic acid) (long chains of short subunit molecules called *nucleotides* linked together in a string). It was possible to imagine that either of the two was the stuff that genes are made of—information might be stored in a sequence of different amino acids, or of different nucleotides, just as different letter combinations make a word on this page. But which one is the stuff of genes, protein or DNA?

This question was answered clearly in a variety of different experiments, all of which shared the same basic design: If you separate the DNA in an individual's chromosomes from the protein there, which of the two materials is able to change another individual's genes?

The Griffith-Avery Experiments

In 1928 a British microbiologist named Fred Griffith made an unexpected observation. He was working with mice infected with the bacterium *Streptococcus pneumoniae* and learned that only bacteria possessing a gene that gave them a slimy covering were able to kill infected mice (such deadly strains are called **virulent**). Griffith found that the slimy coat itself wasn't a poison, because if he first killed the "coated" bacteria, the preparation was harmless to mice. The unexpected observation was a control experiment that Griffith carried out (figure 7.2). When Griffith mixed dead "coated" bacteria (harmless) with live bacteria that lacked the gene to make coats (harmless), the mixture killed mice! When he looked in the blood of the dead mice, he found that the "coatless" live bacteria now had coats—they had somehow acquired the gene they lacked from the dead bacteria. The process of acquiring a gene from another organism is called **transformation.**

The material that transformed the *Streptococcus* bacteria from harmless (without coats) to virulent (with coats) was of course the gene saying how to make the coat. Was this "transforming principle" made of DNA or of protein? The answer came in 1944 when Oswald Avery isolated the key ingredient, a preparation that would readily transform harmless bacteria without coats into virulent, coated ones. What was the ingredient made of? DNA. Simple experiments confirmed Avery's identification: Protein-destroying enzymes had no effect on the preparation's power to transform, while DNA-destroying enzymes rendered the preparation harmless.

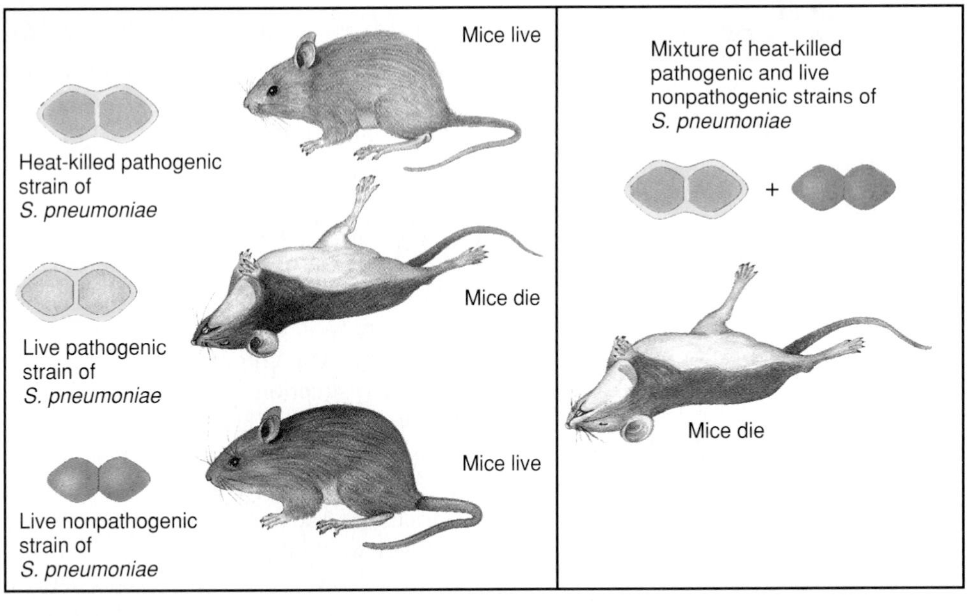

Figure 7.2 Griffith's discovery of transformation.
Transformation, the movement of a gene from one organism to another, provided some of the key evidence that DNA is the genetic material. Griffith found that extracts of dead pathogenic strains of the bacterium *Streptococcus pneumoniae* can "transform" harmless strains, that after being treated with the extract, became able to kill mice. Griffith concluded that the live cells had been "transformed" by the dead ones. Later, Avery analyzed the extract and demonstrated that the material that had passed from dead to living bacteria was DNA.

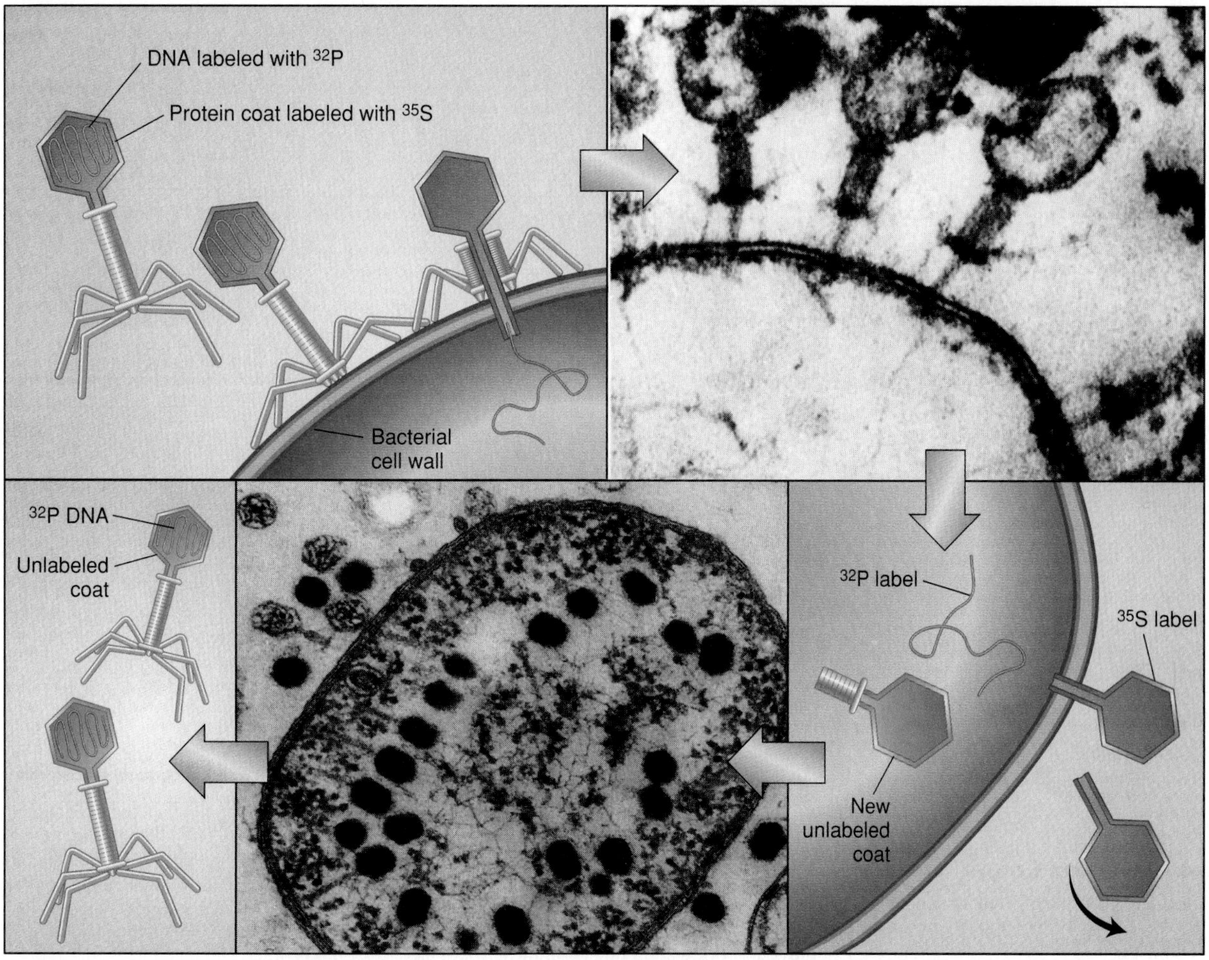

Figure 7.3 The Hershey-Chase experiment.

The experiment that convinced most biologists that DNA was the genetic material was carried out soon after World War II, when radioactive isotopes were first becoming commonly available to researchers. Hershey and Chase used different radioactive labels to "tag" and track protein and DNA. They found that when bacterial viruses inserted their genes into bacteria to guide the production of new viruses, it was DNA and not protein that was inserted. Clearly the virus DNA, not the virus protein, was responsible for directing the production of new viruses.

The Hershey-Chase Experiment

Avery's result was not widely appreciated at first, because most biologists still preferred to think that genes were made of proteins. In 1952, however, a simple experiment carried out by Alfred Hershey and Martha Chase was impossible to ignore. The team studied the genes of viruses that infect bacteria. These viruses attach themselves to the surface of bacterial cells and inject their genes into the interior; once inside, the genes take over the genetic machinery of the cell and order the manufacture of hundreds of new viruses. When mature, the progeny viruses burst out to infect other cells. These bacteria-infecting viruses have a very simple structure: a core of DNA surrounded by a coat of protein.

In this experiment, Hershey and Chase used **radioactive isotopes** to "label" the DNA and protein of the viruses. In one preparation, the viruses were grown so that their DNA contained radioactive phosphorus (^{32}P); in another preparation, the viruses were grown so that their protein coats contained radioactive sulfur (^{35}S). The two preparations were then mixed together, and the mixture was allowed to infect bacteria. After a few minutes Hershey and Chase shook the suspension forcefully to dislodge attacking viruses from the surface of bacteria, used a rapidly spinning centrifuge to isolate the bacteria, and then asked a very simple question: What did the viruses inject into the bacterial cells, protein or DNA? They found that the interiors of the bacterial cells contained the ^{32}P label but not the ^{35}S label. The conclusion is clear and inescapable—the genes that viruses use to specify new generations of viruses are made of DNA and not protein (figure 7.3).

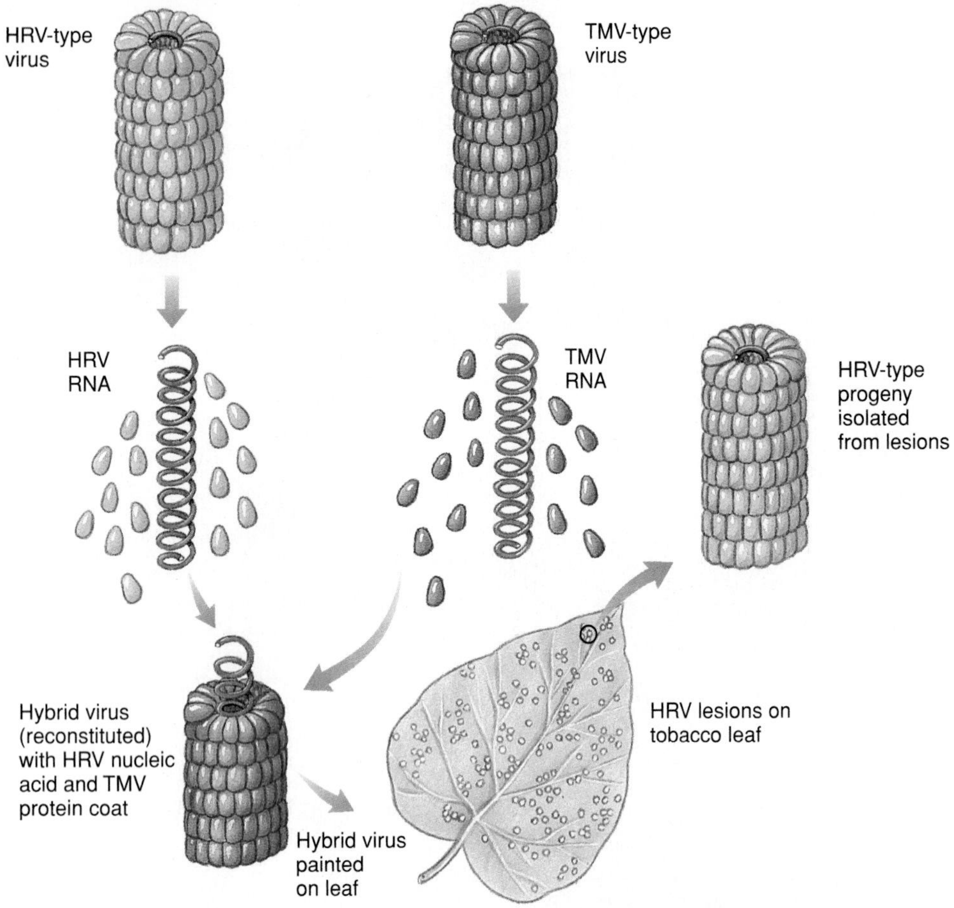

Figure 7.4 Fraenkel Conrat's virus reconstitution experiment.
Both TMV and HRV are plant RNA viruses that infect tobacco plants, causing lesions on the leaves. In this experiment, TMV and HRV were both dissociated into protein and RNA. Then hybrid virus particles were produced by mixing the HRV RNA and the TMV protein. When the reconstituted virus particles were painted onto tobacco leaves, HRV-type lesions developed. From these lesions, normal HRV virus particles could be isolated in great numbers; no TMV viruses could be isolated from the lesions. Thus the RNA (HRV) and not the protein (TMV) contains the information necessary to specify the production of the viruses.

The Fraenkel Conrat Experiment

In a third experiment, carried out in 1957 by Heinz Fraenkel Conrat, the chemical nature of the gene was seen particularly clearly. He isolated viruses from two kinds of plant: TMV (tobacco mosaic virus) from tobacco plants, and HRV (Holmes ribgrass virus) from grass. Both viruses consist of a protein coat wrapped around a single strand of nucleic acid, in this case (as in many viruses, including the AIDS virus) RNA rather than DNA. Fraenkel Conrat found that he could take a virus preparation apart chemically, separating the RNA from the protein, and that when he put the two chemicals back together, they reassembled into infective virus particles. So to find out what genes are made of, Fraenkel Conrat did a mix-and-match experiment: He mixed TMV protein with HRV RNA and then allowed the reassembled viruses to infect plants (figure 7.4). What happened? The infected plants developed the sores characteristic of HRV infection. When the experiment was carried out in reverse, with HRV protein and TMV RNA, the infected plants developed the sores characteristic of TMV infection. Clearly, it was the RNA and not the protein that was directing the production of new viruses in infections by these plant viruses.

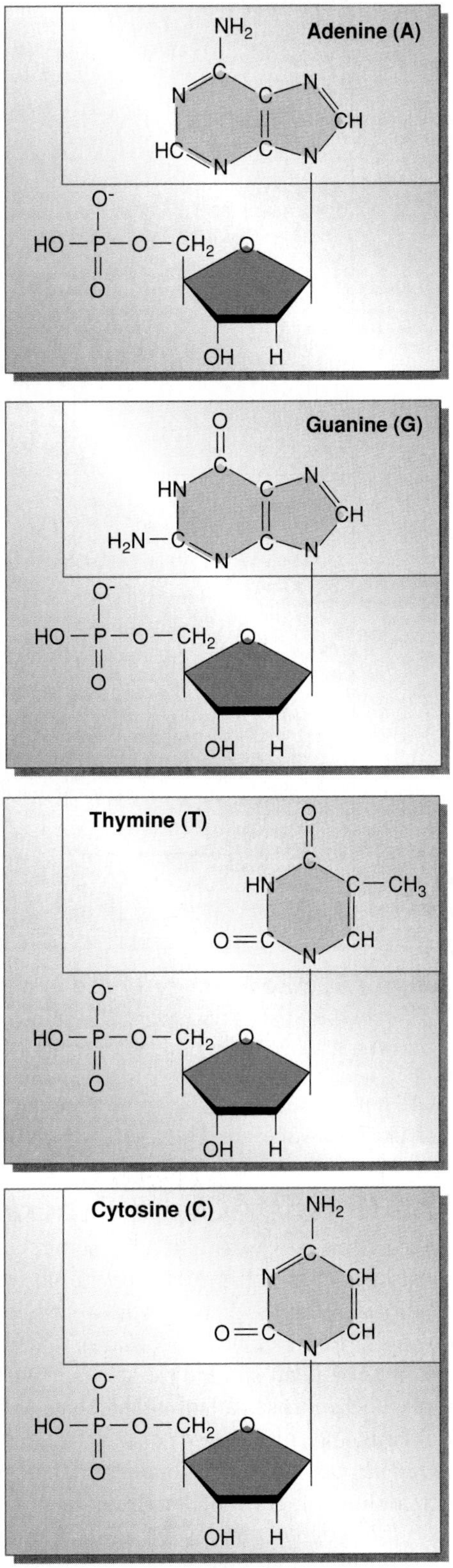

Figure 7.5 The four nucleotide subunits that make up DNA.
The nucleotide subunits of DNA are composed of three elements: a central five-carbon sugar, a phosphate group, and an organic, nitrogen-containing base.

The Structure of DNA

By the early 1950s it was becoming very clear to biologists that genes were in fact made of DNA, and a lot of scientists began to study DNA in earnest. DNA is a long, chainlike molecule made up of subunits called **nucleotides**. Each nucleotide has three parts: a central sugar, a phosphate (PO_4) group, and an organic base. The sugar and the phosphate group are the same in every nucleotide of DNA, but there are four different kinds of bases, two large ones with double-ring structures and two small ones with single rings (figure 7.5). The large bases are called **purines** and their names are **A** (adenine) and **G** (guanine). The small bases are called **pyrimidines** and their names are **C** (cytosine) and **T** (thymine). A key observation, made by biochemist Erwin Chargaff, was that DNA molecules always had equal amounts of purines (A + G) and pyrimidines (T + C). In fact, with slight variations due to imprecision of measurement, the amount of A always equals the amount of T, and the amount of G always equals the amount of C. This observation (A = T, G = C), known as **Chargaff's rule,** suggested that DNA had a regular structure.

The significance of Chargaff's rule became clear in 1953, when British scientists began to study the structure of DNA using X-ray diffraction techniques. In these experiments DNA molecules are bombarded with X-ray beams, and when individual rays encounter atoms their path is bent or diffracted; each atomic encounter creates a pattern on photographic film like the ripples created by tossing a rock into a smooth lake. By analyzing the complex patterns created by many atoms, scientists can work out the locations of all the atoms in the molecule. The first studies, carried out by Rosalind Franklin at Oxford University, suggested that the DNA molecule had the shape of a coiled spring, a form called a **helix.**

Two workers at Cambridge University, Francis Crick and James Watson (an American) learned informally of Franklin's results and, using Tinkertoy models of the bases, quickly deduced the true structure of DNA (figure 7.6): The DNA molecule is a **double helix,** a winding staircase of two strands whose bases face one another. Chargaff's rule is a direct reflection of this structure—every bulky purine on one strand is paired with a slender pyrimidine on the other strand. Specifically, A pairs with T, and G pairs with C (figure 7.7). With this **base pairing,** the molecule keeps a constant thickness because purine–purine and pyrimidine–pyrimidine pairs never occur.

Figure 7.6 The discovery of DNA's structure.
A key breakthrough in our understanding of heredity occurred in 1953, when Watson and Crick deduced the structure of DNA. James Watson was a young American postdoctoral student (he is the one peering up as if afraid their homemade model of the DNA molecule will topple over), and Francis Crick (pointing) was an English scientist.

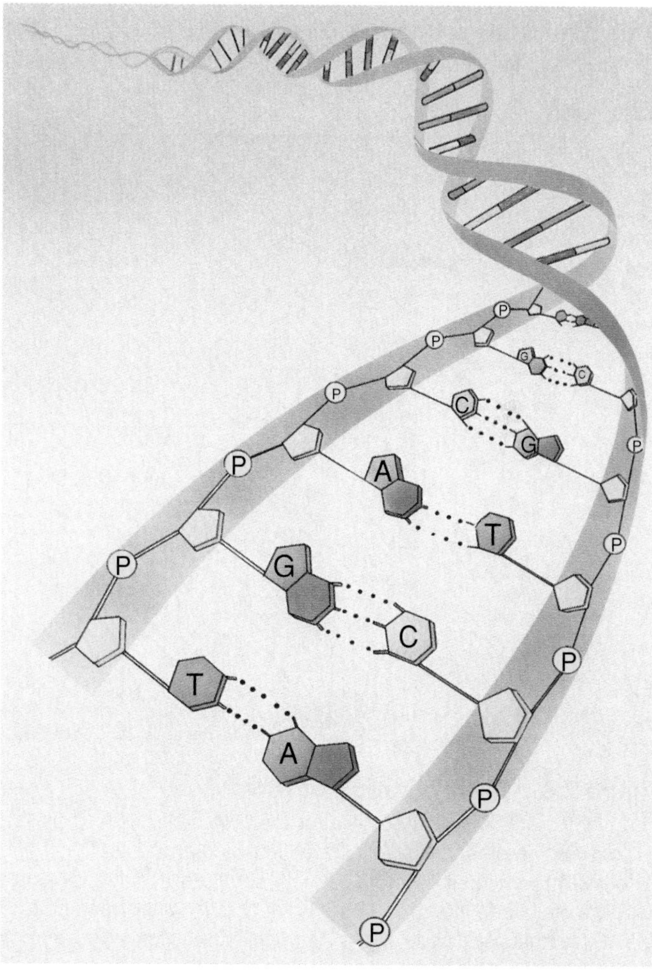

Figure 7.7 Base pairing.
In a DNA duplex molecule, only two base pairs are possible: adenine (A) with thymine (T) and guanine (G) with cytosine (C). A G–C base pair has three hydrogen bonds; an A–T base pair has only two.

How DNA Copies Itself

The attraction that holds the two DNA strands together is the formation of weak hydrogen bonds between the bases that face each other from the two strands. That is why A pairs with T and not C—it can only form hydrogen bonds with T. Similarly, G can form hydrogen bonds with C but not T. In the Watson-Crick model of DNA, the two strands of the double helix are said to be *complementary* to each other. One chain of the helix can have any sequence of bases, of A, T, G, and C, but this sequence completely determines that of its partner in the helix. If the sequence of one chain is ATTGCAT, the sequence of its partner in the double helix must be TAACGTA. Each chain in the helix is a complementary mirror image of the other. This **complementarity** makes it possible for the DNA molecule to copy itself during cell division in a very direct manner. The double helix need only "unzip" and assemble a new complementary chain along each naked single strand by base pairing A with T and G with C (figure 7.8).

The copying of DNA before cell division is called **DNA replication** and is overseen by an enzyme called DNA polymerase. This enzyme reads along the naked single strand and adds the correct complementary nucleotide (A with T, G with C) at each position as it moves, creating a complementary strand. The place where the parent DNA molecule becomes unzipped is called a **replication fork.** At replication forks, the polymerase very actively shuttles up one strand and down the other. Chromosomes each contain a single, very long molecule of DNA, but it is too long to copy conveniently all the way from one end to the other with a single replication fork. Each chromosome is instead copied in segments, each zone of about 100,000 nucleotides having its own replication fork.

The enormous amount of DNA that resides within the cells of your body represents a long series of DNA replications, starting with the DNA of a single cell—the fertilized egg. Living cells have evolved many mechanisms to avoid errors during DNA replication and to preserve the DNA from damage. These mechanisms of **DNA repair** proofread the strands of each daughter cell against one another for accuracy and correct any mistakes. But the proofreading is not perfect. If it were, no mistakes would occur, no variation in gene sequence would result, and evolution would come to a halt.

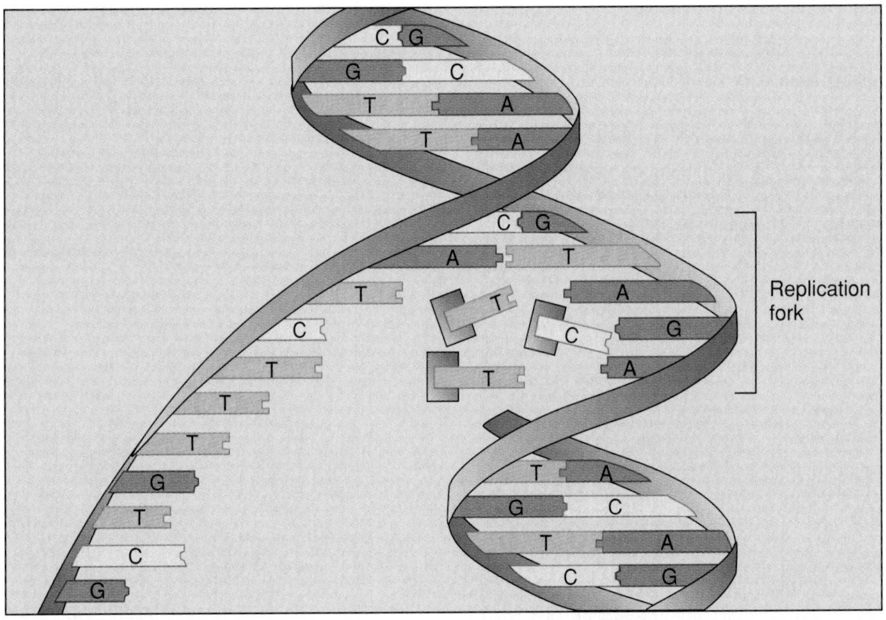

Figure 7.8 The replication of DNA.
Complementarity makes it possible for the DNA molecule to copy itself. This diagram shows a DNA double helix, the bottom portion of which has separated into single strands. The strand on the *right* is shown in the process of re-forming a new double helix. Although not illustrated in this diagram, the other strand is also replicated (but in the opposite direction). In each new molecule of DNA, one strand is "old" DNA from the parent cell and one strand is recently synthesized "new" DNA. This is called semiconservative replication.

7.2 Transcription

The discovery that genes are made of DNA left unanswered the question of how the information in DNA is used. How does a string of nucleotides in a spiral molecule determine if you have red hair? We now know that the information in DNA is arrayed in little blocks, like entries in a dictionary, each block a gene specifying a protein. Proteins are the tools of heredity. Genes control what you are like because the proteins the DNA directs the cell to make determine what the cell will be like.

Making Working Copies of Genes

Just as an architect protects building plans from loss or damage by keeping them safe in a central place and issuing only blueprint copies to on-site workers, so your cells protect their DNA instructions by keeping them safe within a central DNA storage area, the nucleus. The DNA never leaves the nucleus. Instead, "blueprint" copies of particular genes within the DNA instructions are sent out into the cell to direct the assembly of proteins. These working copies of genes are made of ribonucleic acid (RNA) rather than DNA. Recall that RNA is the same as DNA except that the sugars have an extra oxygen atom and T is replaced by a similar nucleotide called U, uracil. The path of information is thus:

$$\text{DNA} \rightarrow \text{RNA} \rightarrow \text{protein}$$

This information path is often called the *central dogma,* because it describes the key organization used by your cells to express their genes (figure 7.9). A cell uses three kinds of RNA in the synthesis of proteins: messenger RNA (mRNA), ribosomal RNA (rRNA), and transfer RNA (tRNA). The specific role of each type of RNA is discussed following.

The use of information in DNA to direct the production of particular proteins is called **gene expression.** Gene expression occurs in two stages: in the first stage, called **transcription,** an mRNA molecule is synthesized from a gene within the DNA; in the second stage, called **translation,** this mRNA is used to direct the production of a protein.

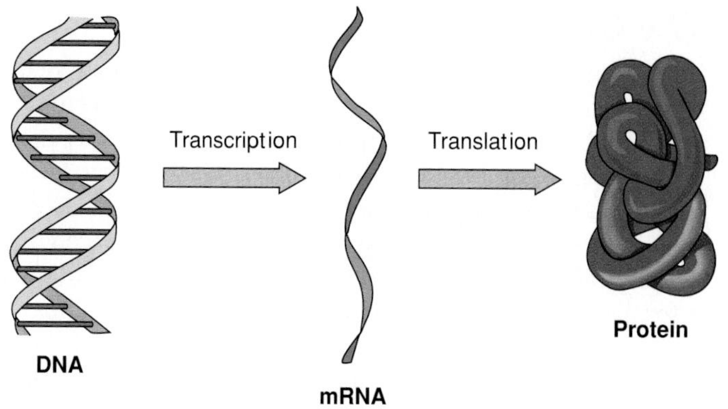

Figure 7.9 The central dogma of gene expression.
Through the production of mRNA (transcription) and the synthesis of proteins (translation), the information contained in DNA is expressed.

The Transcription Process

The RNA copy of a gene used in the cell to produce a protein is called **messenger RNA (mRNA)—**it is the messenger that conveys the information from the nucleus to the cytoplasm. The copying process that makes the mRNA is called transcription—just as monks in monasteries used to make copies of manuscripts by faithfully transcribing each letter, so enzymes within the nuclei of your cells make mRNA copies of your genes by faithfully complementing each nucleotide.

In your cells, the transcriber is a large and very sophisticated protein called **RNA polymerase.** It binds to one strand of a DNA double helix at a particular site called a *promoter* (its shape fits the sequence of bases there) and then moves down the DNA strand like a train on a track. As it goes along the DNA "track," the polymerase pairs each nucleotide with its complementary RNA version (A with U, G with C), building an mRNA chain behind it as it moves down the DNA strand (figure 7.10). Transcription is the production by an RNA polymerase enzyme of an mRNA molecule whose nucleotide sequence is complementary to the gene's DNA nucleotide sequence.

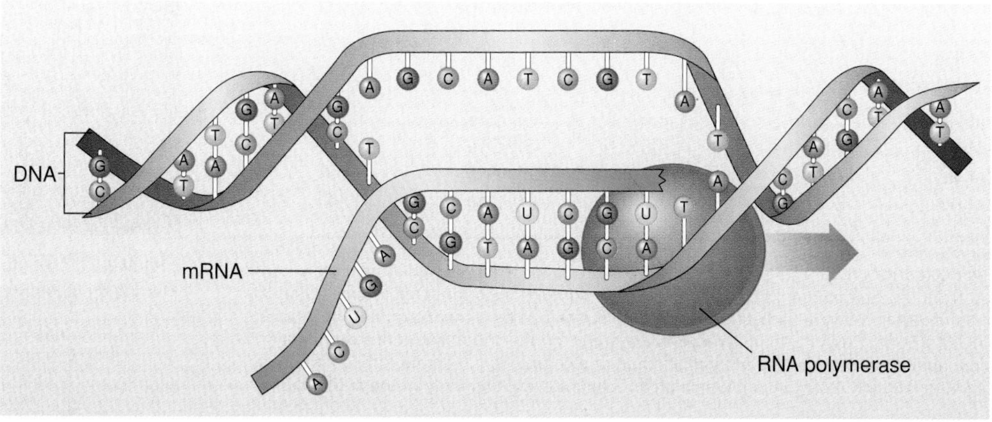

Figure 7.10 Transcription.
One of the strands of DNA functions as a template on which nucleotide building blocks are assembled into mRNA by RNA polymerase as it moves down the DNA strand.

The Genetic Code

The essence of Mendelian genetics is that information determining hereditary traits, traits passed from parent to child, is encoded information. The information is written within the chromosomes in blocks called genes. Genes affect Mendelian traits by directing the production of particular proteins. The essence of gene expression, of using your genes, is reading the information encoded within DNA and using that information to direct the production of the correct protein.

To correctly read a gene, a cell must translate the information encoded in DNA into the language of proteins—that is, it must convert the *order* of the gene's nucleotides into the order of amino acids in a protein. The rules that govern this translation are called the **genetic code.** They are very simple:

1. Each gene is read from a fixed starting position, a nucleotide sequence at its beginning called a promoter site where the RNA polymerase first binds to the DNA.
2. RNA polymerase moves down the DNA in steps that are three nucleotides long.
3. Each three-nucleotide block in the gene corresponds to a particular amino acid.

Special three-nucleotide sequences located at the end of genes say "stop." A three-nucleotide sequence that corresponds to an amino acid is called a **codon.** Biologists worked out which codons correspond to which amino acids by trial-and-error experiments carried out in test tubes. In these experiments investigators used artificial mRNAs to direct the synthesis of proteins in the tube and then looked to see the sequence of amino acids in the proteins. An mRNA that was a string of UUUUUU..., for example, produced a protein that was a string of phenylalanine (PHE) amino acids, telling investigators that the codon UUU corresponded to the amino acid PHE. The entire **genetic code dictionary** is presented in table 7.1. Because at each position of a three-letter codon any of the four different nucleotides (A, U, G, C) may be used, there are 64 different possible three-letter codons ($4 \times 4 \times 4 = 64$) in the genetic code.

The genetic code is universal, the same in practically all organisms. GUC codes for valine in bacteria, in fruit flies (figure 7.11), in eagles, and in your own cells. The only exceptions biologists have ever found to this rule is in the way in which cell organelles (mitochondria and chloroplasts) and a few microscopic protists read the "stop" codons. In every other instance, the same genetic code is employed by all living things.

Table 7.1 The Genetic Code

First Letter	Second Letter: U	C	A	G	Third Letter
U	Phenylalanine	Serine	Tyrosine	Cysteine	U
	Phenylalanine	Serine	Tyrosine	Cysteine	C
	Leucine	Serine	STOP	STOP	A
	Leucine	Serine	STOP	Tryptophan	G
C	Leucine	Proline	Histidine	Arginine	U
	Leucine	Proline	Histidine	Arginine	C
	Leucine	Proline	Glutamine	Arginine	A
	Leucine	Proline	Glutamine	Arginine	G
A	Isoleucine	Threonine	Asparagine	Serine	U
	Isoleucine	Threonine	Asparagine	Serine	C
	Isoleucine	Threonine	Lysine	Arginine	A
		Threonine	Lysine	Arginine	G
	START; Methionine				
G	Valine	Alanine	Aspartate	Glycine	U
	Valine	Alanine	Aspartate	Glycine	C
	Valine	Alanine	Glutamate	Glycine	A
	Valine	Alanine	Glutamate	Glycine	G

A codon consists of three nucleotides read in the sequence shown above. For example, ACU codes threonine. The first letter, *A,* is read from the first column; the second letter, *C,* from the second-letter columns; and the third letter, *U,* from the third-letter column. Each of the codons is recognized by a corresponding anticodon sequence on a tRNA molecule. Some tRNA molecules recognize more than one codon sequence but always for the same amino acid. Most amino acids are encoded by more than one codon. For example, threonine is encoded by four codons (ACU, ACC, ACA, and ACG), which differ from one another only in the third position.

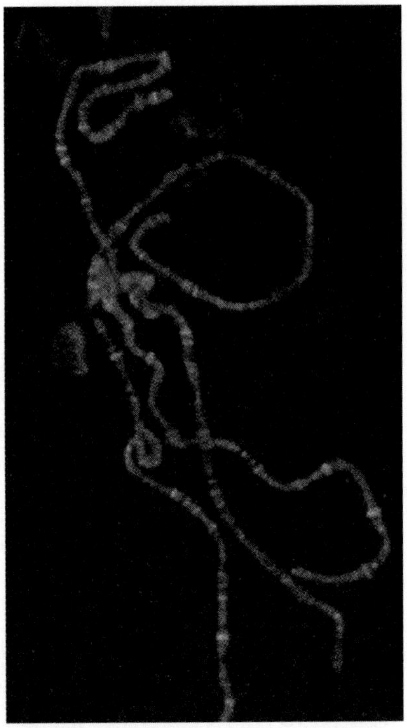

Figure 7.11 Genes on a fruit fly chromosome.

Genes being actively transcribed appear as brighter bands. In each one of them, GUC codes for valine, because the genetic code is universal.

7.3 Translation

The final result of the transcription process is the production of an mRNA copy of a gene. Like a photocopy, the mRNA can be used without damage or wear and tear on the original. After transcription of a gene is finished, the mRNA passes out of the nucleus into the cytoplasm through pores in the nuclear membrane. There, translation of the genetic message occurs. In **translation,** organelles called **ribosomes** use the mRNA produced by transcription to direct the synthesis of a protein.

The Protein-making Factory

Ribosomes are the factories of the cell. Each is very complex, containing over 50 different proteins and several segments of **ribosomal RNA** (rRNA). Just as factories use blueprints to direct the assembly of a car, so ribosomes use mRNA to direct the assembly of a protein.

Ribosomes are composed of two pieces, or subunits, one nested into the other like a fist in the palm of your hand (figure 7.12). The "fist" is the smaller of the two subunits. It contains the rRNA, which has a short nucleotide sequence exposed on the surface of the subunit. This exposed sequence is identical to a sequence called the leader region which occurs at the beginning of all genes. Because of this, an mRNA molecule binds to the exposed rRNA of the small subunit like a fly sticking to flypaper.

The Key Role of tRNA

Directly adjacent to the exposed rRNA sequence is a small pocket or dent in the surface of the ribosome that has just the right shape to bind yet a third kind of RNA molecule, **transfer RNA (tRNA).** It is tRNA molecules that bring amino acids to the ribosome to use in making proteins. tRNA molecules are chains about 80 nucleotides long, folded into a compact shape, with a three-nucleotide sequence jutting out at one end and an amino acid attached to the other.

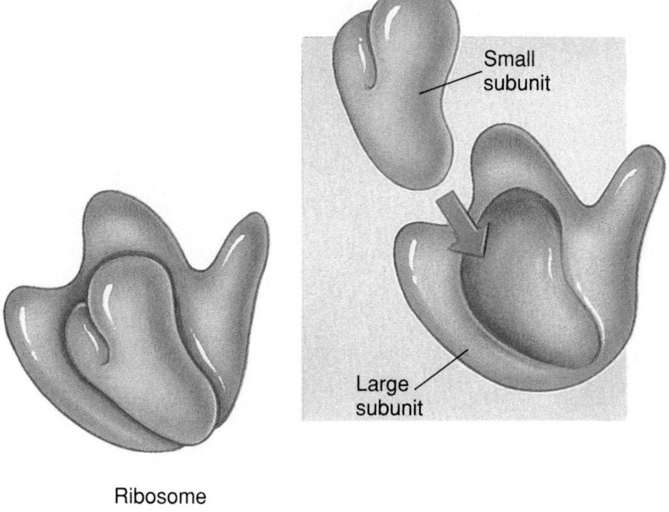

Figure 7.12 A ribosome is composed of two subunits.
The smaller subunit fits into a depression on the surface of the larger one.

The three-nucleotide sequence is very important—called the **anticodon,** it is the complementary sequence to 1 of the 64 codons of the genetic code! Special enzymes match amino acids with their proper anticodons, so that each tRNA carries the amino acid appropriate to its anticodon.

Because the dent in the small subunit is directly adjacent to where the mRNA binds to the rRNA, three nucleotides of the mRNA are positioned directly facing the jutting anticodon of the tRNA (figure 7.13). Like the address on a letter, the anticodon ensures that an amino acid is delivered to its correct "address" on the mRNA where the ribosome is assembling the protein (figure 7.14).

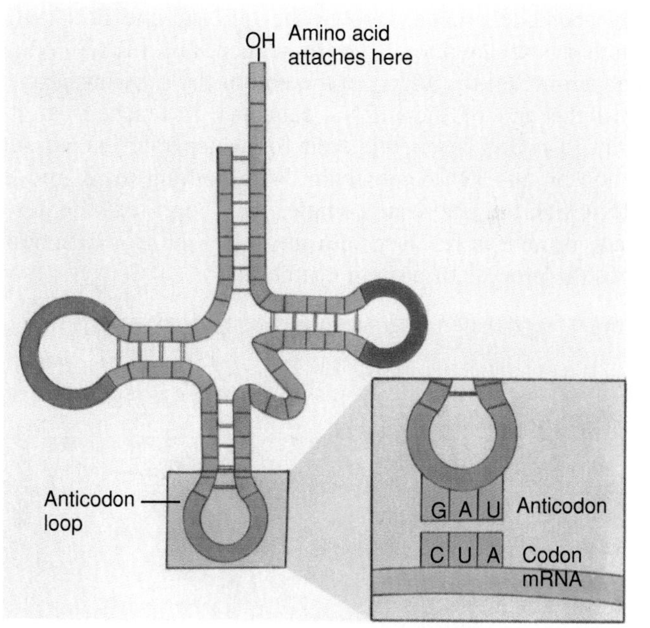

Figure 7.13 Structure of a tRNA molecule.
tRNA is the molecule that "reads" the genetic code and carries the appropriate amino acid into position along the mRNA strand. The anticodon loop matches up to the mRNA codon. The correct amino acid is attached to the opposite end of the tRNA molecule.

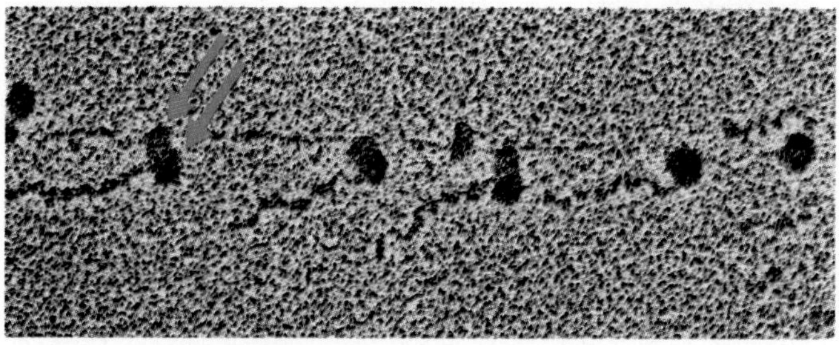

Figure 7.14 Ribosomes assembling proteins.
These ribosomes are reading along an mRNA molecule, assembling proteins that dangle behind them like tadpole tails. Clearly visible (arrows) are the two subunits of each ribosome.

Making the Protein

Once an mRNA molecule has bound to the small ribosomal subunit, the other larger ribosomal subunit binds as well, forming a complete ribosome. The ribosome then begins the process of translation. The mRNA begins to thread through the ribosome like a string passing through the hole in a donut. The mRNA passes through in short spurts, three nucleotides at a time, and at each burst of movement a new three-nucleotide codon is positioned opposite the dent in the small subunit where tRNA molecules bind (figure 7.15).

As each new codon is presented to the tRNA site, the old tRNA paired with the previous codon is passed over to the large ribosomal subunit and the amino acid it carries is attached to the end of a growing amino acid chain. So as the ribosome proceeds down the mRNA, one tRNA after another is selected to match the sequence of mRNA codons. The amino acid is added to the end of the growing protein, until the end of the mRNA sequence is reached. At this point, a codon is encountered for which there is no anticodon on any tRNA molecule. With nothing to fit into the tRNA site, the ribosome complex falls apart, and the newly made protein is released into the cell. Figure 7.16 summarizes the process of protein synthesis.

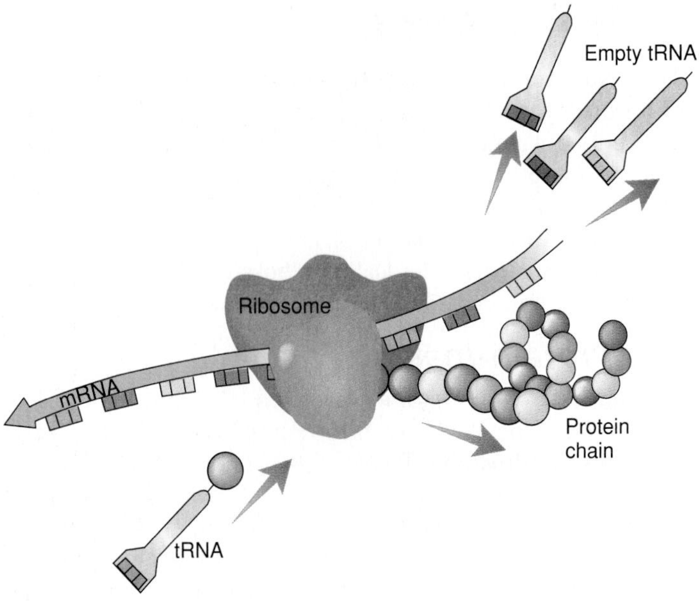

Figure 7.15 Translation.
The mRNA strand acts as a template for tRNA molecules. The appropriate tRNA is selected and positioned by the ribosome, which moves along the mRNA in three-nucleotide steps.

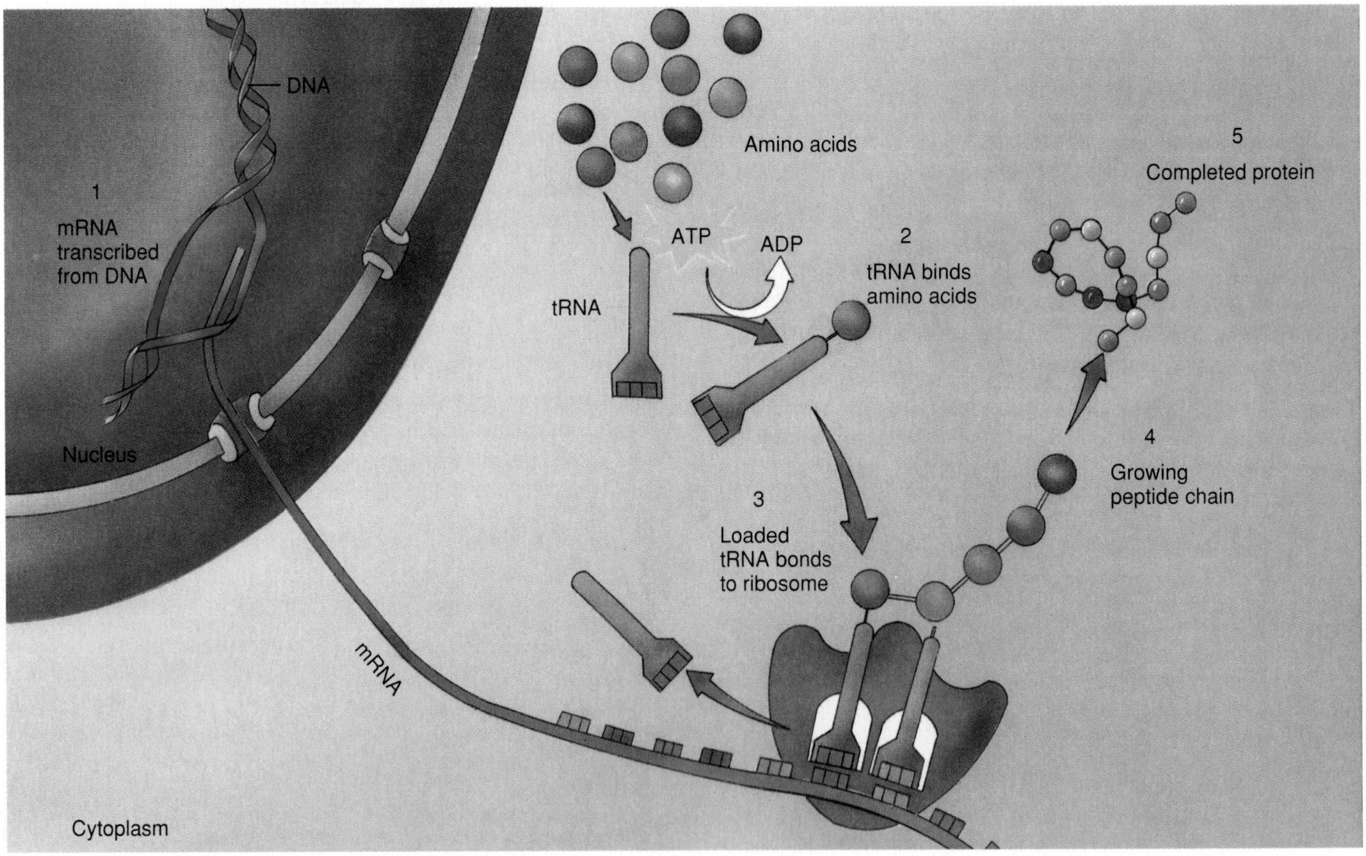

Figure 7.16 An overview of protein synthesis.
The DNA in the nucleus contains the information for the production of cell proteins. First, a complementary copy of the DNA, called mRNA, is made and carried out of the nucleus to a ribosome in the cytoplasm (*1*). At the same time, tRNA binds to an amino acid that corresponds to its anticodon sequence (*2*). Ribosomes bind the loaded tRNAs to their complementary sequences on the strand of mRNA (*3*). tRNA adds its amino acid to the growing peptide chain (*4*), which is released as the completed protein (*5*).

7.4 Regulating Gene Expression

Being able to translate a gene into a protein is only part of the game of gene expression. Every cell must also be able to regulate when particular genes are used. Imagine if every instrument in a symphony played at full volume all the time, all the horns blowing full blast and each drum beating as fast and loudly as it could! No symphony plays that way, because music is more than noise—it is the controlled expression of sound. In the same way, growth and development is the controlled expression of genes, each brought into play at the proper moment to achieve precise and delicate effects. The development of a human being from a single fertilized cell is a biochemical symphony, and the proteins that orchestrate the thousands of delicate changes follow a complex score written in the genes, a score that turns genes on and off like a conductor signaling here a violin, there a drum.

Cells control the expression of their genes by saying *when* individual genes are to be transcribed. At the beginning of each gene are special regulatory sites that act as points of control. Specific regulatory proteins within the cell bind to these sites, turning transcription of the gene off or on.

Turning Genes Off

Many genes are not used often, their transcription turned off except when needed. This is called "negative" control, because the control stops something from happening. In these genes, the regulatory site is located between the place where the RNA polymerase binds to the DNA (called the **promoter** site) and the beginning edge of the gene. When the regulatory protein, called a **repressor,** is bound to its regulatory site, its presence blocks the movement of the polymerase toward the gene, a barrier it cannot pass (figure 7.17*a*). The site where the repressor protein binds is called the **operator.** Imagine if you sat down to eat dinner and someone placed a brick wall between your chair and the table—you could not begin your meal until the wall was removed, any more than the polymerase can begin transcribing the gene until the repressor protein "barrier" is removed.

Turning Genes On

To turn on a gene whose transcription is blocked by a repressor, all that is required is to remove the repressor (figure 7.17*b*). Cells do this by binding special "signal" molecules to the repressor protein; the binding causes the repressor protein to contort into a shape that doesn't fit DNA, and it falls off, removing the barrier to transcription. A specific example demonstrating how repressor proteins work is the set of genes called the *lac* operon in the bacterium *Escherichia coli* (figure 7.18). An **operon** is a segment of DNA containing a cluster of genes that are transcribed as a unit. The operon consists of both protein-encoding (structural) genes and associated regulatory elements—the operator and promoter. When an *E. coli* encounters the sugar lactose, it starts to transcribe the genes that let it break down the lactose to get energy by altering the repressor protein. The lactose binds to the repressor protein, inducing a twist in its shape that causes it to fall from the DNA, and the polymerase is no longer blocked from transcribing the needed genes.

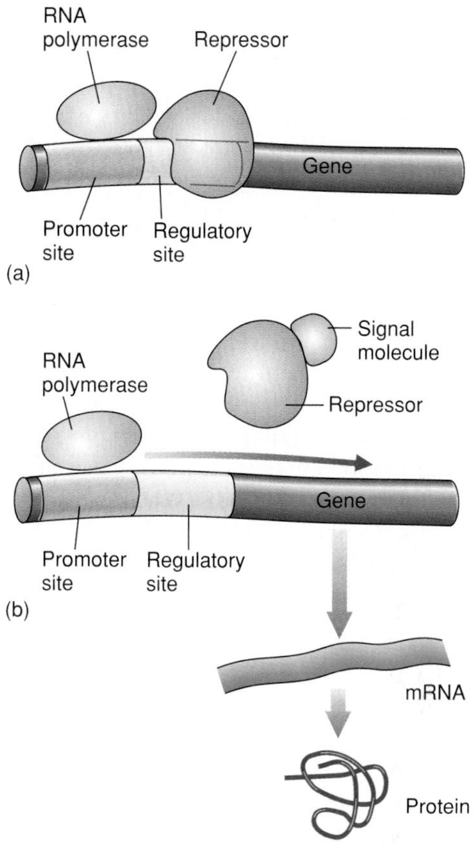

Figure 7.17 How genes are turned off.
Many genes are used only infrequently. The body keeps them from being expressed until needed by the use of a regulatory protein, called a repressor (*a*), that blocks the passage of RNA polymerase. Gene transcription cannot occur until the repressor is removed (*b*).

Activators

Because RNA polymerase binds to a specific promoter site on one strand of the DNA double helix, it is necessary that the DNA double helix unzip in the vicinity of this site in order for the polymerase protein to be able to sit down properly. In many genes, this unzipping cannot take place without the assistance of a regulatory protein called an **activator** that binds to the DNA in this region and helps it unwind. Just as in the case of the repressor protein described previously, cells can turn genes on and off by binding "signal" molecules to the activator protein. These molecules prevent it from binding to the DNA or enable it to do so.

Activators enable a cell to carry out a second level of control. When a bacterium encounters the sugar lactose, it may not be low on energy. Imagine if you had to eat every time you encountered food! When a bacterial cell already has lots of energy, levels of a special "I'm hungry" signal molecule fall. Without being prodded by this signal molecule, the activator protein cannot twist into the proper shape to fit the DNA unwinding site in front of the lactose-using genes; as a result, the genes are not transcribed, even though the repressor protein does not block the polymerase.

Enhancers

A third level of control is exercised by expanding access to the gene. To make the promoters of complexly controlled genes accessible to many regulatory proteins simultaneously, many eukaryotic genes possess special associated sequences called **enhancers.** These enhancer sequences bind specific "signal" proteins that interact with the protein transcription factors that help RNA polymerase find and attach to its binding site at the beginning of the structural gene (figure 7.19). Unlike promoters and activators, which butt right up to the start of a gene, enhancers are usually located far away from the start of the gene, often thousands of nucleotides distant.

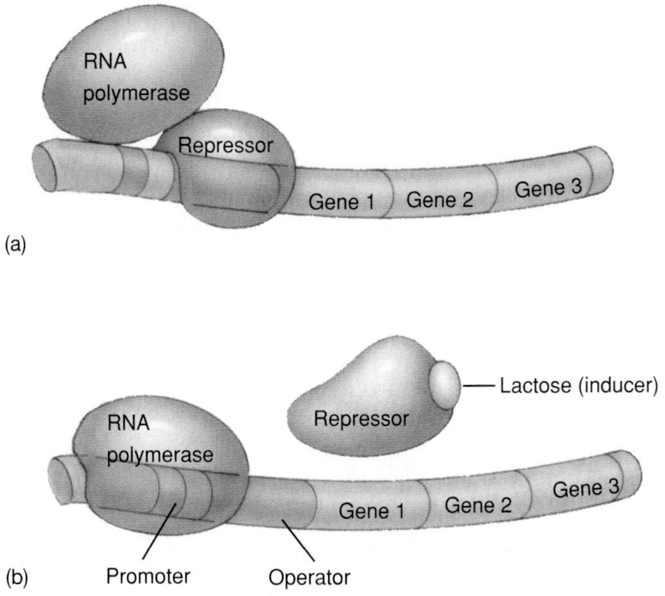

Figure 7.18 How the *lac* operon works.
(*a*) The *lac* operon is shut down ("repressed") when the repressor protein is bound to the operator site. Because promoter and operator sites overlap, polymerase and repressor cannot bind at the same time, any more than two people can sit in one chair. (*b*) The *lac* operon is transcribed ("induced") when lactose binding to the repressor protein changes its shape so that it can no longer sit on the operator site and block polymerase binding.

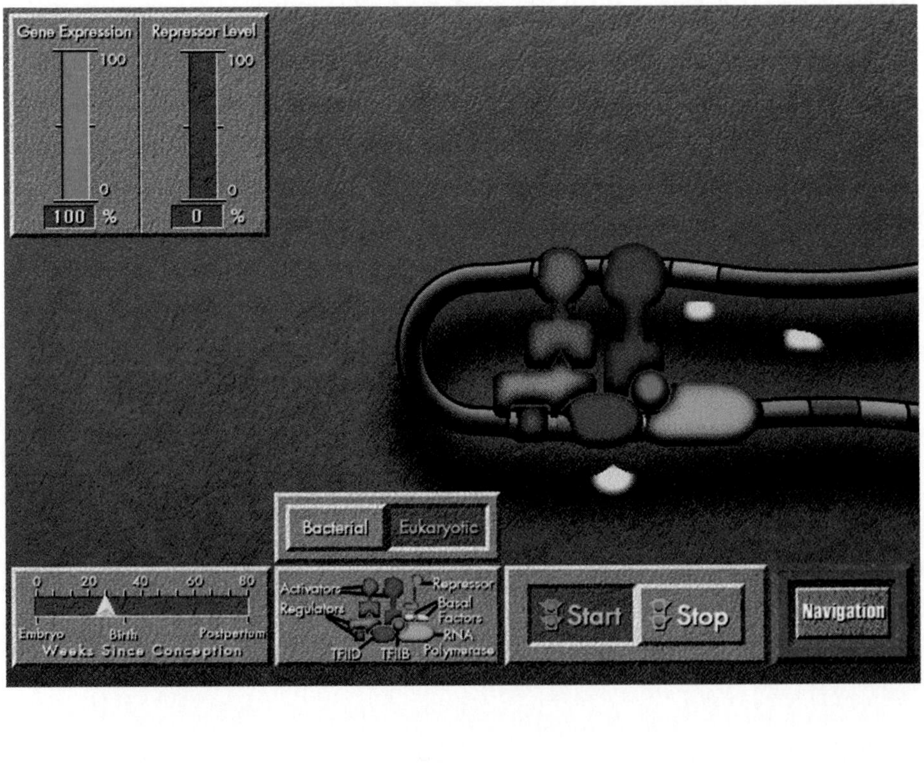

Figure 7.19 Gene regulation in eukaryotes.
The bunched group of proteins in this diagram is a transcription complex aiding the polymerase to transcribe the globin genes. Interacting with the complex are two proteins bound to enhancer sequences. This illustration is a "screen capture" from an interactive CD-ROM exercise that allows you to compare the regulatory strategies employed by bacteria (the *lac* operon) with those employed by eukaryotes (the globin genes). (*Explorations in Cell Biology & Genetics,* Module 16, "Gene Regulation")

Architecture of the Gene

Because of the many levels of control that the cell exercises over transcription, most genes have the same structure that a textbook does, with lots of "front matter" (preface, introduction, table of contents) placed before the actual chapters. A typical gene might have the following layout:

enhancer—//—activator—promoter—repressor—gene

Some genes are more complex than this, with even more regulatory sites nested in the zone in front of the gene, but the general principle is clear: Cells regulate gene transcription by binding proteins to the DNA that enhance or inhibit RNA polymerase.

Introns

While it is tempting to think of a gene as simply the nucleotide version of a protein—an uninterrupted stretch of nucleotides that is read three at a time to make a corresponding chain of amino acids—this simple way of doing things actually occurs only in bacteria. In all organisms that evolved later (that is, the eukaryotes), genes are fragmented. In these more complex genes, the nucleotide sequences encoding the amino acid sequence of the protein (called **exons**) are interrupted frequently by extraneous "extra stuff" called **introns.** In most human genes, the intron material far outweighs the protein-encoding exons. Usually less than 10% of a human gene is exons, and all the rest are noncoding introns. Like cars on a rural highway, exons are scattered here and there within genes, in the correct order but not near one another.

How does the cell manage to make a protein from such a mess? After the mRNA has been copied from the DNA, special enzymes attack it and chop out all the introns! The exons are then stitched together to make the mRNA that would have been transcribed if the introns had never existed, and it is this "processed" mRNA that the ribosomes translate into protein (figure 7.20).

Why this crazy organization? It appears that each exon of a gene encodes a different functional part of the protein. One exon may influence which molecules an enzyme is able to recognize and bind, while a different exon may determine what chemical reaction then takes place. Yet another exon may specify which signal molecules the protein will respond to. Biologists believe that evolution has favored this intron/exon mode of gene organization because it has permitted cells to shuffle exons between genes. This sort of shuffling is a very powerful way to create "new" genes—in one step producing a new kind of functional protein, instead of having to invent each new protein from scratch. Like building models with Tinkertoys or Legos, many different enzymes can be created. Imagine the number of outfits two friends can create by exchanging sweaters, pants, belts, shoes, and hats. The many millions of proteins that occur in human cells appear to have arisen from only a few thousand exons!

Gene Families

Introns are not the only surprise awaiting you in studying eukaryotic genes. Everything we have said in this chapter assumes that chromosomes each carry one copy of each kind of gene—one to make the enzyme that breaks down lactose, for example, and one to encode the protein hemoglobin, which carries oxygen in your blood from lungs to tissues. In fact, this is far from true. Most genes in your cells exist in multiple copies, clusters of almost identical sequences called **multigene families.** Multigene families may contain as few as three or as many as several hundred versions of a gene.

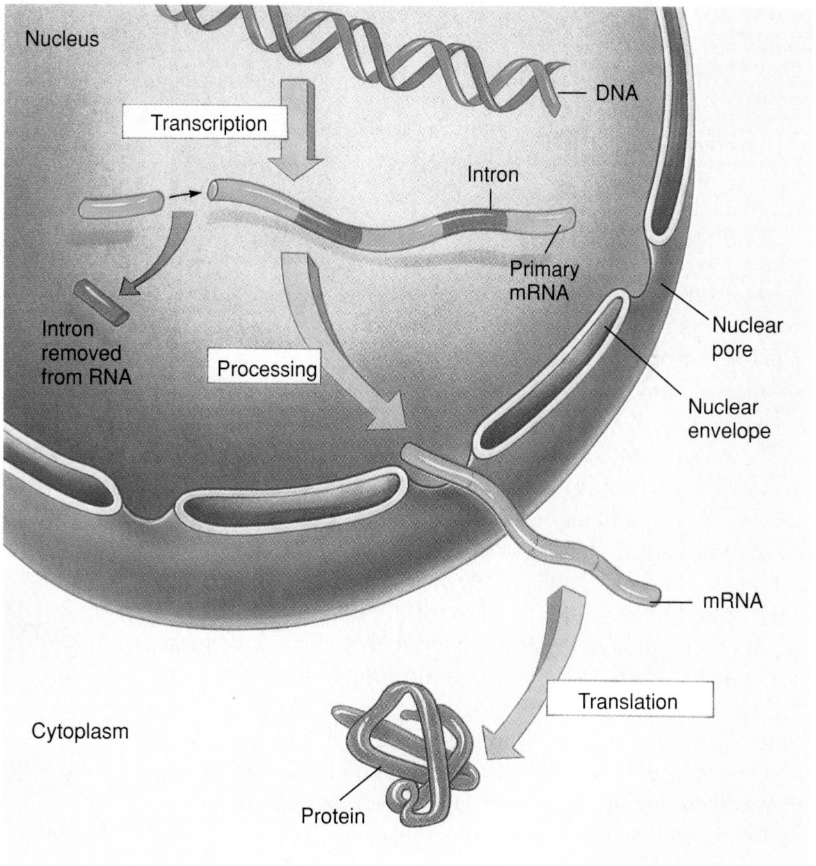

Figure 7.20 Gene processing.
When transcribing DNA into mRNA, the extraneous genetic material (intron) is removed from the mRNA sequence by enzymes before mRNA leaves the nucleus. The edited version of mRNA then enters the cytoplasm, where it participates in translation and protein synthesis.

Transposons: Jumping Genes

Other genes are very unusual in that they are repeated hundreds of thousands of times, scattered randomly about on the chromosomes. These segments, called **transposons,** have the remarkable ability to move about from one chromosomal location to another. Once every few thousand cell divisions, a transposon simply picks up and moves elsewhere, jumping at random to a new location on the chromosome. Transposons appear to be molecular parasites. They are actively transcribed by RNA polymerase but play no functional role in the life of the cell that bears them.

Figure 7.21 Mutation.
Fruit flies normally have one pair of wings, extending from the thorax. This fly is a *bithorax* mutant. Because of a mutation in a gene regulating a critical stage of development, it possesses *two* thorax segments and thus two sets of wings.

Mutation

A change in a cell's genetic message is called a **mutation.** Some mutational changes affect the message itself, altering the sequence of DNA nucleotides. These alterations in the coding sequence are called **point mutations** because they usually involve only one or a few nucleotides. Other classes of mutation involve changes in how the genetic message is organized and are often caused by transposons or jumping genes. When a particular gene is mutated, its function is often destroyed. Thus mutations in genes regulating development can lead to unusual changes in the adult body (figure 7.21).

Point mutations, involving one or a few nucleotides, result from either chemical or physical damage to the DNA or spontaneous pairing errors that occur during DNA replication. The first of these is of particular practical importance because modern industrial societies produce and release into the environment many chemicals capable of damaging DNA, chemicals called **mutagens.** Among the most tragic of these are the tars and other mutagens released into the lungs by the smoking of cigarettes. The mutagens from cigarettes destroy the genes that restrain cell division and lead to cancer in the tissues exposed to the smoke—the lungs. Research has firmly established that cigarette smoking leads to cancer. As you can see in figure 7.22, the more cigarettes you smoke, the greater the likelihood you will develop lung cancer.

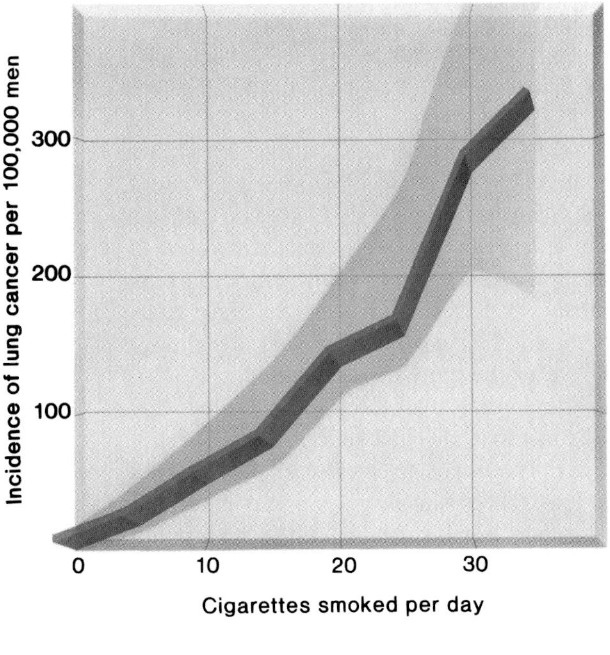

Figure 7.22 Smoking causes cancer.
The annual incidence of lung cancer is clearly related to the number of cigarettes smoked per day. Of every 100 American college students who smoke cigarettes regularly, 1 will be murdered, about 2 will be killed on the roads, and about 25 will be killed by tobacco.

CHAPTER 7

HIGHLIGHTS

7.1 Genes Are Made of DNA

Key Terms

- transformation 138
- nucleotides 141
- double helix 141
- base pairing 141
- complementarity 142

Key Concepts

- A series of critical experiments demonstrated that genes are composed of DNA.
- A key advance was the demonstration by Avery that the active ingredient in Griffith's transforming extract was DNA.
- DNA is a double helix, in which A on one strand pairs with T on the other, and G similarly pairs with C.

7.2 Transcription

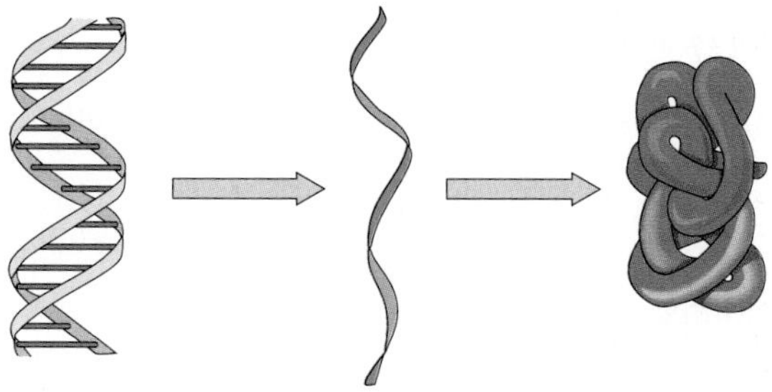

Key Terms

- transcription 143
- mRNA 143
- RNA polymerase 143
- genetic code 144
- codon 144

Key Concepts

- DNA serves as a template on which mRNA is assembled in a process called transcription.
- The genetic information is encoded in DNA in three-nucleotide units called codons.
- The genetic code specifies which codons correspond to each amino acid.
- The beginning of a gene, where the RNA polymerase binds the DNA, is the promoter.

7.3 Translation

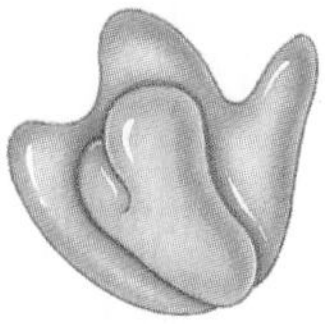

Key Terms

- translation 145
- ribosome 145
- tRNA 145
- anticodon 145

Key Concepts

- mRNA serves as the template guiding assembly of proteins.
- A ribosome moves along the mRNA, adding an amino acid to the end of a growing protein chain as it passes each codon.

7.4 Regulating Gene Expression

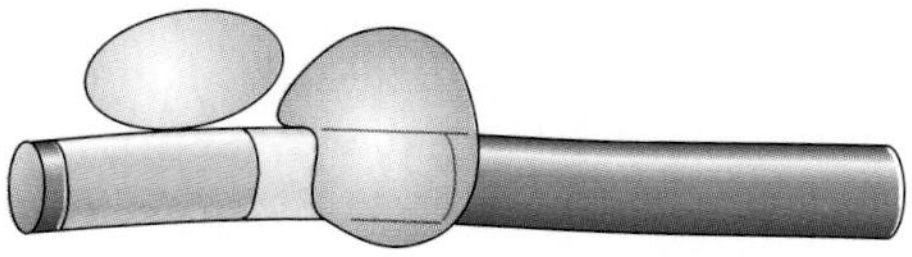

Key Terms

- promoter 147
- repressor 147
- operon 147
- enhancer 148
- intron 149
- multigene family 149
- mutation 150

Key Concepts

- Most genes are regulated by controlling the rate at which they are transcribed.
- Bacteria shut off genes by attaching a repressor protein to or near the promoter, so the polymerase is unable to bind.
- Eukaryotes can use many regulatory proteins simultaneously.

7

CONCEPT REVIEW

1. The Hershey-Chase experiment demonstrated that
   a. virus DNA injected into bacterial cells is apparently the factor involved in directing the production of new virus particles.
   b. RNA is the genetic material of some viruses.
   c. ^{32}P-labeled protein is injected into bacterial cells by viruses.
   d. the transforming principle is the DNA, not the polysaccharide coat.

2. According to Chargaff's rule and the Watson-Crick model of DNA, if you know the base sequence of one strand is AATTCG, then the sequence of the complementary strand must be
   a. AATTCG.
   b. TTGGAC.
   c. TTAACG.
   d. TTAAGC.

3. The sequence of DNA nucleotides that encodes the amino acid sequence of an enzyme is a
   a. gene.
   b. polypeptide.
   c. protein.
   d. phenotype.

4. DNA makes RNA by
   a. translation.
   b. replication.
   c. transcription.
   d. repair.

5. RNA makes proteins by
   a. translation.
   b. replication.
   c. transcription.
   d. repair.

6. The codon is found on transfer RNA.
   a. true
   b. false

7. Which amino acid is specified by the codon CUC? (Consult table 7.1.)
   a. lysine
   b. alanine
   c. proline
   d. leucine

8. In the *lac* region of the bacterium *Escherichia coli's* chromosome, the repressor protein binds at the
   a. activator protein site.
   b. promoter site.
   c. operator site.
   d. transcription unit.

9. Transposons can jump from one chromosome to another.
   a. true
   b. false

10. The transforming principle in the Griffith-Avery experiment proved to be ________.

11. DNA has the shape of a ________.

12. Erwin Chargaff observed that the amount of adenine in all DNA molecules is equal to the amount of ________ and that the amount of guanine is equal to the amount of ________.

13. The ________ is the cell site of protein synthesis.

14. ________ is the molecule that carries amino acids during translation.

15. The genetic code consists of ________ triplet codons.

16. Special sequences of eukaryotic genes called ________ make the promoters of complexly controlled genes accessible to many regulatory proteins simultaneously.

17. A regulatory protein responsible for blockage of DNA transcription is called a ________.

18. Exons alternate with ________ on chromosomes.

19. A chemical capable of damaging DNA is called a ________.

Answers to the Concept Review questions appear in Appendix B.

CHALLENGE YOURSELF

1. From an extract of human cells growing in tissue culture, you obtain a white, fibrous substance. How would you distinguish whether it was DNA, RNA, or protein?
2. Why is hydrogen bonding important to the structure of DNA?
3. Why must the DNA molecule become unzipped in order to function?
4. If every cell in an organism contains a full complement of the genes in that organism, and if each gene codes for an enzyme or other protein, why don't all cells in an organism synthesize the same proteins and therefore have the same structure and function?
5. The nucleotide sequence of a hypothetical eukaryotic gene is shown below:

   TAC ATA CTA GTT AC<u>G</u> TCG CCC GGA AAT ATC

   If a mutation in this gene changed the fifteenth nucleotide (underlined) from guanine to thymine, what effect do you think it would have on the expression of this gene?

FOR FURTHER READING

Cohen, L. "Diet and Cancer." *Scientific American,* November 1987, 42–48. Diet appears to play an important role in cancer, particularly breast cancer, which is associated with high levels of dietary fat.

Crick, F. H. C. "The Discovery of the Double Helix Was a Matter of Selecting the Right Problem and Sticking to It." *Chronicle of Higher Education* 35 (October 5, 1988): B1–2. Francis Crick's own recollections of the hectic days when he and James Watson deduced that the structure of DNA is a double helix.

Doolittle, W. F. "The Origin and Function of Intervening Sequences in DNA—A Review." *American Naturalist* 130 (December 1987):915–28. A thoughtful and comprehensive review of the many ideas concerning why introns exist and what they do.

Nowak, R. "Mining Treasures from 'Junk DNA'." *Science* 263 (February 4, 1994): 608–10. If "junk DNA" really is junk, why do we have so much of it?

Steitz, J. "Snurps." *Scientific American,* June 1988, 56–63. Small nuclear ribonuclear proteins (snurps) help remove introns from mRNA transcripts. They were discovered less than 10 years ago and are now an active subject of research.

Tjian, R. "Molecular Machines That Control Genes." *Scientific American*, February 1995, 54–61. Elaborate protein complexes that assemble on DNA control the activities of our genes.

Todorov, I. "How Cells Maintain Stability." *Scientific American,* December 1990, 66–75. How cells shut down their protein-making factories in times of stress and start them back up again.

TECHNOLOGY LINKS

The Living World Home Page
http://www.wcbp.com/biology/tlw

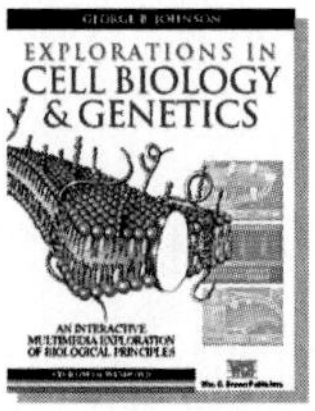

Explorations in Cell Biology & Genetics CD
#15 Reading DNA
#16 Gene Regulation

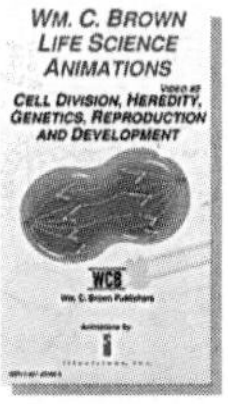

Life Science Animations Videotape 2
#15 DNA Replication
#16 Transcription of a Gene
#17 Protein Synthesis

8

Gene Technology

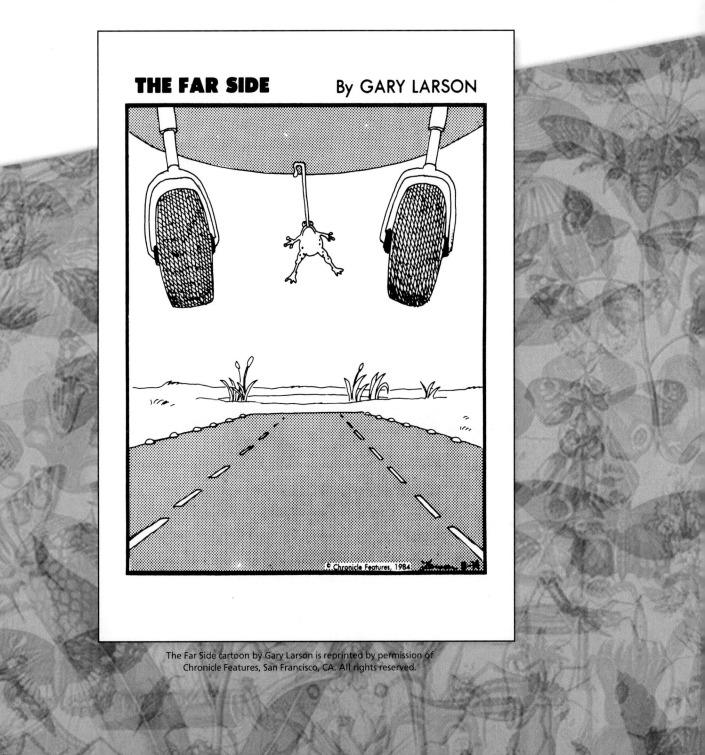

CHAPTER OUTLINE

Figure 8.1 A gene transportation system.
Genetic engineers often insert eukaryotic genes into small circular segments of bacterial DNA called plasmids. The plasmid shown in this electron micrograph was used in the first successful transfer of an animal gene into a bacterium.

Chimera is a mythical creature with the head of a lion, the body of a goat, and the tail of a serpent. No chimera ever existed in nature. But human beings have made them—not the lion-goat-snake variety, but chimeras of much greater importance for the future.

The first "chimera" was created in the San Francisco laboratory of geneticists Stanley Cohen and Herbert Boyer in 1973. They isolated from African clawed toads the gene that encodes toad ribosomal RNA (rRNA) and then stitched that gene into the plasmid shown in figure 8.1. They then transferred the plasmid into *E. coli* bacteria, which began to busily churn out toad rRNA within their cells. No such creature, part toad and part bacterium, had ever existed on earth before.

This experiment ushered in a new age in biology. In the less than 20 years since, the ability to *transfer* DNA from one creature to another has caused a revolution that today affects not only the science of biology but every aspect of society.

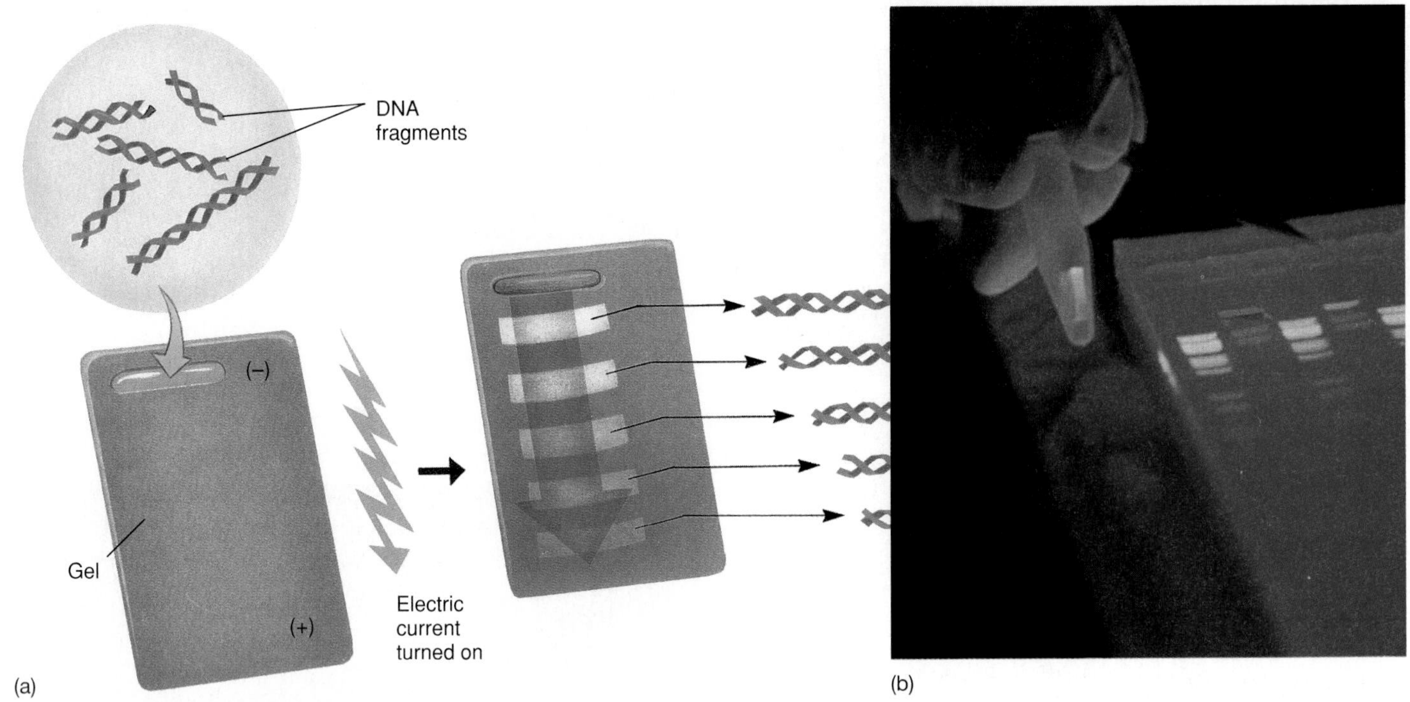

Figure 8.2 Using gel electrophoresis to "see" genes.
Gel electrophoresis allows investigators to separate a complex mixture of DNA fragments according to how long the individual fragments are. (*a*) When placed on one side of a gel, the fragments will migrate different distances when an electric current is applied. The distance a fragment moves relates to its size, larger fragments moving slower and thus traveling shorter distances during the period of time the current is applied. (*b*) The fragments show up as separate bands when viewed under fluorescent light.

8.1 Genetic Engineering

Moving genes from one organism to another is often called **genetic engineering.** Many of the gene transfers have placed eukaryotic genes into bacteria. The bacterial cells are converted by the transfer into tiny factories that produce prodigious amounts of the protein encoded by the eukaryotic gene. Other gene transfers have moved genes from one animal or plant to another. Scientists carrying out these gene transfers use a technique called **gel electrophoresis** to visualize the gene fragments. In figure 8.2, a collection of DNA fragments has been separated by electric current into easily visualized bands.

Genetic engineering is having a major impact on medicine. Most of the insulin used to treat diabetes is now obtained from bacteria that contain a human insulin gene. In late 1990, the first transfers of genes from one human to another were carried out in attempts to correct the effects of defective genes in a rare genetic disorder called "severe combined immune deficiency" (figure 8.3).

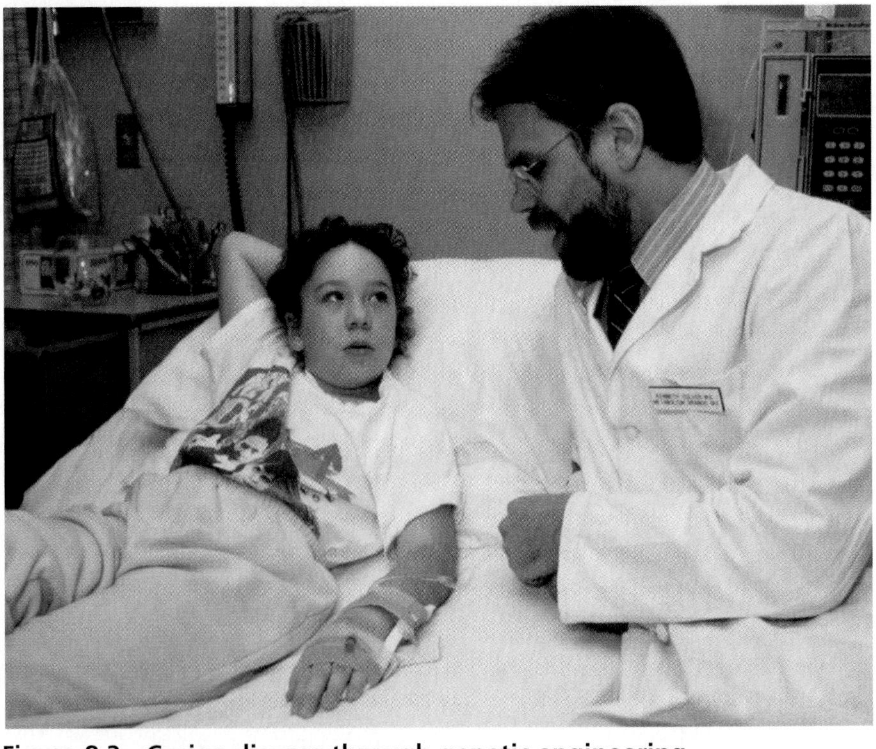

Figure 8.3 Curing disease through genetic engineering.
One of two young girls who were the first humans "cured" of a hereditary disorder by transferring into their bodies healthy versions of the gene they lacked. The transfer was successfully carried out in 1990, and the girls remain healthy.

Performing Genetic Surgery

The first stage in any genetic engineering experiment is to chop up the "source" DNA in order to get a copy of the gene you wish to transfer. This first stage is the key to successful transfer of the gene, and learning how to do it is what has led to the genetic revolution. The trick is in how the DNA molecules are cut. The cutting must be done in such a way that the resulting DNA fragments have "sticky ends" that can later be inserted into another chromosome.

This special form of molecular surgery is carried out by **restriction enzymes,** which are special enzymes that bind to specific short sequences (typically four to six nucleotides long) on the DNA. These sequences are very unusual in that they are symmetrical—the two strands of the DNA duplex have the same nucleotide sequence, running in opposite directions! One of these sequences, for example, is CTTAAG. Try writing down the sequence of the opposite strand: it is GAATTC—the same sequence, written backwards.

What makes the DNA fragments "sticky" is that most restriction enzymes do not make their incision in the center of the sequence; rather, the cut is made to one side. In the sequence written above, for example, the cut is made after the first nucleotide, C/TTAAG. This produces a break, with short single strands of DNA dangling from each end. Because the two single-stranded ends are complementary in sequence, they could pair up and heal the break, with the aid of a sealing enzyme—*or* they could pair with *any other DNA fragment cut by the same enzyme,* because all would have the same single-stranded sticky ends (figure 8.4)! Any gene in any organism cut by the enzyme that attacks CTTAAG sequences can be joined to any other with the aid of a sealing enzyme called a **ligase,** which re-forms the bonds between the sugars and phosphates of DNA. Fragments of mouse and human DNA cleaved by the same enzyme can be joined to one another just as easily as can two bacterial fragments (figure 8.5). This is because they have the same complementary sequences at their ends, and only the ends are involved in rejoining.

Hundreds of different restriction enzymes are known, recognizing a wide variety of four to six nucleotide sequences. Each kind of enzyme attacks only one sequence and always cuts at the same place. By trying one after another, biologists can almost always find a sequence that cuts out the gene they seek. The short sequence the enzyme attacks occurs by chance on both ends of the gene and not within it. The restriction enzymes are the basic tools of genetic engineering.

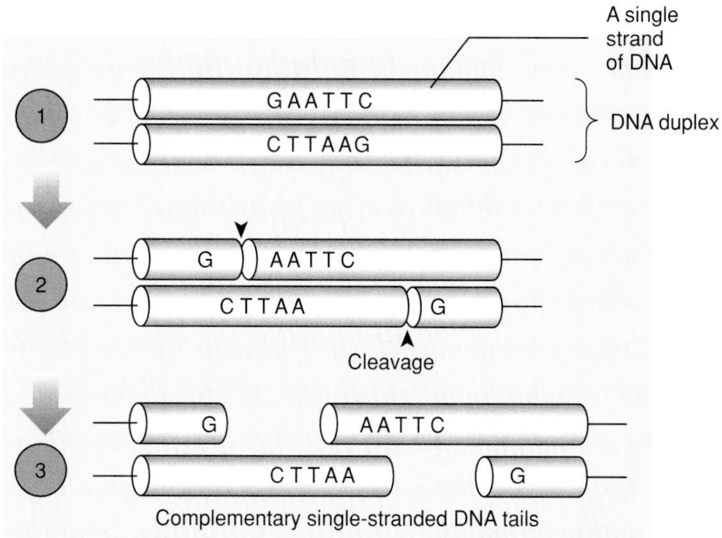

Figure 8.4 How restriction enzymes produce DNA fragments with sticky ends.

Fragments cut by the same enzyme have the same single-stranded ends. (*1*) The same recognition sequence occurs on both strands of the DNA duplex. (*2*) One restriction enzyme, called *Eco*RI, always cleaves sequence GAATTC at the same spot, after the first G. Because the same sequence occurs on both strands, both are cut. But the position of the G is not the same on the two strands; the sequence runs the opposite way on the other strand. As a result, single-stranded tails are produced. (*3*) Because of the symmetry of the sequence, the single-stranded tails are complementary to each other, or "sticky."

Figure 8.5 Larger than life—a hybrid mouse produced by genetic engineering.

These two mice are genetically identical, except that the large one has one extra gene, encoding a potent human growth hormone not normally present in mice. The gene was added to a mouse chromosome using sticky ends similar to those illustrated in figure 8.4 and is now a stable part of the mouse' s genetic endowment.

Courtesy Ralph L. Brinster, University of Pennsylvania, School of Veterinary Medicine.

Transferring the Gene

Every gene transfer presents unique problems, but all share four distinct stages:

1. Cleaving DNA—cutting up the source chromosome.
2. Recombining DNA—placing the DNA fragments into vehicles that can infect the target cells.
3. Cloning—infecting target cells with DNA-bearing vehicles and allowing infected cells to reproduce.
4. Screening—selecting the particular infected cells that have received the gene of interest.

Perhaps the best way to see how a gene transfer experiment works is to follow one through from beginning to end. Let's go back and watch as Cohen and Boyer transfer a toad gene into bacterial cells (figure 8.6):

Stage 1: Cleaving DNA

Step 1. Cohen and Boyer first isolated chromosomal DNA from an adult African clawed toad, *Xenopus laevis,* and treated this DNA with a DNA-cutting enzyme called *Eco*RI. This is the enzyme that attacks the CTTAAG sequence described earlier. The result was a large number of DNA fragments, one of which was the rRNA gene they sought to transfer. Every fragment has the same single-stranded sequences dangling from its two ends. The fragments are usually separated by **electrophoresis,** which permits their relative size to be estimated, as illustrated in figure 8.2.

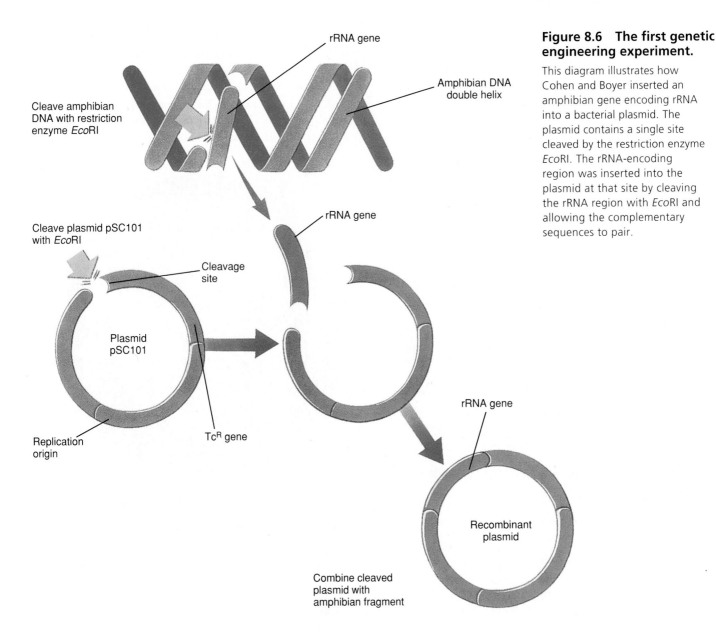

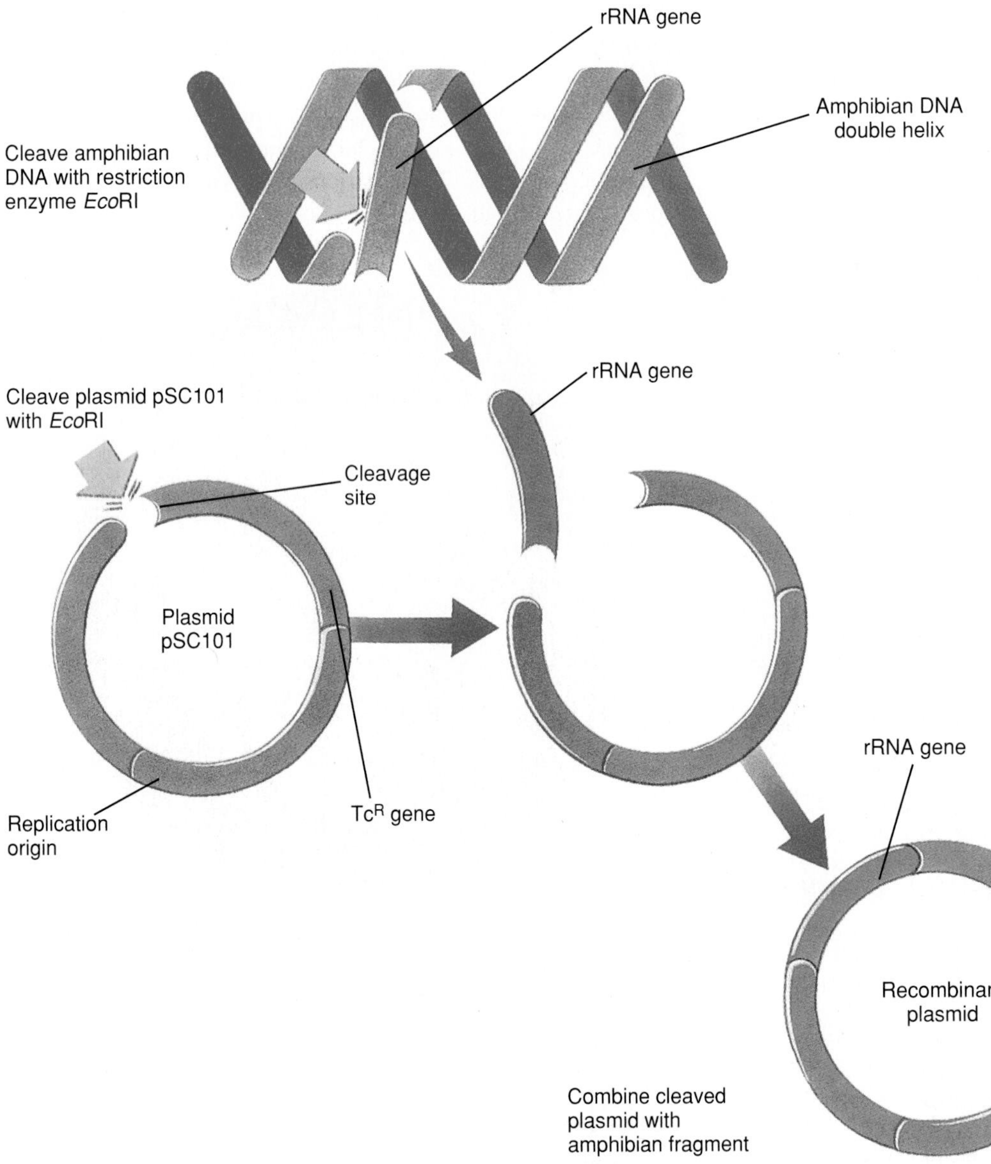

Figure 8.6 The first genetic engineering experiment.

This diagram illustrates how Cohen and Boyer inserted an amphibian gene encoding rRNA into a bacterial plasmid. The plasmid contains a single site cleaved by the restriction enzyme *Eco*RI. The rRNA-encoding region was inserted into the plasmid at that site by cleaving the rRNA region with *Eco*RI and allowing the complementary sequences to pair.

Step 2. The investigators then cut up the DNA of the circular bacterial chromosome with the same enzyme and isolated a fragment that possessed two critical genes: the gene that replicates the DNA and a gene making the cell resistant to the antibiotic tetracycline (a chemical that can kill bacteria). Because there was no CTTAAG sequence within the fragment, its two ends were able to loop around and stick to each other, forming a tiny circle of DNA called a **plasmid,** which is able to replicate outside of the main bacterial chromosome. This plasmid was to be their vehicle, the carrier of the toad rRNA gene.

Stage 2: Recombining DNA

The recombinant DNA was produced by simply mixing together the DNA fragments and plasmids made in stage 1. It is as easy for the ends of the plasmid to stick to the ends of the toad DNA fragments as to each other, because they all have the same short sticky tails.

Stage 3: Cloning

Step 1. The plasmid-toad fragment mixture was added to a growing culture of bacteria, under conditions that encourage bacteria to take up plasmids into their cells.

Step 2. The investigators then added the antibiotic tetracycline to the growing culture. The only cells not killed by the antibiotic were those that had become resistant to tetracycline because they had taken up the plasmid. Thus, this step eliminated all bacterial cells without plasmids. Every cell that survived this step had been infected by one of the plasmids generated by stage 2. Each cell was then isolated and allowed to reproduce, forming a colony of millions of identical cells all containing the plasmid. Thousands of these colonies were obtained; together, they constituted a **clone library.**

Stage 4: Screening

The screening step was the most laborious part of the experiment. Each clone was tested to see if the rRNA gene was present by attempting to pair its DNA with some pure toad rRNA. The toad rRNA is called a **probe** because it is used to detect the presence of genes of similar sequence. Only cells carrying the toad rRNA gene would contain DNA that could form complementary base pairs with the toad rRNA probe (figure 8.7), and so stick to it. Patient searching eventually uncovered a clone whose cells contained DNA that stuck to the toad rRNA probe. These are the cells we seek—bacterial cells that carry the toad gene.

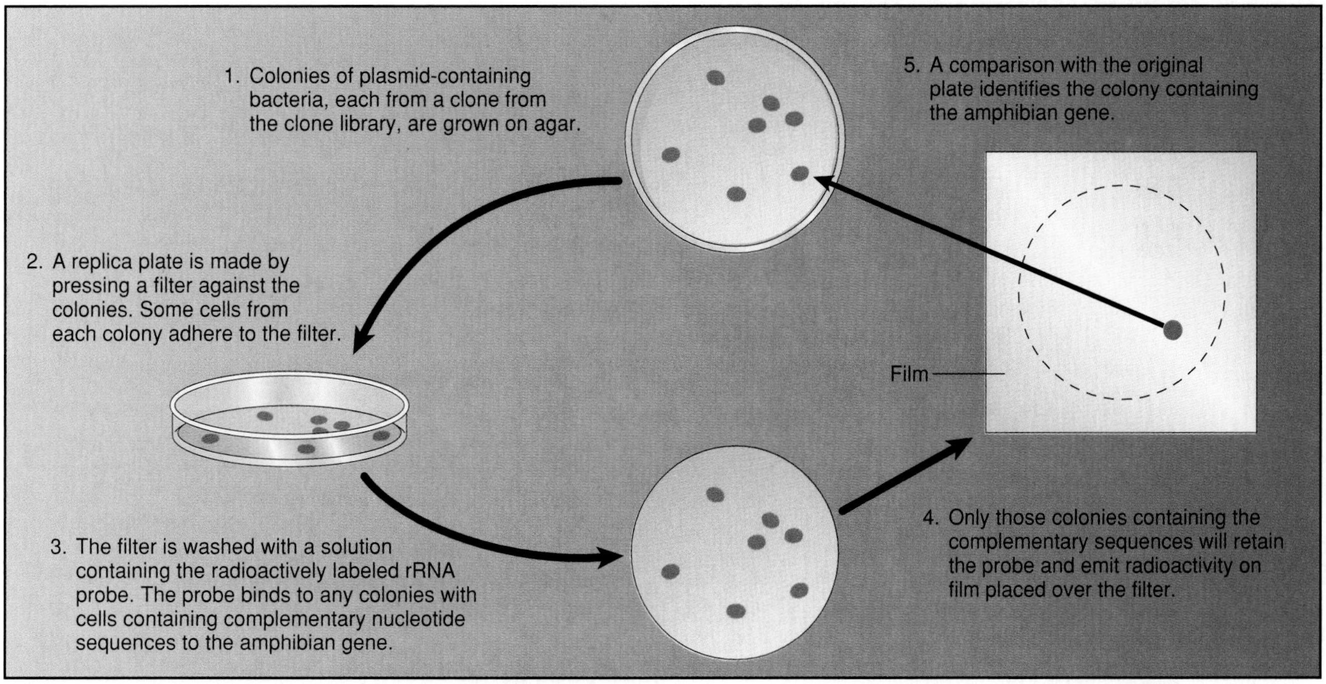

1. Colonies of plasmid-containing bacteria, each from a clone from the clone library, are grown on agar.
2. A replica plate is made by pressing a filter against the colonies. Some cells from each colony adhere to the filter.
3. The filter is washed with a solution containing the radioactively labeled rRNA probe. The probe binds to any colonies with cells containing complementary nucleotide sequences to the amphibian gene.
4. Only those colonies containing the complementary sequences will retain the probe and emit radioactivity on film placed over the filter.
5. A comparison with the original plate identifies the colony containing the amphibian gene.
Film

Figure 8.7 Screening a clone library for rRNA.

Each of the colonies on these bacterial culture plates represents millions of cells descended from a single cell—a **clone.** To test whether the amphibian rRNA gene was present in any particular clone, Cohen and Boyer looked for colonies whose cells contained DNA with a sequence complementary to amphibian rRNA. Pressing the plate against a filter caused some cells from each colony to adhere to the filter. The filter was then washed with a solution containing radioactively labeled amphibian rRNA. Only those colonies containing the amphibian rRNA gene would retain the label.

PCR Amplification

Often, instead of inserting a gene into bacteria in order to generate many copies, and so produce large amounts of the protein the gene encodes, researchers use a technique called **PCR,** or the **polymerase chain reaction** (figure 8.8). To carry out this gene-multiplying procedure, researchers first identify and synthesize short, single-stranded sequences of nucleotides called **primers,** which occur on either side of the DNA region to be amplified. A solution containing DNA, primers, free nucleotides, and an enzyme called DNA **polymerase** is then heated to about 98°C. This high temperature disrupts the hydrogen bonds holding the DNA duplex together, causing it to unravel into single strands. As the solution cools, the primers bind to their complementary sequence near the region of interest on the single strands of DNA. Using the primer as a starting point, the DNA polymerase then proceeds down the strand of DNA, adding nucleotides. When it is done, what used to be the primer is now lengthened into a complementary copy of the entire single-stranded fragment. Because both strands behave this way, there are now two copies of the original fragment. This process is repeated many times, resulting in the desired amplification of the DNA region of interest.

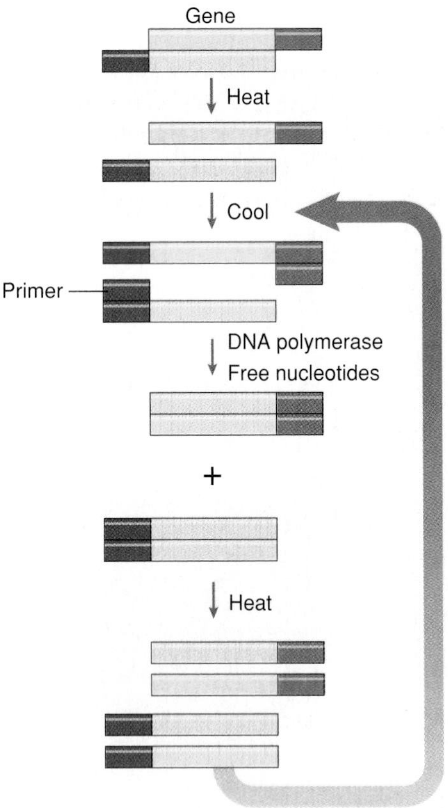

Figure 8.8 The polymerase chain reaction (PCR).
PCR has revolutionized many aspects of biology, because it allows the investigation of minute samples of DNA. In criminal investigations, DNA "fingerprints" can be prepared from cells in a tiny speck of dried blood or at the base of a single human hair.

Formation of cDNA

Virtually every nucleotide within the transcribed portion of a bacterial gene participates in a code specifying an amino acid. However, as you will recall from chapter 7, eukaryotic genes are encoded in segments called *exons* separated from one another by numerous nontranslated sequences called *introns.* The entire gene is transcribed by RNA polymerase, producing what is called the **primary transcript.** Before a eukaryotic gene can be translated, the introns must be cut out of this primary transcript. The fragments that remain (the exons) are then stitched together to form the **processed mRNA,** which is eventually translated in the cytoplasm. When transferring eukaryotic genes into bacteria, it is desirable to transfer DNA that has already been processed this way, instead of the raw eukaryotic DNA, because bacteria lack the enzymes to carry out the processing. To do this, genetic engineers first isolate from the cytoplasm the processed mRNA corresponding to a particular gene (figure 8.9). The cytoplasmic mRNA has *only* exons, properly stitched together. An enzyme called reverse transcriptase is then used to make a DNA version of the mRNA that has been isolated. Such a version of a gene is called copy DNA or **cDNA.** The single strand of cDNA can then be used as a template for the synthesis of a complementary strand. In this way, a gene lacking introns is synthesized that can be used by bacteria to produce proteins.

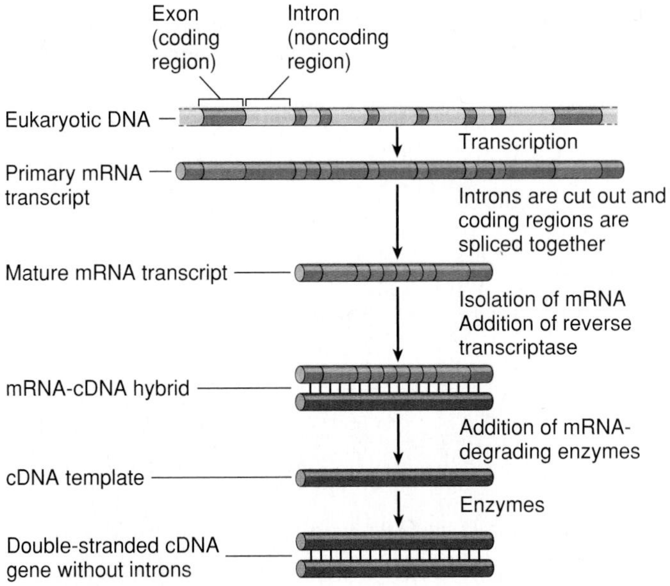

Figure 8.9 Producing an intron-free version of a eukaryotic gene.
The primary transcript of the gene contains numerous introns that are cut out during processing, producing the processed mRNA found in the cytoplasm. The enzyme reverse transcriptase is then used to make a DNA strand complementary to the processed mRNA. Finally, that newly made strand of DNA is used as a template for the enzyme DNA polymerase, which assembles a complementary DNA strand along it, producing a double-stranded DNA version of the gene that is free of introns.

8.2 Transforming Agriculture

One of the greatest impacts of genetic engineering on society has been the successful manipulation of the genes of crop plants to make them more resistant to disease, frost, and other forms of stress; to improve their nutritional balance and protein content; and to make them resistant to herbicides (chemicals that kill plants).

The T_i Plasmid

The key to the great progress that has occurred in plant genetic engineering in recent years was the discovery of a suitable way to transport DNA from one plant to another. For years, genetic engineering in plants was not possible because, unlike bacteria, plants have few viruses that can carry out this critical role. Said simply, genetic engineers lacked a suitable vehicle to carry out stage 3 of the gene transfer process.

The breakthrough came with the use of an unusual bacterial plasmid as a vehicle, a plasmid responsible for a kind of plant cancer that causes large bulbous tumors called crown galls. Named the **T_i plasmid,** this plasmid easily infects broadleaf crop plants such as tomatoes, tobacco, and soybeans, and when it has infected a plant cell, it proceeds to stitch itself into the plant cell's chromosome. To make a genetic engineering vehicle, scientists removed the tumor-causing genes from the T_i plasmid. The vacant space in the now-harmless plasmid could then be filled by DNA introduced by the investigator, and this DNA then carried by the T_i plasmid vehicle into the chromosomes of a target plant.

Unfortunately, bacteria carrying the T_i plasmid do not infect the cereal plants such as corn, rice, and wheat, all of which play very important roles in feeding an increasingly hungry world. Researchers have developed alternative methods to introduce new genes into these plants. In 1991 they succeeded in "shooting" genes into wheat by coating microscopic pellets with gene-bearing DNA (figure 8.10)!

Figure 8.10 Shooting genes into cells.
This DNA particle gun fires tungsten pellets coated with DNA into plant cells such as the ones on the culture plate this experimenter is holding.

Herbicide Resistance

A major recent advance has been the creation of crop plants that are resistant to the herbicide **glyphosate,** the active ingredient in Roundup, a powerful biodegradable herbicide. Roundup is used in orchards and agricultural fields to control weeds. Growing plants need to make a lot of protein, and glyphosate stops them from making protein by destroying an enzyme that manufactures a critical amino acid. Humans are unaffected by glyphosate because we don't make this amino acid anyway—we obtain it from plants we eat! To make crop plants resistant to this powerful plant-killer, genetic engineers screened thousands of organisms until they found a species of bacteria that could make the amino acid in the presence of glyphosate. They then isolated the gene encoding the resistant enzyme (figure 8.11). Using the T_i plasmid as a vehicle, they successfully introduced the resistance gene into noncereal crop plants and, more recently, "shot" it into wheat.

Agriculture is being revolutionized by herbicide-resistant crops for two reasons. First, it lowers the cost of producing a crop, because a crop resistant to Roundup does not need to be weeded—the farmer simply treats the field by spraying it with Roundup, and all growing plants die except the crop, which is resistant. Also, the creation of glyphosate-resistant crops is of major benefit to the environment. Glyphosate is quickly broken down in the environment, which makes its use a great improvement over most commercial herbicides. Perhaps even more important, much of the tragic loss of fertile topsoil to erosion, one of the greatest environmental challenges facing our country today, would not occur if crop land were not intensively plowed to remove weeds.

Figure 8.11 Genetically engineered herbicide resistance.
Cultivation of crops to remove weeds is not necessary if the weeds can be killed by herbicides that do not kill the crop. All four of these petunia plants were exposed to equal doses of a herbicide. The two on *top* were genetically engineered to be resistant to glyphosate, the active ingredient in the herbicide, whereas the two dead ones on the *bottom* were not.

Pest Resistance

Other important efforts of genetic engineers in agriculture have involved making crops resistant to insect pests without spraying with pesticides, a great saving to the environment. Consider cotton. Its fibers are a major source of raw material for clothing throughout the world, yet the plant itself can hardly survive in a field because of the many insects that attack it. Over 40% of the chemical insecticides used today are employed to kill insects that eat cotton plants. The world's environment would greatly benefit if these thousands of tons of insecticide were not needed. Biologists are now in the process of producing cotton plants that are resistant to attack by insects, so that insecticides will not be needed.

One successful approach uses a kind of soil bacterium that secretes enzymes that attack and kill the larvae (caterpillars) of butterflies and other insects, serious pests of some crops. When the genes producing these enzymes were inserted into the chromosomes of tomatoes, the plants began to manufacture the new enzyme, which made them highly toxic to tomato hornworms, one of the most serious pests on commercial tomato crops (figure 8.12). Any hornworms that took a bite of an altered tomato plant quickly died.

Many important plant pests attack roots. To combat these pests, genetic engineers are introducing the insect-killing enzyme isolated from soil bacteria into different kinds of bacteria, ones that colonize the roots of crop plants. Any insect eating such roots consume the bacteria and so are lethally attacked by the enzyme.

Nitrogen Fixation

All of the nitrogen plants use to make their proteins and DNA must be obtained by them from the soil. Where does this nitrogen in soil come from? Bacteria living within the roots of certain plants put it there. They "fix" nitrogen by converting nitrogen gas from the atmosphere into a form that plants can use. Engineering major crops to carry out **nitrogen fixation** for themselves has become the "Holy Grail" of genetic engineers, a difficult quest that many seek. The problem has not been discovering and isolating the necessary genes or getting them into crop plants. Instead, when the nitrogen-fixing genes from bacteria are introduced into plants, they do not seem to function properly in their new host. Researchers are working all over the world to find a way around this difficulty.

Farm Animals

A very important advance in agriculture has been the introduction of **bovine growth hormone (BGH)** into the diet of dairy cows, greatly improving milk production. Instead of extracting this hormone at great expense from the brains of dead cows, the gene has been introduced into bacteria, which produce the hormone so cheaply that it is practical to add it as a supplement to the cows' diet. So-called BGH milk is in fact no different from milk produced by any other dairy cow—the cow simply produces more of it. Milk from cows whose production had been enhanced by BGH went on sale for the first time to the general public in 1994, despite loud criticism from public-interest groups opposed to genetic engineering.

Extra copies of the gene encoding this growth hormone have also been introduced directly into the chromosomes of both cows and pigs to increase their weight. Still underway, these attempts promise to create new breeds of very large and fast-growing cattle and pigs (figure 8.13).

The human version of this same growth hormone is now being tested as a potential treatment for disorders such as dwarfism, in which the pituitary gland fails to make adequate amounts of growth hormones.

Figure 8.12 A successful experiment in pest resistance. Both of these tomato plants were exposed to destructive hornworms under laboratory conditions. The nonengineered plant on the *left* has been completely eaten, while the engineered plant shows virtually no signs of damage.

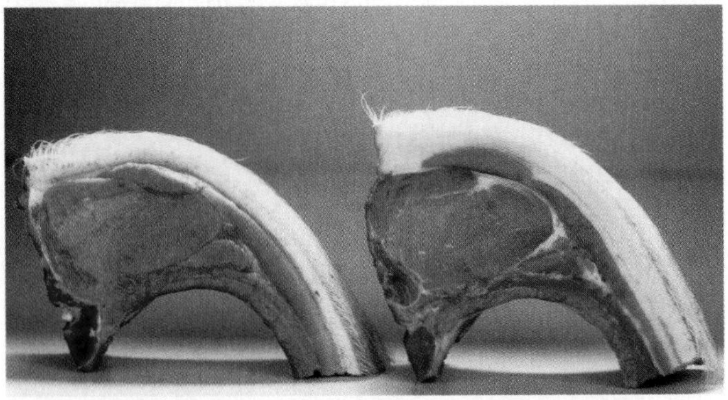

Figure 8.13 Making leaner bacon—genetically. *Left,* meat from a hog that was injected with bovine growth hormone genes. *Right,* meat from a normal hog. Both animals weighed approximately the same, but the specimen on the *left* showed significantly reduced fat content.

8.3 Advances in Medicine

Much of the excitement about genetic engineering has focused on its potential to improve medicine—to aid in curing and preventing illness. Major advances have been made in the production of proteins used to treat illness, in the creation of new vaccines to combat infections (figure 8.14), and in the replacement of defective genes.

Making "Magic Bullets"

Many genetic defects occur because our bodies fail to make critical proteins. *Diabetes* is such an illness. The body is unable to control levels of sugar in the blood because a critical protein, **insulin,** cannot be made. These failures can be overcome if the body can be supplied with the protein it lacks. The donated protein is in a very real sense a pharmaceutical drug, a "magic bullet" to combat the body's inability to regulate itself.

Until recently, the principal problem with using regulatory proteins as drugs was in manufacturing the protein. Proteins that regulate the body's functions are typically present in the body in very low amounts, and this makes them difficult and expensive to obtain in quantity. With genetic engineering techniques, the problem of obtaining large amounts of rare proteins has been largely overcome. The genes encoding the medically important proteins are now introduced into bacteria. Because the host bacteria can be grown cheaply in bulk, large amounts of the desired human protein can be easily isolated. In 1982, the U.S. Food and Drug Administration approved the use of human insulin produced from genetically engineered bacteria, the first commercial product of genetic engineering.

Today hundreds of pharmaceutical companies around the world are busy producing other medically important proteins using these genetic engineering techniques. These products include **anticoagulants** (proteins involved in dissolving blood clots), which are effective in treating heart attack patients, and **factor VIII,** a protein that promotes blood clotting. A deficiency in factor VIII leads to hemophilia, an inherited disorder discussed in chapter 6, which is characterized by prolonged bleeding. For a long time, hemophiliacs received blood factor VIII that had been isolated from donated blood. Unfortunately, some of the donated blood had been infected with viruses such as HIV and hepatitis B, which were then unknowingly transmitted to those people who received blood transfusions. Today the use of genetically engineered factor VIII eliminates the risks associated with blood products obtained from other individuals.

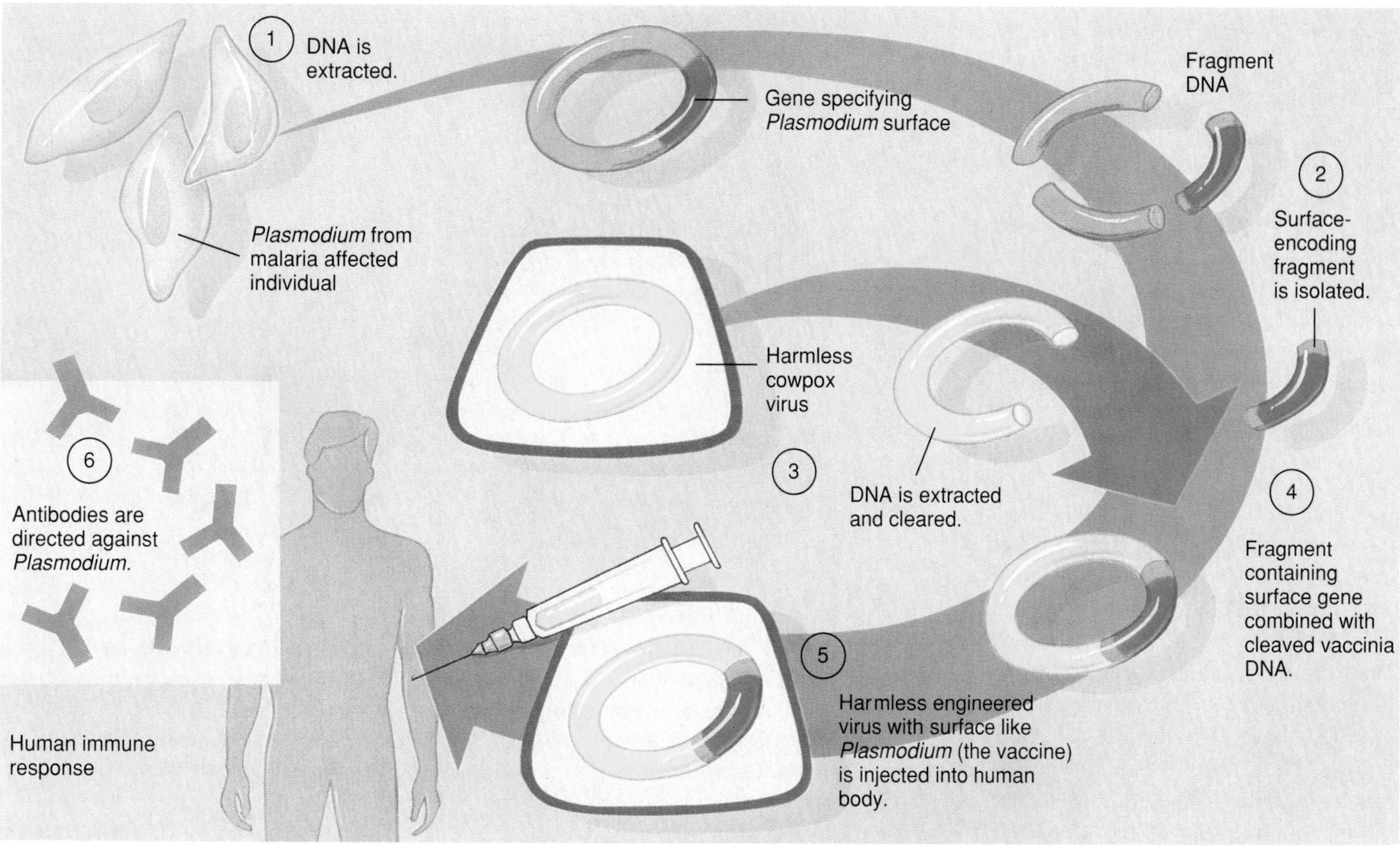

Figure 8.14 How scientists are attempting to construct a piggyback vaccine for malaria.
The organism (*Plasmodium*) that causes malaria is removed from an affected individual and its DNA is extracted. The extracted DNA is fragmented. At the same time, the DNA is isolated from harmless cowpox virus. The two extracts are combined, forming a harmless virus that has been engineered to "look" like malaria to the body. The body responds by forming antibodies against the vaccine, thus conferring resistance to malaria.

Piggyback Vaccines

In figure 8.14 you saw how genetic engineering is being used in a major attempt to create a vaccine that will protect people against malaria, a disease caused by a tiny protozoan and for which there is currently no effective protection. Transmitted by mosquito bites, malaria affected between 300 and 500 million people in 1993, and 1.5 to 2.7 million of them died.

Vaccines are a way to teach our bodies how to defend themselves against disease agents they have not yet met. A harmless version of a disease-causing microbe is injected into the body so that the immune system will develop defenses against the disease. The injected version serves as a model for the body, which carefully examines its surface and uses what it learns to craft defensive proteins called antibodies.

Traditionally, vaccines have been prepared by killing the disease agent (or sometimes by rendering it weak and unable to grow). That way, injecting it into your body doesn't make you sick. This "weakening" process was used in preparing the first Salk polio vaccines. The problem with this approach is that any failure in the weakening or killing process results in introducing the disease into the very patients seeking protection. This danger is one of the reasons rabies vaccines are administered only when an animal suspected of carrying rabies has actually bitten someone.

Now, with the advent of genetic engineering techniques, a new and much safer approach has become possible: inserting the gene encoding the microbe's surface protein into the DNA of a harmless virus. As was illustrated in figure 8.14, the harmless virus now carries the added gene along with it "piggyback" wherever it goes. The modified but still harmless virus becomes an effective and safe **piggyback vaccine.** Injected into a human body, the surfaces of the modified virus display the piggyback surface protein like battle flags. Responding to this challenge, the immune system makes antibodies that attack any cells displaying that protein. As a result, the body is thereafter protected against infection by the microbe from which the piggyback gene was obtained. Any cells of this microbe that invade the body are immediately attacked by antibodies already trained to recognize their surface!

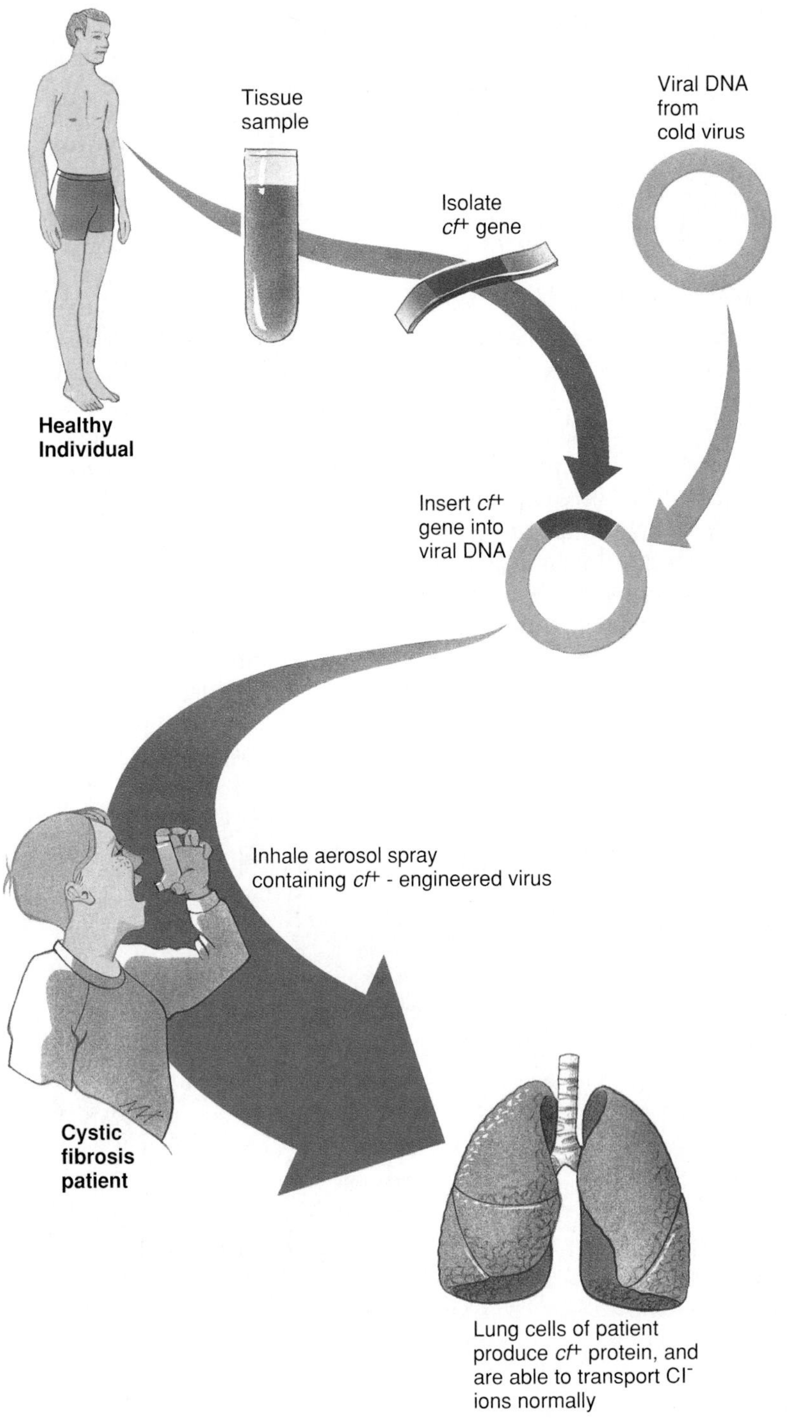

Healthy Individual
Tissue sample
Viral DNA from cold virus
Isolate *cf*+ gene
Insert *cf*+ gene into viral DNA
Inhale aerosol spray containing *cf*+ - engineered virus
Cystic fibrosis patient
Lung cells of patient produce *cf*+ protein, and are able to transport Cl⁻ ions normally

Figure 8.15 Using genetic engineering to combat cystic fibrosis.
One of the most serious human genetic disorders, cystic fibrosis, results from the failure of the body's cells to transport chloride ions (Cl^-) effectively. Using genes from a normal, healthy individual and splicing them into the genes of a cold virus, an engineered virus has been produced that transfers copies of the healthy genes into the individual afflicted with cystic fibrosis.

Among the many other vaccines now being manufactured in this way are ones directed against the herpes virus and hepatitis B virus, the latter the cause of a highly infectious and very dangerous disease.

Human Gene Therapy

In 1990 researchers first attempted to combat genetic defects by the transfer of human genes. Many so-called genetic diseases occur when a person lacks a particular gene because both chromosomes' copies are either defective or missing. One obvious way to cure such disorders is to give the person a working copy of the gene, but until the advent of genetic engineering, **gene therapy** had been impossible. Now, with the techniques of genetic engineering, it is being attempted as a way of combating a variety of genetic disorders such as **cystic fibrosis** (figure 8.15) and muscular dystrophy. Among the first successful attempts has been the transfer of an enzyme-encoding gene into the bone marrow of two girls (one of them is pictured in figure 8.3) suffering from a rare blood disorder caused by the lack of this enzyme.

Human gene transfer is also providing a new and powerful weapon against cancer. All humans possess in their bodies tiny amounts of a protein called TNF (tumor necrosis factor), which attacks and kills cancer cells. Recently, the gene encoding this protein was added by genetic engineers to a kind of white blood cell that is one of the body's guards against cancer. These cells are very skillful at seeking out cancer cells, but like a wooden arrow shot at an army battle tank, they are not very effective at harming them. Armed with this new TNF weapon, however, these engineered white blood cells become more like guided missiles zeroing in on cancer cells with an explosive payload.

The ability of genetic engineers to produce cells that manufacture large quantities of desired proteins has led to the commercial availability of many new drugs within the last few years, some of which are listed in table 8.1. This list grows longer by the day.

Table 8.1 Genetically Engineered Drugs

Product	Effects and Uses
Colony-Stimulating Factors	Stimulate white blood cell production; used to treat infections and immune system deficiencies
Erythropoietin	Stimulate red blood cell production; used to treat anemia in individuals with kidney disorders
Growth Factors	Stimulate differentiation and growth of various cell types; used to aid wound healing
Human Growth Factors	Used to treat dwarfism
Interferons	Disrupt the reproduction of viruses; used to treat some cancers
Interleukins	Activate and stimulate white blood cells; used to treat wounds, HIV infections, cancer, immune deficiencies

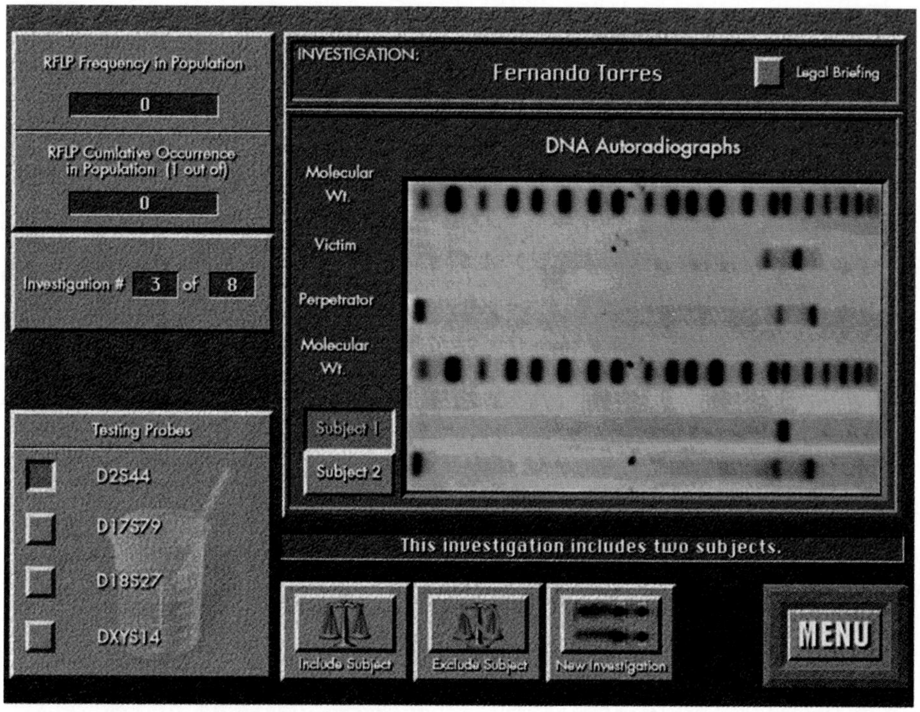

Figure 8.16 You be the judge.

In this interactive CD-ROM exercise you analyze the DNA evidence presented in real courtroom trials, attempting to ascertain the guilt or innocence of the suspect in each instance. The exercise selects a case at random from its library of real cases and presents you with physical evidence (DNA samples from victim, perpetrator, and suspect) and DNA probes to use in comparing suspect and perpetrator. By using a variety of probes, a clear statistical picture usually emerges. (*Explorations in Cell Biology & Genetics,* Module 14, "DNA Fingerprinting")

Genetic engineering is also playing an increasing role in our courtrooms. In the interactive exploration described in figure 8.16 you can be the judge, comparing gene profiles to assess guilt or innocence.

The Human Genome Project

The promise of genetic engineering in fighting disease is bright, and many efforts are underway to increase its usefulness. One of the most significant is a concerted effort to make available to researchers, on a dial-a-gene basis, all of the genes of the human body. That is, quite literally, genetic engineers are attempting to catalogue, locate, read, and make available for potential transfer every gene on each of your 23 different chromosomes. Biologists call the entire collection of genes within human cells the **human genome,** and this effort to locate and write down the nucleotide sequence of each and every human gene is called the **Human Genome Project.** It is one of the largest scientific projects ever undertaken: the human genome contains some 3 trillion nucleotides!

As a first stage in this enormous undertaking, scientists have tackled smaller targets, both to refine their techniques and to learn something of what to expect with the more complex human genome. In figure 8.17 you see some of the early fruits of this labor, reported in 1995—the entire genome sequence of the bacterium *Haemophilus influenzae*. All 1,830,137 nucleotides of the bacterial genome have been sequenced, a single circle of DNA that encodes 1,743 genes. In a very real sense, this genomic sequence is a recipe for life, containing all the information needed to specify a living organism.

Eukaryotes present a much bigger problem, simply because their genomes are so much larger, but progress is rapid. The 12-million-nucleotide sequence of yeast, a single-celled eukaryote, is expected to be completed by 1996. Work on the human genome, a vastly greater project, moves quickly as well. The complete restriction map of the human genome was completed in 1995. This key step divides the human genome into small fragments of manageable size, and orders them on the chromosomes, so that investigators can know just what portion of a chromosome they are sequencing as they examine each fragment. Although a large task indeed, sequencing of the entire human genome seems only a matter of time.

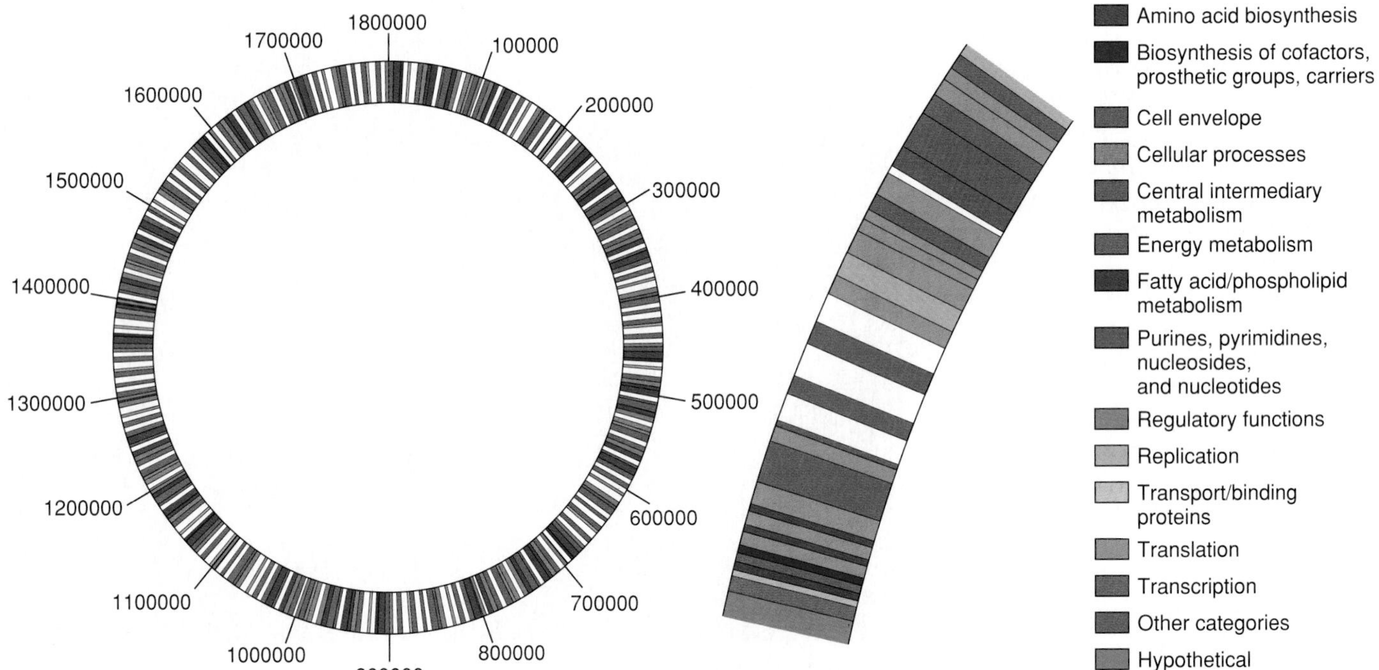

Sources:
Dr. J. Craig Venter and
Dr. Anthony Kerlavage

Figure 8.17 A recipe for life.

The complete nucleotide sequence of the bacterium *Haemophilus influenzae* was reported in 1995. The circle illustrates the 1,830,137 base pairs that encode the 1,743 different genes of a living *H. influenzae* cell. Each band represents a different gene. An enlargement of the band indicates 16 different functional classes to which the genes have been assigned.

CHAPTER 8

HIGHLIGHTS

8.1 Genetic Engineering

Key Terms

- restriction enzyme 157
- plasmid 159
- clone library 159
- probe 159
- PCR 160
- cDNA 160

Key Concepts

- DNA is cut by restriction enzymes into fragments with "sticky" ends.
- Any two fragments of DNA cut by the same restriction enzyme can be joined together, whatever the source of the DNA.
- DNA fragments can be introduced into bacteria, carried on plasmids or viruses. A bacterial colony that grows from such a cell is called a clone.
- Probes such as cDNA can be used to screen bacterial colonies for the presence of a gene you are attempting to transfer.

8.2 Transforming Agriculture

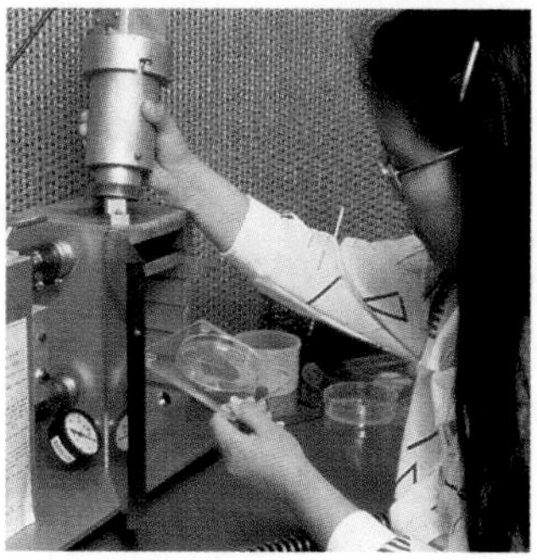

Key Terms

- T_i plasmid 161
- glyphosate 161
- nitrogen fixation 162
- BGH 162

Key Concepts

- It is difficult to transfer genes into plants because of their tough cell walls—few viruses can penetrate them.
- Today genes are inserted into plant cells using the T_i plasmid or are shot with a gun.
- Major advances have been made in transferring genes resistant to herbicides and pests, as well as genes encoding hormones that increase milk production and farm animal size.

8.3 Advances in Medicine

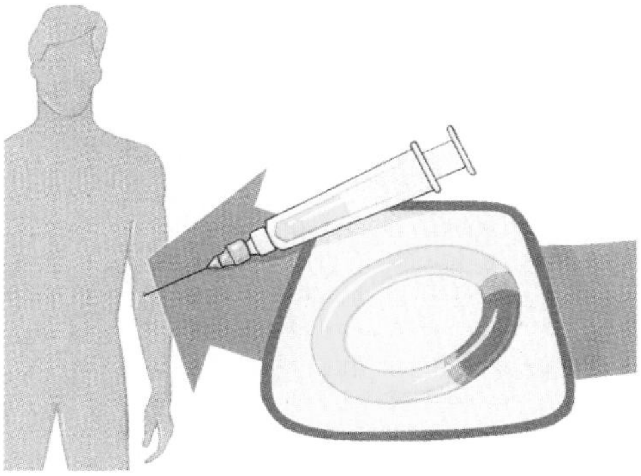

Key Terms

- factor VIII 163
- piggyback vaccine 164
- gene therapy 165
- cystic fibrosis 165
- Human Genome Project 166

Key Concepts

- By transferring human genes into bacteria, it has become possible to produce commercial quantities of rare proteins, making many new drugs practical.
- Effective new vaccines are being made by introducing genes specifying surface proteins of pathogenic microbes into harmless infective agents.
- Transfer of healthy *cf* genes into cystic fibrosis patients may soon cure this hereditary disorder.
- Rapid progress is being made in sequencing the entire human genome.

8

CONCEPT REVIEW

1. Restriction enzymes are invaluable tools for genetic engineering because

a. they cut both strands of DNA at specific sites.

b. "sticky" ends from very different DNA sources can be easily joined together.

c. they are tools of the meat industry.

d. a and b.

2. In gel electrophoresis, larger DNA fragments move _________ distances when an electric current is applied.

a. longer

b. shorter

3. In creating recombinant DNA, the sticky ends of source DNA fragments are joined to the sticky ends of _________ produced by the cutting up of a circular bacterial chromosome.

a. clones

b. antibiotics

c. plasmids

d. restriction enzymes

4. A clone library is

a. a collection of different lines of bacterial cells, each containing a plasmid carrying one series of different DNA fragments.

b. a collection of books about clones.

c. the particular bacterial line that contains the DNA fragment of interest.

d. a series of restriction enzymes that can be used to produce different clones.

5. PCR is a technique used to

a. create clone libraries.

b. transfer genes.

c. chop up source DNA.

d. amplify a region of DNA.

6. A version of DNA that is made from processed RNA is called

a. a plasmid.

b. cDNA.

c. a probe.

d. mDNA.

7. T_i is a virus that infects host cells.

a. true

b. false

8. In piggyback vaccines, a plasmid containing DNA from the disease-causing microbe and from a harmless virus is inserted into a human host. Why is the human body able to form antibodies against the vaccine versus being infected with the disease?

a. The plasmid contains the gene that codes for the microbe's surface proteins but is otherwise harmless.

b. A virulent form of the disease is introduced into the body.

c. PCR was used to create many copies of the gene that encodes the microbe's surface protein.

d. The plasmid acts as a "magic bullet."

9. The transfer of working copies of genes from normal, healthy individuals to afflicted individuals is known as

a. a piggyback vaccine.

b. the Human Genome Project.

c. human gene therapy.

d. DNA fingerprinting.

10. ________ is the last of the four stages of genetic engineering.

11. The enzyme used to create cDNA from mRNA is called _________.

12. _________ is a plasmid used to introduce desirable pieces of DNA into new cells.

13. Lack of _________ leads to dwarfism in animals.

14. Some plants have been genetically engineered to produce _________ that are highly toxic to insect pests.

15. TNF is an abbrevation for the _________.

16. The effort to locate and write down the nucleotide sequence of every human gene is called the _________.

Answers to the Concept Review questions appear in Appendix B.

CHALLENGE YOURSELF

1. A major focus of genetic engineering has been the attempt to produce large quantities of scarce human metabolites by placing the appropriate human genes into bacteria. Human insulin is now manufactured this way. However, if we attempt to use this approach to produce human hemoglobin (β-globin), the experiment does not work. Even if the proper clone from a clone library is identified, the fragment containing the β-globin gene is successfully incorporated into a plasmid, and *E. coli* are successfully infected with the chimeric plasmid, no β-globin is produced by the infected cells. Why doesn't the experiment work?

2. In Michael Crichton's 1990 novel *Jurassic Park,* genetic engineers re-create dinosaurs from samples of fossilized dinosaur DNA. The story relies in part on discoveries and technology that already exist. For example, mosquitoes and other blood-sucking insects dating from the age of the dinosaurs have been found preserved in amber, researchers have been successful in obtaining and analyzing DNA contained in 18-million-year-old fossils of leaves, and transgenic animals have been produced after foreign genes were inserted into the eggs from which they developed. Given these successes, why hasn't the *Jurassic Park* experiment been completed in reality?

FOR FURTHER READING

Beardsley, T. "Vital Data." *Scientific American*, March 1996, 100–105. An up-to-date account of progress in the Human Genome Project.

Lee, T. *The Human Genome Project: Cracking the Genetic Code of Life*. New York: Plenum Press, 1991. A detailed book devoted to a discussion of one of the most intensive scientific undertakings of all time.

Neufield, P. J., and N. Colman. "When Science Takes the Stand." *Scientific American,* May 1990, 46–53. DNA and other evidence is increasingly being applied to the solution of criminal cases but must be used with caution.

Old, L. "Tumor Necrosis Factor." *Scientific American,* May 1988, 59–75. One of the few cancer cures that works (sometimes), this small protein plays an important role in the body's defense against cancer.

Srix, G. "A Recombinant Feast: New Bioengineered Crops Move Toward Market." *Scientific American,* March 1995, 38–40. A current progress report on the growing role of genetic engineering in modern agriculture.

Various authors. "The Human Genome Project." *Science,* 265 (September 30, 1994). An entire issue devoted to the Human Genome Project, including a spectacular wall chart of human chromosomes with all known loci labeled.

Vines, G. "Gene Tests: The Patient's Dilemma." *New Scientist,* November 12, 1994, 40–44. Is genetic testing truly necessary, or will it catapult us into becoming a eugenic society?

TECHNOLOGY LINKS

The Living World Home Page
http://www.wcbp.com/biology/tlw

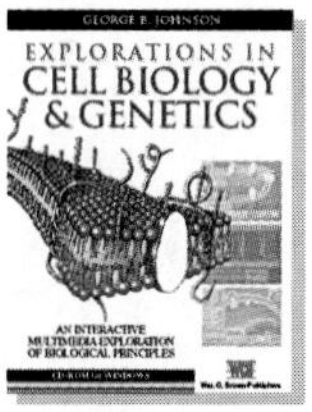

Explorations in Cell Biology & Genetics CD
#11 Constructing a Genetic Map
#14 DNA Fingerprinting
#17 Making a Restriction Map

9

Evolution and Natural Selection

CHAPTER OUTLINE

9.1 Darwin
- Darwin's Voyage on the HMS *Beagle*
- Darwin and Malthus
- Natural Selection
- How Darwin Came to Write His Book

9.2 The Evidence for Evolution
- The Fossil Record
- The Molecular Record
- Comparing Organisms

9.3 How Populations Evolve
- Genes Within Populations
- Why Do Allele Frequencies Change?
- Forms of Selection

9.4 Adaptation: Evolution in Action
- Sickle-Cell Anemia
- Industrial Melanism
- An Overview of Natural Selection

9.5 How Species Form
- The Species Concept
- Isolating Mechanisms
- Adaptive Radiation
- Does Evolution Occur in Spurts?

Highlights

Figure 9.1 Charles Darwin and his eldest son, William. This daguerreotype was made in 1842, the same year Darwin began writing *On the Origin of Species*.

Biologists believe that the great diversity of life on earth—bacteria, elephants, and roses—is the result of a long process of **evolution,** the progressive change that occurs in organisms' characteristics through time. Long ago, dinosaurs walked the earth; now their descendants, the birds, fly over it. In 1859 the English naturalist Charles Darwin (1809–82) (figure 9.1) first suggested an explanation for why evolution occurs. His theory was that the struggle to survive and leave offspring would be more often won by those best able to respond to the challenges of living, a process he called **natural selection.** Over time, this "pruning" process would cause the species to change—to evolve. In the century since, biologists have become convinced Darwin was right.

9.1 Darwin

The son of a moderately wealthy country doctor, Darwin had a troubled education, spending more time outdoors than in school. As a medical student in Edinburgh, he hated dissection (cutting up dead bodies) and for two years continually skipped lectures to spend time collecting beetles instead! In desperation, his father sent him to Cambridge University to train for the ministry. There, in 1831, a professor recommended the 22-year-old Darwin for a post as an unpaid naturalist on a five-year naval voyage to survey the coasts of South America on the HMS *Beagle* (figure 9.2). After interviewing, Darwin was offered the position. His father at first refused to let him go, and Darwin regretfully declined. But his uncle (and the father of his future wife) interceded at the last moment, and Darwin was off on a voyage that would change forever how we think of ourselves. The course followed around the world by the HMS *Beagle* is shown in figure 9.3. During his long journey, Darwin spent much of his time ashore, experiencing firsthand the biological richness of the tropical forests, the extraordinary fossils of Patagonia, and the remarkable life-forms on the **Galápagos Islands** off the west coast of South America.

Figure 9.2 A replica of the HMS *Beagle*.
Charles Darwin set forth on the HMS *Beagle* in 1831 at the age of 22.

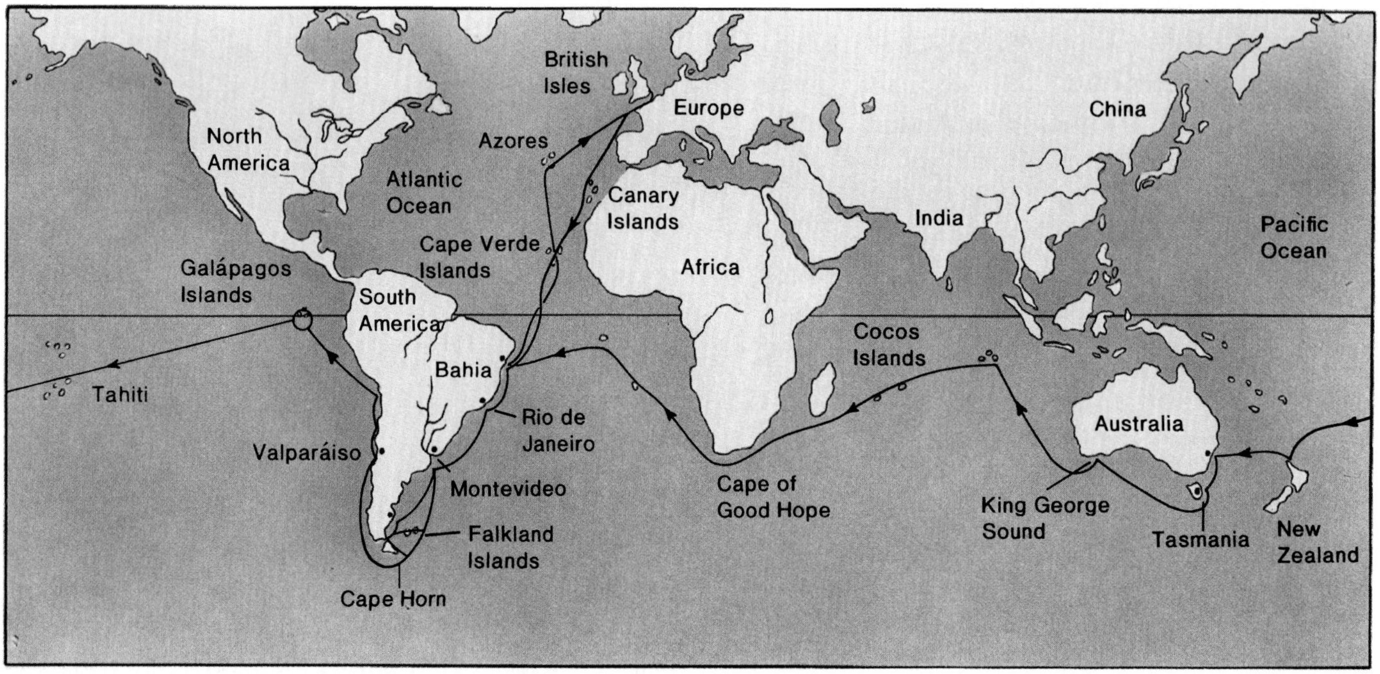

Figure 9.3 The five-year voyage of the HMS *Beagle*.
Although the ship sailed around the world, most of the time was spent exploring the coasts and coastal islands of South America, such as the Galápagos Islands. Darwin's studies of the animals of these islands played a key role in the eventual development of his theory of evolution by means of natural selection.

Darwin's Voyage on the HMS Beagle

When the HMS *Beagle* set sail, Darwin did not believe that species evolved. He, like most people of his time, thought that species had always been the same. But during his five years on the ship, Darwin saw things that made him change his mind, things he could explain in no other way than by evolution. For example, in rich fossil deposits in southern South America, he saw fossils of extinct armadillos that were directly related to the armadillos still living in the same area and yet different from those elsewhere. Why would there be living and fossil organisms in the same place, directly related to one another, unless one had given rise to the other?

Repeatedly, Darwin saw patterns in the differences among related species that suggested to him that the species had changed gradually as they migrated from one place to another. On the Galápagos Islands (figure 9.4), over a thousand kilometers off the coast of Ecuador, Darwin saw more than a half dozen species of finches, all related but each specialized to catch food in a different way. Some of these finches had heavy bills that they used to crack open tough seeds, others had slender bills for catching insects, and one even used a twig to probe for insects like a woodpecker! There were no other small birds on the islands. These finches more closely resembled South American finches than any others (figure 9.5), so perhaps an ancestral finch had been blown over the sea from South America long ago. But if all of these finches were the descendants of a small flock of one species, why did Darwin see many different finch species on the islands? The birds must have changed after they arrived!

Figure 9.5 The ancestor of Darwin's finch?
(*a*) One of Darwin's finches, the medium ground finch. (*b*) The blue-black grassquit, which is found in grasslands along the Pacific Coast, from Mexico to Chile. This species may be the ancestor of Darwin's finches.

Figure 9.4 A view of the Galápagos Islands.
Since Darwin's time, much of the natural habitat of the larger islands has been destroyed by human habitation. Goats, for example, have drastically altered the vegetation.

In a more general sense, Darwin was struck by the fact that all the plants and animals of the Galápagos Islands resembled those of the nearby coast of South America. If each one of these plants and animals had been created independently and simply placed on the Galápagos Islands, why did they not resemble the plants and animals of a similar habitat of faraway Africa, for example? Why did they resemble those of the environmentally different but adjacent South American coast instead? Darwin felt that the simplest explanation was that the ancestors of the animals and plants of the Galápagos Islands must have migrated to the islands from South America long ago and then changed during the years they lived in their new island home. Descent with modification. Evolution.

Darwin and Malthus

It is one thing to observe the results of evolution but quite another to understand how it happens. Darwin's great achievement lies in his perception that evolution occurs because of natural selection. Of key importance to the development of Darwin's insight was his study of Thomas Malthus's *Essay on the Principles of Population.* In his book, Malthus pointed out that populations of plants and animals (including human beings) tend to increase geometrically. A geometric progression is one in which the elements progress by a constant factor, for example, 2, 6, 18, 54, and so forth; in this example, each number is three times the preceding one.

Virtually any kind of animal or plant, if it could reproduce unchecked, would cover the entire surface of the world within a surprisingly short time. In fact, this does not occur; instead, populations of species remain more or less constant year after year, because death intervenes and limits population numbers.

Natural Selection

Sparked by Malthus's ideas, Darwin saw that although every organism has the potential to produce more offspring than are able to survive, only a limited number actually survive and produce their own offspring. Combining this observation with what he had seen on the voyage of the HMS *Beagle,* as well as with his own experiences in breeding domestic animals, Darwin made the key association: *those individuals that possess superior physical, behavioral, or other attributes are more likely to survive than those that are not so well endowed.* By surviving, they have the opportunity to pass on their favorable characteristics to their offspring. Because these characteristics will increase in the population, the nature of the population as a whole gradually changes. Darwin called this process **natural selection.**

How Darwin Came to Write His Book

When Darwin returned from his voyage at the age of 27, he undertook what proved to be a life-long study of plants and animals. He did not write ***On the Origin of Species*** as soon as he got back. Instead, he immersed himself in other studies. Among the books he wrote then are studies of geology and of how Pacific islands are built by coral reefs. He devoted eight years to the study of barnacles, a group of marine animals! A few years after his return, he married his first cousin, Emma, and started a family.

During these years Darwin and others analyzed the specimens he had collected on the voyage of the HMS *Beagle,* and he began to formulate his ideas about evolution. He finally wrote them down in 1842 in a preliminary manuscript that he showed only to a few scientists he knew and trusted. In that year, however, a controversial paper called *Vestiges of Natural Creation* was published, arguing that evolution must have occurred in the earth's past. Copies of the book, a best-seller of the time, were publicly burned, and its author was criticized by both government and church. Shrinking from such public controversy, Darwin put his own manuscript aside. For the next 17 years, he went about his life, enlarging and refining his notes on evolution but saying nothing in public. He raised 10 children and published many books on other scientific subjects, but the manuscript on evolution remained in his drawer.

The stimulus that finally brought Darwin's work into print was a letter he received in early 1858. A young English naturalist named Alfred Russel Wallace sent him a short essay from Malaysia that concisely set forth the theory of evolution by means of natural selection. In the essay he sent Darwin, Wallace wrote, "The life of wild animals is a struggle for existence . . . in which the weakest and least perfectly organized must always succumb . . . giv(ing) rise to successive variations departing further and further from the original type."

Wallace asked Darwin to help him get his essay published. Darwin's scientific friends urged him to get his own work ready, and some months later they arranged for Darwin to present an abstract of his manuscript at a public scientific meeting along with Wallace's paper. Neither Darwin nor Wallace actually attended the meeting on July 1, 1858, because Wallace was still in Malaysia and one of Darwin's sons had died two days before of scarlet fever. Little notice was taken of the two papers at the time, and Darwin set to work preparing his book for publication.

Darwin's book appeared 13 months later in November 1859 and caused an immediate sensation. Many people were deeply disturbed by the idea that human beings were closely related to apes. While humans closely resemble apes, they found the possibility that there might be a direct evolutionary link unacceptable. Darwin's arguments were so compelling, however, that within a few years they came to be completely accepted by biologists around the world. Over a century later, the evidence is even more compelling, and biologists agree as to the general correctness of Darwin's theory, although many people who are not biologists still find his ideas unacceptable.

9.2 The Evidence for Evolution

More than a century has now passed since Darwin's death in 1882. During this period, a great deal of additional evidence has accumulated supporting the theory of evolution, much of it far stronger than that available to Darwin and his contemporaries.

The Fossil Record

The most direct evidence that evolution has occurred is that we can see it happening in the fossil record. Fossils of species embedded in old rock are different from fossils in newer rocks. Darwin predicted that fossil hunters would eventually find "missing links" between the great groups of organisms, and in fact many of them have since been found. The fossil history of the vertebrates, for example, is remarkably complete, with fossil links between fishes and amphibians, reptiles and birds (figure 9.6), and every stage between reptiles and mammals.

Figure 9.6 Fossil of a transitional bird, *Archaeopteryx*.
A fossil link between reptiles and birds, this well-preserved fossil of *Archaeopteryx*, about 150 million years old, was discovered within two years of the publication of *On the Origin of Species*. It has many reptilian traits, including teeth and a tail, and many birdlike features, like feathers.

Dating Fossils

Fossils are created when organisms, footprints, or burrows become buried in sand or sediment. Over time, the calcium in bone and other hard tissues becomes mineralized as the sediment is converted to rock. A fossil is any record of prehistoric life—generally taken to mean older than 10,000 years. By dating the rocks in which fossils occur, biologists can get a very good idea of how old the fossils are. Rocks are usually dated by measuring the degree of radioactive decay of certain radioactive atoms among rock-forming minerals. A radioactive atom is one whose nucleus contains so many neutrons and protons that it is unstable and eventually flies apart, creating smaller, more stable atoms of another element. Because the rate of decay of a radioactive element (how many of its atoms undergo decay in a minute) is constant, scientists can use the amount of radioactive decay to date fossils. The older the fossil, the greater the fraction of its radioactive atoms that have decayed.

One of the most widely employed methods of dating fossils fewer than 50,000 years old is the carbon-14 (^{14}C) **radioisotopic dating** method. Most carbon atoms have an atomic weight of 12 (^{12}C), but a fixed proportion of the carbon atoms in the atmosphere consists of carbon atoms with an atomic weight of 14 (^{14}C). This proportion is captured by plants in photosynthesis, and it is present also in the carbon molecules of animal bodies, all of which come ultimately from plants. After a plant or animal dies, its ^{14}C gradually decays over time, losing neutrons to form nitrogen-14 (^{14}N). It takes 5,600 years for half of the ^{14}C present in a sample to be converted to ^{14}N by this process. This length of time is called the **half-life** of the isotope. Because the half-life of an isotope is a constant that never changes, the extent of radioactive decay allows you to date a sample. Thus, a sample that had a quarter of its original proportion of ^{14}C remaining would be approximately 11,200 years old (two half-lives).

For fossils older than 50,000 years, there is too little ^{14}C remaining to measure precisely, and scientists instead examine the decay of potassium-40 (^{40}K) into argon-40 (^{40}Ar), which has a half-life of 1.3 billion years.

What the Fossil Record Says

Accurate radioactive dates permit biologists to line up fossils in the order of their age, from oldest to youngest. When this is done, there is striking evidence of progressive change. For example, among the hoofed mammals illustrated in figure 9.7, small bony bumps on the nose of the oldest forms became progressively larger until they were large blunt horns on the most recent forms. During the evolution of horses, the number of toes on the front foot was gradually reduced from four to one. Over a 12-million-year span in the early Jurassic period 200 million years ago, the shells of a kind of coiled oyster became progressively larger, thinner, and flatter (figure 9.8). A particularly striking example of progressive change in the fossil record is the evolution of the whale from hoofed mammals, in which the gradual reduction and eventual loss of the rear legs can be readily seen (figure 9.9).

A host of other examples are known, all illustrating a record of *progressive* change through time. The demonstration of this orderly and progressive change in the fossil record is the strongest evidence that evolution has occurred.

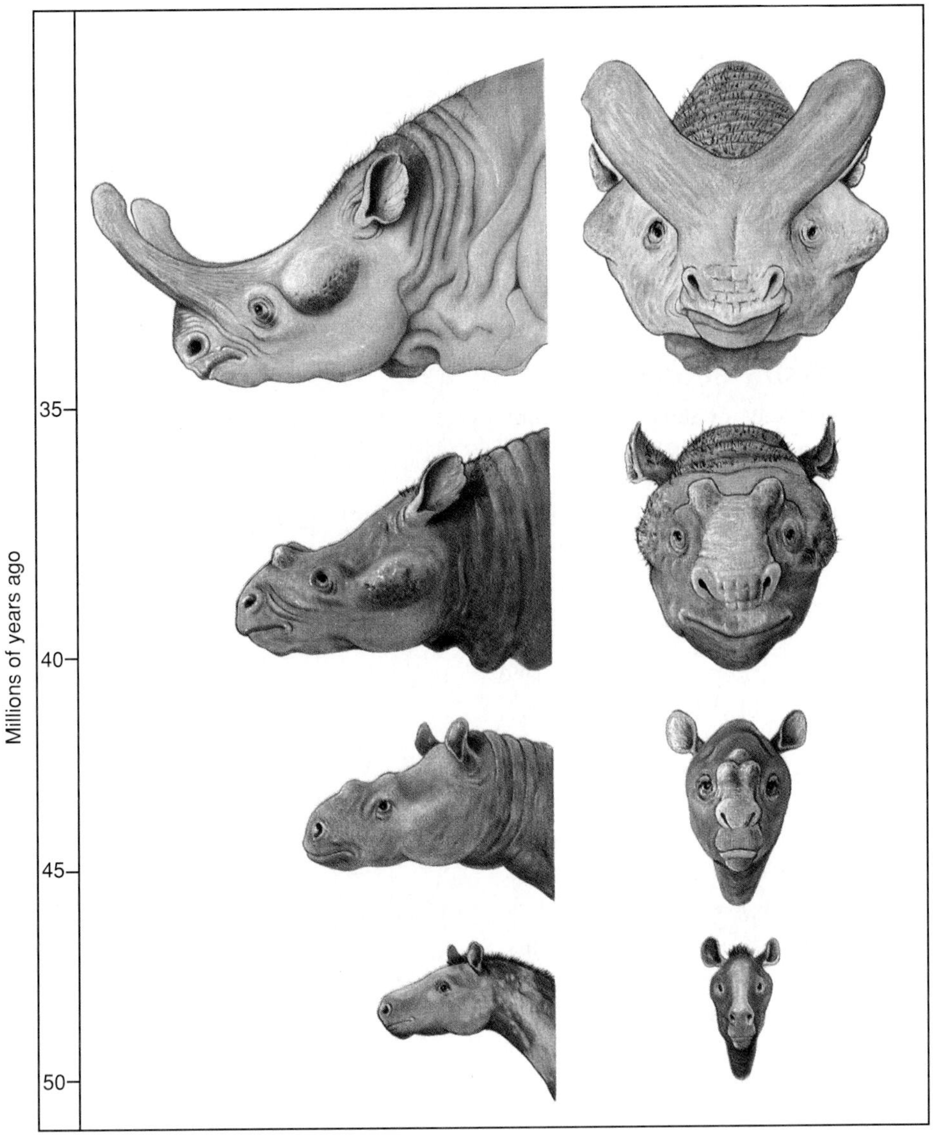

Figure 9.7 Progressive change in the fossil record.
Here you see illustrated changes in a group of hoofed mammals known as titanotheres between about 50 million and 35 million years ago. During this time, the small, bony protuberance located above the nose 50 million years ago evolved into relatively large, blunt horns.

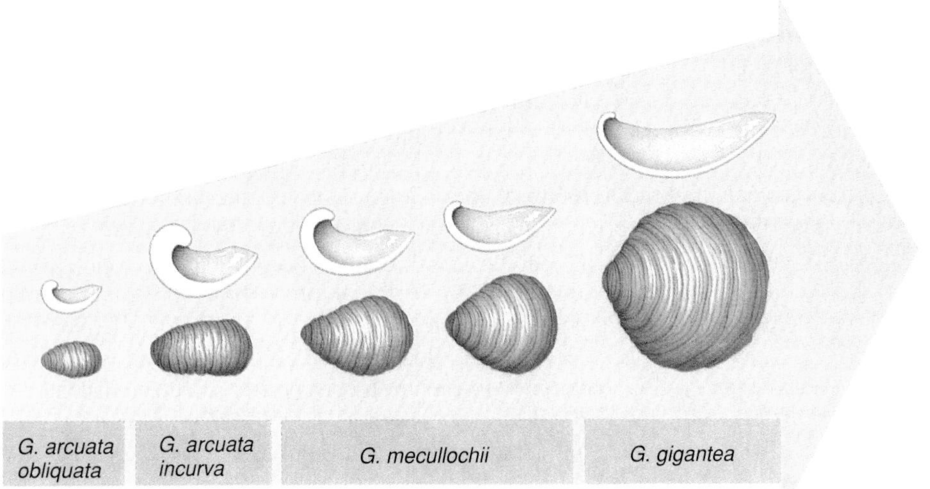

Figure 9.8 Progressive evolution of shell shape in oysters.
During a 12-million-year portion of the early Jurassic period, the shells of a group of coiled oysters became larger, thinner, and flatter. These animals rested on the ocean floor, and the larger, flatter shells may have proven more stable against potentially disruptive water movements.

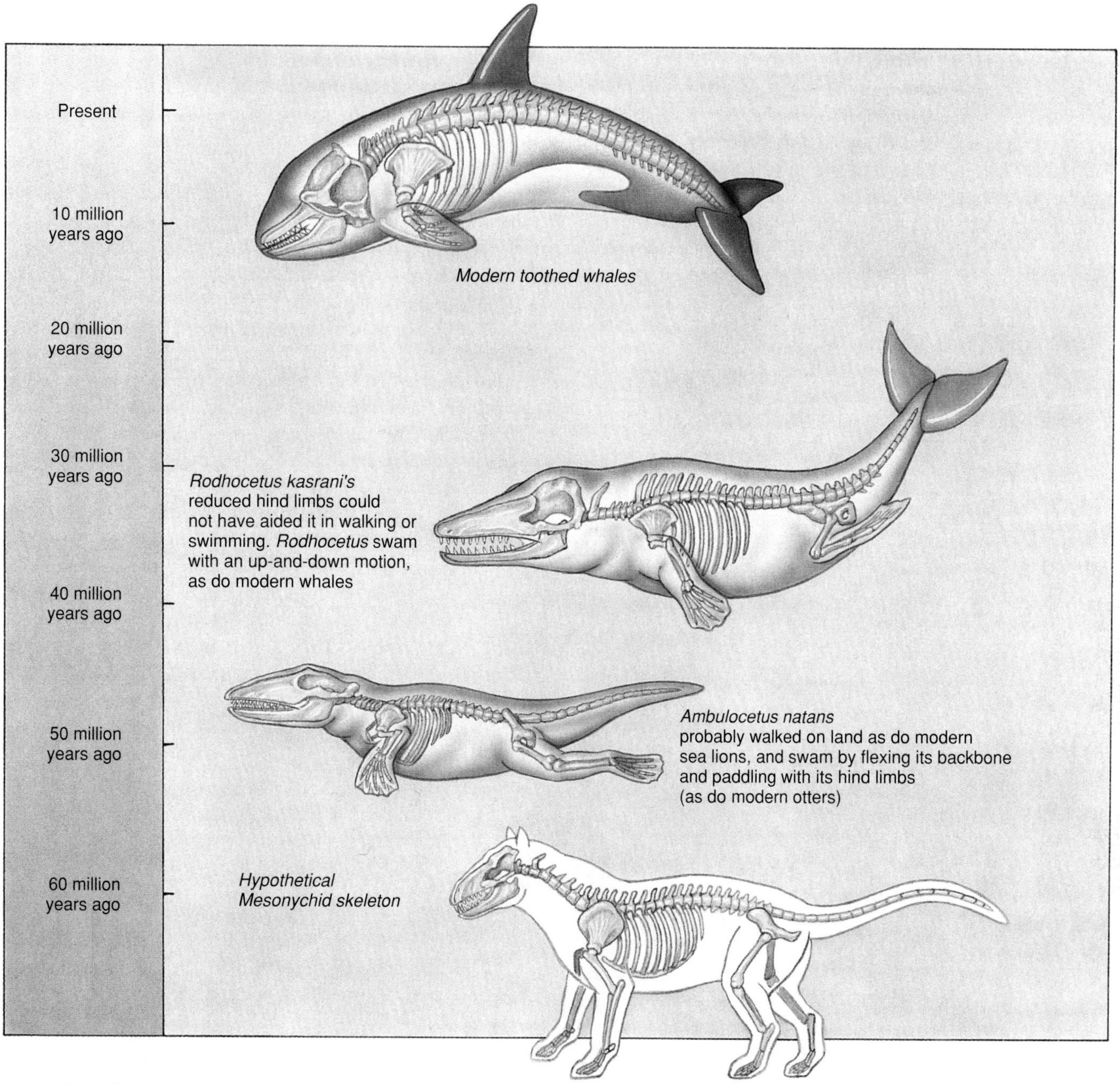
Present
10 million
years ago
20 million
years ago
30 million
years ago
40 million
years ago
50 million
years ago
60 million
years ago
Modern toothed whales
Rodhocetus kasrani's
reduced hind limbs could
not have aided it in walking or
swimming. Rodhocetus swam
with an up-and-down motion,
as do modern whales
Ambulocetus natans
probably walked on land as do modern
sea lions, and swam by flexing its backbone
and paddling with its hind limbs
(as do modern otters)
Hypothetical
Mesonychid skeleton

Figure 9.9 Whales evolved from hoofed mammals.

Whales are thought to have evolved from four-legged land mammals that are also the ancestors of cows. *Ambulocetus natans* and *Rodhocetus kasrani* are recently discovered transitional forms in whale evolution.

The Molecular Record

The picture of progressive evolution painted by the fossil record makes a very strong prediction that can be tested directly. If organisms have changed progressively, then their genes (which, after all, encode the instructions that dictate what they are like) should have changed too. Evolution theory thus predicts that each gene should have to accumulate more and more alterations in its nucleotide sequence over time, as evolution progressively "fixes" one change after another into the genetic instructions.

The Ticking of the Molecular Clock

To test this prediction, biologists first arrange living organisms in terms of their age—that is, how long it has been since each species first appeared in the fossil record. This is usually done by potassium-argon radioactive dating of volcanic rock found near the fossil (figure 9.10). Volcanic rock is used because it was newly formed at the same time as the fossil. Why not simply date the rock in which the fossil rests? Most fossils are found in so-called sedimentary rock, formed by compression of then-existing bits of rock. If dated, sedimentary rock indicates the much older age of the bits from which it was formed.

After the species have been ordered, from newest to oldest, biologists then select a gene and count the number of nucleotide differences between the version carried by one species and that carried by another. If evolution has occurred, then when biologists count the ticks of this **molecular clock,** evolution makes a simple prediction: organisms that are more distantly related should have had time to accumulate a greater number of differences between their genes than have more closely related ones.

When the test is carried out, this is indeed what is seen. The longer the time since species diverged, the greater the number of differences in the nucleotide sequence for the gene encoding the respiratory protein cytochrome *c* (figure 9.11). The same regular pattern is seen in hemoglobin and many other proteins. Time and again, we see direct evidence of progressive change through time. The demonstration of progressive change at the molecular level strongly reinforces the fossil evidence that evolution has occurred.

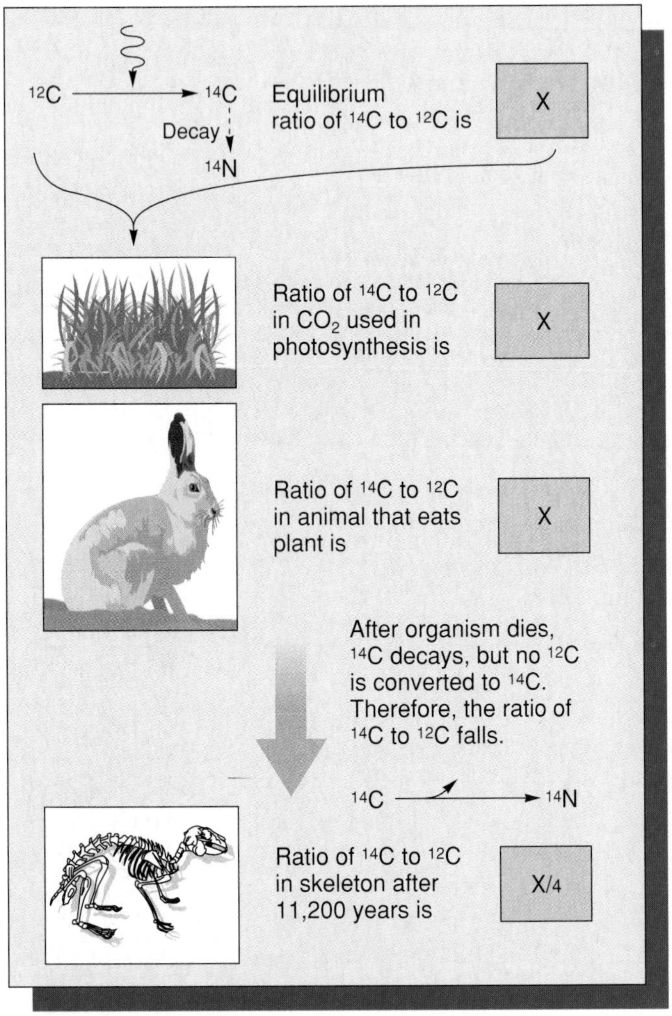

Figure 9.10 Radioactive isotope dating.
This diagram illustrates radioactive dating using carbon-14, a short-lived isotope. Most fossils are dated with much longer-lived isotopes, primarily potassium-40. Called potassium-argon dating, this procedure can be used to accurately date samples many millions of years old.

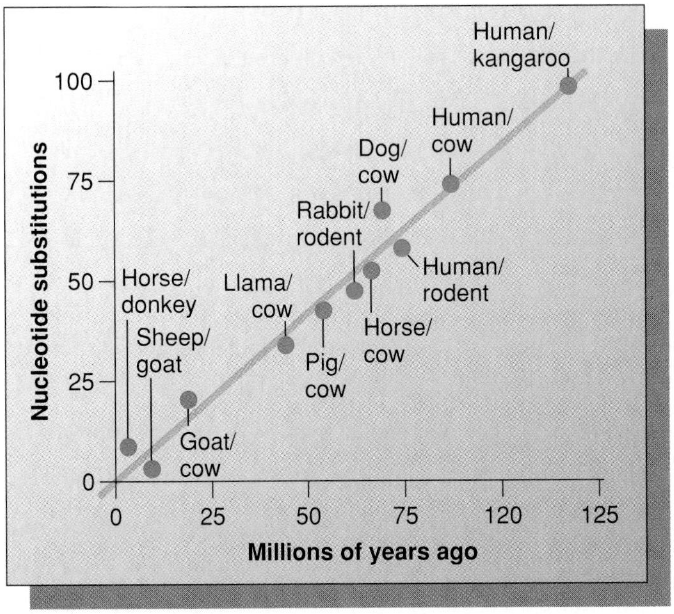

Figure 9.11 The "molecular clock" of cytochrome *c*.
Investigators compared various pairs of organisms and counted the number of nucleotides in the cytochrome c genes that were not the same. Plotting the number of such "substitutions" against the time investigators believed has elapsed since the pair of organisms diverged results in a straight line. This suggests that the cytochrome c gene is evolving at a constant rate.

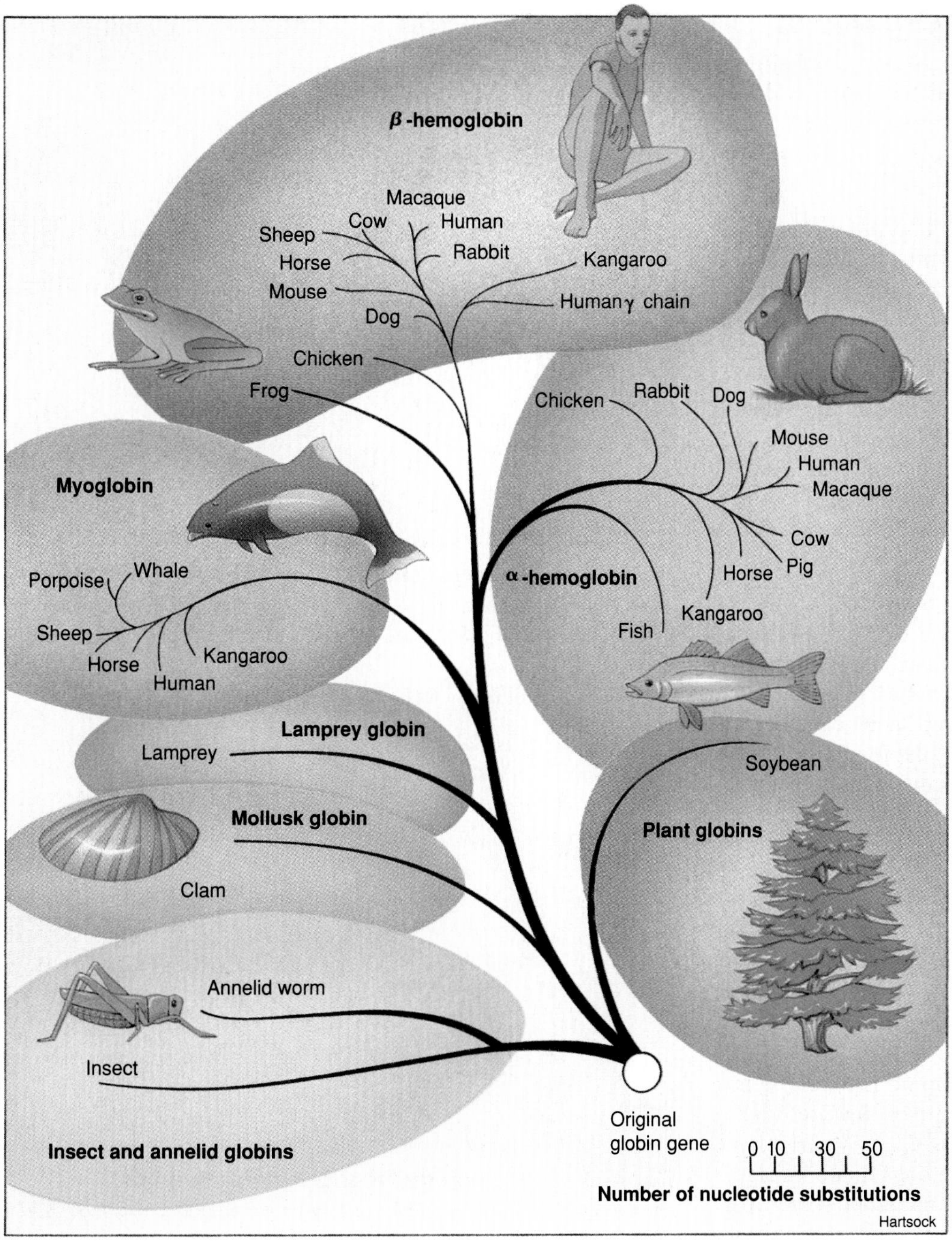

Figure 9.12 An evolutionary tree of the globin gene.
The length of the various lines reflects the number of nucleotide substitutions in the gene. In humans, two of the proteins encoded by versions of the globin gene are α-hemoglobin and β-hemoglobin.

Building Molecular Family Trees

Some genes, like the one encoding hemoglobin, have been so well studied that the entire course of their evolution can be laid out with confidence. By tracing the origin of each particular nucleotide change in the gene, and watching the change then pass down among the species that are descended from the one in which it first occurred, molecular biologists are able to construct a history of the gene, a molecular "family tree." The tree for the globin gene presented in figure 9.12 portrays the evolutionary history of the mammalian hemoglobin gene—not a guess, but a direct observation, recorded in the gene sequences themselves.

Molecular family trees provide strong support of evolution, because they reflect precisely the evolutionary relationships indicated by the fossil record. Whales, dolphins, and porpoises cluster together in the molecular family tree, as do the hoofed animals, and as do the primates. There is no other scientific way to explain these data except as a record of progressive evolutionary change.

Comparing Organisms

Comparing different kinds of organisms has provided yet more evidence for Darwin's theory.

Development Reveals Past Evolution

Much of our evolutionary history can be seen in the way in which human embryos develop. Early in development, we (and all other vertebrate embryos) have gill slits like a fish; at a later stage every human embryo has a long bony tail; and at the fifth month of development, each develops a coat of fine fur! These relict developmental forms suggest strongly that our development has evolved, with new instructions being layered on top of old ones (figure 9.13).

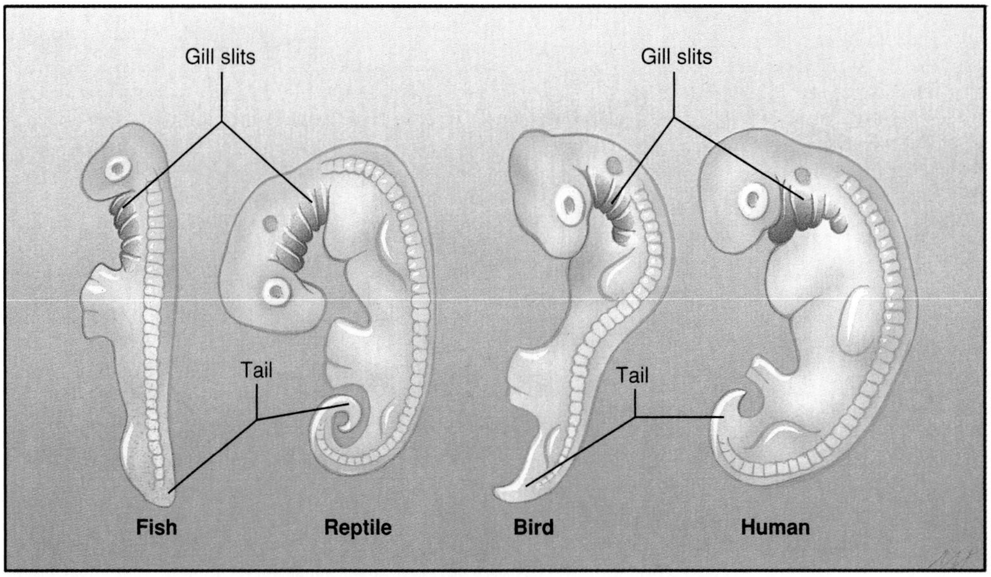

Figure 9.13 Embryos show our early evolutionary history.
These embryos, representing various vertebrate animals, show the primitive features that all share early in their development, such as gill slits and a tail.

Sharing the Same Parts

As vertebrates have evolved, the same bones are sometimes put to different uses, yet they can still be seen, their presence betraying their evolutionary past. For example, the forelimbs of vertebrates are all **homologous structures;** that is, although the structure and function of the bones have diverged, they are derived from the same body part present in a common ancestor. You can see in figure 9.14 how the bones of the forelimb have been modified in one way in the wings of bats, in another way in the fins of porpoises, and in yet other ways in the legs of frogs, horses, and humans.

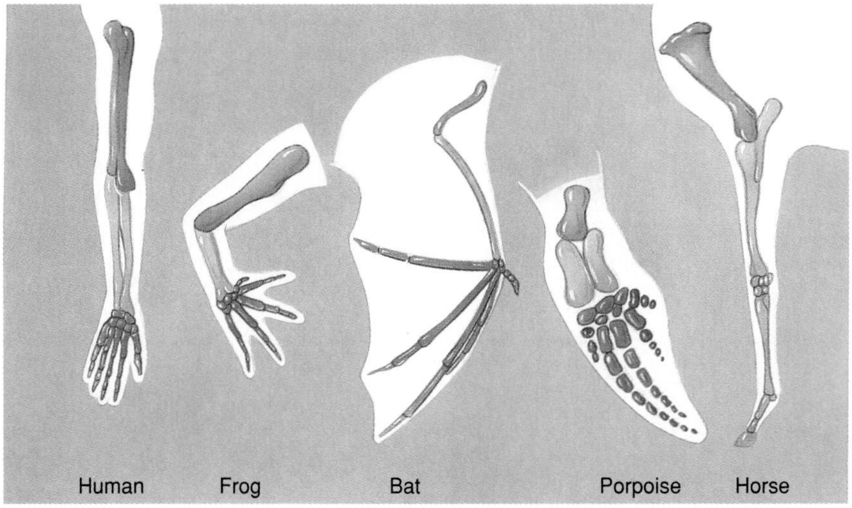

Figure 9.14 Homology among vertebrate limbs.
Homologies among the forelimbs of four mammals and a frog show the ways in which the proportions of the bones have changed in relation to the particular way of life of each organism. Although considerable differences can be seen in form and function, the same basic bones are present in each forelimb.

Not all similar features are homologous. Sometimes features come to resemble each other as a result of parallel evolution of separate lineages. These are called **analogous structures.** For example, the marsupial mammals of Australia evolved in isolation from placental mammals, but similar selective pressures have generated very similar kinds of animals (figure 9.15). Similarly, flippers of penguins and dolphins are analogous structures, originating as two very different structures in two ancestors of distantly related lines and then being modified through natural selection to look the same and serve the same function.

"Leftover" (Vestigial) Organs

Sometimes bones are put to no use at all! In living whales, which evolved from hoofed mammals, the bones of the pelvis that formerly anchored the two hind limbs are all that remain of the rear legs, unattached to any other bones and serving no apparent purpose. Another example of a vestigial organ is the human appendix. In the great apes, our closest relatives, we find an appendix much larger than ours, attached to the gut tube, which functions in digestion, holding bacteria used in digesting the cellulose cell walls of the plants eaten by these primates. The human appendix is a vestigial version of this structure that now serves no function.

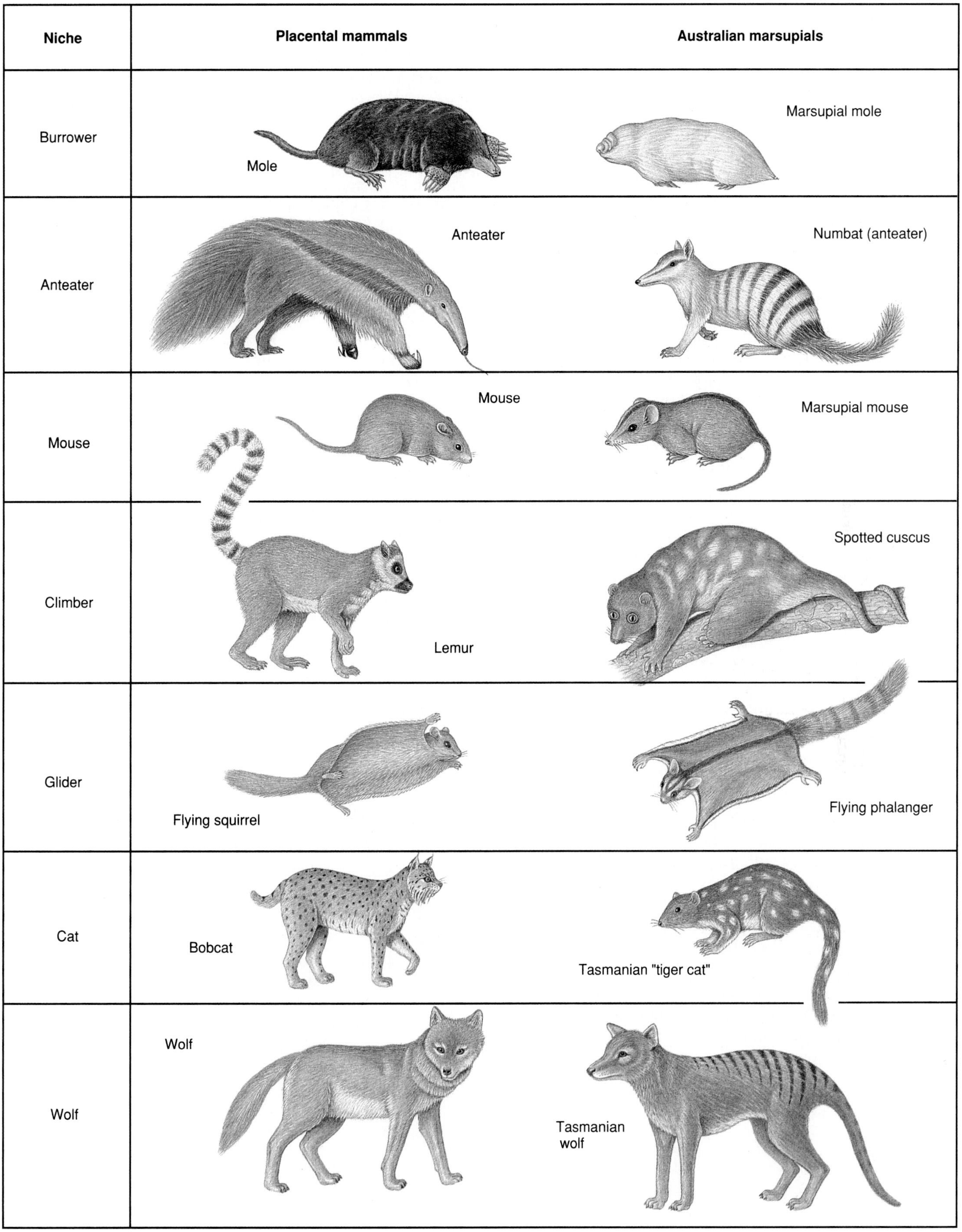
Niche
Placental mammals
Australian marsupials
Burrower
Mole
Marsupial mole
Anteater
Anteater
Numbat (anteater)
Mouse
Mouse
Marsupial mouse
Climber
Lemur
Spotted cuscus
Glider
Flying squirrel
Flying phalanger
Cat
Bobcat
Tasmanian "tiger cat"
Wolf
Wolf
Tasmanian wolf

Figure 9.15 Analogous structures.
The analogous structures of marsupials in Australia (*right column*) and placental mammals in the rest of the world (*left column*) show that the two groups of mammals evolved similarly in response to similar environments.

9.3 How Populations Evolve

The word *evolution* conjures up images of dinosaurs, woolly mammoths frozen in blocks of ice, or Darwin confronting a monkey. Traces of ancient life-forms, now extinct, survive as fossils that help us to piece together the history of life on earth. Thus, evolution is usually interpreted to mean changes in the kinds of animals and plants on the earth, changes that take place over long periods, with new forms replacing old ones. Actually, this kind of evolution is called **macroevolution.** Macroevolution is evolutionary change on a grand scale, encompassing the origin of novel designs, evolutionary trends, new kinds of organisms penetrating new habitats, and major extinction episodes.

Much of Darwin's theory of evolution focuses not on the way in which new species are formed from old ones or macroevolutionary changes but rather on how changes occur within species. A **population** is a group of interbreeding organisms that occupy a given geographic area. As a result of natural selection, a population gradually comes to include more and more individuals with advantageous characteristics, assuming that the characteristics have a genetic basis. In this way, the population evolves. Changes of this sort within populations—the progressive change in allele frequencies—is called **microevolution.** Natural selection is the main process by which microevolutionary change occurs. The essence of Darwin's explanation of evolution is that progressive adaptation by natural selection is responsible for evolutionary changes *within* a species. These changes, when they accumulate, lead to the creation of new kinds of organisms, new species. In short, microevolution drives macroevolution.

Genes Within Populations

From the 1920s onward, scientists began to formulate a comprehensive theory of how **alleles**, alternative forms of a gene, behave in populations and how changes in gene frequencies lead to evolutionary change. Variation within populations puzzled many scientists; dominant alleles were believed to drive recessive alleles out of populations, with selection favoring an optimal form. The solution to the puzzle of why genetic variation persists was developed in 1908 by G. H. Hardy and G. Weinberg. Hardy and Weinberg pointed out that in a large population in which there is random mating, and in the absence of forces that change the proportion of alleles of a given gene, the **allele frequencies,** the original genotype proportions remain constant from generation to generation. Dominant alleles do not, in fact, replace recessive ones. Because their proportions do not change, the genotypes are said to be in **Hardy-Weinberg equilibrium** (figure 9.16). A population that is in Hardy-Weinberg equilibrium is not evolving.

Hardy and Weinberg came to this conclusion by analyzing the frequencies of alleles in successive generations. **Frequency** is defined as the proportion of individuals falling within a category in relation to the total number of individuals being considered. Thus, in a population of 100 cats, with 84 black and 16 white cats, the frequencies of black and white are .84 and .16. By convention, the frequency of the more common of two alleles is designated by the letter p and that of the less common allele by the letter q. Because there are only two alleles, the sum of p and q must always equal one.

In algebraic terms, the Hardy-Weinberg principle is written as an equation. For a gene with two alternative alleles B (frequency p) and b (frequency q), the equation looks like this:

$$(p + q)^2 = p^2 + 2pq + q^2$$

Individuals homozygous for allele *B*	Individuals heterozygous for alleles *B* and *b*	Individuals homozygous for allele *a*

This equation lets us calculate allele frequencies (values of p and q) in a very simple way. For example, in the population of 100 cats, if we assume white color to be a recessive trait, then the 16 white cats represent double-recessive individuals, q^2 in the equation. If $q^2 = .16$, then q must equal .40 (the square root of 16 is 4). Now recall that $p + q = 1$. p must thus equal $1 - .40$, or .60. What proportion of the black cats are heterozygotes? The equation tells us $2pq$ ($2 \times .40 \times .60$), which is .48. This result can be easily visualized in a Punnett square diagram (figure 9.17).

The Hardy-Weinberg theorem is based on certain assumptions. The previous equation is only true if the following four assumptions are met:

1. The size of the population is very large or effectively infinite.
2. Individuals mate with one another at random.
3. All alleles are replaced equally from generation to generation (natural selection is not occurring).
4. There is no input of new copies of any allele from any extraneous source (such as from a nearby population or from mutation).

How valid are the predictions made by the Hardy-Weinberg equation? For many genes, they prove to be very accurate.

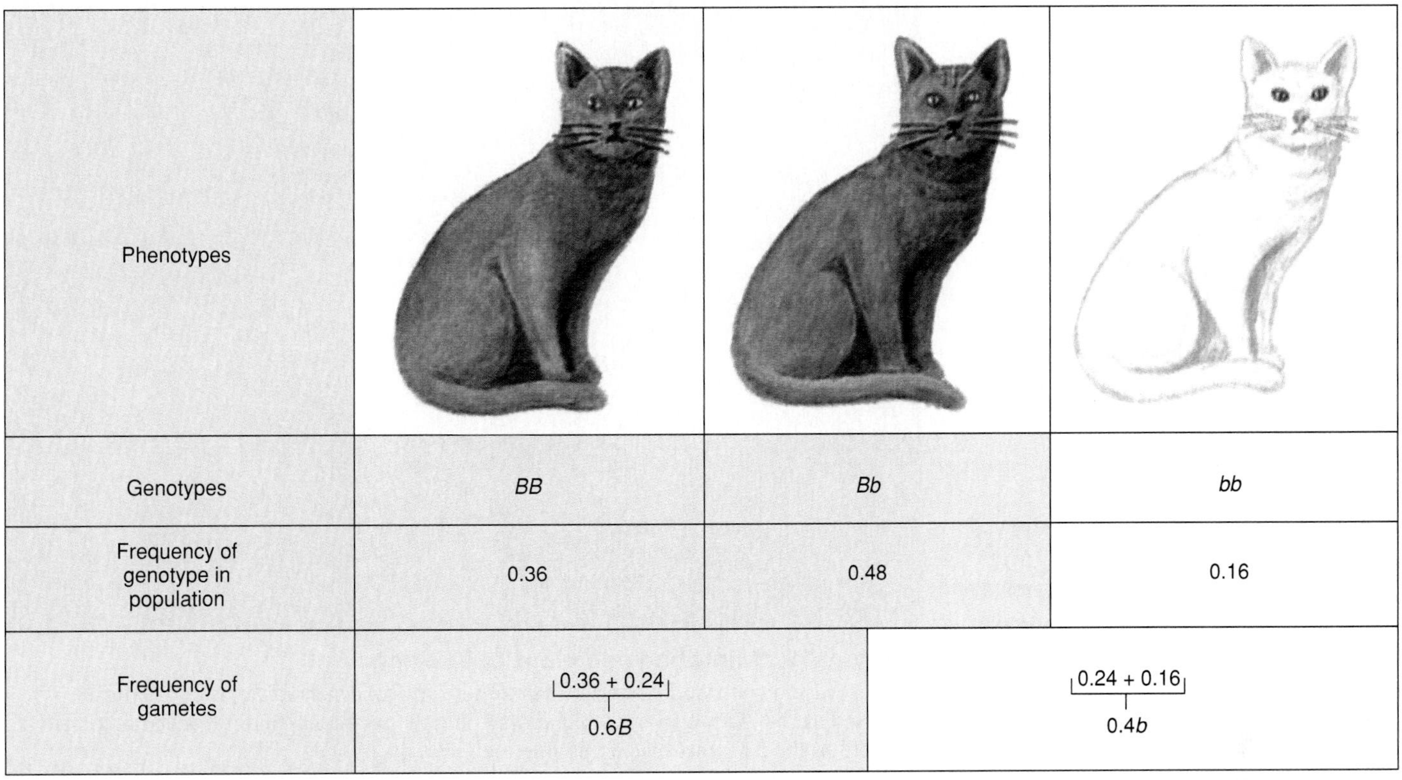

Phenotypes
Genotypes
BB
Bb
bb
Frequency of genotype in population
0.36
0.48
0.16
Frequency of gametes
0.36 + 0.24
0.6B
0.24 + 0.16
0.4b

Figure 9.16 The Hardy-Weinberg equilibrium.
In the absence of factors that alter them, the frequencies of gametes, genotypes, and phenotypes remain constant generation after generation. The example shown here involves a population of 100 cats, in which 16 are white and 84 are black. White cats are *bb*, and black cats are *BB* or *Bb*.

As an example, consider the recessive allele responsible for the serious human disease cystic fibrosis. This allele is present in white North America at a frequency of about 22 per 1,000 individuals, or .022. What proportion of white North Americans, therefore, is expected to express this trait? The frequency of double-recessive individuals (q^2) is expected to be .022 × .022, or 1 in every 2,000 individuals. What proportion is expected to be heterozygous carriers? If the frequency of the recessive allele q is .022, then the frequency of the dominant allele p must be 1 − .022, or .978. The frequency of heterozygous individuals ($2pq$) is thus expected to be 2 × .978 × .022, or 43 in every 1,000 individuals, very close to real estimates.

Most human populations are large and randomly mating and thus are similar to the ideal population envisioned by Hardy and Weinberg. For some genes, however, the calculated predictions do *not* match the actual values. The reasons why explain a great deal about evolution.

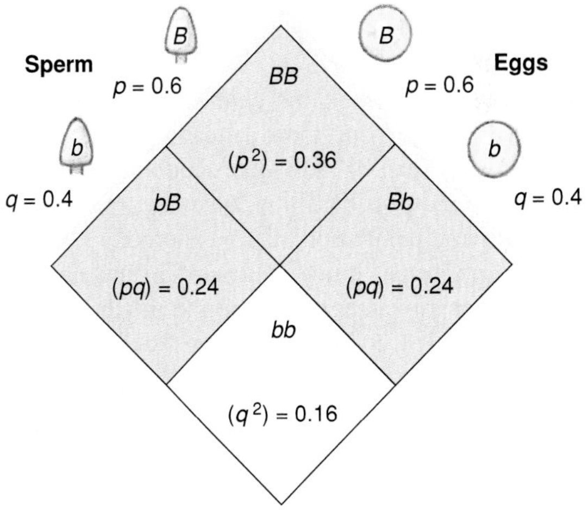

Figure 9.17 A Punnett square analysis.
The potential crosses in this cat population can be determined using a Punnett square analysis.

Why Do Allele Frequencies Change?

Many factors can alter allele frequencies. But only five alter the proportions of homozygotes and heterozygotes enough to produce significant deviations from the proportions predicted by the Hardy Weinberg principle: (1) mutation, (2) migration (including both immigration into and emigration out of a given population), (3) genetic drift (random loss of alleles, which is more likely to occur in small populations), (4) nonrandom mating, and (5) selection (table 9.1).

Mutation

A **mutation** is an error in replication of a nucleotide sequence in DNA. Mutation from one allele to another obviously can change the proportions of particular alleles in a population. But mutation rates are generally too low to significantly alter Hardy-Weinberg proportions of common alleles. Many genes mutate one to ten times per 100,000 cell divisions. Some of these mutations are harmful, while others are neutral or, even rarer, beneficial. This rate is so slow that few populations are around long enough to accumulate significant numbers of mutations.

Figure 9.18 Migration in a plant population.
The yellowish-green cloud around these Monterey pines, *Pinus radiata*, is pollen being dispersed by the wind, a form of migration. The male gametes within the pollen reach the egg cells of the pine passively in this way.

Migration

Migration, defined in genetic terms as the movement of individuals from one population into another, can be a powerful force upsetting the genetic stability of natural populations. Sometimes, migration is obvious, as when an animal moves from one place to another. If the characteristics of the newly arrived animal differ from those already there, and if the newly arrived individual or individuals can adapt to survive in the new area and mate successfully, then the genetic composition of the receiving population may be altered.

Other important kinds of migration are not as obvious. These subtler movements include the drifting of gametes of plants (figure 9.18) or immature stages of marine animals or plants from one place to another. However it occurs, migration can alter the genetic characteristics of populations and prevent the maintenance of the Hardy-Weinberg equilibrium. However, the evolutionary role of migration is more difficult to assess and depends heavily on the selective forces prevailing at the different places where the species occurs.

Table 9.1 Agents of Evolutionary Change

Factor	Description
Mutation	The ultimate source of variation. Individual mutations occur so rarely that mutation alone does not change allele frequency much.
Migration	A very potent agent of change. Migration acts to promote evolutionary change by enabling populations that exchange members to converge toward one another.
Genetic Drift	Statistical accidents. Usually occurs only in very small populations.
Nonrandom Mating	Inbreeding is the most common form. It does not alter allele frequency but decreases the proportion of heterozygotes ($2pq$).
Selection	The only form that produces *adaptive* evolutionary changes. Only rapid for allele frequency greater than .01.

Genetic Drift

In small populations, the frequencies of particular alleles may be changed drastically by chance alone. The individual alleles of a given gene may all be represented in few individuals, and some of them may be accidently lost if those individuals fail to reproduce or die. Allele frequencies appear to change randomly, as if the frequencies were drifting; thus, random loss of alleles is known as **genetic drift.** A series of small populations that are isolated from one another may come to differ strongly as a result of genetic drift.

When one or a few individuals migrate and become the founders of a new, isolated population at some distance from their place of origin, the alleles that they carry are of special significance. Even if these alleles are rare in the source population, they will be a significant fraction of the new population's genetic endowment. This is called the **founder effect.** As a result

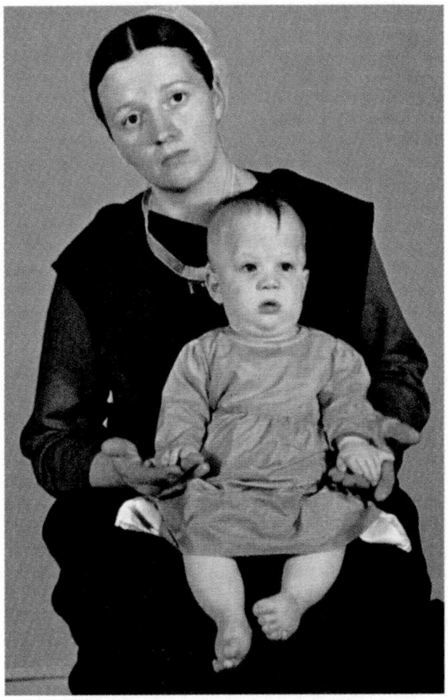

Figure 9.19 A consequence of the founder effect.

This Amish woman is holding her child, who has Ellis-van Creveld syndrome. The characteristic symptoms are short limbs, dwarfed stature, and extra fingers. This disorder was introduced in the Amish community by one of its founders in the eighteenth century and persists to this day because of the reproductive isolation of the Amish.

of the founder effect, rare alleles and combinations often become more common in new, isolated populations. The founder effect is particularly important in the evolution of organisms that occur on oceanic islands, such as the Galápagos Islands, which Darwin visited. Most of the kinds of organisms that occur in such areas were probably derived from one or a few initial founders. In a similar way, isolated human populations are often dominated by the genetic features that were characteristic of their founders, if only a few individuals were involved initially (figure 9.19).

Another ramification of genetic drift is known as the **bottleneck effect,** in which a small founder population becomes the sole source of alleles for a given species. The bottleneck effect can be seen in the current cheetah population. Researchers think that the cheetah population in Africa underwent some sort of crisis about 10,000 years ago, which depleted their numbers considerably. In the last century, another population crisis occurred as cheetahs were almost hunted to extinction. Their decreased numbers have limited cheetahs' genetic variability, with serious consequences. For example, cheetahs are susceptible to a number of fatal diseases. This decrease in genetic variability and its consequences has placed the cheetah population at risk, and scientists fear that the species may become extinct because of lack of allele variation.

Nonrandom Mating

Individuals with certain genotypes sometimes mate with one another either more or less commonly than would be expected on a random basis, a phenomenon known as **nonrandom mating.** One type of nonrandom mating characteristic of many groups of organisms is **inbreeding,** mating with relatives. Inbreeding increases the proportions of individuals that are homozygous, and as a result inbred populations contain more homozygous individuals than predicted by the Hardy-Weinberg principle. For this reason, populations of self-fertilizing plants consist primarily of homozygous individuals, whereas outcrossing plants, which interbreed with individuals different from themselves, have a higher proportion of heterozygous individuals (figure 9.20).

Selection

As Darwin pointed out, some individuals leave behind more progeny than others, and the likelihood they will do so is affected by their inherited characteristics. The result of this process is called **selection** and was familiar even in Darwin's day to breeders of horses and farm animals. In so-called **artificial selection,** the breeder selects for the desired characteristics. In **natural selection,** Darwin suggested the environment plays this role, with conditions in nature determining which kinds of individuals in a population are the most fit and so affecting the proportions of genes among individuals of future populations. This is the key point in Darwin's proposal that evolution occurs because of natural selection: the environment imposes the conditions that determine the results of selection and, thus, the direction of evolution.

Figure 9.20 Outcrossing in action.

This bee is pollinating a desert poppy in eastern Arizona. Flowers pollinated by bees and other insects or animals show more genetic variation (fewer homozygotes) than flowers that are self-pollinating because the pollen is brought to the flower from other individuals the bee has previously visited. In the terms of the Hardy-Weinberg principle, they are less inbred.

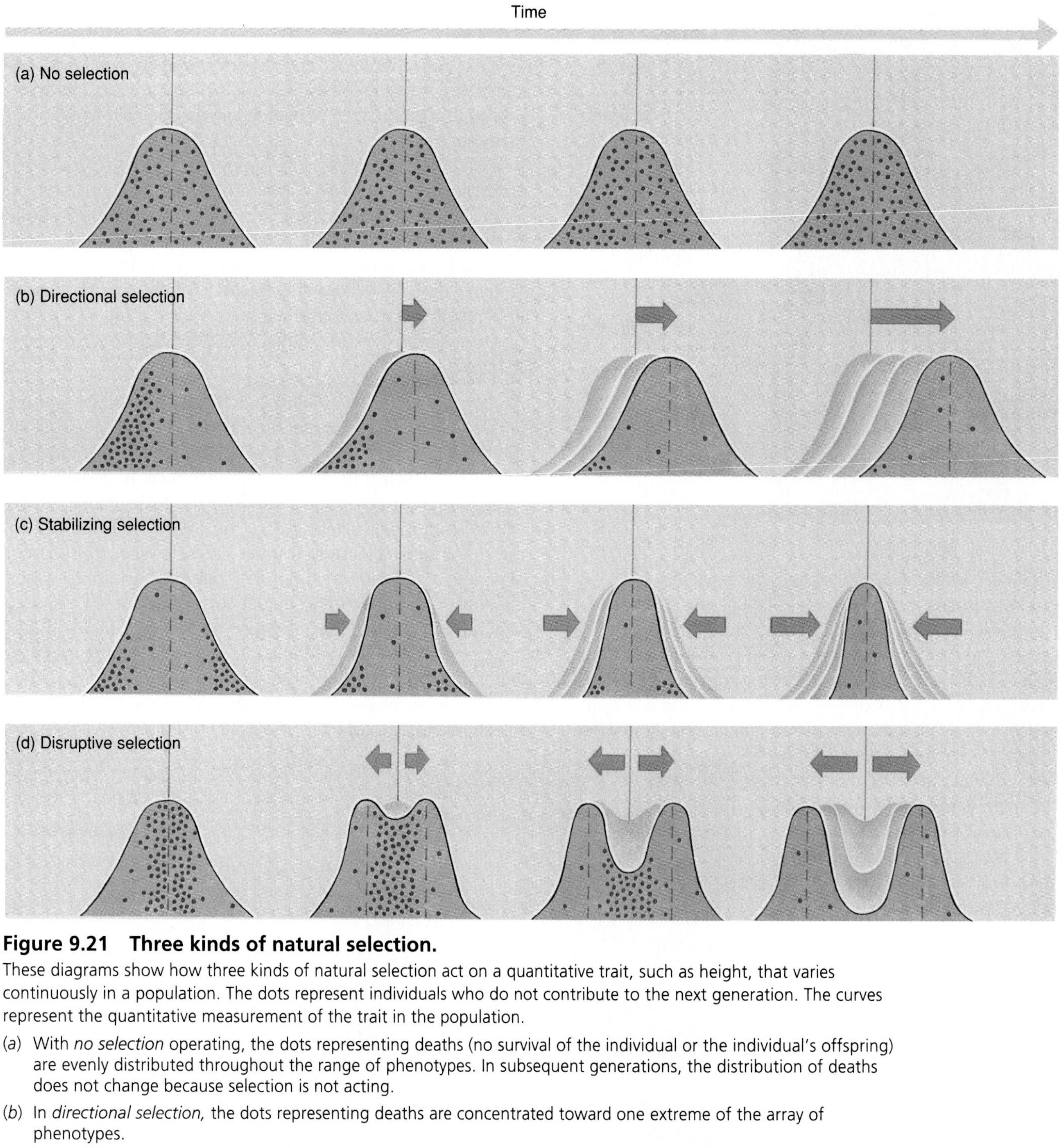

Figure 9.21 Three kinds of natural selection.

These diagrams show how three kinds of natural selection act on a quantitative trait, such as height, that varies continuously in a population. The dots represent individuals who do not contribute to the next generation. The curves represent the quantitative measurement of the trait in the population.

(*a*) With *no selection* operating, the dots representing deaths (no survival of the individual or the individual's offspring) are evenly distributed throughout the range of phenotypes. In subsequent generations, the distribution of deaths does not change because selection is not acting.

(*b*) In *directional selection,* the dots representing deaths are concentrated toward one extreme of the array of phenotypes.

(*c*) In *stabilizing selection,* deaths are scarce in the middle of the range and more concentrated on both ends of the range of phenotypes.

(*d*) In *disruptive selection,* deaths are concentrated in the middle of the range of phenotypes as the extreme forms of the trait are favored.

Forms of Selection

In nature, many traits are affected by more than one gene. Interactions between genes are typically complex. For example, alleles of many different genes play a role in determining human height. In such cases, selection operates on all the genes, influencing most strongly those that make the greatest contribution to the phenotype. How selection changes the population depends on which phenotypes are favored. Three types of natural selection have been identified: (1) directional selection, (2) stabilizing, or balancing, selection, and (3) disruptive selection (figure 9.21).

Directional Selection

When selection acts to eliminate one extreme from an array of phenotypes, the genes determining this extreme become less frequent in the population. The environment dictates which extreme will not be favored. Individuals exhibiting the other extreme of an array of phenotypes will leave more offspring than individuals with the unfavored form of the trait and individuals in the middle of the array. Thus, individuals exhibiting forms of a trait at one extreme become more common (figure 9.22). This form of selection is called **directional selection.**

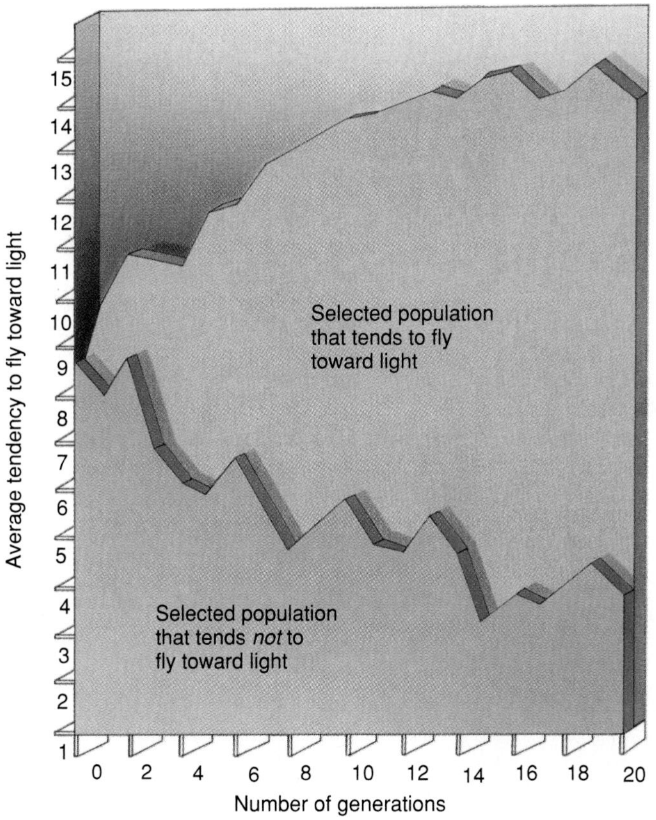

Figure 9.22 Directional selection in action.
In generation after generation, individuals of the fly *Drosophila* were selected for their tendency to fly toward light more strongly than usual (*red curve*). The *blue curve* represents flies selected *against* the tendency to fly toward light. The increasing red curve indicates that when flies with a strong tendency to fly toward light were used as the parents for the next generation, their offspring had a greater tendency to fly toward light. The decreasing blue curve shows the opposite: when flies with a strong tendency to *not* fly toward light were used as the parents for the next generation, their offspring had a reduced tendency to fly toward light.

Stabilizing Selection

When selection acts to eliminate *both* extremes from an array of phenotypes, the frequency of the intermediate type, which is already the most common, is increased. In effect, selection is operating to prevent change away from this middle range of values. In humans, infants with intermediate weight at birth have the highest survival rate (figure 9.23). In ducks and chickens, eggs of intermediate weight have the highest hatching success. This form of selection is called **stabilizing,** or **balancing, selection.**

Disruptive Selection

In some situations, selection acts to eliminate, rather than favor, the intermediate type. For example, in different parts of Africa, the color pattern of the butterfly *Papilio dardanus* is dramatically different, although in each instance it closely resembles the coloring of some other butterfly species that birds do not like to eat. Birds quickly detect and eat butterflies that do not resemble distasteful butterflies, so any intermediate color patterns are eliminated. In this case, selection is acting to eliminate the intermediate phenotypes. This form of selection is called **disruptive selection.** Disruptive selection is far less common than the other two types of selection.

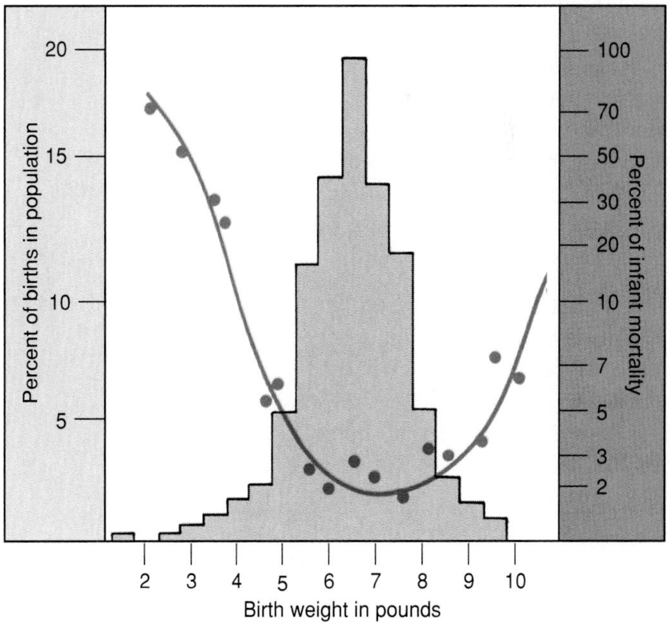

Figure 9.23 Stabilizing selection for birth weight.
The death rate among human babies is lowest at an intermediate birth weight between 7 and 8 pounds. Both larger and smaller babies have a greater tendency to die at or near birth. Stabilizing selection acts to eliminate both extremes from an array of phenotypes and to increase the frequency of the intermediate type.

9.4 Adaptation: Evolution in Action

Darwin's theory of evolution not only states that evolution has occurred, it also suggests a mechanism, natural selection. In the century since Darwin (and Wallace) suggested the pivotal role of natural selection, many examples have been found in which natural selection is clearly acting to change the genetic makeup of species, just as Darwin predicted. Two of the best-studied examples are sickle-cell anemia (a defect in human hemoglobin proteins) and industrial melanism (a darkening of wing pigmentation in moths).

Sickle-Cell Anemia

The first example we will consider is **sickle-cell anemia,** a hereditary disease affecting hemoglobin molecules in the blood. Sickle-cell anemia was first detected in 1910 in Chicago in a blood examination of an individual complaining of tiredness (figure 9.24). It is one of the best understood of all genetic disorders. The disorder arises as a result of a single nucleotide change in the gene encoding *β*-hemoglobin, which causes hemoglobin molecules to clump together (figure 9.25).

Persons homozygous for the sickle-cell genetic mutation in the *β*-hemoglobin gene usually die. This is because the sickled form of hemoglobin does not carry oxygen atoms well and red blood cells that are sickled do not flow smoothly through the tiny capillaries but instead jam up and block blood flow. Heterozygous individuals, who have both a defective and a normal form of the gene, make enough functional hemoglobin to keep their red blood cells healthy.

The Puzzle: Why So Common?

The disorder is now known to have originated in central Africa, where the frequency of the sickle-cell allele is about .12. One in 100 people are homozygous for the defective allele and develop the fatal disorder. Sickle-cell anemia affects roughly two American blacks out of every thousand but is almost unknown among other racial groups.

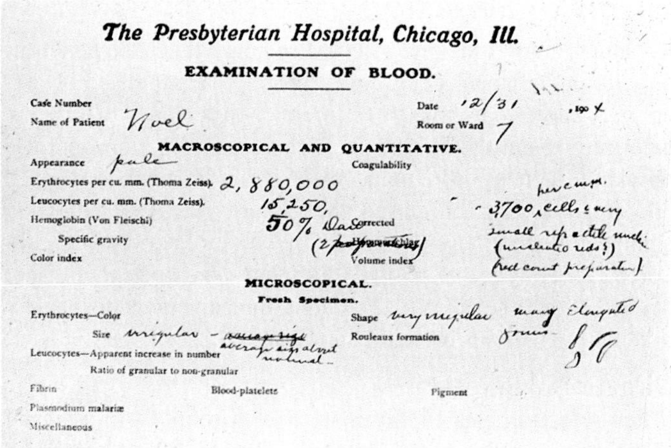

The Presbyterian Hospital, Chicago, Ill.

EXAMINATION OF BLOOD.

Case Number
Name of Patient Noel
Date 12/31, 190 4
Room or Ward 7

MACROSCOPICAL AND QUANTITATIVE.

Appearance pale
Coagulability
Erythrocytes per cu. mm. (Thoma Zeiss) 2,880,000
Leucocytes per cu. mm. (Thoma Zeiss) 15,250
Hemoglobin (Von Fleischl) 50% Dare
Corrected
Specific gravity
Color index
Volume index

MICROSCOPICAL.

Fresh Specimen.

Erythrocytes—Color
Size irregular
Shape very irregular
Rouleaux formation
Leucocytes—Apparent increase in number
Ratio of granular to non-granular
Fibrin
Blood-platelets
Pigment
Plasmodium malariæ
Miscellaneous

Figure 9.24 The first known sickle-cell anemia patient. Dr. Ernest Irons's blood examination report on his patient Walter Clement Noel, December 31, 1904, described his oddly shaped red blood cells.

Figure 9.25 Why the sickle-cell mutation causes hemoglobin to clump. The sickle-cell mutation changes the sixth amino acid in the β-hemoglobin chain (position B6) from glutamic acid (very polar) to valine (nonpolar). The unhappy result of this change is that the nonpolar valine at position B6, protruding from a corner of the hemoglobin molecule, fits nicely into a nonpolar pocket on the opposite side of another hemoglobin molecule. This causes the two molecules to clump together. As each molecule has both a B6 valine and an opposite nonpolar pocket, long chains form. When polar glutamic acid (the normal allele) occurs at position B6, it is not attracted to the nonpolar pocket, and no clumping occurs.

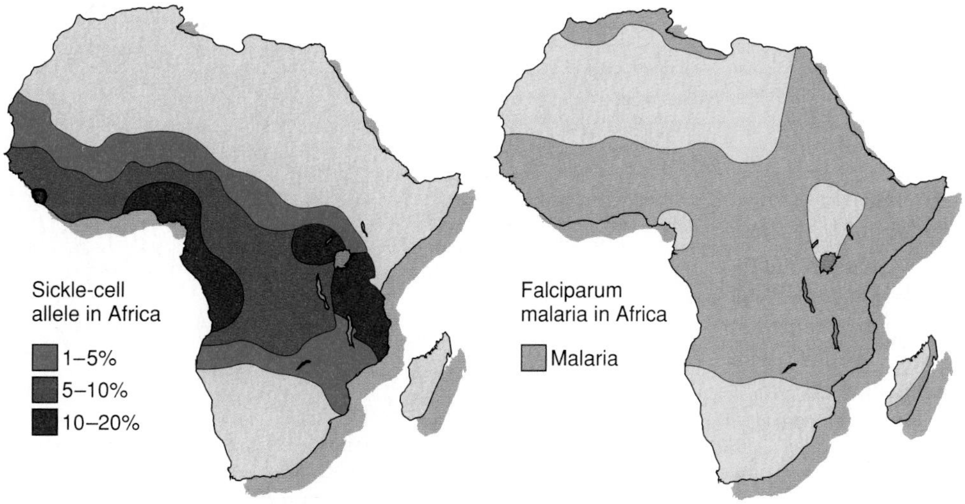

Figure 9.26 How balancing selection maintains sickle-cell anemia.
The diagrams show the frequency of the sickle-cell allele (*left*) and the distribution of falciparum malaria (*right*). Falciparum malaria is one of the most devastating forms of the often-fatal disease. As you can see, its distribution in Africa is closely correlated with that of the allele of the sickle-cell characteristic.

If Darwin is right, and natural selection drives evolution, then why has natural selection not acted against the defective allele in Africa and eliminated it from the human population there? Why is this potentially fatal allele instead very common there?

The Answer: Balancing Selection

The defective allele has not been eliminated from central Africa because people who are heterozygous for the sickle-cell allele are much less susceptible to malaria, one of the leading causes of death in central Africa (figure 9.26). Even though the population pays a high price—the many individuals in each generation who are homozygous for the sickle-cell allele die—the deaths are far fewer than would occur due to malaria if the heterozygous individuals were not malaria resistant. One in 5 individuals are heterozygous and survive malaria, while only 1 in 100 are homozygous and die of anemia. Natural selection has favored the sickle-cell allele in central Africa because the payoff in survival of heterozygotes there more than makes up for the price in death of homozygotes.

Balancing selection is thus acting on the sickle-cell allele: (1) selection tends to eliminate the sickle-cell allele because of its lethal effects on homozygous individuals; (2) selection tends to favor the sickle-cell allele because it protects heterozygotes from malaria. Like an economist balancing a budget, natural selection increases the frequency of an allele in a species as long as there is something to be gained by it, until the cost balances the benefit.

Balancing selection occurs because malarial resistance counterbalances lethal anemia. Malaria is a tropical disease that has not occurred in the United States since the early 1900s, and balancing selection has not favored the sickle-cell allele here. Blacks brought to America several centuries ago from Africa have not gained any evolutionary advantage in all that time from being heterozygous for the sickle-cell allele. There is no benefit to being resistant to malaria if there is no danger of getting malaria anyway. As a result, the selection against the sickle-cell allele in America is not counterbalanced by any advantage, and the allele has become far less common among American blacks than among blacks in central Africa. Thus, *directional* selection has acted to move the frequency of the sickle-cell allele in America in one direction (in this case, towards elimination). Among African Americans, many of whom have lived some 15 generations in a country where malaria has been relatively rare and is now essentially absent, sickle-cell anemia affects only 2 in 1,000 people. Using the Hardy-Weinberg equation, you can calculate that the frequency of the sickle-cell allele is the square root of .002, or approximately .045.

Industrial Melanism

A particularly well-studied case of evolution in action is **industrial melanism,** the darkening of insects in response to industrial pollution.

The Peppered Moth

Industrial melanism has occurred in many insect species, but the best-studied case is the European peppered moth, *Biston betularia.* Until the 1850s, almost all peppered moths in England had light-colored wings (figure 9.27). Dark-winged peppered moths, whose wings contained far more of the dark pigment melanin, were very rare and treasured by British butterfly and moth collectors. Starting around 1850, however, dark-colored peppered moths were found more often, usually in the more heavily industrialized areas of England. Every year after that, more of the dark peppered moths were seen, until after a hundred years, the peppered moth populations near industrial centers were almost 100% dark individuals.

Figure 9.27 Color variants of the peppered moth *Biston betularia.*
Before 1850, almost all moths captured in England were light-colored individuals. Black moths had a dominant allele that was present in populations, although very rare. The trunks of trees at that time were covered with light-colored lichens.

The Concealment Hypothesis

Darwin's theory that evolution occurs because of natural selection suggests a hypothesis to explain the replacement of light-colored moths by dark ones. The dark moths are most common in industrial regions where tree trunks are darkened almost black by the soot of pollution. Perhaps when dark moths sit on soot-darkened bark, they escape being eaten by birds because their dark color makes it hard for the birds to see them against the dark background. Light-colored moths, on the other hand, would stand out against a dark background like this and so would be ready prey for hungry birds. Darwin's theory thus suggests that natural selection is favoring the dark form of peppered moth in industrial areas because the dark color aids their survival, providing concealment from birds. Industrial melanism is a term used to describe the evolutionary process in which darker individuals come to predominate over lighter ones as a result of natural selection (figure 9.28).

Figure 9.28 Industrial melanism.
After 1850, black moths became more common in industrial regions of England. Air pollution that was spreading in the industrialized regions had killed many of the light-colored lichens that had previously occurred on tree trunks. Dark moths were much less conspicuous when resting on dark tree trunks than were light moths.

Testing the Hypothesis

To see if natural selection was really the cause of this evolutionary change in moth color, the concealment hypothesis was tested by the British ecologist H. B. D. Kettlewell in a series of field experiments. First, he reared populations of peppered moths in his laboratory. He marked the undersides of their wings with a dot of paint where birds couldn't see it, so he could recognize them later. When he had produced large numbers of each type of moth, he released equal numbers of light and dark individuals in two locations: woods near the city of Birmingham, which was heavily polluted, and woods in a rural area in Dorset, which was unpolluted. Rings of traps were set up around the woods to see which moths survived.

The result of this experiment was clear-cut (figure 9.29). In polluted Birmingham, two-thirds of the moths that survived to be trapped were dark, while in rural Dorset two-thirds of the survivors were light. The moths that matched the color of the tree trunks were surviving better! Many subsequent experiments confirmed the result of this experiment. Concealed observers could even see the birds passing by dark moths on polluted tree trunks to attack more conspicuous light-colored ones. Thus, the results of the experiment supported the hypothesis that natural selection was indeed the cause of industrial melanism, the dark form being favored in industrial areas.

Dozens of other species of moths have changed in the same way as the peppered moth in industrial areas throughout Europe and North America, with dark forms becoming more common as industrialization spread. In the second half of the twentieth century, as pollution has been brought under better control, the trends are being reversed. Dark peppered moths are becoming rarer. These changes provide clear evidence that allele frequencies in populations change in response to specific factors in the environment.

An Overview of Natural Selection

These case histories of sickle-cell anemia and industrial melanism are but two of many well-documented cases that provide clear evidence that evolutionary changes can be produced by natural selection. All of these changes share the same fundamental characteristic: The change makes the species better adapted to the environment in which it is living. In every case, *the environment dictates the direction and extent of the change.*

This is the key lesson we have learned about evolution: Evolution is not predictable; instead, it is directed by the environment. Just as a football coach has a team try a variety of plays but keeps in the team's game plan only those plays that work, so a species keeps only those changes that "work." The species doesn't decide which changes to keep, any more than a football coach does. The pattern of success determines the outcome.

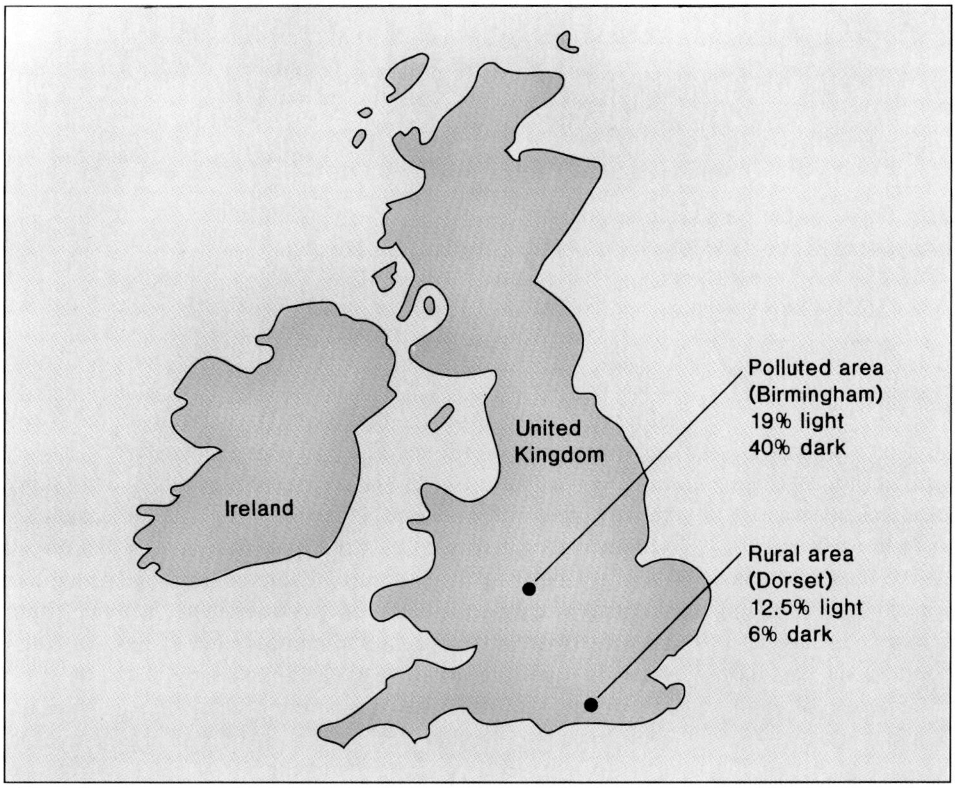

Figure 9.29 Results of Kettlewell's industrial melanism study.
Kettlewell found more dark moths in areas affected by industrial pollution and more light moths in areas with little industrial pollution.

Figure 9.30 Great genetic variation can occur within a single species.
"Looking different" does not mean two animals are members of different species. All of the different breeds of dogs are members of the same species, *Canis familiaris,* and all can breed successfully with each other. Dog breeders engaging in artificial selection are responsible for the great differences among dog breeds.

9.5 How Species Form

A key aspect of Darwin's theory of evolution is his proposal that adaptation (microevolution) leads ultimately to species formation (macroevolution). The way natural selection leads to the formation of new species has been thoroughly documented by biologists, who have observed the stages of the species-forming process, or speciation, in many different plants, animals, and microorganisms. Speciation usually involves progressive change: first, local populations become increasingly specialized; then, if they become different enough, natural selection may act to keep them that way.

The Species Concept

How to define a species was and still is the topic of much debate. Originally, a species was defined as all the individuals in a group able to breed with one another and produce progeny that were still of that species. Thus all dogs, although many appear very dissimilar, are considered members of the same species (figure 9.30). Even if two different-looking individuals appeared among the progeny, they were still considered to belong to the same species. Others define a species as a group of individuals that has its own distinctive role in nature, occupies a particular type of habitat, or displays different activities. The fact that some species of trees, groups of mammals, and fishes can form hybrids with one another, even though they may not do so in nature, complicates the situation further. A **species,** therefore, is generally defined as a group of organisms that is unlike other such groups and that does not integrate extensively with other groups in nature.

Species formation is generally considered to be the final stage of a long evolutionary process: (1) First, local populations become adapted to the particular set of circum-

Figure 9.31 Ecological races of the seaside sparrow.
Subspecies of the seaside sparrow, *Ammodramus maritimus*, are quite local in distribution, and some of them are even in danger of extinction because of the alteration of their habitats. The widespread subspecies *Ammodramus maritimus maritimus* (*1*) is the most common. *A. m. fisheri* (*2*) occurs along the gulf coast, and the Cape Sable seaside sparrow, *A. m. mirabilis* (*3*), occurs in a small area of southwestern Florida. The dusky seaside sparrow, *A. m. nigrescens* (*4*), occurred only near Titusville, Florida. The last individual, a male, died in captivity in 1987.

stances that they face. (2) When they become different enough, they are considered to be ecological races. (3) Then, natural selection may act to reinforce the differences between two races by favoring changes that discourage hybrids. These sorts of changes are called isolating mechanisms. (4) Finally, the two races become incapable of interbreeding and are considered separate species.

Ecological Races

Because natural selection favors reproductive success, which usually means increased survival, it continually molds and shapes a species to improve the "fit" between the species and its environment. When a species lives in several different kinds of environments, selection acts in each case to improve the local population, and this makes the separate populations progressively more different from one another, each becoming better suited to the particular challenges of living where it does. Over time, if their localities are different enough, the local populations can become quite distinct, forming what biologists call **ecological races.** Ecological races are populations of the same species that differ genetically because they are adapted to very different living conditions. Members of different ecological races are not yet so different as to be different species, but they have taken the first step on that road (figure 9.31).

Isolating Mechanisms

Often ecological races continue to diverge, becoming more and more different from each other as natural selection favors different survival strategies in different environments. Eventually a point is reached when the races are so different that biologists consider them separate species. Very often, selection favors changes called **isolating mechanisms,** which prevent the new species from breeding with each other, particularly when the hybrids produced by such matings are ill-suited to either environment. At this point, two species are said to be **reproductively isolated.** Three examples of isolating mechanisms are geographic isolation (figure 9.32), ecological isolation (figure 9.33), and behavioral isolation (figure 9.34).

The reproductive isolation between species is created and maintained by two basic kinds of isolating mechanisms: **prezygotic** isolating mechanisms, which prevent the formation of a zygote, and **postzygotic** isolating mechanisms, which prevent the proper development or functioning of zygotes after they are formed. There are many ways that species achieve prezygotic and postzygotic isolation, the most common of which are outlined in table 9.2.

Table 9.2	Mechanisms for Reproductive Isolation
Geographic isolation	species occur in different areas, which are often separated by a physical barrier such as a river or mountain range.
Ecological isolation	species occur in the same area but they occupy different habitats. Survival of hybrids is low because they are not adapted to either environment of their parents.
Temporal isolation	species reproduce in different seasons or at different times of the day.
Behavioral isolation	species differ in their mating rituals.
Mechanical isolation	structural differences between species prevent mating.
Prevention of gamete fusion	gametes of one species function poorly with the gametes of another species or within the reproductive tract of another species.

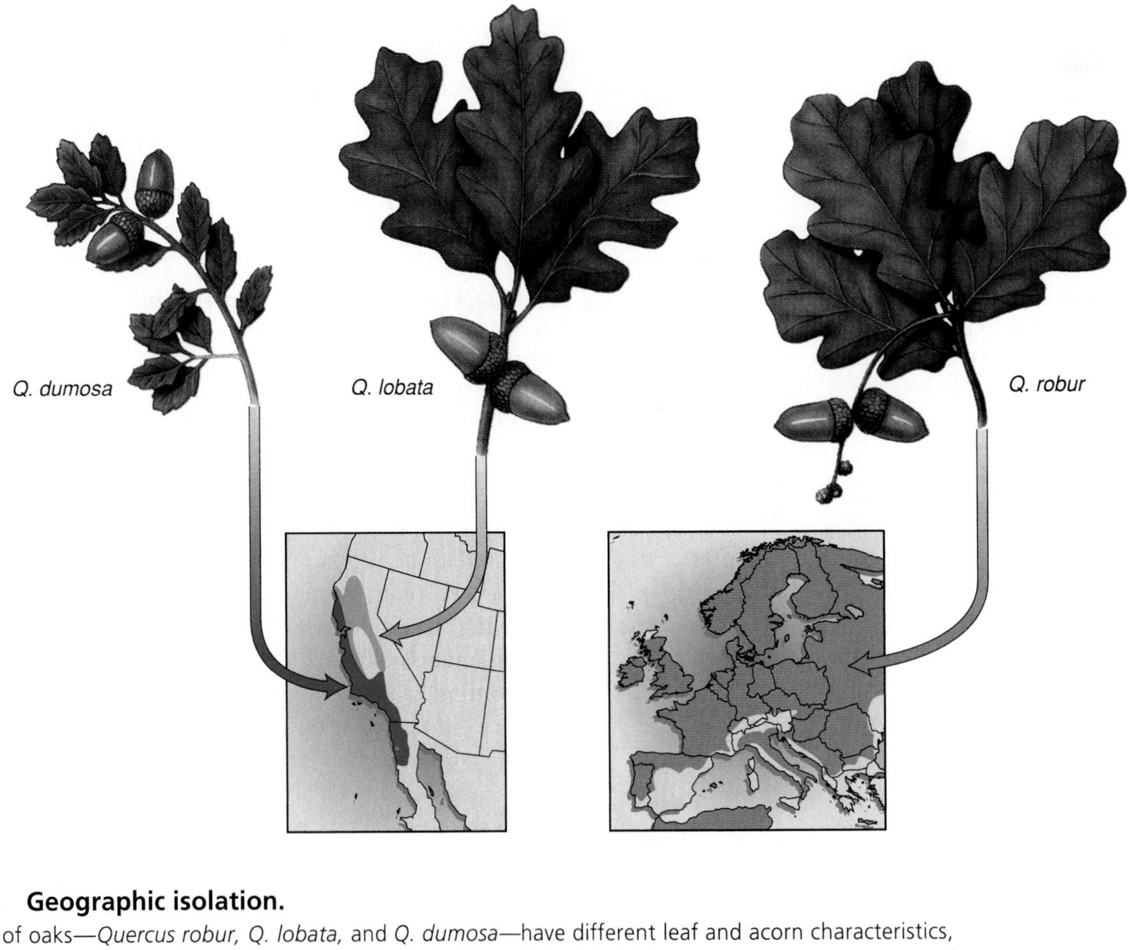

Figure 9.32 Geographic isolation.
Three species of oaks—*Quercus robur, Q. lobata,* and *Q. dumosa*—have different leaf and acorn characteristics, yet all of them could hybridize with one another. Geographic distance, however, keeps the species separate: *Q. lobata* and *Q. dumosa* are found in different habitats in California, while *Q. robur* occurs in Europe.

Figure 9.33 Ecological isolation.
These closely related species of oak—*Quercus lobata* (*a*) and *Quercus dumosa* (*b*)—occur in different habitats and are thus adapted to different environmental pressures.

Figure 9.34 Behavioral isolation.
Males and females of closely related species of Hawaiian *Drosophila* engage in different forms of courtship and mating rituals. (*a*) *Drosophila silvestris.* After approaching the female from the rear, the male lunges forward while vibrating his wings. Then, with his head under the wings of the female, he raises his forelegs up and over the female's abdomen. (*b*) *D. heteroneura.* A male with extended wings approaches a female in a typical courtship posture. (*c*) *D. heteroneura*. Two males interlock antennae as part of the aggressive behavior involved in territorial defense. (*d*) *D. clavisetae.* In this species, the males raise their abdomens up and over their backs and spray a chemical over the female. This chemical acts as a mating signal to attract the female.

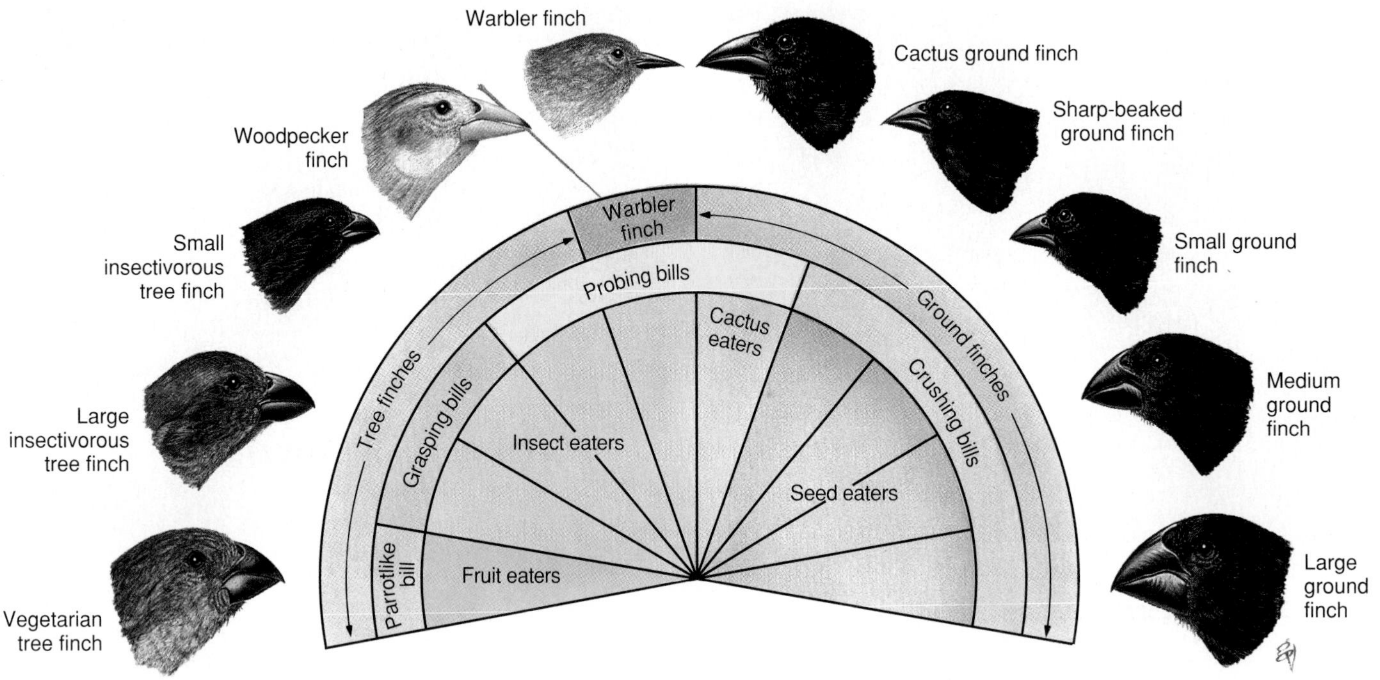

Figure 9.35 Darwin's finches.
Ten species of Darwin's finches from Isla Santa Cruz, one of the Galápagos Islands, showing differences in bills and feeding habits. The bills of several of these species resemble those of different distinct families of birds on the mainland. This condition presumably arose when the finches evolved new species in habitats where other kinds of birds normally occur on the mainland and which are not available to finches there. The woodpecker finch uses cactus spines to probe in crevices of bark and rotten wood for food. All of these birds are thought to have been derived from a single common ancestor, a finch like that shown in figure 9.5*a*.

Adaptive Radiation

One of the most visible manifestations of evolution is the existence of clusters of closely related species. These species often have evolved relatively recently from a common ancestor. The phenomenon by which they change, coming to occupy a series of different habitats within a region, is called **adaptive radiation.** Such clusters are often particularly impressive on islands, in series of lakes, or in other sharply discontinuous habitats. One example of a cluster of species that has undergone adaptive radiation to fill a wide range of habitats is Darwin's finches (figure 9.35).

Another even more striking example occurs among the fruit flies of the Hawaiian Islands. More than a third of all the world's species of *Drosophila,* over 500 species, are found on these islands! They are thought to have arisen from a single common ancestor that reached the first island to form. The other islands arose from the sea one after another, and as each new island arose it was invaded by the groups of *Drosophila* present on the older islands. The result has been wave after wave of species formation, one of the most remarkable examples of macroevolution in action anywhere on earth.

Does Evolution Occur in Spurts?

While biologists accept Darwin's theory that natural selection leads to evolution, they are currently engaged in a lively discussion about the *rate* at which macroevolutionary changes occur. Darwin's theory makes no prediction about this, but most biologists have assumed that species formation is a gradual process that goes on all the time. This hypothesis of ongoing gradual evolutionary change is called **gradualism.**

Recently, some biologists have suggested that the underlying assumption of gradualism, that macroevolution proceeds at a uniform pace, is incorrect. They argue instead that macroevolution occurs in fits and starts. In their view, major environmental upheavals in the past, by creating new challenges to living, have been responsible for the formation of most new species. Because such drastic environmental changes occur very infrequently—separated by quiet periods lasting tens of millions of years—these biologists argue that species formation should occur in spurts, too, separated by long periods of environmental equilibrium in which little change in the environment occurs and few new species are formed. Their proposal, called the hypothesis of **punctuated equilibrium,** predicts that the fossil record will be very discontinuous. Is it? The fossil record of many animals appears "jerky," as this hypothesis predicts. However, the fossil record of other species seems to exhibit progressive change. Because this hypothesis goes to the heart of how macroevolution occurs, it continues to be a focus of interest, and discussion continues.

CHAPTER 9

HIGHLIGHTS

Key Terms | Key Concepts

9.1 Darwin

Key Terms

evolution 171
natural selection 171
Galápagos Islands 172
On the Origin of Species 174

Key Concepts

- In 1831 Darwin began a trip around the world, closely observing the plants and animals he saw.
- In 1859 Darwin published *On the Origin of Species,* in which he proposed that the mechanism of evolution was natural selection.

9.2 The Evidence for Evolution

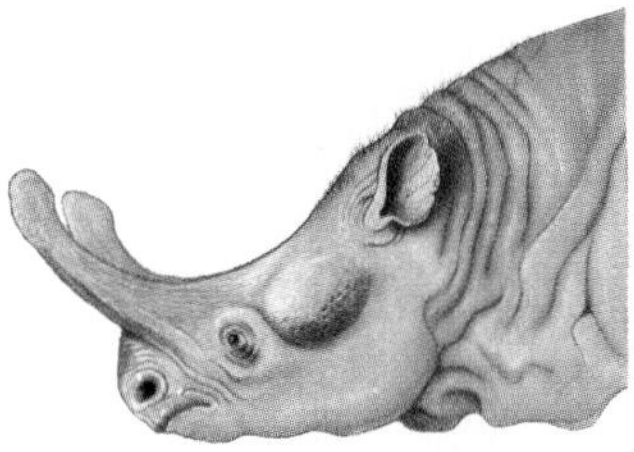

Key Terms

fossil 175
radioisotopic dating 175
molecular clock 178
homologous structure 180

Key Concepts

- If we date fossils, and order them by age, progressive changes are seen. This is direct evidence that evolution has occurred.
- The progressive accumulation of molecular differences and comparisons of living organisms provide additional strong evidence that evolution has occurred.

9.3 How Populations Evolve

Key Terms

microevolution 182
allele frequency 182
Hardy-Weinberg equilibrium 182
directional selection 187
balancing selection 187

Key Concepts

- In a population not undergoing significant evolutionary change, two alleles present in frequencies p and q will be distributed among the genotypes in the proportions $p^2 + 2pq + q^2$, the Hardy-Weinberg equilibrium.
- Allele frequencies change in nature due to mutation, migration, drift, nonrandom mating, and selection.

9.4 Adaptation: Evolution in Action

Key Terms

sickle-cell anemia 188
industrial melanism 190

Key Concepts

- A mutation in hemoglobin causes a condition known as sickle-cell anemia. This recessive mutation is common in central Africa because it renders heterozygous individuals resistant to malaria.
- Vegetation darkened by industrial soot has favored the evolution of darker moths, better concealed from their predators.

9.5 How Species Form

Key Terms

species 192
ecological race 193
isolating mechanism 194
punctuated equilibrium 196

Key Concepts

- Microevolution leads to macroevolution. Adaptation to local habitats leads to divergence and the evolution of ecological races.
- Isolating mechanisms then reinforce the differences, leading to reproductive isolation and species formation.

9

CONCEPT REVIEW

1. Which of the following is *not* one of Darwin's pieces of evidence that evolution occurs?

a. Species on oceanic islands show strong similarities to species living on the nearest mainland.

b. The earth was created in 4004 B.C.

c. Progressive changes in characteristics of plants and animals can be seen in successive rock layers.

d. The plants and animals of each continent are distinctive.

2. Darwin was greatly influenced by an essay written by _________, which pointed out that population growth is geometric while increase in food is arithmetic.

a. Emma Wedgewood

b. Alfred R. Wallace

c. James Usher

d. Thomas Malthus

3. Darwin believed that the major driving force in evolution was

a. natural selection.

b. scientific creation.

c. uniformitarianism.

d. molecular biology.

4. Organisms that are more distantly related have had more time to accumulate genetic differences.

a. true

b. false

5. Which of the following is *not* a factor that causes change in the proportions of homozygous and heterozygous individuals in a population?

a. mutation

b. migration

c. genetic drift

d. random mating

e. selection against a specific allele

6. If you came across a population of plants and discovered a surprisingly high level of homozygosity, what would you predict about their mating system?

a. Their pollen is dispersed by the wind.

b. They probably reproduce asexually.

c. They are predominantly outcrossing.

d. They are predominantly self-fertilizing.

7. Heterozygotes for the sickle-cell allele have less resistance to malaria than individuals who do not have the allele.

a. true

b. false

8. Which type of selection acts to eliminate both extremes from an array of phenotypes?

a. directional

b. stabilizing

c. disruptive

d. intensive

9. Which is *not* a prezygotic isolating mechanism?

a. geographical isolation

b. ecological isolation

c. seasonal isolation

d. hybrid sterility

e. mechanical isolation

10. The arm of a human and the fin of a porpoise are considered to be _________ structures, made of the same bones, which have been modified in size and shape for different uses.

11. The _________ record offers the most direct line of evidence supporting evolution.

12. In the process of dating fossils, a common isotope that changes over time is _________-14.

13. By studying changing genes through time, biologists have studied _________ clocks.

14. The human _________ is an example of a vestigial organ without an apparent function.

15. Genetic change within populations, which is the result of adaptation, is called _________.

16. Over time, local populations can become quite distinct, forming what biologists call ecological _______ .

Answers to the Concept Review questions appear in Appendix B.

CHALLENGE YOURSELF

1. Imagine that you sit on the Supreme Court and are hearing a case in which it is argued that creation science should be taught in public schools alongside evolution as a legitimate alternative scientific explanation of biological diversity. What is the best case that lawyers might make for and against this proposition? How would you vote and why?

2. Will a dominant allele that is lethal be removed from a large population as a result of natural selection? What factors might prevent this from happening? What if the lethal allele is recessive?

3. In a large, randomly mating population with no forces acting to change gene frequencies, the frequency of homozygous recessive individuals for the characteristic of extra-long eyelashes is 90 per 1,000 or .09. What percent of the population carries this desirable trait but displays the dominant phenotype, short eyelashes? Would the frequency of the extra-long eyelash allele increase, decrease, or remain the same if long-lashed individuals preferentially mated with each other and no one else?

FOR FURTHER READING

Darwin, C. R. *On the Origin of Species by Means of Natural Selection, or the Preservation of Favoured Races in the Struggle for Life*. New York: Cambridge University Press, 1975 reprint. One of the most important books of all time, Darwin's long essay is still comprehensible and interesting to modern readers.

Darwin, C. R. *The Voyage of the* Beagle. Garden City. N.Y.: Natural History Press, 1962 reprint. Darwin's own account of his observations and adventures during the famous five-year voyage he took in his twenties.

Gillis, A. "Getting a Picture of Human Diversity." *Bioscience* 44, (January 1994): 8–11. A report on how population geneticists are using variation in human genes to track the history of *Homo sapiens*.

Gould, S. J. *Ever Since Darwin*. New York: W. W. Norton, 1977. An entertaining and insightful collection of essays on evolution and Darwinism.

Malthus, T. R. *An Essay on the Principle of Population, or A View of Its Past and Present Effects on Human Happiness; with an Inquiry into Our Prospects Respecting the Future Removal or Mitigation of the Evils Which It Occasions*. New York: Cambridge University Press, 1992. One of the many scholarly presentations and analyses of Malthus's famous discourse on the control of populations.

O'Brien, S., D. E. Wildt, and M. Bush. "The Cheetah in Genetic Peril." *Scientific American,* May 1986, 84–92. Evidence for an ancient population bottleneck in cheetahs, which has put the survival of the species in doubt.

Raup, D. "The Role of Extinction in Evolution." *Proc. Nat. Acad. Sci. USA* 91 (July 1994): 6758–63. A very interesting reevaluation of the role extinction should play in evolutionary theory.

Webb, G. *The Evolution Controversy in America*. Lexington, KY: University Press of Kentucky, 1994. A recent update on the ongoing battle between creationists and evolutionists.

Weiner, J. *The Beak of the Finch: Evolution in Real Time*. New York: Alfred A. Knopf, 1994. A highly recommended account of ongoing studies of evolution in action among Darwin's Galápagos finches.

TECHNOLOGY LINKS

The Living World Home Page
http://www.wcbp.com/biology/tlw

10

How We Name Living Things

CHAPTER OUTLINE

Figure 10.1 Carolus Linnaeus (1707–78).
This Swedish biologist devised the system of naming organisms that is still in use today.

Our world is populated by some 10 million different kinds of organisms. In order to talk about them and study them, it is necessary to give them names, just as it is necessary that people have names. Of course you cannot remember the name of every kind of organism, any more than the governor of California can recall the name of every person living in the state. In order to do his job, the governor groups the U.S. citizens living in his or her state into categories of various kinds—those who live in Los Angeles County, for example, or other counties like San Francisco, Riverside, or San Diego. Among those state residents who live in Los Angeles County, some live in Hollywood and others live in Watts or other neighborhoods. Among those who live in Hollywood, some live on Vine Street and others on other streets. Finally, among those living on Vine Street, a particular family lives in the house numbered 222, in apartment #3. Thus the governor can identify and deal with a particular city resident out of millions by categorizing the person by county, city, borough, neighborhood, street, house, and apartment. We call this kind of multilevel grouping of individuals **classification.** The classification scheme biologists use was invented by Carolus Linnaeus (figure 10.1) over 230 years ago.

10.1 The Classification of Organisms

Organisms were first classified more than 2,000 years ago by the Greek philosopher Aristotle, who categorized living things as either plants or animals. He classified animals as either land, water, or air dwellers, and he divided plants into three kinds based on stem differences. This simple classification system was expanded by the Greeks and Romans, who grouped animals and plants into basic units such as cats, horses, and oaks. Eventually, these units began to be called **genera** (singular, **genus**), the Latin word for "group." Starting in the Middle Ages, these names began to be systematically written down, using Latin, the language used by scholars at that time. Thus, cats were assigned to the genus *Felis*, horses to *Equus*, and oaks to *Quercus*—names that the Romans had applied to these groups. For genera that were not known to the Romans, new names were invented.

The Invention of the Linnaean System

The classification system of the Middle Ages, called the polynomial system, was used virtually unchanged for hundreds of years, until it was replaced about 200 years ago by the **binomial system** introduced by Linnaeus.

The Polynomial System

Until the mid-1700s, biologists usually added a series of descriptive terms to the name of the genus when they wanted to refer to a particular kind of organism, which they called a **species.** These phrases, starting with the name of the genus, came to be known as **polynomials** (*poly,* "many"; *nomial,* "name"), strings of Latin words and phrases consisting of up to 12 or more words. This would be like the mayor of New York referring to a particular citizen as "Brooklyn resident: Democrat, male, Asian American, middle income, Protestant, elderly, likely voter, short, bald, heavyset, wears glasses, works in the Bronx selling shoes." As you can imagine, these polynomial names were cumbersome. Even more worrisome, the names were altered at will by later authors, so that a given organism really did not have a single name that was its alone.

The Binomial System

A much simpler system of naming animals, plants, and other organisms stems from the work of the Swedish biologist Carolus Linnaeus (1707–78). Linnaeus devoted his life to a challenge that had defeated many biologists before him—cataloging all the different kinds of organisms. In the 1750s he produced several major works that, like his earlier books, employed the polynomial system. But as a kind of shorthand, Linnaeus in these books also included a two-part name for each species. These two-part names, or **binomials** (*bi* is the Latin prefix for "two") have become our standard way of designating species. For example, he designated the willow oak *Quercus phellos* and the red oak *Quercus rubra* (figure 10.2), even though he also included the polynomial name for these species. This naming is like the mayor of New York referring to the Brooklyn resident as Sylvester Kingston—we also use binomial names for ourselves, our so-called "given" and "family" names.

Figure 10.2 How Linnaeus named two species of oaks.
(*a*) Willow oak, *Quercus phellos.* (*b*) Red oak, *Quercus rubra.* Although they are clearly oaks (members of the genus *Quercus*), these two species differ sharply in the shapes and sizes of their leaves and in many other features, including their overall geographic distributions.

Taxonomy

A group of organisms at a particular level in a classification system is called a **taxon** (plural, taxa), and the branch of biology that identifies and names such groups of organisms is called **taxonomy.** Taxonomists are in a real sense detectives, biologists who must utilize clues of appearance and behavior to identify and assign names to organisms.

By agreement among taxonomists throughout the world, no two organisms can have the same name. So that no one country is favored, a language spoken by no country—Latin—is used to make the names. Because the scientific name of an organism is the same anywhere in the world, this system provides a standard and precise way of communicating, whether the language of a particular biologist is Chinese, Arabic, Spanish, or English. This is a great improvement over the use of common names, which often vary from one place to the next. As you can see in figure 10.3, corn in Europe refers to the plant Americans call wheat; a bear is a large placental omnivore in the United States but a koala (a vegetarian marsupial) in Australia; and a robin is a very different bird in Europe and North America.

Also by agreement, the first word of the binomial name is the genus to which the organism belongs. This word is always capitalized. The second word refers to the particular species and is not capitalized. The two words together are called the **scientific name** and are written in italics. The system of naming animals, plants, and other organisms established by Linnaeus has served the science of biology well for nearly 230 years.

Figure 10.3 Common names make poor labels.
The common names corn (*a*), bear (*b*), and robin (*c*) bring clear images to our minds (photos on *right*), but the images would be very different to someone living in Europe or Australia (photos on *left*). There, the same common names are used to label very different species.

Higher Categories

Like the governor of California, a biologist needs more than two categories to classify all the world's living things. In our example, the governor might employ seven categories (county, city, borough, neighborhood, street, house, apartment), and so did Linnaeus (figure 10.4). To make his universal catalog of organisms more useful, Linnaeus grouped the genera with similar properties into a cluster he called a **family** (figure 10.5), and he placed similar families into the same **order.** Orders with common properties were placed into the same **class,** and classes with similar characteristics into the same **phylum** (plural, phyla). Botanists (that is, those who study plants) used to call plant phyla "divisions," but that practice is falling out of fashion. Finally, the phyla were assigned to one of several great groups, the **kingdoms.** Biologists currently recognize six kingdoms: two kinds of bacteria (Archaebacteria and Eubacteria), a largely unicellular group of eukaryotes (Protista), and three multicellular groups (Fungi, Plantae, and Animalia).

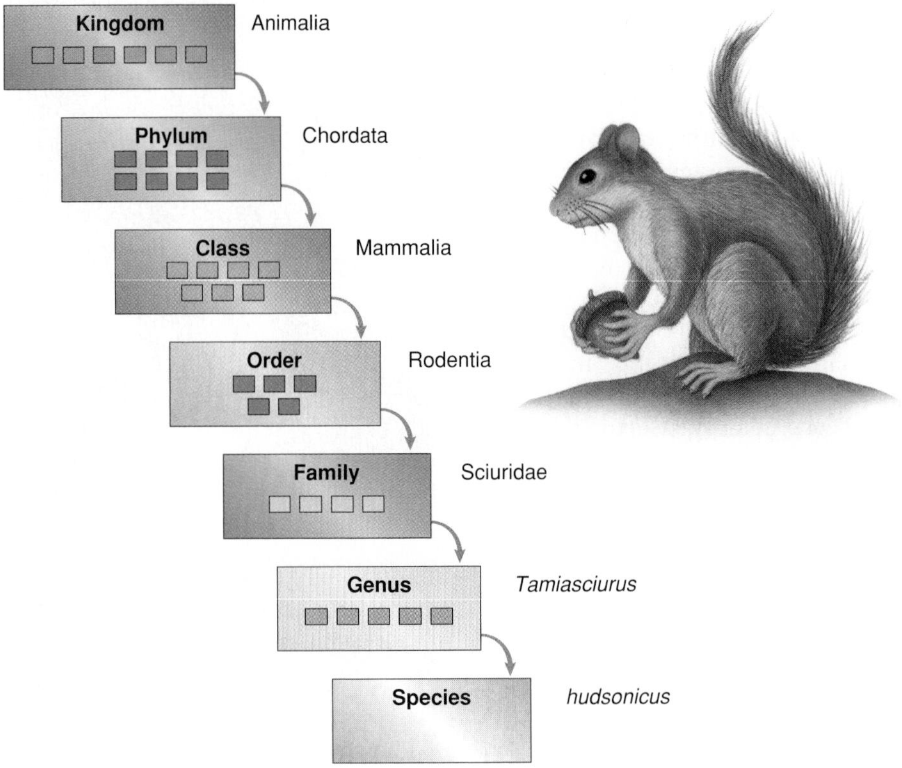

Figure 10.4 The hierarchical system used to classify an organism.
This "box-within-a-box" diagram clearly demonstrates how the different taxa are inclusive of one another in this classification of a squirrel.

Figure 10.5 Diversity within a family: squirrels.
All of these squirrels have the same classification up until their genus. This classification is kingdom Animalia, phylum Chordata, class Mammalia, order Rodentia, and family Sciuridae. (*a*) *Tamiasciurus hudsonicus*, or red squirrel, is an avid tree-dweller. (*b*) *Eutamias townsendii*, or Townsend's chipmunk, is a member of a genus of diurnal, brightly colored, very active ground-dwelling squirrels. (*c*) *Marmota flaviventris*, the yellow-bellied marmot, is a large, playful squirrel that lives in burrows.

In order to remember the seven categories in their proper order, it may prove useful to memorize a phrase such as "**k**indly **p**ay **c**ash **o**r **f**urnish **g**ood **s**ecurity" (**k**ingdom–**p**hylum–**c**lass–**o**rder–**f**amily–**g**enus–**s**pecies).

Each of the categories at every level in the Linnaean system of classification is loaded with information (figure 10.6). Just as knowledge of your zip code may reveal a great deal about your potential spending habits and voting preferences, so knowledge of how an organism is classified implies both a series of characteristics shared by members of the group and also a series of organisms that belong to it (table 10.1).

Take, for example, a honeybee, and look at the seven levels of its classification: The name of the species (level 1) is *Apis mellifera*. Its genus name (level 2) *Apis* is a member of the family Apidae (level 3). All members of this family are bees, some solitary, others living in hives as *A. mellifera* does. Knowledge of its order (level 4), Hymenoptera, gives you a more general kind of information about *A. mellifera*. This order also includes wasps and hornets, and knowing that *A. mellifera* belongs to it reveals that it is like these others in many respects. The honeybee is likely to be able to sting, for example, and may live in colonies. At an even more general level, the class (level 5) of *A. mellifera,* which is Insecta, tells us that the honeybee is an insect, possessing as all insects do three major body segments, with wings and three pairs of legs attached to the middle segment. Its phylum (level 6), Arthropoda, tells us that the honeybee is an arthropod, with a hard cuticle of chitin and jointed appendages. Its kingdom (level 7), Animalia, tells us that *A. mellifera* is a multicellular heterotroph whose cells lack cell walls.

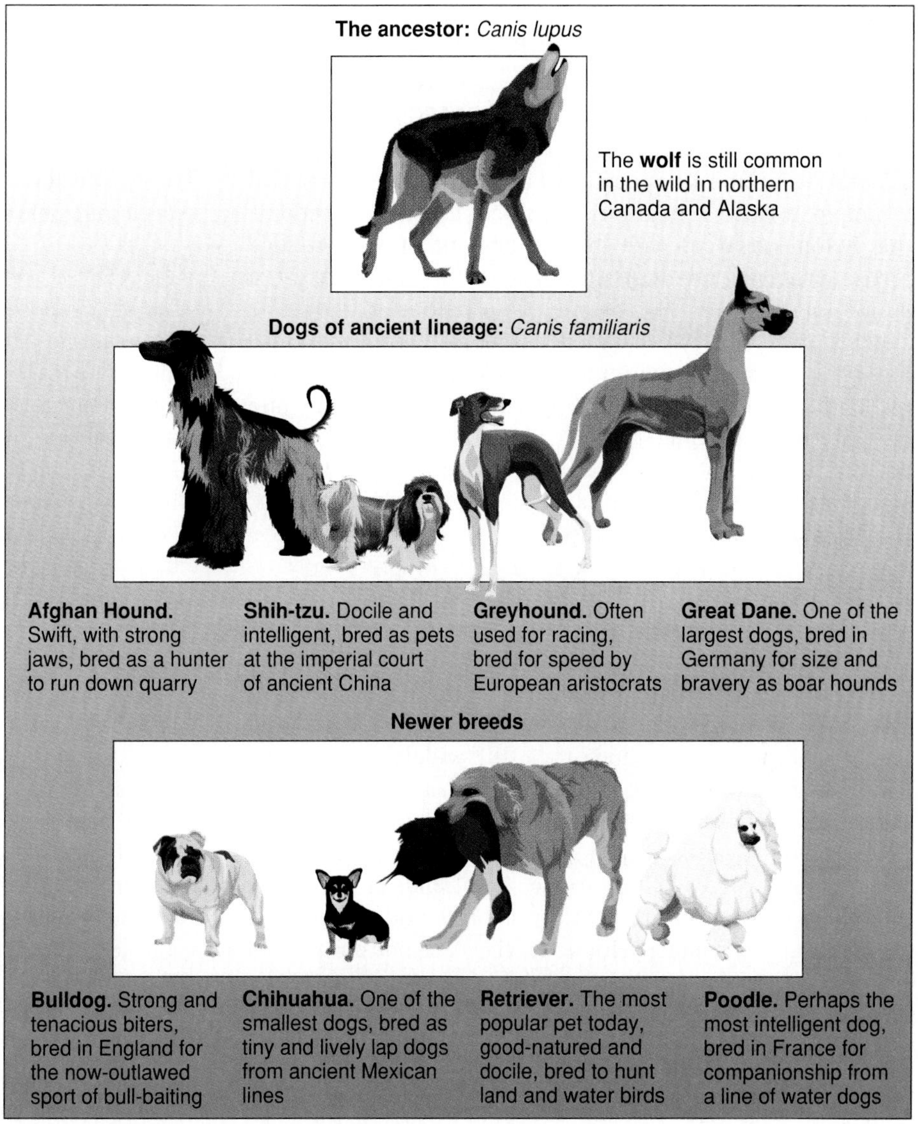

The ancestor: Canis lupus
The wolf is still common in the wild in northern Canada and Alaska
Dogs of ancient lineage: Canis familiaris
Afghan Hound. Swift, with strong jaws, bred as a hunter to run down quarry
Shih-tzu. Docile and intelligent, bred as pets at the imperial court of ancient China
Greyhound. Often used for racing, bred for speed by European aristocrats
Great Dane. One of the largest dogs, bred in Germany for size and bravery as boar hounds
Newer breeds
Bulldog. Strong and tenacious biters, bred in England for the now-outlawed sport of bull-baiting
Chihuahua. One of the smallest dogs, bred as tiny and lively lap dogs from ancient Mexican lines
Retriever. The most popular pet today, good-natured and docile, bred to hunt land and water birds
Poodle. Perhaps the most intelligent dog, bred in France for companionship from a line of water dogs

Figure 10.6 Diversity within a species: dogs.

Domestic dogs are descended from the wolf, but hundreds of different breeds of *Canis familiaris* have been produced by controlled breeding and selection. All of them are fully able to interbreed.

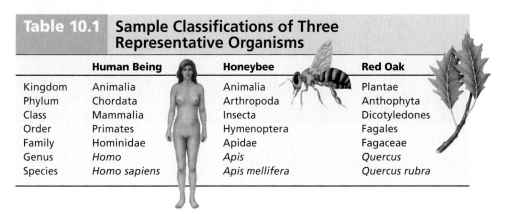

Table 10.1 Sample Classifications of Three Representative Organisms

	Human Being	Honeybee	Red Oak
Kingdom	Animalia	Animalia	Plantae
Phylum	Chordata	Arthropoda	Anthophyta
Class	Mammalia	Insecta	Dicotyledones
Order	Primates	Hymenoptera	Fagales
Family	Hominidae	Apidae	Fagaceae
Genus	*Homo*	*Apis*	*Quercus*
Species	*Homo sapiens*	*Apis mellifera*	*Quercus rubra*

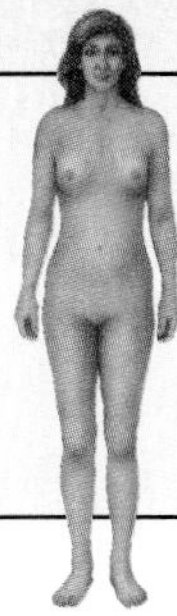

10.2 What Is a Species?

The basic biological unit in the Linnaean system of classification is the species. John Ray (1627–1705), an English clergyman and scientist, was one of the first to propose a general definition of species. In about 1700 he suggested a simple way to recognize a species: all the individuals that belong to it can breed with one another and produce fertile progeny. By Ray's definition, the progeny of a single mating were all considered to belong to the same species, even if they contained different-looking individuals, as long as these individuals could interbreed. All domestic cats are one species (they can all interbreed), while carp are not the same species as goldfish (they cannot interbreed). The donkey you see in figure 10.7 is not the same species as the horse, because when they interbreed, the progeny—mules—are sterile.

The Biological Species Concept

With Ray's observation, the species began to be regarded as an important biological unit that could be cataloged and understood, the task that Linnaeus set himself a generation later. Where information was available, Linnaeus used Ray's species concept, and it is still widely used today. Thus, all the different breeds of dogs you saw in figure 10.6 are *Canis familiaris,* and all can breed successfully with one another, regardless of how different they may appear. Coyotes and wolves belong to the same genus as dogs, *Canis,* and can often produce fertile hybrids with dogs when they come into contact. However, they are not considered members of the same species as dogs because many of the hybrid offspring are *not* fertile.

When the evolutionary ideas of Darwin were joined to the genetic ideas of Mendel in the 1920s to form the field of population genetics, it became desirable to define the category species more precisely. The definition that emerged, the so-called **biological species concept,** was stated by the American evolutionist Ernst Mayr as follows: Species are "groups of actually or potentially interbreeding natural populations which are reproductively isolated from other such groups." In other words, hybrids between species occur rarely in nature, whereas individuals that belong to the same species are able to interbreed freely.

Problems with the Biological Species Concept

The biological species concept works fairly well for animals, where strong barriers to hybridization between species exist, but very poorly for the other kingdoms. The problem is that the biological species concept assumes that organisms regularly **outcross,** that is, interbreed with individuals other than themselves, individuals with a different genetic makeup than their own. Animals regularly outcross, and so the concept works well for animals. However, outcrossing is less common in the other five kingdoms. In bacteria and many protists, fungi, and plants, **asexual reproduction,** reproduction without sex, predominates. These species clearly cannot be characterized in the same way as outcrossing animals and plants—they do not interbreed with one another, much less with the individuals of other species.

Complicating matters further, the reproductive barriers that are the key element in the biological species concept, while common among animal species, are not typical of other kinds of organisms. In fact, there are essentially no barriers to hybridization between the species in many groups of trees and other plants. Even among animals, fish species are able to form fertile hybrids with one another, though they may not do so in nature.

In practice, biologists today recognize species in different groups in much the way they always did, because they differ from one another in their visible features. Within animals, the biological species concept is still widely employed, while among plants and other kingdoms it is not.

Horse Donkey Mule

Figure 10.7 Ray's definition of a species.
According to Ray, donkeys and horses are not the same species. Even though they produce very robust offspring (mules) when they mate, the mules are sterile.

Figure 10.8 Ecotypes in plants.
Shown here are two forms of bishop pine, *Pinus muricata,* in Marin County, just north of San Francisco, California. (*a*) This tree is growing in a flat field behind Inverness Ridge, where it is protected from the ocean winds. (*b*) This tree, growing in an exposed place, has been dwarfed and shaped by the winds.

On the Way to Becoming Species

Within the units that are classified as species, the populations that occur in different places may be more or less distinct from one another. Called **ecological races,** or **ecotypes,** these populations have become adapted to the local environment where they live. When the ranges of two ecotypes come into contact, individuals with intermediate or mixed characteristics are usually common. This occurs because the two populations are interfertile. By contrast, when two *species* of animal or outcrossing plant occur together, intermediate types are not usually found, although the two species may hybridize occasionally.

Ecotypes occur in both animals and plants, although they were first studied in plants. As every gardener knows, plants of the same species may differ greatly in appearance depending on the places where they are grown (figure 10.8). Before a plant character can be used to describe a species, it is first necessary to test whether differences in the character are induced by the environment or truly reflect the genetic makeup of the individual. In the 1920s and 1930s, the great Swedish botanist Gote Turesson devised a simple way to test whether differences between ecotypes of plants growing in different parts of Sweden were genetically determined or caused by environmental factors. He simply dug up the plants wherever they were growing and replanted all of the different ecotypes in a common experimental garden outside his laboratory at the University of Lund, Sweden. In nearly every case, he found that the unique characteristics of individual ecotypes (height, leaf size, degree of hairiness, flowering time, pattern of branching) were maintained when the plants were grown in the common environment at Lund. Therefore, the characteristics had a genetic basis and reflected differences in the genes between the ecotypes.

Ecotypes, populations of a species adapted to a particular place, are the first stage in the formation of new species. As isolated populations become more different from one another, they may eventually come to exploit different resources in different ways. The populations may occur in different habitats, have different feeding habits, or otherwise tend to avoid coming into contact. Evolution then tends to favor genetic changes that further separate their niches. When individuals do hybridize, the gametes may not form offspring that are able to survive, or the offspring may be sterile. Such changes are called **isolating mechanisms.** If the process continues long enough, the populations may become so different that they are considered distinct species.

Most animal species are separated by combinations of isolating mechanisms. Some of these factors prevent the species from hybridizing in the first place; others limit the success of the interspecific hybrids once they are formed. For example, two related species of animal may occur in different habitats, produce their gametes at different times of the year, have different mating behaviors, and produce inviable embryos if hybridization does take place.

How Many Kinds of Species Are There?

Since the time of Linnaeus, about 1.5 million species have been named. This is a far greater number of organisms than Linnaeus suspected when he was developing his system of classification. But the actual number of species in the world is undoubtedly much greater, judging from the very large numbers that are still being discovered. Some scientists estimate that at least 10 million species exist on earth. At least two-thirds of these—more than 6 million species—occur in the tropics. Considering that no more than half a million tropical species have been named, our knowledge of these organisms is obviously very limited.

10.3 Inferring Phylogeny

After naming and classifying some 1.5 million organisms, what have biologists learned? One very important advantage of being able to identify particular species of plants, animals, and other organisms is that individuals of species that are useful to humans as sources of food and medicine can be identified. For example, if you cannot tell the fungus *Penicillium* from *Aspergillus,* you have little chance of producing the antibiotic penicillin. One species of periwinkle plant produces a drug effective against Parkinson's disease, but other closely related species do not, thus knowing which species you are dealing with is critical for obtaining the drug. In a thousand ways, just having names for organisms is of immense importance in our modern world.

Taxonomy also enables us to glimpse the evolutionary history of life on earth. The more similar two taxa are, the more closely related they are likely to be, for the same reason that you are more like your brothers and sisters than like strangers selected from a crowd. By looking at the differences and similarities between organisms, biologists can attempt to reconstruct the tree of life, inferring which organisms evolved from which other ones, in what order, and when. The reconstruction and study of evolutionary trees is called **systematics.**

How to Build a Family Tree

A branching tree that reflects the evolutionary relationships among species is called a **phylogeny.** How do taxonomists reconstruct the phylogeny of a group of organisms? If you were considering a collection of plants, how would you go about determining the family tree of their evolutionary past?

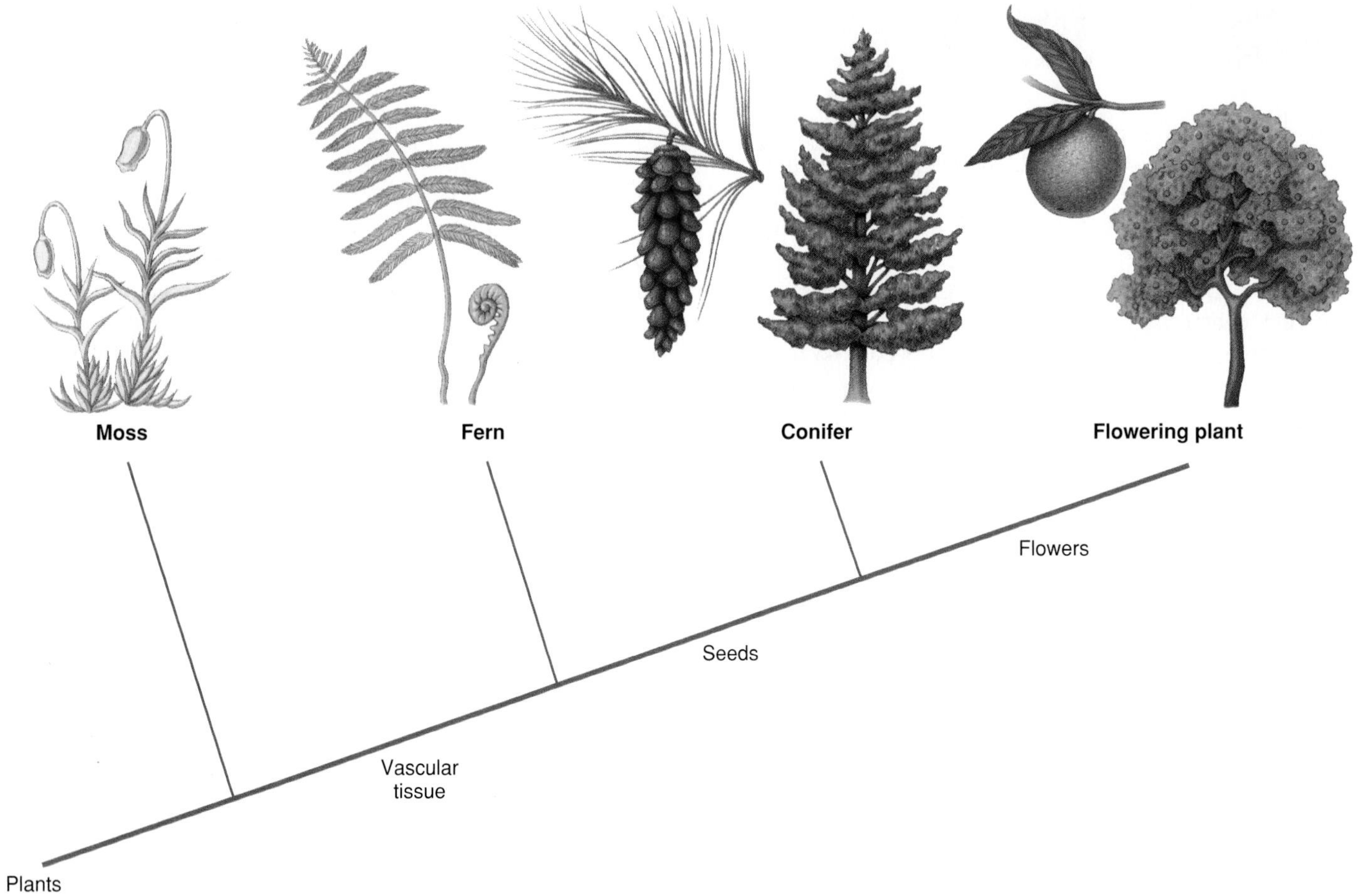

Figure 10.9 A plant cladogram.
The derived characters between the cladogram branch points are shared by all plants above the branch point and are not present in any plants below it.

Cladistics

The simplest and most objective way to construct a phylogeny is to focus on key characters that some of the plants share because they have inherited them from a common ancestor. A **clade** is a group of organisms related by descent, and this approach to constructing a phylogeny is called **cladistics.** Cladistics infers phylogeny (that is, builds family trees) according to similarities derived from a common ancestor, so-called **derived characters.** The key to the approach is being able to identify morphological, physiological, or behavioral traits that differ among the organisms being studied and can be attributed to a common ancestor. By examining the distribution of these traits among the organisms, it is possible to construct a **cladogram** (figure 10.9), a branching diagram that represents the phylogeny.

Cladograms are not true family trees. They do not convey direct information about ancestors and descendants—who came from whom. Instead, they convey comparative information about *relative* relationships. Organisms that are closer together on a cladogram simply share a more recent common ancestor than those that are farther apart. Because the analysis is comparative, it is necessary to have something to anchor the comparison to, some solid ground against which the comparisons can be made. To achieve this, each cladogram must contain an **outgroup,** a rather different organism (but not *too* different) to serve as a baseline for comparisons among the other organisms being evaluated, the **ingroup.** For example, in figure 10.9 the mosses are the outgroup to the clade of plants that have vascular tissue.

The cladogram in figure 10.10 suggests that humans are more closely related to gorillas than to tigers and that tigers, humans, and gorillas as a group are more closely related to each other than to the lizard. The lower the position of a branch on the cladogram, the more distant is the most recent common ancestor of the taxa being compared and the less closely related they are.

Cladistics is a relatively new approach in biology and has become popular among students of evolution. This is because it does a very good job of portraying the *order* in which a series of evolutionary events have occurred. The

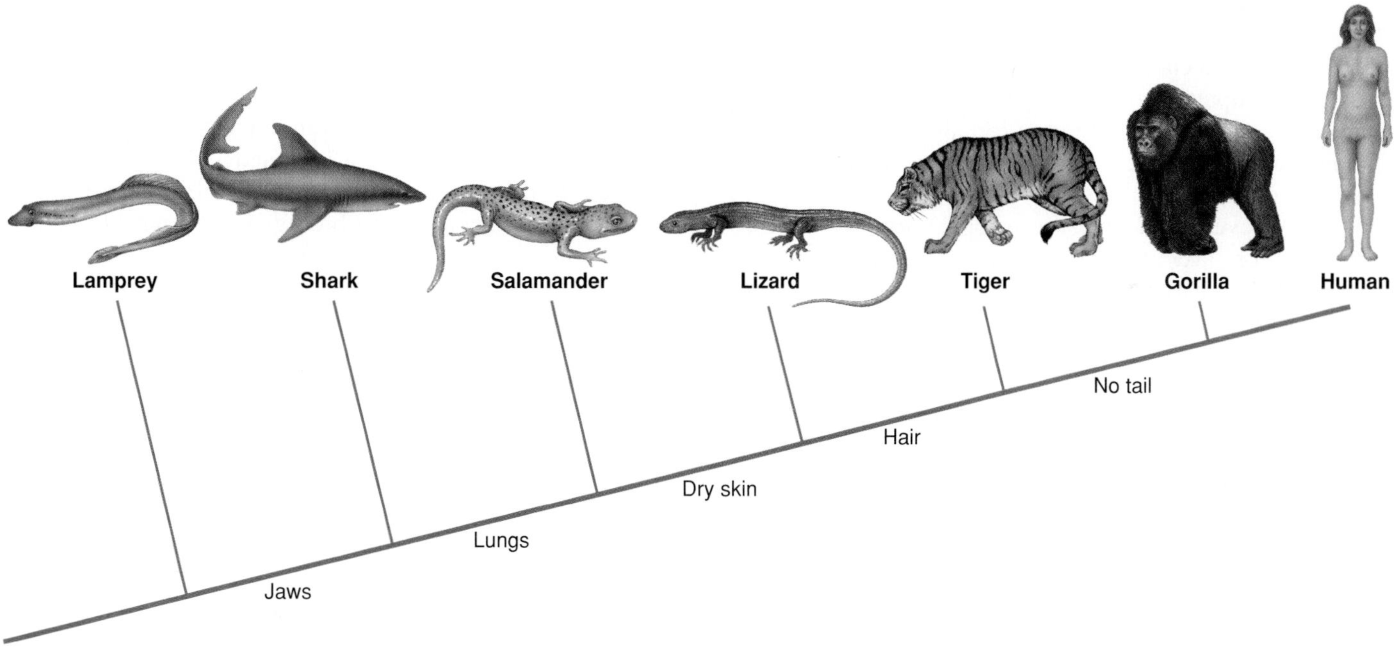

Figure 10.10 An animal cladogram.
The derived characters listed along the main branch of this cladogram are shared by all the animals to the right of the branch point and are not present in any organisms to the left.

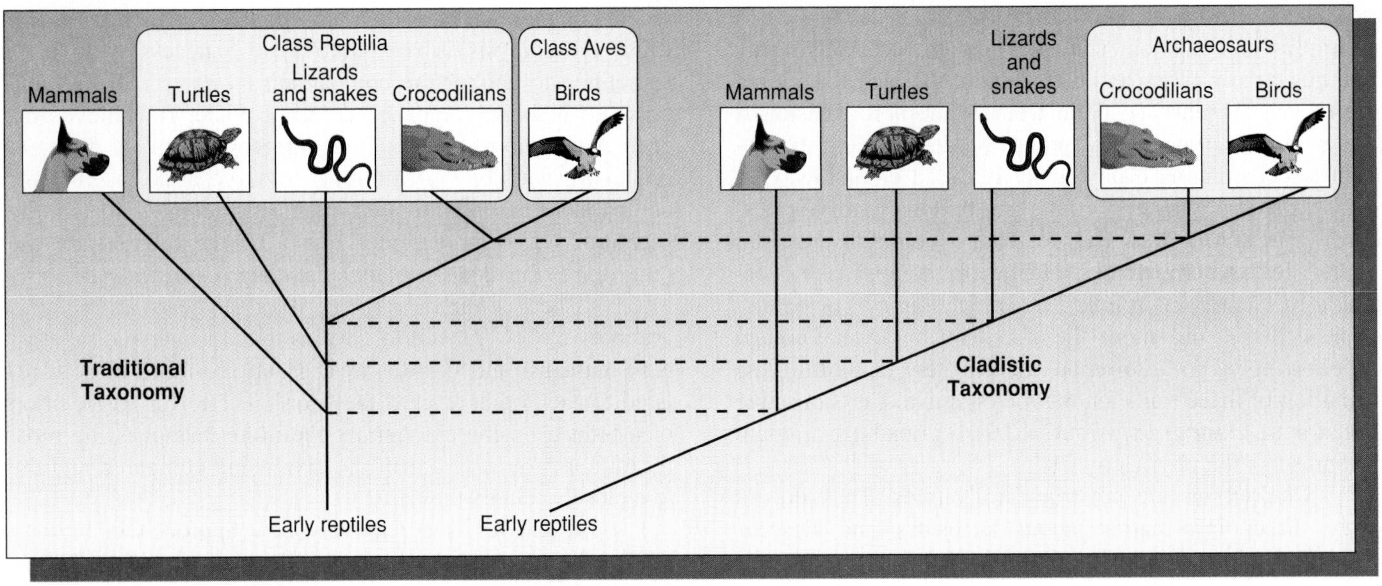

Figure 10.11 Two ways to classify terrestrial vertebrates.
Traditional taxonomic analyses place birds in their own class (Aves) because birds have evolved several unique adaptations that separate them from the crocodiles. Cladistic analyses, however, place crocodiles and birds together (as archaeosaurs) because they share many derived characters.

great strength of a cladogram is that it can be completely objective. A computer fed the data will generate exactly the same cladogram time and again. In fact, most cladistic analyses involve many characters, and computers are required to make the comparisons.

However, sometimes it is necessary to "weight" characters, or take into account the variation in the "strength" of a character—the size or location of a fin, the effectiveness of a lung. For example, let's say that the following are five unique events that occurred on November 22, 1963: (1) my cat was declawed, (2) I had a wisdom tooth pulled, (3) I sold my first car, (4) President Kennedy was assassinated, and (5) I passed physics. Without weighting the events, each one is assigned equal importance. In a nonweighted cladistic sense, they are equal (all happened to you only once, and on that day), but in a practical, real-world sense, they certainly are not. One event, Kennedy's death, had a far greater impact and importance than the others. Because evolutionary success depends so critically on just such high-impact events, modern cladistics attempts to weight the evolutionary significance of the characters being studied, when information is available.

Traditional Taxonomy

Weighting characters lies at the core of **traditional taxonomy.** In this approach, phylogenies are constructed and organisms are classified based on a vast amount of information about the morphological characteristics and biology of the organism gathered over a long period of time. The large amount of information about the biological significance of traits permits a knowledgeable weighting of characters. In traditional taxonomy, the full observational power and judgment of the biologist is brought to bear—and also any biases he or she may have. For example, in classifying the terrestrial vertebrates, traditional taxonomists place birds in their own class (Aves), giving great weight to the characters that made powered flight possible, such as feathers. However, a cladogram of vertebrate evolution (figure 10.11) lumps birds in among the reptiles with crocodiles. This accurately reflects their true ancestry but ignores the immense evolutionary impact of a derived character such as feathers.

In practice, traditional taxonomy is the better approach when a great deal of information is available (you can give a character due evolutionary weight because you have a basis to make the judgment), while cladistics is the better approach when little information is available about how the character affects the life of the organism. DNA sequence comparisons, for example, lend themselves well to cladistics—you have a great many derived characters (differences in sequence that arise during the phylogeny and are inherited by all groups that arise afterwards) but little or no idea of what impact the sequence differences have on the organism.

CHAPTER 10

HIGHLIGHTS

10.1 The Classification of Organisms

Key Terms

- classification 201
- genus 202
- binomial system 202
- taxonomy 203

Key Concepts

- Linnaeus invented the binomial system for naming species, a vast improvement over complex polynomial names.
- Linnaeus placed organisms into seven hierarchical categories: kingdom, phylum, class, order, family, genus, and species.

10.2 What Is a Species?

Key Terms

- biological species concept 206
- ecotype 207
- isolating mechanism 207

Key Concepts

- The biological species concept defines species as groups that cannot successfully interbreed.
- This concept works well for animals and outcrossing plants but poorly for other organisms.
- Ecological races are an intermediate stage in species formation.

10.3 Inferring Phylogeny

Key Terms

- systematics 208
- phylogeny 208
- cladistics 209
- derived character 209
- outgroup 209

Key Concepts

- Taxonomy is of great practical importance.
- Systematics is the study of the evolutionary relationships among a group of organisms.
- Cladistics builds family trees by clustering together those groups that share an ancestral "derived" character.
- Traditional taxonomy classifies organisms based on a large amount of information available, giving due weight to the evolutionary significance of certain characters.

10

CONCEPT REVIEW

1. List the taxonomic categories from most inclusive to least inclusive.
   a. order
   b. kingdom
   c. species
   d. phylum
   e. family
   f. genus
   g. class

2. To which kingdom, phylum, and class do you belong?
   a. Mammalia
   b. Arthropoda
   c. Anthophyta
   d. Animalia
   e. Chordata
   f. Primates

3. Donkeys and horses are not the same species because the hybrids they produce are
   a. ecological races.
   b. fertile.
   c. sterile.
   d. fertile, but nonreproducing.

4. Populations of the same species that have become adapted to the local environments in which they live are called
   a. families.
   b. common names.
   c. hybrids.
   d. ecotypes.

5. About _________ species of organisms have been named.
   a. 1.5 million
   b. 10 million
   c. 2 billion
   d. 10,000

6. How many species of organisms are estimated to live on the earth?
   a. 1.5 million
   b. 10 million
   c. 2 billion
   d. 10,000

7. Organisms within a clade share a more recent _________ than organisms outside the clade.
   a. ingroup
   b. outgroup
   c. common ancestor
   d. species

8. The derived character that is shared only by gorillas and humans is
   a. hair.
   b. lungs.
   c. no tail.
   d. jaws.

9. In constructing phylogenies, phylogenetic systematists and modern cladists often _________ the characters that have evolutionary significance.
   a. weight
   b. ignore
   c. randomize
   d. create

10. _________ devised the binomial system of nomenclature.

11. The statement that species are groups of interbreeding natural populations that are reproductively isolated from other such groups is called the _________.

12. At least _________ of the total number of estimated species of organisms on earth occur in the tropics.

13. _________ is the study of the evolutionary relationships among organisms.

14. A branching tree that reflects the evolutionary relationships among organisms is called a _________.

15. A group of organisms related by descent is called a _________.

16. The shared derived character among conifers and flowering plants is _________.

17. In a cladogram of terrestrial vertebrates, birds are most closely related to _________ because the two groups share many derived characters.

Answers to the Concept Review questions appear in Appendix B.

CHALLENGE YOURSELF

1. What are some of the advantages of the binomial system of naming organisms?
2. Do you think a better system of classification would be to group all photosynthetic eukaryotes, regardless of whether or not they are single-celled, as plants and to group all nonphotosynthetic ones as animals? Present arguments in favor of doing so and other arguments in favor of the system of classification used in this text.
3. How is a species best defined? Would this definition apply to all organisms in all kingdoms? If not, how would you amend your definition to include the other organisms?
4. What is the current assessment of the number of living species on our planet? What criticisms might you have for this projected number and why? What problems might be encountered in the immediate future to come up with a more accurate estimate of the number of living species than the number currently suggested?

FOR FURTHER READING

Gould, S. J. "What Is a Species?" *Discover* 12 (1992): 40–44. One famous evolutionist's view on a question that biologists are still debating a century after Darwin.

Knowlton, N. "Confusion of Kingdoms." *Nature* 359 (October 1992): 488ff. How many kingdoms are there, really? Depends on who you talk to.

Margulis, L. *The Illustrated Five Kingdoms: Guide to the Diversity of Life on Earth.* New York: Harper Collins, 1994. A marvelous account of the diversity of organisms, beautifully illustrated and presented in a logical fashion.

May, R. "How Many Species Inhabit the Earth?" *Scientific American.* (October 1992): 42–48. A good guess at how many species exist now, how they got there, and what the implications are for conservation.

Wayne, R., and J. Gittleman. "The Problematic Red Wolf." *Scientific American,* (July 1995): 36–39. Is the red wolf a species or a long-established hybrid of the grey wolf and the coyote?

Woese, C., O. Kandler, and M. Wheelis. "Towards a Natural System of Organism: Proposal for the Domains Archaea, Bacteria, and Eucarya." *Proc. Nat. Acad. Sci. US* 87 (1990): 4576–79. Conveys much of the excitement of the new taxonomy being generated from molecular analysis; an alternative to six kingdoms.

TECHNOLOGY LINKS

The Living World Home Page
http://www.wcbp.com/biology/tlw

11

The First Single-Celled Creatures

Childhood innocence

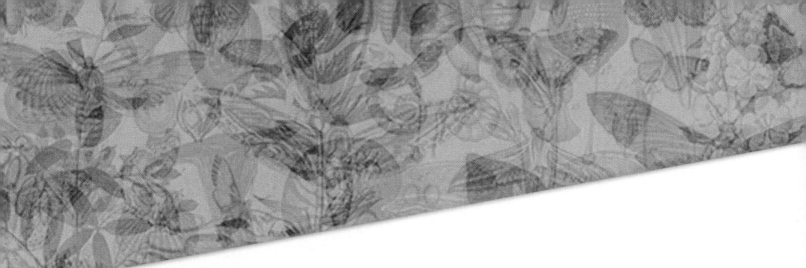

CHAPTER OUTLINE

11.1 Bacteria
- The Structure of a Bacterium
- How Bacteria Reproduce
- Comparing Bacteria (Prokaryotes) to Eukaryotes

11.2 Kinds of Bacteria
- Bacterial Lifestyles

11.3 Viruses
- The Structure of Viruses
- How Viruses Gain Entry into Cells
- How Viruses Reproduce
- Viruses Are Important Agents of Disease

Highlights

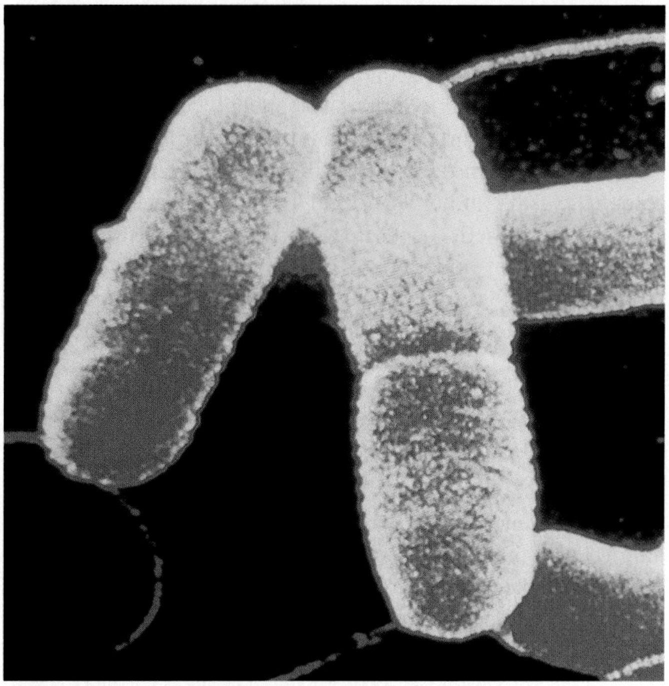

Figure 11.1 A look toward the past.
These bacterial cells are dividing. Bacteria are the most ancient form of life on earth and remain among the most numerous. There are few places on earth where bacteria are not found.

The earliest living organisms of which we have any record are bacteria (figure 11.1). The early stages of life on earth appear to have been rife with evolutionary experimentation. Some of the early bacteria were able to grow in an oxygen-free environment and produced methane as a result of their metabolic activities. Another group of bacteria, called *cyanobacteria,* produced oxygen as a result of its photosynthetic activities. When cyanobacteria appeared at least 3 billion years ago, they played the decisive role in increasing the concentration of free oxygen in the earth's atmosphere from below 1% to the current level of 21%.

11.1 Bacteria

Judging from fossils in ancient rocks, bacteria have been plentiful on earth for over 3.5 billion years. From the diverse array of early living forms, a few became the ancestors of the great majority of organisms alive today. Several forms, including methane-producing bacteria and cyanobacteria, have survived locally or in unusual habitats, but others probably became extinct millions or even billions of years ago. The fossil record indicates that eukaryotic cells, being much larger than bacteria and exhibiting elaborate shapes in some cases, did not appear until about 1.5 billion years ago. Therefore, for at least 2 billion years—nearly half the age of the earth—bacteria were the only organisms that existed.

Today bacteria are the simplest and most abundant form of life on earth. In a spoonful of farmland soil, 2.5 billion bacteria may be present. In 1 hectare (about 2.5 acres) of wheat land in England, the weight of bacteria in the soil is approximately equal to that of 100 sheep!

It is not surprising, then, that bacteria occupy a very important place in the web of life on earth. They play a key role in cycling minerals within the earth's ecosystems. In fact, photosynthetic bacteria were in large measure responsible for the introduction of oxygen into the earth's atmosphere. Bacteria are responsible for some of the most deadly animal and plant diseases, including many that infect humans. Bacteria are our constant companions, present in everything we eat and on everything we touch.

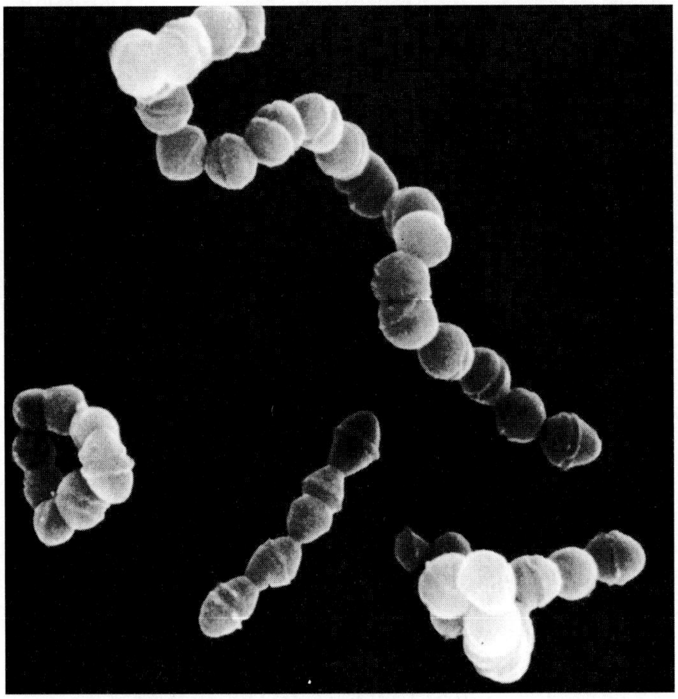

Figure 11.2 Spherical-shaped bacteria.
Bacterial cells are simple in form. Each of the spheres (cocci) adhering in chains in this micrograph is an individual bacterial cell of the genus *Streptococcus.*

The Structure of a Bacterium

The essential character of bacteria can be conveyed in a simple sentence: **Bacteria** are small, simply organized, single cells that lack an organized nucleus. Therefore, they are prokaryotes; their single DNA chromosome is not confined by a nuclear membrane in a nucleus, as in the cells of eukaryotes. Too tiny to see with the naked eye, a bacterial cell is usually simple in form, either rod-shaped (bacilli) like the bacteria of figure 11.1, spherical (cocci) like the bacteria in figure 11.2, or spirally coiled (spirilla). A few kinds of bacteria aggregate into stalked structures or filaments.

The bacterial cell's plasma membrane is encased within a cell wall (figure 11.3). Among members of the more advanced of the two bacterial kingdoms, the cell wall is made of a network of polysaccharide molecules linked together by protein cross-links. Many species of bacteria have only this cell wall, but in others, the cell wall is covered with an outer membrane layer composed of large molecules of lipopolysaccharide (a chain of sugars with lipids attached to it). Bacteria are commonly classified by the presence or absence of this membrane as **gram-positive** (no outer membrane) or **gram-negative** (possess an outer membrane). The name refers to the Danish microbiologist Hans Gram, who developed a staining process that colors the cell wall purple. In bacteria with an outer membrane, the stain cannot reach the cell wall and so the cells do not develop a purple color. The membranes of gram-negative bacteria make them resistant to antibiotics that attack the bacterial cell wall. That is why penicillin, which targets the protein cross-links of the eubacterial cell wall, is effective only against gram-positive bacteria. Outside of the cell wall and membrane, many bacteria possess a gelatinous layer called a **capsule.**

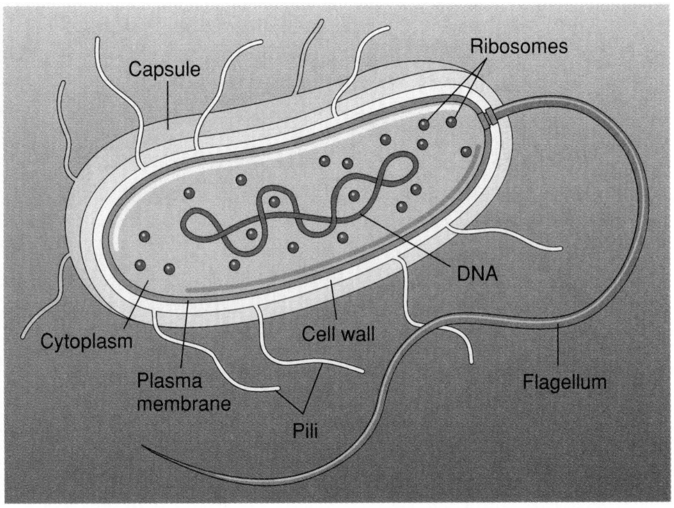

Figure 11.3 The structure of a bacterial cell.
Bacterial cells are small and lack interior organization. The plasma membrane is encased within a rigid cell wall.

Many kinds of bacteria possess threadlike **flagella,** long strands of protein that may extend out several times the length of the cell body. Bacteria swim by twisting these flagella in a corkscrew motion. Flagella may be distributed all over the cell body or be confined to one or both ends of the cell. Some bacteria also possess shorter, thicker outgrowths called **pili,** which act as docking cables, helping the cell to attach to surfaces or other cells.

When exposed to harsh conditions (dryness or high temperature), some bacteria form thick-walled **endospores** around their chromosomes and a small bit of cytoplasm. These endospores are highly resistant to environmental stress and may germinate to form new active bacteria even after centuries. The formation of endospores is the reason that the bacterium responsible for botulism, *Clostridium botulinum,* sometimes persists in cans and bottles if the containers have not been heated at a high enough temperature to kill the spores.

How Bacteria Reproduce

Bacteria, like all other living cells, grow and divide. Bacteria reproduce using a process called **binary fission,** in which an individual cell simply increases in size and divides in two. Following replication of the bacterial DNA, the cell membrane and cell wall grow inward and eventually divide the cell by forming a new wall from the outside toward the center of the old cell.

Some bacteria can pass plasmids from one cell to another in a process called **conjugation.** Recall from chapter 8 that a plasmid is a small, circular fragment of DNA that replicates outside the main bacterial chromosome. In bacterial conjugation, the pilus of one cell, called the donor cell, forms a conjugation bridge that links the donor cell to the recipient cell. The plasmid within the donor cell is then mobilized for transfer. The plasmid in the donor cell then begins to replicate its DNA, passing the replicated copy out across the bridge and into the recipient cell (figure 11.4).

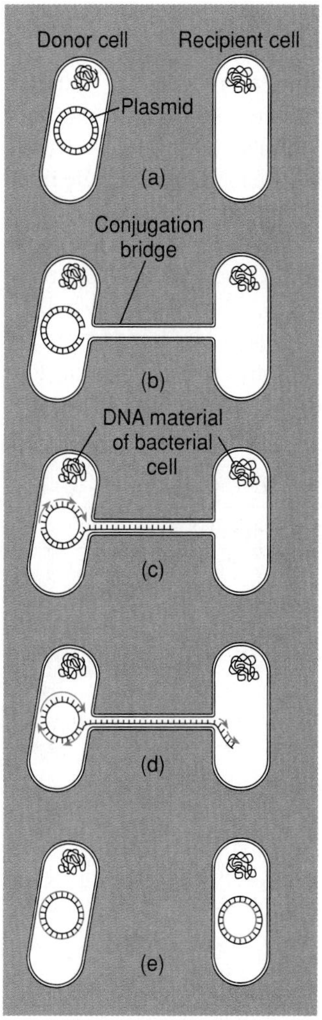

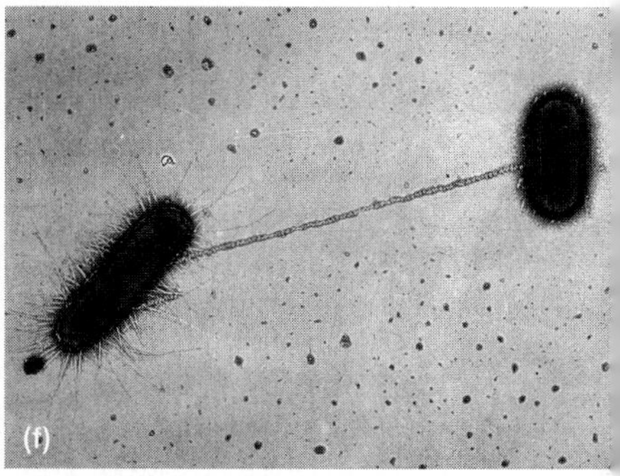

Figure 11.4 Bacterial conjugation.
(*a*) Donor cells contain a plasmid that recipient cells lack.
(*b*) The long pilus connecting the two cells is called a conjugation bridge.
(*c*) Across the bridge, the plasmid replicates a copy of its chromosome. One strand is passed across, the other serving as a template to build a replacement.
(*d*) When the single strand enters the recipient cell, it serves as a template to assemble a double-stranded duplex.
(*e*) When the process is complete, both cells contain a complete copy of the circular plasmid DNA molecule.
(*f*) Photo of the conjugation bridge.

Comparing Bacteria (Prokaryotes) to Eukaryotes

Bacteria differ from eukaryotes in seven key respects:

1. **Internal compartmentalization.** The cytoplasm of bacteria has very little internal organization. Unlike that of eukaryotes, it contains no internal compartments and no internal membrane system. Bacteria are prokaryotes—that is, they possess no cell nucleus.
2. **Cell size.** Most bacterial cells are only about 1 micrometer in diameter, whereas most eukaryotic cells are well over 10 times that size.
3. **Unicellularity.** All bacteria are fundamentally single-celled. Even though some bacteria may adhere together in a matrix of form filaments, their cytoplasm is not directly interconnected, and their activities are not integrated and coordinated, as is the case in multicellular eukaryotes.
4. **Chromosomes.** Bacteria do not possess chromosomes in which proteins are complexed with the DNA, as eukaryotes do. Instead, their DNA exists as a single circle in the cytoplasm.
5. **Cell division.** Cell division in bacteria takes place by binary fission (see chapter 4). The cells simply pinch in two. In eukaryotes, microtubules pull chromosomes to opposite poles during the cell division process, called mitosis.
6. **Flagella.** Bacterial flagella are simple, composed of a single fiber of protein that is spun like a propeller to drive the cell through the water. Eukaryotic flagella are more complex structures, with a 9 + 2 arrangement of microtubules, that whip back and forth rather than rotate.
7. **Metabolic diversity.** Bacteria possess many metabolic abilities that eukaryotes do not. For example, bacteria perform several different kinds of anaerobic and aerobic photosynthesis, while eukaryotes carry out only one. Only bacteria can obtain their energy from oxidizing inorganic compounds (so-called chemoautotrophs). Only bacteria have the ability to fix atmospheric nitrogen.

11.2 Kinds of Bacteria

Although small and internally simple, all bacteria that live today are diverse in their internal chemistry and in many of the details of how they are assembled (table 11.1). If you were to gauge success by numbers, then bacteria are the most successful of all organisms—there are more bacteria in a spoonful of dirt than all the vertebrates on earth. There are over 4,000 named species of bacteria, and undoubtedly many thousands more awaiting discovery. During the 2 billion years they evolved alone on earth, bacteria adapted to many kinds of environments, including some you might consider harsh. They invaded waters that were very salty, very acidic or alkaline, and very hot or cold—anywhere that would support life, they thrived. Today they are found in hot springs where the temperatures exceed 78°C (180°F) and have been recovered living beneath 430 meters of ice in Antarctica!

Early in the history of life, bacteria split into two kinds, so different in structure and metabolism that biologists assign them to entirely separate kingdoms. One branch produced the **archaebacteria** ("ancient bacteria"), its only survivors today confined to extreme environments that may resemble habitats on the early earth. This is the branch from which eukaryotes are thought to have evolved. The other branch produced the **eubacteria** ("true bacteria"), which include nearly all the 4,800 kinds of bacteria that live today.

Table 11.1 Bacteria

Major Group	Typical Examples		Key Characteristics
			Archaebacteria
Archaebacteria	Methanogens, halobacteria	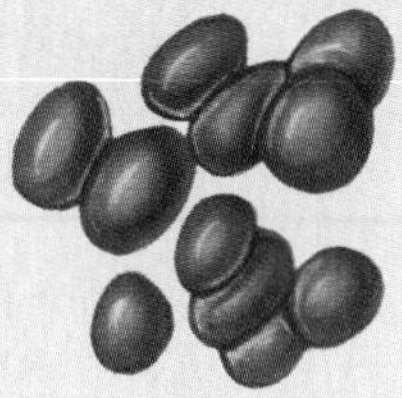	Bacteria that are not members of the kingdom Eubacteria; probably the oldest form of life on earth; anaerobic, with unusual cell walls; some produce methane; others reduce sulfur
			Eubacteria
Actinomyces	*Streptomyces, Actinomyces*	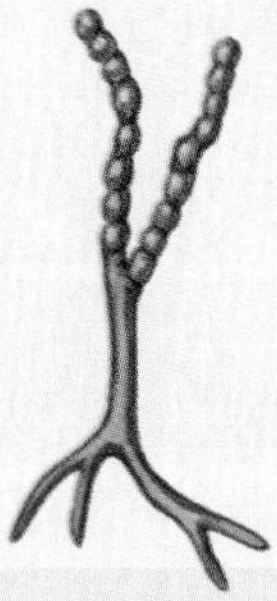	Gram-positive bacteria; form branching filaments and produce spores; often mistaken for fungi; produce many commonly used antibiotics, including streptomycin and tetracycline; one of the most common types of soil bacteria; also common in dental plaque
Chemoautotrophs	Sulfur bacteria, *Nitrobacter, Nitrosomonas*	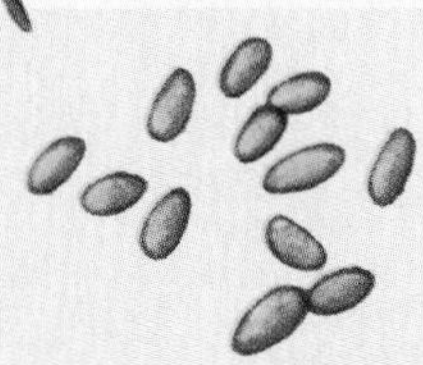	Bacteria able to obtain their energy from inorganic chemicals; most extract chemical energy from reduced gases such as H_2S (hydrogen sulfide), NH_3 (ammonia), and CH_4 (methane); play a key role in the nitrogen cycle
Cyanobacteria	*Anabaena, Nostoc*	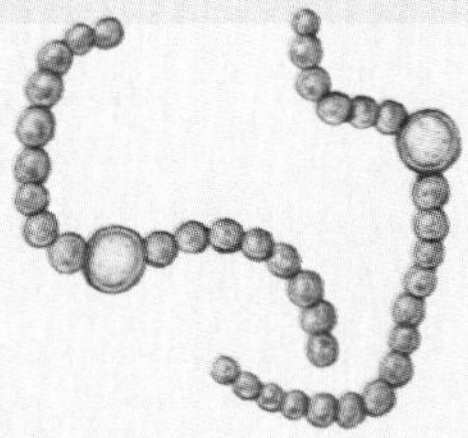	A form of photosynthetic bacteria common in both marine and freshwater environments; deeply pigmented; often responsible for "blooms" in polluted waters
Enterobacteria	*E. coli, Salmonella, Vibrio*	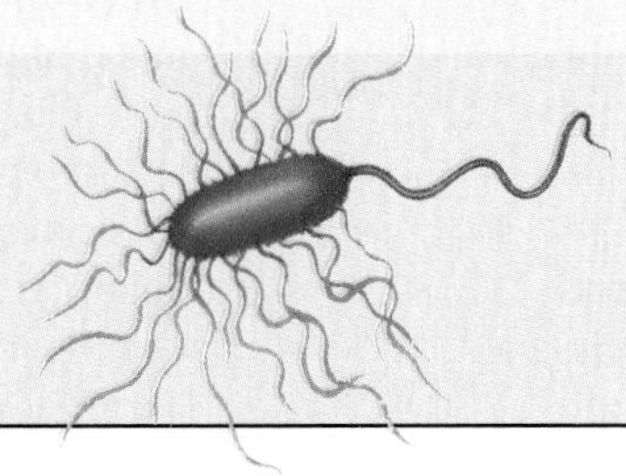	Gram-negative rod-shaped bacteria; do not form spores; usually aerobic heterotrophy; many important diseases are caused by enterics, including bubonic plague and cholera

Archaebacteria and eubacteria differ in four fundamental ways:

1. **Cell wall.** Both kinds of bacteria have cell walls outside the plasma membrane that strengthen the cell. The cell walls of eubacteria are constructed of carbohydrate-protein complexes called peptidoglycans, which link together to create a strong mesh that gives the eubacterial cell wall great strength. The cell walls of archaebacteria lack peptidoglycans.
2. **Plasma membranes.** All bacteria have plasma membranes with a lipid-bilayer architecture (see chapter 3). The plasma membranes of eubacteria and archaebacteria, however, are made of very different kinds of lipids.
3. **Gene translation machinery.** Eubacteria possess ribosomal proteins and an RNA polymerase that are distinctly different from those of eukaryotes. However, the ribosomal proteins and RNA of archaebacteria are very similar to those of eukaryotes.
4. **Gene architecture.** The genes of eubacteria are not interrupted by introns as are those of eukaryotes, while at least some of the genes of archaebacteria do possess introns.

Table 11.1 (continued)

Major Group	Typical Examples		Key Characteristics
			Eubacteria
Gliding and budding bacteria	Myxobacteria, *Chondromyces*		Gram-negative bacteria; exhibit gliding mobility by secreting slimy polysaccharides over which masses of cells glide; some groups form upright multicellular structures carrying spores called fruiting bodies
Pseudomonads	*Pseudomonas*	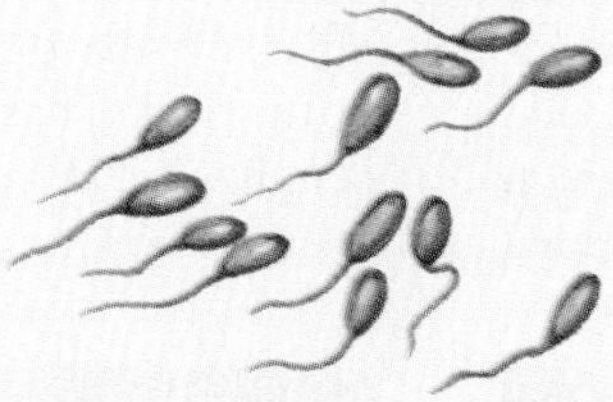	Gram-negative heterotrophic rods with polar flagella; very common form of soil bacteria; also contain many important plant pathogens
Rickettsias and chlamydias	*Rickettsia, Chlamydia*	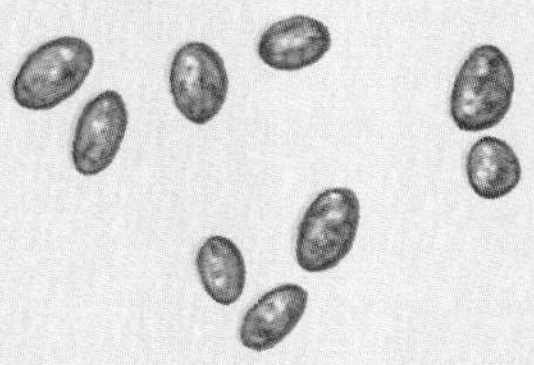	Small gram-negative intracellular parasites; the *Rickettsia* life cycle involves both mammals and arthropods such as fleas and ticks; *Rickettsia* are responsible for many fatal human diseases, including typhus (*Rickettsia prowazekii*) and Rocky Mountain spotted fever; chlamydial infections are one of the most common sexually transmitted diseases
Spirochaetes	*Treponema*	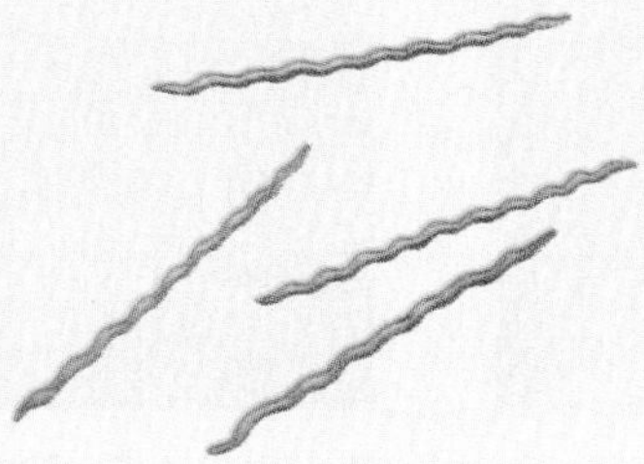	Long, coil-shaped cells; common in aquatic environments; a parasitic form is responsible for the disease syphilis

Bacterial Lifestyles

Archaebacteria

The archaebacteria that survive today are mostly **methanogens,** bacteria that use hydrogen (H_2) gas to reduce carbon dioxide (CO_2) to methane (CH_4). Methanogens are strict anaerobes, poisoned by oxygen gas. They live in swamps and marshes, where other microbes have consumed all the oxygen. The methane that they produce bubbles up as "marsh gas." Methanogens also live in the gut of cows and other herbivores that live on a diet of cellulose, converting the CO_2 produced by that process to methane gas. The few surviving archaebacteria that are not methanogens live in unusually harsh environments, such as the very salty Dead Sea and the Great Salt Lake (over 10 times saltier than seawater). **Thermoacidophiles** favor hot, acidic springs such as the sulfur springs of Yellowstone National Park (figure 11.5), where the water is nearly 80°C, with a very acid pH of 2 or 3.

Eubacteria

Almost all bacteria living today are eubacteria, members of the kingdom Eubacteria. Many are heterotrophs that power their lives by consuming organic molecules, while others are photosynthetic, gaining their energy from the sun. **Cyanobacteria** are among the most prominent of the photosynthetic bacteria. We have already discussed the critical role that the members of this ancient phylum have played in the history of the earth by generating the oxygen in our atmosphere. Cyanobacteria usually have a mucilaginous sheath, which is often deeply pigmented. Nitrogen fixation occurs in almost all cyanobacteria, within specialized cells called **heterocysts,** enlarged cells that occur in many of the filamentous members of this phylum (figure 11.6). These cells begin to form when available nitrogen falls below a certain threshold; when nitrogen is abundant, their formation is inhibited. In **nitrogen fixation,** atmospheric nitrogen is converted to a form in which it can be used by living organisms.

There are numerous phyla of nonphotosynthetic bacteria. Some of these bacteria are **chemoautotrophs,** able to obtain their energy from inorganic chemicals. Most bacteria, however, are **heterotrophs,** obtaining their energy from organic material formed by other organisms. Organic material must be broken down if it is to be used again by organisms. Bacteria and fungi play the leading role in breaking down organic molecules formed by biological processes, thereby making the nutrients in these molecules available once more for recycling. Decomposition is just as indispensable to the continuation of life on earth as is photosynthesis.

Bacteria cause many diseases in humans (table 11.2), including cholera, diphtheria, and leprosy. Among the most serious of bacterial diseases is **tuberculosis (TB),** a disease of the respiratory tract caused by the bacterium *Mycobacterium tuberculosis* (figure 11.7). TB is a leading cause of death throughout the world. Spread through the air, tuberculosis is quite infectious. In most instances, a person becomes infected by inhaling tiny droplets of moisture that contain the bacteria. TB was a major health risk in the United States until the discovery of effective drugs to suppress it in the 1950s. The appearance of drug-resistant strains in the 1990s has raised serious concern within the medical community, as the incidence of TB has increased. The search is on for new types of anti-TB drugs.

Figure 11.5 Thermoacidophiles live in hot springs.
These archaebacteria growing in Sulfide Spring, Yellowstone National Park, Wyoming, are able to tolerate high acid levels and very hot temperatures.

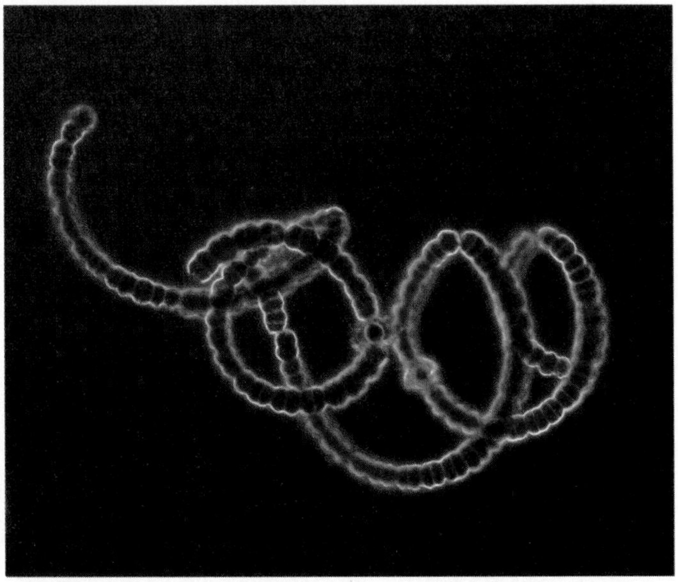

Figure 11.6 The cyanobacterium *Anabaena*.
Individual cells adhere in filaments. The larger cells (areas on the filament that seem to be bulging) are heterocysts, specialized cells in which nitrogen fixation occurs. These organisms exhibit one of the closest approaches to multicellularity among the bacteria.

Table 11.2 Major Bacterial Diseases of Humans

Disease	Pathogen	Vector/Reservoir	Epidemiology
Cholera	*Vibrio cholerae*	Human feces, plankton	Causes severe diarrhea that can lead to death by dehydration if untreated. There is a 50% peak mortality if the disease goes untreated. A major killer in times of crowding and poor sanitation; over 100,000 died in Rwanda in 1994 of cholera.
Diptheria	*Cornybacterium diptheriae*	Humans	Acute inflammation and lesions of mucous membranes. Spread through contact with infected individual. Vaccine available.
Hansen's disease (leprosy)	*Mycobacterium leprae*	Humans, feral armadillos	Chronic infection of the skin, worldwide incidence about 10 to 12 million, especially in Southeast Asia. Spread through contact with infected individuals.
Lyme disease	*Borrelia bergdorferi*	Ticks, deer, rodents	Spread through bite of infected tick. Lesion followed by malaise, fever, fatigue, pain, stiff neck, and headache.
Plague	*Yersinia pestis*	Fleas of rodents	Killed one-fourth of the population of Europe in the seventeenth century; endemic in wild rodent populations of western United States today.
Tuberculosis	*Mycobacterium tuberculosis*	Humans	An acute bacterial infection of the lungs, lymph, and meninges. Its incidence is on the rise, complicated by the development of new strains of the bacteria that are resistant to antibiotics. Tuberculosis took 3.1 million lives in 1995, the most of any disease.
Typhus	*Rickettsia prowazekii*	Lice, rat fleas, humans	Historically a major killer in times of crowding and poor sanitation; transmitted from human to human through the bite of infected lice and fleas. Typhus has a peak untreated mortality of 70%.

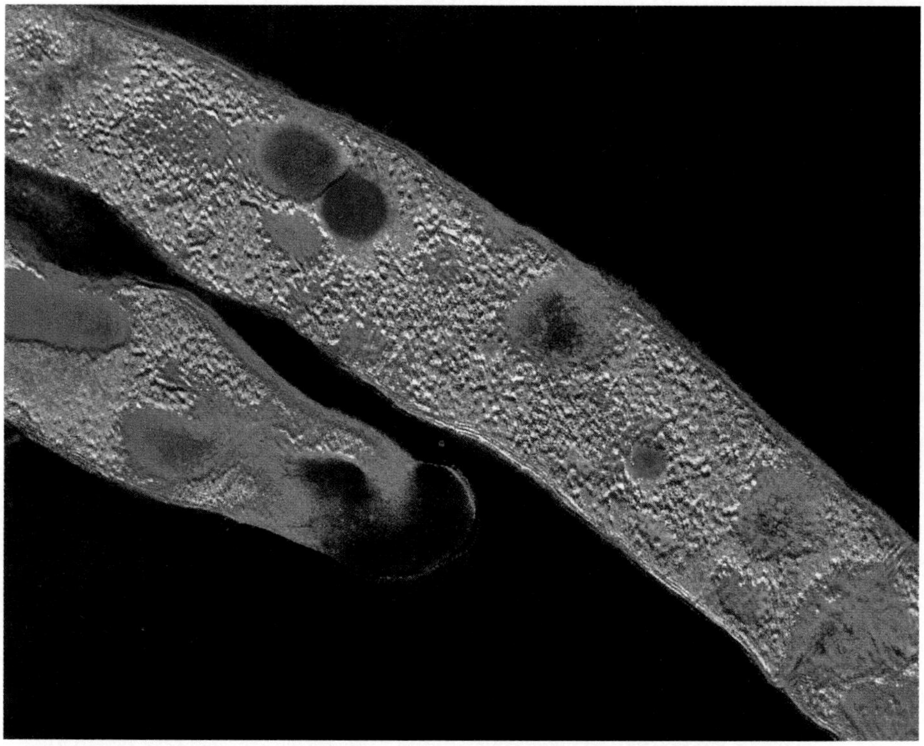

Figure 11.7 ***Mycobacterium tuberculosis.***
This rod-shaped, filament-forming bacterium is responsible for tuberculosis in humans.

11.3 Viruses

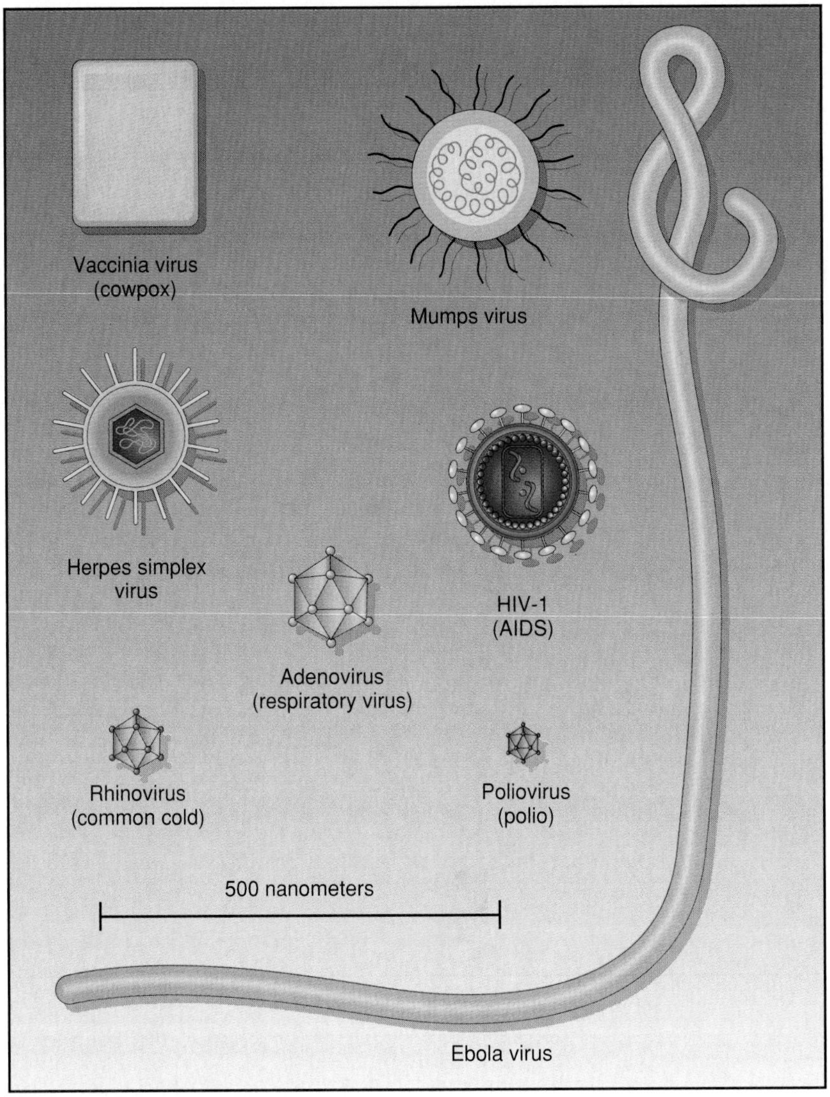

Figure 11.8 Virus sizes.
At the scale shown here, a human hair would be 8 meters thick. Thus, as you can see, viruses are extremely small particles.

The border between the living and the nonliving is very clear to a biologist. Living organisms are cellular and able to grow and reproduce independently, guided by information encoded within DNA. As discussed earlier in this chapter, the smallest creatures living on earth today that satisfy these criteria are bacteria, cells so simple in design that they are thought to resemble the first organisms to evolve as life began on earth. Even simpler than bacteria are viruses (figure 11.8). As you will learn in this section, viruses are so simple that they do not satisfy the criteria for "living."

Viruses arise as fragments of the genome of a host species, and they possess only a portion of the properties of organisms. Viruses are literally "parasitic" chemicals, segments of DNA (or sometimes RNA copies of DNA) wrapped in a protein coat. They cannot reproduce on their own, and for this reason they are not considered alive by biologists. They can, however, reproduce within cells, often with disastrous results to the host organism. For this reason viruses have had, and continue to have, a major impact on the living world.

Biologists first began to suspect the existence of viruses near the end of the last century, when European scientists attempting to isolate the infectious agent responsible for hoof-and-mouth disease in cattle concluded that it could not be a bacterium. The scientists were forced to this startling conclusion by the results of a simple but compelling experiment. When they tried to strain the infectious agent through fine-pored porcelain filters normally used to capture bacteria from growth medium, the infectious agent passed through the filter with no difficulty. The only conclusion possible was that the infectious agent must be much smaller than a typical bacterial cell. Investigating the agent further, the scientists found that the agent could not multiply in solution—it could only reproduce itself within living host cells that it infected. The infecting agents were called viruses.

For many years after their discovery, viruses were regarded as very primitive forms of life, very tiny cells that were perhaps the ancestors of bacteria. We now know this view to be totally incorrect. The true nature of viruses was discovered in 1933, when the biologist Wendell Stanley prepared an extract of a plant virus called *tobacco mosaic virus* (TMV) and attempted to purify it. To his great surprise, the purified TMV preparation precipitated (that is, separated from solution) in the form of crystals. This was surprising because precipitation is something that only chemicals do—the TMV virus was acting like a chemical off the shelf rather than an organism. Stanley concluded that TMV is best regarded as just that—chemical matter rather than a living organism.

With Stanley's experiments, biologists began to understand the true nature of viruses for the first time. Within a few years, scientists disassembled the TMV virus and found that Stanley was right. TMV was not cellular but rather chemical. Each particle of TMV virus is in fact a mixture of two chemicals: RNA and protein. The TMV virus has the structure of a twinkie, a tube made of an RNA core surrounded by a coat of protein. Later workers were able to separate the RNA from the protein and purify and store each chemical. Then, when they reassembled the two components, the reconstructed TMV particles were fully able to infect healthy tobacco plants and so clearly *were* the virus itself, not merely chemicals derived from it. From these experiments and a host of others carried out on other viruses, a general picture has emerged: Viruses are chemical assemblies that can infect cells and replicate within them. They are not alive.

The Structure of Viruses

All organisms appear to have viruses, and in every case the basic structure is the same, a core of nucleic acid surrounded by protein. There is considerable difference, however, in the details. In figure 11.9 you can compare the structure of plant, animal, and bacterial viruses—they are clearly quite different from one another. Many plant viruses like TMV (figure 11.9*a*) have a core of RNA, and some animal viruses like HIV (figure 11.9*b*) do too, but most viruses have DNA in place of RNA. Like TMV, most viruses form a protein sheath, or **capsid,** around their nucleic acid core. The only exceptions are viroids, tiny naked RNA molecules that infect some plant cells. In addition, many viruses form a membranelike **envelope,** rich in proteins, lipids, and glycoprotein molecules, around the capsid.

The smallest viruses are only about 17 nanometers in diameter; the largest ones measure up to 1,000 nanometers, big enough to be barely visible with the light microscope. Viruses are so small that they are smaller than many of the molecules in a cell. Most viruses can be detected only by using the higher resolution of an electron microscope.

Most viruses are more complex than TMV. Several different segments of DNA or RNA may be present in each virus particle, along with many different kinds of protein. Bacterial viruses, called **bacteriophages,** are among the most complex viruses known, built like a lunar lander with head, neck, tail, base plate, and tail fibers (figure 11.9*c*). Most animal and plant viruses are much simpler in appearance than bacteriophage, being either **helical** (rod-shaped with capsid proteins winding around the core in a helix like a winding staircase) or **isometric** (with the capsid proteins forming a sphere of 20 equal triangular facets like a geodesic dome). The geodesic design of isometric viruses like HIV is the most efficient way to assemble subunits to form an external shell with maximum internal capacity.

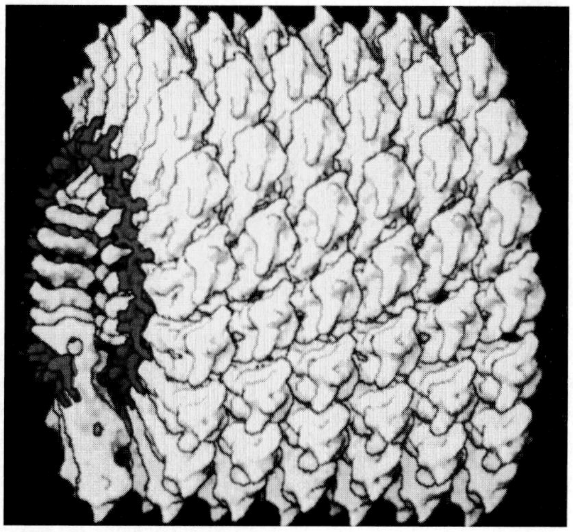
(a)

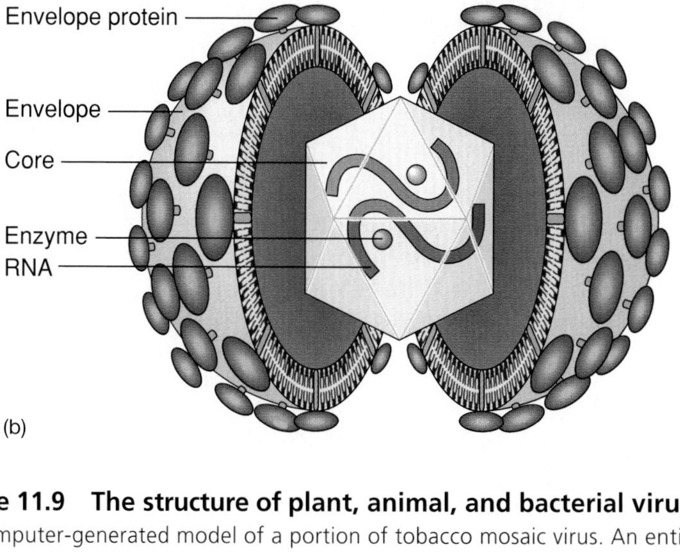

(b)

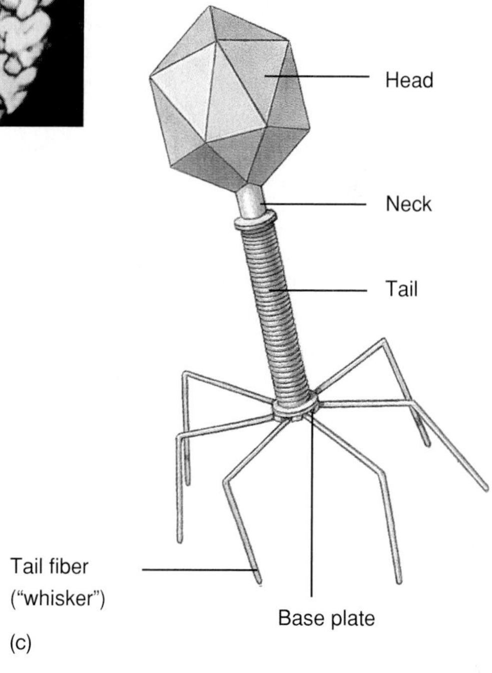

(c)

Figure 11.9 The structure of plant, animal, and bacterial viruses.
(*a*) Computer-generated model of a portion of tobacco mosaic virus. An entire virus consists of 2,130 identical protein molecules (the yellow knobs), which form a cylindrical coat around a single strand of RNA (red in this model).
(*b*) Structure of the AIDS virus. The RNA core is held within a core capsid that is encased by a protein envelope.
(*c*) Structure of the bacterial virus T4 looks for all the world like a lunar lander.

How Viruses Gain Entry into Cells

Before any virus can reproduce, it must first infect a living cell. How does a virus get into a cell? Bacterial viruses punch a hole in the bacterial cell wall and inject their DNA inside. Plant viruses like TMV enter plant cells through tiny rips in the cell wall at points of injury. Animal viruses enter their host cells by endocytosis, a process described in chapter 4. In effect, the cell's plasma membrane dimples inward, surrounding and engulfing the virus particle. Imagine dropping a brick on a feather pillow and you will get some idea of how the animal virus sinks into the membrane.

Before any virus can be engulfed by an animal cell membrane, the virus must first bind to the cell membrane. What does it bind to? From the envelope of the animal virus protrude spikes of *glycoprotein* (a protein with sugars attached to it) and lipid that bind to specific receptor molecules on the cell membrane, usually also glycoproteins. Animal viruses infect only those cells with *receptor proteins* on their surface that match the virus's own surface glycoprotein. HIV infects only white blood cells; polio virus enters only certain spinal nerve cells; hepatitis virus goes only to the liver; and rabies virus infects the brain.

Mammals are protected from virus infection by their immune systems, which produce antibodies that identify and label the virus particles for removal. When mutations arise in the virus gene encoding the glycoprotein, the changes often make it harder for antibodies to recognize the virus. That is why new strains of flu continually arise to which people are not resistant. Like a fugitive changing disguises, the constantly mutating flu virus keeps changing its coat.

Mutations in the gene encoding the receptor protein may also enable the virus to bind to a receptor protein that it earlier failed to recognize. This is thought to be how HIV became a human pathogen. A mutation occurred in the HIV gene encoding glycoprotein gp120 so that the altered form was able for the first time to recognize and bind to the human white blood cell receptor protein **CD4** (figure 11.10). Once able to bind these cells, HIV enters and destroys them, disrupting the human immune system and producing the disease AIDS.

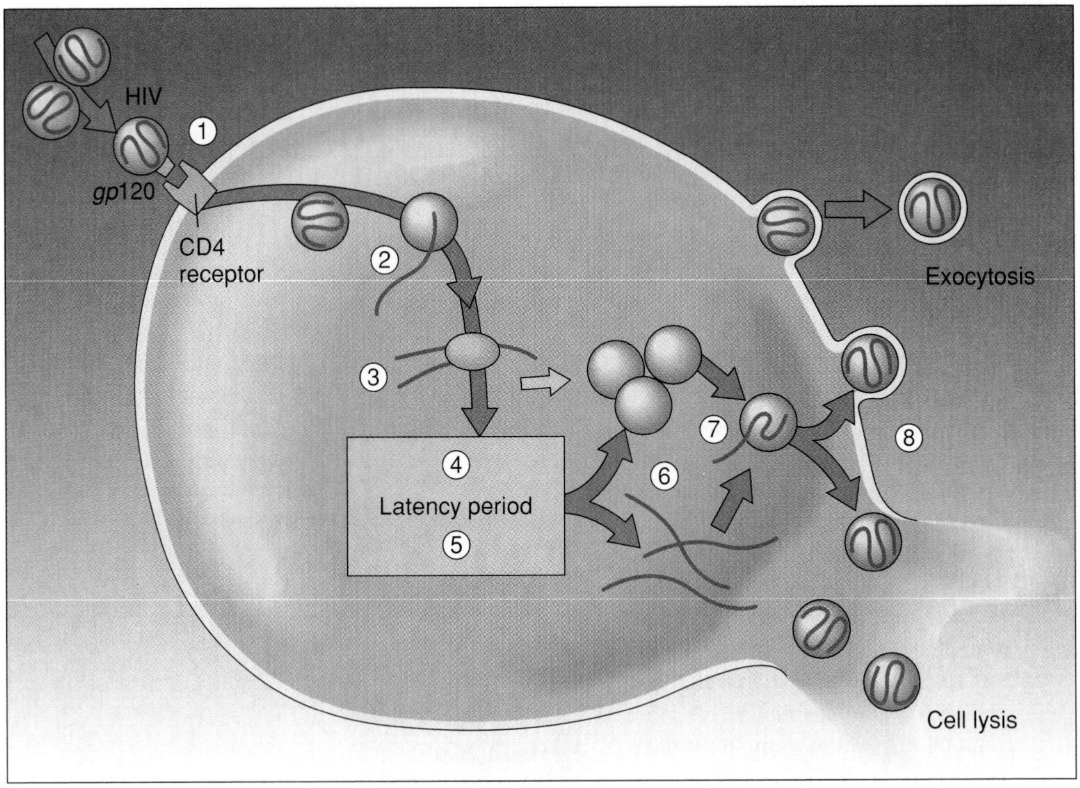

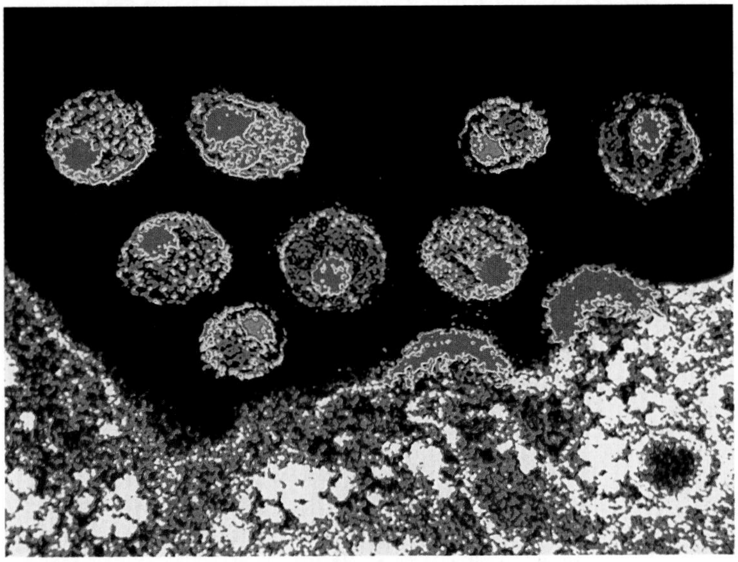

Figure 11.10 The HIV infection cycle.
Top: (*1*) The gp120 glycoprotein of HIV attaches to the CD4 receptor protein on the surface of a T4 white blood cell, initiating endocytosis. (*2*) The viral RNA is released into the cell's cytoplasm. (*3*) A DNA copy is made of the virus RNA. (*4*) The HIV DNA undergoes a long latency period. It is not clear whether the HIV DNA becomes incorporated into the host chromosome or not. (*5*) After a period of some eight to ten years, the virus initiates active replication. (*6*) Both HIV RNA and HIV proteins are made. (*7*) Complete HIV particles are assembled. (*8*) Some T4 cells lyse, releasing free HIV, whereas other HIV particles exit the T4 cell by exocytosis, as shown in the micrograph (*bottom*).

Figure 11.11 Lytic and lysogenic cycles of a bacteriophage.
In the lytic cycle, the bacteriophage exists as viral DNA, free in the bacterial host cell's cytoplasm; the viral DNA directs the production of new viral particles by the host cell. In the lysogenic cycle, the bacteriophage DNA is integrated into the large, circular DNA molecule of the host bacterium. Bacteriophages are much smaller relative to their hosts than illustrated in this diagram.

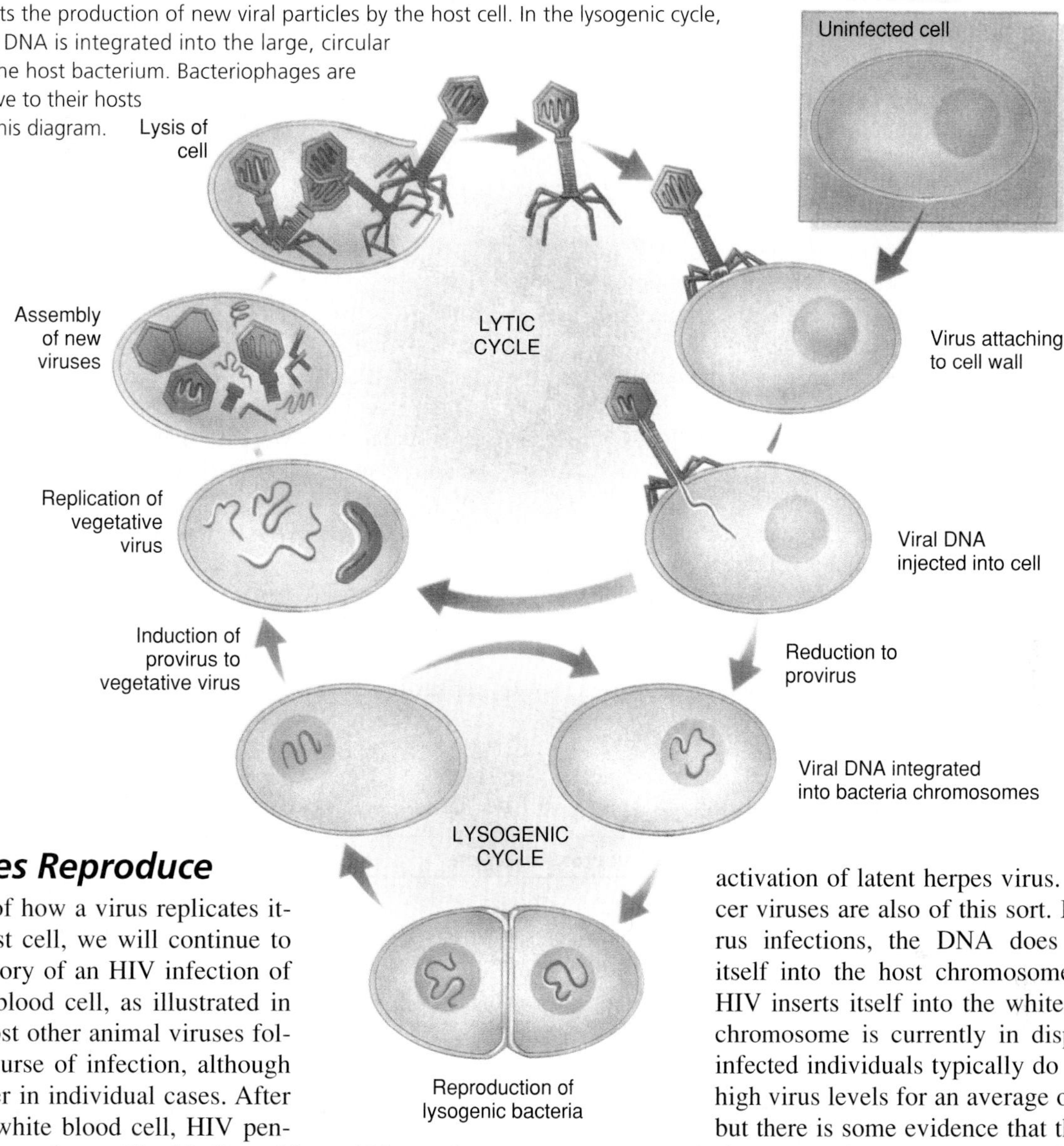

How Viruses Reproduce

As an example of how a virus replicates itself within a host cell, we will continue to examine the history of an HIV infection of a human white blood cell, as illustrated in figure 11.10. Most other animal viruses follow a similar course of infection, although details may differ in individual cases. After docking with a white blood cell, HIV penetrates the cell's membrane. The binding of its gp120 protein to the cell's CD4 receptor triggers endocytosis (called **receptor-mediated endocytosis**). Endocytosis is described in chapter 4. The cell membrane invaginates inward to form a deep cavity around the virus particle. The virus enters the cell within the resulting membrane-bounded vesicle, which soon releases the virus into the cell cytoplasm.

Once it is within the host cell, the HIV virus sheds its envelope and capsid. This leaves a double strand of virus RNA floating in the cytoplasm, along with a virus enzyme that was also within the virus shell. This enzyme, called **reverse transcriptase,** manufactures a double strand of DNA complementary to the virus RNA. In some virus infections, the DNA then inserts itself into the host cell chromosome, where it remains latent until some future event causes it to be excised and to resume activity. Fever blisters are caused by activation of latent herpes virus. Some cancer viruses are also of this sort. In other virus infections, the DNA does not insert itself into the host chromosome. Whether HIV inserts itself into the white blood cell chromosome is currently in dispute. HIV-infected individuals typically do not exhibit high virus levels for an average of 10 years, but there is some evidence that the immune system succeeds in suppressing an ongoing infection over this long period, until its resources eventually become depleted and the infection escapes control.

After a nonlatent virus is transcribed into DNA, or when a latent virus becomes active, the genes of the virus are transcribed into proteins that take over part of the host cell's machinery, directing it to produce many copies of the virus. Newly assembled virus particles leave the cell by exocytosis (budding out), and eventually the host cell ruptures ("lyses"), releasing thousands of additional virus particles. These newly released virus particles are then free to infect other white blood cells, continuing their life cycle and the infection.

The replication of a bacteriophage follows the same general scheme as an animal virus, except that entry is gained by injecting the virus nucleic acid into the cell (figure 11.11) rather than by endocytosis.

Viruses Are Important Agents of Disease

Animal viruses are responsible for many serious communicable diseases, including AIDS, ebola fever, and flu (table 11.3). A variety of childhood diseases are also caused by viruses, including chicken pox, measles, and mumps.

Poliomyelitis, caused by polio virus, was a widespread crippling disease through the first half of the twentieth century, until effective vaccines were developed in the 1950s. With the widespread use of these vaccines, **polio** is now a rare disease in industrialized countries, although it remains a serious and common disease in the tropics and elsewhere in the less developed parts of the world.

Colds are viral infections of the respiratory tract. More than 200 different kinds of viruses that cause colds have been identified, about a third of them RNA viruses called rhinoviruses.

One of the most deadly of viral diseases is ebola, caused by a filamentous virus that attacks connective tissue (figure 11.12). Ebola outbreaks have been confined to local regions of Africa, the latest in Gabon, West Africa, in February of 1996.

Among the most devastating diseases of all time is **influenza,** caused by a virus found in wild ducks living in central and northern Asia. Commonly known as the flu, influenza is a viral infection of the respiratory tract characterized by chills, fever, and muscular aches. When the virus is inhaled, it comes in contact with the cells of the upper respiratory tract and penetrates them. There, the viruses reproduce and spread to new cells. If the body cannot marshal an immune defense, the infection can be fatal. In the flu epidemic of 1918, 22 million people died worldwide in 18 months! Most varieties of the influenza virus are readily recognized by our immune systems, so that flu infections lead only to mild symptoms before the virus is brought under control. However, the influenza virus mutates readily, constantly changing its surface proteins, and occasionally a variant arises that is more difficult for our immune systems to recognize. Vaccines are the best protection against such variants.

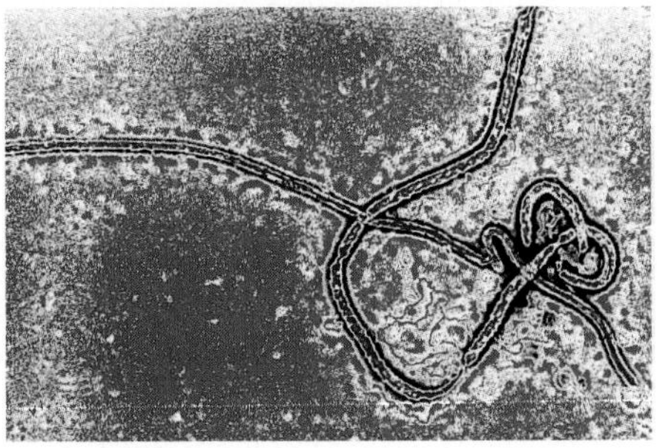

Figure 11.12 The ebola virus. This lethal filamentous virus has an unusual structure, looking like a hose twisted at one end.

Table 11.3 Major Viral Diseases of Humans

Disease	Pathogen	Vector/Reservoir	Epidemiology
AIDS	HIV	STD	Destroys immune defenses, resulting in death by infection or cancer. Four million cases worldwide by 1995.
Ebola fever	Filoviruses	Unknown	Acute hemorrhagic fever; virus attacks connective tissue, leading to massive hemorrhaging and death; peak mortality of up to 90%.
Hepatitis B (viral)	Hepatitis B virus (HBV)	Humans	Highly infectious through contact with infected body fluids. Approximately 1% of U.S. population infected. Vaccine available, no cure. Can be fatal.
Influenza	Influenza viruses	Humans, ducks	Historically a major killer (22 million died in 18 months in 1918–19); wild Asian ducks are a major reservoir. The ducks are not affected by the flu virus, which shuffles its antigen genes while multiplying within them, leading to new flu strains.
Polio	Poliovirus	Humans	Acute viral infection of the CNS that can lead to paralysis and is often fatal. Prior to the development of Salk's vaccine in 1954, 60,000 people a year contracted the disease in the United States alone.
Yellow fever	Flavivirus	Humans, mosquitoes	Spread from individual to individual by mosquito bites; a notable cause of death during the construction of the Panama Canal. If untreated, this disease has a peak mortality rate of 60%.

CHAPTER 11

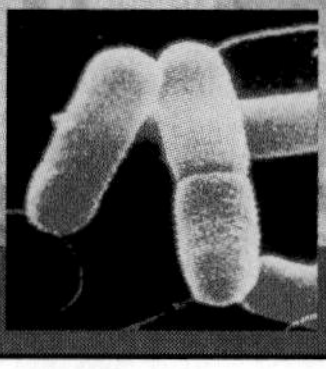

HIGHLIGHTS

	Key Terms	Key Concepts
11.1 Bacteria	bacteria 216 binary fission 217 conjugation 217	• Bacteria are the most ancient form of life on earth. • Bacteria are small, simply organized, single cells that lack an organized nucleus. • Bacteria reproduce by binary fission. • Some bacteria transfer plasmids from one cell to another in a process called conjugation.
11.2 Kinds of Bacteria	archaebacteria 218 eubacteria 218 cyanobacteria 220 nitrogen fixation 220 heterotroph 220	• There are two kingdoms of bacteria, archaebacteria and eubacteria. • The two kingdoms differ in fundamental aspects of their structure and are found living in very different places. • Bacteria are responsible for much of the world's photosynthesis and all of its nitrogen fixation.
11.3 Viruses	envelope 223 receptor protein CD4 224 receptor-mediated endocytosis 225	• Viruses are not organisms, as they cannot reproduce outside of cells. • Every virus has the same basic structure: a core of nucleic acid encased within a sheath of protein. • Animal viruses enter cells by endocytosis, while bacterial viruses inject their nucleic acid into host bacterial cells.

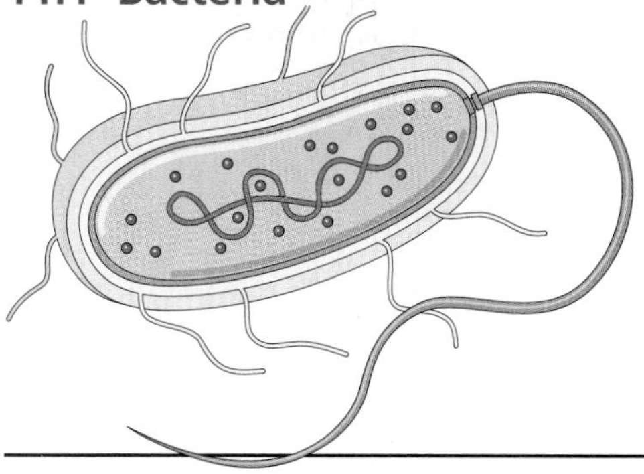

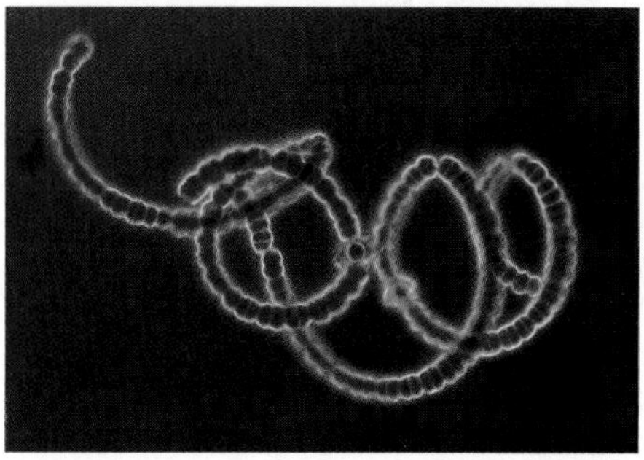

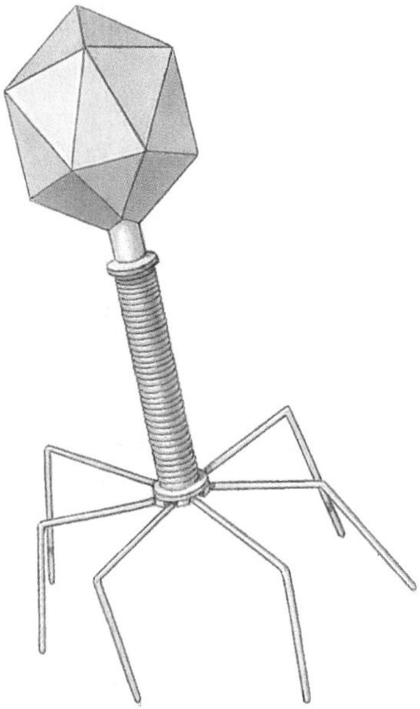

11

CONCEPT REVIEW

1. Bacteria have been on the earth for over _________ billion years, while eukaryotes did not appear in the fossil record until _________ billion years ago.
   a. 4
   b. 3.5
   c. 2.5
   d. 1.5

2. Prokaryotes, including bacteria, lack
   a. DNA.
   b. cells.
   c. an organized nucleus.
   d. cytoplasm.

3. Cell division in bacteria is achieved by
   a. binary fission.
   b. mitosis.
   c. meiosis.
   d. flagella.

4. Eukaryotes are thought to have evolved from the branch of bacteria called
   a. eubacteria.
   b. archaebacteria.

5. In eubacteria, the genes are not interrupted by
   a. introns.
   b. exons.
   c. restriction enzymes.
   d. "stop" codons.

6. Which three of the following diseases are caused by bacteria?
   a. tetanus
   b. AIDS
   c. tuberculosis
   d. cholera
   e. herpes

7. Stanley's evidence that TMV was chemical rather than cellular was that TMV
   a. did not contain DNA or RNA.
   b. could reproduce on its own.
   c. did not contain any protein.
   d. crystallized in solution.

8. The genetic information of viruses is encoded in a molecule of
   a. RNA.
   b. DNA.
   c. either RNA or DNA, but not both.
   d. either RNA or protein, but not both.

9. Once HIV enters the cell and copies its RNA, it undergoes a _________ period of 8 to 10 years before lysing the cell or exiting the cell by exocytosis.
   a. lysogenic
   b. latency
   c. replication
   d. lytic cycle

10. Bacteria that do not have an outer membrane beyond the cell wall are called _________.

11. _________ is the branch of bacteria that includes nearly all of the 4,800 kinds of bacteria that live today.

12. _________ are archaebacteria that are poisoned by oxygen and live in the guts of cows and other herbivores.

13. In cyanobacteria, nitrogen fixation occurs in specialized cells called _________.

14. Viruses cannot _________ on their own.

15. TMV stands for _________.

16. _________ are viruses that infect bacterial cells.

17. Influenza is a viral infection of the _________ tract.

Answers to the Concept Review questions appear in Appendix B.

CHALLENGE YOURSELF

1. In what ways might the early, self-replicating particles that gave rise to the first organisms have resembled, or differed from, viruses?
2. Justify the assertion that life on earth could not exist without bacteria.
3. Why are there no effective antibiotics against viral infections?
4. Why do we say that bacteria are alive and viruses are not?

FOR FURTHER READING

Calwell, J. C., and P. Caldwell. "The African AIDS Epidemic." *Scientific American,* March 1996, 62–68. Nearly 25% of the population in parts of Africa is HIV-positive. Could lack of circumcision in men in this region contribute to the high incidence of AIDS?

Calwell, M. "Blessed with Resistance." *Discover,* January 15, 1994, 46–48. Researchers have identified some individuals who appear to be naturally resistant to the AIDS virus.

Chin, G. J., and J. Marx. "Resistance to Antibiotics." *Science* (April 15, 1994): 359–93. One of several excellent articles in this issue dealing with the current status of antimicrobial drugs in medicine, particularly the development of the disturbing resistance to drugs that appears to be on the increase.

Dale, J. *Molecular Genetics of Bacteria,* New York: John Wiley & Sons, 1994. Much of what we understand about genetics comes from experimentation with bacteria, and almost all of biotechnology is dependent upon the workings of bacteria.

Fenchel, T., and B. J. Finlay. "The Evolution of Life Without Oxygen." *American Scientist* 82 (January/February 1994): 22–29. The role of the prokaryote in eukaryotic evolution is examined.

Garret, L. *The Coming Plague.* New York: Farrar, Straus, and Giroux, 1994. Contains a frightening account of the development and spread of the ebola virus.

Iseman, M. D. "Evolution of Drug-Resistant Tuberculosis: A Tale of Two Species." *Proc. Nat. Acad. Sci. USA* 91 (March 29, 1994): 2428–29. Drug-resistant diseases may be the biggest threat facing modern medicine. This article discusses the mechanisms by which the tuberculosis pathogen is becoming significantly dangerous.

Roizman, B., ed. *Infectious Diseases in an Age of Chance.* Washington, D.C.: National Academy Press, 1995. The impact of human ecology and behavior on disease transmission, an issue of increasing concern because of the potential for rapid worldwide transmission of usually localized outbreaks of tropical diseases.

TECHNOLOGY LINKS

The Living World Home Page
http://www.wcbp.com/biology/tlw

Physiological Concepts of Life Science Videotape
#13 Life Cycle of HIV

12

Advent of the Eukaryotes

"C'mon! Look at these fangs! . . . Look at these claws! . . . You think we're supposed to eat just honey and berries?"

CHAPTER OUTLINE

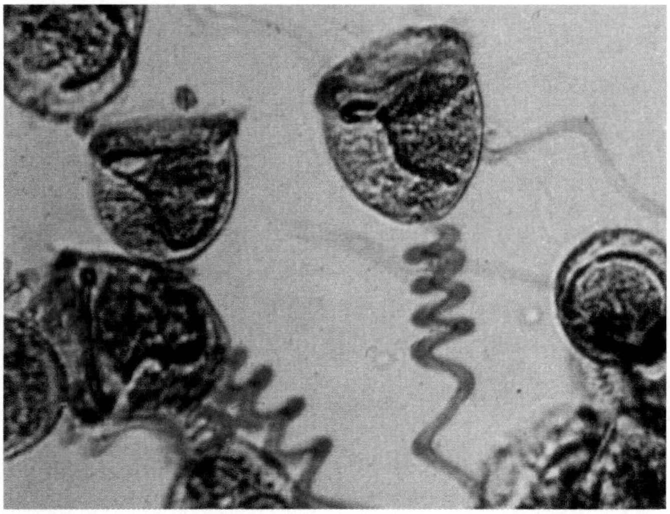

Figure 12.1 Eukaryotes, a new kind of organism. The hallmark of eukaryotes is complexity. These *Vorticella* (a ciliate) are heterotrophic protists that feed on bacteria and have retractable stalks.

For more than half of the long history of life on earth, all life was microscopic in size. The biggest organisms that existed for over 2 billion years were single-celled bacteria fewer than 6 micrometers thick. The first evidence of a different kind of organism is found in tiny fossils in rock 1.5 billion years old. These fossil cells are much larger than bacteria (some as big as 60 micrometers in diameter) and have internal membranes and what appear to be small, membrane-bounded structures. Many have elaborate shapes, and some exhibit spines or filaments. These new, larger fossil organisms mark one of the most important events in the evolution of life, the appearance of a new kind of organism, the eukaryote (figure 12.1). Flexible and adaptable, the eukaryotes rapidly evolved to produce all of the diverse large organisms that populate the earth today, including ourselves—indeed, all organisms other than bacteria are eukaryotes.

12.1 Endosymbiosis

What was the first eukaryote like? We cannot be sure, but a good model is *Pelomyxa palustris,* a single-celled, nonphotosynthetic organism that appears to represent an early stage in the evolution of eukaryotic cells (figure 12.2). The cells of **Pelomyxa** are much larger than a bacterial cell and contain a complex system of internal membranes. Although they resemble some of the largest early fossil eukaryotes, these cells are unlike any other eukaryote: *Pelomyxa* lacks mitochondria and does not undergo mitosis. Its nuclei divide somewhat as do those of bacteria, by pinching apart into two daughter nuclei, around which new membranes form. Although *Pelomyxa* cells lack mitochondria, two kinds of bacteria living within them may play the same role that mitochondria do in all other eukaryotes. This primitive eukaryote is so distinctive that it is assigned a phylum all its own, Caryoblastea.

Biologists know very little of the origin of *Pelomyxa,* except that in many of its fundamental characteristics it resembles the archaebacteria far more than the eubacteria. Because of this general resemblance, it is widely assumed that the first eukaryotic cells were nonphotosynthetic descendants of archaebacteria.

What about the wide gap between *Pelomyxa* and all other eukaryotes? Where did mitochondria come from? Most biologists agree with the theory of **endosymbiosis,** which proposes that mitochondria originated as symbiotic, aerobic (oxygen-requiring) eubacteria (figure 12.3). Symbiosis (Greek, *syn,* together with + *bios,* life) means living together in close association. The aerobic eubacteria that became mitochondria are thought to have been engulfed by ancestral eukaryotic cells much like *Pelomyxa* early in the history of eukaryotes. The most similar eubacteria to mitochondria today are the nonsulfur purple bacteria, which are able to carry out oxidative metabolism (described in chapter 5). Before they had acquired these bacteria, the host cells were unable to carry out the Krebs cycle or other metabolic reactions necessary for living in an atmosphere that contained increasing amounts of oxygen. The engulfed bacteria became the interior portion of the mitochondria we see today.

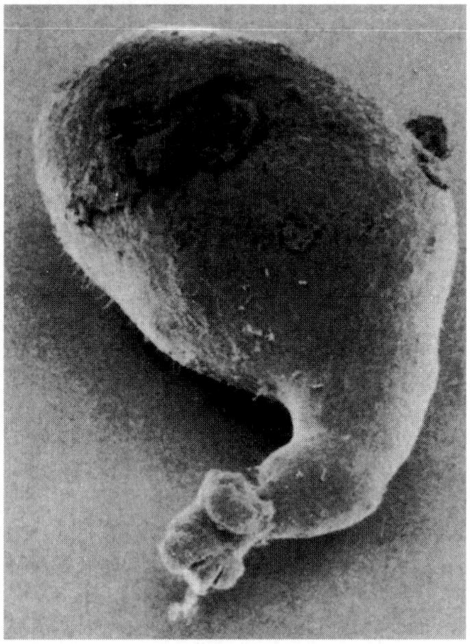

Figure 12.2 ***Pelomyxa palustris.***
This unique, amoeba-like protist lacks mitochondria and may represent a very early stage in the evolution of eukaryotic cells.

Mitochondria are sausage-shaped organelles 1 to 3 micrometers long, about the same size as most eubacteria. Mitochondria are bounded by *two* membranes. The outer membrane is smooth and was apparently derived from the endoplasmic reticulum of the host cell. The inner membrane is folded into numerous layers, resembling the folded membranes of nonsulfur purple bacteria; embedded within this membrane are the proteins that carry out oxidative metabolism.

During the billion and a half years in which mitochondria have existed as endosymbionts within eukaryotic cells, most of their genes have been transferred to the chromosomes of the host cells—but not all. Each mitochondrion still has its own genome, a circular, closed molecule of DNA similar to that found in eubacteria, on which is located genes encoding the essential proteins of oxidative metabolism. These genes are transcribed within the mitochondrion, using mitochondrial ribosomes that are smaller than those of eukaryotic cells, very much like bacterial ribosomes in size and structure. Mitochondria divide by simple fission, just as bacteria do, replicating and sorting their DNA much as bacteria do. However, nuclear genes direct the process, and mitochondria cannot be grown outside of the eukaryotic cell, in cell-free culture.

The theory of endosymbiosis has had a controversial history but has now been accepted by all but a few biologists. The evidence supporting the theory is so extensive that in this text we will treat it as established.

What of mitosis, the other typical eukaryotic process that *Pelomyxa* lacks? The mechanism of mitosis, now so common among eukaryotes, did not evolve all at once. Traces of very different, and possibly intermediate, mechanisms survive today in some of the eukaryotes. In fungi and some groups of protists, for example, the nuclear membrane does not dissolve and mitosis is confined to the nucleus. When mitosis is complete in these organisms, the nucleus divides into two daughter nuclei, and only then does the rest of the cell divide. This separate nuclear division phase of mitosis does not occur in most protists, or in plants or animals. We do not know if it represents an intermediate step on the evolutionary journey to the form of mitosis that is characteristic of most eukaryotes today or if it is simply a different way of solving the same problem. There are no fossils in which we can see the interiors of dividing cells well enough to be able to trace the history of mitosis.

Endosymbiosis Is Not Rare

Many eukaryotic cells contain other endosymbiotic bacteria in addition to mitochondria. Plants and algae contain chloroplasts, bacteria-like organelles that were apparently derived from symbiotic photosynthetic bacteria. The chloroplast body is bounded by two membranes that resemble those of mitochondria and were apparently derived in the same fashion. Chloroplasts are larger than mitochondria, with a more complex system of internal membranes and a larger circle of DNA. As in mitochondria, the gene translation machinery of the chloroplast closely resembles that of bacteria.

While all mitochondria are thought to have arisen from a single symbiotic event, it is difficult to be sure for chloroplasts. Three biochemically distinct classes of chloroplast exist, each resembling a different bacterial ancestor. Red algae possess pigments similar to those of cyanobacteria, plants and green algae more closely resemble the photosynthetic bacteria *Prochloron,* while brown algae and other photosynthetic protists resemble a third group of bacteria. This diversity of chloroplasts has led to the widely held belief that eukaryotic cells acquired chloroplasts by symbiosis at least three different times. Recent comparisons of chloroplast DNA sequences, however, suggest a single origin of chloroplasts, followed by very different postendosymbiotic histories. In each of the three main lines, different genes became relocated to the nucleus, lost, or modified. Chloroplasts seem to have had a rich evolutionary history *after* their entry into eukaryotic cells.

Mitochondria and chloroplasts are not the only eukaryotic organelles that appear to have an endosymbiotic origin. Centrioles, organelles associated with the assembly of microtubules, resemble in many respects spirochete bacteria, and they contain bacteria-like DNA involved in the production of their structural proteins. While the case for endosymbiotic spirochetes is still sketchy, this is an area of active research.

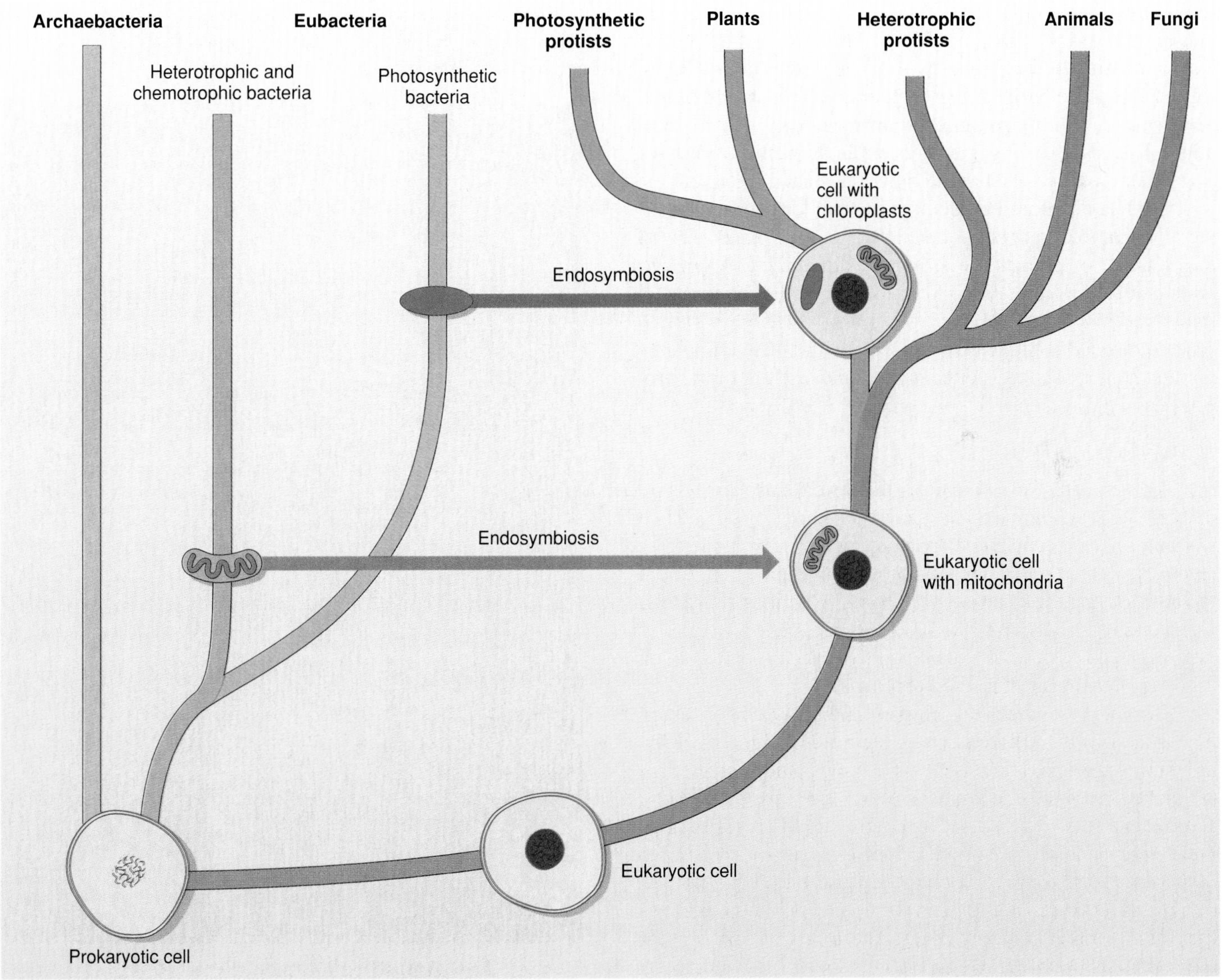

Figure 12.3 The theory of endosymbiosis.
Scientists propose that ancestral eukaryotic cells engulfed aerobic eubacteria, which then became mitochondria in the eukaryotic cell. Chloroplasts may also have originated this way, with eukaryotic cells engulfing photosynthetic eubacteria.

12.2 The Protists

Between the bacteria and the most primitive eukaryotes, the **protists,** there is a gap in the evolutionary record. Microscopic fossils tell us nothing about how the many fundamental changes in internal cell structure evolved. There are no microscopic "missing links." The evolution of the first eukaryotes could not have happened quickly, because they differ from bacteria in so many different complex ways. No other organisms that are transitional between bacteria and eukaryotes survive, however.

The Most Diverse Kingdom

Of the six kingdoms of organisms, the kingdom Protista is by far the most diverse. Among the protists are the simplest of the eukaryotes and also many very complex organisms. Few protists are large; most are unicellular and too small to see with the naked eye. All single-celled eukaryotes (except yeasts) are protists (figure 12.4). However, as you will learn, **multicellularity** has evolved in many lines (figure 12.5). Sometime early in the evolution of protists the *flagellum* evolved, with its complex arrangement of 9 + 2 microtubules. This same arrangement occurs in all protists that have flagella or cilia, as well as in all other flagellated eukaryotes.

The kingdom Protista is a catchall kingdom, containing all eukaryotes that are not animals, fungi, or plants. The 14 major phyla of protists are strikingly different from one another and are with a few exceptions only distantly related (table 12.1). What unites the protists as a kingdom is the absence of special features that characterize the three complexly multicellular kingdoms. For example, protists do not form embryos, and they do not develop complex sexual organs.

Algae and Protozoa

In older systems of classification, there were only three kingdoms—bacteria, animals, and plants. Taxonomists considered anything that was capable of photosynthesis to be a plant, including the green photosynthetic protists, or **algae.** Fungi were included among the plants as well, along with the protists that seemed to resemble fungi, the **molds.** Single-celled eukaryotes that ingested their food "like animals" were considered small, very simple animals and called **protozoa.**

These schemes of classification have been abandoned, and in the six-kingdom system now widely used by taxonomists, algae, slime and water molds, and protozoa are all included among the protists. Why? Because to consider some groups of protists to be plants simply because they have independently acquired chloroplasts in the course of their evolution, or to be animals because they have not acquired chloroplasts and can move, or to be fungi because they don't, misses the key point. These protists differ from other organisms in many significant respects, and it is very misleading to lump them together with unrelated organisms on the basis of a single trait that they acquired independently.

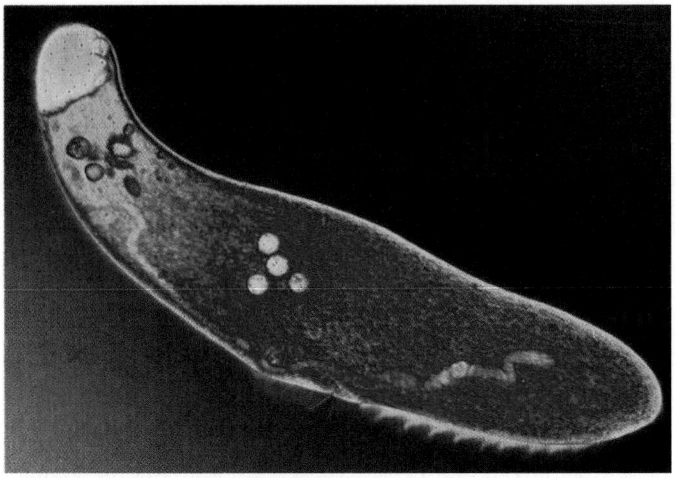

Figure 12.4 A single-celled protist.
Although multicellularity has evolved many times among the protists, most protists are microscopic, single-celled organisms. This *Paramecium* is typical in that its single cell has a complex interior organization.

Figure 12.5 A multicellular protist.
Surrounding this diver are massive "groves" of giant kelp, a kind of brown algae. Kelp groves contain some of the largest organisms on earth.

Table 12.1 The Protists

Group	Phyla	Typical Examples		Key Characteristics	Approximate Number of Living Species
Sarcodina			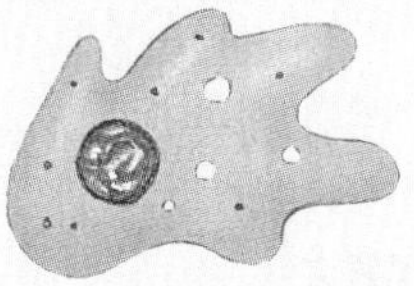	No permanent locomotor apparatus; heterotrophs	13,500
Amoebas	Rhizopoda	*Amoeba*		Move by pseudopodia	
Forams	Foraminifera	Forams		Rigid shells; move by protoplasmic streaming	
Algae			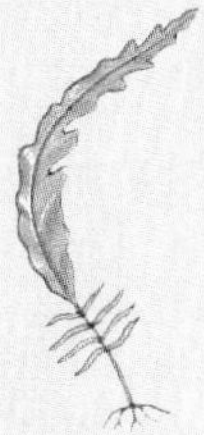	Photosynthetic; multicellular or largely multicellular	12,500
Red	Rhodophyta			Chlorophyll *a* + red pigment	
Brown	Phaeophyta	Kelp		Chlorophyll *a* + chlorophyll *c*	
Green	Chlorophyta	*Chlamydomonas*		Chlorophyll *a* + chlorophyll *b*	
Diatoms	Bacillariophyta	"Diatomaceous earth"	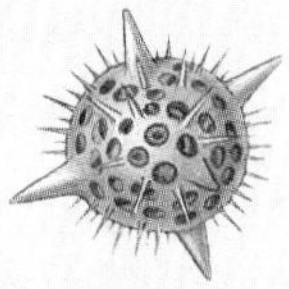	Double shell of silica; photosynthetic; unicellular	11,500
Flagellates				Locomotor flagella	9,500
Dinoflagellates	Dinoflagellata	Red tides	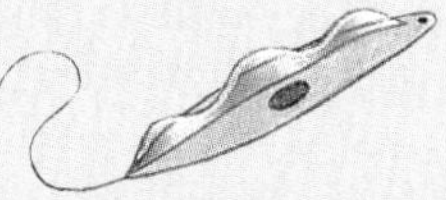	Photosynthetic; unicellular; two flagella	
Zoomastigotes	Zoomastigina	Trypanosomes		Heterotrophic; unicelluluar	
Euglenoids	Euglenophyta	*Euglena*		Some photosynthetic, others heterotrophic; unicellular	
Ciliates	Ciliophora	*Paramecium*	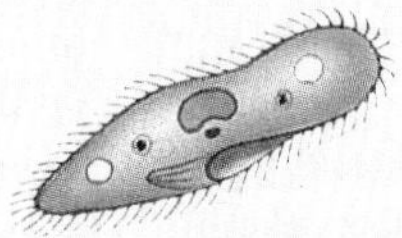	Many cilia; two nuclei per cell; fixed cell shape; heterotrophic; unicellular	8,000
Sporozoans	Sporozoa	*Plasmodium*	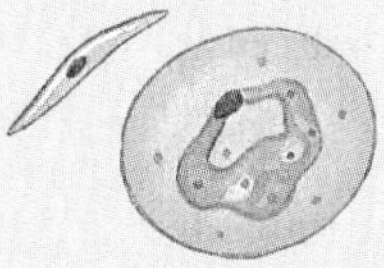	Nonmotile; spore-forming; unicellular parasites	3,900
Molds				Heterotrophs with restricted mobility; cell walls made of carbohydrate	1,150
Cellular slime molds	Acrasiomycota		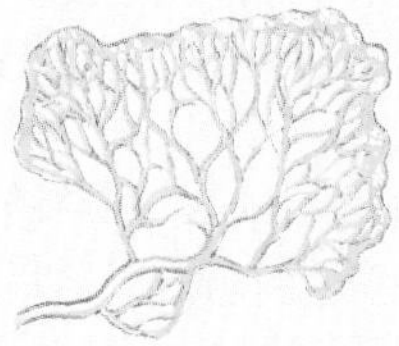	Colonial aggregations of individual cells; most closely related to amoebas	
Plasmodial slime molds	Myxomycota			Stream along as a multinucleate mass of cytoplasm	
Water molds	Oomycota	Rusts and mildew		Terrestrial and freshwater	

12.3 The Evolution of Sex

Of all the differences between bacteria and eukaryotes, a profound one is the capacity for sexual reproduction among eukaryotes. In **sexual reproduction,** two different parents contribute gametes to form the offspring. Gametes are usually formed by meiosis, discussed in chapter 6. In most eukaryotes, the gametes are haploid (have a single copy of each chromosome), and the offspring produced by their fusion are diploid (have two copies of each chromosome). In this section we examine sexual reproduction among the eukaryotes and how it evolved.

Life Without Sex

Sexual reproduction is not the only way that eukaryotes can reproduce. Consider, for example, a sponge. A sponge can reproduce by simply fragmenting its body. Each small portion grows and gives rise to a new sponge. This is an example of **asexual reproduction,** reproduction without forming gametes. In asexual reproduction, the offspring are genetically identical to the parents, barring mutation. Some protists such as the green algae exhibit a true sexual cycle, but only transiently (figure 12.6). The majority of protists reproduce asexually most of the time. The fusion of two haploid cells to create a diploid zygote, the essential act of sexual reproduction, occurs only under stress.

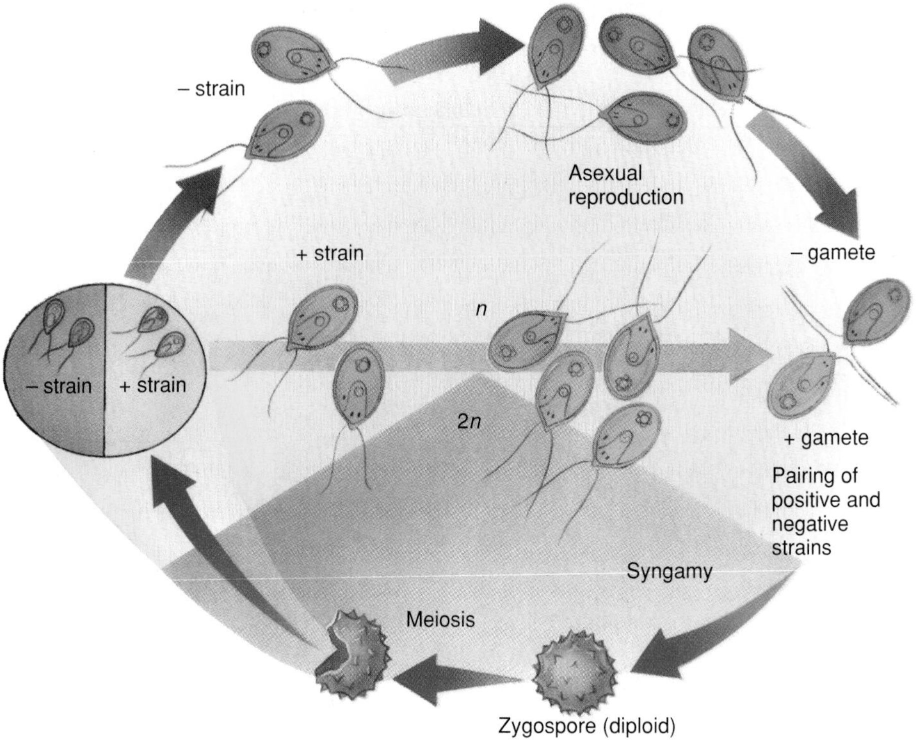

Figure 12.6 The life cycle of *Chlamydomonas.*
Individual cells of *Chlamydomonas* (a microscopic, biflagellated alga, phylum Chlorophyta) are haploid and divide asexually, producing identical copies of themselves. At times, such haploid cells act as gametes, fusing to produce a zygote, as shown in the lower right-hand side of the diagram. The zygote develops a thick, resistant wall, becoming a zygospore. Meiosis takes place, ultimately resulting in the release of four haploid individuals.

The development of an adult from an unfertilized egg is a form of asexual reproduction called **parthenogenesis.** Parthenogenesis is a common form of reproduction among insects. Among bees, for example, fertilized eggs develop into females, while unfertilized eggs become males. Some lizards, fishes, and amphibians reproduce by parthenogenesis, an unfertilized egg undergoing mitosis without cell cleavage to produce a diploid cell, which then undergoes development as if it were a zygote produced by sexual union of two gametes.

Many plants and marine fishes undergo a form of sexual reproduction that does not involve partners. In **self-fertilization,** one individual provides both male and female gametes. Mendel's peas discussed in chapter 6 produced their F_2 generations by "selfing." Why isn't this asexual reproduction (after all, there is only one parent)? This is considered to be sexual rather than asexual reproduction because the offspring are not genetically identical to the parent. During the production of the gametes, considerable genetic reassortment occurs—that is why Mendel's F_2 pea plants were not all the same!

Why Sex?

If reproduction without sex is so common among eukaryotes today, it is a fair question to ask why sex occurs at all. Evolution is the result of changes that occur at the level of *individual* survival and reproduction, and it is not immediately obvious what advantage is gained by the progeny of an individual that engages in sexual reproduction. Indeed, the segregation of chromosomes that occurs in meiosis tends to disrupt advantageous combinations of genes more often than it assembles new, better-adapted ones. Because all the progeny could maintain a parent's successful gene combinations if the parent employed asexual reproduction, the widespread use of sexual reproduction among eukaryotes raises a puzzle: Where is the benefit from sex that promoted the evolution of sexual reproduction?

How Sex Evolved

In attempting to answer this question, biologists have looked more carefully at where sex first evolved—among the protists. Why do many protists form a diploid cell in response to stress? Biologists think this occurs because only in a diploid cell can certain kinds of chromosome damage be repaired effectively, particularly double-strand breaks in DNA. Such breaks are induced, for example, by desiccation—drying out. The early stages of meiosis, in which the two copies of each chromosome line up and pair with each other, seems to have evolved originally as a mechanism for repairing doublestrand damage to DNA by using the undamaged version of the chromosome as a template to guide the fixing of the damaged one. In yeasts, mutations that inactivate the system that repairs double-strand breaks of the chromosomes also prevent crossing over. Thus it seems likely that sexual reproduction and the close association between pairs of chromosomes that occurs during meiosis first evolved as mechanisms to repair chromosomal damage by using the second copy of the chromosome as a template.

Why Sex Is Important

One of the most important evolutionary innovations of eukaryotes was the invention of sex. Sexual reproduction provides a powerful means of shuffling genes, quickly generating different combinations of genes among individuals. Genetic diversity is the raw material for evolution. In many cases the pace of evolution appears to be geared to the level of genetic variation available for selection to act upon—the greater the genetic diversity, the more rapid the evolutionary pace. Programs for selecting larger domestic cattle and sheep, for example, proceed rapidly at first but then slow as all of the existing genetic combinations are exhausted; further progress must then await the generation of new gene combinations. The genetic recombination produced by sexual reproduction has had an enormous evolutionary impact because of its ability to rapidly generate extensive genetic diversity.

Sexual Life Cycles

Many protists are haploid all their lives, but with few exceptions, animals and plants are diploid at some stage of their lives. That is, the body cells of most animals and plants have two sets of chromosomes, one from the male and one from the female parent. The production of haploid gametes by meiosis, followed by the union of two gametes in sexual reproduction, is called the **sexual life cycle.**

Eukaryotes are characterized by three major types of sexual life cycles (figure 12.7).

1. In the simplest of these, found in many algae, the zygote formed by the fusion of gametes is the only diploid cell. This sort of life cycle is said to have **zygotic meiosis,** because in algae the zygote undergoes meiosis. Here the haploid cells occupy the major portion of the life cycle; the zygote undergoes meiosis immediately after it is formed.
2. In most animals, the gametes are the only haploid cells. They exhibit **gametic meiosis,** because in animals meiosis produces the gametes. Here the diploid zygote occupies the major portion of the life cycle.
3. Plants exhibit **sporic meiosis,** because in plants the spore-forming cells undergo meiosis. In plants there is a regular **alternation of generations** between a haploid phase and a diploid phase. The diploid phase produces spores that give rise to the haploid phase, and the haploid phase produces gametes that fuse to give rise to the diploid phase.

The genesis of sex, then, involved meiosis and fertilization with the participation of two parents. We have previously said that bacteria lack true sexual reproduction, although in some groups, two bacteria do pair up in conjugation and exchange parts of their genome. The invention of true sexual reproduction among the protists has no doubt contributed importantly to their tremendous diversification and adaptation to an extraordinary range of ways of life, as we shall see in the following section.

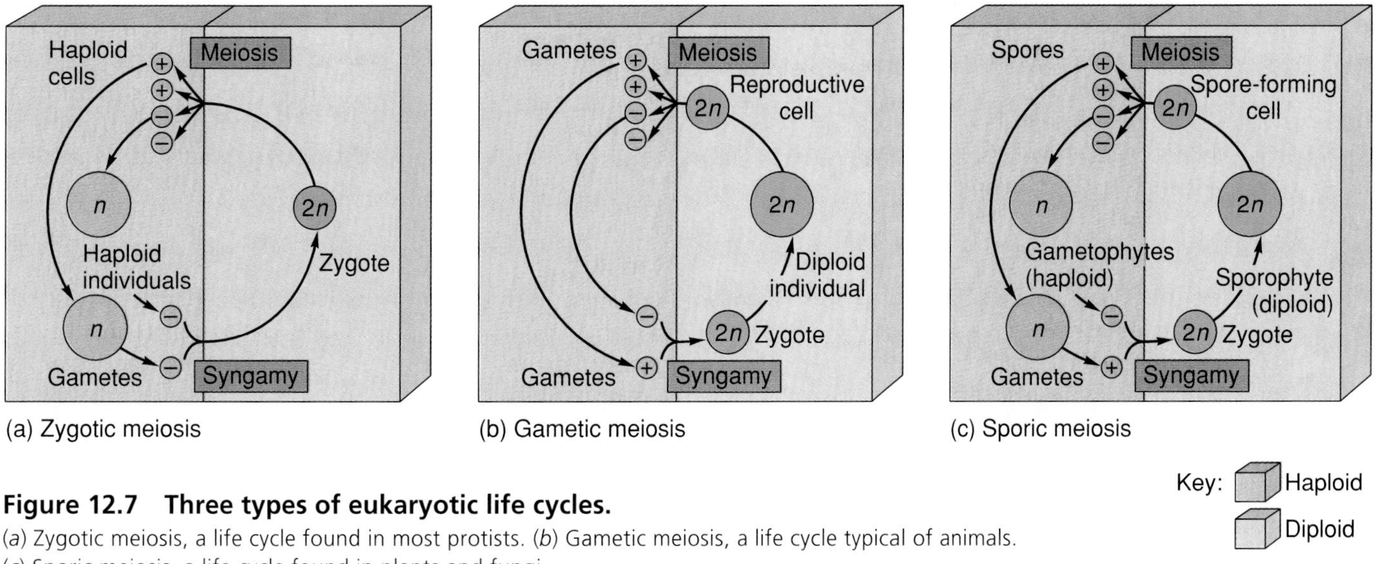

Figure 12.7 Three types of eukaryotic life cycles.
(*a*) Zygotic meiosis, a life cycle found in most protists. (*b*) Gametic meiosis, a life cycle typical of animals. (*c*) Sporic meiosis, a life cycle found in plants and fungi.

12.4 Kinds of Protists

The diversity among the protists is so great that biologists have not yet sorted out the evolutionary relationships among them. Scientists do not yet know enough to construct a general phylogeny such as those described in chapter 10. Lacking a clear evolutionary picture, biologists group the many kinds of protists according to shared characters that seem to be of particular importance. This text separates the protists into seven general groups, distinguished from one another largely by their mode of locomotion. Within these seven groups are 14 major phyla of protists and many minor ones.

Sarcodina

The largest of the seven general groups of protists, the Sarcodina, are distinguished by having no permanent locomotor apparatus. They are all heterotrophs and contain two major phyla, the amoebas and the forams.

Amoebas

Amoebas, members of the phylum Rhizopoda, lack flagella and cell walls. There are several hundred species. They move from place to place by **pseudopodia** (Greek, *pseudo,* false + *podium,* foot), flowing projections of cytoplasm that extend outward (figure 12.8). An amoeba puts a pseudopod forward and then flows into it. Amoebas are abundant in soil, and many are parasites of animals. Reproduction in amoebas occurs by simple *fission* into two daughter cells of equal volume. They do undergo mitosis but lack meiosis and any form of sexuality.

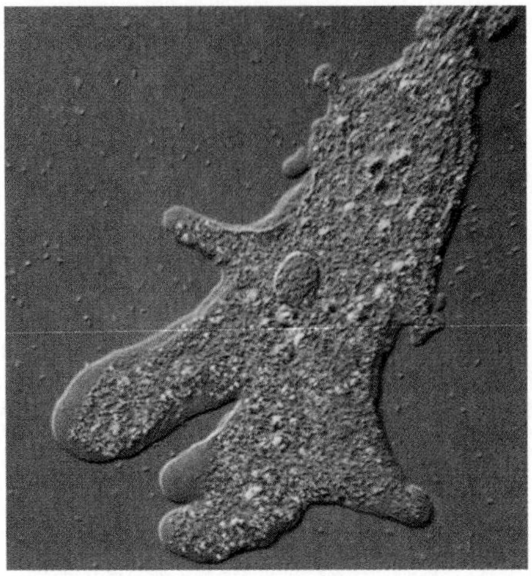

Figure 12.8 An amoeba.
Amoeba proteus is a relatively large amoeba. The projections are pseudopodia; an amoeba moves simply by flowing into them.

Figure 12.9 A representative of phylum Foraminifera.
A living foram, showing the podia, thin cytoplasmic projections that extend through pores in the calcareous test, or shell, of the organism.

Forams

Forams, members of the phylum Foraminifera, possess rigid shells and move by **cytoplasmic streaming.** They are marine protists with pore-studded shells called **tests** that may be as big as several centimeters in diameter. There are several hundred species of forams. Their shells, built largely of calcium carbonate, are often brilliantly colored—vivid yellow, bright red, or salmon pink—and may have many chambers arrayed in a spiral shape resembling a tiny snail. Long, thin cytoplasmic projections called **podia** extend out through the pores in the tests and are used for swimming and capturing prey (figure 12.9). The life cycle of forams is complex, involving alternation between haploid and diploid generations. Limestone is often rich in forams—the White Cliffs of Dover, the famous landmark on the southern England seacoast, is made up almost entirely of foram tests.

Algae

Algae are photosynthetic protists, many of them multicellular. There are three major kinds, distinguished by the kind of chlorophyll pigment they contain: green, red, or brown. Their chloroplasts appear to have arisen by a single endosymbiotic event, before the differences in chlorophyll evolved.

Green Algae

Green algae, members of the phylum Chlorophyta, are of special interest because the ancestor of true plants was a member of this group. Green algae chloroplasts are similar to plant chloroplasts, and like them they contain chlorophylls *a* and *b*. Green algae are an extremely varied group of more than 7,000 species, mostly mobile and aquatic like *Chlamydomonas,* but a few (like *Chlorella*) are immobile in moist soil or on tree trunks. Most green algae are microscopic and unicellular, but some, like *Ulva* (sea lettuce), are large and multicellular. Among the most elaborate is *Volvox,* a hollow sphere made up of tens of thousands of cells (figure 12.10). The two flagella of each cell beat in time with all the others to rotate the colony, which has reproductive cells at one end. *Volvox* borders on true multicellularity.

Red Algae

Red algae, members of the phylum Rhodophyta, possess red pigments called phycobilins that give them their characteristic color (figure 12.11). Almost all of the 4,000 species of red algae are multicellular and live in the sea, where they grow more deeply than any other photosynthetic organism. Red algae have complex bodies made up of interwoven filaments of cells. The laboratory media *agar* is made from the cell walls of red algae. Their life cycle is complex, usually in-

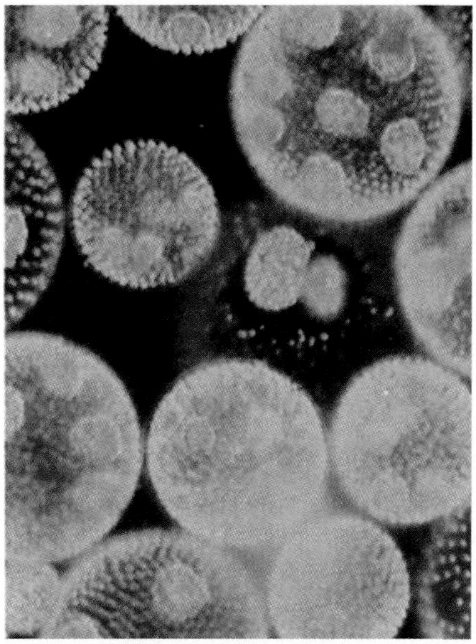

Figure 12.10 Green algae.
The green algae *Volvox* aggregate into hollow balls of cells that function as highly integrated colonies, so integrated that some biologists regard them as multicellular.

Figure 12.11 Red algae.
These red algae have their cellulose cell walls heavily impregnated with calcium carbonate, the same material of which oyster shells are made. Because they are hard and occur on coral reefs, they are called coralline algae.

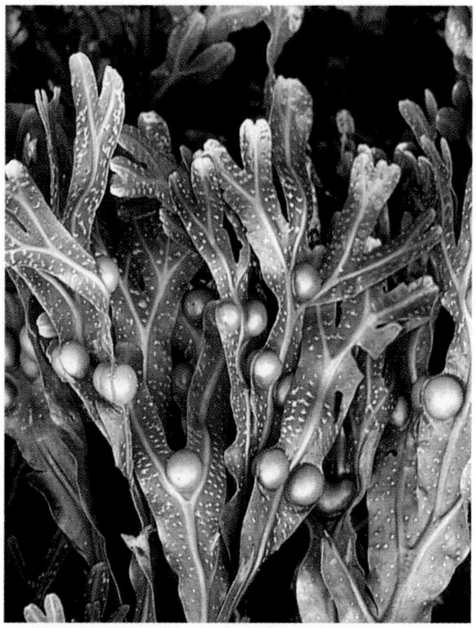

Figure 12.12 Brown algae.
Like all brown algae, this bladder wrock is multicellular. Some brown algae, called kelp, are among the largest of all multicellular organisms, often growing taller than a house.

volving alternation of generations. None of the red algae have flagella or centrioles, suggesting that red algae may be one of the most ancient groups of eukaryotes.

Brown Algae

Brown algae, members of the phylum Phaeophyta, contain the longest, fastest-growing, and most photosynthetically productive living things—giant kelp, with individuals over 100 meters long (figure 12.12). The 1,500 species of brown algae are all multicellular and almost all marine. They are the most conspicuous seaweeds in the ocean. The larger brown algae have flattened blades, stalks, and anchoring bases and often contain complex internal plumbing like that of plants. The life cycle employs an alternation of generations, with the large individuals we see being the sporophyte (diploid) generation.

Diatoms

Diatoms, members of the phylum Bacillariophyta, are photosynthetic unicellular protists with a unique double shell of silica. Like tiny oysters, their shells resemble small boxes with lids, one half fitting inside the other. They are sometimes called golden algae, although they do not resemble the three phyla of true algae in any important aspects. There is one major phylum.

Diatoms are abundant in both oceans and lakes. There are over 11,500 species, of two sorts: some with radial symmetry (those in figure 12.13 look like tiny wheels) and others with bilateral (two-sided) symmetry. The shells of fossil diatoms form thick deposits that are mined commercially as "diatomaceous earth," which is used as an abrasive or to make paint sparkle. Diatoms move by protoplasmic streaming along a groove in their shell. Individuals are diploid and usually reproduce asexually by separating the halves of the shell, each half then growing another half to match it. When individuals get too small because of repeated division, they slip out of their shells, grow to full size, and then regenerate a new set of shells. Under stress, diatoms undergo meiosis and reproduce sexually.

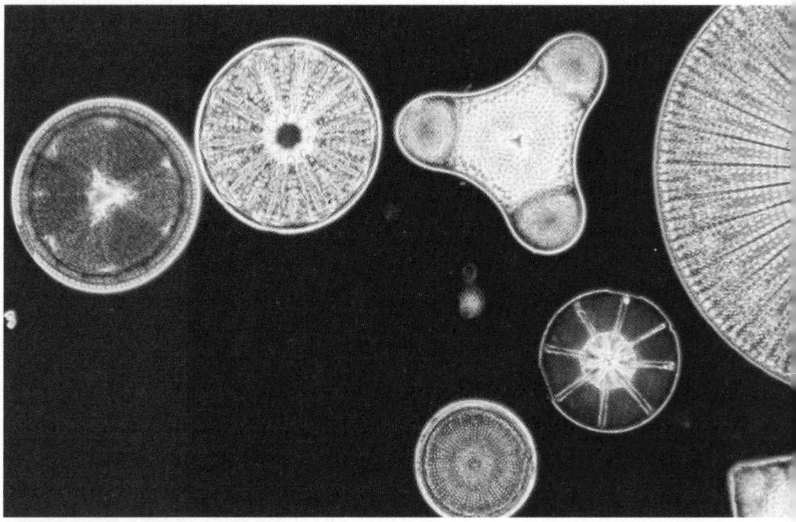

Figure 12.13 Diatoms.
Several different kinds of diatoms with radial symmetry.

Flagellates

Flagellates are protists that move by means of locomotory **flagella.** There are three major phyla: photosynthetic dinoflagellates, heterotrophic zoomastigotes, and euglenoids (in many respects specialized zoomastigotes, some of which are acquired chloroplasts). Some of these flagellates are heterotrophic and others photosynthetic.

Dinoflagellates

Dinoflagellates, members of the phylum Dinoflagellata, are unicellular photosynthetic protists, most with two flagella of unequal length. The dinoflagellates are very unusual protists that do not appear to have any close relatives. There are about 1,000 species. Some occur in fresh water, but most are marine. Luminous dinoflagellates produce the twinkling light sometimes seen in tropical seas at night. Most dinoflagellates have a stiff coat of cellulose, often encrusted with silica, giving them unusual shapes. Their flagella are unique, unlike those of any other phylum: one beats in a groove circling the body like a belt, the other in a groove perpendicular to it (figure 12.14). Their beating rotates the body like a top. A few dinoflagellates produce powerful toxins such as the poisonous "red tides," which are population explosions of such dinoflagellates (figure 12.15). Dinoflagellates reproduce by splitting in half. Their form of mitosis is unique—their chromosomes remain condensed, distributed along the sides of channels containing bundles of microtubules that run through the nucleus.

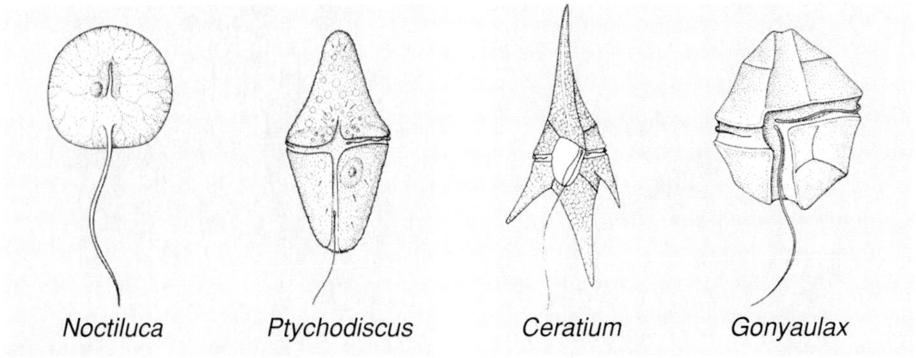

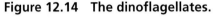

Figure 12.14 The dinoflagellates.
Noctiluca, Ptychodiscus, Ceratium, and *Gonyaulax. Noctiluca,* which lacks the heavy cellulosic armor characteristic of most dinoflagellates, is one of the bioluminescent organisms that causes the waves to sparkle in warm seas at certain times of year.

Figure 12.15 Red tide.
Red tides are caused by population explosions of dinoflagellates. The pigments in the dinoflagellates or, in some cases, other organisms, are responsible for the color of the water.

Zoomastigotes

Zoomastigotes, members of the phylum Zoomastigina, are of special interest because the ancestor of all animals appears to have been a member of this group. The choanoflagellates are the group from which sponges, and probably all other animals, were derived. Zoomastigotes are unicellular, heterotrophic protists that are highly variable in form. There are several thousand species. All have at least one flagellum, and some species have thousands. Most reproduce only asexually. Members of one group, the trypanosomes, are important pathogens of human beings (figure 12.16). Some zoomastigotes live symbiotically in the guts of termites and provide the enzymes the termites use to digest wood (much as bacteria aid cattle and horses in digesting grass).

Euglenoids

Euglenoids, members of the phylum Euglenophyta, are freshwater protists with two flagella that clearly illustrate the folly of attempting to classify protists as tiny animals or plants. About one-third of the 1,000 known species have chloroplasts and are photosynthetic; the others lack chloroplasts, ingest their food, and are heterotrophic. In the dark, many photosynthetic euglenoids reduce the size of their chloroplasts (they may appear to disappear!) and become heterotrophs until put back in light. Euglenoids are very closely related to zoomastigotes, and many taxonomists merge the two phyla. *Euglena,* after which the phylum is named, has a protein scaffold called the pellicle inside the cell membrane and can change shape (figure 12.17). Two flagella are attached at the base of a flask-shaped opening called the reservoir, which is located at the anterior end of the cell. One of the flagella is long and has a row of very fine, short, hairlike projections along one side. A second, shorter flagellum is located within the reservoir but does not emerge from it. A light-sensitive organ, the stigma, helps them move toward light. Reproduction is by mitotic cell division; no sexual reproduction is known in this group.

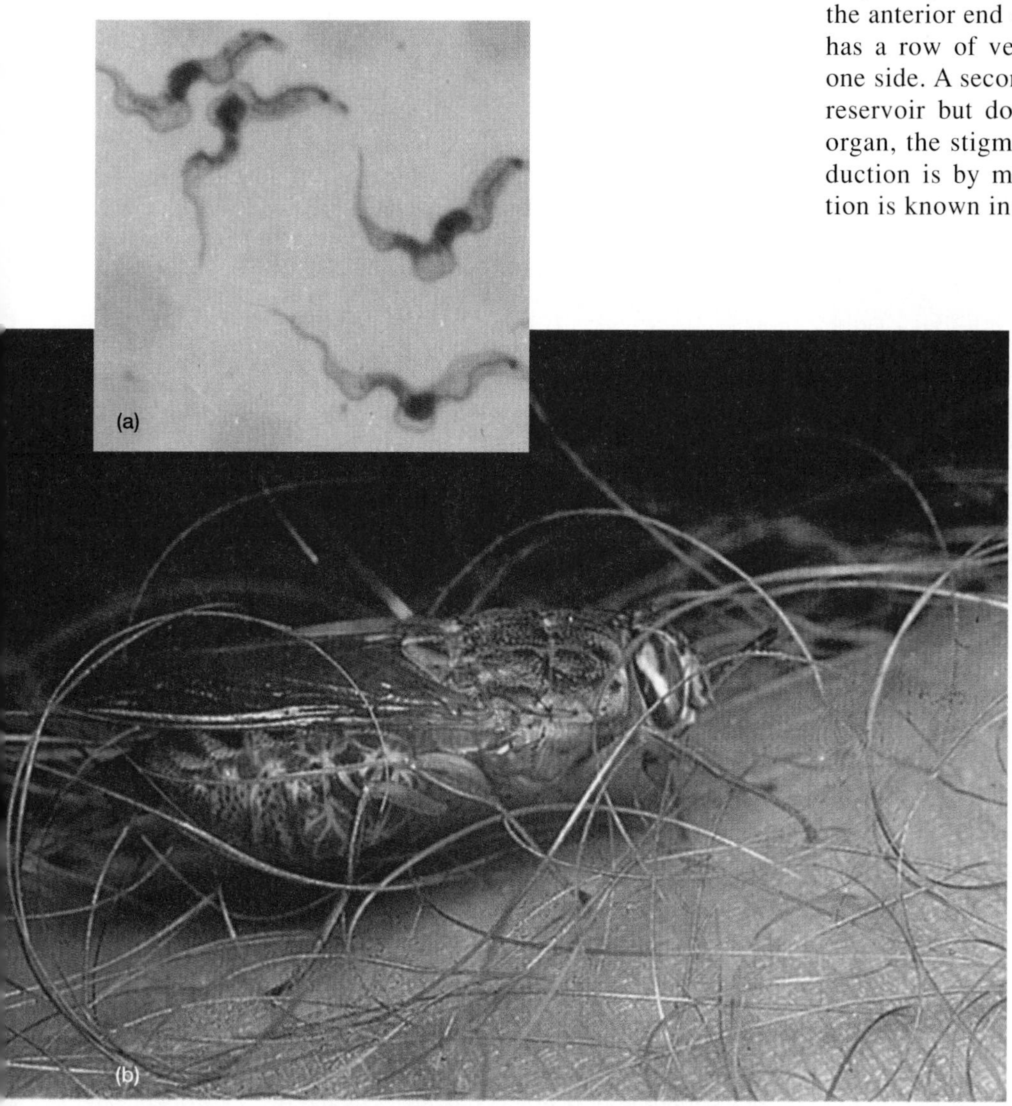

Figure 12.16 Sleeping sickness—a protist's disease.
(*a*) *Trypanosoma*, the protists that cause sleeping sickness, among red blood cells. The undulating, changeable shapes of the trypanosomes are visible in this photograph. (*b*) A tsetse fly, shown here sucking blood from a human arm, in Tanzania, East Africa. Tsetse flies transmit the trypanosomes that cause sleeping sickness.

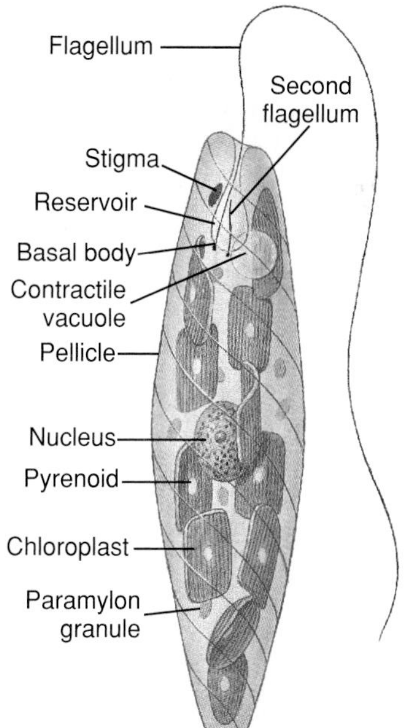

Figure 12.17 Diagram of *Euglena*.
Starch forms around pyrenoids; paramylon granules are areas where food reserves are stored.

Ciliates

The ciliates are very complex and unusual unicellular heterotrophs, with **cilia,** a fixed cell shape, and two nuclei per cell. Ciliates are so different from other eukaryotes (they even use the genetic code differently!) that many taxonomists argue they should be placed in a separate kingdom of their own. There is only one major phylum.

Ciliates, members of the phylum Ciliophora, all possess large numbers of cilia, usually arrayed in long rows down the body or in spirals around it. About 8,000 species have been named. Ciliates have a pellicle, which makes the body wall tough but flexible. The body interior is extremely complex, inspiring some biologists to consider ciliates organisms without cell boundaries rather than unicellular. *Paramecium,* a typical ciliate, has a complex digestive process, with a gullet ("mouth") and intake channel for bacteria and food particles, which are then enclosed in membrane bubbles and digested by enzymes. Reproduction is usually by mitosis, with the body splitting in half. Cells divide asexually for about 700 generations and then die if sexual reproduction has not occurred. Sexual reproduction is by conjugation, duplicate nuclei being exchanged between different mating types across a conjugation bridge (figure 12.18).

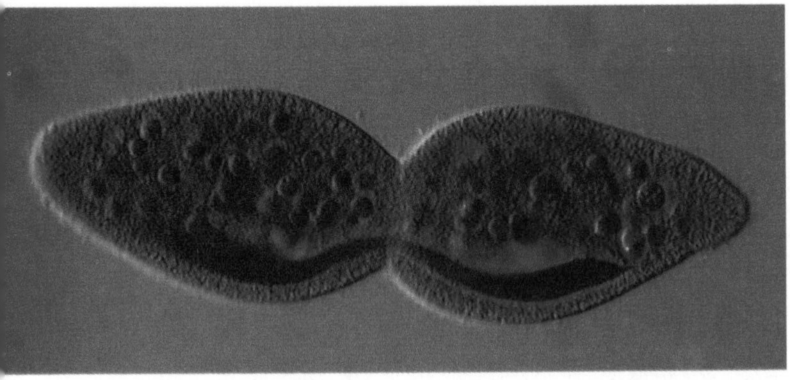

(a)

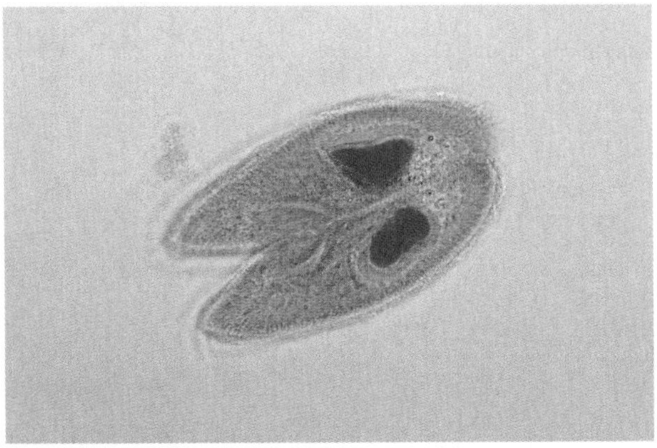

(b)

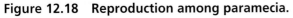

Figure 12.18 Reproduction among paramecia.
When a mature *Paramecium* divides (*a*), two complete individuals result. When two mature cells fuse (*b*), the process is called conjugation.

Molds

Molds are heterotrophs with restricted mobility that are sometimes confused with fungi, although they do not resemble them in any significant respect. For example, the cell walls of molds are made of carbohydrate, like those of all protists, while fungal walls are made of chitin. Also, molds carry out normal mitosis, while fungal mitosis is unusual. There are three major phyla of molds, each quite different in their structure and life cycles. None of the mold phyla are related.

Cellular Slime Molds

Cellular slime molds, members of the phylum Acrasiomycota, are more closely related to amoebas than any other phylum, but they are able to aggregate in times of stress into mobile, multicellular colonies called **slugs.** There are 70 named species, the best known of which is *Dictyostelium discoideum. Dictyostelium* has been the subject of intensive research by biologists interested in the basic mechanisms of development. The complex developmental cycle of this mold is presented in figure 12.19. *Dictyostelium* is basically a unicellular scavenger. When deprived of food, thousands of individual *Dictyostelium* amoebas come together into a slug that moves to a new habitat. There, the colony differentiates into a base, a stalk, and a swollen tip that develops **spores.** Each of these spores, when released, becomes a new amoeba, which begins to feed and so restarts the life cycle.

Plasmodial Slime Molds

Plasmodial slime molds, members of the Myxomycota, are a group of about 500 bizarre species that stream along as a **plasmodium,** a nonwalled multinucleate mass of cytoplasm that looks like an oozing mass of slime. Plasmodia can flow around obstacles and even pass through the mesh in cloth. As they move, they engulf and digest bacteria and other organic material. A plasmodial slime mold contains many nuclei, but these are not separated by cell walls. All of the nuclei undergo mitosis at the same time, in coordinated fashion. If the plasmodium begins to dry or starve, it migrates away rapidly, and then stops and often divides into many small mounds, each of which produces a spore-laden structure. Within the spores, meiosis occurs, producing haploid gametes. Spores germinate when favorable conditions return, the gametes fusing to re-form a diploid plasmodium, in which mitosis continues to occur.

Water Molds

Water molds, members of the phylum Oomycota, are the rusts and mildews that are often seen in moist environments. There are 580 named species, all of which either parasitize living organisms or feed on dead organic matter. Oomycetes are unusual in that their spores are motile, with two flagella, one pointed forward, the other backward. Many oomycetes are important plant pathogens, including *Phytophthora infestans,* which causes late blight in potatoes. This mold was responsible for the Irish potato famine of 1845–47, during which about 400,000 Irish people starved to death.

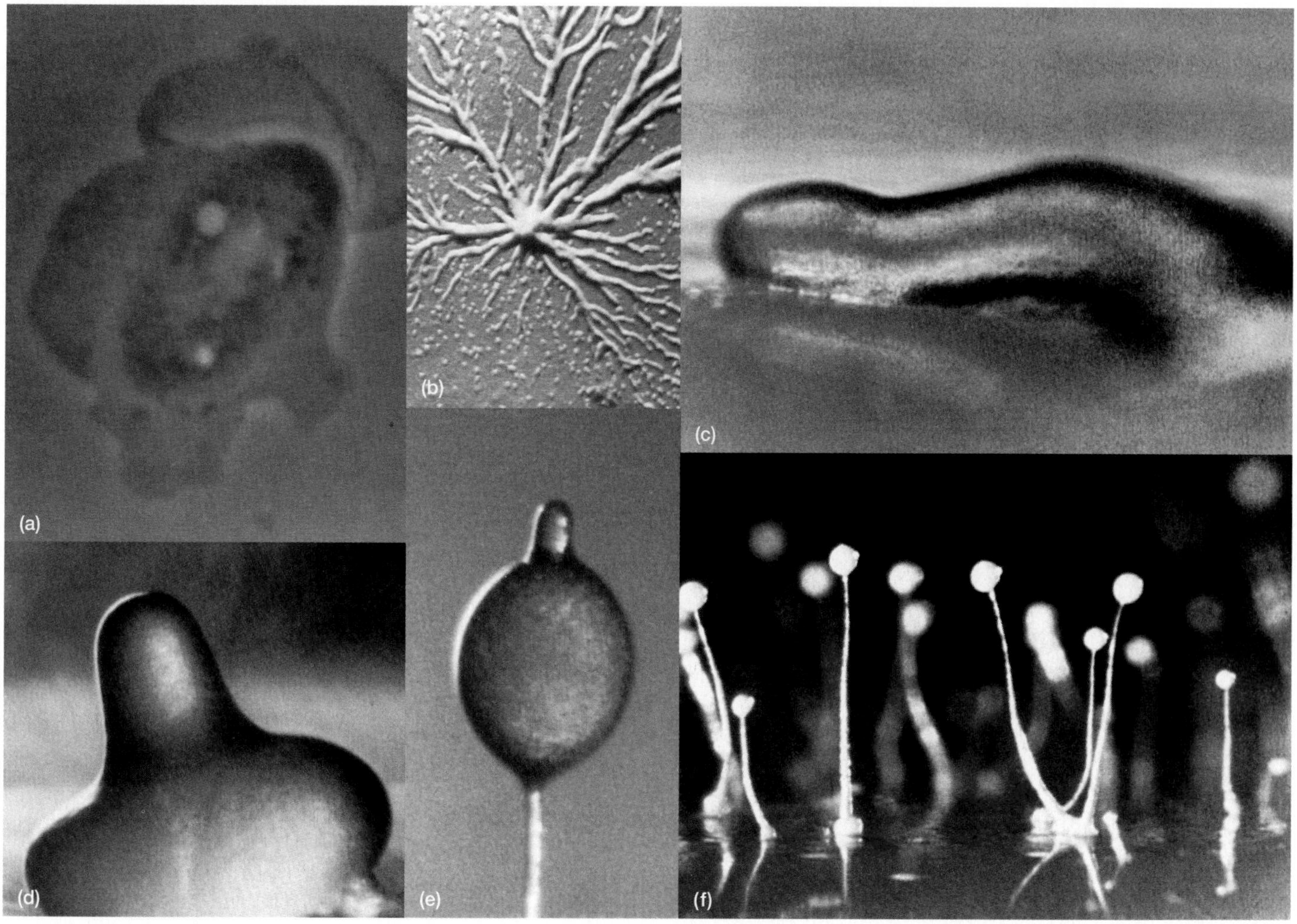

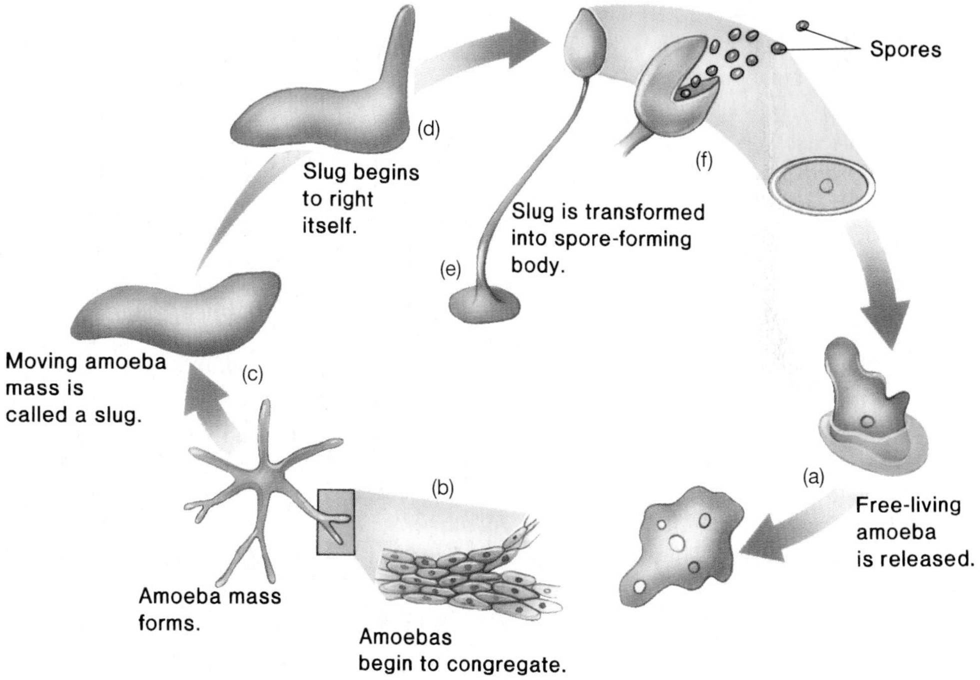

Figure 12.19 Development in *Dictyostelium discoideum.*

In this cellular slime mold (phylum Acrasiomycota), (*a*) germinating spores form amoebas. (*b*) The amoebas aggregate and move toward a fixed center. (*c*) They form a multicellular slug 2 to 3 millimeters long that migrates toward light. (*d*) The slug stops moving and begins to differentiate into a spore-forming body, called a sorocarp. (*e*) The differentiated head of a sorocarp. (*f*) Within the sorocarps, the amoebas become encysted as spores.

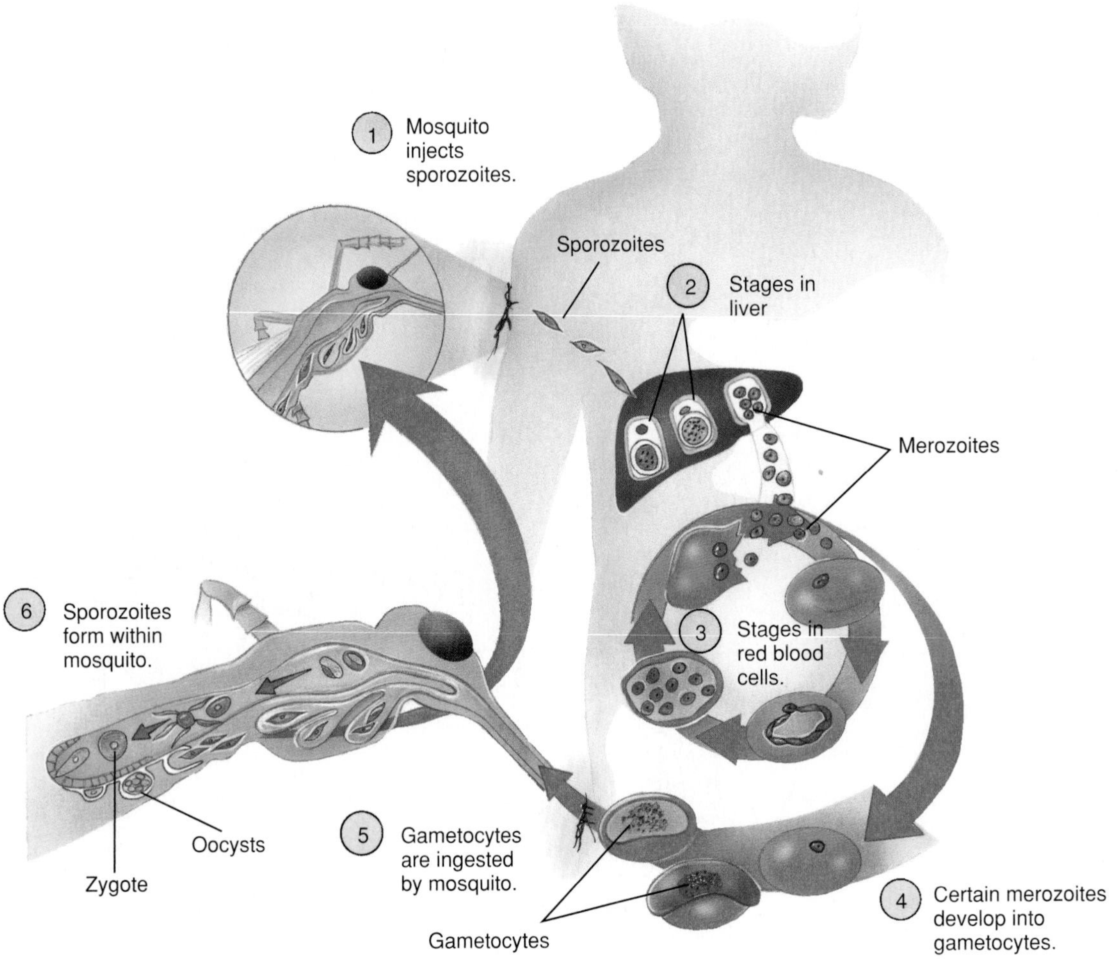
1
Mosquito injects sporozoites.
Sporozoites
2
Stages in liver
Merozoites
3
Stages in red blood cells.
4
Certain merozoites develop into gametocytes.
Gametocytes
5
Gametocytes are ingested by mosquito.
6
Sporozoites form within mosquito.
Oocysts
Zygote

Figure 12.20 ***Plasmodium* is the sporozoan that causes malaria.**
Plasmodium has a complex life cycle that alternates between mosquitoes and mammals.

Sporozoans

Sporozoans are nonmotile, spore-forming, unicellular *parasites* of animals, all members of one phylum. They are responsible for many diseases in humans and domestic animals.

Sporozoans, members of the phylum Sporozoa, infect animals with small spores that are transmitted from host to host. All sporozoans possess a unique arrangement of microtubules and other organelles clustered at one end of the cell, but they are nonmotile, having no flagella. There are 3,900 described species of sporozoans, the best known of which is the malaria-causing parasite *Plasmodium.*

Sporozoans have complex life cycles that involve both asexual and sexual phases. Sexual reproduction involves an alternation of haploid and diploid generations. Both haploid and diploid individuals can also divide rapidly by mitosis, thus producing a large number of small, infective individuals. Sexual reproduction involves the fertilization of a large female gamete by a small, flagellated male gamete. The zygote that results soon becomes a thick-walled cyst called an oocyst, which is highly resistant to drying out and other unfavorable environmental factors. Within the oocyst, meiotic divisions produce infective haploid spores.

An alternation between different hosts often occurs in the life cycles of sporozoan (figure 12.20). Sporozoans of the genus *Plasmodium* are spread from person to person by mosquitoes of the genus *Anopheles;* at least 65 different species of this genus are involved. The sporozoan life cycle stages—called sporozoites, merozoites, and gametocytes—each produce different antigens, and they are sensitive to different antibodies. When a mosquito inserts its proboscis into a human blood vessel, it injects about a thousand sporozoites. They travel to the liver within a few minutes, where they are no longer exposed to antibodies circulating in the blood. If even one sporozoite reaches the liver, it will multiply rapidly there and still cause malaria. The number of malaria parasites increases roughly eightfold every 24 hours after they enter the host's body.

Malaria is one of the most serious diseases in the world. Between 300 and 500 million people are affected by it at any one time, and approximately 1.5 to 2.7 million of them, mostly children, die each year. Efforts to eradicate malaria have focused on (1) the elimination of the mosquito vectors, (2) the development of drugs to poison the parasites once they have entered the human body, and (3) the development of vaccines.

CHAPTER 12

HIGHLIGHTS

12.1 Endosymbiosis

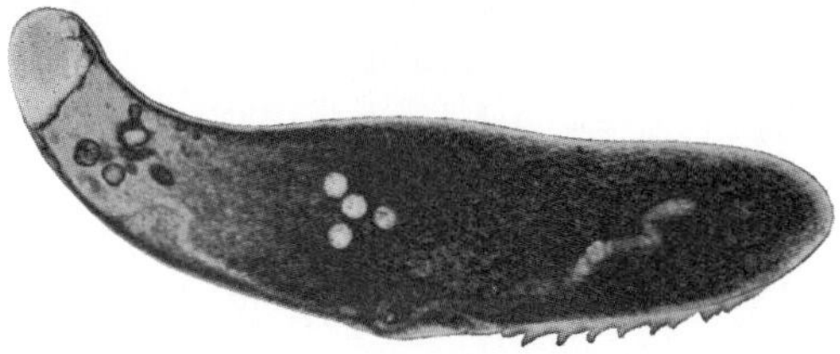

Key Terms

- endosymbiosis 232
- *Pelomyxa* 232
- mitochondria 232

Key Concepts

- The theory of endosymbiosis, accepted by almost all biologists, proposes that mitochondria and chloroplasts were once aerobic eubacteria that were engulfed by ancestral eukaryotes.
- There is some suggestion that centrioles may also have an endosymbiotic origin.

12.2 The Protists

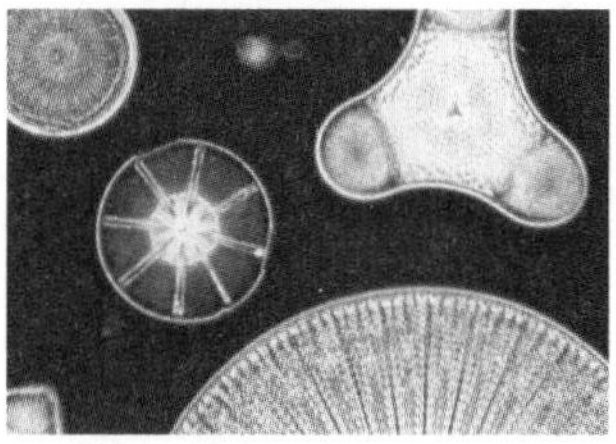

Key Terms

- protist 234
- multicellularity 234
- algae 234

Key Concepts

- Of the six kingdoms, Protista is by far the most diverse.
- Protista is a catchall kingdom containing all eukaryotes that are not fungi, plants, or animals.
- Multicellularity has evolved among the protists many times.

12.3 The Evolution of Sex

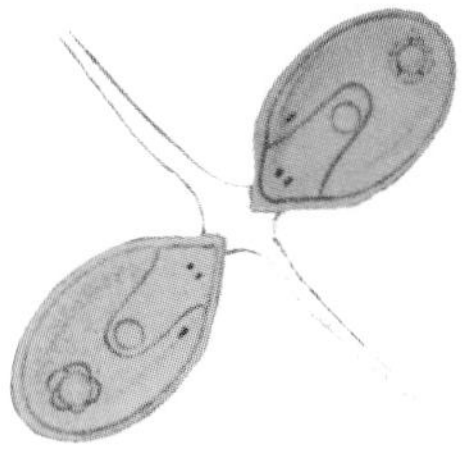

Key Terms

- asexual reproduction 236
- sexual life cycle 237
- alternation of generations 237

Key Concepts

- Reproduction without sex is the rule among protists, which typically resort to sexual reproduction only in times of stress.
- Sex appears to have first evolved as a mechanism to repair damage to DNA.

12.4 Kinds of Protists

Key Terms

- pseudopodia 238
- cytoplasmic streaming 238
- flagella 240
- cilia 242
- spores 242

Key Concepts

- The 14 major phyla of protists can be sorted into seven general groups, based largely on their mode of locomotion.
- Algae are photosynthetic protists, many of which are multicellular.
- Molds are heterotrophs with restricted mobility.

12

CONCEPT REVIEW

1. Endosymbiosis of a _________ bacteria most likely gave rise to the chloroplasts of eukaryotic cells of plants and some algae.
   a. chemoautotrophic
   b. *Pelomyxa*
   c. methanogenic
   d. photosynthetic

2. Mitochondria, chloroplasts, and possibly _________ are eukaryotic cell organelles thought to have arisen by endosymbiosis.
   a. DNA
   b. Golgi bodies
   c. centrioles
   d. lysosomes

3. The kingdom Protista contains all eukaryotes that are not
   a. viruses or bacteria.
   b. animals, fungi, or plants.
   c. photosynthetic.
   d. animals or plants.

4. Parthenogenesis is the development of an adult from _________ and is a common form of asexual reproduction among insects.
   a. an unfertilized egg
   b. a fertilized egg
   c. larvae
   d. a zygote

5. In zygotic meiosis, the _________ cells occupy the major portion of the life cycle.
   a. zygotic
   b. diploid
   c. haploid
   d. sporic

6. Which of the following types of eukaryotes does *not* carry out meiosis?
   a. *Dictyostelium*
   b. *Chlamydomonas*
   c. paramecia
   d. sporozoans

7. Which of the following groups of algae have chloroplasts identical to those found in plants?
   a. euglenoids
   b. red algae
   c. brown algae
   d. green algae

8. *Trypanosoma* is a zoomastigote responsible for the disease in humans called
   a. sleeping sickness.
   b. malaria.
   c. red tides.
   d. tuberculosis.

9. The laboratory medium agar is made from the cell walls of
   a. diatoms.
   b. red algae.
   c. brown algae.
   d. *Euglena*.

10. Mitochondria are sausage-shaped organelles in eukaryotic cells that are thought to have arisen by _________.

11. In _________ reproduction, no gametes are formed.

12. Sexual reproduction seems to have evolved as a mechanism to repair damaged _________ during the close association between pairs of chromosomes during meiosis.

13. Red tides are population explosions of protozoans called _________.

14. When the unicellular form of *Dictyostelium* is deprived of _________, amoebas come together, form a slug, and move to a new area.

15. _________ is the group of protists that is believed to have given rise to animals and certainly gave rise to sponges.

16. Flagella are common among the various phyla of algae except one. Which phylum of algae lacks flagellated cells at any stage of the life cycle? ____________________

Answers to the Concept Review questions appear in Appendix B.

CHALLENGE YOURSELF

1. If plants were derived from green algae, why don't we classify green algae as plants in this text?
2. If mitochondria and chloroplasts originated as symbiotic bacteria, what would you suggest were the characteristics of the organism in which they became symbiotic?
3. What are the advantages of sexual reproduction?
4. Coral reefs are one of the most productive communities on earth even though tropical waters are often poor in nutrients. Why? Why do you think coral reefs are only found in tropical and subtropical waters?

FOR FURTHER READING

Corliss, J. "An Interim Utilitarian ("user friendly") Hierarchical Classification and Characterization of the Protists." *Acta Protozoologica* 33 (1994): 1–51. The title says it all: a recent comprehensible treatment of an often confusing group of organisms.

Dyer, B. D., and R. A. Obar. *Tracing the History of Eukaryotic Cells*. New York: Columbia University Press, 1994. A good overview of the problems involved in this fascinating area.

Lambrecht, F. "Trypanosomes and Hominid Evolution." *BioScience* 35 (1985): 640–46. A fascinating article that charts the probable effects of sleeping sickness in determining the course of human history.

Lee, J. J., S. H. Hunter, and E. C. Bovee, eds. *An Illustrated Guide to the Protozoa*. Lawrence, Ks.: Society of Protozoologists, 1985. Outstanding visual impression of the diversity of protists.

Margulis, L. *Symbiosis in Cell Evolution*. San Francisco: W. H. Freeman, 1980. Outstanding treatment of the origin of eukaryotic cells by serial symbiosis.

Maynard-Smith, J. *The Evolution of Sex*. New York: Cambridge University Press, 1978. An important viewpoint on the origin of sex that clearly outlines the issues currently being argued.

McDermott, J. "A Biologist Whose Heresay Redraws Earth's Tree of Life." *Smithsonian,* August 1989, 72–80. Lynn Margulis's innovative approach to endosymbiosis has revolutionized the way we think about organisms.

Oaks, S., et al. *Malaria—Obstacles and Opportunities*. Washington, D.C.: National Academy Press, 1991. The report of a high-level committee of scientists on the recent worldwide resurgence of malaria, which is staging a dramatic comeback in many countries where it was thought to be under control. Malaria already kills more humans each year than any other communicable disease, and the numbers are rising.

TECHNOLOGY LINKS

The Living World Home Page
http://www.wcbp.com/biology/tlw

Life Science Animations
Videotape 4
#45 Life Cycle of Malaria

13

Evolution of Multicellular Life

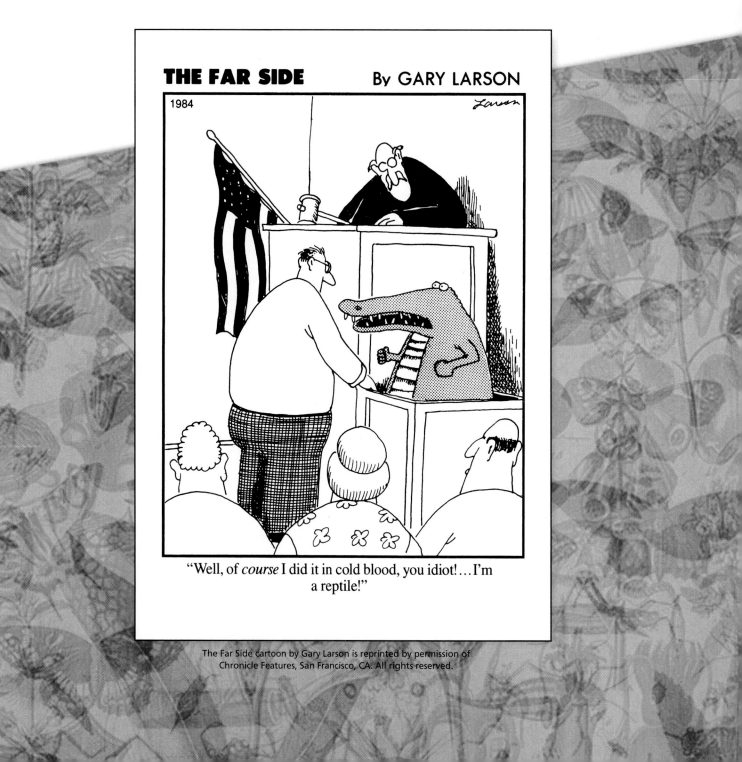

"Well, of *course* I did it in cold blood, you idiot!...I'm a reptile!"

CHAPTER OUTLINE

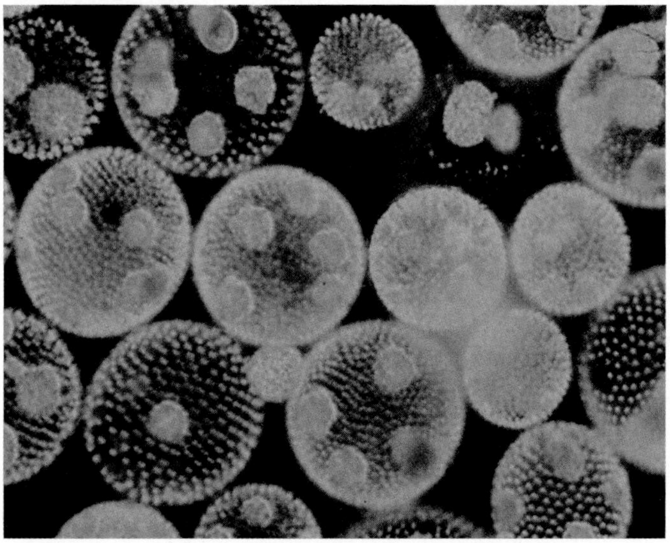

Figure 13.1 Approaching multicellularity.
Individual, motile, unicellular green algae are united in the protist *Volvox* as a hollow colony of cells that moves by the beating of the flagella of its individual cells. Some species of *Volvox* have cytoplasmic connections between the cells that help coordinate colony activities. *Volvox* is a highly complex form of colony that has many of the properties of multicellular life.

Archaebacteria and eubacteria are unicellular (individuals composed of a single cell), as are most protists. The unicellular body plan is a tremendously successful one, comprising about half the biomass on earth. Bacteria have continued their evolution for over 3.5 billion years, using nearly every energy source conceivable. Similarly, modern unicellular protists are not only numerous but extraordinarily diverse. Among the protists, a second and very different way of living evolved some 1.5 billion years ago—multicellularity (figure 13.1).

13.1 Colonial Protists

Advantages of Multicellularity

A single cell has limits. It can only be so big without encountering serious surface-to-volume problems. Said simply, as a cell becomes larger, there is too little surface area for so much volume. The evolution of multicellular individuals composed of many cells solved this problem (figure 13.2). **Multicellularity** is a condition in which an organism is composed of many cells, permanently associated with one another, that integrate their activities. The key advantage of multicellularity is that it allows specialization—distinct types of cells, tissues, and organs can be differentiated within an individual's body, each with a different function. With such functional "division of labor" within its body, a multicellular organism can possess cells devoted specifically to protecting the body, others to moving it about, still others to seeking mates and prey, and yet others to carry on a host of other activities. This allows the organism to function on a scale and with a complexity that would have been impossible for its unicellular ancestors. In just this way, a small city of 50,000 inhabitants is vastly more complex and capable than a crowd of 50,000 people in a football stadium—each city dweller is specialized in a particular activity that is interrelated to everyone else's, rather than just being another body.

Figure 13.2 Kelp, a multicellular protist.
Multicellularity has evolved in many lines of the kingdom Protista. One of these, representative of the phylum Phaeophyta, is a gigantic kelp.

Colonies and Aggregates

A **colonial organism** is a collection of cells that are permanently associated but in which little or no integration of cell activities occurs. Some bacteria are colonial, their cell walls adhering to one another. In a few of the colonial bacteria, cells are held together within a common sheath. Others form filaments, sheets, or three-dimensional aggregates of cells. However, little or no integration of cell activities occurs in any of these colonies. Thus while such bacteria may properly be considered colonial (living together), none are truly multicellular.

Many protists also form colonial assemblies, consisting of many cells with little differentiation or integration. In some protists the distinction between colonial and multicellular is blurred. *Volvox,* the colonial protist illustrated in figure 13.1, is a good example of this. In the green algae *Volvox*, individual motile cells aggregate into a hollow bowl of cells that moves by a coordinated beating of the flagella of the individual cells—like scores of rowers all pulling their oars in concert. A few cells near the rear of the moving colony are reproductive cells, but most are relatively undifferentiated. Some species have cytoplasmic connections between the cells that might permit coordination of some activities. Is *Volvox* multicellular? Is coordination of flagella evidence of integration? To return to our analogy comparing the inhabitants of a small city to a crowd in a football stadium, imagine the stadium crowd were carrying out a "wave," in which segments of the stadium stand and yell, one after another like falling dominos—does this indicate that the football watchers are in fact truly integrated? No. In this sense, *Volvox* is more properly considered colonial—although the distinction is a difficult one.

An **aggregation** is a more transient collection of cells that come together for a period of time and then separate. Cellular slime molds, for example, are unicellular organisms that spend most of their lives moving about and feeding as single-celled amoebas (see chapter 12). They are common in damp soil and on rotting logs, where they move about, ingesting bacteria and other small organisms. When the individual amoebas exhaust the supply of bacteria in a given area and are near starvation, however, all of the individual organisms in that immediate area aggregate into a large moving mass of cells called a slug. By moving to a different location, the aggregation increases the chance that food will be found. In the new habitat, the aggregation differentiates into a base, a stalk, and a swollen tip. Within this tip spores are produced that can pass to distant locations (where there may be more food). Each of these spores, if it falls into a suitably moist habitat, releases a new amoeba, which begins to feed, and the aggregation cycle is started again.

13.2 Multicellularity

True multicellularity, in which the activities of the individual cells are coordinated and the cells themselves are in contact, occurs only in eukaryotes and is one of their major characteristics. Multicellularity has evolved many times among the protists. Three groups of protists have independently attained true but simple multicellularity—the brown algae (phylum Phaeophyta), green algae (phylum Chlorophyta), and red algae (phylum Rhodophyta) (figure 13.3). In **simple multicellular organisms,** individuals are composed of many cells that interact with one another and coordinate their activities, although in no case is individual cell specialization complex.

Simple multicellularity does not imply small size or limited adaptability. Some marine algae grow to be enormous. An individual kelp, one of the brown algae, may grow to tens of meters in length—some taller than a redwood! Red algae grow at great depths in the sea, far below where kelp or other algae are found. Not all algae are multicellular. Green algae, for example, include many kinds of multicellular organisms but an even larger number of unicellular ones.

Figure 13.3 Multicellularity in algae.
Three phyla of algae have become multicellular independently and differ greatly in their photosynthetic pigments and structure. Some red algae (phylum Rhodophyta) are shown here growing on sponges. Members of this phylum occur at greater depths in the sea than the members of either of the other phyla.

Complex Multicellularity

The algae, structurally simple multicellular organisms, fill the evolutionary gap between unicellular protists and complex multicellular organisms (animals, plants, and fungi). In **complex multicellular organisms,** individuals are composed of many highly specialized kinds of cells that coordinate their activities. These organisms fall into three kingdoms:

1. **Plants.** Multicellular green algae were almost certainly the direct ancestors of the plants (see chapter 14) and were themselves considered plants in the last century. However, green algae are basically aquatic and much simpler in structure than plants and are considered protists in the six-kingdom system used widely today.
2. **Animals.** Animals arose from a unicellular protist ancestor. Several groups of animal-like protists have been considered to be tiny animals in the past century ("protozoa"), including flagellates, ciliates, and amoebas. The simplest (and seemingly most primitive) animals today, the sponges, seem clearly to have evolved from a kind of flagellate.
3. **Fungi.** Fungi also arose from a unicellular protist ancestor, one different from the ancestor of animals. Certain protists, including slime molds and water molds, have been considered fungi ("molds"), although they are in fact protists and are not thought to resemble ancestors of fungi. The true protist ancestor of fungi is as yet unknown. This is one of the great unsolved problems of taxonomy.

Two key characteristics of the complex multicellular organisms distinguish them from simple multicellular organisms like marine algae: **cell specialization** and **intercell coordination.** In an animal, plant, or fungus, the body of an individual possesses different kinds of cells that possess very different structures and are coordinated in complex ways.

Perhaps the most important characteristic of complex multicellular organisms is cell specialization. If you think about it, having a variety of different sorts of cells within the same individual implies something very important about the genes of the individual: *different cells are using different genes!* The process whereby a single cell (in humans, a fertilized egg) becomes a multicellular individual with many different kinds of cells is called **development.** The cell specialization that is the hallmark of complex multicellular life is the direct result of cells developing in different ways by activating different genes.

A second key characteristic of complex multicellular organisms is intercell coordination, the adjustment of a cell's activity in response to what other cells are doing. The cells of all complex multicellular organisms communicate with one another with chemical signals called hormones. In some organisms like sponges there is relatively little coordination between the cells; in other organisms like humans almost every cell is under complex coordination.

There are three kingdoms of complexly multicellular organisms: Fungi, Plantae, and Animalia. With the exception of yeasts among the fungi, all members of these three great kingdoms are composed of individuals that are complexly multicellular. In the remainder of this chapter we consider the fungi. In the next chapter, we examine plants. Then in the following four chapters, we explore in detail how animals have evolved.

13.3 Fungi

Of all the bewildering variety of organisms that live on earth, perhaps the most unusual, the most peculiarly different from ourselves, are the fungi. Mushrooms and toadstools are fungi, multicellular creatures that grow so rapidly in size that they seem to appear overnight on our lawns. At first glance, a mushroom looks like a funny kind of plant growing up out of the soil, and that is how taxonomists used to classify fungi, as members of the plant kingdom. The three phyla of fungi used to be called "divisions" because that is what taxonomists used to call plant phyla (some still do, although most have abandoned the term, applying the term "phyla" to all six kingdoms). Scientists who studied fungi, called mycologists, were traditionally members of botany departments. However, when you look more closely, fungi turn out to have nothing in common with plants except that they are multicellular and grow in the ground. As you will see, the more you examine fungi, the more unusual they seem.

Figure 13.4 A mycelium.
A mycelium is a dense, interwoven mat of fungal hyphae. Most of the body of a fungus is occupied by its mycelium. This fungal mycelium is growing through leaves on the forest floor in Maryland.

A Fungus Is Not a Plant

A good way to begin your examination of fungi is to focus on a mushroom and ask: What is different between this mushroom and a plant the same size, say a pea plant? The list of significant differences includes the following:

1. **Fungi are heterotrophs.** Perhaps most obviously, a mushroom is not green. Virtually all plants are photosynthesizers, while no fungi carry out photosynthesis. Instead, fungi obtain their food by secreting digestive enzymes onto whatever they are attached to and then absorbing into their bodies the organic molecules that are released by the enzymes.
2. **Fungi have filamentous bodies.** Fungi are basically filamentous in their growth form (that is, their body consists of long, slender filaments), even though these filaments may be packed together to form complex structures like the mushroom. Plants, in contrast, are basically three-dimensional.
3. **Fungi have nonmotile sperm.** Some plants have motile sperm with flagella. No fungi do.
4. **Fungi have cell walls made of chitin.** The cell walls of fungi are built of polysaccharides (that is, chains of sugars) and chitin, the same tough material that a crab shell is made of. The cell walls of plants are made of cellulose, also a strong building material. Chitin, however, is far more resistant to microbial degradation than is cellulose.
5. **Fungi have nuclear mitosis.** Mitosis in fungi is different from plants or any other eukaryote in one key respect: the nuclear envelope does not break down and re-form. Instead, all of mitosis takes place *within* the nucleus. A spindle apparatus forms there, dragging chromosomes to opposite poles of the *nucleus* (not the cell, as in all other eukaryotes).

You could build a much longer list, but already the take-home lesson is clear: fungi are not like plants at all! Their many unique features are strong evidence that fungi are not closely related to any other group of organisms.

The Body of a Fungus

Fungi exist mainly in the form of slender filaments, barely visible with the naked eye, called **hyphae** (singular, **hypha**). A hypha is basically a long string of cells. The walls dividing one cell from another are called **septa** (singular, **septum**). The presence of septa is another of the ways in which fungi differ fundamentally from all other multicellular organisms, for the septa rarely form a complete barrier! From one fungal cell to the next, cytoplasm flows, streaming freely down the hypha through openings in the septa.

The main body of a fungus is not the mushroom, which is a temporary reproductive structure, but rather the extensive network of fine hyphae that penetrate the soil, wood, or flesh in which the fungus is growing. A mass of hyphae is called a **mycelium** (plural, **mycelia**) and may contain many meters of individual hyphae (figure 13.4). This body organization creates a unique relationship between the fungus and its environment. All parts of the fungal body are metabolically active, secreting digestive enzymes and actively attempting to digest and absorb any organic material with which the fungus comes in contact.

Because of cytoplasmic streaming, it is meaningless to ask how many nuclei a typical fungal cell contains. There are many nuclei connected by the shared cytoplasm of a fungal mycelium. None of them (except for reproductive cells) are isolated in any one cell; all of them are linked cytoplasmically with every cell of the mycelium. Indeed, the entire concept of multicellularity takes on a new meaning among the fungi, the ultimate communal sharers among the multicellular organisms.

How Fungi Reproduce

Fungi reproduce both asexually and sexually. All fungal nuclei except for the zygote are haploid. Often in the sexual reproduction of fungi, individuals of different "mating type" must participate, much as two sexes are required for human reproduction. Sexual reproduction is initiated when two hyphae of genetically different mating types come in contact, and the hyphae fuse. What happens next? In animals and plants, when the two haploid gametes fuse, the two haploid nuclei immediately fuse to form the diploid nucleus of the zygote. As you might by now expect, fungi handle things differently. In most fungi, the two nuclei do not fuse immediately. Instead, they remain unmarried inhabitants of the same house, coexisting in a common cytoplasm for most of the life of the fungus! A fungal hypha that has nuclei within it derived from two genetically different individuals is called a **heterokaryon** (Greek, *heteros,* other + *karyon,* kernel or nucleus). A fungal hyphae in which all the nuclei are genetically similar is said to be a **homokaryon** (Greek, *homo,* one).

When reproductive structures are formed in fungi, complete septa form between cells, the only exception to the free flow of cytoplasm between cells of the fungal body. There are three kinds of reproductive structures: (1) **gametangia,** within which gametes form; (2) **sporangia,** within which sexual spores form; and (3) **conidia,** a form of asexual multinucleate spore.

Spores are a common means of reproduction among the fungi. They are well suited to the needs of an organism anchored to one place. They are so small and light that they may remain suspended in the air for long periods of time, and they are so long that air currents can carry individual spores great distances. When a spore or conidium lands in a suitable place, it germinates and begins to divide, soon giving rise to a new fungal hypha.

Figure 13.5 A mushroom.
The oyster mushroom, *Pleurotus ostreatus,* immobilizes nematodes, which the fungus uses as a source of food.

How Fungi Obtain Nutrients

All fungi obtain their food by secreting digestive enzymes into their surroundings and then absorbing back into the fungus the organic molecules produced by this **external digestion.** Many fungi are able to break down the cellulose in wood, cleaving the linkages between glucose subunits and then absorbing the glucose molecules as food. That is why fungi are so often seen growing on dead trees.

Just as some plants like the Venus flytrap are active carnivores, so some fungi are active predators. For example, the edible oyster fungus *Pleurotus ostreatus* attracts tiny roundworms known as nematodes that feed on it—and secretes a substance that anesthetizes them (figure 13.5). When the worms become sluggish and inactive, the fungal hyphae envelop and penetrate their bodies and absorb their contents, a rich source of nitrogen (always in short supply in natural ecosystems). Other fungi are even more active predators, snaring or trapping prey or firing projectiles into nematodes, rotifers, and other small animals that come near.

Kinds of Fungi

Fungi are an ancient group of organisms at least 400 million years old. There are 77,000 described species, in four groups (table 13.1), and many more awaiting discovery. Many fungi are harmful because they decay, rot, and spoil many different materials as they obtain food and because they cause serious diseases in animals and particularly in plants. Other fungi, however, are extremely useful. The manufacture of both bread and beer depends on the biochemical activities of yeasts, single-celled fungi that produce abundant quantities of carbon dioxide and ethanol. Fungi are used on a major scale in industry to convert one complex organic molecule into another; many commercially important steroids are synthesized in this way.

The three fungal phyla, distinguished from one another primarily by their mode of sexual reproduction, are the zygomycetes, the ascomycetes, and the basidiomycetes. The evolutionary relationships among the three phyla and the imperfect fungi are not clear, although the zygomycetes, which we examine first, are the simplest.

Table 13.1 Fungi

Phylum	Typical Examples	Key Characteristics	Approximate Number of Living Species
Ascomycota	Yeasts, truffles, morels 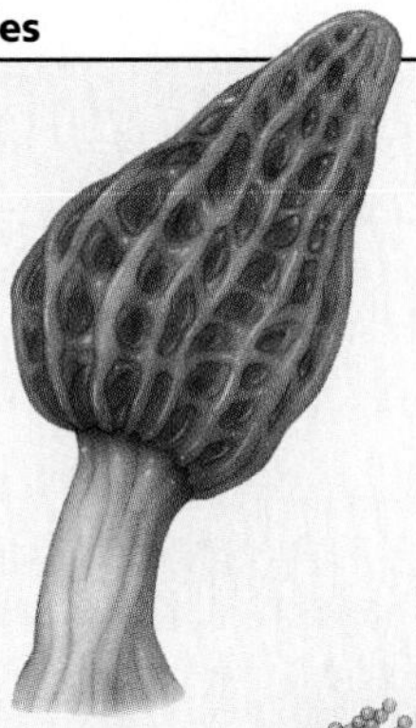	Develop by sexual means; spores are formed inside a sac called an ascus; asexual reproduction is also common	30,000
Imperfect fungi	*Aspergillus, Penicillium* 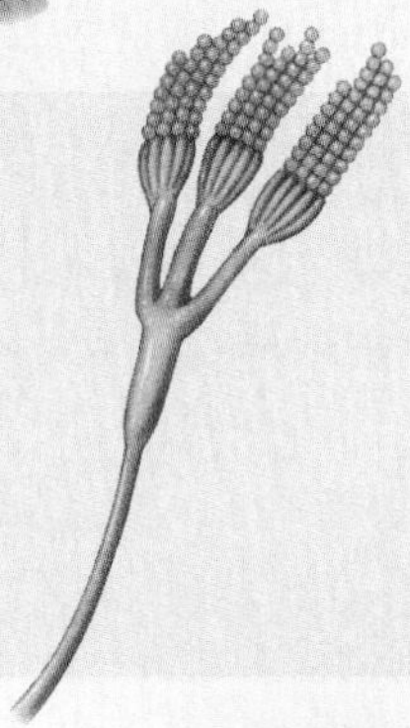	Sexual life cycle has not been observed; most are thought to be ascomycetes that have lost the ability to reproduce sexually	17,000
Basidiomycota	Mushrooms, toadstools, rusts 	Develop by sexual means; spores are borne on club-shaped structures called basidia; the terminal hyphal cell that produces spores is called a basidium; asexual reproduction occurs occasionally	16,000
Zygomycota	*Rhizopus* (black bread mold) 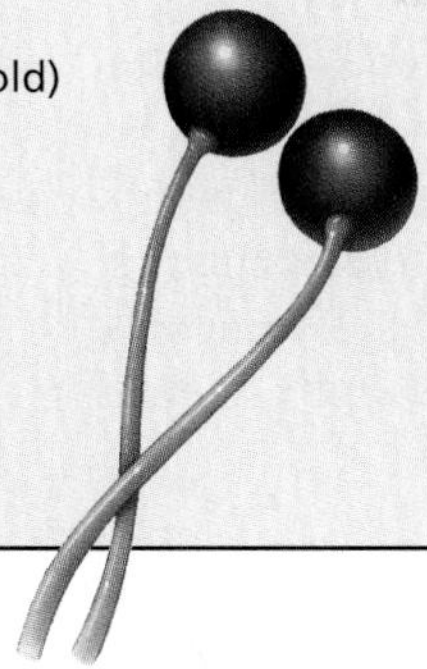	Develop sexually and asexually; multinucleate hyphae lack septa except for reproductive structures; fusion of hyphae leads directly to formation of a zygote, which divides by meiosis just before it germinates	665

Zygomycetes

The **zygomycetes,** members of the phylum Zygomycota, are unique among the fungi in that the fusion of hyphae does not produce a heterokaryon. Instead, the two nuclei fuse and form a single diploid nucleus. Just as the fusion of sperm and egg produce a zygote in plants and animals, so this fusion produces a zygote. The name zygomycetes means "fungi that make zygotes."

Zygomycetes are the exception to the rule among fungi, and there are not many kinds of them, only about 600 named species (less than 1% of the named fungi). Included among them are some of the most frequent bread molds (the so-called black molds) and many microscopic fungi found on decaying organic material.

Reproduction among the zygomycetes is typically asexual. A cell at the tip of a hypha becomes walled off by a complete septum, forming an erect stalk tipped by a sporangium within which haploid spores are produced (figure 13.6). These spores are shed into the wind and blown to new locations, where they germinate and attempt to start new mycelia (figure 13.7). Sexual reproduction is unusual but frequently occurs in time of stress. It leads to the production of a particularly sturdy and resistant structure called a **zygosporangium.** The zygosporangium is a very effective survival mechanism, a resting structure that allows the organism to remain dormant for long periods of time when conditions are not favorable.

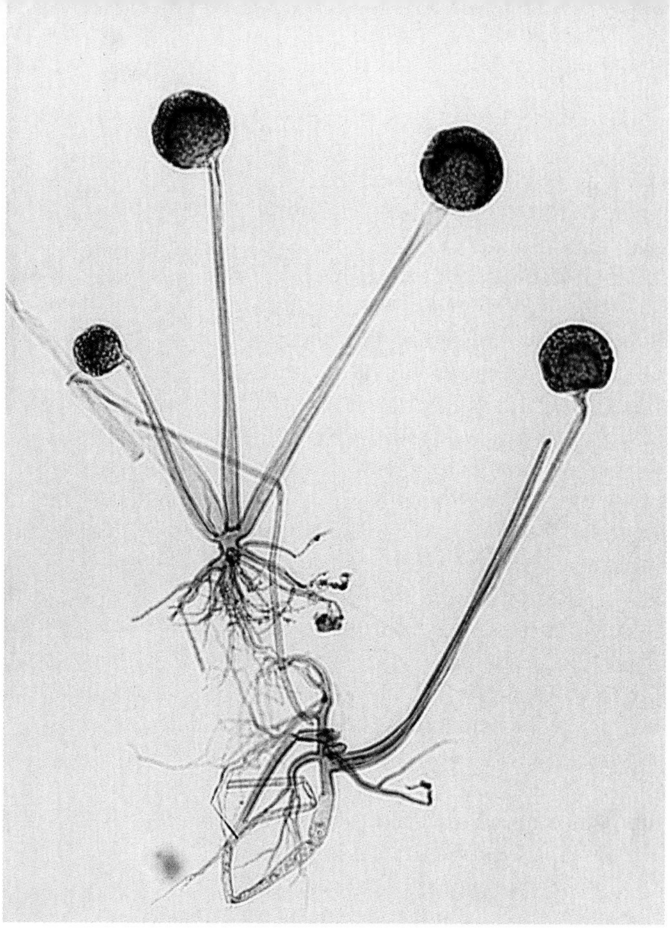

Figure 13.6 ***Rhizopus*, a zygomycete.**
The spore-bearing structures of *Rhizopus* are about a centimeter tall.

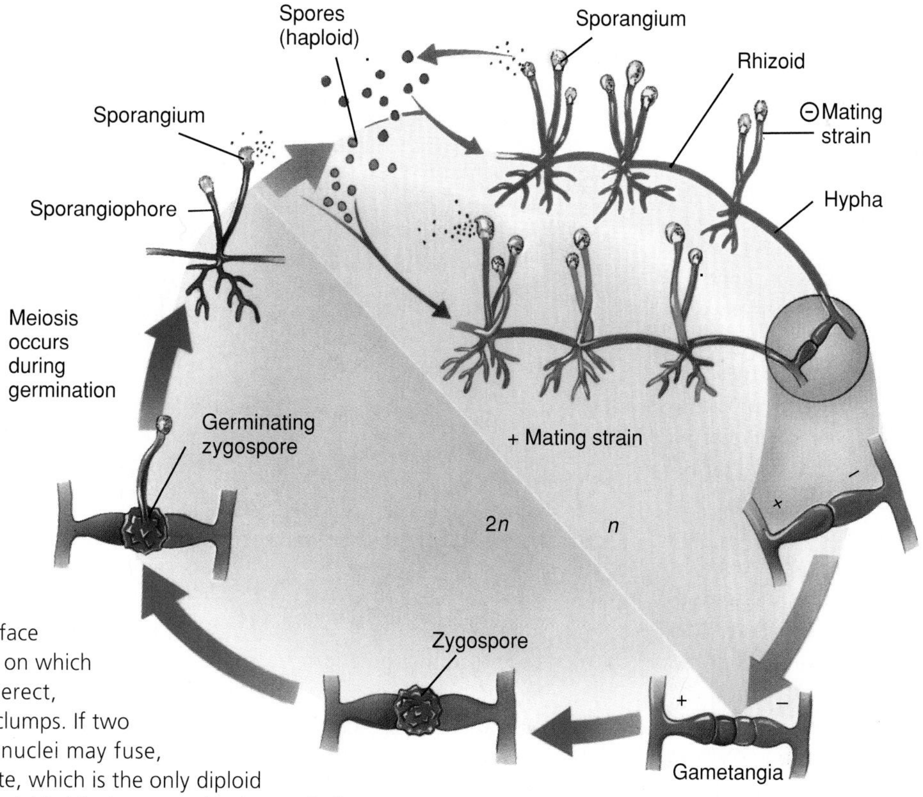

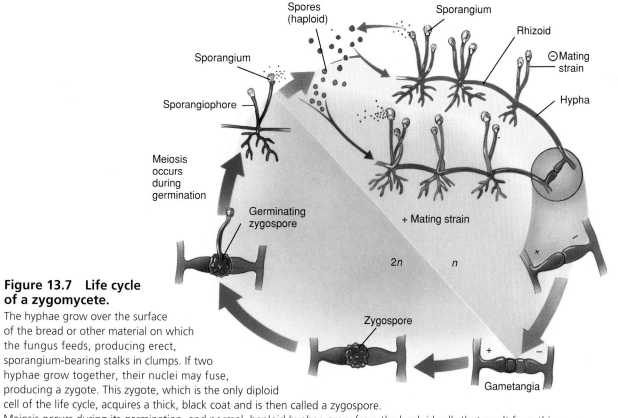

Figure 13.7 Life cycle of a zygomycete.
The hyphae grow over the surface of the bread or other material on which the fungus feeds, producing erect, sporangium-bearing stalks in clumps. If two hyphae grow together, their nuclei may fuse, producing a zygote. This zygote, which is the only diploid cell of the life cycle, acquires a thick, black coat and is then called a zygospore. Meiosis occurs during its germination, and normal, haploid hyphae grow from the haploid cells that result from this process.

Ascomycetes

The second phyla of fungi, the **ascomycetes,** phylum Ascomycota, is by far the largest of the three fungal phyla, with about 30,000 named species and many more being discovered each year (figure 13.8). Among the ascomycetes are such familiar and economically important fungi as yeasts, morels, and truffles, as well as molds such as *Neurospora* (a historically important organism in genetic research) and many of the most serious plant fungal pathogens such as Dutch elm disease and chestnut blight.

Reproduction among the ascomycetes is usually asexual, just as it is among the zygomycetes. However, unlike zygomycetes, which completely lack septa in their hyphae, the hyphae of ascomycetes possess septa dividing one cell from another. The septa are incomplete and have a central large pore in them, so the flow of cytoplasm up and down the hypha is not impeded. Asexual reproduction occurs when the tips of hyphae become fully isolated from the rest of the mycelium by a complete septum, forming asexual spores called conidia, each often containing several nuclei. When one of these conidia is released, air currents carry it to another place, where it may germinate to form a new mycelium.

It is important not to get confused by the number of nuclei in conidia. These multinucleate spores are *haploid,* not diploid, because there is only one version of the genome (the set of ascomycete chromosomes) present, while in a diploid cell there are two genetically different sets of chromosomes present. The actual number of nuclei is not what's important—it's the number of different genomes. Imagine a library: the multiple nuclei of asexual conidia are like multiple copies of the same book.

Although not common, sexual reproduction does occur. In fact, the ascomycetes are named for a characteristic sexual reproductive structure, the **ascus,** which forms when two different hyphae come into contact and fuse (figure 13.9). The ascus is a microscopic cell within which the zygote is formed. The zygote is the only diploid nucleus of the ascomycete life cycle. When a mature ascus bursts, individual spores may be thrown as far as 30 centimeters. Considering how small the ascus is (only 10 micrometers long), this is truly an amazing distance. On the same scale, this would be like you hitting a baseball 1.25 kilometers, 10 times longer than Babe Ruth's longest home run!

Figure 13.8 Representative ascomycetes.
(*a*) A morel, *Morchella esculenta,* a delicious, edible ascomycete (phylum Ascomycota) that appears in early spring in the northern, temperate woods (especially under oaks). (*b*) A cup fungus, phylum Ascomycota, in the rain forest of the Amazon Basin.

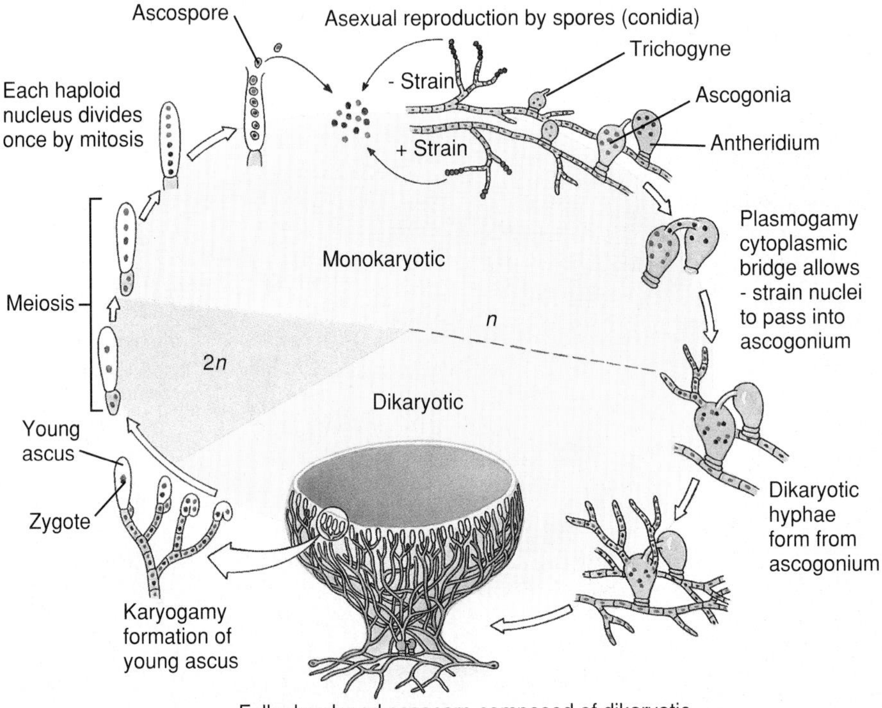

Figure 13.9 Life cycle of an ascomycete.
Asexual reproduction takes place by means of conidia, spores cut off by septa at the ends of modified hyphae. Sexual reproduction occurs when the female gametangia, called ascogonia, fuse with the male gametangium, or antheridium, through a structure called the trichogyne. Following zygote formation, the ascocarp develops.

Basidiomycetes

Members of the third phyla of fungi, Basidiomycota, contain the most familiar of the fungi among their 16,000 named species—the mushrooms, toadstools, puffballs, and shelf fungi (figure 13.10). Many mushrooms are used as food, but others are deadly poisonous. Some species are cultivated as crops—the button mushroom *Agaricus campestris* is grown in more than 70 countries, producing a crop in 1995 with a value of over $15 billion. Also among the **basidiomycetes** are the rusts and smuts, many of them responsible for important plant diseases.

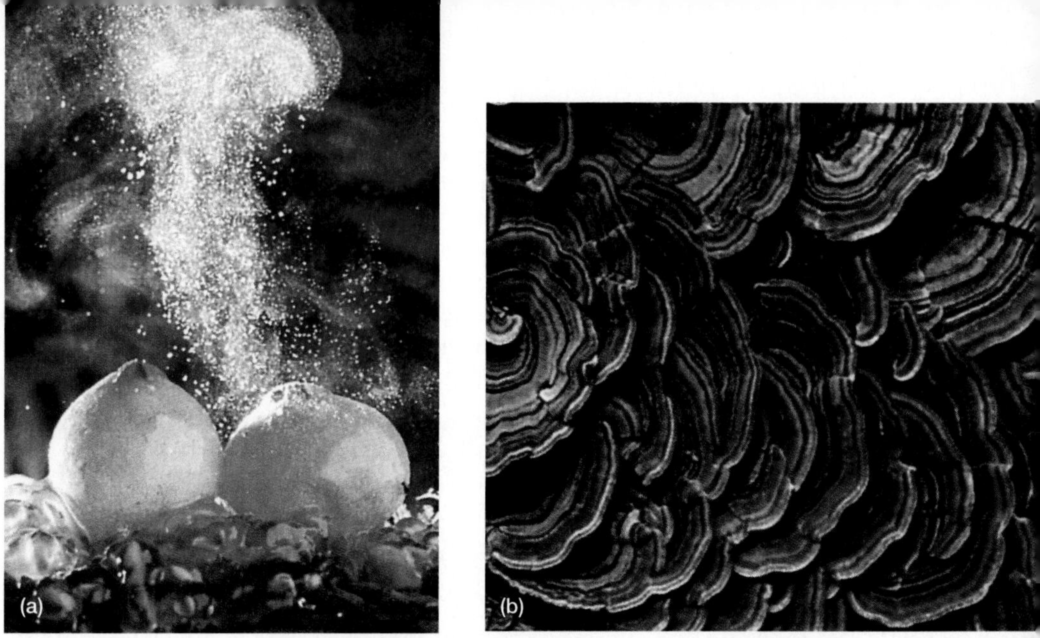

Figure 13.10 Representative basidiomycetes.
(*a*) An earth star, *Geastrum saccatus.* Basidia form within puffballs, which release hundreds of thousands or even millions of basidiospores when mature. (*b*) A shelf fungus, *Trametes versicolor,* growing on a tree trunk. Basidia line the inner surface of tubes that are on the lower surface of many shelf fungi, although some have gills.

There are many technical differences between basidiomycetes and ascomycetes, most of them involving nuances of structure, but one major difference stands out: When hyphae fuse in sexual reproduction, the result in ascomycetes is typically a heterokaryon with several nuclei in each cell, while the result in basidiomycetes is always a **dikaryon,** with precisely two nuclei (one of each mating type) within each cell.

The life cycle of a basidiomycete starts with the production of a hypha from a germinating spore (figure 13.11). These hyphae lack septa at first, just as in zygomycetes. Eventually, however, septa are formed between each of the nuclei—but as in ascomycetes, there are holes in these cell separations, allowing cytoplasm to flow freely between cells. These hyphae grow, forming complex mycelia, and when hyphae of two different mating types fuse, the dikaryon that results goes on to form a dikaryotic mycelium.

The two nuclei in each cell of a dikaryotic hypha can coexist together for a very long time without fusing. Unlike the other two fungal phyla, asexual reproduction is infrequent among the basidiomycetes; they almost always reproduce sexually.

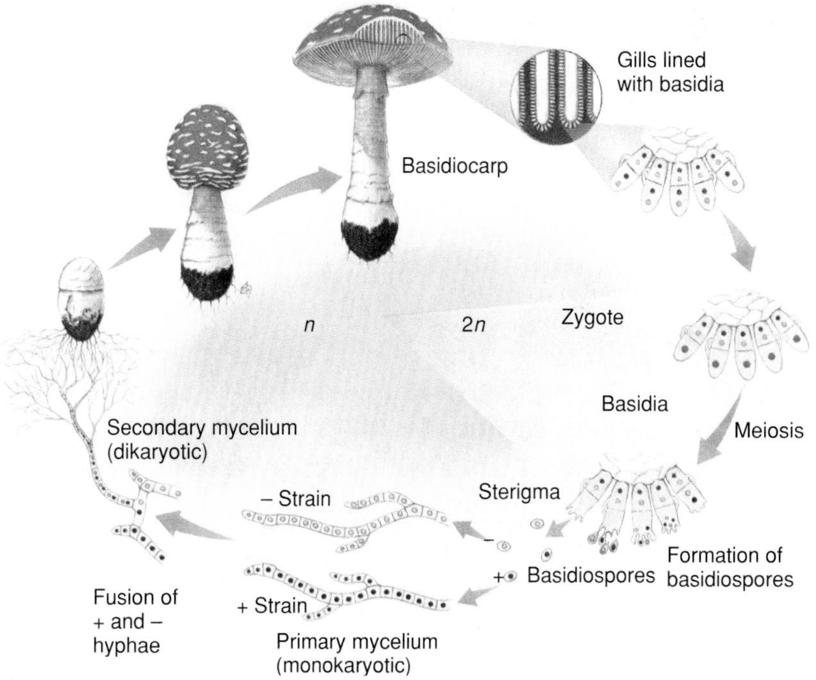

Figure 13.11 Life cycle of a basidiomycete.
Basidiomycetes usually reproduce sexually, with the fusion of nuclei in the basidia to produce a zygote. Meiosis follows syngamy and produces spores that eventually form a basidiocarp.

In sexual reproduction, zygotes (the only diploid cells of the life cycle) form when the two nuclei of dikaryotic cells fuse. This occurs within a club-shaped reproductive structure called the **basidium,** similar to the ascus. Around the many thousands of basidia in which this starts to occur, the mycelium forms a complex structure made up of dikaryotic hyphae called the basidiocarp, or **mushroom.** Meiosis occurs in each basidium, forming haploid spores. The basidia occur in a dense layer on the underside of the cap of the mushroom, where the surface is folded like an accordion. It has been estimated that a mushroom with an 8-centimeter cap produces as many as 40 million spores per hour!

Yeasts

Yeast is the generic (general) name given to unicellular fungi. Although single-celled, yeasts appear almost certainly to have been derived from multicellular ancestors. Some yeasts have been derived from each of the three phyla of fungi, most of them from the ascomycetes. There are about 250 named species of yeasts, including *Saccharomyces cerevisiae,* or baker's yeast, used for thousands of years in the production of bread, beer, and wine (figure 13.12). Other yeasts are important pathogens, including *Candida,* a common source of vaginal infection.

Just as in ascomycetes, most of yeast reproduction is asexual and takes place by cell fission or budding (the formation of a small cell from a portion of a larger one). Sexual reproduction among yeasts occurs when two yeast cells fuse. The new cell containing two nuclei functions as an ascus, with meiosis of the fused nuclei producing four ascospores, which develop directly into new yeast cells.

Figure 13.12 The use of yeasts.
Baking bread involves the metabolic activities of the yeasts, single-celled ascomycetes. Wine making also uses the metabolic properties of yeasts, and the same species, *Saccharomyces cerevisiae,* is usually employed for both processes. In baking bread, yeasts generate carbon dioxide; in making beer or wine, yeasts are used to produce alcohol.

The Imperfect Fungi

In addition to these three phyla of fungi, which, as we have seen, differ primarily in their mode of sexual reproduction, there are some 17,000 described species of fungi in whom sexual reproduction has not been observed. These cannot be formally assigned to one of the three sexually reproducing phyla and so are grouped for convenience as the so-called **imperfect fungi** (figure 13.13). The imperfect fungi are fungi that have lost the ability to reproduce sexually. Most of them appear to be ascomycetes, although some basidiomycetes are also included—you can tell by features of the hyphae and asexual reproduction.

Many of the imperfect fungi are of great economic importance. Some species of *Penicillium* are sources of the well-known antibiotic penicillin, while other species of this genus contribute the characteristic flavors and aromas to cheeses such as Roquefort and Camembert. Species of *Aspergillus* are used for fermenting soy sauce and for the commercial production of citric acid. Most of the fungi that cause skin diseases, including athlete's foot and ringworm, are also imperfect fungi.

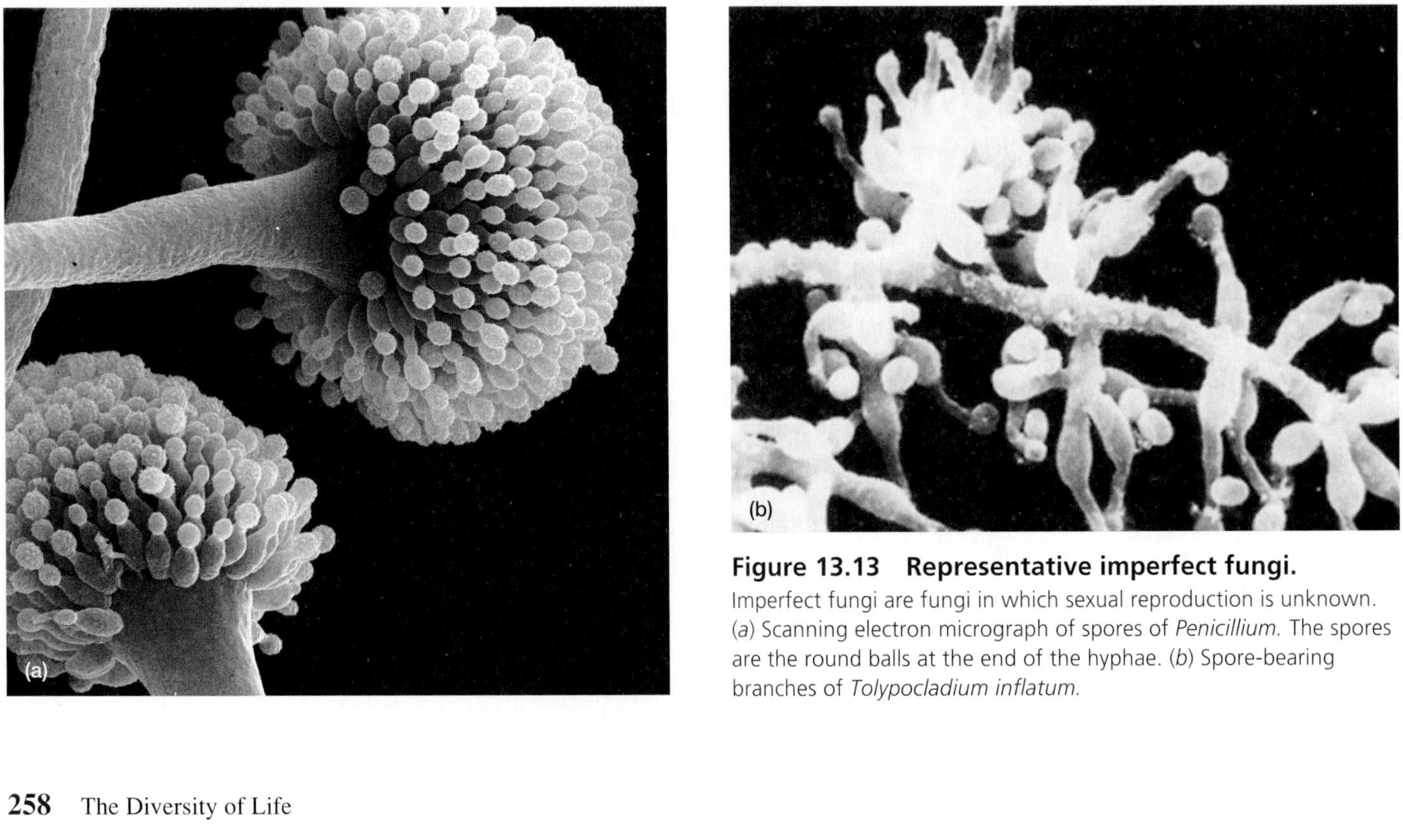

Figure 13.13 Representative imperfect fungi.
Imperfect fungi are fungi in which sexual reproduction is unknown. (*a*) Scanning electron micrograph of spores of *Penicillium.* The spores are the round balls at the end of the hyphae. (*b*) Spore-bearing branches of *Tolypocladium inflatum.*

13.4 Fungal Associations

By now you will be convinced that fungi are very unlike the other multicellular organisms—algae, animals, and plants. But you should not conclude from this that fungi have little to do with other organisms other than to consume their dead bodies. The truth is just the opposite: fungi are involved in a variety of intimate symbiotic associations with algae and plants that play very important roles in the biological world. Recall from our discussion of endosymbiosis in chapter 12 that mutualism is a form of symbiosis in which each partner benefits. Among the fungi, these symbiotic associations typically involve a sharing of abilities between a heterotroph (the fungus) and a photosynthesizer (the algae or plant). The fungus contributes the ability to absorb minerals and other nutrients very efficiently from the environment; the photosynthesizer contributes the ability to use sunlight to power the building of organic molecules. Alone, the fungus has no source of food, the photosynthesizer no source of nutrients. Together, each has access to both food and nutrients, a partnership in which both participants benefit.

Lichens

A **lichen** is a symbiotic association between a fungus and a photosynthetic partner. Ascomycetes are the fungal partners in all but 20 of the 15,000 different species of lichens that have been characterized. Most of the visible body of a lichen consists of its fungus, but interwoven between layers of hyphae within the fungus are cyanobacteria, green algae, or sometimes both. Enough light penetrates the translucent layers of hyphae to make photosynthesis possible. Specialized fungal hyphae envelop and sometimes penetrate the photosynthetic cells, serving as highways to collect and transfer to the fungal body the sugars and other organic molecules manufactured by the photosynthetic cells. The fungus transmits special biochemical signals that direct the cyanobacteria or green algae to produce metabolic substances that they would not if growing independently of the fungus. Indeed, the fungus is not able to grow or survive without its photosynthetic partner. Many biologists characterize this particular symbiotic relationship as one of slavery rather than cooperation, a controlled parasitism of the photosynthetic organism by the fungal host.

The durable construction of the fungus, combined with the photosynthetic abilities of its partner, has enabled lichens to invade the harshest of habitats, from the tops of mountains to dry, bare rock faces in the desert (figure 13.14). In such harsh, exposed areas, lichens are often the first colonists, breaking down the rocks and setting the stage for the invasion of other organisms. A key component of such primary succession, as it is called, are lichens with cyanobacteria that are able to fix atmospheric nitrogen. The activities of these lichens introduce usable nitrogen into the environment in the form of ammonia or organic molecules, where it can be used by other pioneering organisms. Without this nitrogen, the other organisms could not survive.

Lichens are able to survive drying or freezing by converting to a condition we might call suspended animation. When moisture and warmth return, the lichen recovers quickly and resumes its normal metabolic activities, including photosynthesis. In harsh environments, the growth of the lichen may be extremely slow. Some high-mountain lichens covering an area no larger than your fist actually appear to be thousands of years old and therefore are among the oldest living things on earth.

Lichens are extremely sensitive to pollutants in the atmosphere, because they absorb substances dissolved in rain and dew readily. This is why lichens are generally absent in and around cities—they are acutely sensitive to sulfur dioxide produced by automobile traffic and industrial activity. Such pollutants destroy their chlorophyll molecules and thus decrease photosynthesis and upset the physiological balance between the fungus and the algae or cyanobacteria. Degradation or destruction of the lichen results. Lichens are now disappearing even from national parks and other formally remote areas as they are affected increasingly by industrial pollution and automobile exhausts.

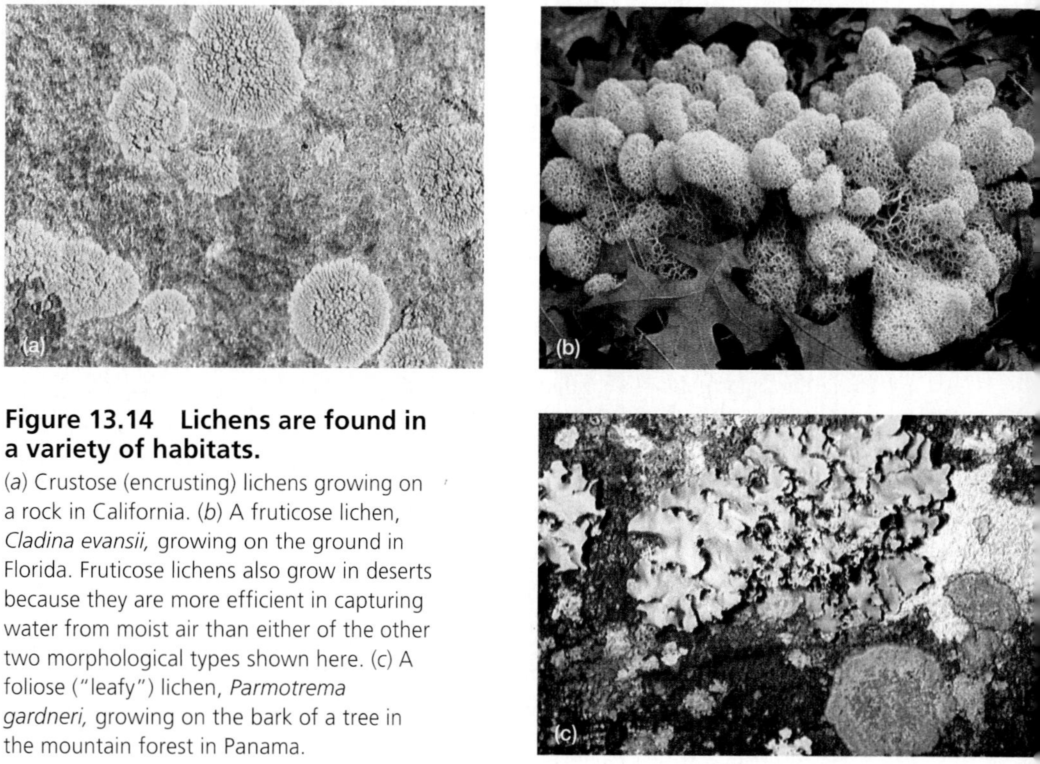

Figure 13.14 Lichens are found in a variety of habitats.
(*a*) Crustose (encrusting) lichens growing on a rock in California. (*b*) A fruticose lichen, *Cladina evansii,* growing on the ground in Florida. Fruticose lichens also grow in deserts because they are more efficient in capturing water from moist air than either of the other two morphological types shown here. (*c*) A foliose ("leafy") lichen, *Parmotrema gardneri,* growing on the bark of a tree in the mountain forest in Panama.

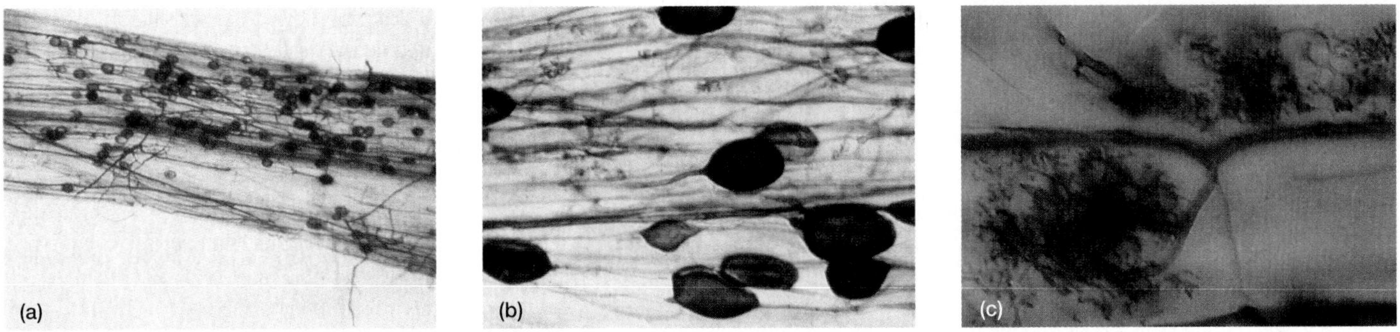

Figure 13.15 Mycorrhizae.
The zygomycete *Glomus versiforme* is shown here growing in the roots of leeks, *Allium porrum.* (*a*) General view of the leek root, showing vesicles (saclike structures) photographed with a dissecting microscope. The root has been squashed in fluid (×16). (*b*) Vesicles in leek root (×63). (*c*) Arbuscules in leek root (×63). Arbuscules are branching structures characteristic of young infections, with vesicles predominating later.

Mycorrhizae

The roots of about 80% of all kinds of plants are involved in symbiotic associations with certain kinds of fungi. In fact, it has been estimated that fungi account for as much as 15% of the total weight of the world's plant roots! Associations of this kind are called **mycorrhizae** (from the Greek *myco,* "fungus" + *rhizos,* "roots") (figure 13.15). In a mycorrhiza, filaments of the fungus act as superefficient root hairs, projecting out from the epidermis, or outermost cell layer, of the terminal portions of the root. The fungal filaments aid in the direct transfer of phosphorus and other minerals from the soil into the roots of the plant, while the plant supplies organic carbon to the symbiotic fungus. Because both partners benefit substantially, the association represents a fine example of mutualism (figure 13.16).

Figure 13.16 How mutualism aids plant growth.
Mycorrhizae aid plant growth by helping the plant roots absorb nutrients from the soil. Soybeans without mycorrhizae (*far left*) grow far more slowly than soybeans with different strains of mycorrhizae (*center* and *right*).

Endomycorrhizae

In almost all mycorrhizae, involving perhaps more than 200,000 species of plants, the fungal hyphae of the mycelium actually penetrate the outer cells of the plant root, as well as extending far out into the soil. These are called **endomycorrhizae.** You might be surprised to learn that all of this symbiotic activity is being carried out worldwide by a very small group of fungi—only 30 species of zygomycetes are known to be involved in these hundreds of thousands of associations!

The earliest fossil plants often have endomycorrhizal roots, which are thought to have played an important role in the invasion of land by plants. The soils of that time would have completely lacked in organic matter other than that contributed to beach sands by dead marine life, and mycorrhizal plants are particularly successful in such infertile soils. The most primitive vascular plants surviving today continue to depend strongly on endomycorrhizae.

Ectomycorrhizae

In perhaps 10,000 species of plants, the mycorrhizae do not physically penetrate the plant root, wrapping around it instead. These are called **ectomycorrhizae.** In marked contrast to endomycorrhizae, these nonpenetrating ectomycorrhizae represent highly specialized relationships, in which a particular species of plant has become associated with a particular fungus, usually a basidiomycete (although some are ascomycetes). At least 5,000 species of fungi have been identified in different ectomycorrhizae, most restricted to a single species of plant. These sorts of mycorrhizae are important because they involve many commercially significant trees in temperate regions, including pines, firs, oaks, beeches, and willows.

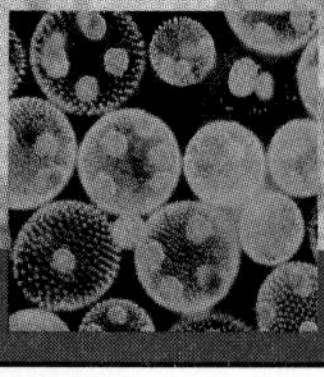

CHAPTER 13

HIGHLIGHTS

13.1 Colonial Protists

Key Terms

multicellularity 250
colonial organisms 250
aggregation 250

Key Concepts

- Multicellular organisms are composed of many cells that integrate their activities.
- Colonial organisms are collections of cells that are permanently associated but in which little integration of cell activities occurs.
- Aggregations are transient collections of cells.

13.2 Multicellularity

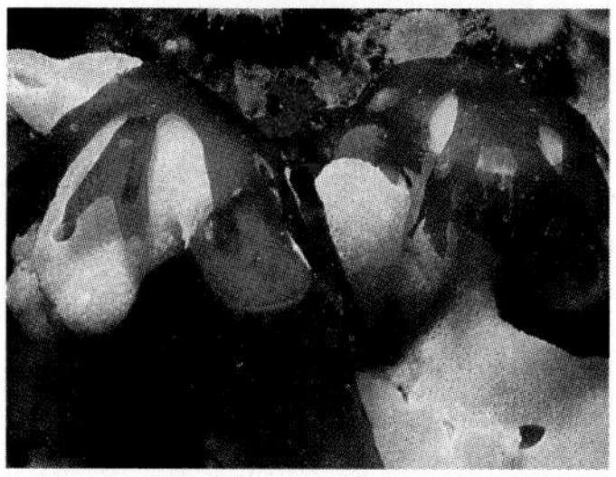

Key Terms

simple multicellular organisms 251
complex multicellular organisms 251
cell specialization 251
intercell coordination 251

Key Concepts

- Simple multicellular organisms lack complex cell specialization.
- Complex multicellular organisms are characterized by cell specialization and intercell coordination.

13.3 Fungi

Key Terms

hyphae 252
mycelium 252
heterokaryon 253
spores 253

Key Concepts

- Fungi are multicellular heterotrophs with filamentous bodies.
- Long strings of fungal cells, called hyphae, form a mass called a mycelium.
- The cells of fungi are interconnected, sharing cytoplasm and nuclei.

13.4 Fungal Associations

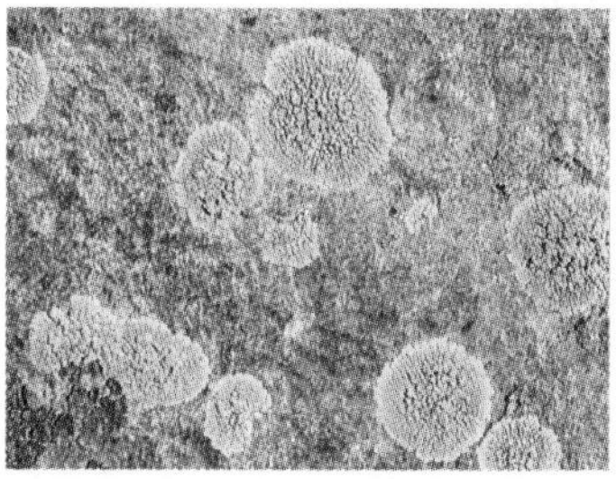

Key Terms

lichen 259
mycorrhizae 260

Key Concepts

- Lichens are symbiotic associations between a fungus and a photosynthetic partner.
- Mycorrhizae are symbiotic associations between a fungus and the roots of a plant.

13

CONCEPT REVIEW

1. The key advantage of multicellularity is that
   a. it allows the organism to be motile.
   b. it allows the formation of specialized tissues.
   c. photosynthesis is possible.
   d. sexual reproduction is possible.

2. Which of the following is not a "complex" multicellular organism?
   a. plants
   b. algae
   c. fungi
   d. fish

3. A fungus is not a plant for a variety of reasons. Which of the following characteristics does not distinguish fungi from plants?
   a. The sperm of fungi do not have flagella.
   b. The cell walls of fungi are made of chitin.
   c. Fungi do not carry out photosynthesis.
   d. Fungi are multicellular.

4. A mass of fungal filaments is called a
   a. hyphae.
   b. mycelium.
   c. septum.
   d. colony.

5. Members of the three phyla of fungi reproduce
   a. asexually only.
   b. sexually only.
   c. both asexually and sexually.
   d. either asexually or sexually, but not both.

6. Which of the following statements describing how fungi eat is *not* true?
   a. Fungi secrete a sticky substance that can trap flies.
   b. Fungi can fire projectiles into animals such as rotifers.
   c. Fungi secrete digestive enzymes into their surroundings.
   d. Fungi can feed on dead trees.

7. Included among the zygomycetes is the black bread mold
   a. zygosporangium.
   b. yeast.
   c. *Neurospora.*
   d. *Rhizopus.*

8. A mushroom is a complex structure of dikaryotic hyphae called the
   a. basidiocarp.
   b. ascus.
   c. zygote.
   d. mycorrhizae.

9. Which of the following organisms can symbiotically associate with a fungus to form a lichen?
   a. plants
   b. mycorrhizae
   c. green algae
   d. coral

10. Simple _________ has evolved in three groups of protists: the brown algae, green algae, and red algae.

11. The slender filaments that make up the body of a fungus are called _________.

12. Fungi are often growing on dead trees because they are able to break down the _________ in the wood and absorb glucose molecules as food.

13. _________ reproduction in the zygomycetes is unusual and frequently occurs in times of _________, when conditions are not favorable.

14. The fungal phylum _________ is the largest of the fungal phyla, with about 30,000 species named.

15. _________ is a member of the imperfect fungi and is a source of a well-known antibiotic.

16. Symbiotic associations between plants and fungi, called _________, aid in the uptake of _________ by plants.

Answers to the Concept Review questions appear in Appendix B.

CHALLENGE YOURSELF

1. If fungi have been so successful as to become an entire kingdom, why do you suppose the protist that first gave rise to the fungi has not persisted?

2. What is there about the way fungi live in nature that helps make them particularly valuable in industrial processes?

3. Many antibiotics, including penicillin, are derived from fungi. Why do you think the fungi produce these substances? Of what use are they to the fungi? In addition, fungi often secrete substances into the food that they are attacking that make these foods unpalatable or even poisonous. What kind of advantage would these substances provide the fungi?

4. What role did mycorrhizae probably play in plants' invasion of land hundreds of millions of years ago?

FOR FURTHER READING

Cooke, W. B. *The Fungi of Our Moldy Earth*. Berlin: J. Cramer, 1986. Description of techniques for collecting, preparing, isolating, and identifying fungi, in addition to a discussion of their ecology and classification.

Hale, M. E. *The Biology of Lichens*. 3d ed. Baltimore, Md.: University Park Press, 1983. A concise account of all aspects of the biology of lichens.

McKnight, K. H., and V. B. McKnight. *A Field Guide to Mushrooms of North America*. Princeton, N.J.: Peterson Field Guide Series, 1987. Excellent identification guide to the common edible and poisonous species of the United States and Canada.

Moore-Landecker, E. *Fundamentals of Fungi*. Englewood Cliffs, N.J.: Prentice Hall, 1990. A good basic text for the study of mycology.

Newhouse, J. R. "Chestnut Blight." *Scientific American* (July 1990): 106–11. Biological control is now being used in an effort to stop the ravages of this fungus.

Simon, L., J. Bousquet, R. C. Levesque, and M. LaLonde. "Origin and Diversification of Endomycorrhizal Fungi and Coincidence with Vascular Land Plants." *Nature* (May 6, 1993): 67–69. The critical relationship between fungi and plant roots is discussed in this very technical article.

Sommer, R. "Why I Continue to Eat Corn Smut." *Natural History* (January 1995): 18–21. A treatise on the fungus that causes "corn smut," the bane of farmers but balm to the gastronauts.

Sternberg, S. "The Emerging Fungal Threat." *Science* (December 9, 1994): 1632–34. Gone are the days when fungal infections meant just ringworm or athlete's foot. In today's increasing population of immunocompromised individuals, fungal infections can mean death.

TECHNOLOGY LINKS

The Living World Home Page
http://www.wcbp.com/biology/tlw

14

Rise of the Flowering Plants

Metamorphosis

CHAPTER OUTLINE

Figure 14.1 A flowering plant.
This water lily is one of about 266,000 species of plants living today.

Until about 440 million years ago, nothing lived on the surface of the earth. Scientists are not sure why it took life so long to reach terrestrial habitats, but they suspect that intense solar radiation may have made the land surface uninhabitable. With the advent of photosynthesis in the earth's oceans, oxygen gas (O_2) began to accumulate in the atmosphere, leading to the development of a layer of ozone (O_3) high in the atmosphere that shielded the surface from much of the sun's ultraviolet radiation. Soon after the appearance of significant amounts of oxygen in the earth's atmosphere, plants (figure 14.1) and fungi invaded the land.

14.1 The Origin of Plants

Plants are complex multicellular organisms that are terrestrial **autotrophs**—that is, they occur almost exclusively on land and feed themselves by photosynthesis. The name *autotroph* comes from the Greek, *autos,* self + *trophos,* feeder. Today, plants are the dominant organisms on the surface of the earth. An estimated 266,000 species are now in existence, covering every part of the terrestrial landscape except the extreme polar regions and the highest mountain tops (figure 14.2). In this section we examine how plants adapted to life on land.

Adapting to Terrestrial Living

The green algae that were probably the ancestors of today's plants are aquatic organisms that cannot survive on land. Before their descendants could live on land, they had to overcome three environmental challenges. First, they had to absorb minerals from the rocky surface. Second, they had to find a means of conserving water. Third, they had to develop a way to reproduce on land.

Absorbing Minerals

Plants require relatively large amounts of six inorganic minerals: nitrogen, potassium, calcium, phosphorus, magnesium, and sulfur. Each of these minerals constitutes 1% or more of a plant's dry weight. Algae absorb these minerals from water, but where is a plant on land to get them? The first plants seem to have developed a special relationship with fungi that was a key factor in their ability to absorb minerals in terrestrial habitats. Within the roots of many early fossil plants like *Cooksonia* and *Rhynia* can be seen fungi, living intimately within and among the root cells. As you may recall from chapter 13, these kinds of symbiotic associations are called **mycorrhizae.** In plants with mycorrhizae, the fungi enable the plant to take up phosphorus, zinc, copper, and other nutrients from rocky soil, while the plant supplies organic molecules to the fungus.

Conserving Water

One of the key challenges to living on land is to avoid drying out. To solve this problem, plants have a watertight outer covering called a **cuticle.** The covering is formed from a waxy substance that is impermeable to water. Like the wax on a shiny car, the cuticle prevents water from entering or leaving the stem or leaves. Water enters the plant only from the roots, while the cuticle prevents water loss to the air. Passages do exist through the cuticle, in the form of specialized pores called **stomata** (singular, **stoma**) in the leaves and sometimes the green portions of the stems (figure 14.3). Stomata, which occur on at least some portions of all plants except liverworts, allow carbon dioxide to pass into the plant bodies for photosynthesis and allow water and oxygen gas to pass out of them. The cells that border stomata expand and contract, thus controlling the loss of water while allowing the entrance of carbon dioxide. In most plants, water enters through the roots and exits through stomata on the underside of the leaves.

Figure 14.2 There are many kinds of plants.
These four very different plants represent 4 of the 12 phyla of plants that live on earth today. Many other phyla originally existed but are now extinct.

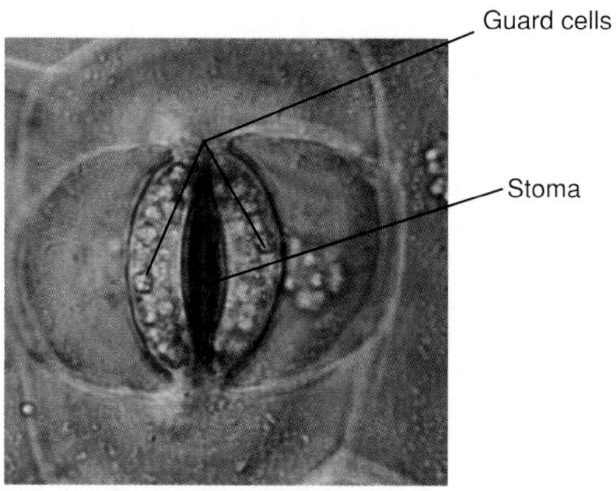

Figure 14.3 A stoma.
A stoma is a passage through the cuticle. Water passes out through the stoma, and carbon dioxide enters by the same portal. The cells flanking the stoma are called guard cells.

Reproducing on Land

To reproduce sexually on land, it is necessary to pass gametes from one individual to another, and because plants cannot move about, it is necessary that the gametes avoid drying out while they are transferred by wind or insects. In the first plants, the eggs were surrounded by a jacket of cells, and a film of water was required for the sperm to swim to the egg and fertilize it. Today, mosses still reproduce this way. However, soon after mosses evolved, changes occurred in the plant life cycle that favored the development of **spores,** which as you will recall from chapter 13 are asexual reproductive cells very resistant to drying out.

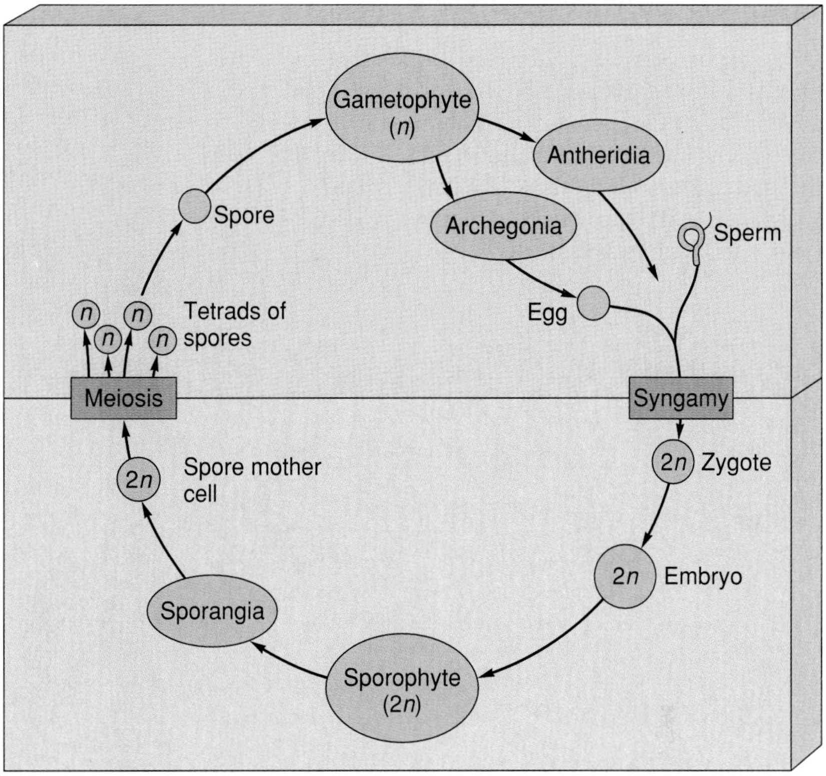

Figure 14.4 Generalized plant life cycle.
In a plant life cycle, there is alternation of generations, in which a diploid generation alternates with a haploid one. Gametophytes, which are haploid (n), alternate with sporophytes, which are diploid ($2n$). Antheridia (male) and archegonia (female), which are the sex organs (gametangia), are produced by the gametophyte, and they, in turn, respectively produce sperm and eggs. The sperm and egg ultimately come together to produce the first diploid cell of the sporophyte generation, the zygote. Meiosis takes place within the sporangia, the spore-producing organs of the sporophyte, resulting in the production of the spores, which are haploid and are the first cells of the gametophyte generation.

Changing the Life Cycle

Among many algae, haploid cells occupy the major portion of the life cycle (see chapter 12). The zygote formed by the fusion of gametes is the only diploid cell, and it immediately undergoes meiosis to form haploid cells again. In early plants, this meiosis became delayed, so that for a significant portion of the life cycle, cells were diploid. This resulted in an **alternation of generations,** in which a diploid generation alternates with a haploid one (figure 14.4). Botanists call the diploid generation the **sporophyte** because it forms haploid spores by meiosis. The haploid generation is called the **gametophyte** because it forms haploid gametes by mitosis. When you look at an early plant, you see largely gametophyte tissue—the sporophytes are smaller brown structures attached to or enclosed within the tissues of the gametophyte (figure 14.5*a*). When you look at plants that evolved later, the vascular plants, you see largely sporophyte tissue. The gametophytes are always much smaller than the sporophytes and are often enclosed within their tissues (figure 14.5*b*).

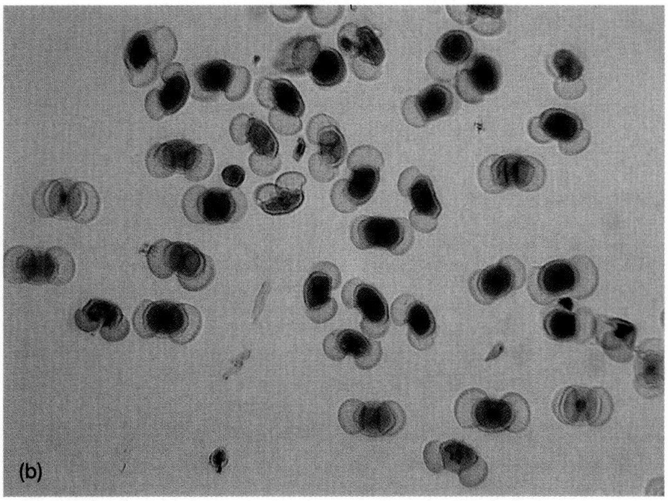

Figure 14.5 Gametophytes of a nonvascular and a vascular plant.
(*a*) A nonvascular plant: the flattened, somewhat circular plants growing on the forest floor in New Zealand are the gametophytes of a tree fern; like most fern gametophytes, they are bisexual and photosynthetic. (*b*) A vascular plant: pine gametophytes (pollen grains) are just large enough to be visible with the naked eye.

Evolution of a Vascular System

Once plants became established on land, many other features developed gradually that aided their evolutionary success in this new, demanding habitat. For example, in the first plants there was no fundamental difference between the aboveground and the belowground parts. Later, roots and shoots with specialized structures evolved, each suited to its particular environment.

One of the most important structural changes in the gradual adaptation of plants to the demands of living on land involved developing better ways of moving water around the body of the plant. In order for a plant to grow high into the air, a relatively efficient plumbing system is required to carry water up from the roots to the leaves and to carry carbohydrates down from the leaves to the roots. These plumbing systems, consisting of specialized strands of hollow cells connected

Table 14.1

Phylum	Typical Examples	Key Characteristics	Approximate Number of Living Species
Anthophyta (flowering plants)	Oak trees, corn, wheat, roses 	Flowering; also called angiosperms; characterized by ovules that are fully enclosed by the carpel; fertilization involves two sperm nuclei; one forms the gamete, the other fuses with polar bodies to form endosperm for the seed; after fertilization, carpels and the fertilized ovules (now seeds) mature to become fruit	235,000
Bryophyta (mosses)	*Polytrichum, Sphagnum* (peat moss) 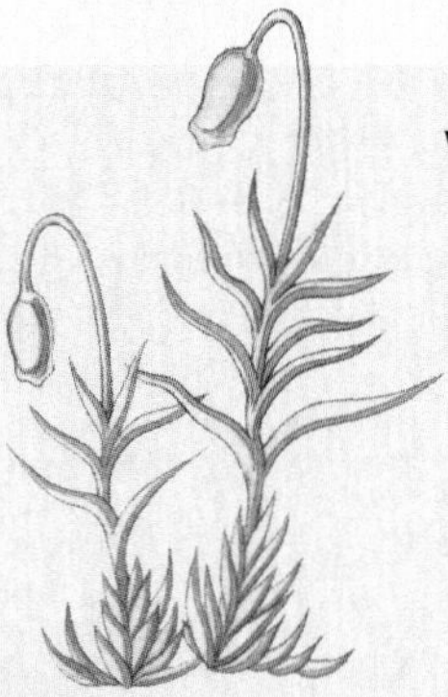	Without vascular tissues; lack true roots and leaves; live in moist habitats and obtain nutrients by osmosis and diffusion; two other phyla, Hepaticophyta (liverworts) and Anthocerophyta (hornworts), are also considered bryophytes and make up 40% of all bryophyte species	16,600
Pterophyta (ferns)	*Azolla, Sphaeropteris* (tree ferns) 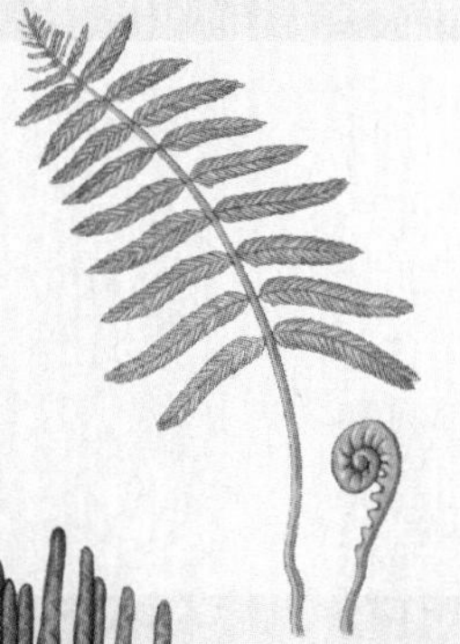	Seedless vascular plants; haploid spores germinate into free-living haploid individuals; two minor phyla, Sphenophyta (horsetails) and Psilophyta (whisk ferns) contain 21 additional species	12,000
Lycophyta (lycopods)	*Lycopodium,* (club mosses) 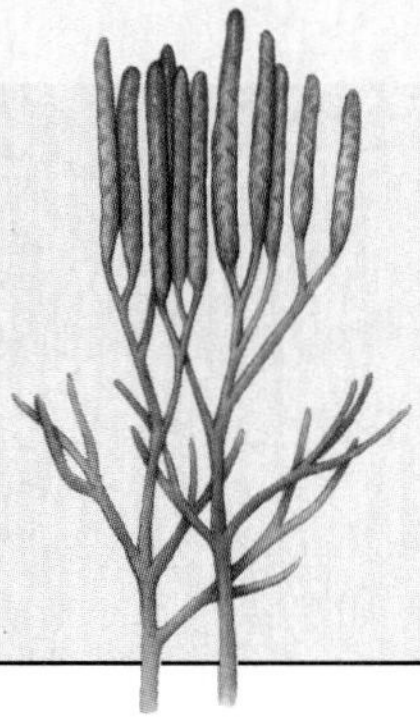	Seedless vascular plants similar in appearance to mosses, but diploid; found in moist habitats	1,000

end to end like a pipeline, are called **vascular systems** (Latin, *vasculum,* vessel or duct). In most vascular plants today they run from near the tip of a plant's roots all the way up the stem and into the leaves. Not all plants have efficient vascular systems. Some plants have none, others simple, inefficient ones, and still others highly sophisticated ones.

Of the 12 phyla of living plants (table 14.1), 9 are vascular plants. The first vascular plants of which we have any complete fossils appeared about 410 million years ago. These plants, members of the extinct phylum Rhyniophyta, had no leaves. The plant body was little more than a simple branching axis with spore-bearing sporangia at the tips of the branches. Other ancient vascular plants evolved leaves, which first appeared as simple extensions out from the stem; in most living plants, leaves form on the branches. Vascular plants proved to be phenomenally successful, and they are by far the most common kind of plant alive today.

Table 14.1 (continued)

Phylum	Typical Examples	Key Characteristics	Approximate Number of Living Species
Coniferophyta (conifers)	Pines, spruce, fir, redwood, cedar 	Gymnosperms; ovules within a carpel but partially exposed at time of pollination; flowerless; seeds are dispersed by the wind; sperm lack flagella; leaves are needlelike or scalelike; most species are evergreens and live in dense stands; among the most common trees on earth	550
Cycadophyta (cycads)	Cycads, sago palms 	Gymnosperms; very slow growing, palmlike trees; sperm have flagella but reach vicinity of egg by a pollen tube	100
Gnetophyta (shrub teas)	Mormon tea, *Welwitschia* 	Gymnosperms; nonmotile sperm; shrubs and vines	70
Ginkgophyta (ginkgo)	Ginkgo trees 	Gymnosperms; fanlike leaves that are dropped in winter (deciduous); seeds fleshy and ill-scented; motile sperm	1

Plants with No Vascular System: Liverworts and Hornworts

The first successful land plants had no vascular system—no tubes or pipes to transport water and nutrients throughout the plant. This greatly limited the maximum size of the plant body because all materials had to be transported by osmosis and diffusion. Only two phyla of living plants (also called "divisions" by some botanists), the **liverworts** (phylum Hepaticophyta) and the **hornworts** (phylum Anthocerophyta), completely lack a vascular system. The word "wort" meant "herb" in medieval Anglo-Saxon when these plants were named. Liverworts are the simplest of all living plants (figure 14.6). About 6,000 species of liverworts and 100 species of hornworts survive today, usually growing inconspicuously in moist and shady places.

Plants with Simple Vascular Systems: Mosses

A third phylum of plants, the **mosses** (phylum Bryophyta), were the first plants to evolve strands of specialized cells that conduct water and carbohydrates, so-called **vascular tissue.** In many mosses, a central strand of vascular tissue conducts water up the stem of the gametophyte. Simple in design, the moss vascular system is composed of conducting cells without specialized wall thickenings—like soft pipes, they cannot carry water very high. In the plants that evolved later, the vascular tissue is made up of specialized conducting cells with strengthened walls, able to raise water to great heights without ballooning out. This is why vascular plants are much taller than mosses. Because their vascular systems are simple, mosses are usually grouped by botanists with the liverworts and hornworts as "nonvascular" plants. Today about 10,000 species of mosses grow in moist places all over the world (figure 14.7). In mosses, as in liverworts and hornworts, the sporophytes are borne on the gametophytes, from which the spores derive their food (figure 14.8). Moss spores take 6 to 18 months to develop and are generally elevated on a stalk. The life cycle of a moss is illustrated in figure 14.9.

Plants with Specialized Vascular Systems: Vascular Plants

The remaining nine phyla of plants, which have efficient vascular systems made of highly specialized cells, are called **vascular plants.** Figure 14.10 shows a leaf at high magnification with the surface layers of cell removed; you can clearly see a network of veins crisscrossing the leaf, so that no cell is far from a vein. Vascular plants often grow to great heights—a tree can be more than 50 meters (163 feet) tall, with a mass of many tons. Today, more than 250,000 species of vascular plants exist on earth.

Figure 14.6 A simple plant.
Liverworts are the simplest of all living plants. In this liverwort, *Marchantia*, the sporophytes are borne within the tissues of the umbrella-shaped structures that arise from the surface of the flat, green, creeping gametophyte. These particular structures develop archegonia within their tissues; another similar structure, borne on different plants of *Marchantia,* produces the antheridia.

Figure 14.7 Mosses love water.
Mosses often completely dominate wet places such as this small waterfall. They are also very conspicuous at high altitudes and latitudes, where their ability to withstand drought helps them survive during the long winters.

Figure 14.8 Moss sporophytes.
Most of a moss is sporophyte. In this hair-cup moss, *Polytrichum*, the small leaves at the base belong to the gametophyte, while each of the large yellowish-brown stalks, with the capsule at its summit, is a sporophyte. Although moss sporophytes may be green and carry out a limited amount of photosynthesis when they are immature, they are soon completely dependent, in a nutritional sense, on the gametophyte.

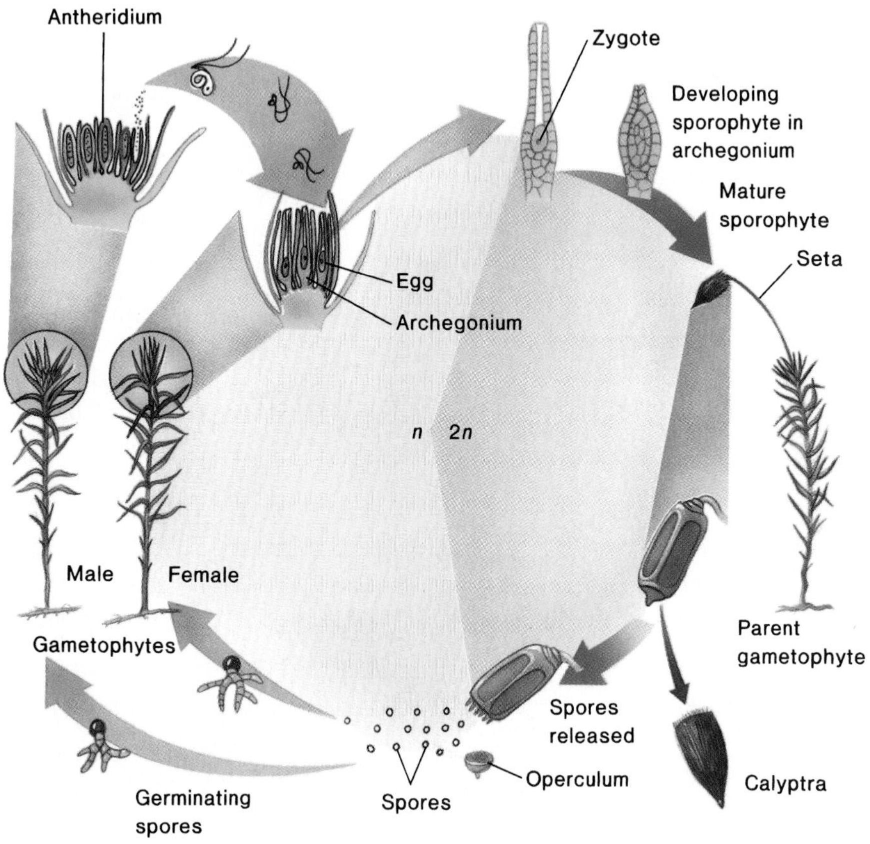

Figure 14.9 The life cycle of a moss.

On the gametophytes, which are haploid, sperm are released from each antheridium. They then swim through free water to an archegonium and down its neck to the egg. Fertilization takes place there; the resulting zygote develops into a sporophyte, which is diploid. The sporophyte grows out of the archegonium and differentiates into a slender, basal stalk, or seta, which has a wollen capsule at its apex. The capsule is covered, at least at first, with a cap formed from the swollen archegonium. The sporophyte grows on the gametophyte and eventually produces spores as a result of meiosis. The spores are shed from the capsule. The spores germinate, giving rise to gametophytes. The gametophytes initially are threadlike; they grow along the ground. Ultimately, buds form on them, from which leafy gametophytes arise.

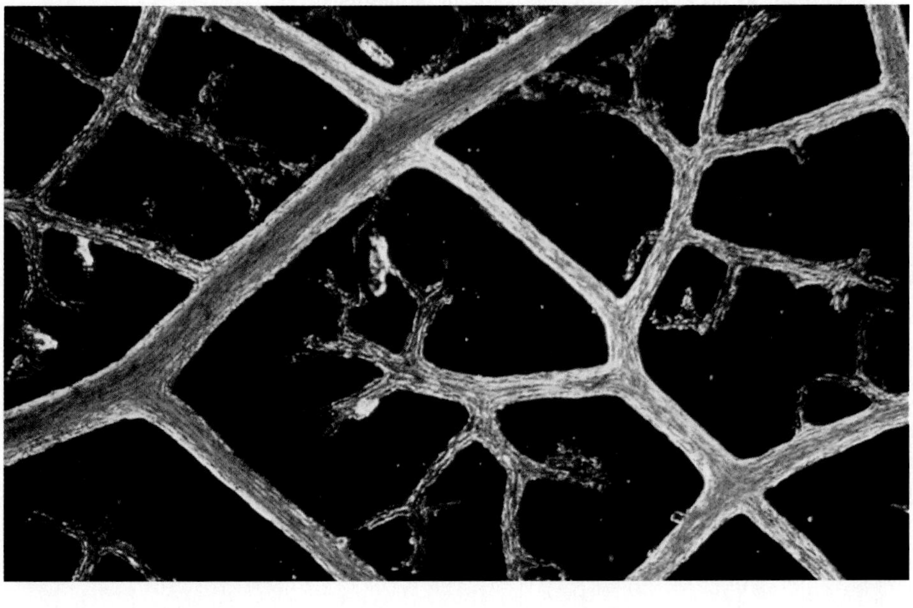

Figure 14.10 The vascular system of a leaf.

The veins of a vascular plant contain strands of specialized cells for conducting carbohydrates, as well as water containing dissolved minerals. The veins run from the tips of the roots to the tips of the shoots and throughout the leaves, as shown in this greatly enlarged photomicrograph of a cleared leaf.

14.2 General Features of Vascular Plants

The first vascular plant appeared approximately 430 million years ago, but only incomplete fossils have been found. The first vascular plants for which we have relatively complete fossils, the extinct phylum Rhyniophyta, lived 410 million years ago. Among them is the oldest known vascular plant, *Cooksonia,* with branched, leafless shoots that form spores at their tips.

Vascular plants are distinguished by several features:

1. **Dominant sporophyte.** The life cycle is dominated by the sporophyte (diploid) generation. The sporophyte individual tends to be much larger than the gametophyte and nutritionally independent of it.
2. **Specialized conducting tissue.** Water- and nutrient-conducting cells have reinforced cell walls to withstand considerable hydrostatic pressure. Within the stems of early vascular plants are easily seen vascular bundles, made up of parallel threads of these conducting cells.
3. **Specialized body form.** It is among vascular plants that we first see the traditional plant architecture of roots, stems, and leaves (figure 14.11). In the earliest vascular plants, there was no differentiation of the plant body into roots and shoots.

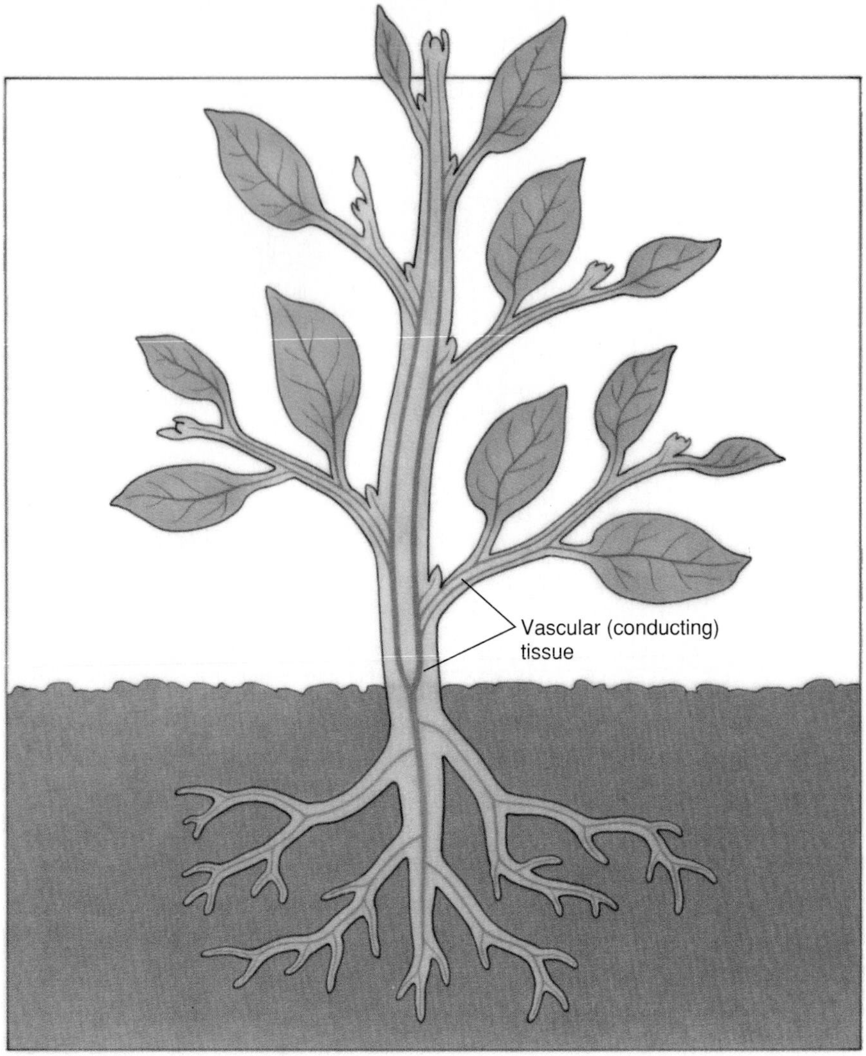

Figure 14.11 The architecture of a plant.
The basic architecture of a vascular plant consists of roots, stems, and leaves. This body design is determined largely by the need to transport water through the plant. Water passes continuously in through the roots, up through the stems, and out through the leaves. At the same time, sugar (carbohydrate) molecules, manufactured as a result of photosynthesis in the leaves and green stems, pass through a parallel conducting system to all parts of the plant, where they are used for growth.

The two types of conducting systems in the earliest vascular plants, still found largely unchanged in vascular plants today, are sieve elements and tracheary elements. Both are elongated cells that occur end to end in strands, like sections of bamboo. **Sieve elements** are relatively soft-walled cells that conduct carbohydrates away from the areas where they are made. They are the characteristic cell type of a kind of tissue called **phloem. Tracheary elements** are hard-walled cells that transport water and dissolved minerals up from the roots. They are the characteristic cell type of a kind of tissue called **xylem.**

Early vascular plants grew by cell division at the tips of the stem and roots. Imagine stacking dishes—the stack can get taller but not wider! This sort of growth pattern is called **primary growth.** The region of growth is called the **apical** (at the tip) **meristem.** A meristem is an actively dividing zone of cells, and growth at the tip of the plant tends to make the plant body taller but not broader.

This simple approach to growth was quite successful. During the so-called Age of Coal, when much of the world's fossil fuel was formed, the lowland swamps that covered Europe and North America were dominated by lycophyte trees, which grew to heights of 10 to 35 meters (33 to 115 feet). The pace of evolution was rapid during this period, for the world's climate was changing, growing dryer and colder. As the world's swamplands began to dry up, the lycophyte trees vanished, disappearing abruptly from the fossil record. They were replaced by tree ferns, with heights of more than 20 meters (66 feet) and trunks 30 centimeters (12 inches) thick. Like the lycophytes, the trunks of tree ferns were formed entirely by primary growth.

Figure 14.12 The evolution of secondary growth.
Extending across vast areas of Asia and North America are the northern forests, such as the one pictured, which consist of tall, thick-trunked trees. The existence of these forests is possible due to the evolutionary advance of secondary growth, growth around the plant's periphery.

About 380 million years ago, vascular plants developed a new pattern of growth, in which a cylinder of meristem cells beneath the bark divides, producing new cells in regions around the plant's periphery. This growth is called **secondary growth.** Secondary growth makes it possible for a plant to increase in diameter. Only after the evolution of secondary growth could vascular plants become thick-trunked and therefore tall. Redwood trees today reach heights of up to 117 meters (384 feet) and trunk diameters in excess of 11 meters (36 feet). This evolutionary advance made possible the evolution of the tall forests that today cover northern North America (figure 14.12). You are familiar with the product of plant secondary growth as **wood.** The growth rings so visible in cross sections of trees (figure 14.13) are zones of secondary growth (spring–summer) spaced by zones of little growth (fall–winter).

Figure 14.13 Growth rings in wood.
This photo shows a cross section of a ponderosa pine (*Pinus ponderosa*). The dark area in the center is heartwood, a nonliving portion of the stem in which no water transport occurs. The lighter part of the wood is sapwood, in which active water conduction takes place.

Seedless Vascular Plants

The earliest vascular plants lacked seeds (figure 14.14), and four of the nine phyla of modern-day vascular plants do not have seeds. These four phyla of living seedless vascular plants, two of which are illustrated in figure 14.15, have free-swimming sperm that require the presence of free water for fertilization.

By far the most abundant of the four phyla of seedless vascular plants contains the **ferns,** with about 12,000 living species. Ferns are found throughout the world, although they are much more abundant in the tropics than elsewhere. Many are small, only a few centimeters in diameter, but some of the largest plants that live today are also ferns. Descendants of the Age of Coal tree ferns, they can have trunks more than 24 meters (79 feet) tall and leaves up to 5 meters (16 feet) long!

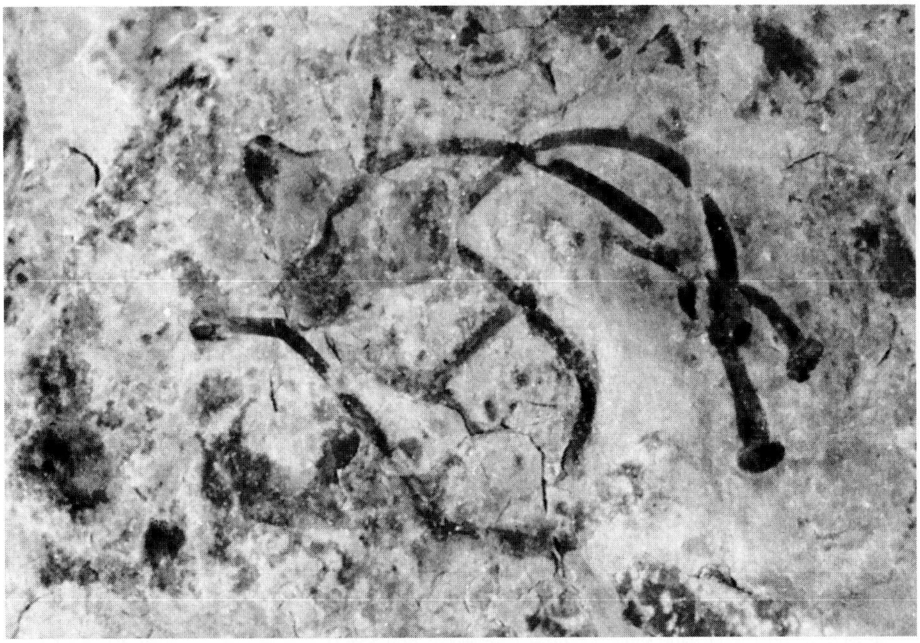

Figure 14.14 The earliest vascular plant.
The earliest vascular plant of which we have complete fossils is *Cooksonia.* This fossil, 410 million years old, consists of little more than a branched stem with spore-producing sporangia at the tips.

Figure 14.15 Seedless vascular plants.
The earliest vascular plants lacked seeds. (*a*) The club moss *Lycopodium lucidulum* (division Lycophyta). Although superficially similar to the gametophytes of mosses, club moss plants are sporophytes. Meiosis occurs in the sporangia of the cones, ultimately resulting in the shedding of haploid spores. These spores grow into gametophytes that, in *Lycopodium,* bear both antheridia and archegonia. (*b*) A tree fern in the forests of Malaysia (division Pterophyta). The ferns are by far the largest group of spore-producing vascular plants.

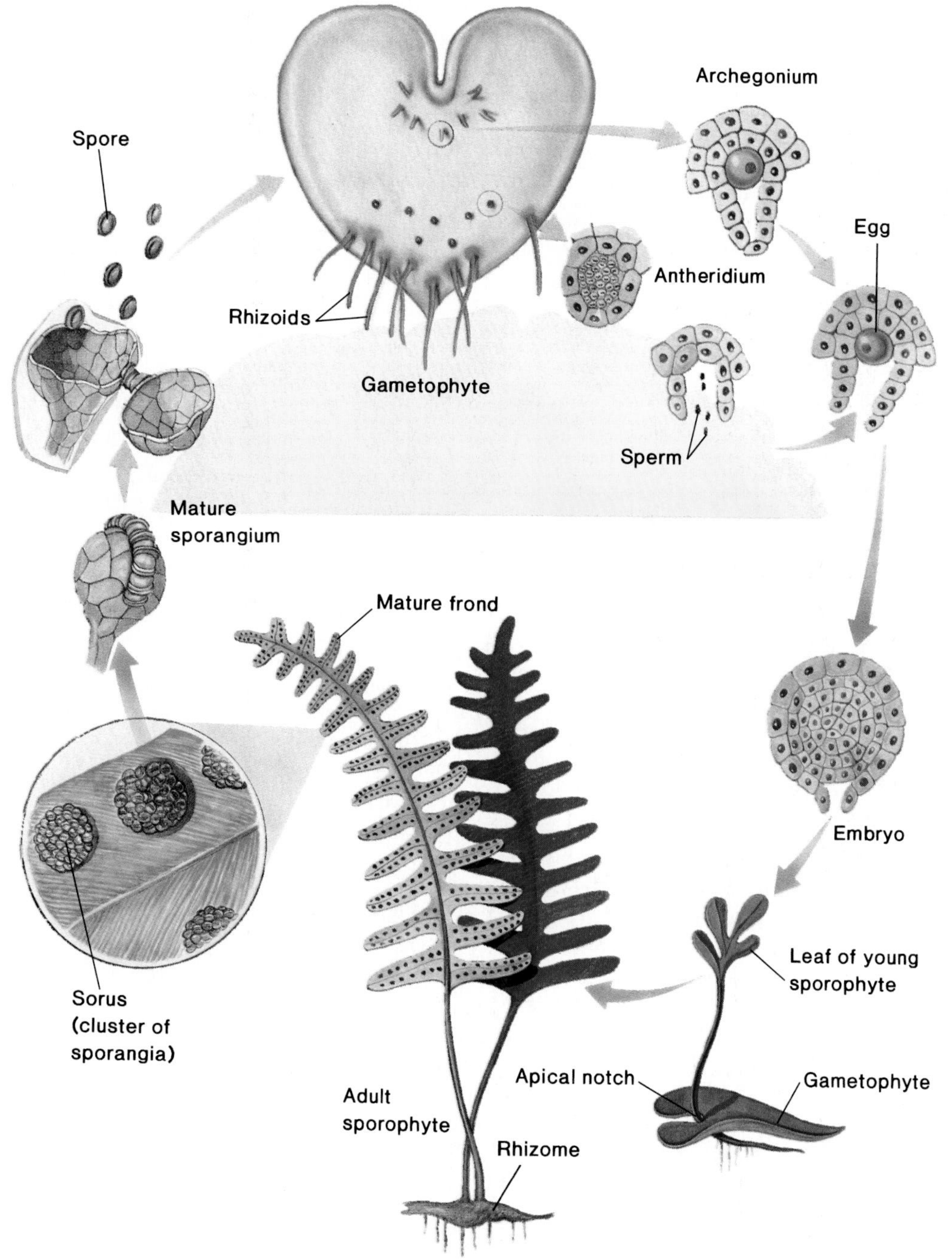

Figure 14.16 Fern life cycle.

The gametophytes, which are haploid, grow in moist places. Rhizoids (anchoring structures) project from their lower surface. Eggs develop in an archegonium, and sperm develop in an antheridium, both located on the gametophyte's lower surface near the apical notch, the region of the most rapid cell division. The sperm, when released, swim through free water to the mouth of the archegonium, entering and fertilizing the single egg. Following the fusion of egg and sperm to form a zygote—the first cell of the diploid sporophyte generation—the zygote starts to grow within the archegonium. Eventually, the sporophyte becomes much larger than the gametophyte—it is known as a fern plant. Most ferns have more or less horizontal stems, called rhizomes, that creep along below the ground. On the sporophyte's leaves, called fronds, occur clusters of sporangia (called sori; singular, sorus), within which meiosis occurs and spores are formed. The release of these spores, which is explosive in many ferns, and their germination lead to the development of new gametophytes.

The Life of a Fern

In ferns, the life cycle of plants begins a revolutionary change that culminates later with seed plants. Nonvascular plants like mosses are made up largely of gametophyte (haploid) tissue. Vascular seedless plants like ferns have both gametophyte and sporophyte individuals, each independent and self-sufficient. The gametophyte produces eggs and sperm; when these swim through water and combine in a fertilized egg, the zygote grows into a sporophyte. The sporophyte bears and releases haploid **spores** that float to the ground and germinate, growing into haploid gametophytes. The fern gametophytes are small, thin, heart-shaped photosynthetic plants, usually no more than a centimeter in length, that live in moist places. The fern sporophytes are much larger and more complex, with long vertical leaves called **fronds.** When you see a fern, you are almost always looking at the sporophyte. The fern life cycle is described in figure 14.16.

14.3 The Advent of Seeds

We now examine a key evolutionary advance among the vascular plants, the development of a protective cover for the embryo. A **seed** is a plant embryo with a durable, watertight cover. The seed is a crucial adaptation to life on land because it protects the embryonic plant from drying out when it is at its most vulnerable stage. The evolution of the seed was a critical step in the domination of the land by plants.

Seed Plants

The change in the life cycle of vascular plants in favor of the sporophyte (diploid) generation, which began among the ferns, reaches its full force with the advent of the seed plants. In seed plants, the gametophyte generation has become even more reduced than in ferns, to the point that gametophytes develop within the tissue of the sporophyte individuals and are entirely dependent upon them for nutrients and water. **Pollen grains,** which are actually tiny male gametophytes, are only a few cells large, surrounded by a thick protective wall. Pollen grains (containing the sperm) are transported by wind, insects, or other animals to the vicinity of the **ovule** (the egg), the female gametophyte; this process is called **pollination.** The pollen grain then cracks open and sprouts, or germinates, and the pollen tube, containing the sperm cells, grows out, transporting the sperm directly to the egg. Thus there is no need for free water in the pollination and fertilization process.

Botanists generally agree that all seed plants are derived from a single common ancestor. There are five living phyla. In four of them, collectively called the **gymnosperms** (Greek, *gymnos,* naked + *sperma,* seed), the ovules are not completely enclosed by sporophyte tissue at the time of pollination. Gymnosperms were the first seed plants. From gymnosperms evolved the fifth group of seed plants, called **angiosperms** (Greek, *angion,* vessel + *sperma,* seed), phylum Anthophyta. Angiosperms are the most recently evolved of all the plant phyla. Angiosperms differ from all gymnosperms in that their ovules are completely enclosed by a vessel of sporophyte tissue called the **carpel** at the time they are pollinated.

The Structure of a Seed

A seed contains a sporophyte embryo surrounded by a protective seed coat (figure 14.17). Seeds are the way in which plants, anchored by their roots to one place in the ground, solve the problem of dispersing their progeny to new locations. The hard cover of the seed (formed from the tissue of the parent plant), protects the seed from drying out during its travel through the air. Many seeds have devices to aid in carrying them farther. Most species of pine, for example, have seeds with thin flat wings attached. These wings help catch air currents, which carry the seeds to new areas.

Once a seed has fallen to the ground, it may lie there, dormant, for many years. When conditions are favorable, however, and particularly when moisture is present, the seed will germinate and begin to grow into a young tree (figure 14.18). Many seeds have abundant food stored in them to provide a ready source of energy for the new plant as it starts its growth. Called endosperm, this food source is discussed later in the chapter.

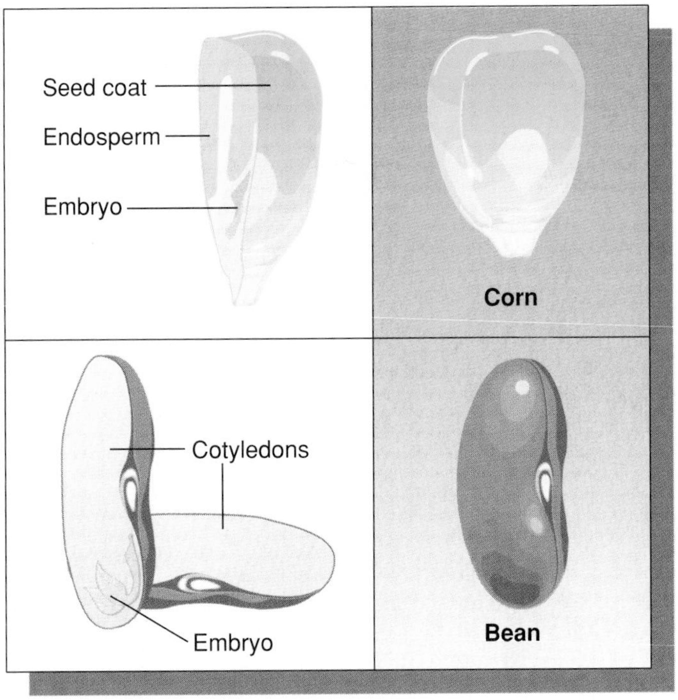

Figure 14.17 Basic structure of seeds.
A seed is a sporophyte (diploid) embryo surrounded by a protective coat. This seed coat, formed of sporophytic tissue from the parent, protects the embryo—the dormant young plant of the next sporophyte generation.

The advent of seeds had an enormous influence on the evolution of plants. Seeds have greatly improved the adaptation of plants to living on land in at least four respects:

1. **Dispersal.** Most important, seeds facilitate the migration and dispersal of plant offspring into new habitats.
2. **Dormancy.** Seeds permit plants to postpone development when conditions are unfavorable, as during a drought, and to remain dormant (figure 14.19) until conditions improve.
3. **Germination.** By making the reinitiation of development dependent upon environmental factors such as temperature, seeds permit the course of embryonic development to be synchronized with critical aspects of the plant's habitat, such as the season of the year.
4. **Nourishment.** The seed offers the young plant nourishment during the critical period just after germination, when the seedling must establish itself.

Figure 14.18 Seeds allow plants to bypass the dry season.
When it does rain, seeds can germinate, and plants can grow rapidly to take advantage of the relatively short periods when water is available. (*a*) This desert tree has tough seeds (*b*) that only germinate after they are cracked. Rains will leach out the chemicals in the seed coats that inhibit germination, and the hard coats of the seeds may be cracked when they are being washed down along arroyos in temporary floods.

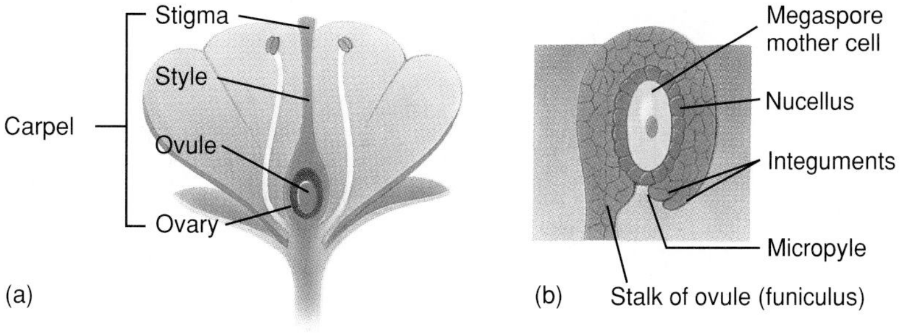

Figure 14.19 The key to dormancy.
Seeds can remain dormant because of the impermeable covering provided by the seed coat. Both gymnosperms and angiosperms have seeds that develop from the carpel. These diagrams show the development of angiosperm seeds as an example. In (*a*), the position of the carpel in an angiosperm flower is indicated. In (*b*), the ovary matures to form the fruit, while the outer covering matures to form the seed coat.

Gymnosperms

The first plants to evolve seeds were gymnosperms. Members of the most familiar of the four phyla of gymnosperms—pine, spruce, hemlock, cedar, redwood, yew, cypress, and fir trees—are trees that produce their seeds in **cones** and for this reason are called **conifers** (phylum Coniferophyta) (figure 14.20). The seeds (ovules) of conifers develop on scales within the cones and are exposed at the time of pollination. Most of the conifers have needlelike leaves, an evolutionary adaptation for retarding water loss. Conifers are often found growing in moderately dry regions of the world, including the vast taiga forests of the northern latitudes. Many are very important as sources of timber and pulp.

Figure 14.20 A conifer.
This Engelmann spruce, *Picea engelmannii,* common in the Rocky Mountains, is typical of the conifers that form extensive forests at high altitudes and latitudes.

There are about 550 living species of conifers. The tallest living vascular plant, the coastal sequoia (*Sequoia sempervirens*), found in coastal California and Oregon, is a conifer and reaches 100 meters (330 feet). The biggest redwood, however, is the mountain sequoia redwood species (*Sequoiadendron gigantea*) of the Sierra Nevadas. The largest individual tree is nicknamed after General Sherman of the Civil War, and it stands more than 80 meters (260 feet) tall while measuring 20 meters (60 feet) around its base. Another much smaller type of conifer, the bristlecone pines in Nevada, may be the oldest trees in the world—about 5,000 years old.

The other three gymnosperm phyla, much less widespread, are cycads (phylum Cycadophyta), ginkgoes (phylum Ginkgophyta), and gnetophytes (phylum Gnetophyta) (figure 14.21). Cycads, the predominant land plant in the early age of dinosaurs, the Jurassic, have short stems and palmlike leaves. They are still widespread throughout the tropics. There is only one living species of ginkgo, the maidenhair tree, which has fan-shaped leaves shed in the autumn. Because ginkgoes are resistant to air pollution, they are commonly planted along city streets. The Gnetophytes are thought to be the most closely related to angiosperms. This phylum contains only three kinds of plants, all unusual. One of them is perhaps the most bizarre of all plants, *Welwitschia,* which grows on the exposed sands of the harsh Namibian Desert of Southwestern Africa. *Welwitschia* acts like a plant standing on its head! Its two beltlike, leathery leaves are generated continuously from their base, splitting as they grow out over the desert sands.

Figure 14.21 The four kinds of gymnosperms.
(*a*) Slash pines, *Pinus elliottii,* in Florida, are representative of the largest division of gymnosperms, the Coniferophyta. (*b*) An African cycad, *Encephalartos kosiensis,* phylum Cycadophyta. The cycads have fernlike leaves and seed-forming cones, like the one shown here. (*c*) Maidenhair tree, *Ginkgo biloba,* the only living representative of the division Ginkgophyta, a group of plants that was abundant 200 million years ago. Among living seed plants, only the cycads and *Ginkgo* have swimming sperm. (*d*) *Welwitschia mirabilis,* phylum Gnetophyta, is found in the extremely dry deserts of southwestern Africa. In *Welwitschia*, two enormous, strap-shaped leaves grow from a circular zone of cell division that surrounds the apex of the carrot-shaped root.

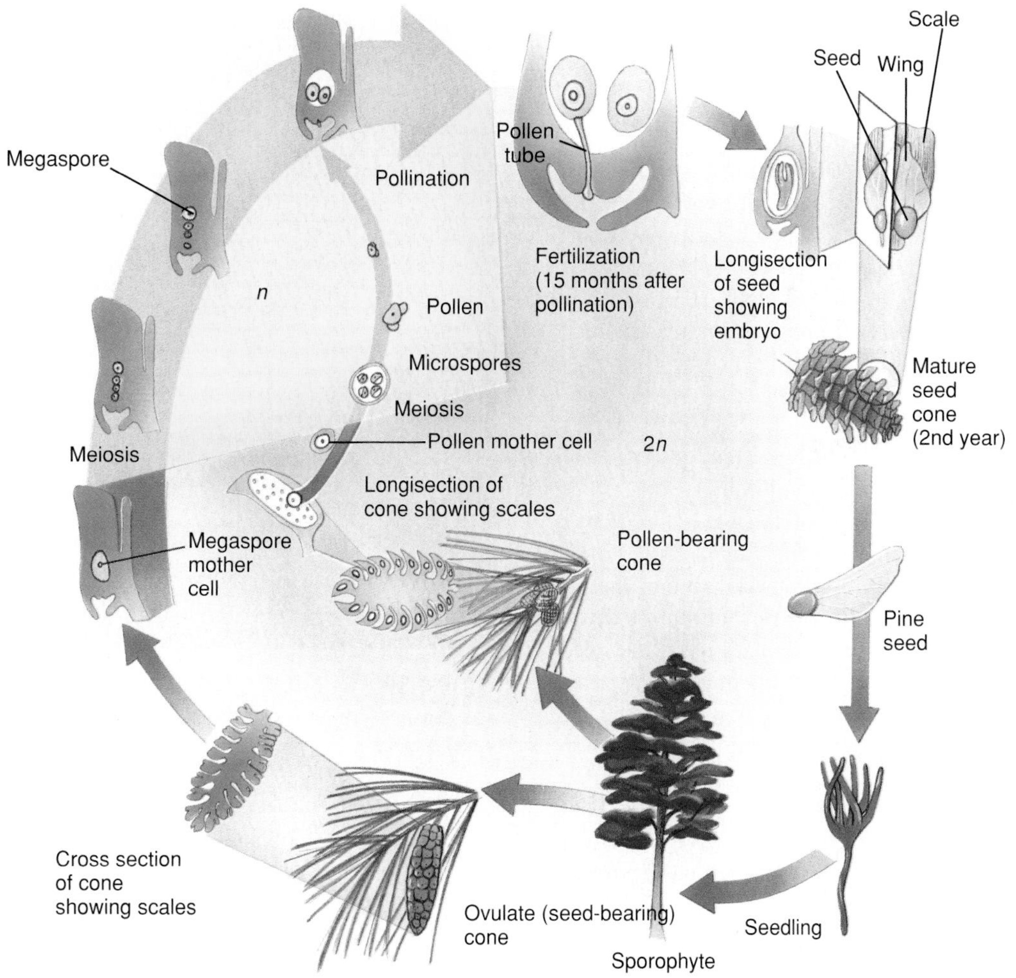

Figure 14.22 Life cycle of a conifer.

In all seed plants, the gametophyte generation is greatly reduced. In conifers such as pine, microsporangia are borne in pairs on the surface of the thin scales of the relatively delicate pollen-bearing cones. A germinating pollen grain is the mature microgametophyte of a pine. Megagametophytes, in contrast, develop within the tissues of the ovule. The familiar seed-bearing cones of pines are much heavier and more substantial structures than the pollen-bearing cones. Two ovules, and ultimately two seeds, are borne on the upper surface of each scale. After a pollen grain has reached a scale, it germinates, and a slender pollen tube grows toward the egg. When the pollen tube grows to the vicinity of the megagametophyte, sperm are released, fertilizing the egg and producing a zygote there. The development of the zygote into an embryo takes place within the ovule, which matures into a seed. Eventually, the seed falls from the cone and germinates, the embryo resuming growth and becoming a new pine tree.

The Life of a Gymnosperm

We will examine conifers as typical gymnosperms. The conifer life cycle is illustrated in figure 14.22. Conifer trees form two kinds of cones. Seed cones contain the female gametophyte, with its egg cells; pollen cones contain pollen grains. Conifer pollen grains are small and light and are carried by the wind to seed cones. Each pollen grain has a pair of air sacs to help carry it in the wind. Because it is very unlikely that any particular pollen grain will succeed in being carried to a seed cone (the wind can take it anywhere), a great many pollen grains are needed to be sure that at least a few succeed in pollinating seed cones. For this reason, pollen grains are shed from their cones in huge quantities, often appearing as a sticky yellow layer on the surfaces of ponds and lakes—and even on windshields.

When a grain of pollen settles down on a scale of a female cone, a slender tube grows out of it down into the scale, delivering the male gamete (the sperm cell) to the female gametophyte containing the egg, or ovum. Fertilization occurs when the sperm cell fuses with the egg, forming a zygote. This zygote is the beginning of the sporophyte generation.

What happens next is the essential improvement in reproduction achieved by seed plants. Instead of the zygote simply growing into an adult sporophyte (a tree—just as you grow directly into an adult from a fertilized zygote—the zygote's growth is delayed. Instead, it forms a seed.

14.4 The Evolution of Flowers

Angiosperms, plants in which the ovum is completely enclosed by sporophyte tissue when it is fertilized, are the most successful of all plants. Ninety percent of all living plants are angiosperms, over 235,000 species, including hardwood trees, shrubs, herbs, grasses, vegetables, and grains—in short, nearly all of the plants that we see every day. Virtually all of our food is derived, directly or indirectly, from angiosperms. In fact, more than half of the calories we consume come from just three species: rice, corn (maize), and wheat.

Rise of the Angiosperms

In a very real sense, the remarkable evolutionary success of the angiosperms is the culmination of the plant kingdom's adaptation to life on land. Angiosperms successfully meet the last difficult challenge posed by terrestrial living: the inherent conflict between the need to obtain nutrients (solved by roots, which anchor the plant to one place) and the need to find mates (solved by making the male gametes very tiny, so they can be carried to other plants). This challenge has never really been overcome by gymnosperms, whose pollen grains are carried passively by the wind on the chance that they might by luck encounter a female cone. Think about how inefficient this is! Angiosperms are also able to deliver their pollen directly, as if in an addressed envelope, from one individual of a species to another. How? *By inducing insects and other animals to carry it for them!* The tool that makes this animal-dictated pollination possible, the great advance of the angiosperms, is the flower (figure 14.23).

Figure 14.23 An angiosperm flower.
This flower of the wild woodland plant *Geranium* shows the five free petals, 10 stamens, and fused carpel typical of angiosperm flowers.

The Flower

Flowers are the reproductive organs of angiosperm plants. A flower is a sophisticated pollination machine. It employs bright colors to attract the attention of insects (or birds or small mammals), nectar to induce the insect to enter the flower, and structures that coat the insect with pollen grains while it is visiting. Then, when the insect visits another flower, it carries the pollen with it into that flower.

The basic structure of a flower consists of four concentric circles (figure 14.24), or **whorls:**

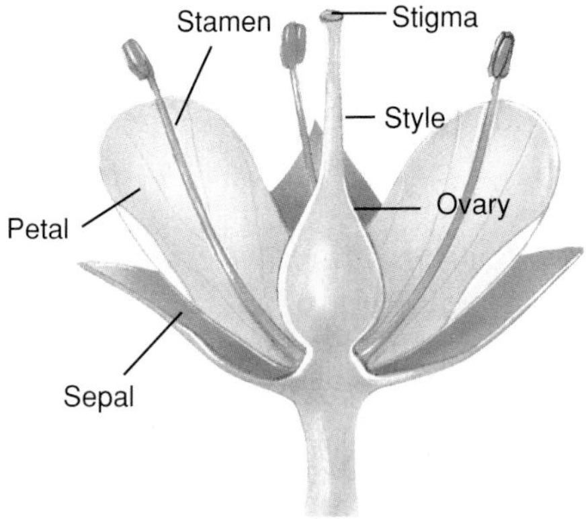

Figure 14.24 An angiosperm flower, diagrammed.
The basic structure of a flower is a series of four concentric circles, or whorls: the sepals, petals, stamens, and the carpel (ovary, style, and stigma).

1. The outermost whorl of the flower is concerned with protecting the flower from physical damage. It is made up of *sepals,* which are in effect modified leaves that protect the flower while it is a bud.
2. The second whorl of the flower is concerned with attracting particular pollinators. It is made up of *petals* that have particular pigments, often vividly colored.
3. The third whorl of the flower is concerned with producing pollen grains. It is made up of *stamens,* which are slender, threadlike filaments with a swollen portion (the *anther*) at the tip that contains the pollen.
4. The fourth and innermost whorl of the flower is concerned with producing eggs. It is made up of the *carpel,* which is sporophyte tissue that completely encases the ovules within which the egg cell develops. The ovules occur in the swollen lower portion of the carpel, called the ovary; usually there is a slender stalk rising from the ovary called the *style,* with a swollen tip called a *stigma* that is receptive to pollen. When the flower is pollinated, a pollen tube grows down from the stigma through the style to the ovary.

Figure 14.25 Red flowers are pollinated by hummingbirds.
This long-tailed hermit hummingbird is extracting nectar from the red flowers of *Heliconia imbricata* in the forests of Costa Rica. Note the pollen on the bird's beak. Hummingbirds of this group obtain nectar primarily at long, curved flowers that more or less match the length and shape of their beaks.

Why There Are Different Kinds of Flowers

If you were to watch insects visiting flowers, you would quickly discover that the visits are not random. Instead, certain insects are attracted by particular flowers. Insects recognize a particular color pattern and odor and search for flowers that look similar. Insects and plants have coevolved (see chapter 27) so that certain insects specialize in visiting particular kinds of flowers. As a result, a particular insect carries pollen from one individual to another *of the same species,* which is the key to successful insect pollination. It is this keying in on particular species that makes insect pollination so effective.

Of all insect pollinators, the most numerous are bees. Bees evolved soon after flowering plants, some 100 million years ago. Today there are over 20,000 species. Bees locate sources of nectar largely by odor at first (that is why flowers smell sweet) and then focus in on the flower's color and shape. Bee-pollinated flowers are usually yellow or blue, and frequently they have guiding stripes or lines of dots to indicate the position in the flower of the nectar (usually in the throat of the flower). While inside the flower, the bee becomes coated with pollen. This coating is far from accidental. Most of the bees visiting flowers actively seek to acquire pollen, which they use as a rich source of protein to feed their larvae.

Many other insects pollinate flowers. Butterflies tend to visit flowers like phlox that have "landing platforms" on which they can perch. These flowers typically have long, slender floral tubes filled with nectar that a butterfly can reach by uncoiling its long proboscis (a hoselike tube extending out from the mouth). Moths, which visit flowers at night, are attracted to white or very pale-colored flowers, often heavily scented, that are easy to locate in dim light.

Red flowers, interestingly, are not typically visited by insects, most of which cannot "see" red as a distinct color. Who pollinates these flowers? Hummingbirds and sunbirds (figure 14.25)! To these birds, red is a very conspicuous color, just as it is to us. Birds do not have a well-developed sense of smell, and do not orient to odor, which is why red flowers often do not smell strongly. Mammals such as nocturnal opossums and bats may visit tree species and even giant blooming cacti to eat the pollen, and, in the process, transport pollen grains from one plant to another.

Some angiosperms have reverted to the wind pollination practiced by their ancestors, notably oaks, birches, and most important the grasses. The flowers of these plants are small, greenish, and odorless.

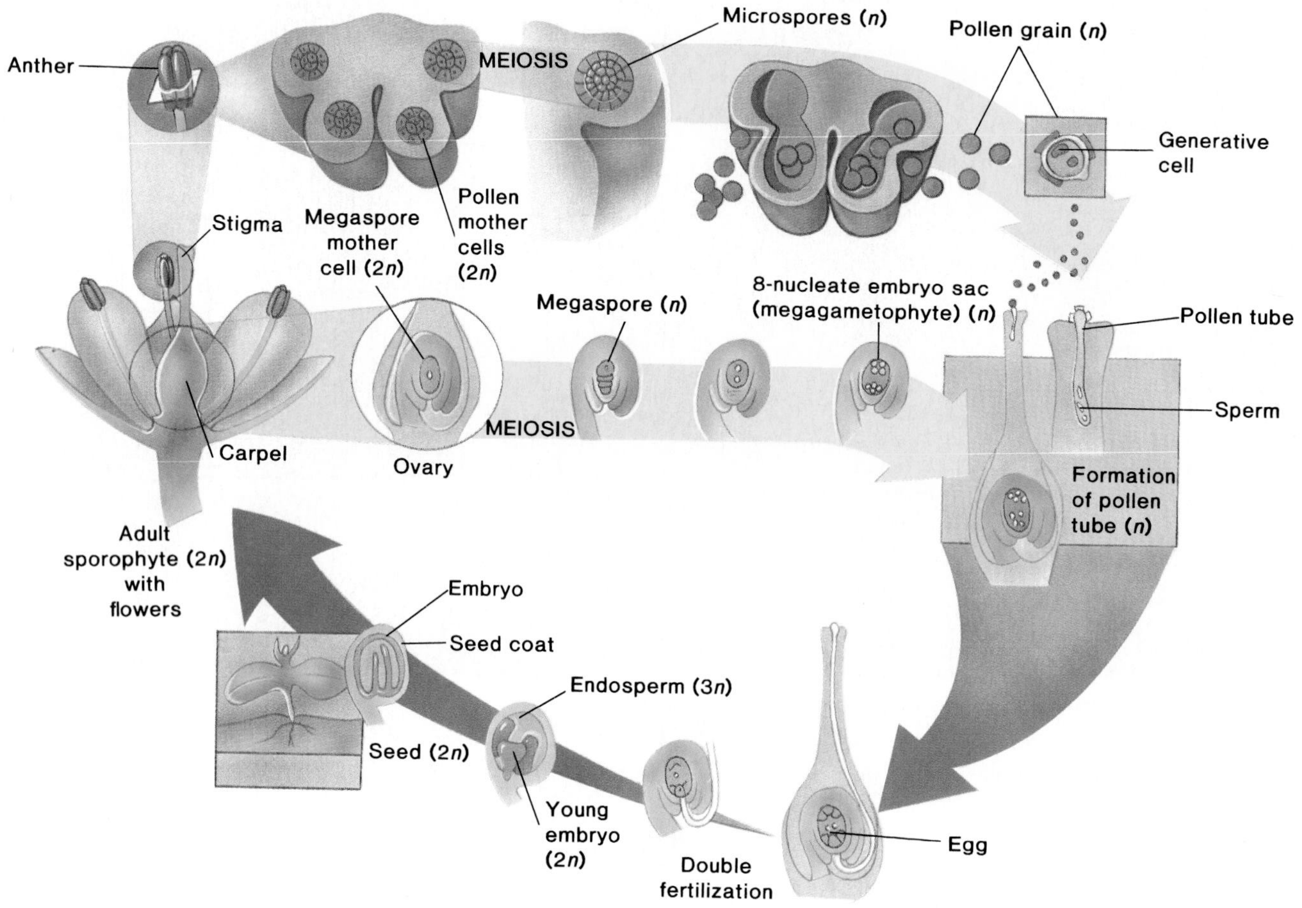

Figure 14.26 Life cycle of an angiosperm.
In angiosperms, as in gymnosperms, the sporophyte is the dominant generation. Eggs form within the megagametophyte, or embryo sac, inside the ovules, which, in turn, are enclosed in the carpels. The carpel is differentiated in most angiosperms into a slender portion, or style, ending in a stigma, the surface on which the pollen grains germinate. The pollen grains, meanwhile, are formed within the sporangia of the anthers and complete their differentiation to their mature, three-celled stage either before or after grains are shed. Fertilization is distinctive in angiosperms, being a double process. A sperm and an egg come together, producing a zygote; at the same time, another sperm fuses with the two polar nuclei, producing the primary endosperm nucleus, which is triploid. Both the zygote and the primary endosperm nucleus divide mitotically, giving rise, respectively, to the embryo and the endosperm. The endosperm is the tissue, unique to angiosperms, that nourishes the embryo and young plant.

Improving Seeds: Double Fertilization

The seeds of gymnosperms often contain food to nourish the developing plant in the critical time immediately after germination, but the seeds of angiosperms have greatly improved on this aspect of seed function. Angiosperms produce a special, highly nutritious tissue called **endosperm** within their seeds. Here is how they do it. In the angiosperm life cycle, the male gametophyte (with haploid gametes) contains *two* sperms, not one (figure 14.26). The first fuses with the egg, as in all sexually reproducing organisms, forming the zygote. The other fuses with two other products of meiosis called polar nuclei to form a triploid (three copies of the chromosomes) endosperm cell. This cell divides much more rapidly than the zygote, giving rise to the nutritive endosperm tissue within the seed. This process of fertilization with two sperm to produce both a zygote and endosperm is called **double fertilization.** Double fertilization occurs only in angiosperms.

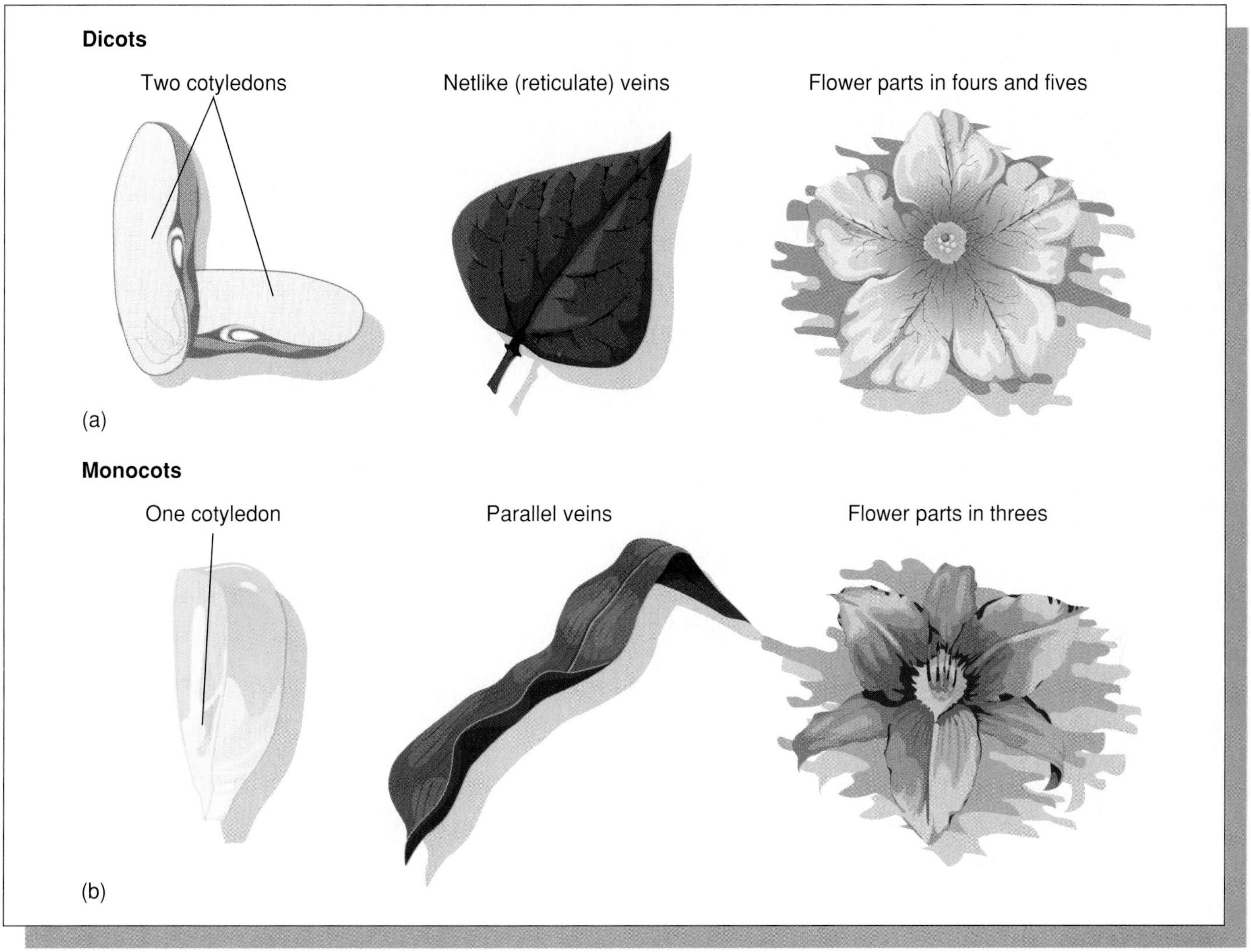

Figure 14.27 Monocots and dicots.
(*a*) Dicots have two cotyledons and netlike (reticulate) veins. Their flower parts occur in fours and fives.
(*b*) Monocots are characterized by one cotyledon, parallel veins, and the occurrence of flower parts in threes.

In some angiosperms, such as the common bean or pea, the endosperm is fully used up by the time the seed is mature. Food reserves are stored in swollen, fleshy leaves. In other angiosperms, such as corn, the mature seed contains abundant endosperm, which is used after germination. Some angiosperm embryos have two seed leaves or cotyledons and are called dicotyledons, or **dicots.** The first angiosperms were like this. Dicots typically have leaves with netlike branching of veins and flowers with four to five parts per whorl. Oak and maple trees are dicots, as are many shrubs.

The embryos of other angiosperms, which evolved somewhat later, have a single seed leaf and are called monocotyledons, or **monocots.** Monocot leaves typically have parallel veins and flowers with three parts per whorl. Grasses, one of the most successful of all plants, are wind-pollinated monocots. The basic characteristics of monocots and dicots are outlined in figure 14.27.

Figure 14.28 A water-dispersed fruit.
This fruit of the coconut, *Cocos nucifera,* is sprouting on a sandy beach. Coconuts, one of the most useful plants for humans in the tropics, have become established on even the most distant islands by drifting in the waves.

Improving Seed Dispersal: Fruits

Just as the mature ovules become seeds, so the mature ovary that surrounds the ovules becomes the **fruit.** A fruit is a mature ripened ovary containing fertilized seeds, surrounded by a carpel. Fruits provide angiosperms with a far better way of dispersing their progeny than simply sending their seeds off on the wind. Instead, just as in pollination, they employ animals. By making fruits fleshy and tasty to animals, angiosperms encourage animals to eat them. The seeds within the fruit are resistant to chewing and digestion. They pass out of the animal with the feces, undamaged and ready to germinate at a new location far from the parent plant.

There are three main kinds of fleshy fruits: berries, drupes, and pomes. In **berries**—examples of which are grapes, tomatoes, and dates—which are typically many-seeded, the fleshy portion of the fruit forms on the inner wall of the carpel. In **drupes**—peaches, olives, plums, and cherries—the inner layer of the fruit is stony and adheres tightly to a single seed. In **pomes**—apples and pears—the fleshy portion of the fruit comes from the petals and sepals of the flower!

Fleshy fruits usually have distinctively colored coverings, often black, bright blue, or red. These colors help attract animals. Berries, for example, are frequently dispersed by birds and sometimes by other vertebrates. Thus a bear eating blueberries is helping the blueberry plant disperse its seeds.

Other kinds of specialized fruits like burdock lack fleshy carpels and are dispersed by attaching themselves to the fur of animals. Coconuts are dispersed by water, colonizing distant islands (figure 14.28). Coconuts are drupes whose outer layer is fibrous rather than fleshy, playing a protective rather than animal-attracting role.

Many plant fruits are specialized for wind dispersal. The small, nonfleshy fruits of the dandelion, for example, have a plumelike modified calyx that lets them be carried long distances on wind currents. Many grasses have dustlike fruit, so light the wind bears them easily. Maples have long wings attached to the fruit. In tumbleweeds, the whole plant breaks off and is blown across open country by the wind, scattering seeds as it moves.

CHAPTER 14

HIGHLIGHTS

14.1 The Origin of Plants

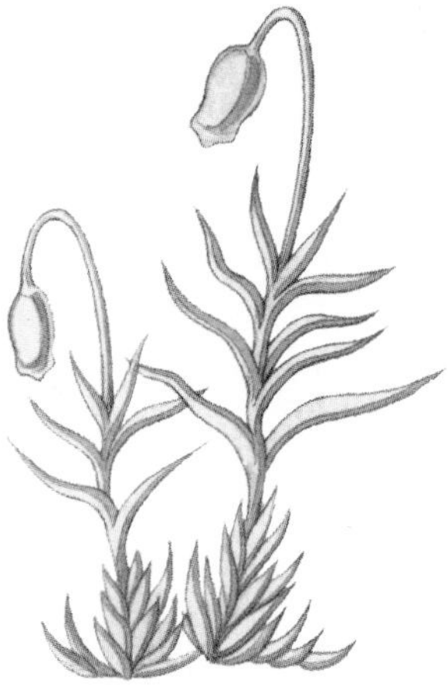

Key Terms

cuticle 266
stomata 266
alternation of generations 267
sporophyte 267
gametophyte 267

Key Concepts

- Plants faced three key challenges in adapting to life on land: absorbing minerals, conserving water, and transferring gametes during sexual reproduction.
- Early plants developed a symbiotic partnership with fungi to aid in absorbing minerals.
- Plants conserve water with a watertight outer covering called a cuticle.
- Plants transfer protected gametes, as spores or pollen.
- Plants exhibit alternation of gametophyte (haploid) and sporophyte (diploid) generations.

14.2 General Features of Vascular Plants

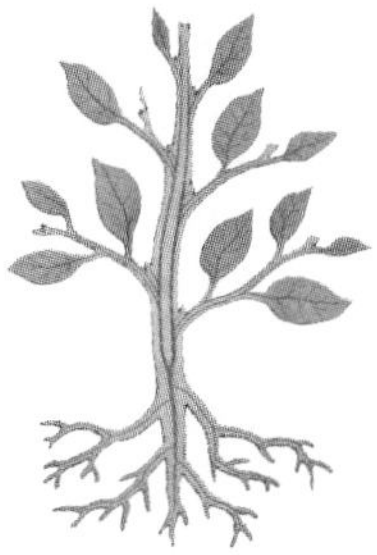

Key Terms

sieve element 272
tracheary element 272
primary growth 272
secondary growth 273

Key Concepts

- Vascular plants possess specialized water-conducting tissues.
- The life cycle of vascular plants is dominated by the sporophyte generation.
- Early vascular plants grew from their tips (primary growth), while later ones also grow in diameter (secondary growth).

14.3 The Advent of Seeds

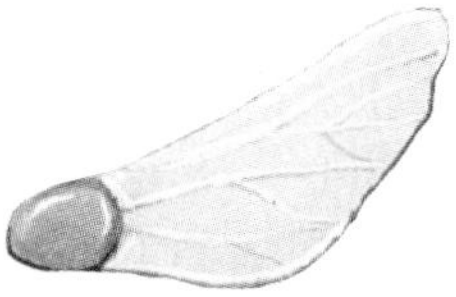

Key Terms

seed 276
gymnosperm 276
cone 278

Key Concepts

- A seed is a plant embryo with a durable, watertight cover.
- In the gymnosperms, the ovules are not completely enclosed within sporophyte tissue when pollinated.
- Seeds often contain nourishment for the germinating seedling.

14.4 The Evolution of Flowers

Key Terms

angiosperm 280
flower 280
endosperm 282
double fertilization 282
dicot 283
monocot 283
fruit 284

Key Concepts

- Angiosperm ovules are completely enclosed by the sporophyte carpel when pollinated.
- Angiosperms, or "flowering plants," are the most successful of all plants.
- Angiosperms are characterized by flowers and fruit, both of which aid dispersal of gametes.

14

CONCEPT REVIEW

1. The evolution of the cuticle in plants was a key adaptation to terrestrial living because it
   a. enhanced absorption of nutrients from soil.
   b. allowed for an alternation of generations.
   c. helped plants to avoid drying out.
   d. enhanced water loss.

2. In early plants, the gametophyte tissue is usually _________ than the sporophyte tissue.
   a. smaller
   b. larger
   c. more vascularized
   d. less visible

3. Which group of plants completely lacks a vascular system?
   a. liverworts
   b. ferns
   c. mosses
   d. cycads

4. Which of the following are found *only* in vascular plants?
   a. a waxy cuticle
   b. carbohydrate molecules
   c. sieve elements
   d. stomata

5. Growth around the plant's periphery, causing it to increase in diameter, is known as ____________ growth.
   a. primary
   b. secondary
   c. tertiary
   d. apical

6. An example of a seedless vascular plant is a
   a. hornwort.
   b. pine tree.
   c. fruit tree.
   d. club moss.

7. Conifers, ginkgoes, cycads, and gnetophytes are collectively called
   a. gymnosperms.
   b. angiosperms.
   c. bryophytes.
   d. seedless vascular plants.

8. Which of the following is *not* a part of the innermost whorl of the flower?
   a. stamens
   b. style
   c. carpels
   d. stigma

9. Which of the following is a unique characteristic of angiosperms?
   a. double fertilization
   b. seeds
   c. leaves
   d. cones

10. The life cycle of plants is characterized by an _________ of diploid and haploid generations.

11. Plant gametes fuse to form a diploid zygote that grows by mitosis to the _________ generation.

12. The oldest known vascular plant is called __________ and had branched, leafless shoots that produced spores.

13. The region of primary growth at the tip of a vascular plant is called the _________.

14. Unlike the gymnosperms, the ovules of angiosperms are completely enclosed by the _________ at the time of pollination.

15. A sperm is haploid. Endosperm is _________.

16. _________ have one cotyledon, parallel veins, and three flower parts per whorl.

17. _________ improve dispersal of seeds in angiosperms because they are fleshy and tasty to animals.

Answers to the Concept Review questions appear in Appendix B.

CHALLENGE YOURSELF

1. Compare and contrast the adaptations to a terrestrial existence of fungi versus plants. Which group of organisms has been more successful, and why?
2. Compare and contrast the alternation of gametophyte and sporophyte generations in mosses, ferns, pines, and flowering plants. Which generation is dominant in each case, if any? Is it haploid or diploid? Which plant groups are more advanced? Are there advantages to having the gametophyte or sporophyte generation dominate?
3. Why do mosses and ferns both require free water to complete their life cycles? At what stage of the life cycle is water required? Do angiosperms also require free water to complete their life cycles? What are the reasons for the difference, if any?
4. Why was the development of seeds so important to the success of angiosperms? How did these characteristics help angiosperms become the dominant photosynthetic organisms on land? In what terrestrial communities are angiosperms *not* dominant? What plant groups are predominant in those areas? Why?

FOR FURTHER READING

Gensel, P. G., and H. N. Andrews. "The Evolution of Early Land Plants." *American Scientist* 75 (1987): 478–89. Excellent, well-illustrated account of the earliest plants.

Hazen-Hammond, S. "The Saguaro Survives: 11,000 Years and Still Growing." *Smithsonian* (January 1996): 76–83. Reports of the death of this charismatic megacactus are premature, despite damage from weather, animals, and vandals.

Heywood, V. H., ed. *Flowering Plants of the World*. Englewood Cliffs, N.J.: Prentice Hall, 1985. An outstanding guide to the families of flowering plants, with excellent illustrations and diagrams.

Mooney, H. A., et al. "Plant Physiological Ecology Today." *BioScience* 37 (1988): 18–67. Nearly an entire issue of *BioScience* devoted to the ways in which plants cope with their environments and scientists study them. Highly recommended.

Norstog, K. "Cycads and the Origin of Insect Pollination." *American Scientist* 75 (1987): 270–79. Studies of one of the most ancient lines of living seed plants suggest that insect pollination was established before the origin of angiosperms.

Raven, R. H., R. F. Evert, and S. E. Eichhorn. *Biology of Plants*. 5th ed. New York: Worth Publishers, 1993. A comprehensive treatment of general botany; a standard text in the field.

Stewart, D. "Green Giants." *Discover*, (April 1990): 60–64. Sequoias are more than majestic trees; they're triumphs of hydraulics, wind resistance, and architecture.

Tyrrell, E. Q., and R. A. Tyrrell. *Hummingbirds: Their Life and Behavior: A Photographic Study of the North American Species*. New York: Crown Publishers, 1985. A magnificently illustrated account of the most highly specialized birds that visit flowers.

TECHNOLOGY LINKS

The Living World Home Page
http://www.wcbp.com/biology/tlw

Life Science Animations
Videotape 5
#46 Journey into a Leaf
#49 How Leaves Change Color and Drop in Fall

15

The Living Plant

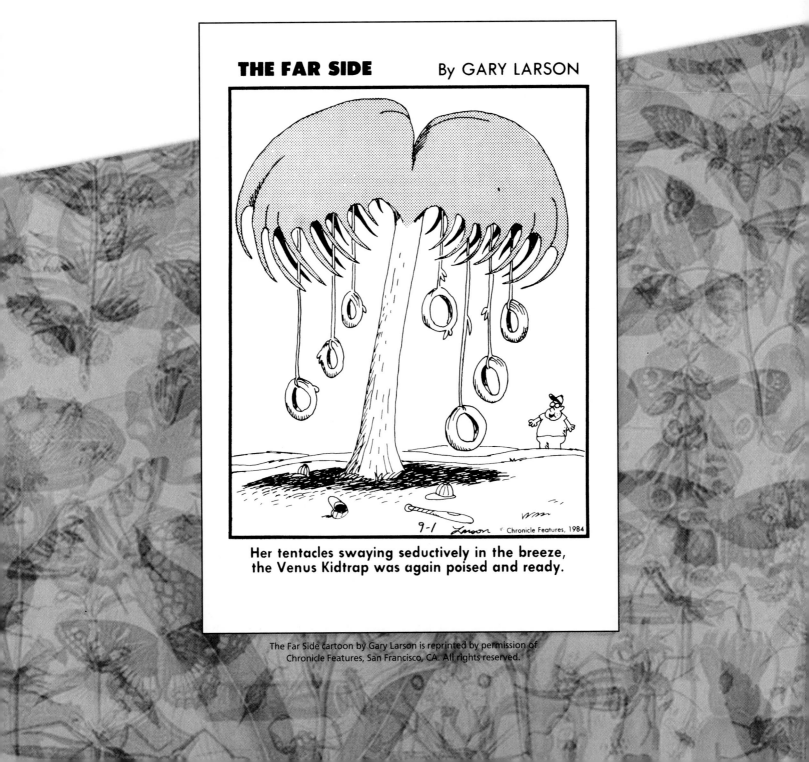

Her tentacles swaying seductively in the breeze, the Venus Kidtrap was again poised and ready.

CHAPTER OUTLINE

Figure 15.1 A plant vibrantly alive.
Each leaf of this plant is covered by a watertight waxy cuticle, causing the water to bead as if on the surface of a freshly waxed car. Light shines through the cuticle and a thin layer of epidermal cells, driving photosynthesis in the chloroplast-containing cells beneath.

At first glance, a plant may not appear vibrantly alive. It does not bound from one place to another like a gazelle, nor does it growl or respond to a caress. But its quiet appearance is deceiving, and its internal structure more complex than you might suspect. Each of the leaves you see in figure 15.1 is in fact a complex structure built of many tissues. This chapter focuses on three facets of the life of plants: (1) how tissues form the structure of the plant body; (2) how this body carries out the essential functions of nutrition and transport; and (3) how flowering plants reproduce.

15.1 The Structure and Function of Plant Tissues

The plants that cover the earth's surface in such bewildering variety all possess the same fundamental architecture. This section is devoted to outlining the basic structure of plants and to analyzing how their aboveground parts and below ground parts differ in growth and form. All parts of plants have an outer covering of protective tissue and an inner matrix of relatively unspecialized tissue, within which is embedded vascular tissue that conducts water, nutrients, and food throughout the plant. Plants cannot move, but they adjust to their environment by growing and changing their form. Although the similarities between a cactus, an orchid, and a pine tree may not at first be obvious, plants have a fundamental unity of structure that is reflected in the construction plan of their respective bodies; in the way they grow, produce, and transport their food; and in how they regulate their development.

Basic Organization of a Vascular Plant

The cells and tissues of vascular plants, the ways in which they grow and develop, and how the body these tissues form carries out the functions of living are the focus of this chapter. We discuss the fundamental differences between the belowground portions of the plant—the roots—and the aboveground portions—the shoot—as well as the structural and functional relationships between them.

A vascular plant is organized along a vertical axis, like a pipe. The part belowground is called the **root;** the part aboveground is called the **shoot** (figure 15.2). Although roots and shoots differ in their basic structure, growth at the tips throughout the life of the individual is characteristic of both. The root penetrates the soil and absorbs water and various ions, which are crucial for plant nutrition. It also anchors the plant. The shoot consists of stem and leaves. The **stem** serves as a framework for the positioning of the **leaves,** where most photosynthesis takes place. The arrangement, size, and other characteristics of the leaves are critically important in the plant's production of food.

A cactus, a rose, and a pine tree all share this same basic structure, although the similarity might not be obvious at first glance. This fundamental unity in structure is reflected not only in the basic organization of the plant body but in the tissues used to construct it, the way it produces and transports food, and the means by which it reproduces, grows, and develops. First, we focus on the tissues, the basic building blocks of the plant body.

Tissue Types in Plants

The organs of a plant—the leaves, stem, and roots—are composed of different combinations of tissues, just as your legs are composed of bone, muscle, and connective tissue. A tissue is a group of similar cells—cells that are specialized in the same way—organized into a structural and functional unit. Most plants have three major tissue types: (1) *ground tissue,* in which the vascular tissue is embedded; (2) *dermal tissue,* the outer protective covering of the plant; and (3) *vascular tissue,* which conducts water and dissolved minerals up the plant and conducts the products of photosynthesis throughout.

Each major tissue type is composed of distinctive kinds of cells, whose structures are related to the functions of the tissues in which they occur. For example, vascular tissue is composed of *xylem,* which conducts water and dissolved minerals, and *phloem,* which conducts carbohydrates (mostly sucrose), which the plant uses as food.

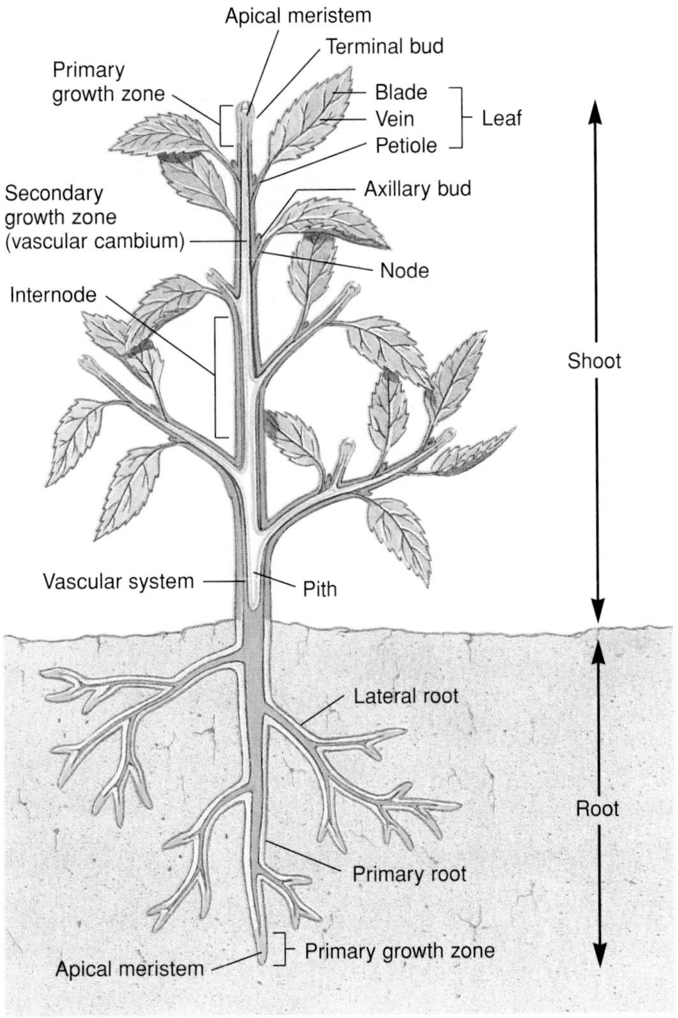

Figure 15.2 The body of a plant.
The plant body consists of an aboveground portion called the shoot (stems and leaves) and a belowground portion called the roots. Elongation of the plant, so-called primary growth, takes place when clusters of cells called the apical meristems divide at the ends of the roots and the stems. Thickening of the plant, so-called secondary growth, takes place in a sheath around the plant, allowing the plant to enlarge in girth like letting out a belt. The yellowish-green areas of the diagram are zones of active elongation.

Plant Cell Types

Meristems

Animals grow all over. As children grow into adults, their torsos grow at the same time their legs do. If, instead, children grew in only one place, with their legs getting longer and longer, they would be growing in a way similar to the way plants grow.

Plants contain growth zones of unspecialized cells called **meristems,** whose only function is to divide. Every time one of these cells divides, one of its two offspring remains in the meristem, whereas the other differentiates into one of the three kinds of plant tissue and ultimately becomes part of the plant body.

In plants, **primary growth** is initiated at the tips by the **apical meristems,** regions of active cell division that occur at the tips of roots and shoots (see figure 15.2). The growth of these meristems results primarily in the extension of the plant body. As it elongates, it forms what is known as the primary plant body, which is made up of the primary tissues.

Growth in thickness, **secondary growth,** involves the activity of the **lateral meristems,** which are cylinders of meristematic tissue. The continued division of their cells results primarily in the thickening of the plant body (see figure 15.2). There are two kinds of lateral meristems: the **vascular cambium,** which gives rise to ultimately thick accumulations of secondary xylem and phloem, and the **cork cambium,** from which arise the outer layers of bark on both roots and shoots.

Ground Tissue

Parenchyma cells are the least specialized and the most common of all plant cell types (figure 15.3); they form masses in leaves, stems, and roots. Parenchyma cells, unlike some other cell types, are characteristically alive at maturity, with fully functional cytoplasm and a nucleus. Most parenchyma cells have only primary cell walls, which are mostly cellulose that is laid down while the cells are still growing.

Collenchyma cells, which are also living at maturity, form strands or continuous cylinders beneath the epidermis of stems or leaf stalks and along veins in leaves. They are usually elongated, with unevenly thickened primary walls, which are their distinguishing feature. Strands of collenchyma provide much of the support for plant organs in which secondary growth has not occurred (figure 15.4).

In contrast to parenchyma and collenchyma cells, **sclerenchyma cells** have tough, thick secondary walls; they usually do not contain living cytoplasm when mature. There are two types of sclerenchyma: **fibers,** which are long, slender cells that usually form strands, and **sclereids,** which are variable in shape but often branched (figure 15.5). Sclereids are sometimes called stone cells because they make up the bulk of the stones of peaches and other "stone" fruits, as well as that of nut shells. Both fibers and sclereids are tough and thick-walled and strengthen the tissues in which they occur.

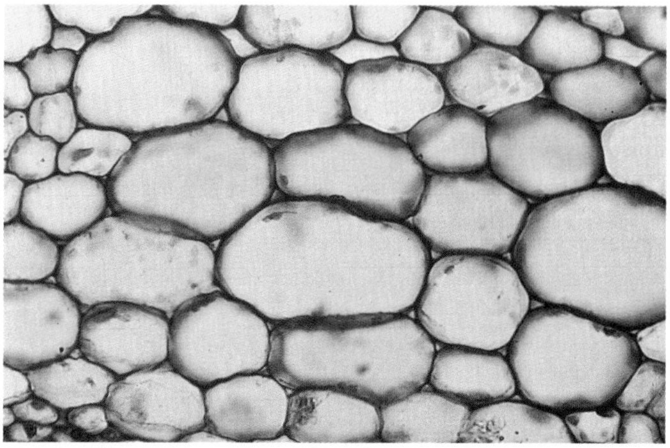

Figure 15.3 Parenchyma cells.
Cross section of parenchyma cells from grass. Only primary cell walls are seen in this living tissue.

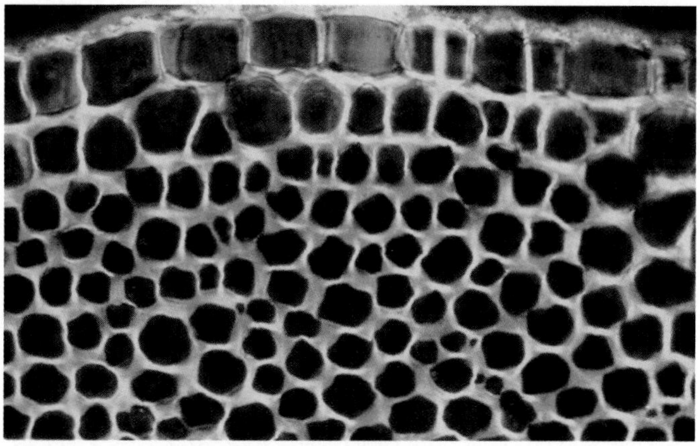

Figure 15.4 Collenchyma cells.
Cross section of collenchyma cells, with thickened side walls, from a young branch of elderberry (*Sambucus*). In other kinds of collenchyma cells, the thickened areas may occur at the corners of the cells or in other kinds of strips.

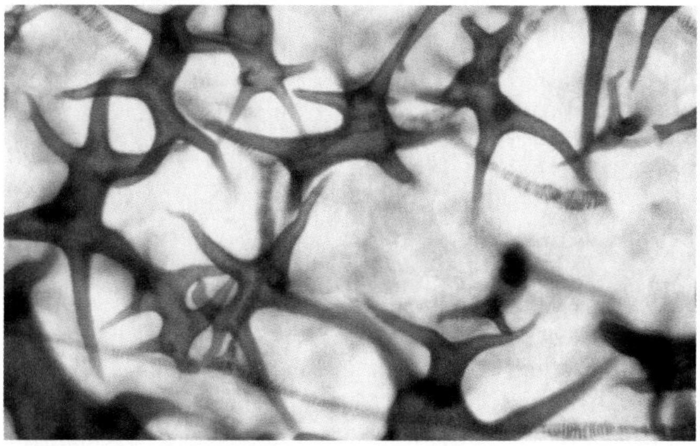

Figure 15.5 Sclereids.
Clusters of sclereids ("stone cells"), stained blue in this preparation, in the pulp of a pear. Such clusters of sclereids give pears their gritty texture.

Dermal Tissue

All parts of the outer layer of a primary plant body are covered by flattened epidermal cells, which are often covered with a thick, waxy layer called the **cuticle.** These are the most abundant cells in the plant epidermis, or skin. They protect the plant and provide an effective barrier against water loss. One type of specialized cell that occurs among the epidermal cells is the guard cell.

Guard cells are paired cells. With the opening that lies between them, they make up the **stomata** (singular, stoma), which occur frequently in the epidermis of leaves and occasionally on other parts of the shoot, such as on stems or fruits (figure 15.6). Oxygen, carbon dioxide, and water pass into and out of the leaves almost exclusively through the stomata, which open and shut in response to such external factors as supply of moisture and light.

Trichomes are outgrowths of the epidermis that occur on the shoot, occurring on the surfaces of stems and leaves. Trichomes vary greatly in form in different kinds of plants. A "fuzzy" or "woolly" leaf is covered with trichomes, which when viewed under the microscope look like a thicket of fibers. Trichomes play an important role in regulating the heat and water balance of the leaf. Much as the hairs of a fur coat provide insulation, so too the trichomes slow heat and water vapor loss from the surface of the leaf. Other kinds of trichomes are glandular, often secreting sticky or toxic substances that may deter potential herbivores.

Other outgrowths of the epidermis occur belowground, on the surface of roots near their tips. Called **root hairs,** these extensions are single cells that keep the root in intimate contact with the particles of soil. Root hairs play an important role in the absorption of water and minerals from the soil.

Vascular Tissue

Vascular plants contain two kinds of conducting or vascular tissue: the xylem and the phloem. **Xylem** is the plant's principal water-conducting tissue, forming a continuous system that runs throughout the plant body. Within this system, water (and the minerals dissolved in the water) passes from the roots up through the shoot in an unbroken stream. When water reaches the leaves, much of it passes into the air as water vapor, through the stomata.

The two principal types of conducting elements in the xylem are **tracheids** and **vessel elements,** both of which have thick secondary walls, are elongated, and have no living cytoplasm at maturity. In conducting elements composed of tracheids, water flows from tracheid to tracheid through openings called pits in the secondary walls. In contrast, vessel elements have not only pits but also definite openings, or perforations, in their end walls by which they are linked together and through which water flows. A linked row of vessel elements forms a vessel (figure 15.7). Primitive angiosperms have only tracheids, but the majority of angiosperms have vessels. Vessels conduct water much more efficiently than do strands of tracheids. In addition to conducting cells, xylem likewise includes fibers and parenchyma cells.

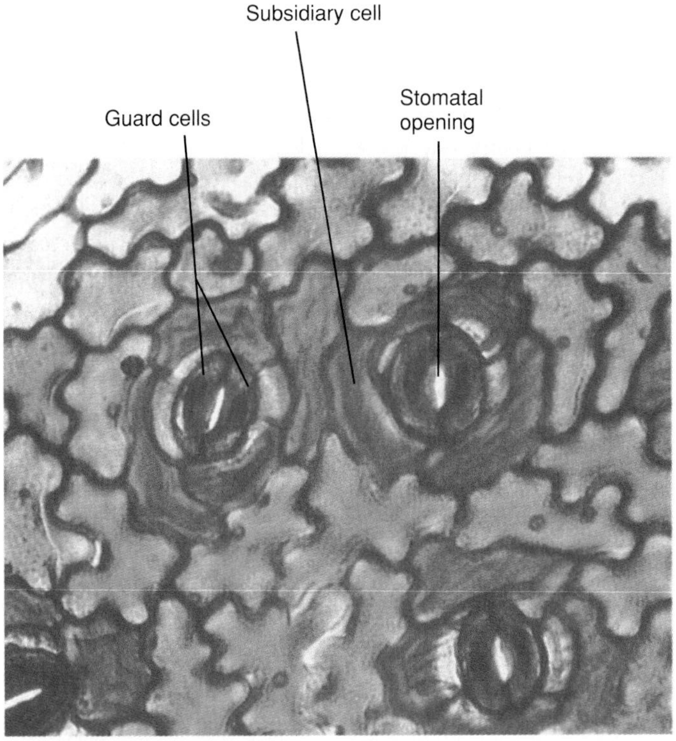

Figure 15.6 Guard cells.
Stomata occur frequently among the epidermal cells of this member of the aralia family (Araliaceae). The epidermis has been peeled off the leaf and stained with a red dye.

Phloem is the principal nutrient-conducting tissue in vascular plants. Different kinds of plants have one of two different kinds of phloem cells: **sieve-tube members** or **sieve cells.** Clusters of pores known as sieve areas occur on both kinds of cells and connect the cytoplasms of adjoining sieve cells and sieve-tube members. Both cell types are living, but their nuclei are lost during maturation.

Angiosperms contain sieve-tube members. In sieve-tube members, the pores in some of the sieve areas are larger than those in others; such sieve areas are called sieve plates. Sieve-tube members occur end to end, forming longitudinal series called sieve tubes. Specialized parenchyma cells known as **companion cells** occur regularly in association with sieve-tube members (figure 15.8). Other vascular plants contain sieve cells. In these, the pores in all of the sieve areas are roughly the same diameter. Sieve cells do not have companion cells. In an evolutionary sense, sieve-tube members clearly are advanced over sieve cells because they are more specialized and presumably more efficient.

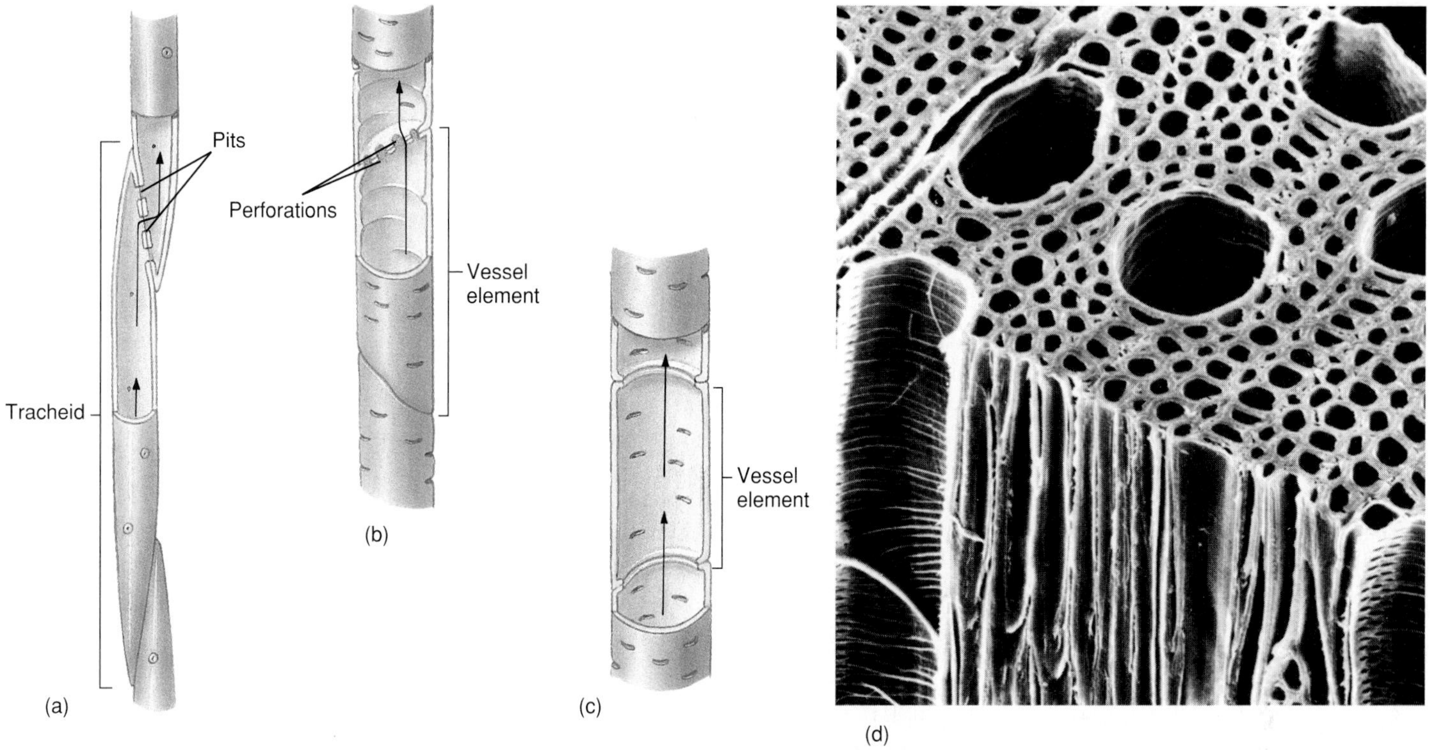

Figure 15.7 Comparison of vessel elements and tracheids.
(*a*) In tracheids the water passes from cell to cell by means of pits. (*b*) In vessel elements, water moves by way of perforations, which may be simple or interrupted by bars. (c) Open-ended vessel elements. (*d*) A scanning electron micrograph of the red maple (*Acer rubrum*), showing the xylem.

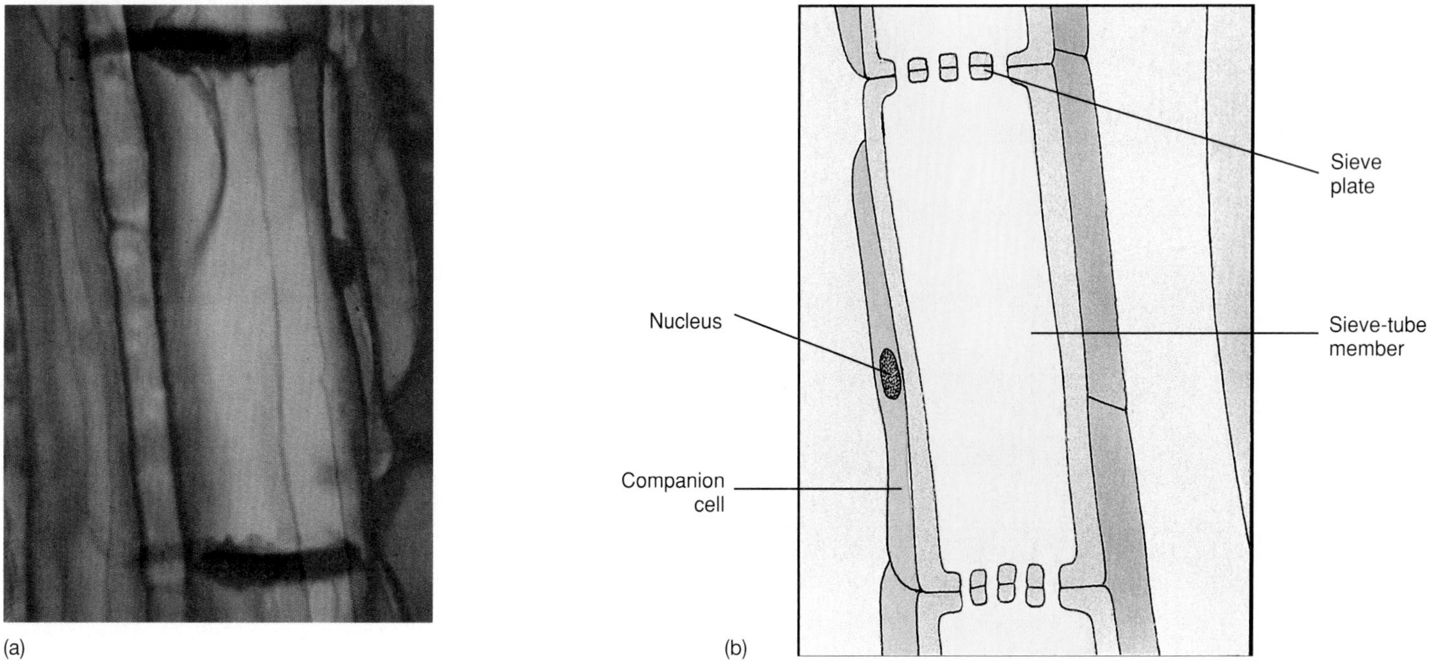

Figure 15.8 Sieve tubes.
(*a*) Sieve-tube member from the phloem of squash (*Cucurbita*), connected with the cells above and below to form a sieve tube. (*b*) In this diagram of (*a*), note the thickened end walls, which are at right angles to the sieve tube. The narrow cell with the nucleus at the left of the sieve-tube member is a companion cell.

The Upper Plant Body: The Shoot

Technically, a plant shoot is that portion of the plant body that lies above the cotyledons, and the plant root is the portion below the cotyledons. In most plants, all of the aboveground parts, including the stem and the leaves, are portions of the shoot. Leaves are usually the most prominent shoot organs and are structurally diverse (figure 15.9).

Leaves

Leaves, outgrowths of the shoot apex, are the light-capturing organs of most plants. Most of the chloroplast-containing cells of a plant are within its leaves, and it is there that the bulk of photosynthesis occurs. The only exceptions to this are found in some plants, such as cacti, whose green stems have largely taken over the function of photosynthesis for the plant. Photosynthesis is conducted mainly by the "greener" parts of plants because they contain more chlorophyll, the most efficient photosynthetic pigment (figure 15.10).

The apical meristems of stems and roots are capable of growing indefinitely under appropriate conditions. Leaves, in contrast, grow by means of **marginal meristems,** which flank their thick central portions. These marginal meristems grow outward and ultimately form the **blade** (flattened portion) of the leaf, while the central portion becomes the midrib. Once a leaf is fully expanded, its marginal meristems cease to function.

In addition to the flattened blade, most leaves have a slender stalk, the **petiole.** Two leaflike organs, the **stipules,** may flank the base of the petiole where it joins the stem. Veins, consisting of both xylem and phloem, run through the leaves. As mentioned in chapter 14, in most dicots, the pattern is net or reticulate venation; in many monocots, the veins are parallel.

A typical leaf contains masses of parenchyma, called **mesophyll** ("middle leaf"), through which the vascular bundles, or veins, run. Beneath the upper epidermis of a leaf are one or more layers of closely packed, columnlike parenchyma cells called **palisade parenchyma.** The rest of the leaf interior, except for the veins, consists of a tissue called **spongy parenchyma** (figure 15.11). Between the spongy parenchyma cells are large intercellular spaces that function in gas exchange and particularly in the passage of carbon dioxide from the atmosphere to the mesophyll cells. These intercellular spaces are connected, directly or indirectly, with the stomata.

Figure 15.9 Leaves.

Angiosperm leaves are stunningly variable. (*a*) Diverse leaves in the herb layer of a Costa Rican rain forest. (*b*) A *compound leaf,* marijuana (*Cannabis sativa*). Such a compound leaf is composed of several leaflets. (*c*) A *simple leaf,* its margin deeply lobed, from the tulip tree (*Liriodendron tulipifera*). A simple leaf has only one leaf element. (*d*) Another compound leaf, from a member of the legume family in the lowland forest of Peru. (*e*) Many unusual arrangements of leaves occur in different kinds of plants. For example, in this miner's lettuce (*Claytonia perfoliata*), an herb of the Pacific states, two leaves are completely fused below each of the clusters of flowers, which seem, therefore, to arise from the center of a single leaf.

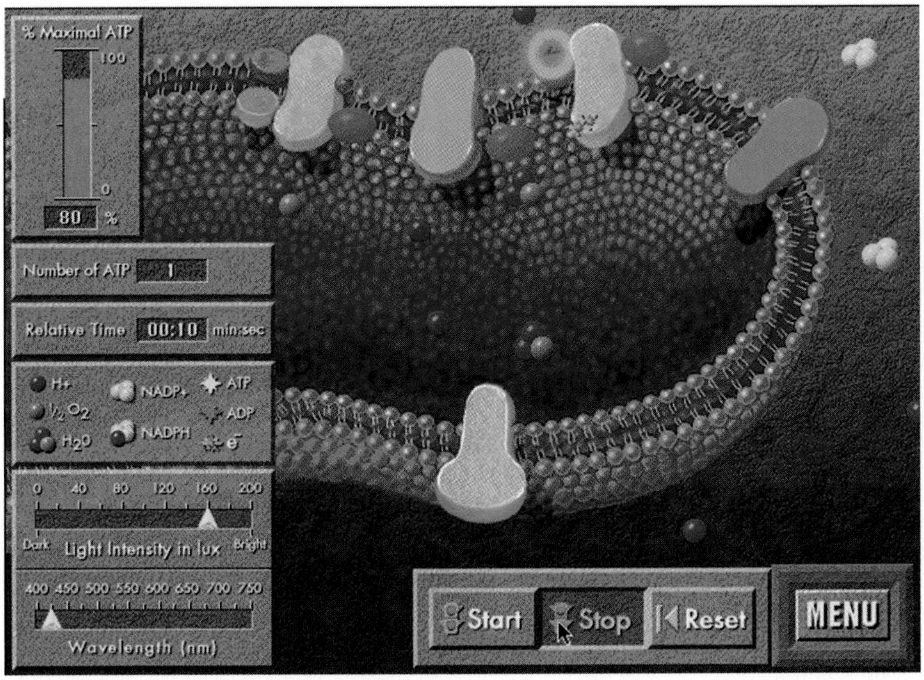

Figure 15.10 Exploring photosynthesis.
This screen capture from an interactive CD-ROM exercise illustrates how the wavelength of light and its intensity affect the output of photosynthesis. Constructing an action spectrum and investigating how much light is enough to optimize photosynthetic yield shows how surprisingly little light is required to drive the photosynthetic machinery at full efficiency. (*Explorations in Cell Biology & Genetics,* Module 9, "Photosynthesis")

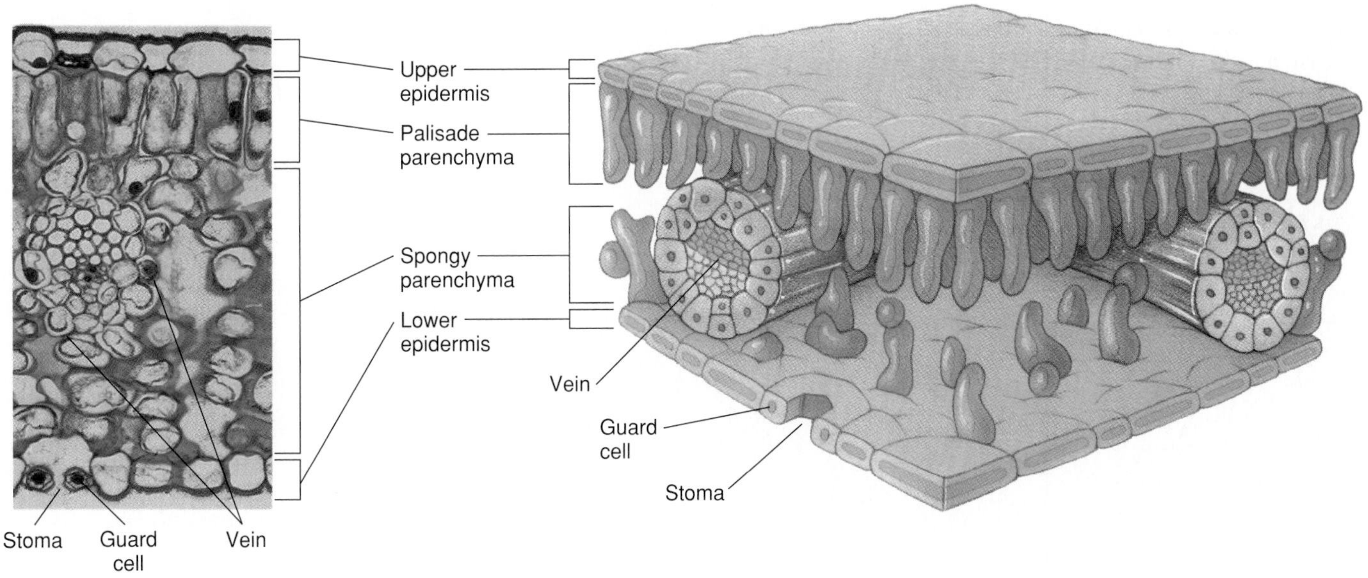

Figure 15.11 A leaf in cross section.
Transection of a lily leaf, showing the arrangement of palisade and spongy parenchyma; a vascular bundle or vein; and the epidermis, with paired guard cells flanking the stoma.

Stems

As mentioned earlier, the stem is that part of the shoot that serves as the framework for the positioning of the leaves. It experiences both primary and secondary growth and also is the source of an economically important plant product—wood.

Primary Growth In the primary growth of a shoot, leaves first appear as leaf primordia (singular, primordium), or rudimentary young leaves, which cluster around the apical meristem, unfolding and growing as the stem itself elongates (figure 15.12). The places on the stem at which leaves form are called nodes. The portions of the stem between these attachment points are called the internodes. As the leaves expand to maturity, a bud, a tiny, undeveloped side shoot, develops in the **axil** of each leaf, the angle between a branch or leaf and the stem from which it arises. These buds, which have their own leaves, may elongate and form lateral branches, or they may remain small and dormant. A hormone diffusing downward from the terminal bud of the shoot continuously suppresses the expansion of the lateral buds. These buds begin to expand when the terminal bud is removed. Therefore, gardeners who wish to produce bushy plants or dense hedges crop off the tops of the plants, thus removing their terminal buds.

Within the soft, young stems, the strands of vascular tissue, xylem and phloem, either occur as a cylinder in the outer portion of the stem, as is common in dicots, or are scattered through it, as is common in monocots (figure 15.13). The vascular bundles contain both primary xylem and primary phloem. At the stage when only primary growth has occurred, the inner portion of the ground tissue of a stem is called the **pith,** and the outer portion is the **cortex.**

Figure 15.12 Where leaves originate.
Scanning electron micrograph of the shoot apex of a silver maple, *Acer saccharinum,* showing a developing shoot during summer, the season of active growth. The apical meristem and leaf primordia are plainly visible.

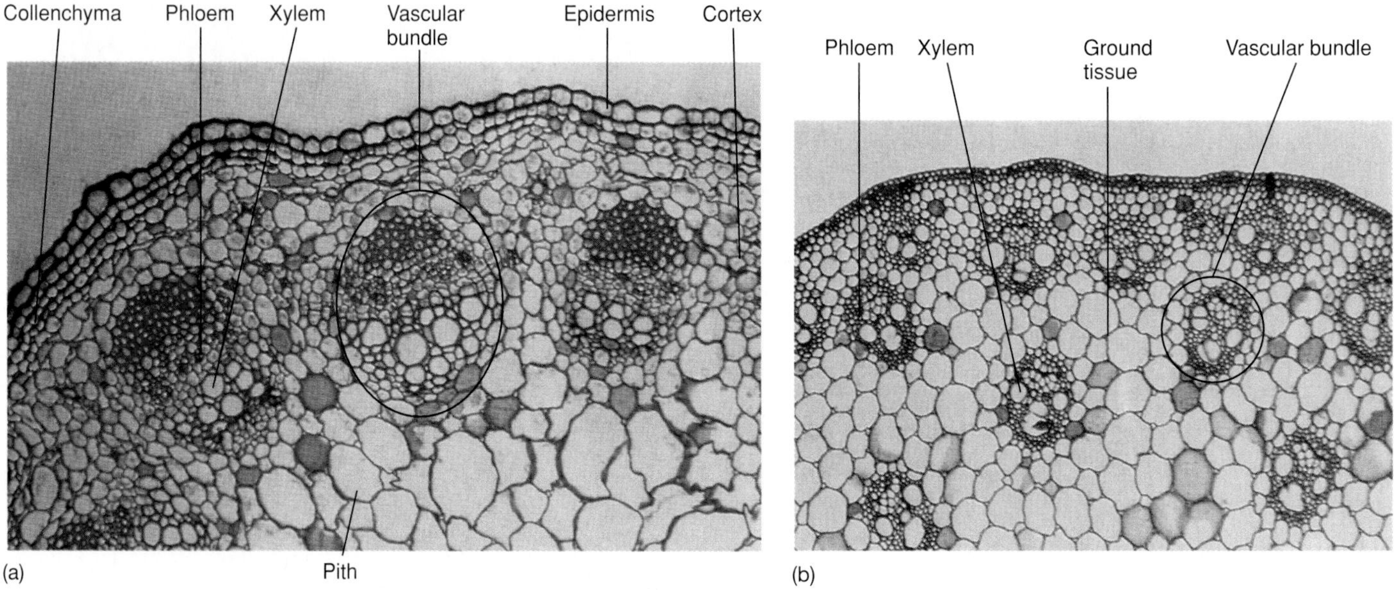

Figure 15.13 A comparison of dicot and monocot stems.
(*a*) Transection of a young stem of a dicot, the common sunflower, *Helianthus annuus,* in which the vascular bundles are arranged around the outside of the stem. (*b*) Transection of the monocot corn, *Zea mays,* with the scattered vascular bundles characteristic of the class.

Secondary Growth In stems, secondary growth is initiated by the differentiation of the **vascular cambium,** which consists of a thin cylinder of actively dividing cells located between the bark and the main stem in mature woody plants. The vascular cambium develops from parenchyma cells within the vascular bundles of the stem, between the primary xylem and the primary phloem (figure 15.14). The cylindrical form of the vascular cambium is completed by the differentiation of some of the parenchyma cells that lie between the bundles. Once established, the vascular cambium consists of elongated, somewhat flattened cells with large vacuoles. The cells that divide from the vascular cambium outwardly, toward the bark, become secondary phloem; those that divide from it inwardly become secondary xylem. The cells of the vascular cambium also divide laterally (side by side), allowing the stem to thicken as the tree or shrub becomes more mature.

While the vascular cambium is becoming established, a second kind of lateral cambium, the cork cambium, develops in the stem's outer layers. The cork cambium usually consists of plates of dividing cells that move deeper and deeper into the stem as they divide. Outwardly, the cork cambium splits off densely packed **cork cells;** they contain a fatty substance and are nearly impermeable to water. Cork cells are dead at maturity. Inwardly, the cork cambium divides to produce a dense layer of parenchyma cells. The cork, the cork cambium that produces it, and this dense layer of parenchyma cells make up a layer called the **periderm,** which is the plant's outer protective covering.

Cork, which covers the surfaces of mature stems or roots, takes the place of the epidermis, which performs a similar function in the younger parts of the plant. The term **bark** refers to all of the tissues of a mature stem or root outside of the vascular cambium. Because the vascular cambium has the thinnest-walled cells that occur anywhere in a secondary plant body, it is the layer at which bark breaks away from the accumulated secondary xylem.

Wood is one of the most useful, economically important, and beautiful products obtained from plants. Anatomically, wood is accumulated secondary xylem. As the secondary xylem ages, its cells become infiltrated with gums and resins, and the wood becomes darker. For this reason, the wood located nearer the central regions of a given trunk, called heartwood, is often darker and denser than the wood nearer the vascular cambium, called sapwood, which is still actively involved in transport within the plant. Because of the way it is accumulated, wood often displays rings. In temperate regions, these rings are annual rings (figure 15.15). In each ring, the light zone is formed of larger cells that form during the spring and summer, when water is plentiful and temperatures warm; the dark zone is denser, formed of thick-walled cells in the fall. The abrupt discontinuity between the layers marks winter, when growth of the dark zone ceases, to be followed in the spring by a new light zone.

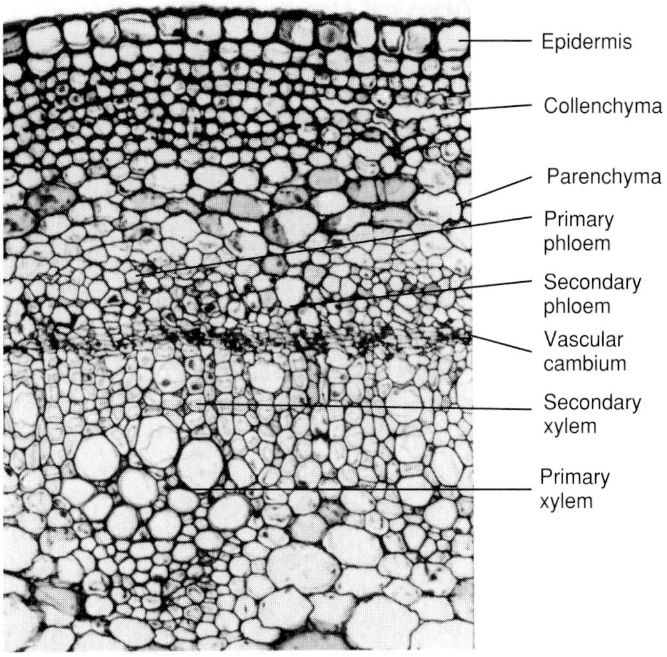

Figure 15.14 Vascular cambium.
Early stage in the differentiation of the vascular cambium in an elderberry stem (*Sambucus canadensis*).

Figure 15.15 Annual rings in a section of pine.
The rings that you see in this section of pine reflect the fact that the vascular cambium of trees divides more actively in the spring and summer, when water is plentiful and temperatures are suitable for growth, than in the fall and winter, when water is scarce and the weather is cold. As a result, layers of larger, thinner-walled cells formed during the growing season alternate with the smaller layers formed during the rest of the year. A count of such annual rings in a tree trunk can be used to calculate the tree's age.

The Lower Plant Body: The Roots

Roots have simpler patterns of organization and development than do stems. Although different patterns exist, the kind of root described here and shown in figure 15.16 is found in many dicots. Roots contain xylem and phloem, just as stems do, but there is no pith in the center of the vascular tissue in most dicot roots. Instead, these roots have a central column of xylem with radiating arms. Alternating with the radiating arms of xylem are strands of primary phloem. Surrounding the column of vascular tissue and forming its outer boundary is a cylinder of cells one or more cell layers thick called the **pericycle.** Branch, or lateral, roots are formed from cells of the pericycle. The outer layer of the root, as in the shoot, is the epidermis. The mass of parenchyma in which the root's vascular tissue is located is the cortex. Its innermost layer—the endodermis—consists of specialized cells that regulate the flow of water between the vascular tissues and the root's outer portion (figure 15.17). The endodermis lies just outside of the pericycle. Endodermis cells are surrounded by a thickened, waxy band called the **Casparian strip.** By the differential passage of minerals and nutrients through endodermis cells, the plant regulates its supply of minerals.

The apical meristem of the root divides and produces cells both inwardly, back toward the body of the plant, and outwardly. Outward cell division results in the formation of a thimblelike mass of relatively unorganized cells, the **root cap,** which covers and protects the root's apical meristem as it grows through the soil.

The root elongates relatively rapidly just behind its tip. Abundant **root hairs,** extensions of single epidermal cells, form above that zone. Virtually all water and minerals are absorbed from the soil through the root hairs, which greatly increase the root's surface area and absorptive powers. In plants with mycorrhizae (see chapter 13), the root hairs are often greatly reduced in number, and the fungal filaments of the mycorrhizae play a role similar to that of the root hairs.

One of the fundamental differences between roots and shoots has to do with the nature of their branching. In stems, branching occurs from buds on the stem surface; in roots, branching is initiated well back of the root apex as a result of cell divisions in the pericycle. The lateral root primordia grow out through the cortex toward the surface of the root, eventually breaking through and becoming established as lateral roots (figure 15.18). Secondary growth in roots, both main roots and laterals, is similar to that in stems.

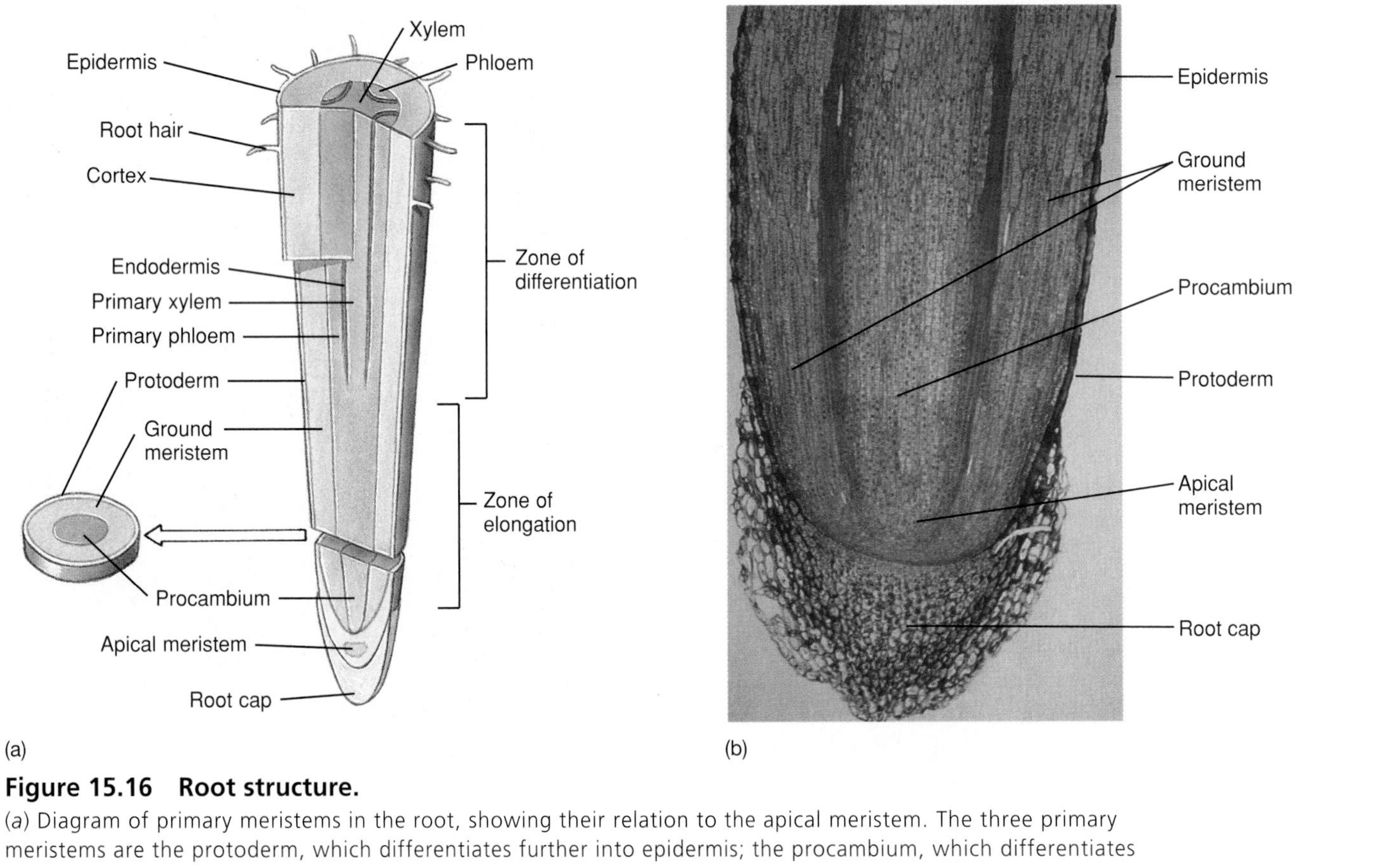

Figure 15.16 Root structure.

(*a*) Diagram of primary meristems in the root, showing their relation to the apical meristem. The three primary meristems are the protoderm, which differentiates further into epidermis; the procambium, which differentiates further into primary vascular strands; and the ground meristem, which differentiates further into ground tissue. (*b*) Median longitudinal section of a root tip in corn, *Zea mays,* showing the differentiation of protoderm, procambium, and ground meristem.

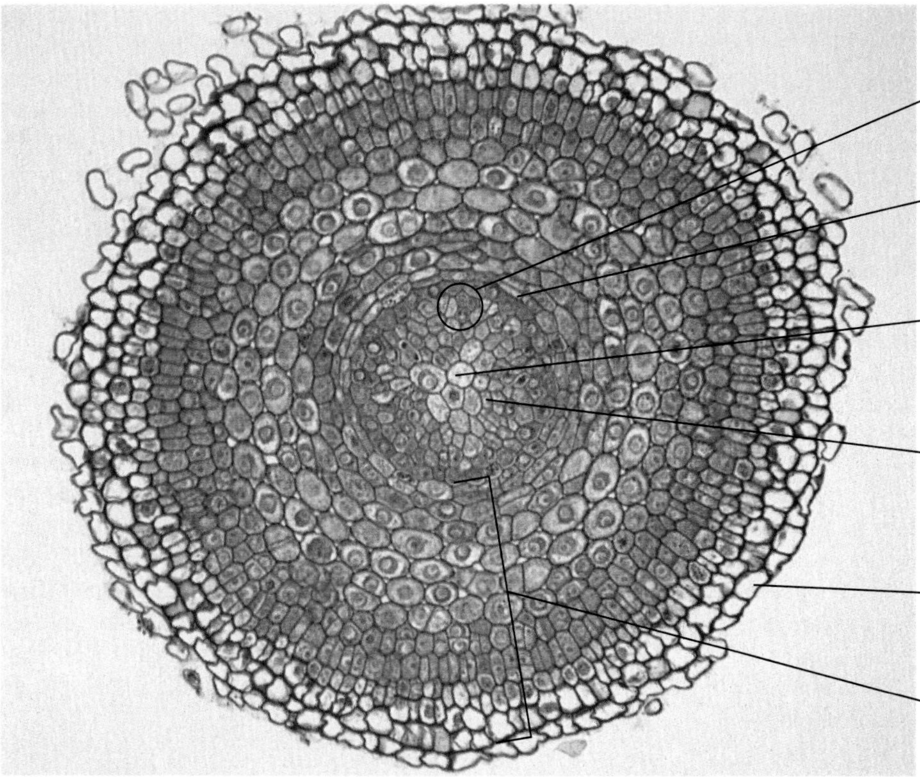

Figure 15.17 A root cross section.
Cross section through a root of a buttercup, *Ranunculus californicus*.

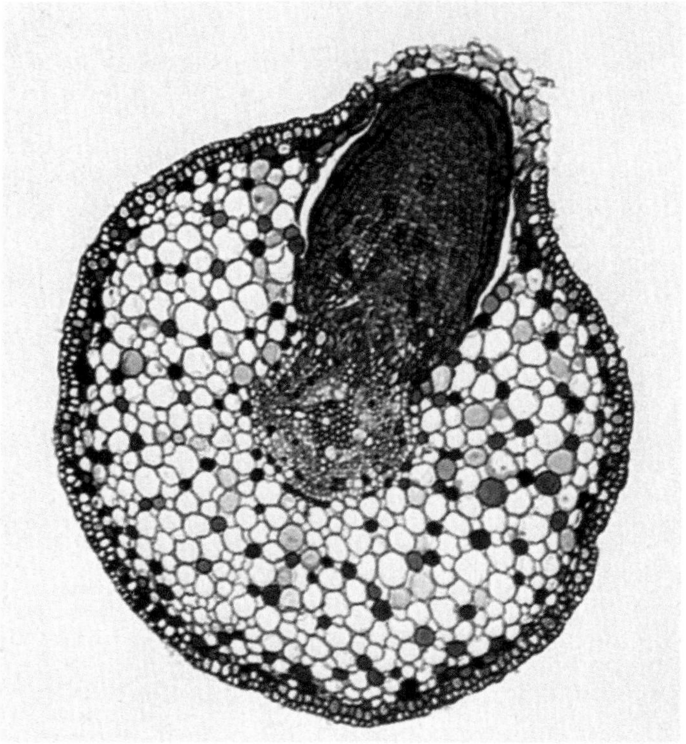

Figure 15.18 Lateral roots.
A lateral root growing out through the cortex of the black willow, *Salix nigra*. Lateral roots originate beneath the surface of the main root, whereas lateral stems originate at the surface.

15.2 Plant Nutrition and Transport

Plants have a conducting system, as humans do, for transporting fluids and nutrients from one part to another and special organs for reproduction and the gathering of energy. And, like humans, plants regulate their growth and organ functioning with hormones, chemicals that act as messengers to coordinate the many activities of the plant body. This section focuses on the internal activities of living plants: (1) the movement of fluids and nutrients within plants; (2) the hormones that control how plants grow; and (3) plant responses to environmental stimuli, such as light, gravity, and touch.

Water Movement

Functionally, a plant is essentially a tube with its base embedded in the ground. At the base of the tube are roots, and at its top are leaves. For a plant to function, two kinds of transport processes must occur: first, the carbohydrate molecules produced in the leaves by photosynthesis must be carried to all of the other living plant cells. To accomplish this, liquid, with these carbohydrate molecules dissolved in it, must move both up and down the tube. Second, nutrients and water in the ground must be taken up by the roots and ferried to the leaves and other plant cells. In this process, liquid moves up the tube. Plants accomplish these two processes by using chains of specialized cells: those of the phloem transport photosynthetically produced carbohydrates up and down the tube, and those of the xylem carry water and minerals upward.

Cohesion-Adhesion-Tension Theory

Many of the leaves of a large tree may be more than 10 stories off the ground. How does a tree manage to raise water so high? If a long, hollow tube, closed at one end, is filled with water and placed, open end down, in a full bucket of water, gravity acts (pushes) on the column of air over the bucket. The weight of the air (at sea level) exerts an amount of pressure that is defined as 1 atmosphere downward on the water in the bucket and thus presses the water up into the tube. But gravity also acts to pull the water within the tube down. The interaction of these two forces determines the water level in the tube. At sea level, the water rises to about 10.4 meters. If the tube is any higher, a vacuum forms in the upper, closed end of the tube and fills with water vapor.

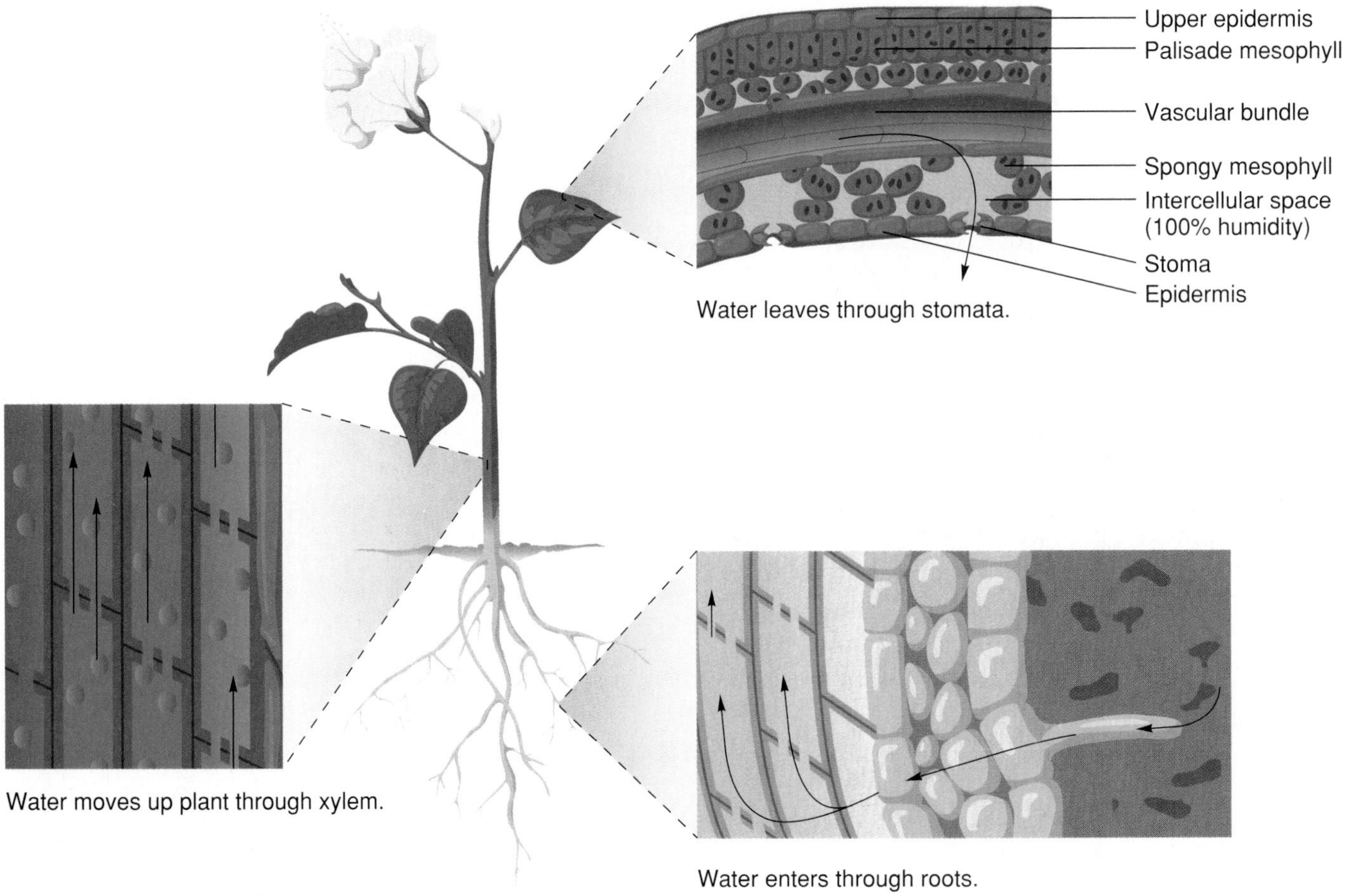

Figure 15.19 Transpiration.
The movement of water upward in the xylem, combined with the entrance of water through the roots, causes water to leave through the stomata in the leaves.

Opening the tube and blowing air across the upper end demonstrates how water rises higher than 10.4 meters in a plant. The stream of relatively dry air causes water molecules to evaporate from the water surface in the tube. The water level in the tube does not fall because, as water molecules are drawn from the top, they are replenished by new water molecules forced up from the bottom. This, in essence, is what happens in plants. The passage of air across leaf surfaces results in the loss of water by evaporation, creating a "pull" at the open upper end of the "tube." Meanwhile, new water molecules entering through the roots of the plant are pushed up by atmospheric pressure. In addition to these pushing and pulling forces, the *adhesion* of water molecules to the walls of the very narrow tubes that occur in plants also helps to maintain water flow to the tops of plants.

A column of water in a tall tree does not collapse simply because of its weight because water molecules have an inherent strength that arises from their tendency to form hydrogen bonds with one another. These hydrogen bonds cause **cohesion** of the water molecules; in other words, a column of water resists separation. This resistance, called tensile strength, varies inversely with the diameter of the column; that is, the smaller the diameter of the column, the greater the tensile strength. Therefore, plants must have very narrow transporting vessels to take advantage of tensile strength.

How the combination of the forces of gravity, tensile strength, and cohesion affect water movement in plants is called the **cohesion-adhesion-tension theory.**

Transpiration

The cohesion-adhesion-tension theory explains the process by which water leaves a plant, a process called **transpiration.**

More than 90% of the water taken in by plant roots is ultimately lost to the atmosphere, almost all of it from the leaves. It passes out primarily through the stomata in the form of water vapor. On its journey from the plant's interior to the outside, a molecule of water first passes into the pockets of air within the leaf by evaporating from the walls of the spongy mesophyll that lines the intercellular spaces. These intercellular spaces open to the outside of the leaf by way of the stomata. The water that evaporates from these surfaces of the spongy mesophyll cells is continuously replenished from the tips of the veinlets in the leaves. Because the strands of xylem conduct water within the plant in an unbroken stream all the way from the roots to the leaves, when a portion of the water vapor in the intercellular spaces passes out through the stomata, the supply of water vapor in these spaces is continually renewed (figure 15.19).

Structural features such as the stomata, the cuticle, and the intercellular spaces in leaves have evolved in response to one or both of two contradictory requirements: minimizing the loss of water to the atmosphere, on the one hand, and admitting carbon dioxide, which is essential for photosynthesis, on the other. How plants resolve this problem is discussed a little later. A consideration of how roots absorb water must come first.

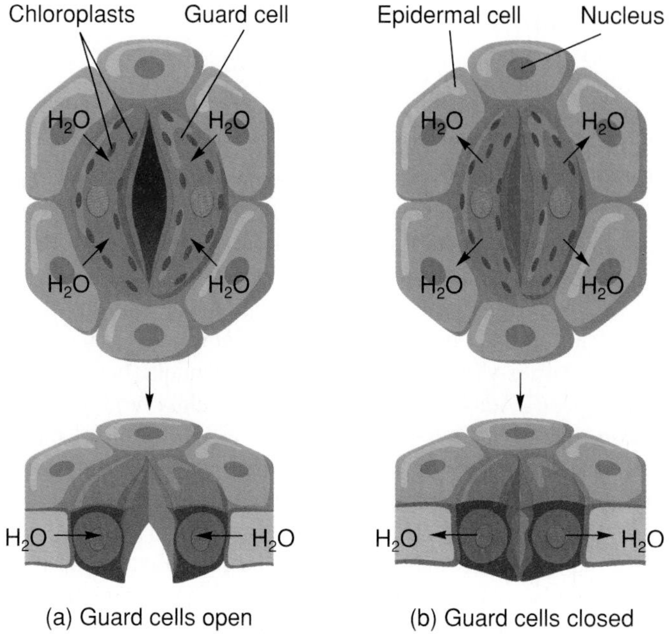

Figure 15.20 How guard cells regulate the opening and closing of stomata.
(a) When guard cells contain a high level of solutes, water enters the guard cells, causing them to bow outward. This bowing opens the stoma. *(b)* When guard cells contain a low level of solutes, water leaves the guard cells, causing them to become flaccid. This flaccidity closes the stoma.

Regulation of Transpiration: Open and Closed Stomata

The only way plants can control water loss on a short-term basis is to close their stomata. Many plants can do this when subjected to water stress. But the stomata must be open at least part of the time so that carbon dioxide, which is necessary for photosynthesis, can enter the plant. In its pattern of opening or closing its stomata, a plant must respond to both the need to conserve water and the need to admit carbon dioxide.

The stomata open and close because of changes in the water pressure of their guard cells. Stomatal guard cells have a distinctive shape—they are thicker on the side next to the stomatal opening and thinner on their other sides and ends. When the guard cells are turgid (plump and swollen with water), they become bowed in shape, thus opening the stomata as wide as possible (figure 15.20).

A number of environmental factors affect the opening and closing of stomata. The most important is water loss. The stomata of plants that are wilted because of a lack of water tend to close. An increase in carbon dioxide concentration also causes the stomata of most species to close. In most plant species, stomata open in the light and close in the dark. Very high temperatures (above 30° to 35°C) also tend to cause stomata to close.

Water Absorption by Roots

Most of the water absorbed by plants comes in through the root hairs, which collectively have an enormous surface area (figure 15.21). The root hairs are **turgid**—plump and swollen with water—because they contain a higher concentration of dissolved minerals than does the water in the soil solution; water, therefore, tends to move into them steadily. Once inside the roots, water passes inward to the conducting elements of the xylem.

Water is not the only substance that enters the roots by passing into the cells of root hairs. Membranes of root hair cells contain a variety of ion transport channels that actively pump specific ions into the plant, even against large concentration gradients. These ions, many of which are plant nutrients, are then transported throughout the plant as a component of the water flowing through the xylem (figure 15.22).

Virtually all of the absorption of water and minerals from the soil takes place through the root hairs. These root hairs greatly increase the surface area and therefore the absorptive powers of the roots. In plants that have ectomycorrhizae (see chapter 13) the root hairs often are greatly reduced in number; the fungal filaments take their place in promoting absorption.

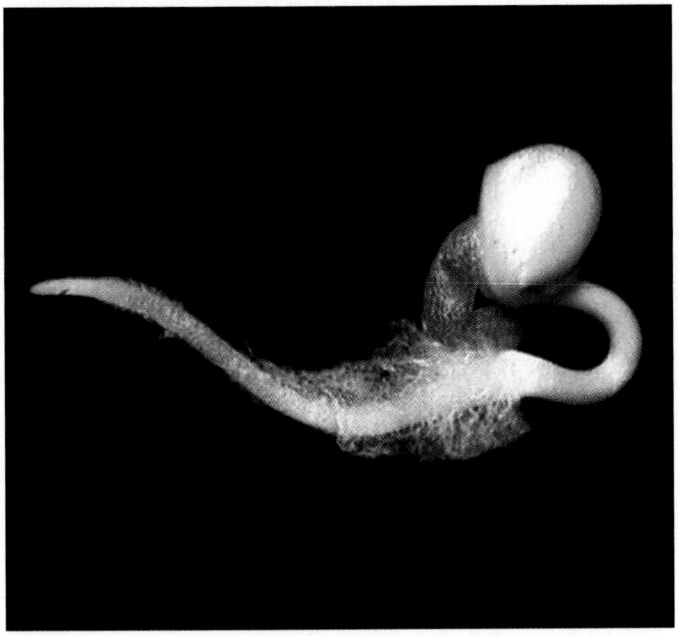

Figure 15.21 Root hairs.
Abundant fine root hairs can be seen in the back of the root apex of this germinating seedling of radish, *Raphanus sativus*.

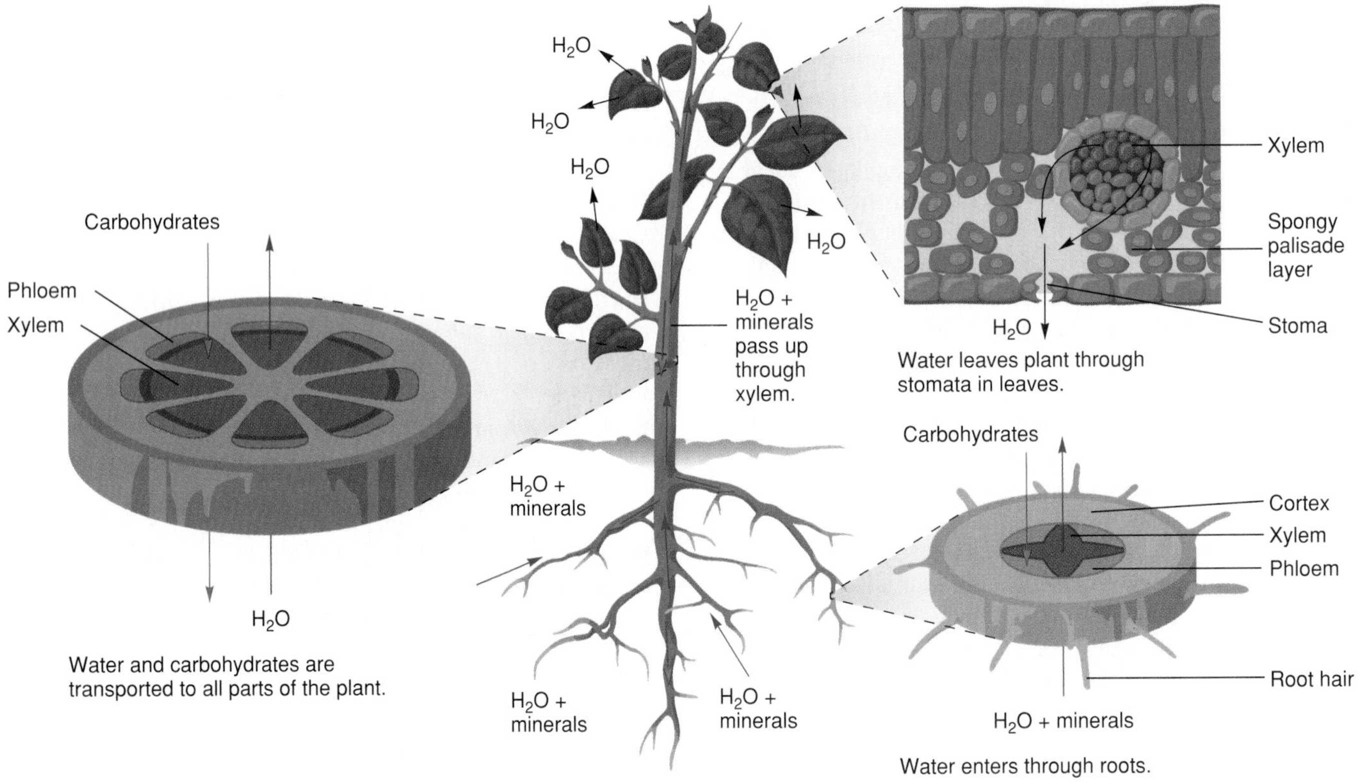

Figure 15.22 The flow of materials into, out of, and within a plant.
Water and minerals enter through the roots of a plant and are transported through the xylem to all parts of the plant body (*blue arrows*). Water leaves the plant through the stomata in the leaves (*black arrows*). Carbohydrates synthesized in the leaves are circulated throughout the plant by the phloem (*red arrows*).

Carbohydrate Transport

Most of the carbohydrates manufactured in plant leaves and other green parts are moved through the phloem to other parts of the plant. This process, known as **translocation,** makes suitable carbohydrate building blocks available at the plant's actively growing regions. The carbohydrates concentrated in storage organs such as tubers, often in the form of starch, are also converted into transportable molecules, such as sucrose, and moved through the phloem.

The pathway that sugars and other substances travel within the plant has been demonstrated precisely by using radioactive isotopes and aphids, a group of insects that suck the sap of plants. Aphids thrust their piercing mouthparts into the phloem cells of leaves and stems to obtain the abundant sugars there. When the aphids are cut off of the leaf, the liquid continues to flow from the detached mouthparts and is thus available in pure form for analysis. The liquid in the phloem contains 10 to 25% dissolved solid matter, almost all of which is sucrose.

Using aphids to obtain the critical samples and radioactive tracers to mark them, researchers have learned that movement of substances in the phloem can be remarkably fast—rates of 50 to 100 centimeters per hour have been measured. This translocation movement is a passive process that does not require the expenditure of energy. The **mass flow** of materials transported in the phloem occurs because of water pressure, which develops as a result of osmosis. First, sucrose produced as a result of photosynthesis is actively "loaded" into the sieve tubes of the vascular bundles. This loading increases the solute concentration of the sieve tubes, so water passes into them by osmosis. An area where the sucrose is made is called a *source;* an area where sucrose is delivered from the sieve tubes is called a *sink.* Sinks include the roots and other regions where the sucrose is being unloaded. There the solute concentration of the sieve tubes is decreased as the sucrose is removed. As a result of these processes, water moves in the sieve tubes from the areas where sucrose is being taken into those areas where it is being withdrawn, and the sucrose moves passively with the water (figure 15.23).

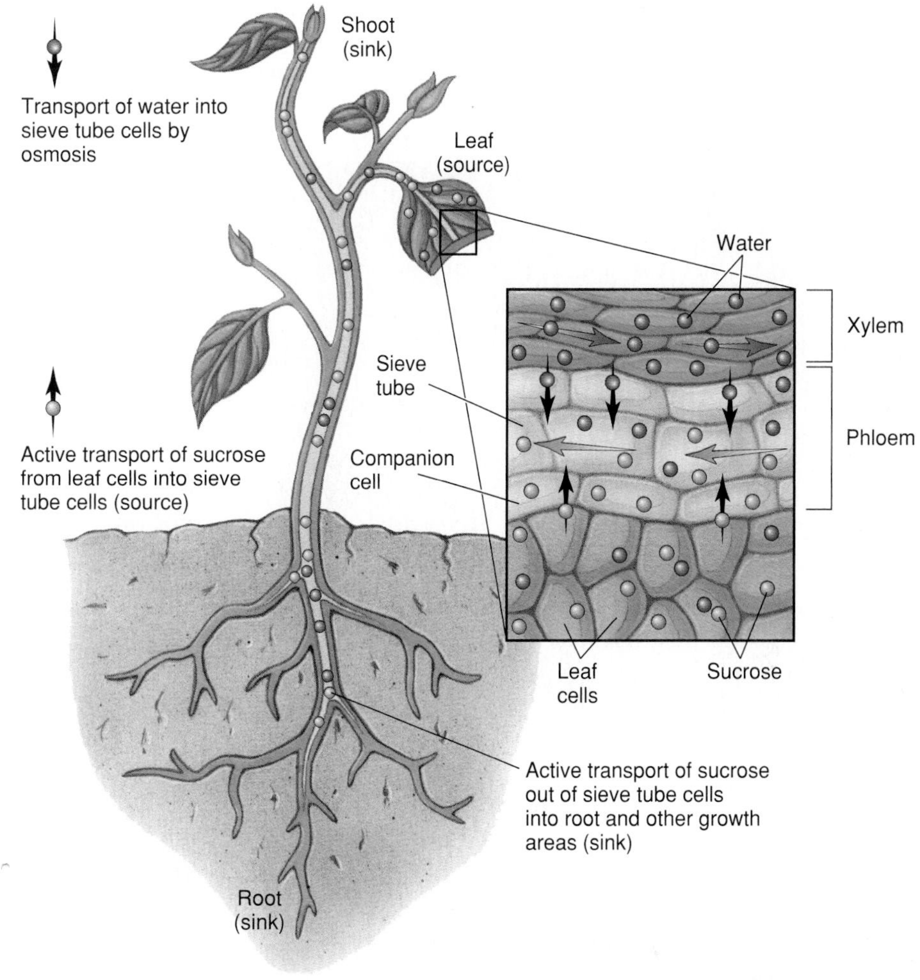

Figure 15.23 Mass flow.
Carbohydrates made in the leaves (the "source") are loaded into sieve tubes, and water follows by osmosis. At points where carbohydrates are needed (the "sinks," such as roots and shoots), the carbohydrates are unloaded, and water again follows by osmosis.

Essential Plant Nutrients

Just as human beings need certain nutrients, such as carbohydrates, amino acids, and vitamins, to survive, plants also need various nutrients to remain alive and healthy. Lack of an important nutrient may slow a plant's growth or make the plant more susceptible to disease or even death.

Nutrients are involved in plant metabolism in many ways (table15.1). *Nitrogen* is an essential part of proteins and of nucleic acids. *Potassium* ions regulate the **turgor pressure** (the pressure within a cell that results from water moving into the cell) of guard cells and therefore the rate at which the plant loses water and takes in carbon dioxide. *Calcium* is an essential component of the middle lamellae, the structural elements laid down between plant cell walls, and it also helps to maintain the physical integrity of membranes. *Magnesium* is a part of the chlorophyll molecule. The presence of *phosphorus* in many key biological molecules such as nucleic acids and ATP has been explored in detail in earlier chapters. *Sulfur* is a key component of an amino acid (cysteine) essential in building proteins.

Some plants are able to use other organisms directly as sources of nitrogen, just as animals do. These are the carnivorous plants. Carnivorous plants have adaptations to lure and trap insects and other small animals. The plants digest their prey with enzymes secreted from various kinds of glands. The Venus flytrap (*Dionaea muscipula*) has sensitive hairs on each side of each leaf, which, when touched, trigger the two halves of the leaf to snap together (figure 15.24*a* and *b*). Pitcher plants attract insects with their bright, flowerlike colors, and once inside the pitchers, the insects slide down into a cavity filled with water and digestive enzymes (figure 15.24*c*).

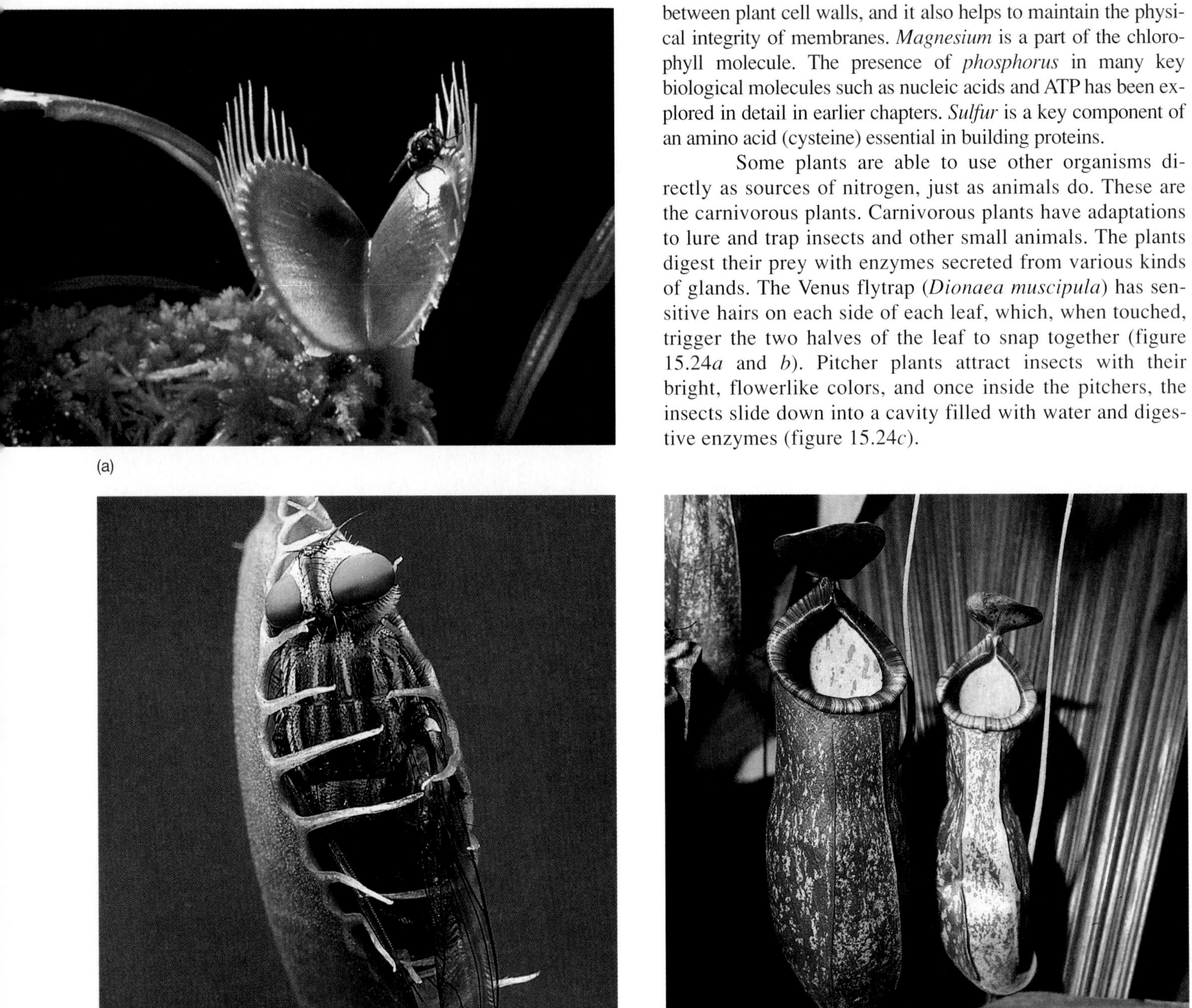

(a)

(b)

(c)

Figure 15.24 Carnivorous plants.

(*a*) Venus flytrap, *Dionaea muscipula,* which inhabits low boggy ground in North and South Carolina. (*b*) A Venus flytrap leaf has snapped together, imprisoning a fly. (c) A tropical Asian pitcher plant, *Nepenthes.* Insects seeking nectar enter the pitchers, which are modified leaves, and are trapped and digested. Complex communities of invertebrate animals and protists inhabit the pitchers.

Table 15.1 Plant Nutrients

Nutrient	Relative Abundance in Plant Tissue (ppm)
Nitrogen	1,000,000
Potassium	250,000
Calcium	125,000
Magnesium	80,000
Phosphorus	60,000
Sulfur	30,000
Chlorine	3,000
Iron	2,000
Boron	2,000
Manganese	1,000
Zinc	300
Copper	100
Molybdenum	1

Note: ppm = parts per million. Parts per million equals units of an element by weight per million units of oven-dried plant material.

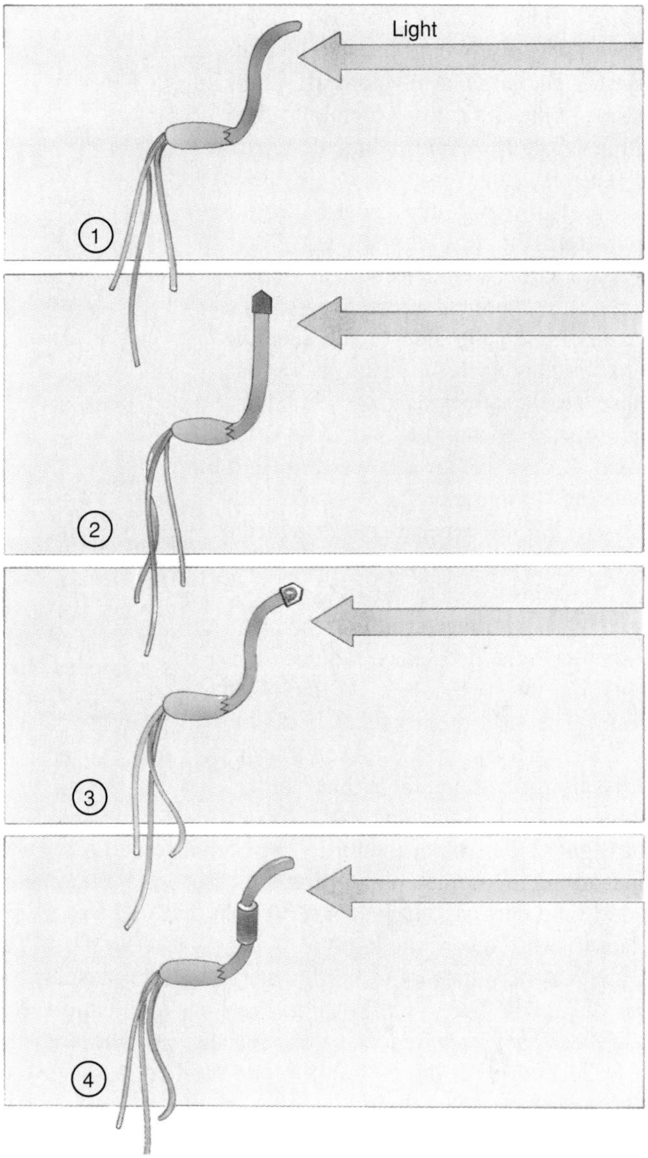

Figure 15.25 The Darwins' discovery of auxin. Charles and Francis Darwin found that a young grass seedling normally bends toward the light (*1*). If they covered the tip of the seedling with a lightproof collar, however, the seedling did not bend toward the light (*2*). When the tip of the seedling was covered with a transparent collar, the bending did occur (*3*). When the Darwins placed the collar below the tip, the seedling again bent toward the light (*4*). From these experiments, the Darwins concluded that, in response to light, an "influence" that causes bending was transmitted from the tip of the seedling to the area below the tip, where bending usually occurs.

Regulating Plant Growth: Plant Hormones

Hormones are chemical substances produced in small, often minute quantities in one part of an organism and then transported to another part of the organism, where they stimulate certain physiological processes and inhibit others. How they act in a particular instance is influenced both by what the hormones themselves are and by how they affect the particular tissue that receives their message.

In animals, hormones are usually produced at definite sites, normally in organs that are solely concerned with hormone production. In plants, on the other hand, hormones are produced in tissues that are not specialized for that purpose but that carry out other, usually more obvious functions.

At least five major kinds of hormones are found in plants: auxin, cytokinins, gibberellins, ethylene, and abscisic acid. Other kinds of plant hormones certainly exist but are less well understood. The study of plant hormones, especially how hormones produce their effects, is today an active and important field of research.

Auxin

In his later years, the great evolutionist Charles Darwin became increasingly devoted to the study of plants. In 1881, he and his son Francis published a book called *The Power of Movement in Plants,* in which they reported their systematic experiments concerning the way in which growing plants bend toward light, a phenomenon known as **phototropism.**

After conducting a series of experiments (figure 15.25), the Darwins hypothesized that, when plant shoots were illuminated from one side, an "influence" that arose in the uppermost part of the shoot was then transmitted downward, causing the shoot to bend. Later, several botanists conducted a series of experiments that demonstrated that the substance causing the shoots to bend was a chemical we call **auxin.** Auxin is now known to regulate cell growth in plants.

How auxin controls plant growth was discovered in 1926 by Frits Went, a Dutch plant physiologist, in the course of studies for his doctoral dissertation. From his experiments, described in figure 15.26, Went was able to show that the substance that flowed into the agar from the tips of the light-grown grass seedlings enhanced cell elongation. This chemical messenger caused the tissues on the side of the seedling into which it flowed to grow more than those on the opposite side. He named the substance that he had discovered auxin, from the Greek word *auxein,* meaning "to increase."

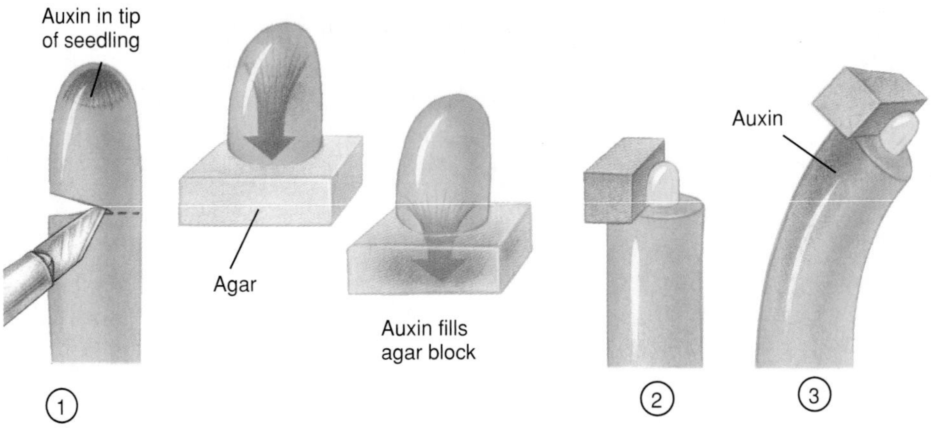

Figure 15.26 Went's demonstration of how auxin affects plant growth. Frits Went, a Dutch plant physiologist, discovered how auxin controls plant growth. Went removed the tips of grass seedlings and put them on agar gel. Auxin flowed from the tips of the seedlings into the agar blocks (*1*). Went then placed these blocks of agar on one side of the ends of grass seedlings that had been grown in the dark and from which the tips had been removed (*2*). The seedlings bent away from the side on which the auxin-filled agar block was placed (*3*). Went concluded that auxin promoted cell elongation and that it accumulated on the side of a grass seedling away from the light.

Went's experiments provided a basis for understanding the responses the Darwins had obtained some 45 years earlier: Grass seedlings bend toward the light because the auxin contents on the two sides of the shoot differ. The side of the shoot that is in the shade has more auxin; therefore, its cells elongate more than those on the lighted side, bending the plant toward the light. Later experiments by other investigators showed that auxin in normal plants migrates from the illuminated side to the dark side in response to light and thus causes the plant to bend toward the light.

Auxin appears to act by increasing the plasticity of the plant cell wall, within minutes of its application. Researchers have reported that the covalent bonds linking the polysaccharides of the cell wall to one another change extensively in response to auxin.

Synthetic auxins are routinely used to control weeds. When applied as herbicides, they are used in higher concentrations than those at which auxin normally occurs in plants. One of the most important of the synthetic auxins used in this way is 2,4-dichlorophenoxyacetic acid, usually known as 2,4-D (figure 15.27). It kills weeds in lawns without harming the grass because 2,4-D affects only broad-leaved dicots. When treated, the weeds literally "grow to death," rapidly depleting all metabolic reserves so that no source of energy remains for transport or other essential functions.

Closely related to 2,4-D is the herbicide 2,4,5-trichlorophenoxyacetic acid (2,4,5-T), which is widely used to kill woody seedlings and weeds. Notorious as the "Agent Orange" of the Vietnam War, 2,4,5-T is easily contaminated with a by-product of its manufacture, dioxin. Dioxin is harmful to people because it is an **endocrine disrupter,** a chemical that interferes with the course of human development. The growing release of endocrine disrupters as by-products of modern chemical manufacturing is a subject of great environmental concern.

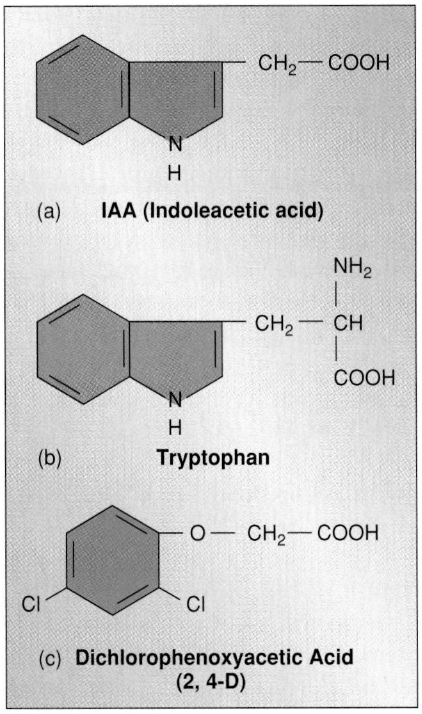

Figure 15.27 Auxins. (*a*) Indoleacetic acid (IAA) is the only known naturally occurring auxin. (*b*) Tryptophan is the amino acid from which plants synthesize IAA. (*c*) Dichlorophenoxyacetic acid (2,4-D), a synthetic auxin, is a widely used herbicide.

Cytokinins

A **cytokinin** is a plant hormone that, in combination with auxin, stimulates cell division in plants and determines the course of differentiation. Substances with these properties are widespread, both in bacteria and in eukaryotes. In vascular plants, most cytokinins seem to be produced in the roots, from which they are then transported throughout the rest of the plant. Cytokinins apparently stimulate cell division by influencing the synthesis or activation of proteins specifically required for mitosis.

Gibberellins

Synthesized in the apical portions of both shoots and roots, **gibberellins** have important effects on stem elongation in plants and play the leading role in controlling this process in mature trees and shrubs (figure 15.28). In these plants, the application of gibberellins characteristically promotes elongation within the spaces between leaf nodes on stems, and this effect is enhanced if auxin is also present. Gibberellins are also involved with many other aspects of plant growth, such as inducing flowering and hastening seed germination.

Ethylene

Ethylene is produced in large quantities during a certain phase of fruit ripening, when the fruit's respiration is proceeding at its most rapid rate. At this phase, complex carbohydrates are broken down into simple sugars, cell walls become soft, and the volatile compounds associated with flavor and scent in the ripe fruits are produced. When applied to fruits, ethylene hastens their ripening.

One of the first lines of evidence that led to the recognition of ethylene as a plant hormone was the observation that gases from oranges caused premature ripening in bananas. Such relationships have led to major commercial uses. Tomatoes are often picked green and then artificially ripened as desired by the application of ethylene. Ethylene is widely used to speed the ripening of lemons and oranges as well. Carbon dioxide produces effects opposite to those of ethylene in fruits, and fruits that are being shipped are often kept in an atmosphere of carbon dioxide if they are not intended to ripen yet.

Genetic engineers, using techniques described in chapter 8, have placed bacterial genes into tomatoes that slow the ripening process. Because until now commercial tomatoes have had to be picked very early in order to get them to market before they become overripe, store-bought tomatoes have typically lacked the taste of "home-grown" tomatoes. Because of their delayed ripening, the genetically engineered tomatoes can be left on the vine longer, greatly improving their taste.

Figure 15.28 The effect of a gibberellin. Although more than 60 gibberellins have been isolated from natural sources, apparently only one kind is active in shoot elongation. As shown in this photo, cabbage plants produce tall flowering shoots when treated with this form of gibberellin.

Abscisic Acid

Abscisic acid is a naturally occurring plant hormone that is synthesized mainly in mature green leaves, fruits, and root caps. The hormone was given its name because it stimulates leaves to age rapidly and fall off (the process of abscission), but evidence that abscisic acid plays an important natural role in this process is scant. However, we do know that abscisic acid suppresses the growth and elongation of buds and promotes aging, counteracting some of the effects of the gibberellins (which stimulate bud growth and elongation) and of auxin (which tends to retard aging).

Plant Responses to Environmental Stimuli

Plants respond to different environmental stimuli in a variety of ways. As discussed earlier in the chapter, plants bend toward light as they grow in response to the environmental stimulus. A host of other plant responses, including flowering, dropping of leaves, and yellowing of leaves due to loss of chlorophyll, are also prompted by various environmental stimuli. *Tropisms,* or movement responses to external stimuli, control patterns of plant growth and thus plant appearance. Three major classes of plant tropisms are considered here: phototropism, gravitropism, and thigmotropism.

Phototropism

Phototropism, the bending of plants toward directional sources of light, was introduced in the discussion of auxin. In general, stems are positively phototropic, growing toward the light, whereas roots are negatively phototropic, growing away from it. The phototropic reactions of stems are clearly of adaptive value because they allow plants to capture greater amounts of light than would otherwise be possible. Auxin is involved in most, if not all, of plants' phototropic growth responses.

Figure 15.29 Tropism guides plant growth.
The branches of this fallen tree are growing straight up because they are gravitropic and also phototropic.

Gravitropism

Another familiar plant response is **gravitropism** which causes stems to grow upward and roots downward. Both of these responses clearly have adaptive significance: stems that grow upward are apt to receive more light than those that do not; roots that grow downward are more apt to encounter a more favorable environment than those that do not. The phenomenon is now called gravitropism because it is clearly a response to gravity. Figure 15.29 illustrates the effect of both phototropism and gravitropism.

Thigmotropism

Still another commonly observed response of plants is **thigmotropism,** a name derived from the Greek root *thigma,* meaning "touch." Thigmotropism is defined as the response of plants to touch. Examples include plant tendrils that rapidly curl around and cling to stems or other objects and twining plants, such as bindweed, that also coil around objects. These behaviors result from rapid growth responses to touch. Specialized groups of cells in the plant epidermis appear to be concerned with thigmotropic reactions, but again, their exact mode of action is not well understood.

Photoperiodism

Essentially all eukaryotic organisms are affected by the cycle of night and day, and many features of plant growth and development are keyed to changes in the proportions of light and dark in the daily 24-hour cycle. Such responses constitute **photoperiodism,** a mechanism by which organisms measure seasonal changes in relative day and night length. One of the most obvious of these photoperiodic reactions concerns angiosperm flower production.

Day length changes with the seasons; the farther from the equator you are, the greater the variation. Plants' flowering responses fall into three basic categories in relation to day length: Long-day plants initiate flowers when days become longer than a certain length and nights become shorter. Short-day plants, on the other hand, begin to form flowers when days become shorter than a critical length and nights become longer. Thus, many spring and early summer flowers are long-day plants, and many fall flowers are short-day plants (figure 15.30). Commercial plant growers use these responses to day length to time flower blooming for specific holidays or occasions.

In addition to long-day and short-day plants, a number of plants are described as day-neutral. Day-neutral plants produce flowers whenever environmental conditions are suitable, without regard to day length.

Phytochromes

Flowering responses to daylight are controlled by several chemicals that interact in complex ways. Although the nature of some of these chemicals has been deduced, how the various chemicals work together to promote or inhibit flowering responses is still being debated.

One class of these chemicals—the **phytochromes**—has been demonstrated to influence plants' flowering responses. Phytochromes are pigment molecules, meaning that they absorb light energy and undergo changes in their chemical structure as a result. In plants, when light in the red wavelengths (about 660 nanometers) strikes the phytochrome molecule P_r, the molecule is converted to an alternative form, P_{fr}. The reaction is reversed when light in the far-red wavelengths (about 730 nanometers) strikes P_{fr}, converting it back to P_r. Daylight contains both red and far-red wavelengths, but night light contains only far-red wavelengths. Thus, during nighttime, any P_{fr} that remains in the plant is converted to P_r. Of the two pigments, only P_{fr} is biologically active or able to exert a biological response on a plant.

Scientists hypothesize that in short-day plants, P_{fr} acts as a flowering inhibitor, while in long-day plants, it acts as a flowering promoter. During seasons of long days and short nights, P_{fr} accumulates in short-day plants during the day and is destroyed at night. However, enough P_{fr} is left over at daybreak to inhibit flowering. As the days get shorter and the nights get longer, however, short-day plants have a chance to "catch up" and destroy larger amounts of the flowering inhibitor P_{fr} during the longer nights. Thus, short-day plants flower during seasons with longer nights and shorter days.

In contrast, during seasons of long days and short nights, P_{fr} accumulates in long-day plants during the day and is destroyed at night, but enough P_{fr} remains at daybreak to promote flowering. As the days shorten and the nights lengthen, however, the long nights provide a longer opportunity for P_{fr} destruction, and flowering in the long-day plants is inhibited.

After many experiments, scientists hypothesized that a hormone they called **florigen** also exerted a flower-inducing response in plants. Another hormone, not yet isolated or named, seemed to induce a flower-inhibiting response in plants. These hormones are produced in the leaves and are transmitted to the bud, where they exert their different effects. In long-day plants, florigen is produced during long days, and the flower-inhibiting hormone is produced during short days. In short-day plants, the flower-inhibiting hormone is produced during long days, and florigen is produced during short days. Although much experimental evidence points to the existence of these hormones, scientists have not yet been able to isolate and study these particular chemical messengers.

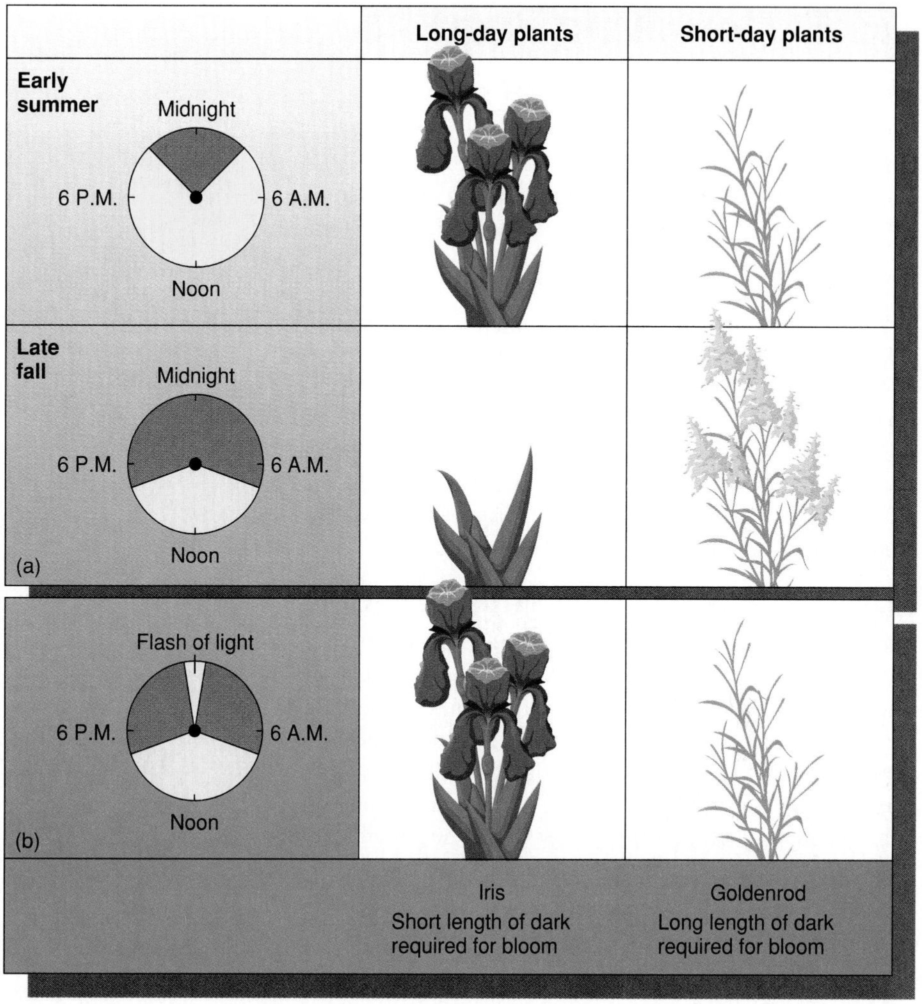

Figure 15.30 Photoperiodism in plants.
(*a*) The iris is a long-day plant that is stimulated by short spring nights to bloom in the spring. The goldenrod is a short-day plant that is stimulated by the long nights of fall to bloom in the fall. (*b*) If the long night of winter is artificially interrupted by a flash of light, the goldenrod will not bloom and the iris will. In each case, the duration of uninterrupted darkness determines when flowering occurs.

Dormancy

Plants respond to their external environment largely by changes in growth rate. Plants' ability to stop growing altogether when conditions are not favorable—to become dormant—is critical to their survival.

In temperate regions, dormancy is generally associated with winter, when low temperatures and the unavailability of water because of freezing make it impossible for plants to grow. During this season, the buds of deciduous trees and shrubs remain dormant, and the apical meristems remain well protected inside enfolding scales. Perennial herbs spend the winter underground as stout stems or roots packed with stored food. Many other kinds of plants, including most annuals, pass the winter as seeds.

15.3 Flowering Plant Reproduction

Flowering plants, or angiosperms, dominate the earth except for the great northern forests, the polar regions, the high mountains, and the driest deserts. Among the features that have contributed to their success are their unique reproductive structures, which include the flower and the fruit. As discussed in chapter 14, flowers bring about the precise transfer of pollen by insects and other animals, which allows plants to exchange gametes with one another, even though each plant is rooted in one place. Fruits play an important role in the dispersal of angiosperms from place to place. Not only were both flowers and fruits key elements in the early success of angiosperms, but their evolution produced most of the striking differences seen among different angiosperms today. This section explains how flowering plants reproduce.

Formation of Angiosperm Gametes

As mentioned in chapter 14, plant life cycles are characterized by an alternation of generations, in which a diploid sporophyte generation gives rise to a haploid gametophyte generation. In angiosperms, the gametophyte generation is very small and is completely enclosed within the tissues of the parent sporophyte. The male gametophytes, or microgametophytes, are **pollen grains.** The female gametophyte, or megagametophyte, is the **embryo sac.** Pollen grains and the embryo sac both are produced in separate, specialized structures of the angiosperm flower.

Like animals, angiosperms have separate structures for producing male and female gametes, but the reproductive organs of angiosperms are different from those of animals in two ways: First, in angiosperms, both male and female structures usually occur together in the same individual flower. Second, angiosperm reproductive structures are not permanent parts of the adult individual. Angiosperm flowers and reproductive organs develop seasonally; these flowering seasons correspond to times of the year most favorable for pollination.

Structure of the Angiosperm Flower

As noted in chapter 14, a typical angiosperm flower is composed of whorls, a circle of parts present at a single level along an axis. The outermost whorl consists of structures called **sepals** (figure 15.31). The sepals protect the other flower whorls and serve as the flower's attachment point to the stalk. The sepals are usually green, although in some angiosperm species, they are brightly colored. All of the sepals together are called the **calyx.**

The second whorl of a flower is composed of **petals.** Most angiosperm petals are vibrantly colored, and in those plants pollinated by animals (such as insects and birds), the petals may have characteristic shapes and features that attract the pollinating animal. All of the petals together are called the **corolla.**

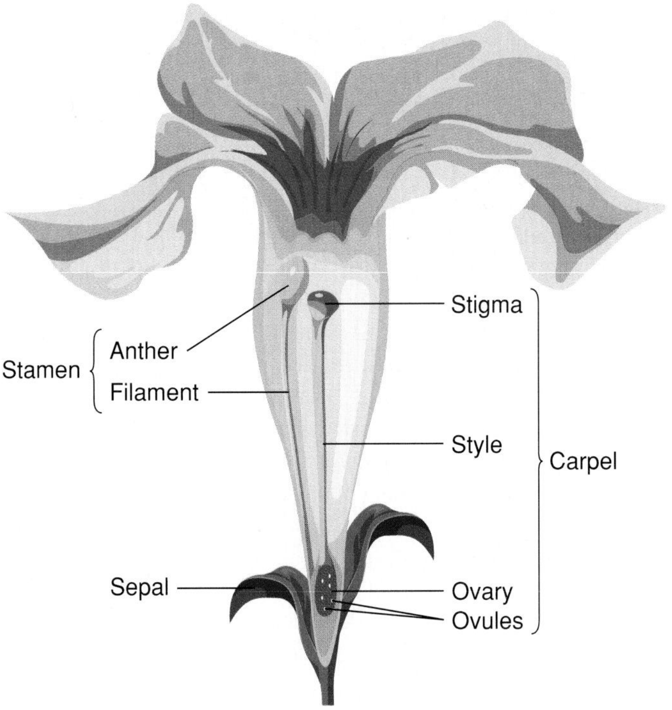

Figure 15.31 Structure of an angiosperm flower. The male reproductive structure is called the stamen and consists of the anther and filament. The female reproductive structure is called the carpel and consists of the stigma, style, and ovary. The ovary encloses the ovules, which contain eggs that will develop into seeds after fertilization.

The third whorl of a flower is composed of the male reproductive structures. The male reproductive structures are the **stamens.** Each stamen consists of a slender stalk to which is attached the **anther.** At the end of each anther are two sacs in which the pollen grains develop. All of the stamens of an angiosperm flower are called the **androecium,** which means "male household."

The fourth whorl of a flower is composed of female reproductive structures, the **carpel.** Each carpel is composed of a **stigma, style,** and **ovary.** The stigma, the top part of the carpel, is covered with a sticky, sugary liquid to which pollen grains adhere during pollination. The liquid also nourishes the pollen grain as it makes its way to the ovary. The style is the elongated portion of the carpel that leads to the ovary. The ovary contains the **ovules.** Within the ovules, the haploid megaspores develop into embryo sacs that will contain the egg. After fertilization, the ovules develop into seeds that give rise to the sporophyte generation. All of the carpels of an angiosperm flower are called the **gynoecium,** meaning "female household."

Pollen Formation

Pollen grains form in the two pollen sacs located in the anther. Each pollen sac contains specialized chambers in which the *microspore mother cells* are enclosed and

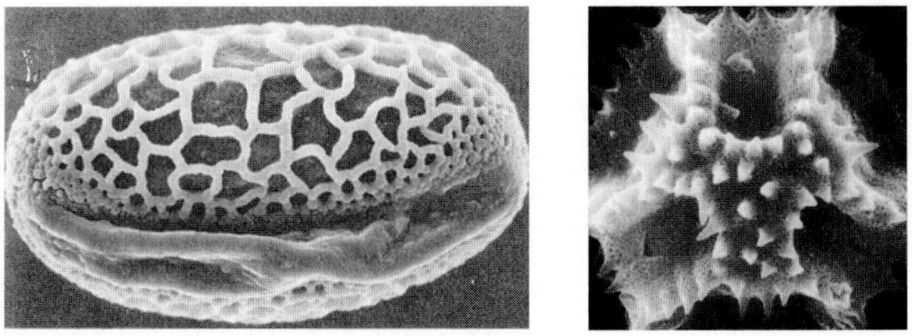

(a) (b)

Figure 15.32 Pollen grains.
(*a*) In the Easter lily, *Lilium candidum,* the pollen tube emerges from the pollen grain through the groove or furrow that occurs on one side of the grain. (*b*) In a plant of the sunflower family, *Hyoseris longiloba,* three pores are hidden among the ornamentation of the pollen grain. The pollen tube may grow out through any one of them.

protected. The microspore mother cells undergo meiosis to form four haploid microspores. Subsequently, mitotic divisions form four pollen grains.

Pollen grain shapes are specialized for specific flower species. As discussed in more detail later in the chapter, fertilization requires that the pollen grain grow a tube that penetrates the style until the ovary is encountered. Most pollen grains have a furrow from which this pollen tube emerges. Some pollen grains have three furrows (figure 15.32).

Egg Formation

Eggs develop in the ovules of the angiosperm flower. Within each ovule is a megaspore mother cell. Each megaspore mother cell undergoes meiosis to produce four haploid megaspores. In most plants, only one of these megaspores, however, survives; the rest are absorbed by the ovule. The lone remaining megaspore undergoes repeated mitotic divisions to produce eight haploid nuclei, which are enclosed within an embryo sac. Within the embryo sac, the eight nuclei are arranged in precise positions. One nucleus is located near the opening of the embryo sac; this nucleus is the egg cell. The other seven are arranged around the embryo sac: two are located in the middle of the embryo sac and are called polar nuclei; two nuclei flank the egg cell; and the other three nuclei are located at the end of the embryo sac, opposite the egg cell.

Pollination

Pollination is the process by which pollen is placed on the stigma. The pollen may be carried to the flower by wind or by animals, or it may originate within the individual flower itself. When pollen from a flower's anther pollinates the same flower's stigma, the process is called *self-pollination.*

In many angiosperms, the pollen grains are carried from flower to flower by insects and other animals that visit the flowers for food or other rewards (figure 15.33) or are deceived into doing so because the flower's characteristics suggest such rewards. A liquid called **nectar,** which is rich in sugar as well as amino acids and other substances, is often the reward sought by animals. Successful pollination depends on the plants attracting insects and other animals regularly enough that the pollen is carried from one flower of that particular species to another.

Figure 15.33 Insect pollination.
This bumblebee, *Bombus*, has become covered with pollen while visiting a flower. The bee will transfer large quantities of the pollen to the next flower it visits.

The relationship between such animals, known as *pollinators,* and the flowering plants has been important to the evolution of both groups. By using insects to transfer pollen, the flowering plants can disperse their gametes on a regular and more or less controlled basis, despite their being anchored to the ground.

For pollination by animals to be effective, a particular insect or other animal must visit plant individuals of the same species. A flower's color and form have been shaped by evolution to promote such specialization. Yellow flowers are particularly attractive to bees, whereas red flowers attract birds but are not particularly noticed by insects. Some flowers have very long floral tubes with the nectar produced deep within them; only the long, slender beaks of hummingbirds or the long, coiled tongues of moths or butterflies can reach such nectar supplies.

In certain angiosperms and all gymnosperms, pollen is blown about by the wind and reaches the stigmas passively. For such a system to operate efficiently, the individuals of a given plant species must grow relatively close together because wind does not carry pollen very far or very precisely, compared with transport by insects or other animals. Because gymnosperms, such as spruces or pines, grow in dense stands, wind pollination is very effective. Wind-pollinated angiosperms, such as birches, grasses, and ragweed, also tend to grow in dense stands. The flowers of wind-pollinated angiosperms are usually small, greenish, and odorless, and their petals are either reduced in size or absent altogether. They typically produce large quantities of pollen.

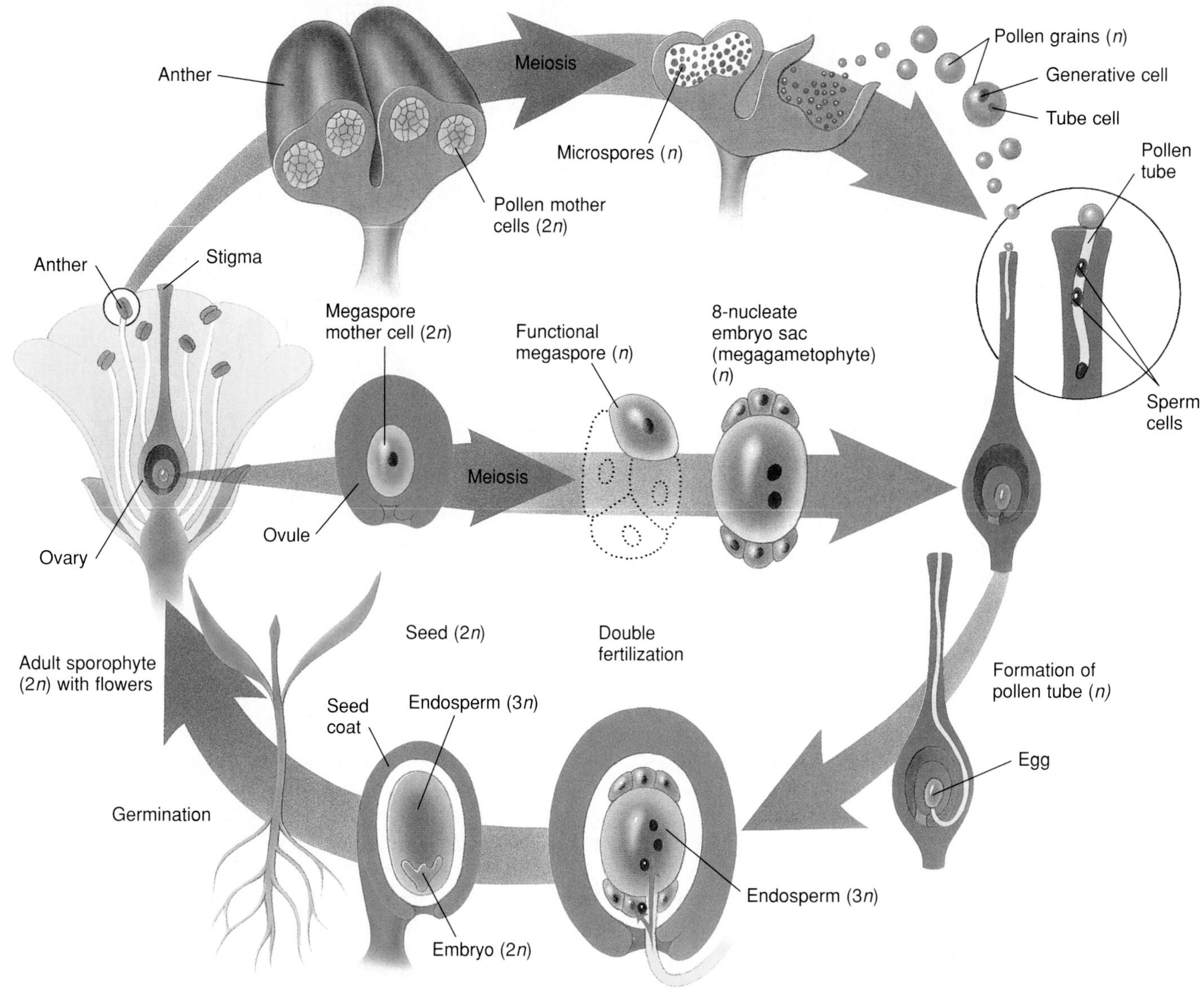

Figure 15.34 Angiosperm life cycle.

Eggs form within the embryo sac, inside the ovules, which, in turn, are enclosed in the carpels. The pollen grains, meanwhile, are formed within the sporangia of the anthers and are shed. Fertilization is a double process. A sperm and an egg come together, producing a zygote; at the same time, another fuses with the polar nuclei to produce the endosperm. The endosperm is the tissue, unique to angiosperms, that nourishes the embryo and young plant.

Fertilization

Fertilization in angiosperms is a complex, somewhat unusual process in which two sperm cells are utilized (figure 15.34). This unique process, called **double fertilization,** was described in detail in chapter 14. Double fertilization results in two key developments: (1) the fertilization of the egg and (2) the formation of a nutrient substance called endosperm that nourishes the embryo. Once fertilization is complete, the embryo develops by dividing numerous times. Meanwhile, protective tissues enclose the embryo, resulting in the formation of the seed. The seed, in turn, is enclosed in another structure called the fruit. These typical angiosperm structures evolved in response to the need for seeds to be dispersed over long distances to ensure genetic variability. Seeds and fruits are tied to the activity of animals, who carry the seeds to new habitats.

Once a pollen grain has been spread by wind, an animal, or self-pollination, it adheres to the sticky, sugary substance that covers the stigma and begins to grow a **pollen tube** which pierces the style. The pollen tube, nourished by the sugary substance, grows until it reaches the ovule in the ovary. Meanwhile, one of the cells within the pollen grain inside the tube divides to form two sperm cells.

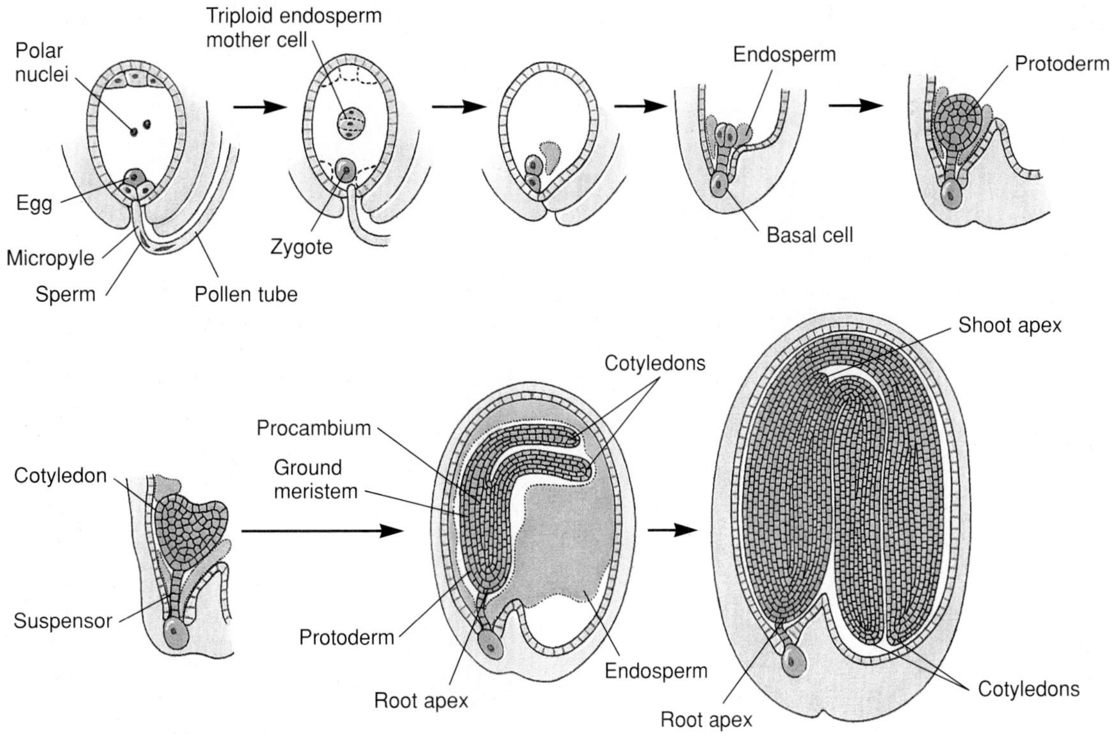

Figure 15.35 Development in an angiosperm embryo.
The zygote first divides into four cells. The basal cell, the one nearest the opening through which the pollen tube entered, undergoes a series of divisions and forms a narrow column of cells called the suspensor. The other three cells continue to divide and form a mass of cells arranged in layers. By about the fifth day of cell division, the principal tissue systems of the developing plant can be detected within this mass.

The pollen tube eventually reaches the embryo sac in the ovule. At the entry to the embryo sac, the tip of the pollen tube bursts and releases the two sperm cells. Simultaneously, the two nuclei that flank the egg cell disintegrate, and one of the sperm cells fertilizes the egg cell, forming a zygote. The other sperm cell fuses with the two polar nuclei located at the center of the embryo sac, forming the triploid ($3n$) primary endosperm nucleus. The primary endosperm nucleus eventually develops into the endosperm, which nourishes the embryo.

Seed Formation

The entire series of events that occur between fertilization and maturity is called *development*. During development, cells become progressively more specialized, or differentiated. The first stage in the development of a plant zygote is active cell division to form an organized mass of cells, the embryo. In angiosperms, the differentiation of cell types within the embryo begins almost immediately after fertilization (figure 15.35). By the fifth day, the principal tissue systems can be detected within the embryo mass, and within another day, the root and shoot apical meristems can be detected.

Early in the development of an angiosperm embryo, a profoundly significant event occurs: the embryo simply stops developing and becomes dormant. In many plants, embryo development is arrested soon after apical meristems and the first leaves, or **cotyledons,** are differentiated. The integuments—the coats surrounding the embryo—develop into a relatively impermeable seed coat, which encloses the quiescent embryo within the seed, together with a source of stored food.

Once the seed coat fully develops around the embryo, most of the embryo's metabolic activities cease; a mature seed contains only about 10% water. Under these conditions, the seed and the young plant within it are very stable. **Germination,** or the resumption of metabolic activities that leads to the growth of a mature plant, cannot take place until water and oxygen reach the embryo, a process that sometimes involves cracking the seed. Seeds of some plants have been known to remain viable for hundreds of years. Environmental factors help ensure that the plant germinates only under appropriate conditions.

Figure 15.36 Animal-dispersed fruits.
(*a*) The bright red berries of this honeysuckle, *Lonicera hispidula,* are highly attractive to birds, just as are red flowers. Birds may carry the berry seeds either internally or stuck to their feet for great distances. (*b*) The spiny fruits of this burgrass, *Cenchrus incertus,* adhere readily to any passing animal, as you will know if you have stepped on them.

Fruits

During seed formation, the flower ovary begins to develop into fruit. Paralleling the evolution of angiosperm flowers, and of equal importance to angiosperm success, has been the evolution of these fruits. Fruits form in many ways and exhibit a wide array of modes of specialization in relation to their dispersal.

Fruits that have fleshy coverings, often black, bright blue, or red, are normally dispersed by birds and other vertebrates. Like the red flowers discussed in relation to pollination by birds, the red fruits signal an abundant food supply (figure 15.36*a*). By feeding on these fruits, the birds and other animals carry seeds from place to place before excreting the seeds as solid waste. The seeds, not harmed by the animal digestive system, thus are transferred from one suitable habitat to another. Other fruits are dispersed by attaching themselves to the fur of mammals or the feathers of birds (figure 15.36*b*). Still other fruits, such as those of mangroves, coconuts, and certain other plants that characteristically occur on or near beaches, swamps, or other bodies of water are regularly spread from place to place in the water.

Germination

What happens to a seed when it encounters conditions suitable for its germination? First, it imbibes water. Seed tissues are so dry at the start of germination that the seed takes up water with great force, after which metabolism resumes. Initially, the metabolism may be anaerobic, but when the seed coat ruptures, aerobic metabolism takes over. At this point, oxygen must be available to the developing embryo because plants, which drown for the same reason people do, require oxygen for active growth. Few plants produce seeds that germinate successfully underwater, although some, such as rice, have evolved a tolerance of anaerobic conditions. Figure 15.37 shows the development of a dicot and monocot from germination to maturity.

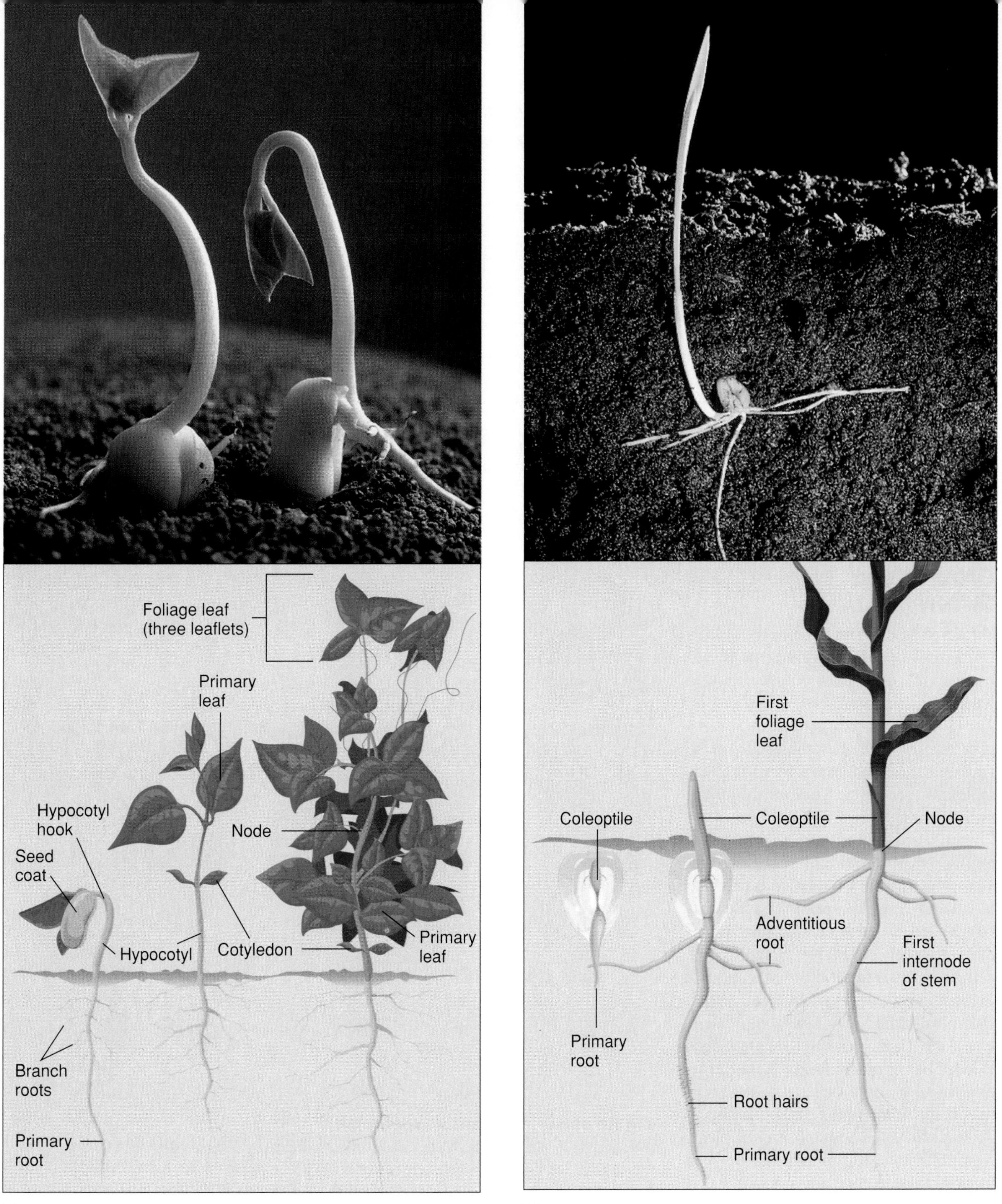

Figure 15.37 Development of angiosperms.
(*a*) Dicot development in a soybean. The two cotyledons of the dicot are pulled up through the soil along with the hypocotyl (the stem below the cotyledons). The cotyledons are the first leaves that perform photosynthesis. As other leaves develop, they take over photosynthesis entirely, and the cotyledons shrivel and fall off the stem. Flowers develop in buds at the nodes. (*b*) Monocot development in corn. Monocots have one single, underdeveloped cotyledon, which does not appear in the development of the mature plant. The coleoptile is a primordial stem; it encloses and protects the shoot and leaves as it pushes its way up through the soil.

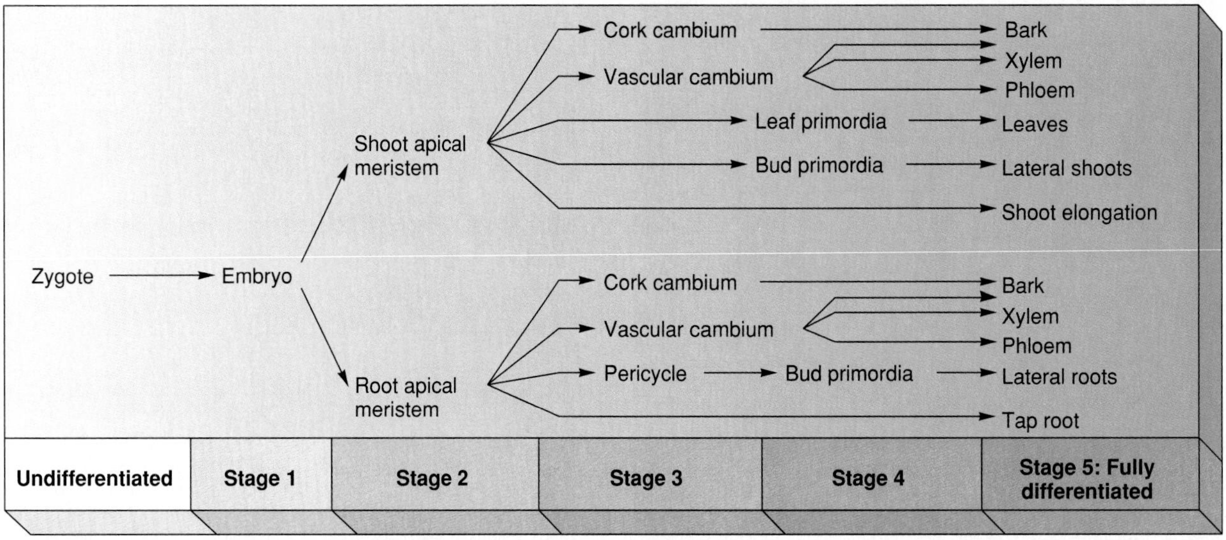

Figure 15.38 Stages of plant differentiation.
As this diagram shows, the different cells and tissues in a plant all originate from the shoot and root apical meristems.

Growth and Differentiation

Once a seed has germinated, the plant's further development depends on the activities of the meristematic tissues, which interact with the environment. The shoot and root apical meristems give rise to all of the other cells of the adult plant. Differentiation, or the formation of specialized tissues, occurs in five stages in plants (figure 15.38).

After a seed germinates, the pattern of growth and differentiation that was established in the embryo is repeated indefinitely until the plant dies. But differentiation in plants, unlike that in animals, is largely reversible. Botanists first demonstrated in the 1950s that individual differentiated cells isolated from mature individuals could give rise to entire individuals. F. C. Steward was able to induce isolated bits of phloem taken from carrots to form new plants, plants that were normal in appearance and fully fertile (figure 15.39). Regeneration of entire plants from differentiated tissue has since been carried out in many plants, including cotton, tomatoes, and cherries. These experiments clearly demonstrate that the original differentiated phloem tissue still contains all of the genetic potential needed for the differentiation of entire plants. No information is lost during plant tissue differentiation, and no irreversible steps are taken.

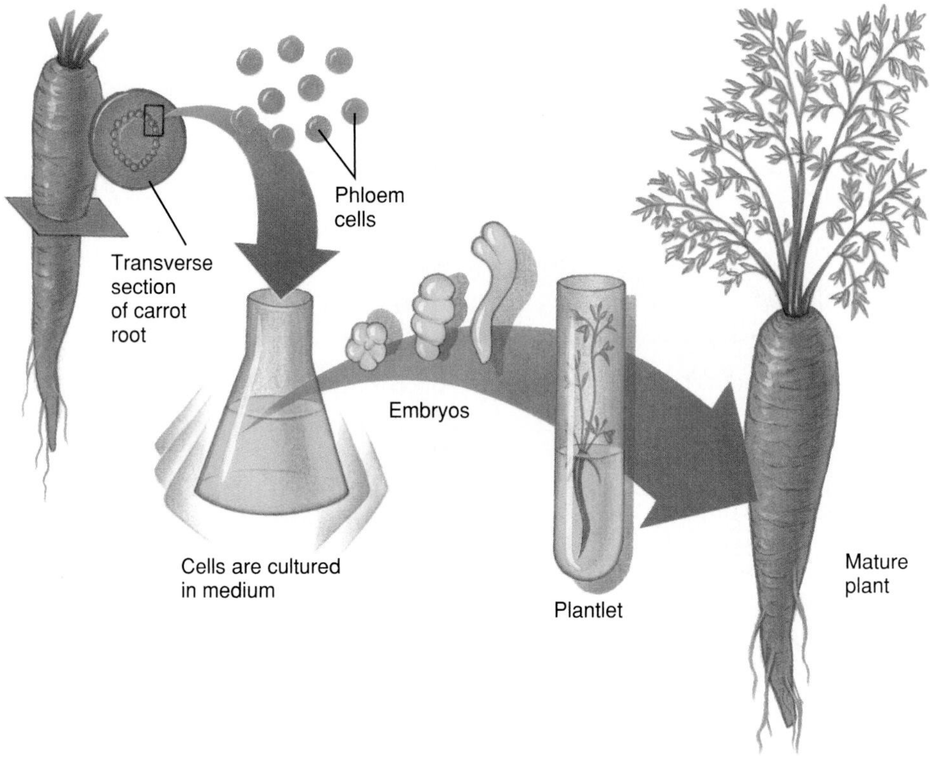

Figure 15.39 Steward's experiment.
Phloem tissue was isolated from carrots in the laboratory of F. C. Steward at Cornell University. The disks of tissue were grown in a flask in which the medium was constantly agitated to bring a fresh supply of nutrients to the masses of undifferentiated cells that soon formed. From the phloem tissue, Steward was able to induce the formation of normal, fully fertile carrot plants.

CHAPTER 15

HIGHLIGHTS

15.1 The Structure and Function of Plant Tissues

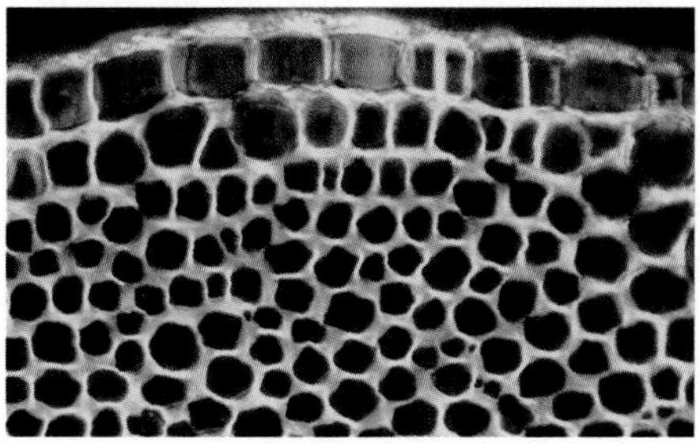

Key Terms

shoot 290
meristems 291
primary growth 291
secondary growth 291
stomata 292
vascular cambium 297

Key Concepts

- A vascular plant is organized along a vertical axis.
- The part aboveground is called the shoot, and encompasses the stem, its branches, and the leaves.
- The part belowground is called the root.
- Plants grow from actively dividing zones called meristems.
- Growth from the tip of the shoot or root is called primary growth.
- Growth in girth is called secondary growth. It takes place in a meristem called the vascular cambium.

15.2 Plant Nutrition and Transport

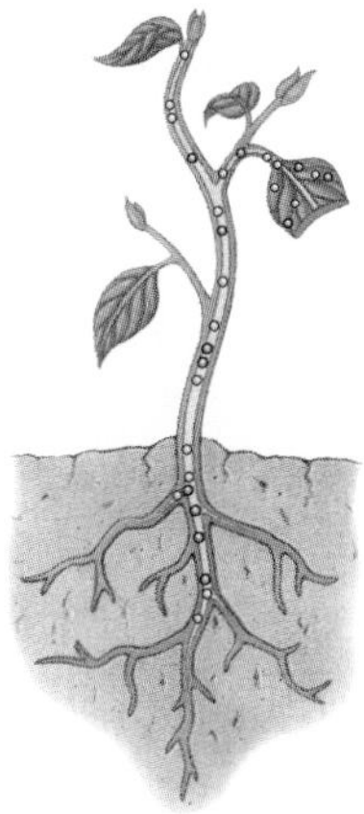

Key Terms

cohesion 301
transpiration 301
translocation 303
turgor pressure 304
phototropism 305
auxin 305

Key Concepts

- Water is held up in the conducting vessels of plants by adhesion to the vessel walls and cohesion of the water molecules to one another.
- Water is pulled up the plant by transpiration, which is evaporation from the surface of leaves.
- Carbohydrates are translocated throughout the plant by mass flow.
- Auxin is a hormone that regulates cell growth in plants.

15.3 Flowering Plant Reproduction

Key Terms

calyx 310
corolla 310
androecium 310
gynoecium 310
double fertilization 312

Key Concepts

- The male and female reproductive structures occupy the third and fourth whorl of a flower.
- If the male and female structures mature at the same time, the flower may self-pollinate.
- If the male and female structures mature at different times, or on different flowers, the pollen grains from the male structures are carried to the female structures of different flowers by insects or by the wind.

15

CONCEPT REVIEW

1. The shoot of a vascular plant consists of
   a. stem and roots.
   b. stem.
   c. stem and leaves.
   d. the apical meristem.

2. Parenchyma cells
   a. are alive at maturity.
   b. have secondary cell walls.
   c. do not contain living cytoplasm when mature.
   d. are a type of dermal tissue in plants.

3. Some 90% of the water taken up by roots is lost to the atmosphere. Which leaf structure accounts for the greatest portion of this water loss?
   a. cuticle
   b. mesophyll
   c. transpiration
   d. stomata

4. Xylem
   a. is the principal water-conducting tissue of plants.
   b. conducts the products of photosynthesis.
   c. is made of sieve cells.
   d. is not found in roots.

5. What takes place within the intercellular spaces between the spongy parenchyma cells in leaves?
   a. water absorption
   b. gas exchange
   c. photosynthesis
   d. production of the periderm

6. Wood consists of accumulated
   a. primary xylem.
   b. primary phloem.
   c. secondary xylem.
   d. secondary phloem.

7. Cells of the _______ regulate the flow of water laterally between the vascular tissues and the cell layers in the outer portions of the root.
   a. periderm
   b. endodermis
   c. pericycle
   d. pith

8. Plants bend toward the light because
   a. cells on the shaded side of the stem elongate more than those on the sunny side.
   b. cells on the sunny side of the stem elongate more than cells on the shaded side.
   c. cells on the sunny side of the stem accumulate auxin.
   d. auxin suppresses lateral bud growth.

9. Dormancy in plants is commonly associated with which two of the following?
   a. warm temperatures
   b. cold temperatures
   c. lack of carbon dioxide
   d. lack of available water

10. Which type of cell division produces the gametes in plants?
    a. meiosis
    b. mitosis

11. Pollen grains on the stigma of a flower reach the egg by forming a
    a. cotyledon.
    b. seed.
    c. pollen tube.
    d. carpel.

12. Fleshy fruits that are brightly colored are often dispersed by
    a. insects.
    b. water.
    c. mammals.
    d. birds.

13. Primary growth is initiated by the _______ meristems, whereas secondary growth involves the activity of the _______ meristems.

14. A special type of _______ cells give pears their gritty texture.

15. In stems, the cylinders that contain both primary xylem and primary phloem are called _______.

16. The process by which water leaves a plant is called _______.

17. Stomata in leaves open and close because of changes in the water pressure of their _______ cells.

18. _________ is the plant hormone involved in fruit ripening.

19. _________ is the response in plants that causes roots to grow downwards.

20. _________ are the most numerous and constant pollinators of insect-pollinated plants.

21. The first step in seed germination occurs when the seed _______.

Answers to the Concept Review questions appear in Appendix B.

CHALLENGE YOURSELF

1. When plant roots are deprived of oxygen, they lose their ability to absorb ions. Why is this? What does this say about the ion absorption process?
2. Why do gardeners often remove many of a plant's leaves after transplanting it?
3. If day length in a particular place at a particular time of year were 10 hours, which would produce flowers: a short-day plant, a long-day plant, both, or neither? Why? Do you think there are any short-day plants in the tropics? Why?
4. When poinsettias are kept inside a house after the holidays, they rarely bloom again. Why do you think this might be, and what might you do to get them to produce flowers a second time?
5. Angiosperms usually produce both pollen and ovules within a single flower or at least on a single plant. Why don't all angiosperms simply self-pollinate? What are the advantages of self-pollination?
6. What role did bees play in the origin of angiosperms and in their subsequent diversification?

FOR FURTHER READING

Chapin, F. S., III. "Integrated Responses of Plants to Stress." *Bioscience* 41 (1991): 29–36. All plants respond to stress of many types in basically the same way, which we are beginning to understand more completely.

Hopkins, W. G. *Introduction to Plant Physiology*. New York: John Wiley & Sons, 1995. An up-to-date treatment of the entire field.

Lipske, M. "Forget Hollywood: These Bloodthirsty Beauties are For Real." *Smithsonian,* December 1992, 49–59. Lively description of carnivorous plants and the biological communities that live and function within and on their leaves.

Proctore, M., and P. Yeo. *The Pollination of Flowering Plants*. New York: Talpinger Publishing, 1973. An excellent introduction to pollination biology, clearly and interestingly presented.

Raven, R. H., R. F. Evert, and S. E. Eichhorn: *Biology of Plants,* 5th ed. New York: Worth Publishers, 1993. A comprehensive treatment of general botany; a standard text in the field.

Sandved, K. B., and G. T. Prance. *Leaves*. New York: Crown Publishers, 1984. This book provides an incredibly beautiful introduction to the diversity of leaves.

Simons, P. "The Secret Feelings of Plants." *New Scientist* 136 (1992): 29–32. A fascinating account of recent investigation into the ways in which plants respond to touch.

TECHNOLOGY LINKS

The Living World Home Page
http://www.wcbp.com/biology/tlw

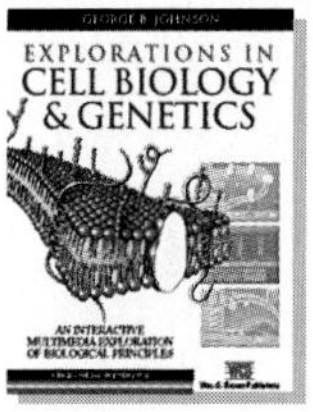

Explorations in Cell Biology & Genetics CD
#9 Photosynthesis

Life Science Animations Videotape 5
#47 How Water Moves Through a Plant
#48 How Food Moves from a Source to a Sink
#50 Mitosis & Cell Division in Plants

16

Evolution of the Animal Phyla

Great moments in evolution.

CHAPTER OUTLINE

Figure 16.1 Desert tarantula on sandstone.
This spider (*Aphonopelma chalcodes*) belongs to the phylum Arthropoda. The arthropods, the mollusks (phylum Mollusca), and our own phylum, Chordata, are the three most successful land-dwelling phyla.

Animals are the eaters of the earth, and all of them are multicellular heterotrophs. They are distinct among multicellular organisms in that animal cells lack rigid cell walls and are usually quite flexible. Many are able to move from place to place in search of food, which they ingest. Animals are very diverse in form, ranging in size from fleas too small to see with the naked eye to enormous whales and giant squids. The animal kingdom includes about 36 phyla, most of which occur in the sea, with far fewer in freshwater and fewer still on land. Members of three phyla, Arthropoda (insects and spiders, figure 16.1), Mollusca (snails), and Chordata (vertebrates), dominate animal life on land.

16.1 Some General Features of Animals

In this chapter we explore the great diversity of the animal kingdom. Most animals reproduce sexually. Animal eggs, which are nonmotile, are much larger than the small, usually flagellated sperm. In animals, the cells formed in meiosis function directly as gametes; they do not divide by mitosis first, as they do in plants and fungi, but rather fuse directly with one another to form the zygote. Consequently, there is no counterpart among animals of the alternation of haploid (gametophyte) and diploid (sporophyte) generations characteristic of plants (see chapter 14). With few exceptions, animals are diploid and the gametes are the only haploid cells in their life cycles.

This chapter describes a series of nine key evolutionary innovations, each exemplified by a major **phylum** (figure 16.2). As you will recall from chapter 10, a phylum is the highest level of biological organization within a kingdom. There are about 36 phyla in the animal kingdom. While many other interesting and important innovations also arose, these nine serve to highlight both the progressive nature of animal evolution and the important role of certain key elements of body architecture. The key characteristics of these nine phyla as well as several additional animal phyla of significance are shown in table 16.1. The exact order in which the major animal phyla evolved is not known with certainty, but the rough outlines of their evolutionary history can be seen in the progressive change brought about by the nine major adaptations you will now explore. In reviewing these nine stages of animal evolution, you will be tracing the evolutionary history of the animals, an evolutionary journey that took place in the sea over 600 million years ago.

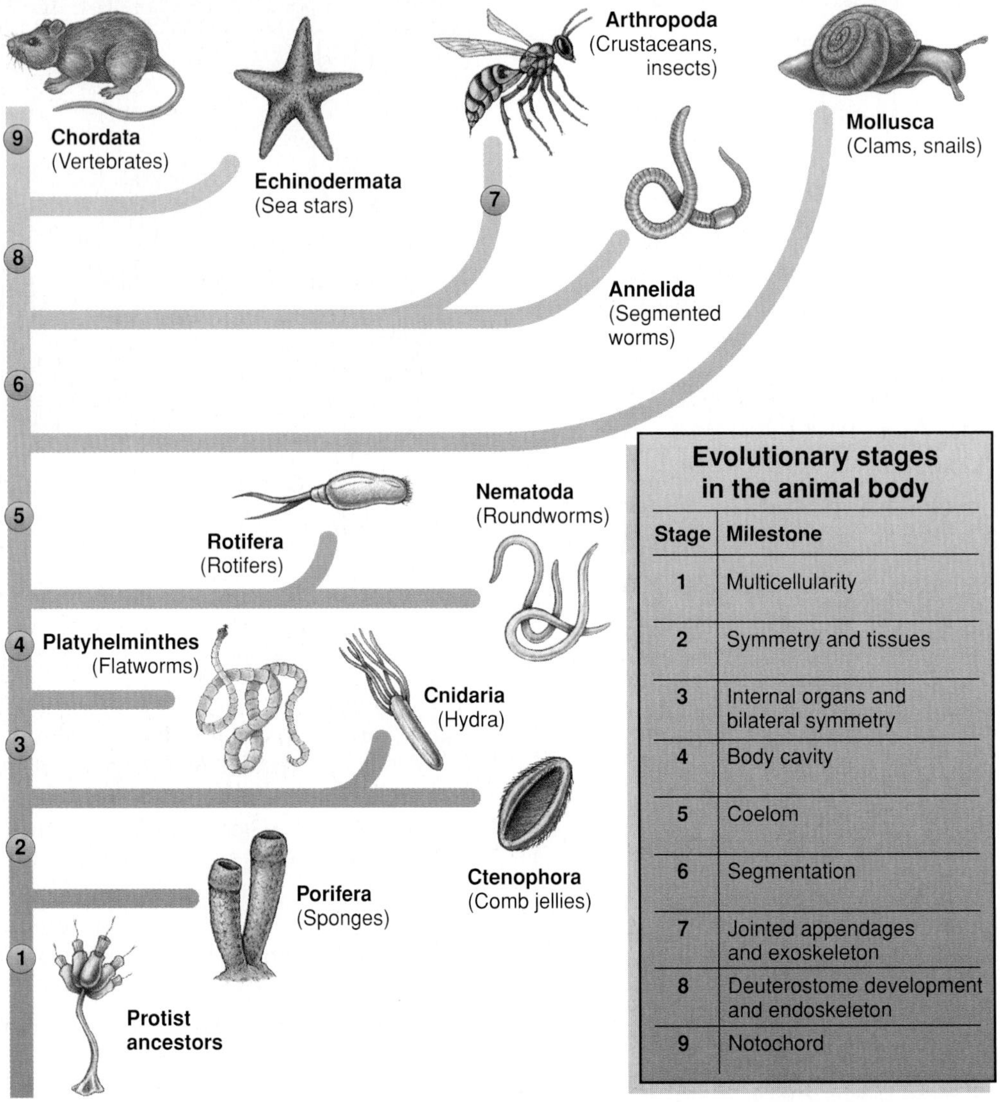

Evolutionary stages in the animal body

Stage	Milestone
1	Multicellularity
2	Symmetry and tissues
3	Internal organs and bilateral symmetry
4	Body cavity
5	Coelom
6	Segmentation
7	Jointed appendages and exoskeleton
8	Deuterostome development and endoskeleton
9	Notochord

Figure 16.2 The animal body: an evolutionary journey.
In this chapter we examine nine key innovations in the body design of animals, using in each case a major phylum to explore the nature of the change.

Table 16.1 The Major Animal Phyla

Phylum	Typical Examples		Key Characteristics	Approximate Number of Named Species
Arthropoda (arthropods)	Beetles, other insects, crabs, spiders		Most successful of all animal phyla; chitinous exoskeleton covering segmented bodies with paired, jointed appendages; many insect groups have wings	1,000,000
Mollusca (mollusks)	Snails, oysters, octopuses, nudibranchs		Soft-bodied coelomates whose bodies are divided into three parts: head–foot, visceral mass, and mantle; many have shells; almost all possess a unique rasping tongue called a radula; 35,000 species are terrestrial	110,000
Chordata (chordates)	Mammals, fish, dinosaurs, birds, amphibians		Segmented coelomates with a notochord; possess a dorsal nerve cord, pharyngeal slits, and a tail at some stage of life; in vertebrates, the notochord is replaced during development by the spinal column; 20,000 species are terrestrial	42,500
Platyhelminthes (flatworms)	*Planaria*, tapeworms, liver flukes	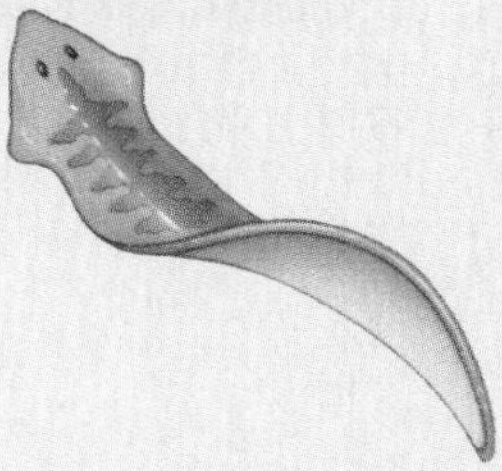	Solid, unsegmented, bilaterally symmetrical worms; no body cavity; digestive cavity has only one opening	15,000
Nematoda (roundworms)	*Ascaris*, pinworms, hookworms, *Filaria*	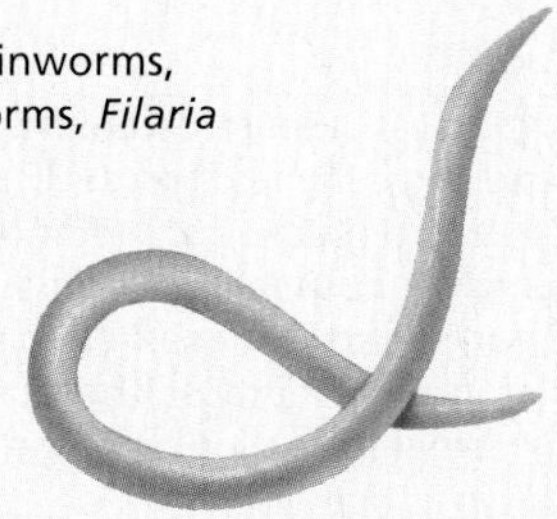	Pseudocoelomate, unsegmented, bilaterally symmetrical worms; tubular digestive tract passing from mouth to anus; tiny; without cilia; live in great numbers in soil; some are important animal parasites	12,000+
Annelida (segmented worms)	Earthworms, polychaetes, leeches, beach tube worms	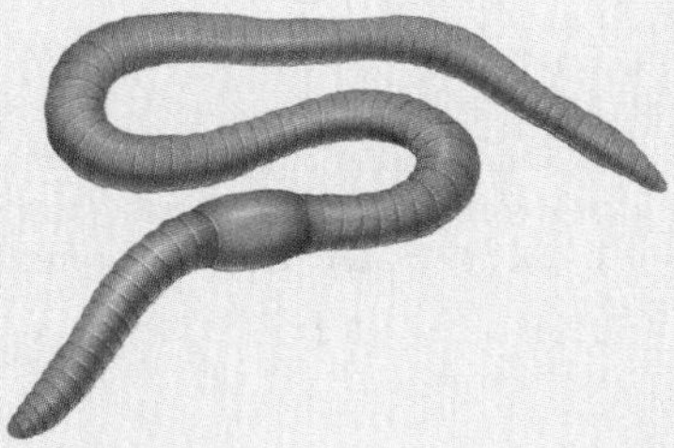	Coelomate, serially segmented, bilaterally symmetrical worms; complete digestive tract; terrestrial forms have bristles called setae on each segment that anchor them during crawling	12,000

continued

Table 16.1 The Major Animal Phyla (continued)

Phylum	Typical Examples	Key Characteristics	Approximate Number of Named Species
Cnidaria (jellyfish)	Jellyfish, hydra, corals, sea anemones 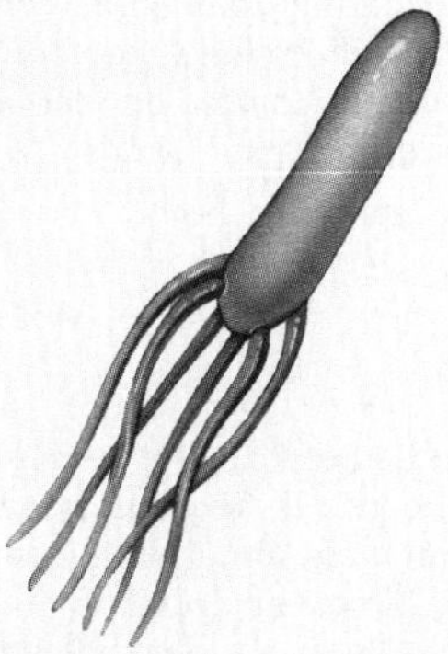	Soft, gelatinous, radially symmetrical bodies whose digestive cavity has a single opening; possess tentacles armed with stinging cells called cnidocytes that shoot sharp harpoons called nematocysts; marine	10,100
Echinodermata (starfish)	Sea stars, sea urchins, sand dollars, sea cucumbers 	Deuterostomes with radially symmetrical adult bodies; endoskeleton of calcium plates; five-part body plan and unique water vascular system with tube feet; able to regenerate lost body parts; marine	6,000
Porifera (sponges)	Barrel sponges, boring sponges, basket sponges, vase sponges 	Asymmetrical bodies; without distinct tissues or organs; saclike body consists of two layers breached by many pores; internal cavity lined with food-filtering cells called choanocytes; mostly marine (150 species live in freshwater)	5,150
Bryozoa (moss animals)	*Bowerbankia, Plumatella,* sea mats, sea moss 	Microscopic, aquatic deuterostomes that form branching colonies, possess U-shaped row of ciliated tentacles for feeding called a lophophore that usually protrudes through pores in a hard exoskeleton; also called "Ectoprocta" because the anus or proct is external to the lophophore	4,000
Rotifera (wheel animals)	Rotifers 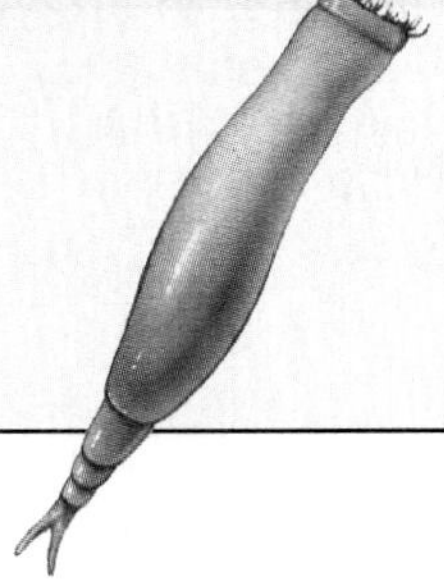	Small, aquatic pseudocoelomates with a crown of cilia around mouth resembling a wheel; almost all live in freshwater	2,000

Table 16.1 The Major Animal Phyla (continued)

Phylum	Typical Examples	Key Characteristics	Approximate Number of Named Species
Minor worms	Velvet worms, acorn worms, arrowworms, giant tube worms 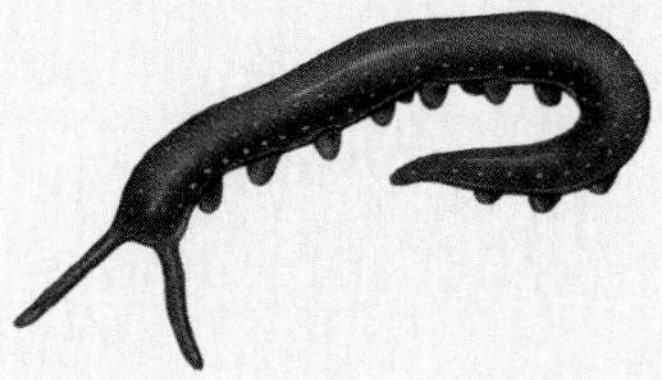	**Chaetognatha** (arrowworms): coelomate deuterostomes; bilaterally symmetrical; large eyes and powerful jaws **Hemichordata** (acorn worms): marine worms with dorsal *and* ventral nerve cords **Onychophora** (velvet worms): protostomes with a chitinous exoskeleton; evolutionary relicts **Pognophora** (tube worms): sessile deep-sea worms with long tentacles; live within chitinous tubes attached to the ocean floor **Rhynchocoela** (ribbon worms): acoelomate, bilaterally symmetrical marine worms with long, extendable proboscis	980
Brachiopoda (lamp shells)	*Lingula* 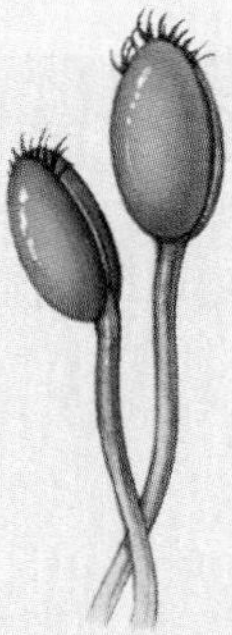	Like bryozoans, possess a lophophore, but within two clamlike shells; more than 30,000 species known as fossils	250
Ctenophora (sea walnuts)	Comb jellies, sea walnuts 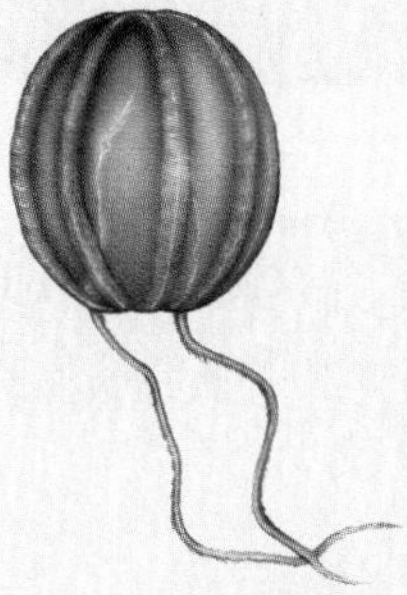	Gelatinous, almost transparent, often bioluminescent marine animals; eight bands of cilia; largest animals that use cilia for locomotion; complete digestive tract with anal pore	100
Phoronida (lophophores)	*Phoronis* 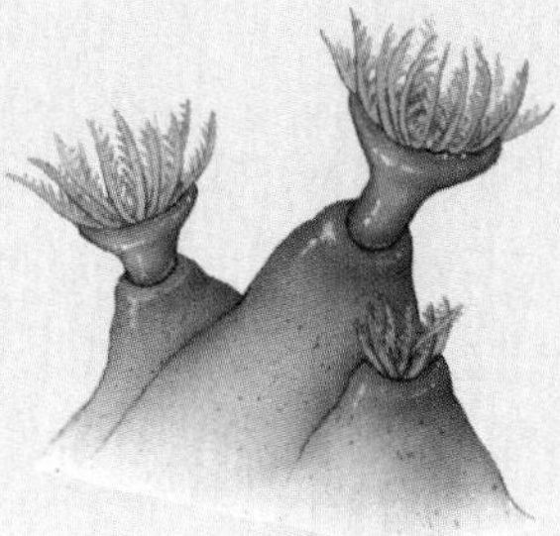	Lophophorate, protostome tube worms; often live in dense populations; unique U-shaped gut, instead of the straight digestive tube of other tube worms	12
Loricifera (sand animals)	*Nanoloricus mysticus* 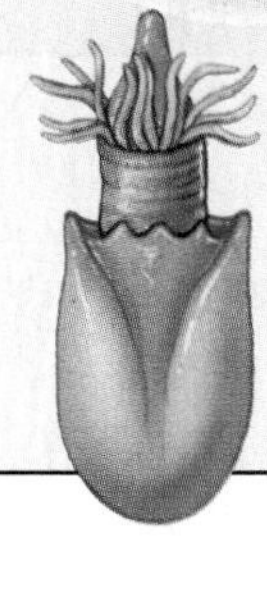	Tiny, bilaterally symmetrical pseudocoelomates that live in spaces between grains of sand; mouthparts possess a unique flexible tube; a recently discovered animal phylum (1983)	6

16.2 The Simplest Animals

The kingdom Animalia consists of two subkingdoms: (1) *Parazoa,* animals that lack a definite symmetry and possess neither tissues nor organs, and (2) *Eumetazoa*, animals that have a definite shape and symmetry, and in most cases tissues organized into organs. The subkingdom Parazoa consists primarily of the sponges, phylum Porifera. The other animals, composing about 35 phyla, belong to the subkingdom Eumetazoa.

Sponges: Animals Without Tissues

Sponges, members of the phylum Porifera, are the simplest complexly multicellular organisms. Their bodies contain a variety of highly specialized cells, although their cells are not organized into tissues. Most sponges completely lack symmetry. Sponges lack organs, their bodies consisting of little more than masses of specialized cells embedded in a gel-like matrix, like chopped fruit in jello. However, sponge cells do possess a key property of animal cells: cell recognition. The ability of a sponge cell to recognize another sponge cell enables a sponge to pass through a fine silk mesh with individual cells separating and then reaggregating on the other side to reform the sponge.

About 5,000 species exist, almost all in the sea (a few live in freshwater). Some are tiny, others more than 2 meters in diameter (figure 16.3). The body of an adult sponge is anchored in place on the seafloor and functions as a water-filtering machine, enabling the sponge to trap protists and tiny animals that live in seawater. The body is shaped like a vase, its outside covered with a skin of flattened cells called epithelial cells that protect the sponge. The body of the sponge is perforated by tiny holes. The name of the phylum, Porifera, refers to this system of pores. Facing into the internal cavity are unique flagellated cells called **choanocytes,** or collar cells. The beating of the flagella of the many choanocytes that line the body cavity of the sponge draws water in through the pores and drives it through the cavity. One cubic centimeter of sponge tissue can propel more than 20 liters of water a day in and out of the sponge body! Why all this moving of water? The sponge is a "filter feeder." The beating of each choanocyte's flagellum draws water down through its collar, made of small hairlike projections resembling a picket fence. Any food particles in the water are trapped in the fence and later ingested by the choanocyte or other cells of the sponge.

The choanocytes of sponges very closely resemble a kind of protist called choanoflagellates, which seem almost certain to have been the ancestors of sponges. Indeed, they may be the ancestors of *all* animals, although it is difficult to be certain that sponges are the direct ancestors of the other more complex phyla of animals.

Key characteristics of the phylum Porifera are summarized in figure 16.4.

(a)

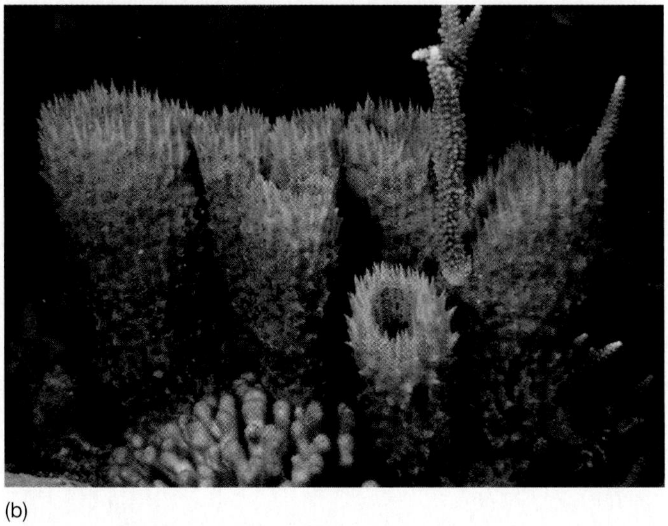

(b)

Figure 16.3 Diversity in sponges.
These two marine sponges are barrel sponges. They are among the largest of sponges, with well-organized forms. Many are more than 2 meters in diameter (*a*), while others are smaller (*b*).

1 Phylum Porifera: Sponges

Key Evolutionary Advance: MULTICELLULARITY. The body of a sponge (phylum Porifera) is complexly **multicellular**—that is, it contains many cells, of several distinctly different types, whose activities are coordinated with each other. The sponge body is **not symmetrical** and has **no organized tissues.**

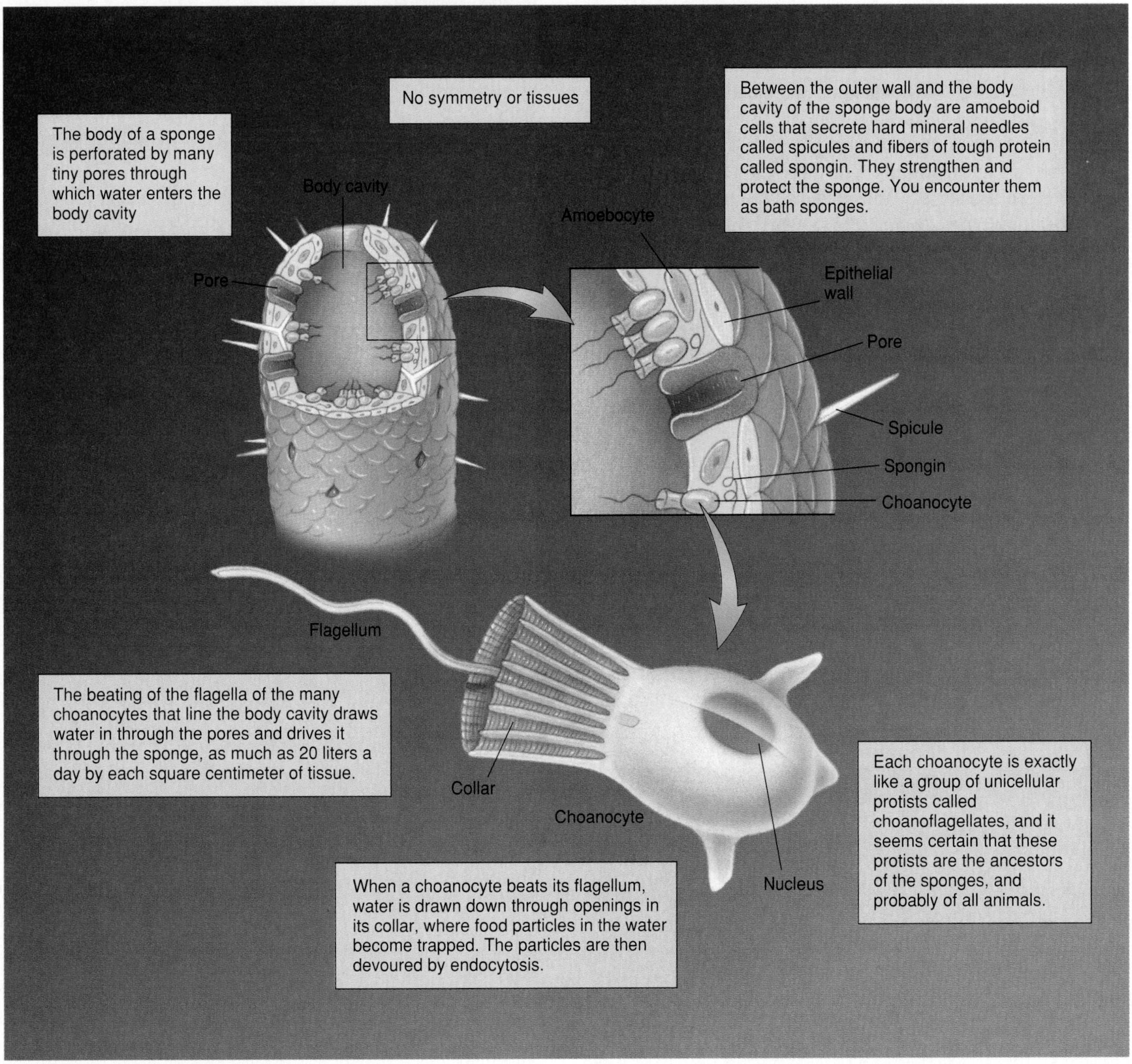

No symmetry or tissues
The body of a sponge is perforated by many tiny pores through which water enters the body cavity
Between the outer wall and the body cavity of the sponge body are amoeboid cells that secrete hard mineral needles called spicules and fibers of tough protein called spongin. They strengthen and protect the sponge. You encounter them as bath sponges.
Body cavity
Amoebocyte
Pore
Epithelial wall
Pore
Spicule
Spongin
Choanocyte
Flagellum
The beating of the flagella of the many choanocytes that line the body cavity draws water in through the pores and drives it through the sponge, as much as 20 liters a day by each square centimeter of tissue.
Collar
Choanocyte
Nucleus
Each choanocyte is exactly like a group of unicellular protists called choanoflagellates, and it seems certain that these protists are the ancestors of the sponges, and probably of all animals.
When a choanocyte beats its flagellum, water is drawn down through openings in its collar, where food particles in the water become trapped. The particles are then devoured by endocytosis.

Figure 16.4 Phylum Porifera: sponges.

Cnidarians: Tissues Lead to Greater Specialization

All animals other than sponges have both symmetry and tissues and thus are eumetazoans. Three distinct cell layers form in the embryos of all eumetazoans: an outer **ectoderm,** an inner **endoderm,** and an in-between **mesoderm.** These three embryonic tissues give rise to the basic body plan, differentiating into the many tissues of the adult body. In general, the outer coverings of the body and the nervous system develop from the ectoderm, the digestive organs and intestines develop from the endoderm, and the skeleton and muscles develop from the mesoderm (in cnidarians mesoderm is poorly developed).

The most primitive eumetazoans to exhibit symmetry and tissues are two **radially symmetrical** phyla, with body parts arranged around a central axis like the petals of a daisy. These two phyla are *Cnidaria* (pronounced ni-DAH-ree-ah), which includes jellyfish, hydra, sea anemones, and corals (figure 16.5), and *Ctenophora* (pronounced tea-NO-fo-rah), a minor phylum that includes the comb jellies. The bodies of all other eumetazoans are marked by a fundamental bilateral symmetry. Even starfish, which are radially symmetrical as adults, are bilaterally symmetrical when young.

All **cnidarians** are carnivores that capture their prey, such as fishes and shellfish, with tentacles that ring their mouths. These tentacles bear unique stinging cells called **cnidocytes,** which occur in no other organism and give the phylum its name. Within each cnidocyte is a small but powerful harpoon called a **nematocyst,** which cnidarians use to spear their prey and then draw the harpooned prey back to the tentacle containing the cnidocyte. The cnidocyte builds up a very high internal osmotic pressure and uses it to push the nematocyst outward so explosively that the barb can penetrate even the hard shell of a crab.

Key characteristics of the phylum Cnidaria are summarized in figure 16.6.

(b)

(c)

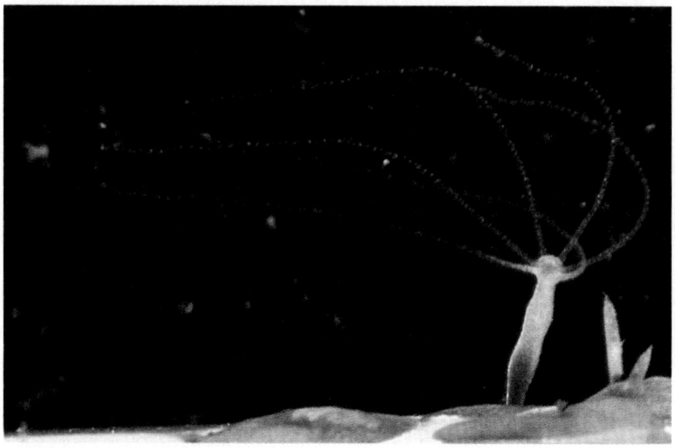

(a)

Figure 16.5 Representative cnidarians.

(*a*) *Hydra.* (*b*) Yellow cup coral. (*c*) Portuguese man-of-war, *Physalia utriculus.* The Portuguese man-of-war is a colony of cnidarians that has adopted the way of life characteristic of the jellyfish. This highly integrated colonial organism can ensnare good-sized fishes by using its painful stings and tentacles, which are sometimes over 15 meters long.

2 Phylum *Cnidaria:* Cnidarians

Key Evolutionary Advances: SYMMETRY and TISSUES. The cells of a cnidarian like *Hydra* are organized into specialized tissues. The interior gut cavity is specialized for **extracellular digestion**—that is, digestion within a gut cavity rather than within individual cells. Unlike sponges, cnidarians are **radially symmetrical,** with parts arranged around a central axis like the petals of a daisy.

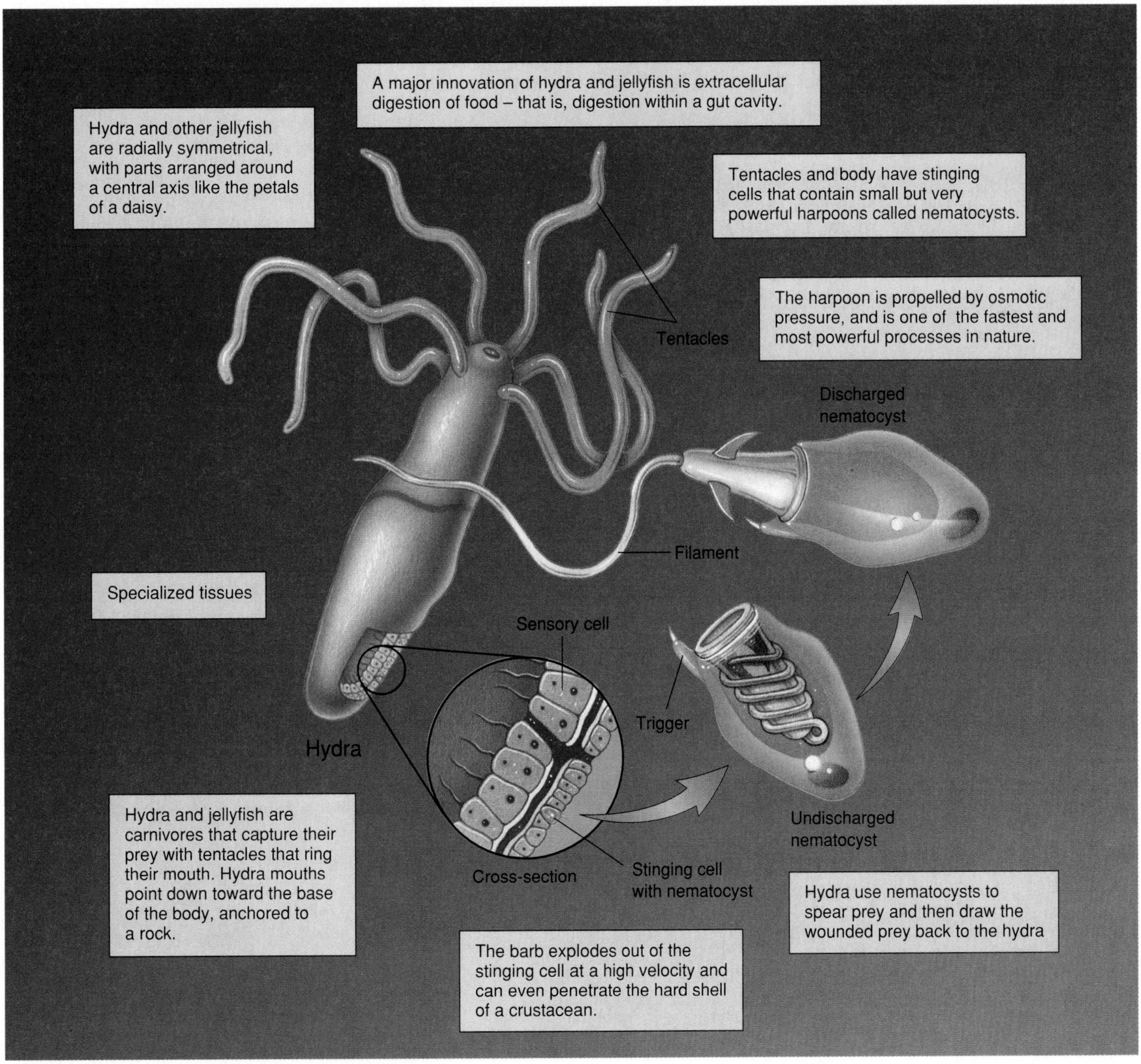

Figure 16.6 Phylum Cnidaria: cnidarians.

Cnidarians have two basic body forms, **medusae** and **polyps** (figure 16.7). Medusae are free-floating, gelatinous, and often umbrella-shaped. Their mouths point downward, with a ring of tentacles hanging down around the edges (hence the radial symmetry). Medusae are commonly called "jellyfish" because of their gelatinous interior or "stinging nettles" because of their nematocysts. Polyps are cylindrical, pipe-shaped animals that are usually attached to a rock. They also exhibit radial symmetry. In polyps, the mouth faces away from the rock and therefore is often directed upward. Many cnidarians exist only as medusae, others only as polyps, and still others alternate between these two phases during the course of their life cycles (figure 16.8).

A major evolutionary innovation among the cnidarians, as compared with sponges, is the **extracellular digestion** of food. Food trapped by a sponge choanocyte is taken directly into that cell, or into a circulating amoeboid cell, by endocytosis. In a cnidarian, food is digested *outside of cells,* in a gut cavity. Extracellular digestion is the same heterotrophic strategy pursued by fungi, except that fungi digest food outside their bodies, while animals digest it within their bodies, in a cavity. This evolutionary advance has been retained by all of the more advanced groups of animals. For the first time it became possible to digest an animal larger than oneself.

The comb jellies (phylum Ctenophora) are transparent relatives of the cnidarians. They propel themselves through the water by beating plates of fused cilia—they are the largest animals that use cilia for locomotion.

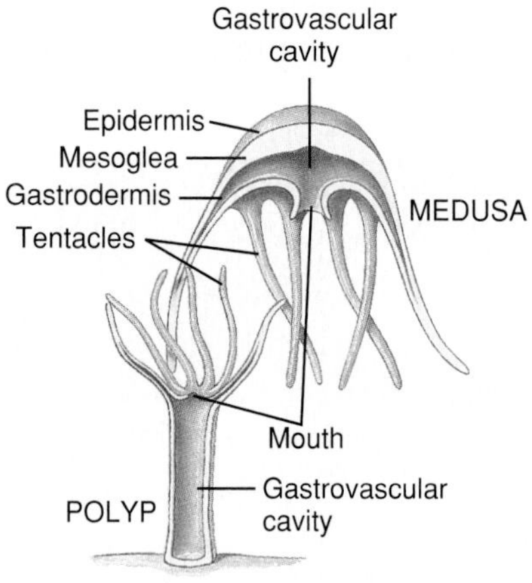

Figure 16.7 The two basic body forms of cnidarians.
The medusa (*above*) and the polyp (*below*) are the two phases that alternate in the life cycles of many cnidarians, but several species (corals and sea anemones, for example) exist only as polyps.

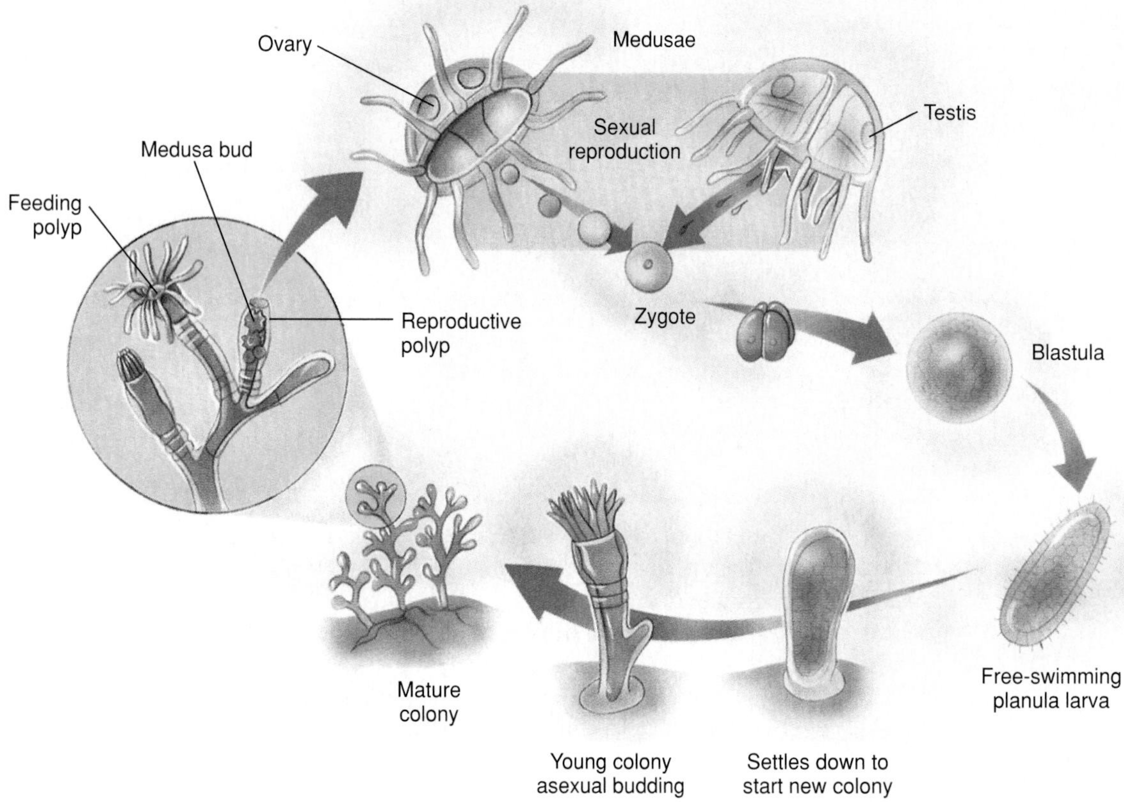

Figure 16.8 The life cycle of *Obelia,* a marine colonial hydroid.
Polyps reproduce asexually by budding, forming colonies. They may also give rise by the formation of specialized buds to medusae, in which gametes are produced. These gametes fuse, producing zygotes that develop into planulae, which, in turn, settle down to produce polyps.

16.3 The Advent of Bilateral Symmetry

All eumetazoans other than cnidarians and ctenophores are **bilaterally symmetrical**—that is, they have a right half and a left half that are mirror images of each other. In looking at a bilaterally symmetrical animal, you refer to the top half of the animal as **dorsal** and the bottom half as **ventral** (figure 16.9). The front is called **anterior** and the back **posterior.** Bilateral symmetry was a major evolutionary advance among the animals because it allows different parts of the body to become specialized in different ways. For example, most bilaterally symmetrical animals have evolved a definite head end, a process called **cephalization.** Animals that have heads are often active and mobile, moving through their environment headfirst, with sensory organs concentrated in front so the animal can test for food, danger, and mates as it enters new surroundings.

Figure 16.9 How radial and bilateral symmetry differ.
(*a*) Radial symmetry is the regular arrangement of parts around a central axis, so that any plane passing through the central axis divides the organism into halves that are approximate mirror images.
(*b*) Bilateral symmetry is reflected in a body form in which the right and left halves of an organism are approximate mirror images.

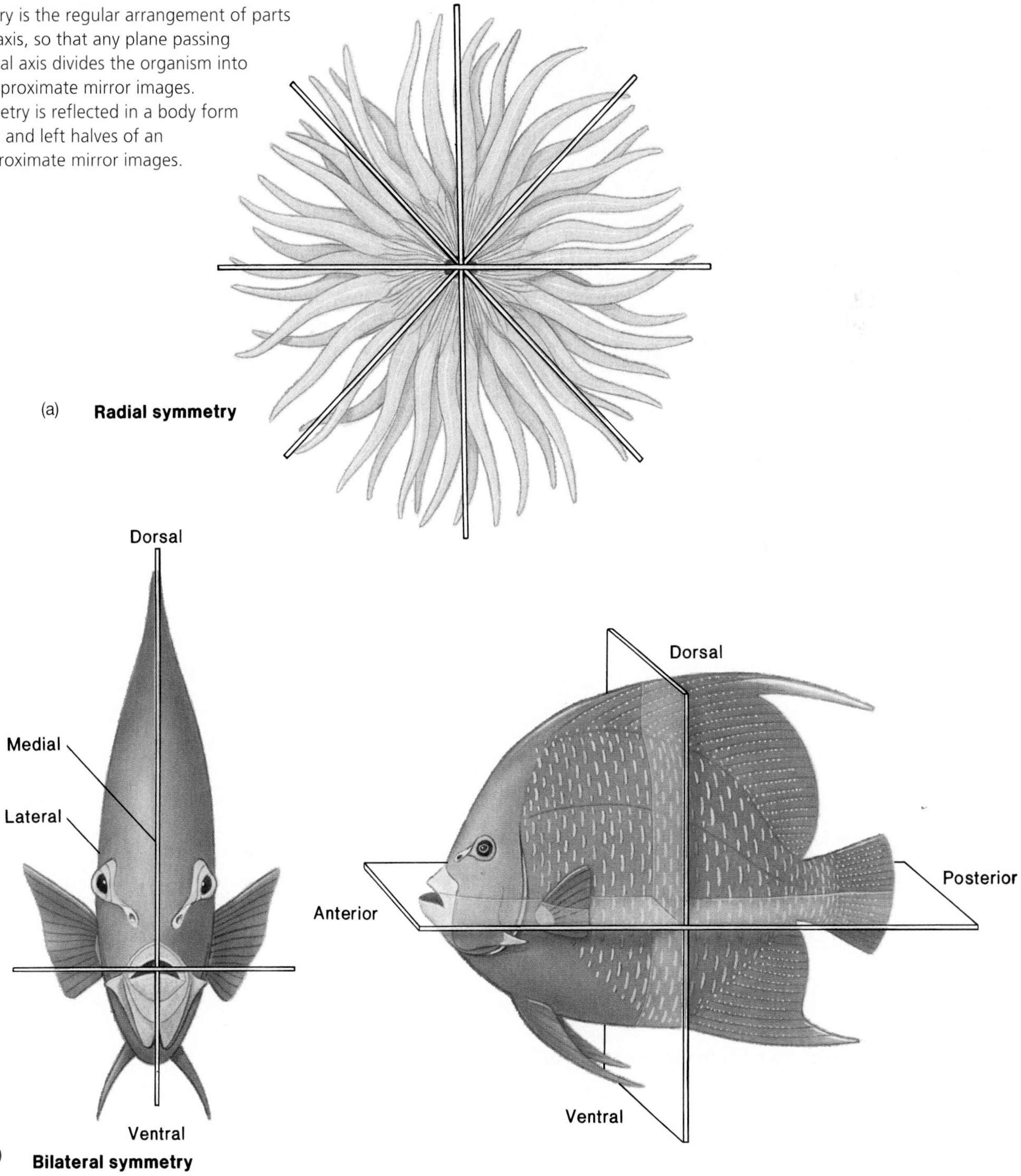

Solid Worms: Internal Organs

The simplest of all bilaterally symmetrical animals are the **solid worms.** By far the largest phylum of these, with about 15,000 species, is Platyhelminthes (pronounced plat-ee-hel-MIN-theeze), which includes the flatworms (figure 16.10).

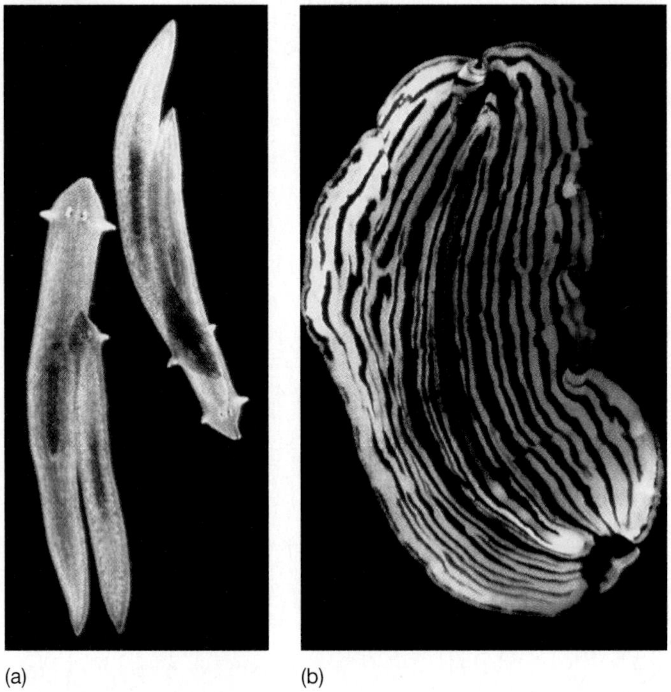

Figure 16.10 Flatworms.
(*a*) A common flatworm, *Planaria.* (*b*) A free-living marine flatworm.

Flatworms are the simplest animals in which organs occur. An organ is a collection of different tissues that function as a unit. The testes and uterus of flatworms are reproductive organs, for example. The dark spots on the head are eyespots that can detect light, although they cannot focus an image like your eyes can.

Solid worms lack any internal cavity other than the digestive tract. Flatworms are soft-bodied animals flattened from top to bottom, like a piece of tape or ribbon. If you were to cut a flatworm in half across its body, you would see that the gut is completely surrounded by tissues and organs (figure 16.11). This solid body construction is called **acoelomate,** meaning without a body cavity.

Flatworms are thin because of their acoelomate body design. They lack any circulatory system, so dissolved substances such as oxygen, carbon dioxide, and nutrients must pass through the solid body by diffusion. Having a thin body shortens the distance these substances must move. Diffusion is also facilitated by the fact that the gut of a flatworm is highly branched—some portion of it runs close to practically all of the flatworm's tissues. The gut has only one opening, the mouth, so that material must move through the gut in two ways, food in and wastes out.

Most species of flatworm are parasitic. Flatworms occur within the bodies of members of almost every other animal phylum. They range in size from free-living flatworms less than 1 millimeter long to tapeworms many meters long.

Key characteristics of the phylum Platyhelminthes are summarized in figure 16.12.

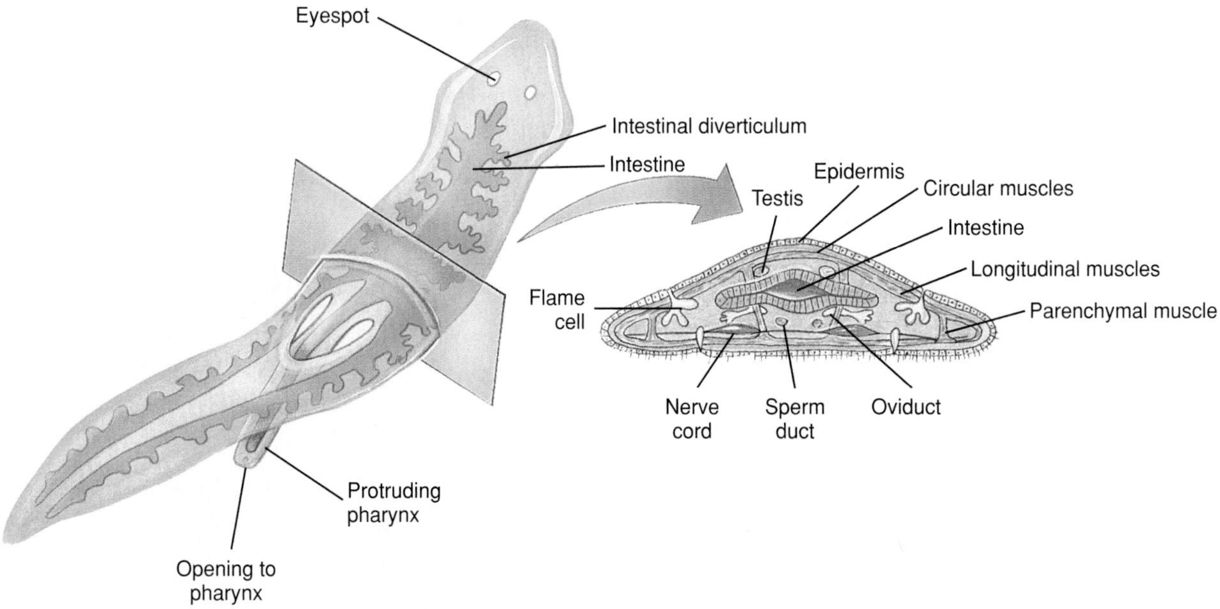

Figure 16.11 Diagram of flatworm anatomy.
The organism shown is *Dugesia,* the familiar freshwater "planaria" of many biology laboratories.

3 *Phylum Platyhelminthes:* Solid Worms

Key Evolutionary Advances: INTERNAL ORGANS and BILATERAL SYMMETRY. The evolution of the mesoderm in acoelomate solid worms such as the flatworms (phylum Platyhelminthes) allowed the formation of digestive and other **organs.** Flatworms were the first animals to be **bilaterally symmetrical** and to have a distinct **head.**

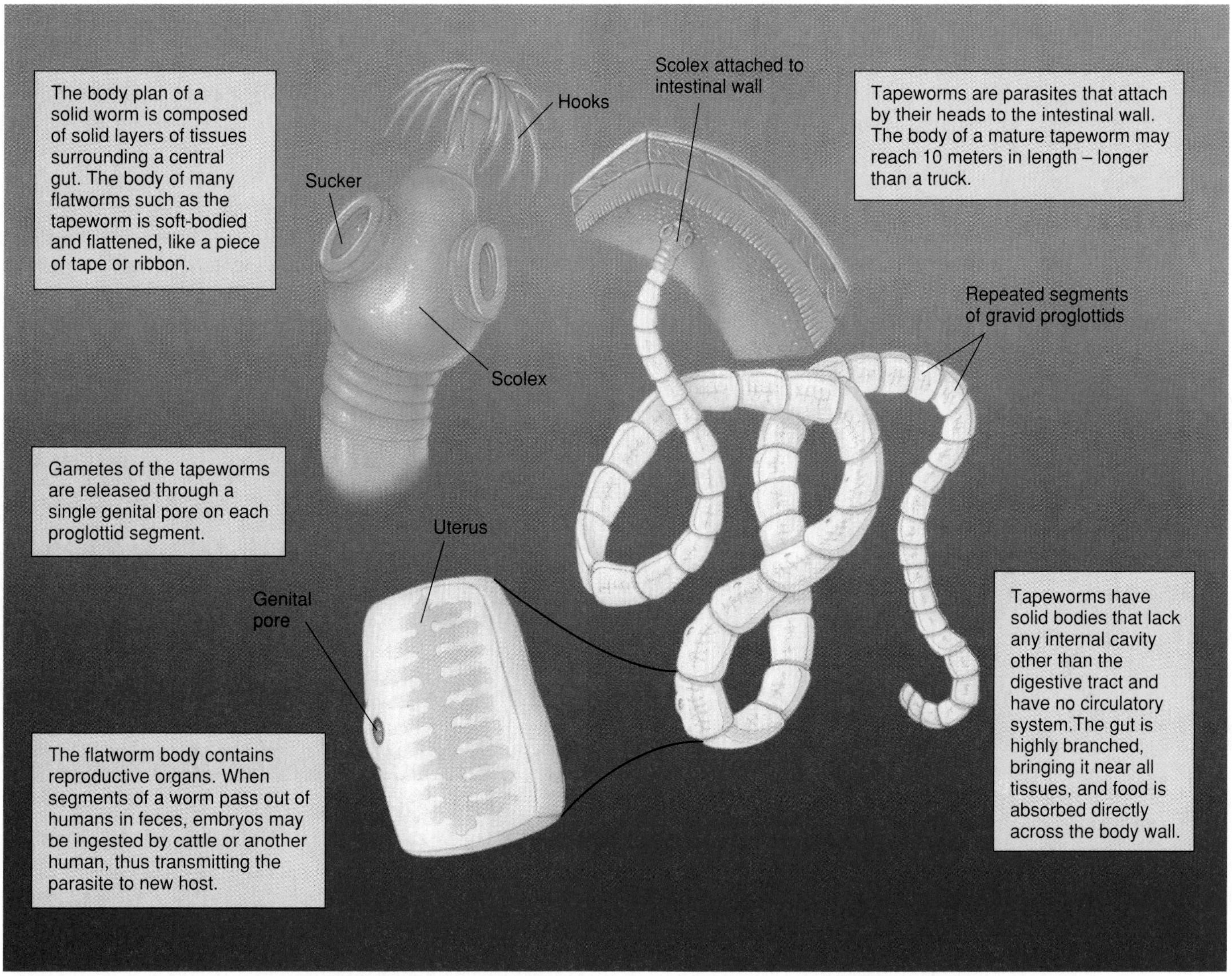

The body plan of a solid worm is composed of solid layers of tissues surrounding a central gut. The body of many flatworms such as the tapeworm is soft-bodied and flattened, like a piece of tape or ribbon.
Hooks
Sucker
Scolex
Scolex attached to intestinal wall
Tapeworms are parasites that attach by their heads to the intestinal wall. The body of a mature tapeworm may reach 10 meters in length – longer than a truck.
Repeated segments of gravid proglottids
Gametes of the tapeworms are released through a single genital pore on each proglottid segment.
Uterus
Genital pore
The flatworm body contains reproductive organs. When segments of a worm pass out of humans in feces, embryos may be ingested by cattle or another human, thus transmitting the parasite to new host.
Tapeworms have solid bodies that lack any internal cavity other than the digestive tract and have no circulatory system.The gut is highly branched, bringing it near all tissues, and food is absorbed directly across the body wall.

Figure 16.12 Phylum Platyhelminthes: solid worms.

16.4 The Advent of a Body Cavity

All bilaterally symmetrical animals other than solid worms have a cavity within their body (figure 16.13) The evolution of an internal body cavity was an important improvement in animal body design for several reasons:

1. **Circulation.** Fluids that move within the body cavity can serve the function of a circulatory system, permitting the rapid passage of materials from one part of the body to another, opening the way to larger bodies.
2. **Movement.** Fluid in the cavity makes the animal's body rigid, permitting resistance to muscle contraction and thus opening the way to muscle-driven body movement.
3. **Organ function.** In a fluid-filled enclosure, body organs can function without being deformed by surrounding muscles. For example, food can pass through a gut suspended within a cavity freely, at a rate not controlled by when the animal moves.

Roundworms: Pseudocoelomates

Among animals, seven phyla are characterized by a body cavity located between the endoderm and the mesoderm. Such a cavity is called a **pseudocoel,** and the animals in which it occurs are **pseudocoelomates.** Only one of the seven phyla includes a large number of species, the phylum Nematoda. There are over 12,000 species of **nematodes,** most of them microscopic animals that live in soil. It has been estimated that a spadeful of rich soil may contain, on the average, 1 million nematodes (figure 16.14)! A layer of muscles extends beneath a flexible thick cover of epidermis and cuticle along the length of the worm. These long muscles push against the cuticle and the pseudocoel, whipping the body from side to side as the animal moves.

A second phylum consisting of animals with a pseudocoelomate body plan is Rotifera, the rotifers. **Rotifers** are common, small, basically aquatic animals that have a crown of cilia at their heads; they range from 0.04 to 2 millimeters long (figure 16.15). About 2,000 species exist throughout the world. Bilaterally symmetrical and covered with chitin, rotifers depend on their cilia for both locomotion and feeding, ingesting bacteria, protists, and small animals.

All pseudocoelomates have a one-way digestive tract, in which food passes in through the mouth, along a digestive tract, and out the anus. This is a major improvement in body design, as it permits far greater specialization of the digestive tract. A one-way digestive tract can function like an assembly line, food being acted on in different ways in each section as it passes down the tract.

Key characteristics of the phylum Nematoda are summarized in figure 16.16.

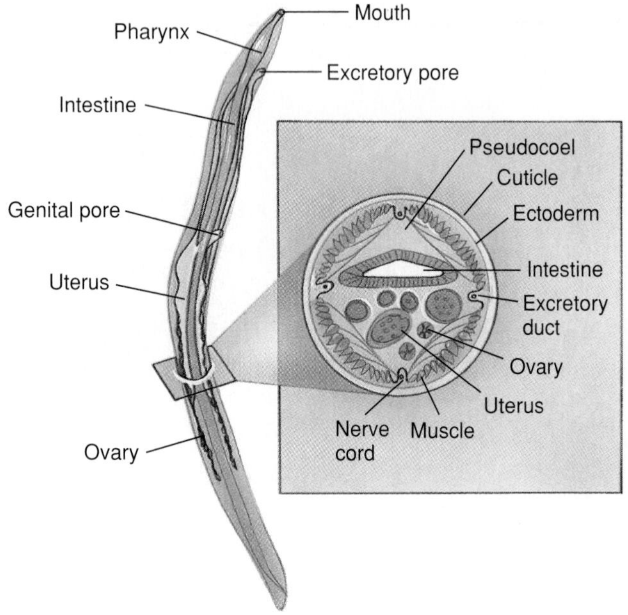

Figure 16.13 An internal body cavity.
In this cross section of a parasitic roundworm, *Ascaris lumbricoides,* the body cavity, called a pseudocoel, is positioned between the intestine derived from the endoderm and muscle tissues derived from the mesoderm.

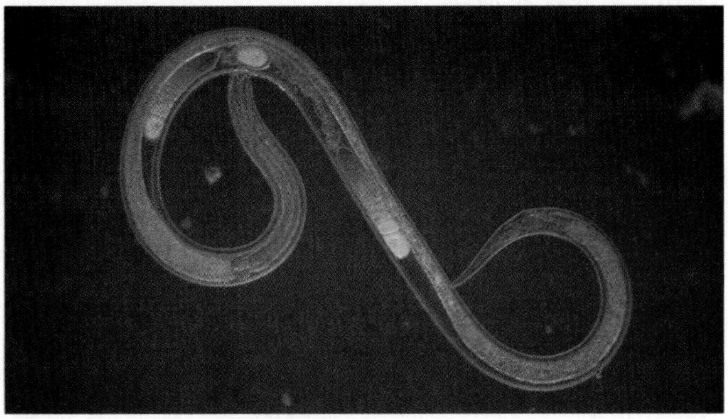

Figure 16.14 Nematodes.
One square meter of ordinary garden, lawn, or forest soil teems with millions of nematodes. Although most are similar in form, they range from about 0.2 millimeter to about 6 millimeters long. A trained nematologist (student of nematodes) must examine slide mounts with a compound microscope to determine which species are present.

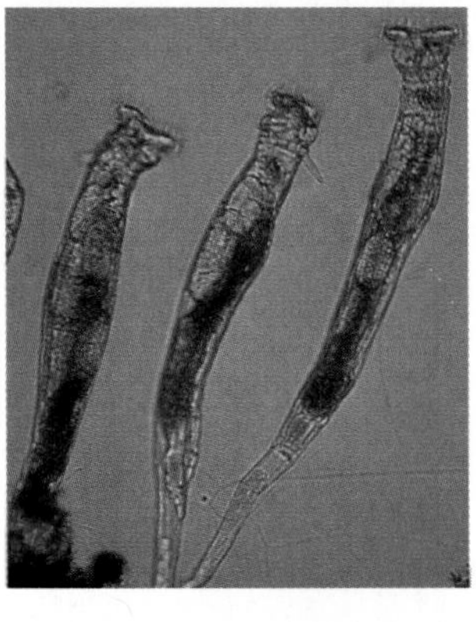

Figure 16.15 Rotifers (phylum Rotifera).
These common aquatic animals depend on their crown of cilia for feeding and locomotion.

4 *Phylum Nematoda:* Roundworms

Key Evolutionary Advance: BODY CAVITY. The major innovation in body design in roundworms (phylum Nematoda) is a **body cavity** between the gut and the body wall. This cavity is the pseudocoelom. This body cavity allows nutrients to circulate throughout the body and prevents organs from being deformed by muscle movements.

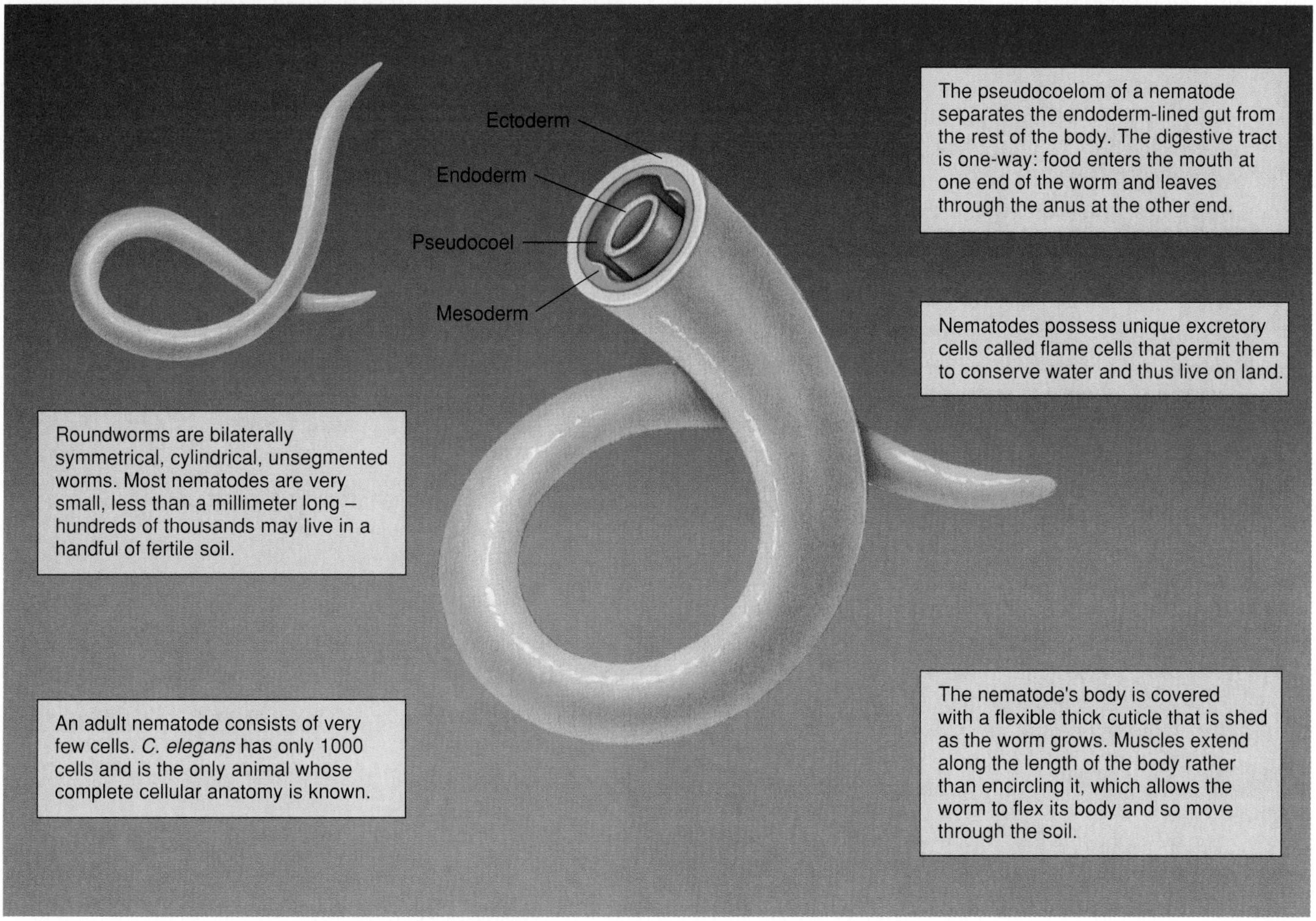

Ectoderm
Endoderm
Pseudocoel
Mesoderm
The pseudocoelom of a nematode separates the endoderm-lined gut from the rest of the body. The digestive tract is one-way: food enters the mouth at one end of the worm and leaves through the anus at the other end.
Nematodes possess unique excretory cells called flame cells that permit them to conserve water and thus live on land.
Roundworms are bilaterally symmetrical, cylindrical, unsegmented worms. Most nematodes are very small, less than a millimeter long – hundreds of thousands may live in a handful of fertile soil.
An adult nematode consists of very few cells. *C. elegans* has only 1000 cells and is the only animal whose complete cellular anatomy is known.
The nematode's body is covered with a flexible thick cuticle that is shed as the worm grows. Muscles extend along the length of the body rather than encircling it, which allows the worm to flex its body and so move through the soil.

Figure 16.16 Phylum Nematoda: roundworms.

Mollusks: Coelomates

Even though acoelomates and pseudocoelomates have proven very successful, a third way of organizing the animal body has evolved that occurs in the bulk of the animal kingdom. This new body design repositions the fluid-filled body cavity so that it develops not between endoderm and mesoderm but rather entirely within the mesoderm. Such a body cavity is called a **coelom,** and animals that possess such a cavity are called **coelomates.**

What is the functional difference between a pseudocoel and a coelom, and why has the latter kind of body cavity been so overwhelmingly more successful? The answer has to do with the nature of animal embryonic development. In animals, development of specialized tissues involves a process called **primary induction,** in which one of the three primary tissues (endoderm, mesoderm, and ectoderm) interacts with another. The interaction requires physical contact. A major advantage of the coelomate body plan is that it allows contact between mesoderm and endoderm, so that primary induction can occur during development. For example, contact between mesoderm and endoderm permits localized portions of the digestive tract to develop into complex, highly specialized regions like the stomach. In pseudocoelomates, mesoderm and endoderm are separated by the body cavity, limiting developmental interactions between these tissues.

The appearance of the coelom does resurrect one old problem that had been solved by pseudocoelomates—circulation. In coelomates, the gut is again surrounded by tissue that presents a barrier to diffusion, just as it was in solid worms. This problem is solved among coelomates by the development of a **circulatory system,** a network of vessels that carries fluids to all parts of the body. The circulating fluid, or blood, carries nutrients and oxygen to the tissues and removes wastes and carbon dioxide. Blood is usually pushed through the circulatory system by contraction of one or more muscular hearts.

The least advanced of the coelomates, the only major phylum of coelomates without segmented bodies, are the Mollusca (figure 16.17). The **mollusks** are the largest animal phylum, except for the arthropods, with over 110,000 species. Mollusks occur almost everywhere. Next to insects, they are the most successful land animal. There are more terrestrial mollusk species (35,000) than there are terrestrial vertebrate species (20,000).

Mollusks include three classes with outwardly different body plans. However, the seeming differences hides a basically similar body design. The body of all mollusks is composed of three distinct segments: a head, a central section that contains the body's organs, and a foot. Wrapped around the visceral mass like a cape is a heavy fold of tissue called the **mantle,** with the gills positioned on its inner surface like the lining of a coat. The **gills** are filamentous projections of tissue, rich in blood vessels, that capture oxygen from the water circulating between the mantle and visceral mass and release carbon dioxide.

(a)

(b)

Figure 16.17 Two classes of mollusks.
(*a*) A bivalve. (*b*) A cephalopod.

The three major classes of mollusks, all different variations upon this same basic design, are gastropods, bivalves, and cephalopods.

1. **Gastropods** (snails and slugs) use the muscular foot to crawl, and their mantle often secretes a single, hard, protective shell. All terrestrial mollusks are gastropods.
2. **Bivalves** (clams, oysters, and scallops) secrete a two-part shell with a hinge, as their name implies (figure 16.17*a*). They filter-feed by drawing water into their shell.
3. **Cephalopods** (octopuses and squids) have modified the mantle cavity to create a jet propulsion system that can propel them rapidly through the water (figure 16.17*b*). Their shell is greatly reduced to an internal structure or is absent.

One of the most characteristic features of mollusks is the **radula,** a rasping, tonguelike organ. With rows of pointed, backward-curving teeth, the radula is used by some snails to scrape algae off rocks. Cephalopods and many other gastropods are active predators that use their radula as a weapon to puncture their prey. The small holes often seen in oyster shells are produced by gastropods that have bored holes to kill the oyster and extract its body for food.

Key characteristics of the phylum Mollusca are summarized in figure 16.18.

5 *Phylum Mollusca:* Mollusks

Key Evolutionary Advance: COELOM. The body cavity of a mollusk like this snail (phylum Mollusca) is a coelom, completely enclosed within the mesoderm. This allows physical contact between the mesoderm and the endoderm, permitting interactions that lead to development of highly specialized organs such as a stomach.

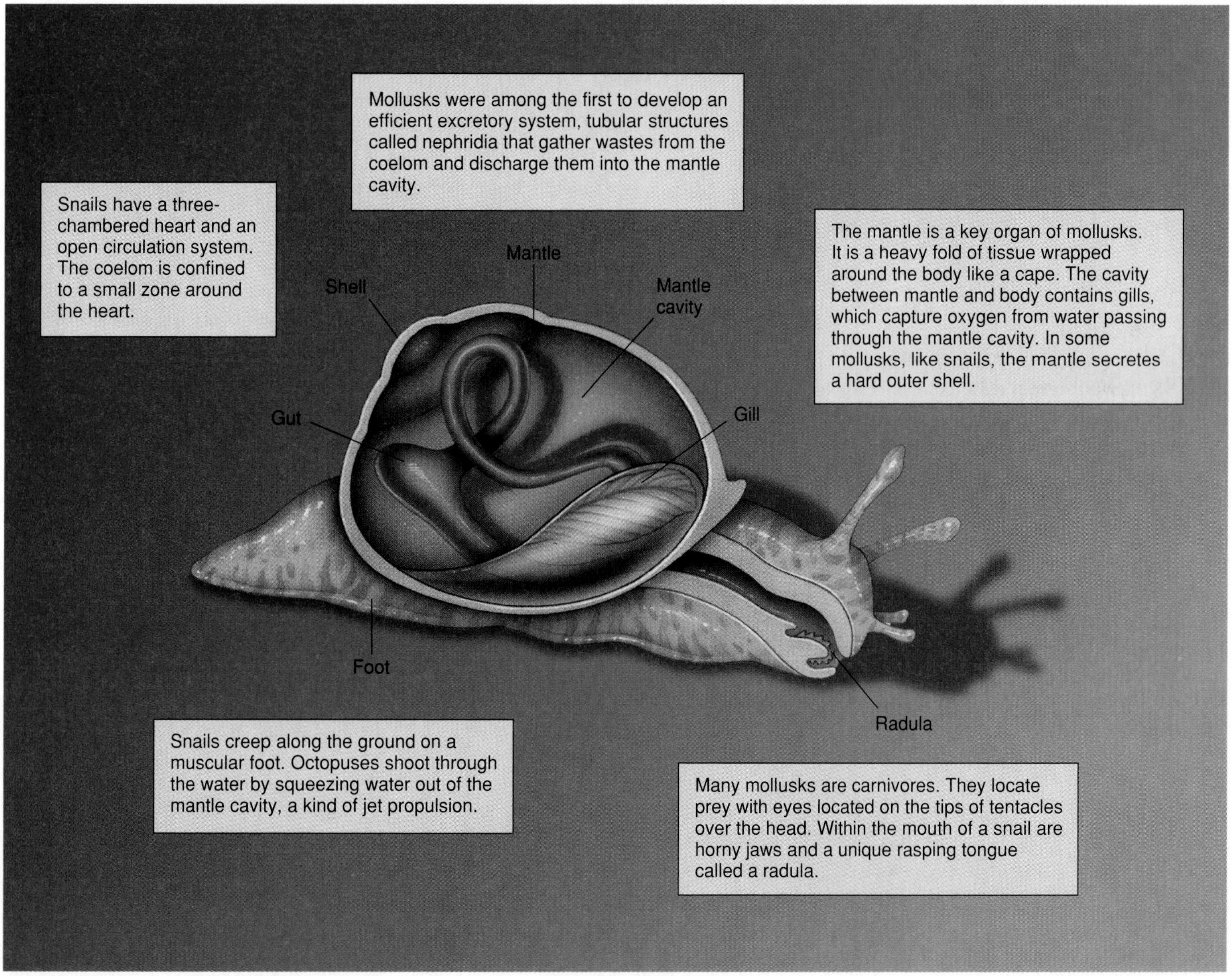

Mollusks were among the first to develop an efficient excretory system, tubular structures called nephridia that gather wastes from the coelom and discharge them into the mantle cavity.
Snails have a three-chambered heart and an open circulation system. The coelom is confined to a small zone around the heart.
Mantle
Shell
Mantle cavity
The mantle is a key organ of mollusks. It is a heavy fold of tissue wrapped around the body like a cape. The cavity between mantle and body contains gills, which capture oxygen from water passing through the mantle cavity. In some mollusks, like snails, the mantle secretes a hard outer shell.
Gut
Gill
Foot
Radula
Snails creep along the ground on a muscular foot. Octopuses shoot through the water by squeezing water out of the mantle cavity, a kind of jet propulsion.
Many mollusks are carnivores. They locate prey with eyes located on the tips of tentacles over the head. Within the mouth of a snail are horny jaws and a unique rasping tongue called a radula.

Figure 16.18 Phylum Mollusca: mollusks.

Annelids: The Rise of Segmentation

One of the early key innovations in body plan to arise among the coelomates was **segmentation,** the building of a body from a series of similar segments. The first segmented animals to evolve were the **annelid worms,** the phylum Annelida (figure 16.19) These advanced coelomates are assembled as a chain of nearly identical segments, like the boxcars of a train. The great advantage of such segmentation is the evolutionary flexibility it offers—a small change in an existing segment can produce a new kind of segment with a different function. Thus, in an annelid worm, some segments are modified for reproduction, some for feeding, and others for eliminating wastes.

Two-thirds of all annelids live in the sea (about 8,000 species), and most of the rest—some 3,100 species—are earthworms. The basic body plan of an annelid is a tube within a tube: the digestive tract is a tube suspended within the coelom, which is itself a tube running from mouth to anus.

1. **Repeated segments.** The body segments of an annelid are visible as a series of ringlike structures running the length of the body, looking like a stack of donuts. The segments are divided from one another internally by partitions, just as walls separate the rooms of a building. In each of the cylindrical segments, the digestive, excretory, and locomotor organs are repeated. The body fluid within the coelom of each segment creates a hydrostatic (liquid-supported) skeleton that gives the segment rigidity, like an inflated balloon. Muscles within each segment play against the fluid in the coelom. Because each segment is separate, each is able to expand or contract independently. This lets the worm body move in ways that are quite complex. When an earthworm crawls on a flat surface, for example, it lengthens some parts of its body while shortening others.
2. **Specialized segments.** The anterior (front) segments of annelids contain the sensory organs of the worm. Some of these are organs sensitive to light, and elaborate eyes with lenses and retinas have evolved in some annelids. A well-developed cerebral ganglion, or brain, is contained in one anterior segment.
3. **Connections.** Because partitions separate the segments, it is necessary to provide ways for materials and information to pass between segments. A circulatory system carries blood from one segment to another, while nerve cords connect the nerve centers or ganglia located in each segment with each other and the brain. The brain can then coordinate the worm's activities.

Segmentation underlies the body organization of all advanced coelomate animals, not only annelids but also arthropods (crustaceans, spiders, and insects) and chordates (mostly vertebrates). Sometimes the segmentation is not obvious when viewing an adult arthropod or vertebrate. In many arthropods, the segments are fused, making it difficult to perceive the underlying pattern. In the adult human, too, segments are not apparent, although they can be clearly seen during embryonic development. For example, vertebrate muscles develop from repeated blocks of tissue called somites that occur in the embryo. Another example of vertebrate segmentation is the vertebral column, which is a stack of very similar vertebrae.

Key characteristics of the phylum Annelida are summarized in figure 16.20.

(a)

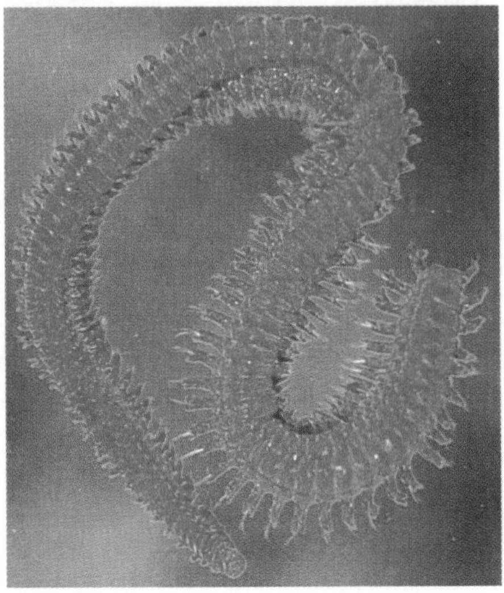

(b)

Figure 16.19 Representative annelids.

(*a*) Earthworms are the terrestrial annelids. This night crawler, *Lumbricus terrestris,* is in its burrow.
(*b*) Shiny bristle worm, *Oenone fulgida,* a polychaete.

6 *Phylum Annelida:* Annelids

Key Evolutionary Advance: SEGMENTATION. Marine polychaetes and earthworms (phylum Annelida) were the first organisms to evolve a body plan based on **repeated body segments.** Most segments are identical and are separated from other segments by partitions.

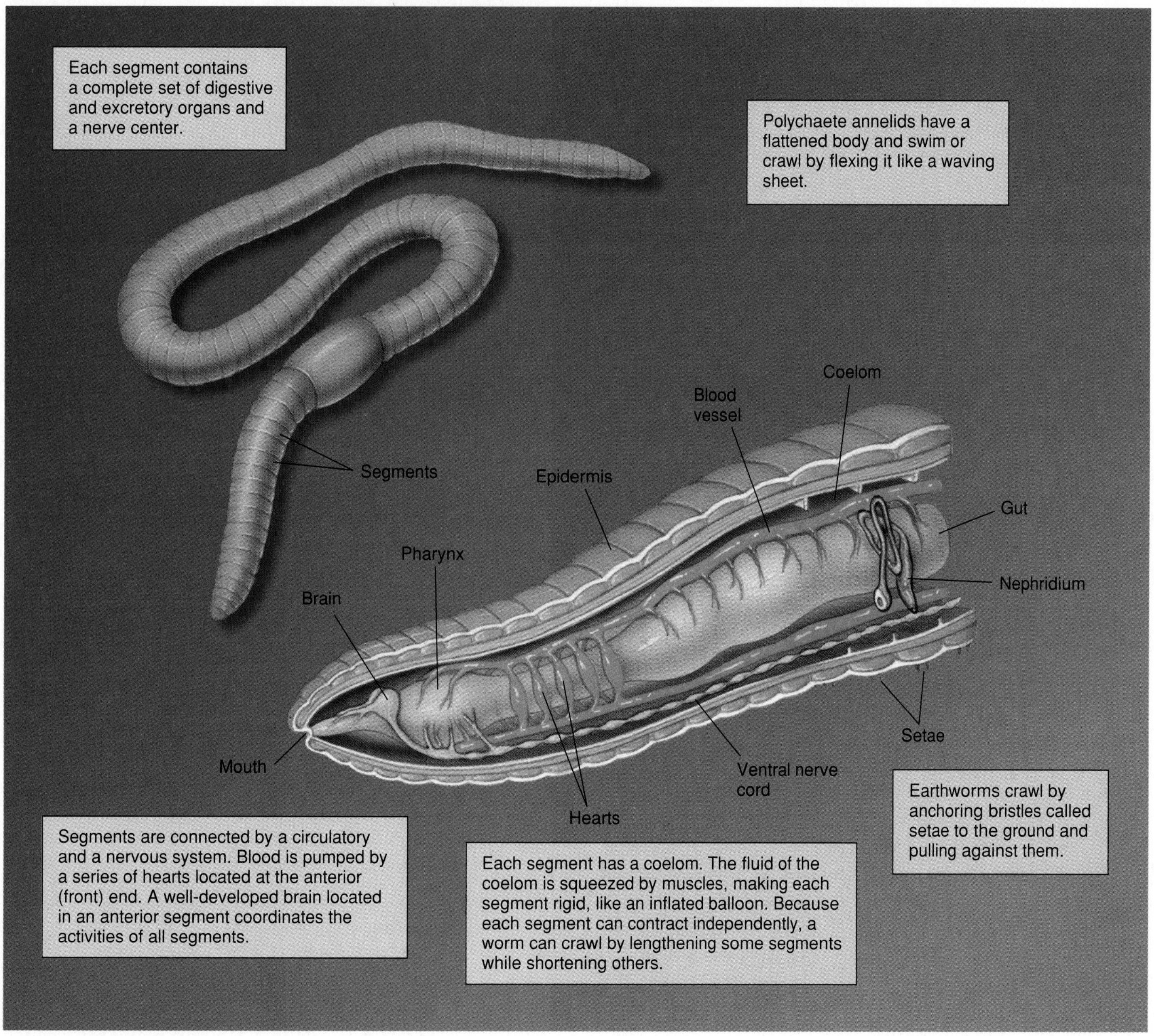

Each segment contains a complete set of digestive and excretory organs and a nerve center.
Polychaete annelids have a flattened body and swim or crawl by flexing it like a waving sheet.
Coelom
Blood vessel
Segments
Epidermis
Gut
Pharynx
Nephridium
Brain
Setae
Mouth
Ventral nerve cord
Hearts
Earthworms crawl by anchoring bristles called setae to the ground and pulling against them.
Segments are connected by a circulatory and a nervous system. Blood is pumped by a series of hearts located at the anterior (front) end. A well-developed brain located in an anterior segment coordinates the activities of all segments.
Each segment has a coelom. The fluid of the coelom is squeezed by muscles, making each segment rigid, like an inflated balloon. Because each segment can contract independently, a worm can crawl by lengthening some segments while shortening others.

Figure 16.20 Phylum Annelida: annelids.

Arthropods: Advent of Jointed Appendages

The evolution of segmentation among the annelids marked the first major innovation in body structure among the coelomates. An even more profound innovation remained. It marks the origin of the body plan characteristic of the most successful of all animal groups, the **arthropods,** phylum Arthropoda (figure 16.21). This innovation was the development of jointed appendages.

The name "arthropod" comes from two Greek words, *arthros,* jointed, and *podes,* feet. All arthropods have jointed appendages. Some are legs, and others may be modified for other uses. To gain some idea of the importance of jointed appendages, imagine yourself without them—no hips, knees, ankles, shoulders, elbows, wrists, or knuckles. Without jointed appendages, you could not walk or grasp an object. Arthropods use jointed appendages as legs and wings for moving, as antennae to sense their environment, and as mouthparts for sucking, ripping, and chewing prey. A scorpion, for example, seizes and tears apart its prey with mouthpart appendages modified as large pincers.

Jointed appendages have proven very successful—about two-thirds of all named species on earth are arthropods. Scientists estimate that a quintillion (a billion billion) insects are alive at any one time—200 million insects for each living human!

The arthropod body plan, however, has one great limitation: Arthropods have a rigid external skeleton, or **exoskeleton,** made of chitin. In any animal, a key function of the skeleton is to provide places for muscle attachment, and in arthropods the muscles attach to the interior surface of the hard chitin shell, which also protects the animal from predators and impedes water loss.

What is the limitation? Chitin is tough but brittle and cannot support great weight. As a result, the exoskeleton must be much thicker to bear the pull of the muscles in large insects than in small ones, so there is a limit to how big an arthropod body can be. That is why you don't see beetles as big as birds or crabs the size of a cow—the exoskeleton would be so thick the animal couldn't move its great weight. In fact, the great majority of arthropod species consist of small animals—mostly about a millimeter in length—but members of the phylum range in adult size from about 80 micrometers long (some parasitic mites) to 3.6 meters across (a gigantic crab found in the sea off Japan). Some lobsters are nearly a meter in length. The largest living insects are about 33 centimeters long, but the giant dragonflies that lived 300 million years ago had wingspans of as much as 60 centimeters (2 feet)!

Because this size limitation is inherent in the body design of arthropods, no arthropods have ever grown to great size. As we will see, to overcome this limitation, a strong, flexible endoskeleton is required.

Arthropod bodies are segmented like those of annelids, from which they almost certainly evolved. Individual segments often exist only during early development, however, and fuse into functional groups as adults. For example, caterpillars have many segments, while butterflies have only three functional body units—head, thorax, and abdomen—each composed of several fused segments.

Figure 16.21 An arthropod.
The most common class of arthropods is Insecta, the insects. More than 70% of all named animal species are insects. The insect in this photograph is the monarch butterfly, *Danaus plexippus*.

Many arthropods, which include the spiders, mites, scorpions, and a few others, lack jaws, or **mandibles,** and are called **chelicerates.** Their mouthparts, known as **chelicerae,** evolved from the appendages nearest the animal's anterior end. The remaining arthropods have mandibles, formed by the modification of one of the pairs of anterior appendages. These arthropods, called **mandibulates,** include the crustaceans, insects, centipedes, millipedes, and a few other small groups.

Key characteristics of the phylum Arthropoda are summarized in figure 16.22.

7 *Phylum Arthropoda:* Arthropods

Key Evolutionary Advances: JOINTED APPENDAGES and EXOSKELETON. Insects and other arthropods (phylum Arthropoda) have a coelom, segmented bodies, and **jointed appendages.** The three body regions of an insect (head, thorax, and abdomen) are each actually composed of a number of segments that fuse during development. All arthropods have a strong **exoskeleton** made of chitin. One class of arthropods, the insects, has evolved **wings,** which permit them to fly rapidly through the air.

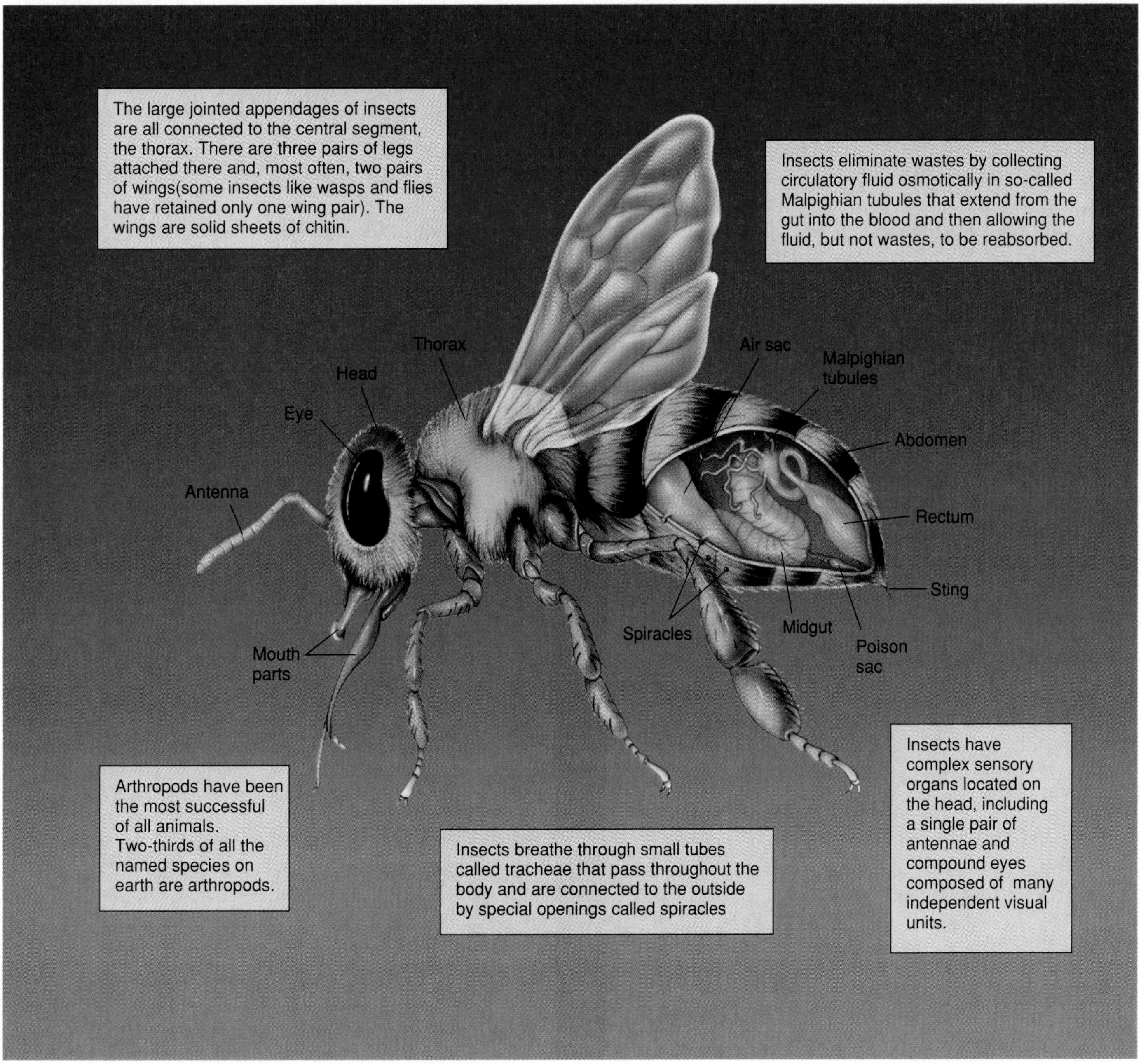

The large jointed appendages of insects are all connected to the central segment, the thorax. There are three pairs of legs attached there and, most often, two pairs of wings(some insects like wasps and flies have retained only one wing pair). The wings are solid sheets of chitin.
Insects eliminate wastes by collecting circulatory fluid osmotically in so-called Malpighian tubules that extend from the gut into the blood and then allowing the fluid, but not wastes, to be reabsorbed.
Thorax
Head
Eye
Antenna
Mouth parts
Air sac
Malpighian tubules
Abdomen
Rectum
Sting
Spiracles
Midgut
Poison sac
Arthropods have been the most successful of all animals. Two-thirds of all the named species on earth are arthropods.
Insects breathe through small tubes called tracheae that pass throughout the body and are connected to the outside by special openings called spiracles
Insects have complex sensory organs located on the head, including a single pair of antennae and compound eyes composed of many independent visual units.

Figure 16.22 Phylum Arthropoda: arthropods.

Figure 16.23 Horseshoe crabs.
These horseshoe crabs, *Limulus,* are emerging from the sea to mate at the edge of Delaware Bay, New Jersey, in early May.

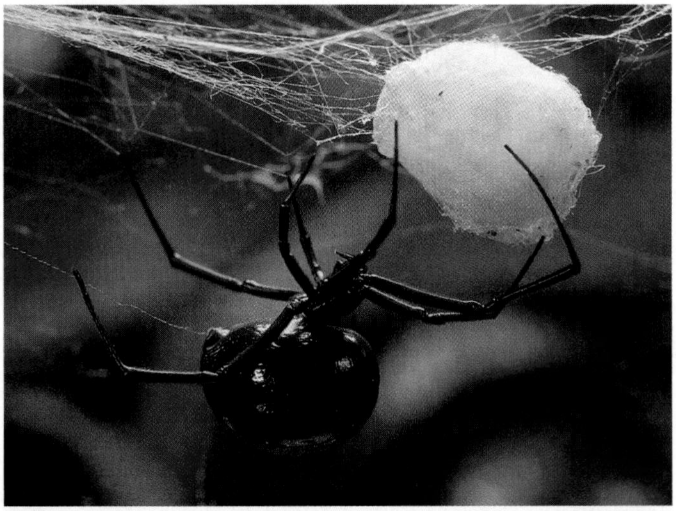

(a)

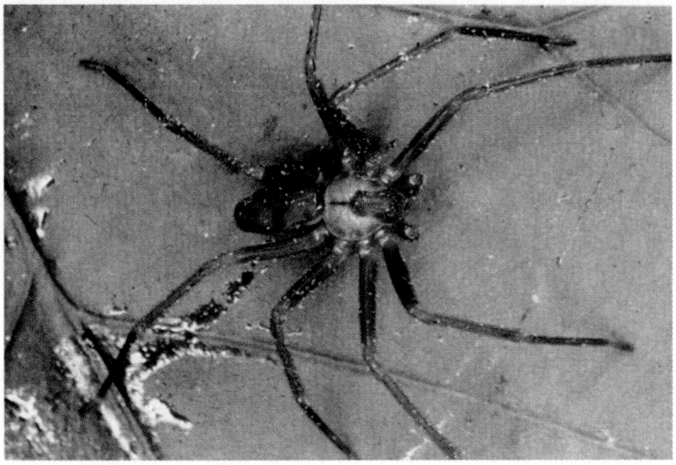

(b)

Figure 16.24 Arachnids.
(*a*) One of the two poisonous spiders in the United States and Canada, the black widow spider, *Latrodectus mactans.* (*b*) The other genus of poisonous spiders of this area, the brown recluse, *Loxosceles reclusa.* Both species are common throughout temperate and subtropical North America, but they rarely bite humans.

Chelicerates

The chelicerate fossil record goes back as far as that of any multicellular animal, about 630 million years. A major group of arthropods, the now-extinct trilobites, was also abundant then, and horseshoe crabs living today (figure 16.23) seem to be directly descended from them. By far the largest of the three classes of chelicerates is the largely terrestrial class Arachnida (figure 16.24), with some 57,000 named species, including the spiders, ticks, mites, scorpions, and daddy longlegs. Most **arachnids** are carnivorous, although mites are largely herbivorous. Ticks are blood-feeding ectoparasites of vertebrates, and some ticks may carry diseases, such as Rocky Mountain spotted fever and Lyme disease.

Mandibulates

The **crustaceans** (subphylum Crustacea) are a large, diverse group of primarily aquatic organisms, including some 35,000 species of crabs, shrimps, lobsters, crayfish, barnacles, water fleas, pillbugs, and related groups (figure 16.25). Often incredibly abundant in marine and freshwater habitats and playing a role of critical importance in virtually all aquatic ecosystems, crustaceans have been called "the insects of the water." Most crustaceans have two pairs of antennae, three pairs of chewing appendages, and various numbers of pairs of legs (figure 16.26). Crustaceans differ from the insects—but resemble millipedes and centipedes—in that they have legs on their abdomen as well as on their thorax. They are the only arthropods with two pairs of antennae.

Figure 16.25 Crustaceans.
(*a*) A freshwater crayfish, *Procambarus*. (*b*) Edible crab, *Cancer pangyrus*. (*c*) A copepod, member of an abundant group of marine and freshwater crustaceans (order Copepoda), most of which are a few millimeters long. Copepods are important components of the plankton. (*d*) Gooseneck barnacle, *Lepas anatifera,* feeding. These are stalked barnacles; many others lack a stalk.

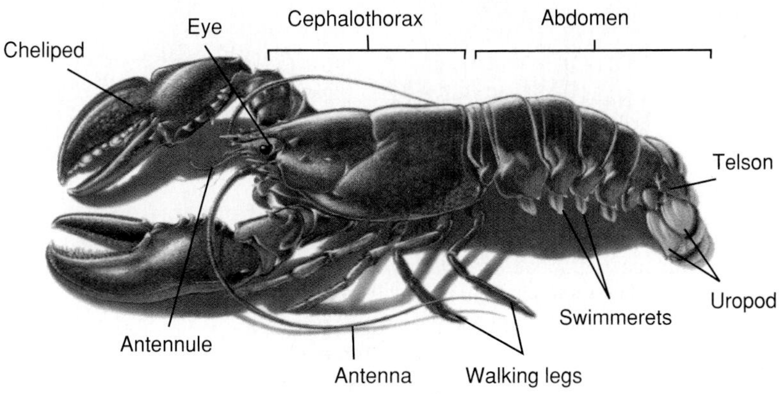

Figure 16.26 Body of a lobster, *Homarus americanus.*
Some of the specialized terms used to describe crustaceans are indicated. For example, the head and thorax are fused together into a cephalothorax. Appendages called swimmerets occur in lines along the sides of the abdomen and are used in reproduction and also for swimming. Flattened appendages known as uropods form a kind of compound "paddle" at the end of the abdomen. Lobsters may also have a telson, or tail spine.

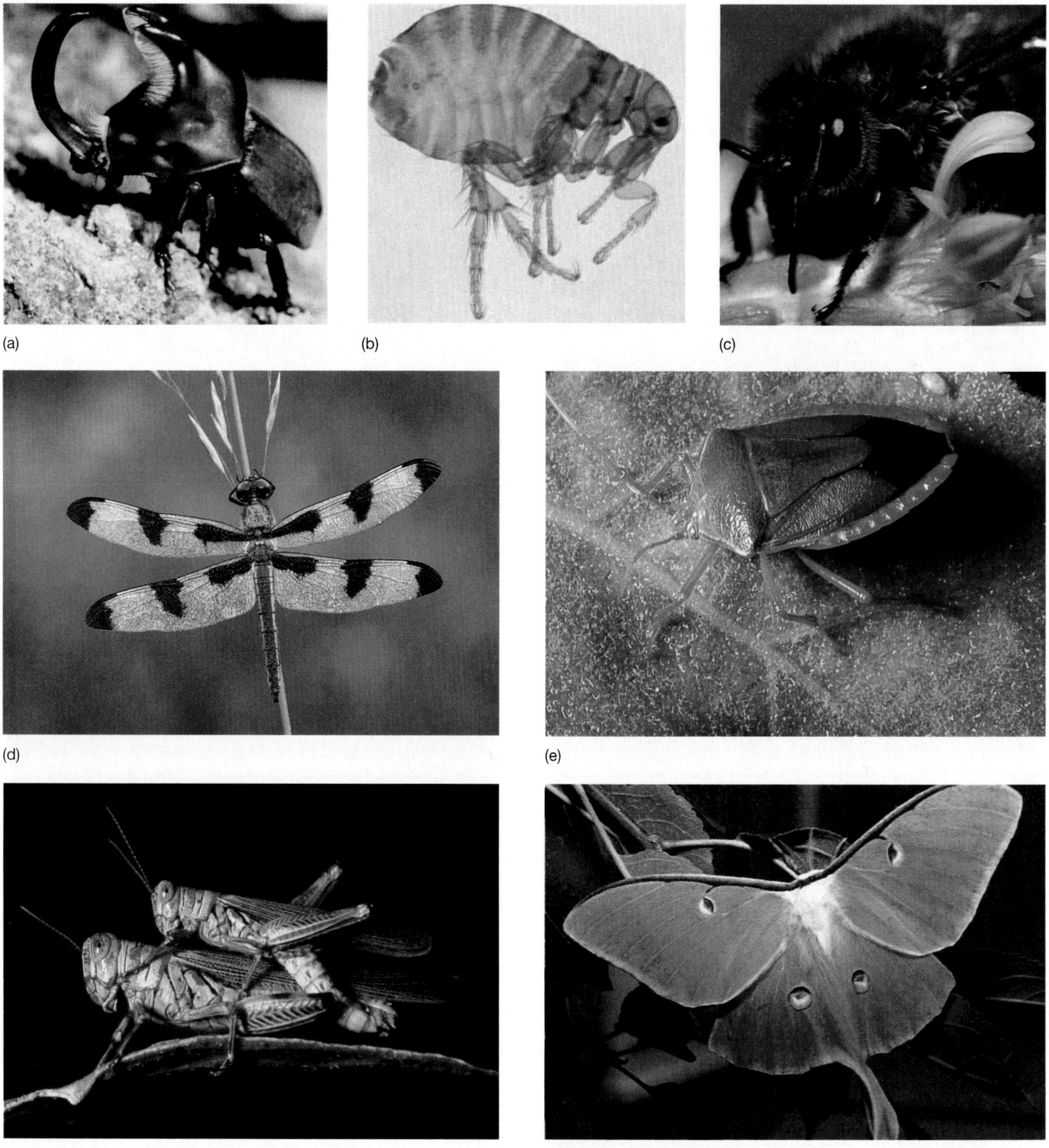

Figure 16.27 Insect diversity.

(*a*) Some insects have a tough exoskeleton, like this South American scarab beetle, *Dilobderus abderus* (order Coleoptera). (*b*) Human flea, *Pulex irritans* (order Siphonaptera), in California. Fleas are flattened laterally, slipping easily through hair. (c) The honeybee, *Apis mellifera* (order Hymenoptera), is a widely domesticated and efficient pollinator of flowering plants. (*d*) This pot dragonfly (order Odonata) has a fragile exoskeleton. (*e*) A true bug, *Edessa rufomarginata* (order Hemiptera), in Panama. (*f*) Copulating grasshoppers (order Orthoptera). (*g*) Luna moth, *Actias luna,* in Virginia. Luna moths and their relatives are the most spectacular insects.

The **insects,** class Insecta, are by far the largest group of arthropods, whether measured in terms of numbers of species or numbers of individuals; as such, they are the most abundant group of eukaryotes on earth (figure 16.27). Although primarily a terrestrial group, insects live in every conceivable habitat on land and in freshwater, and a few have even invaded the sea. More than 70% of all the named animal species are insects, and the actual proportion is undoubtedly much higher because millions of additional forms await detection, classification, and naming. About 90,000 described species are found in the United States and Canada, but the actual number of species in this region approaches 125,000.

Most insects are relatively small, ranging in size from 0.1 millimeter to about 30 centimeters in length. Insects have three body sections:

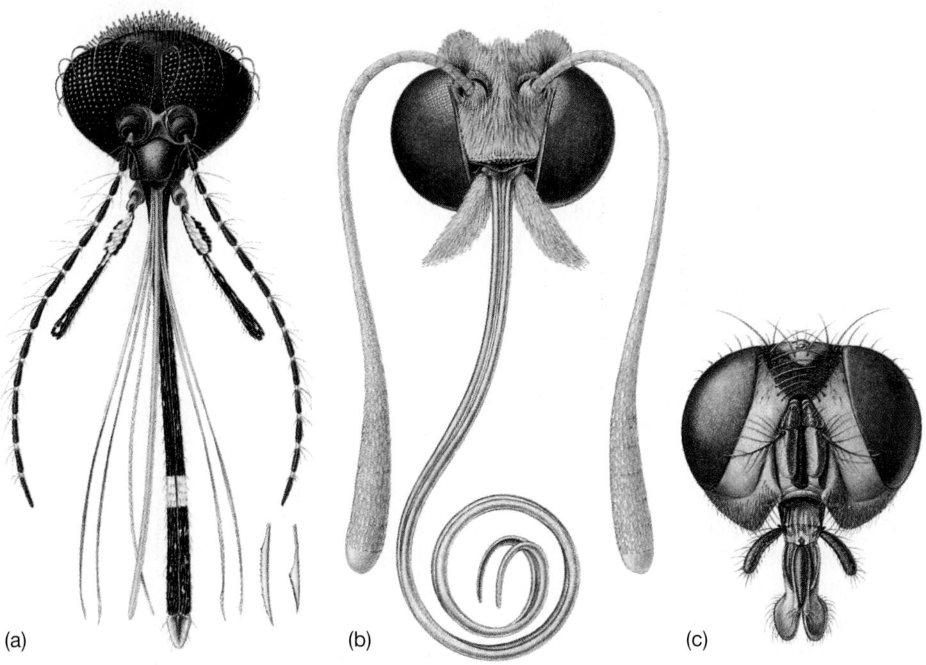

Figure 16.28 Modified mouthparts in three kinds of insects.
(*a*) Mosquito, *Culex;* mouthparts modified for piercing. (*b*) Alfalfa butterfly, *Colias;* mouthparts modified for sucking nectar from flowers. (c) Housefly, *Musca domestica;* mouthparts modified for sopping up liquids.

1. **Head.** The head of insects is very elaborate, with a single pair of antennae and elaborate mouthparts (figure 16.28). Most insects have compound eyes, which are composed of independent visual units.
2. **Thorax.** The thorax consists of three segments, each of which has a pair of legs. Most insects also have two pairs of wings attached to the thorax. In some insects like flies, one of the pairs of wings has been lost during the course of evolution.
3. **Abdomen.** The abdomen consists of up to 12 segments. Digestion takes place primarily in the stomach, and excretion takes place through organs called Malpighian tubules. Most of the water and salts are reabsorbed by the hindgut, and thus the Malpighian tubules constitute an efficient mechanism for water conservation and were a key adaptation facilitating invasion of the land by arthropods.

Millipedes and centipedes are closely related to insects. Centipedes have one pair of legs on each body segment, millipedes two (figure 16.29). The centipedes are all carnivorous and feed mainly on insects. The appendages of the first trunk segment are modified into a pair of venomous fangs. In contrast, most millipedes are herbivores, feeding mainly on decaying vegetation. Millipedes live mainly in damp, protected places, such as under leaf litter, in rotting logs, under bark or stones, or in the soil.

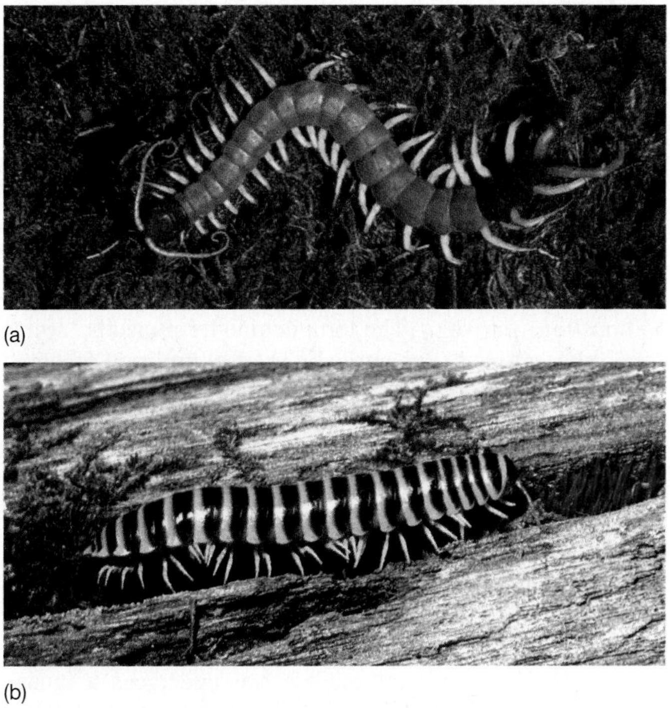

Figure 16.29 Centipedes and millipedes.
Centipedes are active predators, whereas millipedes are sedentary herbivores. (*a*) Centipede, *Scolopendra.* (*b*) Millipede, *Sigmoria,* in North Carolina.

16.5 Redesigning the Embryo

There are two major kinds of coelomate animals. All the coelomates we have met so far have essentially the same kinds of embryos, starting as hollow balls of cells that indent to form a two-layer-thick ball opening to the outside. In mollusks, annelids, and arthropods, the mouth develops from or near this opening, called the blastopore. This same pattern of development is seen in all noncoelomate animals. An animal whose mouth develops in this way is called a **protostome** ("first mouth"). A second kind of coelomate animal developed after the arthropods in which this basic pattern of embryological development is altered. In the echinoderms, chordates, and a few other small, related phyla, the anus develops from or near the blastopore, and the mouth forms later on another part of the embryo. These animals are called **deuterostomes** ("second mouth"). Echinoderms and chordates are clearly related to each other by their shared pattern of embryonic development.

Deuterostomes represent a revolution in embryonic development. While the pattern of cell division in protostomes is spiral, it is radial in deuterostomes. In protostomes, the developmental fate of each cell is fixed, because the chemicals that act as development signals are localized in different parts of the egg. By contrast, deuterostome embryos have identical cells, each of which, if separated from the others at an early enough stage, can develop into a complete organism, because the developmental signals are generated by the chromosomes. In protostomes, positional information is the key to the developmental fate of embryo cells, while in deuterostomes, groups of cells move around during the course of development to form new tissue associations.

(a) (b) (c)

Figure 16.30 Diversity in echinoderms.
(*a*) Sea star, *Oreaster occidentalis* (class Asteroidea), in the Gulf of California, Mexico. (*b*) Feather star (class Crinoidea) on the Great Barrier Reef in Australia. (c) Brittle star, *Ophiothrix* (class Ophiuroidea).

Echinoderms: The First Deuterostomes

The first deuterostomes, marine animals called **echinoderms** in the phylum Echinodermata**,** appeared more than 650 million years ago. The term echinoderm means "spiny skin" and refers to an **endoskeleton** composed of hard, calcium-rich plates called ossicles just beneath the delicate skin. When they are first formed, the plates are enclosed in living tissue, and so are truly an endoskeleton, although in adults they fuse, forming a hard shell. About 6,000 species of echinoderms are living today, almost all of them on the ocean bottom (figure 16.30). Many of the most familiar animals seen along the seashore are echinoderms, including sea stars (starfish), sea urchins, sand dollars, and sea cucumbers.

The body plan of echinoderms undergoes a fundamental shift during development: All echinoderms are bilaterally symmetrical as larvae but become radially symmetrical as adults. Adult echinoderms have a five-part body plan, easily seen in the five arms of a sea star. Its nervous system consists of a central ring of nerves from which five branches arise—while the animal is capable of complex response patterns, there is no centralization of function, no "brain." Some echinoderms like feather stars have 10 or 15 arms, but always multiples of five.

A key evolutionary innovation of echinoderms is the development of a hydraulic system to aid movement. Called a **water vascular system,** this fluid-filled system is composed of a central ring canal from which five radial canals extend out into the arms. From each radial canal tiny vessels extend out through short side branches into thousands of tiny, hollow **tube feet.** At the base of each tube foot is a fluid-filled muscular sac that acts as a valve. When a sac contracts, its fluid is prevented from reentering the radial canal and instead is forced into the tube foot, thus extending it. When extended, the tube foot attaches itself to the ocean bottom, often aided by suckers. The sea star can then pull against these tube feet and so haul itself over the seafloor.

Key characteristics of the phylum Echinodermata are summarized in figure 16.31.

8 *Phylum Echinodermata:* Echinoderms

Key Evolutionary Advances: DEUTEROSTOME DEVELOPMENT and ENDOSKELETON. Echinoderms like sea stars (phylum Echinodermata) are coelomates with a **deuterostome** pattern of development. A delicate skin stretches over an **endoskeleton** made of calcium-rich plates, often fused into a continuous, tough spiny layer.

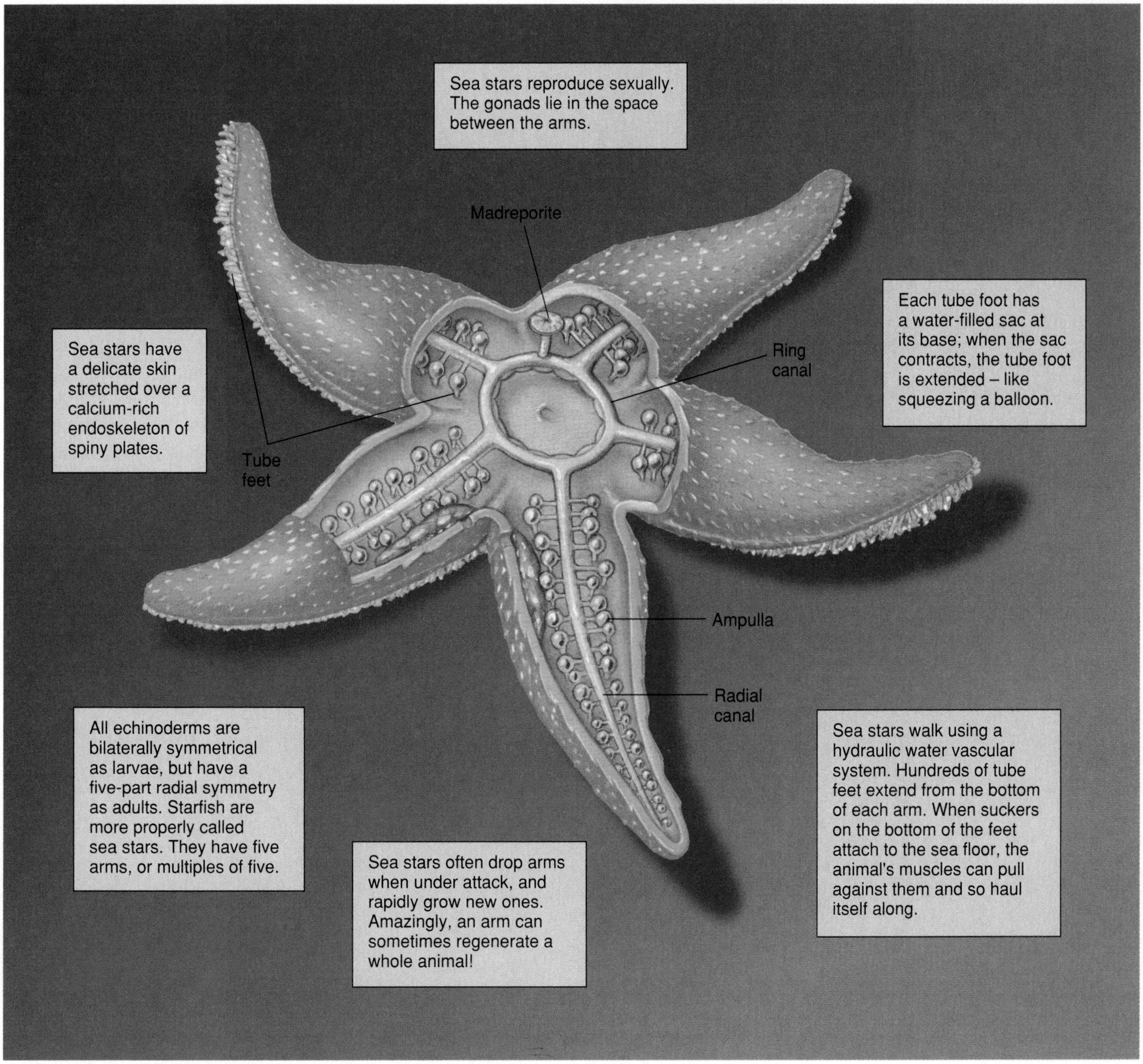

Sea stars reproduce sexually. The gonads lie in the space between the arms.
Madreporite
Each tube foot has a water-filled sac at its base; when the sac contracts, the tube foot is extended – like squeezing a balloon.
Ring canal
Sea stars have a delicate skin stretched over a calcium-rich endoskeleton of spiny plates.
Tube feet
Ampulla
Radial canal
All echinoderms are bilaterally symmetrical as larvae, but have a five-part radial symmetry as adults. Starfish are more properly called sea stars. They have five arms, or multiples of five.
Sea stars often drop arms when under attack, and rapidly grow new ones. Amazingly, an arm can sometimes regenerate a whole animal!
Sea stars walk using a hydraulic water vascular system. Hundreds of tube feet extend from the bottom of each arm. When suckers on the bottom of the feet attach to the sea floor, the animal's muscles can pull against them and so haul itself along.

Figure 16.31 Phylum Echinodermata: echinoderms.

Chordates: Improving the Skeleton

The endoskeleton of echinoderms is functionally similar to the exoskeleton of arthropods, in that it is a hard shell that encases the body, with muscles attached to its inner surface. The second major kind of deuterostome, the **chordates,** employ a very different kind of endoskeleton, one that is truly internal. Members of the phylum Chordata are characterized by a flexible rod called a **notochord** that develops along the back of the embryo. Muscles attached to this rod allowed early chordates to swing their backs back and forth, swimming through the water. This key evolutionary advance, attaching muscles to an internal element, started chordates along an evolutionary path that leads to the vertebrates and for the first time to truly large animals (figure 16.32).

The approximately 42,500 species of chordates are distinguished by three principal features:

1. **Notochord.** A long, stiff rod that forms beneath the nerve cord, between it and the developing gut (which becomes the stomach and intestines) in the early embryo.
2. **Nerve cord.** A single dorsal (along the back) hollow nerve cord, to which the nerves that reach the different parts of the body are attached.
3. **Pharyngeal slits.** A series of slits behind the mouth into the pharynx, which is a muscular tube that connects the mouth to the digestive tract and windpipe.

All chordates have all three of these characteristics at some time in their lives. For example, human embryos have pharyngeal slits, a nerve cord, and a notochord as embryos.

Key characteristics of the phylum Chordata are summarized in figure 16.33.

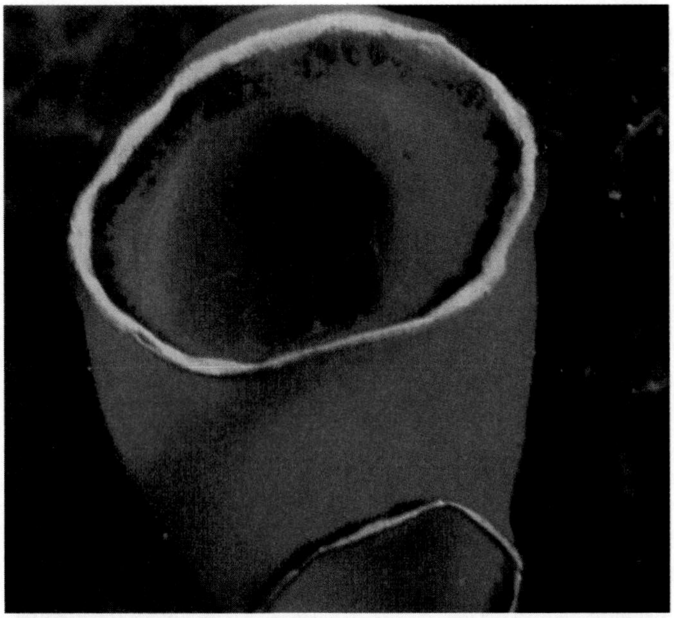

(a)

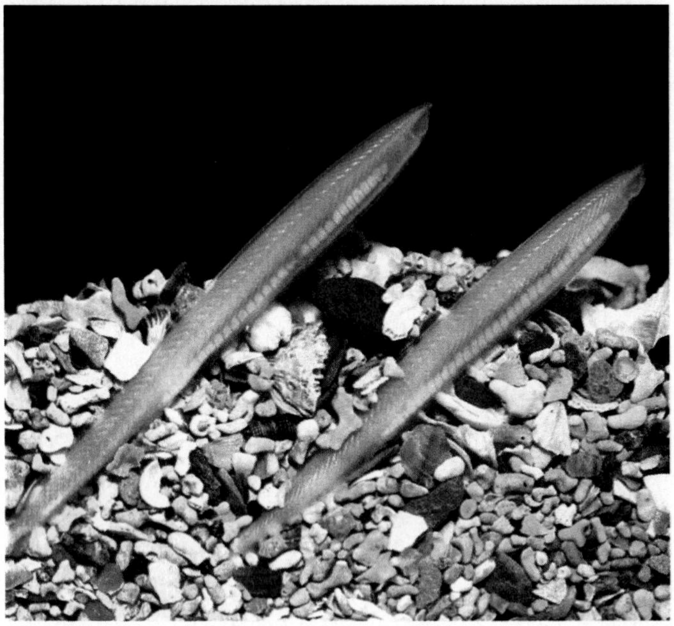

(b)

(c)

Figure 16.32 Diversity in chordates.
(*a*) A beautiful blue and gold tunicate. (*b*) Two lancelets, *Branchiostoma lanceolatum,* partly buried in shell gravel, with their anterior ends protruding. The muscle segments are clearly visible in this photograph. The numerous square, pale yellow objects along the side of the body are gonads, indicating that these are male lancelets. (c) Terrestrial vertebrates are among the most successful animal groups. This tiger, leaping toward a potential dinner, is both large and very mobile, two traits characteristic of many terrestrial vertebrates.

9 *Phylum Chordata:* Chordates

Key Evolutionary Advance: NOTOCHORD. Vertebrates, tunicates, and lancelets are chordates (phylum Chordata), coelomate animals with a flexible rod, the **notochord,** that acts to anchor internal muscles, permitting rapid body movements. Chordates also possess **pharyngeal slits** (relics of their aquatic ancestry) and a dorsal **hollow nerve cord.** In vertebrates, the notochord is replaced during embryonic development by the vertebral column.

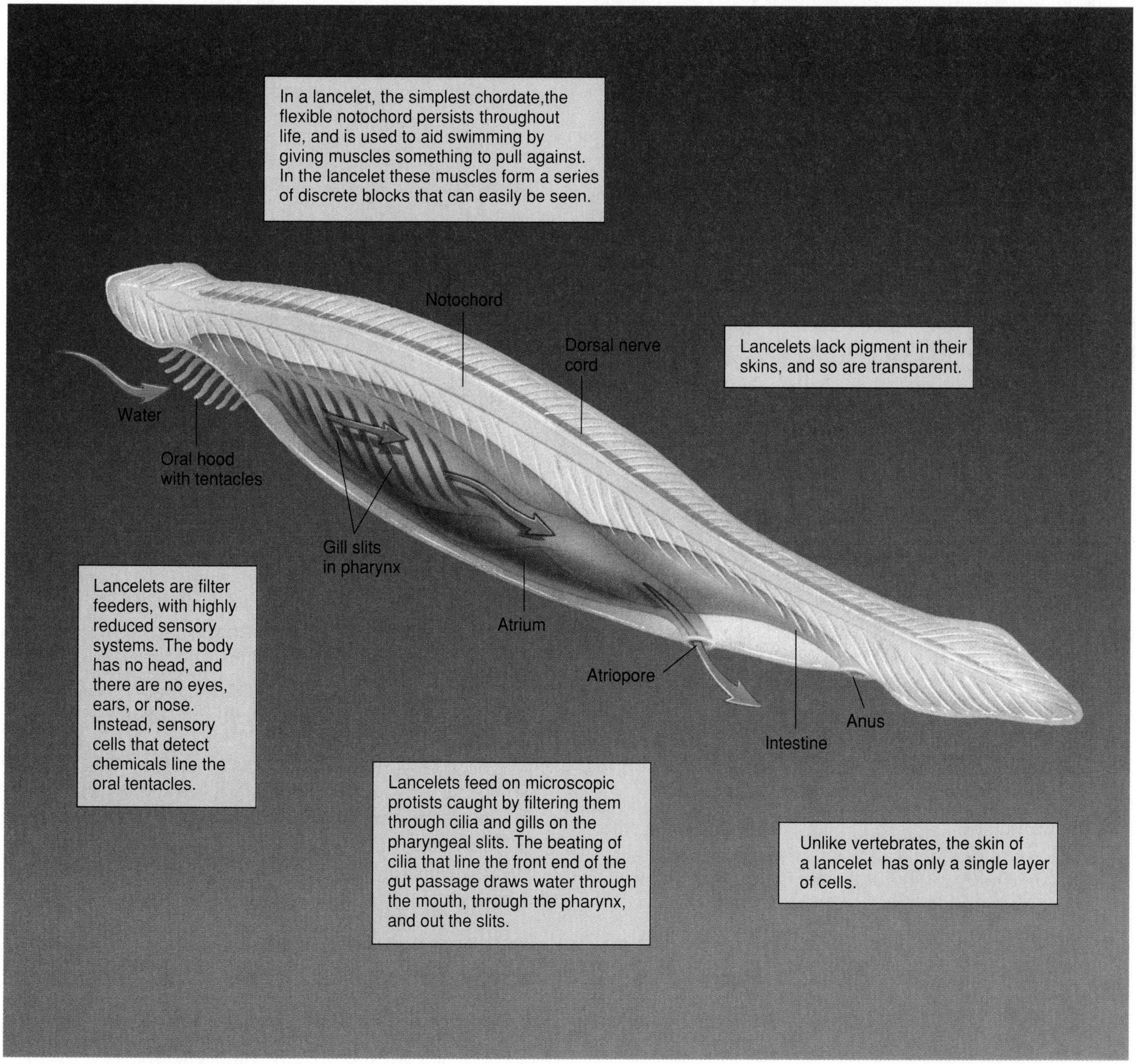

In a lancelet, the simplest chordate,the flexible notochord persists throughout life, and is used to aid swimming by giving muscles something to pull against. In the lancelet these muscles form a series of discrete blocks that can easily be seen.
Notochord
Dorsal nerve cord
Lancelets lack pigment in their skins, and so are transparent.
Water
Oral hood with tentacles
Gill slits in pharynx
Atrium
Atriopore
Intestine
Anus
Lancelets are filter feeders, with highly reduced sensory systems. The body has no head, and there are no eyes, ears, or nose. Instead, sensory cells that detect chemicals line the oral tentacles.
Lancelets feed on microscopic protists caught by filtering them through cilia and gills on the pharyngeal slits. The beating of cilia that line the front end of the gut passage draws water through the mouth, through the pharynx, and out the slits.
Unlike vertebrates, the skin of a lancelet has only a single layer of cells.

Figure 16.33 Phylum Chordata: chordates.

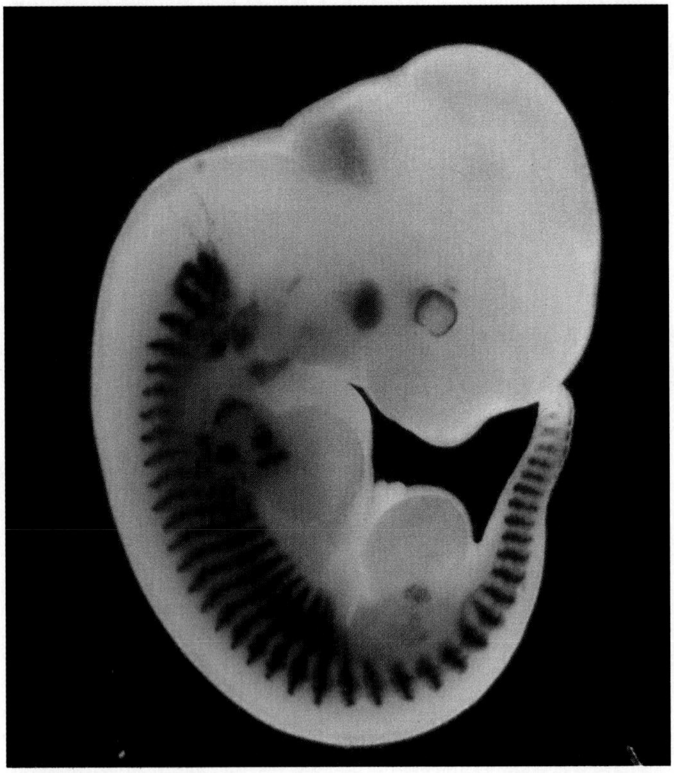

Figure 16.34 A mouse embryo.
At 11.5 days of development, the muscle is already divided into segments called somites (stained dark in this photo), reflecting the fundamentally segmented nature of all chordates.

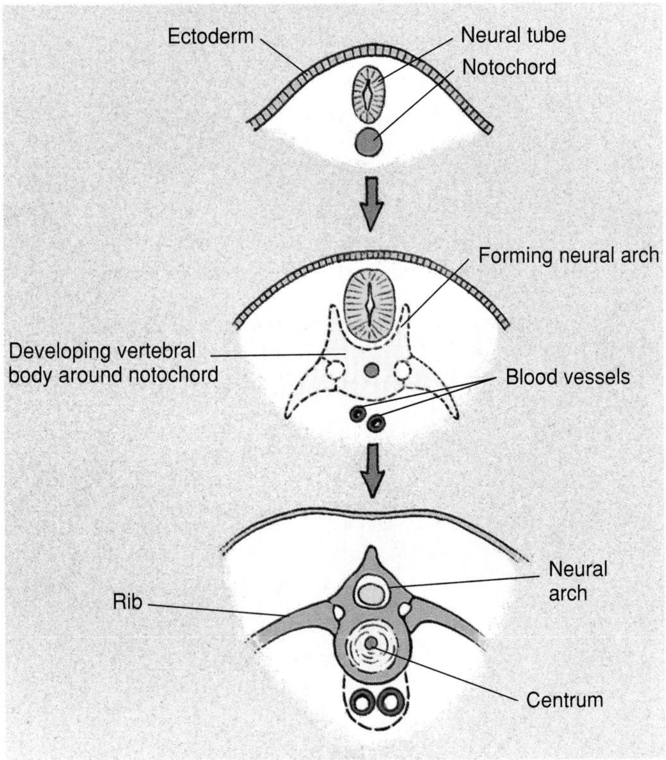

Figure 16.35 Embryonic development of a vertebrate.
During the course of development, the flexible notochord is surrounded and eventually replaced by a bony covering, the vertebral column, that protects the neural tube and provides a strong, flexible rod against which the muscles pull when the animal swims or moves.

Chordates are deuterostome coelomates, whose nearest relations in the animal kingdom are the echinoderms, also deuterostomes. In their body plan, chordates are segmented, and distinct blocks of muscles can be seen clearly in many forms (figure 16.34). Many chordates have jointed appendages. With the exception of tunicates and a small group of fishlike marine animals, the lancelets, all chordates are **vertebrates.** Vertebrates differ from tunicates and lancelets in two important respects:

1. **Backbone.** In vertebrates, the notochord becomes surrounded and then replaced during the course of the embryo's development by a bony vertebral column, a tube of hollow bones called vertebrae that encloses the dorsal nerve cord like a sleeve and protects it (figure 16.35).
2. **Head.** All vertebrates except the earliest fishes have a distinct and well-differentiated head, with a skull and brain. For this reason, the vertebrates are sometimes called the craniate chordates (Greek, *kranion,* skull).

All vertebrates have an internal skeleton made of bone or cartilage against which the muscles work. This endoskeleton makes possible the great size and extraordinary powers of movement that characterize the vertebrates. An interesting feature of vertebrates and other chordates is that they have a tail that extends beyond the anus, at least during their embryonic development; nearly all other animals have a terminal anus.

The endoskeleton of most vertebrates is made of **bone** (in a few, like sharks, the bone is replaced with more flexible cartilage). Bone is a special form of tissue containing fibers of the protein collagen that are coated with a calcium phosphate salt. Bone is formed in two stages. First, collagen is laid down in a matrix of fibers along lines of stress to provide flexibility, and then calcium minerals impregnate the fibers, providing rigidity. The great advantage of bone over chitin as a structural material is that bone is strong without being brittle.

The evolution and characteristics of the major groups of vertebrates are discussed in detail in the next chapter.

CHAPTER 16

HIGHLIGHTS

Key Terms | Key Concepts

16.1 Some General Features of Animals

Key Terms

phylum 322

Key Concepts

- There are about 36 phyla of animals.
- Animal cells lack cell walls.
- All animals are multicellular heterotrophs, and most are mobile.

16.2 The Simplest Animals

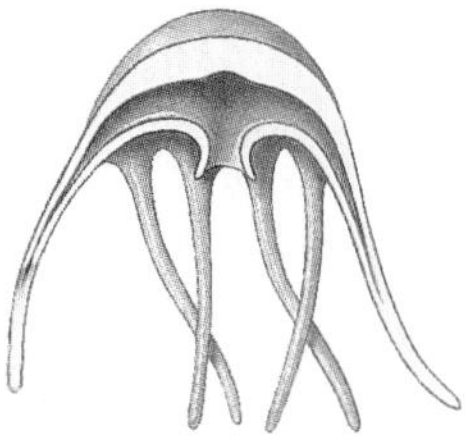

Key Terms

metazoans 326
choanocytes 326
radially symmetrical 328
nematocyst 328

Key Concepts

- Metazoans possess tissues and organs, and their bodies have a definite shape and symmetry.
- Protists called choanoflagellates very closely resemble the choanocytes of sponges and may be the ancestors of all animals.
- The most primitive metazoans exhibit radial symmetry.

16.3 The Advent of Bilateral Symmetry

Key Terms

bilaterally symmetrical 331
acoelomate 332

Key Concepts

- All advanced metazoans are bilaterally symmetrical at some stage of their life cycle.
- Acoelomate animals have a solid body with no internal cavity.
- This severely limits their size, as diffusion is the only way they can transport materials.

16.4 The Advent of a Body Cavity

Key Terms

pseudocoelomates 334
coelom 336
coelomates 336
radula 336
segmentation 338
exoskeleton 340

Key Concepts

- A pseudocoelom is a body cavity between the endoderm and the mesoderm.
- A coelom is a body cavity entirely within the mesoderm. It has the advantage that the primary tissues can interact with one another.
- Segmentation is the building of a body from a series of similar sections. It has the great advantage that different segments can specialize in different ways.
- A rigid exoskeleton of chitin limits body size because chitin is brittle.

16.5 Redesigning the Embryo

Key Terms

protostome 346
deuterostomes 346
endoskeleton 346
notochord 348

Key Concepts

- A notochord is a flexible rod to which muscles attach, allowing early chordates to swing their bodies back and forth.
- The function of the notochord is taken over by the backbone of vertebrates.

16

CONCEPT REVIEW

1. The phylum of animals that contains the largest number of named species is
   a. Chordata.
   b. Mammals.
   c. Annelida.
   d. Arthropoda.

2. Choanocytes of sponges bear a striking resemblance to the _________, members of one group of zoomastigotes.
   a. acoelomates
   b. choanoflagellates
   c. cnidarians
   d. collar cells

3. All animals other than sponges have both _________ and tissues and are called eumetazoans.
   a. multicellularity
   b. choanocytes
   c. symmetry
   d. circulatory systems

4. Cnidarians spear their prey with powerful harpoons called
   a. nematocysts.
   b. cnidocytes.
   c. hydra.
   d. choanocytes.

5. Which of the following is not present in flatworms?
   a. uterus or testes
   b. organs
   c. bilateral symmetry
   d. body cavity

6. Where does a coelom originate?
   a. in the endoderm
   b. between the endoderm and the mesoderm
   c. in the mesoderm
   d. between the mesoderm and the ectoderm

7. Octopuses and squids are in the phylum _________, along with snails and clams.
   a. Echinodermata
   b. Mollusca
   c. Cnidaria
   d. Crustacea

8. Which of the following are not arthropods?
   a. earthworms
   b. crayfish
   c. spiders
   d. butterflies

9. Which two phyla listed here are deuterostomes?
   a. Platyhelminthes
   b. Chordata
   c. Echinodermata
   d. Arthropoda

10. Three characteristics that distinguish the chordates from other animals are
    a. a single, hollow dorsal nerve cord.
    b. a notochord at some time in development.
    c. pharyngeal slits at some time in development.
    d. segmentation.
    e. a nervous system.

11. Unlike tunicates and lancelets, vertebrates have a(n)
    a. nervous system.
    b. coelom.
    c. exoskeleton.
    d. backbone.

12. The subkingdom _________ consists mainly of the sponges, phylum _________.

13. The three distinct layers of cells that form in the embryos of all eumetazoans are the _________, _________, and _________.

14. The process of evolving a definite head area is called _________.

15. _________, a type of roundworm, are usually microscopic and are found in large numbers in soil.

16. Next to arthropods, _________ are the most successful land animals.

17. The key evolutionary advance in annelids is _________.

18. The three body sections in insects are the head, _________, and _________.

19. A sea cucumber is in the phylum _________.

20. Animals in which the mouth develops from or near the blastopore in the embryo are called _________.

21. Tunicates belong to the phylum _________.

Answers to the Concept Review questions appear in Appendix B.

CHALLENGE YOURSELF

1. Why are the most primitive animals encountered in the sea?
2. In what ways is an earthworm more complex than a flatworm?
3. What is the evolutionary advantage of having a shell?
4. What are the key adaptations that facilitated the invasion of the land by arthropods? Why is the phylum so successful?
5. Why is it believed that echinoderms and chordates, which are so dissimilar, are members of the same evolutionary line?

FOR FURTHER READING

Clark, R. B. *Dynamics in Metazoan Evolution: The Origin of the Coelom and Segments.* Oxford: Clarendon Press, 1964. Dated, but classic, treatment of the impact of the coelom on animal evolution.

Colbert, E. H. *Evolution of the Vertebrates: A History of Backboned Animals Through Time.* New York: Wiley-Liss, 1991. A nice, complete, up-to-date discussion of vertebrate phylogeny, including fossil vertebrates and evolution of the group as a whole.

Droser, M., R. Fortey, and X. Li. "The Ordovician Radiation." *American Scientist,* March/April 1996, 122–31. Changes in world climates played a pivotal role in biological diversification.

Grimaldi, D. "Captured in Amber." *Scientific American,* April 1996, 84–91. The exquisitely preserved tissues of insects in 25-million-year-old amber is revealing some genetic secrets of evolution.

Kirchner, W. H., and W. F. Towne. "The Sensory Basis of the Honeybee's Dance Language." *Scientific American,* June 1994, 74–80. The often explicit communication seen in the social insects has fascinated scientists and nonscientists alike for over a century. This discussion looks at how a complex dance communicates so much information about a food source.

Levinton, J. S. "The Big Bang of Animal Evolution." *Scientific American,*

November 1992, 84–91. An interesting discussion of the sudden appearance of large numbers of new species in the fossil record.

Natural History 104, March 1995. An issue devoted almost entirely to fascinating and enlightening articles about spiders: fangs, silks, fights, webs, babies, fossils and—of course—evolution.

Ruppert, E. E. *Invertebrate Zoology,* 6th ed. Fort Worth, TX.: Saunders College Press, 1994. The definitive, classic textbook created by Robert D. Barnes for the field of invertebrate zoology. The best, most comprehensive text on invertebrate zoology available.

Shear, W. A. "One Small Step for an Arthropod: In a Giant Leap in the History of Life, Invertebrates Set Foot on Land More Than 400 Million Years Ago." *Natural History,* March 1993, 46–51. Who were the first terrestrial animals? New evidence uncovered in New York, Scotland, and Wales points to tiny carnivorous arachnids and myriapods.

Vacelet, J., and N. Boury-Esnault. "Carnivorous Sponges." *Nature* 373, (January 26, 1995): 33–35. A shallow-water cave in the Mediterranean yields a microcrustacean-eating sponge previously only encountered in very deep water.

TECHNOLOGY LINKS

The Living World Home Page
http://www.wcbp.com/biology/tlw

17

History of Terrestrial Vertebrates

CHAPTER OUTLINE

Figure 17.1 A terrestrial vertebrate.
This koala is a marsupial mammal. Marsupials evolved near the end of the evolution of vertebrates. Except in isolated Australia, marsupials were eventually replaced by another kind of mammal, placental mammals.

Of all animals, vertebrates are the most familiar to us, both because that is what *we* are and because all other land animals bigger than our fist are vertebrates, too. **Vertebrates** are chordates, members of the great animal phylum Chordata, in which a flexible rod—the notochord—develops early in life along the back of the embryo. In vertebrates, this rod becomes surrounded and then replaced during development by a tube of hollow bones called the vertebral column, spine, or most simply, **backbone.** In this chapter, we trace the evolutionary history of vertebrates, passing from fishes without jaws to amphibians, dinosaurs, and eventually mammals like the koala you see in figure 17.1.

17.1 Overview of Vertebrate Evolution

When scientists first began to study and date fossils, they had to find some way to organize the different time periods from which the fossils came. They divided the earth's past into large blocks of time called **eras.** Eras are further subdivided into smaller blocks of time called **periods,** and some periods, in turn, are subdivided into **epochs,** which can be divided into **ages.**

The Paleozoic Era

Virtually all of the major groups of organisms that survive at the present time, except for the plants, originated at the beginning of the **Paleozoic era,** during or soon after the Cambrian period (590 to 505 M.Y.A.), exclusively in the sea. Thus, the diversification of animal life on earth is basically a marine record, and the fossils from the Paleozoic era all originated in the sea (figure 17.2).

Many of the animal phyla that appeared in the Cambrian period have no living relatives. Their fossils indicate that this was a period of experimentation with different body forms and ways of life, some of which ultimately led to the contem-

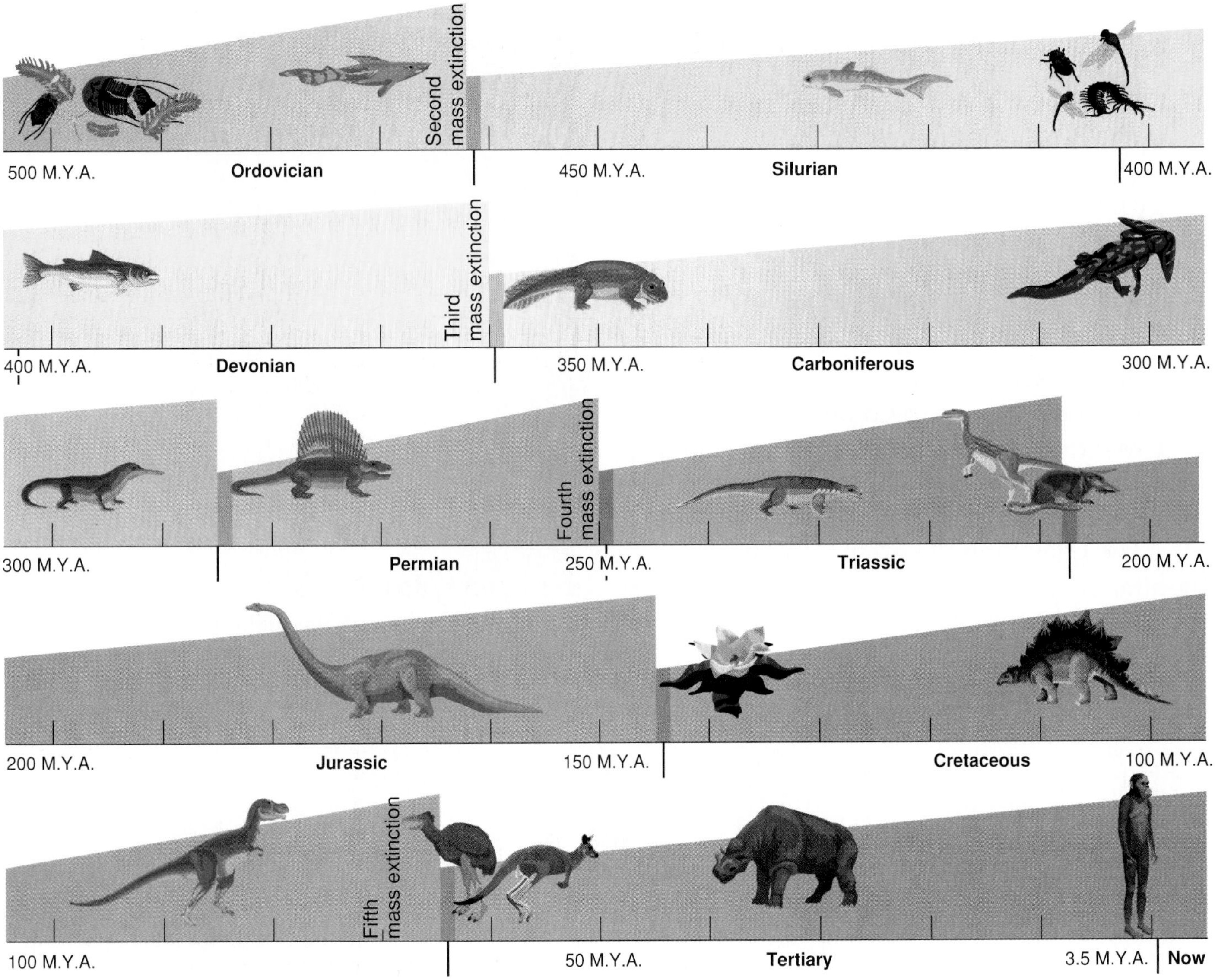

Figure 17.2 An evolutionary timeline.
Vertebrates evolved in the seas about 470 million years ago (M.Y.A.) and invaded land about 100 million years later. Dinosaurs and mammals evolved in the Triassic period about 220 million years ago. Dinosaurs dominated life on land for over 150 million years, until their sudden extinction 65 million years ago left mammals free to flourish.

porary phyla of animals and others to extinction. For example, the trilobites (figure 17.3) appear to be the ancestors of at least one living group, the horseshoe crabs, whereas the ammonites, which were abundant 100 million years ago, have no surviving descendants.

The first vertebrates evolved about 470 million years ago in the oceans—fishes without jaws. They didn't have paired fins either—many of them looked something like a flat hotdog with a hole at one end and a fin at the other. For 100 million years, a parade of different kinds of fishes were the only vertebrates on earth. They became the dominant creatures in the sea, some bigger than cars.

Invasion of the Land

Only a few of the animal phyla that evolved in the Cambrian seas have invaded the land successfully; most others have remained exclusively marine. The first organisms to colonize the land were plants, about 410 million years ago. The ancestors of plants were specialized members of a group of photosynthetic protists known as the green algae. It seems probable that plants first occupied the land in symbiotic association with fungi, as discussed in chapter 13.

The second major invasion of the land, and perhaps the most successful, was by the arthropods, a phylum of hard-shelled animals with jointed legs and a segmented body. This invasion of the land occurred soon after the evolution of the plants.

Vertebrates initiated the third major invasion of the land. The first vertebrates to live on land were the amphibians, represented today by frogs, toads, and salamanders. The earliest amphibians known are from the Devonian period (figure 17.4), and among their descendants are the reptiles, which became the ancestors of the dinosaurs, birds, and mammals.

That all four of the major groups of multicellular organisms—plants, fungi, arthropods, and vertebrates—colonized the land within a few tens of millions of years of one another is probably related to the development of suitable environmental conditions, such as the formation of a layer of ozone (O_3) in the atmosphere.

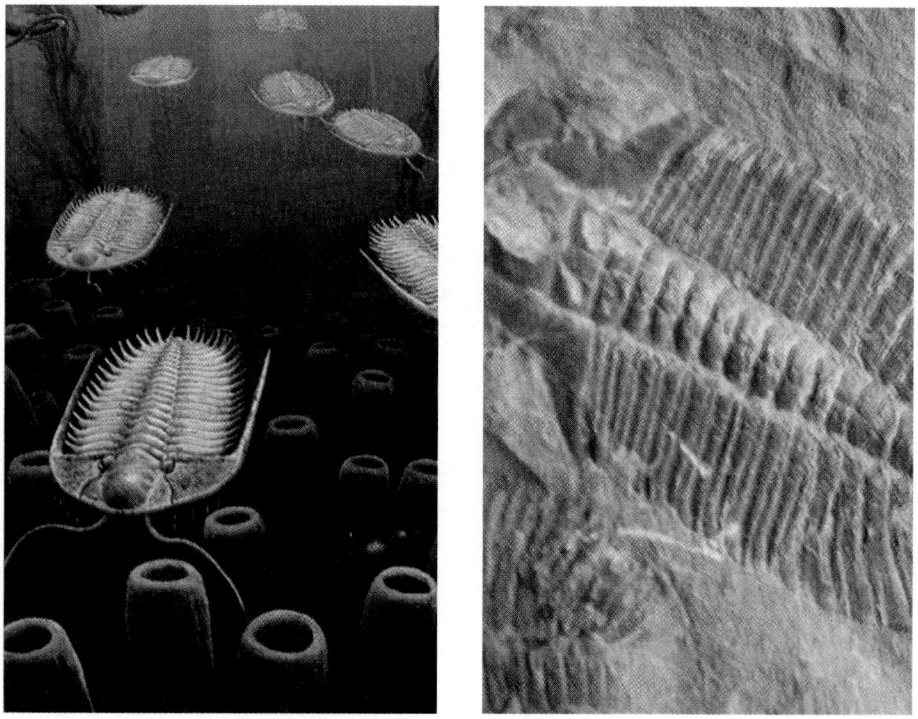

(a) (b)

Figure 17.3 Life in the Cambrian.
(*a*) Trilobites are shown swimming in this reconstruction of a community of marine organisms in the late Cambrian period, 500 to 550 million years ago. Trilobites were early members of the arthropod phylum. On the seafloor is a colony of sponges, members of another ancient animal phylum. (*b*) A fossil trilobite.

Figure 17.4 An early amphibian.
Ichthyostega was one of the first amphibians with teleologically efficient limbs for crawling on land and a relatively advanced ear structure for picking up airborne sounds. Despite these features, *Ichthyostega,* which lived about 350 million years ago, was still quite fishlike in overall appearance.

Figure 17.5 Dinosaurs.
Some of the remarkable diversity of dinosaurs is shown in this famous reconstruction from the Peabody Museum, Yale University. This painting covers a span of approximately 320 million years (reading *right* to *left*) during the later Paleozoic and Mesozoic eras, ending 65 million years ago at the end of the Cretaceous period (*far left*). Throughout this

Mass Extinctions

The history of the Paleozoic era has been marked by periodic major episodes of extinction, called **mass extinctions.** Four of these mass extinctions occurred during the Paleozoic, the first of them near the end of the Cambrian period about 505 million years ago. At that time, most of the existing families of trilobites (see figure 17.3), a very common type of marine arthropod, became extinct. Additional mass extinctions occurred about 438 and 360 million years ago.

The fourth and most drastic mass extinction in the history of life on earth happened during the last 10 million years of the Permian period, marking the end of the Paleozoic era. It is estimated that 96% of all species of marine animals that were living at that time became extinct! All of the trilobites disappeared forever. Brachiopods, marine animals resembling mollusks but with a different filter-feeding system, were extremely diverse and widespread during the Permian; only a few species survived. Bryozoans, marine filter feeders that formed coral-like colonies in oceans throughout the world in the Permian, became rare afterward.

Mass extinctions left vacant many evolutionary opportunities, and for this reason they were followed by rapid evolution among the relatively few plants, animals, and other organisms that survived the extinction. Little is known about the causes of major extinctions. Possible causes include volcanic eruptions or a meteor impact. In the case of the Permian mass extinction, some scientists argue that the extinction was brought on by a gradual accumulation of carbon dioxide in ocean waters. Such an increase would have severely disrupted the ability of animals to carry out metabolism and form their shells. This hypothesis would explain why sedentary marine organisms (with little tolerance for CO_2) were preferentially eliminated, while more mobile organisms (with much higher tolerances for CO_2) survived.

The Mesozoic Era

The **Mesozoic era** (248 to 65 M.Y.A.) was a time of intensive evolution of terrestrial plants and animals. The major evolutionary lines on land had been established earlier, but the evolutionary expansion of these lines, which led to the major groups of organisms living today, occurred during the Mesozoic era, starting with the great evolutionary leap that brought the amphibians onto the land. Frogs, salamanders, and other much larger amphibians that are now extinct were the first vertebrates to live successfully on land—as they still do. Amphibians in turn gave rise to the first reptiles about 300 million years ago. Within 50 million years the reptiles, better suited than amphibians to living out of water, replaced them as the dominant land animal on earth.

With the success of the reptiles, vertebrates truly came to dominate the surface of the earth. Many kinds of reptiles evolved, from those smaller than a chicken to others bigger than a truck. Some flew, and others swam. From among them eventually evolved reptiles, which gave rise to the three

Figure 17.5 (continued)
vast period of time, the remarkable increase in structural complexity and overall diversity of the dinosaurs can be seen, until they abruptly became extinct, giving way to the dominant mammals of the Cenozoic era. Flowering plants can be seen for the first time at the left-hand side of the illustration.

great lines of terrestrial vertebrates: dinosaurs, birds, and mammals. Dinosaurs and mammals appear at about the same time in the fossil record, 220 million years ago, but the dinosaurs quickly won out in the evolutionary competition, and for over 150 million years dinosaurs dominated the face of the earth. (Think of it—over a *million centuries*! If you could look back to that distant time and have each century flash by your eye in a brief minute, it would take a thousand days to see it all.) Over all these long centuries the largest mammal was no bigger than a cat. Dinosaurs reached the height of their diversification and dominance of the land during the Cretaceous and Jurassic periods (figure 17.5).

The Mesozoic era has traditionally been divided into three periods: the Triassic, the Jurassic, and the Cretaceous. Because of the major extinction that ended the Paleozoic, only 4% of species survived into the Mesozoic. All through the Mesozoic, these species evolved into new species, genera, and families. Both on land and in the sea, the number of species of almost all groups of organisms has been climbing steadily for the past 250 million years and is now at an all-time high.

This extended recovery from the great Permian extinction had one interruption. About 65 million years ago at the end of the Cretaceous period, dinosaurs disappeared, along with the flying reptiles called pterosaurs (figure 17.6) and the great marine reptiles. This extinction marks the end of the Mesozoic era. Mammals quickly took their place, becoming in their turn abundant and diverse—as they are today.

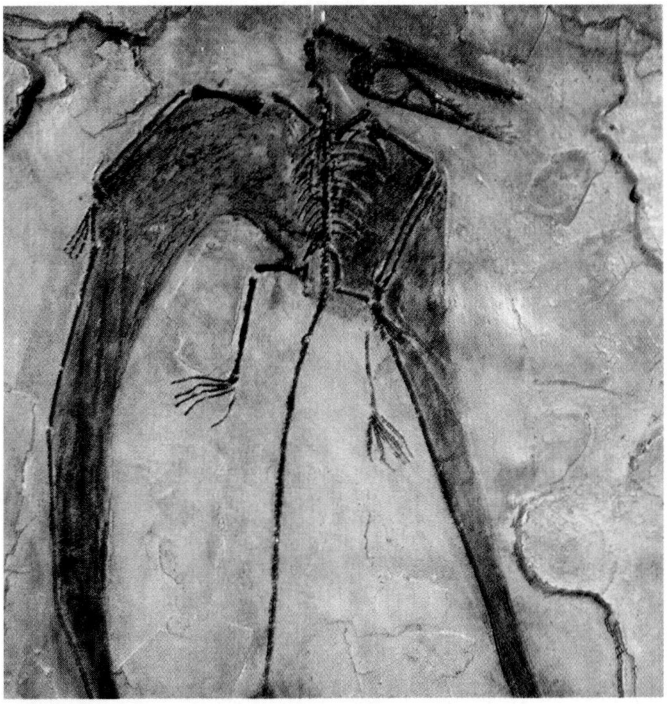

Figure 17.6 An extinct flying reptile.
Pterosaurs, such as the one pictured, became extinct with the dinosaurs about 65 million years ago. Flight has evolved three separate times among vertebrates; however, birds and bats are the only representatives still with us.

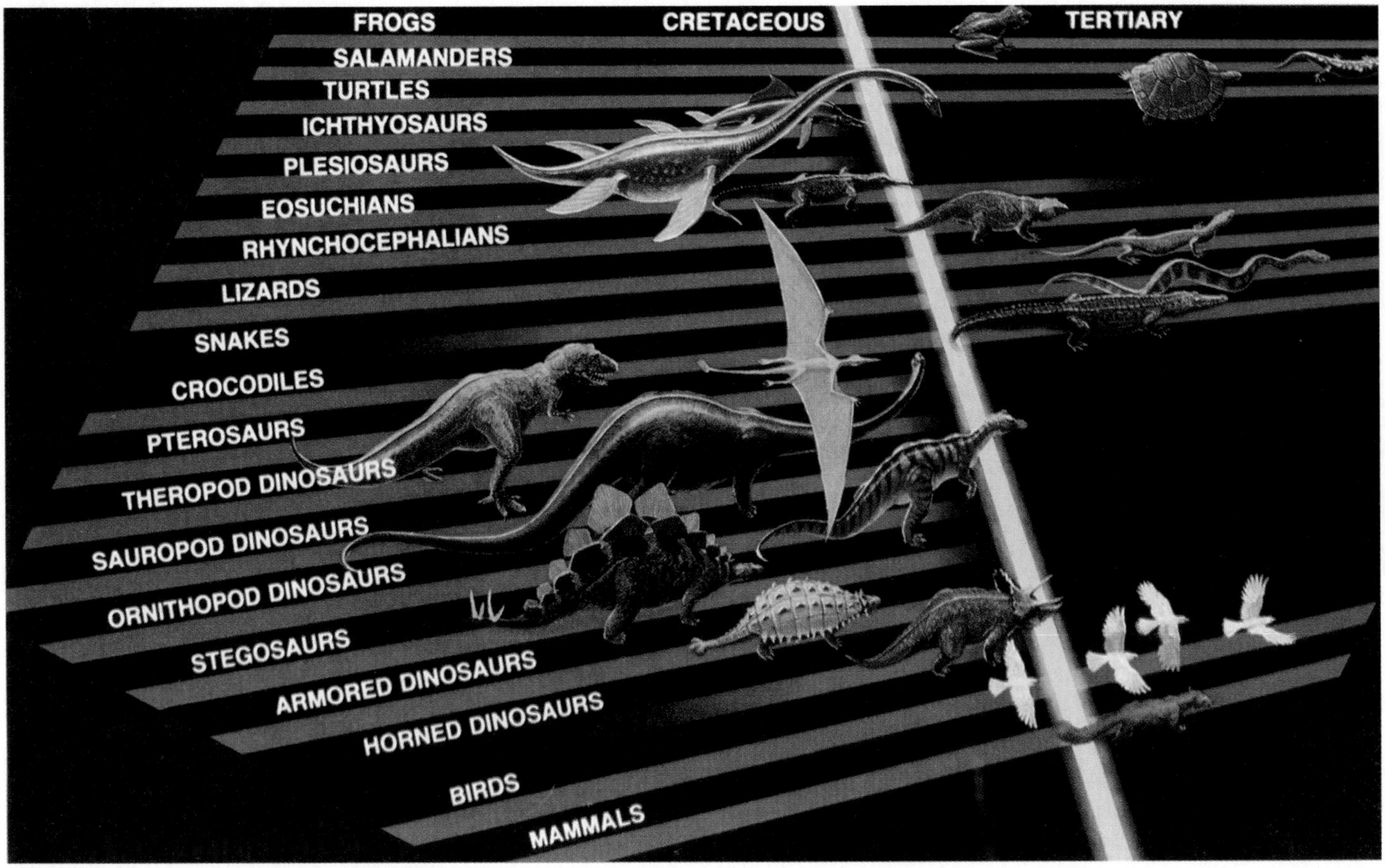

Figure 17.7 Extinction of the dinosaurs.
The dinosaurs became extinct 65 million years ago (*yellow line*) in a major extinction event that also eliminated all the great marine reptiles (plesiosaurs and ichthyosaurs) as well as the largest of the primitive land mammals. The birds and smaller mammals survived and went on to occupy the aerial and terrestrial modes of living in the environment left vacant by the dinosaurs. Crocodiles, small lizards, and turtles also survived, but reptiles never again achieved the diversity of the Cretaceous period.

What Happened to the Dinosaurs?

Dinosaurs disappear abruptly from the fossil record 65 million years ago (figure 17.7), their extinction marking the end of the Mesozoic era. Why? Many explanations have been advanced, including volcanoes and infectious disease. The most widely accepted theory assigns the blame to a comet that hit the earth then. Physicist Luis W. Alvarez and his associates discovered that the usually rare element iridium was abundant in many parts of the world in a thin layer of sediment that marked the end of the Cretaceous period. Iridium is rare on earth but common in meteorites. Alvarez and his colleagues have proposed that if a large meteorite 10 kilometers in diameter struck the surface of the earth 65 million years ago, a dense cloud would have been thrown up. The cloud would have been rich in iridium, and as its particles settled, the iridium would have been incorporated into the layers of sedimentary rock being deposited at the time. By darkening the world, the cloud would have greatly slowed or temporarily halted photosynthesis and driven many kinds of organisms to extinction.

The Alvarez hypothesis has become widely accepted among biologists, although it remains controversial among some. These holdouts argue that it is not certain that dinosaurs became extinct suddenly, as they would have by a meteorite collision, and question whether other kinds of animals and plants show the types of patterns that would have been expected as the result of a meteorite collision. The issue was largely settled in the minds of most scientists by the discovery in the last decade of an impact crater, in the sea just off the coast of the Yucatán peninsula. For hundreds of kilometers in all directions are signs of the meteor's impact, including quartz crystals with shock patterns that could only have been produced by an enormous impact (they are produced, for example, as a by-product of nuclear tests). Large amounts of soot were deposited in rock worldwide at the time of the extinction, indicating very widespread burning. At what date does radio-dating place the Yucatán impact? Sixty-five million years ago.

The Cenozoic Era

The relatively warm, moist climates of the early Cenozoic era (65 M.Y.A. to present) have gradually given way to today's colder and drier climate. The first half of the Cenozoic was very warm, with junglelike forests at the poles. With the extinction of the dinosaurs and many other organisms and this change in climate, new forms of life were able to invade new habitats. Mammals diversified from earlier, small nocturnal forms to many new forms. Most present-day orders of mammals appeared at this time, a period of great diversity.

Then the world's climate began to cool, and ice caps formed at the poles. As glaciation in Antarctica became fully established by about 13 million years ago, regional climates cooled dramatically. A series of so-called ice ages followed, the most recent only a few million years ago. Many very large mammals evolved during the ice ages, including mastodons, mammoths (figure 17.8), saber-toothed tigers, and enormous cave bears.

The Antarctic ice mass that formed as a result of these glaciations has made the world's climate cooler near the poles, warmer near the equator, and drier in the middle latitudes than ever before. In general, forests covered most of the land area of continents, except for Antarctica, until about 15 million years ago, when the forests began to recede rapidly and modern plant communities appeared. During the past several million years, the formation of extensive deserts in northern Africa, the Middle East, and India made migration between Africa and Asia very difficult for organisms of tropical forests. Overall, the Cenozoic era has been characterized by sharp differences in habitat, even within small areas, and the regional evolution of distinct groups of plants and animals (figure 17.9). These factors have facilitated the rapid formation of many new species.

Figure 17.8 Life in downtown Los Angeles 15 thousand years ago.
Trapped in large pools of tar, the skeletons of mammoths have been preserved in the La Brea Tar Pits located in Los Angeles, California.

Figure 17.9 Regional evolution of specialized organisms.
The Cenozoic era was characterized by the evolution of many new species adapted to particular environments, such as this mother and baby three-toed sloth (*Bradypus infuscatus*). Sloths reside in the canopy of the tropical rain forest, as shown here in Panama.

17.2 The Parade of Vertebrates

A series of key evolutionary advances allowed vertebrates to first conquer the sea and then the land. In this section we explore the history of vertebrates (figure 17.10), and in the following section we examine the key evolutionary advances that permitted vertebrates to successfully invade the land. As you will see, this invasion was a staggering evolutionary achievement, involving fundamental changes in many body systems.

Fishes Dominate the Sea

The first vertebrates evolved in the sea about 470 million years ago. They were jawless **fishes** called **agnathans** that could only feed on foods they could suck into their mouths. Most seem to have lived on the ocean bottom, feeding by filtering tiny invertebrates from water passing through their mouths. Their bodies were covered with thick, bony plates so that they were built a bit like tanks with tail fins. These first vertebrates were a great evolutionary success, dominating the world's oceans for about 150 million years, until they were eventually replaced by new kinds of fishes that were hunters.

Only 63 species of jawless fishes exist today, half of them eel-like lamprey parasites and the other half scavenger hagfish. The feeding habits of these living agnathans are specialized—they are parasites on other fish or are scavengers. Indeed, their success in exploiting these particular evolutionary opportunities may be the reason they alone have survived into the modern world as the only representatives of what was once the dominant group of vertebrates.

Lampreys have round mouths that function like suction cups (figure 17.11). Using these specialized mouths, lampreys attach themselves to fish, rasping through the skin of the fish with their tongues and then sucking out the blood through the hole.

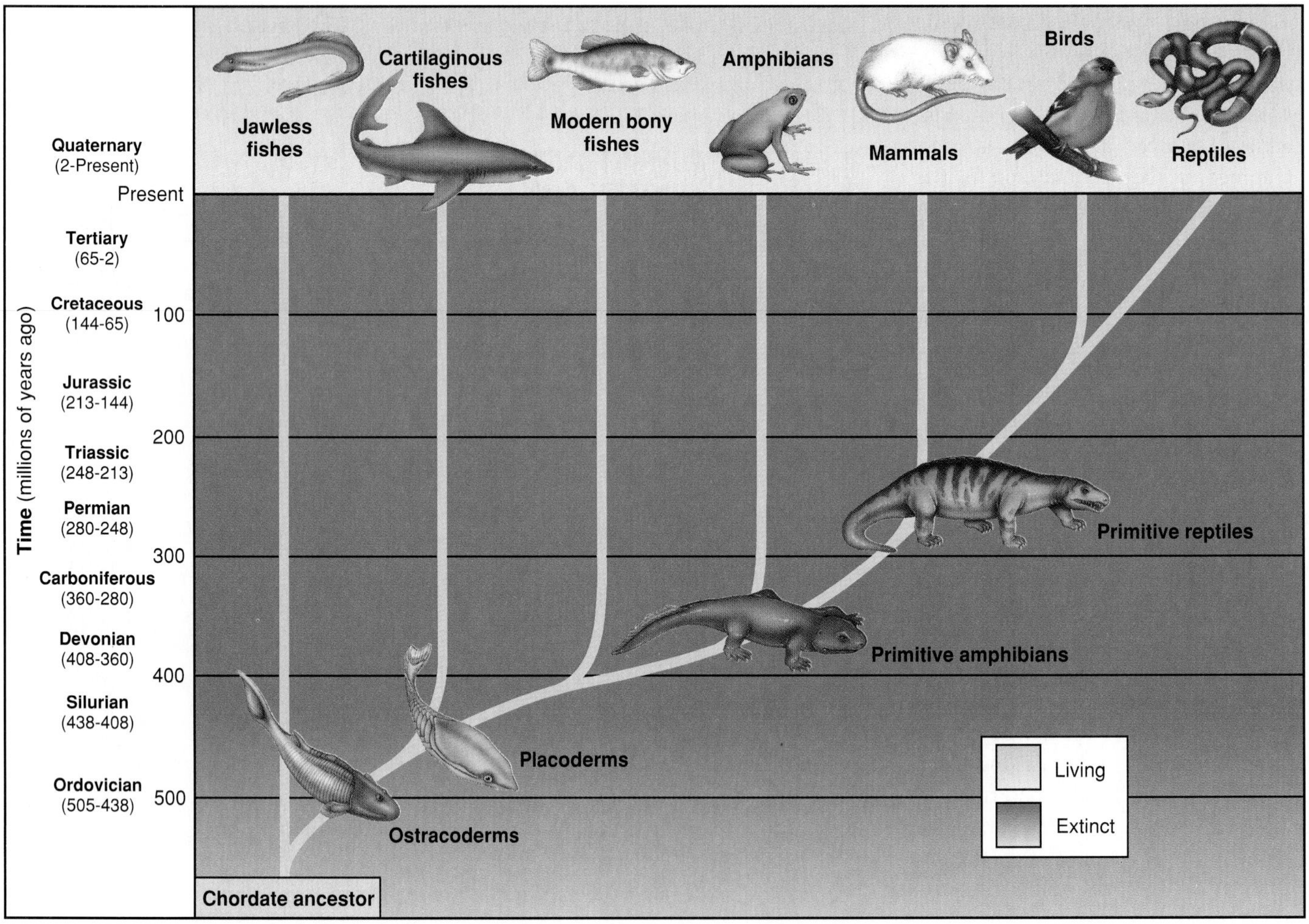

Figure 17.10 Vertebrate family tree.
Most of the kinds of primitive amphibians are now extinct, although two groups survive as frogs and salamanders. Primitive reptiles arose from amphibians and gave rise to mammals and to dinosaurs, which survive today as birds. Ostracoderms and Placoderms are extinct groups of fishes.

The Evolution of Jaws

The evolution of fishes in the sea has been dominated by two questions, two evolutionary challenges to surviving as a predator in water:

1. What is the best way to grab hold of potential prey?
2. What is the best way to pursue prey through water?

The fishes that replaced the jawless ones 360 million years ago were powerful predators with much better solutions to both evolutionary challenges. They solved the problem of grasping prey with strong biting jaws that had jagged, bony edges that served as teeth (figure 17.12). They solved the problem of chasing prey through water with a bony skeleton that could support strong muscles (remember, a muscle needs something to pull against) and paired fins for better control of fast swimming.

The earliest jawed fishes had small bodies covered with protective spines. Later, larger jawed fishes called *placoderms* evolved. These fishes had massive heads armored with heavy bony plates, the largest over 9 meters (30 feet) long! Both spiny fishes and placoderms are extinct now, replaced in turn by fishes that evolved even better ways of moving through the water, the sharks and the bony fishes. The replacement did not happen overnight—the earliest sharks and bony fishes first appear in the fossil record soon after spiny fishes and placoderms do. However, after sharing the seas for 150 million years, the long competition finally ended with the complete disappearance of the less maneuverable early jawed fishes. For the last 250 million years, all fishes swimming in the world's oceans and rivers have been either sharks (and their relatives, the rays) or bony fishes.

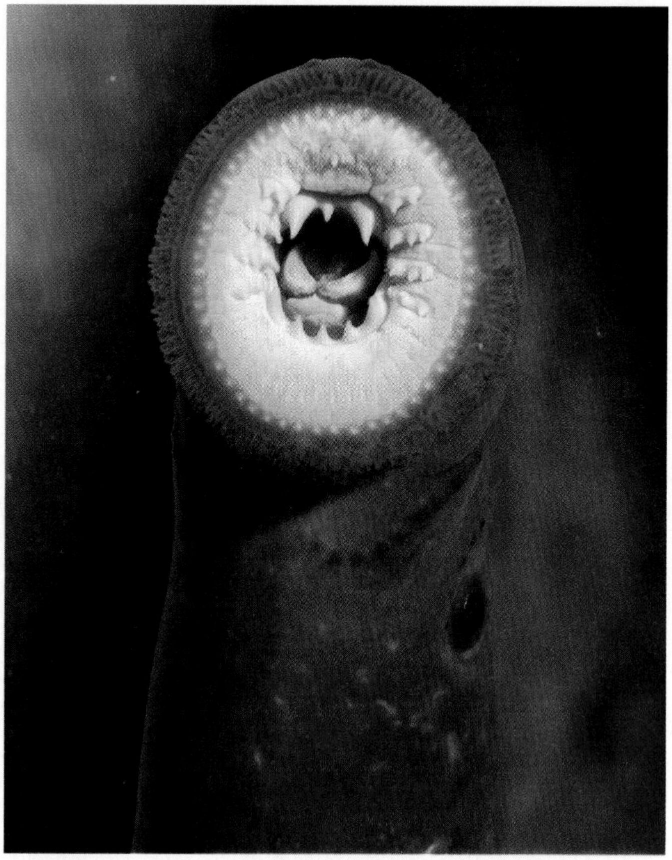

Figure 17.11 Specialized mouth of a lamprey.
Lampreys use their suckerlike mouths to attach themselves to the fish on which they prey. When they have done so, they bore a hole in the fish with their teeth and feed on its blood.

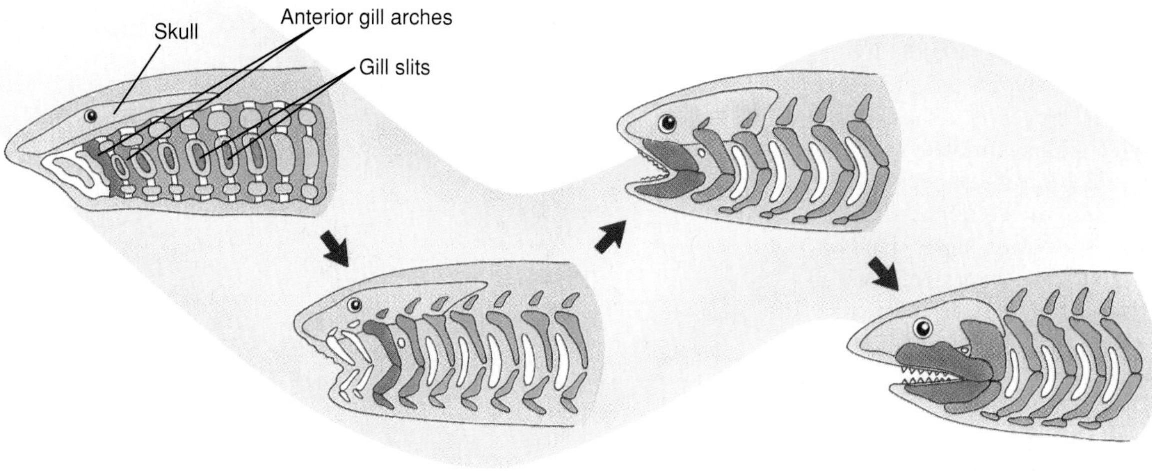

Figure 17.12 Evolution of the jaw.
Jaws evolved from the anterior gill arches of ancient, jawless fishes.

Sharks

The problem of fast and maneuverable swimming was solved in sharks by the replacement of the heavy bone skeleton of the early fishes with a far lighter one made of strong, flexible cartilage. Members of this group, the class Chondrichthyes, are very powerful swimmers, with a back fin, a tail fin, and two sets of paired side fins for controlled thrusting through the water (figure 17.13). Some of the largest sharks filter their food from the water like jawless fishes, but most are predators, their mouths armed with rows of hard, sharp teeth. Today there are about 850 species of sharks and rays.

Bony Fishes

The problem of fast and maneuverable swimming was solved in bony fishes in a very different way. The bony skeleton of their ancestors was retained and a gas-filled sac called a **swim bladder** was added that gave their bodies buoyancy in the water (figure 17.14). A bony fish uses its swim bladder like a child uses an inner tube, to float in the water without expending energy in swimming—a shark sinks if it stops swimming. By adjusting the amount of gas in its swim bladder, a bony fish can rise up and down in the water the same way a submarine does. This simple solution to the challenge of swimming has proven to be a great success. Of the nearly 20,000 living species of fishes in the world today, 18,000 species are bony fishes (class Osteichthyes) with swim bladders (figure 17.15). That's more species than all other kinds of vertebrates combined! In fact, if you could stand in one place and have every vertebrate animal alive today pass by you, one after the other, half of them would be bony fishes. The major groups of fishes are reviewed in table 17.1.

Bony fishes have a number of features that are unique to this class alone. They have a highly developed **lateral line system,** a sensory system that enables them to detect changes in water pressure and thus the movement of predators and prey in the water.

Figure 17.13 Chondrichthyes.
The blue shark is a member of the class Chondrichthyes, which are mainly predators or scavengers and spend most of their time in graceful motion. As they move, they create a flow of water past their gills, from which they extract oxygen.

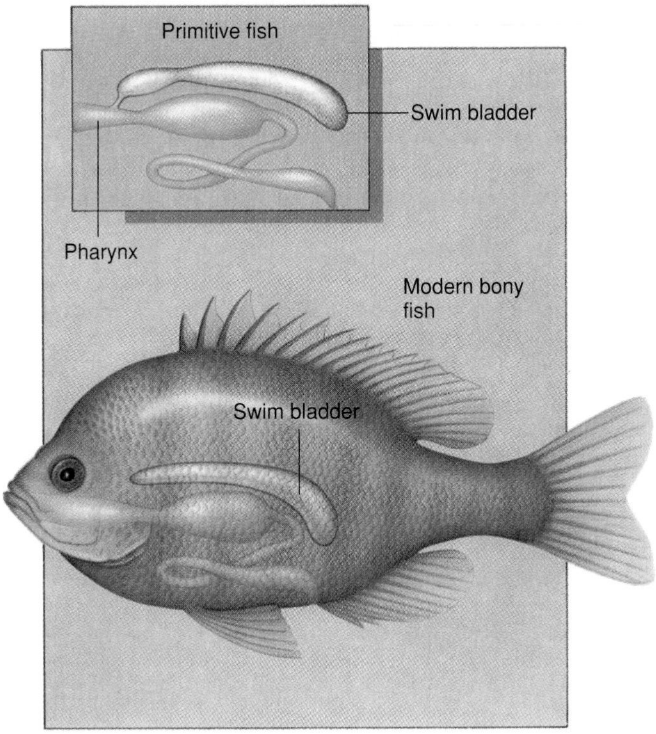

Figure 17.14 Diagram of a swim bladder.
The bony fishes use a swim bladder, which evolved as an outpocketing of the pharynx, to control their buoyancy in water.

Figure 17.15 Osteichthyes.
Bony fishes are extremely diverse, containing more species than all other kinds of vertebrates combined. This Korean angelfish, *Pomacanthus semicircularis,* in Fiji, is one of the many striking fishes that live around coral reefs in tropical seas.

Table 17.1 Major Groups of Fishes

Class	Typical Examples		Key Characteristics	Approximate Number of Living Species
Acanthodii	Spiny fishes		First fishes with jaws; all now extinct; paired fins supported by sharp spines	Extinct
Placodermi	Armored fishes		Jawed fishes with heavily armored heads; often quite large	Extinct
Osteichthyes	Ray-finned fishes		Most diverse group of vertebrates; swim bladders and bony skeletons; paired fins supported by bony rays	18,000
	Lobe-finned fishes		Largely extinct group of bony fishes; ancestral to amphibians; paired lobed fins	7
Chondrichthyes	Sharks, skates, rays		Streamlined hunters; skeleton of cartilage; no swim bladder; internal fertilization	850
Agnatha	Lampreys, hagfishes		Largely extinct group of jawless fishes with no paired appendages; the two surviving groups are parasites	63

Amphibians Invade the Land

The first vertebrates to come onto land from the sea were **amphibians.** Which kind of fishes were their ancestors? Compare a fish and a frog. A difference that commands attention is that frogs have legs and fish don't. Lobe-finned fish evolved 390 million years ago and have sturdy fins with a pattern of bones that bears a remarkable resemblance to that of amphibian limbs. Only seven species survive today, a single species of coelacanth (figure 17.16) and six species of lungfish. Scientists have debated whether the direct ancestor of amphibians was a coelacanth or a lungfish. However, recent evidence has led paleontologists to believe that the direct ancestor was neither a coelacanth nor a lungfish but a third type of lobe-finned fish that is now extinct.

The successful invasion of land by amphibians involved a number of major innovations:

1. *Legs*, which evolved from fins, were necessary to support the body's weight as well as to allow movement from place to place (figure 17.17).
2. *Lungs* were necessary, even though there is far more oxygen in air than water, because the delicate structure of fish gills requires the buoyancy of water to support it.

Figure 17.16 The living coelacanth, *Latimeria chalumnae.*

Discovered in the western Indian Ocean in 1938, this coelacanth represents a group of fishes that had been thought to be extinct for about 70 million years. Scientists who studied living individuals in their natural habitat at depths of 100 to 200 meters observed them drifting in the current and hunting other fishes at night. Some individuals are nearly 3 meters long.

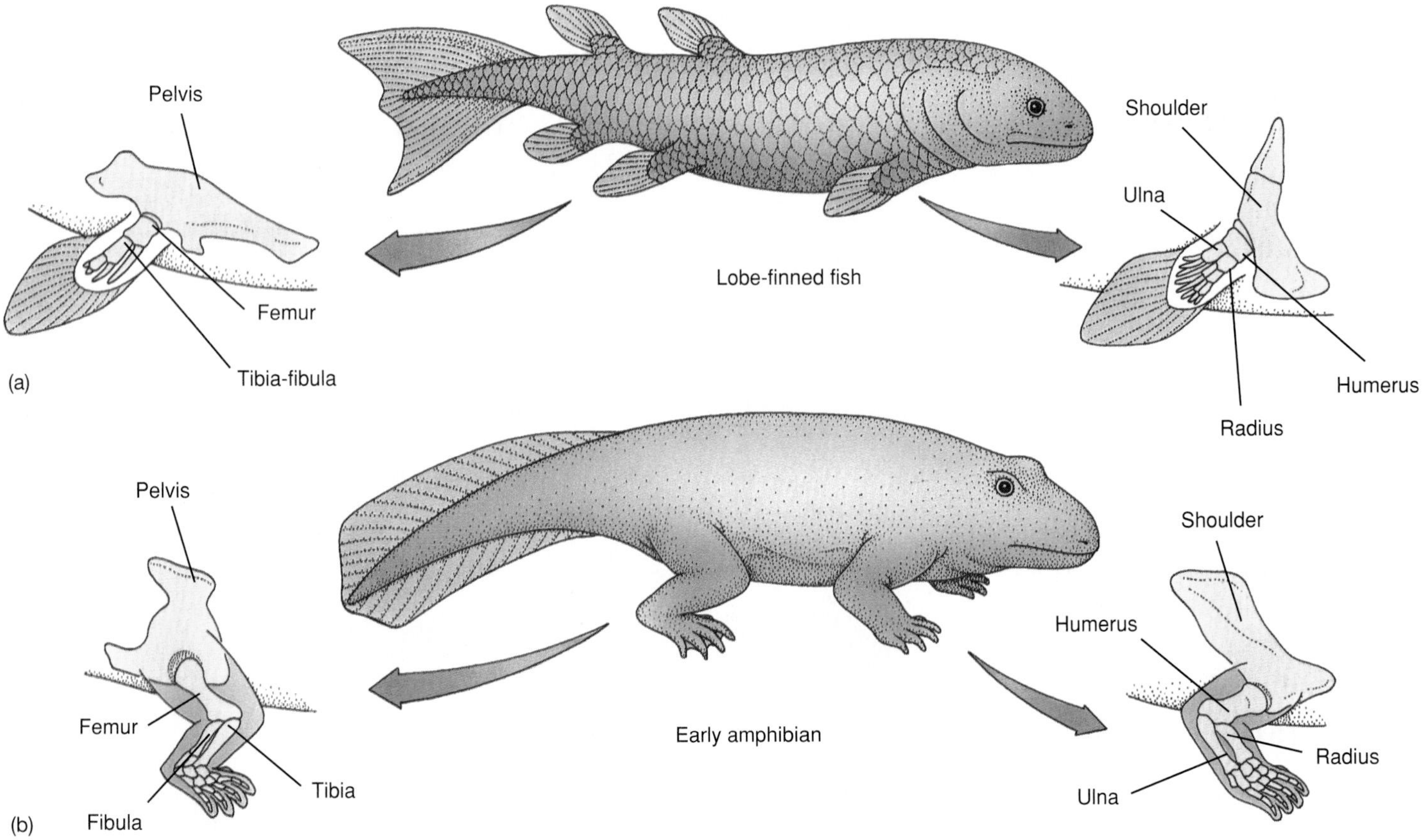

Figure 17.17 Legs evolved from fins.

(*a*) The limbs of a lobe-finned fish. Some lobe-finned fishes could move out onto land. (*b*) The limbs of an early amphibian. As illustrated by their skeletal structure, the legs of primitive amphibians could clearly function on land better than could the fins of lobe-finned fishes.

Table 17.2 Orders of Amphibians

Order	Typical Examples		Key Characteristics	Approximate Number of Living Species
Anura	Frogs, toads		Compact tailless body; large head fused to the trunk; rear limbs specialized for jumping	3,680
Caudata	Salamanders, newts	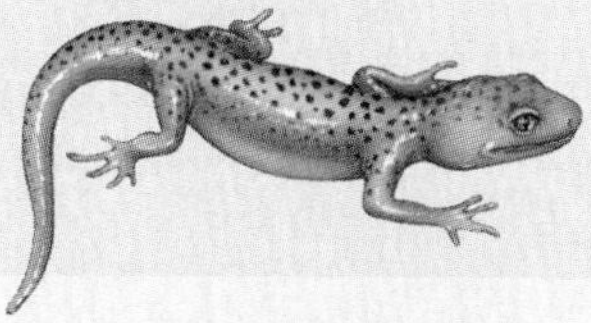	Slender body; long tail and limbs set out at right angles to the body	369
Gymnophiona	Caecilians		Tropical group with a snakelike body; no limbs; little or no tail	160

3. The *heart* had to be redesigned to deliver to the amphibian tissues the greater amounts of oxygen required by muscles for walking.
4. *Reproduction* had to be carried out in water until ways could be found to prevent the eggs from drying out.
5. Most important, ways had to be developed to prevent the *body* itself from drying out!

Amphibians solved these problems only imperfectly, but their solutions worked well enough that amphibians have survived for 350 million years (figure 17.18). Evolution does not insist on perfect solutions, only workable ones.

Amphibians were the dominant land vertebrate for 100 million years. They first became common in the late Paleozoic era, when much of North America was covered by low-land tropical swamps. Amphibians reached their greatest diversity during the mid-Permian period, when 40 families existed. Sixty percent of them were fully terrestrial, with bony plates and armor covering their bodies, and many grew to be very large—some as big as a pony! After the great Permian extinction the terrestrial forms began to decline, and by the time dinosaurs evolved, only 15 families remained, all aquatic. Only two of these families survived the age of the dinosaurs, both aquatic: the anurans (frogs and toads) and the urodeles (salamanders).

Figure 17.18 Class Amphibia.
This red-eyed tree frog, *Agalychnis callidryas*, is a member of the group of amphibians that includes frogs and toads (order Anura).

Approximately 4,200 species of amphibians (class Amphibia) exist today, in 37 different families, all aquatic and all descended from the two aquatic families that survived the age of dinosaurs (table 17.2). All of today's amphibians must reproduce in water and live the early part of their lives there, so amphibians are not completely terrestrial. However, in most habitats, particularly in the tropics, they are often today the most abundant and successful vertebrates to be found.

Reptiles Conquer the Land

Figure 17.19 Class Reptilia.
A river crocodile, *Crocodilus acutus.* Most crocodiles resemble birds and mammals in that they have a four-chambered heart; all other living reptiles have a three-chambered one. Crocodiles, like birds, are related to dinosaurs, rather than to any of the other living reptiles.

If we think of amphibians as the "first draft" of a manuscript about survival on land, then reptiles were the finished book. For each of the five key challenges to living on land, **reptiles** improved on the innovations first attempted by amphibians: (1) legs were arranged to more effectively support the body's weight, allowing reptile bodies to be bigger and to *run;* (2) lungs and (3) heart were altered to make them far more efficient; (4) eggs were encased in watertight covers; and (5) the skin was covered with dry scales to retain water. Reptiles were the first truly *terrestrial* vertebrates.

Today some 6,000 species in the class Reptilia (table 17.3), mostly snakes and lizards, are found in practically every wet and dry habitat on earth (figure 17.19). And *today's* reptiles don't begin to tell the tale. When reptiles first evolved, about 320 million years ago, the world was entering a long, dry period, conditions to which reptiles were very well suited, and thus reptiles quickly diversified. In particular, their ability to retain water allowed reptiles to maintain large bodies in dry conditions, something amphibians could not do. Within 100 million years, the large-body niche was totally taken over by the reptiles—all land vertebrates bigger than a chicken-sized animal were reptiles.

A series of different reptiles dominated life on land, one replacing another as the dominant form.

Pelycosaurs

At first, "sail-back" **pelycosaurs** rose to prominence, the first land vertebrates to be efficient predators. With long, sharp "steak knife" teeth, pelycosaurs like *Dimetrodon* were the first land vertebrates able to kill animals its own size. The sail appears to have helped control body temperature. Pelycosaurs were dominant for 50 million years, at their height composing 70% of all land vertebrates. They died out about 250 million years ago, replaced by their direct descendants, the therapsids.

Therapsids

The **therapsids** had a more upright stance than the sprawling pelycosaurs from which they evolved. Therapsids were the immediate ancestors of the mammals, and paleontologists suspect they may have been warm-blooded. Being warm-blooded (that is, maintaining a constant high body temperature) uses considerable fuel to produce body heat, and so therapsids would have had to eat 10 times more frequently than cold-blooded pelycosaurs.

Thecodonts

The therapsids were in turn replaced by cold-blooded reptiles called **thecodonts.** Most thecodonts resembled crocodiles, but later forms were the first reptiles to be bipedal—to stand on two feet. Cold-blooded thecodonts were able to compete successfully against warm-blooded therapsids because the world's climate grew warm 230 million years ago, and thecodonts had to eat far less.

Dinosaurs

From thecodonts, five great lines of large-bodied reptiles developed: **dinosaurs,** a very diverse group, some of which grew to be larger than houses; **crocodiles,** which have changed little from that time until now; **ichthyosaurs** and **plesiosaurs,** marine reptiles that lived in the oceans; and **pterosaurs,** which were flying reptiles.

Dinosaurs were the most successful land vertebrates ever, dominating life on land for over 150 million years. They represent a significant improvement in body design over thecodonts. Their legs are positioned directly beneath the body, rather than projecting out in a "V" like therapsids and thecodonts. Placing their body's weight directly over the legs allowed some dinosaurs to run with great speed and agility.

At about the same time that dinosaurs evolved, about 220 million years ago, mammals evolved, but while dinosaurs flourished there were no large-bodied mammals. That niche was filled.

Table 17.3 Orders of Reptiles

Class	Typical Examples	Key Characteristics	Approximate Number of Living Species
Ornithischia	Stegosaur 	Dinosaurs with two pelvic bones facing backward, like a bird's pelvis; herbivores, with turtlelike upper beak; legs under body	Extinct
Saurischia	Tyrannosaur 	Dinosaurs with one pelvic bone facing forward, the other back, like a lizard's pelvis; both plant- and flesh-eaters; legs under body	Extinct
Pterosauria	Pterodactyl 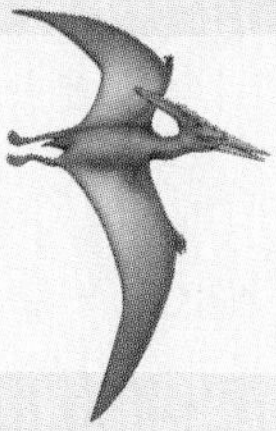	Flying reptiles; wings were made of skin stretched between fourth finger and body; wingspans of early forms typically 60 centimeters; later forms nearly 8 meters	Extinct
Plesiosauria	Plesiosaur 	Barrel-shaped marine reptiles with sharp teeth and large paddle-shaped fins; some had snakelike necks twice as long as their body	Extinct
Ichthyosauria	Ichthyosaur 	Streamlined marine reptiles with many body similarities to sharks and modern fishes	Extinct
Squamata, suborder Sauria	Skink 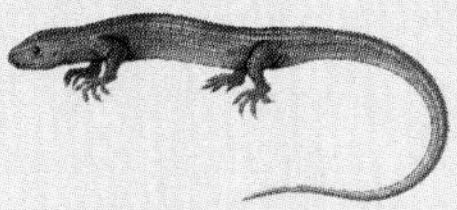	Lizards; limbs set out at right angles to body; anus is in transverse (sideways) slit; most are terrestrial	3,800
Suborder Serpentes	Kingsnake 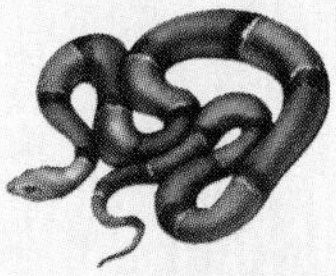	Snakes; no legs; move by slithering; scaly skin is shed periodically; most are terrestrial	3,000
Testudines	Sea turtle 	Ancient armored reptiles with shell of bony plates to which vertebrae and ribs are fused; sharp, horny beak without teeth	250
Crocodylia	Crocodile 	Advanced reptiles with four-chambered heart and socketed teeth; anus is a longitudinal (lengthwise) slit; closest living relatives to birds	25
Rhynchocephalia	Tuatara 	Sole survivors of a once successful group that largely disappeared before the dinosaurs; fused, wedgelike, socketless teeth; primitive third eye under skin of forehead	1

Birds Master the Air

Birds evolved from small bipedal dinosaurs about 150 million years ago, but they were not common until the flying reptiles called pterosaurs became extinct along with the dinosaurs. Unlike pterosaurs, birds are insulated with feathers. Birds are so structurally similar to dinosaurs in all other respects that many scientists consider birds to be simply feathered dinosaurs.

The oldest bird of which there is a clear fossil is *Archaeopteryx* (meaning "ancient wing"), which was about the size of a crow and shared many features with small, bipedal, carnivorous dinosaurs. For example, it had teeth and a long reptilian tail. And unlike the hollow bones of today's birds, its bones were solid. Because of these features, early finds of **Archaeopteryx** were classified as *Compsognathus,* a dinosaur of similar size, until impressions of feathers were discovered on the fossils.

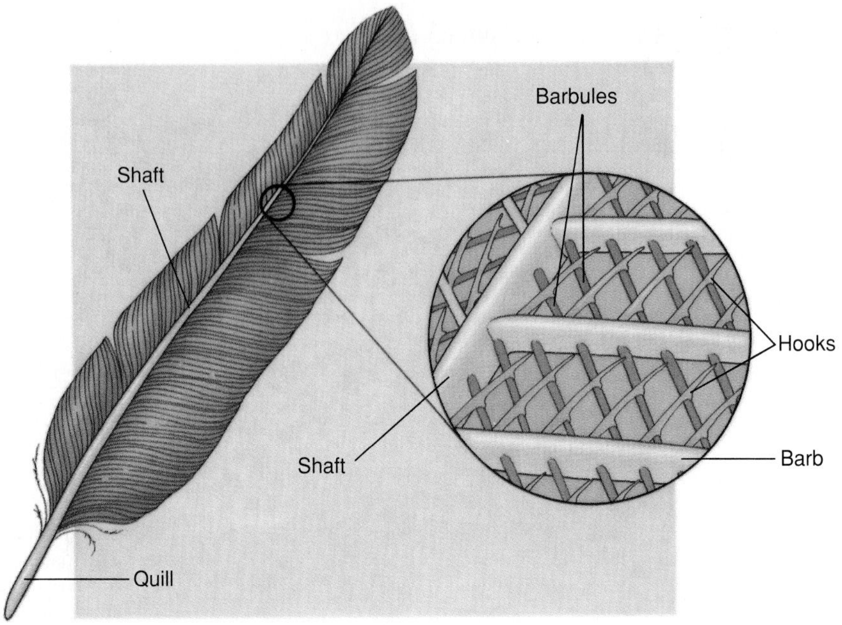

Figure 17.20 Feathers are more complex than they seem.
The barbs off the main shaft of a feather have secondary branches called barbules. The barbules of adjacent barbs are attached to one another by microscopic hooks.

Archaeopteryx had two key avian characteristics:

1. *Feathers,* a structure unique in the animal world. Derived from reptilian scales, feathers are the ideal adaptation for flight because they are lightweight and easily replaced if damaged (figure 17.20);
2. *A well-developed collarbone* (the so-called wishbone), which acts as a torque-absorbing strut linking the shoulders. The wishbone is absolutely essential for flight.

Birds, like mammals, are warm-blooded. They generate enough heat through metabolism to maintain a high body temperature. Many scientists think the later dinosaurs, from which birds evolved, were warm-blooded. Birds maintain body temperatures significantly higher than most mammals, ranging from 40° to 42°C (104° to 108°F). For comparison, your body temperature is 37°C (98°F). The high body temperature permits metabolism in the bird's flight muscles to proceed rapidly, necessary to satisfy the large energy requirements of flight.

Today over 9,000 species of birds (class Aves) occupy a variety of habitats all over the world (figure 17.21). The major orders of birds are reviewed in table 17.4. You can tell a great deal by examining the beaks. For example, carnivorous birds such as hawks have a sharp beak for tearing apart meat, the beaks of ducks are flat for shoveling through mud, and the beaks of finches are short and thick for crushing seeds.

Figure 17.21 Class Aves.
The birds (class Aves) are a large and successful group of about 9,000 species, more than any other class of vertebrates except the bony fishes. Pictured here is the northern saw-whet owl, *Aegolius acadius.* Owls, hawks, eagles, and similar predatory birds are called raptors.

Table 17.4 Major Orders of Birds

Order	Typical Examples	Key Characteristics	Approximate Number of Living Species
Passeriformes	Crows, mockingbirds, robins, sparrows, starlings, warblers	*Songbirds* Well-developed vocal organs; perching feet; dependent young	5,276 (largest of all bird orders; contains over 60% of all bird species)
Apodiformes	Hummingbirds, swifts	*Fast fliers* Short legs; small bodies; rapid wing beat	428
Piciformes	Honeyguides, toucans, woodpeckers	*Woodpeckers or toucans* Grasping feet; chisel-like, sharp bills that can break down wood	383
Psittaciformes	Cockatoos, parrots	*Parrots* Large, powerful bills for crushing seeds; well-developed vocal organs	340
Charadriiformes	Auks, gulls, plovers, sandpipers, terns	*Shorebirds* Long, stiltlike legs; slender, probing bills	331
Columbiformes	Doves, pigeons	*Pigeons* Perching feet; rounded, stout bodies	303
Falconiformes	Eagles, falcons, hawks, vultures	*Birds of prey* Carnivorous; keen vision, sharp, pointed beaks for tearing flesh; active during the day	288
Galliformes	Chickens, grouse, pheasants, quail	*Gamebirds* Often limited flying ability; rounded bodies	268
Gruiformes	Bitterns, coots, cranes, rails	*Marsh birds* Long, stiltlike legs; diverse body shapes; marsh-dwellers	209
Anseriformes	Ducks, geese, swans	*Waterfowl* Webbed toes; broad bill with filtering ridges	150
Strigiformes	Barn owls, screech owls	*Owls* Nocturnal birds of prey; strong beaks; powerful feet	146
Ciconiiformes	Herons, ibises, storks	*Waders* Long-legged; large bodies	114
Procellariiformes	Albatrosses, petrels	*Seabirds* Tube-shaped bills; capable of flying for long periods of time	104
Sphenisciformes	Emperor penguins, crested penguins	*Penguins* Marine; modified wings for swimming; flightless; found only in Southern Hemisphere; thick coat of insulating feathers	18
Dinornithiformes	Kiwis	*Kiwis* Flightless; small; primitive; confined to New Zealand	2
Struthioniformes	Ostriches	*Ostriches* Powerful running legs; flightless; only two toes, very large	1

Mammals Adapt to Colder Times

The **mammals** that evolved about 220 million years ago side by side with the dinosaurs would look strange to you, not at all like modern-day lions and tigers and bears. They share two key characteristics with the mammals of today:

1. Female mammals have mammary glands, which produce milk to nurse the rapidly growing newborns.
2. Among living vertebrates, only mammals have hair, and all mammals do (even whales have a few sensitive bristles on their snout). A hair is a filament composed of dead cells filled with the protein keratin. The primary function of hair is insulation.

Figure 17.22 Class Mammalia.
These meadow mice are placental mammals. There are more species of rodents like these mice than any other kind of mammal. As in the time of the dinosaurs, most mammals living today are smaller than cats.

Monotremes

The first mammals probably most closely resembled today's **monotremes,** which have a pelvic structure similar to early reptiles and lay shelled eggs. When the dinosaurs disappeared, their niche as the large land vertebrate was taken over by such mammals, whose fur-insulated bodies were better suited to the colder climates that became typical then. Monotremes nurtured their young with milk produced by mammary glands (hence the name "mammal").

Marsupials

Only three species of monotreme survive now—monotremes were replaced 100 million years ago by a different kind of mammal that invested even more effort in caring for its young. Called **marsupials,** these mammals did not have shelled eggs. Instead, young were born within the mother's body, and within days after birth transferred to pouches where they could be nursed by the mother's milk in relative protection. About 280 species of marsupials survive today, almost all of them in Australia and New Guinea.

Placental Mammals

Placental mammals (class Mammalia) invest even more care in nurturing their young than marsupials do, carrying the developing young within the mother's body far longer—you were nurtured in this fashion for nine months (figure 17.22). During this long period of protected growth and development, the young are fed not by mother's milk but rather by transferring nutrients directly from the mother's blood supply through a tissue called a *placenta* (that is why they are called "placental mammals").

Today, over 4,000 species of placental mammals occupy all the large-body niches that dinosaurs once claimed, among many others. They range in size from 1.5-gram shrews to 100-ton whales. Almost half of all mammals are rodents—mice and their relatives. Almost one-quarter of all mammals are bats! Mammals have even invaded the seas, as plesiosaurs and ichthyosaurs did so successfully millions of years earlier—79 species of whales and dolphins live in today's oceans. The placental mammals that walked the earth during the ice ages were even larger than today's versions; the world's climate has warmed again in recent times, favoring smaller bodies, which are easier to cool.

Primates, the order to which we belong, are not a major group; there are only 233 known species. Human beings evolved only very recently, less than 2 million years ago. There have been at least three species of humans, but our species, *Homo sapiens,* is the only one that survives today. We are notable among primates for having less hair, walking upright, making tools on a regular basis, and having complicated language. The major orders of mammals are described in table 17.5.

Table 17.5 Major Orders of Mammals

Order	Typical Examples		Key Characteristics	Approximate Number of Living Species
Rodentia	Beavers, mice, porcupines, rats, squirrels		*Small plant-eaters* Chisel-like incisor teeth	1,814
Chiroptera	Bats		*Flying mammals* Primarily fruit- or insect-eaters; elongated fingers; thin wing membrane; nocturnal; navigate by sonar	986
Insectivora	Moles, shrews		*Small burrowing mammals* Insect-eaters; most primitive placental mammals; spend most of their time underground	390
Marsupialia	Kangaroos, koalas		*Pouched mammals* Have abdominal pouch for young	280
Carnivora	Bears, cats, dogs, weasels, wolves		*Carnivorous predators* Teeth adapted for shearing flesh; no native families in Australia	240
Primates	Apes, humans, lemurs, monkeys		*Tree-dwellers* Large brain size; binocular vision; opposable thumb; end product of a line that branched off early from other mammals	233
Artiodactyla	Cattle, deer, giraffes, pigs, sheep		*Hoofed mammals* With two or four toes; herbivores	211
Cetacea	Dolphins, porpoises, whales		*Fully marine mammals* Streamlined bodies; front limbs modified into flippers; no hind limbs; blowholes on top of head; no hair except on muzzle	79
Lagomorpha	Rabbits, hares, pikas		*Rodentlike jumpers* Four upper incisors (rather than the two seen in rodents); hindlegs often longer than forelegs, an adaptation for jumping	69
Pinnipedia	Sea lions, seals, walruses		*Marine carnivores* Feed mainly on fish; limbs modified for swimming	34
Edentata	Anteaters, armadillos, sloths		*Toothless insect-eaters* Many are toothless but have some degenerate, peglike teeth	30
Perissodactyla	Horses, rhinoceroses, zebras		*Hoofed mammals with one or three toes* Herbivorous teeth adapted for chewing	17
Proboscidea	Elephants		*Long-trunked herbivores* Two upper incisors elongated as tusks; largest living land animal	2

17.3 Meeting the Challenge of Life on Land

The Challenge of Obtaining Oxygen

One of the defining challenges faced by vertebrates as they evolved, first in the oceans and then on land, was the need to acquire adequate amounts of oxygen to fuel their metabolism. Vertebrates first evolved in the sea, and they had to obtain from it the oxygen needed for fueling their muscles and growing their tissues. Most of the primitive phyla obtained oxygen by direct diffusion from seawater, which contains about 10 milliliters of dissolved oxygen per liter (figure 17.23*a*). Sponges, cnidarians, many flatworms and roundworms, and some annelid worms all obtain their oxygen by diffusion from surrounding water. Similarly, members of the vertebrate class Amphibia conduct gas exchange by direct diffusion through their moist skin (figure 17.23*b*).

The more advanced marine invertebrates (mollusks, arthropods, and echinoderms) possess special respiratory organs called gills that increase the surface area available for diffusion of oxygen. A **gill** is basically a thin sheet of tissue that waves through the water. Gills can be simple, as in the papulae of echinoderms (figure 17.23*c*), or complex, as in the highly convoluted gills of fish (figure 17.23*e*). In bony fishes, the gills are constructed in a particularly effective way—they are in fact the most efficient oxygen-gathering organs ever to have evolved.

Terrestrial arthropods do not have a single major respiratory organ like a gill. Instead, a network of air ducts called tracheae, branching into smaller and smaller tubes, carries air to every part of the body (figure 17.23*d*). Very small tracheoles are in direct contact with individual cells, and oxygen diffuses from the tracheoles across plasma membranes and into the cells of the body. The openings of tracheae to the outside are through special structures called spiracles, which can be closed and opened. Terrestrial vertebrates, except for some amphibians, do have a single respiratory organ, called the lung (figure 17.23*f*).

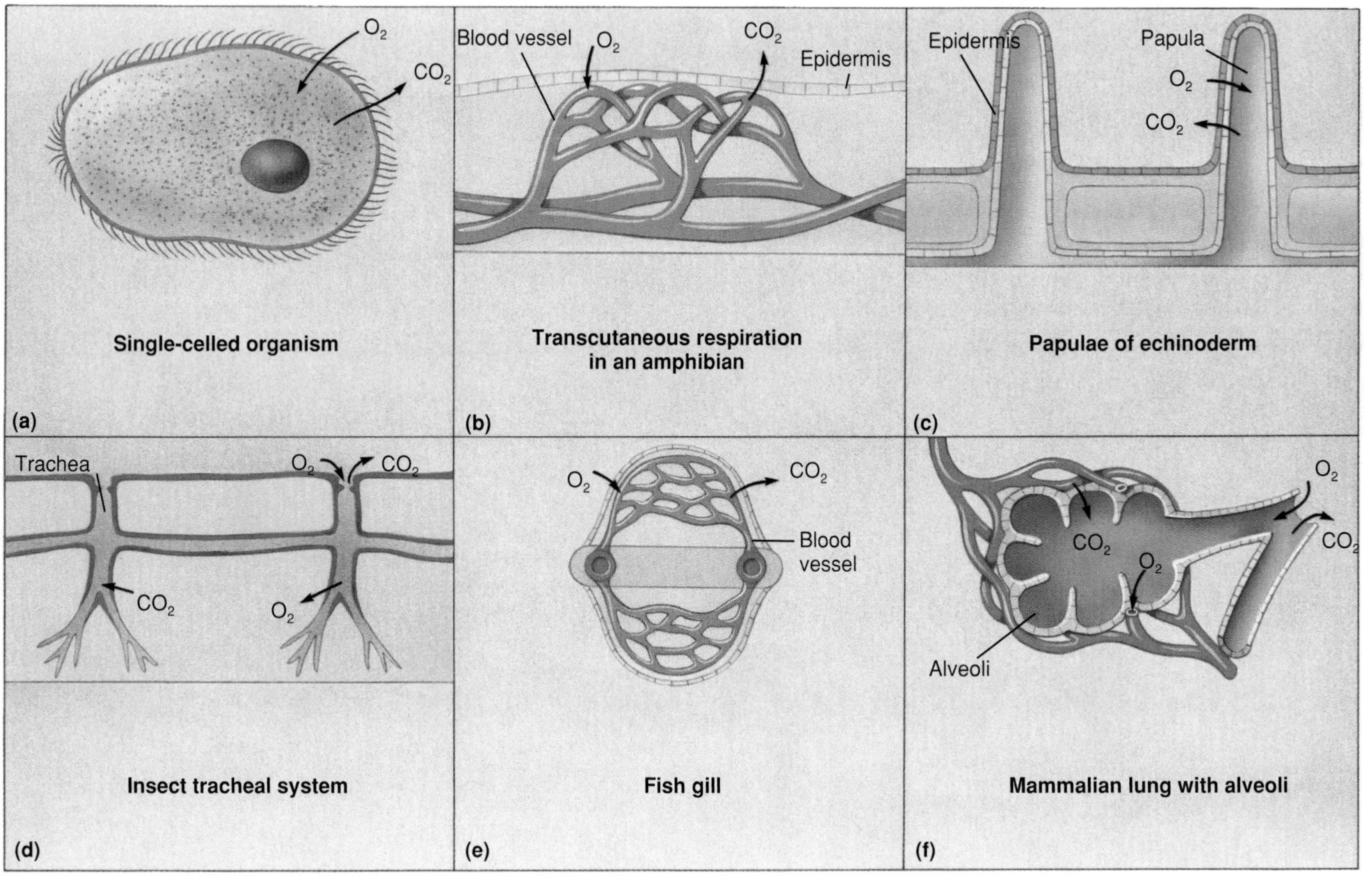

Figure 17.23 Gas exchange in animals.
(*a*) Gases diffuse directly into single-celled organisms. (*b*) Amphibians and many other multicellular organisms respire through their skin (transcutaneous respiration). (*c*) Echinoderms have protruding papulae, which provide an increased respiratory surface. (*d*) Insects respire through tracheae, which open to the outside. (*e*) Fish gills provide a very large respiratory surface and employ countercurrent flow. (*f*) Mammalian lungs provide a large respiratory surface but do not permit countercurrent flow.

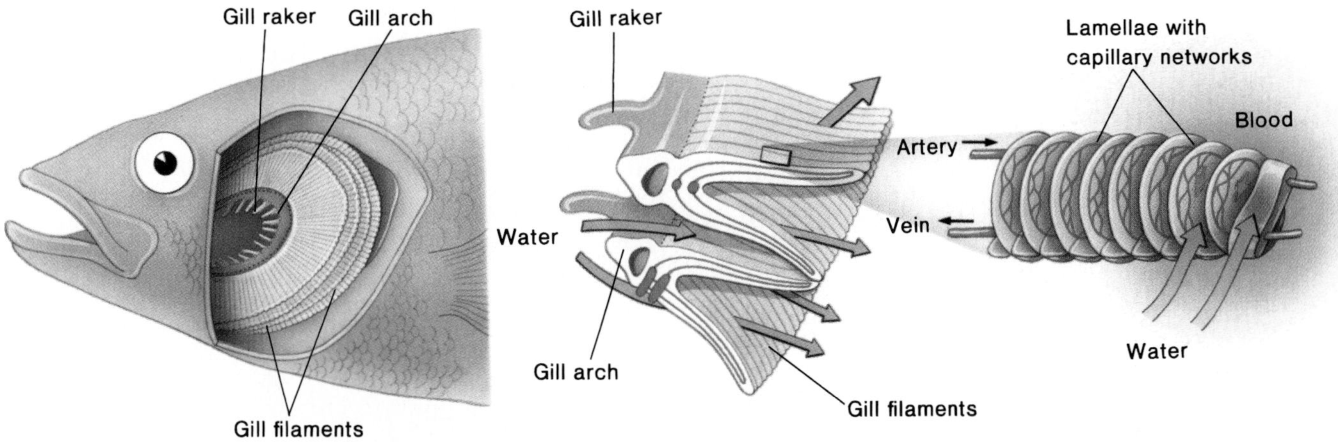

Figure 17.24 Structure of a fish gill.
Water passes from the gill arch over the filaments (from *left* to *right* in the diagram). Water always passes the lamellae in the same direction, which is opposite to the direction the blood circulates across the lamellae. This opposite orientation of blood and water flow is critical to the success of the gill's operation.

Fish Get Oxygen from Water with Gills

Have you ever seen the face of a swimming fish up close? A fish swimming in water continuously opens and closes its mouth, just as if it were trying to eat the water. That is just what the fish is doing, ramming water into its mouth with its throat closed. The rear of its mouth cavity is slit open so that the mouthful of water passes back outside—only a curtain of gills hangs between its mouth and the slit. When the fish gulps water, the water is pushed through the mouth cavity and out the slit—and (and this is the whole point) is forced past the gills on its one-way journey.

This swallowing process, which seems so awkward, is at the heart of a great advance in gill design achieved by the fishes. What is important about the swallowing is that it causes the water to always move past the fish's gills *in the same direction*. The gills of a mollusk are in an internal cavity, also, but in a mollusk's respiratory cavity, the water simply sloshes in and out of the cavity, washing back and forth over the gills. Why is it so much better to always move the water past the gill in the same direction, as a fish does? Because moving the water past the gills in the same direction permits **countercurrent flow,** which is a supremely efficient way of extracting oxygen. Here is how it works:

Each gill is composed of two rows of gill filaments, thin membranous plates stacked one on top of the other and projecting out into the flow of water (figure 17.24). As water flows past the filaments from front to back, oxygen diffuses from the water into blood circulating within the gill filament. Within each filament the blood circulation is arranged so that the blood is carried in the direction opposite the movement of the water, from the back of the filament to the front.

Because of the countercurrent flow, the blood in the fish's gills can build up oxygen concentrations as high as those of the water entering the gill (figure 17.25*a*). If the flow of water and blood had been in the *same* direction (concurrent flow), the highest concentration possible is the halfway point, the average between water and deoxygenated blood (figure 17.25*b*).

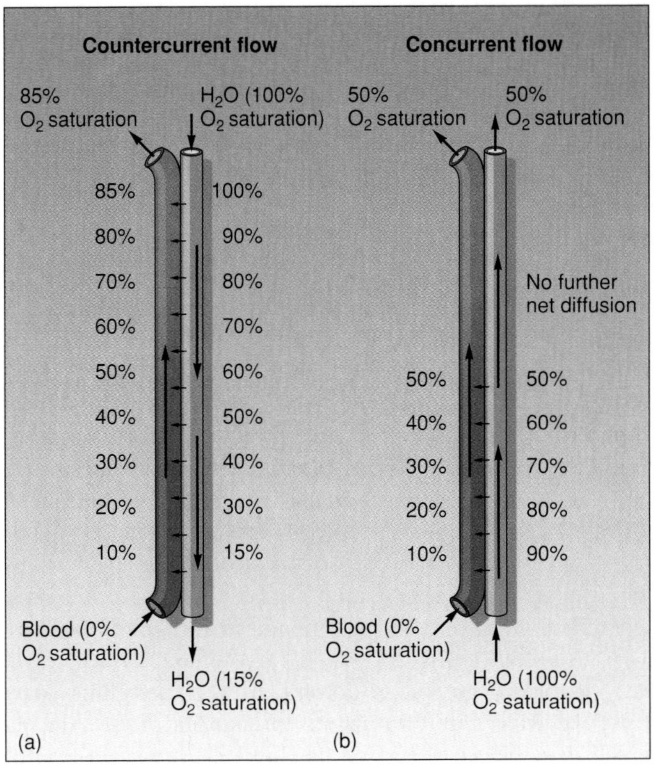

Figure 17.25 Countercurrent flow.
(*a*) Countercurrent flow of water and blood increases the efficiency of oxygen extraction in this example to 85%, compared with 50% with concurrent flow (*b*).

The gills of bony fishes are the most efficient respiratory machines that have ever evolved among organisms. They are able to extract up to 85% of the available oxygen from water.

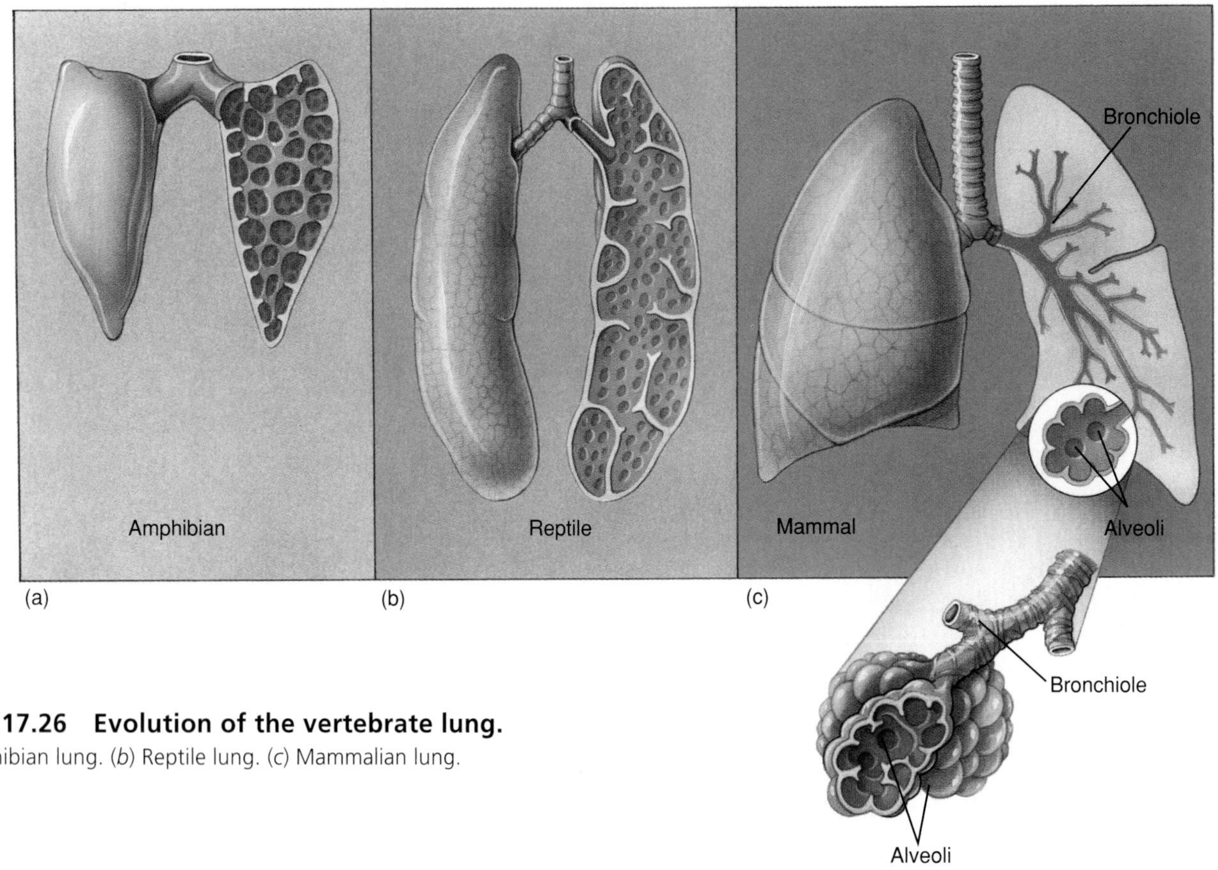

Figure 17.26 Evolution of the vertebrate lung.
(*a*) Amphibian lung. (*b*) Reptile lung. (*c*) Mammalian lung.

Amphibians Get Oxygen from Air with Lungs

One of the major challenges facing the first land vertebrates was obtaining oxygen from air. While there is a lot more oxygen in air than in water (air contains about 210 milliliters of oxygen per liter, over 20 times as much as seawater), all this oxygen doesn't do an aspiring land dweller much good if the machinery for gathering up the oxygen doesn't work—and fish gills, which are superb oxygen-gathering machines in water, won't work in air. The system of delicate membranes upon which gills depend has no skeleton or other means of support and the membranes collapse on top of one another in air, reducing the gill to a soggy lump—that's why a fish dies when kept out of water, literally drowning in air for lack of oxygen.

Unlike a fish, if you lift a frog out of water and place it on dry ground, it doesn't drown. Partly this is because the frog is able to respire through its moist skin, but mainly it is because the frog has lungs. A **lung** is a respiratory organ designed like a bag. The amphibian lung is hardly more than a sac with a convoluted internal membrane (figure 17.26*a*). The air moves into the sac through a tubular passage from the head and then back out again through the same passage. In a resting human, an inhalation adds about 500 milliliters of additional air to the volume already there, about 1,200 milliliters. The air mixes, and then 500 milliliters is exhaled. When each breath is completed, the lung still contains 1,200 milliliters, but the air in the lung is richer in oxygen because of the mixing with the newly breathed air. Because the diffusion surface of the lung is not exposed to fully oxygenated air, but rather to this mixture of fresh and partly depleted air, the respiratory efficiency of lungs is much less than that of gills. But because there is so much more oxygen *in* air, it doesn't have to be. This inefficient arrangement works perfectly well for an amphibian, which doesn't need any more oxygen than simple lungs provide.

Reptiles and Mammals Increase the Lung Surface

Reptiles are far more active than amphibians, so they need more oxygen. But reptiles cannot rely on their skin for respiration the way amphibians can; their dry scaly skin is "watertight" to avoid water loss. Instead, the lungs of reptiles contain many small air chambers, which greatly increase the surface area of the lung available for diffusion of oxygen (figure 17.26*b*).

Because mammals maintain a constant body temperature by heating their bodies metabolically, they have even greater metabolic demands for oxygen than do reptiles. The problem of harvesting more oxygen is solved by increasing the diffusion surface area within the lung even more. The lungs of mammals possess on their inner surface many small chambers called **alveoli,** which are clustered together like grapes (figure 17.26*c*). Each cluster is connected to the main air sac in the lung by a short passageway called a bronchiole. Air within the lung passes through the bronchioles to the alveoli, where all oxygen uptake and carbon dioxide disposal takes place. In more active mammals, the individual alveoli

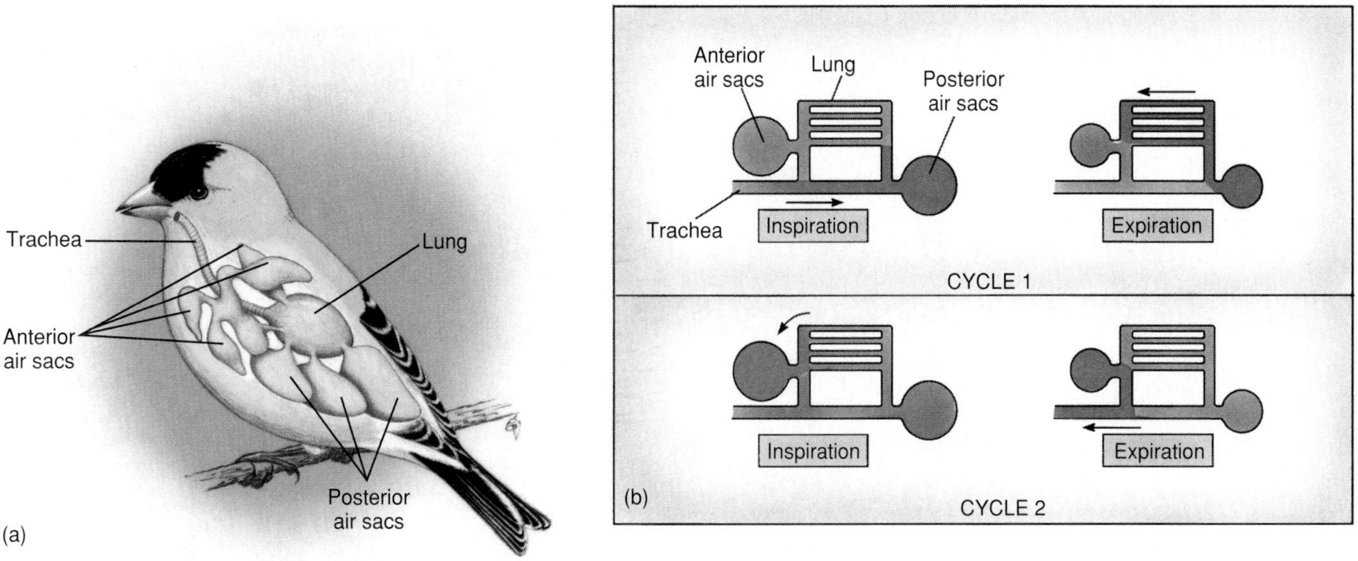

Figure 17.27 How a bird breathes.
(*a*) A bird's respiratory system is composed of the trachea, anterior air sacs, lungs, and posterior air sacs.
(*b*) Breathing occurs in two cycles: in cycle 1, air is drawn from the trachea into the posterior air sacs and then is exhaled into a lung; in cycle 2, the air is drawn from the lung into the anterior air sacs and then is exhaled through the trachea. Passage of air through the lungs is always in the same direction, from posterior to anterior (*right* to *left* here). Because blood circulates in the lung from anterior to posterior, the lung achieves a type of countercurrent flow that is very efficient in picking up oxygen from the air.

are smaller and more numerous, increasing the diffusion surface area even more. Humans have about 300 million alveoli in each of their lungs, for a total surface area devoted to diffusion of about 80 square meters (about 42 times the surface area of the body)!

Birds Perfect the Lung

There is a limit to how much efficiency can be improved by increasing the surface area of the lung, a limit that has already been reached by the more active mammals. When birds evolved, this efficiency just wasn't good enough. Flying creates a respiratory demand for oxygen that exceeds the capacity of the saclike lungs of even the most active mammal. Unlike bats, whose flight involves considerable gliding, most birds beat their wings rapidly as they fly, often for quite a long time. This intensive wing beating uses up a lot of energy quickly, because the wing muscles must contract very frequently. Flying birds thus must carry out very active oxidative respiration within their cells to replenish the ATP expended by their contracting flight muscles, and this requires a great deal of oxygen—more than a saclike mammalian lung can deliver, however clustered its alveoli.

The bird lung copes with the demands of flight by finding a new way to improve the efficiency of the lung, one that does not involve further increases in its surface area. Can you guess what it is? In effect, they do what fishes do! An avian lung is connected to a series of air sacs behind the lung. When a bird inhales, the air passes directly to these air sacs, which act as holding tanks (figure 17.27). When the bird exhales, the air flows from the air sacs forward into the lung and then on through another set of air sacs in front of the lungs and out of the body. What is the advantage of this complicated passage? It creates a unidirectional flow of air through the lungs.

Air flows through the lungs of birds in one direction only, from back to front. This one-way air flow results in two significant improvements: (1) There is no dead volume, as in the mammalian lung, so the air passing across the diffusion surface of the bird lung is always fully oxygenated. (2) Just as in the gills of fishes, the flow of blood past the lung runs in a direction different from that of the unidirectional air flow. It is not opposite, as in fish; instead, the latticework of capillaries is arranged across the air flow, at a 90-degree angle. This is not as efficient as the 180-degree arrangement of fishes, but the blood leaving the lung can still contain more oxygen than exhaled air, which no mammalian lung can do. That is why a sparrow has no trouble flying at an altitude of 6,000 meters on an Andean mountain peak, while a mouse of the same body mass and with a similar high metabolic rate will stand panting, unable even to walk. The sparrow is simply getting more oxygen than the mouse.

Just as fish gills are the most efficient aquatic respiratory machines, so bird lungs are the most efficient atmospheric ones. Both achieve high efficiency by using forms of countercurrent flow. No one knows what the lungs of dinosaurs (the immediate ancestors of birds) were like, but if dinosaurs were indeed warm-blooded, it is likely that they evolved lungs more efficient than those of today's reptiles. As dinosaurs evolved at about the same time as mammals, they may have adapted the same evolutionary strategy of expanding surface area or have adopted the approach seen in their descendants the birds.

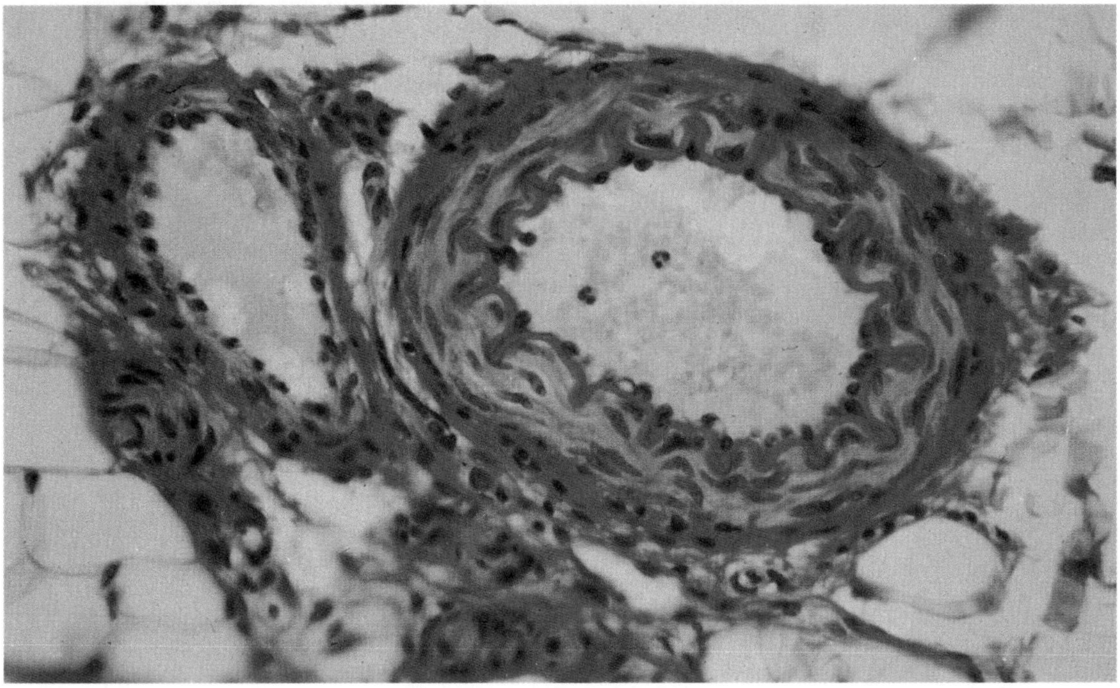

Figure 17.28 A closeup look at blood vessels. In circulatory systems that contain blood vessels, veins (*left*) remove carbon dioxide from tissues, while arteries (*right*) deliver oxygen to muscles and other tissues.

The Challenge of Building a Better Heart

A second critical evolutionary challenge that has served to define what vertebrates are like today has been the need to deliver oxygen efficiently to the body's tissues. The nature of the circulatory system determines how easy or difficult it is to deliver oxygen to muscles and remove carbon dioxide (figure 17.28).

Among simple multicellular animals, the movement of oxygen and carbon dioxide occurs by diffusion (figure 17.29*a*). Some portion of every cell of the organism is exposed to the external liquid environment, enabling individual cells to capture oxygen from that environment and to release carbon dioxide into it.

As multicellular animals have become larger, however, with layers of cells stacked on one another, this simple approach doesn't work—interior cells are too far from the environment for diffusion to be effective. This dilemma was solved with the advent of the body cavity, first seen among pseudocoelomates such as nematodes. Fluids within this body cavity constitute a primitive kind of circulatory system, one that permits materials to pass from one cell to another without leaving the organism (figure 17.29*b*). One cell is able to take up oxygen from the environment, and another distant cell can receive and use that oxygen. This transport of material from one place to another within an organism by passage through an internal fluid is called **circulation.**

A circulatory system may be either closed or open (figure 17.29*c* and *d*). In a *closed system* the fluid is enclosed within **blood vessels** and so is separated from the rest of the body's fluids and does not mix freely with them. Materials pass into and out of the circulating fluid by diffusion. In an *open system,* by contrast, there is no distinction between circulating fluid and body fluid generally—the circulating fluid *is* the body fluid.

Either an open or a closed circulatory system will work, and both function successfully in different phyla of animals. Annelid worms, for example, have a closed circulatory system in which two major blood vessels extend the length of the animal, one on top and the other on the bottom, with branches extending out in each segment to the muscles and organs. All vertebrates have closed circulatory systems. Insects and other arthropods, however, evolved an open circulatory system. A muscled tube within the central body cavity forces the cavity's fluid out into the body through a network of channels and spaces. The fluid then flows back into the central cavity.

Whatever the animal, any circulation system requires not only a system of passageways through which blood can circulate but also a pump to force the fluid through them. Much of the evolution of the vertebrate circulation system has focused on the pump, called a **heart.**

The chordates that were ancestral to the vertebrates had simple tubular hearts, similar to those now seen in lancelets. The pump was little more than a specialized zone of the ventral (lower) artery, more heavily muscled than the rest of the arteries, which beat in simple peristaltic waves. (A peristaltic wave is one that starts at one end of a tube and moves progressively along toward the other. If you take a soft drink straw and squeeze one end, and then move your squeezing fingers down the straw, that's a peristaltic wave.) Peristaltic pumps such as these hearts are not very efficient. Blood is pushed in *both* directions as the heart contracts, and pumping only happens because the as-yet-uncontracted portion of the blood vessel has a larger diameter than the contracted portion, thus less resistance to flow.

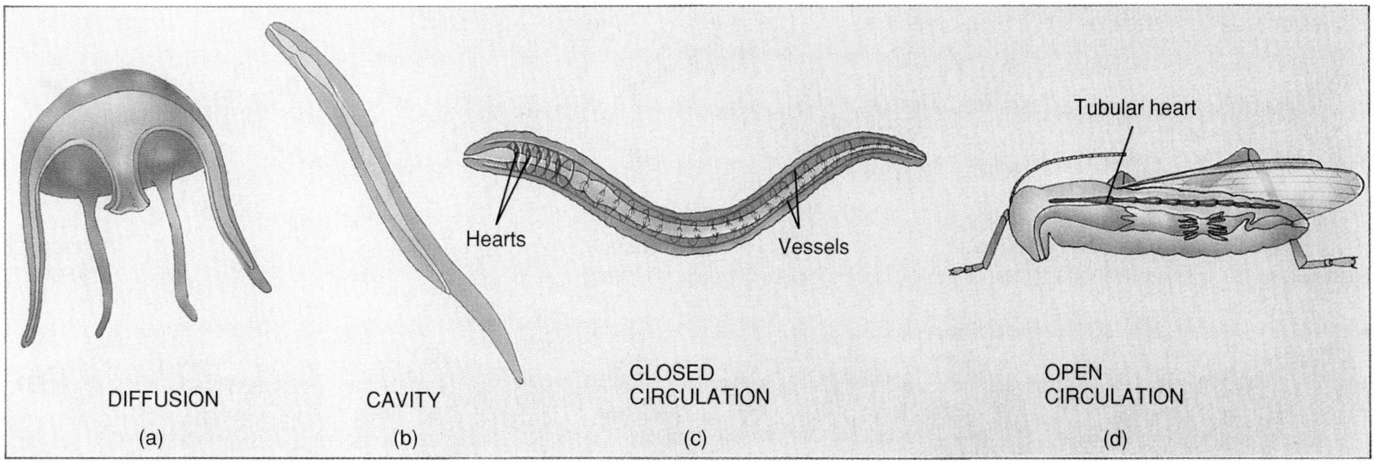

Figure 17.29 Evolution of circulatory systems.
(*a*) In simple diffusion, as in this jellyfish, the organism's internal fluid reaches all parts of the body. This diffusion is a passive process and does not require the organism to move or to expend energy. (*b*) Circulation in the body cavity of a nematode. As the nematode moves, its contracting muscles push fluid back and forth within the body cavity. (*c*) The closed circulatory system of an annelid contains several hearts that pump blood through blood vessels to the body tissues and a second set of blood vessels that pump the blood back from the tissues to the heart. (*d*) Insects have an open circulatory system. The tubular heart pumps blood out to body tissues, from which the blood seeps back rather than traveling in vessels.

The evolution of the heart among vertebrates reflects two great transitions in their history. The first was the shift from filter feeding to active prey capture, which accompanied the evolution of jaws. The second was the shift to greater metabolic activity, which accompanied invasion of the land and control of body temperatures.

Fish Develop a Heart with Chambers

The development of gills by fishes created a serious problem. Because resistance to flow through any tube increases dramatically as the tube's diameter gets smaller, the tiny diameters of the many small blood vessels within fish gills creates an enormous resistance to flow of the blood through the gills, a resistance too great for any peristaltic pump to overcome. This problem was solved by the replacement of the peristaltic pump of early chordates with a pump of very different design: a chamber-pump heart.

The fish heart is basically a tube with four chambers side by side (figure 17.30*a*). The first two chambers (the sinus venosus and the atrium) are collection chambers, the second two (the ventricle and the conus arteriosus) are pumping chambers.

1. **Sinus venosus.** To make movement of blood toward the heart as resistance-free as possible, the first chamber is a large collection chamber. A one-way valve at its entrance prevents blood from flowing back out when the chamber's walls contract.
2. **Atrium.** To deliver the right amount of blood to the pump quickly, the second chamber accepts from the sinus venosus a volume of blood that is just enough to fill the ventricle into which it empties.
3. **Ventricle.** To provide enough force to push the blood through the gills, the third chamber is a thick-walled pump with enough muscle to contract strongly.
4. **Conus arteriosus.** To smooth the pulsations and add still more thrust, the fourth chamber is a second, more elongated pump.

You can see the ancestry of fishes in how their hearts operate. Like the early chordates from which they evolved, the sequence of the heartbeat is peristaltic, starting at the rear and moving to the front. The first of the four chambers to contract is the sinus venosus, which loads the atrium; then the atrium contracts, filling the ventricle with the proper amount of blood; then the ventricle contracts forcefully, slamming blood into the conus arteriosus, which contracts last, smoothing the pulsation. Despite shifts in the relative positions of the chambers in the vertebrates that evolved later, this heartbeat sequence is maintained unchanged in all vertebrates.

The fish heart represents one of the great evolutionary innovations of the vertebrates. Its great advantage is that the blood it delivers to the tissues of the body is fully oxygenated, being pumped first through the gills and from there to the rest of the body, before returning to the heart. This arrangement has one great limitation, however. After passing through the fine network of tiny blood vessels in the gills, the flow of blood has lost much of the force contributed by the contraction of the heart, so the circulation of blood from the gills to the rest of the body is sluggish. This means that oxygen cannot be delivered to body muscles at a high rate. Fish have never evolved a means to overcome this limitation.

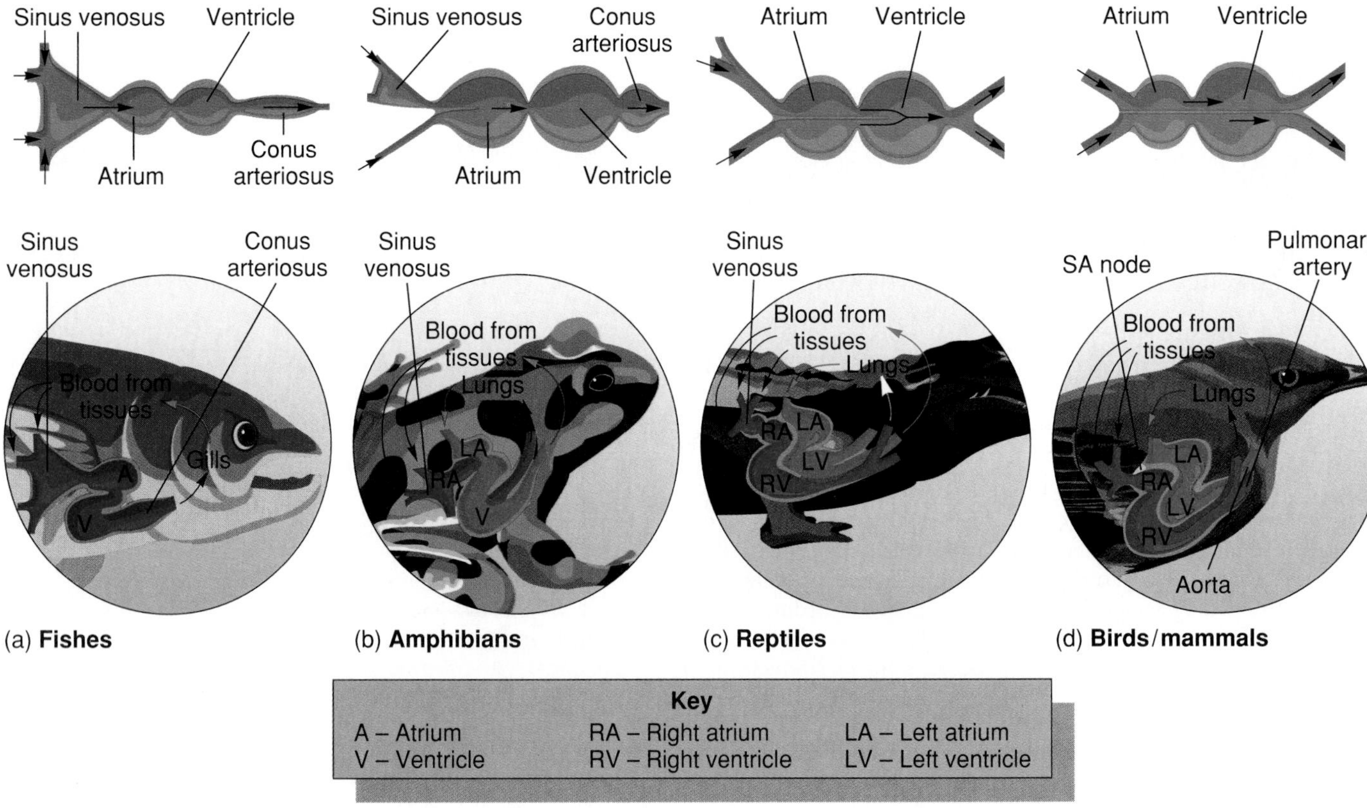

Figure 17.30 Evolution of the vertebrate heart.
(*a*) In fishes, the heart is shaped somewhat like a continuous tube, with no separations between the different chambers. (*b*) Amphibians have the beginning of a separation between the right and left halves of the heart. (*c*) In reptiles, this separation is even more complete, and three distinct chambers are evident. (*d*) In birds and mammals, the separation is complete, and four distinct chambers are visible.

Amphibians Evolve the Pulmonary Vein

When amphibians began to extract oxygen from air using lungs, the inherent limitation of pumping blood through the respiratory organ to the rest of the tissues presented an even greater problem, both because the first lungs were probably quite inefficient and because standing and walking required delivery of more oxygen to the muscles. This problem was solved in amphibians by a redesign of the plumbing—adding another pair of major blood vessels. These vessels, called the **pulmonary veins,** reroute blood after it passes through the lungs, carrying it back to the heart to be pumped again before it is sent out to the tissues of the body.

The great advantage of this arrangement is that oxygenated blood could be pumped to the muscles at much higher pressures. The disadvantage is that the oxygenated blood from the lungs is mixed in the heart with the nonaerated blood being returned to the heart from the rest of the body. Because of this, the amphibian heart pumps out a mixture of aerated and nonaerated blood rather than fully aerated blood. Many of the subsequent modifications in the evolution of the heart have managed to minimize this disadvantage.

The amphibian heart is different from that of fishes in an essential way (figure 17.30*b*). A dividing wall, or **septum,** partitions the entrance to the heart and the exit from it, so that the aerated blood entering from the lungs tends to stay in the side of the heart that will exit toward the rest of the body, and the nonaerated blood entering the heart from the rest of the body tends to stay on the other side of the heart that will exit the heart toward the lungs. The septum at the heart's entrance divides the sinus venosus and atrium completely and partially divides the conus arteriosus. In effect, the changes in the amphibian heart match it to the two circulation paths created by the pulmonary vein, one in which aerated blood is directed through the heart toward the body's tissues, and the other in which nonaerated blood is directed through the heart toward the lungs.

Terrestrial Vertebrates Have a Divided Heart

In the amphibian heart, the separation of the lung and body circulations is still imperfect, with a lot of mixing still occurring in the heart. (It is no accident that amphibians are usually sluggish.) Among reptiles, further extension of the septum dividing the heart greatly reduced the amount of

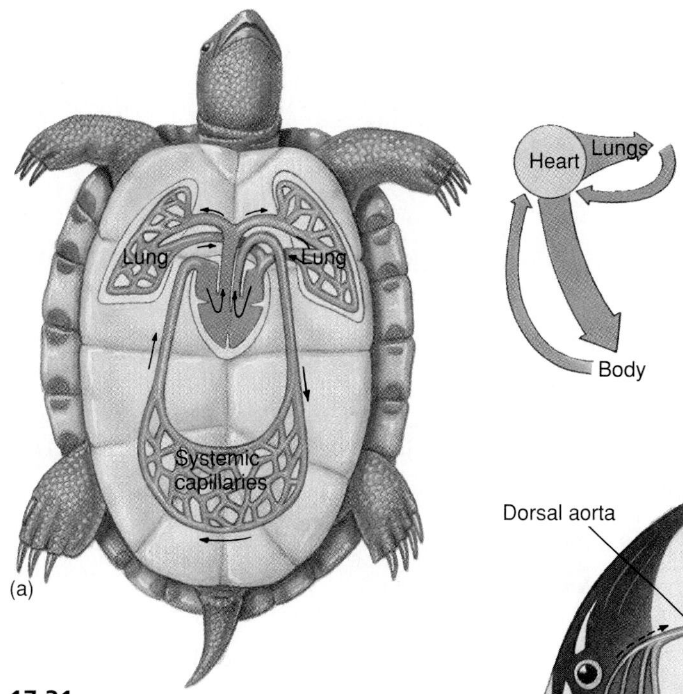

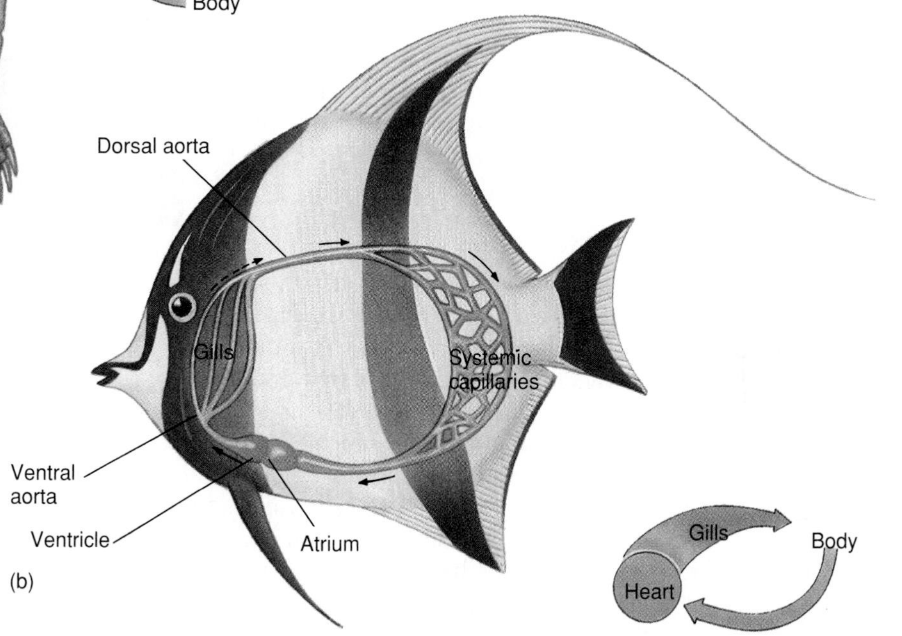

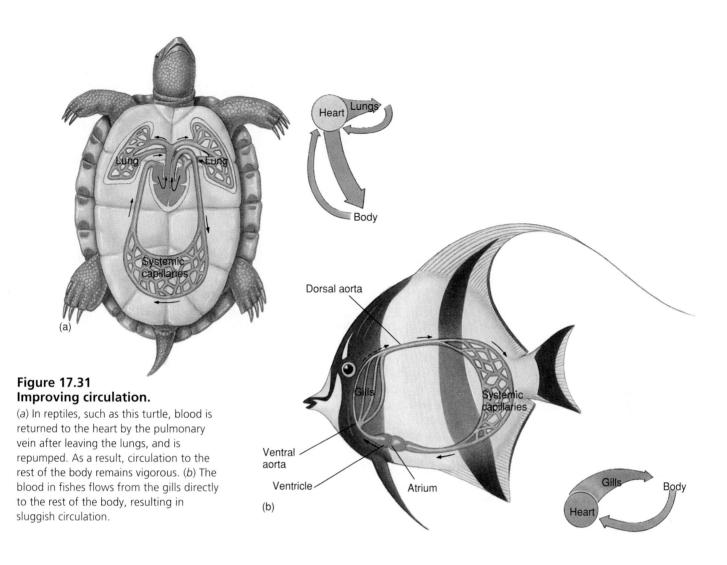

Figure 17.31
Improving circulation.
(*a*) In reptiles, such as this turtle, blood is returned to the heart by the pulmonary vein after leaving the lungs, and is repumped. As a result, circulation to the rest of the body remains vigorous. (*b*) The blood in fishes flows from the gills directly to the rest of the body, resulting in sluggish circulation.

mixing (figure 17.30*c*). The septum dividing the atrium extends well into the ventricle, partially dividing it and so creating a much better separation of flow between aerated and nonaerated blood within the heart. At the other end, the conus arteriosus is divided completely in two, forming the trunks of the two large arteries leaving the heart.

Because the reptilian heart achieves a better separation of aerated and nonaerated blood, it works much more efficiently than the amphibian heart or the fish gill (figure 17.31). The separation within the ventricle is not complete, however, so mixing still occurs, limiting the efficiency of the system.

Only a slight alteration was needed to complete the evolutionary transition of the vertebrate heart begun by the amphibians. This has happened three different times, independently, in the hearts of mammals, birds, and crocodiles. The septum dividing the ventricle is extended even farther until the wall in the center of the heart is complete, dividing the pumping chamber into two parts (figure 17.30*d*). The closing of the ventricular septum finally solved the problem originally created when amphibians developed the pulmonary vein. Mixing no longer takes place within the heart, so blood pumped to the tissues of the body is fully aerated.

The sinus venosus, a major chamber in the fish heart, is reduced in size in amphibians and even further reduced in reptiles. In mammals and birds, the sinus venosus is no longer evident as a separate chamber, but some tissue from it remains and plays a very important role. Throughout the evolutionary history of the heart, the sinus venosus has functioned both as a collection chamber and as the site of origin of the heartbeat. This second function is indispensable, and mammals and birds have retained some of the excitatory tissue of the sinus venosus in the wall of the right atrium—right where it was previously located in the fish heart. Called the **pacemaker,** it is the point of origin of each heartbeat.

The Challenge of Retaining Water

A third key evolutionary challenge facing the vertebrates has been the need to retain water in many different environments. Vertebrates first evolved in water, and even today, no vertebrate can do without it for long. Approximately two-thirds of every vertebrate's body is water, and if the amount of water falls much lower than this, the individual dies. Losing water is called **dehydration.** You might think that dehydration presents no problem to a fish—it lives submerged in water—but you would be wrong. Remember osmosis? Osmosis is the process that causes a net movement of water across membranes toward regions of higher ion concentration. A fish in seawater has within its body only about one-third the salt concentration of the surrounding water, thus water tends to leave its body by osmosis. Marine fishes avoid "drying out" in seawater by reabsorbing water through remarkable organs called kidneys.

Marine Fish Use Kidneys to Retain Water

The vertebrate kidney is thought to have evolved first in freshwater fishes, who take in excess water by osmosis (their bodies are hypertonic, having more salt than the fresh water around them) and need a way of disposing of it, as well as of body wastes (figure 17.32*a*). The kidney is a complex organ made up of thousands of repeating disposal units called **nephrons,** each with the structure of a bent tube (figure 17.33). Blood pressure forces the fluid in blood past a filter at the top of each nephron that retains blood cells, proteins, and other useful large molecules in the blood but allows the water, and the small molecules and wastes dissolved in it, to pass through the filter and into the bent tube part of the nephron. As the filtered fluid passes through the nephron tube, useful sugars and ions are recovered from it by active transport, leaving the water and metabolic wastes dissolved in it (such as urea) behind in a fluid called urine. A freshwater fish drinks little but urinates a lot.

The challenge faced by marine fish is just the opposite: their bodies are hypotonic to their surroundings and lose water continuously by osmosis to the more salty water in which they swim (figure 17.32*b*). To make up for water lost from their bodies, marine fishes drink a lot of water. The nephrons in the kidneys of marine fishes absorb water out of the filtrate, concentrating it so as to lose as little water as possible. Because the seawater they drink contains a lot of salt, marine fishes don't take up ions from the filtrate in their nephrons; instead, they do exactly the opposite. The ion channels in the nephron membranes of marine fishes are reversed in orientation, allowing the nephron to actively excrete into the urine excess salts. A marine fish drinks a lot but urinates little.

Sharks are an exception. Instead of actively pumping ions out of their bodies through their kidneys, sharks' kidneys reabsorb them into their bodies! The ion concentration in their blood is 100 times as high as in mammals and is the same as the ion concentration in salt water. Water does not tend to enter or leave a shark by osmosis, as there is no osmotic difference between the shark's fluids and the saltwater in which it is swimming.

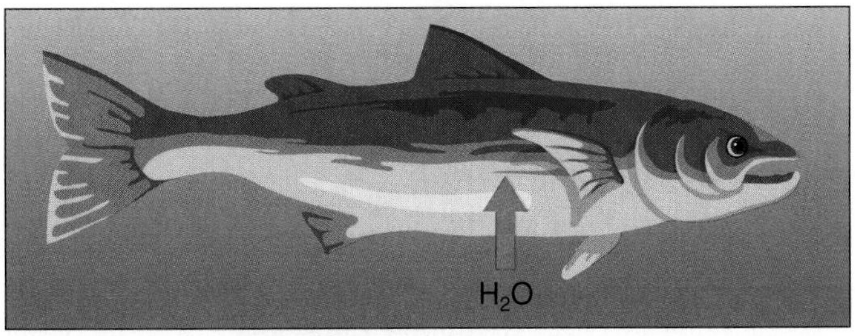

(a) Hypertonic fish in fresh water

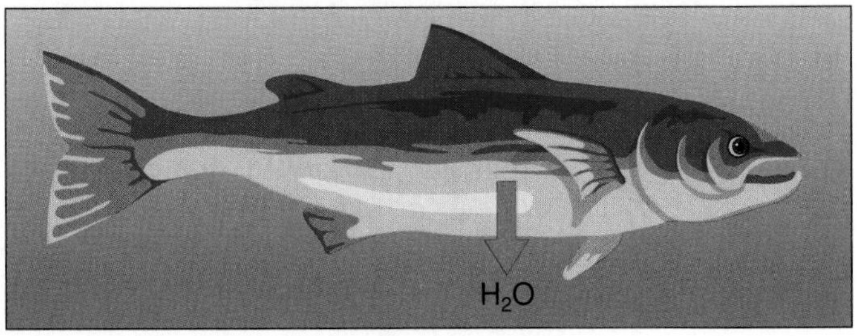

(b) Hypotonic fish in salt water

Figure 17.32 The challenge of living in water.
In fresh water (*a*), the body of a fish contains more salt than the surrounding water, so that water diffuses into the body. In salt water (*b*), the body of a fish contains less salt than surrounding water, and water diffuses out of the body. Fresh water and marine fish have developed a variety of measures to counter these tendencies.

Reptiles and Mammals Concentrate Their Urine

Amphibians have the same kind of kidney as freshwater fishes. They have dilute urine and reabsorb little water. Reptiles living on dry land face water loss not from osmosis but from desiccation. They absorb much more of the water in their kidney filtrate than an amphibian does, producing a far more concentrated urine. This urine, however, cannot become any more concentrated than the reptile's blood plasma; otherwise, the reptile's body water would simply flow into its urine by osmosis while the urine was still in the kidneys.

Mammals and birds do an even better job of water retention than reptiles, because they have evolved kidneys that can remove far more water from the urine. Human urine may be four times as concentrated as blood plasma, and the urine of desert mice over 20 times as concentrated! How do mammalian and avian kidneys avoid the osmotic trap of having body water run back into the urine within the kidney? By a remarkably simple and powerful innovation: bending the nephron tube, producing the loop of Henle (figure 17.34). This bending extends the end segment of the nephron into a region of the kidney with a very high local concentration of urea. This high salt concentration draws water out of the tube by osmosis, leaving behind in the urine only a very highly concentrated urea solution.

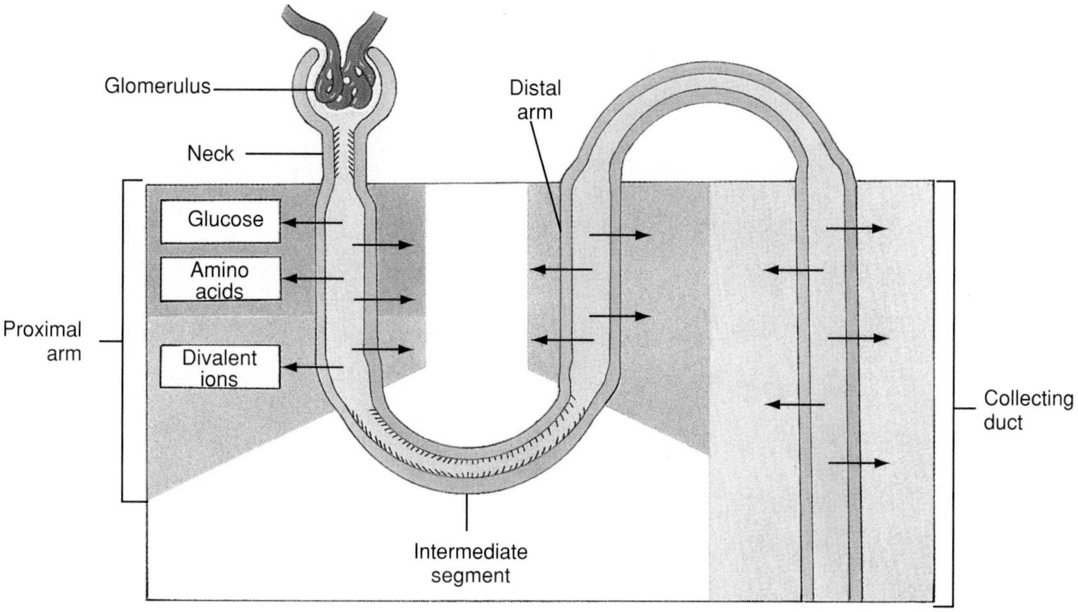

Figure 17.33 Basic organization of the vertebrate nephron.
The nephron tube of the freshwater fish is a basic design that has been retained in the kidneys of marine fishes and terrestrial vertebrates that evolved later. Sugars, small proteins, and divalent ions such as Ca^{++} are recovered at the beginning of the tube (the so-called proximal arm). Ions such as Na^+ and Cl^- are recovered after the initial bend (the so-called distal arm). Mammals and birds achieve much greater water conservation by bending the nephron tube, producing what is known as the loop of Henle (see figure 17.34).

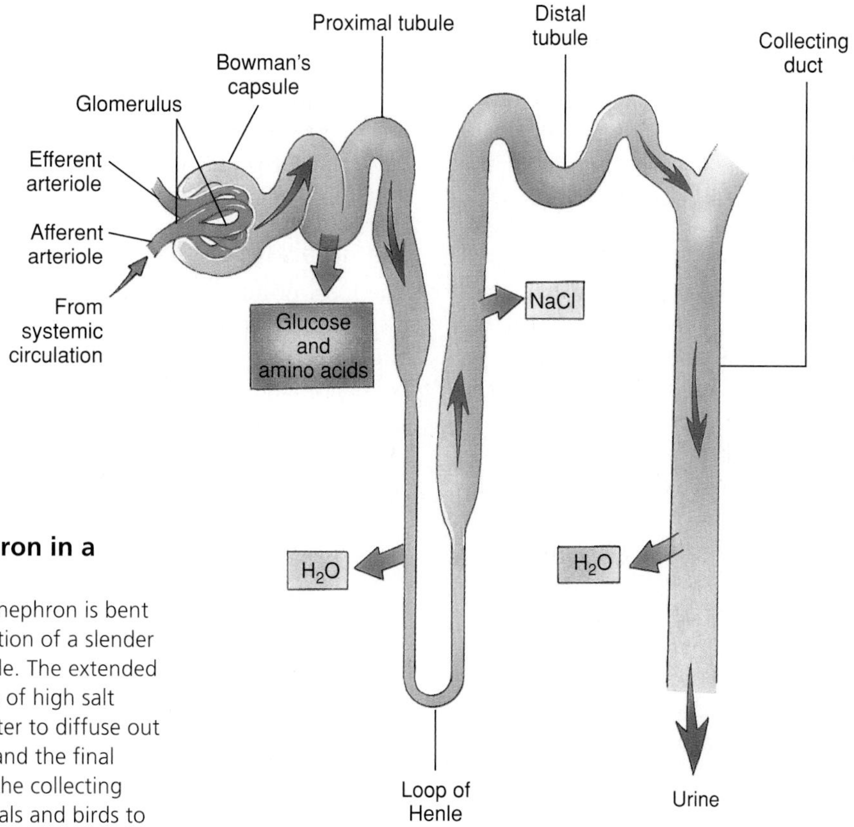

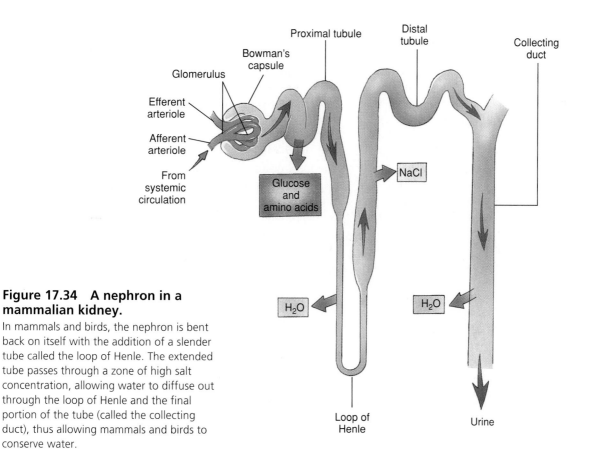

Figure 17.34 A nephron in a mammalian kidney.
In mammals and birds, the nephron is bent back on itself with the addition of a slender tube called the loop of Henle. The extended tube passes through a zone of high salt concentration, allowing water to diffuse out through the loop of Henle and the final portion of the tube (called the collecting duct), thus allowing mammals and birds to conserve water.

Reptiles Evolve a Watertight Skin

Living on land presented vertebrates with a new and different challenge for retaining water: a wet body loses moisture rapidly to air by evaporation, like damp laundry hanging on a clothesline. Loss of water to air blowing over its body is a very serious problem for any amphibian living on dry land. Like a wet towel steaming in the hot sun, water quickly evaporates away from the surface of the amphibian's wet skin. To slow the water loss, amphibians secrete a slippery mucus that retains water—it is what makes their skin feel slimy to the touch. But even with this coating, the skin of an amphibian is not very watertight.

Reptiles evolved a direct and effective way to avoid losing body water to skin evaporation—their skin is not wet! Reptile skin is totally dry. It is covered with tough scales and is quite watertight. This dry skin completely freed reptiles from the necessity of living in a wet environment. Mammals and birds, which evolved from reptiles, also have skin that is dry and impermeable to water.

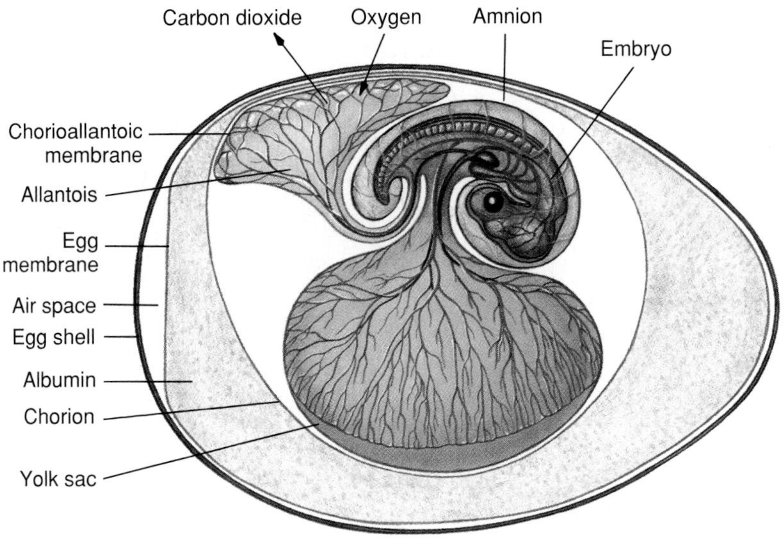

Figure 17.35 The watertight egg.
The amniotic egg is perhaps the most important feature that allows reptiles to live in a wide variety of terrestrial habitats.

Reptiles Evolve a Watertight Egg

The challenge of retaining water has affected another aspect of vertebrate life—reproduction. Loss of water is not a significant problem for the eggs of a fish or an amphibian. Typically the eggs are released by the female into water, fertilized there by sperm released nearby by a male, and go on to develop free in the water. However, for a reptile living on dry land, reproduction presents a serious water-loss problem: There is no pool of water within which to fertilize its eggs. To prevent the eggs and sperm from drying out, fertilization must be internal. To prevent the fertilized eggs from drying out afterwards, the embryos within reptile eggs are each surrounded by a watertight protective membrane called the chorion. The chorion allows oxygen to enter the egg and carbon dioxide to leave. Lying within the chorion is another membrane, the amnion, which encloses the embryo within a watery environment. This kind of watertight, fluid-filled egg is called an **amniotic egg** (figure 17.35). Each egg is provided with a large amount of rich food for the embryo, called yolk. The zygote develops within the egg, eventually achieving the form of a miniature adult before it completely depletes its supply of yolk and leaves the egg to face its fate as an adult.

Amniotic eggs evolved first among the reptiles and represent a key step in the journey of vertebrates onto land. Amniotic eggs are also employed by birds and mammals. A tough shell surrounds and protects the eggs of reptiles and birds. The shell of bird eggs is hard enough to withstand the weight of the adult bird while it sits on the egg, warming it with its body. The higher temperature hastens the development of the bird embryo. The young that hatch from the eggs of most bird species are not able to survive unaided, since their development is still incomplete. The young birds are fed and nurtured by their parents, and they grow to maturity gradually.

Mammals Nourish Their Embryos Internally

The most primitive mammals, the monotremes, lay eggs like the reptiles from which they evolved. The living monotremes consist solely of the duck-billed platypus and the echidna. No other mammals lay eggs.

The young of all other mammals are nourished and protected by their mother, in one of two ways:

1. *Marsupials* nourish their young with milk, giving birth to embryos at a very early stage of development. The tiny animals continue their development nursing in pouches on the mother's body, eventually emerging when they are able to function as young adults.
2. *Placental mammals* retain their young for a much longer period within the body of the mother. To nourish them, they have evolved a specialized, massive network of blood vessels called a placenta, through which nutrients are channeled to the embryo from the mother's blood.

CHAPTER 17

HIGHLIGHTS

17.1 Overview of Vertebrate Evolution

Key Terms

Paleozoic era 356
mass extinctions 358
Mesozoic era 358

Key Concepts

- All major groups of living organisms except plants arose in the early Paleozoic era.
- Arthropods and vertebrates invaded land from the sea soon after plants and fungi.
- Periodic mass extinctions have occurred; the greatest, at the end of the Paleozoic era, rendered 90% of all species extinct.
- In the Mesozoic era reptiles were the dominant terrestrial vertebrate, particularly dinosaurs.
- At the end of the Mesozoic era dinosaurs disappeared, and mammals took their place.

17.2 The Parade of Vertebrates

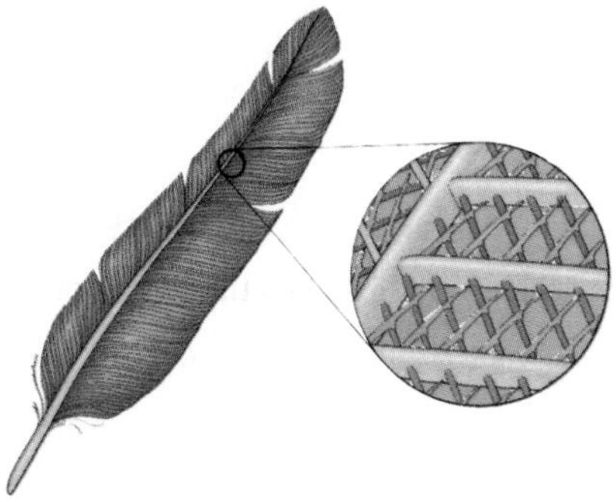

Key Terms

agnathan 362
swim bladder 364
therapsid 368
thecodont 368
dinosaur 368
Archaeopteryx 370
monotreme 372
marsupial 372
placental mammal 372

Key Concepts

- The first vertebrates were jawless fishes.
- The three key innovations in the evolutionary history of fishes were jaws, fins, and the swim bladder.
- Amphibians, the first vertebrates to live on land, evolved legs, lungs, and the pulmonary vein.
- Reptiles were well adapted to living on land, with dry skin and watertight eggs.
- Birds evolved feathers, hollow bones, and very efficient lungs.
- Most present-day mammals are placental mammals.

17.3 Meeting the Challenge of Life on Land

Key Terms

gill 374
countercurrent flow 375
alveoli 376
pulmonary vein 380
septum 380
nephron 382
amniotic egg 384

Key Concepts

- Marine and aquatic vertebrates extract oxygen from water very efficiently with gills.
- Reptiles and mammals perfect the lung by increasing its surface area with alveoli.
- Terrestrial vertebrates evolved a septum within the heart to separate the pulmonary and systemic circulations.
- Reptiles, birds, and mammals conserve water by using kidneys to remove it from urine.

17

CONCEPT REVIEW

1. Place the following periods and eras in order, oldest first.
   a. Permian
   b. Cambrian
   c. Cretaceous
   d. Cenozoic
   e. Devonian

2. The first vertebrates evolved about _________ years ago.
   a. 1.5 billion
   b. 250 million
   c. 470 million
   d. 10 million

3. Which of the following invaded land first?
   a. amphibians
   b. arthropods
   c. plants
   d. dinosaurs

4. Match the organism with the period in which it arose.

a. amphibians	(1) Cambrian
b. chordates	(2) Permian
c. dinosaurs	(3) Devonian
d. flowering plants	(4) Cretaceous

5. During the extinction that marked the end of the Cretaceous period, which of the following did *not* become extinct?
   a. ammonites
   b. pterosaurs
   c. great marine reptiles
   d. smaller plankton

6. Which of the following groups of vertebrates do *not* have past or present representatives with flight?
   a. amphibians
   b. reptiles
   c. birds
   d. mammals

7. To solve the problem of fast and maneuverable swimming, sharks have a cartilaginous skeleton, while Osteichthyes have
   a. heavy bony plates.
   b. swim bladders.
   c. lungs.
   d. suckerlike mouths.

8. Which of the following characteristics are *not* adaptations that contributed to the successful invasion of land by amphibians?
   a. legs
   b. lungs
   c. jaws
   d. pulmonary veins

9. Mammals are unique among all vertebrates because they
   a. do not lay eggs.
   b. have hair.
   c. are warm-blooded.
   d. have a cartilaginous skeleton.

10. The lungs of _________ are connected to anterior and posterior air sacs and are the most complex and efficient lungs among vertebrates.
    a. amphibians
    b. reptiles
    c. birds
    d. mammals

11. Because marine fish continuously lose water by osmosis to their salty environment, they must _________ a lot and excrete excess salts.
    a. urinate
    b. swim
    c. eat
    d. drink

12. _________ are fossil organisms that appear to be the ancestors to horseshoe crabs.

13. The most dramatic extinction event in the history of life on earth happened at the end of the _________ period.

14. A _________ is an eel-like agnathan that attaches itself to fish with its suckerlike mouth.

15. More than half of all vertebrate species are _________.

16. _________ are the group of dinosaurs that were most likely warm-blooded and were the immediate ancestors of mammals.

17. The oldest fossil bird, called _________, had feathers and a well-developed collarbone.

18. The passage of water in the direction opposite to the direction of blood flow in fish gills permits _________, an extremely efficient way of extracting oxygen.

19. A division of the pumping chamber of the heart by a completely extended septum has occurred independently in mammals, birds, and _________.

20. One of the pivotal innovations that led to the success of reptiles is the watertight egg, called an _________ egg.

Answers to the Concept Review questions appear in Appendix B.

CHALLENGE YOURSELF

1. Dinosaurs and mammals both lived throughout the Mesozoic era, a period of more than 150 million years; all this time dinosaurs were the dominant form and mammals a minor group. Both mammals and small reptiles survived the Cretaceous extinction. Why do you suppose reptiles did not go on to become dominant again, rather than mammals?
2. What are the advantages and disadvantages of endothermy compared to ectothermy? How was this thought to have played a role in the rise and fall of some reptile groups?
3. What limits the ability of amphibians to occupy the full range of terrestrial habitats and allows other terrestrial vertebrates to occur in them successfully?
4. Of the 42,500 species of living chordates, twice as many (about 19,000 fishes) live in the sea as on the surface of the land (about 6,000 reptiles and 4,500 mammals). Why do you think there are so many more species of chordates in the sea than on land?

FOR FURTHER READING

Conant, R. *A Field Guide to Reptiles and Amphibians: Eastern and Central North America.* Boston: Houghton-Mifflin, 1991. The standard herpetological field guide used by most field biologists.

Gill, F. B. *Ornithology.* New York: W. H. Freeman, 1990. Excellent overall account of the structure, biology, and other features of the birds.

Gore, R. "Dinosaurs." *National Geographic,* January 1993, 2–54. A marvelous overview of current research into dinosaurs; beautifully illustrated.

Miller, P. "Jane Goodall." *National Geographic,* December 1995, 102–29. Jane Goodall's decades of study show that chimpanzees in the wild are startlingly like us.

Norman, D. *Prehistoric Life: The Rise of the Vertebrates.* New York: Macmillan, 1994. A very readable account of the history of terrestrial vertebrates.

Novacek, M., et al. "Fossils of the Flaming Cliffs." *Scientific American,* December 1994, 60–9. Mongolia's Gobi Desert contains one of the richest assemblages of dinosaur remains ever found.

Rismiller, P. D., and R. S. Seymour. "The Echidna." *Scientific American,* February 1991, 96–103. The secret of the natural history and reproductive behavior of this spiny mammal of Australia and New Guinea is now being explored.

Schwartz, D. "Snatching Scientific Secrets from the Hippo's Gaping Jaws." *Smithsonian,* March 1996, 90–102. A jaw that hears underwater may be part of the sound system used by "river horses" to detect friend and foe.

Sereno, P. "Africa's Dinosaur Castaways." *National Geographic,* July 1996, 106–19. New dinosaur finds in the Sahara include enormous therapod carnivores bigger that *T. rex.*

Tuttle, M. D. "Saving North America's Beleaguered Bats." *National Geographic,* August 1995, 37–57. An engaging account of how conservationists are creating bat sanctuaries in mines and caves to help save these insect-eating mammals.

Tuttle, R. "Apes of the World." *American Scientist,* 78 March 1990, 115–25. A survey of our closest living relatives reveals rich prospects for further study and an urgent need for conservation.

TECHNOLOGY LINKS

The Living World Home Page
http://www.wcbp.com/biology/tlw

Life Science Animations
Videotape 5
#53 Continental Drift and
Plate Tectonics

18

How Humans Evolved

CHAPTER OUTLINE

Figure 18.1 A near relative.
Gorillas, chimpanzees, and humans are so similar that, were they any other group, scientists would probably call them the same genus. Most wild populations of primates are in imminent danger of extinction.

In 1871 Charles Darwin published another groundbreaking book, *The Descent of Man.* In this book, he suggested that humans evolved from the same African ape ancestors that gave rise to the gorilla (figure 18.1) and the chimpanzee. Although little fossil evidence existed at that time to support Darwin's case, numerous fossil discoveries made since then strongly support his hypothesis. Human evolution is the part of the evolution story that most interests people, and it is also the part about which we know the most. In this chapter we follow the evolutionary journey that has led to humans, telling the story chronologically. It is an exciting story, replete with controversy.

18.1 The Evolution of Primates

The idea that humans have evolved from apelike ancestors has proved to be very controversial, as Darwin suspected it would. Many well-meaning people in Darwin's day refused to accept his theory because it viewed humanity as the product of a long evolutionary journey, rather than as the immediate product of divine creation (figure 18.2). In the United States several states passed laws in the 1920s making it illegal to teach Darwin's theory of evolution in the classrooms of public schools (as in the famous Scopes trial of 1925). These laws were on the books and enforced until 1968, when the Supreme Court declared them unconstitutional. Even today, over 20 states have policies permitting local school districts to include creationism (the idea that humans are the direct product of divine creation) as an alternative to evolution in biology courses, although laws actually requiring "equal time" were ruled unconstitutional in 1986. Human evolution thus remains a controversial subject—not because of scientific doubt but because of disagreement about where science stops and religion begins.

Figure 18.2 Darwin greets his monkey ancestor.
In his time, Darwin was often portrayed unsympathetically, as in this drawing from an 1874 publication.

We begin our journey by looking past apes to a more distant past about 80 million years ago. During the age of dinosaurs, a small mammal appeared, no bigger than your fist. This tiny tree shrew lived in trees and survived by catching insects (figure 18.3). It had big eyes and a long, shrewlike nose. These tree shrews were the ancestors of the first primates, the evolutionary line leading to humans.

A distinguishing characteristic of early primates, and of most all mammals in general, is their unique dentition. Humans have four types of teeth: **incisors,** sharp, chisel-shaped, used for biting; **canines,** sharp, pointed teeth; and **premolars** and **molars,** flattened, ridge-surfaced, used for grinding and crushing. The numbers of each of these four types of teeth varies in other types of mammals according to the different specializations and methods of feeding used by the animal. For example, in deer, which are herbivores, grinding teeth predominate because they must pulverize the cell wall of plant tissue before digesting it. In cats, which are carnivores, most of the teeth are sharply pointed and are suited for capturing and killing prey. Because teeth fossilize well, scientists often use the shape, size, and number of teeth to hypothesize about the diet and habitat of fossilized organisms.

Figure 18.3 Where primates come from.
The mouse-sized, insect-eating tree shrew of the early Cenozoic era, reconstructed here, probably resembled the common ancestor of the primates and insectivores, two contemporary orders of mammals.

The Earliest Primates

The small nimble tree shrews gave rise 40 million years ago to larger animals called **primates.** Primates are the line of animals leading to humans. Evolution, selecting for changes that made tree-living insect catchers better and better at stalking their insect prey along slender branches, had in primates produced two distinct improvements:

1. **Grasping fingers and toes.** Unlike the clawed feet of tree shrews and today's squirrels, primates had grasping hands and feet that enabled them to grip limbs, hang from branches, and seize food.
2. **Binocular vision.** Unlike the eyes of shrews and squirrels, which are placed at the sides of the head so that their two fields of vision do not overlap, the eyes of primates are shifted forward to the front of the face, producing overlapping "binocular" vision that lets the brain judge distance precisely.

Other mammals have binocular vision, but only primates have both binocular vision and grasping hands. These changes guided the evolution of increased intelligence that would become the hallmark of the primates.

The first primates were **prosimians** ("before monkeys"). A prosimian looks something like a cross between a squirrel and a cat. By 38 million years ago, prosimians were common in North America, Europe, Asia, and Africa. Only a few prosimians survive today, principally on the island of Madagascar about 480 kilometers (300 miles) off the east coast of Africa. This island, about twice the size of the state of Arizona, is the home of all 24 surviving species of lemurs, prosimians the size of a tomcat with a long tail for balancing. Most lemurs are nocturnal, as were the early prosimians—that is, they hunt insects at night and sleep during the day. The large eyes you see in the tarsier in figure 18.4 are characteristic of nocturnal animals, helping them to see effectively in dim light. Today the tropical rain forests of Madagascar are being rapidly destroyed by an expanding human population—as the lemur's forest home disappears, our oldest living link to our past will soon be extinct in the wild.

Figure 18.4 A prosimian.
The tarsier, a prosimian native to tropical Asia, has binocular vision and large eyes, adaptations to nocturnal living in trees.

Most Primates Are Monkeys

About 36 million years ago, a revolution occurred in how primates lived—they became active during the day. This change had far-reaching effects. Because vision is far more important for daytime hunting, evolution favored many improvements in eye design, including the appearance for the first time of color-sensing cells called cones in the eye—color vision. The improved senses were governed by an expanded brain. We call these new, day-active primates **monkeys.**

Monkeys seem to have replaced prosimians rather rapidly. Feeding mainly on fruits and leaves rather than insects, they were the first primates with an opposable thumb—standing out at an angle from the other digits, it can be bent back against them to grasp an object or tool. Monkeys live in social groups and care for their young for prolonged periods. This long childhood of learning seems to be a necessary part of the development of the large brains of higher primates.

Monkeys first evolved in central Africa and are common there still. Some migrated early on to South America, where they developed in isolation—their descendants in South America today are easy to identify, since unlike African monkeys they grasp objects with long, prehensile (grasping) tails. Others moved to Asia, where they still lack prehensile tails.

The Path to Humanity: Apes

Humans did not evolve from monkeys but from another kind of primate that evolved independently from prosimian ancestors about 25 million years ago—the **apes** (figure 18.5). Together with human beings and their direct ancestors, apes make up a group called the hominoids. Apes have larger brains than monkeys, and the more advanced ones lack tails. With the exception of the gibbon, which has a tail and is relatively small, all living apes are larger than any monkey (figure 18.6). Apes exhibit the most adaptable behavior of any mammal except human beings. Once common, they are rare today, confined to relatively small areas in Africa and Asia. No apes ever occurred in North America.

Studies of ape DNA have told us a great deal about how apes evolved. The most primitive apes, the line leading to gibbons, diverged from other apes about 10 million years ago, while orangutans split off about 8 million years ago. The key split between the hominids (the human line) and the line leading to gorillas and chimpanzees occurred only about 5 million years ago. Because the split was so recent, the genes of humans and chimpanzees have not had time to evolve many differences. Thus human and chimpanzee DNA differs in less than 3% of their nucleotide sequence. Your hemoglobin molecule and that of a chimpanzee differ in only a single amino acid!

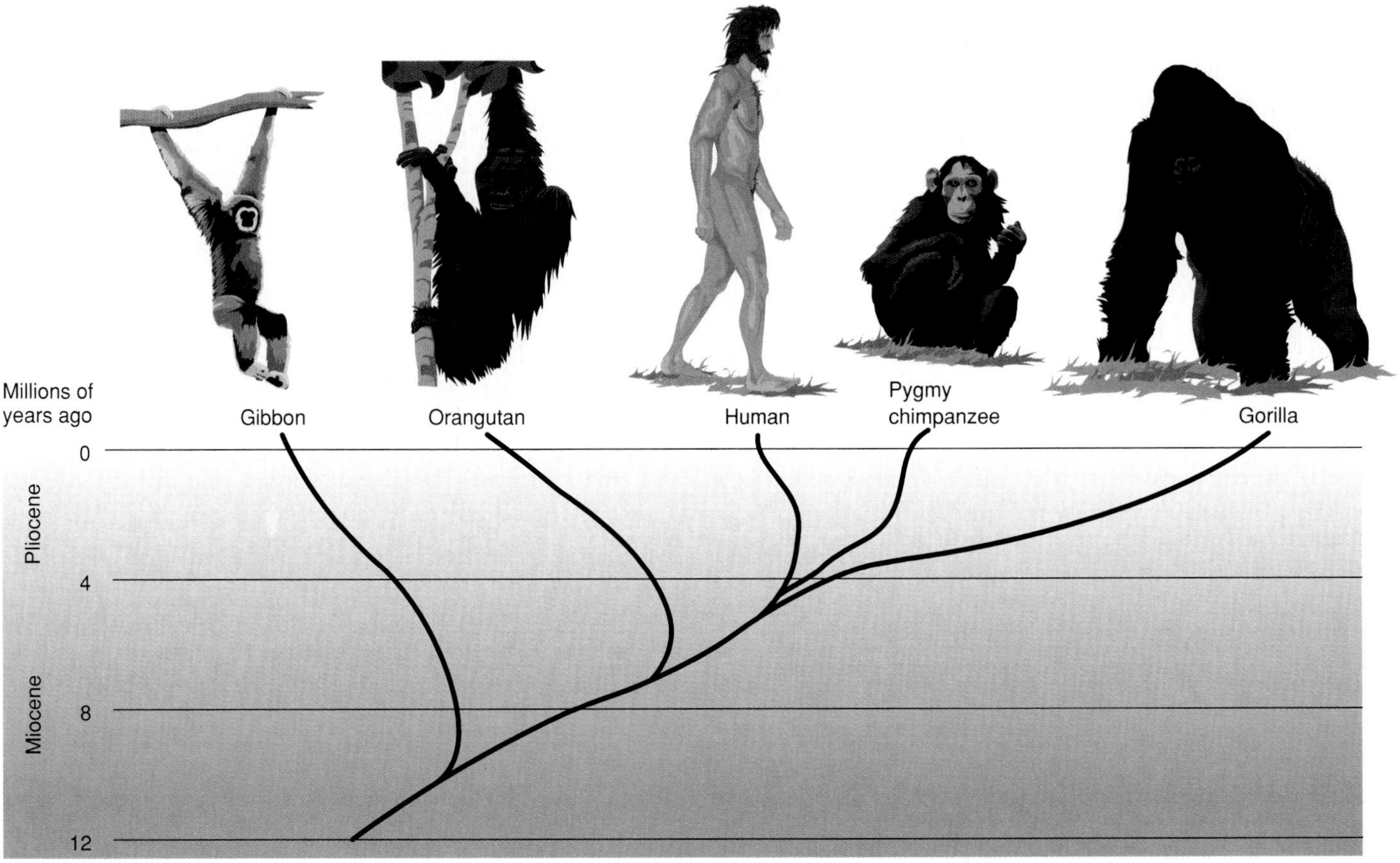

Figure 18.5 The evolution of living hominoids—apes and humans.
Humans are closely related to chimpanzees and gorillas.

(a) (b) (c) (d)

Figure 18.6 The living apes.
(*a*) Gorilla; (*b*) Mueller gibbon; (*c*) chimpanzee; (*d*) orangutan.

18.2 Evolutionary Origins of Humans

Fifteen million years ago, the world's climate began to get cooler, and the great forests were eventually replaced by open grassland savannas. In response to these changes a new kind of ape evolved, our direct ancestor.

Hominids

Assigned to the genus *Australopithecus* ("Southern apes"), these new apes along with humans are classified as **hominids,** meaning of the human line. They exhibit two critical early steps on the path leading to the evolution of humans:

1. **Bipedalism** (walking upright on two feet). Apes have arms longer than their legs and a narrow pelvis. Their spinal cord exits from the back of the skull at about a 45-degree angle to the ground (figure 18.7). Australopithecines have shorter arms (there are no trees to swing from in grassland savannas) and a bowl-shaped pelvis (to center the weight of the body over the legs); and their spinal cord exits from the bottom of the skull approximately at a 90-degree angle (vertical) from the ground (the head is on *top* of the body now, instead of in front of it).
2. **Larger brains** (more brain per pound of body weight). Australopithecines weighed about 18 kilograms (40 pounds) and were about 1 meter (3.5 feet) tall, but their brains were typically far larger than an ape of similar size, occupying a volume of up to 550 cubic centimeters—bigger than a large gorilla's brain of 400 cubic centimeters.

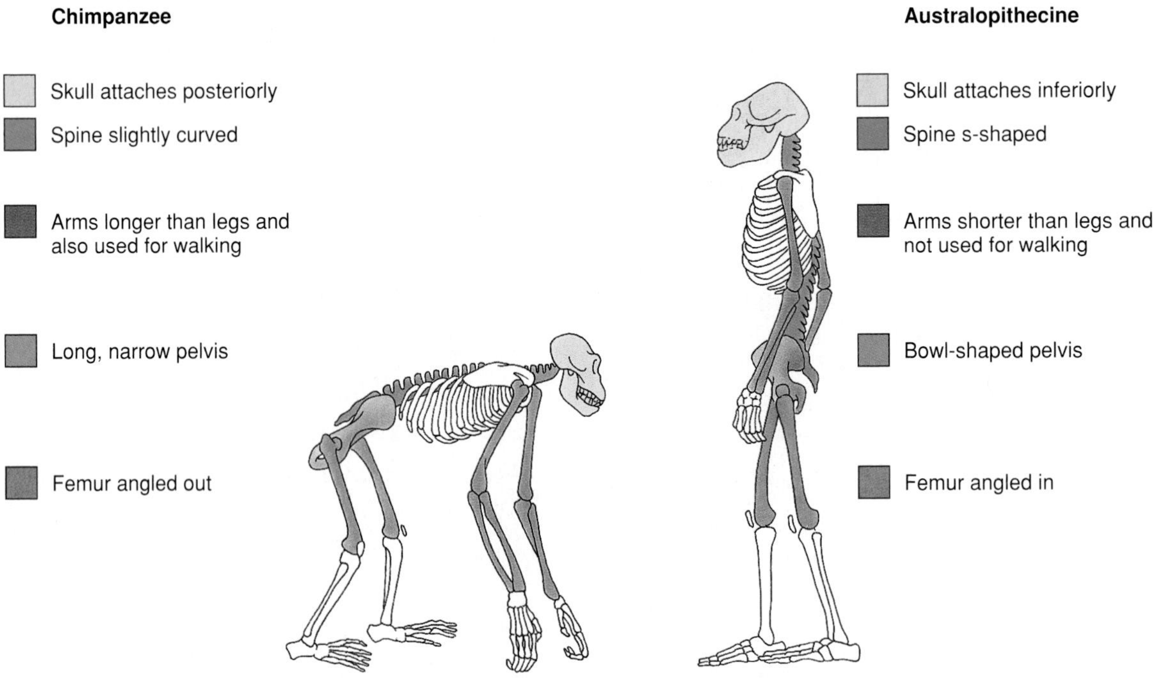

Figure 18.7 Skeletal features of apes and australopithecines compared.
Chimpanzees have arms longer than their legs, and their spinal cord exits from the back of their skull. Early humans were able to walk upright because they had shorter arms, and their spinal cord exited from the bottom of the skull.

Discovery of Australopithecus

The first hominid fossil of the genus ***Australopithecus*** was discovered in 1924 by Raymond Dart, an anatomy professor in Johannesburg, South Africa. One day a mine worker brought into his office an unusual chunk of rock—actually a rock-hard mixture of sand and soil. Picking away at it, Professor Dart uncovered a skull unlike that of any ape he had ever seen, and he was an expert. Beautifully preserved, the skull appeared to be that of a five-year-old individual that still had its milk teeth (figure 18.8). The fossil had a rounded jaw unlike the pointed jaw of apes, and the size of its braincase indicated a brain far larger than any ape of similar size. What riveted Dart's attention was that the skull was embedded among rocks and fossils several million years old! Until that time, the oldest reported hominid fossils had been less than 500,000 years old, so the ancientness of this skull was unexpected and exciting. Nor was the initial impression of great age misleading—we now know that Dart's discovery is fully 2.8 million years old.

Figure 18.8 Fossil and reconstruction of *Australopithecus africanus.*
The original Taung specimen of *A. africanus* (*foreground*) was discovered near Taung, South Africa, in 1924. The skull's reconstruction (*background*) was made by Professor Raymond Dart, who described the find in 1925. The specimen comprises part of a juvenile skull and mandible and an endocast (petrified sediments filling the cranium) of the right half of the brain. The overall configuration is apelike, but the teeth and braincase have several hominid features, and the opening between the vertebral and cranial cavities is set forward beneath the skull, implying an erect, bipedal posture.

Our Evolutionary Tree Has Many Branches

Dart called his find *Australopithecus africanus* (*australo*, southern + *pithecus*, ape), the ape from South Africa. Dart argued right from the first that *Australopithecus* was the direct ancestor of humans, the long-sought "missing link" to the apes. At first, few believed him, but soon the evidence began to mount and eventually the scientific community became convinced. In 1938 a second, stockier kind of *Australopithecus* was unearthed in South Africa. Called *A. robustus*, it had massive teeth and jaws. In 1959, paleoanthropologist Mary Leakey discovered a third kind of *Australopithecus* in East Africa. Named *A. boisei* (after Charles Boise, an American-born business man who contributed to the Leakeys' projects), it was even more stockily built. Nicknamed "nutcracker man" because of its powerful jaws, *A. boisei* had a great bony ridge on the crest of the head to anchor immense jaw muscles—like a Mohawk haircut but of bone. Like the other australopithecines, *A. boisei* was very old—almost 2 million years. In 1989 yet a fourth kind of heavily built australopithecine was reported, a massively boned ancestor of *A. boisei*.

These early hominids lived in the East African Rift Valley, where an over 3,000 kilometer-long fault line splits Ethiopia off from the rest of Africa along a line running from the Red Sea down into Mozambique. Until 5 million years ago, this valley was part of a vast tropical forest that spanned the continent. Like the tropical forests of today, these were rich with species, diverse and teeming with life. Starting about 5 million years ago, these great tropical forests were replaced by open, grassy plains dotted with scattered trees—the African savanna, much as we know it today.

It was into this challenging new environment that australopithecines first appeared. Biologists believe that living in the savanna fostered the evolution of an upright posture, the hallmark of the australopithecines. An individual that walks upright is said to be **bipedal.** Judging from the age of their fossils, many of the australopithecines lived at the same time, sharing the African savannas. This seems to have been a time of great evolutionary experimentation. Nevertheless, all of these early hominids shared three key characteristics:

1. *They were bipedal.* At least four lines of evidence suggest that the early hominids walked upright. First, the upper end of the australopithecine tibia (the "shinbone" below the knee) is shaped to bear more weight than an ape's leg. Second, the femurs (the "thighbone" above the knee) are angled inward, directly below the body, to carry its

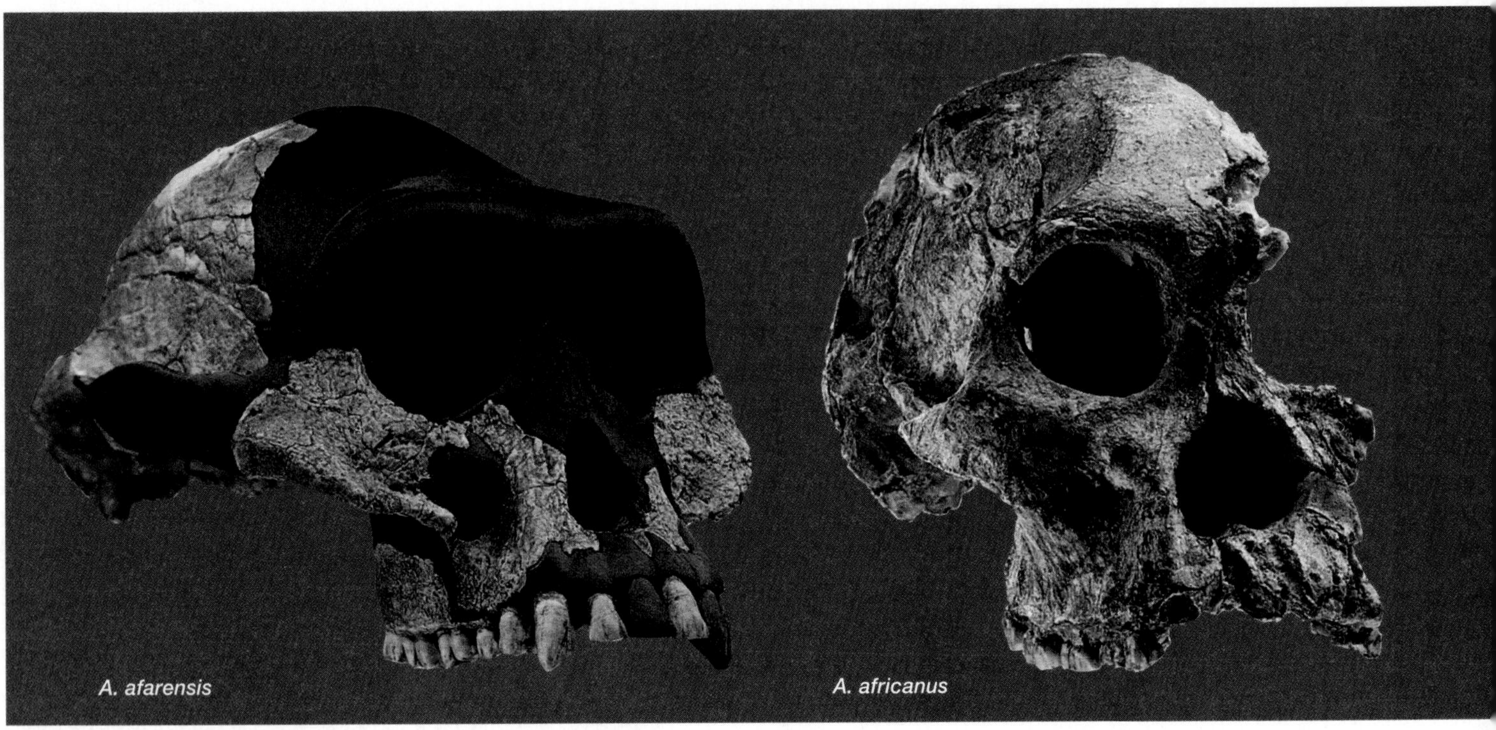

Figure 18.9 Nearly human.
These four skulls, all photographed from the same angle, are among the best specimens available of the key *Australopithecus* species.

weight, rather than angling away from the pelvis like an ape's. Third, the pelvis is bowl-shaped, centering the body weight over the legs, rather than tall and narrow, pitching the body weight forward as an ape's pelvis does. Fourth, the position where the spine attaches to the skull is shifted down underneath the head, rather than being located near the rear of the skull as in an ape.

2. *Their teeth reflect a varied diet.* Apes have the teeth of carnivores, adapted for eating a restricted diet of flesh and soft tissue rather than for grinding. They have large canine (fang) teeth, and their teeth have only a thin layer of tooth enamel. Australopithecines, on the other hand, have the teeth of omnivores, with smaller canines and much thicker tooth enamel. Their teeth are much like human teeth today.
3. *They had large brains.* An adult ape brain is typically about 400 cubic centimeters in volume, while australopithecine brains ranged in size from 400 to 550 cubic centimeters (still much smaller than your brain, which is about 1,350 cubic centimeters).

Different types of australopithecines (figure 18.9) ate different things. *A. robustus* had the strong jaw muscles and flat grinding teeth of a plant-eater. *A. boisei,* with its huge molars and immense jaw muscles anchored to the bony crest of its skull, appears to have been adapted to chewing dry seeds, nuts, and other tough material. The relatively large incisors of *A. africanus* are more typical of carnivores, although its well-developed grinding teeth would have allowed it to eat plants as well.

Lucy: The Most Complete Ancient Hominid

In 1974, anthropologist Don Johanson went to the remote Afar Desert of Ethiopia in search of early human fossils—and hit the jackpot. He found the most complete and best preserved skeleton of a prehuman hominid ever (figure 18.10). Nicknamed "Lucy," the slender australopithecine skeleton was 40% complete, and over 3 million years old, older than any of the more massive australopithecines described previously. The skeleton was assigned the scientific name *A. afarensis.*

The shape of the pelvis indicated that Lucy was a female, and the similarity of her leg bones and knee joint to those of modern humans proved that she walked upright. Her teeth were distinctly humanlike, but the shape of the head resembled that of an ape, and the brain was not larger than an ape's, about 400 cubic centimeters. Apparently, hominids walked upright before they acquired large brains.

Since Johanson's discovery, numerous other specimens of *A. afarensis* have been unearthed. The reconstruction of Lucy illustrated in figure 18.10, prepared for The Living World Education Center of the St. Louis Zoo, was based on a careful analysis of how muscles and ligaments attached to the skull and skeleton.

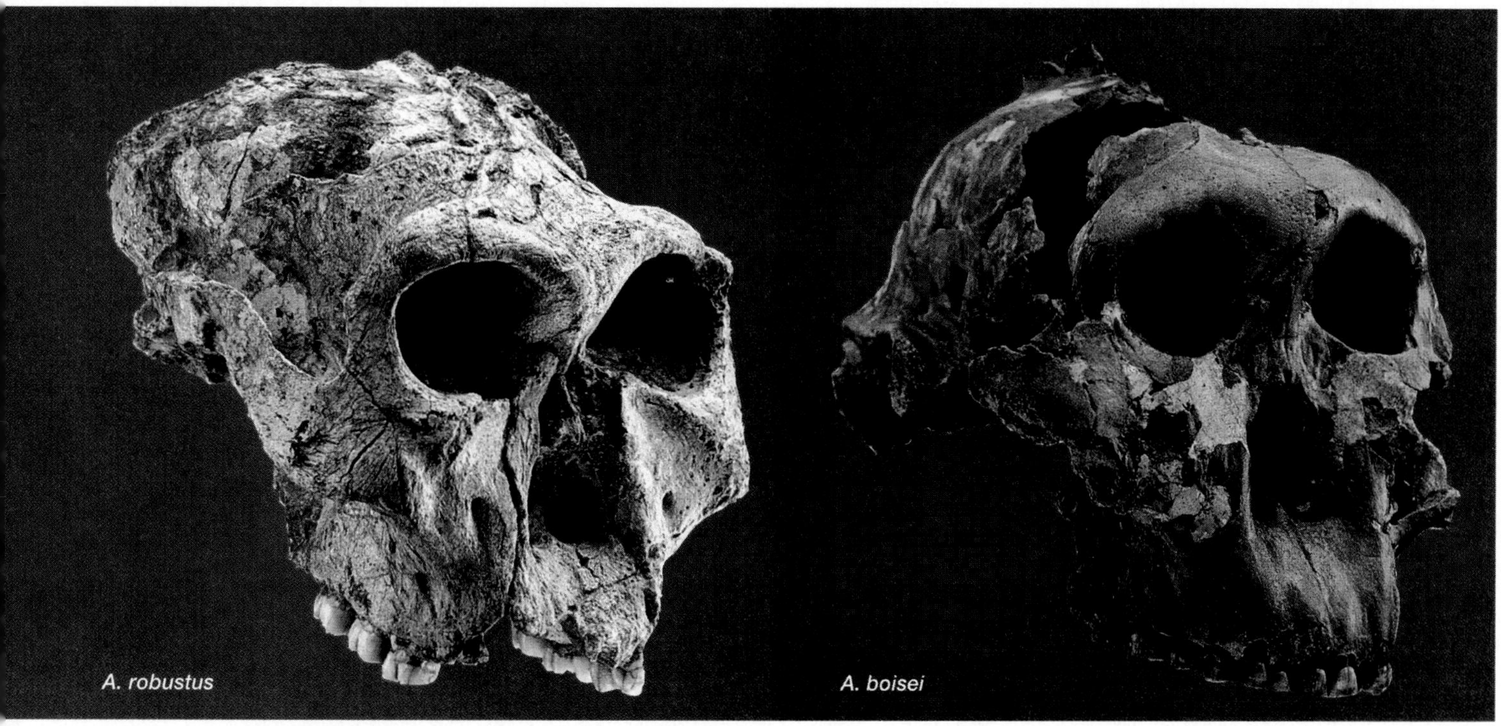

Figure 18.9 (continued).

(a)

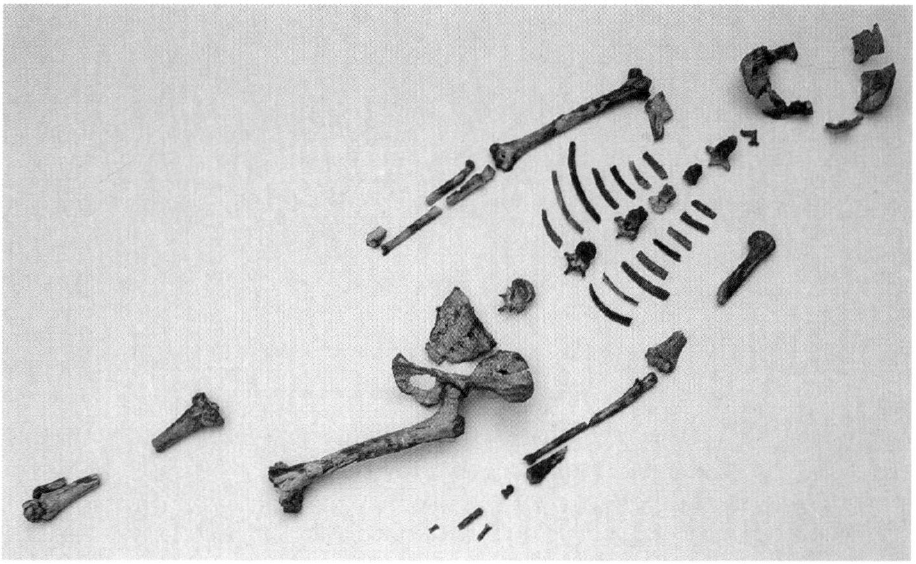

(b)

Figure 18.10 "Lucy."
(*a*) This reconstruction was made by a careful study of muscle attachments to the skull and skeleton. (*b*) Lucy is the most complete skeleton of *Australopithecus* discovered so far.

The Oldest Hominids

In 1994 a fossil was unearthed in Ethiopia (the country where Lucy was found) that had some hominid features and was 4.4 million years old. Named *Ardipithecus ramidus,* these near-apes may be early ancestors of modern apes. In 1995, hominid fossils of nearly the same age, 4.2 million years old, were found in the Rift Valley in Kenya. These individuals were much less apelike than *A. ramidus.* While clearly australopithecine, the fossils were intermediate in many ways between apes and *Australopithecus afarensis.* Named *Australopithecus anamensis,* numerous specimens of this species were found, and most researchers agree that these slightly built *A. anamensis* individuals represent the true base of our family tree, the first members of the genus *Australopithecus,* ancestor to all the others. While there is considerable disagreement among researchers about how to draw the tree—some would put a species on one branch, some on another (figure 18.11)—there is no doubt at all among them that the genus *Australopithecus* was immediately ancestral to our own genus, *Homo,* our link to the past.

The footprints shown in figure 18.12 were made in soft, damp volcanic ash by an early australopithecine, perhaps *A. anamensis,* 3.7 million years ago. The arch, big toe, and heel marks are those of a bipedal hominid. It seems that walking upright has been a feature of hominids from the very beginning.

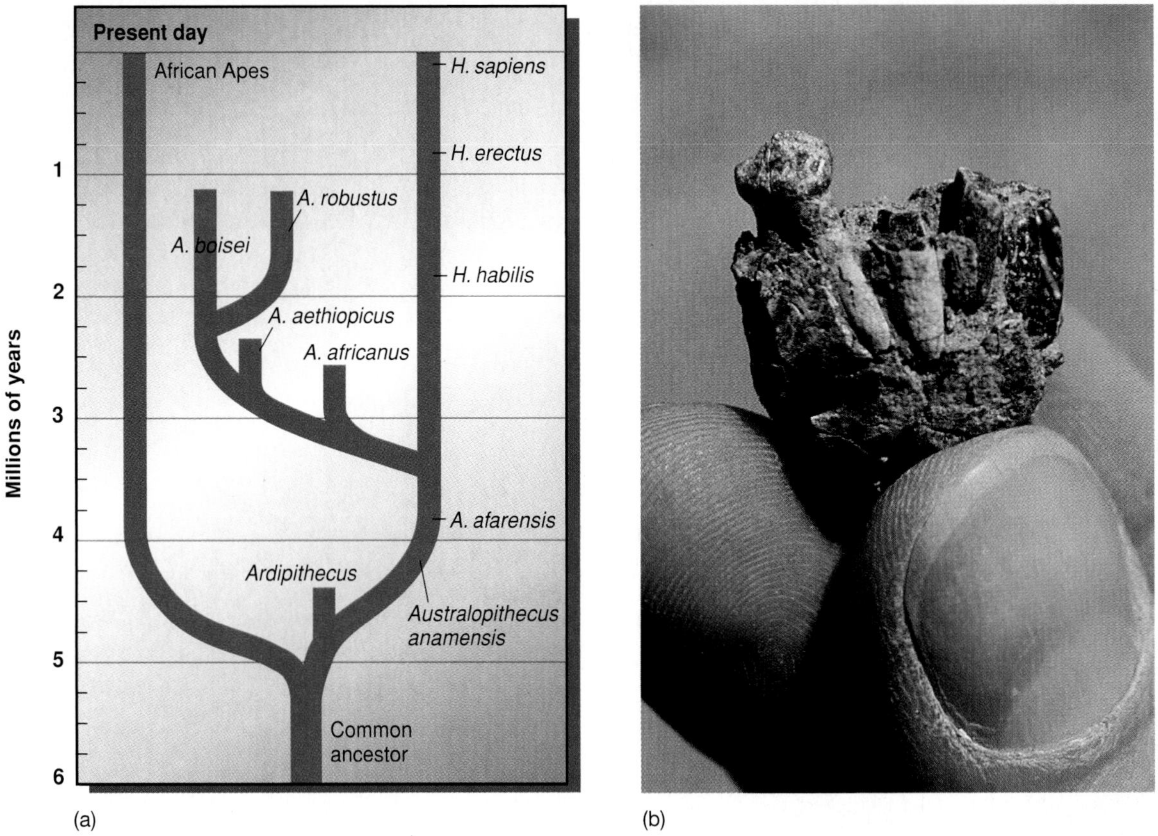

Figure 18.11 The path of human evolution.
There is considerable discussion about the exact path of human evolution. (*a*) The version presented here is the most commonly accepted, although many points are actively disputed. (*b*) The photo shows a tooth from *Ardipithecus ramidus,* discovered in 1994. The name *ramidus* is from the Latin word for "root," as *A. ramidus* is thought to be near the root of the hominoid family tree. The earliest known hominid, at 4.4 million years old, *A. ramidus* was about the size of a chimpanzee and apparently could walk upright.

Figure 18.12 Footprints of two individuals of the genus *Australopithecus.*
These footprints were made in Africa 3.7 million years ago. A mother and a child walking on the beach might leave such tracks. But these tracks are not human. Preserved in volcanic ash, these tracks record the passage of two individuals of the genus *Australopithecus*. They were found at Laetoli, East Africa, in 1978 by a team of workers headed by Mary Leakey. The tracks continue 24 meters before disappearing. Looking at them, we journey into the past.

18.3 The First Humans

We are the third and only surviving species of humans. The first humans, members of the genus *Homo,* evolved from australopithecine ancestors about 2 million years ago, only to be replaced in turn by a second "improved version" of human that moved out of Africa and spread across the earth.

African Origin: Homo habilis

In the early 1960s, more hominid bones were discovered close to the Olduvai Gorge site in Tanzania, where *A. boisei* had been unearthed, and scattered among the bones were stone tools. Although the fossil bones were badly crushed, painstaking reconstruction of the many pieces suggested a skull with a brain volume of about 640 cubic centimeters, much larger than the australopithecine range of 400 to 550 cubic centimeters. Much discussion ensued about whether this fossil was human, until 1972, when paleoanthropologist Richard Leakey discovered a virtually complete skull (figure 18.13). The skull, 1.8 million years old, had a brain volume of 775 cubic centimeters and many of the characteristics of human skulls. Clearly it was human and not australopithecine. Because of its association with tools, this early human was called ***Homo habilis,*** meaning "handy man." Skeletons discovered in 1987 indicate that *H. habilis* was small in stature, like *Australopithecus. H. habilis* lived in Africa for 500,000 years and then became extinct, replaced by a new kind of human with an even larger brain.

Figure 18.13 The Leakeys describe *Homo habilis.*
In 1972 at East Turkana, Kenya, Richard and Mary Leakey's team of researchers discovered the most complete *Homo habilis* skull to date. Debate surrounded its identification and age; some thought it resembled the australopithecines, while others felt it belonged with *Homo.* Initially, the skull was dated at 2.6 million years, making it "the oldest man" ever found. Later its age was revised to 1.8 million years.

Out of Africa: Homo erectus

Our picture of what *H. habilis* was like lacks detail, because it is based on only a few specimens, and some scientists still dispute *H. habilis*'s qualifications as a true human since it had not moved far from its australopithecine roots. There is no such doubt about the species that replaced it, ***Homo erectus.*** Many specimens have been found, and *H. erectus* was without any doubt a true human. The story of how this second species of human was discovered is a fascinating one.

Java Man

After the publication of Darwin's book *On the Origin of Species* in 1859, there was much public discussion about "the missing link," the fossil ancestor common to both humans and apes. Puzzling over the question, a Dutch doctor and anatomist named Eugene Dubois took a very simple approach to the problem—he went to the zoo. Looking at the apes there, he was most drawn to the orangutans, the "old men of the forest," from Java and Borneo. In many anatomical features they seemed to him to resemble what a missing link should look like. So Dubois decided to seek fossil evidence of the missing link in the home country of the orangutan, Java. Closing his practice, Dubois set off for Java in search of ancient humans.

Dubois set up practice in a river village in eastern Java. Digging into a hill that villagers claimed had "dragon bones," he unearthed a skull cap and a thighbone in 1891. He was very excited by his find, informally called **Java man,** for three reasons:

1. The structure of the thigh bone clearly indicated that the individual had long, straight legs and was an excellent walker.
2. The size of the skull cap suggested a *very* large brain, about 1,000 cubic centimeters.
3. Most surprising, the bones seemed as much as 500,000 years old, judged by other fossils Dubois unearthed with them.

The fossil hominid that Dubois had found was far older than any discovered up to that time, and few scientists were willing to accept that it was an ancient species of human. Finally, after years of arguing to an unconvinced audience, Dubois in disgust buried the Java man skull cap and thighbone under the floorboards of his dining room and for 30 years refused to let anyone see them.

Peking Man

Another generation passed before scientists were forced to admit that Dubois had been right all along. In the 1920s a skull was discovered near Peking (now Beijing), China, that closely resembled Java man. Continued excavation at the site eventually revealed 14 skulls, many excellently preserved (figure 18.14). Crude tools were also found, and most important of all, the ashes of campfires. Casts of these fossils were distributed for study to laboratories around the world. The originals were loaded onto a truck and evacuated from Peking at the beginning of World War II, only to disappear into the confusion of history. No one knows what happened to the truck or its priceless cargo. Fortunately, Chinese scientists have excavated numerous additional skulls of **Peking man** since 1949.

Figure 18.14 Peking man excavation site.
A specimen of *Homo erectus* known as Peking man was found near Beijing, China (formerly Peking). The site later revealed a total of 14 skulls, a number of other human bones, tools, campfire ashes, and the bones of horses, bears, elephants, deer, and other animals. These findings indicate that Peking man was social, used tools, and cooked food over campfires.

A Very Successful Species

Java man and Peking man are now recognized as belonging to the same species, *Homo erectus*. *H. erectus* was a lot larger than *H. habilis*—about 1.5 meters tall. It had a large brain, about 1,000 cubic centimeters (figure 18.15), and walked erect. Its skull had prominent brow ridges and, like modern humans, a rounded jaw. Most interesting of all, the shape of the skull interior suggests that *H. erectus* was able to talk.

Where did *H. erectus* come from? It should come as no surprise to you that it came out of Africa. In 1976 a complete *H. erectus* skull was discovered in East Africa. It was 1.5 million years old, a million years older than the Java and Peking finds, and its appearance marked the beginning of the great human expansion. Far more successful than *H. habilis, H. erectus* quickly became widespread and abundant in Africa, and within 1 million years had migrated into Asia and Europe. A social species, *H. erectus* lived in tribes of 20 to 50 people, often dwelling in caves. They successfully hunted large animals, butchered them using flint and bone tools, and cooked them over fires—the site in China contains the remains of horses, bears, elephants, deer, and rhinoceroses.

H. erectus survived for over a million years, longer than any other species of human. These very adaptable humans only disappeared in Africa about 500,000 years ago, as modern humans were emerging. Interestingly, they survived even longer in Asia, until about 250,000 years ago.

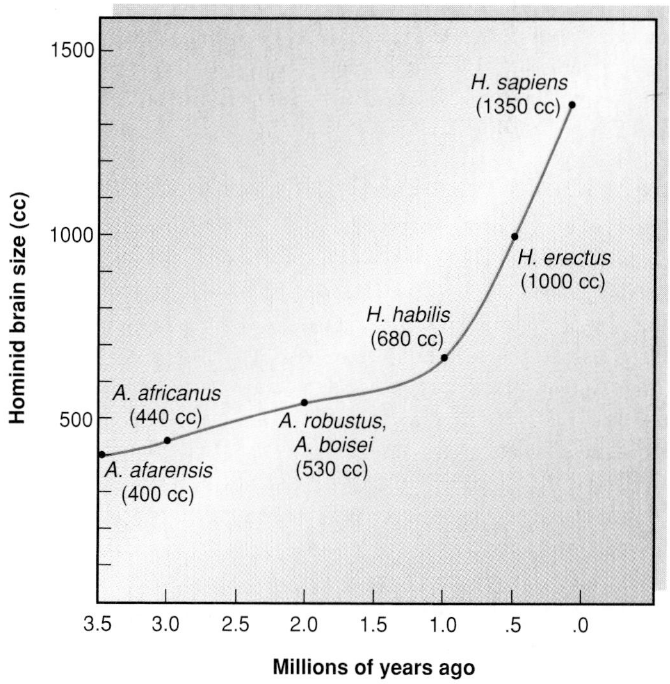

Figure 18.15 Brain size increased as hominids evolved.
Homo erectus had a larger brain than *H. habilis,* which was larger than the australopithecines.

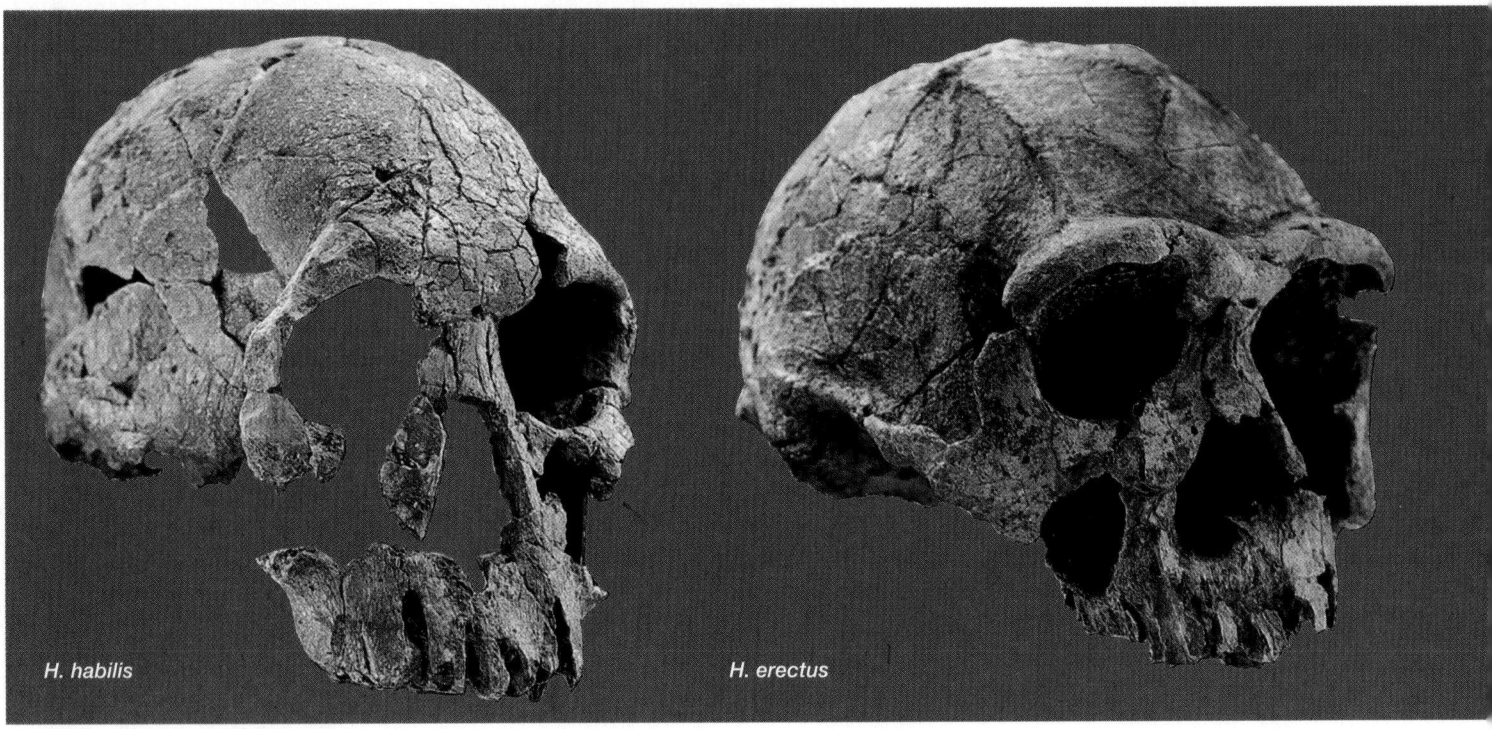

Figure 18.16 Our own genus.
These four skulls illustrate the changes that have occurred during the evolution of the three species of humans.
The Cro-Magnon skull is essentially the same as human skulls today. The skulls were photographed from the same angle.

18.4 Our Own Species

Finally we come to the end of our evolutionary journey, when our own species, ***Homo sapiens*** ("wise man") appeared, about 500,000 to 300,000 years ago (figure 18.16). We are newcomers to the human family—*H. sapiens* has not been around nearly as long as *H. erectus* was. Still, we have changed quite a bit since those first days.

Out of Africa Again: Homo sapiens

The origin of human races is a much-debated point among scientists studying human evolution. Many have argued that the different races evolved from *H. erectus* independently, each adapted to a different place—Orientals in Asia, Caucasians in Europe, aborigines in Australia, and so on. Others have felt it unlikely that the same species would evolve more than once and argue that the races of humans appeared *after* our species evolved from *H. erectus*. Recently, scientists studying DNA within human mitochondria have added fuel to the fire of this controversy. Examining mitochondrial DNA from living humans all over the world, they argue that all the races of humans indeed originated from one *H. sapiens* female ancestor. Where? You guessed it. Africa.

Human races evolved only recently in the evolutionary scale of things, and there has not been enough time for lots of differences to accumulate, so the exact human tree cannot be reliably traced using this approach. However, there are some evident differences, and it turns out that the greatest number of different mitochondrial DNA sequences occur among modern Africans. Since DNA accumulates mutations over time, the oldest DNA should show the largest number of mutations, so this result argues that humans have been living in Africa longer than on any other continent. This suggests that *H. sapiens* evolved in Africa, but such a conclusion is quite controversial, and there is no firm consensus among researchers yet as to where our species originated. Some researchers argue that the mitochondrial data point to the right answer—that our species was born in Africa, and from there spread to all parts of the world, retracing the path taken by *H. erectus* half a million years before (figure 18.17). Other researchers argue that our species evolved from separate but occasionally interbreeding groups of *H. erectus*.

A clearer analysis is possible using chromosomal DNA, segments of which are far more variable than mitochondrial DNA, providing more "markers" to compare. When a variable segment of DNA from human chromosome 12 was analyzed in 1996, a clear picture emerged. A total of 24 different versions of this segment were found. Fully 21 of them were present in human populations in Africa, while three were found in Europeans and only two in Asians and in Americans. This result argues strongly that chromosome 12 has existed in Africa far longer than among non-African humans, strongly supporting an African origin of *H. sapiens*. Recently discovered fossils of early *H. sapiens* from Africa also lend strong support to this hypothesis.

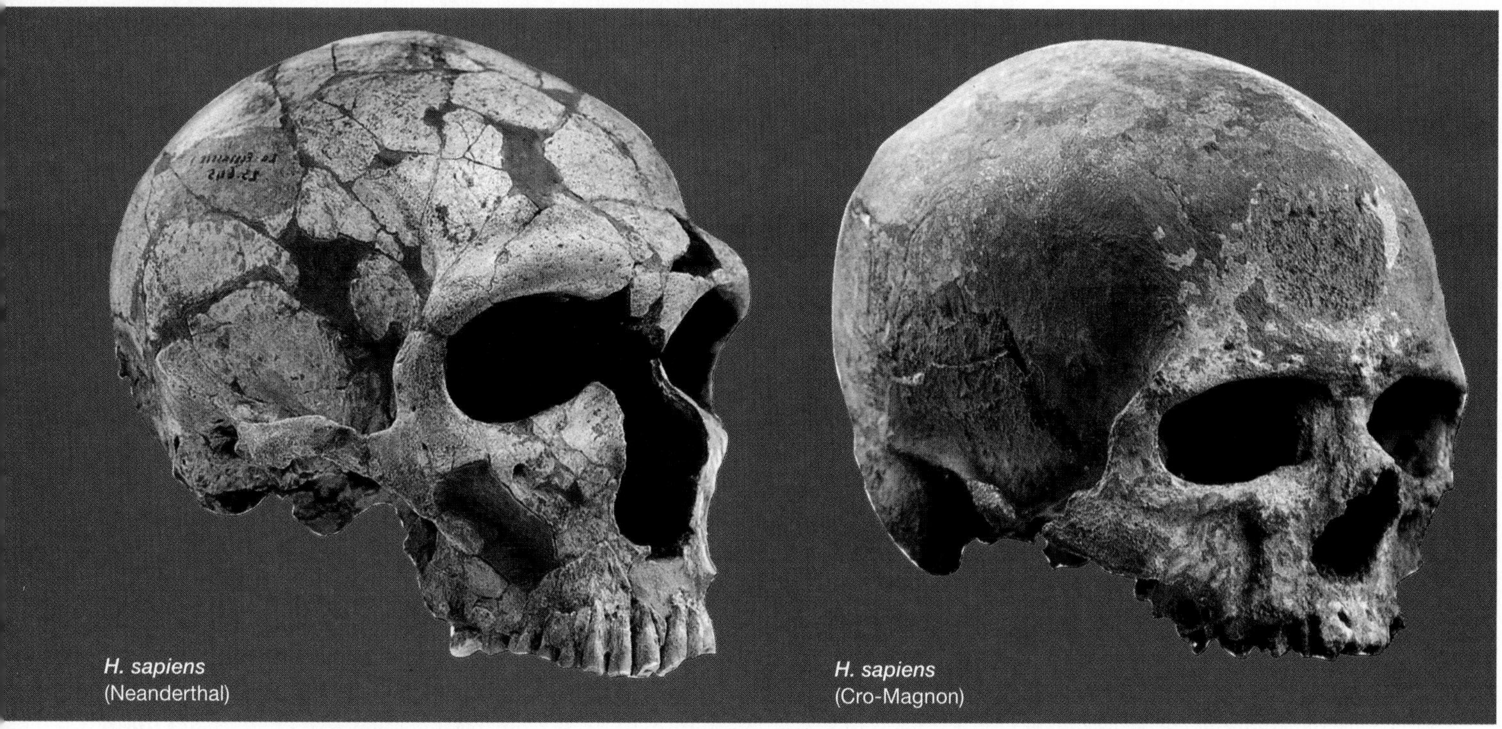

Figure 18.16 (continued).

Neanderthal Man

H. sapiens first appeared in Europe as *H. erectus* was becoming rarer, about 130,000 years ago. Because the first fossils to be found, in 1856, were from the Neander Valley of Germany, these early European humans were called **Neanderthals** ("Thal" means valley in Old German). Compared to ourselves, the European Neanderthals were powerfully built, short, and stocky. Their skulls were massive, with protruding faces and heavy bony ridges over the brows. Their brains were even larger than those of modern humans!

The Neanderthals made diverse tools, including scrapers, spear heads, and hand axes. They lived in huts or caves. Rare at first outside of Africa, they became progressively more abundant in Europe and Asia, and by 70,000 years ago had become common. Neanderthals took care of their injured and sick and commonly buried their dead, often placing food, weapons, and even flowers with the dead bodies. Such attention to the dead strongly suggests that they believed in a life after death. For the first time, we see evidence of the kinds of symbolic thinking characteristic of modern humans.

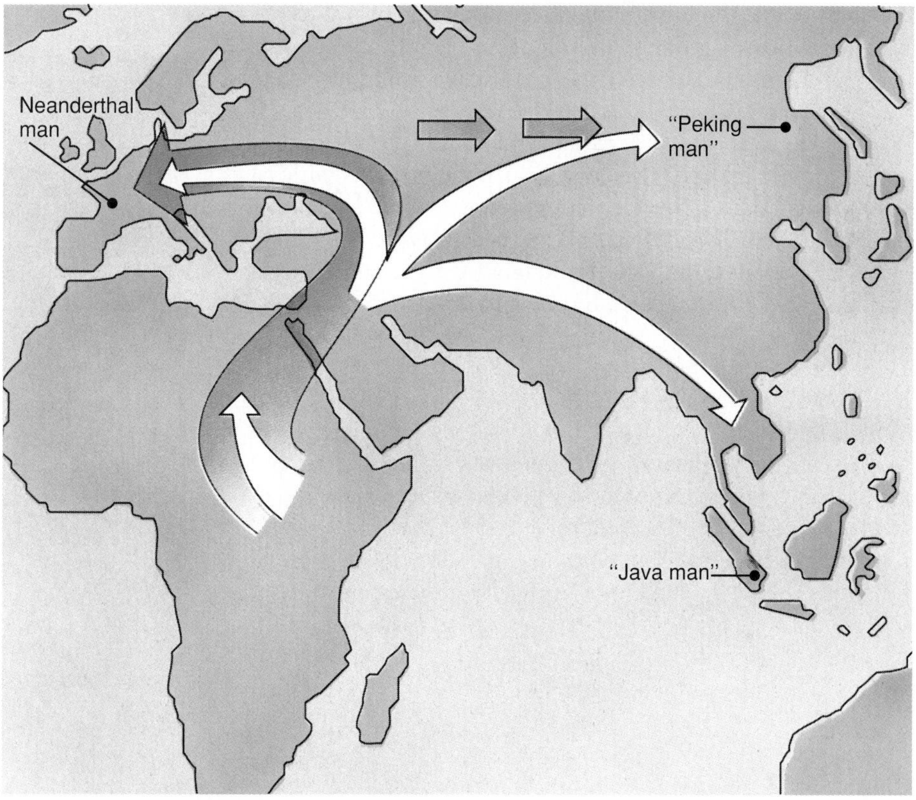

Figure 18.17 Out of Africa—twice.
Evidence now supports the idea that our species first evolved in Africa. *Homo* thus spread from Africa to Europe and Asia twice. First *H. erectus* (*white arrow*) spread as far as Java and China. Later, that species was replaced by *H. sapiens* (*red arrow*) in a second wave of immigration.

Cro-Magnon Man

Fossils of the European Neanderthals abruptly disappear from the fossil record about 34,000 years ago and are replaced by fossils of essentially modern humans, the **Cro-Magnons** (named after the valley in France where their fossils were first discovered). We can only speculate why this sudden replacement occurred, but it was complete all over Europe in a short period. There is some evidence that the Cro-Magnons came from Africa—fossils of essentially modern aspect but as much as 100,000 years old have been found there. Cro-Magnons seem to have replaced the Neanderthals completely in the Middle East by 40,000 years ago, and then spread across Europe, coexisting and possibly even interbreeding with the Neanderthals for several thousand years.

Figure 18.18 Cave painting.
Paintings of animals, and sometimes of hunters, were made by Cro-Magnon people, our immediate ancestors. They were created for a period of about 20,000 years, until about 8,000 to 10,000 years ago, especially in Europe.

The Cro-Magnons used sophisticated stone tools and also made tools out of bones and horns. They had a complex social organization and are thought to have had full language capabilities. They lived by hunting. The world was cooler than it is now—the time of the last great ice age—and Europe was covered with grasslands inhabited by large herds of grazing animals. Pictures of them can be seen in elaborate and often beautiful cave paintings made by Cro-Magnons throughout Europe (figure 18.18).

Humans of modern appearance eventually spread across Siberia to North America, which they reached at least 13,000 years ago, after the ice had begun to retreat and a land bridge still connected Siberia and Alaska. By 10,000 years ago, about 5 million people inhabited the entire world (compared with more than 5.8 billion today).

We Are Unique

We humans are animals and the product of evolution. Our evolution has been marked by a progressive increase in brain size, distinguishing us from other animals in several ways. First, humans are able to make and use tools effectively—a capability that, more than any other factor, has been responsible for our dominant position in the animal kingdom. Second, while not the only animal capable of conceptual thought, we have refined and extended this ability until it has become the hallmark of our species. Lastly, we use symbolic language and can with words shape concepts out of experience. Our language capability has allowed the accumulation of experience, which can be transmitted from one generation to another. Thus we have what no other animal has ever had: cultural evolution. Through culture, we have found ways to change and mold our environment, rather than changing evolutionarily in response to the demands of the environment. We control our biological future in a way never before possible—an exciting potential and frightening responsibility (figure 18.19).

Figure 18.19 The human species.
Humans, unlike other animals, have the ability to control their environment through culture. All other animals are controlled by their environment.

CHAPTER 18

HIGHLIGHTS

	Key Terms	Key Concepts
18.1 The Evolution of Primates 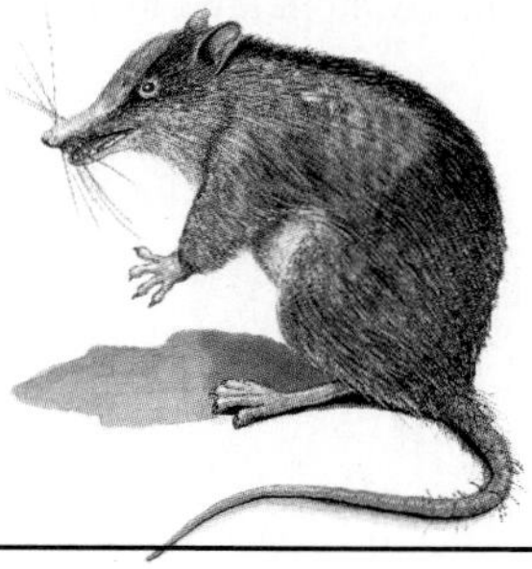	primate 391 prosimian 392 ape 392	• Primates are the order of mammals that contains humans. • Prosimians were the first primates, small and nocturnal. • Prosimians were replaced by their descendants, which are day-active. • Apes also evolved from prosimians and have larger brains.
18.2 The Evolutionary Origins of Humans 	hominid 394 *Australopithecus* 395 bipedal 395	• Two genera are considered hominid (of the human line): *Australopithecus* and *Homo*. • Hominids are characterized by upright walking and large brains. • Many stocky species of *Australopithecus* appear to be side branches on the evolutionary tree. • The slender species like Lucy appear to be our immediate ancestors. • The oldest hominids are 4.2 million years old.
18.3 The First Humans 	*Homo habilis* 400 *Homo erectus* 400	• The first human, *Homo habilis,* appeared 1.8 million years ago. • The second species of human, *H. erectus,* appeared in Africa about 1.5 million years ago and survived for a million years, longer than any other species of human. • *H. erectus* migrated out of Africa to Europe and Asia.
18.4 Our Own Species 	*Homo sapiens* 402 Neanderthals 403 Cro-Magnons 404	• Our species, *Homo sapiens,* evolved in Africa about half a million years ago. • Migrating out of Africa, *H. sapiens* retraced the spread of *H. erectus,* eventually supplanting it. • Early *H. sapiens,* called Neanderthals, actually had bigger brains than Cro-Magnons (modern) *H. sapiens.*

18

CONCEPT REVIEW

1. The ancestors of the first primates were
   a. lemurs. c. mice.
   b. monkeys. d. tree shrews.

2. Prosimians arose about _________ years ago.
   a. 70 million
   b. 40 million
   c. 10 million
   d. 1 million

3. Unlike African and Asian monkeys, South American monkeys have
   a. color vision. c. prehensile tails.
   b. grasping fingers. d. no tails.

4. Historically, apes occurred in North America.
   a. true
   b. false

5. Arrange the following animals in order of their evolutionary distance from humans, the most distant first.
   a. gibbons d. monkeys
   b. gorillas e. orangutans
   c. prosimians

6. Which of the following characteristics were *not* exhibited by australopithecines?
   a. the use of tools
   b. bipedalism
   c. large brains
   d. omnivorous teeth

7. A fossil named _________ seems to be intermediate between apes and *Australopithecus* and may represent the true base of our family tree.
   a. *Homo habilis*
   b. *Ardipithecus ramidus*
   c. *Homo erectus*
   d. "Lucy"

8. _________ survived longer than any other species of human.
   a. *Homo habilis* c. *Homo erectus*
   b. *Homo sapiens* d. Neanderthal man

9. Our own species, *Homo sapiens,* appeared about _________ years ago.
   a. 10 million
   b. 10,000
   c. 100,000
   d. 500,000

10. Neanderthal man was common about 70,000 years ago in _________ and Asia.
    a. Europe
    b. North America
    c. Africa
    d. Australia

11. Arrange the following hominids in the order in which they are thought to have evolved, the most distant first.
    a. *Australopithecus*
    b. Cro-Magnons
    c. *Homo erectus*
    d. *Homo habilis*
    e. Neanderthals

12. Unlike the eyes of squirrels and shrews, the eyes of primates are shifted forward, giving them _________ vision.

13. _________ are apes that have tails and are relatively small.

14. "Lucy" is the most complete fossil hominid of the genus _________.

15. *Homo* _________ replaced *Homo habilis.*

16. Dubois discovered the fossil of __________ man.

17. The scientific name of Neanderthal man is *Homo* _________.

18. Humans crossed the land bridge across Siberia to North America about _________ years ago.

19. _________ evolution affects modern human races as much as biological evolution.

Answers to the Concept Review questions appear in Appendix B.

CHALLENGE YOURSELF

1. Create a hypothesis for why all living lemurs on the earth are on the island of Madagascar.
2. Studies of the DNA of primates have revealed that humans differ from gorillas in only 1.2% of the DNA nucleotide sequences and from chimpanzees in less than 3%. This degree of genetic similarity is the same as is usually seen among "sibling" species (that is, species that have only recently evolved from a common ancestor). Yet humans are assigned not only to a different genus but to a different family! Do you think this is legitimate, or are humans just a rather unusual kind of African ape?
3. Why is it incorrect to state that humans evolved from apes or monkeys?
4. What evidence would solve some unanswered questions about human evolution?
5. Modern humans, *Homo sapiens,* evolved from *Homo erectus* fewer than 1 million years ago. Do you think that evolution of the genus *Homo* is over, or might another species of humans evolve within the next million years? Do you think this would involve the extinction of *H. sapiens?*

FOR FURTHER READING

Blumenschine, R., and J. Cavallo. "Scavenging and Human Evolution." *Scientific American,* October 1992, 90–96. An interesting, if controversial, argument that our early ancestors were better scavengers than hunters.

Gore, R. "Neanderthals." *National Geographic,* January 1996, 2–35. Early humans of our own species were in many ways markedly different from people today.

Johanson, D. "Face-to-Face with Lucy's Family." *National Geographic,* March 1996, 96–115. The discovery of *Australopithecus afarensis*, the immediate ancestor of humans.

Leakey, M. "The Dawn of Humans: The Farthest Horizon." *National Geographic,* September 1995, 38–51. An account of what we know of the first hominids, and how we came to know it.

Milton, K. "Diet and Primate Evolution." *Scientific American,* August 1993, 86–93. The nature of food available in the early angiosperm forests of the Cretaceous and how that has shaped how the human evolutionary line has developed.

Nitecki, M., and D. Nitecki. *Origins of Anatomically Modern Humans*. New York: Plenum Press, 1994. A collection of papers on Paleolithic humans, a very controversial period in human evolution.

Simons, E. "Human Origins." *Science* 245 (September 1989): 1343–50. A thorough review of the skeletal evidence that indicates that all major steps in human evolution took place in Africa.

Thorne, A., and M. Wolpoff. "The Multiregional Evolution of Humans." *Scientific American,* April 1992, 76–83. The argument against African origin of modern *H. sapiens*.

Wood, B. "The Oldest Hominid Yet." *Nature,* September 1994, 280–81. An account of the discovery of *Ardipithecus ramidus*, a primitive early hominid linking hominids with chimpanzees. This important find was first reported by Tim White and others in a more technical companion article on pages 306-12 in the same issue.

TECHNOLOGY LINKS

The Living World Home Page
http://www.wcbp.com/biology/tlw

19

The Human Body

"Pull out, Betty! Pull out! . . . You've hit an artery!"

CHAPTER OUTLINE

19.1 How the Body Is Organized
- Tissues
- Organs
- Organ Systems

19.2 The Principal Tissues
- Epithelium Is Protective Tissue
- Connective Tissue Supports the Body
- Muscle Tissue Lets the Body Move
- Nerve Tissue Conducts Signals Rapidly

19.3 The Human Skeleton
- The Need for a Skeleton
- Organization of the Skeleton
- The Skeleton Is Made of Bone
- Tendons Connect Bone to Muscle

19.4 The Human Muscular System
- Kinds of Muscle
- How Muscles Work

Highlights

Figure 19.1 The human body.
The human body is a highly organized organic machine. Its many organs are composed of different combinations of tissues, much as the engine, transmission, brakes, and other systems of a car are composed of different combinations of materials.

The human body is at the same time both beautiful and incredibly complex (figure 19.1). To understand it, we must first look at how it is put together. Our bodies have the same general architecture as all vertebrates: food flows through a long tube from mouth to anus, which is suspended within an internal body cavity called the *coelom.* Our coelom is divided into two parts: the *thoracic cavity,* which contains the heart and lungs, and the *abdominal cavity,* which contains the stomach, intestines, and liver. The body is supported by an internal scaffold, or skeleton, made up of jointed bones that grow as the body grows. A bony skull surrounds and protects the brain, while a column of bones, the vertebrae, surrounds the spinal cord.

19.1 How the Body Is Organized

Like all animals, your body is composed of cells, over 100 trillion of them. It's difficult to picture how many 100 trillion actually is. One hundred trillion seconds is longer than there has been a United States (310 years!). A line with 100 trillion cars in it would stretch from earth to the sun and back 50 million times! Not all of these 100 trillion cells are the same, of course. If they were, we would not be bodies but amorphous blobs. Our bodies contain over 100 different kinds of cells.

Tissues

Groups of cells of the same type are organized within the body into **tissues,** which are the structural and functional units of the human body. A tissue is a group of cells of the same type that performs a particular function in the body.

Tissues form as the human body develops. Early in development, the growing mass of cells that will become a person differentiates into three fundamental layers of cells: endoderm, mesoderm, and ectoderm. These three kinds of embryonic cell layers in turn differentiate into the more than 100 different kinds of cells in the adult human body.

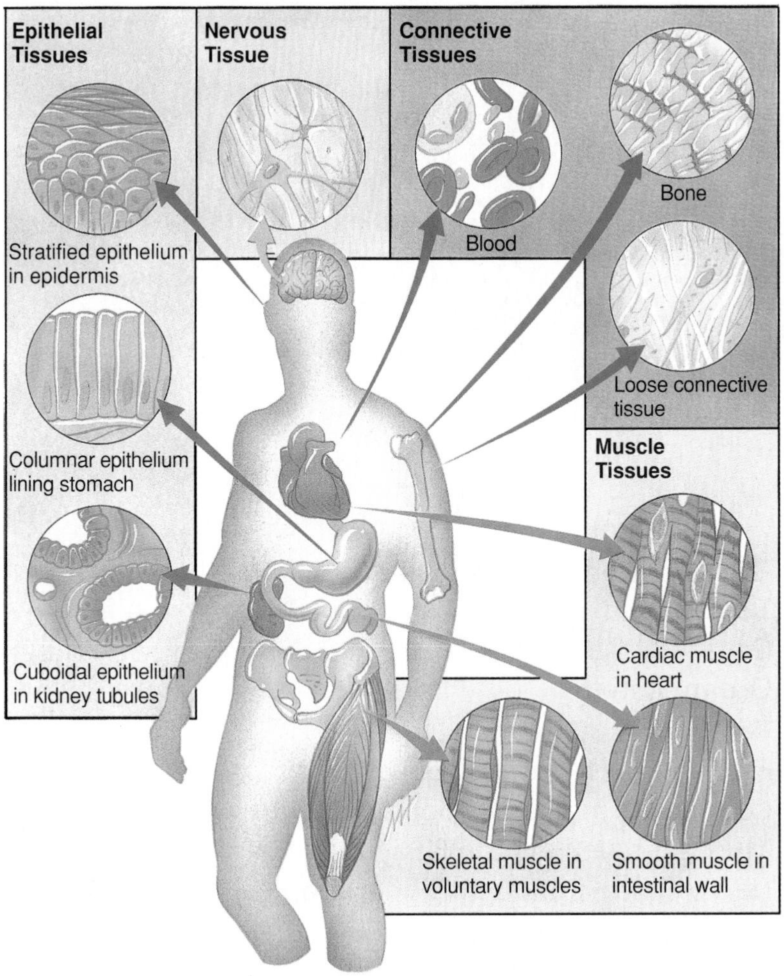

Figure 19.2 Human tissue types.
The four basic classes of tissue are epithelial, nervous, connective, and muscle.

It is possible to assemble many different kinds of tissue from 100 cell types, but biologists have traditionally grouped adult tissues into four general classes: *epithelial, nervous, connective,* and *muscle tissue* (figure 19.2). Of these, connective tissues are particularly diverse. The blood cells flowing through your veins are connective tissue, and so are the bone cells in your skull.

Organs

Organs are body structures composed of several different tissues grouped together into a larger structural and functional unit, just as a factory is a group of people with different jobs who work together to make something. Your heart is an organ. It contains cardiac muscle tissue wrapped in connective tissue and joined to many nerves. All of these tissues work together to pump blood through your body: the cardiac muscles squeeze to push the blood; the connective tissues act as a bag to hold the heart in the proper shape and ensure that the different chambers of the heart contract in the proper order; and the nerves control the rate at which the heart beats. No single tissue can do the job of the heart, any more than one piston can do the job of an automobile engine.

You are probably familiar with many of the major organs of your body. Your lungs are organs that extract oxygen from the air; your stomach an organ that digests food; and your liver an organ that controls the level of sugar and other chemicals in your blood. Organs are the machines of the human body, each built from several different tissues and each doing a particular job. How many others can you name?

Organ Systems

An **organ system** is a group of organs that work together in the human body to carry out an important function. Organizing machines to work together is a familiar concept to most of us. A car, for example, is made of a variety of machines—an engine, a transmission, brakes, wheels—that act together to move the vehicle. In your body, organs work together in the same way. For instance, your digestive system is composed of individual organs that break up the food (teeth), pass the food to the stomach (esophagus), break down the food (stomach), absorb the food (intestine), and expel the solid residue (rectum). If all of these machines do their job right, your body obtains energy and necessary building materials from the food you eat. Figure 19.3 illustrates the relationship between cells, tissues, organs, and organ systems.

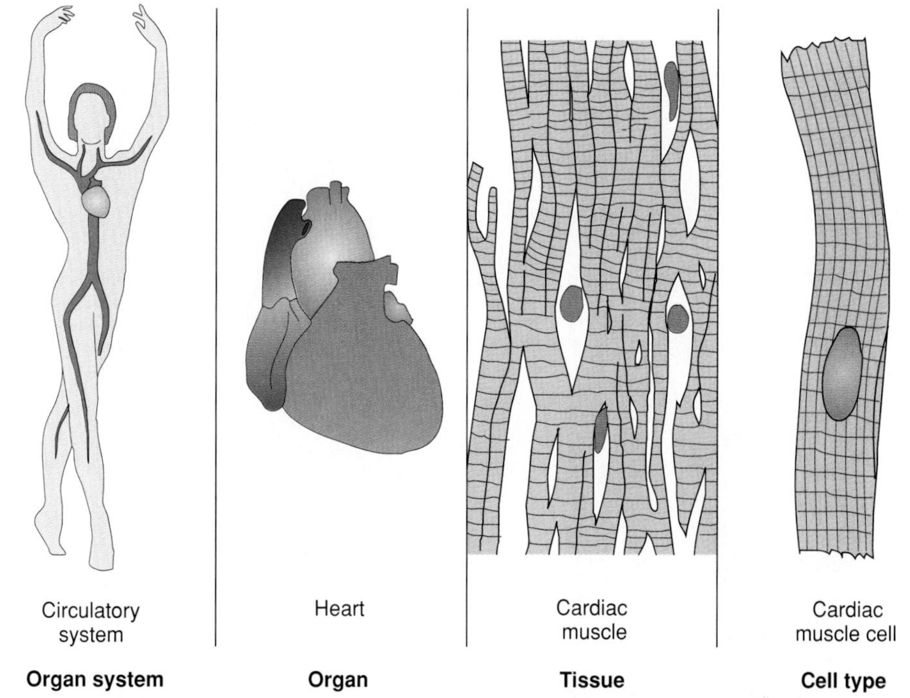

Figure 19.3 Levels of organization within the body.
Similar cell types operate together and form tissues. Tissues functioning together form organs. Several organs working together to carry out a function for the body are called an organ system. The circulatory system is an example of an organ system.

Your body contains 11 principal organ systems (figure 19.4):

1. **Skeletal.** Perhaps the most important feature of the vertebrate body is its bony internal skeleton. The skeletal system protects the body and provides support for locomotion and movement. Its principal components are bones, skull, cartilage, and ligaments. Like arthropods, vertebrates have jointed appendages, the arms, hands, legs, and feet.
2. **Circulatory.** The circulatory system transports oxygen, nutrients, and chemical signals to the cells of the body and removes carbon dioxide, chemical wastes, and water. Its principal components are the heart, blood vessels, and blood. Blood is a cell-rich fluid enclosed within blood vessels.
3. **Endocrine.** The endocrine system coordinates and integrates the activities of the body. Its principal components are the pituitary, adrenal, thyroid, and other ductless glands.
4. **Nervous.** The activities of the body are coordinated by a brain, which receives stimuli, integrates information, and directs the body's activities. Its principal components are nerves, sense organs, brain, and spinal cord.
5. **Respiratory.** The respiratory system captures oxygen and exchanges gases. Its principal components are the lungs, trachea, and other air passageways.
6. **Immune.** The immune system removes foreign bodies from the bloodstream. The principal components are lymphocytes, macrophages, and antibodies.
7. **Digestive.** The digestive system captures soluble nutrients from ingested food. Its principal components are mouth, esophagus, stomach, intestines, liver, and pancreas. Like all vertebrates, the human body has a large body cavity, the coelom, within which are found the large organs of the body. Imagine a balloon full of water that is floating within another, larger balloon, also full of water. Pushing your thumb into the outer balloon doesn't deform the inner one. The fluid within the coelom protects the organs from being deformed by body movements in much the same way.
8. **Urinary.** The urinary system removes metabolic wastes from the bloodstream. Its principal components are the kidneys, bladder, and associated ducts.
9. **Integumentary.** The integumentary system covers and protects the body. Its principal components are skin, hair, nails, and sweat glands.
10. **Muscular.** The muscular system produces movement, both within the body and of its limbs. Its principal components are skeletal muscle, cardiac muscle, and smooth muscle.
11. **Reproductive.** The reproductive system carries out reproduction. Its principal components are testes in males, ovaries in females, and associated reproductive structures.

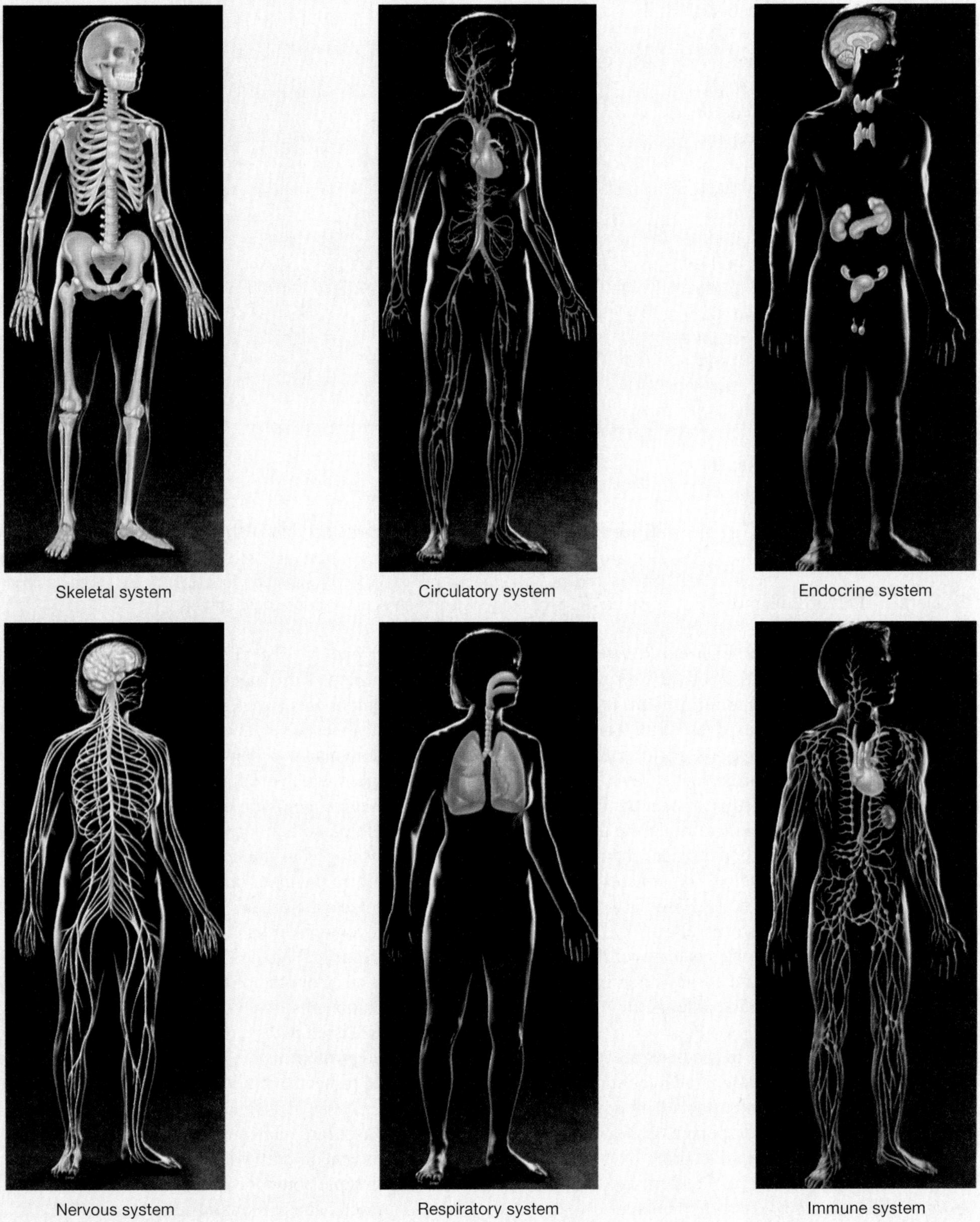

Figure 19.4 Human body organ systems.
Ten of the eleven principal organ systems are shown, including both male and female reproductive systems.
The integumentary system, not pictured, is comprised of the skin and other surface structures of the human body.

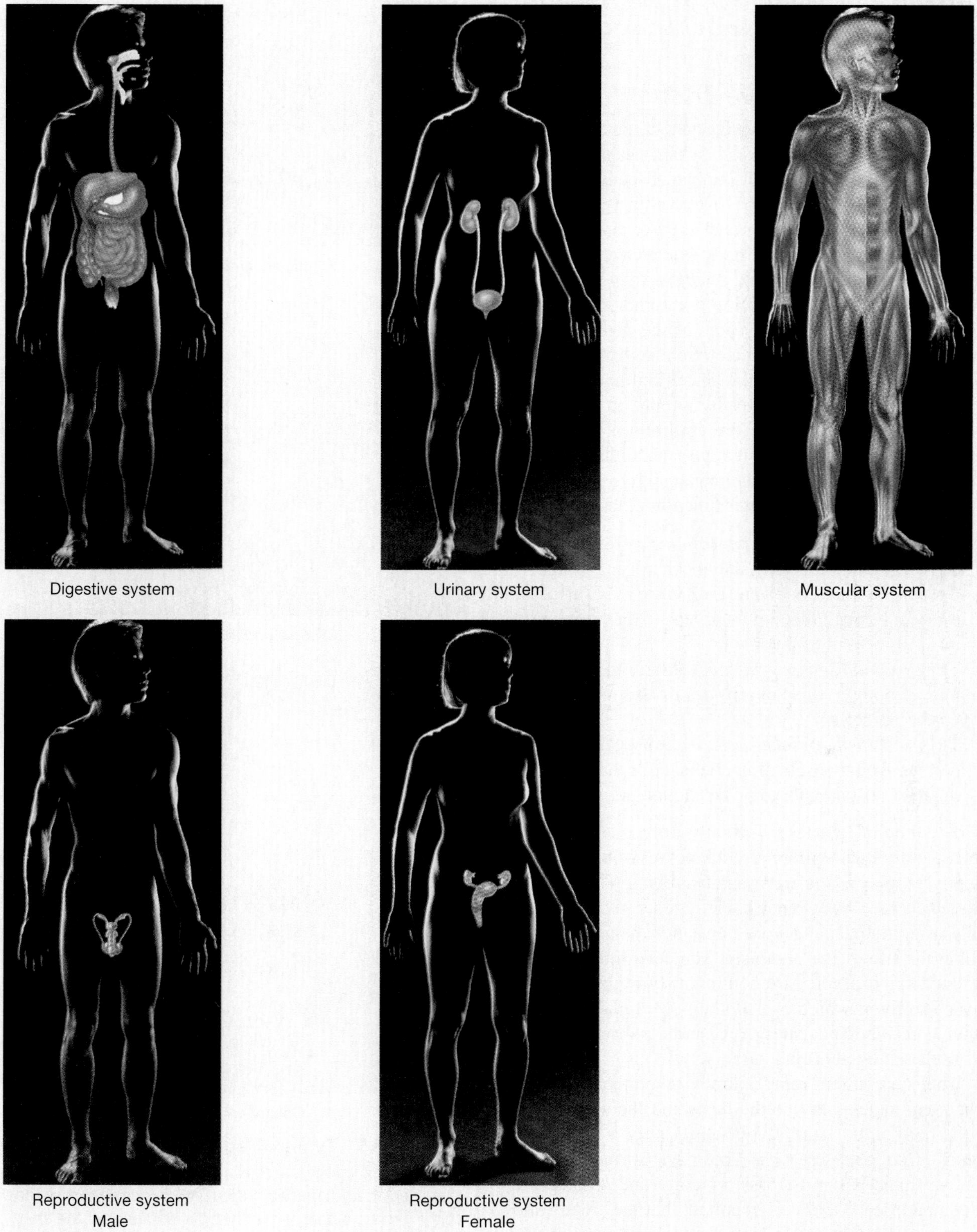

Figure 19.4 (continued).

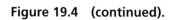

19.2 The Principal Tissues

The human body is composed of over 100 kinds of cells, traditionally grouped into four classes: epithelial, connective, nervous, and muscle.

Epithelium Is Protective Tissue

Epithelial cells are the guards and protectors of the body. They cover its surface and determine which substances enter it and which do not. The organization of the human body is fundamentally tubular, with one tube (the digestive tract) suspended inside another (the body cavity or coelom) like an inner tube inside a tire. The outside of the body is covered with cells (skin) that develop from embryonic *ectoderm* tissue; the body cavity is lined with cells that develop from embryonic *mesoderm* tissue; and the hollow inner core of the digestive tract (the gut) is lined with cells that develop from embryonic *endoderm* tissue. All three kinds of epithelial cells, although different in embryonic origin, are broadly similar in form and function and together are called the **epithelium.** Epithelial cells are in turn classified into three types according to their shapes: squamous, cuboidal, and columnar (figure 19.5).

The body's epithelial layers function in three ways:

1. They *protect the tissues beneath them* from dehydration (water loss) and mechanical damage. Because epithelium encases all the body's surfaces, every substance that enters or leaves the body must cross an epithelial layer.
2. They *provide sensory surfaces*. Many of the human body's sense organs are in fact modified epithelial cells.
3. They *secrete materials*. Most secretory glands are derived from pockets of epithelial cells that pinch together during embryonic development.

Layers of epithelial tissue are usually only one or a few cells thick. Individual epithelial cells possess only a small amount of cytoplasm and have a relatively low metabolic rate. However, they have remarkable regenerative abilities. The cells of epithelial layers are constantly being replaced throughout the life of the organism. The cells lining the digestive tract, for example, are continuously replaced every few days. The liver, which is a football-sized gland formed of epithelial tissue, can readily regenerate substantial portions of itself removed during surgery.

There are three general kinds of epithelial tissue. First, the membranes that line the lungs and the major cavities of the body are a **simple epithelium** only a single cell layer thick. You can see why—these are surfaces across which many materials must pass, entering and leaving the body's compartments, and it is important that the "road" into and out of the body not be too long. Second, the skin, or epidermis, is a **stratified epithelium** composed of more complex epithelial cells several layers thick. Several layers are necessary to provide adequate cushioning and protection and

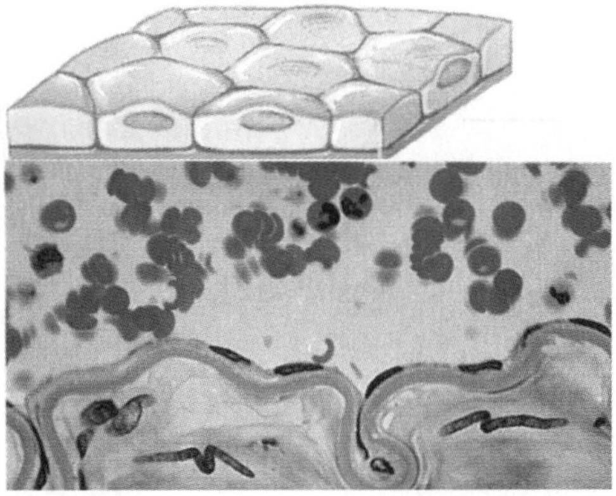

(a) Simple squamous

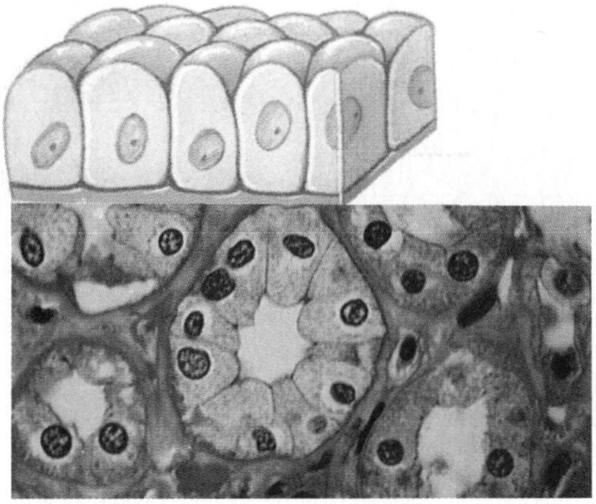

(b) Simple cuboidal

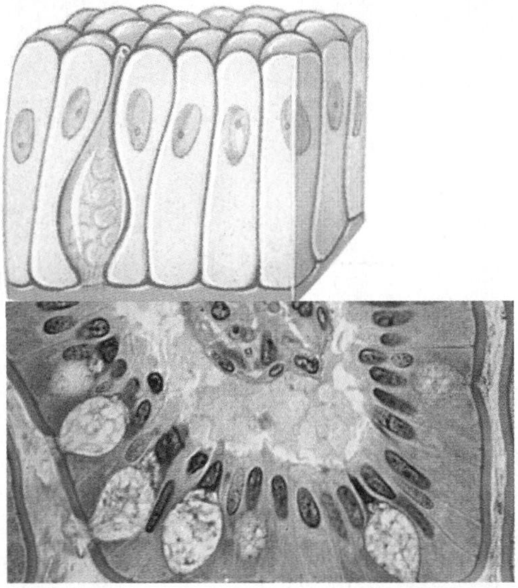

(c) Columnar

Figure 19.5 Types of epithelial cells, based on shape. (*a*) Squamous epithelial cells lining the artery seen here have characteristically flattened nuclei. The round cells above the epithelium are blood cells in the hollow interior of the artery. (*b*) Cuboidal epithelial cells form the walls of these kidney tubules, seen in cross section. (*c*) Columnar epithelial cells form the outer cell layer of this human intestine. Interspersed among the epithelial cells are goblet cells, which secrete mucus.

to enable the skin to continuously replace its cells. Table 19.1 summarizes the characteristics of these two types of epithelial tissues.

The third type of epithelial tissue is found in the **glands** of the body. Endocrine glands secrete hormones into the blood. Exocrine glands (those with ducts that open to the body's outside) secrete sweat, milk, saliva, and digestive enzymes out of the body. Exocrine glands also secrete digestive enzymes into the stomach. If you think about it, the stomach and digestive tract are *outside* the body, since they are the inner canal that passes right through the body. It is possible for a substance to pass all the way through this digestive tract, from mouth to anus, and never enter the body at all. A substance must cross an epithelial layer to truly enter the body.

Table 19.1 Epithelial Tissue

Tissue	Typical Location	Tissue Function
Simple Epithelium		
Squamous	Lining of lungs, capillary walls, and blood vessels	Cells very thin; provides a thin layer across which diffusion can readily occur
Cuboidal	Lining of some glands and kidney tubules; covering of ovaries	Cells rich in specific transport channels; functions in secretion and specific absorption
Columnar	Surface lining of stomach, intestines, and parts of respiratory tract	Thicker cell layer; provides protection and functions in secretion and absorption
Stratified Epithelium		
Squamous	Outer layer of skin; lining of mouth	Tough layer of cells; provides protection
Columnar	Lining of parts of respiratory tract	Functions in secretion of mucus; dense with cilia (small, hairlike projections) that aid in movement of mucus; provides protection

Connective Tissue Supports the Body

The cells of connective tissue provide the body with its structural building blocks and also with its most potent defenses. Derived from the mesoderm, these cells are sometimes densely packed together, and sometimes widely dispersed, just as the soldiers of an army are sometimes massed together in a formation and sometimes widely scattered as guerrillas. **Connective tissue** cells fall into three functional categories: (1) the cells of the immune system, which act to defend the body; (2) the cells of the skeletal system, which support the body; and (3) the blood and fat cells, which store and distribute substances throughout the body.

Immune Connective Tissue

The cells of the immune system roam the body within the bloodstream. They are mobile hunters of invading microorganisms and cancer cells. The two principal kinds of immune system cells are **macrophages,** which engulf and digest invading microorganisms (figure 19.6), and **lymphocytes** (also called white blood cells), which make antibodies or attack virus-infected cells.

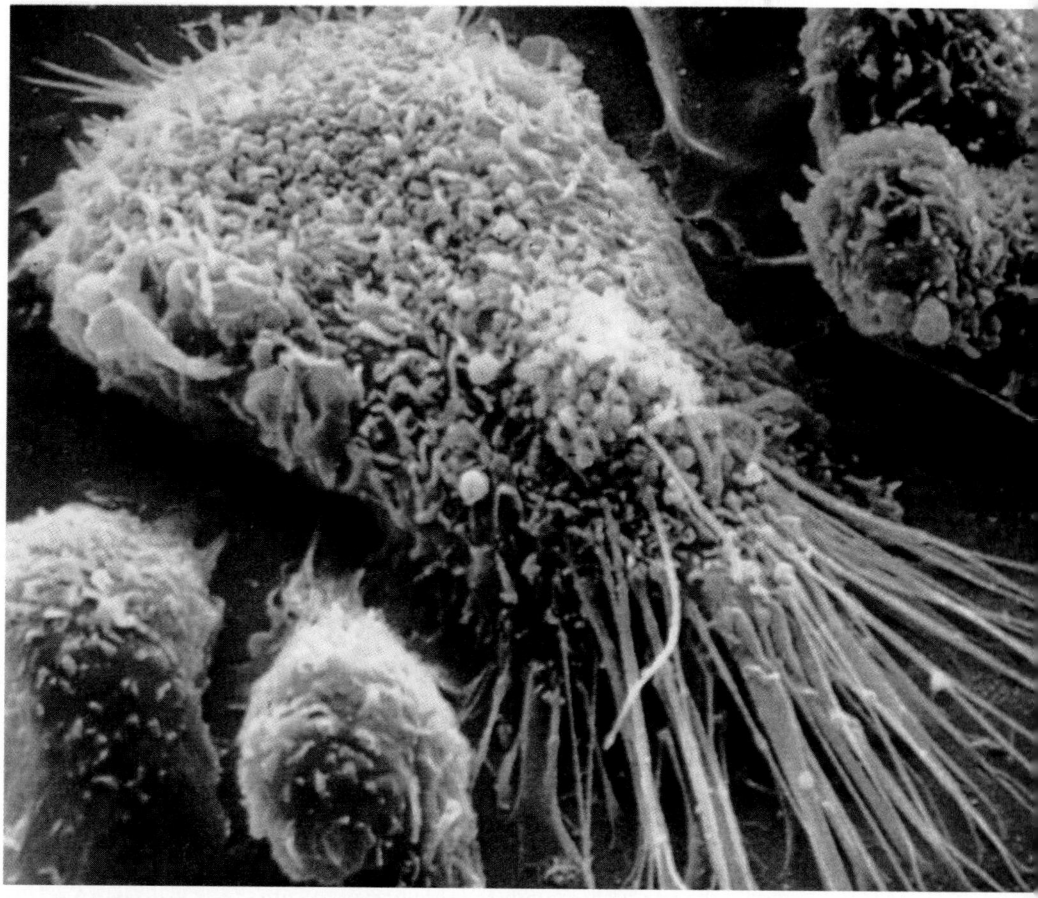

Figure 19.6 A macrophage cell of the immune system.
Here a macrophage reaches out with cytoplasmic extensions to ingest bacterial cells that have infected the bloodstream.

Skeletal Connective Tissue

Three kinds of connective tissue are the principal components of the skeletal system: fibroblasts, cartilage, and bone. All are composed of similar cells and differ mostly in the nature of the material that is laid down between individual cells.

1. **Fibroblasts.** The most common kind of connective tissue in the human body consists of flat, irregularly branching cells called fibroblasts that secrete structurally strong proteins into the spaces between the cells. The many types of proteins give tissues different strengths. The most commonly secreted protein, collagen, is the most abundant protein in the human body: in fact, one-quarter of all the protein in your body is collagen! Fibroblasts are active in wound healing; scar tissue, for example, possesses a collagen matrix.
2. **Cartilage.** In cartilage, the collagen matrix between cells forms in long parallel arrays along the lines of mechanical stress. What results is a firm and flexible tissue of great strength (figure 19.7), just as strands of nylon molecules laid down in long, parallel arrays produce strong, flexible ropes. Cartilage covers the ends of bones that come together in joints, such as the knee, ankle, and elbow.
3. **Bone.** Bone is similar to cartilage, except that the collagen fibers are coated with a calcium phosphate salt, making the tissue rigid (figure 19.8). The structure of bone and the way it is formed are discussed later in this chapter.

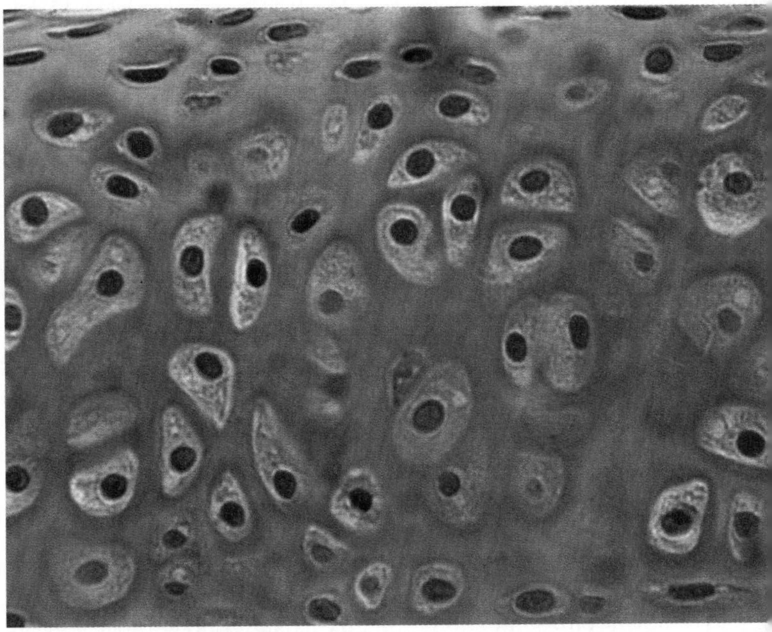

Figure 19.7 Cartilage.
Cartilage forms tissue of great strength and flexibility. Individual fibers are too fine to be seen in this photo.

Storage and Transport Connective Tissue

The third general class of connective tissue is made up of cells that are specialized to accumulate and transport particular molecules. They include the fat-accumulating cells of **adipose tissue** (the tissue that thickens the middle of many overweight people) (figure 19.9). They also include red blood cells, called **erythrocytes** (figure 19.10). About 5 billion erythrocytes are present in every milliliter of your blood. Erythrocytes transport oxygen and carbon dioxide in human blood. They are unusual in that during their maturation they lose most of their organelles, including nucleus, mitochondria, and endoplasmic reticulum. Instead, occupying the interior of each erythrocyte are about 300 million molecules of hemoglobin, the protein that carries oxygen.

The fluid, or **plasma,** in which erythrocytes move is both the "banquet table" and the "refuse heap" of the human body. Practically every substance used by your cells is dissolved in plasma, including inorganic salts like sodium and calcium, body wastes, and food molecules like sugars, lipids, and amino acids. Plasma also contains a wide variety of proteins, including antibodies and particularly albumin, which gives the blood its viscosity. The various types of connective tissue are summarized in table 19.2.

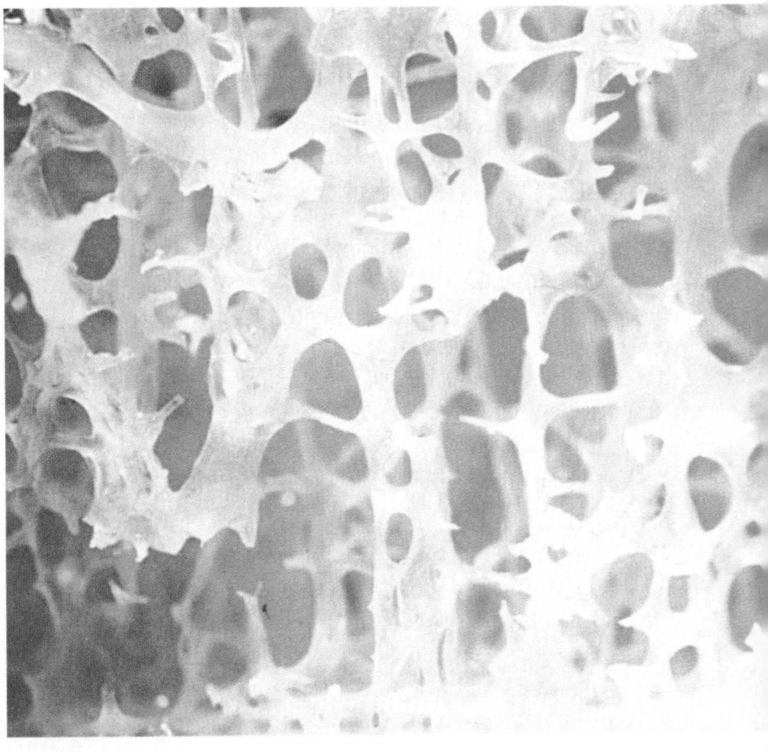

Figure 19.8 The interior of bone.
While we think of bone as hard and solid, the interiors of many bones are composed of a delicate latticework. Bone, like most tissues in your body, is a dynamic structure, constantly renewing itself.

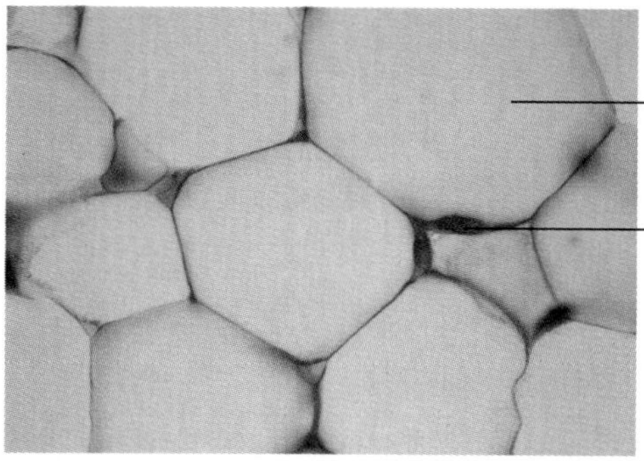

Figure 19.9 Adipose tissue.
Fat is stored in globules of adipose tissue, a type of loose connective tissue. As a person gains or loses weight, the size of the fat globules increases or decreases. A person cannot decrease the number of fat cells by losing weight.

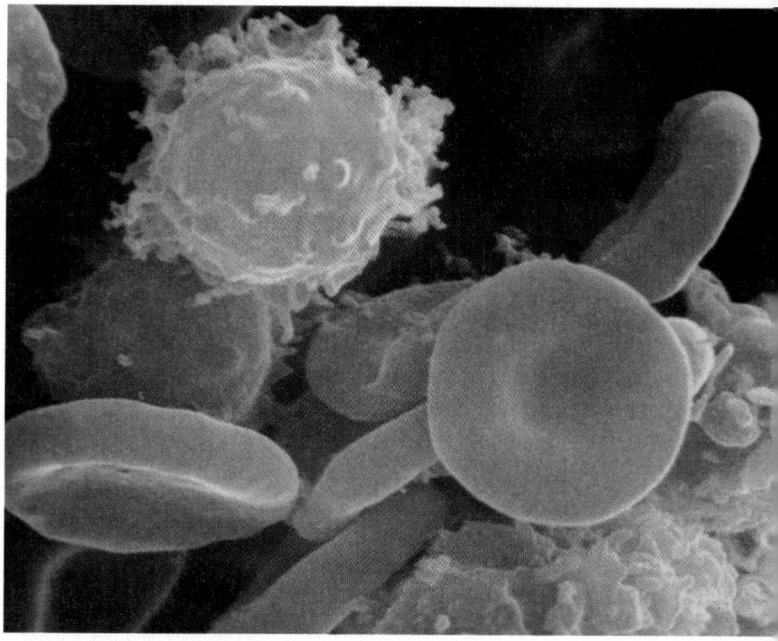

Figure 19.10 Erythrocytes.
Erythrocytes, also called red blood cells, are a form of connective tissue composed of individual cells that are not physically attached to one another. Each red blood cell has the shape of a flattened sphere with a depressed center.

Table 19.2 Connective Tissue

Tissue	Typical Location	Tissue Function	Characteristic Cell Types
Immune System			
White blood cells	Circulatory system	Attack and remove invading microorganisms and virus-infected cells	Macrophages; lymphocytes; mast cells
Skeletal System			
Fibroblasts:			
Loose	Beneath skin and other epithelial tissues	Support; provide a fluid reservoir for epithelium	Fibroblasts
Dense	Tendons; sheath around muscles; kidney; liver; dermis of skin	Provide flexible, strong connections	Fibroblasts
Elastic	Ligaments; large arteries; lung tissue; skin	Enable tissues to expand and then return to normal size	Fibroblasts
Cartilage	Spinal disks; knees and other joints; ear; nose; tracheal rings	Provides flexible support; functions in shock absorption and reduction of friction on load-bearing surfaces	Chondrocytes (specialized fibroblast-like cells)
Bone	Most of skeleton	Protects internal organs; provides rigid support for muscle attachment	Osteocytes (specialized fibroblast-like cells)
Storage and Transport			
Red blood cells	In plasma	Transport oxygen	Red blood cells (erythrocytes)
Adipose tissue	Beneath skin	Stores fat	Specialized fibroblasts (adipocytes)

Muscle Tissue Lets the Body Move

Muscle cells are the workhorses of the human body. The distinguishing characteristic of muscle cells, the thing that makes them unique, is the abundance of contractible protein fibers within them. These fibers, called **microfilaments,** are made of the proteins actin and myosin. All human cells have a fine network of these microfilaments, but muscle cells have many more than other cells. Crammed in like the fibers of a rope, they take up practically the entire volume of the muscle cell. When actin and myosin slide past each other, shortening the fibers, the muscles contract. Like slamming a spring-loaded door, the shortening of all of these fibers together within a muscle cell can produce considerable force. The human body possesses three different kinds of muscle cells: **smooth muscle, skeletal muscle,** and **cardiac muscle** (figure 19.11). In smooth muscle, the microfilaments are only loosely organized. In cardiac and skeletal muscle, the microfilaments are bunched together into fibers called **myofibrils.** Each myofibril contains many thousands of microfilaments, all aligned to provide maximum force when they simultaneously shorten. Cardiac and skeletal muscle are often called striated muscles because the alignment of so many microfilaments gives the muscle myofibril a banded appearance. The different types of muscle tissue are summarized in table 19.3 and later in this chapter.

Table 19.3 Muscle Tissue

Tissue	Typical Location	Characteristic Cell Types
Smooth muscle	Walls of blood vessels, stomach, and intestines	Smooth muscle cells
Cardiac muscle	Walls of heart	Heart muscle cells
Skeletal muscle	Voluntary muscles of the body	Skeletal muscle cells or fibers

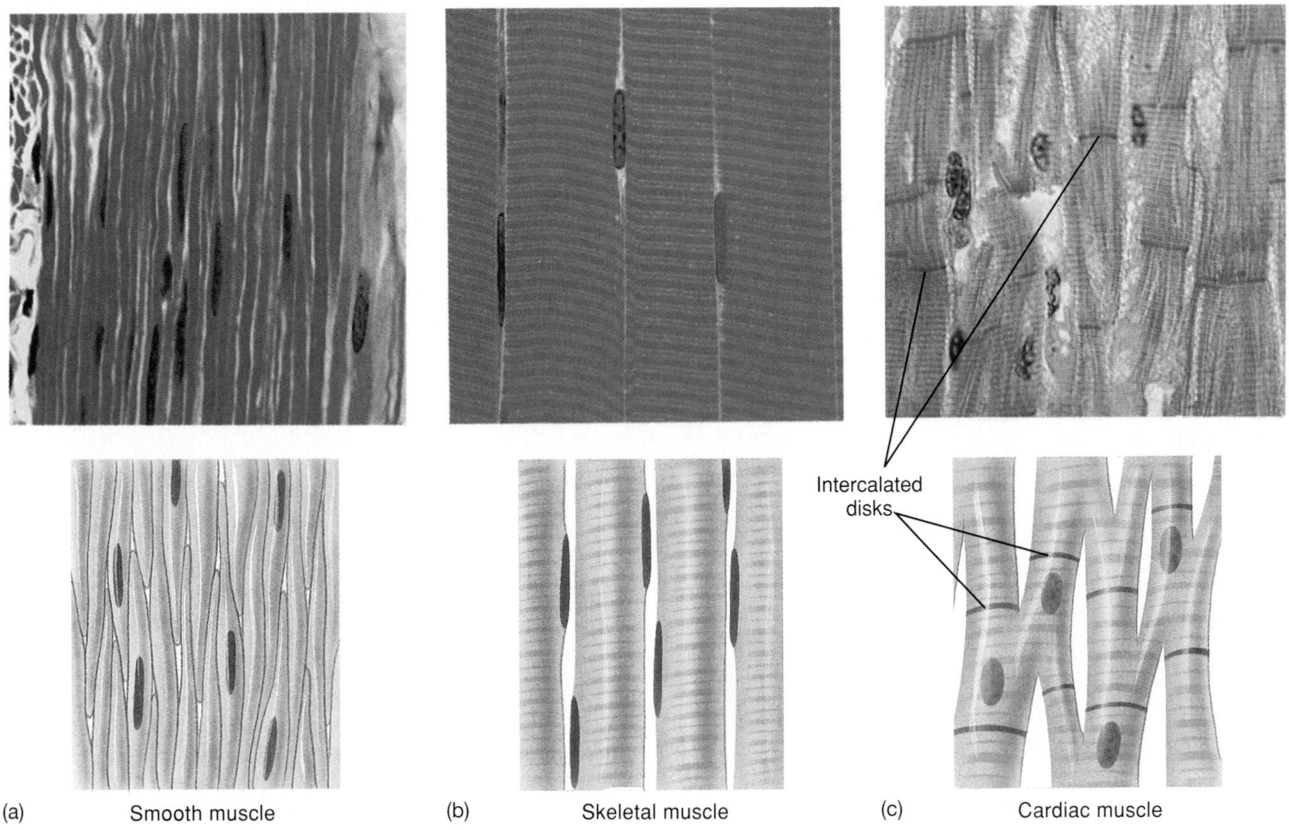

Figure 19.11 Types of muscle.
(*a*) Smooth muscle cells are long and spindle-shaped. (*b*) Skeletal (or striated) muscle cells are formed by the fusion of several muscle cells, end to end, to form a long fiber. (*c*) Cardiac muscle cells are organized into long, branching chains that interconnect, forming a lattice.

Nerve Tissue Conducts Signals Rapidly

Nerve cells carry information rapidly from one organ to another. Nerve tissue, the fourth major class of human tissue, is composed of two kinds of cells: (1) **neurons,** which are specialized for the transmission of nerve impulses, and (2) supporting **glial cells,** which supply the neurons with nutrients, support, and insulation.

Neurons have a highly specialized cell architecture that enables them to conduct signals rapidly throughout the body. Their plasma membranes are rich in ion-selective channels that maintain a voltage difference between the interior and the exterior of the cell, the equivalent of a battery. When ion channels in a local area of the membrane open, ions flood in from the exterior, temporarily wiping out the charge difference. This process, called depolarization, tends to open nearby voltage-sensitive channels in the neuron membrane, resulting in a wave of electrical activity that travels down the entire length of the neuron as a nerve impulse. This process is described in detail in chapter 23.

Each neuron is composed of three parts: (1) a **cell body,** which contains the nucleus; (2) threadlike extensions called **dendrites,** which act as antennae, bringing nerve impulses to the cell body from other cells or sensory systems; and (3) a single, long extension called an **axon,** which carries nerve impulses away from the cell body, often for considerable distances (figure 19.12). For example, a single neuron that activates the muscles in your thumb may have its cell body in your spinal cord and an axon that extends all the way across your shoulder and down your arm to your thumb. Single neuron axons more than a meter long are common.

The body contains many different kinds of neurons. Some neurons are tiny and have only a few projections, others are bushy and have more projections, and still others have extensions that are meters long. Neurons are not normally in direct contact with one another. Instead, a tiny gap called a **synapse** separates them. Neurons communicate with other neurons by passing chemical signals called **neurotransmitters** across the gap. Special receptor proteins present on the far side of the synapse respond to the arrival of the neurotransmitters by starting a new nerve impulse in the receiving neuron.

The nerves of the human body appear as fine white threads when viewed with the naked eye, but they are actually composed of bundles of axons. Like a telephone trunk cable, nerves include large numbers of independent communication channels—bundles composed of hundreds of axons, each connecting a nerve cell to a muscle fiber. In addition, the nerve contains numerous supporting glial cells bunched around the axons. It is important not to confuse a nerve with a neuron. A nerve is made up of the axons of many neurons, just as a cable is made of many wires.

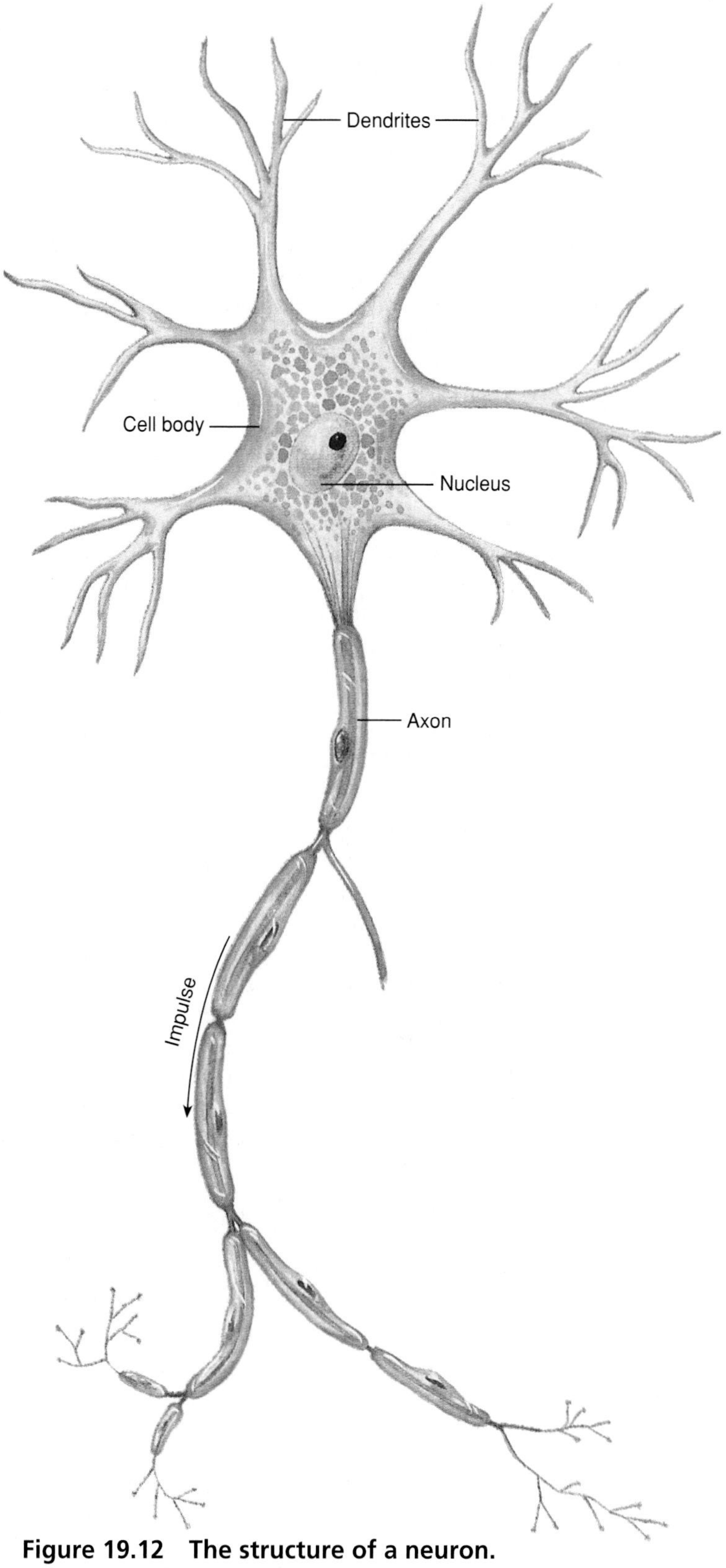

Figure 19.12 The structure of a neuron.
A nerve impulse is received by the dendrites and then passed to the cell body and out through the axon to a muscle fiber or another neuron.

19.3 The Human Skeleton

If you've ever tried to lift a heavy suitcase or push a car, you are familiar with the basic problem posed by movement: working against gravity. The reason you cannot toss a car into the air with your little finger is because gravity is pulling down on the car far harder than your finger can push up. To move the car, you have to lift with a greater force than gravity is exerting. All motion must meet this requirement. Your body uses the chemical energy of ATP to supply that force, using it to alter the length of myofibrils within muscle cells, causing the cells to shorten. When many muscle cells shorten all at once, they can exert great force (figure 19.13).

Figure 19.13 Energy drives muscle contraction. A weight lifter overcomes the force of gravity when lifting heavy weights through the simultaneous contraction of numerous muscle cells.

The Need for a Skeleton

With muscles alone, your body could not move—it would simply pulsate as its muscles contracted and relaxed in futile cycles. For a muscle to produce movement, it must direct its force against another object. Humans and other vertebrates are able to move because the opposite ends of their muscles are attached to an internal scaffold of bone, a **skeleton,** so that the muscles have something to pull against.

Organization of the Skeleton

The human skeleton is made up of 206 individual bones. If you saw them as a pile of bones jumbled together, it would be hard to make any sense of them. To understand the skeleton, it is necessary to group the 206 bones according to their function and position in the body. The 80 bones of the **axial skeleton** support the main body axis, while the remaining 126 bones of the **appendicular skeleton** support the arms and legs (figure 19.14). These two skeletons function more or less independently—that is, the muscles controlling the axial skeleton (postural muscles) are managed by the brain separately from those controlling the appendages (manipulatory muscles).

The Axial Skeleton

The axial skeleton is made up of the skull, backbone, and rib cage. Of the skull's 28 bones, only 8 form the cranium, which encases the brain; the rest are facial bones and middle ear bones. An additional bone, the hyoid bone, supports the tongue but is not really part of the skull.

The skull is attached to the upper end of the backbone, which is also called the **spine,** or vertebral column. The spine is made up of 26 vertebrae, stacked one on top of the other to provide a flexible column surrounding and protecting the spinal cord. Curving forward from the vertebrae are 12 pairs of ribs, attached at the front to the breastbone, or sternum, and forming a protective cage around the heart and lungs.

The Appendicular Skeleton

The 126 bones of the appendicular skeleton are attached to the axial skeleton at the shoulders and hips. The shoulder, or **pectoral girdle,** is composed of two large, flat shoulder blades, each connected to the top of the breastbone by a slender, curved collarbone. The arms are attached to the pectoral girdle; each arm and hand contains 30 bones. The collarbone is the most frequently broken bone of the body. Can you guess why? Because if you fall on an outstretched arm, a large component of the force is transmitted to the collarbone.

The **pelvic girdle** forms a bowl that provides strong connections for the legs, which must bear the weight of the body. Each leg and foot contains a total of 30 bones. The bones of the adult human skeleton are summarized in table 19.4.

Table 19.4 Bones of the Adult Skeleton

Axial Skeleton	
Skull (8 cranial bones, 14 facial bones, and 6 middle ear bones)	28 bones
Hyoid	1 bone
Vertebral column	26 bones
Thoracic cage (24 ribs, 1 sternum)	25 bones
Appendicular Skeleton	
Pectoral girdle	4 bones
Upper limbs	60 bones
Pelvic girdle	2 bones
Lower limbs	60 bones
Total	206 bones

Figure 19.14
Axial and appendicular skeletons.
The axial skeleton is shown in the darker shade, and the appendicular skeleton is shown lighter.

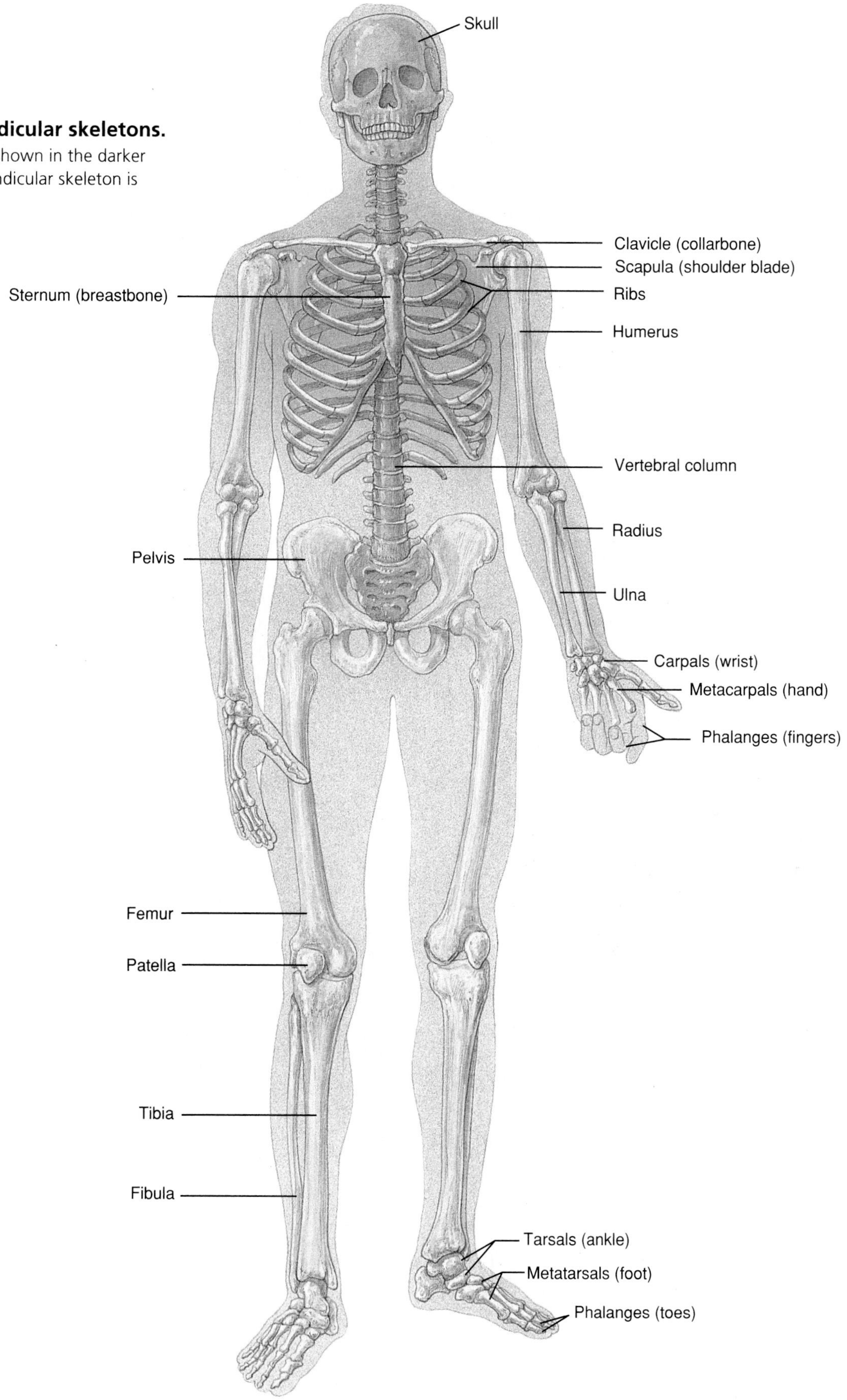

The Skeleton Is Made of Bone

The human skeleton is strong because of the structural nature of bone. Bone is produced by coating collagen fibers with a calcium phosphate salt, making a material that is strong without being brittle. To understand how coating collagen fibers with calcium salts makes such an ideal structural material, consider fiberglass. Fiberglass is composed of glass fibers embedded in epoxy glue. The individual fibers are rigid, giving great strength, but they are also brittle. The epoxy glue, on the other hand, is flexible but weak. The composite, fiberglass, is both rigid and strong because when stress causes an individual fiber to break, the crack runs into glue before it reaches another fiber. The glue distorts and reduces the concentration of the stress—in effect, the glue spreads the stress over many fibers.

The construction of bone is similar to that of fiberglass: small, needle-shaped crystals of a calcium phosphate mineral, hydroxyapatite, surround and impregnate collagen fibrils within bone. No crack can penetrate far into bone because any stress that breaks a hard hydroxyapatite crystal passes into the collagenous matrix, which dissipates the stress. The hydroxyapatite mineral provides rigidity, whereas the collagen "glue" provides flexibility.

Most of us think of bones as solid and rocklike. But actually, bone is a dynamic tissue that is constantly being reconstructed throughout your life. The outer layer of bone is very dense and compact and so is called **compact bone.** The interior is less compact, with a more open lattice structure, and is called **spongy bone.** Red blood cells form in the red marrow of spongy bone (figure 19.15). New bone is formed

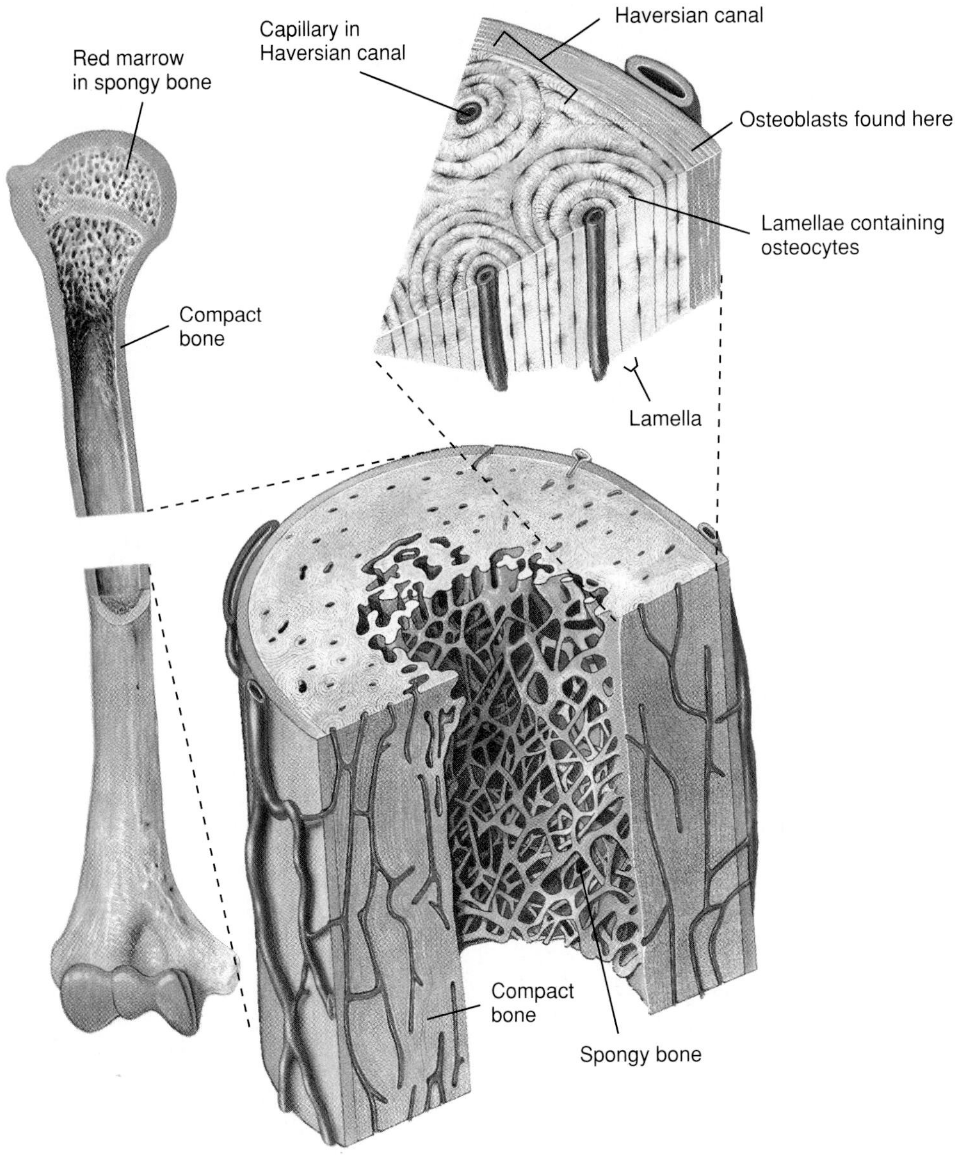

Figure 19.15 The structure of bone.
Some parts of bones are dense and compact, giving the bone strength. Other parts are spongy, with a more open lattice; it is in the red marrow that most red blood cells are formed.

in two stages: First, collagen is secreted by cells called **osteoblasts,** which lay down a matrix of fibrils along lines of stress. Then calcium minerals impregnate the fibrils. Bone is laid down in thin, concentric layers, like layers of paint on an old pipe. The layers form as a series of tubes around a narrow central channel called a **Haversian canal** (figure 19.16), which runs parallel to the length of the bone. The many Haversian canals within a bone, all interconnected, contain blood vessels and nerves that provide a lifeline to its living, bone-forming cells.

When bone is first formed in the embryo, osteoblasts use the cartilage skeleton as a template for bone formation. Later, new bone is formed along lines of stress. That is why long-distance runners must slowly increase the distances they attempt, to allow their bones to strengthen along lines of stress; otherwise, stress fractures can cripple them.

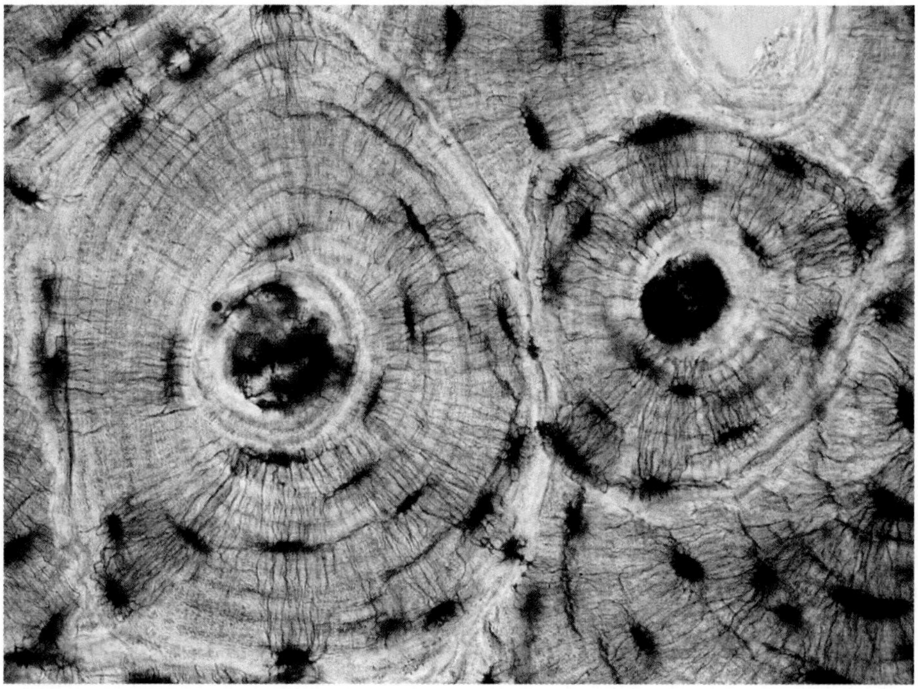

Figure 19.16 Compact bone.
The large, circular structures are Haversian canals.

Tendons Connect Bone to Muscle

Muscles are attached to bones by straps of dense connective tissue called **tendons.** Bones pivot about flexible connections called **joints,** pulled back and forth by the muscles attached to them. Each muscle pulls on a specific bone. One end of the muscle, the *origin,* is attached by a tendon to a bone that remains stationary during a contraction. This provides an object against which the muscle can pull. The other end of the muscle, the *insertion,* is attached to a bone that moves if the muscle contracts.

Muscles can only pull, not push, because their myofibrils contract rather than expand. For this reason, the muscles in your movable joints are attached in opposing pairs, called flexors and extensors. When the **flexor** muscle of your leg contracts, the lower leg is moved closer to the thigh. When the **extensor** muscle of your leg contracts, the lower leg is moved in the opposite direction, farther away (figure 19.17).

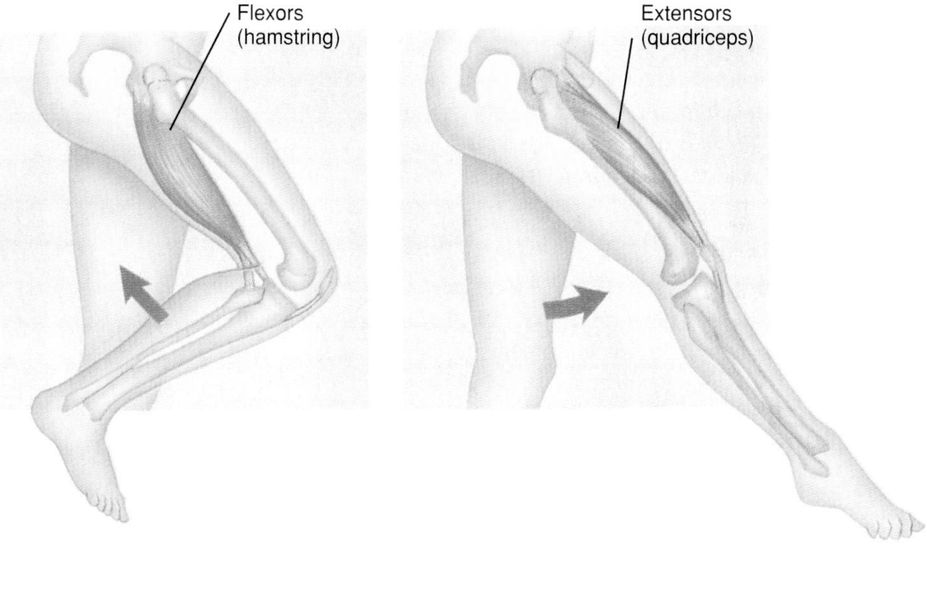

Figure 19.17 Flexor and extensor muscles.
Limb movement is always the result of muscle contraction, never muscle extension. Muscles that retract limbs are called flexors; those that extend limbs are called extensors. Thus, you bend your leg back by contracting your hamstring muscle, a flexor muscle, and you extend your leg by contracting your quadriceps, an extensor muscle.

19.4 The Human Muscular System

Muscle cells are the workhorses of the human body. The key property of muscle cells is the relative abundance of the protein filaments actin and myosin within them, which enable a muscle cell to contract. These protein filaments are present as part of the cytoskeleton of all eukaryotic cells, but they are far more abundant in muscle cells.

Kinds of Muscle

Food moves through your intestines because of the rhythmic contractions of smooth muscle. Your body is able to move because your skeletal muscles pull your bones with considerable force. Your heart pumps because of the contraction of another kind of muscle, cardiac muscle. These three kinds of muscle together form the human muscular system.

Smooth Muscle

Smooth muscle cells are long and spindle-shaped, each containing a single nucleus. The interiors of smooth muscle cells are packed with **myofibrils** each composed of thousands of actin and myosin microfilaments, called **myofilaments.** However, the individual myofilaments are not aligned into orderly assemblies as they are in skeletal and cardiac muscles. Smooth muscle tissue is organized into sheets of cells. In some tissues, smooth muscle cells contract only when they are stimulated by a nerve or hormone. Examples are the muscles that line the walls of many of your blood vessels and those that make up the iris of your eye. In other smooth muscle tissue, such as that found in the wall of the gut, the individual cells contract spontaneously, leading to a slow, steady contraction of the tissue.

Skeletal Muscle

Skeletal muscles move the bones of the skeleton. Some of the major muscles are shown in figure 19.18. Skeletal muscle cells are produced during development by the fusion of several cells at their ends to form a very long fiber. Each of these muscle cell fibers still contains all the original nuclei, pushed out to the periphery of the cytoplasm. Each skeletal muscle is a tissue made up of numerous individual muscle cell fibers that act as a unit.

Skeletal muscles are specialized for rapid, strong contraction. In contrast, smooth muscles are specialized for slow, maintained contraction. Imagine a large raft being towed upstream by many small canoes, with each canoe bound to the raft by its own towline. This is analogous to the contraction of smooth muscles—each muscle cell participates individually in the contraction of the muscle. Now imagine placing all the rowers in one galley where they row in concert, pulling the raft far more effectively. This is analogous to the contraction of skeletal muscle, in which numerous muscle cells pool their resources.

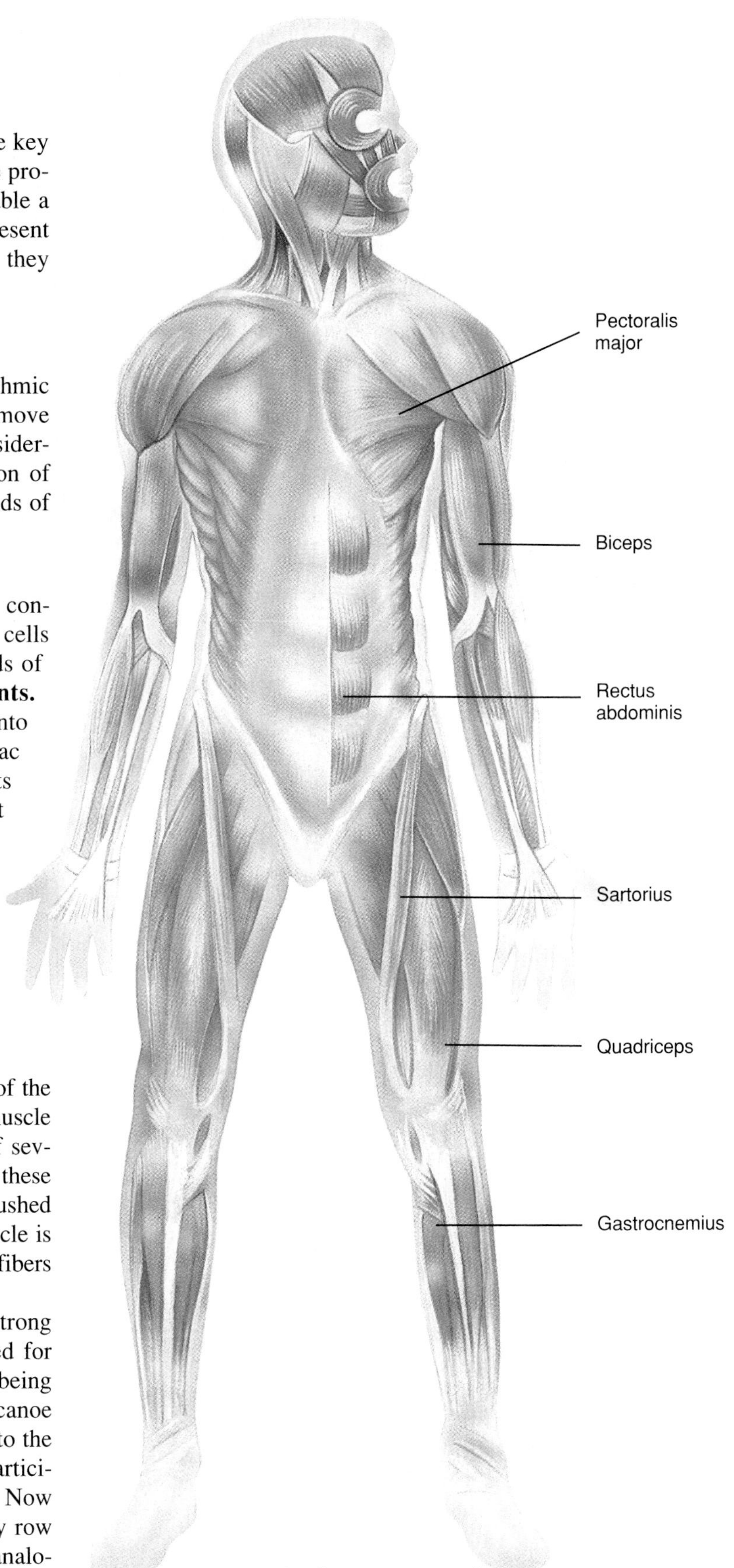

Figure 19.18 The human muscular system. The major muscles in the human body are labeled.

Cardiac Muscle

Your heart is composed of striated muscle fibers arranged very differently from the fibers of skeletal muscle. Instead of very long, multinucleate cells running the length of the muscle, heart muscle is composed of chains of single cells, each with its own nucleus. Chains of cells are organized into fibers that branch and interconnect, forming a latticework. This lattice structure is critical to the way heart muscle functions. Each heart muscle cell is coupled to its neighbors electrically by tiny holes called *gap junctions* that pierce the plasma membranes in regions where the cells touch each other. Heart contraction is initiated at one location by the opening of transmembrane channels that depolarize the membrane; a wave of electrical depolarization then passes from cell to cell across the gap junctions, causing the heart to contract in an orderly pulsation.

How Muscles Work

Far too fine to see with the naked eye, the individual myofilaments of your muscles are only 6 nanometers thick. Each is composed of long, threadlike filaments of the proteins actin and myosin. An **actin filament** consists of two strings of actin molecules wrapped around one another, like two strands of pearls loosely wound together. A **myosin filament** is also composed of two strings of protein wound about each other, but a myosin filament is 10 times longer than an actin filament, and the myosin strings have a very unusual shape. One end of a myosin filament consists of a very long rod, while the other end consists of a double-headed globular region, or "head." In electron micrographs, a myosin filament looks like a two-headed snake. This odd structure is the key to how muscles work.

How Myofilaments Contract

Look at the diagram of myosin and actin in a myofilament (figure 19.19), and focus on the myosin heads. When a myofilament contracts, the heads of the myosin filaments move first. Like flexing your hand downward at the wrist, the heads bend backward and inward. This moves them closer to their rodlike backbones and several nanometers in the direction of the flex. In itself, this myosin head-flex accomplishes nothing—but the myosin head is attached to the actin filament! As a result, the actin filament is pulled along with the myosin head as it flexes, causing the actin filament to slide by the myosin filament in the direction of the flex. As one after another myosin head flexes, the myosin in effect "walks" step by step along the actin. Each step uses a molecule of ATP to recock the myosin head before each flex (figure 19.20).

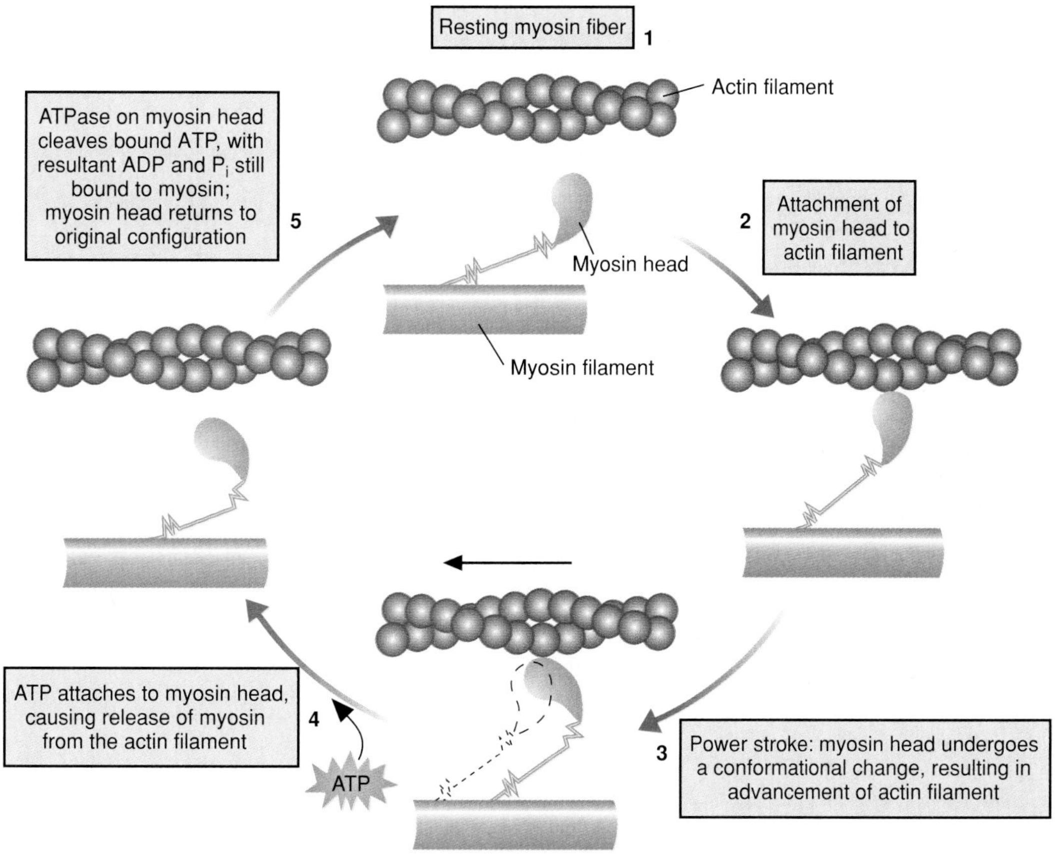

Figure 19.19 The mechanisms of myofilament contraction.
Actin is pulled along myosin (to the *left* in this diagram). The myosin heads bind to the actin filament and undergo a conformational change in shape, which causes the actin filament to be "dragged" along the myosin filament. An ATP molecule attaches to the myosin head, which recocks the mechanism, returning the myosin head to its extended position.

How does this sliding of actin past myosin lead to myofilament contraction and muscle cell movement? Within each myofilament, the actin is anchored at one end, at a position in striated muscle called the Z line (figure 19.21). Because it is tethered like this, the actin cannot simply move off, any more than a chained dog can run down the street. Instead, the actin pulls the anchor with it! As actin moves past myosin, it drags the Z line toward the myosin. The secret of muscle contraction is that each myosin is interposed between two pairs of actin filaments—the myofilament is attached at *both* ends to Z lines by actin. One moving to the left and the other to the right, the two pairs of actin molecules drag the Z lines toward each other as they slide past the myosin core. As the Z lines are pulled closer together, the cell membranes to which they are attached move toward one another, and the cell contracts.

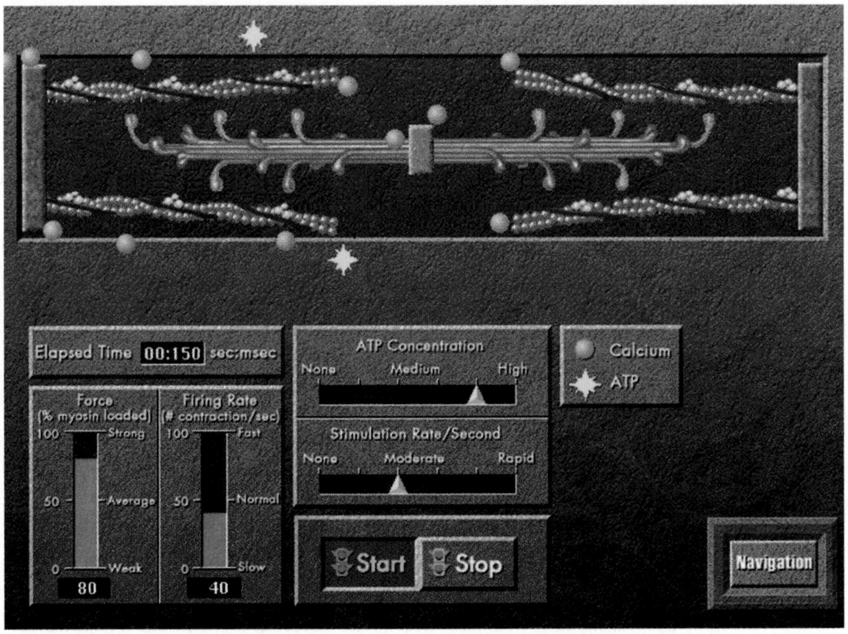

Figure 19.20 Muscle contraction requires ATP.
This is a screen capture from an interactive CD-ROM that explores how the availability of ATP influences the frequency and strength of muscle contraction. (*Explorations in Human Biology,* Module 4, " Muscle Contraction")

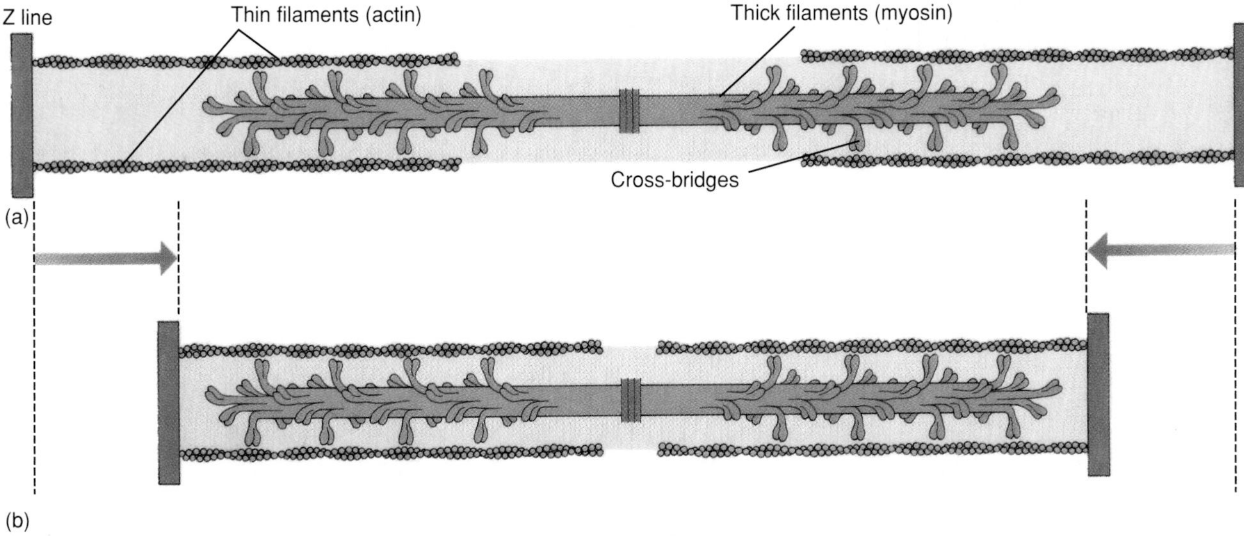

Figure 19.21 The interaction of actin and myosin filaments.
The heads on the two ends of the myosin filament are oriented in opposite directions. Thus, as the right-hand end of the myosin filament "walks" along the actin filaments, pulling them and their attached Z line leftward toward the center, the left-hand end of the same myosin filament "walks" along the actin filaments, pulling them and their attached Z line rightward toward the center. The result is that both Z lines move toward the center—and contraction occurs.

How Nerves Signal Muscles to Contract

In vertebrate skeletal muscle, contraction is initiated by a nerve impulse. The nerve fiber is embedded in the surface of the muscle fiber, forming a **neuromuscular junction.** When a signal reaches the end of a neuron, at the point where the neuron almost touches the muscle cell (the **motor end plate**), the neuron releases the chemical acetylcholine into the tiny gap separating neuron from muscle (figure 19.22). The acetylcholine passes across the gap to the muscle cell membrane, binds to receptor proteins there, and so causes ion channels in the motor end plate to open. The muscle membrane is said to be depolarized, because opening these channels allows ions to move freely in and out, wiping out any differences there might be in ion concentration between inside and outside.

How does depolarization of the muscle membrane cause the muscle cell to contract? The endoplasmic reticulum of the skeletal muscle cell, which is called the **sarcoplasmic reticulum,** wraps around each myofibril like a sleeve (illustrated on the next page in figure 19.23). Within it are embedded numerous protein channels that permit calcium ions to move across the membrane. In resting muscle, calcium ions are actively pumped out of the myofibril's cytoplasmic space, out across the sarcoplasmic reticulum, by an ATP-driven ion pump. This concentrates all the calcium ions of the resting cell within the intracellular space of the sarcoplasmic reticulum system. When the muscle membrane is depolarized, the wave of depolarization passes to the sarcoplasmic reticulum, opening calcium channels and releasing concentrated calcium into the myofibril's cytoplasmic space.

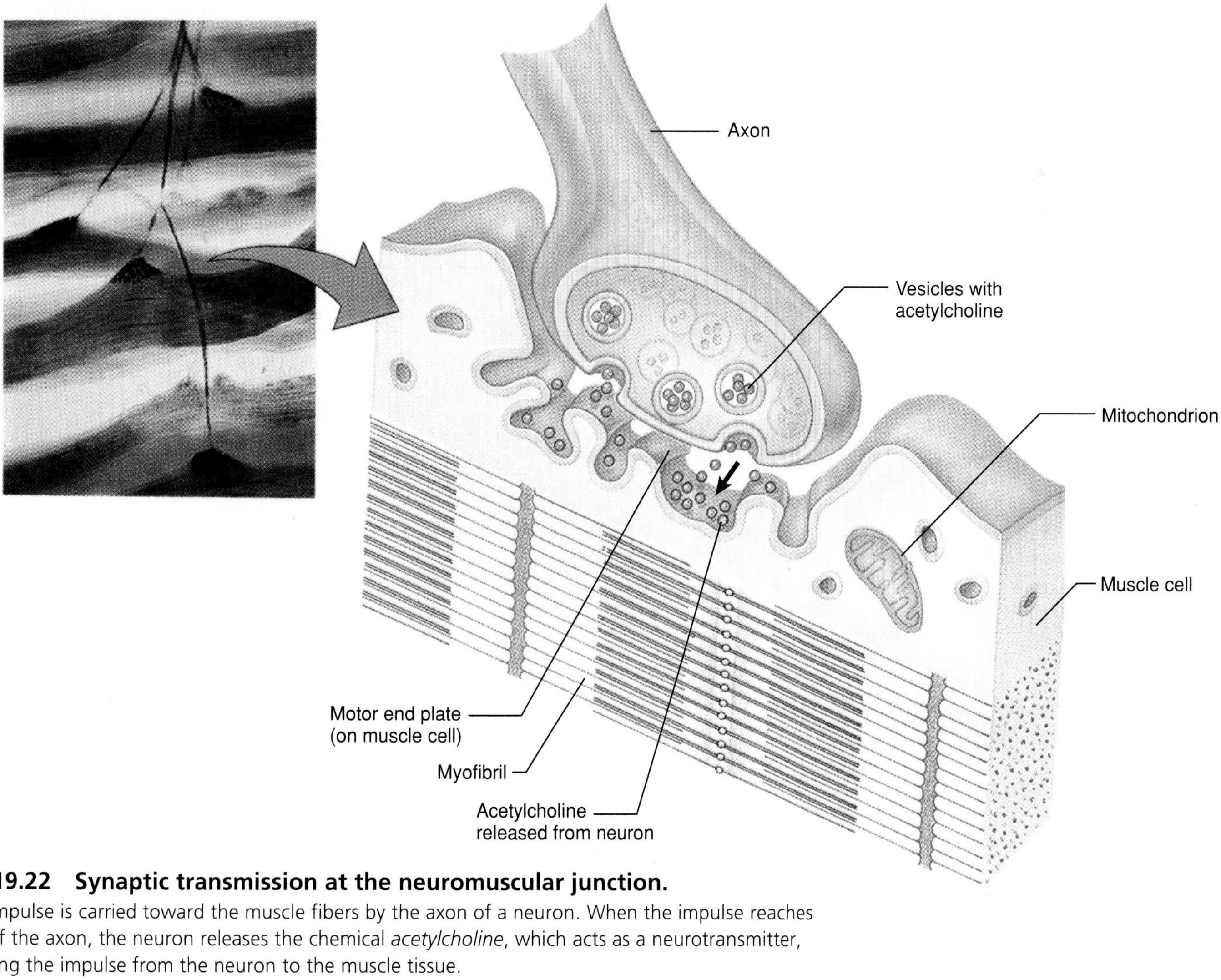

Figure 19.22 Synaptic transmission at the neuromuscular junction.
A nerve impulse is carried toward the muscle fibers by the axon of a neuron. When the impulse reaches the end of the axon, the neuron releases the chemical *acetylcholine*, which acts as a neurotransmitter, transmitting the impulse from the neuron to the muscle tissue.

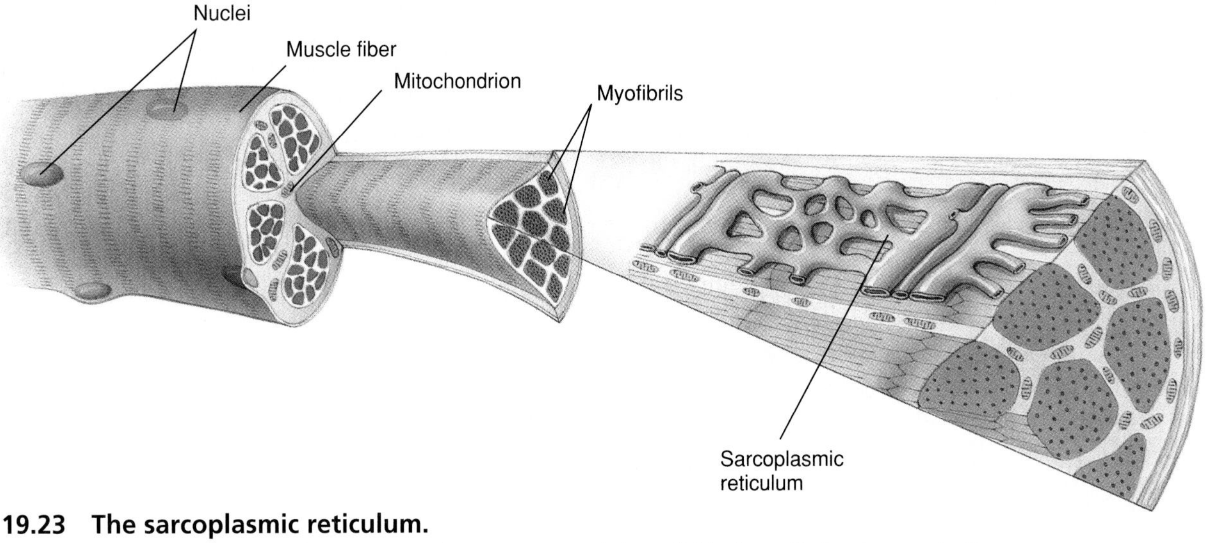

Figure 19.23 The sarcoplasmic reticulum.
The sarcoplasmic reticulum is a system of membranes that wraps around the individual myofibrils of a muscle fiber.

Calcium acts as a trigger to initiate contraction of the myofibril. Here is how:

1. In resting muscle, myosin filament heads are not able to interact with actin, because the sites on actin where the myosin heads must make contact are covered by the protein tropomyosin.
2. Calcium causes the tropomyosin to reposition to a new location, where it does not cover the myosin binding sites on actin (figure 19.24).
3. When this repositioning has occurred, the myosin heads attach to actin and, using ATP energy, move along the actin in a stepwise fashion to shorten the myofibril.

Thus, the release of calcium by the nerve's stimulation of the sarcoplasmic reticulum acts as a switch, triggering the contraction of the myofibril.

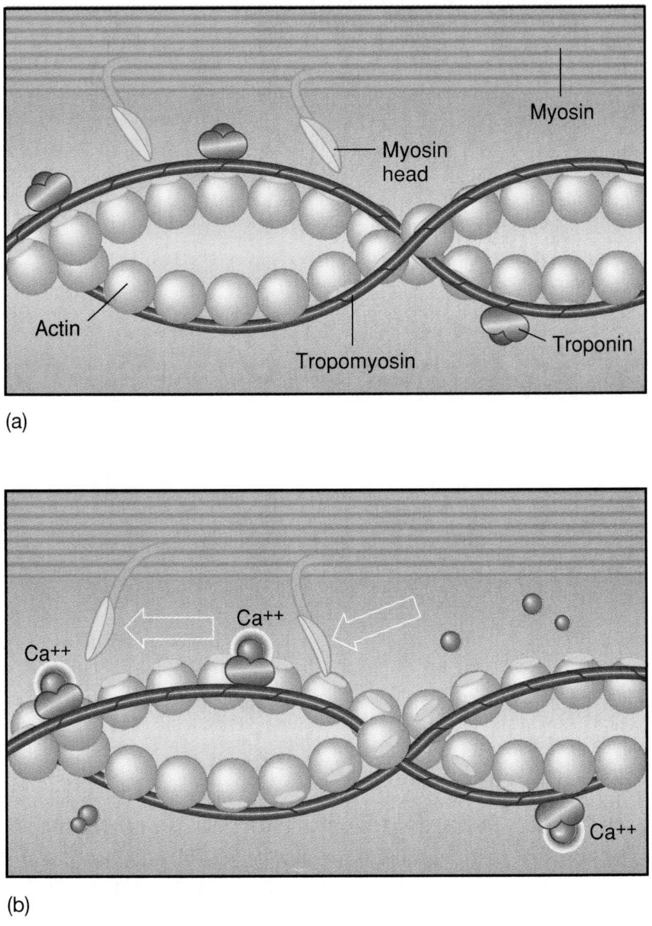

Figure 19.24 How calcium controls muscle contraction.
(*a*) When the muscle is at rest, a long filament composed of the molecule tropomyosin blocks the myosin binding sites of the actin molecule. Without actin's ability to form links with myosin at these sites, muscle contraction cannot occur. (*b*) When calcium ions bind to another protein, troponin, the resulting complex displaces the filament of tropomyosin, exposing the myosin binding sites of actin. Cross-links form between actin and myosin, and contraction occurs.

CHAPTER 19

HIGHLIGHTS

	Key Terms	Key Concepts
19.1 How the Body Is Organized 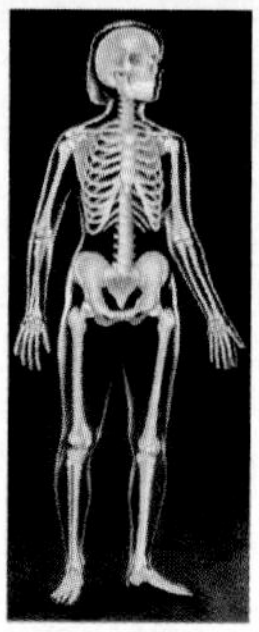	tissues 410 organs 410 organ system 411	• The human body is organized like that of all vertebrates in general structure. • The human body's 100 trillion cells are organized into tissues, which are the actual structural and functional units of the body.
19.2 The Principal Tissues 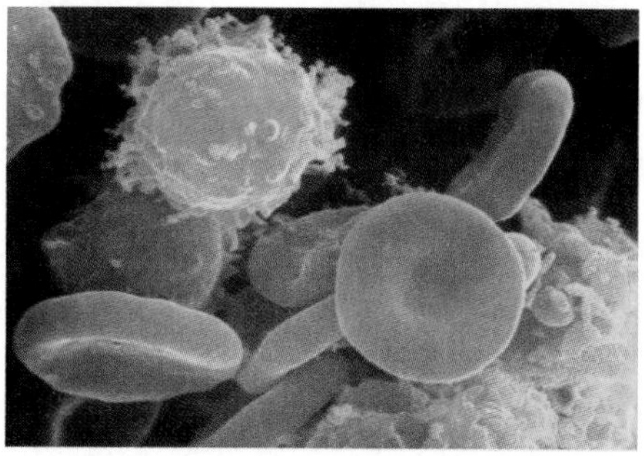	epithelium 414 connective tissue 415 fibroblasts 416 bone 416 adipose tissue 416 neurons 419	• Adult tissues are grouped into four general classes: epithelial, connective, nervous, and muscle tissue. • The outermost of the principal tissues, epithelium, protects the tissues beneath it from dehydration. • Connective tissue supports the body structurally, defends it with the immune system, and transfers materials via the blood.
19.3 The Human Skeleton 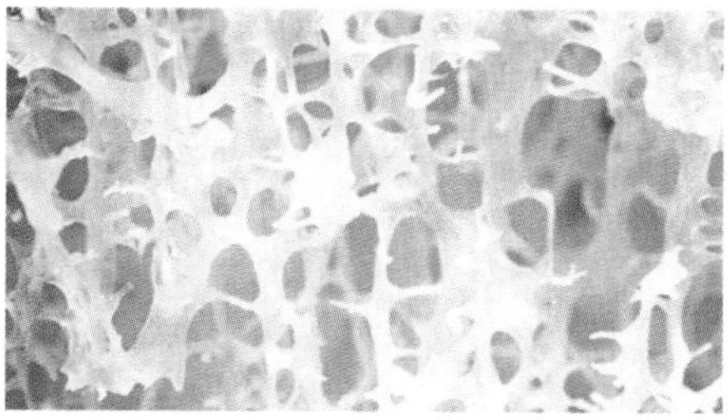	axial skeleton 420 appendicular skeleton 420 Haversian canal 423 tendons 423	• The skeleton protects the internal organs of the body and provides a strong and rigid base against which muscles can pull. • Bone is a dynamic tissue, constantly growing and renewing itself.
19.4 The Human Muscular System 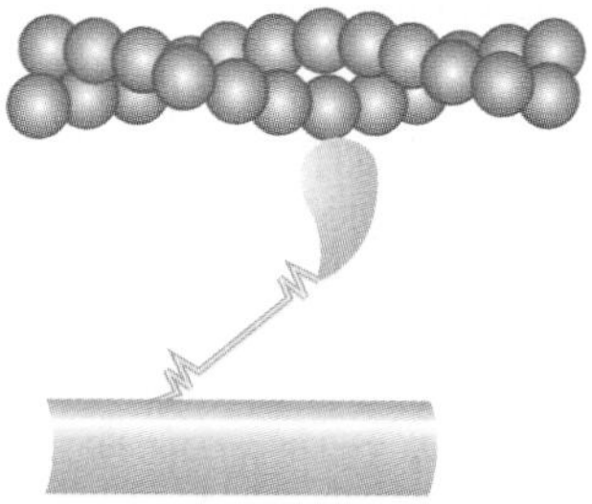	myofibrils 424 actin filament 425 myosin filament 425 motor end plate 427 sarcoplasmic reticulum 427	• Muscle cells do the actual work of movement. • Muscle cells contain large amounts of the protein filaments actin and myosin. • Muscle cells contract when myosin "walks" along the actin filament, driven by the cleavage of ATP.

19

CONCEPT REVIEW

1. An organ consists of several types of
   a. organ systems.
   b. organisms.
   c. tissues.
   d. organs.

2. Which of the following is *not* a principal organ system of the human body?
   a. circulatory system
   b. urinary system
   c. muscular system
   d. lymphatic system
   e. secretory system

3. Each of the following is a function of epithelial tissue except
   a. movement.
   b. protection.
   c. sensory ability.
   d. secretion.

4. Which of the following is *not* a cell type of the connective tissue?
   a. lymphocytes
   b. fat cells
   c. columnar cells
   d. erythrocytes

5. The erythrocyte
   a. fights infection.
   b. carries oxygen.
   c. coagulates the blood.
   d. dissolves substances.

6. Each of the following is a type of muscle tissue except
   a. skeletal.
   b. smooth.
   c. cardiac.
   d. glial.

7. The impulse within a nerve cell travels from
   a. dendrites to axon to cell body.
   b. dendrites to cell body to axon.
   c. cell body to axon to dendrites.
   d. cell body to dendrites to axon.

8. The vertebral column belongs to the _________ skeleton.
   a. appendicular
   b. axial

9. Select the *incorrect* statement about human bone tissue.
   a. Osteoblasts secrete collagen.
   b. The Haversian canal contains blood vessels.
   c. Bone is a type of connective tissue.
   d. The interior of bones is hollow.

10. Select the *incorrect* statement about cardiac muscle.
    a. It is multinucleated.
    b. It composes the heart.
    c. The fibers are striated.
    d. It is composed of chains of cells.

11. Calcium in the muscle cell attaches to
    a. the neurotransmitter.
    b. actin.
    c. myosin.
    d. troponin.

12. The human body consists of about 100 _________ cells.

13. Antibodies are part of the _________ system.

14. _________ tissue covers the surfaces of the body.

15. Neurons have two kinds of projections from the cell body, the _________ and the _________.

16. The human body consists of _________ individual bones.

17. Among myofilaments, the protein _________ moves past myosin during muscle contraction.

18. When a nerve signal reaches the neuromuscular junction, the chemical _________ is released into the gap between the end of the neuron and the muscle tissue.

19. The source of energy for muscle contraction is _________.

Answers to the Concept Review questions appear in Appendix B.

CHALLENGE YOURSELF

1. Chemotherapy is frequently used as a cancer treatment, designed to kill the rapidly dividing malignant cells. Why do you think this treatment typically causes patients to lose their hair and often interferes with the function of their gastrointestinal tract?

2. Is most of bone tissue a living or nonliving substance? Why?

3. In vertebrates, contraction of cardiac muscle in the heart and smooth muscle in the stomach and intestines is initiated by the muscles themselves, independent of the nervous system; in contrast, contraction of the skeletal muscles controlling movements of the jaw and fingers, for example, is initiated by impulses from the nervous system. What do you think are the advantages of having different mechanisms for initiating contractions in muscles such as these?

FOR FURTHER READING

Alexander, R. M. *Bones: The Unity of Form and Function.* New York: Macmillan, 1994. A beautiful overview, including some nice color photographs and illustrations.

Caplan, A. "Cartilage." *Scientific American,* October 1984, 84–97. An interesting account of the many roles played by cartilage in the vertebrate body.

Cohen, C. "The Protein Switch of Muscle Contraction." *Scientific American,* November 1975, 36–45. How proteins associated with myofilaments interact with Ca^{++} ions to trigger contraction.

Key, G. *Comparative Anatomy of the Vertebrates,* 7th ed. St. Louis: Times Mirror/Mosby, 1992. A very readable introduction to the comparative anatomy of the vertebrates.

National Geographic Society. *The Incredible Machine.* Washington, D.C.: National Geographic Society, 1986. A series of outstanding articles on the human body, focusing on its major organ systems. Beautifully illustrated and fun to read.

Rogers, M. "The Nature of Muscles." *National Wildlife,* October/November 1990, 34–41. How efficient are the muscles of larger animals compared to smaller ones? You may be surprised!

Rosenfeld, A. "There's More to Skin Than Meets the Eye." *Smithsonian,* May 1988, 159–80. An entertaining and informative account of the many tasks carried out by the body's largest organ.

"The Skin." *American Health.* January/February 1994, 86–87. A nice overview article about the skin, focusing on its many discrete functions. Includes clear illustrations.

TECHNOLOGY LINKS

The Living World Home Page
http://www.wcbp.com/biology/tlw

Explorations in Human Biology CD
#4 Muscle Contraction

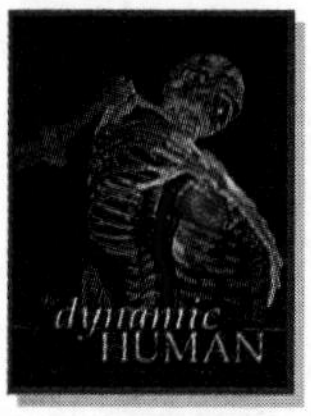

The Dynamic Human CD
Anatomical Orientation
Skeletal System
Muscular System

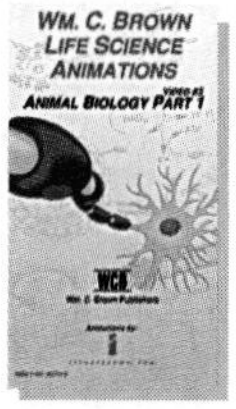

Life Science Animations Videotape 3
#29 Levels of Muscle Structure
#30 Sliding Filament Model of Muscle Contraction
#31 Regulation of Muscle Contraction

20

Circulation and Respiration

The real reason dinosaurs became extinct

CHAPTER OUTLINE

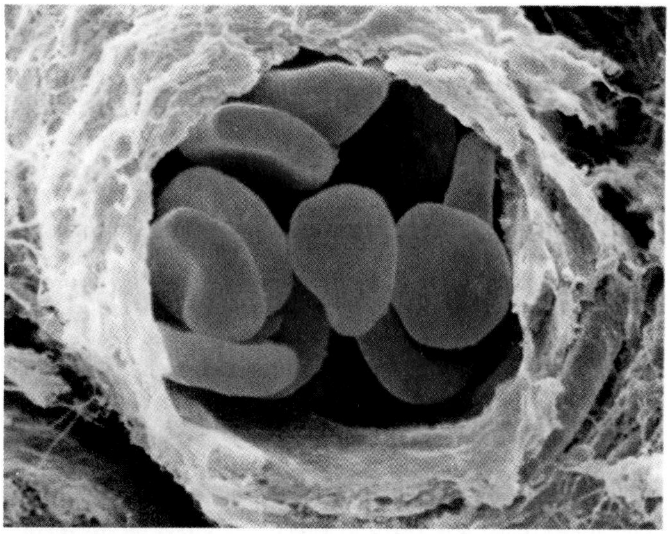

Figure 20.1 A blood vessel.
This ruptured tube is a blood vessel full of red blood cells, which move through the blood vessels, transporting oxygen and carbon dioxide from one place to another.

Every cell in your body must acquire the energy it needs for living from other molecules. Like residents of a city whose food is imported from farms in the countryside, the cells of your body need trucks to carry the food, highways for the trucks to travel on, and a way to cook the food when it arrives. In your body, the trucks are your blood, the highways your blood vessels (figure 20.1), and the fuel oxygen molecules. Remember from chapter 5 that your cells obtain their energy by "burning" sugars like glucose, using up oxygen and generating carbon dioxide. The same blood that carries food and oxygen to the cells also takes away the "garbage," carbon dioxide.

The organ system responsible that provides the transportation, the trucks and highways, is called the *circulatory system,* while the organ system that acquires the oxygen fuel and disposes of the carbon dioxide waste is called the *respiratory system.* We discuss the functions of these two organ systems in this chapter.

20.1 The Circulatory System

The human circulatory system is made up of three elements: (1) the **heart**, a muscular pump that pushes blood through the body; (2) the **blood vessels,** a network of tubes through which the blood moves; and (3) the **blood,** which circulates within these vessels. The plumbing part of the circulatory system, the heart and blood vessels, is sometimes called the **cardiovascular system** (figure 20.2).

Blood leaves the heart through vessels known as **arteries.** From the arteries the blood passes into a network of smaller arteries called **arterioles.** From these, it is eventually forced through the **capillaries,** a fine latticework of very narrow tubes (from the Latin, *capillus,* "a hair"). While passing through the capillaries, the blood exchanges gases and metabolites (glucose, vitamins, hormones) with the cells of the body. After traversing the capillaries, the blood passes into a third kind of vessel, the **venules,** or small veins. A network of venules empties into larger **veins** that collect the circulating blood and carry it back to the heart (figure 20.3).

The capillaries have a much smaller diameter than the other blood vessels of the body. Blood leaves the human heart through a large artery, the aorta, a tube that has a radius about 1 centimeter (about the same as your thumb), but when it reaches the capillaries it passes through vessels with an average radius of only 8 micrometers, a reduction in radius of some 1,250 times! This decrease in size of blood vessels has a very important consequence. When a fluid such as blood flows through a tube, it meets a frictional resistance as the fluid passes over the walls of the tube—the narrower the tube, the greater the resistance. The capillary network presents a high resistance to flow, a resistance that must be overcome by the strength of the heartbeat.

Because it is a closed loop, every part of the cardiovascular system must have the same overall flow rate (about 5 liters per minute in adult humans). Flow rate (volume per unit time) is not the same as velocity (distance moved per unit time). Even though the velocity of flow in individual capillaries is much less than in arteries, the capillary network has the same overall flow rate as arteries because there are so very many more capillaries.

Arteries: Highways from the Heart

Arteries carry blood away from the heart, but an artery is more than simply a pipe. Because blood comes from the heart, not in a smooth flow, but rather in pulses, slammed into the artery in great big slugs as the heart forcefully ejects its contents with each contraction, the artery has to be able to *expand;* otherwise there would be no place for the added fluid to go. An artery, then, is designed as an expandable tube, with its walls made up of three layers of tissue (figure 20.4*a*). The innermost thin layer is composed of endothelial cells. Surrounding them is a thick layer of smooth muscle and elastic fibers, which in turn is encased within an envelope of protective connective

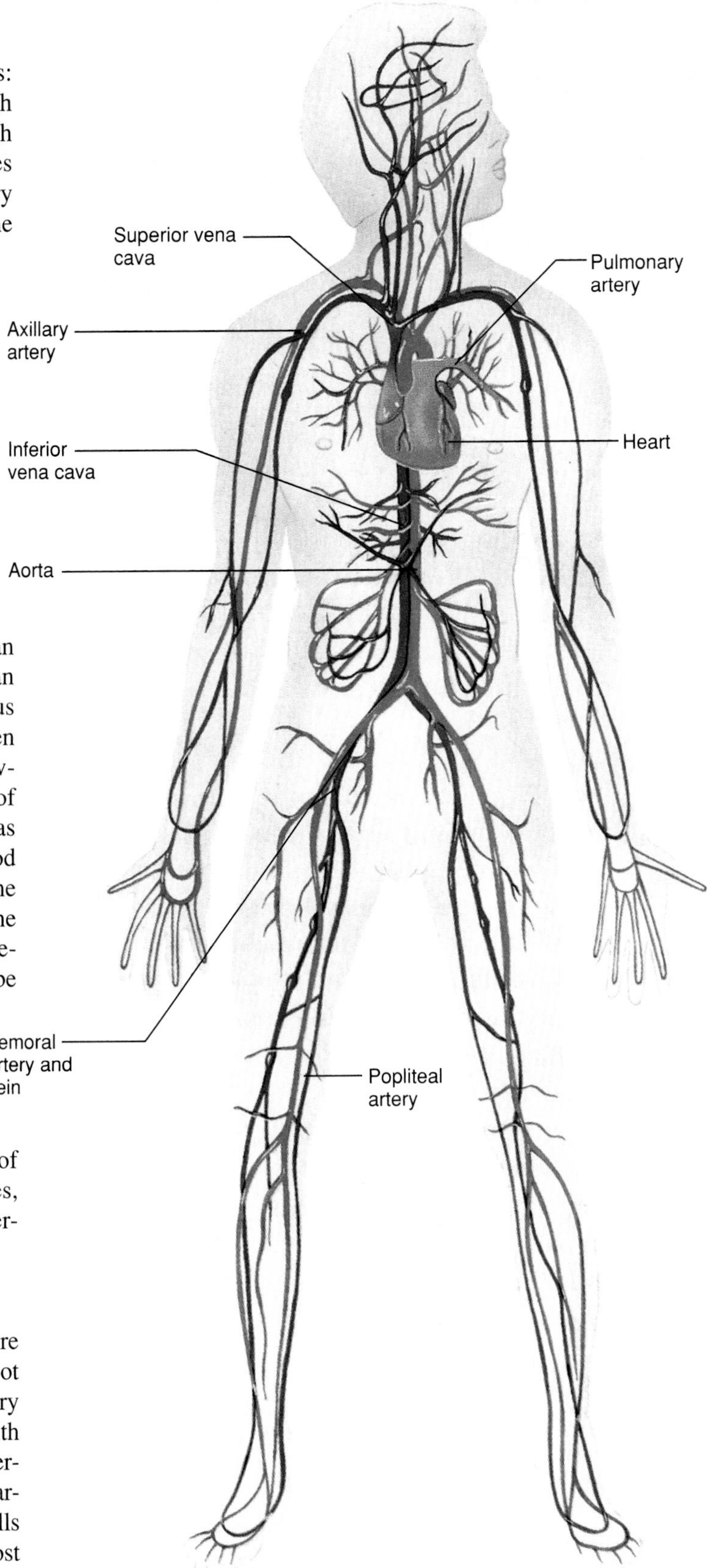

Figure 20.2 The cardiovascular system.
The heart and blood vessels comprise the cardiovascular system. Some of the major veins and arteries of the human body are labeled.

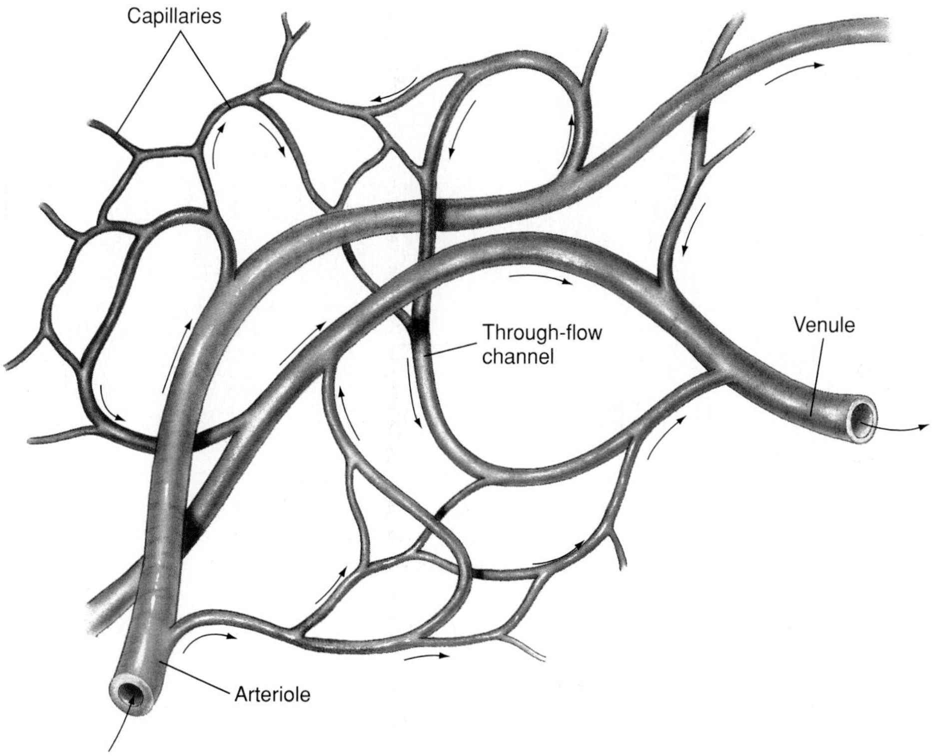

Figure 20.3 The capillary network connects arteries with veins.
Through-flow channels connect arterioles directly to venules. Branching from these through-flow channels is a network of finer channels, the capillaries. Most of the exchange between the body tissues and the red blood cells occurs while they are in this capillary network. Through-flow channels allow the blood to bypass the capillary network when it is needed elsewhere.

tissue. Because this sheath and envelope are elastic, the artery is able to expand its volume considerably when the heart contracts, shoving a new volume of blood into the artery—just as a tubular balloon expands when you blow more air into it. The steady contraction of the smooth muscle layer strengthens the wall of the vessel against overexpansion.

Arterioles differ from arteries in two ways. They are smaller in diameter, and the muscle layer that surrounds an arteriole can be relaxed under the influence of hormones to enlarge the diameter. When the diameter increases, the blood flow also increases, an advantage during times of high body activity. Most arterioles are also in contact with nerve fibers. When stimulated by these nerves, the muscle lining of the arteriole contracts, constricting the diameter of the vessel. Such contraction limits the flow of blood to the extremities during periods of low temperature or stress. You turn pale when you are scared or cold because the arterioles in your skin are constricting. You blush for just the opposite reason. When you overheat or are embarrassed, the nerve fibers connected to muscles surrounding the arterioles are inhibited, which relaxes the smooth muscle and causes the arterioles in the skin to expand, bringing heat to the surface for escape.

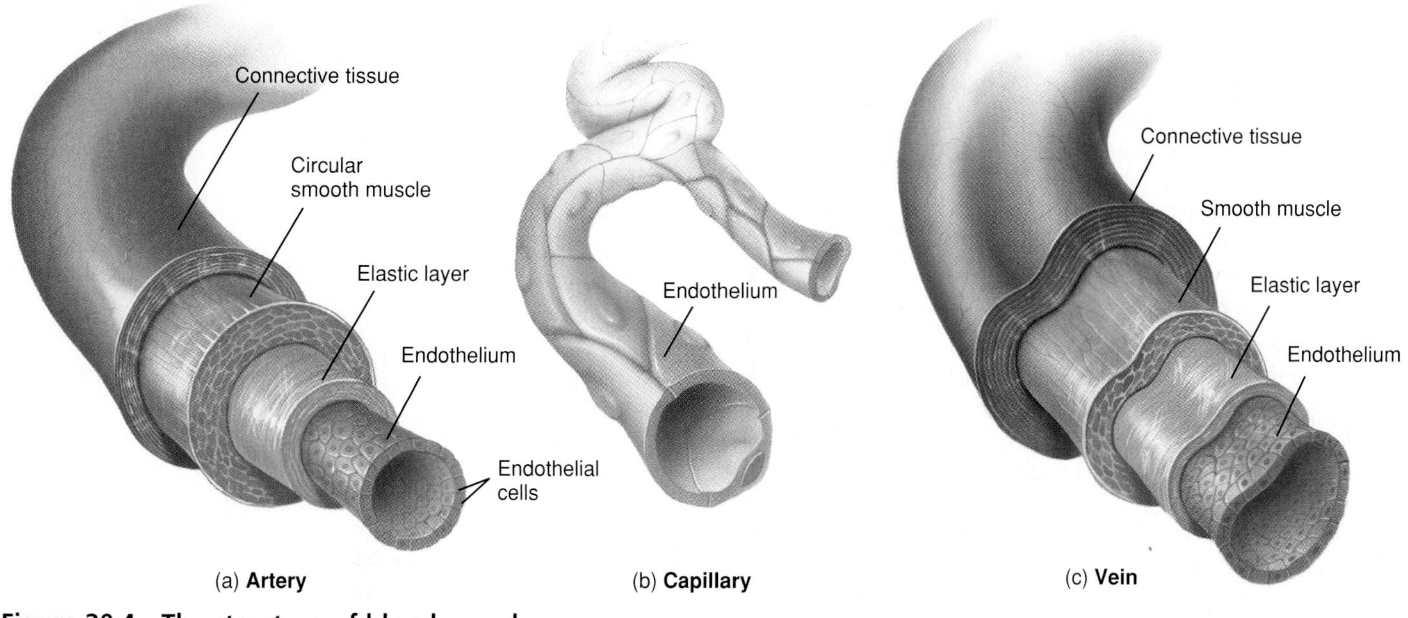

Figure 20.4 The structure of blood vessels.
(*a*) Arteries, which carry blood away from the heart, are expandable and are composed of layers of tissue. (*b*) Capillaries are simple tubes whose thin walls facilitate the exchange of materials between the blood and the cells of the body. (*c*) Veins, which transport blood back to the heart, do not need to be as sturdy as arteries. The walls of veins have thinner muscle and elastic layers than arteries, and they collapse when empty.

Capillaries: Where Exchange Takes Place

Capillaries are where oxygen and food molecules are transferred from the blood to the body's cells and where waste carbon dioxide is picked up. In order to facilitate this back-and-forth traffic, capillaries have thin walls across which gases and metabolites pass easily. Capillaries have the simplest structure of any element in the cardiovascular system. They are built like a soft-drink straw, simple tubes with walls only one cell thick (figure 20.4*b*). The average capillary is about 1 millimeter long and connects an arteriole with a venule. All capillaries are very narrow, with an internal diameter of about 8 micrometers, just bigger than the diameter of a red blood cell (5 to 7 micrometers). This design is critical to the function of capillaries. By bumping against the sides of the vessel as they pass through, the red blood cells are forced into close contact with the capillary walls, making exchange easier (figure 20.5).

No cell of your body is more than 100 micrometers from a capillary. At any one moment, about 5% of your blood is in your capillaries, a network that amounts to several thousand miles in overall length. If all the capillaries in your body were laid end to end, they would extend across the United States! Individual capillaries have high resistance to flow because of their small diameters. However, the total cross-sectional area of the extensive capillary network (that is, the sum of all the diameters of all the capillaries, expressed as area) is greater than that of the arteries leading to it. As a result, the blood pressure is actually far lower in the capillaries than in the arteries. This is important, because the walls of capillaries are not strong, and they would burst if exposed to the pressures that arteries routinely withstand.

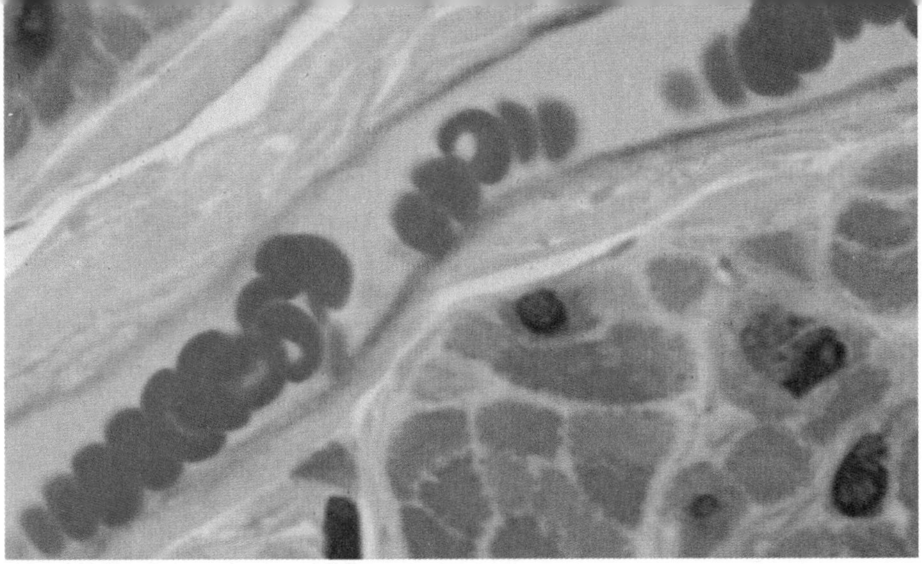

Figure 20.5 Red blood cells within a capillary.
The red blood cells in this capillary pass along in single file. Many capillaries are even narrower than the one shown here. However, red blood cells can pass through capillaries narrower than their own diameter, pushed along by the pressure of the pumping heart.

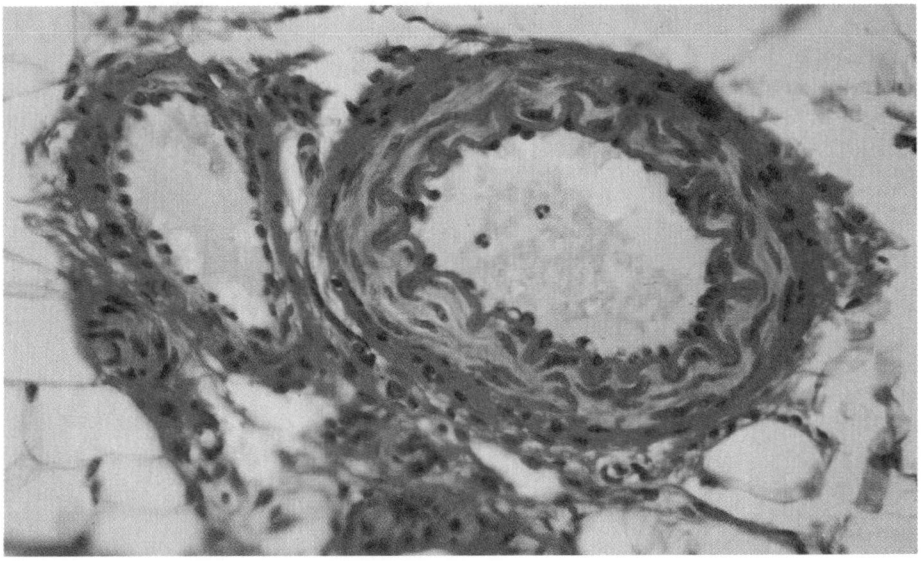

Figure 20.6 Veins and arteries.
The vein (*left*) has the same general structure as the artery (*right*) but much thinner layers of muscle and elastic fiber. An artery retains its shape when empty, but a vein collapses.

Veins: Returning Blood to the Heart

Veins are vessels that return blood to the heart. Veins do not have to accommodate the pulsing pressures that arteries do, because much of the force of the heartbeat is weakened by the high resistance and great cross-sectional area of the capillary network. For this reason, the walls of veins have much thinner layers of muscle and elastic fiber (figure 20.4*c*). An empty artery is still a hollow tube, like a pipe, but when a vein is empty, its walls collapse like an empty balloon (figure 20.6).

Because the pressure of the blood flowing within veins is low, it becomes important to avoid any further resistance to flow, lest there not be enough pressure to get the blood back to the heart. Because a wide tube presents much less resistance to flow than a narrow one, the internal passageway of veins is often quite large, requiring only a small pressure difference to return blood to the heart. The diameters of the largest veins in the human body, the venae cavae, which lead into the heart, are fully 3 centimeters, wider than your thumb! Veins also have unidirectional valves that aid the return of blood, by preventing it from flowing backward.

The Lymphatic System: Recovering Lost Fluid

The cardiovascular system is very leaky. Fluids are forced out across the thin walls of the capillaries by the pumping pressure of the heart. Although this loss is unavoidable—the circulatory system could not do its job of gas and metabolite exchange without tiny vessels with thin walls—it is important that the loss be made up. In your body, about 3 liters of fluid leave your cardiovascular system in this way each day, more than half the body's total supply of about 5.6 liters of blood! To collect and recycle this fluid, the body utilizes a second circulatory system called the **lymphatic system** (figure 20.7). Open-ended lymphatic capillaries gather up liquid from the spaces surrounding cells and carry it through a series of progressively larger vessels to two large lymphatic vessels, which resemble veins. These lymphatic vessels drain into veins in the lower part of the neck through one-way valves. Once within the lymphatic system, this fluid is called **lymph.**

Before being drained into the veins for recirculation, lymph passes through filters called lymph nodes that remove bacteria and other debris from the lymph.

Fluid is driven through the lymphatic system when its vessels are squeezed by the movements of the body's muscles. The lymphatic vessels contain a series of one-way valves that permit movement only in the direction of the neck (figure 20.8).

The lymphatic system has several other important functions. It returns proteins to the circulation, transports fats absorbed from the intestine, and carries bacteria and dead blood cells to the lymph nodes and spleen for destruction.

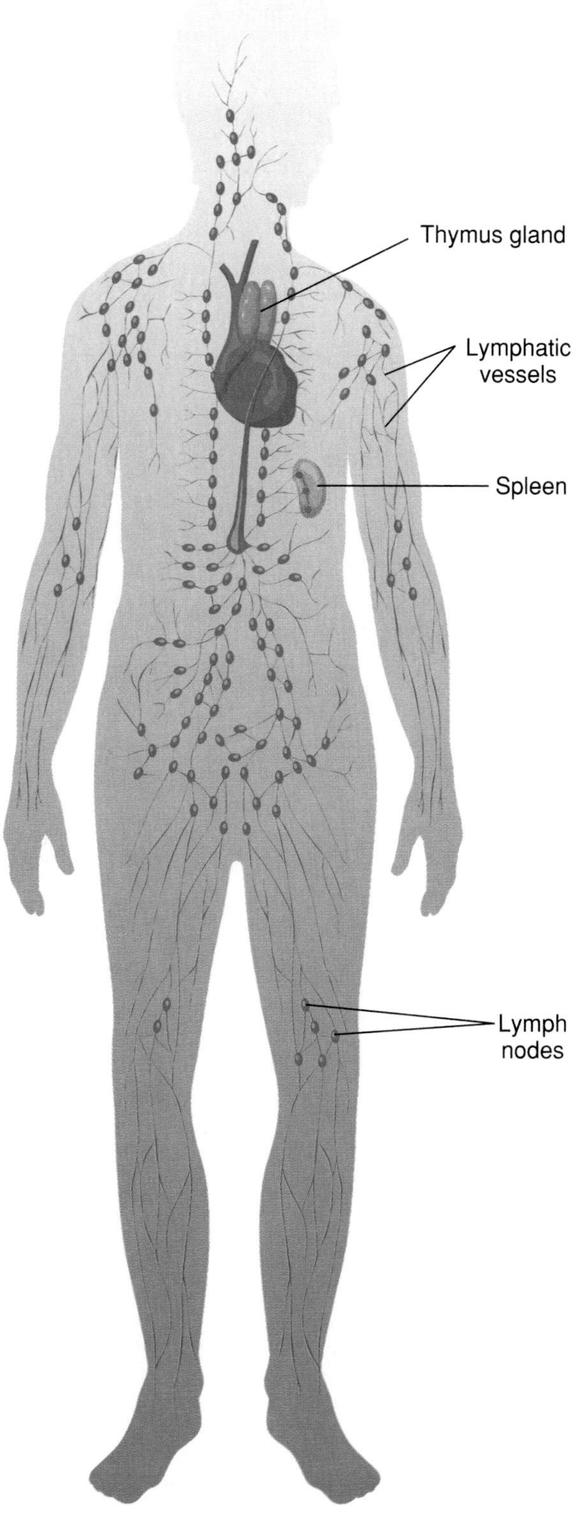

Figure 20.7 The human lymphatic system.
The lymphatic system consists of lymphatic vessels and capillaries, lymph nodes, and lymphatic organs, including the spleen and thymus gland.

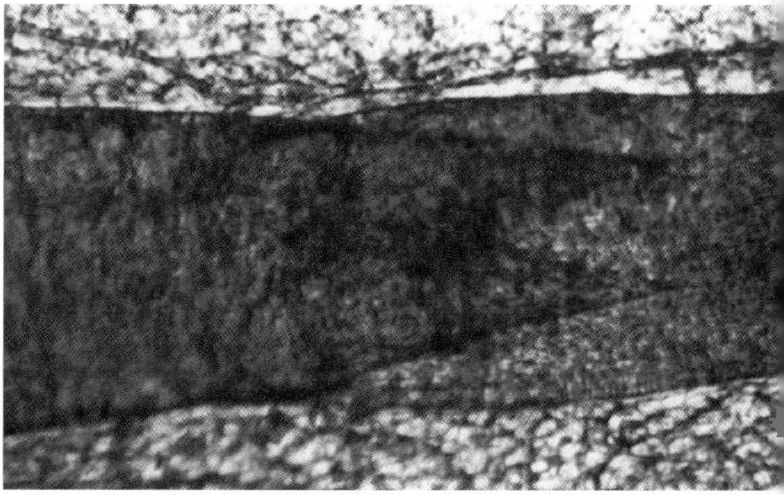

Figure 20.8 A lymphatic vessel valve.
Flow from left to right is not retarded because such flow tends to force open the valve. Flow from right to left is prevented because such flow tends to force the valve closed.

20.2 Blood

About 5% of your body mass is composed of the blood circulating through the arteries, veins, and capillaries of your body. This blood is composed of a fluid called **plasma,** together with several different kinds of cells that circulate within that fluid.

Blood Plasma: The Blood's Fluid

Blood plasma is a complex solution of water with three very different sorts of substances dissolved within it:

1. **Metabolites and wastes.** If the circulatory system is the highway of the vertebrate body, the blood contains the traffic traveling on that highway. Dissolved within its plasma are glucose, vitamins, hormones, and wastes that circulate between the cells of the body.
2. **Salts and ions.** Like the seas in which life arose, plasma is a dilute salt solution. The chief plasma ions are sodium, chloride, and bicarbonate. In addition, trace amounts of other salts, such as calcium and magnesium, as well as metallic ions, including copper, potassium, and zinc, are present in plasma. The composition of the plasma is not unlike that of seawater.
3. **Proteins.** Blood plasma is 90% water. Passing by all the cells of the body, blood would soon lose most of its water to them by osmosis if it did not contain as high a concentration of proteins as the cells it passes. Water is not lost to the cells because blood plasma contains proteins. Some of these are antibody proteins that are active in the immune system. More than half the amount of protein that is necessary to balance the protein content of the cells of the body consists of a single protein, **serum albumin,** which circulates in the blood as an osmotic counter force. Human blood contains 46 grams of serum albumin per liter—that's over half a pound of it in your body. Starvation, and protein deficiency diseases such as kwashiorkor, produce swelling of the body because the body's cells take up water from the albumin-deficient blood.

Blood cell		Life span in blood	Function
Erythrocyte		120 days	O_2 and CO_2 transport
Leukocytes Neutrophil	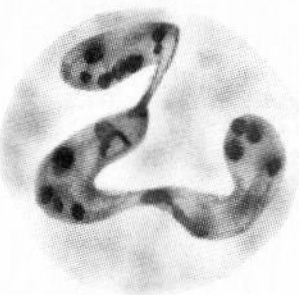	7 hours	Immune defenses
Eosinophil	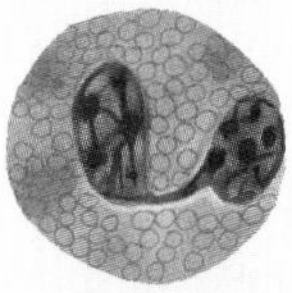	Unknown	Defense against parasites
Basophil	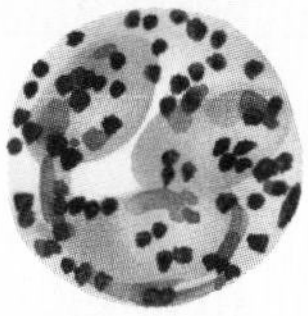	Unknown	Inflammatory response
Monocyte	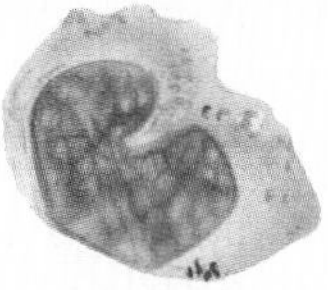	3 days	Immune surveillance (precursor of tissue macrophage)
B-lymphocyte	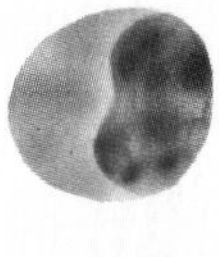	Unknown	Antibody production (precursor of plasma cells)
T-lymphocyte	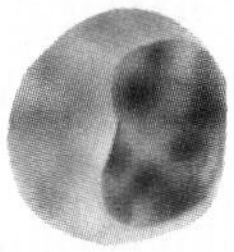	Unknown	Cellular immune response
Platelets	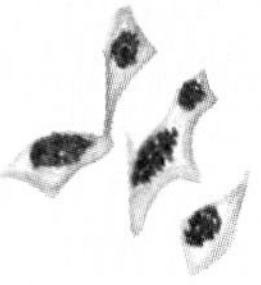	7-8 days	Blood clotting

Figure 20.9 Types of blood cells.
Erythrocytes, leukocytes (neutrophils, eosinophils, basophils, monocytes, and lymphocytes), and platelets are the three principal types of blood cells in humans.

Blood Cells: Cells That Circulate Through the Body

Although blood is liquid, nearly half of its volume is actually occupied by cells. The fraction of the total volume of the blood that is occupied by cells is referred to as the blood's **hematocrit.** In humans, the hematocrit is usually about 45%. The three principal types of cells in the blood are erythrocytes, leukocytes, and cell fragments called platelets (figure 20.9).

Erythrocytes Carry Hemoglobin

Each milliliter of blood contains about 5 billion **erythrocytes,** also called **red blood cells.** Each erythrocyte is a flat disk with a central depression on both sides, something like a doughnut with a hole that doesn't go all the way through (figure 20.10). Erythrocytes carry oxygen to the cells of the body. Almost the entire interior of an erythrocyte is packed with hemoglobin, a protein that binds oxygen in the lungs and delivers it to the cells of the body.

Mature human erythrocytes function like boxcars rather than trucks. Like a vehicle without an engine, erythrocytes contain neither a nucleus nor the machinery to make proteins. Because they lack a nucleus, these cells are unable to repair themselves and therefore have a rather short life; any one erythrocyte lives only about four months. New erythrocytes are constantly being synthesized and released into the blood by cells within the soft interior marrow of bones.

Leukocytes Defend the Body

Less than 1% of the cells in human blood are **leukocytes,** also called **white blood cells.** Leukocytes are larger than red blood cells. They contain no hemoglobin and are essentially colorless. There are several kinds of leukocytes, each with a different function. All of these functions, however, help defend the body against invading microorganisms and other foreign substances, as you will see in chapter 22. Unlike other blood cells, leukocytes are not confined to the bloodstream; they are mobile soldiers that also migrate out into the fluid surrounding cells.

Platelets Help Blood to Clot

Certain large cells within the bone marrow, called **megakaryocytes,** regularly pinch off bits of their cytoplasm. These cell fragments, called **platelets,** contain no nuclei. Entering the bloodstream, they play a key role in blood clotting. In a clot, a gluey mesh of **fibrin** protein fibers sticks platelets together to form a mass that plugs the rupture in the blood vessel. The clot provides a tight, strong seal, much as the inner lining of a tubeless tire seals punctures. The fibrin that forms the clot is made in a series of reactions that start when circulating platelets first encounter the site of an injury. Responding to chemicals released by the damaged blood vessel cells, the platelets release a protein factor into the blood that starts the clotting process.

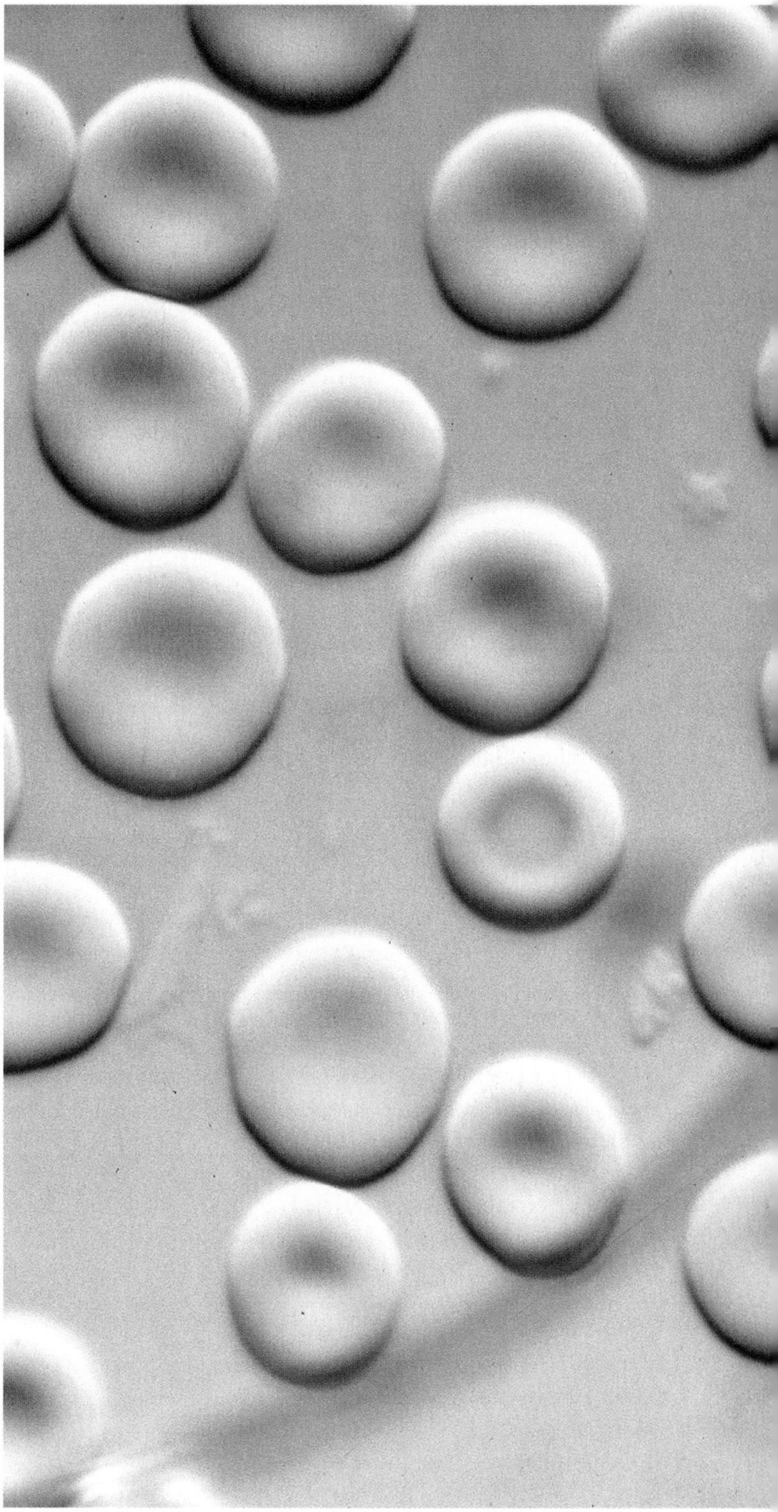

Figure 20.10 Human erythrocytes (red blood cells). Human erythrocytes lack nuclei, which gives them a characteristic collapsed appearance, like a pillow upon which someone has sat (×1,000).

20.3 The Heart and How It Works

During the course of vertebrate evolution, hearts have gotten bigger and thus are able to pump harder. Within any one class of vertebrate, the size of the heart is a constant fraction of the body mass. In mammals, for example, the heart is very close to 0.6% of the total body mass, whether the mammal is a 0.10-kilogram shrew or a 100-kilogram human.

The human heart, like that of all mammals and birds, is really two separate pumping systems operating together within a single unit. One of these pumps blood to the lungs, while the other pumps blood to the rest of the body. A cut through the heart from top to bottom clearly shows its *double* pump organization. The left side has two connected chambers, and so does the right, but the two sides are not connected with one another.

Circulation Through the Heart

Let's follow the journey of blood through your heart, starting with the entry of oxygen-rich blood into the heart from the lungs (figure 20.11). Oxygenated blood from the lungs enters the left side of the heart, emptying directly into the **left atrium** through large vessels called the **pulmonary veins.** From the atrium, blood flows through an opening into the adjoining chamber, the **left ventricle.** Most of this flow, roughly 80%, occurs while the heart is relaxed. When the heart starts to contract, the atrium contracts first, pushing the remaining 20% of its blood into the ventricle.

After a slight delay, the ventricle contracts. The walls of the ventricle are far more muscular than those of the atrium, and thus this contraction is much stronger. It forces most of the blood out of the ventricle in a single strong pulse. The blood is prevented from going back into the atrium by a large, one-way valve, the **mitral valve,** whose flaps are pushed shut as the ventricle contracts. Strong fibers that prevent the flaps from moving too far when closing are attached to their edges. These fibers operate the same way as the chain on a screen door—the door can be opened only as far as the slack in the chain permits.

Prevented from reentering the atrium, the blood within the contracting left ventricle takes the only other passage out. It moves through a second opening that leads into a large blood vessel called the **aorta.** The aorta is separated from the left ventricle by a one-way valve, the **aortic valve.** Unlike the mitral valve, the aortic valve is oriented to permit the flow of the blood *out* of the ventricle. Once this outward flow has occurred, the aortic valve closes, preventing the reentry of blood from the aorta into the heart. Many arteries branch from the aorta, carrying oxygen-rich blood to all parts of the body. The first to branch are the coronary arteries,

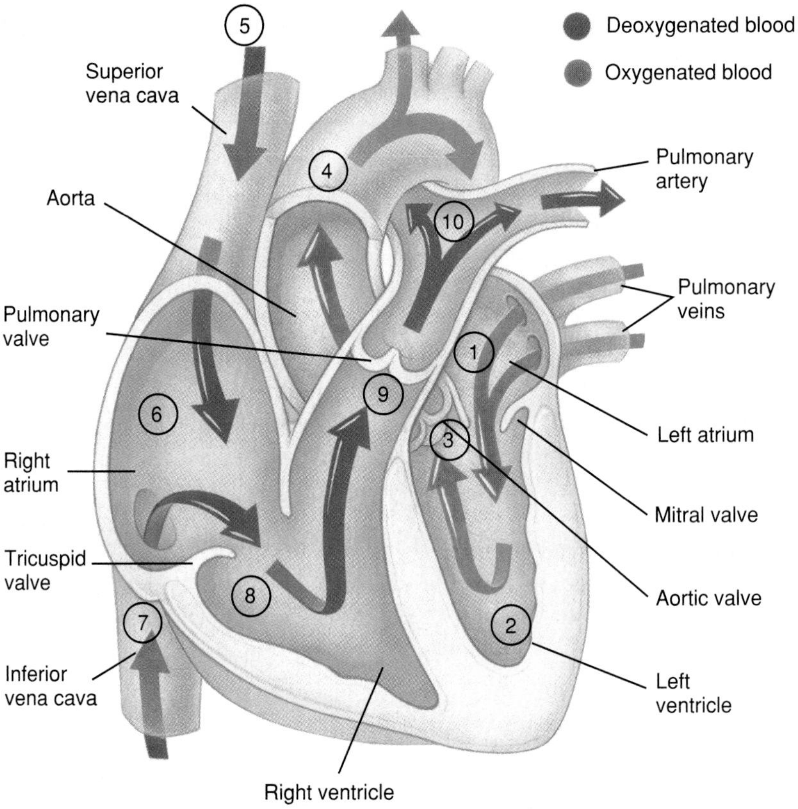

Figure 20.11 The path of blood through the human heart.
(*1*) Oxygenated blood from the lungs enters the left atrium of the heart by way of the pulmonary veins. This blood then enters the left ventricle (*2*), from which it passes into the aorta (*3*) to circulate throughout the body and deliver oxygen to the tissues (*4*). When gas exchange has taken place at the tissues, veins return blood to the heart (*5*). After entering the right atrium by way of the superior and inferior venae cavae (*6, 7*), deoxygenated blood passes into the right ventricle (*8*) and then through the pulmonary valve (*9*) to the lungs by way of the pulmonary artery (*10*).

which carry freshly oxygenated blood to the heart itself; the muscles of the heart do not obtain their supply of blood from within the heart.

The blood that flows into the arterial system eventually returns to the heart after delivering its cargo of oxygen to the cells of the body. In returning it passes through a series of progressively larger veins, ending in two large veins that empty into the right side of the heart. The **superior vena cava** drains the upper body, the **inferior vena cava** the lower body. The largest blood vessels in the body, they both empty deoxygenated blood into the right atrium of the heart.

The right side of the heart is similar in organization to the left side. Blood passes from the **right atrium** into the **right ventricle** through a one-way valve, the **tricuspid valve.** It passes out of the contracting right ventricle through a second valve, the **pulmonary valve,** into the **pulmonary arteries,** which carry the deoxygenated blood to the lungs. The blood then returns from the lungs to the left side of the heart with a new cargo of oxygen, which is pumped to the rest of the body.

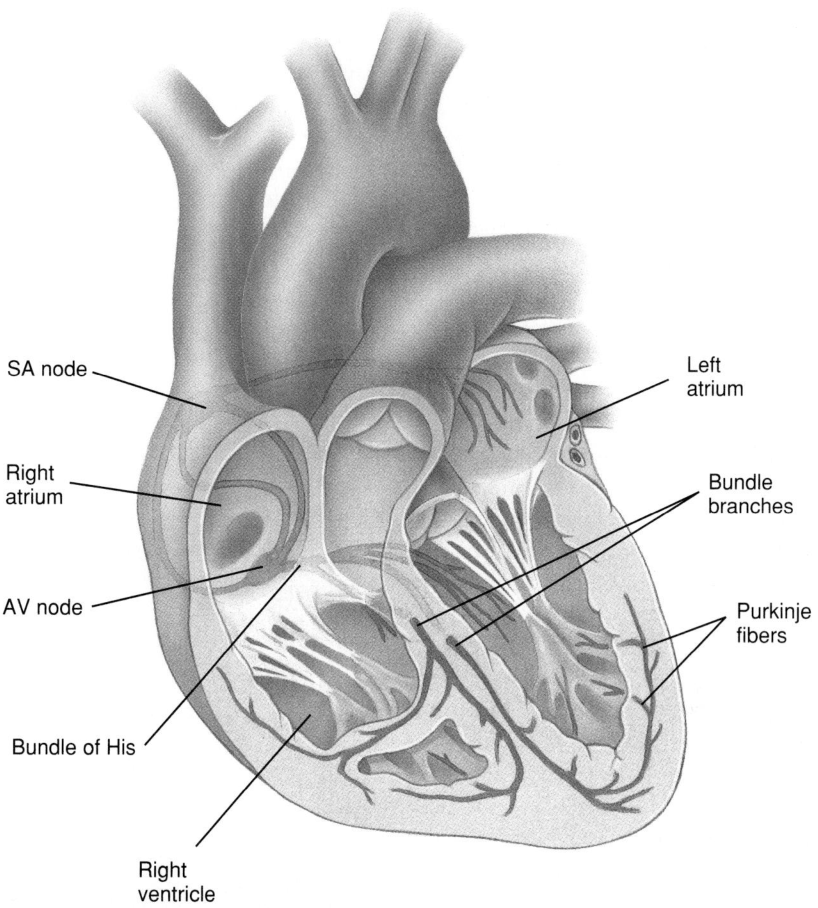

Figure 20.12 How the heart contracts.
Contraction of the human heart is initiated by a wave of depolarization that begins at the SA node. After passing over the right and left atria and causing their contraction, the wave of depolarization reaches the AV node, from which it passes to the ventricles. The depolarization is conducted rapidly over the surface of the ventricles by a set of fibers called Purkinje fibers, which make up the bundle of His and branch throughout the ventricle.

How the Heart Contracts

The contraction of the heart consists of a carefully orchestrated series of muscle contractions. First, the atria contract together, followed by the ventricles. Contraction is initiated by the **sinoatrial (SA) node,** a small cluster of excitatory cardiac muscle cells embedded in the upper wall of the right atrium (figure 20.12). The cells of the SA node act as a pacemaker for the rest of the heart. Their membranes spontaneously depolarize with a regular rhythm that determines the rhythm of the heart's beating. Each depolarization initiated within this pacemaker region passes quickly from one heart muscle to another in a wave that envelops the left and the right atria almost simultaneously.

But the wave of depolarization does not immediately spread to the ventricles. There is a pause before the lower half of the heart starts to contract. The reason for the delay is that the atria of the heart are separated from the ventricles by connective tissue, and connective tissue cannot propagate a depolarization wave. The depolarization would not pass to the ventricles at all except for a slender connection of cardiac muscle cells known as the **atrioventricular (AV) node,** which connects across the gap to a strand of specialized muscle known as the **bundle of His.** Bundle branches divide into fast-conducting **Purkinje fibers,** which initiate the almost simultaneous contraction of all the cells of the right and left ventricles about 0.1 seconds after the atria contract. This delay permits the atria to finish emptying their contents into the corresponding ventricles before those ventricles start to contract.

Monitoring the Heart's Performance

As you can see, the heartbeat is not simply a squeeze-release, squeeze-release cycle but rather a little play in which a series of events occur in a predictable order. The simplest way to monitor heartbeat is to listen to the heart at work, using a stethoscope. The first sound you hear, a low-pitched *lub,* is the closing of the mitral and tricuspid valves at the start of ventricular contraction. A little later, you hear a higher-pitched *dub,* the closing of the pulmonary and aortic valves at the end of ventricular contraction. If the valves are not closing fully, or if they open incompletely, turbulence is created within the heart. This turbulence can be heard as a **heart murmur.** It often sounds like liquid sloshing.

A second way to examine the events of the heartbeat is to monitor the blood pressure. During the first part of the heartbeat, the atria are filling. At this time the pressure in the arteries leading from the left side of the heart out to the tissues of the body decreases slightly. This low pressure is referred to as the **diastolic** pressure. During the contraction of the left ventricle, a pulse of blood is forced into the systemic arterial system, immediately raising the blood pressure within these vessels. The high blood pressure produced in this pushing period, which ends with the closing of the aortic valve, is referred to as the **systolic** pressure. Normal blood pressure values are 70 to 90 diastolic and 110 to 130 systolic. When the inner walls of the arteries accumulate fats, as they do in the condition known as **atherosclerosis,** the diameters of the passageways are narrowed. If this occurs, the systolic blood pressure is elevated.

A third way to monitor a heartbeat is to measure the waves of depolarization. Because the human body basically consists of water, it conducts electrical currents rather well. A wave of membrane depolarization passing over the surface of the heart generates an electrical current that passes in a wave throughout the body. The magnitude of this electrical pulse is tiny, but it can be detected by sensors placed on the skin. A recording of these impulses is called an electrocardiogram (figure 20.13).

The evolution of multicellular organisms has depended critically on the ability to circulate materials throughout the body efficiently. Vertebrates carefully regulate the operation of their circulatory systems and are able to integrate their body activities. Indeed, the metabolic demands of different vertebrates have shaped the evolution of circulatory adaptations (figure 20.14).

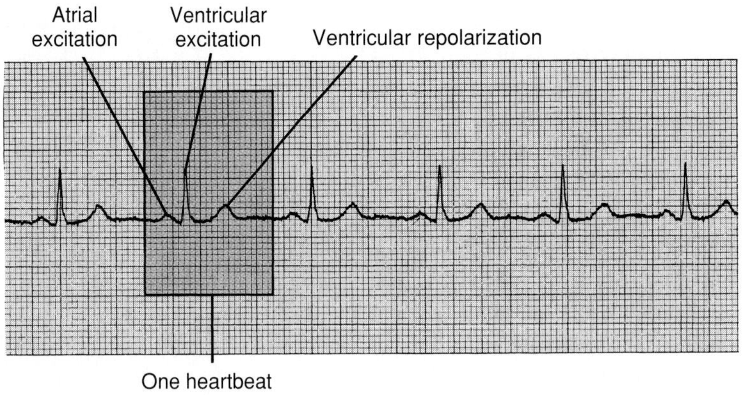

Figure 20.13 An electrocardiogram.
The chart shows the three successive impulses of a normal heartbeat. The first depolarization represents the atrial excitation. A tenth of a second later, a much stronger wave reflects both the excitation of the ventricles and relaxation of the atria. Finally, perhaps 0.2 seconds later, a third wave results from the relaxation, or repolarization, of the ventricles.

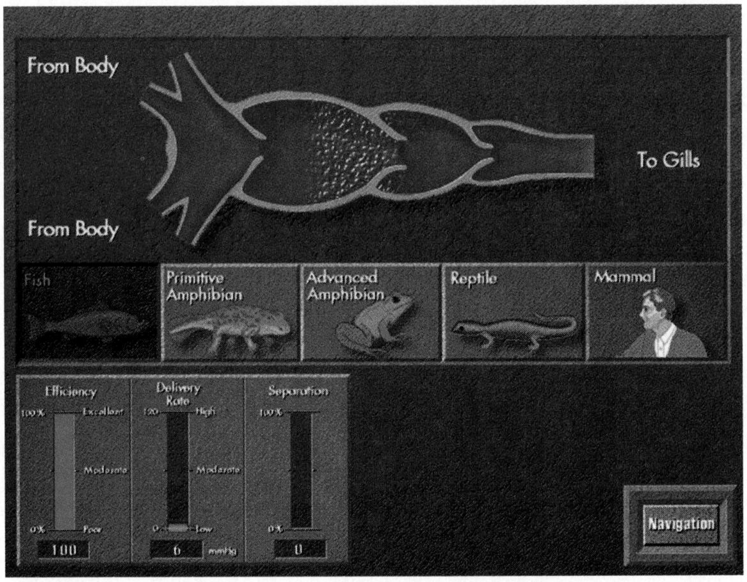

Figure 20.14 Evolution of the vertebrate heart.
The basic vertebrate heart, which evolved in fishes, is a four-chambered pump. This screen capture from a CD-ROM interactive exercise demonstrates how the evolution of the pulmonary vein increased overall delivery of oxygenated blood to the body, and how subsequent extension of a septum within the heart further increased efficiency, eventually leading to the double-pump heart we see in humans today. (*Explorations in Human Biology,* Module 5, "Evolution of the Heart")

20.4 The Respiratory System

All animals obtain the energy that powers their lives by oxidizing molecules rich in energy-laden carbon–hydrogen bonds. This oxidative metabolism requires a ready supply of oxygen. The evolution of oxygen-gathering mechanisms among the vertebrates has favored changes that increase this efficiency. The most efficient mechanism to evolve in water is the gill of bony fishes; the most efficient mechanism to evolve in air is the two-cycle lung of birds. The oxygen-gathering mechanism of mammals, although less efficient than birds, adapts them well to their terrestrial habitat.

Like all mammals, humans obtain the oxygen they need for metabolism from air which is about 21% oxygen gas. The uptake of oxygen from air and the simultaneous release of carbon dioxide together are called **respiration.** The organ system responsible for carrying out respiration, the **respiratory system,** is composed of large, saclike organs called **lungs** and the passages that lead to them (figure 20.15). A pair of lungs is located in your chest, or **thoracic,** cavity. The two lungs hang free within the cavity, connected to the rest of the body only at one position, where the lung's blood vessels and air tube enter. This air tube is called a **bronchus.** It connects each lung to a long tube called the **trachea,** which passes upward and opens into the rear of the mouth.

Air normally enters through the nostrils, where it is warmed. In addition, the nostrils are lined with hairs that filter out dust and other particles. As the air passes through the nasal cavity, an extensive array of cilia further filters and moistens the air (figure 20.16). Mucous secretory ciliated cells in the bronchi also trap foreign particles and carry them upward, where they can be swallowed. The air then passes through the back of the mouth, entering first the **larynx** (voice box) and then the trachea. From there, it passes down through the bronchus to the lungs. Because the air crosses the path of food at the back of the throat, a special flap called the epiglottis covers the trachea whenever food is swallowed, to keep it from "going down the wrong pipe."

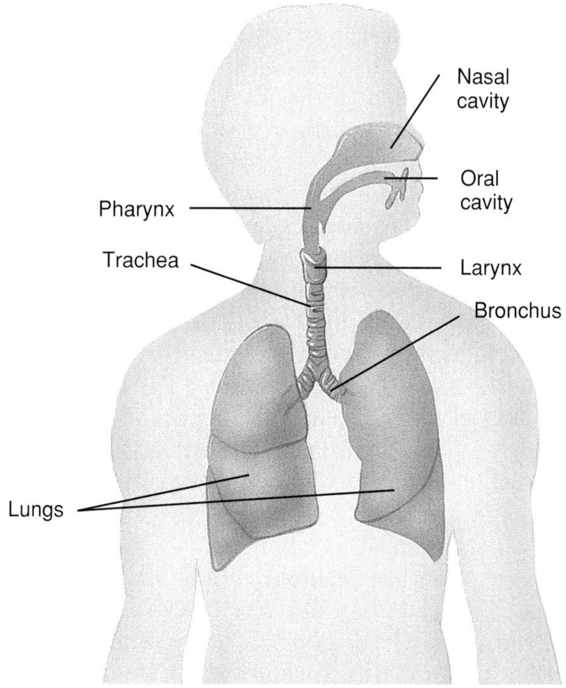

Figure 20.15 The human respiratory system.
The respiratory system consists of the lungs and the passages that lead to them.

Figure 20.16 Cilia line the respiratory tract.
Cilia, shown above at magnification ×3,500, cover the epithelial lining of the nasal cavity, trachea, and bronchi. These cilia filter and moisten air that is inhaled.

The Lungs

Your lungs are large organs that fill your chest cavity. The interior of each lung is not an open cavity, like a bag. Instead, it is subdivided into millions of small chambers called **alveoli,** clustered together like bunches of grapes (figure 20.17). The alveoli greatly increase the diffusion surface of the lung. The total surface devoted to diffusion in your lungs is as much as 80 square meters, an area the size of an average classroom! The alveoli are surrounded by so many capillaries, it is as if blood were flowing over them in a continuous sheet. Alveoli are connected to the bronchi by a branching network of tubes called **bronchioles,** some of which are surrounded by smooth muscle and are sensitive to levels of oxygen.

The human respiratory apparatus is simple in structure and functions as a one-cycle pump. The thoracic cavity is bounded on its sides by the ribs and on the bottom by a thick layer of muscle, the **diaphragm,** which separates the thoracic cavity from the abdominal cavity. Each lung is covered by a very thin, smooth membrane called the **pleural membrane.** A second pleural membrane lines the interior of the thoracic cavity, into which the lungs hang. Within the cavity, the weight of the lungs is supported by water, the **interpleural fluid.**

The interpleural fluid not only supports the lungs, but it also plays another important role by permitting an even application of pressure to all parts of the lung. You can visualize the pleural membranes as a system of two balloons of different sizes, one nested inside the other, with the space between them completely filled with a very thin film of water (figure 20.18). The inner balloon opens out to the atmosphere, so two forces act upon it: air pressure from the atmosphere pushes it outward, and water pressure from the interpleural fluid pushes it inward.

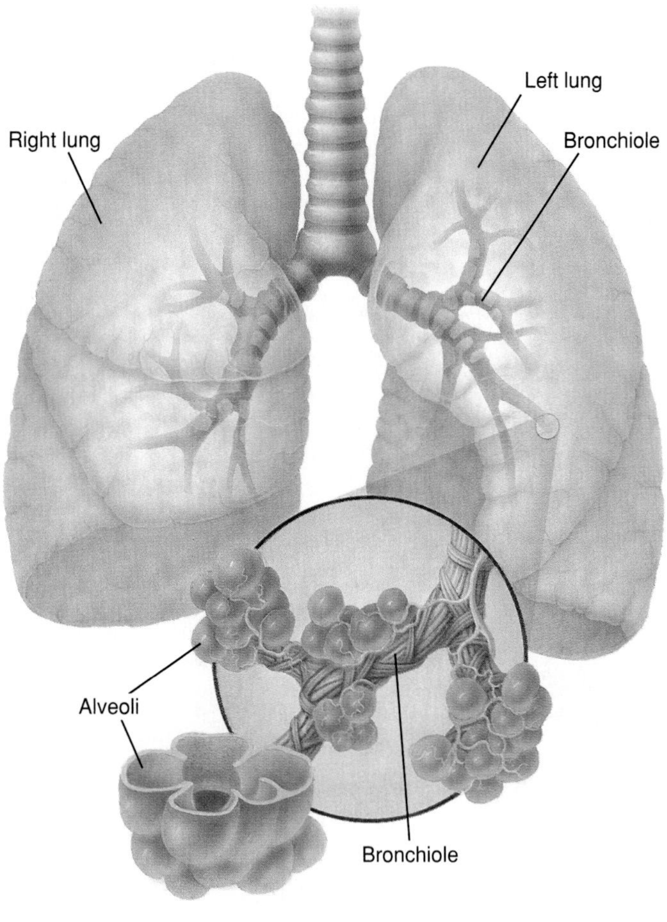

Figure 20.17 Internal structures of the lung.
Gas exchange occurs through the alveoli into the bloodstream.

The Mechanics of Breathing

The active pumping of air in and out through the lungs is called **breathing.** During *inhalation,* muscular contraction causes the walls of the chest cavity to expand so that the rib cage moves outward and upward. The diaphragm is dome-shaped when relaxed but moves downward as it flattens during this contraction (figure 20.19*a*). In effect, we have enlarged the outer balloon by pulling it in all directions. This expansion causes the interpleural fluid pressure to decrease to a level less than that of the air pressure within the inner balloon. Since fluids are incompressible, the wall of the inner balloon is pulled out. As the inner balloon expands, its internal air pressure tends to decrease, and since the inner balloon is connected to the atmosphere, air rapidly moves in to equalize pressure.

During *exhalation,* the ribs and diaphragm return to their original resting position. In doing so, they exert pressure on the fluid. This pressure is transmitted uniformly by the fluid over the entire surface of the lung (the inner balloon), forcing air from the inner cavity back out to the atmosphere (figure 20.19*b*). A typical breath at rest moves about 0.5 liters of air, called the **tidal volume.** The extra amount that can be forced into and out of the lung is about 4.5 liters in men and 3.1 liters in women. The air remaining in the lung after such a maximal expiration is the residual volume, typically about 1.2 liters.

When each breath is completed, the lung still contains a volume of air, the **residual volume.** In human lungs this volume is about 1,200 milliliters. Each inhalation adds from 500 milliliters (resting) to 3,000 milliliters (exercising) of additional air. Each exhalation removes approximately the same volume as inhalation added, reducing the air volume in the lungs once more to about 1,200 milliliters. Because the diffusion surfaces of the lungs are not exposed to fully oxygenated air, but rather to a mixture of fresh and partly oxygenated air, the respiratory efficiency of your lungs is far from maximal. A bird, for example, whose lungs do not retain a residual volume, is able to achieve far greater respiratory efficiency.

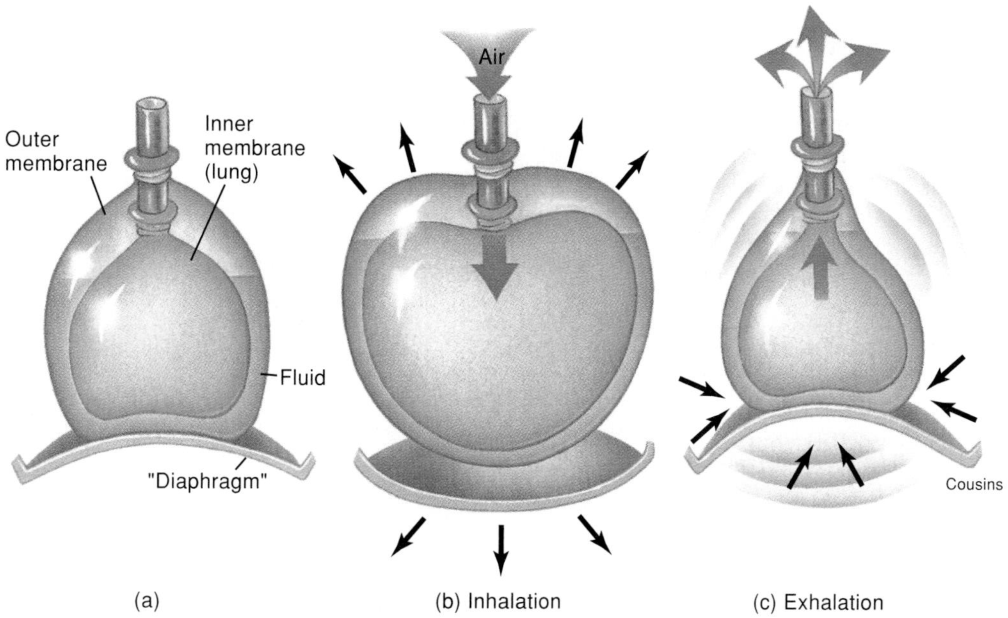

Figure 20.18 A simple experiment that shows how you breathe.
(*a*) Your lung has the structure of a balloon within a balloon, with a thin layer of fluid between the two balloons. When the outer balloon is pulled downward and outward, as shown in (*b*), reducing the air pressure inside the inner balloon, the inner balloon expands as air moves into it (inhalation). When the outer balloon "relaxes," or recoils (*c*), air is forced out of the inner balloon (exhalation).

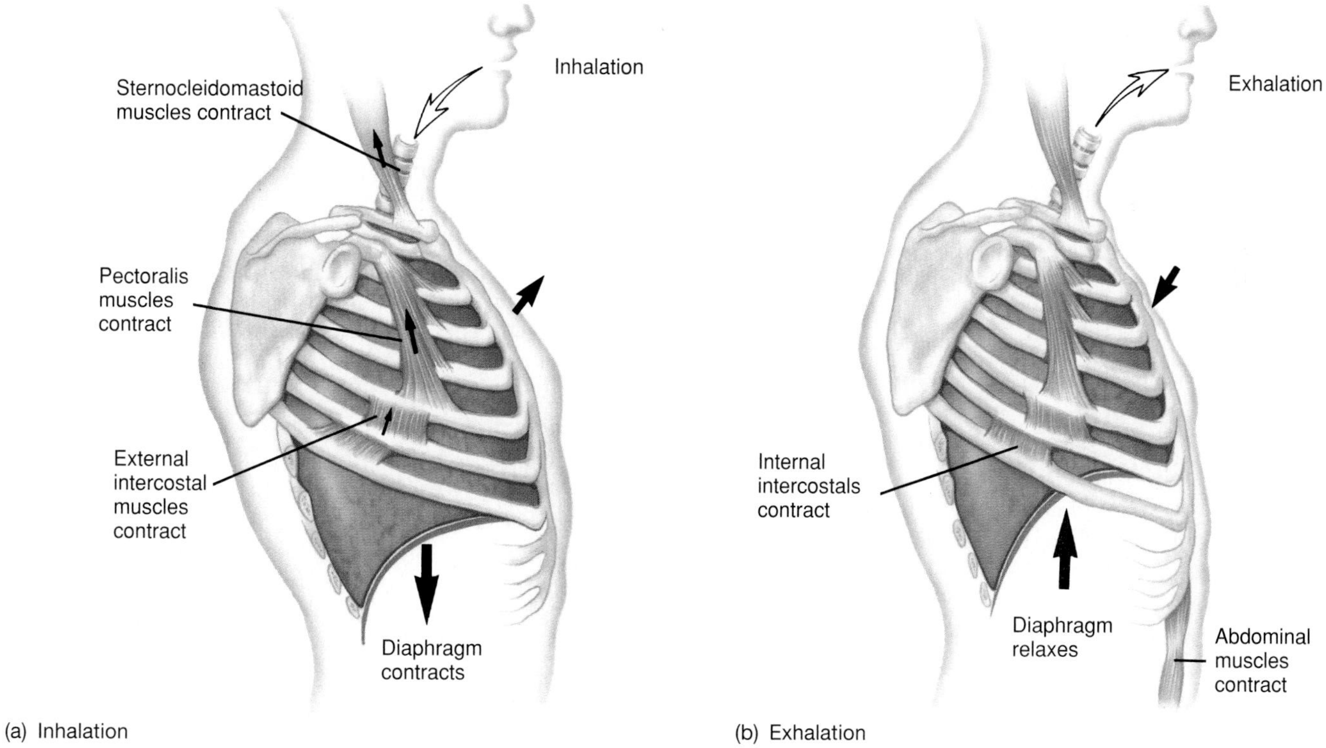

Figure 20.19 How a human breathes.
(*a*) During inhalation, the diaphragm contracts, and the chest cavity expands downward and outward. This reduces the air pressure inside the lungs, and air from outside the body flows into the lungs. (*b*) During exhalation, muscles used to inhale relax, and abdominal muscles contract, forcing air out of the lungs.

How Respiration Works: Gas Exchange

When oxygen has diffused from the air into the moist cells lining the inner surface of the lung, its journey has just begun. Passing from these cells into the bloodstream, the oxygen is carried throughout the body by the circulatory system, described earlier in this chapter. It has been estimated that it would take a molecule of oxygen three years to diffuse from your lung to your toe if transport depended only on diffusion, unassisted by a circulatory system.

Oxygen moves within the circulatory system carried piggyback on the protein **hemoglobin.** Hemoglobin molecules contain iron, which combines with oxygen in a reversible way (figure 20.20). Hemoglobin is manufactured within red blood cells and gives them their color. Hemoglobin never leaves these cells, which circulate in the bloodstream like ships bearing cargo. Oxygen binds to hemoglobin within these cells as they pass through the capillaries surrounding the alveoli of the lungs. Carried away within red blood cells, the oxygen-bearing hemoglobin molecules later release their oxygen molecules to metabolizing cells at distant locations in the body.

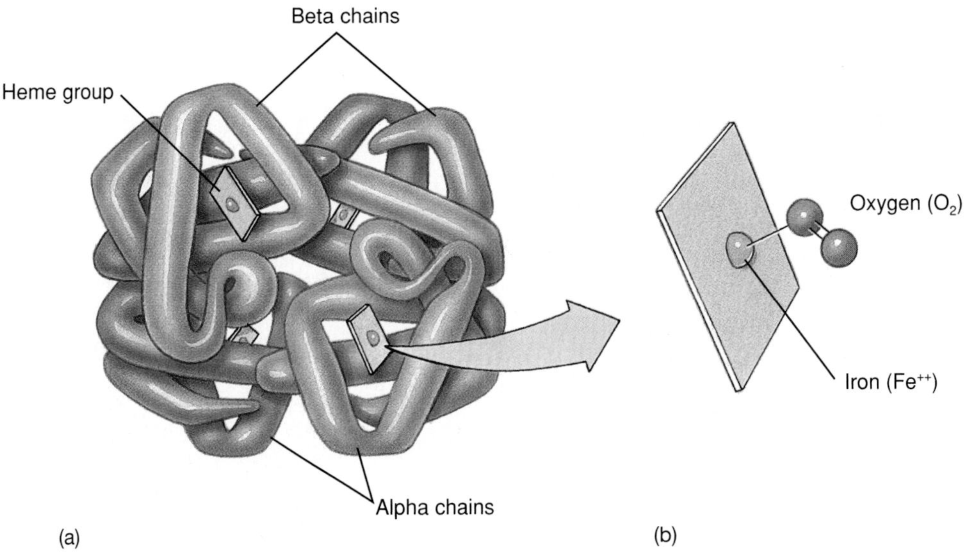

Figure 20.20 The hemoglobin molecule.
(*a*) The hemoglobin molecule is actually composed of four protein chain subunits: two copies of the "alpha chain" and two copies of the "beta chain." Each chain is associated with a heme group, and each heme group (*b*) has a central iron atom, which can bind to a molecule of oxygen.

O_2 Transport

Hemoglobin molecules act like little oxygen (O_2) sponges, soaking oxygen up within red blood cells and causing more to diffuse in from the blood plasma. At the high O_2 levels that occur in the blood supply of the lung, most hemoglobin molecules carry a full load of oxygen atoms. Later, in the tissues, the O_2 levels are much lower, so that hemoglobin gives up its bound oxygen.

In tissue, the presence of carbon dioxide (CO_2) causes the hemoglobin molecule to assume a different shape, one that gives up its oxygen more easily. This speeds up the unloading of oxygen from hemoglobin even more. The effect of CO_2 on oxygen unloading, called the **Bohr effect,** is of real importance, since CO_2 is produced by the tissues at the site of cell metabolism. For this reason the blood unloads oxygen more readily within those tissues undergoing metabolism and generating CO_2.

CO_2 Transport

At the same time the red blood cells are unloading oxygen, they are also absorbing CO_2 from the tissue. Only a tiny fraction of the CO_2 the blood carries is actually dissolved in the plasma, and about another one-fifth is carried by hemoglobin molecules within red blood cells. All the rest of the CO_2 is carried within the cytoplasm of the red blood cells. How do the red blood cells hold all this CO_2 in? To keep CO_2 from diffusing out of the red blood cells back into the plasma where CO_2 levels are low, an enzyme, carbonic anhydrase, combines CO_2 molecules with water to form carbonic acid (H_2CO_3), which dissociates into **bicarbonate ions** that do not diffuse outward. This soaking up of CO_2 by red blood cells removes large amounts of CO_2 from the blood plasma, facilitating the diffusion of more CO_2 into it from the surrounding tissue. The facilitation is critical to CO_2 removal, since the difference in CO_2 concentration between blood and tissue is not large (only 5%).

The red blood cells carry their cargo of bicarbonate ions back to the lungs. The lower CO_2 concentration in the air inside the lungs causes the carbonic anhydrase reaction to proceed in the reverse direction, releasing gaseous CO_2 which diffuses outward from the blood into the alveoli. With the next exhalation, this CO_2 leaves the body. The one-fifth of the CO_2 bound to hemoglobin also leaves because hemoglobin has a greater affinity for oxygen than for CO_2 at low CO_2 concentrations. The diffusion of CO_2 outward from the red blood cells causes the hemoglobin within these cells to release its bound CO_2 and take up oxygen instead. The red blood cells, with their newly bound oxygen, then start the next respiratory journey. Figure 20.21 presents a diagram of oxygen and carbon dioxide transport.

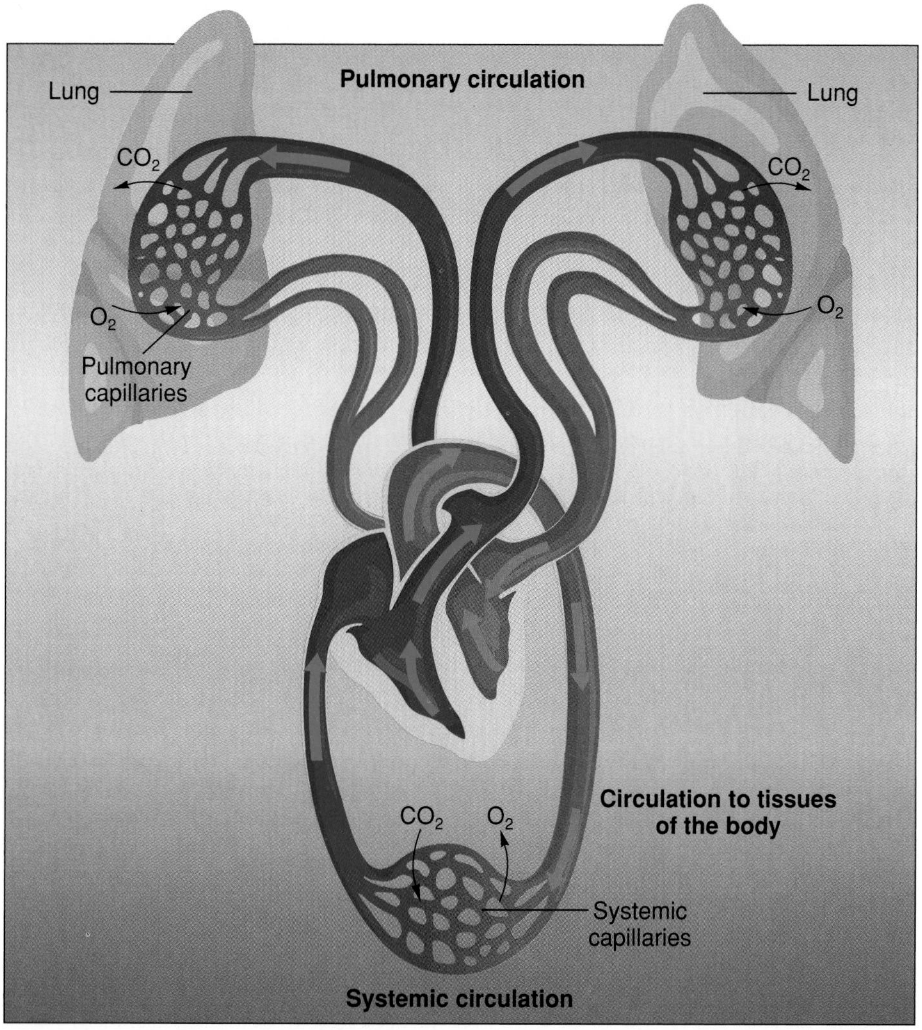

Figure 20.21 The respiratory path.

Oxygen levels are higher in the lungs than in the blood, so oxygen diffuses into the blood and binds to hemoglobin. In body tissues, oxygen levels are lower because oxygen is used up in metabolic processes, so oxygen diffuses from the blood into the tissues. Carbon dioxide, a waste product of metabolic processes, is removed from the tissues in a manner similar but opposite to oxygen delivery.

NO Transport

Hemoglobin also has the ability to hold and release the gas nitric oxide (NO). Nitric oxide, although a noxious gas in the atmosphere, has an important physiological role in the body, acting on many kinds of cells to change their shape and function. For example, in blood vessels the presence of NO causes the blood vessels to expand because it relaxes the surrounding muscle cells. Thus, blood flow and blood pressure are regulated by the amount of NO released into the bloodstream.

Hemoglobin carries NO in a special form called super nitric oxide. In this form, NO has acquired an extra electron and is able to bind to an amino acid, called cysteine, present in hemoglobin. In the lungs, hemoglobin that is dumping CO_2 and picking up O_2 also picks up NO as super nitric oxide. In blood vessels at the tissues, hemoglobin that is releasing its O_2 and picking up CO_2 can do one of two things with nitric oxide. To increase blood flow, hemoglobin can release the super nitric oxide as NO into the blood, making blood vessels expand because NO acts as a relaxing agent. Or, hemoglobin can trap any excesses of NO on its iron atoms left vacant by the release of oxygen, causing blood vessels to constrict. When the red blood cells return to the lungs, hemoglobin dumps its CO_2 and the regular form of NO bound to the iron atoms. It is then ready to pick up O_2 and super nitric oxide and continue the cycle.

How the Brain Controls Breathing

Did you ever stop to think how often you take a breath, without once consciously thinking about it? You took your first breath within moments of being born, and since then have repeated the process over 200 million times. Every breath is initiated by a respiratory control center in the brain, which sends "contract" nerve signals to the diaphragm and, for deeper breathing, to the muscles of the rib cage. These contractions regularly expand the chest cavity. Other neurons then act to inhibit the stimulation of these muscles so that they relax and the body exhales. These signals from the brain are not subject to voluntary control; we cannot consciously stop breathing for more than a minute or so. You breathe more slowly when you sleep, more rapidly when you run. You need less oxygen when you sleep because many of your muscles are not being used, and you need more oxygen when you run because you are using muscles intensively.

The brain knows when to speed up breathing by keeping track of muscle use. When you run, rapid muscle contraction adds increased amounts of CO_2 to your blood. In a simple feedback mechanism, special sensory cells in the brain called chemoreceptors detect that rise in CO_2 and transmit nerve impulses to the brain's respiratory center, which signals the muscles of the diaphragm and rib cage to produce more rapid and deeper breathing. Other chemoreceptors in the aorta and carotid artery are also sensitive to greatly reduced levels of oxygen.

20.5 Lung Cancer and Smoking

Of all the diseases to which humans are susceptible, none is more feared than cancer. One in every four deaths in the United States is caused by cancer. The American Cancer Society estimates that 554,740 people will die of cancer in 1996. About 30% of these—158,700 people—will die of lung cancer. About 140,000 cases of lung cancer were diagnosed each year in the 1980s, and 90% of these persons died within three years. Lung cancer is one of the leading causes of death among adults in the world today. What has caused lung cancer to become a major killer of Americans in this century? To solve this mystery, we must first look more closely at cancer.

The Nature of Cancer

Cancer is a growth disorder of cells. It starts when an apparently normal cell begins to grow in an uncontrolled and invasive way. The result is a ball of cells, called a **tumor,** that constantly expands in size (figure 20.22). When this ball remains a hard mass, it is called a **sarcoma** if connective tissue, such as muscle, is involved or a **carcinoma** if epithelial tissue, such as skin, is involved. Often cancer cells develop abnormally and can no longer recognize the tissue where they originated; such cells leave the tumor mass and spread throughout the body, forming new tumors at distant sites. The spreading cells are called **metastases.** As a lung cancer tumor grows within the lung, for example, its cells invade the surrounding tissues and eventually break through into the lymph and blood vessels (figure 20.23). Once the cancer cells have done this, they spread rapidly through the body, lodging and growing at many locations, particularly the brain. Death soon follows.

The search for a cause of cancers such as lung cancer has uncovered a host of environmental factors that appear to be associated with cancer. For example, the incidence of cancer per 1,000 people is not uniform throughout the United States. Rather, it is centered in cities and in the Mississippi Delta, suggesting that pollution and pesticide runoff may contribute to cancer (figure 20.24). When the many environmental factors associated with cancer are analyzed, a clear pattern emerges: most cancer-causing agents, or **carcinogens,** share the property of being potent **mutagens.** A mutagen is a chemical or radiation that damages DNA, destroying or changing genes (a change in a gene is called a mutation). The conclusion that cancer is caused by mutation is now supported by an overwhelming body of evidence.

What sort of genes are being mutated? In the last several years researchers have found that mutation of only a few genes is all that is needed to transform normally dividing cells

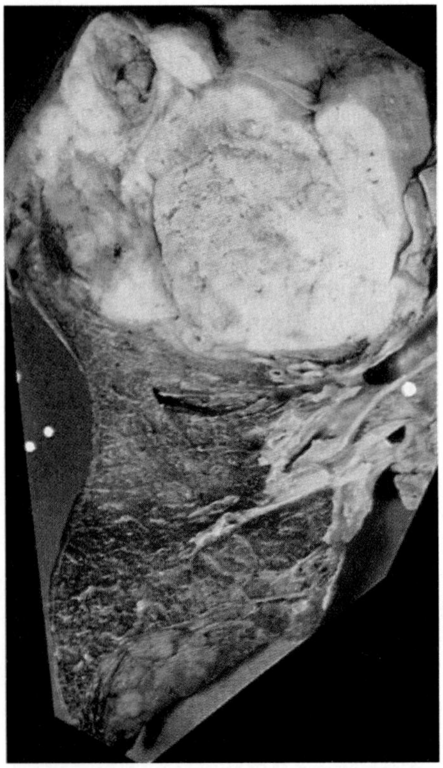

Figure 20.22 A cancerous lung.
The bottom half of this lung is normal; the top half has been taken over completely by a cancerous tumor. As you might imagine, a lung in this condition does not function well, but the difficulty in breathing does not cause death. Rather, death usually results when the cancer spreads to other tissues throughout the body.

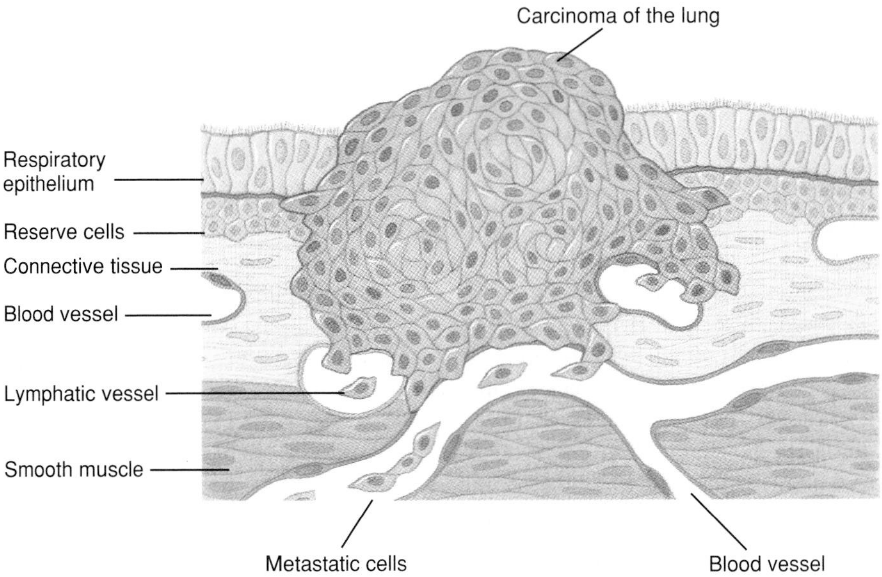

Figure 20.23 Portrait of cancer.
The ball of cells is a carcinoma, developing from epithelial cells lining the interior surface of a human lung. As the mass of cells grows, it invades surrounding tissues, eventually penetrating lymphatic vessels and blood vessels, both of which are plentiful within the lung. These vessels carry metastatic cancer cells throughout the body, where they lodge and grow, forming new foci of cancerous tissue.

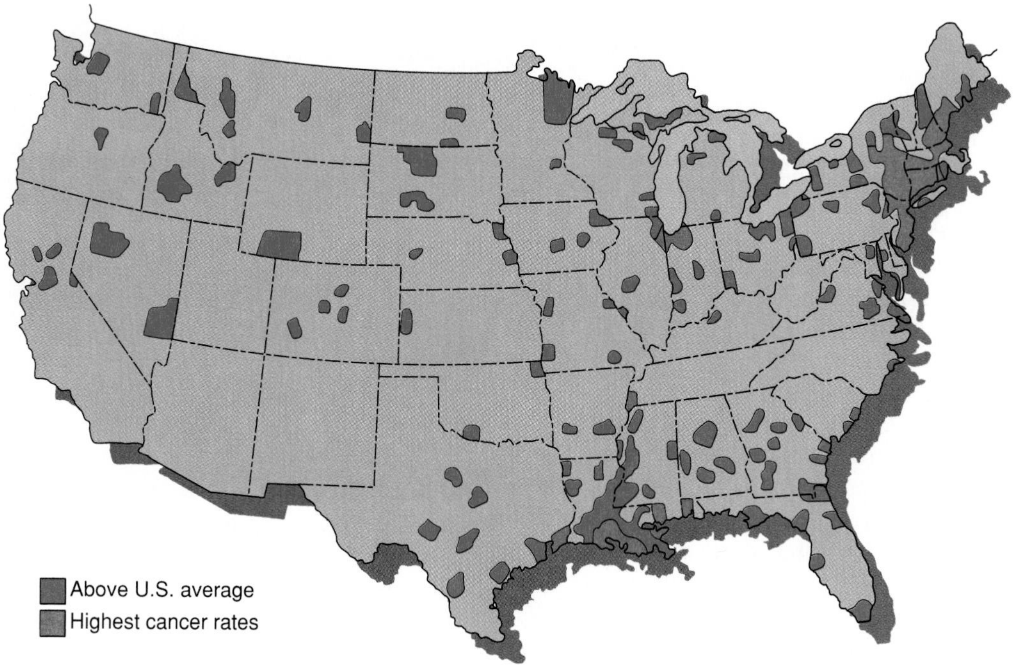

Figure 20.24 Cancer in the United States.
The incidence of cancer per 1,000 people is not uniform throughout the United States. It is centered in cities where chemical manufacturing is common and in the Mississippi Delta. This suggests that pollution and pesticide runoff may contribute to the development of cancer.

into cancerous ones. Identifying and isolating these cancer-causing genes, investigators have learned that all are involved with regulating cell proliferation (how fast cells grow and divide). The emerging picture is that cancer is caused by damage to genes that normally restrict cell proliferation. In effect, the "don't grow" control that normally prevents all of your cells from dividing unceasingly is inactivated. Your cells guard against unwanted proliferation with a variety of safety backup systems, so more than one gene needs to be damaged before the controls are lost. Thus, cancer is usually a disease of old people—because it takes many years of mutations before all the gene "safety-switches" are knocked out by random mutations.

What Causes Lung Cancer?

If cancer is caused by damage to growth-regulating genes, damage often created by exposure to mutagens, what then has led to the rapid increase in **lung cancer** in the United States in this century? What might the mutagen be? The essential clue lies in the fact that women began to "catch" cancer much later than men. As late as 1920, lung cancer was a rare disease. But American men began smoking in large numbers at the turn of the century, and 20 years later (it takes time for mutations occurring at random to "hit" several genes) the incidence of lung cancer among American men started to rise, reaching its current high level in the 1960s (figure 20.25). For the last 30 years, the incidence of lung cancer in American men has held steady at about 1 per 1,000 per year. However, as late as 1963, when lung cancer among males was near current levels, this disease was still rare in females. In the United States that year, only 6,588 females died of lung cancer. Because of social mores, significant

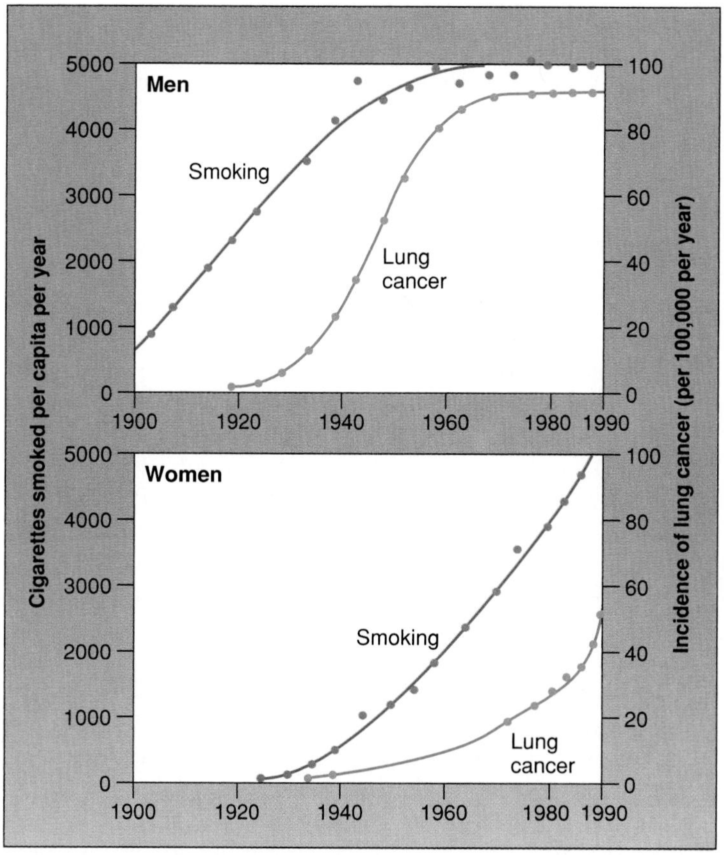

Figure 20.25 Incidence of lung cancer in men and women.
As smoking increased in men in the early 1900s, the incidence of lung cancer also increased. Women followed suit years later, and in 1990, more than 49,000 women died of lung cancer.

numbers of American women did not smoke until 1945, when after World War II many social conventions changed—and 20 years later the rise in lung cancer began among American women. Every year since, the incidence of lung cancer among American women has risen, until today lung cancer death rates among women are approaching those of males. The conclusion is inescapable: *With the 20-year lag for mutational buildup, the incidence of lung cancer tracks the incidence of cigarette smoking, in both sexes.* Of the estimated 177,000 Americans who will get lung cancer in 1996, about 87% of them will be smokers.

Smoking is a popular pastime in the United States. Twenty-nine percent of the U.S. population smokes. American smokers consumed over 500 billion cigarettes in 1995. These cigarettes emit in their tobacco smoke some 3,000 chemicals, among them vinyl chloride, benzo (*a*) pyrenes, and nitroso-*nor*-nicotine, all potent mutagens. Smoking introduces these mutagens to the tissues of your lungs. Among smokers, the current rate of deaths resulting from lung cancer is 180 per 100,000, or about 2 of each 1,000 smokers *each year*.

In the face of these facts, why do so many people continue to smoke? Because the nicotine in cigarette smoke is an addictive drug. Researchers have identified the receptor on central nervous system neurons that it binds to, and they have demonstrated that smoking leads to a decrease in the brain's population of these receptors, leading to a craving for more cigarettes. This mechanism of nicotine addiction is very similar to that of cocaine addiction; once a person becomes addicted, it is difficult to quit. About half of those who try eventually succeed. Because your life is at stake, it is well worth the effort.

Clearly, an effective way to avoid lung cancer is not to smoke. Life insurance companies have computed that, on a statistical basis, smoking a single cigarette lowers your life expectancy 10.7 minutes (more than the time it takes to smoke the cigarette!). Every pack of 20 cigarettes bears an unwritten label: "*The price of smoking this pack of cigarettes is 3-1/2 hours of your life*" (figure 20.26). Smoking a cigarette is very much like going into a totally dark room with a person who has a gun and standing still. The person with the gun cannot see you, does not know where you are, and shoots once in a random direction. A hit is unlikely, and most shots miss. As the person keeps shooting, however, the chance of eventually scoring a hit becomes more likely. In the same way, every time an individual smokes a cigarette, mutagens are being shot at his or her genes, and sooner or later cancer-causing genes will be hit. Nor do statistics protect any one individual: nothing says the first shot will not hit. Older people are not the only ones who die of lung cancer (figure 20.27).

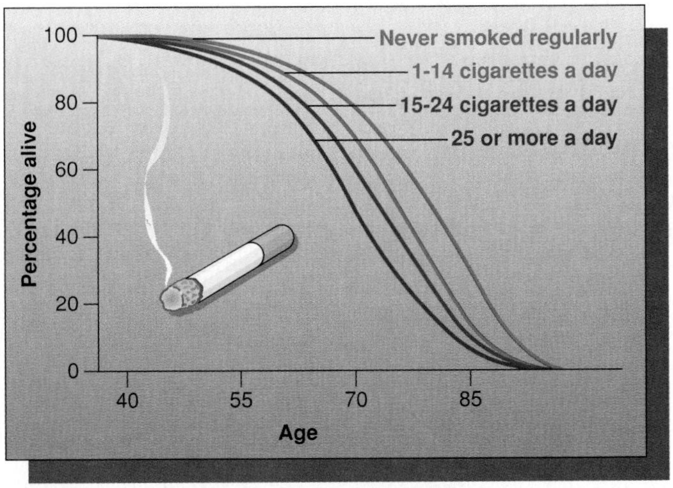

Figure 20.26 Tobacco kills one smoker in two.
Studies have shown a clear relationship between cigarette smoking and reduced life expectancy. Smoking a single cigarette has been estimated to lower life expectancy by 10.7 minutes.
Source: Data from *New Scientist*, October 15, 1994.

Figure 20.27 A teenage smoker.
Older people are not the only ones who will suffer from lung cancer as a consequence of smoking cigarettes.

CHAPTER 20

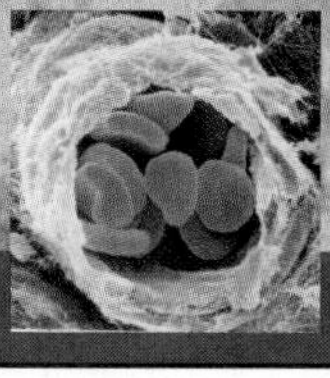

HIGHLIGHTS

20.1 The Circulatory System

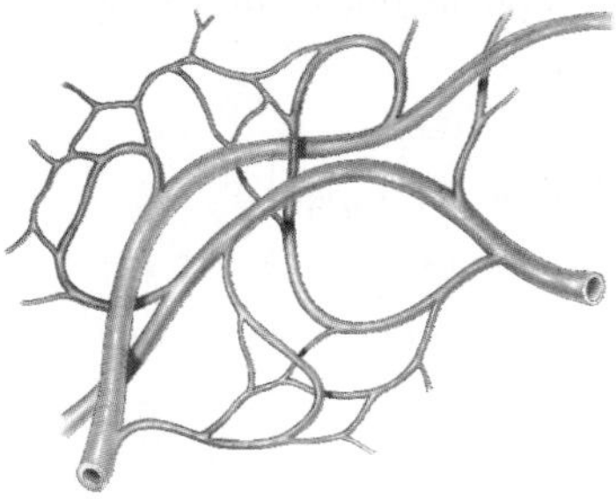

Key Terms

cardiovascular system 434
arteries 434
capillaries 434
veins 434
lymphatic system 437

Key Concepts

- Blood is pumped out from the heart through arteries and returns through veins.
- Materials pass in and out of the blood as it passes through a network of tiny tubes called capillaries, which connects arterial to venous circulation.
- Fluid that passes out of the capillaries is recovered by the lymphatic system.

20.2 Blood

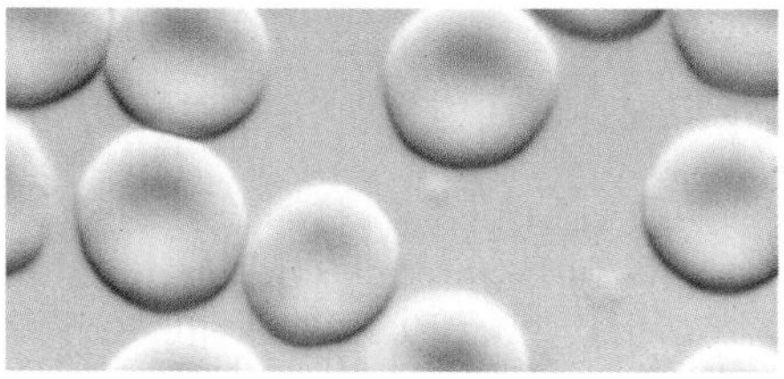

Key Terms

plasma 438
erythrocytes 439
leukocytes 439
platelets 439

Key Concepts

- Blood is composed of a fluid called plasma, white blood cells of the immune system, and red blood cells that carry oxygen.
- Dissolved in plasma are many kinds of molecules, including a considerable amount of serum albumin protein, which osmotically balances the blood.

20.3 The Heart and How It Works

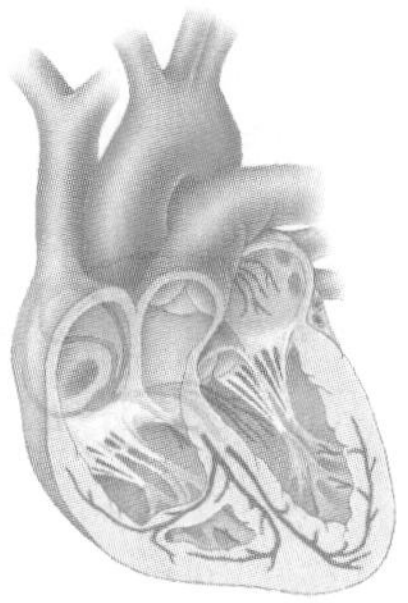

Key Terms

atrium 440
pulmonary veins 440
ventricle 440
sinoatrial (SA) node 441
diastolic 442
systolic 442
atherosclerosis 442

Key Concepts

- The heart is a double pump.
- The right side pumps oxygen-poor blood to the lungs and then returns it via pulmonary veins to the left side.
- The left side pumps oxygen-rich blood to the body and then returns it to the right side.
- The contraction of the heart starts in the upper wall of the right atrium and spreads as a wave across the heart.

20.4 The Respiratory System

Key Terms

bronchus 443
trachea 443
alveoli 444
diaphragm 444
Bohr effect 446
bicarbonate ions 446

Key Concepts

- Respiration is the uptake of oxygen gas from air and the simultaneous release of carbon dioxide.
- Expansion of the chest cavity draws air into the lungs; relaxation forces it back out.
- CO_2 is transported largely as bicarbonate ion.

20.5 Lung Cancer and Smoking

Key Terms

metastases 448
carcinogens 448
lung cancer 449

Key Concepts

- Cancer is unrestrained cell proliferation caused by damage to growth-regulating genes.
- Cigarette smoking is the principal cause of lung cancer.
- Smoking produces lung cancer by introducing carcinogens into the lungs.

20

CONCEPT REVIEW

1. Select the smallest type of blood vessel.
   a. artery
   b. capillary
   c. vein
   d. venule

2. Which of the following statements is *not* true about arteries?
   a. They carry blood away from the heart.
   b. They are capable of expanding.
   c. They contain many valves.
   d. They deliver blood to arterioles.

3. Each is a function of the lymphatic system except
   a. delivering oxygen to cells.
   b. returning fluid to veins.
   c. absorbing fat.
   d. carrying bacteria to lymph nodes.

4. Blood plasma is about _________ water.
   a. 25%
   b. 50%
   c. 75%
   d. 90%

5. Erythrocytes are packed with
   a. fibrinogen.
   b. serum albumin.
   c. hemoglobin.
   d. platelets.

6. Which valve is between the left atrium and ventricle?
   a. tricuspid
   b. mitral
   c. aortic
   d. pulmonary

7. Blood pumped out of the left ventricle moves into the
   a. aorta.
   b. pulmonary veins.
   c. right ventricle.
   d. superior vena cava.

8. Arrange the following in the order in which air contacts them during breathing, starting from the mouth and nose.
   a. bronchus
   b. alveoli
   c. larynx
   d. hemoglobin
   e. trachea

9. Which of the following statements is *not* true about the diaphragm?
   a. It separates two body cavities.
   b. It is a thick layer of muscle.
   c. It is located inside the lungs.
   d. It is part of the human respiratory apparatus.

10. Air is transported through the alveoli walls by
    a. active transport.
    b. diffusion.
    c. osmosis.
    d. endocytosis.

11. A rise in carbon dioxide in the blood _________ the breathing rate.
    a. decreases
    b. increases

12. Air normally enters your body through the nostrils, which are filled with hairs. What is the function of these hairs?
    a. They allow you to breathe while you eat.
    b. They slow and regulate the passage of air.
    c. They moisten the air.
    d. They filter out dust and other particles.

13. Arterioles have a thick middle layer of _________ muscle.

14. No cell of your body is more than _________ away from a capillary.

15. _________ capillaries collect fluid that leaks from the cardiovascular system.

16. The function of _________ is to keep the blood plasma in osmotic equilibrium with the cells of the body.

17. For a blood pressure reading of 110/70, 110 is the _________ pressure.

18. Alveoli are connected to bronchi by a network of tubes called _________.

19. Special chemoreceptors in the _________ detect levels of CO_2 in your blood.

20. _________ are agents that cause cancer.

Answers to the Concept Review questions appear in Appendix B.

CHALLENGE YOURSELF

1. People who appear to have drowned can often be revived, in some cases after being underwater for as long as half an hour. In every case of full recovery after extended submergence, however, the person has been submerged in very cold water. Try to explain this observation.

2. Why can't humans or other mammals breathe water as well as air? In other words, what is it about the lung as a gas exchange organ that makes it unable to function effectively when filled with water?

3. The evidence associating lung cancer with smoking is overwhelming, and as you have learned in this chapter, we now know in considerable detail the mechanism whereby smoking induces cancer—it introduces powerful mutagens into the lungs, which cause mutations to occur; when a growth-regulating gene is mutated by chance, cancer results. The process is no more mysterious than the fact that death results from shooting shotguns at random in crowded football stadiums. In light of this, why do you suppose that cigarette smoking is still legal?

FOR FURTHER READING

Bartecchi, C. E., et al. "The Global Tobacco Epidemic." *Scientific American,* May 1995, 44–51. Cigarette smoking has stopped declining in the United States and is rising elsewhere in the world.

Gillis, A. "As Good as Blood? The Creation of Blood Substitutes Has Been Harder than Scientists Predicted, but Recent Improvements Are Encouraging." *BioScience* 43 (September 1993): 517–20. The construction of a synthetic version of something as complex as a red blood cell is understandably quite challenging.

Long, P. "A Town with a Golden Gene." *Health* 8 (January/February 1994): 60ff. In Limone, Italy, a fortuitous mutation in a gene is providing several generations of a family with resistance to heart disease via regulated levels of HDL. HDLs and their role in heart disease are discussed along with the study of the Italian villagers.

Mestrel, R. "The Secret Life of the Nose." *New Scientist* 138 (April 17, 1993): 13. Poorly functioning or nonfunctioning noses play a critical role in difficulties such as asthma, lung problems, and breathing problems.

Saul, H. "Fine Young Slobs?" *New Scientist* 142 (April 13, 1994): 24–25. Are bad diet and sedentary lifestyle in children and adolescents increasing their risk for heart disease and other cardiovascular problems? Health professionals believe they are.

Stone, R. "New Radon Study: No Smoking Gun." *Science* 263 (January 28, 1994): 465. The last decade has seen rising alarm in a potential new killer: radon gas. How much of a link is there between this allegedly invisible, odorless, deadly gas and lung cancer?

Travis, J. "Helping Premature Lungs Breathe Easier." *Science* 261 (July 23, 1993): 261. One of the highest causes of infant mortality in premature babies is respiratory distress, caused principally because the immature lung cannot yet secrete enough surfactant, a chemical that decreases surface tension in the lungs allowing for gas transfer. Researchers are looking for a cost-effective way to reproduce a key protein required for the synthesis of surfactant.

Willensky, D. "Circulatory Systems." *American Health* 13 (May 1994): 104–5. A nice overview article discussing the structure and function of the circulatory system, with illustrations.

TECHNOLOGY LINKS

The Living World Home Page
http://www.wcbp.com/biology/tlw

Explorations in Human Biology CD
#5 Evolution of the Heart
#6 Smoking and Cancer

The Dynamic Human CD
Cardiovascular System
Lymphatic System
Respiratory System

Life Science Animations Videotape 4
#37 Blood Circulation
#38 Production of Electrocardiogram
#39 Common Congenital Defects of the Heart

21

The Path of Food Through the Body

THE FAR SIDE By GARY LARSON

"Just think ... Here we are, the afternoon sun beating down on us, a dead, bloated rhino underfoot, and good friends flying in from all over. ... I tell you, Frank, this is the best of times."

CHAPTER OUTLINE

Figure 21.1 Food gives us energy and nutrients. The digestive system breaks down food molecules so that our cells can harvest the energy stored within the chemical bonds.

The human body is a living machine that uses energy to move, to dream, to grow, and to perform countless other activities you rarely think about. Flicking your eyelashes takes energy, and so does loving someone. We obtain the energy that fuels our lives from the food we eat. The carbohydrates, fats, and proteins that we use as food are rich in high-energy chemical bonds. We use this chemical energy to make ATP, the molecule that carries energy from one place to another in our bodies. In addition, food gives us many *materials* that we need to build our bodies: calcium for bones, amino acids for making proteins, vitamins and ions and a host of other molecules. Every molecule in your body, every hair and bone and cell, is built from materials that you eat (figure 21.1). In this chapter we discuss how the body uses food and water, the raw materials of life.

21.1 Diet: What We Need to Eat and Why

The food you eat provides both a source of energy and a supply of raw materials that your body is not able to manufacture for itself.

Calories for Energy

An optimal **diet** contains more carbohydrates than fats and also a significant amount of protein, as the "pyramid of nutrition" recommended by the federal government indicates (figure 21.2). Fats have a far greater number of energy-rich carbon–hydrogen bonds and thus a much higher energy content per gram than carbohydrates or proteins. That makes fats a very efficient way to store energy. When you eat, your body has only two choices about what to do with the energy it harvests from your food: either it is metabolized by the muscles and other cells of the body, or it is converted into fat and stored in fat cells. Thus we have the simple equation:

$$\text{food} - \text{exercise} = \text{fat}$$

Carbohydrates are obtained primarily from cereals and grains, breads, fruits, and vegetables. On the average, carbohydrates contain 4.1 **calories** per gram; fats, by comparison, contain 9.3 calories per gram, over twice as much. Dietary fats are obtained from oils, margarine, and butter and are abundant in fried foods, meats, and processed snack foods, such as potato chips and crackers. Protein, like carbohydrates, has 4.1 calories per gram and can be obtained from dairy products, poultry, fish, meat, and grains (figure 21.3).

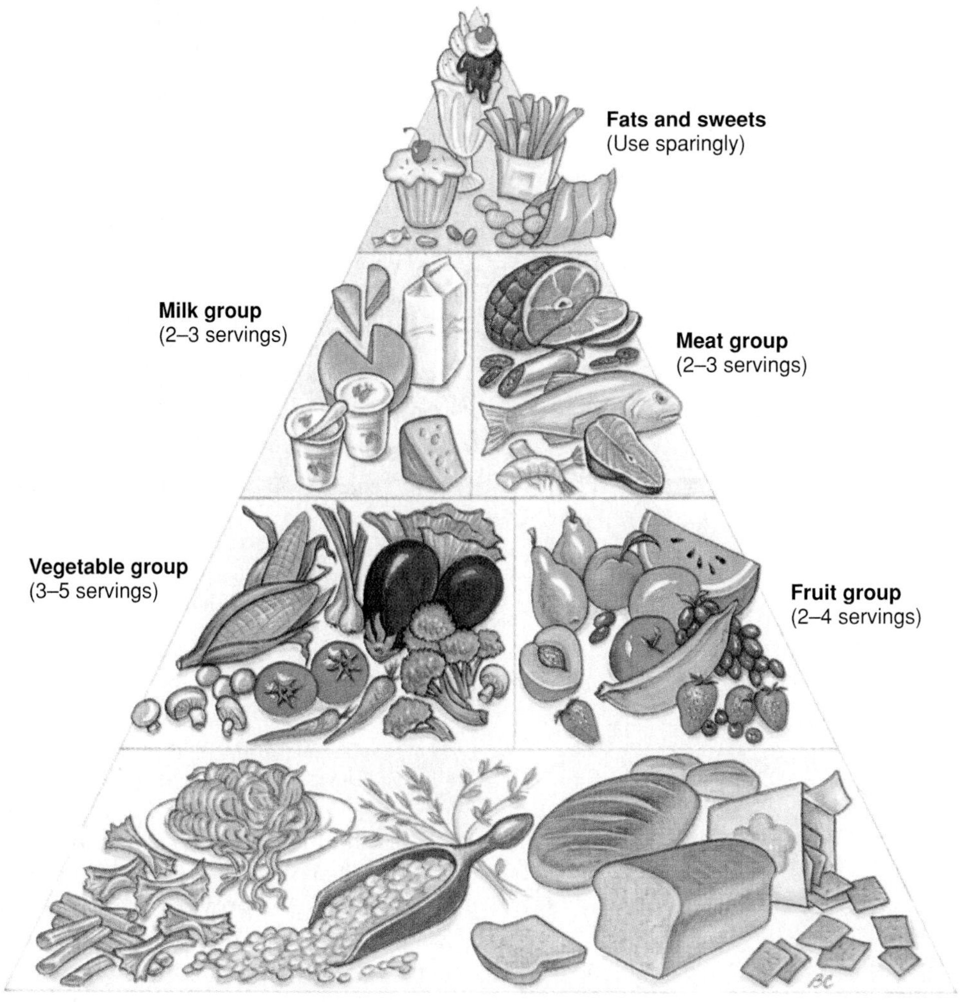

Figure 21.2 The pyramid of nutrition.
A healthy diet uses more foods at the base of the pyramid and fewer at the top.
Source: U.S. Department of Agriculture, U.S. Department of Health and Human Services

Your body uses carbohydrates for energy. Your body uses fats to construct cell membranes and other cell structures, to insulate nervous tissue, and to provide energy. Fats also contain certain fat-soluble vitamins that are essential for proper health. Proteins are used as building materials for cell structures, enzymes, hemoglobin, hormones, and muscle and bone tissue.

In wealthy countries such as those of North America and Europe, obesity (being significantly overweight, see figure 21.4 on next page) is common, the result of habitual overeating and high-fat diets. In the United States, about 30% of middle-aged women and 15% of middle-aged men are classified as overweight (that is, more than 20% heavier than the average person of the same sex and height). Being overweight is highly correlated with coronary heart disease and many other disorders.

One essential characteristic of food is simply bulk, its content of undigested fiber. The large intestine of humans has evolved as an organ adapted to process food that has a relatively high fiber content. Presumably, our ancestors ate a high-fiber diet. Diets that are low in fiber, now common in the United States, result in a slower passage of food through the colon. This low dietary fiber content is thought to be associated with levels of colon cancer in the United States that are among the highest in the world.

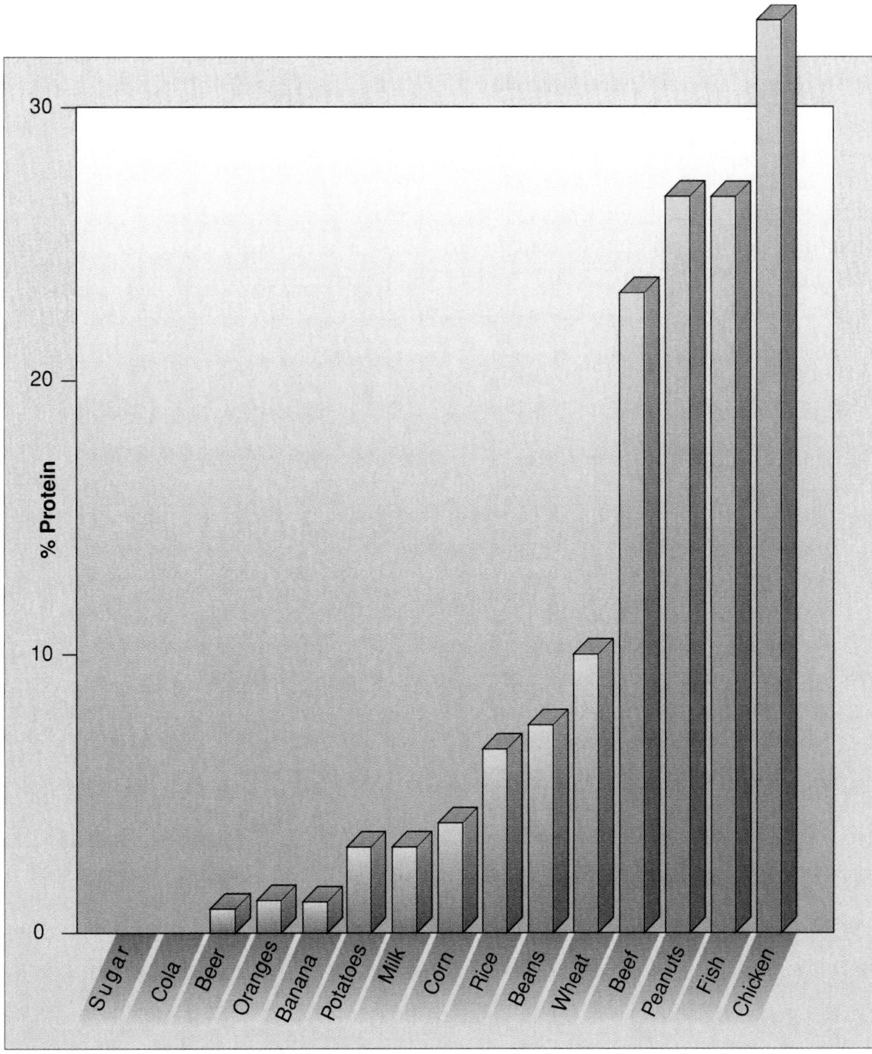

Figure 21.3 The protein content of a variety of common foods.
Humans and many other vertebrates must eat proteins to obtain the amino acids they are unable to synthesize.

Essential Substances for Growth

Over the course of their evolution, many animals lose the ability to manufacture certain substances they need, substances that often play critical roles in their metabolism. They lose this ability for the same reason a child wastes food—there's already plenty of it around, so why worry? Mosquitoes and many other blood-sucking insects, for example, cannot manufacture cholesterol, but they obtain it in their diet because human blood is rich in cholesterol. Humans are unable to manufacture 8 of the 20 amino acids used to make proteins, presumably because they were common in the diet of our ancestors: lysine, tryptophan, threonine, methionine, phenylalanine, leucine, isoleucine, and valine. These amino acids, called **essential amino acids,** must therefore be obtained from proteins in the food we eat. For this reason, it is important to eat so-called complete proteins, that is, ones containing all the essential amino acids. Humans (and all other vertebrates) have also lost the ability to synthesize certain polyunsaturated fats that provide backbones for the many kinds of fats our bodies manufacture.

One food people commonly eat as a source of protein is beef. Recently in Britain, however, the appearance of bovine spongiform encephalopathy (BSE), know as mad-cow disease, has caused schools to ban beef from their menus! Mad-cow disease is caused by an infectious protein called a **prion.** Prions have no nucleic acid that can be denatured by heat or other food-processing methods, such as pasteurization, sterilization, freezing, or drying. People who eat beef infected with mad-cow disease run the risk of contracting prions. In humans, prions can cause Creutzfeldt-Jakob disease (CJD), which affects the brain. Indeed, the deaths of four dairy farmers and two teenagers from CJD in Britain prompted the British government to destroy thousands of prion-infected cattle in 1996.

Vitamins

Essential organic substances that are used in trace amounts are called **vitamins.** Humans require at least 13 different vitamins (table 21.1). Many vitamins are required cofactors for cellular enzymes. Humans, monkeys, and guinea pigs, for example, have lost the ability to synthesize ascorbic acid (vitamin C) and will develop the potentially fatal disease called scurvy—characterized by weakness, spongy gums, and bleeding of the skin and mucous membranes—if vitamin C is not supplied in their diets. All other mammals are able to synthesize ascorbic acid. Energy, proper vitamins, and exercise all interact to determine proper nutrition (figure 21.5).

Trace Elements

In addition to supplying energy and essential organic compounds, the food that you consume must also supply your body with *essential minerals* such as calcium and phosphorus, as well as a wide variety of **trace elements,** which are minerals required in very small amounts. Among the trace elements are *iodine* (a component of thyroid hormone), *cobalt* (a component of vitamin B_{12}), *zinc* and *molybdenum* (components of enzymes), *manganese,* and *selenium.* All of these, with the possible exception of selenium, are also essential for plant growth; we obtain them directly from plants that we eat or indirectly from animals that have eaten plants or plant-eaters.

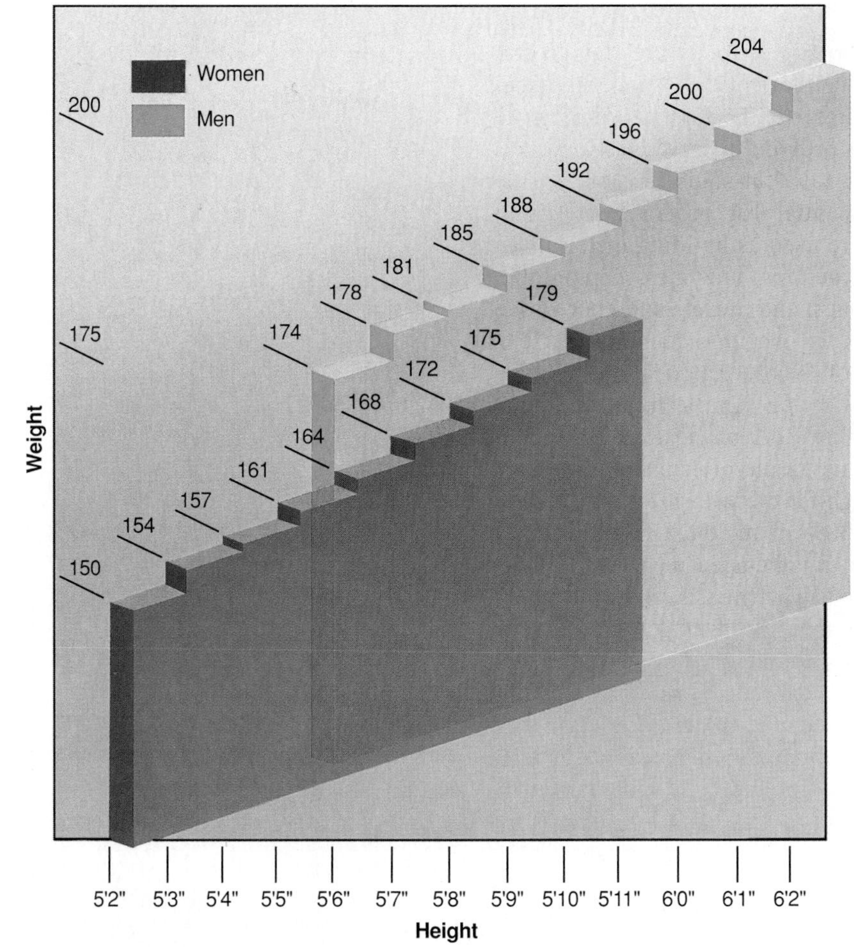

Figure 21.4 Are you overweight?

Obesity is usually defined as being more than 20% heavier than the average person of the same sex and height. This chart presents the weight values at which obesity begins in Americans for a variety of heights.

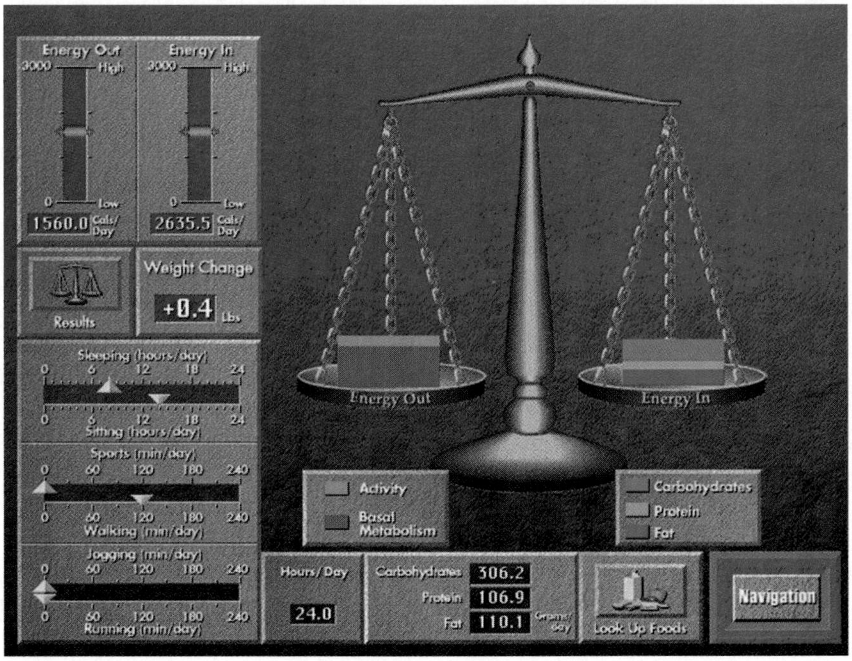

Figure 21.5 How diet and exercise influence weight.

Diet and exercise interact to determine whether we gain or lose weight. This figure is a screen capture from an interactive exercise that allows you to hypothetically alter both the kinds of food you eat, how much of each you consume, and the amount of exercise you perform. You can investigate how different diets influence the "burn fat/make fat" decision. You will discover that the key decision depends on levels of blood glucose, not ATP. (For example, a high-calorie diet can actually cause you to *lose* weight by burning fat if that diet is very low in carbohydrates.) (*Explorations in Human Biology,* Module 7, "Diet and Weight Loss")

Table 21.1 Major Vitamins

Vitamin	Function	Dietary Source	Recommended Daily Allowance (milligrams)	Deficiency Symptoms
Vitamin A	Used in making visual pigments, maintenance of epithelial tissues	Green vegetables, carrots, milk products, liver	1	Night blindness, flaky skin
B-Complex Vitamins				
B_1	Coenzyme in CO_2 removal during cellular respiration	Meat, grains, legumes	1.5	Beriberi, weakening of heart, edema
B_2 (riboflavin)	Part of coenzymes FAD and FMN, which play metabolic roles	In many different kinds of foods	1.8	Inflammation and breakdown of skin, eye irritation
B_3 (niacin)	Part of coenzymes NAD^+ and $NADP^+$	Liver, lean meats, grains	20	Pellagra, inflammation of nerves, mental disorders
B_5 (pantothenic acid)	Part of coenzyme-A, a key connection between carbohydrate and fat metabolism	In many different kinds of foods	5 to 10	Rare: fatigue, loss of coordination
B_6 (pyridoxine)	Coenzyme in many phases of amino acid metabolism	Cereals, vegetables, meats	2	Anemia, convulsions, irritability
B_{12} (cyanocobalamin)	Coenzyme in the production of nucleic acids	Red meat, dairy products	0.003	Pernicious anemia
Biotin	Coenzyme in fat synthesis and amino acid metabolism	Meat, vegetables	Minute	Rare: depression, nausea
Folic acid	Coenzyme in amino acid and nucleic acid metabolism	Green leafy vegetables, whole grain products	0.4	Anemia, diarrhea
Vitamin C	Important in forming collagen, cement of bone, teeth, connective tissue of blood vessels; may help maintain resistance to infection	Fruit, green leafy vegetables	45	Scurvy, breakdown of skin, blood vessels
Vitamin D (calciferol)	Increases absorption of calcium and promotes bone formation	Dairy products, cod liver oil	0.01	Rickets, bone deformities
Vitamin E (tocopherol)	Protects fatty acids and cell membranes from oxidation	Margarine, seeds, green leafy vegetables	15	Rare
Vitamin K	Essential to blood clotting	Green leafy vegetables	0.03	Severe bleeding

21.2 The Human Digestive System

Our bodies obtain the energy needed for growth and activity by using the energy of sugars, fatty acids, and amino acids to make ATP. Eating the body of an animal or plant, however, does not in itself provide you with a rich source of these small molecules, because few organisms contain significant concentrations of free sugars, fatty acids, and amino acids. Instead, they contain starches, fats, and proteins, which are long chains built of the simple molecules, like a necklace is built of pearls. Your body cannot extract energy from these big molecules without first breaking them down. Imagine trying to obtain building materials to construct a new house by consuming old houses—you would find very few loose bricks lying around and few boards. To build the house, you would first have to disassemble the old ones to get the building materials you need. In just the same way, the bodies of the organisms we eat have all their "building materials" tied up in large molecules, which must be broken down into small molecules before we can extract any energy from them. This process is called **digestion,** and the system of organs that carries it out is called the **digestive system** (figure 21.6).

Enzymes that break up proteins into amino acids are called *proteases;* enzymes that break up starches and other carbohydrates into sugars are called *amylases;* and enzymes that break up fats and other lipids into small fatty acids are called *lipases*. Other enzymes break down DNA and RNA into nucleotides. Most of these digestive enzymes cannot tolerate high acid concentrations, so the vertebrate digestive process is carried out in two places: digestion by acids takes place first, in the stomach, and then the food moves to the small intestine, where the acid is neutralized and the digestive enzymes dismantle the large molecules. The products of digestion (free amino acids, sugars, and fatty acids) then pass across the wall of the small intestine into the bloodstream.

Digesting proteins—breaking them up into amino acids—presents a particularly tough problem, because almost all proteins are either folded into tight balls or wound together into tough fibers. Protein-digesting enzymes cannot attack these balls and fibers, because they cannot get at individual protein strands. The human body solves this problem by carrying out digestion in two steps: first, hydrochloric acid (HCl) is used to unfold large proteins into single polypeptide strands, and then a battery of highly specific enzymes attack the strands, as well as starches and fats.

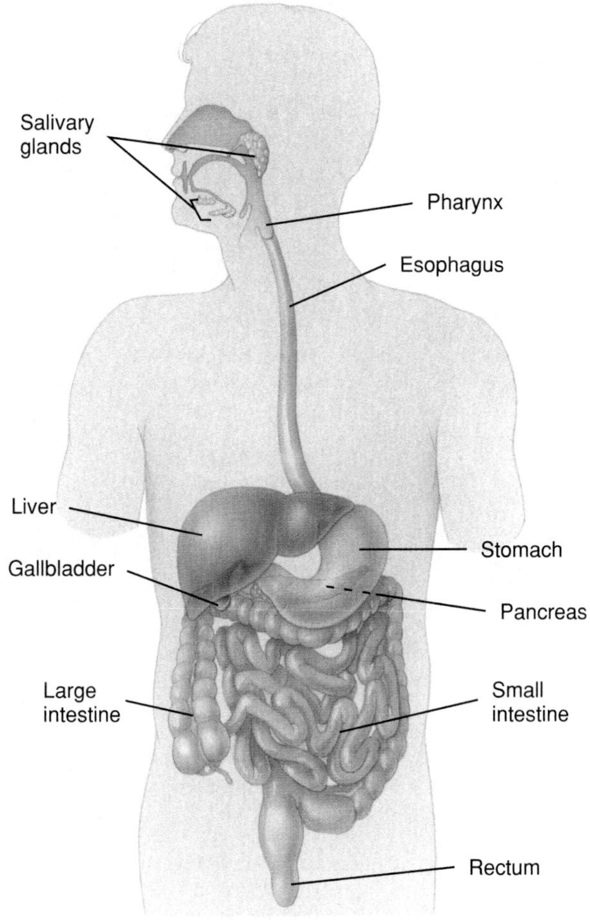

Figure 21.6 The human digestive system.
The human digestive system consists of a tubular gastrointestinal tract and accessory digestive organs.

Where It All Begins: The Mouth

Food is taken in through the **mouth.** Because we humans are omnivores, eating both animal and plant food regularly, our **teeth** are structurally intermediate between the pointed cutting and ripping teeth characteristic of meat-eaters (carnivores) and the flat, grinding teeth characteristic of plant-eaters (herbivores) (figure 21.7). Our teeth and tongue start the mechanical breakdown of the food particles.

Within the mouth, the **tongue** mixes the food with a mucous solution called **saliva.** Saliva is secreted into your mouth by three pairs of salivary glands located above and below the jaw and in the tongue. Saliva moistens and lubricates the food so that it is swallowed more readily and does not abrade the tissue it passes on its way down to the stomach. Saliva also contains **amylase** enzymes, which start breaking down starch into free sugars. This is the first digestive process that occurs as food passes through the digestive system. However, salivary amylase is not essential because similar enzymes released into the small intestine from the pancreas can also digest carbohydrates very effectively.

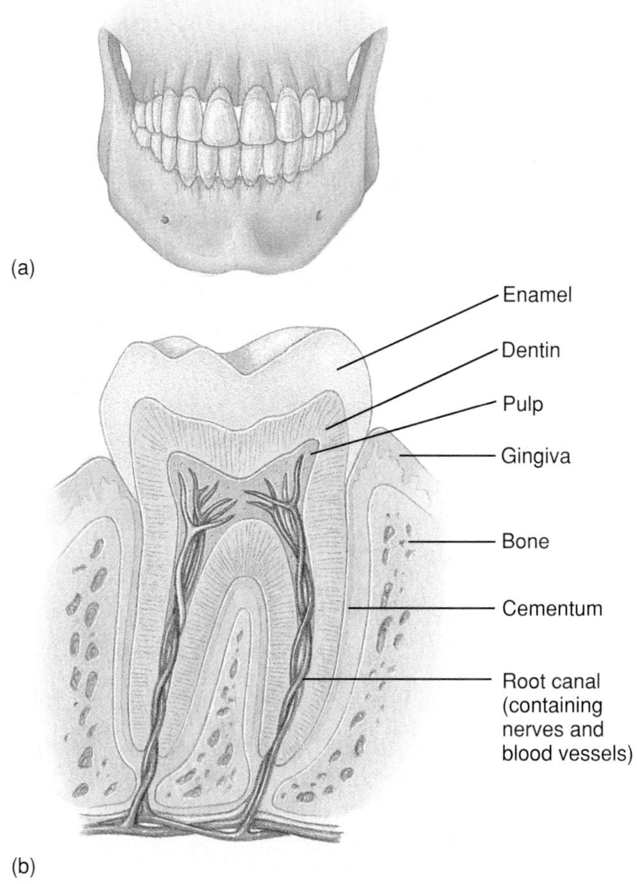

Figure 21.7 Human teeth.

(*a*) The front six teeth on the upper and lower jaws are cuspids and incisors. The remaining teeth, running along the sides of the mouth, are grinders called premolars and molars. Hence, humans are carnivores in the front of their mouths and herbivores in the back. (*b*) Each tooth is alive, with a central pulp containing nerves and blood vessels. The actual chewing surface is a hard enamel layered over the softer dentin, which forms the body of the tooth.

The Journey of Food to the Stomach

After passing through an opening at the back of the mouth, the food enters the **esophagus,** a tube that connects the mouth to the stomach. No digestion takes place in the esophagus. Its role is simply to move the food down to the stomach (figure 21.8). In adult humans, the esophagus is about 25 centimeters (10 inches) long, and its lower end opens into the stomach. The lower two-thirds of the esophagus is wrapped in smooth muscle. Successive waves of contraction of these sheets of muscle move food down through the esophagus to the stomach. Such rhythmic waves of muscular contraction in the walls of a tube are called **peristalsis.** Because the movement of food through the esophagus is primarily caused by these peristaltic contractions, humans can swallow even if they are upside down.

The exit of food from the esophagus to the stomach is controlled by a muscular ring called a **sphincter.** When the sphincter is contracted, it keeps food in the stomach. If the sphincter does not close properly, food backs up into the esophagus, and the stomach acid that comes with it causes a burning sensation called "heartburn" (though it actually has nothing to do with the heart). This problem often arises during pregnancy, when the digestive organs are displaced far upward.

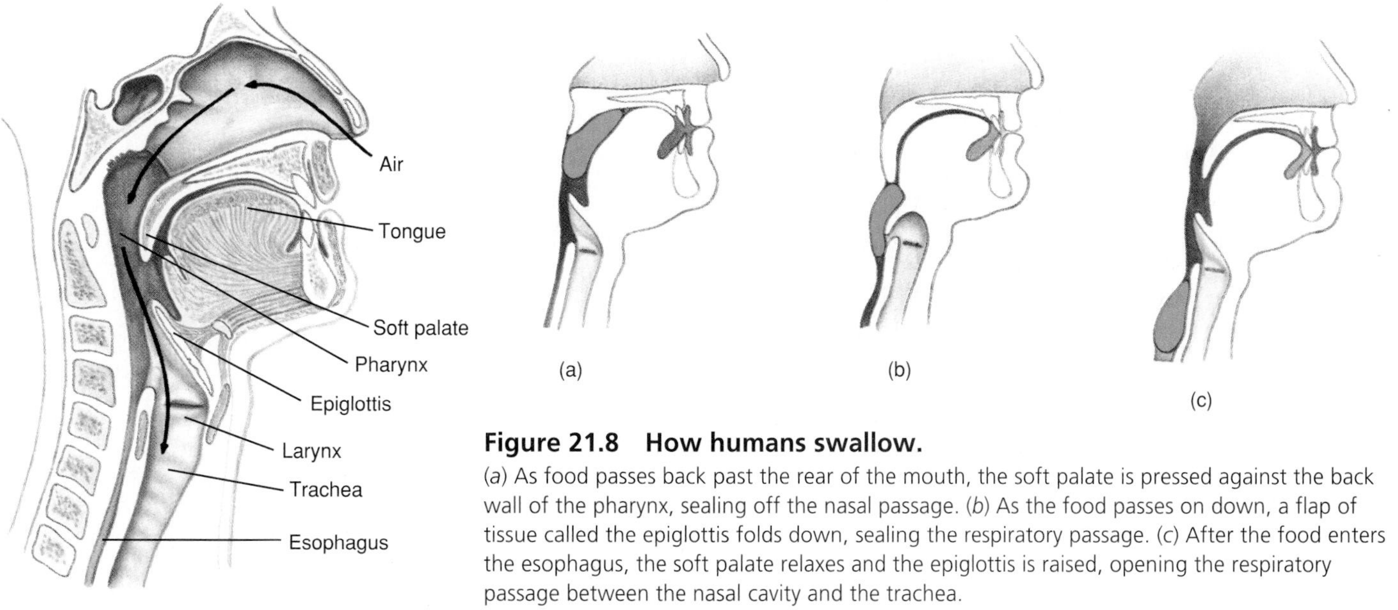

Figure 21.8 How humans swallow.

(*a*) As food passes back past the rear of the mouth, the soft palate is pressed against the back wall of the pharynx, sealing off the nasal passage. (*b*) As the food passes on down, a flap of tissue called the epiglottis folds down, sealing the respiratory passage. (*c*) After the food enters the esophagus, the soft palate relaxes and the epiglottis is raised, opening the respiratory passage between the nasal cavity and the trachea.

Preliminary Digestion: The Stomach

The **stomach** is a saclike portion of the digestive tract (figure 21.9), located below the lungs just beneath the diaphragm. The stomach has several functions: *temporary storage, mechanical breakdown of food,* and the *unraveling of proteins.* The stomach organizes your body's digestive process. It collects ingested food, partly digests the protein, and feeds its contents in a controlled fashion into the primary digestive organ, the small intestine.

The epithelial lining of the stomach is the source of the "digestive juices," a combination of HCl and digestive enzymes that, when mixed with food, forms a semisolid material called *chyme*. The lining of the upper stomach is dotted with deep depressions called *gastric pits* (figure 21.10). In a gastric pit the epithelial membrane is folded inward, the inner walls acting as exocrine glands. The inner walls of the pit contain two kinds of secreting cells, parietal cells, which secrete HCl, and chief cells, which secrete the protein-digesting enzyme

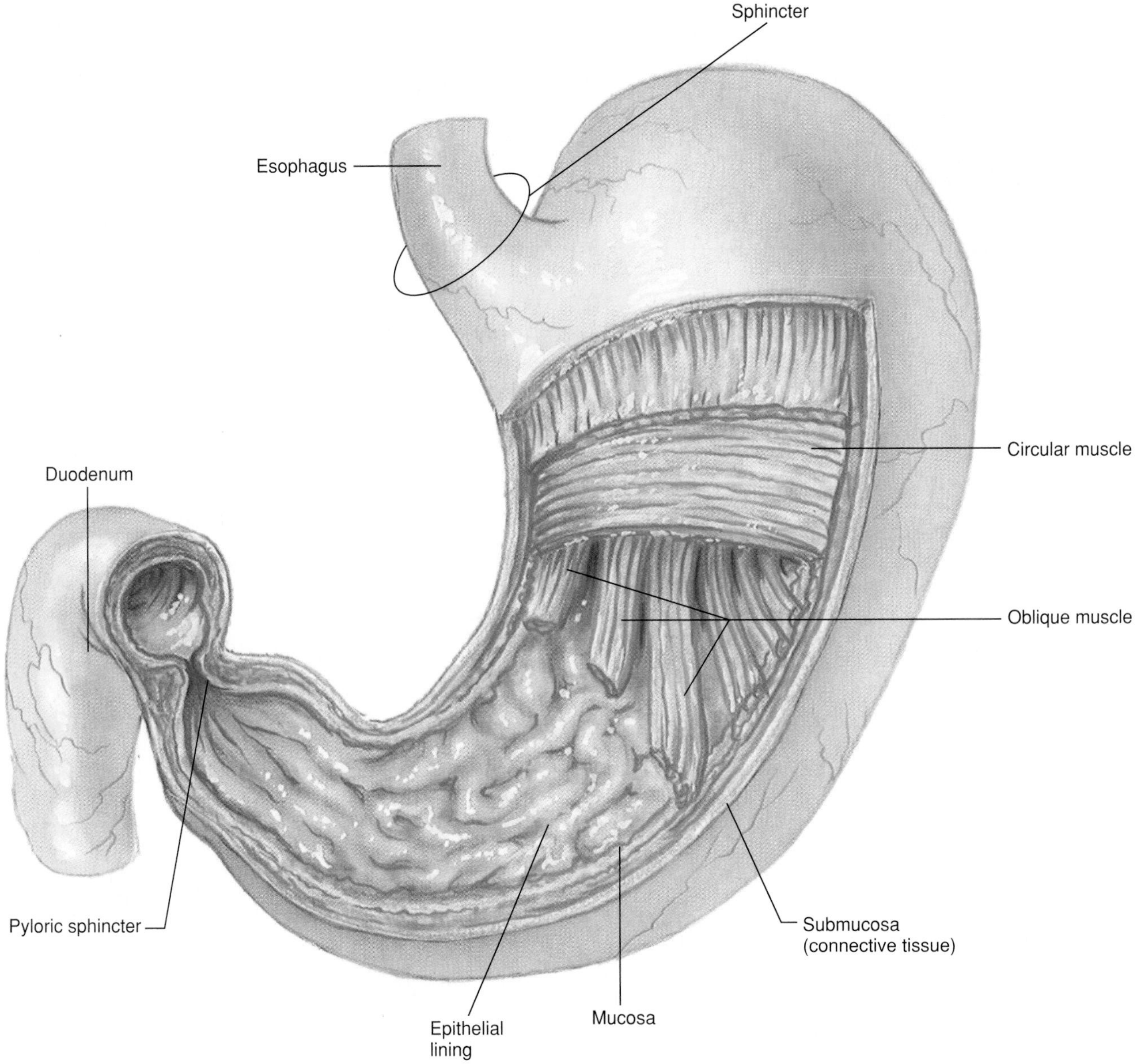

Figure 21.9 The major regions and structures of the stomach.
Entrance of food into the stomach from the esophagus is controlled by the sphincter. The epithelial walls of the stomach are lined with gastric pits, which contain glands that secrete hydrochloric acid and the enzyme pepsinogen. A band of smooth muscle called the pyloric sphincter controls the entrance to the duodenum, the upper part of the small intestine.

pepsin. Actually, they secrete an inactive form of the enzyme called pepsinogen, which is activated to pepsin when it reaches the stomach and acid in the stomach cleaves off a terminal fragment. This is a very clever way of ensuring that the pepsin does not attack the cells of the gastric pit!

Your stomach secretes about 2 liters of HCl every day, creating a very concentrated acid solution. This solution is about 3 million times more acidic than the bloodstream. The HCl unfolds proteins, even the tough fibers of connective tissue, because low pH values (between 1.5 and 2.5) disrupt the attraction between the carboxyl and amino side groups of proteins, which is responsible for holding the polypeptide chains of a protein in their three-dimensional shape. Pepsin then cleaves the polypeptide chains into several long fragments, so the protein cannot refold in the small intestine after the acid is neutralized. Stomach acid does not carry out any other digestive activity. It has a very limited ability to break proteins into amino acids, or carbohydrates into their constituent sugars, and it does not attack fats at all.

The acidic solution within the stomach also kills most of the bacteria that are ingested with the food. The few bacteria that survive the stomach and enter the intestine intact are able to grow and multiply there, particularly in the large intestine. In fact, most vertebrates harbor thriving colonies of bacteria within their intestines, and bacteria are a major component of feces. Bacteria that live within the digestive tract of cows and other grazers play a key role in the ability of these mammals to digest cellulose.

It is important that your stomach not produce *too* much acid. If it did, your body could not neutralize the acid later in the small intestine, a step essential for the final stage of digestion. Production of acid is controlled by hormones. These hormones are produced by endocrine cells scattered within the walls of the stomach. The hormone **gastrin** regulates the synthesis of HCl by the parietal cells of the gastric pits, permitting HCl to be made only when the pH of the stomach is higher than about 1.5. Some people greatly overproduce gastrin, which in turn causes their stomach to produce too much acid. The excessive acid can burn a hole called an **ulcer** in the walls of the small intestine. Overproduction of gastrin is rare, however. Most ulcers are caused by acid secreted by an ulcer-causing bacterium, *Helicobacter pylori.*

Only some water and a few substances such as aspirin and alcohol are absorbed into the blood through the wall of the stomach. All other products of digestion are absorbed in the intestine.

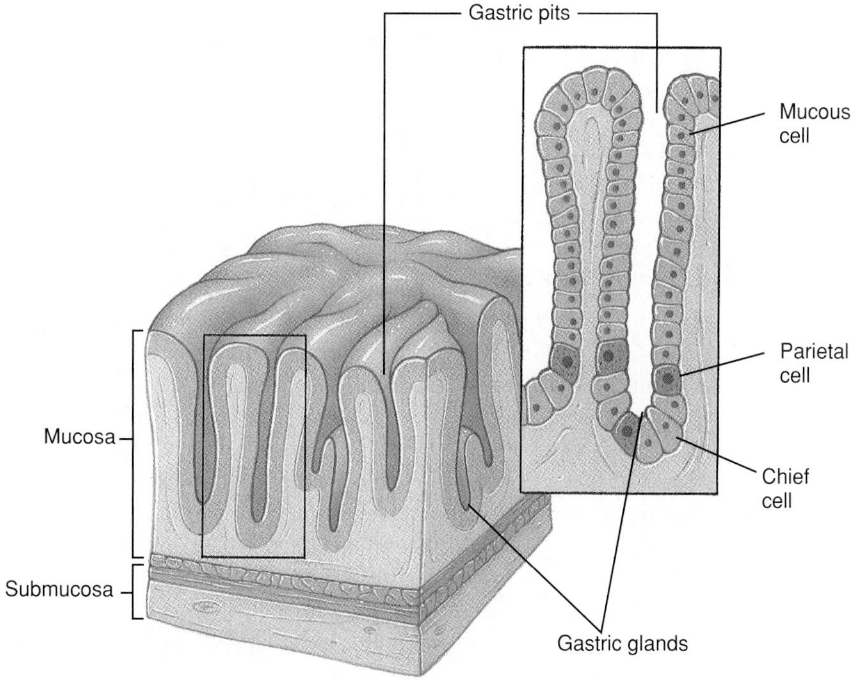

Figure 21.10 Gastric pits.
Gastric pits are deep invaginations of the stomach epithelium down into the underlying mucosa, at the base of which parietal and chief cells secrete hydrochloric acid and a protein that is cleaved into the enzyme pepsin.

Digestion and Absorption: The Small Intestine

The digestive tract exits from the stomach into the **small intestine,** where the breaking down of large molecules into small ones occurs. Only relatively small portions of food are introduced into the small intestine at one time, to allow time for acid to be neutralized and enzymes to act. The small intestine is the true digestive vat of the human body. Within it, carbohydrates are broken down into sugars, proteins into amino acids, and fats into fatty acids. Once these small molecules have been produced, they pass across the epithelial wall of the small intestine into the bloodstream.

Some of the enzymes necessary for these digestive processes are secreted by the cells of the intestinal wall. Most, however, are made in a large gland called the **pancreas,** situated near the junction of the stomach and the small intestine. It is one of the body's major exocrine (secreting through ducts) glands.

The pancreas has three types of cells, each secreting different types of molecules:

1. **Digestive enzymes.** Cells of the pancreas secrete a host of different enzymes that break down fats, carbohydrates, and proteins.
2. **Bicarbonate.** Special cells secrete a base, bicarbonate, that neutralizes the acid, HCl, entering the small intestine from the stomach. Production of bicarbonate is very important, since most of the digestive enzymes secreted by the pancreas will not work in acid solution.
3. **Insulin.** Groups of cells called islets of Langerhans, scattered throughout the pancreas, act as endocrine (ductless) glands, producing the hormone insulin which acts in the liver to regulate the level of glucose in the blood. We discuss the liver later in this chapter.

The pancreas sends bicarbonate and digestive enzymes into the small intestine through a duct that empties into its initial segment, the **duodenum.** Your small intestine is approximately 6 meters long—unwound and stood on its end, it would be far taller than you are! Only the first 25 centimeters, about 4% of the total length, is the duodenum. It is within this initial segment, where the pancreatic enzymes and detergents enter the small intestine, that digestion occurs.

Much of the food energy our bodies harvest is obtained from fats. Before fats can be digested by enzymes, they must be made soluble in water. This process is carried out by a collection of detergent molecules known as *bile salts* secreted into the duodenum from the **liver.** Bile salts solve a key problem of digestion, which is that fat is not soluble in water. How can the body use fats as food, if fat is not soluble in the blood, which carries foods to the cells of the body? Bile salts solve this problem by acting as a superdetergent. They combine with fats to form microscopic droplets in a process called emulsification. These tiny droplets remain suspended in water practically indefinitely.

All the rest of the small intestine (96% of its length) is called the **ileum.** The ileum is devoted to absorbing water and the products of digestion into the bloodstream. The lining of the small intestine is covered with fine fingerlike projections called **villi** (singular, **villus**), each too small to see with the naked eye (figure 21.11*a*). In turn, each of the cells covering a villus is covered on its outer surface by a field of cytoplasmic projections called **microvilli** (figure 21.11*b* and *c*). Both kinds of projections greatly increase the absorptive surface of the lining of the small intestine. The average surface area of the small intestine of an adult human is about 300 square meters, more than the surface of many swimming pools!

The amount of material passing through the small intestine is startlingly large. Per day, an average human consumes about 800 grams of solid food, and 1,200 milliliters of water, for a total volume of about 2 liters. To this amount is added about 1.5 liters of fluid from the salivary glands, 2 liters from the gastric secretions of the stomach, 1.5 liters from the pancreas, 0.5 liters from the liver, and 1.5 liters of intestinal secretions. The total adds up to a remarkable 9 liters—more than 10% of the total volume of your body! However, although the flux is great, the *net* passage is small. Almost all these fluids and solids are reabsorbed during their passage through the small intestine—about 8.5 liters across the walls of the small intestine and 0.35 liters across the wall of the large intestine. Of the 800 grams of solid and 9 liters of liquid that enter the digestive tract each day, only about 50 grams of solid and 100 milliliters of liquid leave the body as feces. The fluid absorption efficiency of the digestive tract thus approaches 99%, very high indeed.

Concentration of Solids: The Large Intestine

The **large intestine,** or **colon,** is much shorter than the small intestine, approximately 1 meter long. No digestion takes place within the large intestine, and only about 4% of fluid absorption occurs there. The large intestine is not convoluted, lying instead in three relatively straight segments, and its inner surface does not possess villi. As a consequence, the large intestine has only one-thirtieth the absorptive surface area of the small intestine. Although sodium and vitamin K are absorbed across its walls, the primary function of the large intestine is to act as a refuse dump. Within it, undigested material, including large amounts of plant fiber and cellulose, is compacted and stored. Many bacteria live and actively divide within the large intestine, where they play a role in the processing of undigested material into the final excretory product, **feces.**

The final segment of the digestive tract is a short extension of the large intestine called the **rectum.** Compact solids within the colon pass into the rectum as a result of the peristaltic contractions of the muscles encasing the large intestine. From the rectum, the solid material passes out of the body through the **anus.**

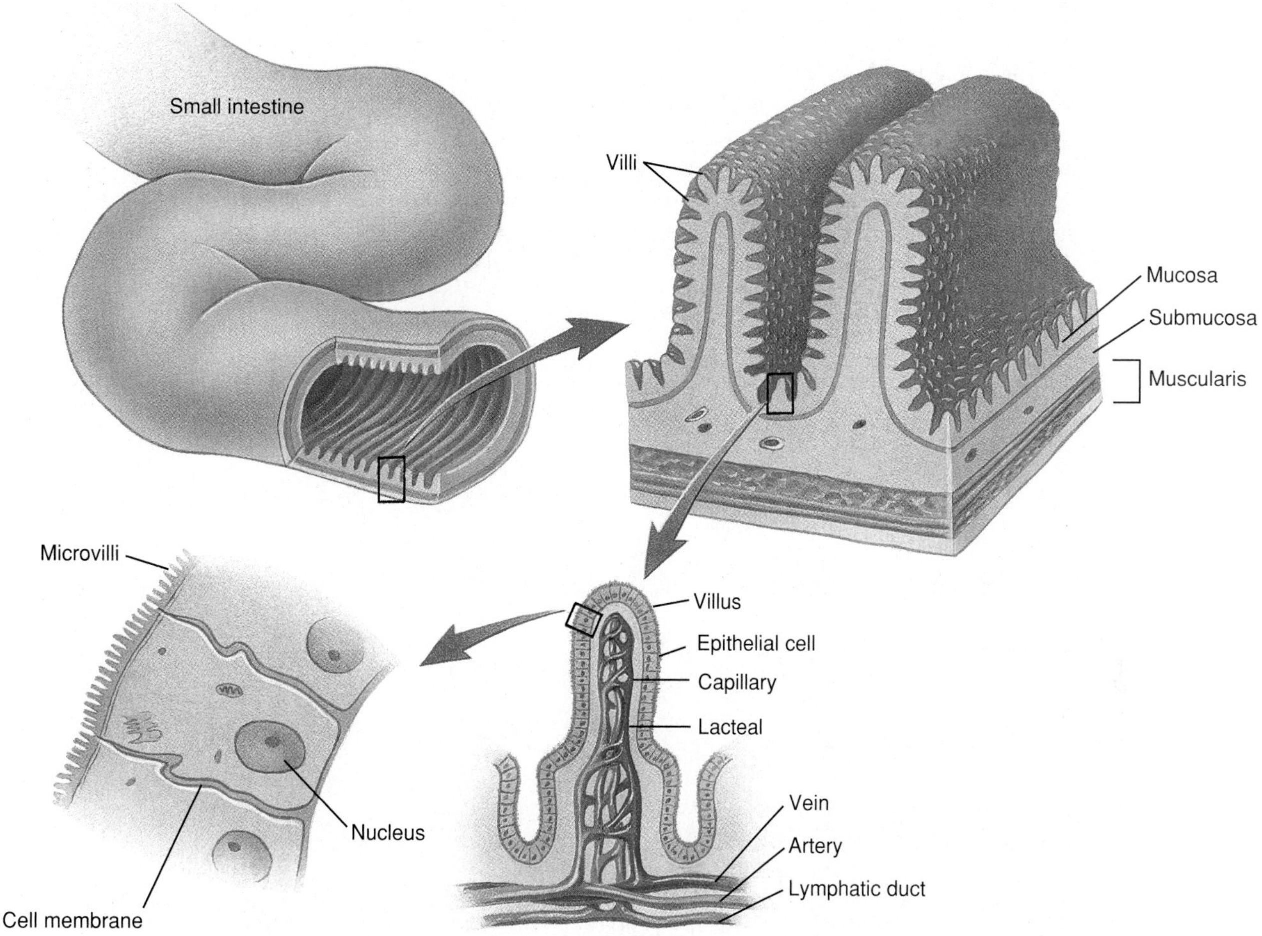

(a)

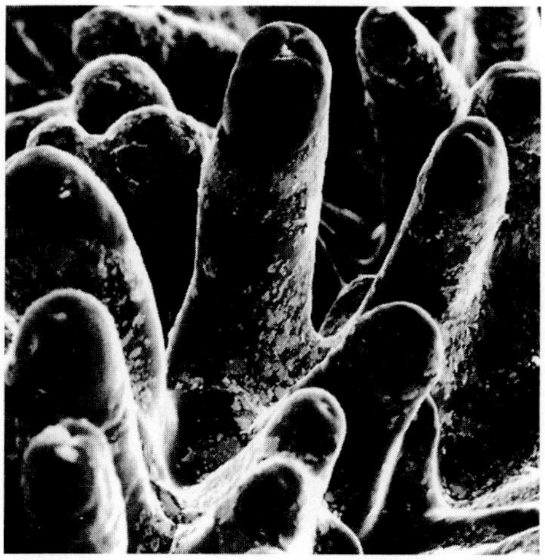

(b)

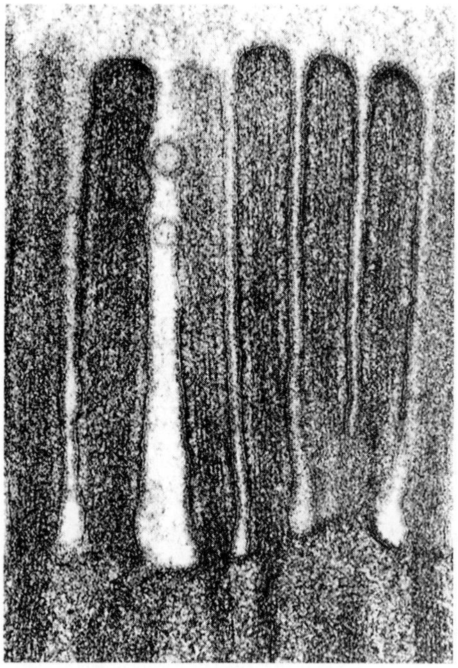

(c)

Figure 21.11 The small intestine.
(*a*) Cross section of the small intestine with details showing villi structure. (*b*) Microvilli, shown in a scanning electron micrograph, are very densely clustered, giving the small intestine an enormous surface area, which is very important for efficient absorption. (*c*) Intestinal microvilli as shown in a transmission electron micrograph.

21.3 Using the Products of Digestion: The Liver

The **liver** is the body's principal metabolic factory, turning the building materials (sugars, amino acids, and fatty acids) arriving from the digestive tract in the bloodstream into substances that are used by the different cells of the body. Blood from the small intestine, rich with the products of digestion, is collected into the portal vein, which carries it to the liver. After flowing through the many fine passages of the liver, the blood is re-collected into the hepatic vein, which carries it back to the heart to be circulated to the rest of the body (figure 21.12).

The liver is the largest internal organ in your body. In an adult human it weighs about 1.5 kilograms (over 3 pounds) and is the size of a small football. The liver carries out a wide variety of metabolic functions: it supplies quick energy, metabolizes alcohol, makes proteins, stores vitamins and minerals, regulates blood clotting, monitors the production of cholesterol, and detoxifies poisons. It also produces the detergent bile salts already mentioned.

One of the liver's most important functions is to regulate the amount of glucose circulating in the blood. It is very important that your blood maintain a relatively constant level of glucose, because the different tissues of the body depend upon it for energy. Brain cells, for example, can store very little glucose, and they lack the enzymes to convert fat or amino acids into glucose. Brain cells are thus totally dependent on blood plasma for glucose and cease to function if the level of glucose in the blood falls much below normal values.

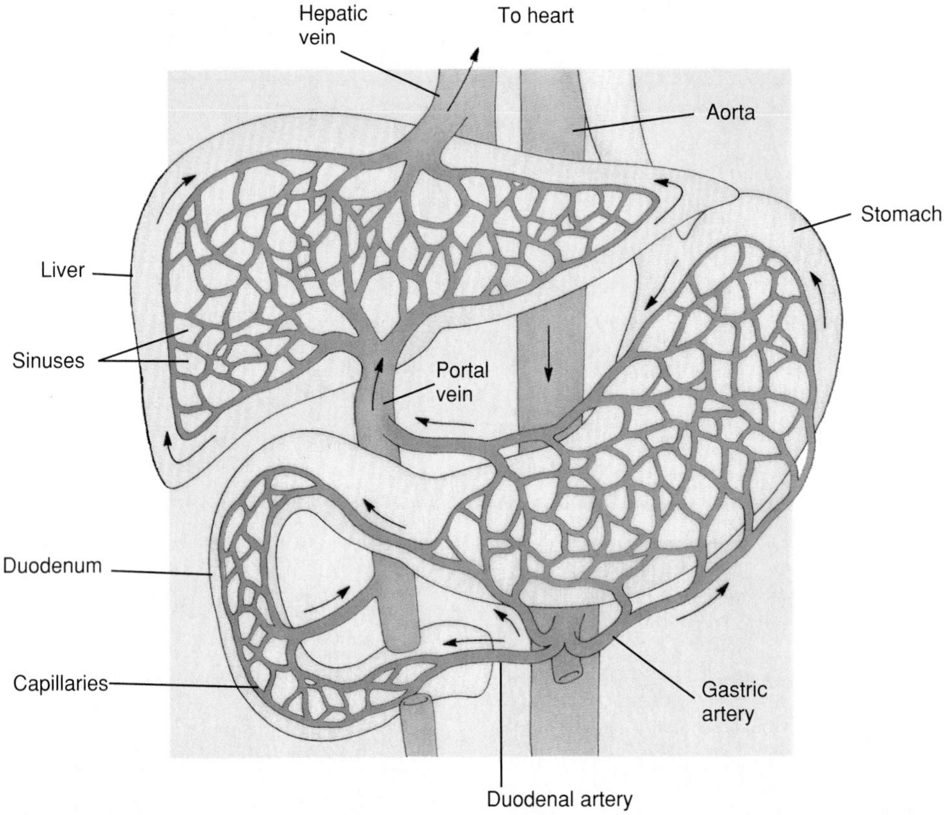

Figure 21.12 Hepatic-portal circulation.
Blood from the stomach and small intestine, rich with the metabolites of digestion, is collected into the portal vein, which carries it to the liver. After flowing through the sinuses of the liver, the liver-processed blood is re-collected into the hepatic vein, which carries it back toward the heart.

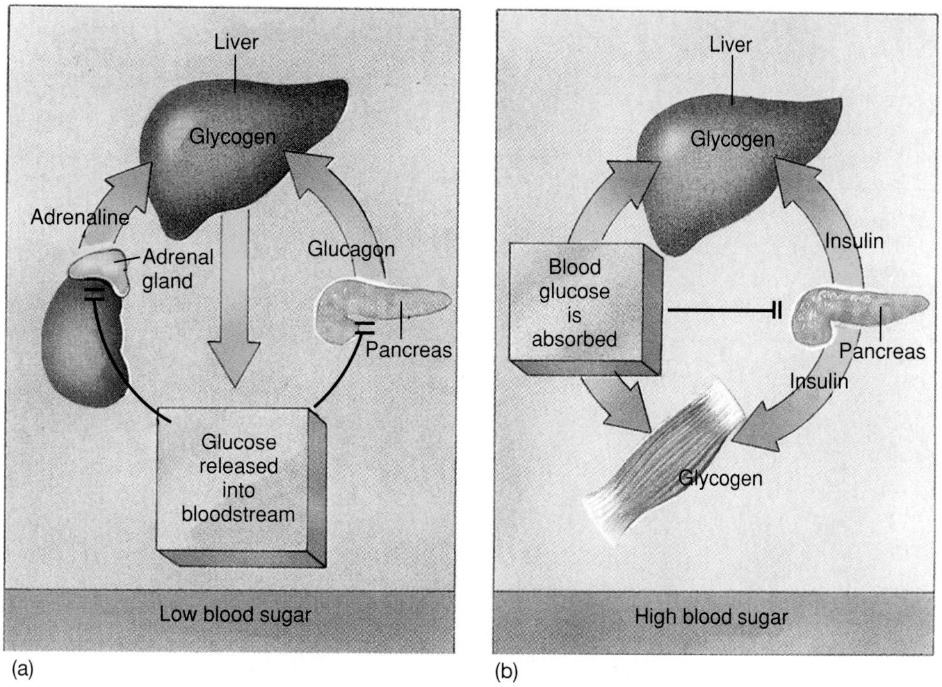

Figure 21.13 Control of blood glucose levels.
(*a*) When blood glucose levels are low, cells within the pancreas release the hormone glucagon into the bloodstream; other cells within the adrenal gland release the hormone adrenaline into the bloodstream. When they reach the liver, these two hormones act to increase the liver's breakdown of glycogen to glucose. (*b*) When blood glucose levels are high, other cells within the pancreas produce the hormone insulin, which stimulates the liver and muscles to convert blood glucose into glycogen.

Regulating Blood Glucose Levels

The need for a constant supply of glucose in the blood presents a potential problem. Most humans eat sporadically. Most of us eat three meals a day. Food thus enters our digestive systems at intervals separated by long periods of fasting. Much of the food is digested relatively quickly, and the metabolites such as glucose and amino acids pass through the lining of the small intestine into the bloodstream right away. Do you see the problem? Unless something is done, the level of glucose in the blood would rise quickly right after a meal and then fall rapidly during a period of starvation. However, these abrupt swings in blood glucose levels are avoided in a very logical way: by establishing a *reservoir,* or metabolic bank, in the liver.

Here is how the liver regulates blood glucose levels: When excessive amounts of glucose are present in the blood passing through the liver, as occurs soon after a meal, the liver converts the excess glucose into a starchlike chain of glucose subunits called **glycogen.** This process is stimulated by the hormone **insulin,** produced by the pancreas. The liver stores some of this glycogen, and the rest is stored in the muscles, where it is easily available to fuel muscle contractions. When blood glucose levels decrease, as they do during sleep or between meals, the glucose deficit in the blood plasma is made up from the glycogen reservoir, stimulated by the hormone glucagon. The human liver stores enough glycogen to supply glucose to the bloodstream for about 10 hours of fasting. If fasting continues, the liver begins to convert amino acids into glucose to maintain the glucose level in the blood. The liver, your body's metabolic reservoir, thus acts much like a bank, making deposits and withdrawals in the "currency" of glucose molecules (figure 21.13).

The liver has a limited storage capacity. When its glycogen reservoir is full, it continues to remove excess glucose molecules from the body by converting them to fat, which is stored elsewhere in the body. That is why overeating frequently results in the deposit of fat around the stomach or on the hips.

Also like a bank, the liver exchanges currencies, converting surplus amounts of molecules such as amino acids and fats to glucose and storing it as glycogen. Excess amino acids that may be present in the blood are converted to glucose by liver enzymes. The first step in this conversion is the removal of the amino group (NH_3) from the amino acid, a process called *deamination.* Unlike plants, animals cannot reuse the nitrogen from these amino groups and must excrete it as nitrogenous waste. The product of amino acid deamination, ammonia (NH_4), combines with carbon dioxide to form urea. The urea is released by the liver into the bloodstream, where—as you will learn next—the kidneys subsequently remove it. Figure 21.14 on the following page summarizes all of the organs of the digestive system and their respective functions.

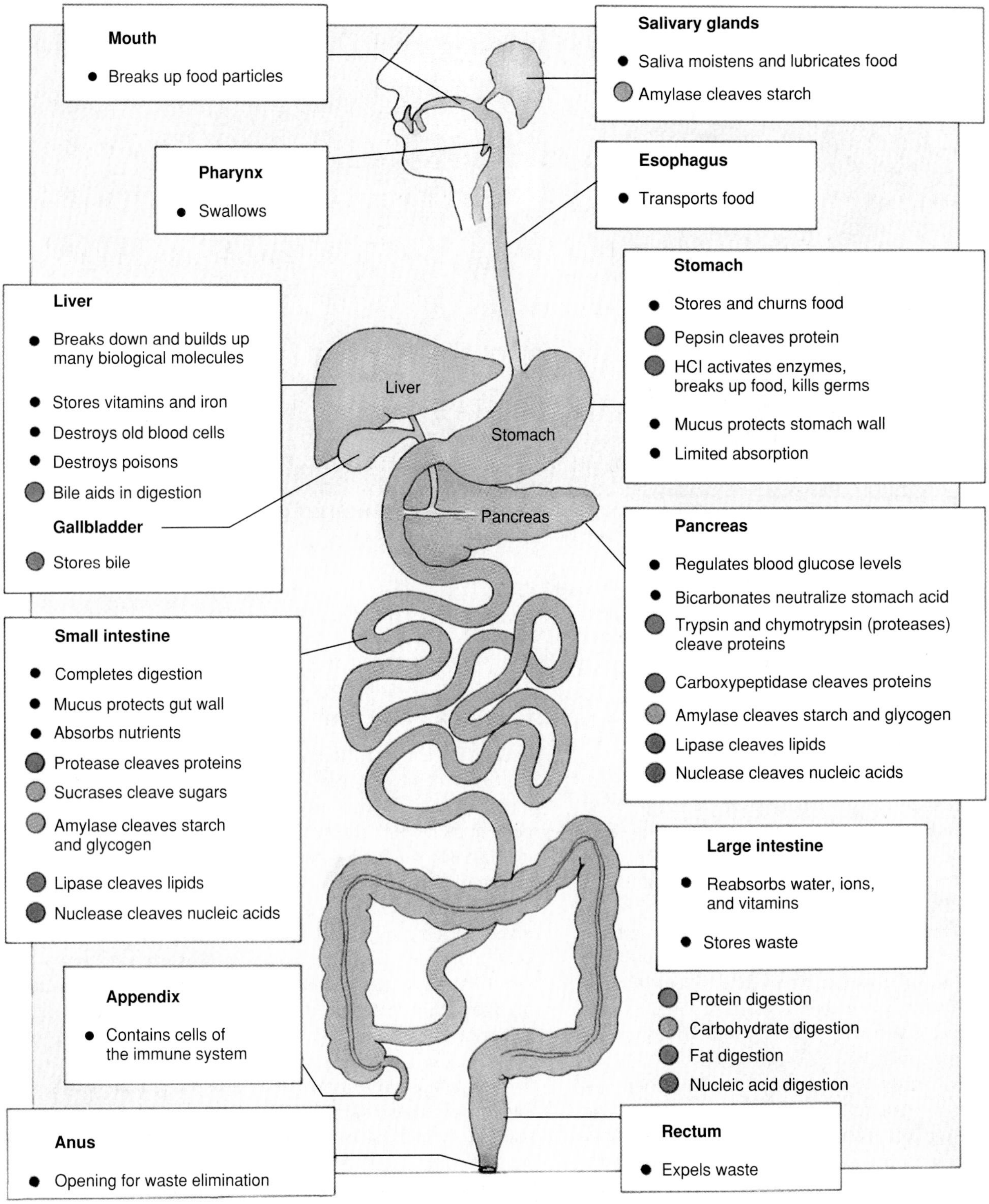

Mouth
Breaks up food particles
Salivary glands
Saliva moistens and lubricates food
Amylase cleaves starch
Pharynx
Swallows
Esophagus
Transports food
Stomach
Stores and churns food
Pepsin cleaves protein
HCl activates enzymes, breaks up food, kills germs
Mucus protects stomach wall
Limited absorption
Liver
Breaks down and builds up many biological molecules
Stores vitamins and iron
Destroys old blood cells
Destroys poisons
Bile aids in digestion
Gallbladder
Stores bile
Liver
Stomach
Pancreas
Pancreas
Regulates blood glucose levels
Bicarbonates neutralize stomach acid
Trypsin and chymotrypsin (proteases) cleave proteins
Carboxypeptidase cleaves proteins
Amylase cleaves starch and glycogen
Lipase cleaves lipids
Nuclease cleaves nucleic acids
Small intestine
Completes digestion
Mucus protects gut wall
Absorbs nutrients
Protease cleaves proteins
Sucrases cleave sugars
Amylase cleaves starch and glycogen
Lipase cleaves lipids
Nuclease cleaves nucleic acids
Large intestine
Reabsorbs water, ions, and vitamins
Stores waste
Appendix
Contains cells of the immune system
Protein digestion
Carbohydrate digestion
Fat digestion
Nucleic acid digestion
Anus
Opening for waste elimination
Rectum
Expels waste

Figure 21.14 The organs of the digestive system and their functions.
The human digestive system contains some dozen different organs that act on the food you consume, starting with the mouth and ending with the anus. All of these organs must work properly in order for the body to properly obtain nutrition.

21.4 How the Body Eliminates Wastes: The Kidneys

If you live an average life span of 70 years, you will in the course of your life eat some 22 tons of food and drink over 7,250 gallons of fluid. Do you ever wonder what happens to it all? We don't just "burn it all up." Every atom in those tons of food and gallons of water still exists, either in your body or eliminated from it. Your body eliminates all of this excess material in one of two ways: respiration and excretion.

Respiration

Almost all the energy in the food you eat is in carbon–hydrogen bonds, and when this energy has been extracted, the carbon and hydrogen atoms that are left behind are combined with oxygen to form CO_2 and water. Most of this CO_2 is eliminated from the body through the lungs as you exhale.

Excretion

But what of other atoms in your food? What of the nitrogen atom in each amino acid of all the proteins in every hamburger you ever ate? And what about the phosphorous and nitrogen atoms in each link of the DNA in every cell of every steak and salad? And the sodium atoms in the salt you consume with every french fry? Of the long list of chemicals we each consume every day of our lives, almost all eventually leave our bodies by a general process called **excretion.** This waste removal is conducted by the kidneys and associated organs, known as the **urinary system** (figure 21.15).

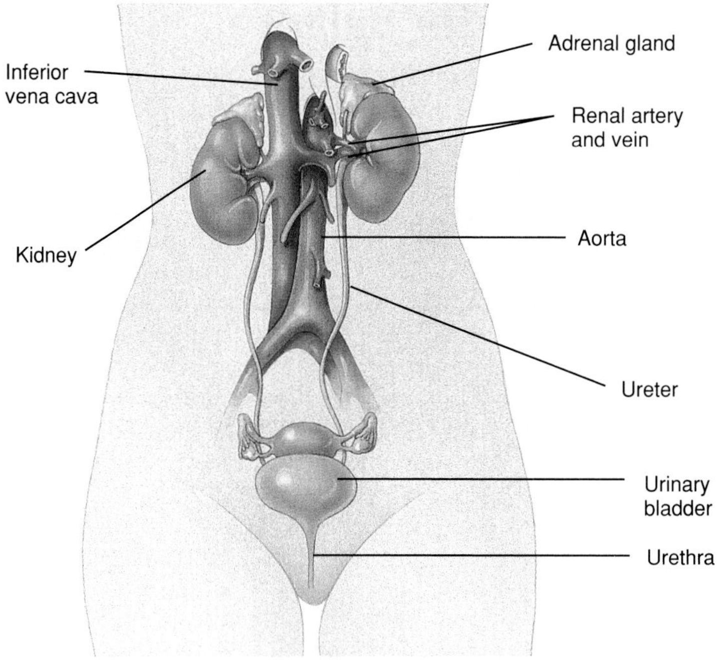

Figure 21.15 The female urinary system.
The urinary system consists of the kidneys, the ureter, which transports urine from the kidneys to the urinary bladder, and the urethra.

Disposing of Nitrogen Wastes

Your body eliminates wastes with a pair of bean-shaped, reddish brown **kidneys,** each the size of a small fist, located in the lower back region. The outer layer of the kidney in mammals is called the renal cortex, while the inner region is known as the renal medulla. A series of converging tubes leads from the medulla to the ureter, which transports urine from each kidney to the urinary bladder.

A major function of the kidneys is the elimination from the body of a variety of potentially harmful substances that animals eat, drink, or inhale. Among the most important are the waste products produced from the proteins we eat. Enzymes in the liver break down the amino acid components of proteins by removing their amino groups, producing ammonia (NH_3), which is extremely toxic. Getting rid of this very dangerous waste product of metabolism is a key task performed by the kidneys and liver working together. First, the liver links pairs of amino groups in a complicated series of chemical reactions to CO_2 to form a largely harmless molecule called **urea** (figure 21.16). Urea can safely be transported and excreted at high concentrations. The urea is carried by the bloodstream to the kidneys, where it is excreted as a principal component of urine. In this process the kidneys receive a flow of about 2,000 liters of blood each day—more than the volume of a car! Because your body actually holds only 5.6 liters of blood, you can see that your blood goes through the kidneys for cleaning many times during the day (about 350 times, or once every four minutes).

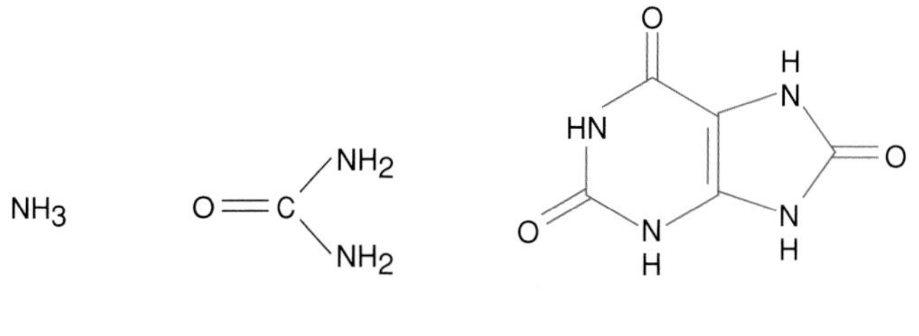

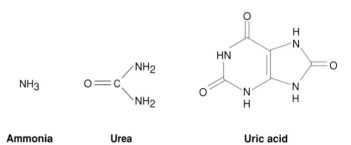

Figure 21.16 Nitrogenous wastes.
When amino acids are metabolized, the immediate by-product is ammonia, which is quite toxic. Humans convert ammonia to urea, which is less toxic. Birds and terrestrial reptiles convert it instead into uric acid, which is insoluble in water.

21.5 Water Balance

Our bodies constantly lose water. Approximately two-thirds of your body is water, and if the amount falls much lower than this you will die. Vertebrates evolved in the sea, and no vertebrate is able to survive much drying out. Why not simply be watertight? If the human body were coated in plastic, like a sandwich in a sandwich bag, our bodies would lose no water. Can you think of why your body is not coated like that? The reason we are not "Glad wrapped" is that many essential body functions *require* the loss of water. We use the evaporation of water from our skins to cool our bodies when they overheat. We use water to carry wastes from the body in urine. Our bodies remain healthy as long as they take in as much *water* as they expend in carrying out these necessary activities.

The average person drinks over a liter of fluid every day, which over a lifetime is enough to fill a tanker truck (figure 21.17). Unless our bodies lose an equal amount in sweat and urine, we would swell up like a balloon. Sweating plays an important role in controlling body temperature (figure 21.18), while urine is a vehicle for discharging urea and excess salts from the body. Your blood pressure is determined by the difference between the amount of fluid you take in and what you excrete—your body constantly varies the amount of water it contributes to urine so as to keep the blood pressure constant.

A major function of the kidneys is water conservation. Your body has a higher concentration of water than does the surrounding air, so it tends to lose water to the air by evaporation. To minimize this loss of water, your body tries to recapture as much water as possible from its urine, because the water in urine is lost from the body.

Another function of the kidneys is to regulate the osmotic composition of the blood—how much water and salt it contains. The proper operation of the many organ systems of the body requires that the osmotic concentration of the blood—the concentration of solutes dissolved within it—be kept within narrow bounds. If you had just drunk a large amount of water, your blood would become diluted unless you could somehow get rid of the extra water. The kidneys remove that excess water by increasing urine production.

Your body also monitors the levels of salt in your blood. When levels of sodium ion fall below normal, your brain directs the adrenal gland to send out hormones that cause the kidneys to extract more sodium ions from the water passing through them. On the other hand, if levels of salt in the bloodstream rise too high, hormone levels are decreased so that more salt is excreted via the urine. In the total absence of hormone, humans may excrete up to 25 grams of salt a day in their urine.

Figure 21.17 A lot of liquid.
During a lifetime, a person might drink as much fluid as it would take to fill this tanker truck.

Figure 21.18 Excretion through perspiration.
Perspiring, or sweating, plays an important role in controlling the temperature of your body.

Structure of the Kidney

The kidney is a complex structure composed of roughly 1 million structures called **nephrons** (figure 21.19), each of which is composed of three elements:

1. **Filter.** The filtration device at the top of each nephron is called a **Bowman's capsule.** Within each capsule an arteriole enters and splits into a fine network of vessels called a **glomerulus.** The walls of these capillaries act as a filtration device. Blood pressure forces fluid through the capillary walls. These walls withhold proteins and other large molecules in the blood, while passing water, small molecules, ions, and urea, the primary waste product of metabolism.
2. **Tube.** The Bowman's capsule is connected to a long narrow tube called a renal tubule, which is bent back on itself in its center, called the **loop of Henle.** This long hairpin loop is a reabsorption device. Like the mammalian small intestine, it extracts from the filtrate passing through the tube molecules useful to the body, such as glucose and a variety of ions. Otherwise, these useful molecules would be lost in the urine.
3. **Duct.** The tube empties into a large collection tube called a **collecting duct.** The collecting duct operates as a water conservation device, reclaiming water from the urine so that it is not lost from the body. Human urine is four times as concentrated as blood plasma—that is, the collecting ducts remove much of the water from the filtrate passing through the kidney. Your kidneys achieve this remarkable degree of water conservation by a simple but superbly designed mechanism: they bend the duct back alongside the nephron tube and make the duct permeable to urea. This greatly increases the local salt (urea) concentration in the tissue surrounding the tube, causing water in urine to pass out of the tube by osmosis. The salty tissue sucks up water from the urine like blotting paper, passing it on to blood vessels that carry it out of the kidneys and back to the bloodstream.

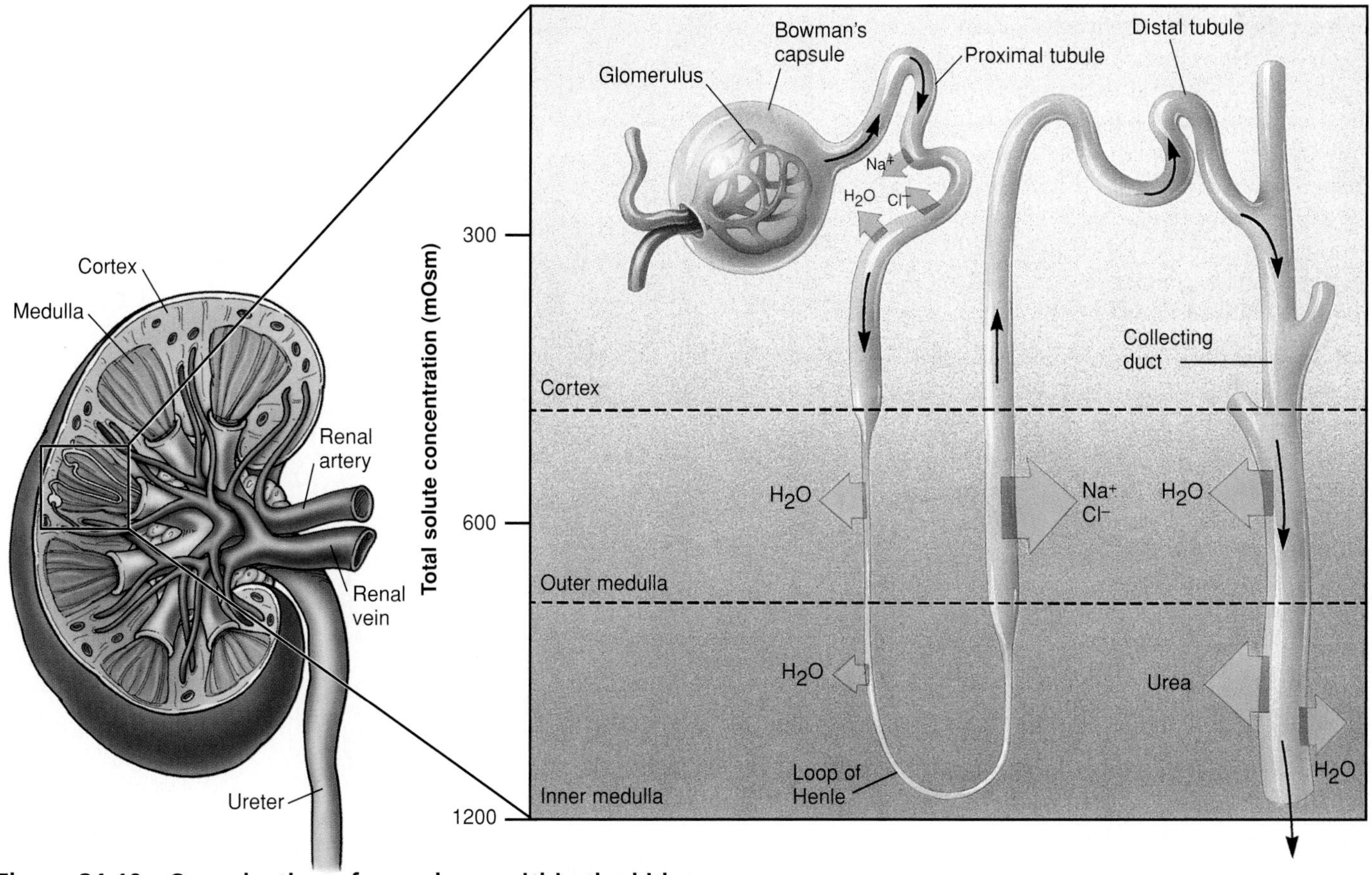

Figure 21.19 Organization of a nephron within the kidney.
The glomerulus is enclosed within a filtration device called a Bowman's capsule. Blood pressure forces liquid from blood through the glomerulus and into the proximal segment of the tubule, where glucose and small proteins are reabsorbed from the filtrate. The filtrate then passes through a double-loop arrangement consisting of the loop of Henle and the collecting duct, which act to remove water from the filtrate. The water is then collected by blood vessels and transported out of the kidney to the systemic (body) circulation.

The Kidney at Work

The formation of urine within your kidneys involves the movement of several kinds of molecules between nephrons and the capillaries that surround them. Five steps are involved: pressure filtration, reabsorption of water, selective reabsorption of ions and nutrients, tubular excretion, and further reabsorption of water.

Pressure Filtration

Driven by the blood pressure, small molecules are pushed across the thin walls of the glomerulus to the inside of the Bowman's capsule. Blood cells and large molecules like proteins cannot pass through, and as a result the blood that enters the glomerulus is divided into two paths: nonfilterable blood components that are retained and leave the glomerulus in the bloodstream and filterable components that pass across and leave the glomerulus in the urine. This filterable stream is called the **glomerular filtrate.** It contains water, nitrogenous wastes (principally urea), nutrients (principally glucose and amino acids), and a variety of ions.

Reabsorption of Water

Filtrate from the glomerulus passes down the descending arm of the loop of Henle. The walls of this portion of the tube are impermeable to either salts or urea but are freely permeable to water. Because (for reasons we discuss later) the surrounding tissue has a high concentration of urea, water passes out of the descending arm by osmosis, leaving behind a more concentrated filtrate.

Selective Reabsorption

At the turn in the loop, the walls of the tubule become permeable to salts and other nutrients, like sugars and amino acids, but much less permeable to water. As the concentrated filtrate passes up this ascending arm, these nutrients pass out into the surrounding tissue, where they are carried away by blood vessels (figure 21.20). In the upper region of the ascending arm are active transport channels that pump out salt (NaCl). Left behind in the filtrate is the urea that initially passed through the glomerulus as nitrogenous waste. The urea concentration is becoming very high within the tubule.

Tubular Excretion

In the ascending loop, substances are also added to the urine by a process called tubular excretion. This active transport process excretes into the urine other nitrogenous wastes such as uric acid and ammonia, as well as excess hydrogen ions.

Further Reabsorption of Water

The tubule then empties into a collecting duct that passes back through the tissue of the kidney. Unlike the tubule, the lower portions of the collecting duct are permeable to urea, some of which diffuses out into the surrounding tissue (that is why the tissue surrounding the descending arm of the loop of Henle has a high urea concentration). A high urea concentration in the tissue results, causing even more water to pass outward from the filtrate by osmosis. The filtrate that is left after salts, nutrients, and water have been removed is **urine.**

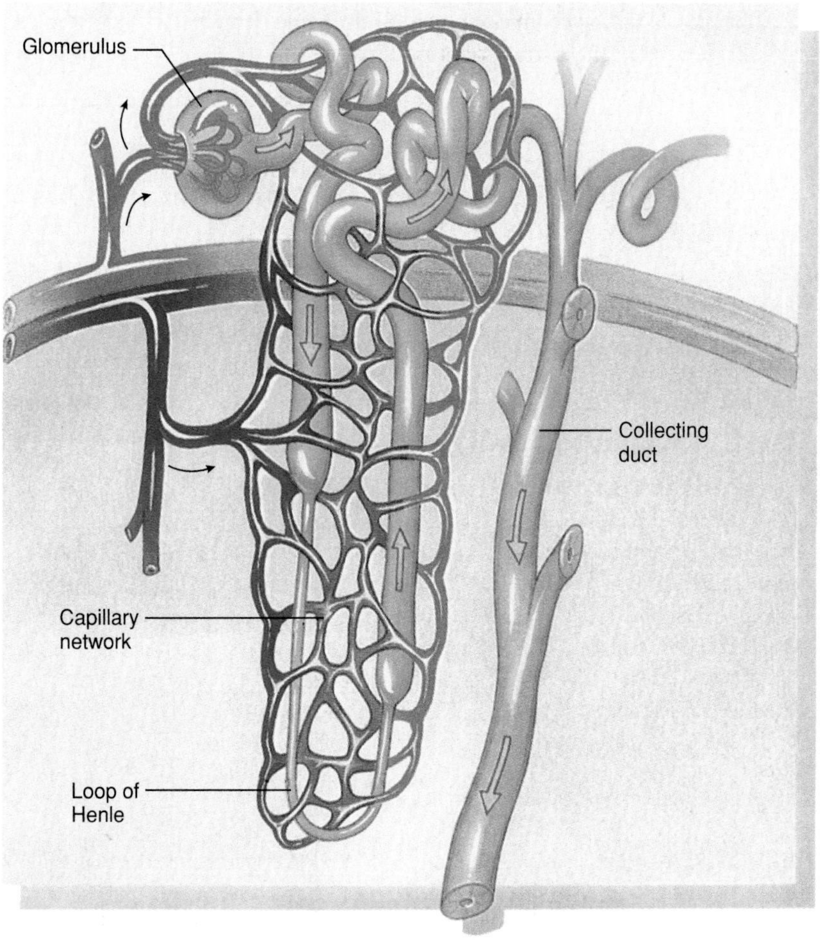

Figure 21.20 The collecting duct.
After the filtrate passes through the loop of Henle, in close proximity to a network of capillaries, it passes out of the nephron, emptying into the collecting duct. The collecting duct bends back past the loop of Henle, where some of the urea from the filtrate passes out into the tissue.

CHAPTER 21

HIGHLIGHTS

Key Terms | Key Concepts

21.1 Diet: What We Need to Eat and Why

Key Terms

diet 456
calories 456
essential amino acid 457
vitamin 458
trace element 458

Key Concepts

- Ingested calories are either metabolized by the body or stored as fat.
- Healthy diets contain those necessary substances we cannot manufacture, such as essential amino acids, vitamins, and trace elements.

21.2 The Human Digestive System

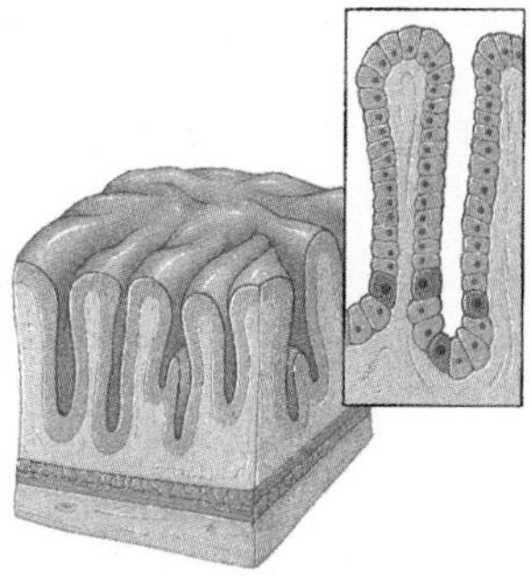

Key Terms

digestion 460
peristalsis 461
gastrin 463
duodenum 464
villi 464

Key Concepts

- Digestion is the breaking down of macromolecules into small components that can be metabolized.
- Digestion is carried out by enzymes.
- Protein digestion is aided by acid, which opens up folded proteins so that enzymes can attack them.
- Acid predigestion occurs in the stomach; most enzymatic digestion occurs in the duodenum, the initial portion of the small intestine.

21.3 Using the Products of Digestion: The Liver

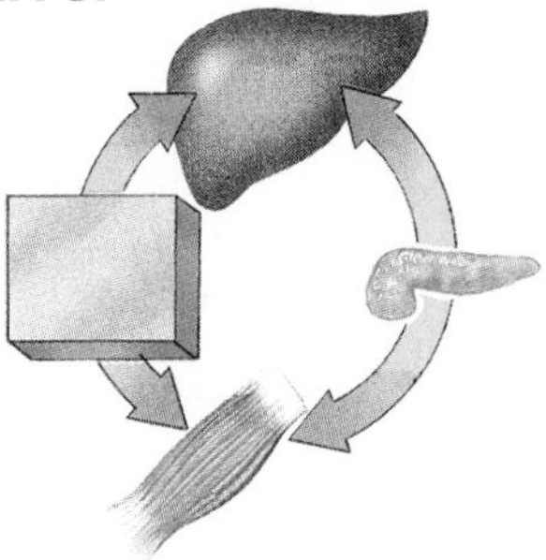

Key Terms

liver 466
glycogen 467
insulin 467

Key Concepts

- The liver, the body's largest internal organ, monitors metabolism.
- The liver regulates levels of glucose in the blood by storing excess glucose as glycogen and then converting it to fat.
- The hormone insulin stimulates this conversion of glucose to glycogen.
- When blood glucose levels fall, another hormone called glucagon stimulates the breakdown of glycogen to glucose.

21.4 How the Body Eliminates Wastes: The Kidneys

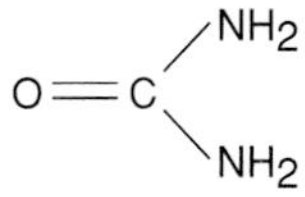

Key Terms

excretion 469
kidney 469
urea 469

Key Concepts

- Kidneys cleanse the blood of many harmful substances.
- Among the most important of these are the nitrogen by-products of protein metabolism.

21.5 Water Balance

Key Terms

nephron 471
glomerulus 471
loop of Henle 471
collecting duct 471
urine 472

Key Concepts

- A major function of the kidneys is water conservation.
- A second key function of the kidneys is to regulate how much salt the blood contains.
- Mammals concentrate their urine by removing water from it; this allows them to retain the water when the urine is discarded.

21

CONCEPT REVIEW

1. Which of the following amino acids is *not* essential for humans?
   a. arginine
   b. leucine
   c. lysine
   d. valine

2. Each of the following is a final product of human digestion except
   a. amino acids.
   b. fatty acids.
   c. free sugars.
   d. starch.

3. Starchy foods such as potatoes begin being digested in the
   a. mouth.
   b. esophagus.
   c. stomach.
   d. duodenum.

4. Protein-rich foods, such as steak, are broken down in the stomach by the combined action of (pick two)
   a. low pH.
   b. high pH.
   c. bile.
   d. amylase.
   e. pepsin.

5. Most of the enzymes that complete the breakdown of food items into simple sugars, amino acids, and fatty acids are secreted by which of the following?
   a. large intestine
   b. gastric pits
   c. liver
   d. pancreas

6. The main function of the large intestine is
   a. digestion.
   b. absorption.
   c. compaction.
   d. secretion.

7. Select the incorrect statement about the liver.
   a. The liver is the largest internal organ of the human body.
   b. The liver regulates the level of sugar in the blood.
   c. The liver receives blood from the heart via the hepatic vein.
   d. The liver acts on amino acids.

8. The human body excretes nitrogen mainly as
   a. ammonia.
   b. urea.
   c. uric acid.
   d. water.

9. The glomerulus is a
   a. cup.
   b. network of capillaries.
   c. tubule.
   d. all of the above.

10. Which of the following is recovered in the collecting duct of the nephron?
    a. glucose
    b. NaCl
    c. water
    d. proteins

11. Approximately _________ nephrons are found in each kidney.
    a. 100
    b. 10,000
    c. 100,000
    d. 1,000,000

12. _________ essential amino acids must be obtained from our diet.

13. A band of muscle called the _________ controls the entrance to the duodenum from the stomach.

14. Food moves down the esophagus by rhythmic muscular contractions called _________.

15. The very numerous _________ that cover the epithelial wall of the small intestine greatly increase the surface area for absorption of digested foods.

16. By the action of the hormone insulin, the liver stores excess glucose as the molecule _________.

17. The _________ ion is secreted into the small intestine by the pancreas to neutralize stomach acid.

18. Pressure _________ occurs inside the Bowman's capsule.

19. The loop of _________ is part of the tube of the nephron.

Answers to the Concept Review questions appear in Appendix B.

CHALLENGE YOURSELF

1. While some animals are capable of synthesizing specific vitamins and amino acids, others cannot synthesize these compounds and must obtain them in their diet. What are the relative advantages and disadvantages of an animal that possesses the biochemical machinery to manufacture specific nutrients that other animals cannot manufacture?

2. You are adrift on a life raft on the ocean with no source of freshwater, and you recall that marine bony fishes and some marine birds satisfy their bodies' needs for water by drinking seawater. Should you drink seawater? Note that human kidneys can remove up to about 6 grams of sodium ions from the blood for every liter of urine they produce, and that a liter of seawater contains about 12 grams of sodium ions. Explain your answer.

FOR FURTHER READING

Allison, D., ed. *Handbook of Assessment Methods for Eating Behaviors and Weight Related Problems: Measures, Theory, and Research*. Thousand Oaks, Calif.: Sage Publications, 1995. Technical, but an important seminal contribution to the diagnosis and psychology of eating disorders.

Alper, J. "Ulcers as an Infectious Disease." *Science* 260 (April 9, 1993): 159–60. Peptic, gastric, and duodenal ulcers are the plague of many. Previously believed to be stress or diet-related, recent evidence points to a bacterial culprit: *Helicobacter pylori.*

American Dietetic Association. "Position of the American Dietetic Association: Vegetarian Diets." *Journal of the American Dietetic Association* 93 (November 1993): 1317–19. It is critical for vegetarians to get not only a good source of protein in their diet but a steady supply of the essential nutrients often unavailable to the "casual vegetarian."

Brown, L., and E. Pollitt. "Malnutrition, Poverty, and Intellectual Development." *Scientific American*, February 1996, 38–43. A poor diet influences mental development in many unexpected ways.

O'Brien, C. "Lucky Break for Kidney Disease Gene." *Science* 264 (June 24, 1994): 1844. Identification of the gene apparently responsible for the development of kidney disease offers a wealth of possible treatment—including gene therapy.

Pain, S. "Beating Bacteria with a Taste for Teeth." *New Scientist* 137 (February 27, 1993): 11. The Actinomycetes are the bacteria responsible for plaque formation and subsequent decay. What strategies (besides conventional oral hygiene) can be employed to fight these bacteria?

Service, R. "Stalking the Start of Colon Cancer." *Science* 263 (March 1994): 1559–60. Colon cancer is one of the deadliest varieties there is—and it's a disease that's strongly heritable.

Sokolov, R. "Pyramid Power: The USDA Has Abandoned the Four Basic Food Groups, and Confusion Reigns." *Natural History* 103 (January 1994): 72–78. Remember what you learned in grade school about the four food groups? Well, you can forget about it. There are at least six of them now. This is a good summary of the most recent attempt by the USDA to define dietary guidelines—a must-read for everyone, not just biology students.

TECHNOLOGY LINKS

The Living World Home Page
http://www.wcbp.com/biology/tlw

Explorations in Human Biology CD
#7 Diet and Weight Loss

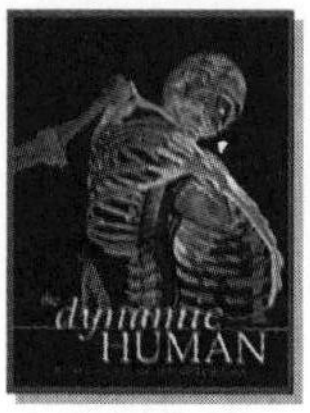

The Dynamic Human CD
Digestive System
Urinary System

Life Science Animations Videotape 3
#33 Peristalsis
#34 Digestion of Carbohydrates
#35 Digestion of Proteins

22

How the Body Defends Itself

CHAPTER OUTLINE

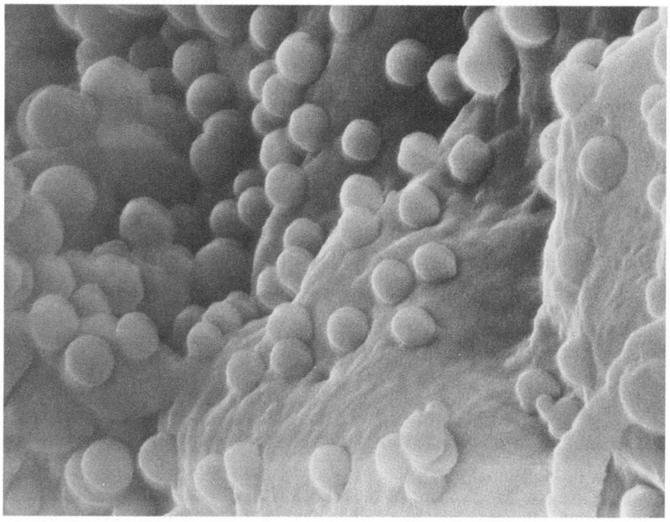

Figure 22.1 HIV, the virus that causes AIDS.
Large numbers of spherical virus particles are budding out from the surface of an infected white blood cell. All of these particles will be fully able to infect neighboring cells.

When you think of how animals defend themselves, it is natural to think of armor, of dinosaurs covered like tanks with heavy plates, of turtles and clams and armadillos. Humans in the Middle Ages emulated this approach by actually wearing body armor. However, even armor offers no protection against the most dangerous enemies the human body faces—our distant relatives, the microbes. Every human body offers a feast in nutrients for tiny, single-celled creatures, as well as a warm, sheltered environment in which they can grow and reproduce. We live in a world awash with microbes, and no human being can long withstand their onslaught unprotected. We survive because the human body has a variety of very effective defenses against this constant attack. The AIDS virus is fatal precisely because it disables the body's protective mechanism (figure 22.1).

22.1 Skin: The First Line of Defense

Your body is defended from infection the same way knights defended medieval cities. "Walls and moats" make entry difficult; "roaming patrols" attack strangers; and "sentries" challenge anyone wandering about and call patrols if a proper "ID" is not presented.

1. **Walls and moats.** The outermost defense of the human body is the **skin,** an efficient barrier to penetration by microbes. The lungs also have important barriers that protect their delicate alveoli from invasion.
2. **Roaming patrols.** When infection threatens, the initial response of your body is to mount a **cellular counterattack,** using a battery of cells and chemicals that kill microbes. These defenses act very rapidly after the onset of infection.
3. **Sentries.** Your body is also guarded by mobile cells that patrol the bloodstream, scanning the surfaces of every cell they encounter. They are part of the **immune system.** One kind of immune cell aggressively attacks and kills any cell identified as foreign, whereas the other type marks the foreign cell or virus for elimination by the roaming patrols.

Skin is the outermost layer of the human body and provides its first defense against invasion by microbes. Skin is our largest organ, comprising some 15% of our total weight. One square centimeter of skin from your forearm (about the size of a dime) contains 200 nerve endings, 10 hairs and muscles, 100 sweat glands, 15 oil glands, 3 blood vessels, 12 heat-sensing organs, 2 cold-sensing organs, and 25 pressure-sensing organs. Skin has three distinct layers: an outer **epidermis,** a lower **dermis,** and an underlying **subcutaneous layer.** Cells of the outer epidermis are continually being worn away and replaced by cells moving up from below—in one hour your body loses and replaces approximately 1.5 million skin cells!

Epidermis Is the "Bark" of the Human Body

The epidermis of skin is from 10 to 30 cells thick, about as thick as this page. The outer layer, called the **stratum corneum,** is the one you see when you look at your arm or face. Cells from this layer are continuously subjected to damage. They are abraded, injured, and worn by friction and stress during the body's many activities. They also lose moisture and dry out. The body deals with this damage not by repairing cells but by replacing them. Cells from the stratum corneum are shed continuously, replaced by new cells produced deep within the epidermis. The cells of the innermost **basal layer** are among the most actively dividing cells of the vertebrate body. New cells formed there migrate upward, and as they move they manufacture keratin protein, which makes them tough. Each cell eventually arrives at the outer surface and takes its turn in the stratum corneum, living there for about a month before it is shed and replaced by a newer cell. Persistent dandruff (psoriasis) is a chronic skin disorder in which new cells reach the epidermal surface every three or four days, about eight times faster than normal.

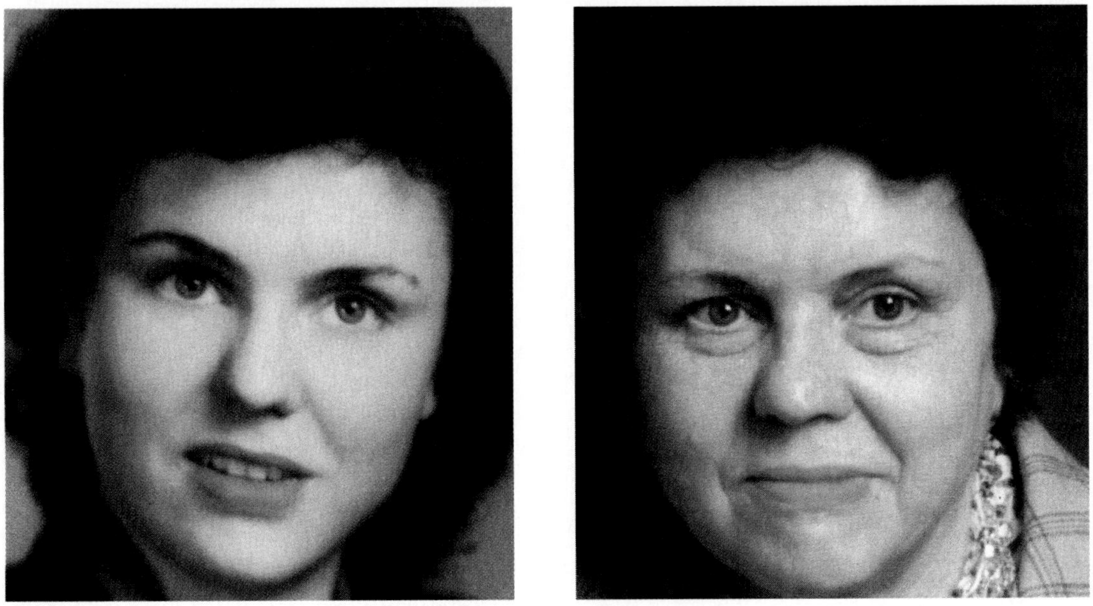

(a) (b)

Figure 22.2 Skin changes as we age.

For the most part, skin ages gradually, but on the face, the changes are more dramatic. (*a*) At 19 this woman's face appears youthful and smooth. (*b*) Forty years later, the production of skin oil is much less; her face appears less smooth and elastic and begins to exhibit wrinkles.

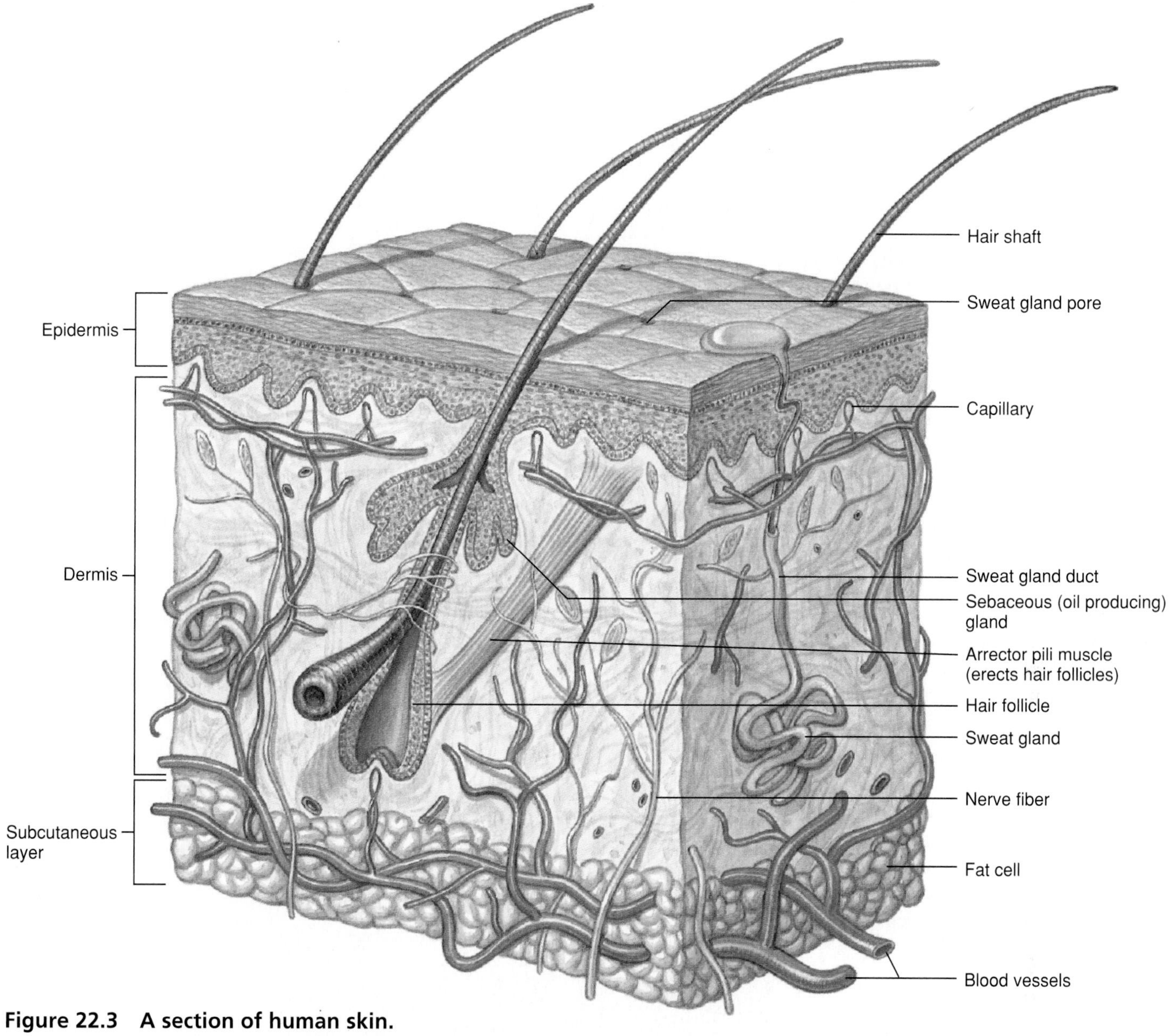

Figure 22.3 A section of human skin.
The skin defends the body by providing a barrier and sweat and oil glands whose secretions make the skin's surface acidic enough to inhibit the growth of microorganisms.

The Lower Skin Provides Support

The dermis of the skin is from 15 to 40 times thicker than the epidermis. It provides structural support for the epidermis, as well as a matrix for the many specialized cells residing within the skin. The wrinkling that occurs as we grow older occurs here (figure 22.2). The leather used to manufacture belts and shoes is derived from thick animal dermis. The layer of subcutaneous tissue below the dermis is composed of fat-rich cells that act as shock absorbers and provide insulation, which conserves body heat.

The Battle Is on the Surface

Your skin not only defends your body by providing a nearly impermeable barrier, but it also reinforces this defense with chemical weapons. The oil and sweat glands within the epidermis, for example (figure 22.3), make the skin's surface very acidic, which inhibits the growth of many microbes. Sweat also contains the enzyme lysozyme, which attacks and digests the cell walls of many bacteria. Cells lining the bronchi and bronchioles secrete a layer of sticky mucus that traps microorganisms before they can reach the warm, moist lungs (ideal breeding grounds for microbes). Cilia on the cells continually sweep this mucus upward, where it can be swallowed. Thus, potential invaders are carried out of the lungs like bound prisoners to be executed by gastric HCl.

Our surface defenses are very effective, but they are occasionally breached. Through breathing, eating, or cuts and nicks, bacteria and viruses now and then enter our bodies. When these invaders reach deeper tissue, a second line of defense comes into play, a cellular counterattack.

22.2 Cellular Counterattack: The Second Line of Defense

When the body's interior is invaded, a host of cellular and chemical defenses swing into action. Four are of particular importance: (1) cells that kill invading microbes; (2) proteins that kill invading microbes; (3) the inflammatory response, which speeds defending cells to the point of infection; and (4) the temperature response, which elevates body temperature to slow the growth of invading bacteria.

Cells That Kill Invading Microbes

The most important counterattack to infection is mounted by white blood cells, which attack invading microbes. These cells patrol the bloodstream and await invaders within the tissues. The three basic kinds of killing cells are macrophages, neutrophils, and natural killer cells. Each uses a different tactic to kill invading microbes.

1. **Foot soldiers.** Patrolling white blood cells called **macrophages** (Greek,"big eaters") kill bacteria one at a time by ingesting them (figure 22.4), much as an amoeba ingests a food particle. Although some macrophages are anchored within particular organs, particularly the spleen, most patrol the byways of the body, circulating in the blood, lymph, and fluid between cells. Macrophages are among the most actively mobile cells of the human body.
2. **Kamikazes.** Other white blood cells called **neutrophils** act like kamikazes. They release chemicals (identical to household bleach) to "neutralize" the entire area, killing any bacteria in the neighborhood—and themselves in the process. A neutrophil is like a grenade tossed into an infection. It kills everything in the vicinity. Macrophages, by contrast, kill only one invading cell at a time, but they live to keep on doing it.
3. **Internal security patrol.** A third kind of white blood cell, called **natural killer cells,** do not attack invading microbes but rather the body cells that are infected by them. Natural killer cells puncture the membrane of the target cell, allowing water to rush in and causing the cell to swell and burst (figure 22.5). Natural killer cells are particularly effective at detecting and attacking body cells that have been infected with viruses. They are also one of the body's most potent defenses against cancer, which they kill before the cancer cell has a chance to develop into a tumor.

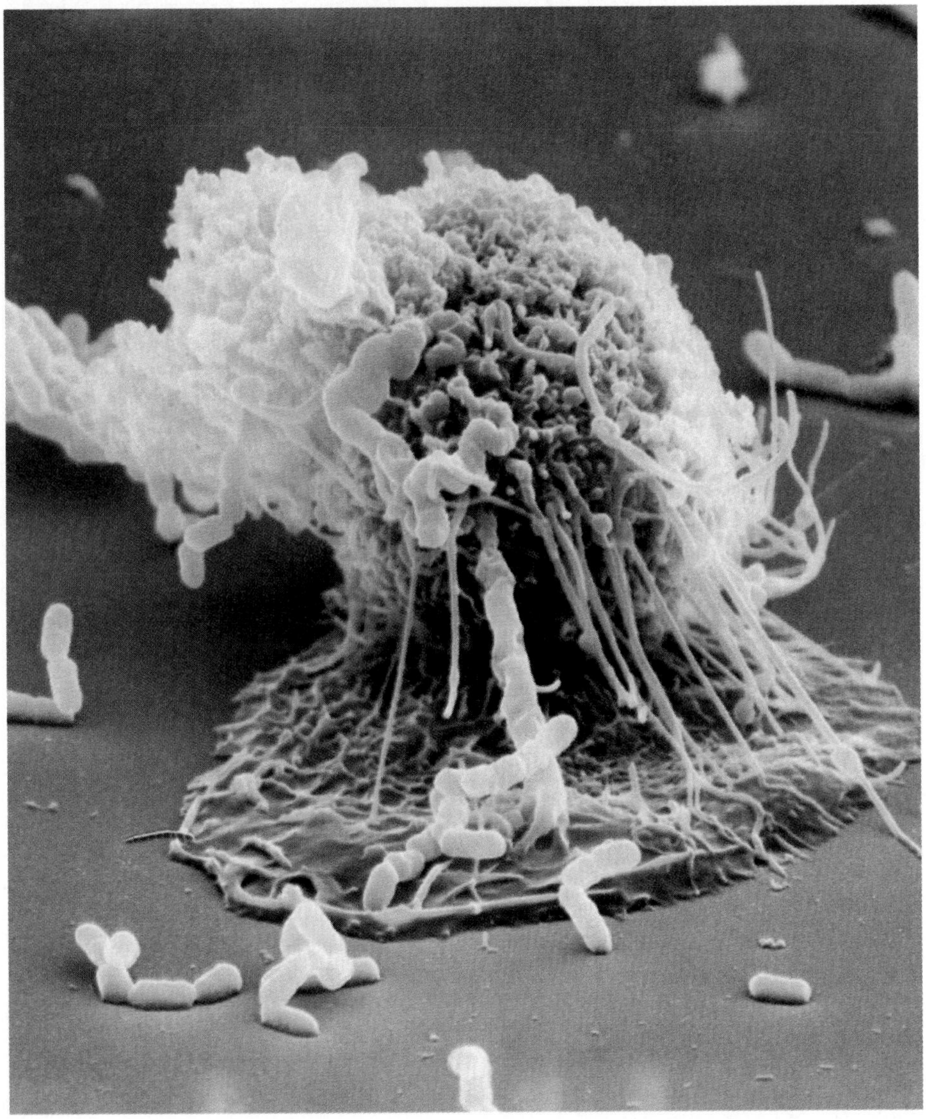

Figure 22.4 A macrophage in action.
In this scanning electron micrograph, a macrophage is "fishing" with long, sticky cytoplasmic extensions. Bacterial cells unfortunate enough to come in contact with the extensions are drawn back to the macrophage and engulfed.

Proteins That Kill Invading Microbes

The human body also employs a very effective chemical defense to complement its cellular defenses, called the **complement system.** It consists of special complement proteins that circulate in the blood plasma in an inactive state. Their defensive activity is triggered when they encounter the cell walls of bacteria or fungi. On detecting a microbial cell wall, the complement proteins interact with one another to form a membrane attack complex (MAC). The MAC inserts itself into the pathogen's cell membrane, forming a hole (figure 22.6). Like a dagger through the heart, this wound is fatal to the invading cell—water rushes in causing the cell to swell and burst.

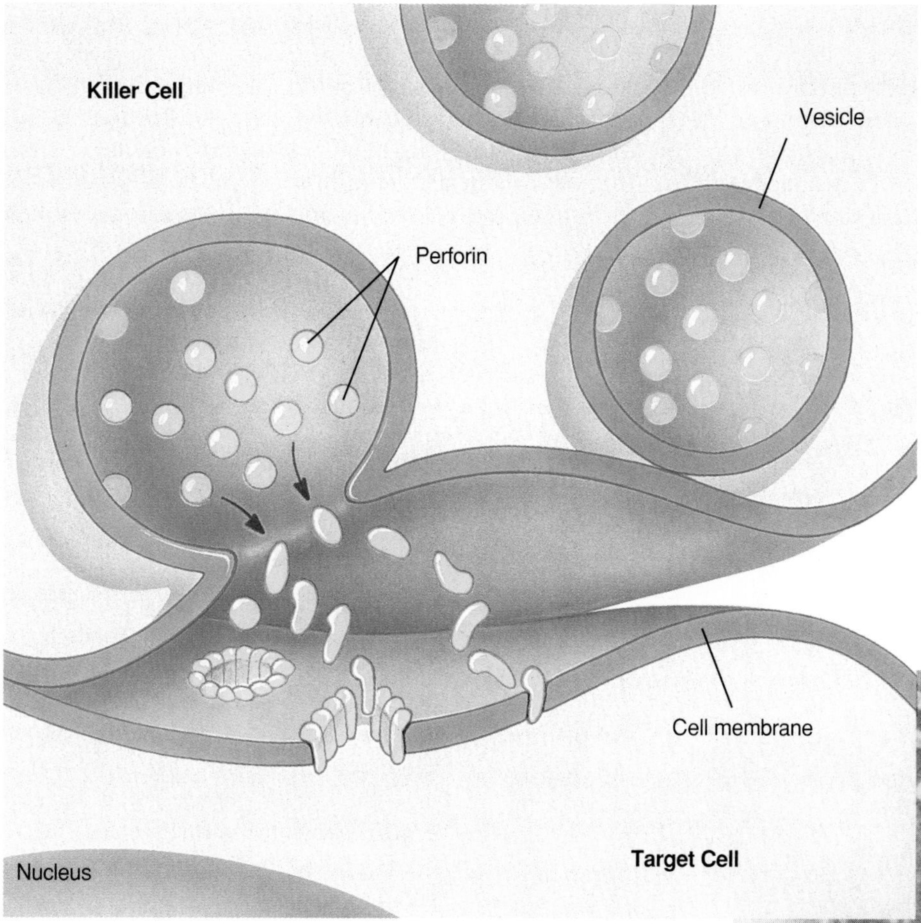

(a)

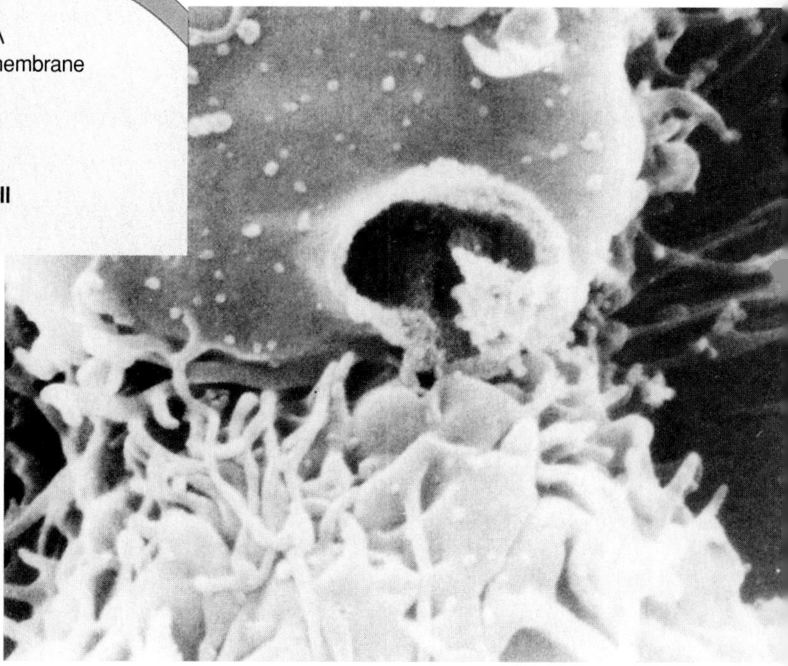

(b)

Figure 22.5 How killer cells kill target cells.
(*a*) The initial event is the tight binding of the killer cell to the target cell. Binding initiates a chain of events within the killer cell in which vesicles loaded with perforin molecules move to the outer cell membrane and expel their contents into the intercellular space over the target. The perforin molecules insert into the membrane like boards on a picket fence to form a pore that admits water and ruptures the cell. (*b*) Death of a tumor cell. A killer cell has attacked this cancer cell, punching a hole in its cell membrane. Water has rushed in, making it balloon out. Soon it will burst.

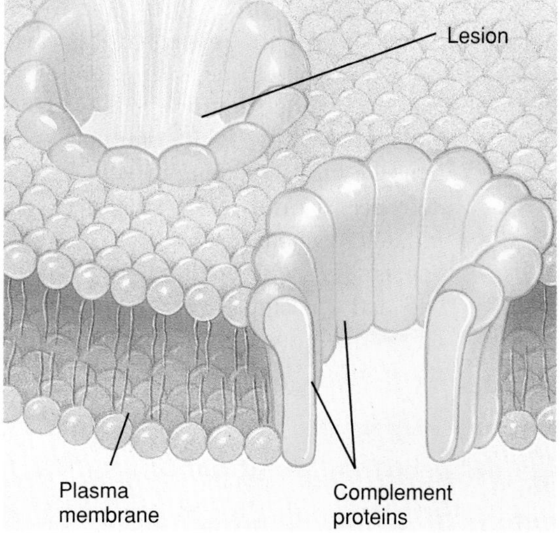

Figure 22.6 How complement creates a hole in a cell.
The complement proteins form a transmembrane channel resembling the perforin-lined lesion made by killer cells. There are some differences between the two. The complement proteins are free-floating in blood plasma, and they attach to the invading microbe directly. The perforin molecules are produced in killer cells and poke holes in infected body cells, not the actual microbe. But the final results of the two defenses are very similar.

The Inflammatory Response

The aggressive cellular and chemical counterattacks to infection are made more effective by the **inflammatory response.** Infected or injured cells release chemical alarm signals that cause blood vessels to expand, both increasing the flow of blood to the site of infection or injury and, by stretching their thin walls, making the capillaries more permeable. This produces the redness and swelling so often associated with infection. The increased blood flow through larger, leakier capillaries promotes the migration of macrophages and neutrophils to the site of infection, where they can attack invading microbes. Neutrophils arrive first, spilling out chemicals that kill the microbes (as well as tissue cells in the vicinity and themselves). Macrophages then clean up the remains of all the dead cells. This counterattack takes a considerable toll; the pus associated with infections is a mixture of dead or dying neutrophils, broken down tissue cells, and dead pathogens.

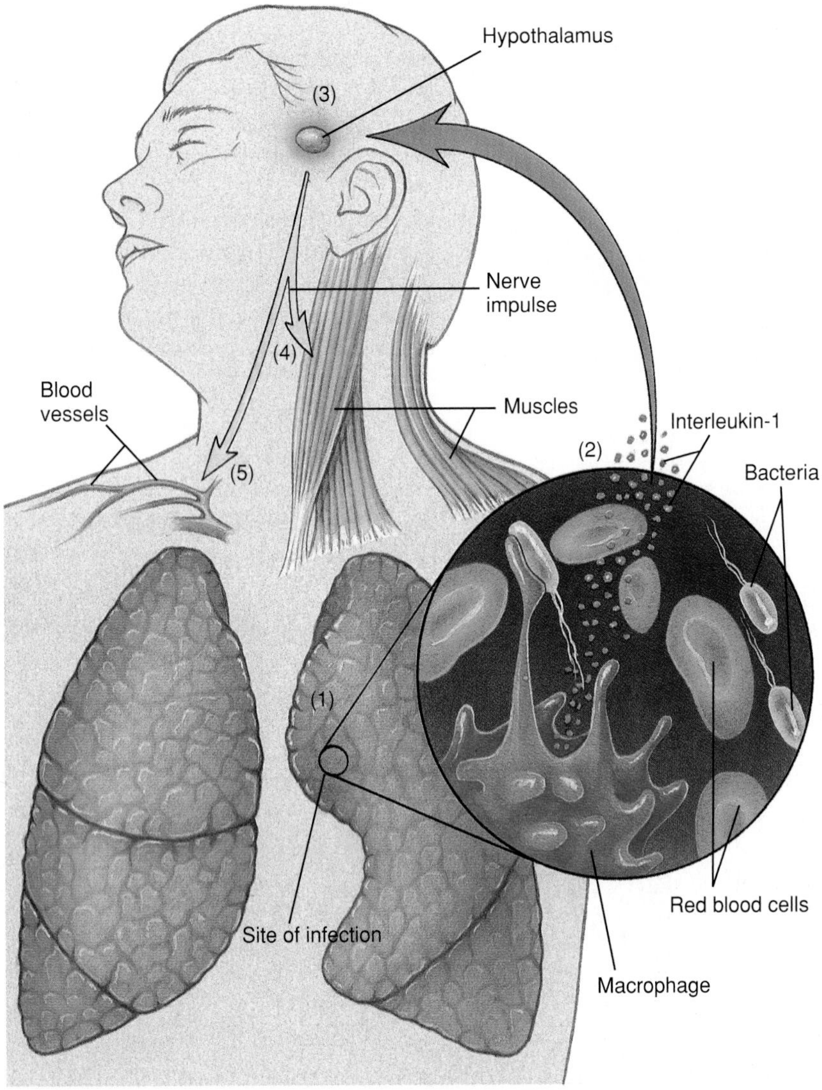

Figure 22.7 How an infection causes fever.
At the site of infection (*1*), macrophages release a "message" molecule, interleukin-1, which passes through the bloodstream (*inset, 2*) to the brain. There it stimulates the hypothalamus (*3*), the body's thermostat, triggering it to set a higher temperature. To raise the body's temperature to the new setting, the brain sends nerve impulses (*4*) to muscles, ordering them to contract; as a result, the body shivers, producing heat. Other nerve impulses (*5*) order blood vessels near the skin to constrict, minimizing heat loss. The higher temperature aids the immune response and inhibits the growth of invading microorganisms. When the immune system begins to make headway against the infection, reducing the numbers of invading microbes, production of interleukin-1 by macrophages stops. The fever "breaks," and body temperature soon falls to normal values.

The Temperature Response

Bacteria do not grow well at high temperatures. Thus when macrophages initiate their counterattack, they increase the odds in their favor by sending a message to the brain to raise the body's temperature. The cluster of brain cells that serves as the body's thermostat responds to the chemical signal by boosting the body temperature several degrees above the normal value of 37°C (98.6°F). The higher-than-normal temperature that results is called a **fever** (figure 22.7). While fever is quite effective at inhibiting microbial growth, very high fevers are dangerous because excessive heat can inactivate critical cellular enzymes. In general, temperatures greater than 39.4°C (103°F) are considered dangerous; those greater than 40.6°C (105°F) are often fatal.

The second line of your body's defenses, with both chemical and cellular weapons, provides it with a sophisticated defense against microbial infection. Only occasionally do bacteria or viruses overwhelm these defenses. When this happens, they face yet a third line of defense, more difficult to evade than any they have encountered. It is the immune system, the most elaborate of your body's defenses. Unlike other defenses, the immune system remembers previous encounters with potential invaders, and if they reappear, the immune system is ready for them.

22.3 The Immune System: The Third Line of Defense

Our immune system is not localized to any one place in the body nor is it controlled by any one organ. Rather, it is composed of a host of individual cells, an army of defenders that rush to the site of an infection to combat invading microorganisms. These cells, the white blood cells mentioned in chapter 11, arise in the bone marrow and circulate in blood and lymph. They are very numerous—of the 100 trillion cells of your body, 2 in every 100 are white blood cells! Macrophages are white blood cells, as are neutrophils and natural killer cells. Three other kinds of white blood cells are the key players in the immune defense. They are **T cells** (which attack and kill invading cells), **B cells** (which label invaders for later destruction by macrophages), and **helper T cells** (which control the other two kinds). Although not bound together, these white blood cells exchange information and act in concert as a functional, integrated system.

To understand how this third line of defense works, imagine you have just come down with the flu. Influenza viruses enter your body in small water droplets inhaled into your respiratory system. If they avoid becoming ensnared in the mucus lining the respiratory membranes (first line of defense), and avoid consumption by macrophages (second line of defense), the viruses infect and kill mucous membrane cells. You feel sick because large numbers of the cells lining your respiratory tract are dying.

At this point macrophages initiate the immune defense. Macrophages inspect the surfaces of all cells they encounter. Every cell in your body carries special marker proteins on its surface called MHC ("major histocompatibility") proteins, which inform the macrophage that the cell belongs to you and not some other individual. The MHC proteins are different for each individual, much as fingerprints are. By comparing the MHC proteins on the surface of other cells with those on its own, macrophages can identify foreign cells.

Macrophages also identify foreign cells in cooperation with T and B cells, using cell surface receptor proteins on their surfaces to detect foreign proteins. We will discuss this recognition process shortly.

Macrophages that encounter pathogens—either a foreign cell such as a bacterial one, which lacks proper MHC proteins, or a virus-infected body cell with telltale virus proteins stuck to its surface—respond by secreting a chemical alarm signal, and it is this alarm signal that calls up the third line of defense. The alarm signal is a protein called **interleukin-1** (Latin for "between white blood cells"). This protein activates helper T cells, a special chemical message that only particular cells can receive. The helper T cells are the "generals" of the immune system. They respond to the interleukin-1 alarm broadcast by macrophages by simultaneously initiating two different parallel lines of immune system defense: the cytotoxic response carried out by T cells and the antibody response carried out by B cells (figure 22.8).

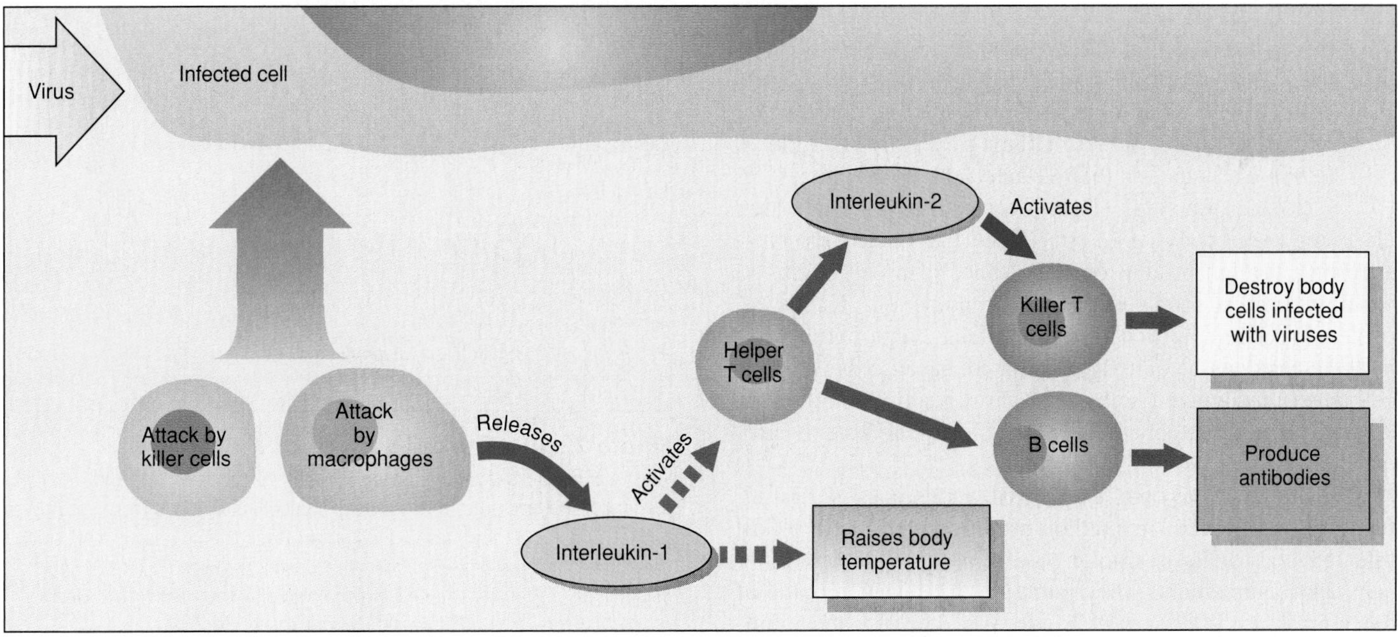

Figure 22.8 The immune response.
The attacks by killer cells and macrophages occur simultaneously, but the macrophage attack leads to further activation of the "specific" immune system, which involves T cells and B cells.

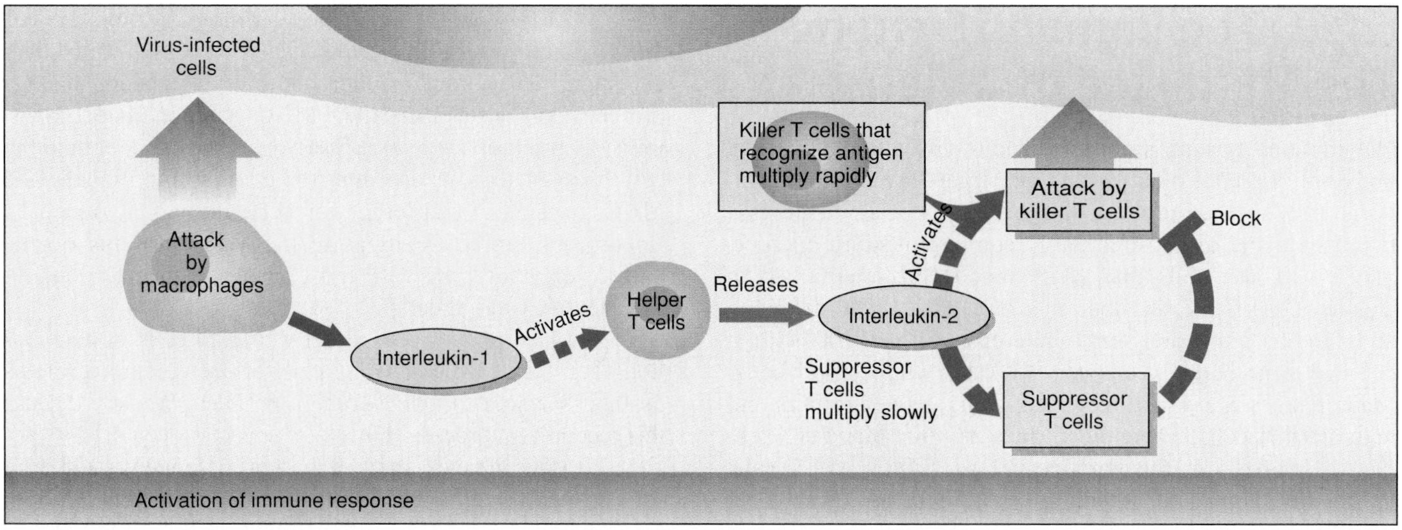

Figure 22.9 The T cell immune defense.
After being activated by the interleukin-1 alarm, helper T cells release interleukin-2, which stimulates the multiplication of killer T cells. Body cells that have been infected by the antigen are destroyed, and as the infection subsides, suppressor T cells "turn off" the immune response.

T Cells: Attack on Infected Body Cells

The awakening of helper T cells by the interleukin-1 alarm signal causes them to immediately unleash a very potent attack directed against the virus, which now resides within the cells of your respiratory tract. The decision is a difficult one, not taken until the first two lines of defense have been breached. The only way to eliminate the virus now is to kill the infected cells that harbor it. Using a second chemical signal called **interleukin-2,** the helper T cells call into action a class of cell assassins known as **killer T cells,** which recognize and destroy body cells infected with the virus (figure 22.9). Infected body cells display little bits of viral protein on their surface, and it is these telltale traces that the killer T cells zero in on, helped in their ID check by macrophages.

How can a single T cell deal with a major infection involving many body cells? Any T cell that detects the presence of a foreign protein on the surface of a cell begins to multiply rapidly, soon forming large numbers of T cells capable of recognizing that foreign protein. Large numbers of infected cells can be quickly eliminated, because the single T cell able to recognize the invading virus is amplified in number to form a large clone of identical T cells, all able to carry out the attack.

Any of your body's cells that bears traces of viral infection is destroyed. The method used by killer T cells to kill infected body cells is similar to that used by natural killer cells and complement—they puncture the cell membrane of the infected cell. Cancer cells are also detected and eliminated by this screening.

How do killer T cells recognize viral proteins? Your body makes millions of different types of T cells. Each type bears a single, unique kind of receptor protein on its cell membrane (figure 22.10), a receptor able to bind a particular viral or bacterial protein. Early in your life, when you were still a fetus, any cells with receptors that recognize your own body proteins were eliminated; the other millions sit like so many bear traps in the bloodstream, waiting (after being "set" by interleukin-2) to encounter any cell bearing a protein that is not "you" and destroy it.

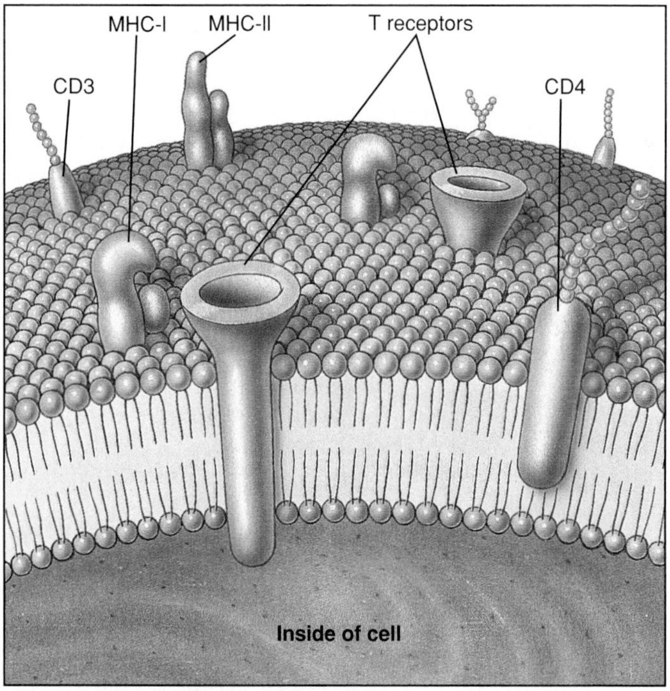

Figure 22.10 The cell surface of a T cell.
Within the lipid bilayer are embedded three types of proteins: *glycoproteins*, markers that identify the cell as being of a particular type (the glycoprotein labeled CD3 identifies the cell as a T cell, and the CD4 marker is characteristic of a particular kind of T cell); *MHC proteins* that identify the cell as belonging to that individual; and *T receptor proteins* that fit particular foreign molecules.

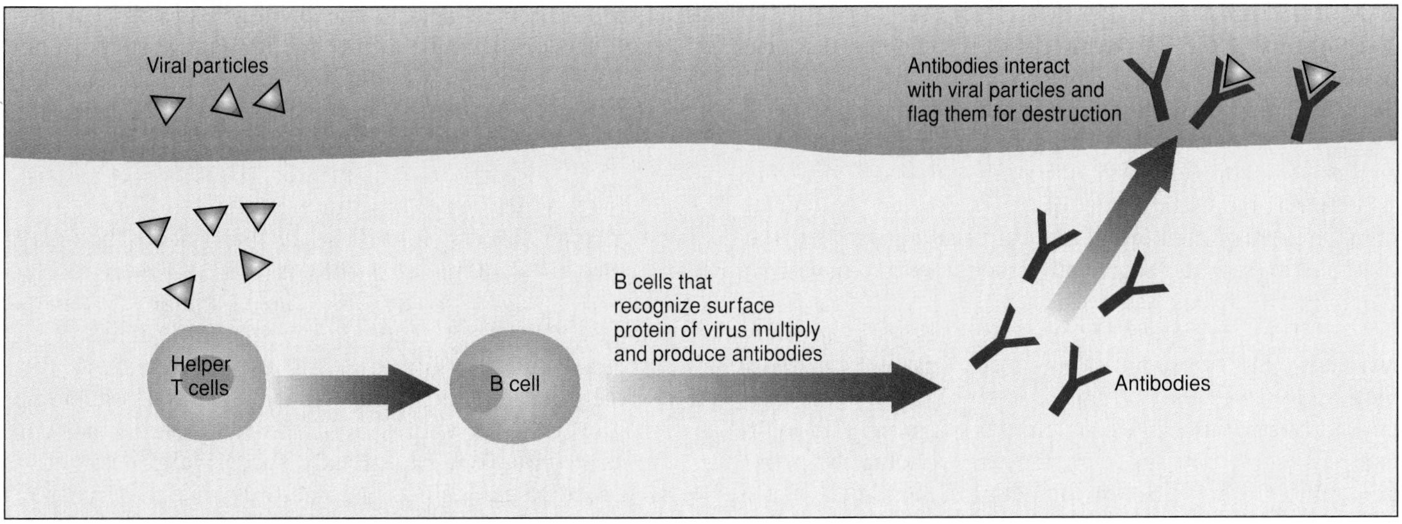

Figure 22.11 The B cell immune defense.
Invading viral particles are bound by B cells, which interact with helper T cells and are activated to divide. The multiplying B cells secrete antibodies which bind to invading microbes and tag them for destruction by macrophages.

B Cells: Attack on Invading Microbes

The interleukin-2 released by helper T cells also simultaneously activates the second kind of defensive white blood cell, the B cell (figure 22.11). The B cell portion of the immune defense is aimed directly against the invading microbe. Like killer T cells, B cells have receptor proteins on their surface, one type of receptor for each type of B cell. B cells recognize invading microbes much as killer T cells recognize infected cells, but unlike killer T cells they do not go on the attack themselves. Rather, they mark the pathogen for destruction by mechanisms that have no "ID check" system of their own. Early in the immune response, the markers placed by B cells alert complement proteins to attack the cells carrying them. Later in the immune response, the markers placed by B cells activate macrophages and natural killer cells.

The way B cells do their marking is simple and foolproof: When a B cell encounters a foreign microbe with a surface protein that matches its own type of receptor, it simply sticks its own receptor protein onto the microbe. The B cell receptor acts as a flag to attract the attention of macrophages and natural killer cells. The B cell is stimulated by its encounter to divide repeatedly, forming a large population of cells with the same receptor as the invading microbe. Finally, all these B cells stop growing and go into the business of making lots and lots of receptor protein, which they secrete into the bloodstream. We call these released receptor proteins **antibodies.** Flooding through the body, these antibody proteins (figure 22.12) attach themselves to any other pathogen cells that might be present, flagging those cells for destruction.

The B cell defense is very powerful because it amplifies the reaction to an initial pathogen encounter a millionfold. It is also a very long-lived defense in that a few

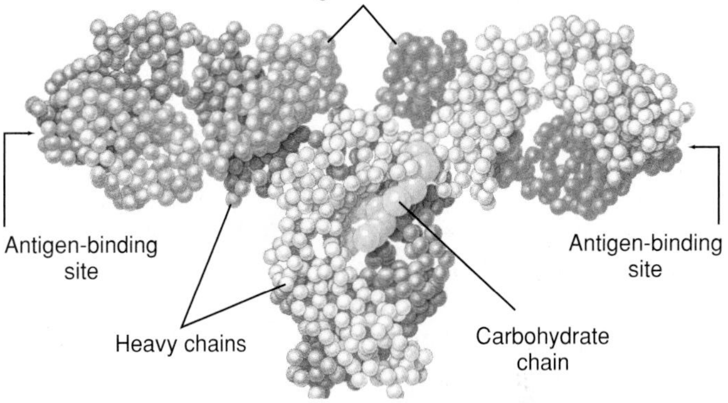

Figure 22.12 An antibody molecule.
In this molecular model of an antibody molecule, each amino acid is represented by a small sphere. Each molecule consists of four protein chains, two identical "light" (red) and two identical "heavy" (blue). The four protein chains wind around one another to form a Y shape. Foreign molecules, called antigens, bind to the arms of the Y.

of the multiplying B cells do not become antibody producers. Instead they become a line of **memory B cells** that continue to patrol your body's tissues, circulating through your blood and lymph for a long time—sometimes for the rest of your life. If the pathogen they are specialized to recognize ever appears again, they are ready, a host of memory cells able to recognize that particular pathogen and start the production of a new generation of antibody-producing cells directed against it. There is no waiting until an infection is well underway and macrophages sound the alarm—the antibodies generated by the new cells cause macrophages to destroy the pathogen *before* you become ill. You are not even aware of the battle going on in your body, and you are said to be **immune** to the infection.

Vaccination

Vaccination is the introduction into your body of a dead or disabled pathogen or, more commonly these days, of a harmless microbe with pathogen proteins displayed on its surface. The vaccination triggers an immune response against the pathogen, without an infection ever occurring. Afterwards, the bloodstream of the vaccinated person contains circulating memory B cells directed against that specific pathogen. The vaccinated person is said to have been "immunized" against the disease.

Through genetic engineering, scientists are now routinely able to produce "piggyback" vaccines made of harmless viruses that contain in their DNA a single gene cut out of a pathogen, a gene encoding a protein normally exposed on the pathogen's surface. By splicing the pathogen gene into the DNA of the harmless host, that host is induced to display the protein on its surface. The harmless virus displaying the pathogen protein is like a sheep in wolf's clothing, unable to hurt you but raising alarm as if it could. Your body responds to its presence by making an antibody directed against the pathogen protein, an antibody that acts like an alarm to the immune system, should that pathogen ever visit your body. Scientists are currently attempting to construct a vaccine against AIDS using this technique (figure 22.13).

If the activities of memory B cells provide such an effective defense against future infection, why can you catch some diseases like flu more than once? The reason you don't stay immune to flu is that the flu virus has evolved a way to evade the human immune system—it changes. The genes encoding the surface proteins of the flu virus mutate very rapidly. Thus the shapes of these surface proteins alter swiftly. Your memory B cells do not recognize viruses with altered surface proteins as being the same viruses it has already successfully defeated or been vaccinated against, because the memory B cell's receptors no longer "fit" the new shape of the flu surface proteins. When the new version of flu virus invades your body, you are back where you started and need to mount an entirely new immune defense.

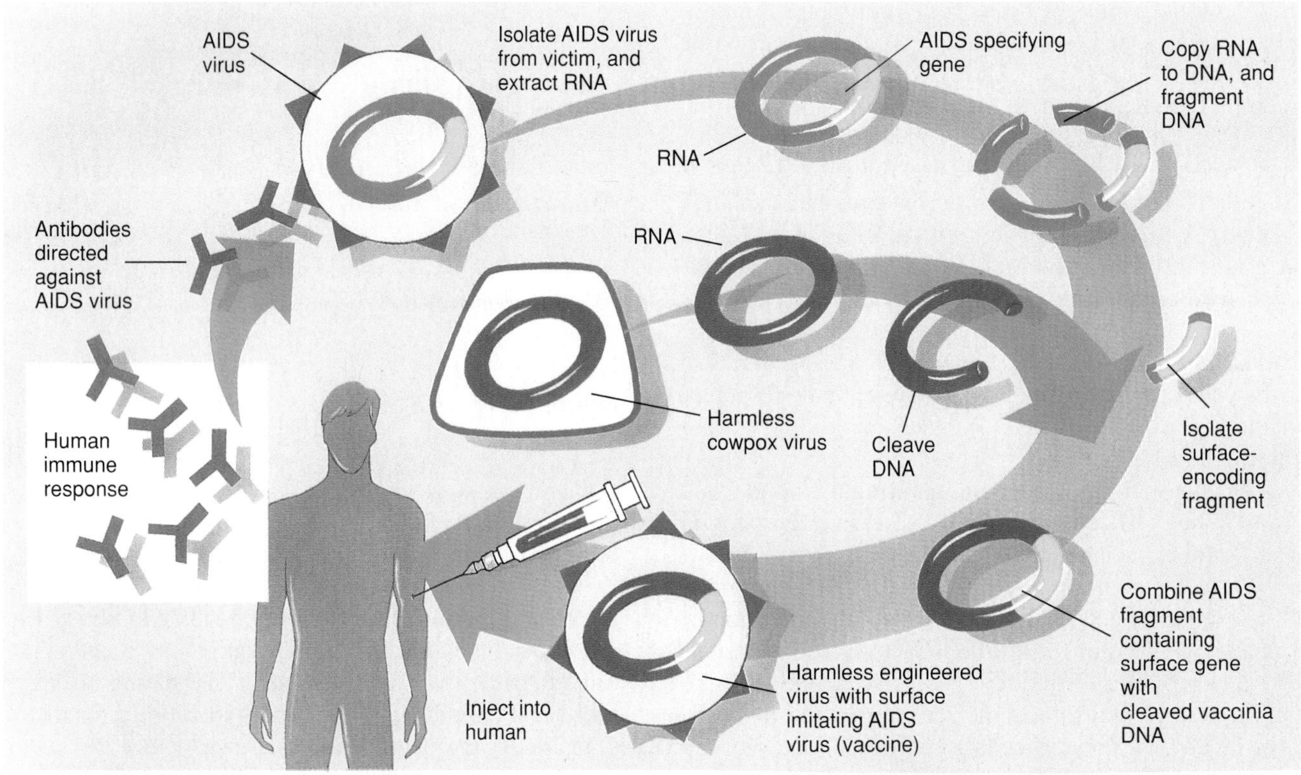

Figure 22.13 Constructing an AIDS vaccine.
Of the several genes of the human immunodeficiency virus (HIV), one would be selected that encodes a surface feature of the virus. All of the other HIV genes would be discarded. This one gene would not in itself be harmful to humans; it simply would encode the HIV surface protein known as glycoprotein 120. This one gene would be inserted into the DNA of a harmless vaccinia cowpox virus, resulting in a harmless virus whose surface imitates the AIDS virus. Persons injected with the vaccinia virus would not become ill (the vaccinia virus is harmless), but they would develop antibodies directed against the infecting virus surface. Because the surface contains the HIV protein gp120, the new antibodies would protect the infected person against any subsequent exposure to the HIV virus.

Sometimes the new mutations cause the flu virus surface proteins to assume shapes our immune system does not readily recognize. When this happened in 1918, over 22 million Americans and Europeans died in 18 months (figure 22.14). Less profound changes in flu virus surface proteins occur periodically, resulting in new strains of flu for which we are not immune. Attempts to improve our defenses against infection are among the most active areas of research today.

One of the most intensive efforts in the history of medicine is currently underway to develop an effective vaccine against HIV, the virus responsible for AIDS. The HIV virus has nine genes, encoding a variety of proteins. Initial efforts have focused on producing a subunit vaccine containing the HIV *env* (envelope) gene, which encodes the protein recognized by receptors on the surface of helper T cells. Unfortunately, these vaccines do not appear to be effective.

New efforts are targeting the small *nef* gene. In 1995 researchers learned that mutations in this gene can render the HIV virus nonvirulent, apparently by slowing replication of the virus. This suggests that an "attenuated" vaccine might be successful where a subunit one was not, the vaccine being the full virus in a nonvirulent *nef*-mutated form. Work continues at a fever pitch.

Unfortunately for attempts to develop an AIDS vaccine, the HIV virus mutates even more rapidly than the flu virus. Even vaccines that work in the laboratory, such as an attenuated version of the chimpanzee form of HIV, are not effective outside the laboratory, where new strains of HIV are encountered. This high mutation rate remains the single biggest obstacle to developing a successful AIDS vaccine.

As you can see, the immune system, often aided by vaccination, can respond in a variety of different ways to different kinds of pathogens (figure 22.15). However, as discussed in the next section, its ability to function normally and efficiently is often disrupted.

Figure 22.14 The flu epidemic of 1918 killed 22 million in 18 months.
With 25 million Americans alone infected during the influenza epidemic, it was hard to provide care for everyone. The Red Cross often worked around the clock.

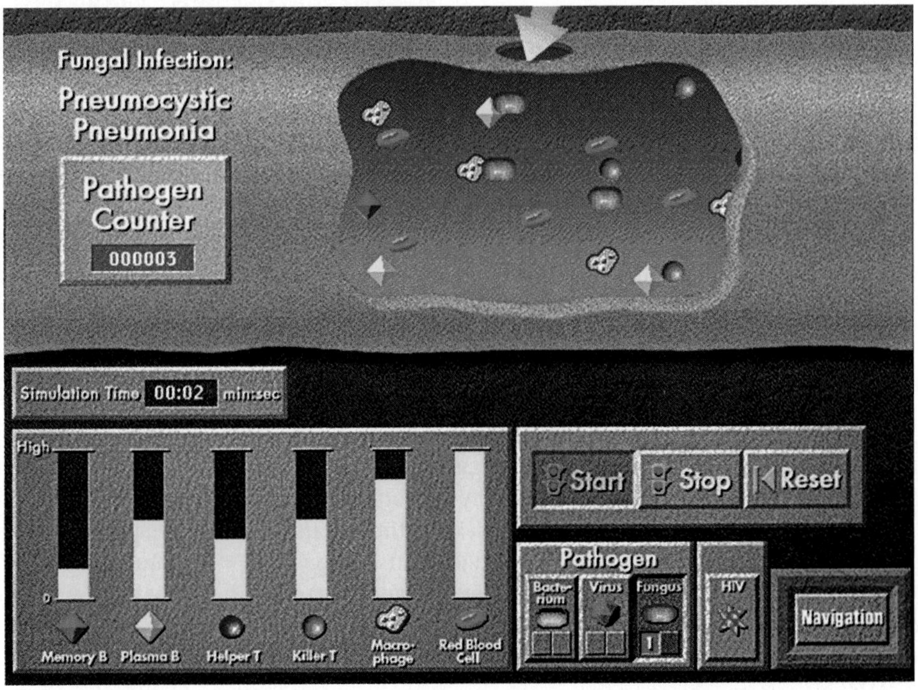

Figure 22.15 Many cells are involved in the immune response.
Infection by a pathogen such as the fungus *Pneumocystis* initiates a chain of events that alters the population of many white blood cells. This is a screen capture from an interactive CD-ROM exploration of the immune response. The exploration allows you to infect an individual with a virus (HIV), a microbe (*Pneumocystis*), or a cancer (Kaposi's sarcoma) and investigate the changes in the number of each cell type as the immune response progresses. (*Explorations in Human Biology,* Module 12, "Immune Response")

22.4 Immune System Failure

Although the immune system is one of the most sophisticated systems of the human body, it is still not perfect. Many of the major diseases we face, and some minor irritations as well, reflect failure of the immune system.

Autoimmune Diseases

The ability of killer T cells and B cells to distinguish cells of your own body—"self" cells—from nonself cells is the key ability of the immune system that makes your body's third line of defense so effective. In certain diseases, this ability breaks down, and the body attacks its own tissues. Such diseases are called **autoimmune diseases.**

Multiple sclerosis is an autoimmune disease that usually strikes people between the ages of 20 and 40. In multiple sclerosis, the immune system attacks and destroys the sheath of myelin that insulates motor nerves (like the rubber covering electrical wires). Degeneration of the myelin sheath interferes with transmission of nerve impulses, until eventually they cannot travel at all. Voluntary functions, such as movement of limbs, and involuntary functions, such as bladder control, are lost, leading finally to paralysis and death. Scientists do not know what stimulates the immune system to attack myelin.

Figure 22.16 The house dust mite *Dermatophagoides*.
This tiny animal causes an allergic reaction in many people.

Another autoimmune disease is type 1 diabetes, in which cells are unable to take in glucose because the pancreas fails to produce insulin (recall from chapter 21 that insulin plays a key role in the liver's regulation of levels of glucose in the blood). Type 1 diabetes is thought to result from an immune attack on the insulin-manufacturing cells of the pancreas. Again, no one knows why the attack occurs. Other autoimmune diseases are rheumatoid arthritis (an immune system attack on the tissues of the joints), lupus (in which the connective tissue and kidneys are attacked), and Graves disease (in which the thyroid is attacked).

Allergies

Although your immune system provides very effective protection against fungi, parasites, bacteria, and viruses, sometimes it does its job too well, mounting a major defense against a harmless substance. Such an immune response is called an **allergy.** Hay fever, sensitivity to even tiny amounts of plant pollen, is a familiar example of an allergy. Many people are allergic to proteins released from the feces of a minute mite that lives on grains of house dust (figure 22.16). The dust that the mite calls home is present in mattresses and pillows, and the mite goes out on foraging expeditions and consumes the dead skin cells that many of us shed in large quantities daily. Many people sensitive to feather pillows are in reality allergic to the mites that are residents of the feathers.

What makes an allergic reaction uncomfortable, and sometimes dangerous, is the involvement of antibodies attached to a kind of white blood cell called a **mast cell.** It is the job of the mast cells in an immune response to initiate an inflammatory response. When they encounter something that matches their antibody, mast cells release histamines and other chemicals that cause capillaries to swell. **Histamines** also increase mucus production by cells of the mucous membranes, resulting in runny noses and nasal congestion (all the symptoms of hay fever). Most allergy medicines relieve these symptoms with antihistamines, chemicals that block the action of histamines.

Asthma is a form of allergic response in which histamines cause the narrowing of air passages in the lungs. People who have asthma have trouble breathing when exposed to substances to which they are allergic.

22.5 AIDS: Immune System Collapse

AIDS (acquired immunodeficiency syndrome) was first recognized as a disease in 1981. By the end of 1995, more than 300,000 Americans had already died of AIDS, and more than 1.5 million other Americans were thought to be infected with **HIV** (human immunodeficiency virus), the virus that causes the disease (figure 22.17). The World Health Organization estimates that 40 million people will be infected with HIV by the year 2000. HIV apparently evolved from a very similar virus that infects chimpanzees in Africa when a mutation arose that allowed the virus to recognize a human cell surface receptor called **CD4.** This receptor is present in the human body on certain immune system cells, notably macrophages and helper T cells. It is the identity of these immune system cells that leads to the devastating nature of the disease.

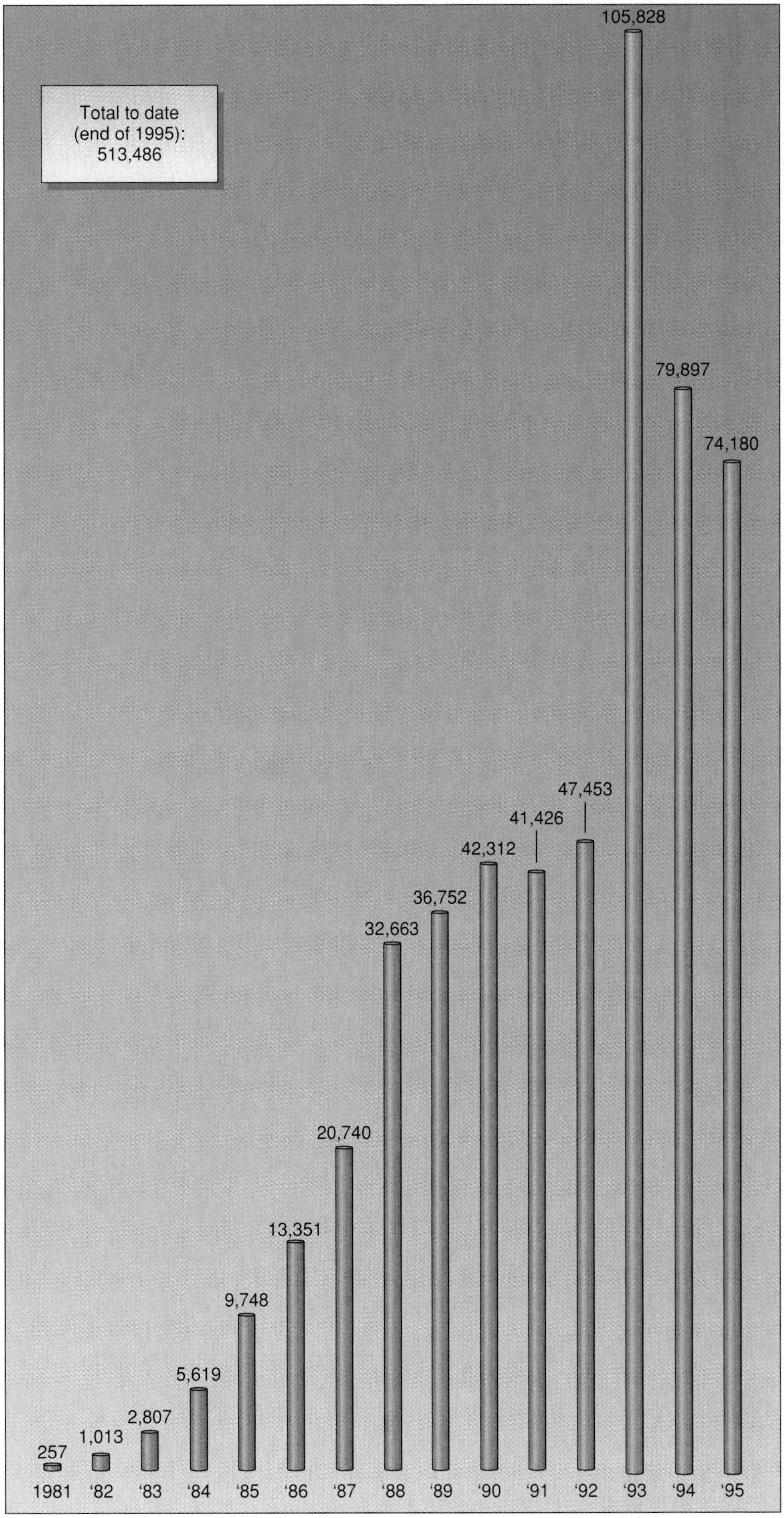

Figure 22.17 The AIDS epidemic in the United States.

The U.S. Centers for Disease Control (CDC) reports that 74,180 new AIDS cases were reported in 1995, with a total of 513,486 cases in the United States and over 310,000 deaths. Over 1.5 million other individuals are thought to be infected with the HIV virus in the United States and 14 million worldwide. The 100,000th AIDS case was reported in August 1989, eight years into the epidemic; the next 100,000 cases took just 26 months; the third 100,000 cases took barely 19 months (May 1993), and the fourth 100,000 took only 13 months (June 1994). The extraordinarily high numbers seen in 1993 reflect an expansion of the definition of what constitutes an AIDS case.

Source: Data from U.S. Center for Disease Control and Prevention, Atlanta, GA.

How AIDS Attacks the Immune System

HIV attacks and cripples the immune system. It invades macrophages, which shelter it from immune defenses and carry it throughout the body (figure 22.18). It also attacks helper T cells, killing ever larger numbers of them as the infection progresses (figure 22.19). When the number of helper T cells in the blood has fallen below 200 (a normal count is 800 to 1,000), a person is said to have AIDS. Without helper T cells to activate and direct B cells and killer T cells, the immune response cannot occur. The body is overwhelmed by pathogens and cancers that it normally would defeat. Pneumocystic pneumonia, a common cause of AIDS death, is a fungal infection that normal immune systems easily combat. Kaposi's sarcoma, another major cause of AIDS death, is a form of cancer that is rare except in very old people because normal immune surveillance easily detects and removes it. Scientists believe that practically everyone infected with HIV will eventually develop AIDS, although the time between infection and onset of AIDS can be 10 years or longer. During this time, the infected person is fully able to transmit the HIV virus to others.

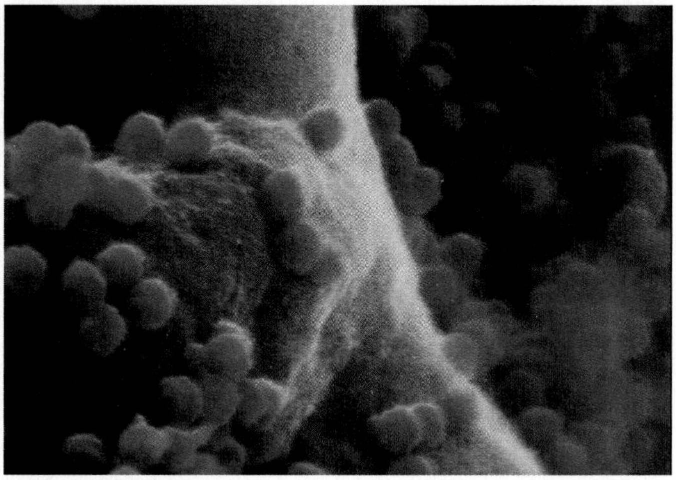

Figure 22.18 HIV viruses infecting a macrophage.
Each blue sphere is an HIV virus.

How HIV Is Transmitted

There is no cure for AIDS. The only way to protect yourself is to avoid exposure to HIV. HIV is a fragile virus that cannot exist for long outside the body. You can get HIV only by taking into your body the HIV-infected blood cells or body fluids of infected individuals. Most of the virus is present within macrophages rather than as free virus particles, and because both semen and vaginal secretions are rich in macrophages, it is easy to contract HIV through sexual intercourse with an infected person. Worldwide, most HIV infections are spread this way. The virus can be transmitted to either partner during sexual intercourse (the frequency of AIDS worldwide is about the same in men and women), and because macrophages can cross any mucous membrane easily, the virus can be transmitted through either vaginal or oral sex. Use of a condom (so called "safe sex") greatly reduces, but does not eliminate, the risk of getting HIV.

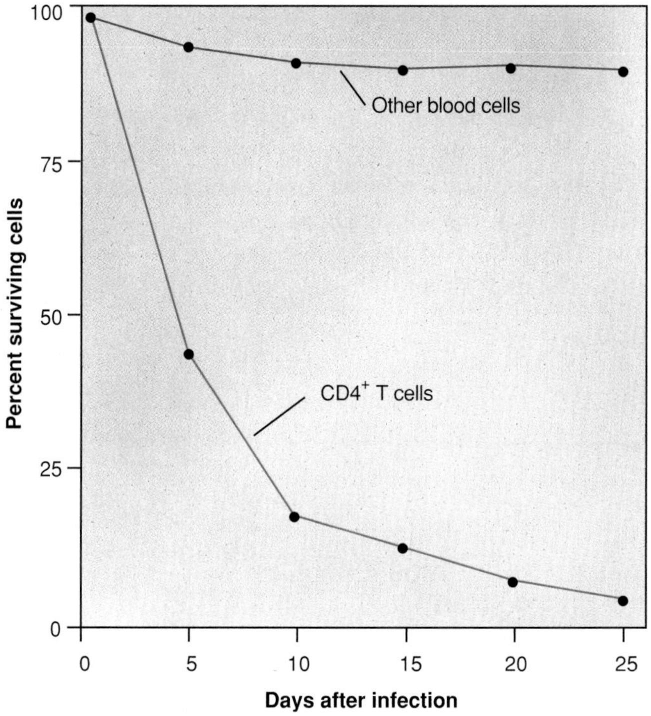

Figure 22.19 Survival of T cells in culture after exposure to AIDS virus.
The virus has little effect on the number of other kinds of blood cells (blue), but it causes the number of white blood cells with CD4 receptors to decline dramatically (red).

Because blood contains many macrophages, it provides a ready vehicle for transmitting AIDS. People who inject intravenous drugs often become infected with HIV by sharing or reusing hypodermic syringes that have become contaminated with HIV-containing blood. The majority of HIV infections in the United States in the late 1980s were transmitted in this manner.

HIV is not transmitted through the air, on toilet seats, or in any other condition where a macrophage would not survive. It cannot be contracted through casual contact, such as shaking hands, sharing food, or drinking from the same water fountain as an infected person, for the simple reason that macrophages cannot be transmitted by such contact. Although HIV is found in saliva, tears, and urine, it occurs there only in very low concentrations, too few virus particles to successfully initiate an infection. That is why you cannot catch HIV from the small amount of saliva exchanged when kissing. Biting insects such as mosquitoes and ticks do not transmit HIV because they do not transmit macrophages. AIDS is not an easy disease to catch. Only exposure to infected needles or unprotected sexual contact with infected individuals commonly lead to AIDS.

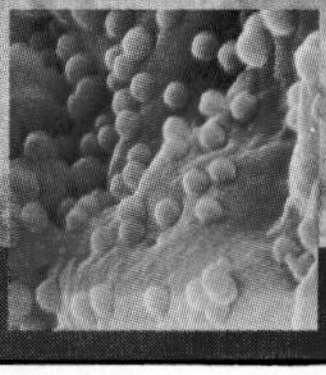

CHAPTER 22

HIGHLIGHTS

Key Terms | Key Concepts

22.1 Skin: The First Line of Defense

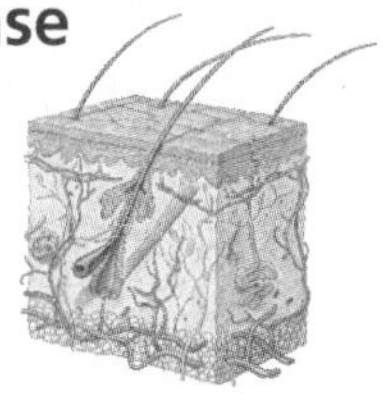

Key Terms

epidermis 478
dermis 478
stratum corneum 478

Key Concepts

- Skin offers an efficient barrier to penetration by microbes.
- Skin's surface is very acidic, inhibiting the growth of microbes.
- Sweat contains enzymes that attack and digest the cell walls of bacteria.

22.2 Cellular Counterattack: The Second Line of Defense

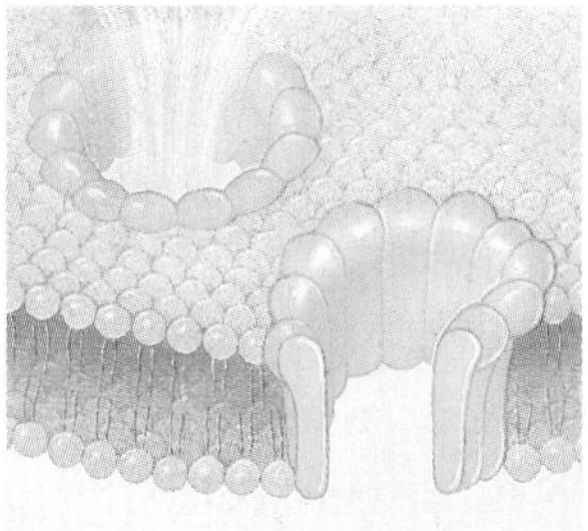

Key Terms

macrophage 480
neutrophil 480
complement system 481
inflammatory response 482
fever 482

Key Concepts

- Macrophages are white blood cells that patrol the bloodstream, ingesting bacteria by phagocytosis.
- Neutrophils kill all the cells at a site of infection.
- Complement proteins insert into a pathogen's cell membrane, causing the cell to rupture like a punctured balloon.
- The body expands blood vessels and raises temperature in response to infection to aid the body's defenses.

22.3 The Immune System: The Third Line of Defense

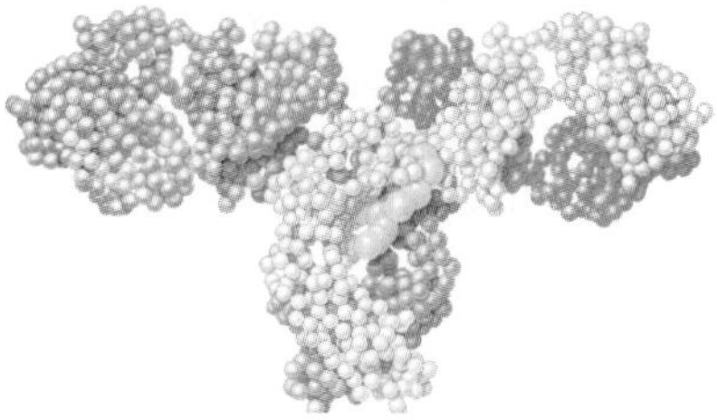

Key Terms

T cell 483
B cell 483
helper T cell 483
interleukin-1 483
antibody 485
vaccination 486

Key Concepts

- Macrophages release chemicals that initiate the immune response by activating helper T cells.
- Helper T cells in turn activate T cells, which destroy infected body cells, and B cells, which label microbes for destruction by macrophages.
- Residual T cells and B cells so speed future immune responses that immunity results.

22.4 Immune System Failure

Key Terms

autoimmune disease 488
allergy 488

Key Concepts

- When the body's immune system attacks its own tissues, autoimmune disease results.
- Allergies are inappropriate immune responses to harmless substances.

22.5 AIDS: Immune System Collapse

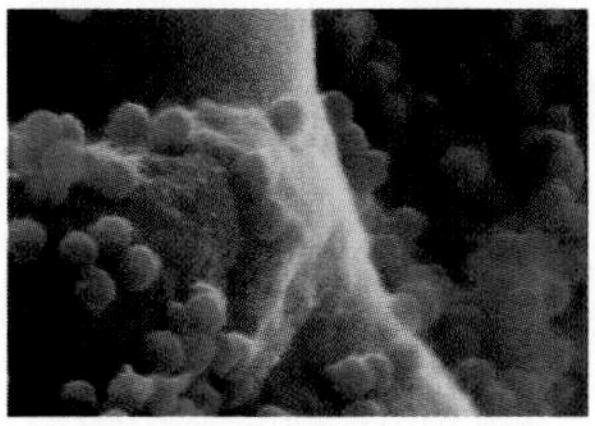

Key Terms

HIV 489
CD4 489

Key Concepts

- AIDS is a disease that destroys the immune response by killing helper T cells and macrophages.
- Over 300,000 Americans have died of AIDS since 1981.
- AIDS is caused by HIV, which is transmitted in body fluids, typically via sexual intercourse or by infected needles during drug use.

22

CONCEPT REVIEW

1. Your body loses and replaces approximately _________ skin cells every hour.
   a. 1,000
   b. 1.5 million
   c. 1 billion
   d. 1.5 billion

2. Which of the following statements does *not* describe the skin?
   a. It is the largest human body organ.
   b. Its oil and sweat glands make the skin's surface very acid.
   c. It secretes gastric HCl, which inhibits the growth of many microbes.
   d. It contains sweat glands that secrete an enzyme that attacks the cell walls of many bacteria.

3. White blood cells that kill bacteria by digesting them are called
   a. natural killer cells.
   b. macrophages.
   c. T cells.
   d. lymphocytes.

4. Membrane attack complexes that form holes in the pathogen's membrane are part of the defense called the
   a. complement system.
   b. immune response.
   c. inflammatory response.
   d. temperature response.

5. The inflammatory response results in _________, which promotes the migration of macrophages and neutrophils to the site of infection.
   a. fever
   b. the specific immune response
   c. increased blood flow
   d. the secretion of lysozyme

6. Which of the following cell types in the immune response controls other types of cells?
   a. helper T cell
   b. B cell
   c. cytotoxic T cell
   d. mast cell

7. The alarm signal protein secreted by macrophages in the immune response is called
   a. a histamine.
   b. an antibody.
   c. MHC.
   d. interleukin-1.

8. Cells called _________ allow you to be immune to infections for a long time, sometimes the rest of your life, after the initial infection.
   a. killer T cells
   b. memory B cells
   c. helper T cells
   d. antigens

9. Which of the following is *not* an autoimmune disease?
   a. AIDS
   b. type I diabetes
   c. type II diabetes
   d. multiple sclerosis

10. The AIDS virus is remarkably effective at short-circuiting the immune response because it infects
    a. helper T cells.
    b. B cells.
    c. mast cells.
    d. stem cells.

11. Biting insects such as mosquitoes do not transmit AIDS because
    a. the only way to get AIDS is through unprotected sex.
    b. they do not transfer blood.
    c. they do not transmit macrophages.
    d. HIV is transmitted through the air.

12. The _________ is the middle layer of the skin.

13. _________ is an enzyme in the sweat of the body.

14. Normal internal human body temperature is _______ °C.

15. B cells secrete protective molecules called _________.

16. _________ is the introduction into your body of a dead pathogen to trigger an immune response against the pathogen.

17. Multiple sclerosis develops from the degeneration of the _________ around neurons.

18. HIV can recognize the surface receptor on human cells called _________.

Answers to the Concept Review questions appear in Appendix B.

CHALLENGE YOURSELF

1. Do you think a virus or other pathogen that invades a vertebrate host, avoids the host's immune defenses entirely, reproduces quickly within the host, and leads to the rapid death of the host is well adapted or poorly adapted to that host? Explain.

2. Why might attempting to bring down a slight fever actually be counterproductive?

3. AIDS is a virus that destroys the human immune system by killing helper T cells, which are necessary to activate the immune response. The African green monkeys from which the AIDS virus is thought to have arisen do not suffer from AIDS. How do you imagine they have escaped this?

FOR FURTHER READING

Boon, T. "Teaching the Immune System to Fight Cancer." *Scientific American,* March 1993, 82–89. Can antigens on the surface of cancerous tumors be used to sensitize the immune system to the development of subsequent tumors?

Brunetta, L. "Cancer and the Immune Response: A Chemical Link?" *Technology Review* 97 (January 1994): 12–13. Inflammation and infection have been linked to developing cancer, but no one really knew why. Researchers now find that when the immune system is active (such as during inflammation or infection), elevated levels of nitrous oxide are produced by the body. The nitric oxide, in turn, has been shown to cause damage to DNA and hasten cell death—perhaps this is the link!

Caldwell, J., and P. Caldwell. "The African AIDS Epidemic." *Scientific American,* March 1996, 62–68. In parts of Africa nearly 25% of the population is infected with the HIV virus as a result of heterosexual transmission of the virus.

Calwell, M. "Blessed with Resistance." *Discover,* January 1994, 46–48. Epidemiologists have encountered a small population of individuals in Nairobi who appear to be resistant to AIDS, despite repeated exposure to the virus.

Cohen, J. "Vaccines Get a New Twist." *Science,* 264 (April 22, 1994): 503–5. Everyone knows you get vaccines to *prevent* disease. Scientists are now experimenting with using them to *treat* disease.

Glausiusz, J. "The Secret Healing Power of Sharks." *Discover,* January 1994, 86. Dogfish shark epithelial cells have apparently yielded an antibiotic effective against bacteria, fungi, and protozoa. Now in the process of being synthesized commercially, this drug may offer some powerful resistance to immunosuppressed patients.

Hoffman, M. "AIDS: Solving the Molecular Puzzle." *American Scientist* 82 (March/April 1994): 171–77. Despite more than a decade of solid, intensive, well-funded research, researchers are still not all that close to understanding exactly how the HIV virus actually *causes* AIDS.

Smith, K. A. "Interleukin-2." *Scientific American,* March 1990, 50–57. The first hormone of the immune system to be recognized, it helps the body to mount a defense against microorganisms by triggering the multiplication of only those cells that attack an invader.

TECHNOLOGY LINKS

The Living World Home Page
http://www.wcbp.com/biology/tlw

Explorations in Human Biology CD
#12 Immune Response
#13 AIDS

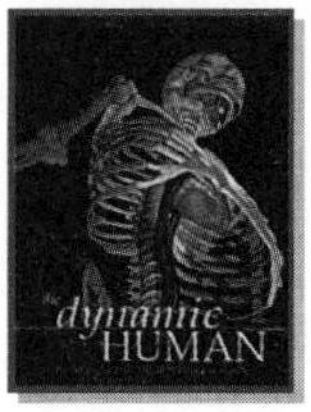

The Dynamic Human CD
Lymphatic System

Life Science Animations Videotape 4
#41 B-Cell Immune Response
#42 Structure and Function of Antibodies
#43 Types of T Cells

23

The Nervous System

"The picture's pretty bleak, gentlemen. ... The world's climates are changing, the mammals are taking over, and we all have a brain about the size of a walnut."

CHAPTER OUTLINE

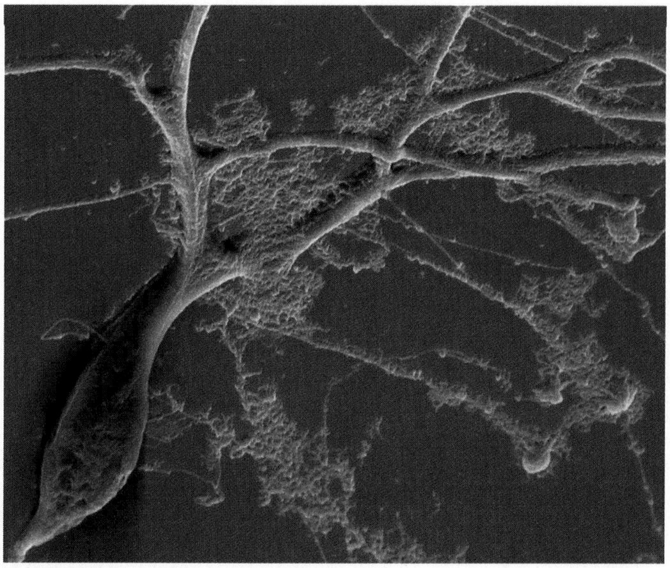

Figure 23.1 A human nerve cell.
The cell body is at the *lower left.* The branching network of fibers extending up from the cell body is made up of dendrites, which carry signals to the cell body. The nervous system controls and integrates the body's activities.

Humans, like all other animals except sponges, use a network of nerve cells (figure 23.1) to gather information about the body's condition and the external environment, to process and integrate that information, and to issue commands to the body's muscles and glands. Just as telephone cables run from every compartment of a submarine to the conning tower, where the captain controls the ship, so bundles of nerve cells called nerves connect every part of your body to its command and control center, the brain and spinal cord. Your body is run just like a submarine, with status information about what is happening in organs and outside the body flowing into the command center, which analyzes the data and issues commands to glands and muscles.

23.1 Organization of the Nervous System

The human nervous system has three elements (figure 23.2):

1. **Central nervous system (CNS):** *central processing.* The brain and spinal cord act as the central control region of the nervous system, processing information and issuing commands.
2. **Motor nervous system:** *output.* Nerves transmit commands from the CNS to voluntary muscles like your leg muscles (figure 23.3), to involuntary muscles like those carrying out peristalsis in your intestines, and to endocrine (hormone-producing) glands like the pituitary gland in your head.
3. **The sensory nervous system:** *input.* A variety of sensory receptors gather information about the environment and about your body's condition. Then sensory nerves transmit this information to the CNS.

The motor and sensory nervous systems are quite distinct and separate. **Motor nerves** (*efferent*) carry information out and away from the CNS, and **sensory nerves** (*afferent*) carry information in toward the CNS. However, because together the motor and sensory nerves make up all the nerve pathways of the body outside the brain and spinal cord, the two nervous systems are sometimes referred to collectively as the *peripheral nervous system.*

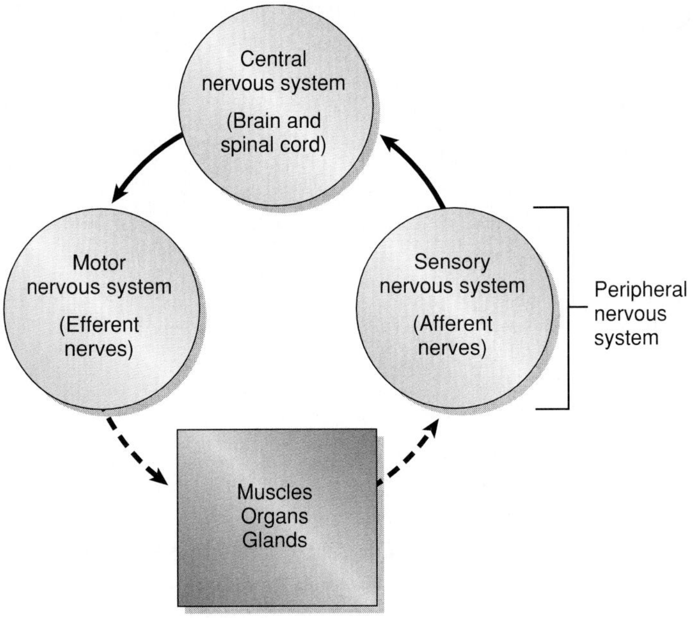

Figure 23.2 How the human nervous system is organized.
The central nervous system, consisting of the brain and spinal cord, issues commands via the motor nervous system and receives information from the sensory nervous system. The motor and sensory nervous systems together make up the peripheral nervous system.

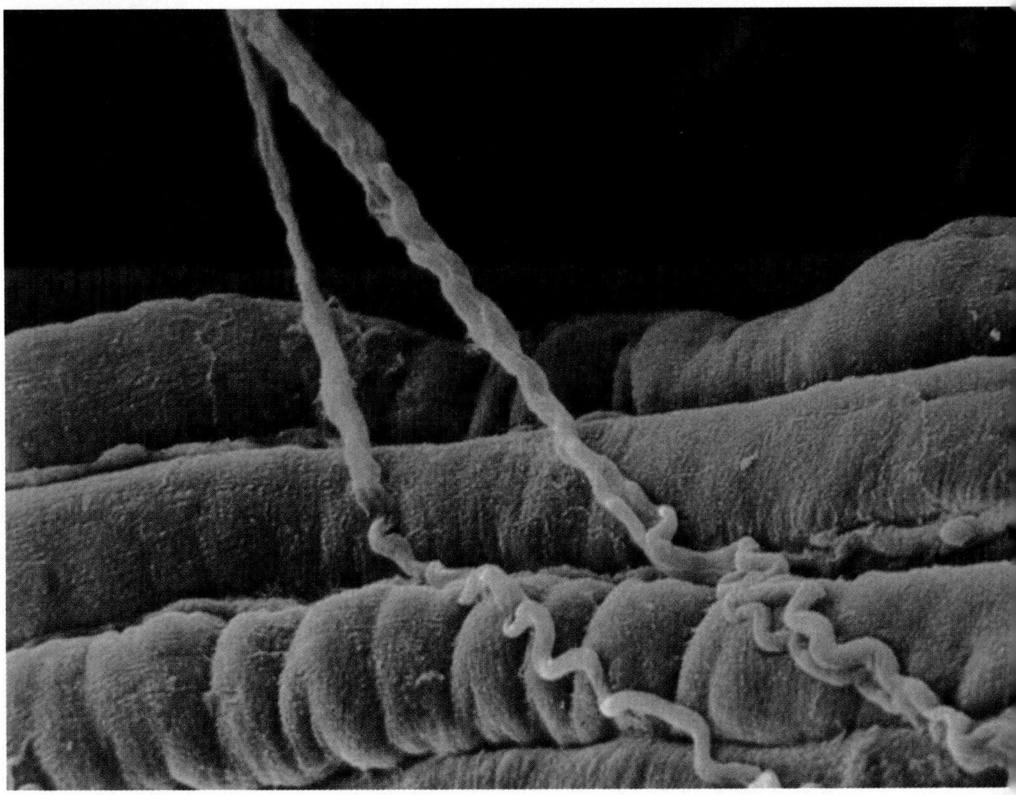

Figure 23.3 Motor nerves transmit commands to individual leg muscles.
The slender, twisted threads are motor nerves, and the long, thick strands at the bottom are leg muscle fibers. Often a nerve carries many independent fibers; these branch off to establish contact with different target muscles.

23.2 The Central Nervous System

The structure and function of the vertebrate brain has long been the subject of scientific inquiry. Despite ongoing research, scientists are still not sure how the brain performs many of its functions. For instance, scientists continue to look for the mechanism the brain employs to store memories, and they do not understand how some memories can be "locked away," only to surface in times of stress. The brain is the most complex vertebrate organ ever to evolve, and it can perform a bewildering variety of complex functions.

In a developing human embryo, the central nervous system starts out as a tube. The front part of the tube develops three bulges that form the brain. The rest of the tube forms the long, straight spinal cord. The brain has three main parts: a center for association and higher thought called the **cerebrum;** a center for information processing composed of the **thalamus** and the **hypothalamus;** and a center for motor coordination located in the **cerebellum** and **brain stem.** The **spinal cord** extends from the brain stem and consists of sensory and motor nerve tracts in the cervical, thoracic, and lumbar regions of the spine (figure 23.4).

The motor pathways can be further subdivided into the **voluntary nervous system,** which relays commands to skeletal muscles, and the **autonomic nervous system,** which stimulates glands and relays commands to the smooth muscles of the body. The voluntary nervous system can be controlled by conscious thought. You can, for example, command your hand to move. The autonomic nervous system, by contrast, cannot be controlled by conscious thought. You cannot, for example, tell the smooth muscles in your digestive tract to speed up their action. The central nervous system issues commands over both voluntary and autonomic systems, but you are conscious of only the voluntary commands.

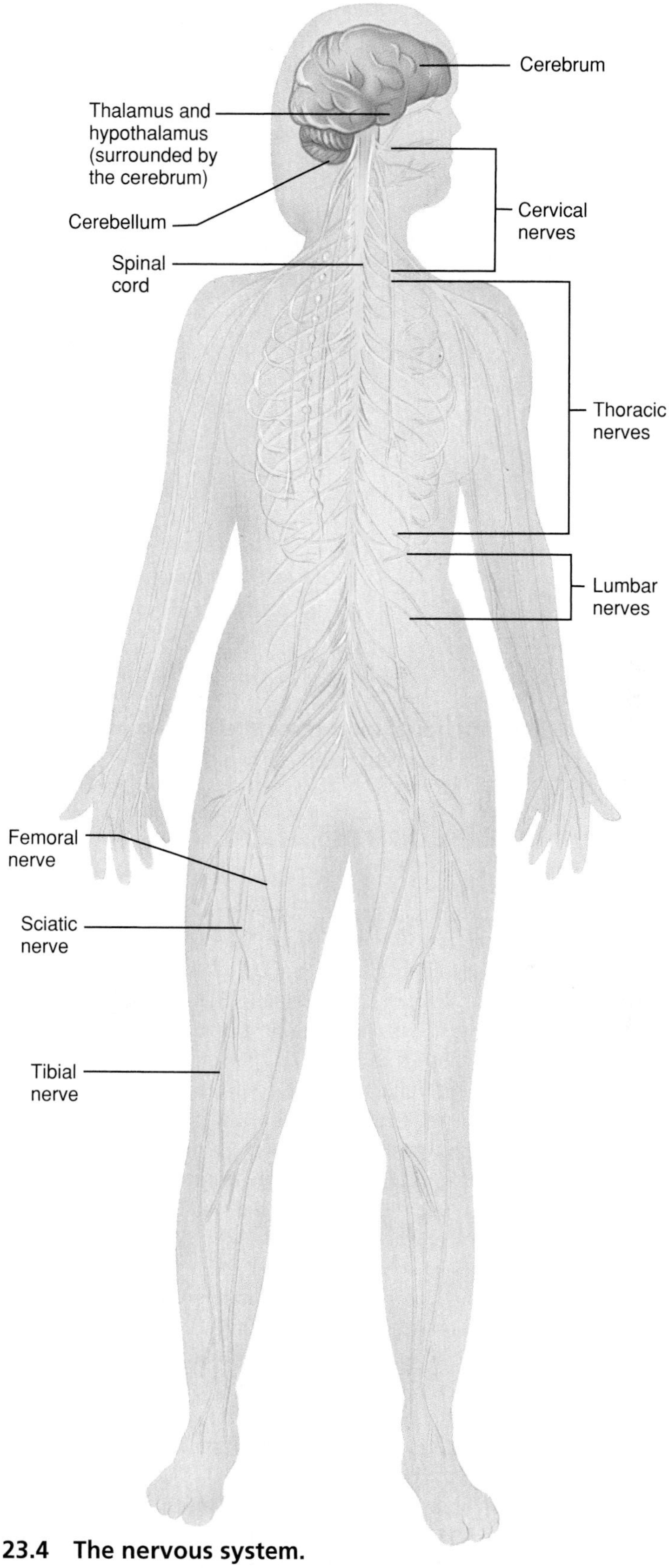

Figure 23.4 The nervous system.
The brain is colored pink and the spinal cord and nerves are colored yellow.

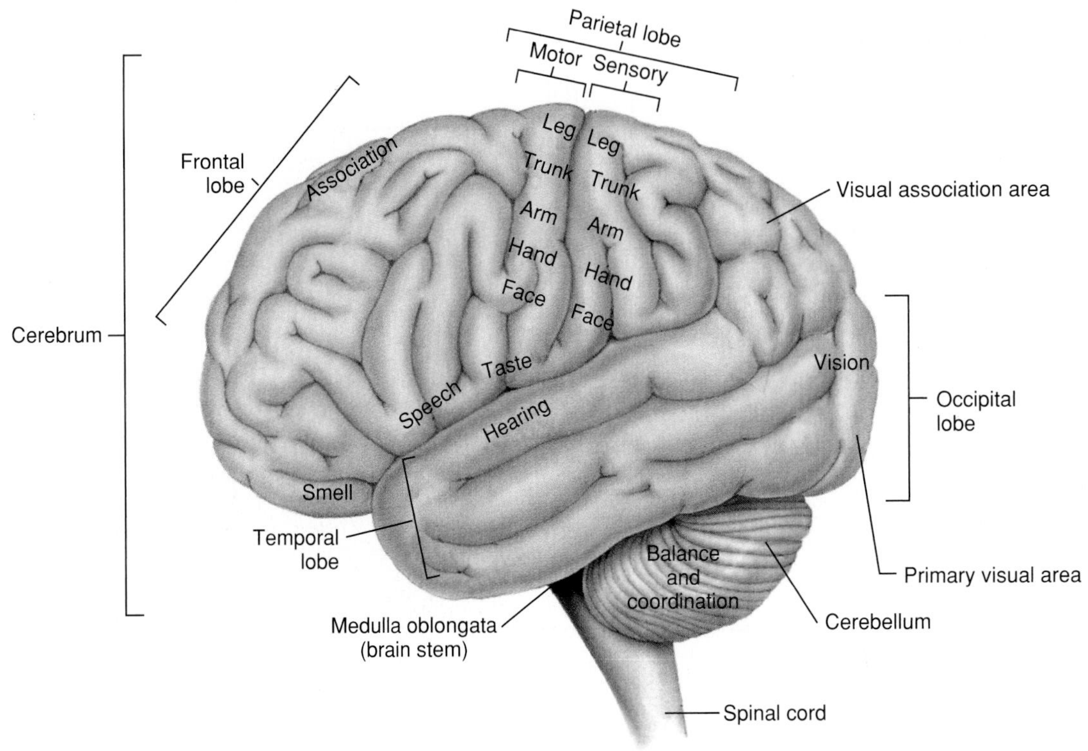

Figure 23.5 The major functional regions of the human brain.
Specific areas of the cerebral cortex are associated with different regions and functions of the body.

The Cerebrum Is the Control Center of the Brain

About 85% of the weight of the human brain is made up of the **cerebrum.** The cerebrum is the large rounded area of the brain divided by a groove into right and left halves called cerebral hemispheres. It functions in language, conscious thought, memory, personality development, vision, and a host of other activities we call "thinking and feeling." Figure 23.5 shows general areas of the brain and the functions they control. The cerebrum, which looks like a wrinkled mushroom, is positioned over and surrounding the rest of the brain, like a hand holding a fist. Much of the neural activity of the cerebrum occurs within a thin, gray outer layer only a few millimeters thick called the **cerebral cortex** (*cortex* is Latin for "bark of a tree"). This layer is gray because it is densely packed with neuron cell bodies. The human cerebral cortex contains the cell bodies of more than 10 billion nerve cells, roughly 10% of all the neurons in the brain. The wrinkles in the surface of the cerebral cortex increase its surface area (and number of cell bodies) threefold. Underneath the cortex is a solid white region of myelinated nerve fibers that shuttles information between the cortex and the rest of the brain (figure 23.6).

The right and left cerebral hemispheres are linked by a bundle of neurons called a **tract.** This tract serves as an information highway, telling each half of the brain what the other half is doing. Because these tracts cross over, each half of the brain controls muscles and glands on the opposite side

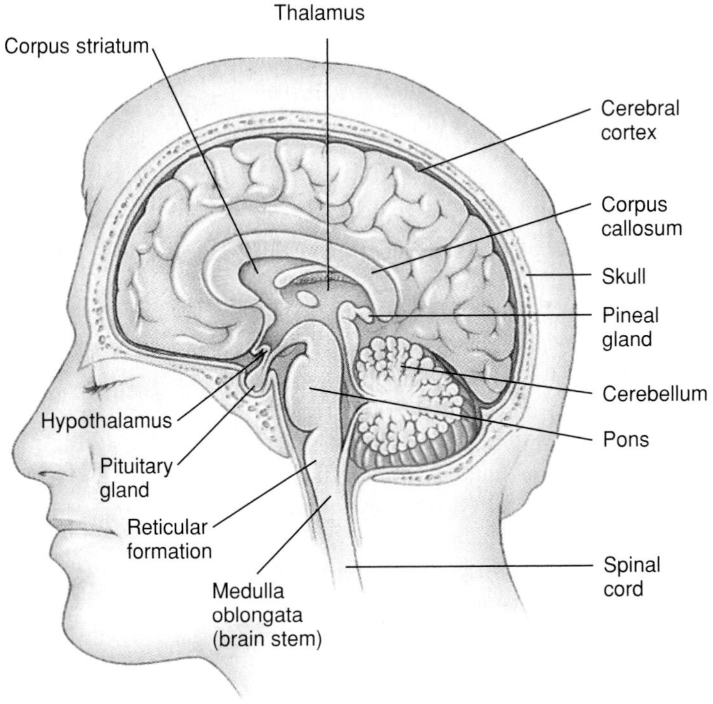

Figure 23.6 A section through the human brain.
The cerebrum occupies most of the brain. Only its outer layer, the cerebral cortex, is visible in surface view.

of the body. In general, the left brain is associated with language, speech, and mathematical abilities while the right brain is associated with intuitive, musical, and artistic abilities.

Researchers have found that the two sides of the cerebrum can operate as two different brains. For instance, in some people the tract between the two hemispheres has been cut by accident or surgery. In laboratory experiments, one eye of an individual with such a "split brain" is covered and a stranger is introduced. If the other eye is then covered instead, the person does not recognize the stranger who was just introduced!

Sometimes blood vessels in the brain are blocked by blood clots, causing a disorder called a **stroke.** During a stroke, circulation to an area in the brain is blocked and the brain tissue dies. A severe stroke in one side of the cerebrum may cause paralysis of the other side of the body.

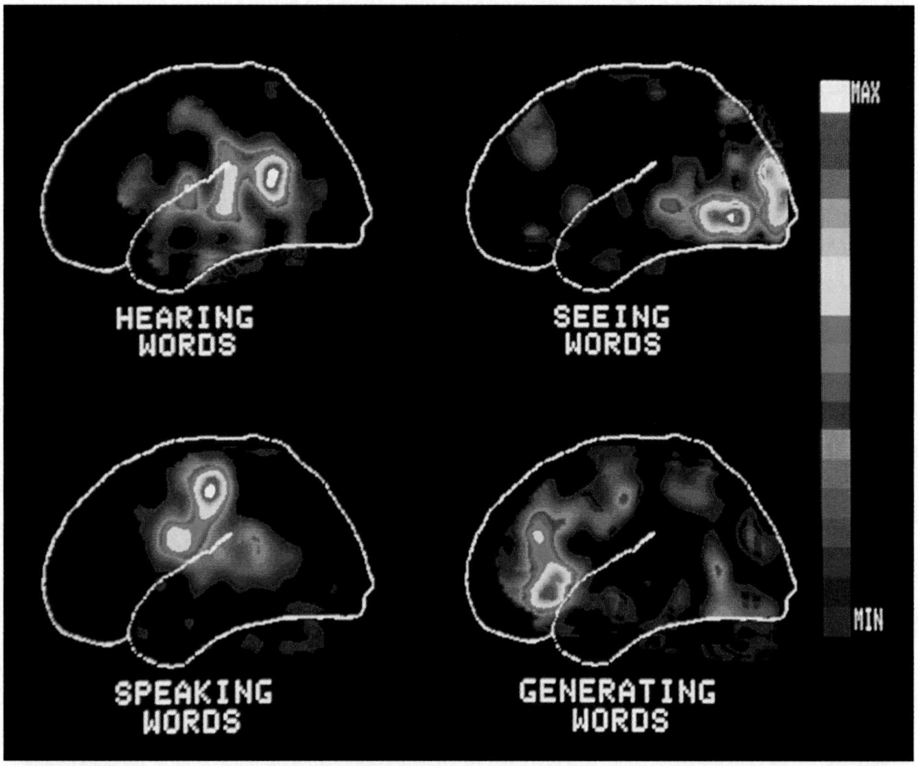

Figure 23.7 Different brain regions control various activities.
This illustration shows how the brain reacts in human subjects asked to listen to a spoken word, to read that same word silently, to repeat the word out loud, and then to speak a word related to the first. Regions of white, red, and yellow show the greatest activity. Compare this to figure 23.5 to see how regions of the brain are mapped.

The Thalamus and Hypothalamus Process Information

Beneath the cerebrum are the thalamus and hypothalamus, important centers for information processing. The **thalamus** is the major site of sensory processing in the brain. Auditory (sound), visual, and other information from sensory receptors enters the thalamus and then is passed to the sensory areas of the cerebral cortex. The thalamus also controls balance. Information about posture, derived from the muscles, and information about orientation, derived from sensors within the ear, combines with information from the cerebellum and passes to the thalamus. The thalamus processes the information and channels it to the appropriate motor center on the cerebral cortex. Different areas of the cerebral cortex control different activities (figure 23.7).

The **hypothalamus** integrates all the internal activities. It controls centers in the brain stem that in turn regulate body temperature, blood pressure, respiration, and heartbeat. It also directs the secretions of the brain's major hormone-producing gland, the pituitary gland. The hypothalamus is linked by an extensive network of neurons to some areas of the cerebral cortex. This network, along with parts of the thalamus and the hypothalamus, is called the **limbic system** (figure 23.8). The operations of the limbic system are responsible for many of the most deep-seated drives and emotions of vertebrates, including pain, anger, sex, hunger, thirst, and pleasure.

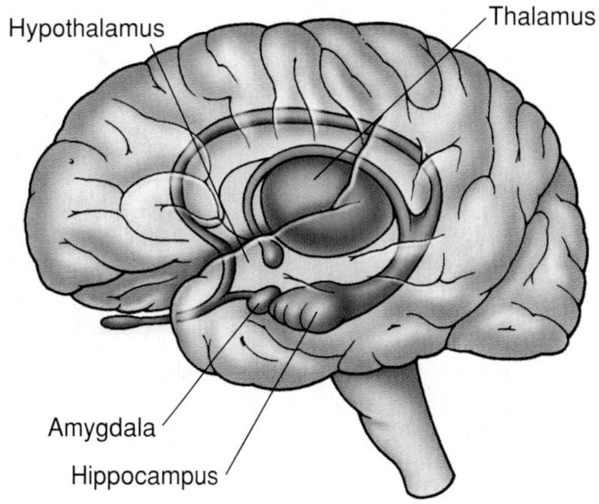

Figure 23.8 The limbic system.
The hippocampus and the amygdala are the major components of the limbic system, which controls our most deep-seated drives and emotions.

The Cerebellum Coordinates Muscle Movements

Extending back from the base of the brain is a structure known as the **cerebellum.** The cerebellum controls balance, posture, and muscular coordination. This small, cauliflower-shaped structure, while well developed in humans and other mammals, is even better developed in birds. Birds perform more complicated feats of balance than we do, because they move through the air in three dimensions. Imagine the kind of balance and coordination needed for a bird to land on a branch, stopping at precisely the right moment without crashing into it.

The Brain Stem Controls Vital Body Processes

The base of the brain, called the **brain stem** or sometimes the **medulla oblongata,** connects the rest of the brain to the spinal cord. This stalklike structure contains nerves that control your breathing, swallowing, digestive processes, the beating of your heart, and the diameter of your blood vessels. A network of nerves called the **reticular formation** runs through the brain stem and connects to other parts of the brain. Their widespread connections make these nerves essential to consciousness, awareness, and sleep. One part of the reticular formation filters sensory input, enabling you to sleep through repetitive noises such as traffic yet awaken instantly when a telephone rings.

The Spinal Cord Carries Information to and from the Brain

The **spinal cord** is a cable of neurons extending from the brain down through the backbone. The gray nerve-cell bodies form a column in the center of the cord, surrounded by a sheath of axons and dendrites, which make the outer edges of the cord white because they are coated with myelin. The spinal cord is surrounded and protected by the backbone, through which spinal nerves pass out to the body (figure 23.9). Messages from the body and the brain run up and down the spinal cord, an information highway.

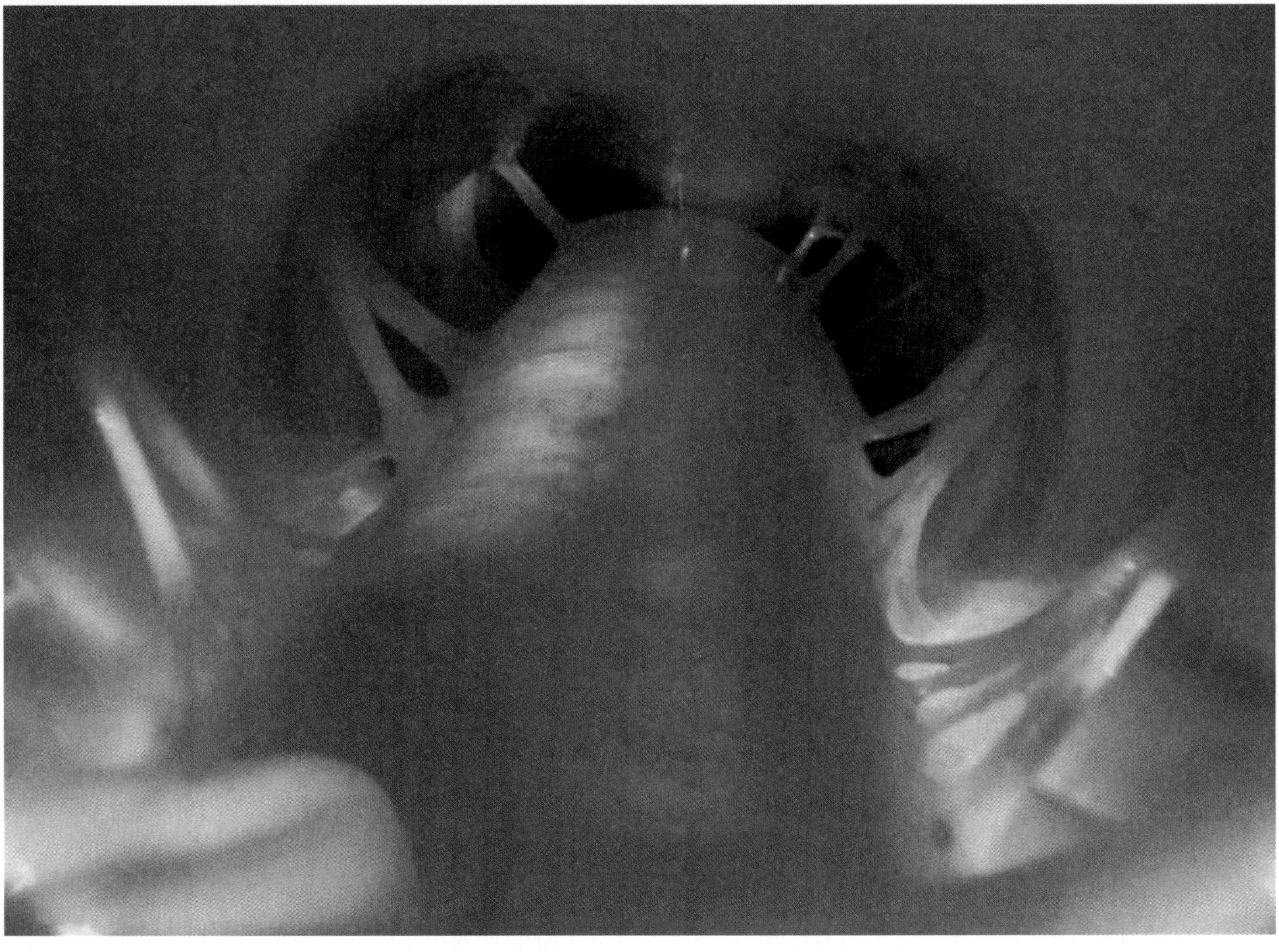

Figure 23.9 A view down the human spinal cord.
Pairs of spinal nerves can be seen extending out from the spinal cord. Along these nerves, the brain and spinal cord communicate with the body.

23.3 The Motor Nervous System

Motor neurons carry information from the central nervous system to muscles and glands. For example, if your eyes see a runaway car speeding toward you, the CNS sends messages through motor neurons to glands that secrete the hormone adrenaline. The adrenaline increases your heartbeat and breathing rate. The CNS also sends messages through motor neurons to many muscles, which contract and get your body out of there—fast!

In each segment of the spine, motor nerves extend out of the spinal cord to the muscles. Motor nerves from the spine control most of the muscles below the head. This is why injuries to the spinal cord often paralyze the lower part of the body. A muscle is paralyzed and cannot move if its motor neurons are damaged.

Reflexes Enable You to Act Quickly

The motor neurons of your body have been wired to enable your body to act particularly quickly in time of danger—even before you are consciously aware of the threat. These sudden, involuntary movements are called reflexes. A **reflex** produces a rapid motor response to a stimulus because the sensory neuron bringing information about the threat passes the information directly to a motor neuron. The escape reaction of a fly about to be swatted is a reflex. One of the most frequently used reflexes in your body is blinking, a reflex that protects your eyes. If anything, such as an insect or a cloud of dust, approaches your eye, the eyelid blinks closed even before you realize what has happened. The reflex occurs before the cerebrum is aware the eye is in danger.

Because they involve passing information between few neurons, reflexes are very fast. Many reflexes never reach the brain. The "danger" nerve impulse travels only as far as the spinal cord and then comes right back as a motor response. Most reflexes involve a single connecting neuron, called an **interneuron,** between the sensory neuron and the motor neuron. A few, like the knee-jerk reflex (figure 23.10), are monosynaptic reflex arcs. In these, the sensory neuron synapses directly with a motor neuron in the spine—there is no interneuron intermediary between them. If you step on something sharp, your leg jerks away from the danger: the prick causes nerve impulses in sensory neurons, which pass up the spinal cord directly to motor neurons, which cause your leg muscles to contract, jerking your leg up.

The Autonomic Nervous System

Some motor neurons are active all the time, even when you are asleep. These neurons carry messages from the CNS that keep the body going even when it is not active. These neurons are called the **autonomic nervous system.** The word *autonomic* means involuntary. The autonomic nervous system carries messages to muscles and glands that usually work without our noticing.

The autonomic nervous system is the command network the CNS uses to maintain the body's homeostasis. Using it, the CNS regulates heartbeat and controls muscle contractions in the walls of your blood vessels. It directs the muscles that control blood pressure, breathing, and the movement of food through the digestive system. It also carries messages that help stimulate glands to secrete tears, mucus, and digestive enzymes.

The autonomic nervous system is composed of two elements that act in opposition to one another:

1. The **sympathetic nervous system.** One division of the autonomic nervous system, called the *sympathetic,* dominates in times of stress. It controls the "fight-or-flight" reaction, increasing blood pressure, heart rate, breathing rate, and blood flow to the muscles. It consists of a network of long extensions (called axons) extending out from motor neurons within the spine; these axons extend to a cluster of neuron cell bodies called ganglia in the immediate vicinity of an organ. It also consists of short motor neurons extending from the ganglia to the nearby organ.

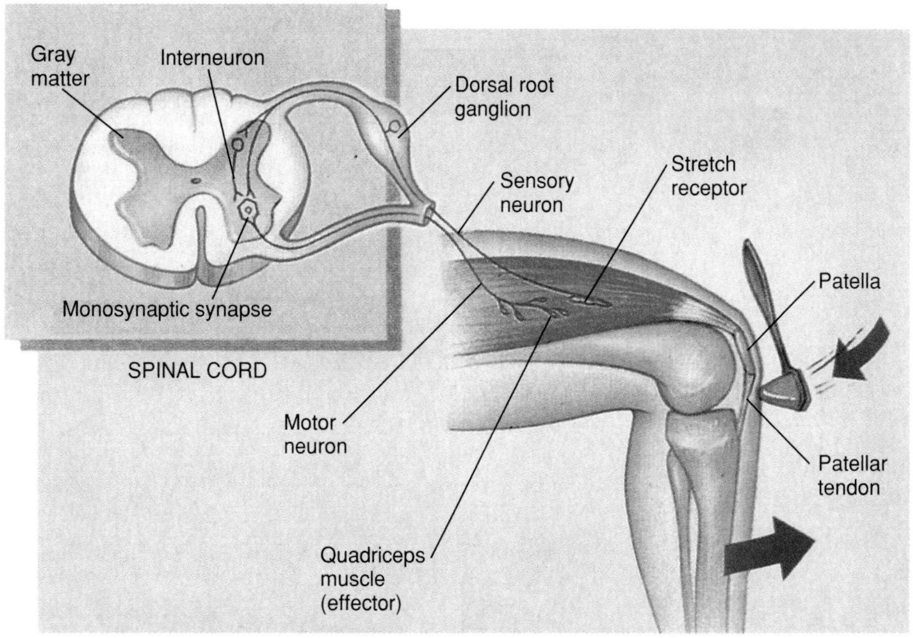

Figure 23.10 The knee-jerk reflex.
The most famous involuntary response, the knee jerk, is produced by activating stretch receptors in the quadriceps muscle. (The hammer hits the patellar tendon, which anchors it.) A signal travels up a sensory neuron to the spine, where the sensory neuron stimulates a motor neuron, which sends a signal to the quadriceps muscle to contract.

2. The **parasympathetic nervous system.** Another division, called the *parasympathetic,* has the opposite effect. It conserves energy by slowing the heartbeat and breathing rate and by promoting digestion and elimination. It consists of a network of short motor axons extending out from the spine to ganglia located near the spine. It also consists of long motor neurons extending from the ganglia directly to each target organ.

Most glands, smooth muscles, and cardiac muscles constantly get input from *both* the sympathetic and parasympathetic systems. The CNS controls activity by varying the ratio of the two signals to either stimulate or inhibit the organ (figure 23.11).

Although the autonomic nervous system can carry out its tasks automatically, on "autopilot," it is not completely independent of voluntary control. For instance, although it carries out breathing without your noticing, you can decide to stop breathing for a short time. However, anytime your voluntary override of the autonomic nervous system so disturbs your body's homeostasis that life is threatened, the brain intervenes to override your voluntary control—you lose consciousness, and the autonomic nervous system takes over again and restores normal functions. That is why you cannot hold your breath indefinitely.

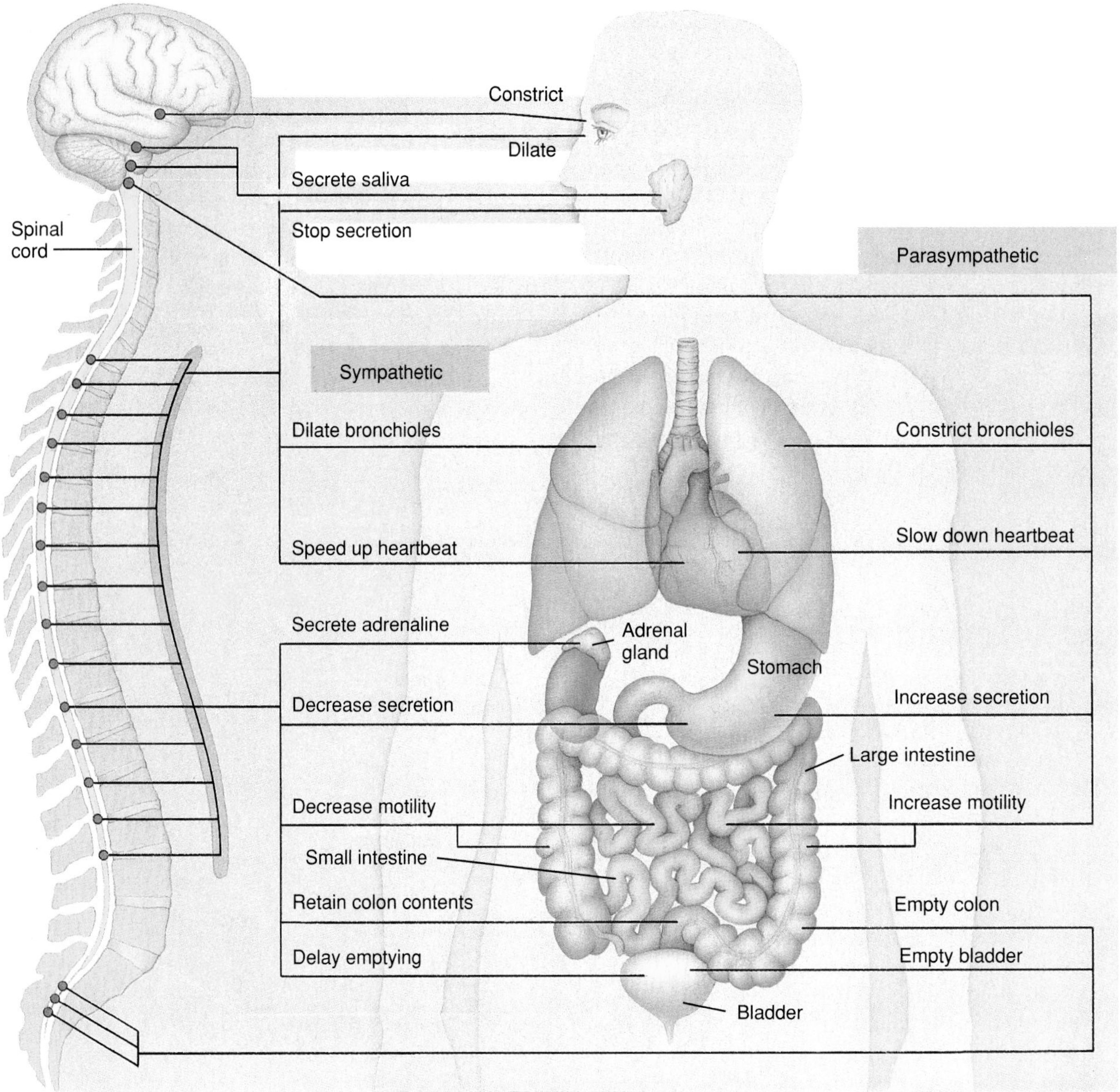

Figure 23.11 How the sympathetic and parasympathetic nervous systems interact.
A nerve path runs from both of the systems to every organ indicated except the adrenal gland.

23.4 How Nerves Work

The basic structural unit of the nervous system, whether central, motor, or sensory, is the nerve cell, or **neuron**.

Neurons

Neurons make up the brain and spinal cord, the motor neurons, and the sensory neurons. All neurons have the same basic structure (figure 23.12). Short, slender branches called **dendrites** extend from one end of a neuron's cell body. Dendrites are input channels. Nerve impulses travel inward along them, toward the **cell body.** Projecting out from the other end of the cell body is a single, long, tubelike extension called an **axon.** Axons are output channels. Nerve impulses travel outward along them, away from the cell body, toward other neurons or to muscles or glands.

Most neurons are unable to survive alone for long; they require the nutritional support provided by companion **neuroglial cells.** More than half the volume of the human nervous system is composed of supporting neuroglial cells. In many neurons, including the motor neurons that extend from brain to muscle, impulse transmission along the very long axons is facilitated by neuroglial **Schwann cells.** These cells envelop the axon at intervals with a sheath of fatty material called myelin, which acts as an electrical insulator. The **myelin sheath** is interrupted at intervals, leaving uninsulated gaps called **nodes of Ranvier** where the axon is in direct contact with the surrounding fluid. The nerve impulse jumps from node to node, speeding its travel down the axon.

When a neuron is "at rest," not carrying an impulse, active transport channels in the neuron's cell membrane transport sodium ions (Na^+) out of the cell and potassium ions (K^+) in. This sodium-potassium pump was described in chapter 4. Sodium ions cannot easily move back into the cell once they are pumped out, so the concentration of sodium ions builds up outside the cell. Similarly, potassium ions accumulate inside the cell, although not as densely since many potassium ions are able to diffuse out through open channels. The effect of this is to make the outside of the neuron more positive than the inside. The cell membrane is said to be "polarized."

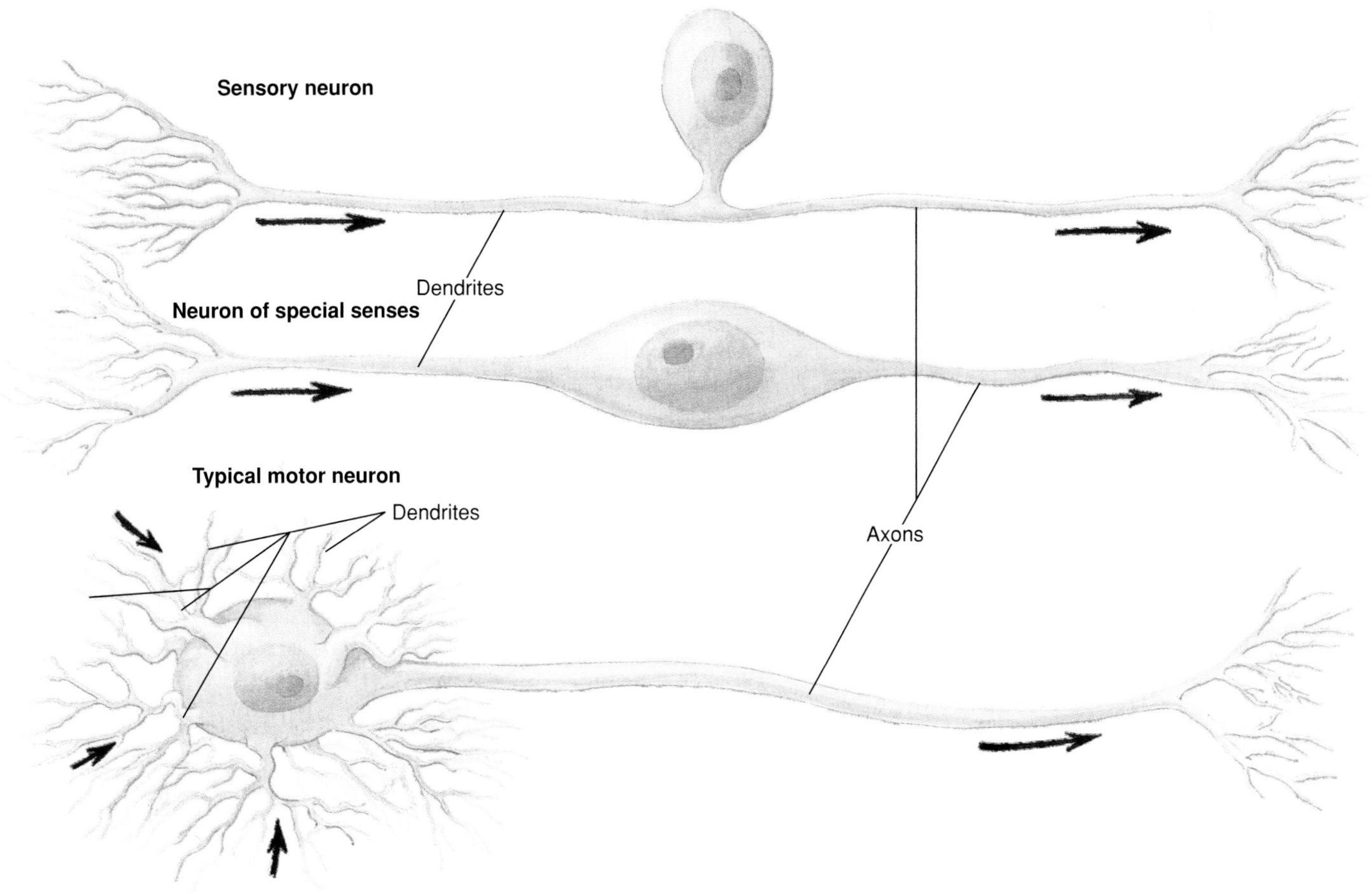

Figure 23.12 Three different types of neurons.
A sensory neuron, a neuron of special senses (found in the eyes, nose, and ears), and a typical motor neuron all have the same basic structure.

The Nerve Impulse

Neurons are constantly expending energy to pump sodium ions out of the cell, in order to maintain a positive charge on the exterior of the cell and a negative charge on the interior. The net negative charge of most proteins within the cell also adds to this charge difference. This charge separation is called the **resting potential.** Using sophisticated instruments, scientists have been able to measure the voltage difference between the neuron interior and exterior as −70 millivolts (millionths of a volt). The resting potential is the starting point for a nerve impulse.

A nerve impulse travels along the axon and dendrites as an *electrical current* generated by ions moving in and out of the neuron through **voltage-gated channels** (that is, protein channels in the neuron membrane that open and close in response to an electrical voltage). The impulse starts when pressure or other sensory inputs disturb a neuron's cell membrane, causing sodium channels on a dendrite to open. As a result, sodium ions flood into the neuron from outside, and for a brief moment the inside of the membrane is "depolarized," becoming more positive than the outside in that immediate area of the dendrite.

The sodium channels in the small patch of depolarized membrane remain open for only about half a millisecond. However, if the change in voltage is big enough, it causes nearby voltage-gated channels to open, which starts a wave of depolarization moving down the neuron, as the opening of the gated channels causes nearby voltage-gated channels to open, like a chain of falling dominoes. This moving local reversal of voltage is called an **action potential.** When the action potential has passed, the voltage-gated sodium channels snap closed again and the resting potential is restored (figure 23.13).

The depolarization and restoration of the resting potential takes only about 5 milliseconds. Fully 100 such cycles could occur, one after another, in the time it takes to say the word *nerve*.

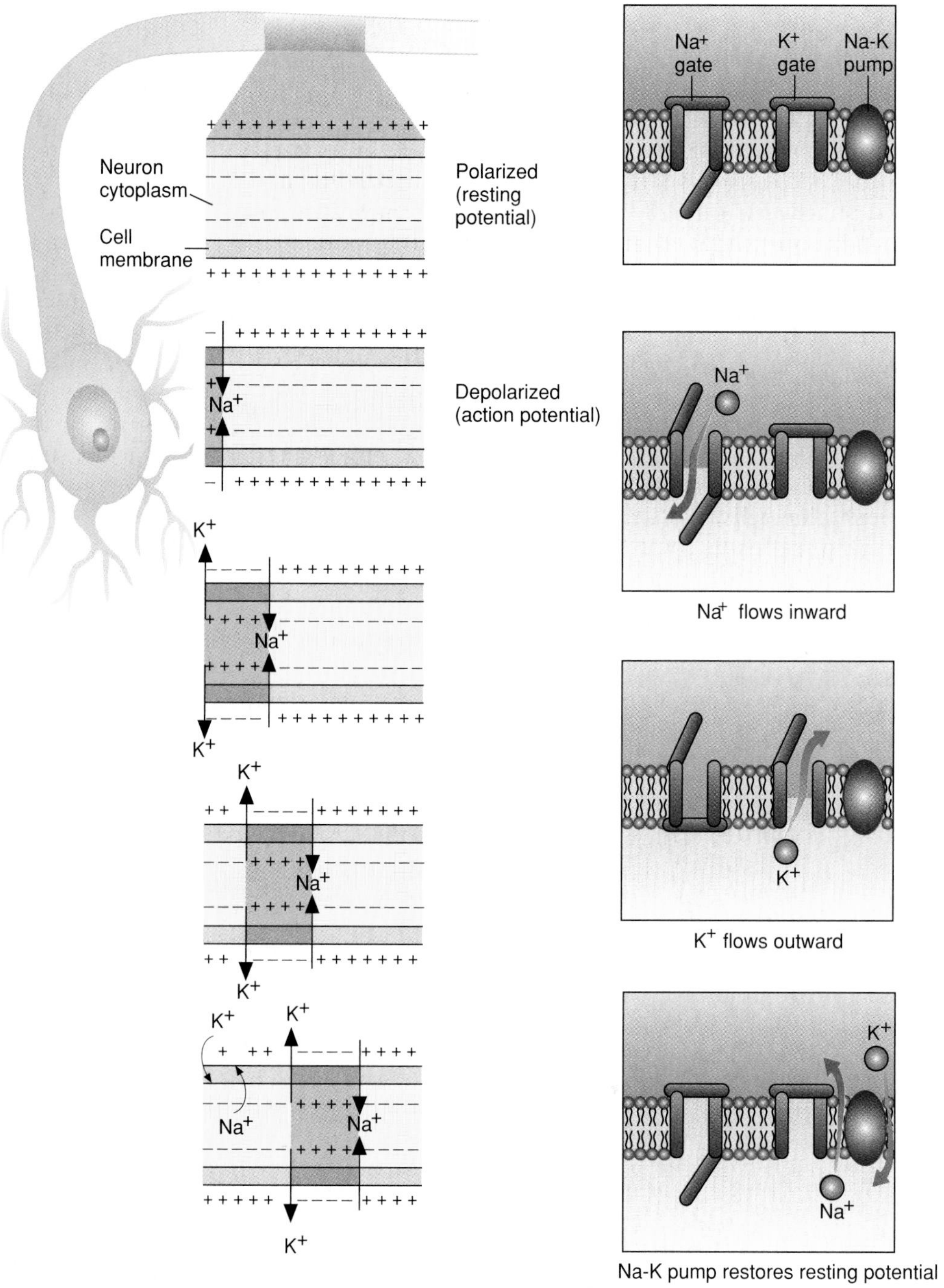

Figure 23.13 Transmission of an action potential.
At the resting membrane potential, the membrane is said to be polarized (charged), with the inside of the membrane negatively charged compared to the outside of the membrane. This polarization is established because of an uneven distribution of Na^+ and K^+ across the membrane. Na^+ is concentrated on the outside of the membrane and K^+ is concentrated on the inside. As the membrane depolarizes, Na^+ channels open and Na^+ flows into the cell, which causes the inside of the cell to become more positive. The K^+ channels open and K^+ flows out of the cell, which begins to repolarize the cell, restoring the inside of the cell to a negative charge. The sodium-potassium pump restores the resting membrane potential.

The Synapse

A nerve impulse can travel only so far along a neuron. Eventually it reaches the end of the axon, usually positioned very close to another neuron or to a muscle cell or gland. Axons, however, do not actually make direct contact with other neurons or with target tissue. Instead, a narrow gap, 10 to 20 nanometers across, separates the axon tip and the target neuron or tissue. This junction of an axon with another cell is called a **synapse.** The membrane on the near (axon) side of the synapse is called the **presynaptic membrane;** the membrane on the far (receiving) side of the synapse is called the **postsynaptic membrane** (figure 23.14).

When a nerve impulse gets to the end of an axon, its message must cross the synapse if it is to continue. Messages do not "jump" across synapses. Instead, they are carried across by chemical messengers called **neurotransmitters.** These chemicals are packaged in tiny sacs, or vesicles, at the tip of the axon. When a nerve impulse arrives at the tip, it causes the sacs to release their contents into the synapse. The neurotransmitters diffuse across the synapse and bind to receptors in the membrane of the cell on the other side, passing the signal to that cell by causing special ion channels in the postsynaptic membrane to open (figure 23.15). Because these channels open when stimulated by a chemical (in this case, a neurotransmitter) they are said to be chemically gated.

Why go to all this trouble? Why not just wire the neurons directly together? For the same reason that the wires of your house are not all connected but instead are separated by a host of switches. When you turn on one light switch, you don't want every light in the house to go on, the toaster to start heating, and the television to come on! If every neuron in your body were connected to every other neuron, it would be impossible to move your hand without moving every other part of your body at the same time. Synapses are the control switches of the nervous system.

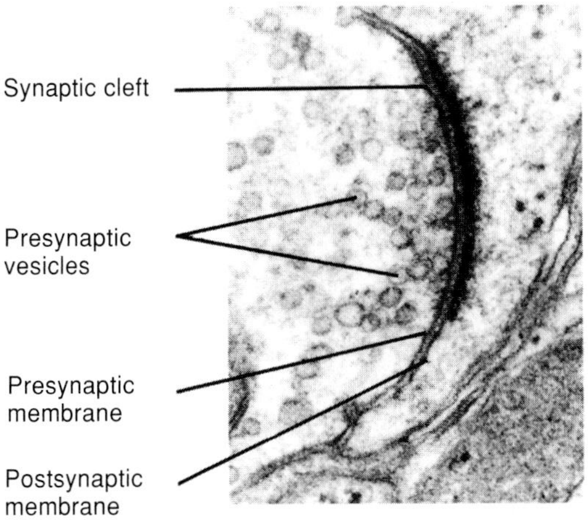

Figure 23.14 A synapse between two neurons.
This micrograph clearly shows the space between the presynaptic and postsynaptic membranes, which is called the synaptic cleft.

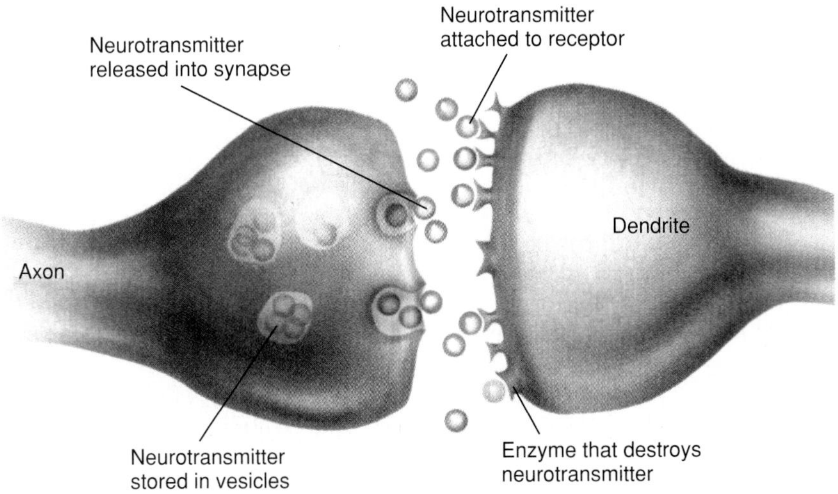

(a)

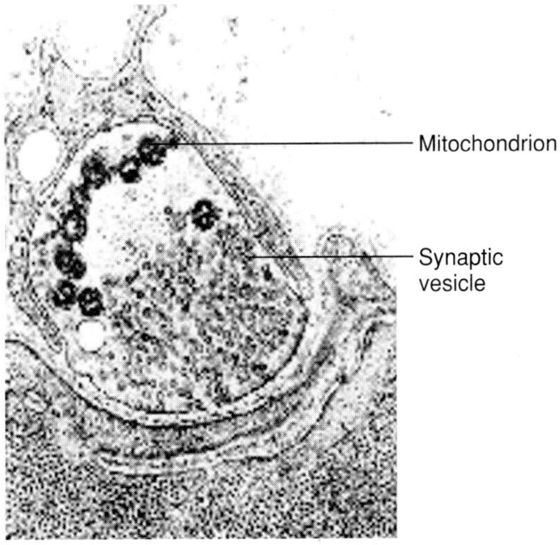

(b)

Figure 23.15 Events at the synapse.
(*a*) When a nerve impulse reaches the end of an axon, it releases a neurotransmitter into the synaptic spaces. In this illustration the neurotransmitter is acetylcholine. When acetylcholine is released into the synapse, acetylcholine molecules diffuse across the synapse and bind to receptors on the dendrite, initiating an impulse in the next neuron. Enzymes destroy the neurotransmitter molecules to prevent continuous stimulation of the postsynaptic neuron.
(*b*) A transmission electron micrograph of the tip of an axon filled with synaptic vesicles.

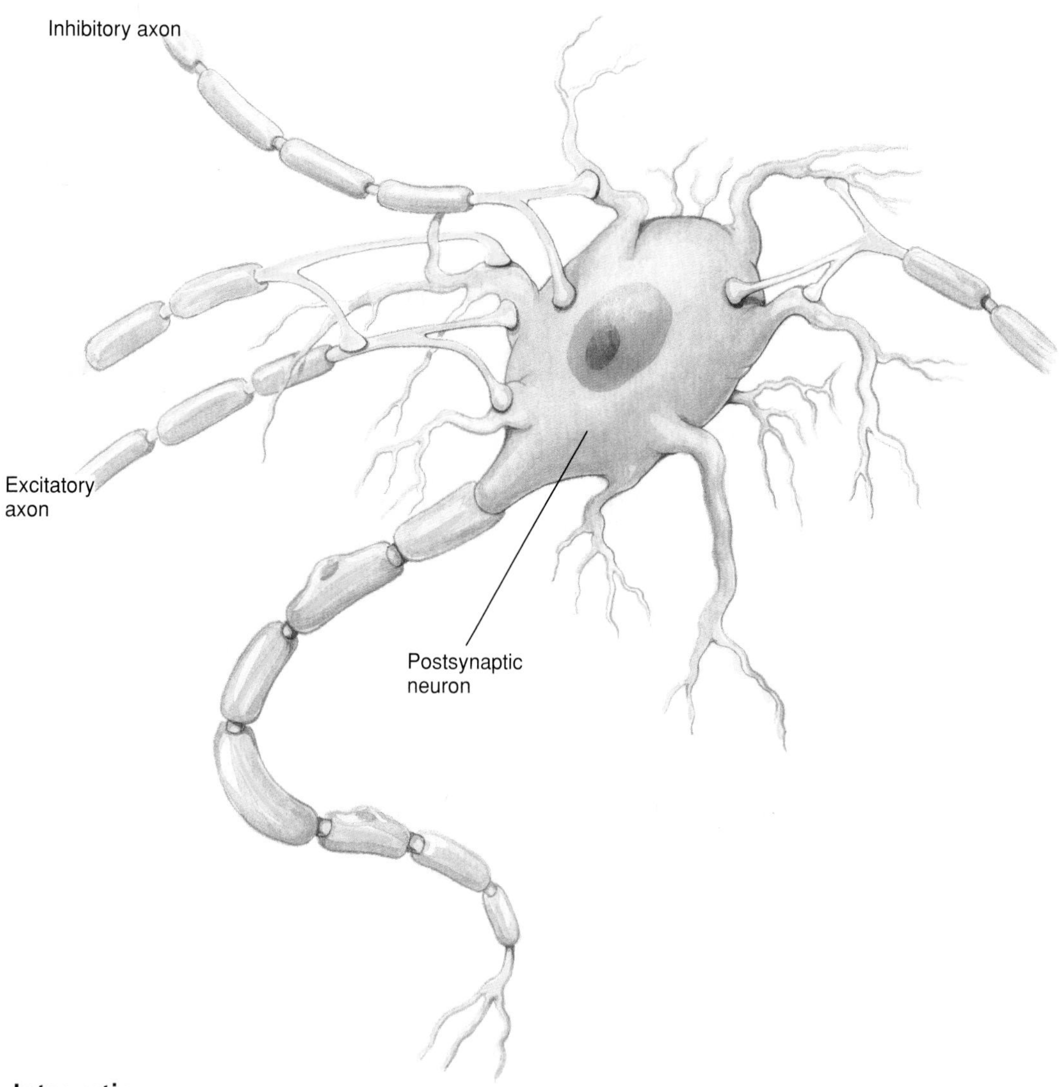

Figure 23.16 Integration.
Many different axons synapse with the cell body of the postsynaptic neuron illustrated here. Each axon contributes to the likelihood that an action potential will be realized and a nerve impulse will be sent down the axon extending below.

Integration

The human nervous system uses dozens of different kinds of neurotransmitters, each recognized by specific receptors on receiving cells. They fall into two general classes, depending on whether they excite or inhibit passage across the synapse.

In an *excitatory synapse,* the receptor protein is a chemically-gated sodium channel, meaning that a sodium channel through the membrane is opened by a chemical, the neurotransmitter. On binding with a neurotransmitter whose shape fits it, the sodium channel opens, allowing sodium ions to flood inward. If enough sodium ion channels are opened by neurotransmitters, an action potential begins in the receiving cell.

In an *inhibitory synapse,* the receptor protein is a chemically-gated potassium channel. Binding with its neurotransmitter opens the potassium channel, leading to the exit of positively charged potassium ions and a more negative interior in the receiving cell. This inhibits the start of an action potential, because the negative voltage change inside means that even more sodium ion channels must be opened to get a domino effect started among voltage-gated sodium channels, and so start an action potential.

An individual nerve cell can possess both kinds of synaptic connections to other nerve cells. When signals from both excitatory and inhibitory synapses reach the body of a neuron, the excitatory effects (which cause less internal negative charge) and the inhibitory effects (which cause more internal negative charge) interact with one another. The result is a process of **integration** in which the various excitatory and inhibitory electrical effects tend to cancel or reinforce one another (figure 23.16). Neurons often receive many inputs. A single motor neuron in the spinal cord may have as many as 50,000 synapses on it!

Neuromodulators

Your body sometimes deliberately prolongs the transmission of a signal across a synapse by slowing the destruction of neurotransmitters. It does this by releasing into the synapse special long-lasting chemicals called **neuromodulators.** Some neuromodulators aid the release of neurotransmitters into the synapse; others inhibit the reabsorption of neurotransmitters so that they remain in the synapse; still others delay the breakdown of neurotransmitters after their reabsorption, leaving them in the tip to be released back into the synapse when the next signal arrives. All neuromodulators "hot-wire" the synapse by prolonging the time neurotransmitters spend there.

Mood, pleasure, pain, and other mental states are determined by particular groups of neurons in the brain that use special sets of neurotransmitters and neuromodulators. Mood, for example, is strongly influenced by the neurotransmitter serotonin. Many researchers think that depression results from a shortage of serotonin. Prozac, the world's best-selling antidepressant, inhibits the reabsorption of serotonin, increasing the amount in the synapse by slowing its removal (figure 23.17).

Drug Addiction

When a cell of your body is exposed to a chemical signal for a prolonged period, it tends to lose its ability to respond to the stimulus with its original intensity. (You are familiar with this loss of sensitivity—when you sit in a chair, how long are you aware of the chair?) Nerve cells are particularly prone to this loss of sensitivity. If receptor proteins within synapses are exposed to high levels of neurotransmitter molecules for prolonged periods, that nerve cell often responds by inserting fewer receptor proteins into the membrane. This feedback is a normal part of the functioning of all neurons, a simple mechanism that has evolved to make the cell more efficient by adjusting the number of "tools" (receptor proteins) in the membrane "workshop" to suit the workload.

The drug cocaine is a neuromodulator that causes abnormally large amounts of neurotransmitters to remain in the synapses for long periods of time. The central nervous system adjusts to the increased firing that results by producing fewer receptors in the postsynaptic membrane. The result is **addiction** (figure 23.18). The decreased number of receptors creates a less-sensitive nerve pathway. Physiologically, the only way a person can then maintain normal functioning is to continue to take the drug. Thus you can see that addiction is not simply a psychological state to be overcome by willpower. It is a physiological dependence caused by changes in neurons.

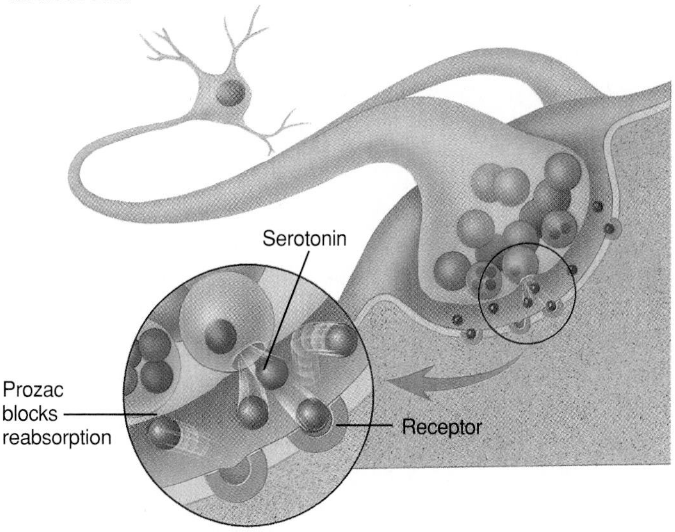

Figure 23.17 Drugs alter transmission of impulses across the synapse.
Depression can result from a shortage of the neurotransmitter serotonin. The antidepressant drug Prozac works by blocking reabsorption of serotonin in the synapse, making up for the shortage.

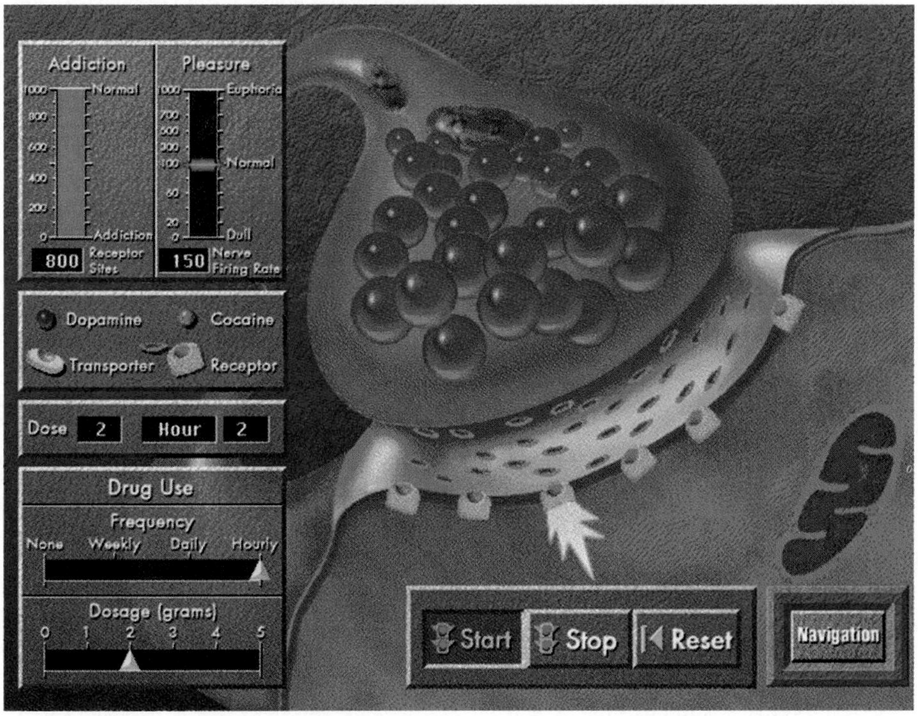

Figure 23.18 Drug addiction.
Addictive drugs typically increase pleasure by causing neurons in the brain to fire more often. Addiction is the brain's way of attempting to turn down the volume. This is a screen capture from an interactive CD-ROM exercise in which the population of postsynaptic receptors of the neurotransmitter decreases in response to cocaine, which prolongs neurotransmitter survival in the synapse. (*Explorations in Human Biology,* Module 10, "Drug Addiction")

23.5 The Sensory Nervous System

Did you ever wonder what it would be like not to know anything about what is going on around you? Imagine if you couldn't hear, or see, or feel, or smell. After a while, a human goes mad if completely deprived of sensory input. Our senses are our bridge to experience, the way our body relates to everything around it.

The **sensory nervous system** tells the central nervous system what is happening. Sensory neurons carry impulses to the CNS from more than a dozen different types of sensory cells that detect changes outside and inside your body. Called **sensory receptors,** these specialized sensory cells detect many different things, including changes in blood pressure, strain on ligaments, and smells in the air. Particularly complex sensory receptors, made up of many cell and tissue types, are called **sensory organs.** Your eyes and ears are sensory organs, and so are the taste buds in your mouth.

A sunset, a symphony, the motion of the dancer in figure 23.19—to the brain they are all the same, a pattern of arriving nerve impulses. How does the brain know whether an incoming nerve impulse is light, sound, or pain? This information is built into the "wiring"—into which neurons interact while passing the information to the CNS and into the location in the brain where the information is sent. The brain "knows" it is responding to light because the message from a sensory neuron is wired to light receptor cells. That is why when you press your fingertips gently against the corners of your eyes, you "see stars"—the brain treats any impulse from the eyes as light, even though the eye received no light.

Sensory Receptors

The path of sensory information to the CNS is a simple one, composed of three stages:

1. **Stimulation.** A physical stimulus impinges on a sensory receptor.
2. **Transduction.** The sensory receptor initiates the opening or closing of ion channels in a sensory neuron.
3. **Transmission.** The sensory neuron conducts a nerve impulse along an afferent pathway to the CNS.

All sensory receptors are able to initiate nerve impulses by opening or closing **stimulus-gated channels** within sensory neuron membranes. Except for visual photoreceptors, these channels are Na^+ channels that depolarize the membrane and so start an electrical signal. The channels are opened by chemical or mechanical stimulation, often a disturbance such as touch, heat, or cold. The receptors differ from one another in the nature of the environmental input that triggers the opening of the channel. Your body contains many sorts of receptors, each sensitive to a different aspect of the body's condition or to a different quality of the external environment.

Traditionally, human sensory receptors are grouped into two classes. Those that sense information relating to the body's internal condition are known as **interoceptors.** Those that sense the exterior world are known as **exteroceptors.** You sense your body's temperature with interoceptors, while you see with exteroceptors. You are not consciously aware of many of your body's interoceptors, while most exteroceptors have direct effects on awareness.

Figure 23.19 A ballet dancer.
Continuous sensory input allows this dancer to practice her art. She feels her arms and legs in their proper positions; her eyes determine her position on stage; her ears help her to integrate her movements with the orchestra; and her inner ears provide her with a keen sense of balance.

Sensing Your Body's Condition

Sensory receptors inside the body inform the CNS about the condition of the body. Much of this information passes to a coordinating center in the brain, the hypothalamus, the part of the brain responsible for maintaining the body's homeostasis—that is, keeping the body's internal environment constant. The human body uses a variety of different sensory receptors, each responding to different aspects of the internal or external environment.

1. **Temperature change.** Two kinds of nerve endings in the skin are sensitive to changes in temperature, one stimulated by cold, the other by warmth. By comparing information from the two, the CNS can learn what the temperature is and if it is changing.
2. **Blood chemistry.** Receptors in the walls of arteries sense CO_2 levels in the blood. The brain uses this information to regulate the body's respiration rate, increasing the rate of breathing when CO_2 levels rise above normal.
3. **Pain.** Damage to tissue is detected by special nerve endings within tissues, usually near the surface, where damage is most likely to occur. When these nerve endings are physically damaged or deformed, the CNS responds by reflexively withdrawing the body segment and often by changing heartbeat and blood pressure as well.
4. **Muscle contraction.** Buried deep within muscles are sensory receptors called stretch receptors. In each, the end of a sensory neuron is wrapped around a muscle fiber (figure 23.20): when the muscle is stretched, the fiber elongates, stretching the spiral nerve ending and causing repeated signals to be sent to the brain. From these signals the brain can determine the rate of change of muscle length at any given moment. The CNS uses this information to control movements that require the combined action of several muscles, such as those that carry out breathing or walking.
5. **Blood pressure.** Blood pressure is sensed by neurons called baroreceptors with highly branched nerve endings within the walls of major arteries. When blood pressure increases, the stretching of the arterial wall causes the sensory neuron to increase the rate at which it sends signals to the CNS, while when it decreases, the rate of firing of the sensory neuron goes down (figure 23.21). Thus the frequency of impulses provides the CNS with a continuous measure of blood pressure.
6. **Touch.** Touch is sensed by pressure receptors buried below the surface of the skin. There are a variety of different types, some specialized to detect rapid changes in pressure, others to measure the duration and extent to which pressure is applied, and still others sensitive to vibration.

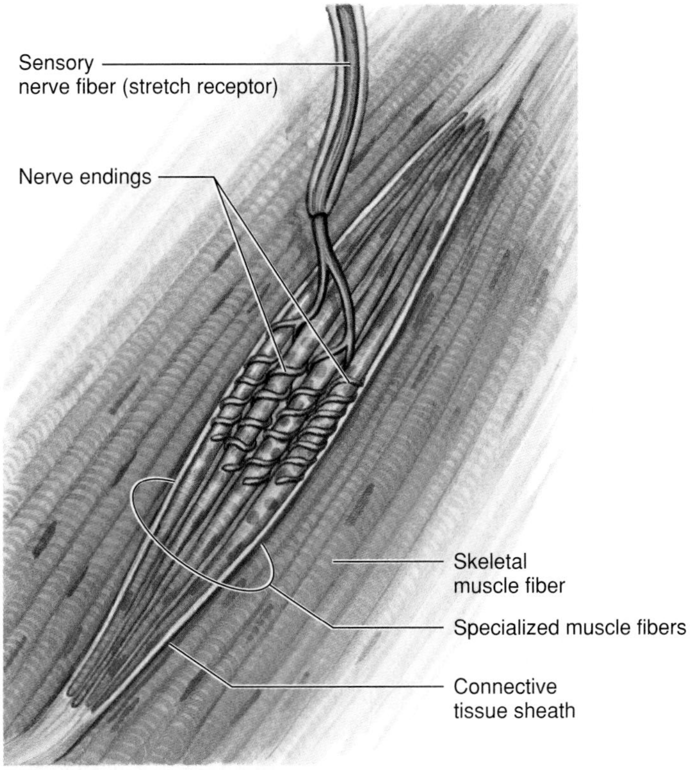

Figure 23.20 A stretch receptor embedded within skeletal muscle.
Stretching the muscle elongates the specialized muscle fibers, which deforms the nerve endings, causing them to send a nerve impulse out along the nerve fiber.

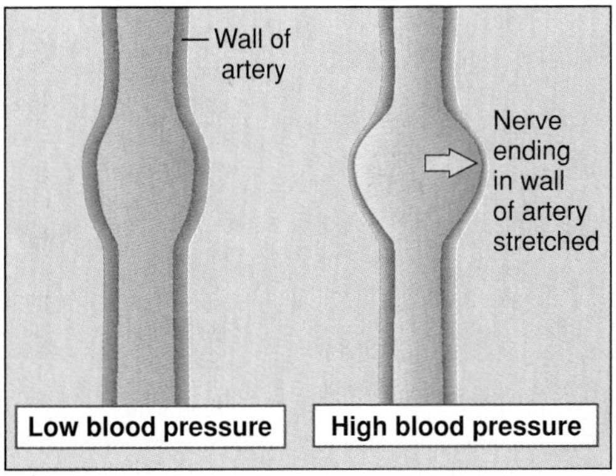

Figure 23.21 How a baroreceptor works.
A network of nerve endings covers a region where the wall of the artery is thin. High blood pressure causes the wall to balloon out there, stretching the nerve endings and causing them to fire impulses.

Sensing Gravity: Balance

Receptors in the joints, tendons, and muscles detect changes in the relative positions of parts of the body. Other receptors in the ear are sensitive to the position of the head. Together, input from all these receptors informs the brain where the body is in three dimensions. This knowledge enables you to move freely and maintain your balance.

1. **Balance.** In order to keep the body's balance, the brain needs a frame of reference, and the reference point it uses is gravity. The **otolith** sensory receptors that detect gravity are located in a series of hollow chambers within the inner ear. To illustrate how these receptors work, imagine a pencil standing in a glass. No matter which way you tip the glass, the pencil will roll along the rim, applying pressure to the lip of the glass. If you want to know the direction the glass is tipped, you need only ask where on the rim pressure is being applied. Using gravity receptors, the brain is able to determine its vertical position.
2. **Motion.** The brain senses motion in a way similar to that used to determine its vertical position, by employing a receptor in which fluid deflects cilia in a direction opposite that of the motion. Within the inner ear are three fluid-filled **semicircular canals,** each oriented in a different plane at right angles to the other two so that motion in any direction can be detected. Protruding into the canal are groups of cilia from sensory cells. The cilia from each cell are arranged in a tentlike assembly called a cupula (figure 23.22), which is pushed by moving ear fluid in a direction opposite that of the head's movement. Because the three canals are oriented in all three planes, movement in any plane is sensed by at least one of them, and the brain is able to analyze complex movements by comparing the sensory inputs from each canal.

The brain uses its ability to detect gravity and head movements to balance the body in space. However, it does not have a speedometer. The semicircular canals detect when the head *changes* speed or direction, but they do not react if the body moves in a straight line because when the head is moving in a straight line, the fluid in the canals does not move. That is why traveling in a car or airplane at constant speed gives no sense of motion. The semicircular canals detect movement only if the car or airplane changes speed or turns—when the motion changes direction, the fluid tends to keep going in its original direction and so bends the cilia within the semicircular canals.

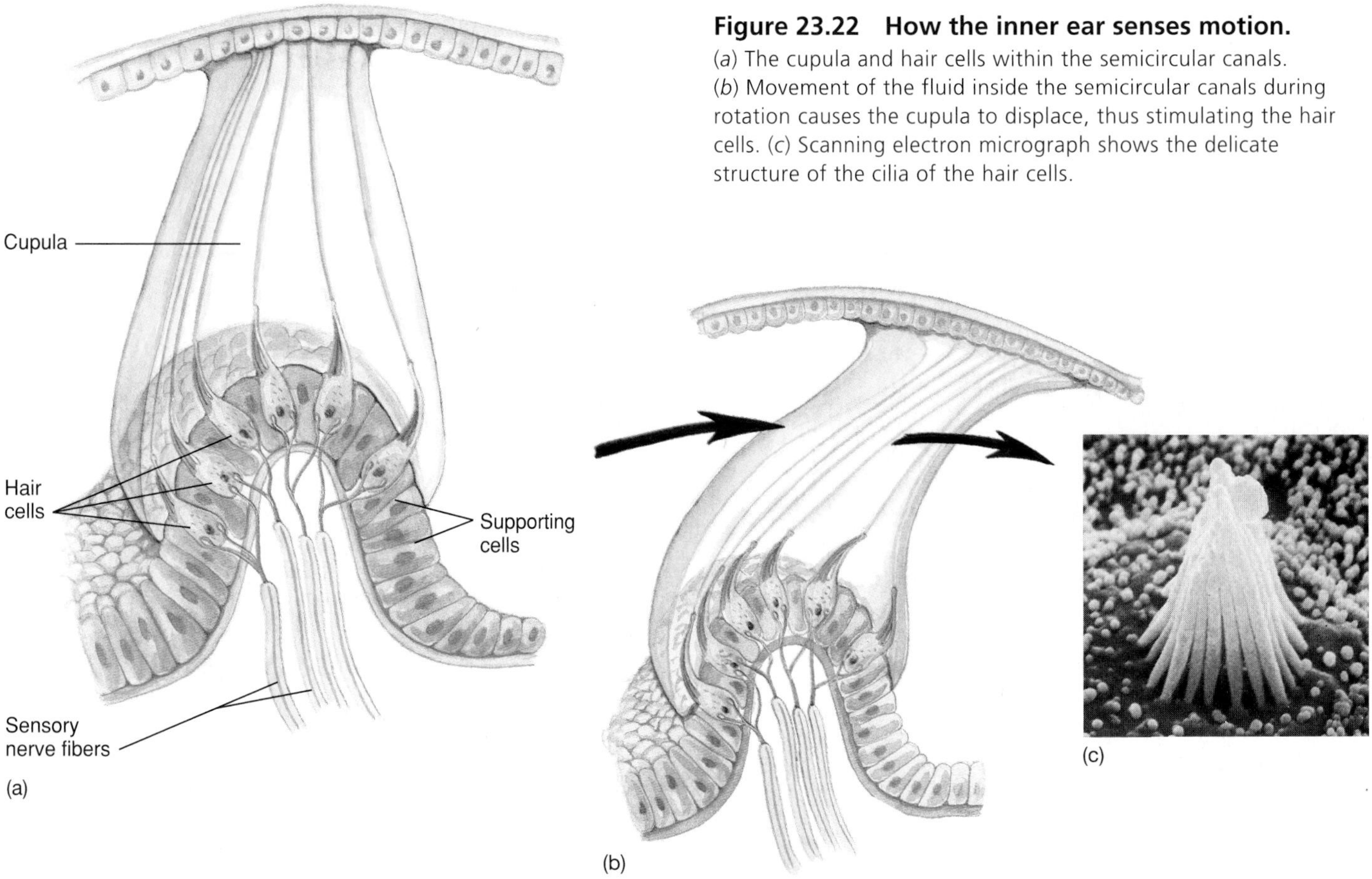

Figure 23.22 How the inner ear senses motion. (*a*) The cupula and hair cells within the semicircular canals. (*b*) Movement of the fluid inside the semicircular canals during rotation causes the cupula to displace, thus stimulating the hair cells. (*c*) Scanning electron micrograph shows the delicate structure of the cilia of the hair cells.

The Problem of Sensing the Exterior World

Sensing the nature of objects in your external environment presents a complex problem. Imagine an object some distance away. The information that any of your sensory receptors can obtain about that object is limited by the nature of the stimulus available.

1. **Attention.** Some sensory systems provide only enough information for you to determine that the object is present, indicating little or nothing about where it is located. Smell sometimes works in this fashion.
2. **Location.** Other sensory systems provide you information about the direction of the object. They permit you to locate the object by moving toward it. Sound works in this fashion for humans. If an object emits sound, you can determine quite accurately the direction of the object by moving your head—but not the distance. A bat, which has far more sophisticated sound-processing equipment than you do, estimates distance by bouncing a sound wave off distant objects and timing how long it takes the sound wave to return.
3. **Imaging.** You have one sensory system that provides information about both direction and distance—vision. With both kinds of information, your central nervous system is able to construct a three-dimensional image of the object and its surroundings.

The human repertoire of sensory systems is not perfect, and it does not allow us to discriminate all potential cues from the environment. Dogs, for example, have far better olfactory senses than we do, and eagles have far more sensitive vision. Other vertebrates use entirely different senses to construct three-dimensional images in the dark. Rattlesnakes can "see" heat. They possess a pair of pit organs, one located on each side of the head between the eye and the nostril. Using them, the snake can locate and strike a motionless warm animal in total darkness by perceiving the thermal radiation emanating from the body of the prey. Bats use high-frequency sound to detect prey in the dark (figure 23.23).

Figure 23.23 Using ultrasound to locate a moth.
This bat is emitting high-frequency "chirps" as it flies. It then listens for the sound's reflection against the moth. By timing how long it takes for a sound to return, the bat can "see" the moth even in total darkness.

Sensing Chemicals: Taste and Smell

Although humans are not as good at sensing chemicals as most mammals, we are able to detect many of the chemicals in the air and in the food we eat.

1. **Taste.** Embedded within the surface of the tongue are taste receptors, specialized to detect four classes of chemicals: sweet (sugars), sour (acids), bitter (complex organics), and salty (salts) (figure 23.24*a*). Sensitivity to each type of chemical varies on different parts of the tongue. Each receptor is actually a specialized sensory cell located in structures called taste buds (figure 23.24*b*). When the tongue encounters a chemical, information from the sensory cells passes to sensory neurons, which transmit the signals to the brain.
2. **Smell.** In your nose are chemically sensitive neurons whose cell bodies are embedded within the epithelium of the nasal passage (figure 23.25). When they detect chemicals, these sensory neurons transmit information to a location in the brain where smell information is processed and analyzed. In many vertebrates (dogs are a familiar example) these neurons are far more sensitive than in humans.

Smell as well as taste is very important in telling us about our food. That is why when you have a bad cold and your nose is stuffed up, your food has little taste. Other receptors also play a role. Thus the "hot" sensation of foods such as chili peppers is detected by pain receptors, not chemical receptors.

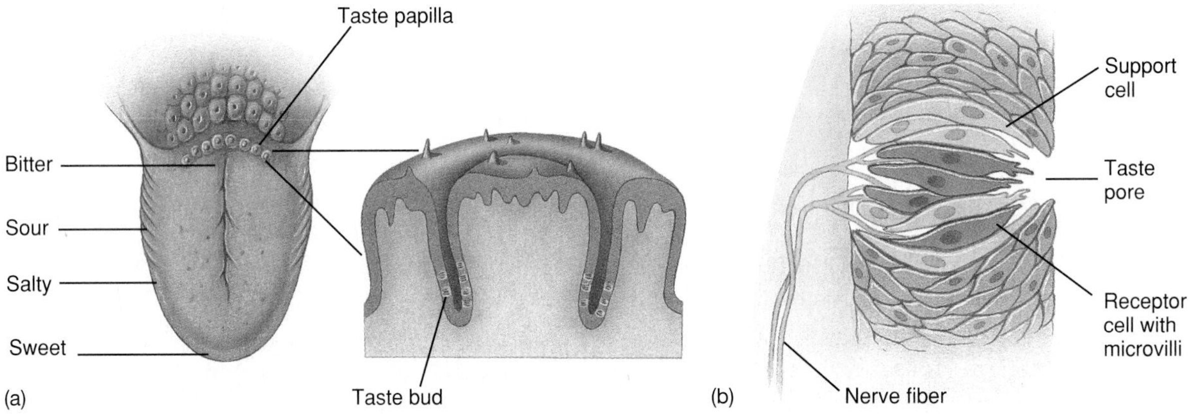

Figure 23.24 Taste.
(*a*) Taste buds in human beings are sensitive to four kinds of chemicals (bitter, sour, salty, and sweet), depending on their location on the tongue. Groups of taste buds are typically organized in sensory projections called papillae. (*b*) Individual taste buds are bulb-shaped collections of chemical receptor cells that open out into the mouth through a taste pore.

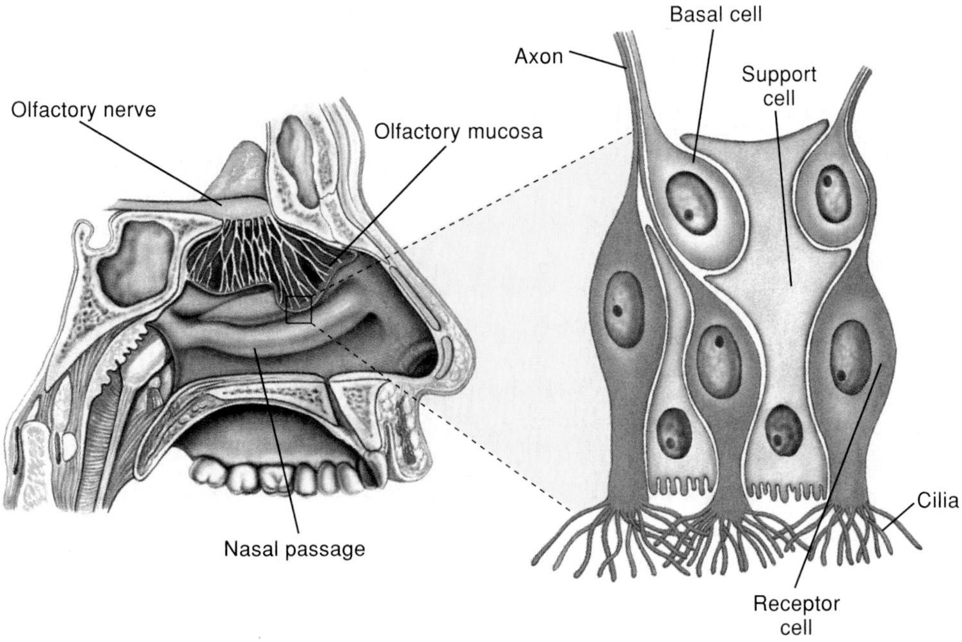

Figure 23.25 Smell.
Humans smell by using receptor cells located in the lining of the nasal passage. The receptor cells are neurons. Axons from these sensory neurons project back through the olfactory nerve directly to the brain.

Sensing Sounds: Hearing

When you hear a sound, you are detecting the air vibrating—waves of pressure in the air beating against your ear, pushing a membrane called the **eardrum** in and out. On the other side of the eardrum are three small bones that act as a lever system to increase the force of the vibration. They transfer the amplified vibration across a second membrane to fluid within the inner ear. The chamber of the inner ear is shaped like a tightly coiled snail shell and is called the **cochlea** (figure 23.26), from the Latin name for "snail." It is connected to the throat by the eustachian tube in such a way that there is no difference in air pressure between the middle ear and the outside. That is why your ears sometimes "pop" when landing in an airplane—the pressure is equalizing between the two sides of the eardrum.

The sound receptors within the cochlea are hair cells that rest on a membrane that runs up and down the middle of the curving chamber, separating it into two halves like a wall. The hair cells do not project into the fluid filling the cochlea; instead, they are covered by a second membrane. When a sound enters the cochlea, the sound waves cause this membrane "sandwich" to vibrate, bending the hairs pressed against the outer membrane and causing them to send nerve impulses to sensory neurons that travel to the brain.

Sounds of different frequencies cause different parts of the membrane to vibrate, and thus fire different sensory neurons—the identity of the sensory neuron being fired tells the CNS what the frequency of the sound is. The intensity of the sound is determined by how *often* the neurons fire. Our ability to hear depends upon the flexibility of the membranes within the cochlea. Humans cannot hear low-pitched sounds, below 20 vibrations (or cycles) per second, although some vertebrates can. As children, we can hear high-pitched sounds, up to 20,000 cycles per second, but this ability decreases as we get older. Other vertebrates can hear sounds at far higher frequencies. Dogs readily hear sounds of 40,000 cycles per second and so can respond to a high-pitched dog whistle when it seems silent to a human observer.

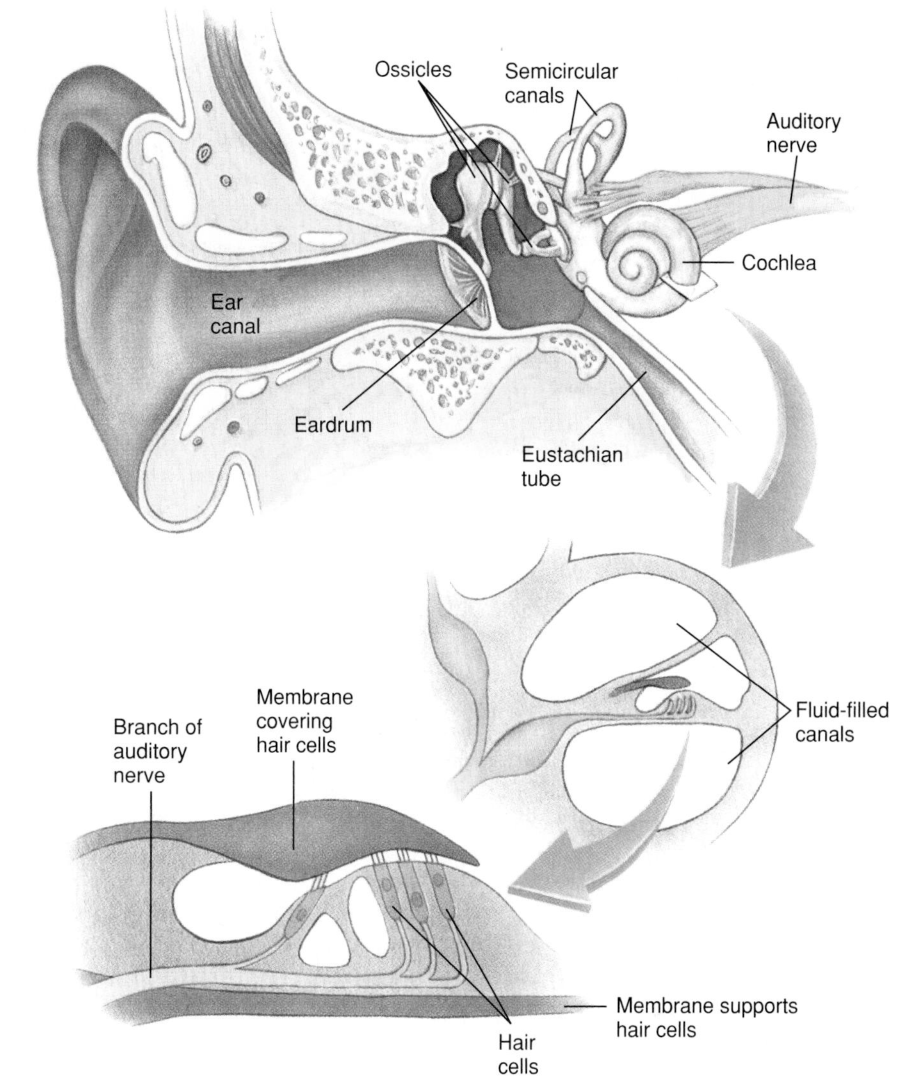

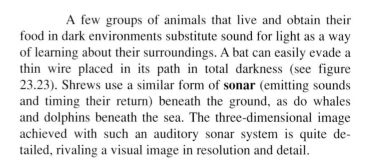

Figure 23.26 Structure and function of the human ear.
Sound waves passing through the ear canal beat on the eardrum, pushing a set of three small bones, or ossicles, against an inner membrane. This sets up a wave motion in the fluid filling the canals within the cochlea. When the sound wave beats against the sides of the canals, the membrane covering the hair cells moves back and forth against the hair cells, which causes associated neurons to fire impulses.

A few groups of animals that live and obtain their food in dark environments substitute sound for light as a way of learning about their surroundings. A bat can easily evade a thin wire placed in its path in total darkness (see figure 23.23). Shrews use a similar form of **sonar** (emitting sounds and timing their return) beneath the ground, as do whales and dolphins beneath the sea. The three-dimensional image achieved with such an auditory sonar system is quite detailed, rivaling a visual image in resolution and detail.

Sensing Light: Vision

No other stimulus provides as much detailed information about the environment as light. Vision, the perception of light, is carried out by a special sensory apparatus called an eye. All the sensory receptors described to this point have been chemical or mechanical ones. Eyes contain sensory receptors called rods and cones that respond to photons of light. The light energy is absorbed by pigments in the rods and cones, which respond by triggering nerve impulses in sensory neurons. Humans have extremely good eyesight. We see in color and can see fine details and movement. Birds are the only animals with better eyesight than humans.

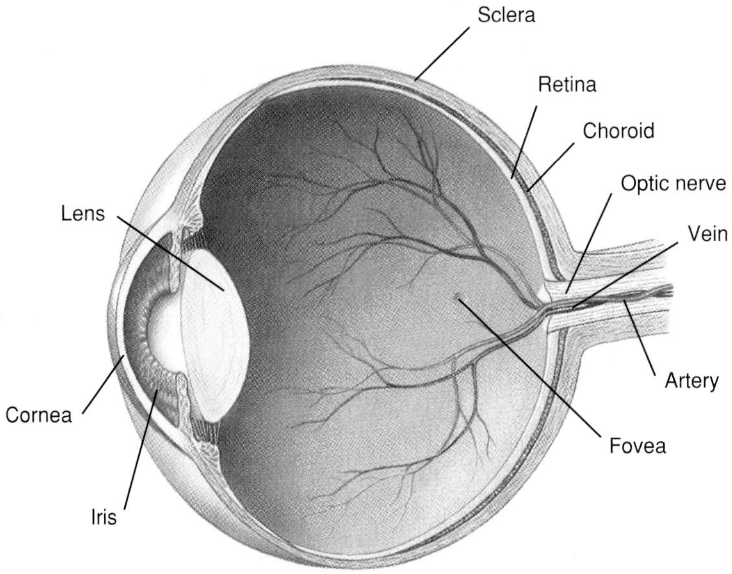

Figure 23.27 The structure of the human eye.
Light passes through the transparent cornea and is focused by the lens on the rear surface of the eye, the retina, at a particular location called the fovea. The retina is rich in photoreceptors.

Structure of the Human Eye

The human eye works like a lens-focused camera (figure 23.27). Light first passes through a transparent protective covering, the **cornea,** which begins to focus the light onto the rear of the eye. The beam of light then passes through the **lens,** which completes the focusing. The lens is a fat disc, somewhat resembling a flattened balloon. It is attached by suspending ligaments to **ciliary muscles.** When these muscles contract, they change the shape of the lens, and thus the point of focus on the rear of the eye. The amount of light entering the eye and reaching the lens is controlled by a shutter, called the **iris,** between the cornea and the lens. The transparent zone in the middle of the iris, the **pupil,** gets larger in dim light and smaller in bright light.

The light that passes through the pupil is focused by the lens onto the back of the eye. An array of light-sensitive receptor cells lines the back surface of the eye, called the **retina.** The retina is the light-sensing portion of the eye. It contains about 1 billion light-sensitive receptor cells called **rods** and **cones,** which, when stimulated by light, generate nerve impulses that travel to the brain along a short, thick nerve pathway called the optic nerve (figure 23.28). Rods are receptor cells that are extremely sensitive to light, and they can detect various shades of gray even in dim light. However, they cannot distinguish colors, and because they do not detect edges well, they produce poorly defined images. Cones are receptor cells that detect color and are sensitive to edges so that they produce sharp images. The center of the retina contains a tiny pit densely packed with some 3 million cones. This area produces the sharpest image, which is why we tend to move our eyes so that the image of an object we want to see clearly falls on this area.

The lens of the human eye is constructed to filter out short-wavelength light. This solves a difficult optical problem: any uniform lens bends short wavelengths more than it does longer ones, a phenomenon known as chromatic aberration. Consequently, these short wavelengths cannot be brought into focus simultaneously with longer wavelengths. Unable to focus the short wavelengths, your eye (and the eyes of all vertebrates) eliminates them. Insects, whose eyes do not focus light, are able to see these lower, ultraviolet wavelengths quite well and often use them to locate food or mates.

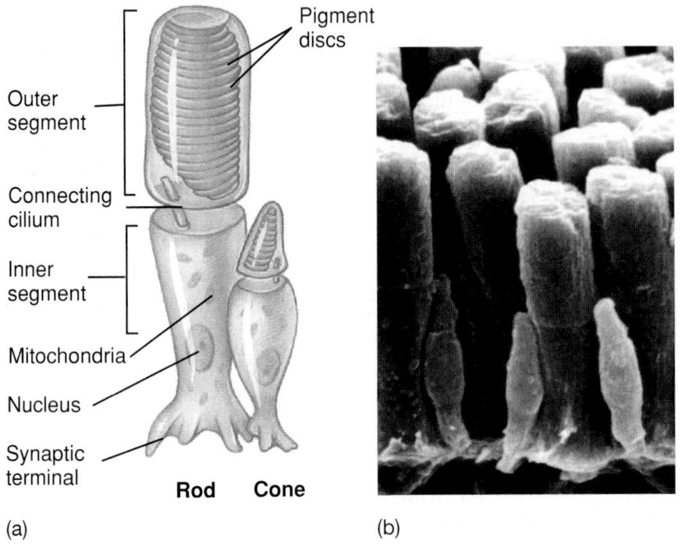

Figure 23.28 Rods and cones.
(*a*) The broad tubular cell on the *left* is a rod. The shorter, tapered cell next to it is a cone. (*b*) Electron micrograph of rods and cones.

How Rods and Cones Work

Light is the most useful stimulus for us to learn about the external environment. Because light travels in a straight line and arrives virtually instantaneously, visual information can be used to determine both the direction and the distance of an object. Many vertebrates have eyes that are more light-sensitive than ours—an owl, for example, has very large eyes that gather in far more light than human eyes do, enabling owls to see well in very dim light (although not in total darkness).

A rod or cone cell in your eye is able to detect a single photon of light. How can it be so sensitive? The primary sensing event of vision is the absorption of a photon of light by a pigment. The pigments in rods and cones are made from plant pigments called carotenoids. Humans cannot make carotenoids, and we must obtain them in our diet. That is why eating carrots is said to be good for night vision—the orange color of carrots is due to the presence of carotenoids called carotenes. The visual pigment in the human eye is a fragment of carotene called ***cis*-retinal.** The pigment is attached to a protein called **opsin** to form a light-detecting complex called **rhodopsin.**

When it receives a photon of light, the pigment undergoes a change in shape. This change in shape must be large enough to alter the shape of the opsin protein attached to it. When light is absorbed by the *cis*-retinal pigment, the linear end of the molecule rotates sharply upward, straightening out that end of the molecule (figure 23.29). The new form of the pigment is referred to as ***trans*-retinal.** This radical change in the pigment's shape induces a change in the shape of the protein opsin to which the pigment is bound, initiating a chain of events that leads to the generation of a nerve impulse.

Each rhodopsin activates several hundred molecules of a protein called transducin. Each of these activates several hundred molecules of an enzyme whose product stimulates sodium channels in the photoreceptor membrane at a rate of about 1,000 per second. This cascade of events allows a single photon to have a large effect on the receptor.

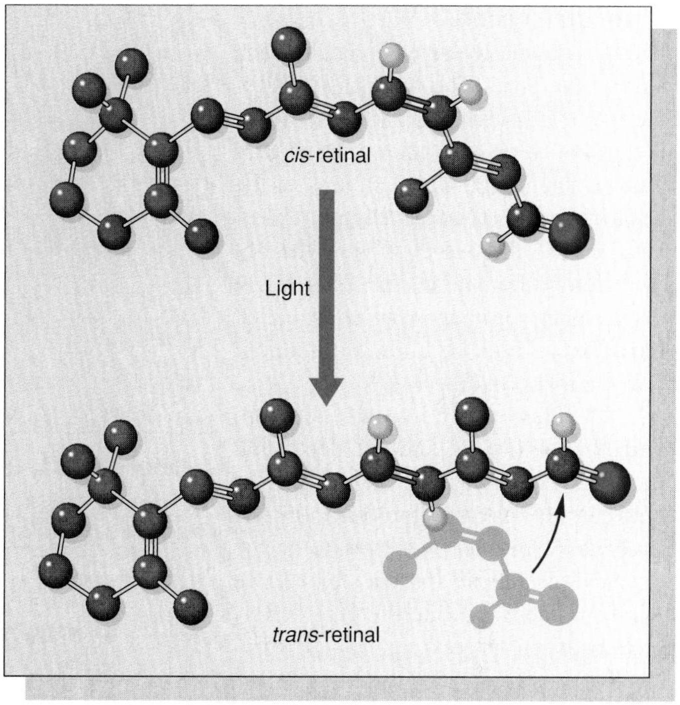

Figure 23.29 Absorption of light.
When light is absorbed by *cis*-retinal, the pigment undergoes a change in shape and becomes *trans*-retinal. This shape change initiates a chain of events that leads to the generation of a nerve impulse.

Color Vision

Three kinds of **cone cells** provide us with color vision. Each possesses a different version of the opsin protein (that is, one with a distinctive amino acid sequence and thus a different shape). These differences in shape affect the flexibility of the attached retinal pigment, shifting the wavelength at which it absorbs light (figure 23.30). In rods, light is absorbed at 500 nanometers. In cones, the three versions of opsin absorb light at 455 nanometers (blue absorbing), 530 nanometers (green absorbing), or 625 nanometers (red absorbing).

There is no reason in principle why additional kinds of cones could not be generated by other alterations in *cis*-retinal, which would make them absorb light of other wavelengths (say, yellow), but they are not needed. By comparing the relative intensities of the signals from the three cones, the brain can calculate the intensity of yellow light.

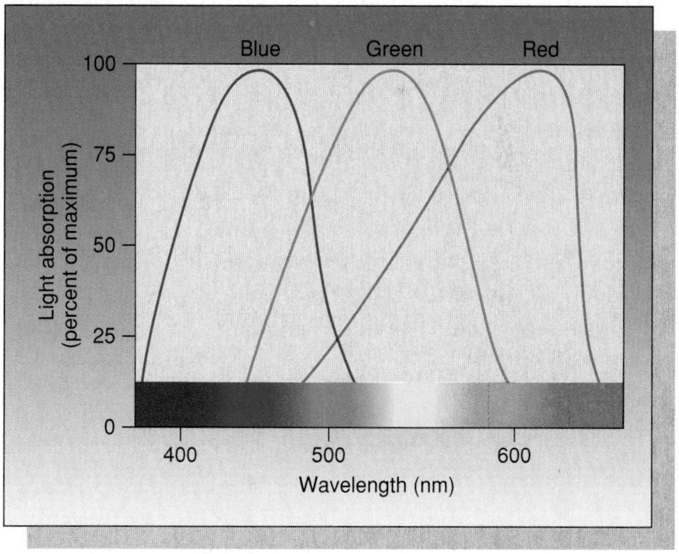

Figure 23.30 Color vision.
The absorption spectrum of *cis*-retinal is shifted in cone cells from the 500 nanometers characteristic of rod cells. The amount of the shift determines what color the cone absorbs: a shift down to 455 nanometers yields blue absorption; a shift up to 530 nanometers yields green absorption; and a shift farther up to 625 nanometers yields red absorption.

Binocular Vision

Humans have two eyes, one on each side of the head. The field of vision of the two eyes overlaps—each sees about a third of what the other sees (figure 23.31). The image that each sees of the same object, however, is slightly different, because the two eyes view the object from different angles. This slight displacement of images (an effect called **parallax**) permits the brain to figure out how far away an object is.

Interestingly, we are not born with this ability to perceive depth and distance. By comparing the differences between the images provided by the two eyes with what we know to be the physical distance to specific objects, we learn as babies to interpret the difference between the two images as representing that amount of distance. Depth perception is learned by trial and error, the brain exploring and remembering how physical distances to an object compare with the amount of difference between the images.

The path of light through each eye is the reverse of what you might expect. The rods and cones are at the rear of the retina, not the front. Light passes through several layers of ganglion and bipolar cells before it reaches the rods and cones (figure 23.32). The retina contains about 3 million cones, most of them located in the central region of the retina called the **fovea,** and approximately 1 billion rods, very few of which are present in the fovea. The eye forms its sharpest image in the fovea region.

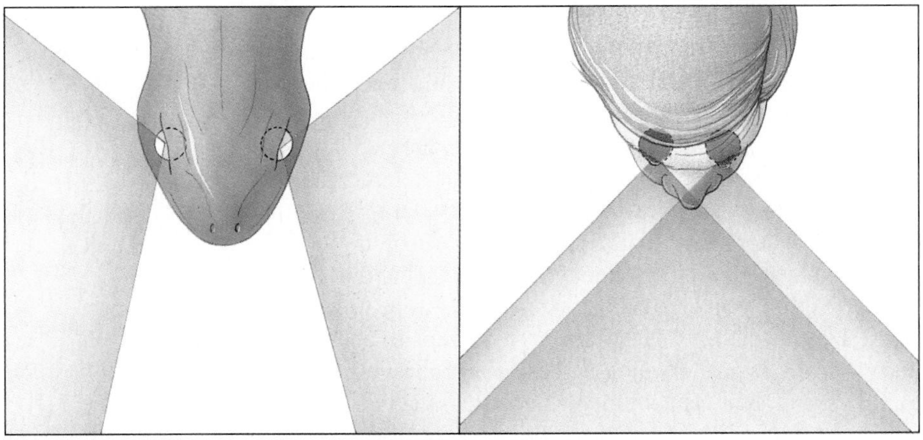

Figure 23.31 Binocular vision.
When the eyes are located on the sides of the head (as on the *left*), the two vision fields do not overlap and binocular vision does not occur. When both eyes are located toward the front of the head (as on the *right*) so that the two fields of vision overlap, depth can be perceived.

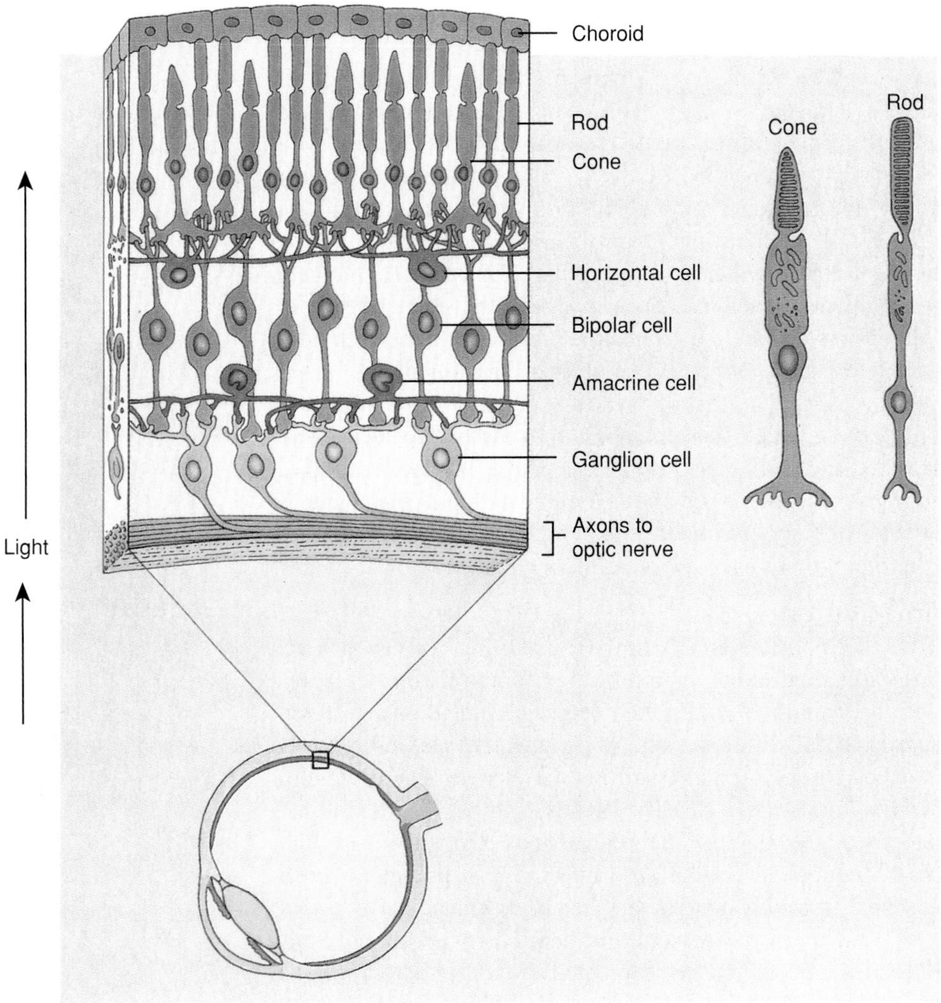

Figure 23.32 Structure of the retina.
Note that the rods and cones are at the rear of the retina, not the front. Light passes through four other types of cells in the retina before it reaches the rods and cones.

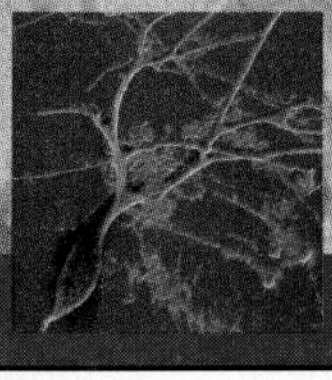

CHAPTER 23

HIGHLIGHTS

	Key Terms	Key Concepts

23.1 Organization of the Nervous System

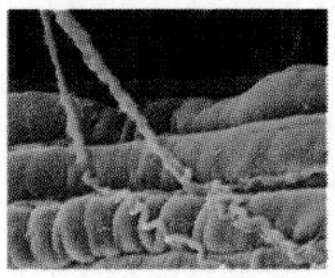

Key Terms

- central nervous system 496
- motor nerve 496
- sensory nerve 496

Key Concepts

- The nervous system is composed of sensory nerves, which carry signals to the central nervous system, and motor nerves, which carry signals away from it.

23.2 The Central Nervous System

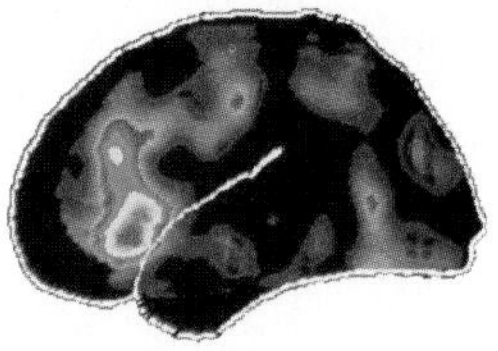

Key Terms

- voluntary nervous system 497
- autonomic nervous system 497
- cerebral cortex 498
- hypothalamus 499

Key Concepts

- The central nervous system is composed of the brain and spinal cord.
- Most of the mass of the brain is the cerebrum; its surface is the site of conscious thought.
- Other parts of the brain integrate information and coordinate body activities.

23.3 The Motor Nervous System

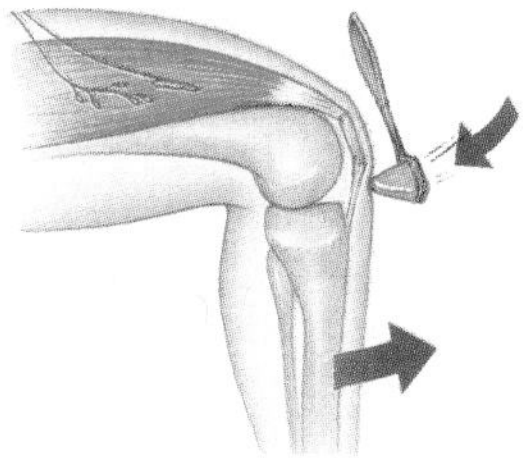

Key Terms

- reflex 501
- interneuron 501
- sympathetic nervous system 501
- parasympathetic nervous system 502

Key Concepts

- Voluntary motor nerves send commands from the brain to skeletal muscles.
- Involuntary motor nerves constitute the autonomic nervous system.
- The autonomic nervous system is composed of two antagonistic elements; the balance of the two determines the degree of stimulation.

23.4 How Nerves Work

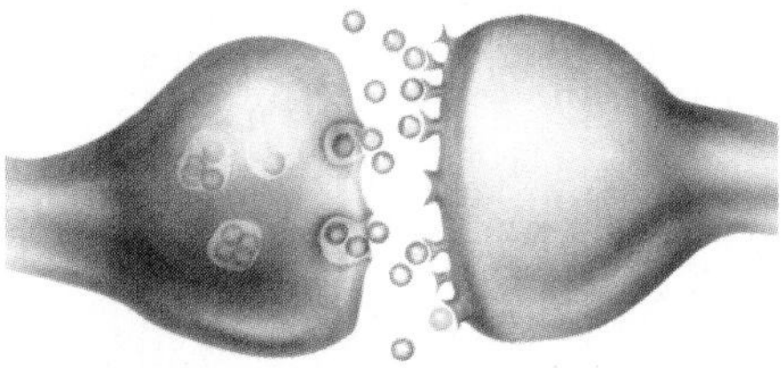

Key Terms

- resting potential 504
- voltage-gated channels 504
- action potential 504
- synapse 505
- neurotransmitter 505
- integration 506
- neuromodulator 507

Key Concepts

- Nerves are composed of the axons of many neurons, bundled together.
- Neurons expend ATP to pump out Na^+ and so polarize the cell membrane, generating a resting potential.
- A nerve impulse is a propagating depolarization of the cell membrane.
- Nerve impulses cross synapses via chemicals called neurotransmitters.

23.5 The Sensory Nervous System

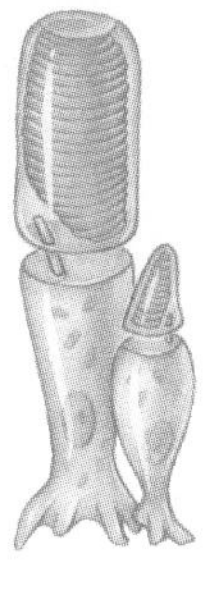

Key Terms

- stimulus-gated channel 508
- interoceptor 508
- retina 514
- rod 514
- cone 514
- rhodopsin 515

Key Concepts

- Interoceptors of many kinds monitor the body's internal condition.
- Sensing the exterior world focuses on three sorts of stimuli: chemicals, sound, and light.
- Both sound and light can be used to form three-dimensional images.

23

CONCEPT REVIEW

1. The sensory nervous system consists of _________ nerves.
   a. afferent
   b. efferent

2. Select the largest region of the human brain.
   a. cerebellum
   b. cerebrum
   c. hypothalamus
   d. thalamus

3. The hindbrain is devoted primarily to
   a. sensory integration.
   b. integrating visceral activities.
   c. processing of visual information.
   d. coordinating motor reflexes.

4. The hypothalamus controls
   a. respiration.
   b. sensory integration.
   c. language.
   d. conscious thought.

5. The parasympathetic branch of the autonomic nervous system _________ the rate of heartbeat.
   a. decreases
   b. increases

6. A reflex does not involve the
   a. sensory nerve.
   b. motor nerve.
   c. cerebrum.
   d. spinal cord.

7. At the synapse, the action potential moves along the neuron in which order?
   a. axon—dendrite—synapse
   b. axon—synapse—dendrite
   c. dendrite—axon—synapse
   d. synapse—axon—dendrite

8. Schwann cells, which envelop the axon at intervals along its length, act as
   a. nerves.
   b. tracts.
   c. electrical insulators.
   d. dendrites.

9. Hair cells in the otolith membrane detect
   a. light.
   b. vibrations.
   c. fluid movement.
   d. gravity.

10. The ability to hear often decreases with age because the
    a. cilia degenerate.
    b. hair cells stiffen.
    c. flexibility of the basilar membrane changes.
    d. tympanic membrane breaks.

11. If light is turned off in a room, the _________ of the retina become more important to vision.
    a. cones
    b. rods

12. _________ ions move back into the nerve cell easily once they are pumped out.

13. The wave of depolarization that moves down a neuron is called an _________.

14. The specialized chemicals that carry out communication across synaptic junctions are called _________.

15. Blockage of blood in the brain can cause a disorder called a _________.

16. There are _________ types of sensory cells for taste.

17. The _________ are three fluid-filled structures that detect motion changes.

18. The first part of the eye to intercept light is the _________.

Answers to the Concept Review questions appear in Appendix B.

CHALLENGE YOURSELF

1. Ouabain is a drug that poisons the sodium-potassium pump. What would be the *immediate* effect of ouabain on a neuron's ability to generate action potentials? Would the *long-term* effect be different? Explain.
2. If a nerve impulse can jump from node to node along a myelinated axon, why can't it jump from the presynaptic cell to the postsynaptic cell across a synaptic cleft?
3. Most people at one time or another have sensed an oncoming storm by detecting the increase in humidity in the air. What sort of receptors detect humidity? Why do you suppose that hot days seem so much hotter when the air is humid?
4. Why does traveling in a car offer very little sense of motion?

FOR FURTHER READING

Benson, D. *The Neurology of Thinking*. New York: Oxford University Press, 1994. Cognition has its own special branch within the realm of neuroscience. The question of awareness is hotly debated in science (and nonscience) today. How can cognition be described in strict clinical terms?

Byrne, R. *The Thinking Ape: Evolutionary Origins of Intelligence.* New York: Oxford University Press, 1995. What is intelligence, who has it, and how did they get it?

Fay, R., and A. Popper, eds. *Comparative Hearing: Mammals.* New York: Springer-Verlag, 1994. You and your dog or cat are both mammals, so how come your pet hears someone at the front door minutes before you do?

Finger, S. *Origins of Neuroscience: A History of Explorations into Brain Function.* New York: Oxford University Press, 1994. A thorough look at one of the most fascinating areas of research in human biology.

Freedman, D. *Brainmakers: How Scientists Are Moving Beyond Computers to Create a Rival to the Human Brain.* New York: Simon and Schuster, 1994. Artificial intelligence and the human brain are explored in this interesting look at what was only science fiction a generation ago.

Holloway, M. "Rx for Addiction." *Scientific American,* March 1991, 94–104. A look at the molecular mechanisms underlying drug addiction.

Mestel, R. "We're All Connected." *Discover,* January 1994, 88. An interesting look at innervation of the skin and at how the nervous system and the immune system may interact in skin tissue during times of stress to produce such phenomena as hives or rashes.

Pines, M., ed. *Seeing, Hearing, and Smelling the World: New Findings Help Scientists Make Sense of Our Senses.* Chevy Chase, Md: Howard Hughes Medical Institute, 1995. The latest experimental roundup of data contributing to our understanding of senses and sensation.

Swerdlow, J. "Quiet Miracles of the Brain." *National Geographic,* June 1995, 2–41. New research reveals the brain's flexibility and leads to ingenious treatments for age-old disorders.

Winson, J. "The Meaning of Dreams." *Scientific American,* November 1990, 86–96. This challenging article argues that dreams reflect a fundamental aspect of how mammals process their memories.

TECHNOLOGY LINKS

The Living World Home Page
http://www.wcbp.com/biology/tlw

Explorations in Human Biology CD
#8 Nerve Conduction
#9 Synaptic Transmission
#10 Drug Addiction

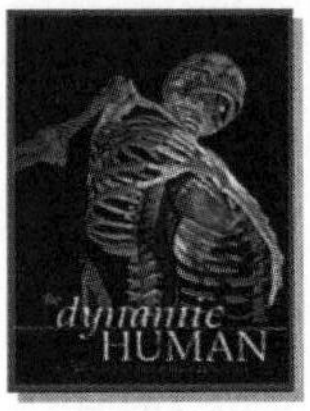

The Dynamic Human CD
Nervous System

Life Science Animations Videotape 3
#22 Formation of Myelin Sheath
#25 Reflex Arcs

Physiological Concepts of Life Science Videotape
#6 Conduction of Nerve Impulses

24

Chemical Signaling Within the Body

CHAPTER OUTLINE

Figure 24.1 Our bodies are complex machines.
Our activities must be coordinated for integration. The nervous system and the endocrine system provide this integration through nervous impulses and chemical messages (hormones), respectively.

The tissues and organs of your body carry out a multitude of activities—capturing oxygen and digesting hot dogs, walking and singing, sweating and seeing far into the distance (figure 24.1). All of this activity must be coordinated, or else we would be unable to walk and eat a hot dog at the same time. It is this coordination that makes the human body an *organism,* rather than a smoothly functioning set of organs. As we discussed in chapter 23, the central nervous system uses chemical messengers as well as nerve signals to coordinate the body's organ systems. The three major classes of chemical messengers are (1) *second messengers,* which work only within cells, (2) *neurotransmitters* released by axons, whose effects last only a short time, and (3) *hormones,* stable, long-lasting chemicals released from glands. A hormone is a chemical signal produced in one part of the body that is stable enough to be transported in active form far from where it is produced and that typically acts at a distant site.

24.1 The Neuroendocrine System

There are three big advantages to using chemical messengers rather than electrical signals (nerves) to control body organs. First, chemical molecules can spread to all tissues via the blood (imagine trying to wire every cell with its own nerve!). Second, chemical signals can persist much longer than electrical ones, a great advantage for hormones controlling slow processes like growth and development. Third, many different kinds of chemicals can act as hormones, so different hormone molecules can be targeted at different tissues.

Hormones are produced by glands, most of which are controlled by the central nervous system. Because these glands are completely enclosed in tissue rather than having ducts that empty to the outside, they are called **endocrine glands** (from the Greek *endon,* within). Hormones are secreted from them directly into the bloodstream (this is in contrast to **exocrine glands,** which, like sweat glands, have ducts). Your body has a dozen major endocrine glands that together make up the endocrine system. While motor nerves regulate the body's movements, the endocrine system is the body's chief means of regulating its ongoing activities.

The *endocrine system* and the *motor nervous system* are the two main routes the central nervous system uses to issue commands to the organs of the body. The two are so closely linked that they are often considered a single system—the **neuroendocrine system.** The hypothalamus can be considered the main switchboard of the neuroendocrine system. Your hypothalamus is continually checking conditions inside your body. Are you too hot or too cold? Are you running out of fuel? Is your blood pressure too high? If your internal environment starts to get out of balance, your hypothalamus has several ways to set things right again. For example, if it wants to speed up your heart rate, it can send a nerve signal to the medulla, or it can use a chemical command, causing the adrenal gland to produce the hormone adrenaline, which also speeds up the heart rate. Which command the hypothalamus uses depends on the desired duration of the effect. A chemical message is typically far longer lasting than a nerve signal.

How the endocrine system is regulated by the brain was until recently one of the great mysteries of medicine. It has long seemed likely that the hypothalamus issues commands to a nearby gland, the pituitary, which in turn sends chemical signals to the various hormone-producing glands of the body. But how does the hypothalamus issue its commands? The pituitary, which receives them, is located very close by, suspended from the hypothalamus by a short stalk (figure 24.2)—but no nerves connect the pituitary to the hypothalamus or to any other part of the brain. In the 1930s researchers discovered a network of tiny blood vessels that spans the short distance between the hypothalamus and the pituitary, suggesting that perhaps a chemical message passed from the hypothalamus to the pituitary. It was not easy to verify this hypothesis, because very little of any such message could be expected to be present in any one brain at one time. However, after concentrating the hypothalamus glands from 1 million pigs, the first of these hormones was isolated in 1969, a short peptide called thyrotropinreleasing hormone (TRH). The release of TRH from the hypothalamus triggers the pituitary to release a hormone called thyrotropin, which travels to the thyroid and causes the thyroid gland to release thyroid hormones.

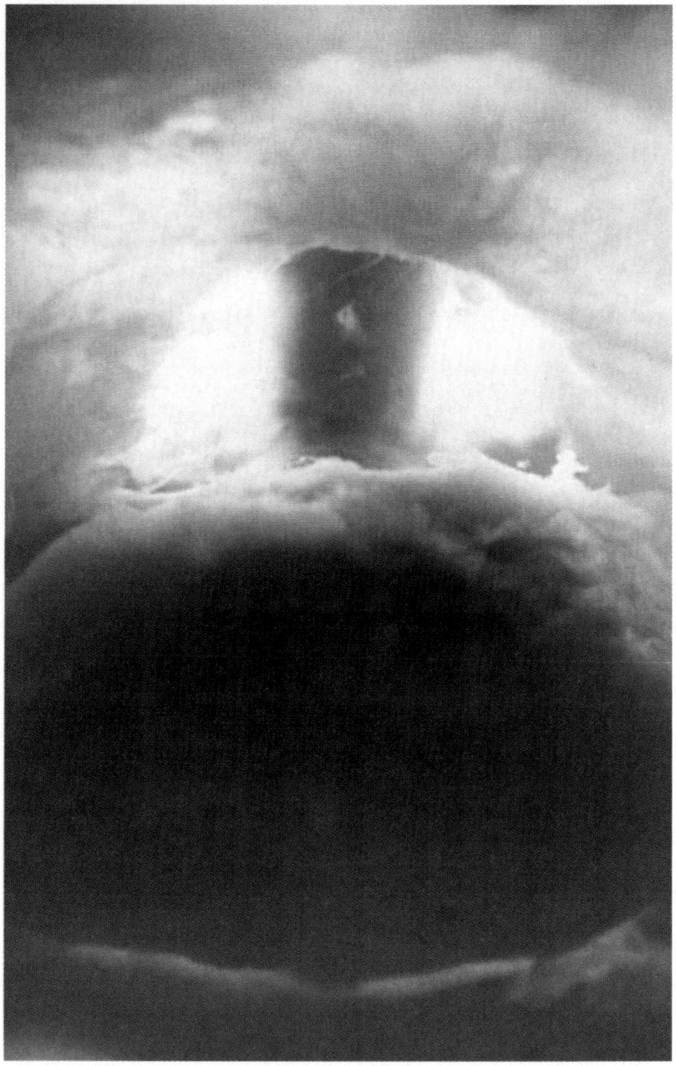

Figure 24.2 The pituitary gland.
The pituitary gland, which hangs by a short stalk from the hypothalamus, regulates the hormone production of many of the body's endocrine glands. Here, it is enlarged 30 times.

Six other hypothalamic regulatory hormones have since been isolated, which together govern the pituitary. The CNS regulates the body's hormones through a chain of command. Each of the seven "releasing" hormones made by the hypothalamus causes the pituitary to synthesize a corresponding pituitary hormone, which travels to a distant endocrine gland and causes that gland to begin producing its particular endocrine hormone.

24.2 How Hormones Work

The key reason why hormones are effective messengers within the body is because a particular hormone can be directed at a specific target cell. How does the target cell recognize that hormone, ignoring all others? Within the plasma membrane of the target cell are embedded receptor proteins that match the shape of the potential signal hormone like a hand fits a glove. As you recall from chapter 23, nerve cells have highly specific receptors within their synapses, each receptor shaped to "respond" to a different neurotransmitter molecule. Cells that the body has targeted to respond to a particular hormone have within their plasma membranes receptor proteins shaped to fit that hormone and no other. Thus, chemical communication within your body involves *two* elements: a molecular signal (the hormone) and a protein receptor on target cells. The system is highly specific because each protein receptor has a shape that only a particular hormone fits (figure 24.3).

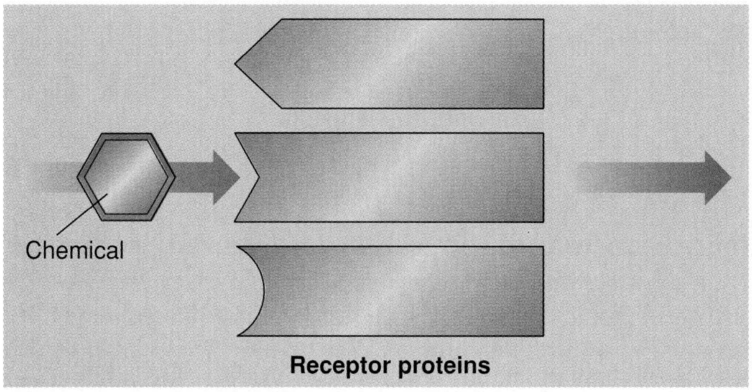

Figure 24.3 Specificity.
A key advantage of chemical signals is that they can be directed to a specific receptor protein on a target cell.

The path of communication taken by a hormonal signal within the human body can be visualized as a series of simple steps (figure 24.4):

1. **Issuing the command.** The hypothalamus of the central nervous system controls the release of many hormones. It chemically signals the pituitary gland, which is located very close to it, to release hormones into the bloodstream.
2. **Transporting the signal.** While hormones can act on an adjacent cell, most are transported throughout the body by the bloodstream.
3. **Hitting the target.** When a hormone encounters a cell with a matching receptor, called a target cell, the hormone binds to that receptor.
4. **Having an effect.** When the hormone binds to the receptor protein, the protein responds by changing shape, which triggers a change in cell activity.

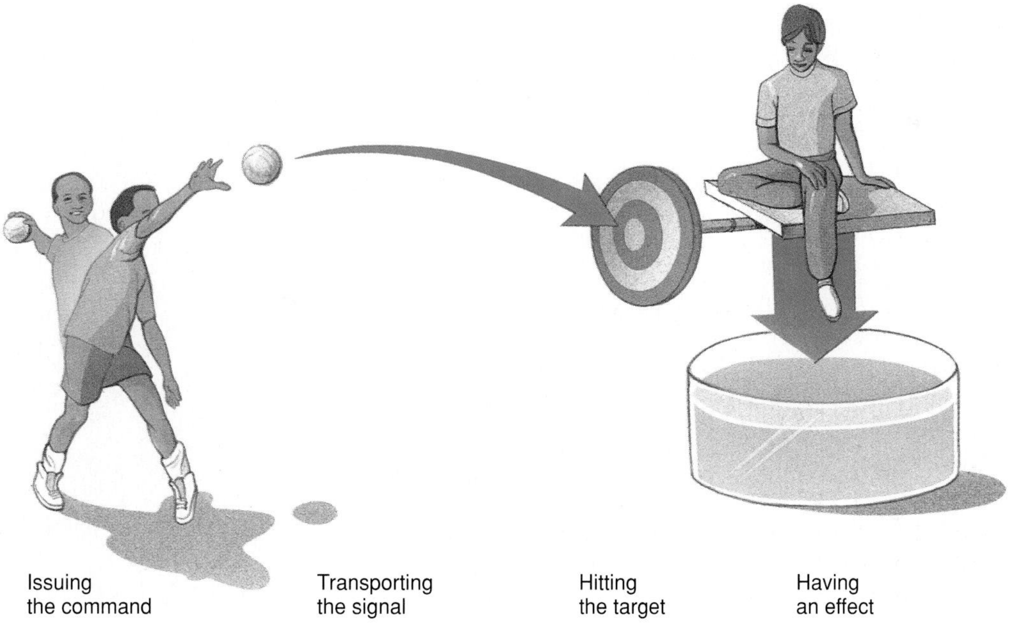

Figure 24.4 The steps of hormonal communication.
Generally, a part of the neuroendocrine system issues a command in the form of a chemical messenger (hormone); the hormone is transported to the target cell via the bloodstream; the hormone reaches the target cell and binds with the cell receptor; and the hormone receptor complex triggers changes in the target cell.

Steroid Hormones Enter Cells

Some protein receptors designed to recognize hormones are located in the cytoplasm of the target cell. The hormones in these cases are lipid-soluble molecules, typically **steroid hormones.** The chemical shape of these molecules resembles chicken wire (figure 24.5). They pass across the lipid bilayer of the cell membrane and bind to receptors within the cytoplasm. This complex of receptor and hormone then binds to the DNA in the nucleus and causes a change in the activity of a particular gene. The change in this gene's activity is responsible for the effect of the hormone (figure 24.6).

The steroids that weight lifters and other athletes sometimes use to "bulk up" turn on genes and thus trick their muscle cells into adding growth. **Anabolic steroids** are synthetic compounds that resemble the male sex hormone testosterone. Their injection into muscles causes the muscle cells to produce more protein, resulting in bigger muscles and increased strength. However, anabolic steroids have many dangerous side effects in both men and women, including liver damage, heart disease, and psychological disorders. Some experts also believe that steroid use can be linked to cancer. Anabolic steroids are illegal, and for many sports, athletes are tested for their presence. Fortunately, many athletes have stopped using anabolic steroids because of the side effects and the negative publicity afforded to the abuse of these compounds.

All steroid hormones are manufactured from cholesterol, a complex molecule composed of four rings. The hormones that promote the development of secondary sexual characteristics are steroids. They include cortisone and testosterone, as well as the hormones estrogen and progesterone, which control the female reproductive system and are discussed in the next chapter.

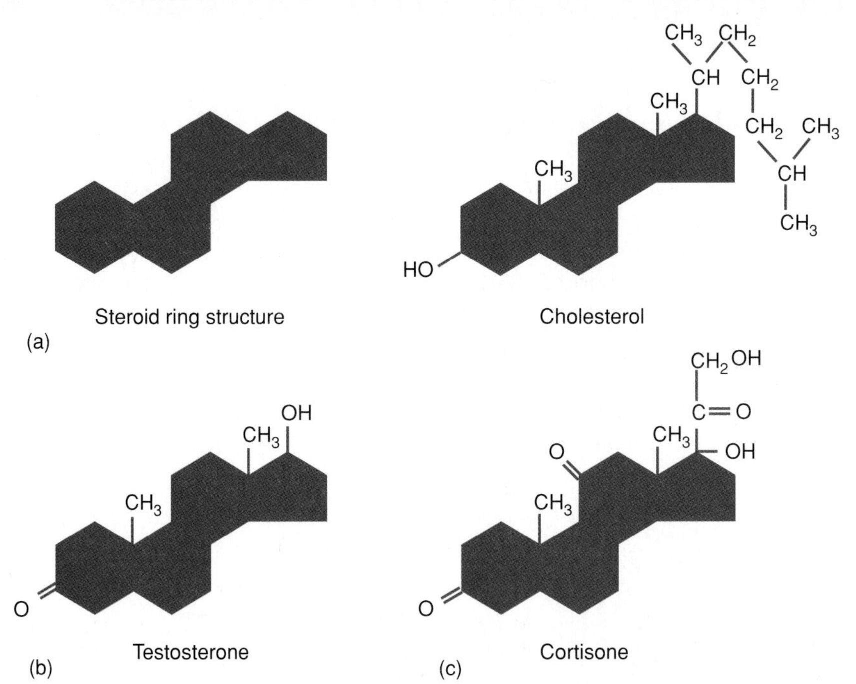

(a)
Steroid ring structure
Cholesterol
HO
(b)
Testosterone
(c)
Cortisone
OH
O

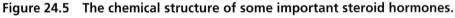

Figure 24.5 The chemical structure of some important steroid hormones.
(*a*) Many steroid hormones are similar in structure to the blood lipid cholesterol.
(*b*) Testosterone stimulates development of the male genital tract; (*c*) cortisone promotes the breakdown of muscle proteins in metabolism.

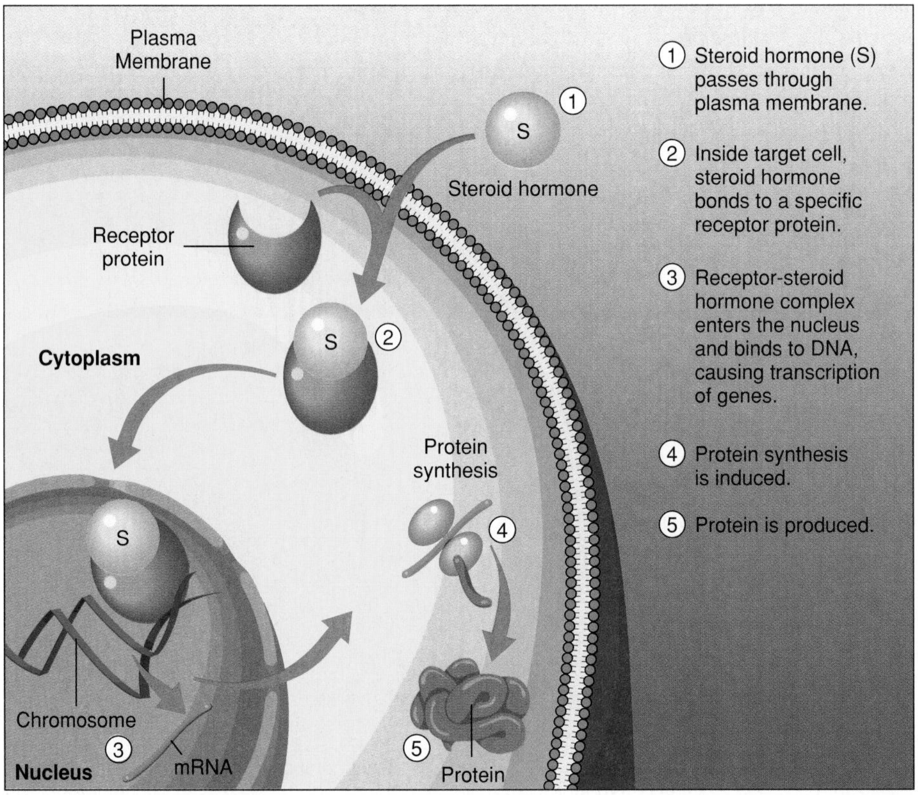

Plasma Membrane
Steroid hormone
Receptor protein
Cytoplasm
Protein synthesis
Chromosome
mRNA
Nucleus
Protein
① Steroid hormone (S) passes through plasma membrane.
② Inside target cell, steroid hormone bonds to a specific receptor protein.
③ Receptor-steroid hormone complex enters the nucleus and binds to DNA, causing transcription of genes.
④ Protein synthesis is induced.
⑤ Protein is produced.

Figure 24.6 How steroid hormones work.
Steroid hormones (*S*) are lipid soluble and thus readily pass through the plasma membrane of cells into the cytoplasm. There they bind to specific receptor proteins, forming a complex that enters the nucleus and binds to specific regulatory sites on chromosomes. The binding activates the genes regulated by the site and thus results in the production of specific proteins.

Peptide Hormones Act at the Cell Surface

Other hormone receptors are embedded within the cell membrane, with their recognition region directed outward from the cell surface. The hormones that bind to these receptors are typically short peptide chains (although some, like insulin, are full-sized proteins). The binding of the **peptide hormone** to the receptor triggers a change in the cytoplasmic end of the receptor protein. This change then triggers events within the cell cytoplasm, usually through intermediate within-cell signals called **second messengers,** which greatly amplify the original signal (figure 24.7).

How does a second messenger amplify a hormone's signal? Second messengers act by activating enzymes (figure 24.8). One of the most common second messengers is cyclic AMP. Cyclic AMP is made from ATP by an enzyme that removes two phosphate units, forming AMP; the ends of the AMP join, forming a circle. A single hormone molecule binding to a receptor in the cell membrane can result in the formation of many second messengers in the cytoplasm. Each second messenger can, in turn, activate many molecules of a certain enzyme. Sometimes each of these enzymes in turn activates many other enzymes. Thus, second messengers enable each hormone molecule to have a tremendous effect inside the cell, far greater than if the hormone had simply entered the cell and sought out a single target.

The hormone insulin provides a well-studied example of how peptide hormones achieve their effects within target cells. Most human cells have insulin receptors in their membranes—typically only a few hundred but in tissues involved in glucose metabolism far more. A single liver cell, for example, may have 100,000 of them. When insulin binds one of these insulin receptors, the receptor protein changes its shape, goading an adjacent signal-modulating protein on the cell interior called a G protein to activate the production of cyclic AMP. The cyclic AMP in turn activates a variety of cellular enzymes, in a cascading series of events that greatly amplifies the strength of the original signal.

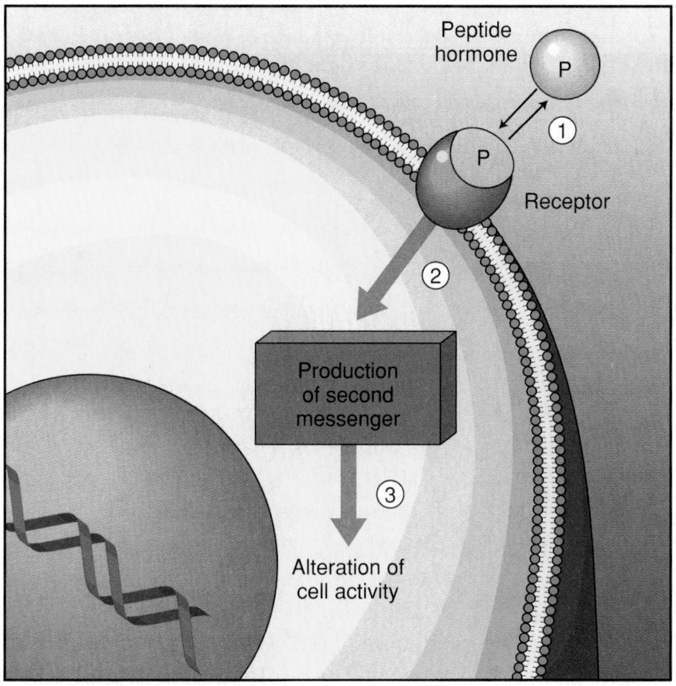

Figure 24.7 How peptide hormones work.
(*1*) The peptide hormone binds with its membrane receptor. (*2*) The hormone-receptor combination triggers a series of biochemical reactions that produces the second messenger. (*3*) The second messenger triggers a series of reactions that lead to altered cell functions.

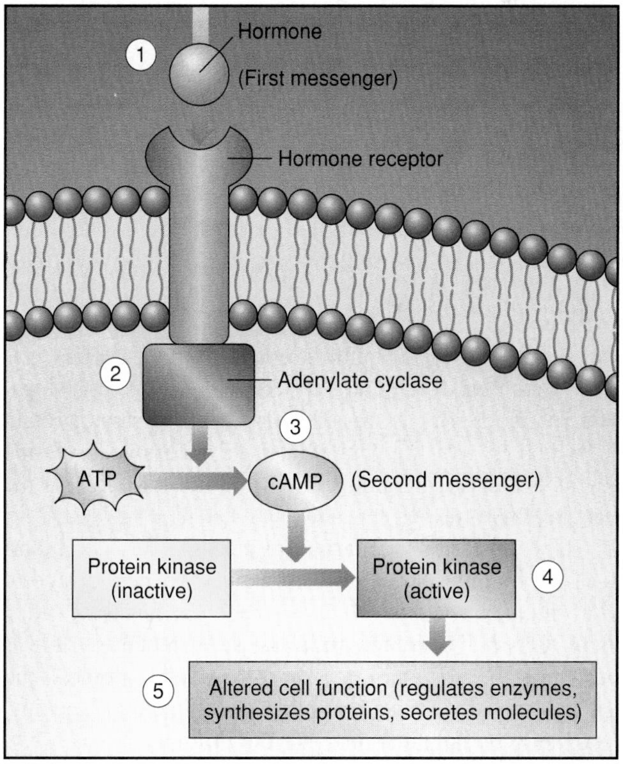

Figure 24.8 How second messengers work.
(*1*) Peptide hormones bind to specific receptor sites on the surface of target cells. (*2*) The hormone-receptor complex activates adenylate cyclase. (*3*) Adenylate cyclase converts ATP into cyclic AMP (cAMP) in the cell cytoplasm. Cyclic AMP acts as a second messenger and activates enzymes called protein kinases. (*4*) Protein kinases catalyze a wide variety of actions. (*5*) Depending on the nature of the first messenger, a variety of altered cell functions may result. Because of the presence of a second messenger, the effect on the cell is greatly amplified.

24.3 The Major Endocrine Glands

The human body has about a dozen major endocrine glands, shown in figure 24.9 and table 24.1 (pp. 528–29).

The Pituitary: The Master Gland

The **pituitary gland,** located in a bony recess in the brain below the hypothalamus, is where nine major hormones are produced. Because many of these hormones act principally to influence other endocrine glands, it was fashionable until relatively recently to regard the pituitary as a "master gland" orchestrating the endocrine system. But in fact, as shown here, that role is reserved for the hypothalamus, because the hypothalamus controls the pituitary, just as a general controls the orders a captain issues to the troops. The pituitary remains, however, one of the most important endocrine glands.

The pituitary is actually two glands. The back portion, or *posterior lobe,* regulates water conservation and milk letdown and uterine contraction in women; the front portion, or *anterior lobe,* regulates the other endocrine glands (figure 24.10).

The Posterior Pituitary

The role of the posterior pituitary first became evident in 1912, when a remarkable medical case was reported: a man who had been shot in the head developed a surprising disorder—he began to urinate every 30 minutes, unceasingly. The bullet had lodged in his pituitary gland, and subsequent research demonstrated that surgical removal of the pituitary also produces these unusual symptoms. Pituitary extracts were shown to contain a substance that makes the kidneys conserve water, and in the early 1950s the peptide hormone **vasopressin** (also called antidiuretic hormone, **ADH**) was isolated. Vasopressin regulates the kidney's retention of water. When vasopressin is missing, the kidneys cannot retain water, which is why the bullet caused excessive urination (and why excessive alcohol, which inhibits vasopressin secretion, has the same effect).

The posterior pituitary also releases a second hormone of very similar structure—both are short peptides composed of nine amino acids—but very different function, called **oxytocin.** Oxytocin initiates milk release in mothers. Here is how it works: Sensory receptors in the mother's nipples send messages to the hypothalamus when stimulated by sucking, causing the hypothalamus to make the pituitary release oxytocin. The oxytocin travels in the bloodstream to the breasts, where it stimulates contraction of the muscles around the ducts into which the mammary glands secrete milk. Oxytocin also stimulates uterine contraction in women during childbirth, which is why the uterus of a nursing mother returns to its normal size after its extension during pregnancy more quickly than does the uterus of a mother who does not nurse her baby. Oxytocin and vasopressin are both synthesized inside neurons within the hypothalamus, and they are transported down nerve axons to the posterior pituitary.

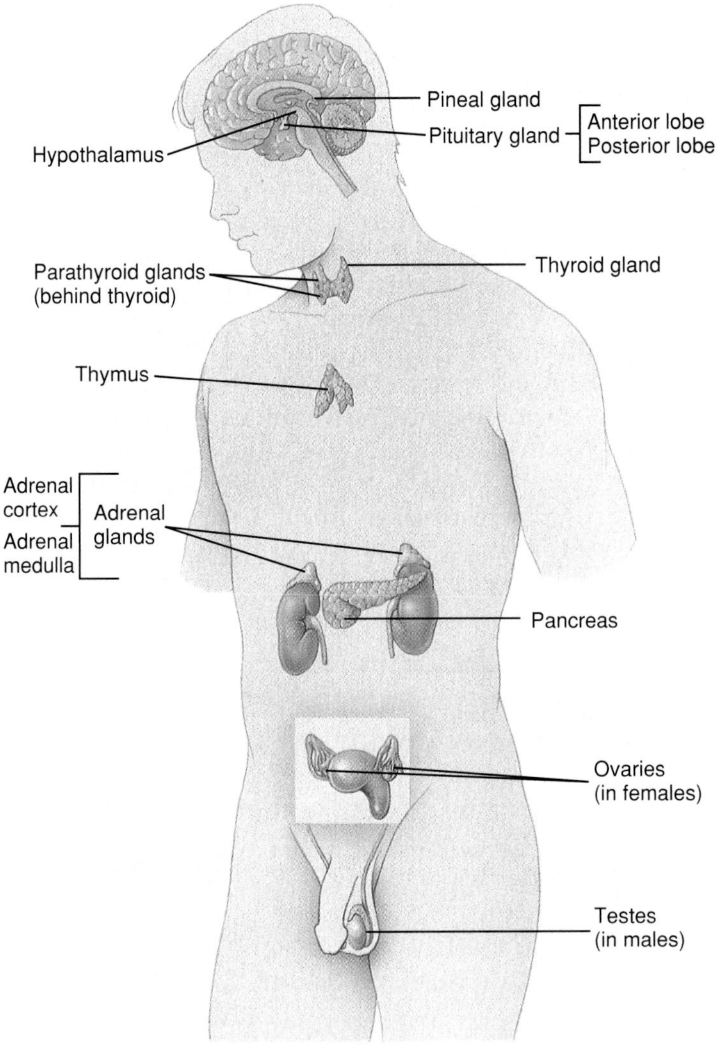

Figure 24.9 Major glands of the human endocrine system. Hormone-secreting cells are clustered in endocrine glands.

The Anterior Pituitary

The key role of the anterior pituitary first became understood in 1909, when a 38-year-old South Dakota farmer was cured of the growth disorder acromegaly by the surgical removal of a pituitary tumor. Acromegaly is a form of giantism in which the jaw begins to protrude and the features thicken. It turned out that giantism is almost always associated with pituitary tumors. Robert Wadlow, born in Alton, Illinois, in 1928, grew to a height of 8 feet, 11 inches, and weighed 475 pounds before he died from infection at age 22—the tallest human being ever recorded. Skull X-rays showed he had a pituitary tumor. Pituitary hormones have also proven to be the cause of several other well-known cases of giantism, such as the 8-foot, 2-inch Irish giant Charles Byrne, born in 1761; his skeleton, preserved in the Royal College of Surgeons, London, shows the effects of a pituitary tumor.

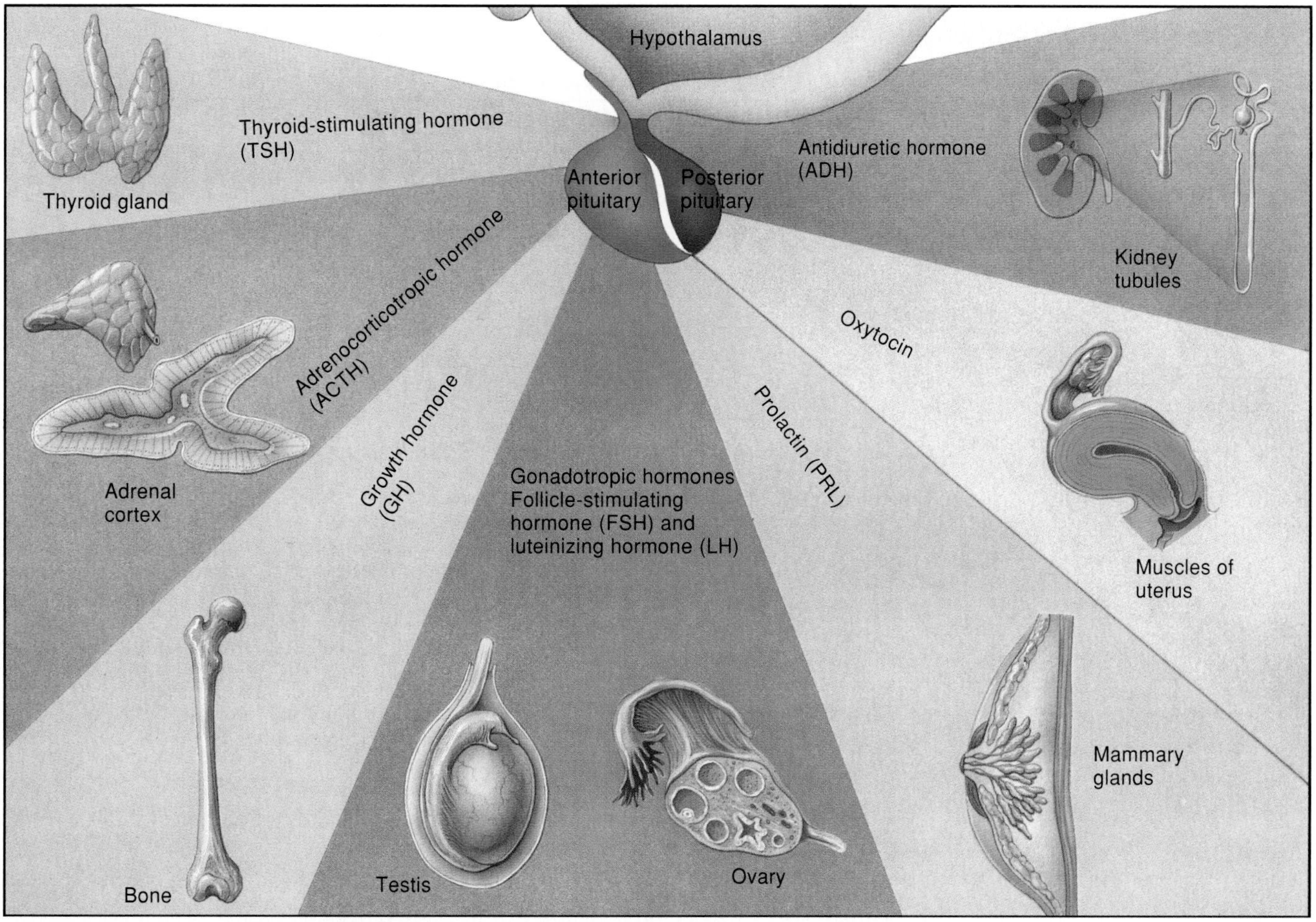

Figure 24.10 The role of the pituitary.
Interactions between the anterior and posterior lobes of the pituitary (two distinct glands) and various organs of the human body are shown in this diagram.

Why did removal of the pituitary tumor cure the South Dakota farmer? Pituitary tumors produce giants because the tumor cells produce prodigious amounts of a growth-promoting hormone. This **growth hormone (GH),** a long peptide of 191 amino acids, is normally produced in only minute amounts by the anterior pituitary gland and usually only during periods of body growth, such as infancy and puberty.

We now know that the anterior pituitary gland produces seven major peptide hormones, each controlled by a particular releasing signal secreted from cells in the hypothalamus. All seven of these hormones have major endocrine roles, as well as poorly understood roles in the central nervous system. These seven hormones are listed here:

1. **Thyroid-stimulating hormone (TSH).** TSH stimulates the thyroid gland to produce the thyroid hormone thyroxin, which in turn stimulates oxidative respiration.
2. **Luteinizing hormone (LH).** LH plays an important role in the female menstrual cycle, discussed in chapter 25. It also stimulates the male gonads to produce testosterone, which initiates and maintains the development of male secondary sexual characteristics (those external features not involved directly in reproduction).
3. **Follicle-stimulating hormone (FSH).** FSH is significant in the female menstrual cycle, and, in males, it stimulates cells in the testes to produce a hormone that regulates development of the sperm.
4. **Adrenocorticotropic hormone (ACTH).** ACTH stimulates the adrenal gland to produce a variety of steroid hormones. Some of these hormones regulate the production of glucose from fat; others regulate the balance of sodium and potassium ions in the blood; and still others contribute to the development of male secondary sexual characteristics.
5. **Somatotropin, or growth hormone (GH).** GH stimulates the growth of muscle and bone throughout the body.
6. **Prolactin.** Prolactin stimulates the breasts to produce milk.
7. **Melanocyte-stimulating hormone (MSH).** In reptiles and amphibians, melatonin stimulates color changes in the epidermis. The function of this hormone in humans is still poorly understood.

Table 24.1 The Principal Endocrine Glands

Endocrine Gland and Hormone	Target	Principal Actions
Posterior Lobe of Pituitary		
Oxytocin (OT)	Uterus Mammary glands	Stimulates contraction of uterus Stimulates ejection of milk
Vasopressin (antidiuretic hormone, ADH)	Kidneys	Conserves water; increases blood pressure
Anterior Lobe of Pituitary		
Growth hormone (GH) (somatotropin)	General	Stimulates growth by promoting protein synthesis and breakdown of fatty acids
Prolactin (PRL)	Mammary glands	Sustains milk production after birth
Thyroid-stimulating hormone (TSH)	Thyroid gland	Stimulates secretion of thyroid hormones
Adrenocorticotropic hormone (ACTH)	Adrenal cortex	Stimulates secretion of adrenal cortical hormones
Follicle-stimulating hormone (FSH)	Gonads	In females, stimulates ovarian follicle, stimulates secretion of estrogen; in males, stimulates production of sperm cells
Luteinizing hormone (LH)	Ovaries and testes	Stimulates ovulation and corpus luteum formation in females; stimulates secretion of testosterone in males
Thyroid Gland		
Thyroid hormone (thyroxine, T_4, and triiodothyronine, T_3)	General	
Calcitonin	Bone	Lowers blood calcium level by inhibiting release of calcium from bone
Parathyroid Glands		
Parathyroid hormone (PTH)	Bone, kidneys, digestive tract	Increases blood calcium level by stimulating bone breakdown; stimulates calcium reabsorption in kidneys; activates vitamin D
Adrenal Medulla		
Epinephrine (adrenaline) and norepinephrine (noradrenaline)	Skeletal muscle, cardiac muscle, blood vessels	Initiates stress responses; increases heart rate, blood pressure, metabolic rate; dilates blood vessels; mobilizes fat; raises blood glucose level
Adrenal Cortex		
Aldosterone	Kidney tubules	Maintains proper balance of sodium and potassium ions
Cortisol	General	Adaptation to long-term stress; raises blood glucose level; mobilizes fat
Pancreas		
Insulin	General	Lowers blood glucose level; increases storage of glycogen in liver
Glucagon	Liver, adipose tissue	Raises blood glucose level; stimulates breakdown of glycogen in liver

Table 24.1 (continued)

Endocrine Gland and Hormone	Target	Principal Actions
Ovary		
Estrogen	General; female reproductive structures	Stimulates development of secondary sex characteristics in females and growth of sex organs at puberty; prompts monthly preparation of uterus for pregnancy
Progesterone	Uterus, breasts	Completes preparation of uterus for pregnancy Stimulates development of breasts
Testes		
Testosterone	General; male reproductive structures	Stimulates development of secondary sex characteristics in males and growth spurt at puberty; Stimulates development of sex organs; stimulates sperm production
Pineal Gland		
Melatonin	Hypothalamus	Function not well understood; may help control onset of puberty in humans
Thymus		
Thymosin	White blood cells	Promotes production and maturation of white blood cells
Hypothalamus		
Thyrotropin-releasing hormone (TRH)	Anterior pituitary	Stimulates TSH release from anterior pituitary
Corticotropin-releasing hormone (CRH)	Anterior pituitary	Stimulates ACTH release from anterior pituitary
Gonadotropin-releasing hormone (GnRH)	Anterior pituitary	Stimulates FSH and LH release from anterior pituitary
Prolactin-releasing hormone (PRH)	Anterior pituitary	Stimulates PRL release from anterior pituitary
Growth hormone-releasing hormone (GHRH)	Anterior pituitary	Stimulates GH release from anterior pituitary
Prolactin-inhibiting hormone (PIH)	Anterior pituitary	Inhibits PRL release from anterior pituitary
Growth hormone-inhibiting hormone (GHIH)	Anterior pituitary	Inhibits GH release from anterior pituitary

The Thyroid: A Metabolic Thermostat

The **thyroid gland** is shaped like a shield (its name comes from *thyros,* the Greek word for shield). It lies just below the Adam's apple in the front of the neck. The thyroid makes several hormones, the two most important of which are **thyroxin,** which increases metabolic rate and promotes growth, and **calcitonin,** which stimulates calcium uptake by bones.

Thyroxine performs several important functions in the human body. Without adequate thyroxin, growth is retarded. Children with underactive thyroid glands are not able to carry out carbohydrate breakdown and protein synthesis at normal rates, a condition called cretinism that results in stunted growth. Mental retardation can also result, because thyroxin is needed for normal development of the central nervous system. Thyroxin contains iodine, and if the amount of iodine in your diet is too low, the thyroid cannot make adequate amounts of thyroxin and will grow larger in a futile attempt to manufacture more. The greatly enlarged thyroid gland that results is called a goiter (figure 24.11). This need for iodine in your diet is why iodine is added to table salt.

Calcitonin plays a key role in maintaining proper calcium levels in the body. If levels of calcium in your blood become too high, calcitonin stimulates calcium deposition in bone, thus lowering calcium levels in the bloodstream. As you will see, another hormone has an even more critical role in maintaining the body's levels of calcium.

The Parathyroids: Builders of Bone

The **parathyroid glands** are four small glands attached to the thyroid. Small and unobtrusive, they were ignored by researchers until well into this century. The first suggestion that the parathyroids produce a hormone came from experiments in which they were removed from dogs: the concentration of calcium in the dogs' blood plummeted to less than half the normal level. However, if an extract of the parathyroid gland was administered, calcium levels returned to normal. If an excess was administered, calcium levels in the blood became *too* high, and the bones of the dogs were literally dismantled by the extract. It was clear that the parathyroid glands were producing a hormone that acted on calcium uptake into and out of the bones.

The hormone produced by the parathyroids is **parathyroid hormone (PTH).** It is one of only two hormones in your body that is absolutely essential for survival (the other is aldosterone, a hormone produced by the adrenal glands, discussed in the next section). PTH regulates the level of calcium in your blood. Recall that calcium ions are the key actors in muscle contraction—by initiating calcium release, nerve impulses cause muscles to contract. You cannot live without the muscles that pump your heart and drive your body, and these muscles cannot function if calcium levels are not kept within narrow limits.

Figure 24.11 A goiter.
This condition is caused by a lack of iodine in the diet.

PTH acts as a fail-safe to make sure calcium levels never fall too low (figure 24.12). PTH is released into the bloodstream, where it travels to the bones and acts on the osteoclast cells within bones, stimulating them to dismantle bone tissue and release calcium into the bloodstream. PTH also acts on the kidneys to resorb calcium ions from the urine and leads to activation of vitamin D, necessary for calcium absorption by the intestine. A diet deficient in vitamin D leads to poor bone formation, a condition called rickets. When PTH is synthesized by the parathyroids in response to falling levels of calcium in the blood, the body is essentially sacrificing bone to keep calcium levels within the narrow limits necessary for proper functioning of muscle and nerve.

The Adrenals: Two Glands in One

Your body has two **adrenal glands,** one located just above each kidney. Each adrenal gland is composed of two parts: an inner core called the **medulla,** which produces the peptide hormones adrenaline (also called epinephrine) and norepinephrine, and an outer shell called the **cortex,** which produces the steroid hormones cortisol and aldosterone.

The Adrenal Medulla: Emergency Warning Siren

The medulla releases adrenaline and norepinephrine in times of stress. These hormones act as emergency signals that stimulate rapid deployment of body fuel. The "alarm" response these hormones produce throughout the body is identical to the individual effects achieved by the sympathetic nervous system, but it is much longer lasting. Among the effects of these hormones are an accelerated heartbeat, increased blood pressure, higher levels of blood sugar, and increased blood flow to the heart and lungs. These hormones can thus be thought of as extensions of the sympathetic nervous system.

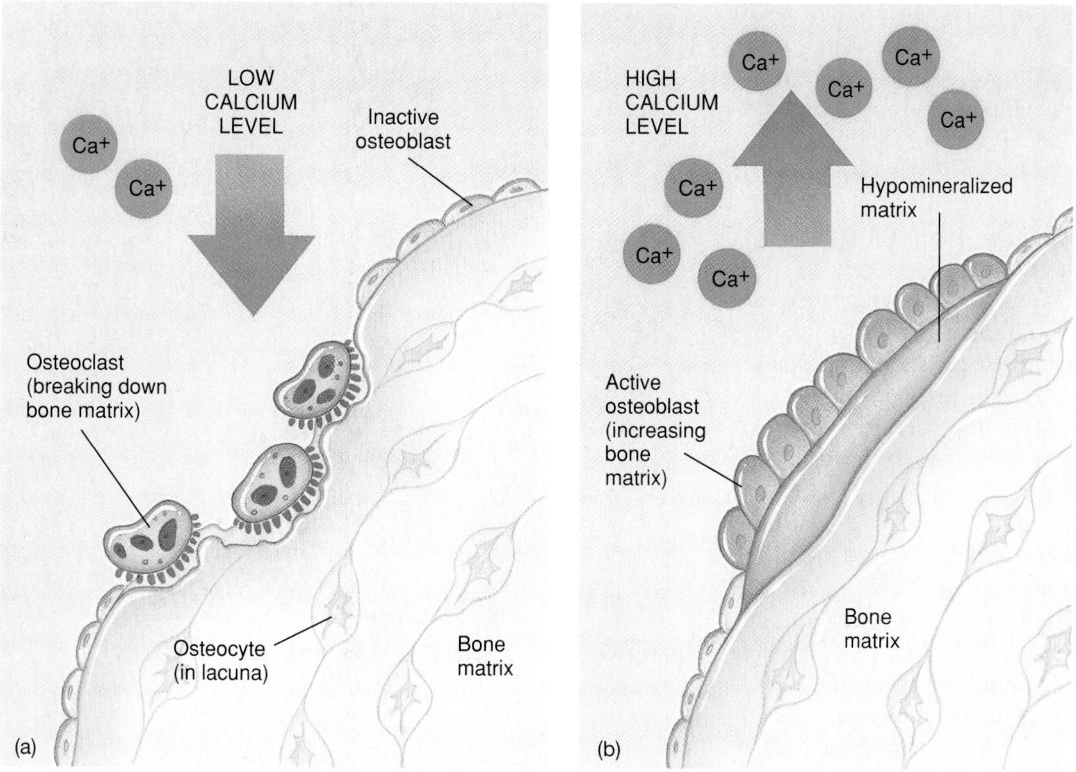

Figure 24.12 Maintenance of proper calcium levels in the blood.
(*a*) When calcium levels in the blood become too low, the parathyroid gland produces additional amounts of PTH, a hormone that stimulates the osteoclast cells to break down calcium in bone. This raises the calcium levels in the blood. (*b*) Conversely, abnormally high levels of calcium in the blood trigger the thyroid gland to secrete calcitonin, which promotes the activity of osteoblasts to remove calcium from the blood and deposit it in bone.

The Adrenal Cortex: Maintaining the Proper Amount of Salt

The adrenal cortex produces the steroid hormone cortisol. Cortisol (also called hydrocortisone) acts on many different cells in the body to maintain nutritional well-being. It stimulates carbohydrate metabolism and reduces inflammation. Synthetic derivatives of this hormone, such as prednisone, have widespread medical use as anti-inflammatory agents. The ability of many cortisol-derived steroids to stimulate muscle growth has also led to the abuse of anabolic steroids by athletes.

The adrenal cortex also produces aldosterone. Aldosterone acts primarily in the kidney to promote the uptake of sodium and other salts from the urine. Sodium ions play crucial roles in nerve conduction and many other body functions. Aldosterone is, with PTH, one of the two endocrine hormones essential for survival. That is why removal of the adrenal glands is invariably fatal.

The Pancreas: The Body's Dietitian

The **pancreas** gland is located behind the stomach and is connected to the front end of the small intestine by a narrow tube. It secretes a variety of digestive enzymes into the digestive tract through this tube, and for a long time it was thought to be solely an exocrine gland. In 1869, however, a German medical student named Paul Langerhans described some unusual clusters of cells scattered throughout the pancreas. In 1893, doctors concluded that these clusters of cells, which came to be called islets of Langerhans, produced a substance that prevented diabetes mellitus. **Diabetes mellitus** is a serious disorder in which affected individuals are unable to take up glucose from the blood, even though their levels of blood glucose become very high. Such individuals lose weight and literally starve. Breaking down their fat reserves results in the production of acids called ketones, which lower the pH of the blood. Affected individuals may eventually suffer brain damage and even death. Diabetes is the leading cause of blindness among adults, and it accounts for one-third of all kidney failures. It is the seventh leading cause of death in the United States.

The substance produced by the islets of Langerhans, which we now know to be the peptide hormone *insulin,* was not isolated until 1922. Two young doctors working in a Toronto hospital injected an extract purified from beef pancreas glands into a 13-year-old boy, a diabetic whose weight had fallen to 29 kilograms (65 pounds) and who was not expected to survive. The hospital record gives no indication of the historic importance of the trial, only stating, "15 cc of MacLeod's serum. 7-1/2 cc into each buttock." With this single injection, the glucose level in the boy's blood fell 25%—his cells were taking up glucose. A more potent extract soon brought levels down to near normal.

This was the first instance of successful insulin therapy. We now know that the islets of Langerhans in the pancreas produce two hormones that interact to govern the levels of glucose in the blood (figure 24.13). These hormones are *insulin* and *glucagon.* Insulin is a storage hormone, designed to put away nutrients for leaner times. It promotes the accumulation of glycogen in the liver and triglycerides in fat cells. When you eat, beta-cells in the islets of Langerhans secrete insulin, causing your body to store away glucose to be used later. When body activity causes the level of glucose in the blood to fall as it is used up as fuel, other cells in the islets of Langerhans, called alpha-cells, secrete glucagon, which causes liver cells to release stored glucose and fat cells to break down triglycerides. The two hormones thus work together to keep glucose levels in the blood within narrow bounds (figure 24.14).

Twelve million Americans, and over 100 million people worldwide, have **diabetes.** We now know that there are *two* kinds of diabetes mellitus. About 10% of affected individuals suffer from type I diabetes, a hereditary autoimmune disease in which the immune system attacks the islets of Langerhans, resulting in abnormally low insulin secretion. Called juvenile onset diabetes, this type usually develops before age 20. Affected individuals can be treated by daily injections of insulin. Active research on the possibility of transplanting islets of Langerhans holds much promise as a lasting treatment for this form of diabetes.

In type II diabetes, the number of insulin receptors on the target tissue is abnormally low, while the level of insulin in the blood is often higher than normal. This form of diabetes usually develops in people over 40 years of age. It is almost always a consequence of excessive weight; in the United States, 90% of those who develop Type II diabetes are obese. Cells of these individuals, sated with food, adjust their appetite for glucose downward by reducing their sensitivity to insulin. As a drug addict's neurons reduce their number of neurotransmitter receptors after continued exposure to a drug, the obese individual's cells reduce their number of insulin receptors. To compensate, the pancreas pumps out ever more insulin, and, in some people, the insulin-producing cells are unable to keep up with the ever-heavier workload and stop functioning. Type II diabetes is usually treatable by diet and exercise, and most affected individuals do not need daily injections of insulin.

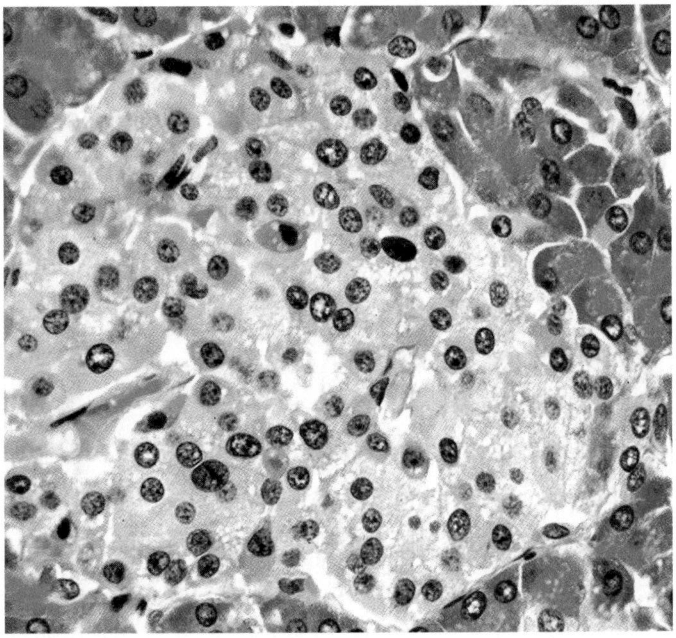

Figure 24.13 Islets of Langerhans.
Glucagon and insulin are produced by clumps of cells within the pancreas called islets of Langerhans. The islet in this photomicrograph is the lighter-colored area.

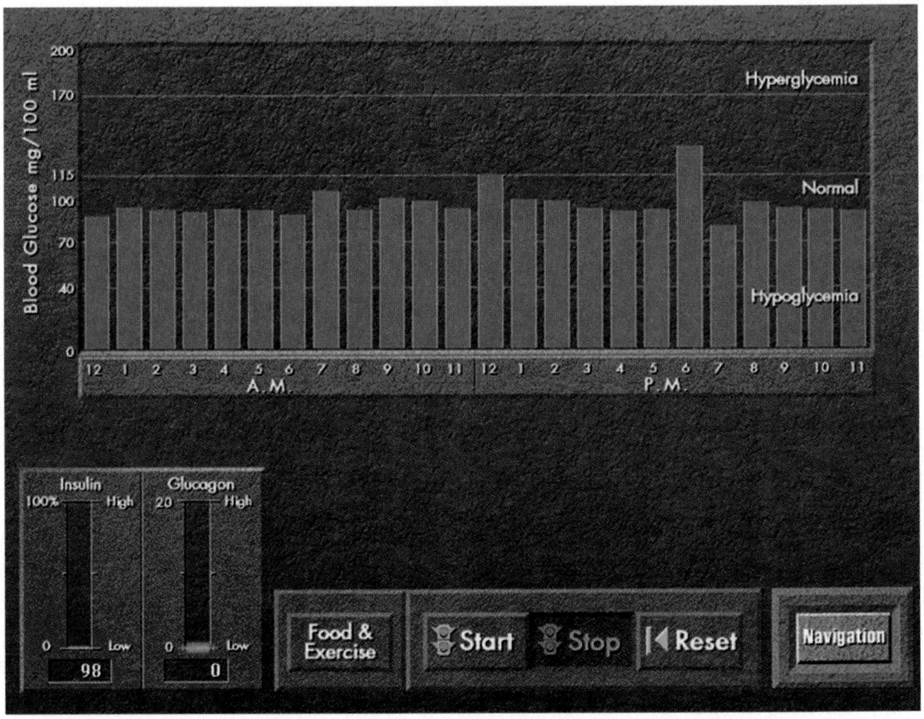

Figure 24.14 Blood glucose levels are kept within a narrow range.
Note on this chart that the sharp rise in blood glucose after lunch and after dinner are quickly reduced to the normal level. This is a screen capture from an interactive CD-ROM that allows the user to vary food consumption and watch how the hormones insulin and glucagon interact to keep blood glucose levels within a narrow range. (*Explorations in Human Biology,* Module 11, "Hormone Action")

CHAPTER 24

HIGHLIGHTS

Key Terms | **Key Concepts**

24.1 The Neuroendocrine System

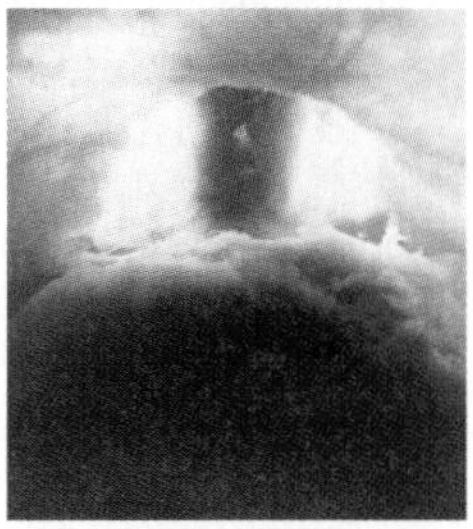

Key Terms

endocrine gland 522
neuroendocrine system 522

Key Concepts

- Many of the hormone-producing endocrine glands are under the direct control of the central nervous system.
- The hypothalamus produces releasing hormones that cause the pituitary to release particular hormones that circulate via the bloodstream to their target tissues.

24.2 How Hormones Work

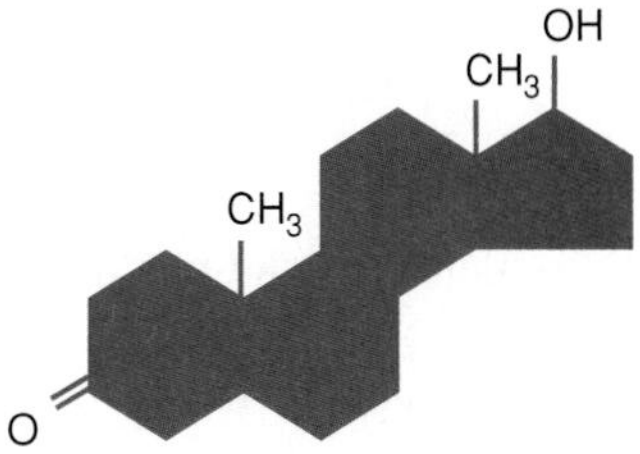

Key Terms

steroid hormone 524
anabolic steroid 524
peptide hormone 525
second messenger 525

Key Concepts

- Steroid hormones enter cells and interact with receptors located in the cytoplasm or nucleus.
- Peptide hormones interact with receptors located on the cell surface.
- Hormones often initiate a cascading series of events in the cell that greatly increases the power of the signal.

24.3 The Major Endocrine Glands

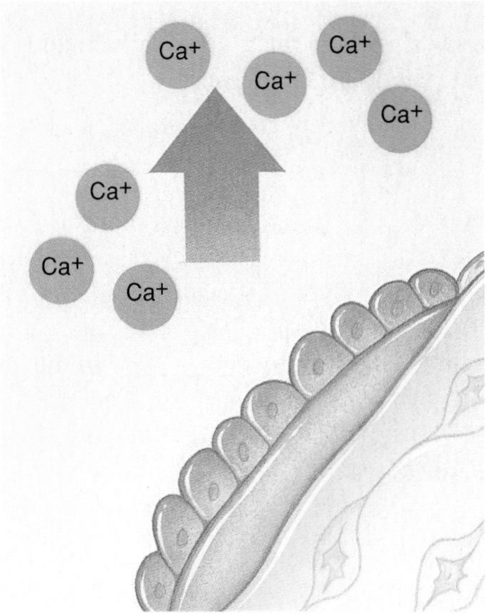

Key Terms

pituitary gland 526
diabetes mellitus 531

Key Concepts

- The anterior pituitary gland produces seven so-called pituitary hormones.
- Each of these hormones is released by a particular releasing hormone from the hypothalamus and circulates to a particular target tissue.

24

CONCEPT REVIEW

1. Hormones are
   a. enzymes.
   b. regulatory chemicals.
   c. produced at one place in the body but exert their influence at another.
   d. all of the above.
   e. b and c.

2. An example of a peptide hormone is
   a. cortisone.
   b. progesterone.
   c. insulin.
   d. testosterone.

3. Which of the following is *not* one of the seven principal pituitary hormones?
   a. thyroid hormone
   b. melanocyte-stimulating hormone
   c. somatotropin
   d. luteinizing hormone

4. Noradrenaline and adrenaline are hormones that produce an "alarm" response. They are produced by the adrenal gland, which is located
   a. in the brain.
   b. just above the kidneys.
   c. in the neck.
   d. in the testes or ovaries.

5. The adrenal cortex controls the uptake of sodium and other salts within the kidney by regulating the production of the hormone
   a. aldosterone.
   b. insulin.
   c. glucagon.
   d. antidiuretic hormone.

6. Type II diabetes
   a. results in abnormally low insulin secretion.
   b. is hereditary.
   c. is most effectively treatable with injections of insulin.
   d. is almost always a consequence of excessive weight.

In questions 7–16, match the hormones in the left-hand column with their effects in the right-hand column.

7. insulin	a. ovulation
8. glucagon	b. secondary sex characteristics
9. ADH	c. reaction to stress
10. TSH	d. sodium balance
11. ACTH	e. reabsorption of water
12. aldosterone	f. milk production
13. epinephrine	g. stimulates adrenal gland
14. prolactin	h. stimulates thyroid gland
15. testosterone	i. increase sugar in the blood
16. LH	j. decreases sugar in the blood

17. Most endocrine glands are under the control of the _________ system.

18. The _________ sends signals to the pituitary gland and thus controls the neuroendocrine system.

19. Hormone molecules are peptides or _________.

20. ADH is secreted by the _________ pituitary gland.

21. The _________ is the outer layer of the adrenal gland.

22. The parathyroids control the level of _________ in the blood.

23. The _________ secretes insulin.

Answers to the Concept Review questions appear in Appendix B.

CHALLENGE YOURSELF

1. Why do you suppose the brain goes to the trouble of synthesizing releasing hormones rather than simply directing the production of the pituitary hormones immediately?

2. Why do you suppose steroid hormones do not employ second messengers when so many peptide hormones do?

3. Two different organs, A and B, are sensitive to a particular hormone. The cells in both organs have identical receptors for the hormone, and hormone-receptor binding produces the same intracellular second messenger in both organs. However, the intracellular effects of the hormone on A and B are very different. Explain.

FOR FURTHER READING

Atkinson, M., and N. MacLaren. "What Causes Diabetes?" *Scientific American,* July 1990, 62–71. For insulin-dependent diabetic patients, the answer is an autoimmune ambush of the body's insulin-producing cells. Why the attack begins and persists is now becoming clear.

Davis, D. L., and H. L. Bradlow. "Can Environmental Estrogen Cause Breast Cancer?" *Scientific American,* October 1995, 166–72. Nonnaturally occurring estrogens ("xenoestrogens") in the environment may be contributing to the phenomenal rates of breast cancer in women.

Dayton, L. "When Hormones Say It's Time to Fight." *New Scientist* 137 (January 9, 1993): 16. Fight, run, or be frightened: adrenergic receptors respond to adrenaline by preparing the body to do something, *quick.*

Dranov, P. "Tired? Wired? It Could Be Your Thyroid." *American Health* 13 (May 1994): 90–93. Over- or underactive thyroid glands account for medical treatment for at least 11 million Americans. There are probably more suffering from some kind of thyroid problem, as many symptoms of thyroid malfunction are misdiagnosed, if they are even diagnosed at all.

Goldman, B. "The Essence of Attraction." *Health* 8 (March/April 1994): 40–42. Human pheromones, primitive responses, desire, and an entrepreneur who is bottling the stuff.

Hoberman, J., and C. Yesalis. "The History of Synthetic Testosterone." *Scientific American,* February 1995, 76–81. Long banned in sports, these anabolic steroids greatly enhance athletic performance.

Kalin, N. "The Neurobiology of Fear." *Scientific American,* May 1993, 94–99. The hormonal response to anxiety, fear, and stress is examined, along with the localized regions of the nervous system in which the activity occurs.

Lacy, P. "Treating Diabetes with Transplanted Cells." *Scientific American*, July 1995, 50–58. Implanting islet cells of the pancreas can potentially cure diabetes—if a way can be found to avoid immune attacks.

Levay, S., and D. Hamer. "Evidence for a Biological Influence in Male Homosexuality." *Scientific American,* May 1994, 44–49. Nature or nurture again? More scientists look at the hypothalamus and make some conclusions favoring nature.

Skerret, P. "Sperm Busters." *Technology Review* 97 (January 1994): 19–20. An injectable form of birth control for men that operates by blocking gonadotropin-releasing hormones in the pituitary is not far off!

TECHNOLOGY LINKS

The Living World Home Page
http://www.wcbp.com/biology/tlw

Explorations in Human Biology CD
#11 Hormone Action

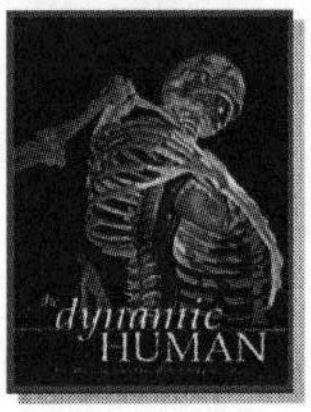

The Dynamic Human CD
Endocrine System

Life Science Animations Videotape 3
#28 Peptide Hormone Action (cAMP)

Physiological Concepts of Life Science Videotape
#10 Action of Steroid Hormone on Target Cells

25

Human Reproduction and Development

"It's still hungry . . . and I've been stuffing worms into it all day."

CHAPTER OUTLINE

Figure 25.1 Sex is more than academic.
In today's society it is vital that young adults understand not only the basic facts of human reproduction but also the consequences of their sexual behavior.

Few subjects are of more direct concern to students than sex. Many students must make important decisions about sex, and all need to be well informed. The subject of this chapter is thus of far more than academic interest (figure 25.1). Sexual reproduction first evolved among marine organisms, long before the appearance of vertebrates. Seawater is not a hostile environment for gametes or for young organisms. Humans and other mammals, because they live on dry land, fertilize gametes internally and nourish their developing young within the mother's body. In this chapter, we review the functioning of the human reproductive system and the course of development from fertilization to birth.

25.1 The Human Reproductive System

The human reproductive system (figure 25.2), like that of all vertebrates, joins egg cells and sperm cells to form a zygote. The zygote develops first into an embryo, then a fetus, and finally, after about nine months, into a child. The egg is fertilized by the sperm within the female, and the zygote develops into a mature fetus there.

Males

The male gamete, or **sperm,** is highly specialized for its role as a carrier of genetic information. Produced after meiosis, the sperm cells have 23 chromosomes instead of the 46 found in all other cells of the human body. Unlike these other cells, sperm do not complete their development successfully at 37°C (98.6°F), the normal human body temperature. The sperm-producing organs, the **testes,** move during the course of fetal development out of the body proper and into a sac called the **scrotum** (figure 25.3). The scrotum, which hangs between the legs of the male, maintains the testes at a temperature about 3°C cooler than that of the rest of the body.

Male Gametes Are Formed in the Testes

The two testes are composed of several hundred compartments, each packed with large numbers of tightly coiled tubes called **seminiferous tubules.** Within these tubes sperm production takes place (figure 25.4). The number of sperm that is produced is truly incredible. A typical adult male produces several hundred million sperm each day of his life! Those that are not ejaculated from the body are broken down, and their materials are resorbed and recycled.

The testes contain cells that secrete the male sex hormone **testosterone.** Sperm and all the cells of the testes also require a combination of the pituitary hormones FSH and LH for their normal function.

After a sperm cell is manufactured within the testes, it is delivered to a long, coiled tube called the **epididymis,** where it matures. The sperm cell is not motile when it arrives in the epididymis, and it must remain there for at least 18 hours before its motility develops. From there, the sperm is delivered to another long tube, the **vas deferens,** where it is stored. When sperm are delivered during intercourse, they travel through a tube from the vas deferens to the **urethra,** where the reproductive and urinary tracts join, emptying through the penis.

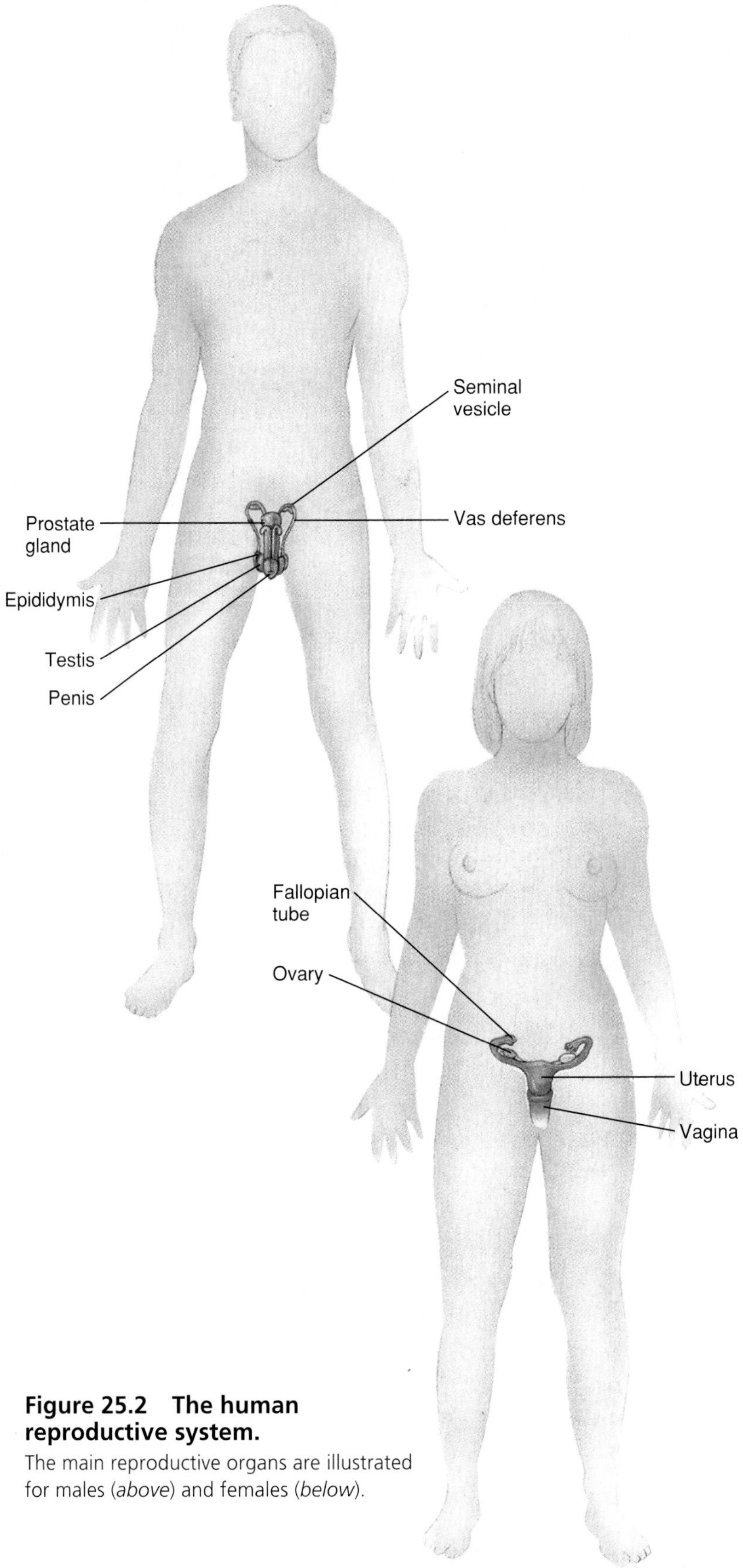

Figure 25.2 The human reproductive system.
The main reproductive organs are illustrated for males (*above*) and females (*below*).

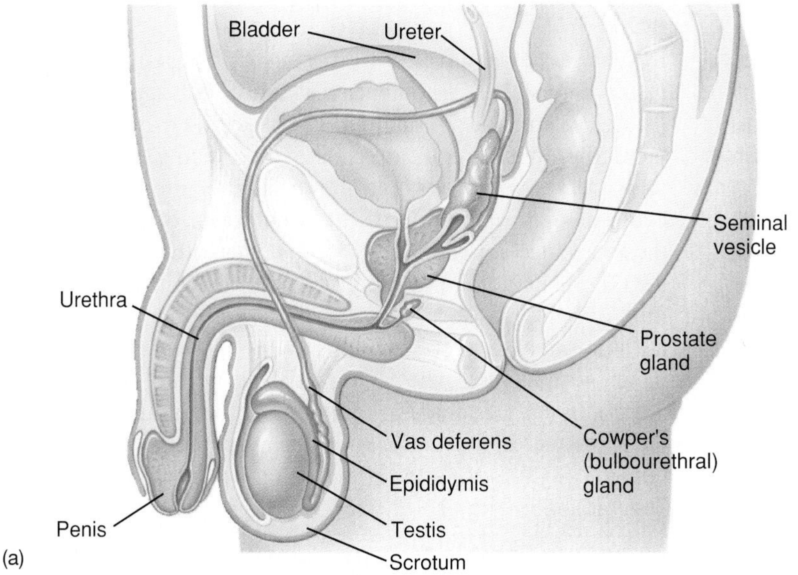

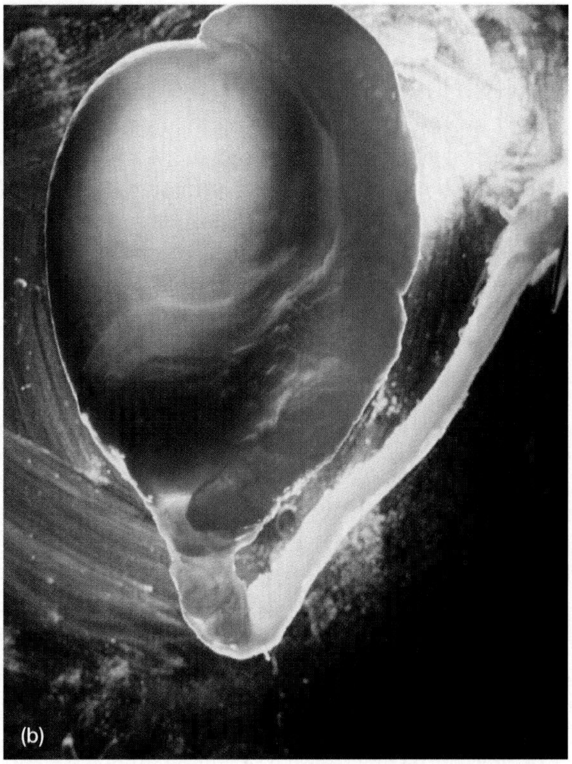

Figure 25.3 The male reproductive system.
(*a*) Organization of the male reproductive organs. (*b*) A human testis. The testis is the darker sphere in the center of the photograph; within it, sperm are formed. Cupped above the testicle is the epididymis, a highly coiled passageway within which sperm complete their maturation. Extending away from the epididymis is a long tube, the vas deferens.

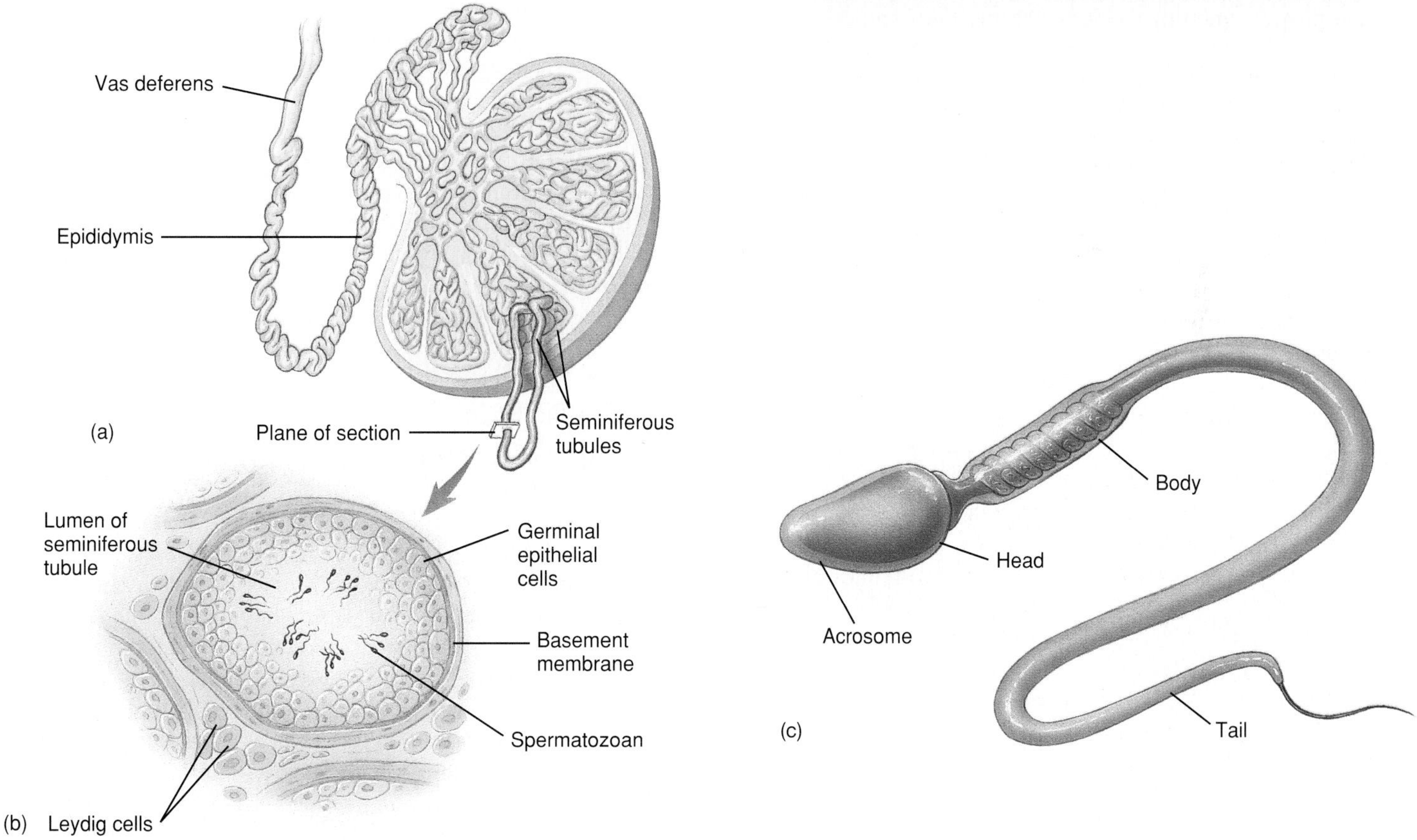

Figure 25.4 The interior of the testis, site of sperm production.
Within the seminiferous tubules of the testis (*a*), cells develop into sperm (*b*). (*c*) Each sperm possesses a long tail coupled to a head, which contains a nucleus. The tip, or acrosome, contains enzymes to help the sperm cell digest a passageway for fertilization.

Male Gametes Are Delivered by the Penis

The **penis** is an external tube containing two long cylinders of spongy tissue side by side. Below and between them runs a third cylinder of spongy tissue that contains in its center a tube called the urethra, through which both semen (during ejaculation) and urine (during urination) pass (figure 25.5). Why this unusual design? The penis is designed to inflate! The spongy tissues that make up the three cylinders is riddled with small spaces between the cells, and when nerve impulses from the CNS cause the arterioles leading into this tissue to expand, blood collects within these spaces. Like blowing up a balloon, this causes the penis to become erect and rigid. Continued stimulation by the CNS is required for this erection to be maintained.

Erection can be achieved without any physical stimulation of the penis, but physical stimulation is required for semen to be delivered. Stimulation of the penis, as by repeated thrusts into the vagina of a female, leads first to the mobilization of the sperm. In this process, muscles encircling the vas deferens contract, moving the sperm along the vas deferens into the urethra. Eventually, the stimulation leads to the violent contraction of the muscles at the base of the penis. The result is **ejaculation,** the forceful ejection of about 5 milliliters of semen. Semen is a collection of secretions from the prostate and other glands that provides metabolic energy sources for the sperm. Within this small 5-milliliter volume are several hundred million sperm. Because the odds against any one individual sperm cell successfully completing the long journey to the egg and fertilizing it are extraordinarily high, successful fertilization requires a high sperm count. Males with fewer than 20 million sperm per milliliter are generally considered sterile.

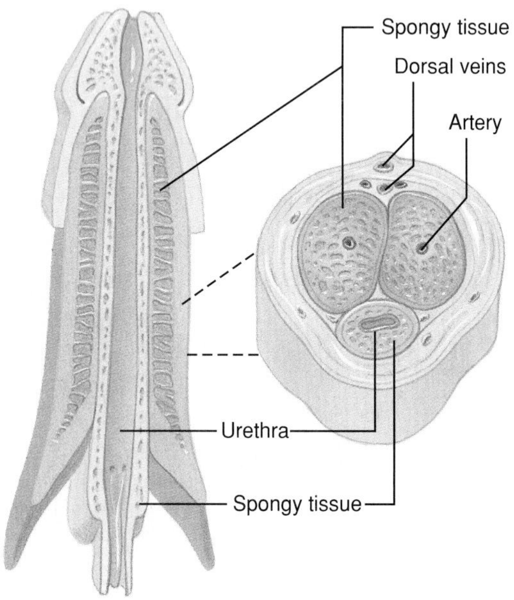

Figure 25.5 Structure of the penis.
(*Left*) Longitudinal section; (*right*) cross section.

Females

In females, eggs develop from cells called **oocytes,** located in the outer layer of compact masses of cells called **ovaries,** located within the abdominal cavity. Unlike males, whose gamete-producing cells are constantly dividing, females have at birth all the oocytes that they will ever produce. At each cycle of egg maturation, called **ovulation,** one of a few of these oocytes initiates development; the others remain in a developmental holding pattern. It has been suggested that this long maintenance period is the reason that developmental abnormalities crop up with increasing frequency in pregnancies of women who are over 35 years old. The oocytes are continually exposed to mutation throughout life, and after 35 years, the odds of a harmful mutation having occurred become high enough to increase significantly the incidence of fetal abnormalities.

Only One Female Gamete Matures Each Month

At birth, a female's ovaries contain some 2 million oocytes, all of which have begun the first meiotic division. At this stage they are called **primary oocytes.** Each primary oocyte is poised to develop further, but it does not continue on with meiosis. Instead, it waits to receive the proper developmental "go" signal, and until a primary oocyte receives this signal, its meiosis is arrested in prophase of the first meiotic division. In fact, very few ever do receive the awaited signal, which turns out to be the hormone FSH we met in the previous chapter.

With the onset of puberty, females mature sexually. At this time the release of FSH initiates the resumption of the first meiotic division in a few oocytes, but a single oocyte soon becomes dominant, the others regressing. Approximately every 28 days after that, another oocyte matures, although the exact timing may vary from month to month. Only about 400 of the approximately 2 million oocytes a woman is born with mature during her lifetime. When they mature, the egg cells are called **ova** (singular, **ovum**), the Latin word for "egg."

Fertilization Occurs As an Egg Journeys to the Uterus

When the ovum is released at ovulation, it is swept by the beating of cilia into one of the **fallopian tubes,** which lead away from the ovary into the **uterus** (figure 25.6). Smooth muscles lining the fallopian tubes contract rhythmically, moving the egg down the tube to the uterus in much the same way that food is moved down through your intestines, pushing it along by squeezing the tube behind it. The journey of the egg through the fallopian tube is a slow one, taking from five to seven days to complete. If the egg is unfertilized, it loses its capacity to develop within a few days. Any egg that arrives at the uterus unfertilized can never become so. For this reason the sperm cannot simply lie in wait within the uterus. To fertilize an egg successfully, a sperm must make its way far up the fallopian tube, a long passage that few survive.

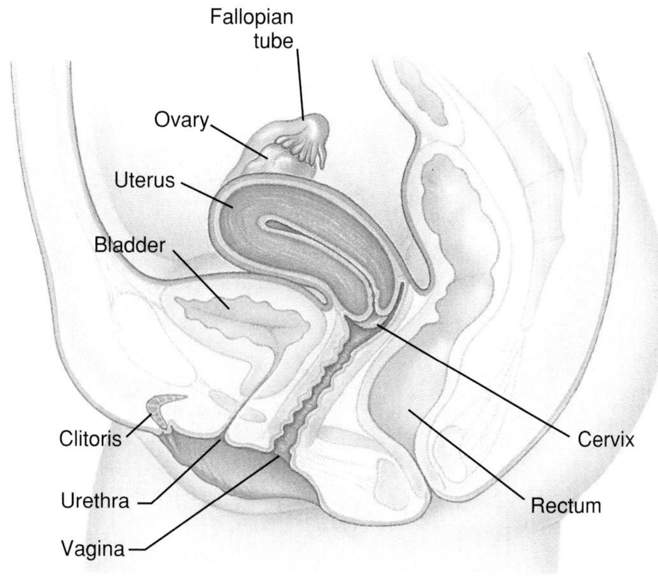

Figure 25.6 The female reproductive system.

The organs of the female reproductive system are specialized to produce gametes and to provide a site for embryo development if the gamete is fertilized.

Sperm are deposited within the vagina, a thin-walled muscular tube about 7 centimeters long that leads to the mouth of the uterus. This opening is bounded by a muscular ring called the **cervix.** The uterus is a hollow, pear-shaped organ about the size of a small fist. Its inner wall, the **endometrium,** has two layers. The outer of these layers is shed during menstruation, while the one beneath it generates the next layer. Sperm entering the uterus swim up to and enter the fallopian tube. They swim upward against the current generated by the tube's contractions, which are carrying the ovum downward toward the uterus.

When a sperm succeeds in fertilizing an egg high in the fallopian tube, the fertilized egg—now an embryo—continues on its journey down the fallopian tube. When it reaches the uterus, the new embryo attaches itself to the endometrial lining and starts the long developmental journey that eventually leads to the birth of a child (figure 25.7).

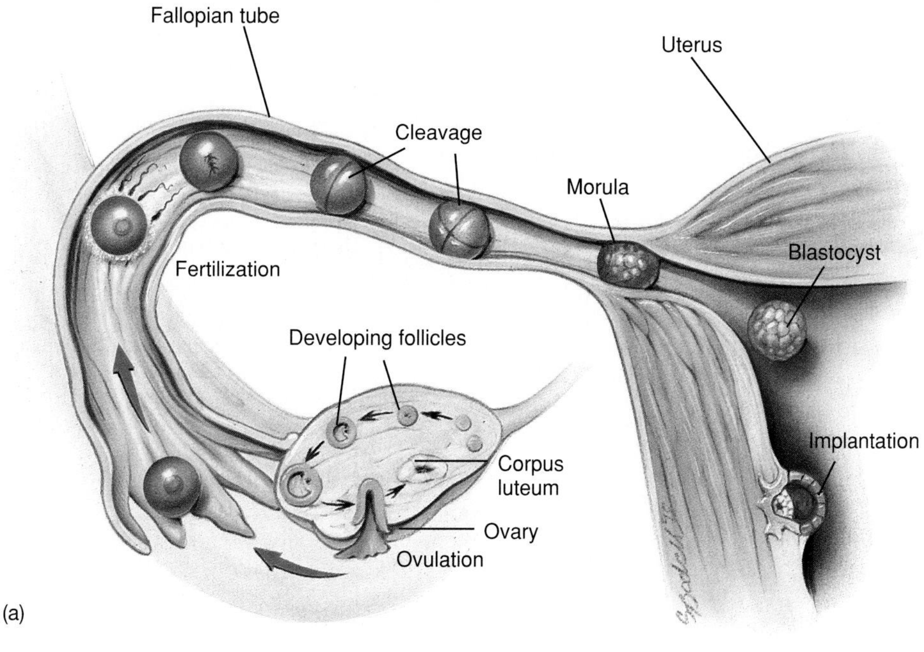

(a)

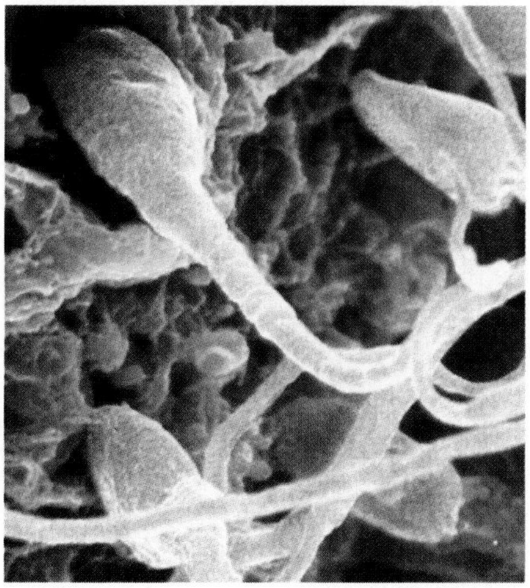

(b)

Figure 25.7 The journey of an ovum.

Produced within a follicle (*a*) and released at ovulation, an ovum is swept up into a fallopian tube and carried down by waves of contraction of the tube walls. Fertilization occurs within the tube, by sperm journeying upward. Several mitotic divisions occur while the fertilized ovum continues its journey down the fallopian tube. The fertilized ovum implants itself within the wall of the uterus, where it continues its development. (*b*) Human sperm fertilizing an ovum. Only the heads and a portion of the long, slender tails of these sperm are shown in this scanning electron micrograph.

25.2 The Female Reproductive Cycle

The female reproductive cycle is composed of two distinct phases, the follicular phase and the luteal phase. These phases are coordinated by a family of hormones.

Sex Hormones

Hormones play many roles in human reproduction. Sexual development, delayed in mammals, is initiated by hormones that coordinate simultaneous sexual development in many kinds of tissues. The production of gametes is another closely orchestrated process, involving a series of carefully timed developmental events. Successful fertilization initiates yet another developmental "program," in which the female body prepares itself for the many changes of pregnancy.

Production of the sex hormones that coordinate all these processes is coordinated by the hypothalamus, which sends releasing hormones to the pituitary, directing it to produce particular sex hormones (figure 25.8). Feedback plays a key role in regulating these activities of the hypothalamus. When target organs receive a pituitary hormone, they begin to produce a hormone of their own, which circulates back to the hypothalamus, shutting down production of the pituitary hormone.

Triggering the Maturation of an Egg

The first, or **follicular,** phase of the reproductive cycle is when the egg develops within the ovary. This development is carefully regulated by hormones. The pituitary, after receiving a chemical signal from the hypothalamus, starts the cycle by secreting **follicle-stimulating hormone (FSH),** which binds to receptors on the surface of cells surrounding the egg (the oocyte and its surrounding mass of tissue is called a **follicle**) and triggers resumption of meiosis. Normally, only a few eggs at any one time have developed far enough to respond immediately to FSH. FSH levels then fall. Because FSH levels are reduced before other eggs ripen to maturity, only a few eggs ripen in every cycle.

The fall of FSH levels is achieved by a feedback command to the pituitary. FSH does not itself carry out this feedback—instead, it sends another hormone as a messenger. FSH not only triggers final egg development, it also causes the ovary to start producing the female sex hormone **estrogen.** Rising levels of estrogen in the bloodstream feed back to the hypothalamus, which responds to the rising estrogen by commanding the pituitary to cut off the further production of FSH. This "shuts the door" on further FSH-induced egg development. The rise in estrogen levels signals the completion of the follicular phase of the reproductive cycle.

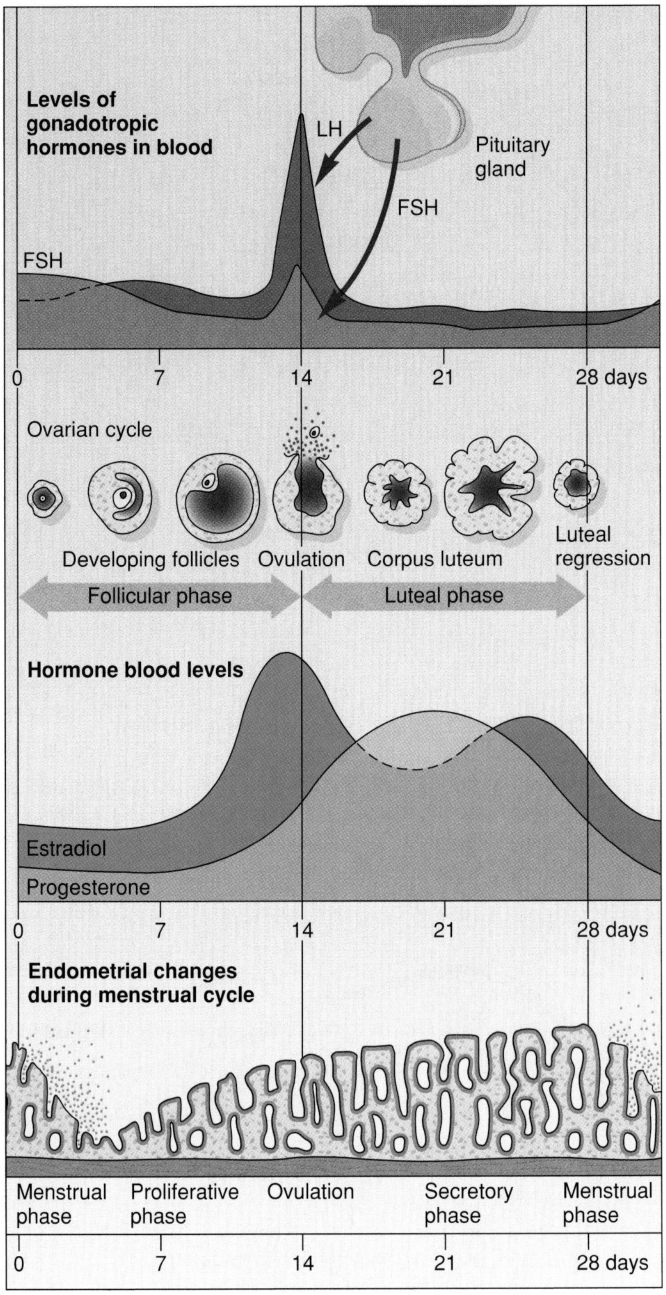

Figure 25.8 The human menstrual cycle.
The growth and thickening of the uterine (endometrial) lining is governed by levels of the hormone progesterone; menstruation, the sloughing off of this blood-rich tissue, is initiated by lower levels of progesterone.

Preparing the Body for Fertilization

The second, or **luteal,** phase of the cycle follows smoothly from the first. The hypothalamus responds to estrogen not only by shutting down the pituitary's FSH production but also by causing the pituitary to begin secreting a second hormone, called **luteinizing hormone (LH).** LH is the hormone that causes ovulation, sending the now-mature egg on its journey towards fertilization (figure 25.9). LH is carried in the bloodstream to the developing follicle, where it inhibits further estrogen production and causes the wall of the follicle to burst. The egg within the follicle is released into one of the fallopian tubes extending from the ovary to the uterus.

After the egg's release and departure, the ruptured follicle repairs itself, filling in and becoming yellowish. In this condition it is called the **corpus luteum,** which is simply the Latin phrase for "yellow body." The corpus luteum soon begins to secrete a hormone, **progesterone,** which inhibits FSH (a backup for estrogen in preventing further ovulations). Progesterone is the body's signal to prepare the uterus for fertilization. If fertilization occurs, the corpus luteum continues to produce progesterone for several weeks. The rising levels of progesterone initiate the many physiological changes associated with pregnancy. Among them are thickening of the walls of the uterus in preparation for the implantation of the developing embryo.

However, if fertilization does *not* occur soon after ovulation, production of progesterone slows and eventually ceases, marking the end of the luteal phase. The decreasing levels of progesterone cause the thickened layer of blood-rich tissue to be sloughed off, a process that results in the bleeding associated with menstruation. **Menstruation,** or "having a period," usually occurs about midway between successive ovulations, although its timing varies widely for individual females.

At the end of the luteal phase, neither estrogen nor progesterone is being produced. In their absence, the pituitary can again initiate production of FSH, thus starting another reproductive cycle. Each cycle begins immediately after the preceding one ends. A cycle usually occurs every 28 days, or a little more frequently than once a month, although this varies in individual cases. The Latin word for month is *mens,* which is why the reproductive cycle is called the **menstrual cycle,** or monthly cycle.

Two other hormones, both secreted by the pituitary, are important in the female reproductive system. For the first couple of days after childbirth, the mammary glands produce a fluid called colostrum, which contains protein and lactose but little fat. Then milk production is stimulated by the hormone **prolactin,** usually by the third day after delivery. When the infant suckles at the breast, the hormone **oxytocin** is released, initiating milk release. Earlier, in combination with chemicals released from the uterus, oxytocin initiates labor and delivery.

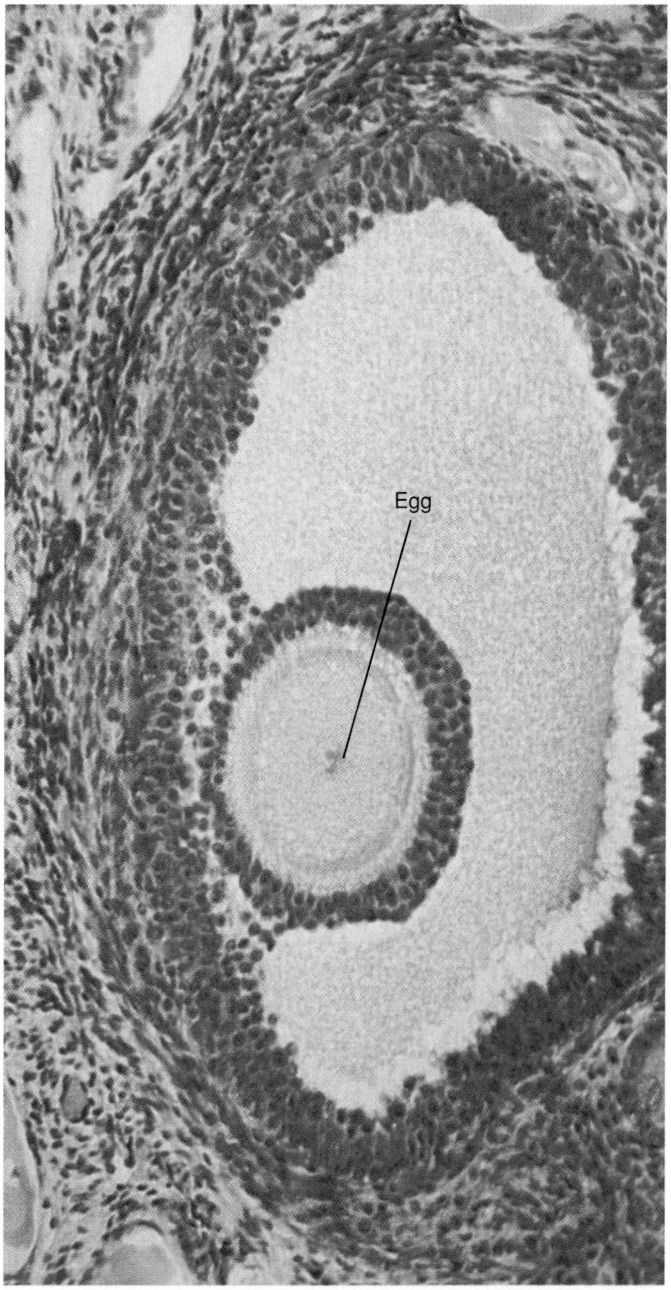

Figure 25.9 A mature egg within an ovarian follicle.
In each menstrual cycle, a few follicles are stimulated to grow under the influence of FSH, but only one achieves full maturity.

25.3 Embryonic Development

Development in humans is a complex, dynamic process, a symphony of cell movement and change that starts with the formation of a single diploid cell from sperm and egg. From this single cell will arise 100 trillion others as the individual the cell is destined to become develops and grows to adulthood. The developmental journey is begun with a series of rapid cell divisions that creates a ball of cells. Some of these cells then migrate to the inside to form a more or less spherical structure that possesses the three primary tissue layers. The development of the specific tissues of the human body follows. Then organs are formed. Only after some three months, when this pattern of development is complete, does growth begin in earnest, a process that takes another six months. A human baby is 266 days old when it is born. We now review what happens during this long developmental period. The most crucial events of development occur very early in pregnancy, many of them before a woman knows she is pregnant.

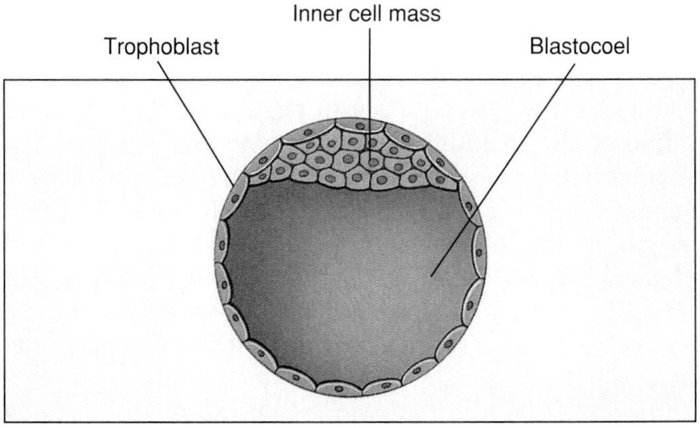

Figure 25.10 A human blastocyst.
A human blastocyst is composed of a sphere of cells, the trophoblast, surrounding a cavity, the blastocoel, and an inner cell mass. The inner cell mass will become the developing embryo. The trophoblast contains enzymes that dissolve an opening into the lining of the uterus, in which the blastocyst implants.

Cleavage: Setting the Stage for Development

The first major event in human embryonic development is the rapid division of the zygote into a larger and larger number of smaller and smaller cells, becoming first 2 cells, then 4, then 8, and so on. The first of these divisions occurs about 30 hours after union of the egg and the sperm, and the second, 30 hours later. During this period of division, called **cleavage,** the overall size of the embryo does not increase. The resulting tightly packed mass of about 32 cells is called a **morula,** and each individual cell in the morula is referred to as a **blastomere.**

During this period, the embryo continues its journey down the mother's fallopian tube. On about the sixth day, the embryo reaches the uterus, attaches to the uterine lining, and penetrates into the tissue of the lining. The embryo now begins to grow rapidly, initiating the formation of the membranes that will later surround, protect, and nourish it. One of these membranes, the **amnion,** will enclose the developing embryo, whereas another, the **chorion,** will interact with uterine tissue to form the **placenta,** which will nourish the growing embryo.

The cells of the morula continue to divide without an overall increase in size, each cell secreting a fluid into the center of the cell mass. Eventually, a hollow ball of 500 to 2,000 cells is formed, surrounding a fluid-filled cavity called the **blastocoel.** Within the ball is an inner cell mass concentrated at one pole that goes on to form the developing embryo. This is the **blastocyst** (figure 25.10). The outer sphere of cells is called the trophoblast. The trophoblast develops into the membrane surrounding the embryo (the chorion) and a complex series of membranes known as the placenta, which connects the developing embryo to the blood supply of the mother.

Gastrulation: The Onset of Developmental Change

Ten to 11 days after fertilization, certain groups of cells move inward from the surface of the cell mass in a carefully orchestrated migration called **gastrulation** (figure 25.11). First, the lower cell layer of the blastula cell mass differentiates into **endoderm,** one of the three primary tissues, and the upper layer into **ectoderm.** Just after this differentiation, much of the **mesoderm** and endoderm arises by the invagination of cells that move from the upper layer of the cell mass *inward,* along the edges of a furrow that appears at the midline of the embryo. The site of this invagination appears as a slit on the surface of the embryo, called the **primitive streak** (figure 25.12).

How can cells move within a cell mass? Apparently the migrating cells creep over the stationary ones by a series of microfilament contractions. The migrating cells move as a single mass because they adhere to one another. How do cells "know" to which other cell to adhere? Within an adhering cell, genes have been expressed that cause the synthesis of molecules on the cell surface that adhere to similar molecules on the surfaces of the other adhering cells.

During gastrulation, about half of the cells of the blastocyst cell mass move into the interior of the human embryo. This movement largely determines the future development of the embryo. By the end of gastrulation, distribution of cells into the three primary tissue types has been completed. The ectoderm is destined to form the epidermis and neural tissue. The mesoderm is destined to form the connective tissue, muscle, and vascular elements. The endoderm forms the lining of the gut and its derivative organs.

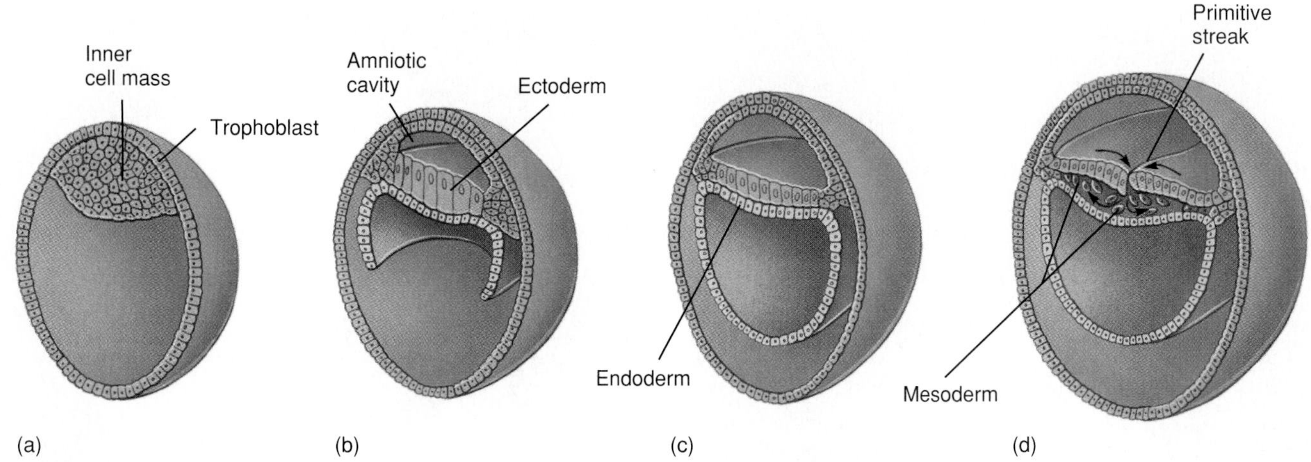

Figure 25.11 Gastrulation.
The amniotic cavity forms within the inner cell mass (*a*), and in its base, layers of ectoderm and endoderm differentiate (*b* and *c*). A primitive streak develops, through which cells destined to become mesoderm migrate into the interior (*d*).

Neurulation: Determination of Body Architecture

In the third week of embryonic development, the three primary cell types begin their development into the tissues and organs of the body. As in all vertebrates, this begins with the formation of two characteristic vertebrate features, the notochord and the hollow dorsal nerve cord. This stage in development is called **neurulation.**

The first of these two structures to form is the **notochord,** a flexible rod. Soon after gastrulation is complete, it forms from mesoderm tissue along the midline of the embryo, below its dorsal surface.

After the notochord has been formed, the dorsal nerve cord forms from the region of the ectoderm that is located above the notochord and later differentiates into the spinal chord and brain. Formation of the dorsal nerve cord starts when a layer of ectodermal cells situated above the notochord invaginates inward, forming a long groove called the **neural groove** along the long axis of the embryo. Then the edges of this groove move toward each other and fuse, creating a long, hollow tube that runs beneath the surface of the embryo's back.

While the dorsal nerve cord is forming from ectoderm, the rest of the basic architecture of the human body is being rapidly determined by changes in the mesoderm. On either side of the developing notochord, segmented blocks of tissue form. Ultimately, these blocks, or **somites,** give rise to the muscles, vertebrae, and connective tissues. As development continues, more and more somites are formed. Within another strip of mesoderm that runs alongside the somites, many of the significant glands of the body, including the kidneys, adrenal glands, and gonads, develop. The remainder of the mesoderm layer moves out and around the inner endoderm layer of cells and eventually surrounds it entirely. As a result, the mesoderm forms a hollow tube within the ectoderm. The space within this tube is called the **coelom.** It contains the endoderm layers that ultimately form the lining of the stomach and gut.

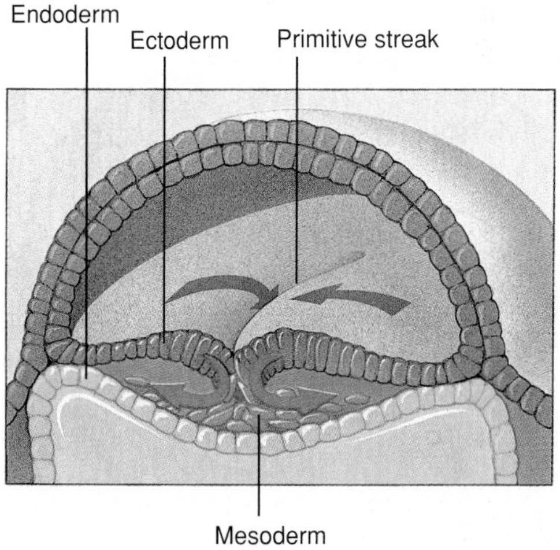

Figure 25.12 Development of a primitive streak.
As the layers of ectoderm and endoderm develop, an indentation, or primitive streak, develops in the ectoderm. This signals the migration of cells to the area between the ectoderm and endoderm, where the mesoderm will develop.

By the end of the third week, over a dozen somites are evident, and the blood vessels and gut have begun to develop. At this point the embryo is about 2 millimeters (less than a tenth of an inch) long.

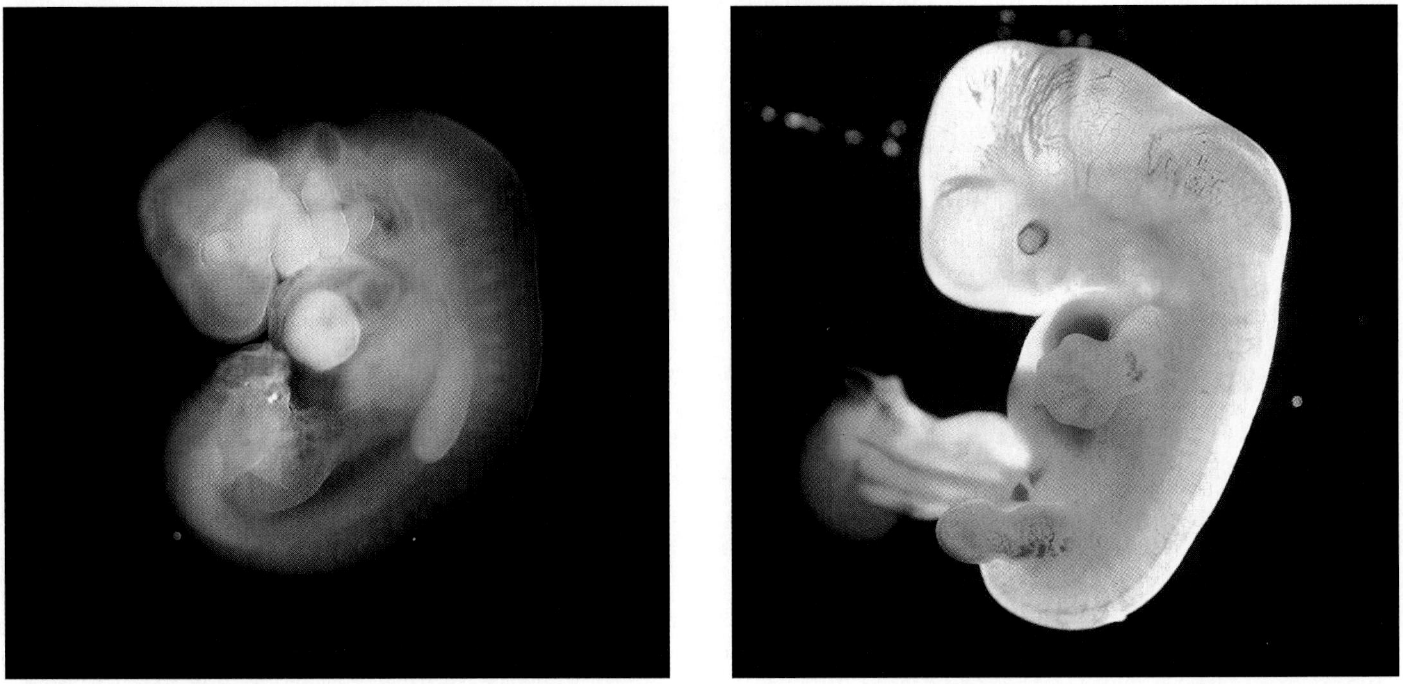

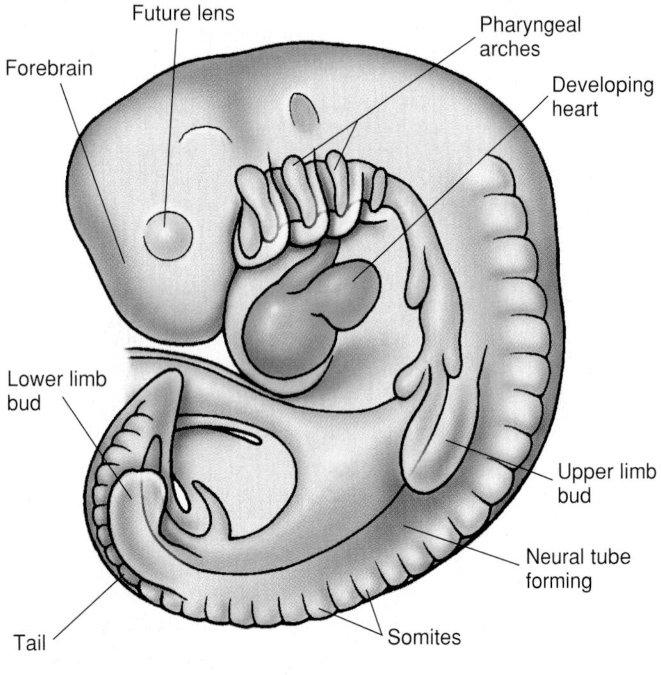

(a)

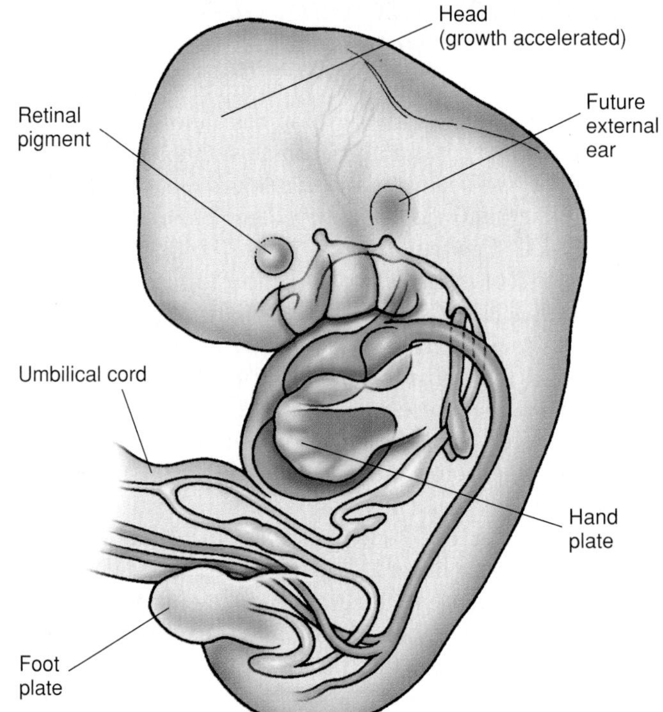

(b)

Figure 25.13 The developing human.
(*a*) Four weeks; (*b*) seven weeks; (*c*) three months; and (*d*) four months.

25.4 Fetal Development

In the fourth week of pregnancy, the body organs begin to form, a process called **organogenesis** (figure 25.13*a*). The eyes form, and the heart begins to pulsate, develops four chambers, and begins a rhythmic beating. At 70 beats per minute, the little heart is destined to beat more than 2.5 billion times during a lifetime of about 70 years. More than 30 pairs of somites are visible by the end of the fourth week, and the arm and leg buds have begun to form. The embryo more than doubles in length during this week, reaching about 5 millimeters.

By the end of the fourth week of pregnancy, the developmental scenario is far advanced, although most women are not yet aware that they are pregnant at this

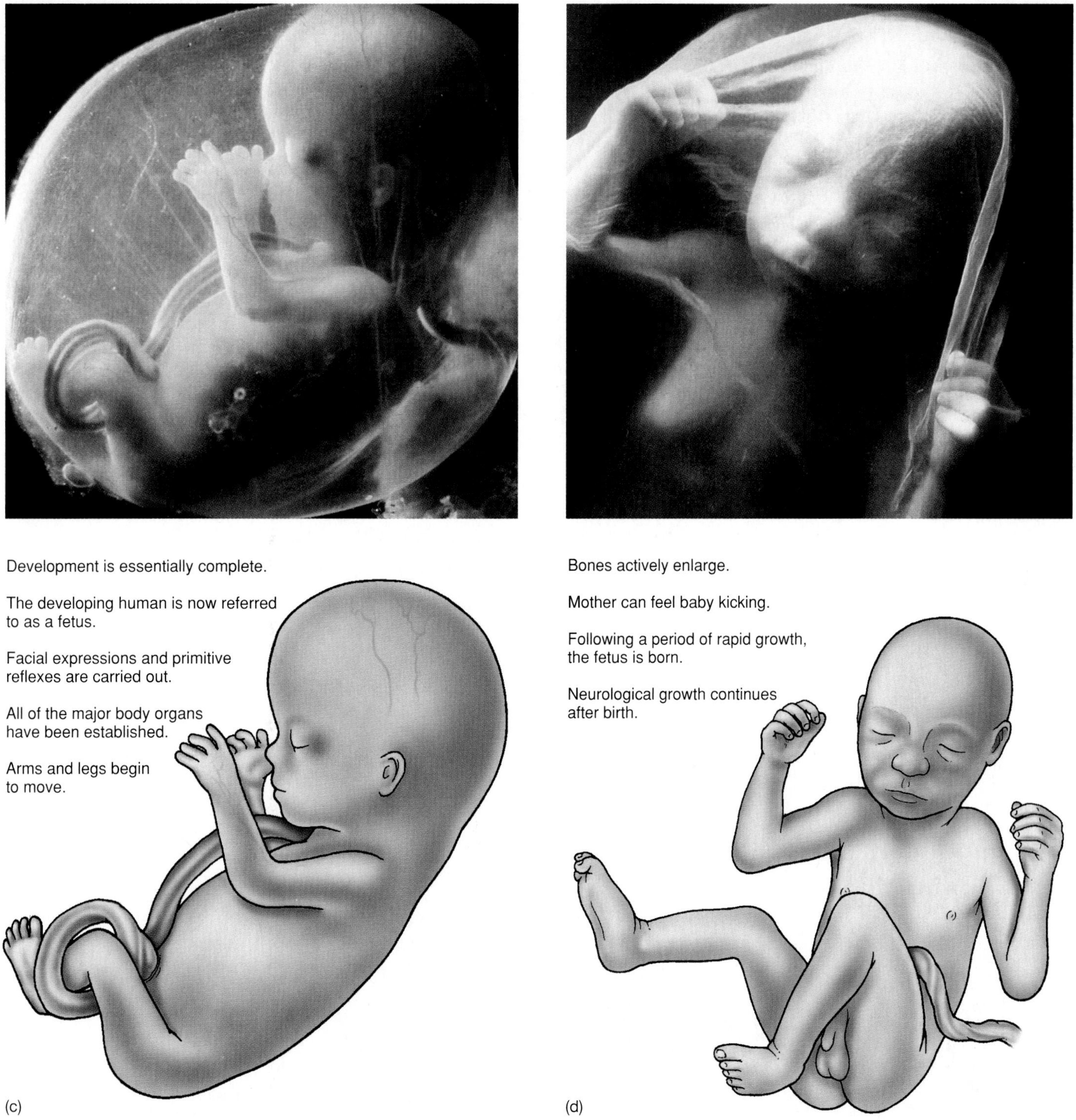
Development is essentially complete.
The developing human is now referred to as a fetus.
Facial expressions and primitive reflexes are carried out.
All of the major body organs have been established.
Arms and legs begin to move.
(c)
Bones actively enlarge.
Mother can feel baby kicking.
Following a period of rapid growth, the fetus is born.
Neurological growth continues after birth.
(d)

Figure 25.13 (continued).

stage. Early pregnancy is a very crucial time in development because the proper course of events can be interrupted easily. Alcohol use by pregnant women during the first months of pregnancy is one of the leading causes of birth defects, producing **fetal alcohol syndrome,** in which the baby is born with a deformed face and often severe mental retardation. One in 250 newborns in the United States is affected with fetal alcohol syndrome. Obviously, women likely to become pregnant should avoid alcohol. Also during the first months of pregnancy, contraction of rubella (German measles) by the mother can upset organogenesis in the developing embryo. Most spontaneous abortions (miscarriages) occur during this period.

The Second Month: Morphogenesis

During the second month of pregnancy, **morphogenesis** takes place (figure 25.13*b*). The miniature limbs of the embryo assume their adult shapes. The arms, legs, knees, elbows, fingers, and toes can all be seen—as well as a short, bony tail. The bones of the embryonic tail, an evolutionary reminder of our past, later fuse to form the coccyx. Within the body cavity, the major internal organs are evident, including the liver and pancreas. By the end of the second month, the embryo has grown to about 25 millimeters in length—it is 1 inch long. It weighs perhaps a gram and is beginning to look distinctly human.

The Third Month: Completion of Development

Development of the embryo is essentially complete. From this point on, the developing human is referred to as a **fetus** rather than an embryo. What remains is essentially growth. The nervous system and sense organs develop during the third month. The embryo begins to show facial expressions and carries out primitive reflexes such as the startle reflex and sucking. By the end of the third month, all of the major organs of the body have been established and the arms and legs begin to move (figure 25.13*c*).

The Second Trimester: The Fetus Grows in Earnest

The second trimester is a time of growth. In the fourth and fifth months of pregnancy, the fetus grows to about 175 millimeters in length (almost 7 inches long), with a body weight of about 225 grams. Bone formation occurs actively during the fourth month. During the fifth month, the head and body become covered with fine hair. This downy body hair, called **lanugo,** is another evolutionary relic and is lost later in development. By the end of the fourth month, the mother can feel the baby kicking; by the end of the fifth month, she can hear its rapid heartbeat with a stethoscope.

In the sixth month, growth accelerates. By the end of the sixth month, the baby is over 0.3 meters (1 foot) long and weighs 0.6 kilograms (about 1.5 pounds)—and most of its prebirth growth is still to come. At this stage, the fetus cannot yet survive outside the uterus without special medical intervention.

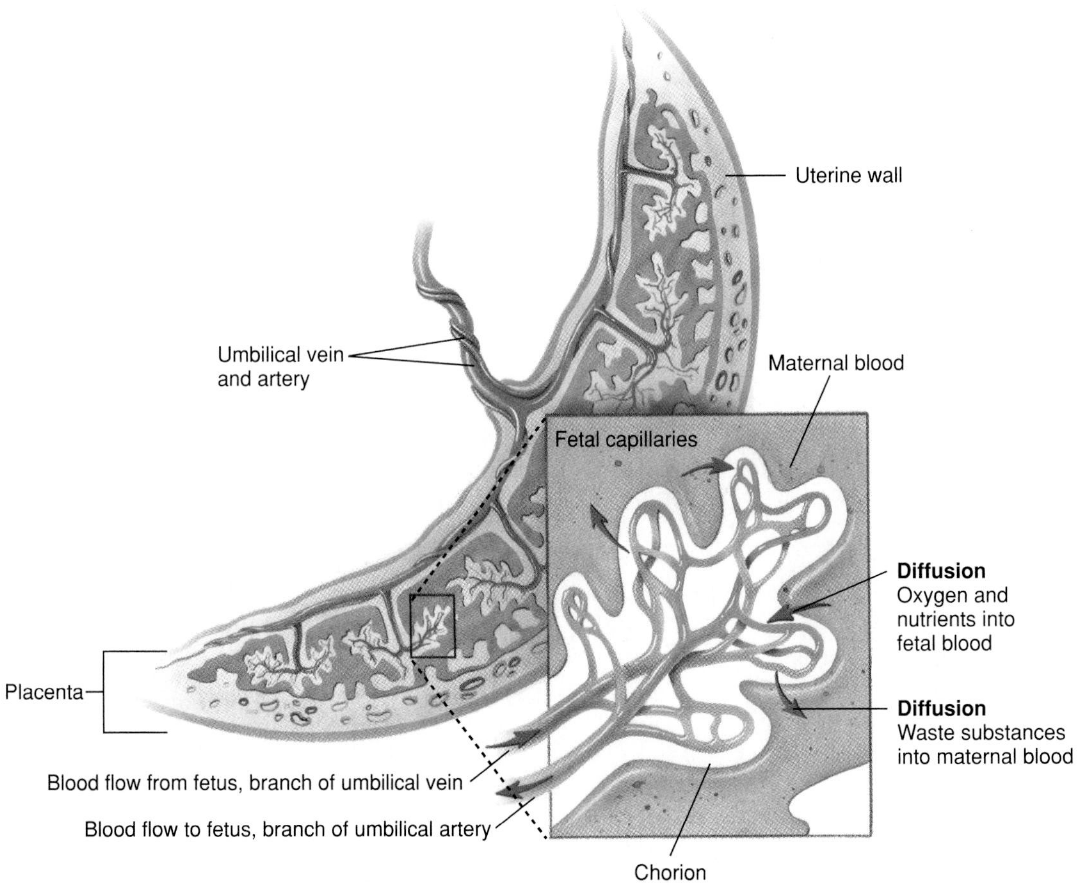

Figure 25.14 Structure of the placenta.

Oxygen and nutrients enter the fetal blood from the maternal blood by diffusion. Waste substances enter the maternal blood from the fetal blood, also by diffusion.

The Third Trimester: The Pace of Growth Accelerates

The third trimester is a period of rapid growth. In the seventh, eighth, and ninth months of pregnancy the weight of the fetus more than doubles. This increase in bulk is not the only kind of growth that occurs. Most of the major nerve tracts are formed within the brain during this period, as are new brain cells.

All of this growth is fueled by nutrients provided by the mother's bloodstream, passing into the fetal blood supply within the placenta (figure 25.14). The undernourishment of the fetus by a malnourished mother can adversely affect this growth and result in severe retardation of the infant. Retardation resulting from fetal malnourishment is a severe problem in many underdeveloped countries, where poverty and hunger are common.

By the end of the third trimester, the neurological growth of the fetus is far from complete and, in fact, continues long after birth. But by this time the fetus is able to exist on its own. Why doesn't the fetus continue to develop within the uterus until its neurological development is complete? What's the rush to get out and be born? Because physical growth is continuing as well, and the fetus is about as large as it can get and still be delivered through the pelvis without damage to mother or child. As any woman who has had a baby can testify, it is a tight fit. Birth takes place as soon as the probability of survival is high. For better or worse, the infant is then a person.

Postnatal Development

Growth continues rapidly after birth (figure 25.15). Babies typically double their birth weight within a few months. Different organs grow at different rates, however. The reason that adult body proportions are different from infant ones is that different parts of the body grow or cease growing at different times. At birth, the developing nervous system is generating new nerve cells at an average rate of more than 250,000 per minute. Then, about six months after birth, this astonishing production of new neurons ceases permanently. Because both jaw and skull continue to grow at the same rate, the proportions of the head do not change after birth. That is why a young human fetus seems so incredibly adultlike.

Figure 25.15 A mother nursing her infant.
Infants grow very rapidly during the first year after birth. During this time, the nursing mother needs adequate nutrition to enable her to support her growing baby.

25.5 Birth Control

Not all couples want to initiate a pregnancy every time they have sex, yet sexual intercourse may be a necessary and important part of their emotional lives together. Some religious groups believe that sexual intercourse has only a reproductive function and thus should be limited to situations in which pregnancy is desired—among married couples wishing to have children. Most couples, however, do not limit sexual relations to procreation, and among them unwanted pregnancy presents a real problem. The solution to this dilemma is to find a way to avoid reproduction without avoiding sexual intercourse, an approach that is commonly called **birth control** or contraception.

Several different birth-control methods are currently available (figure 25.16). These methods differ from one another in their effectiveness and in their acceptability to different couples (table 25.1).

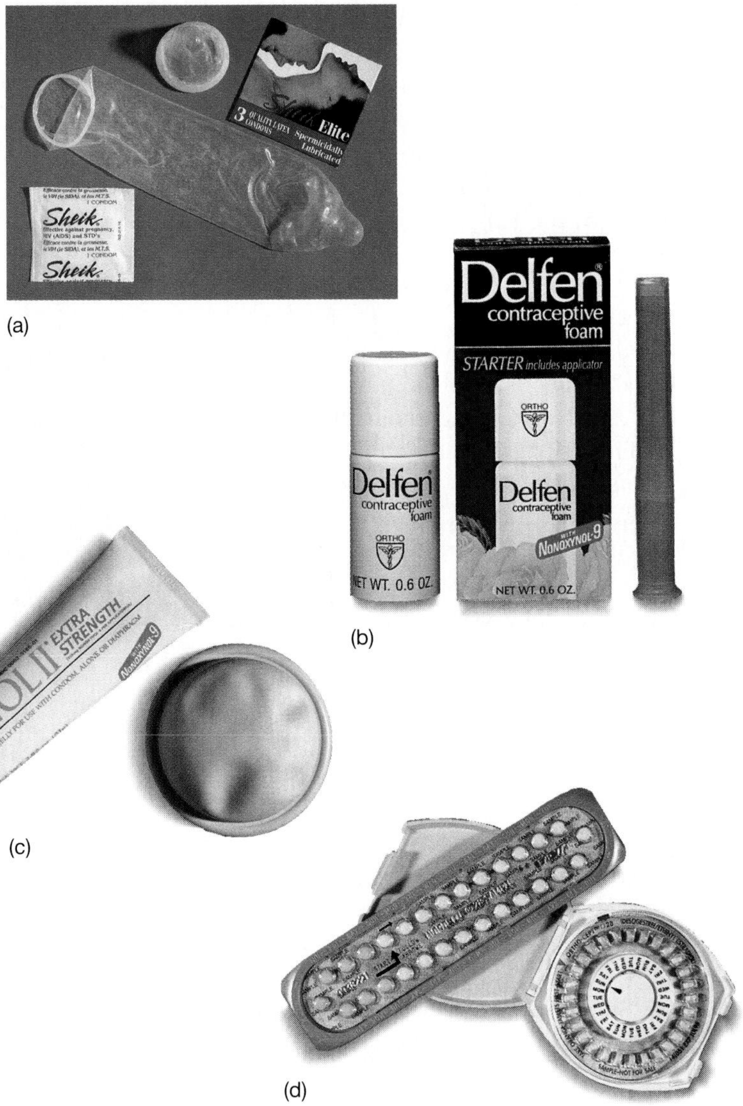

Figure 25.16 Four common birth-control methods. (*a*) Condom; (*b*) foam; (*c*) diaphragm and spermicidal jelly; and (*d*) oral contraceptives.

Abstinence

The simplest and most reliable way to avoid pregnancy is not to have sex at all. Of all methods of birth control, this is the most certain—and the most limiting, since it denies a couple the emotional support of a sexual relationship.

A variant of this approach is to avoid sex only on the two days preceding and following ovulation, because this is the period during which successful fertilization is likely to occur. The rest of the sexual cycle is relatively "safe" for intercourse. This approach, called the **rhythm method,** is satisfactory in principle but difficult in application because ovulation is not easy to predict and may occur unexpectedly. The effectiveness of the rhythm method is low; the failure rate is estimated to be about 40% (that is, 40 pregnancies per 100 women practicing the rhythm method per year). This is to be compared with a pregnancy rate of 90% for totally unprotected sex.

Another variant of this approach is to have only incomplete sex—withdraw the penis before ejaculation, a procedure known as **coitus interruptus.** This requires considerable willpower, often destroys the emotional bonding of intercourse, and is not as reliable as it might seem. Prematurely released sperm can be secreted by the penis within its lubricating fluid, and a second sexual act may transfer sperm ejaculated earlier. The failure rate of this approach is estimated at 25%.

Prevention of Egg Maturation

Since about 1960, a widespread form of birth control in the United States has been the daily ingestion of hormones, or **birth-control pills.** These pills contain estrogen and progesterone, either taken together in the same pill or sequentially taken in separate pills. These hormones shut down production of the pituitary hormones FSH and LH, fooling the body into acting as if ovulation had already occurred, when in fact it has not. The ovarian follicles do not ripen in the absence of FSH, and ovulation does not occur in the absence of LH. Taken correctly (at the same time each day, so that the "swing" between pills is not too great), birth-control pills provide a very effective means of birth control, with failure rates of less than 10%. There is no conclusive evidence of any serious side effects for the great majority of women.

Surgical implants of capsules under the skin that slowly release birth-control hormones, over periods ranging from six months to up to five years, are even more effective, with failure rates of less than 1%.

Table 25.1 Nonsurgical Methods of Birth Control

Device	Action	Failure Rate*	Advantages	Disadvantages
Intrauterine device (IUD)	Small plastic or metal device placed in the uterus; prevents implantation; some contain copper; others release hormones	1–5	Convenient; highly effective; infrequent replacement	Can cause excess menstrual bleeding and pain; risk of perforation, infection, expulsion, pelvic inflammatory disease, and infertility; not recommended for those who eventually intend to conceive or are not monogamous; dangerous during pregnancy
Oral contraceptive	Hormones (progesterone analog alone or in combination with other hormones) primarily prevent ovulation	1–5 depending on type	Convenient; highly effective; provide significant noncontraceptive health benefits, such as protection against ovarian and endometrial cancers	Must be taken regularly; possible minor side effects, which new formulations have reduced; not for women with cardiovascular risks (mostly smokers over age 35)
Condom	Thin sheath for penis collects semen; "female condoms" sheath vaginal walls	3–15	Easy to use; effective; inexpensive; protects against some sexually transmitted diseases	Requires male cooperation; may diminish spontaneity; may deteriorate on the shelf
Diaphragm	Soft rubber cup covers entrance to uterus, prevents sperm from reaching egg, and holds spermicide	4–25	No dangerous side effects; reliable if used properly; provides some protection against sexually transmitted diseases and cervical cancer	Requires careful fitting; some inconvenience associated with insertion and removal; may be dislodged during intercourse
Cervical cap	Miniature diaphragm covers cervix closely, prevents sperm from reaching egg, and holds spermicide	Probably similar to that of a diaphragm	No dangerous side effects; fairly effective; can remain in place longer than diaphragm	Problems with fitting and insertion; comes in limited number of sizes
Foams, creams, jellies, vaginal suppositories	Chemical spermicides inserted in vagina before intercourse prevent sperm from entering uterus	10–25	Can be used by anyone who is not allergic; protect against some sexually transmitted diseases; no known side effects	Relatively unreliable; sometimes messy; must be used five to ten minutes before each act of intercourse
Implant (Norplant)	Capsules surgically implanted under skin slowly release hormone that blocks ovulation	0.3	Very safe, convenient, and effective; very long lasting (five years); may have nonreproductive health benefits like those of oral contraceptives	Irregular or absent periods; minor surgical procedure needed for insertion and removal; some scarring may occur
Injectable contraceptive (Depo-Provera)	Injection every three months of a hormone that is slowly released and prevents ovulation	1	Convenient and highly effective; no serious side effects other than occasional heavy menstrual bleeding	Animal studies suggest it may cause cancer, though new studies in humans are mostly encouraging; occasional heavy menstrual bleeding

*Failure rate is expressed as pregnancies per 100 actual users per year.
Source: Data from the American College of Obstetricians and Gynecologists.

Prevention of Embryo Implantation

The insertion of a coil or other irregularly shaped object into the uterus is an effective means of birth control, since the irritation in the uterus prevents the implantation of the descending embryo within the uterine wall. Such **intrauterine devices (IUDs)** are very effective, with a failure rate of less than 4%. Their high degree of effectiveness, like surgically implanted hormones, undoubtedly reflects their being "no brainers"—once they are inserted, they can be forgotten. The great disadvantage of this method is that many women attempting to use IUDs cannot; the device causes them cramps, pain, and sometimes bleeding.

A controversial birth-control pill developed in France called **RU486,** not currently readily available in the United States, can be taken within the first few months of pregnancy. RU486 is a progesterone antagonist that blocks the effect of progesterone, which is needed to maintain the endometrium. The endometrium regresses and menstruation ensues, bringing the implanted embryo with it. Because it can be considered an "abortant," many women find RU486 objectionable.

Sperm Blockage

If sperm is not delivered to the uterus, fertilization cannot occur. One way to prevent the delivery of sperm is to encase the penis within a thin rubber bag, or **condom.** In principle, this method is easy to apply and foolproof, but in practice, it proves to be less effective than you might expect, with a failure rate of up to 15%. Many males do not favor the use of condoms because they tend to decrease their sensory pleasure. Nevertheless, condoms are the most commonly employed form of birth control in the United States. Condoms are also used widely as a means of so-called safe sex to avoid contracting AIDS or other sexually transmitted diseases (figure 25.17).

A second way to prevent the entry of sperm is to place a cover over the cervix. The most commonly employed type of cover is a rubber dome called a **diaphragm,** which is inserted immediately before intercourse. Because the dimensions of individual cervices vary, diaphragms must be fitted by a physician. Failure rates average up to 20% for diaphragms, perhaps because of the propensity to insert them carelessly or incorrectly.

Sperm Destruction

A third general approach to birth control is to remove or destroy the sperm after ejaculation. Washing out the vagina immediately after intercourse is difficult to do well, as it involves a rapid dash to the bathroom immediately after ejaculation and a very thorough washing. Not surprisingly, this method's failure rate is estimated at over 40%. Sperm can be destroyed within the vagina with **spermicidal jellies, sponges,** and **foams.** These require application immediately before intercourse. The failure rate varies widely, from 3 to 22%.

Surgical Intervention

A completely effective means of birth control is the surgical removal of a portion of the tube through which the gametes are delivered to the reproductive organs. In males, such an operation is called a **vasectomy.** It involves the removal of a portion of the vas deferens, the tube through which sperm travel to the penis. The operation is simple and can be carried out in a physician's office. In females, the comparable operation involves the removal of a section of each of the two fallopian tubes through which the egg travels to the uterus. Since these tubes are located within the abdomen, the operation, called a **tubal ligation,** is more difficult than a vasectomy.

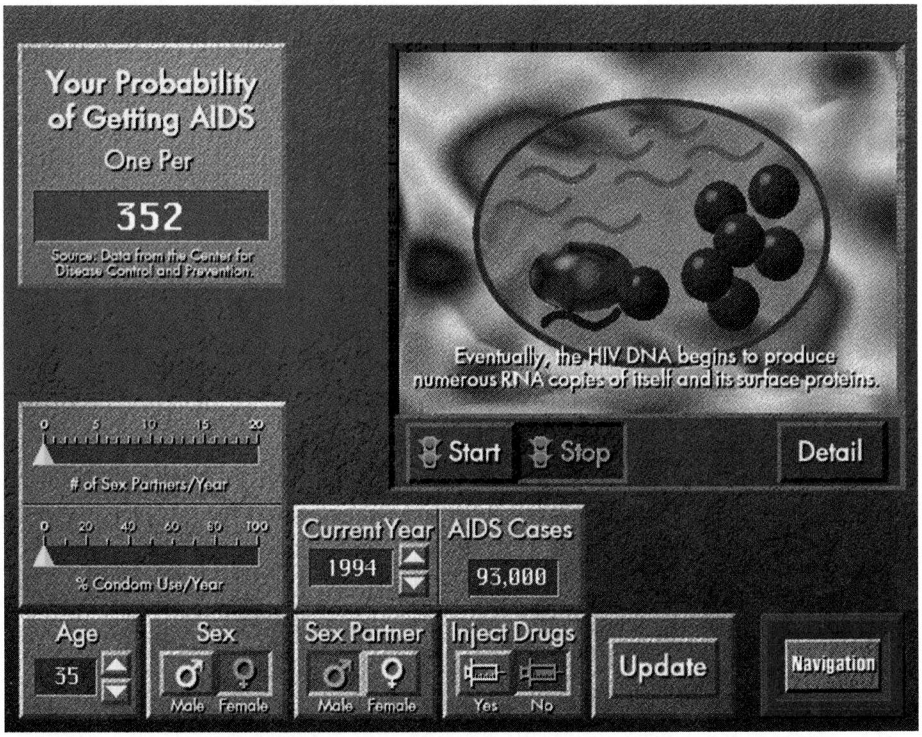

Figure 25.17 Using condoms reduces the risk of AIDS.
This screen capture from an interactive CD-ROM exploration allows you to explore the risk factors associated with AIDS by varying key aspects of behavior, such as using or failing to use condoms when having sex, postponing sex until marriage, and other personal choices that dramatically influence the probability of contracting HIV. (*Explorations in Human Biology*, Module 13, "AIDS")

CHAPTER 25

HIGHLIGHTS

Key Terms | **Key Concepts**

25.1 The Human Reproductive System

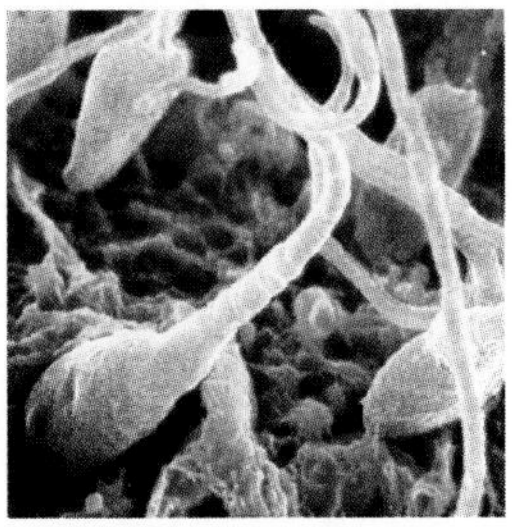

Key Terms

sperm 538
oocytes 540
fallopian tubes 540
endometrium 541

Key Concepts

- Male gametes (sperm) are produced in the testes in very large numbers.
- At least 20 million sperm must be delivered during intercourse for fertilization to be likely.
- Fertilization occurs in the fallopian tubes within a few days after egg release.
- When the developing embryo reaches the uterus, it embeds within the endometrium.

25.2 The Female Reproductive Cycle

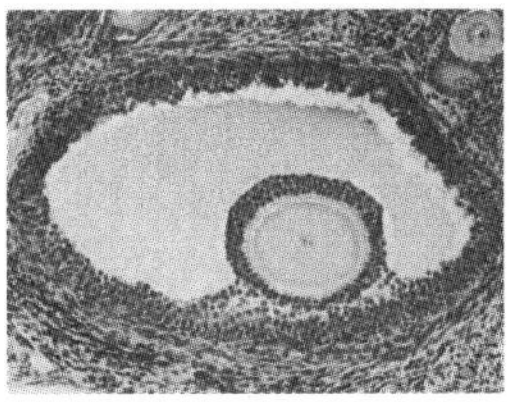

Key Terms

follicular phase 542
follicle 542
luteal phase 543
menstruation 543

Key Concepts

- The hypothalamus regulates the reproductive cycle with a battery of hormones.
- A reproductive cycle takes about 28 days and ends with expulsion of the endometrium.

25.3 Embryonic Development

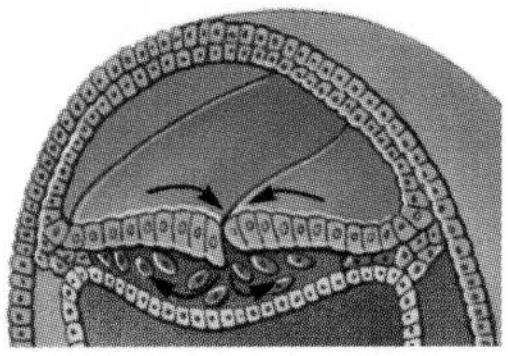

Key Terms

cleavage 544
blastocyst 544
gastrulation 544
neurulation 545

Key Concepts

- The fertilized egg cleaves into a ball of cells called a blastomere.
- Cell movements during gastrulation form a hollow embryo with three primary tissues.
- By three weeks after fertilization, the body has a neural cord and somites.

25.4 Fetal Development

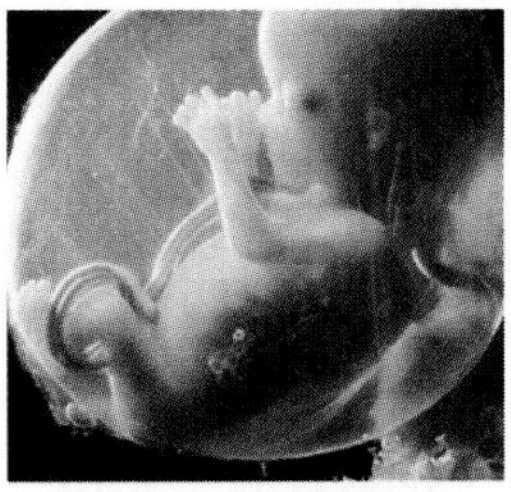

Key Terms

organogenesis 546
fetal alcohol syndrome 547
morphogenesis 548

Key Concepts

- The major organs of the body form in the fourth week.
- The limbs of the body assume their adult shape in the second month.
- Development is essentially complete eight weeks after fertilization.

25.5 Birth Control

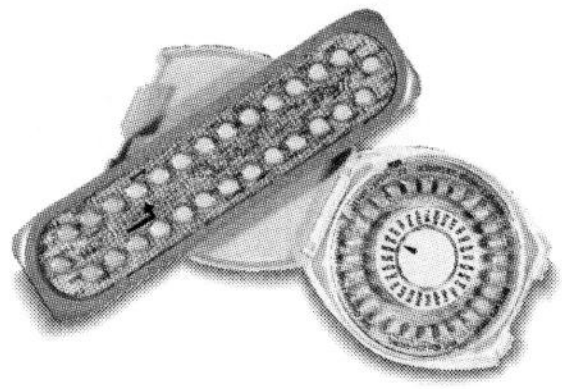

Key Terms

rhythm method 550
IUD 552
RU486 552
condom 552

Key Concepts

- The most effective nonsurgical methods of birth control involve using hormones or other chemicals to prevent egg maturation or embryo implantation.

25

CONCEPT REVIEW

1. Select the incorrect statement about male gametes.
   a. Sperm begin developing before puberty.
   b. Sperm do not develop successfully at 37°C.
   c. Sperm are made in the seminiferous tubules.
   d. Sperm are capable of movement.

2. The spongy tissue in the penis permits
   a. sperm cell production.
   b. sperm cell storage.
   c. erection.
   d. urination.

3. The most common site of fertilization is the
   a. ovary.
   b. vagina.
   c. fallopian tube.
   d. oocyte.

4. Rising levels of estrogen _________ the secretion of FSH.
   a. increase
   b. decrease

5. Ovulation is triggered by the increase of
   a. estrogen.
   b. progesterone.
   c. FSH.
   d. LH.

6. After ovulation, the _________ begin(s) to secrete progesterone to prepare the uterus for fertilization.
   a. ovaries
   b. corpus luteum
   c. pituitary
   d. hypothalamus

7. During the period of cleavage, the embryo is a tightly packed mass of cells called a
   a. morula.
   b. blastocyst.
   c. blastomere.
   d. gastrula.

8. In gastrulation, cells from the upper layer of the embryo _________ to form the mesoderm and much of the endoderm.
   a. migrate
   b. divide
   c. move outward
   d. invaginate

9. Segmented blocks of tissue called _________ on either side of the developing notochord give rise to the muscles, vertebrae, and connective tissue.
   a. mesoderm
   b. somites
   c. neural grooves
   d. spines

10. Which of the following does *not* occur during the first three months of pregnancy?
    a. organogenesis
    b. formation of the placenta
    c. formation of the major nerve tracts within the brain
    d. development of the limb buds

11. By the end of the third trimester, the neurological growth of the fetus is complete.
    a. true
    b. false

12. Which of the following birth-control methods involves sperm blockage?
    a. abstinence
    b. use of a condom
    c. birth-control pill
    d. use of a spermicide

13. The _________ is the storage site for sperm cells.

14. At birth, a human female's ovaries contain about _________ million oocytes.

15. The _________ is a muscular tube about 7 centimeters long that leads to the uterus.

16. _________ is the hormone that stimulates milk production after birth, and _________ is the hormone that initiates milk release once the infant suckles at the breast.

17. The _________ is the site of invagination of cells during gastrulation of the embryo.

18. The _________ is the primary layer that forms neural tissue.

19. The _________ is the primary layer that forms the digestive system.

20. The _________ is the primary layer that forms connective tissue.

21. Coitus interruptus is the withdrawal of the penis before _________.

22. A _________ is the surgical removal of a portion of the vas deferens.

Answers to the Concept Review questions appear in Appendix B.

CHALLENGE YOURSELF

1. Relatively few kinds of animals have both male and female sex organs in the same animal, while most plants do. Propose an explanation for this.
2. How is the location of the testes in the scrotum an advantage for sperm cell production?
3. In mammals, female sexual development is under *negative* control; that is, the absence of the *SRY* gene product and the absence of testosterone result in the development of female structures. Negative control makes it possible for embryos of either sex to develop inside female parents, which have no *SRY* gene product and very low levels of testosterone. If female sexual development were triggered by *positive* controls and male sexual development were under negative controls, how might the development of embryos inside female parents be affected?

FOR FURTHER READING

Aral, S., and K. Holmes. "Sexually Transmitted Diseases in the AIDS Era." *Scientific American,* February 1991, 52–59. Gonorrhea, syphilis, and other infections still exact a terrible toll.

Campbell, K., and J. Wood. *Human Reproductive Ecology: Interactions of Environment, Fertility, and Behavior*. New York: Academy of Sciences, 1994. A comprehensive compilation of papers on the extensive subject of human reproduction.

Dutton, G. "Boosting Fertility." *Popular Science* 242 (April 1993): 34. The revolutionary artificial reproduction technique of encapsulating embryos is discussed.

Greene, W. "AIDS and the Immune System." *Scientific American,* September 1993, 98–105. This feature article discusses how the virus is transmitted and how it "programs" the immune system to replicate the virus—with the subsequent consequences.

Holmes, B. "Genetic Engineering in the Womb." *New Scientist* 145 (March 4, 1995): 16. Why wait until a baby is born to try to correct genetic defects? Scientists have already met with some success in in utero gene transfer.

Michael, R. *Sex in America: A Definitive Survey*. Boston: Little, Brown, 1994. A very definitive survey: a detailed look at sexual behavior in the United States (includes statistics).

Munson, P., D. Counts, and G. Cutler. "Patterns of Human Growth." *Science* 268 (April 21, 1995): 442–47. Human infant development is explored in detail.

Nowak, M., and A. McMichael. "How HIV Defeats the Immune System." *Scientific American*, August 1995, 58–65. The HIV virus appears to undergo a continuous—and dangerous—evolution in the body.

Paabo, S. "The Y Chromosome and the Origin of All of Us (Men)." *Science* 268 (May 26, 1995): 1141–42. Why are there Y chromosomes? Are men really necessary? The origin of man and population genetics are explored under these blanket questions.

"Special Issue: Frontiers in Biology: Development." *Science* 266 (October 28, 1994). An entire issue dedicated to current issues and topics of research in developmental biology.

TECHNOLOGY LINKS

The Living World Home Page
http://www.wcbp.com/biology/tlw

Explorations in Human Biology CD
#3 Life Span and Life Style
#13 AIDS

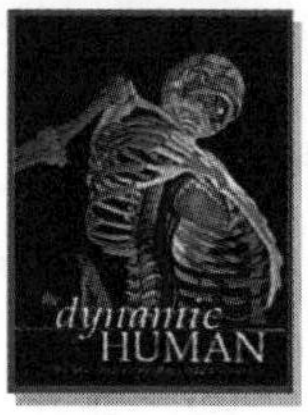

The Dynamic Human CD
Reproductive System

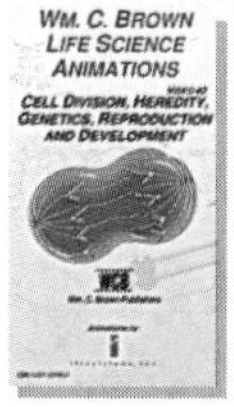

Life Science Animations Videotape 2
#19 Spermatogenesis
#20 Oogenesis
#21 Human Embryonic Development

26

Ecosystems

THE FAR SIDE By GARY LARSON

"And see this ring right here, Jimmy? ... That's another time when the old fellow miraculously survived some big forest fire."

CHAPTER OUTLINE

Figure 26.1 View of the United States from space.
Global environmental destruction has become increasingly obvious when the earth is viewed from outer space. The lights you see are large cities; increasingly, the world's population is centered in cities such as these. Large concentrations of people create unique problems in efficiently distributing and using resources.

You cannot listen to a newscast or read a newspaper today without encountering news about the environment—much of it bad. We are told that the earth's climate is warming because of the greenhouse effect, that industrial chemicals leaking into the atmosphere are destroying its ozone and exposing us to dangerous ultraviolet radiation, and that the world's tropical rain forests are being destroyed so quickly that in 20 years little will remain. Such issues are important to every one of us, because we will all have to live in the world we seem to be destroying (figure 26.1). There is no place else to go.

As you will learn in these last three chapters, today's problems have taught us that we *share* the environment with all the other organisms of the earth, a sharing that is complex and on which we all depend. Together the earth's organisms weave an intricate tapestry of interactions that defines the environment.

26.1 What Is an Ecosystem?

The word **ecology** was coined in 1866 by the great German biologist Ernst Haekel to describe the study of how organisms interact with their environment. It comes from the Greek words *oikos* (house, place where one lives) and *logos* (study of). Our study of ecology, then, is a study of the house in which we live. Do not forget this simple analogy built into the word "ecology"—most of our environmental problems could be avoided if we treated the world in which we live the same way we treat our own homes. Would you pollute your own house?

More specifically, ecology is the study of the interactions of living organisms with one another and with their physical environment (soil, water, weather, and so on). Ecologists, the scientists who study ecology, view the world as a patchwork quilt of different environments, all bordering on and interacting with one another. Consider for a moment a patch of forest, the sort of place a deer might live. Ecologists call the collection of creatures that live in a particular place a **community**—all the animals, plants, fungi, and microorganisms that live together in a forest, for example, are the forest community. Ecologists call the place where a community lives its **habitat**—the soil, and the water flowing over it, are key components of the forest habitat. A habitat is a mailing address, the community its residents. The sum of these two, community and habitat, is an ecological system, or **ecosystem.** An ecosystem is a self-sustaining collection of organisms and their physical environment (figure 26.2).

Figure 26.2 A forest ecosystem.
The redwood forest ecosystem of coastal California and southwestern Oregon is dominated by redwoods (*Sequoia sempervirens*), although many other kinds of plants and animals live beneath and within them.

The Inhabitants of Ecosystems

Imagine that you could fence off a square kilometer of forest and then go in and collect every single creature. What would you get? Well, starting with the animals, you might capture a bear and some deer, as well as an assortment of raccoons, squirrels, rabbits, chipmunks, and moles. Lizards might dart among the leaves, while snakes and toads lie quiet. You would catch all sorts of birds, from vultures to warblers to hummingbirds. If there were a lake, you would find bass or perch and a variety of turtles and frogs. And this collection, a small zoo in itself, would be only a small sampling of the members of the animal kingdom. In addition, the soil contains immense numbers of worms and other organisms, and the forest is alive with different kinds of spiders and insects—ants, ticks, chiggers, mosquitoes, and thousands of different kinds of beetles.

And what of the plants? A North American forest community is typically dominated by one or a few kinds of trees, such as the pine forests of the southeast or the oak and hickory forests of the northeast. Within the forest other kinds of trees occur less frequently, as well as a variety of shrubs, bushes, grasses, herbs, and other plants. Then there are the mushrooms and other fungi, which obtain their food by attaching themselves to plants or animals, secreting digestive chemicals onto them, and then absorbing the rich nutrients that result. A fungus does outside what you do in your stomach. Many grow as fine threads throughout the forest floor.

Any forest pond or stream contains countless creatures too small to see with the naked eye: algae that clump together to form pond scum, and single-celled predators tearing around in every drop of water. And in both water and soil live bacteria, the most numerous of all the creatures of the forest. A single teaspoon of the forest soil is home to billions of bacteria. They play many critical roles, including changing atmospheric nitrogen gas into a form organisms can use and decomposing dead organisms. Without bacteria, the bodies of all the organisms that ever died in the forest would soon be stacked to the sky.

If you collected and removed every single organism, all the animals and plants and fungi and microscopic creatures and bacteria, your square kilometer would be bare, stripped of life. What would remain would be the nonliving surroundings—the ground and rocks and water, the wind that blows over it, the rain and sunlight that fall upon it. This portion of the forest ecosystem is the **physical habitat.** The minerals and water that a physical habitat contains, and the weather to which it is exposed, are what largely determine which kinds of organisms are able to live there.

A forest ecosystem, then, is a rich collection of animals, plants, and other organisms, along with the soil, light, and water they share. Standing in the forest, one tree is surrounded by all of this, the house in which it lives.

26.2 Energy Flows Through Ecosystems

A flower blooming, a squirrel running along a tree branch, a worm burrowing through the soil—every motion of even the tiniest creatures in an ecosystem uses energy. Of all the factors that influence which types of organisms live in an ecosystem and how many of each, none is more important than the flow of energy.

The Path of Energy: Who Eats Whom in Ecosystems

Energy flows into the biological world from the sun, which shines a constant beam of light on our earth. Life exists on earth because some of that continual flow of light energy can be captured and transformed into chemical energy through the process of photosynthesis and used to make organic molecules. These organic molecules are what we call food. Living organisms use the energy in food to make new materials for growth, to repair damaged tissues, to reproduce, and to do myriad other things that require energy, like turning the pages of this text.

You can think of all the organisms in an ecosystem as chemical machines fueled by energy captured in photosynthesis. The organisms that first capture the energy, the **producers,** are plants (and some bacteria and algae), which produce their own energy-storing molecules by carrying out photosynthesis. All other organisms in an ecosystem are **consumers,** obtaining their energy-storing molecules by consuming plants or other animals. Ecologists assign every organism in an ecosystem to a trophic, or feeding, level, depending on the source of its energy. A **trophic level** is composed of those organisms within an ecosystem whose source of energy is the same number of consumption "steps" away from the sun. Thus, a plant's trophic level is 1, while animals that graze on plants are in trophic level 2, and animals that eat these grazers are in trophic level 3 (figure 26.3). Higher trophic levels exist for animals that eat higher on the food chain.

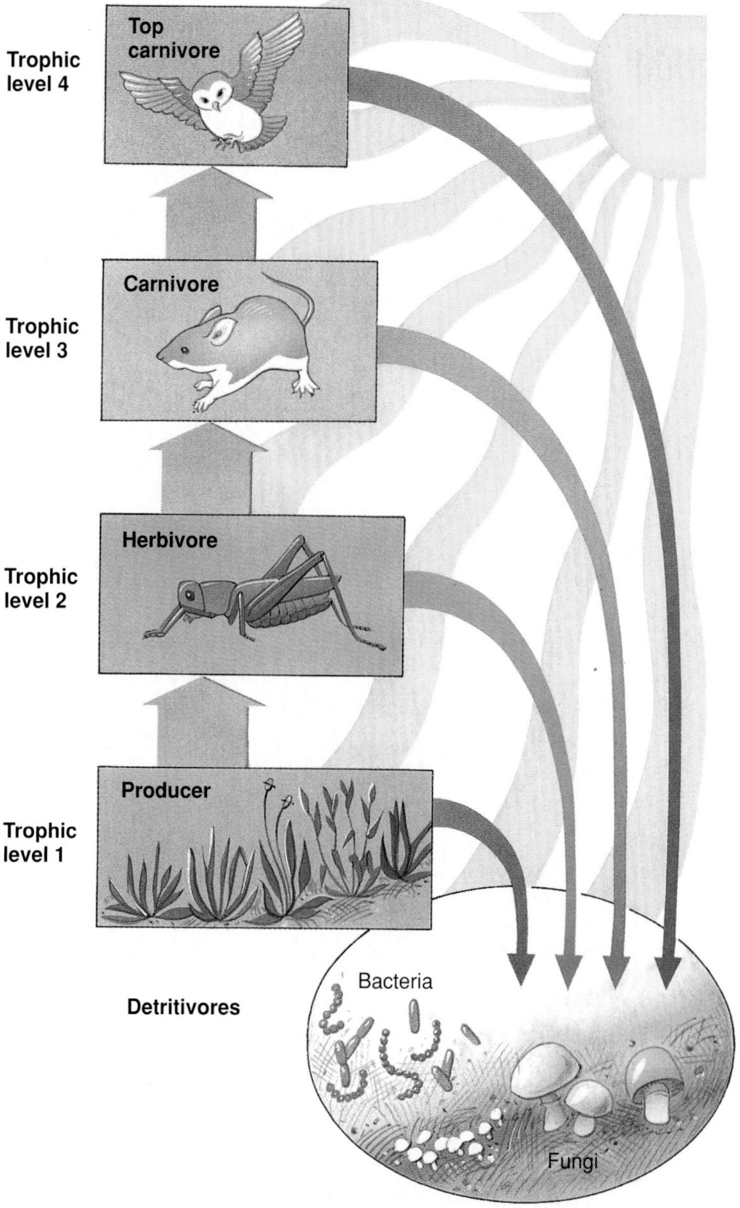

Figure 26.3 Trophic levels within an ecosystem.
Ecologists assign all the members of a community to various trophic levels based on feeding relationships. Producers, such as photosynthetic plants, obtain their energy directly from the sun and are assigned to trophic level 1. Animals that eat plants (herbivores) are at trophic level 2. Animals that eat plant-eating animals (carnivores) are at trophic level 3 and higher. Detritivores make use of all trophic levels for food.

Producers

The lowest trophic level of any ecosystem is occupied by the producers—green plants in most land ecosystems (and, usually, algae in freshwater). Plants use the energy of the sun to build energy-rich sugar molecules. They also absorb nitrogen and other key substances from the air and soil and build them into biological molecules. It is important to realize that plants consume as well as produce. The roots of a plant, for example, do not carry out photosynthesis—there is no sunlight underground. Roots obtain their energy the same way you do, by using energy-storing molecules produced elsewhere (in this case, in the leaves of the plant).

Herbivores

At the second trophic level are **herbivores,** animals that eat plants. They are the *primary consumers* of ecosystems. Deer and horses are herbivores, and so are rhinoceroses, chickens, and caterpillars. Most herbivores rely on "helpers" to aid in the digestion of cellulose, a structural material found in plants. A cow, for instance, has a thriving colony of bacteria in its gut that digests cellulose for it. So does a termite. Humans cannot digest cellulose, because we lack these bacteria—that is why a cow can live on a diet of grass and you cannot.

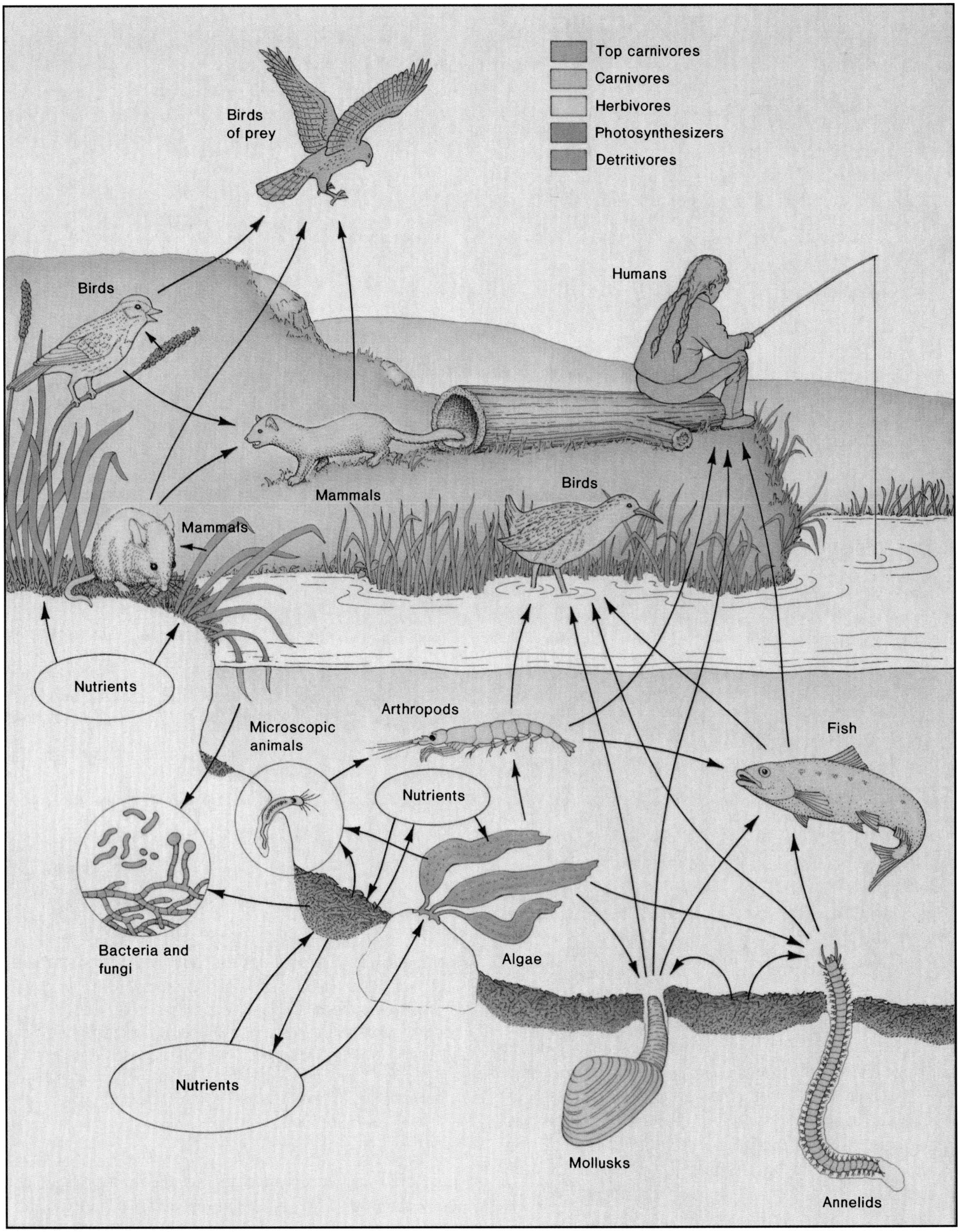

Figure 26.4 A typical food web.
A food web is much more complicated than a linear food chain. The path of energy passes from one trophic level to another and back again in complex ways. The arrows in this figure show the complex relationships among the trophic levels (color coded) along a freshwater lake.

Carnivores

At the third trophic level are animals that eat herbivores, called **carnivores** (meat-eaters). They are the *secondary consumers* of ecosystems. Tigers and wolves are carnivores, and so are mosquitoes and blue jays. Some animals, like bears and humans, eat both plants and animals and are called **omnivores.** They use the simple sugars and starches stored in plants as food and not the cellulose.

Many complex ecosystems contain a fourth trophic level, composed of animals that consume other carnivores. They are called tertiary consumers, or top carnivores. A weasel that eats a blue jay is a tertiary consumer. Only rarely do ecosystems contain more than four trophic levels, for reasons that will become clear in a moment.

Detritivores

In every ecosystem there is a special class of consumers called **detritivores,** or decomposers. They obtain their energy from the organic wastes and dead bodies that are produced at all trophic levels. Bacteria and fungi are the principal decomposers in land ecosystems, but worms and insects are detritivores too, as are vultures.

How Long Can a Food Chain Be?

Food energy passes through an ecosystem from one trophic level to another. When the path is a simple linear progression from producers to herbivores to carnivores, like the links of a chain, it is called a **food chain.** In most ecosystems, however, the path of energy is not a simple linear one, because individual animals often feed at several trophic levels. This creates a complicated path of energy called a **food web** (figure 26.4).

How much energy passes through an ecosystem? **Primary productivity** is the total amount of light energy converted to organic compounds in a given area per unit of time. An ecosystem's **net primary productivity** is the total amount of energy fixed by photosynthesis per unit of time, minus that which is expended by the metabolic activities of ecosystem organisms. The total weight of all ecosystem organisms, called the ecosystem's **biomass,** increases as a result of the ecosystem's net productivity. Some ecosystems, such as cornfields or cattail swamps, have a high net primary productivity. Others, such as tropical rain forests, also have a relatively high net primary productivity, but a rain forest has a much larger biomass than a cornfield. Consequently, a rain forest's net primary productivity is much lower in relation to its biomass.

When a plant uses the chemical energy from sunlight to make structural molecules such as cellulose, it loses a lot of the energy as heat. In fact, only about half of the energy captured by the plant ends up stored in its molecules. The other half of the energy is lost. This is the first of many such losses as the energy passes through the ecosystem. When a herbivore uses plant molecules to make herbivore molecules, only about 10% of the energy present in the plant molecules ends up in the herbivore's molecules. Ninety percent of the remaining energy is lost. And when a carnivore eats the herbivore, 90% of what little remains is then lost in making carnivore molecules. At each trophic level, the energy stored by the organism is about a tenth that of the level below it (figure 26.5).

Trophic efficiency is a law we cannot repeal, and as we seek ways to maintain ever-larger human populations we should ponder the lion. It takes a very large population of zebras and wildebeests to support a small population of lions on the Serengeti Plain of Africa. Unlike the lion, humans can choose to eat either meat or plants. The choice makes a difference to our future, because we eat steak and hamburger only at a great cost in energy. It takes 10 pounds of grain to build 1 pound of human tissue if we eat the grain, but it takes 100 pounds of grain to build 1 pound of human tissue if a cow eats the grain and we eat the cow. Humanity, to survive, may have to learn to eat lower on the food chain.

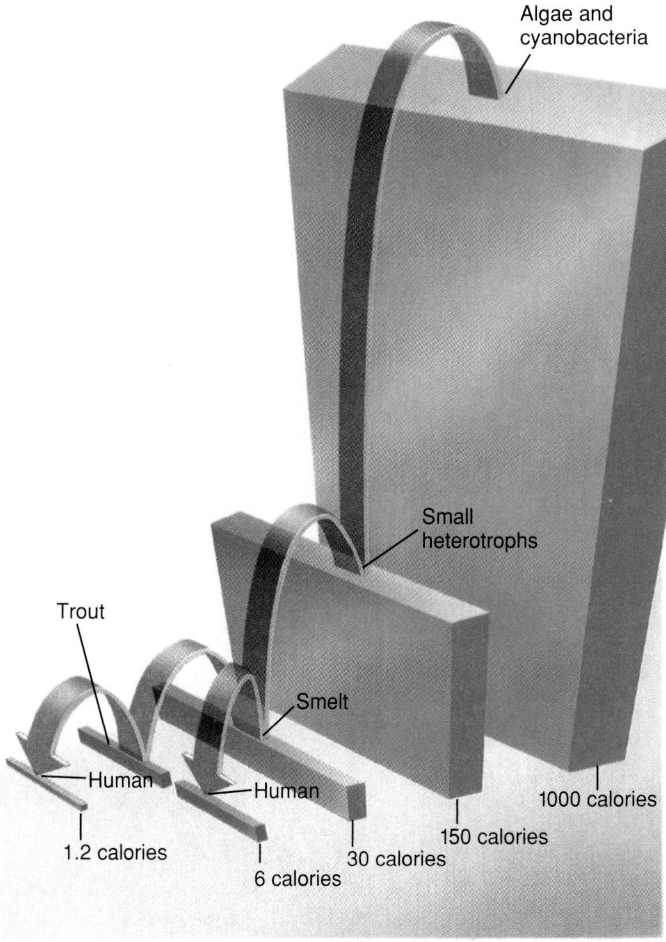

Figure 26.5 Energy loss in an ecosystem.
In a classic study of Cayuga Lake in New York, the path of energy was measured precisely at all points in the food web. This diagram summarizes the results. Photosynthetic algae and cyanobacteria fix the energy of the sun; animal plankton (small heterotrophs) feed on them; and both are consumed by smelt. The smelt are eaten by trout, with about a 10-fold loss in fixed energy; the amount of biomass available in smelt is at least 10 times greater than that available in trout, which we humans prefer to eat.

26.3 Materials Cycle Within Ecosystems

Unlike energy, which flows through the earth's ecosystems in one direction (from the sun to producers to consumers), the physical components of ecosystems are passed around and reused within ecosystems. Ecologists speak of such constant reuse as recycling, or, more commonly, **cycling.** Materials that are constantly recycled include all the inorganic (noncarbon) chemicals that make up the soil, water, and air. While many of them are important, the proper cycling of three materials is particularly critical to the health of any ecosystem: water, carbon, and soil nutrients (nitrogen plus phosphorus).

The paths of water, carbon, and soil nutrients as they pass from the environment to living organisms and then back to the environment form closed circles, or cycles. In each cycle, the inorganic substance resides for a time in an organism and then returns to the nonliving environment. Often, nutrients pass from one organism to the body of another that feeds on primary ones. In almost all cases, only a tiny fraction of the material is tied up in living organisms at any one time.

The Water Cycle

Of all the nonliving components of an ecosystem, water has the greatest influence on the living portion. The availability of water in large measure determines the biological richness of an ecosystem—how many different kinds of creatures live there and how many of each.

Water cycles within ecosystems in two ways: the environmental water cycle and the organismic water cycle.

The Environmental Water Cycle

In the environmental water cycle, water vapor in the atmosphere condenses and falls to the earth's surface as rain or snow. Heated there by the sun, it reenters the atmosphere by **evaporation** from lakes, rivers, and oceans (figure 26.6).

The Organismic Water Cycle

In the organismic water cycle, surface water does not return directly to the atmosphere. Instead, it is taken up by the roots of plants. After passing through the plant, the water reenters the atmosphere through tiny openings in the leaves, evaporating from their surface. This evaporation from leaf surfaces is called **transpiration.** Transpiration is also driven by the sun: the sun's heat creates wind currents that draw moisture from the plant by passing air over the leaves.

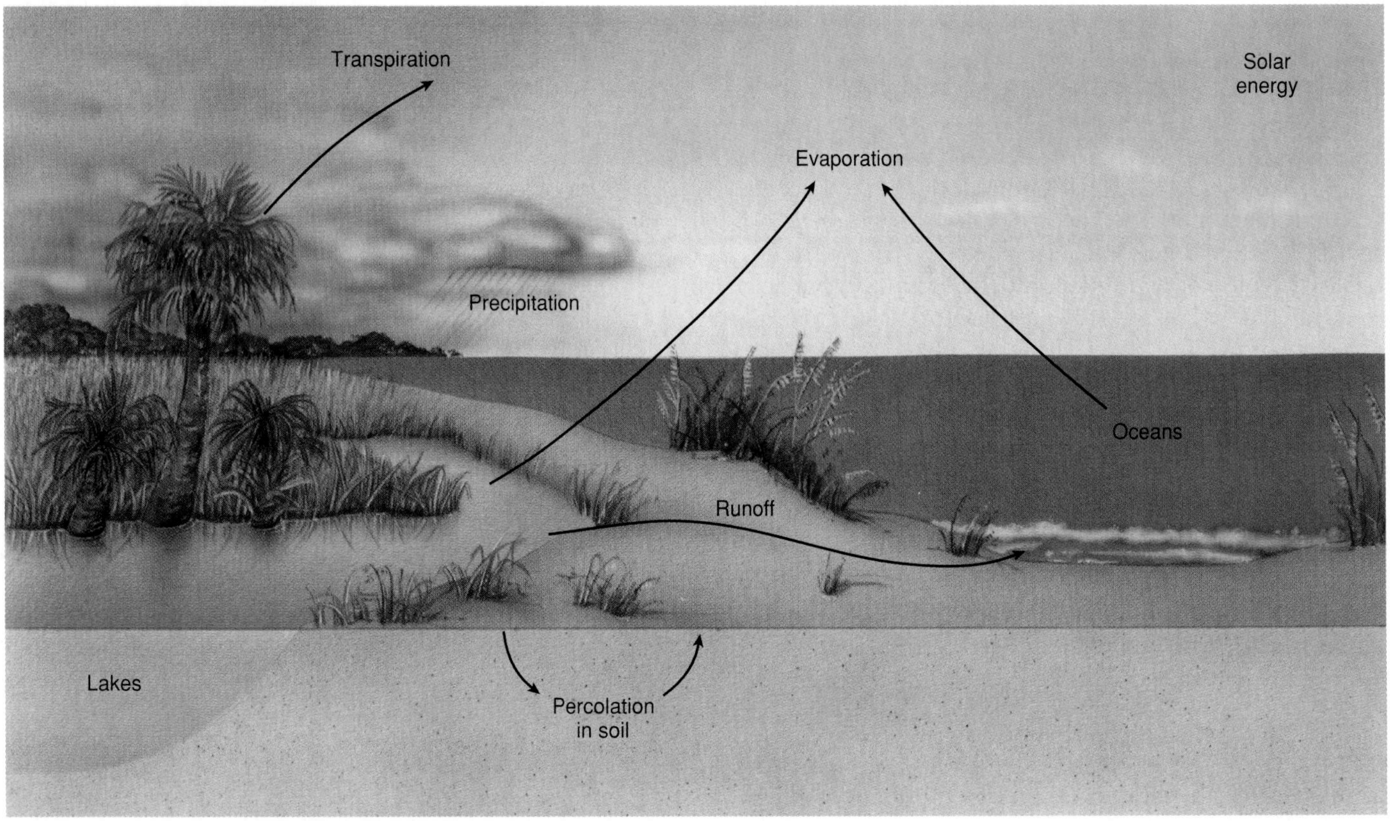

Figure 26.6 The environmental water cycle.
Precipitation on land eventually makes its way to the ocean via groundwater, lakes, and finally, rivers. Solar energy causes evaporation, adding water to the sky. Plants give off excess water through transpiration, also adding water to the atmosphere. Atmospheric water falls as rain or snow over land and oceans, completing the water cycle.

Breaking the Cycle

In very dense forest ecosystems, such as tropical rain forests, more than 90% of the moisture in the ecosystem is taken up by plants and then transpired back into the air. Because so many plants in a rain forest are doing this, the vegetation is the primary source of local rainfall. In a very real sense, these plants create their own rain: the moisture that travels up from the plants into the atmosphere falls back to earth as rain.

Where forests are cut down, the organismic water cycle is broken, and moisture is not returned to the atmosphere. Water drains off to the sea instead of rising to the clouds and falling again on the forest. As early as the late 1700s, the great German explorer Alexander von Humbolt reported that stripping the trees from a tropical rain forest in Columbia prevented water from returning to the atmosphere and created a semiarid desert. It is a tragedy of our time that just such a transformation is occurring in many tropical areas, as tropical and temperate rain forests are being clear-cut or burned in the name of "development" (figure 26.7).

Groundwater

Much less obvious than the surface waters seen in streams, lakes, and ponds is the groundwater, which occurs in permeable, saturated, underground layers of rock, sand, and gravel called aquifers. In many areas, groundwater is the most important water reservoir; for example, in the United States, more than 96% of all freshwater is groundwater. Groundwater flows much more slowly than surface water, anywhere from a few millimeters to as much as a meter or so per day. In the United States, groundwater provides about 25% of the water used for all purposes and provides about 50% of the population with drinking water. Rural areas tend to depend on groundwater almost exclusively, and its use is growing at about twice the rate of surface water use.

Because of the greater rate at which groundwater is being used, the increasing chemical pollution of groundwater is a very serious problem. Pesticides, herbicides, and fertilizers are key sources of groundwater pollution. Recharging the aquifers is an important strategy for conservation, but the purity of the water there initially is also an important consideration. Because of the large volume of water, its slow rate of turnover, and its inaccessibility, removing pollutants from aquifers is virtually impossible.

Figure 26.7 Burning (clear-cutting) forests.
The high density and large size of plants in a forest translate into great quantities of water being transpired to the atmosphere, creating rain over the forests. In this way rain forests perpetuate the wet climate that supports them. Tropical deforestation will permanently alter the climate in these areas, creating arid zones.

The Carbon Cycle

The earth's atmosphere contains plentiful carbon, present as carbon dioxide (CO_2) gas. This carbon cycles between the atmosphere and living organisms, often being locked up for long periods of time in fossil organisms. The cycle is begun by plants that use carbon dioxide in photosynthesis as the raw material for building organic molecules—in effect, they trap the carbon atoms of carbon dioxide within the living world. The carbon atoms are returned to the atmosphere's pool of carbon dioxide through respiration, combustion, and erosion (figure 26.8).

Respiration

Most of the organisms in ecosystems respire—that is, they extract energy from organic food molecules by stripping away the carbon atoms and combining them with oxygen to form carbon dioxide. Plants respire, as do the herbivores that eat the plants and the carnivores that eat the herbivores. All of these organisms use oxygen to extract energy from food, and carbon dioxide is what is left when they are done.

Combustion

A lot of carbon is tied up in wood, and it may stay trapped there for many years, only returning to the atmosphere when the wood is burned. Sometimes the duration of the carbon's visit to the organic world is long indeed. Plants that become buried in sediment, for example, may be gradually transformed by pressure into coal or oil. The carbon originally trapped by these plants is only released back into the atmosphere when the coal or oil (called **fossil fuels**) is burned.

Erosion

Very large amounts of carbon are present in seawater as dissolved carbon dioxide. Substantial amounts of this carbon are extracted from the water by marine organisms, which use it to build their calcium carbonate shells. When these marine organisms die, their shells sink to the ocean floor, become covered with sediments, and form limestone. Eventually, as the ocean recedes and the limestone becomes exposed to weather and erodes, the carbon is returned to the cycle.

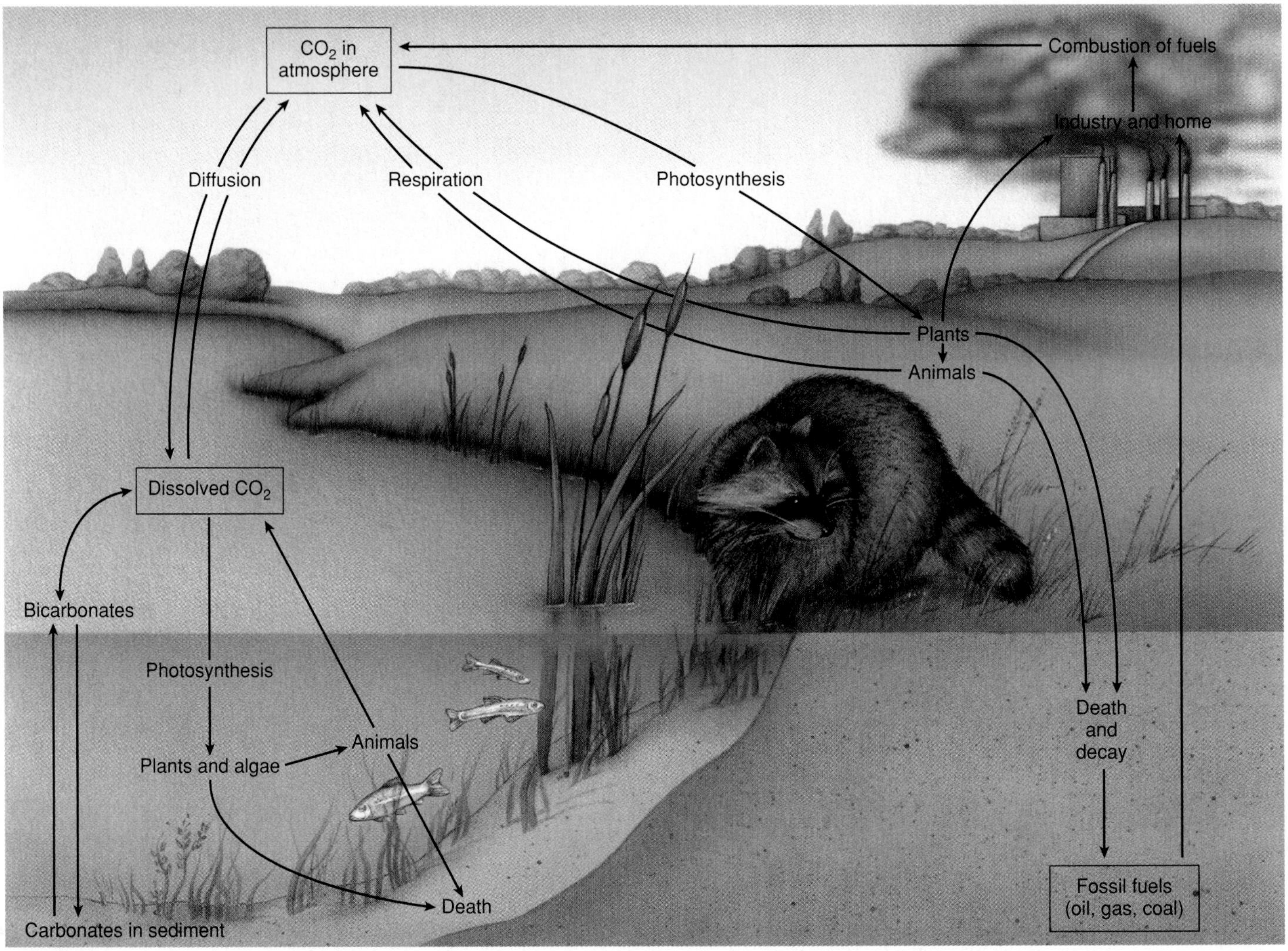

Figure 26.8 The carbon cycle.
Carbon from the atmosphere and from water is fixed by photosynthetic organisms and returned through respiration, combustion, or erosion.

Nutrient Cycles

The two key soil nutrients in ecosystems are nitrogen and phosphorus, both of which are often in very limited supply. The fertility of farmland is determined in large degree by the availability of nitrogen and phosphorus in the soil, and the same is true of natural ecosystems.

The Nitrogen Cycle

Organisms contain a lot of nitrogen (a principal component of protein) and so does the atmosphere, which is 79% nitrogen gas (N_2). However, the chemical connection between these two reservoirs is very delicate, because most living organisms are unable to utilize the nitrogen gas so plentifully available in the air surrounding them. The two nitrogen atoms of nitrogen gas are bound together by a particularly strong "triple" covalent bond that is very difficult to break. Luckily, a few bacteria can break the nitrogen triple bond and bind its nitrogen atoms to hydrogen (forming "fixed" nitrogen, ammonia [NH_3]) in a process called **nitrogen fixation.** Bacteria evolved the ability to do this early in the history of life, before photosynthesis had introduced oxygen gas into the earth's atmosphere, and that is still the only way the bacteria are able to do it—even a trace of oxygen poisons the process. In today's world, awash with oxygen, these bacteria live encased within bubbles called cysts that admit no oxygen or within special airtight cells in nodules of tissue on the roots of beans, aspen trees, and a few other plants. Figure 26.9 shows how bacteria make needed nitrogen available to other organisms, a process called the nitrogen cycle.

The growth of plants in ecosystems is often severely limited by the availability of "fixed" nitrogen in the soil, which is why farmers fertilize fields. This agricultural practice is a very old one, known even to primitive societies—the Indians instructed the pilgrims to bury fish, a rich source of fixed nitrogen, with their corn seeds. Today most fixed nitrogen added to soils by farmers is not organic but instead is produced in factories by industrial rather than bacterial nitrogen fixation, a process that accounts for a prodigious 30% of the entire nitrogen cycle.

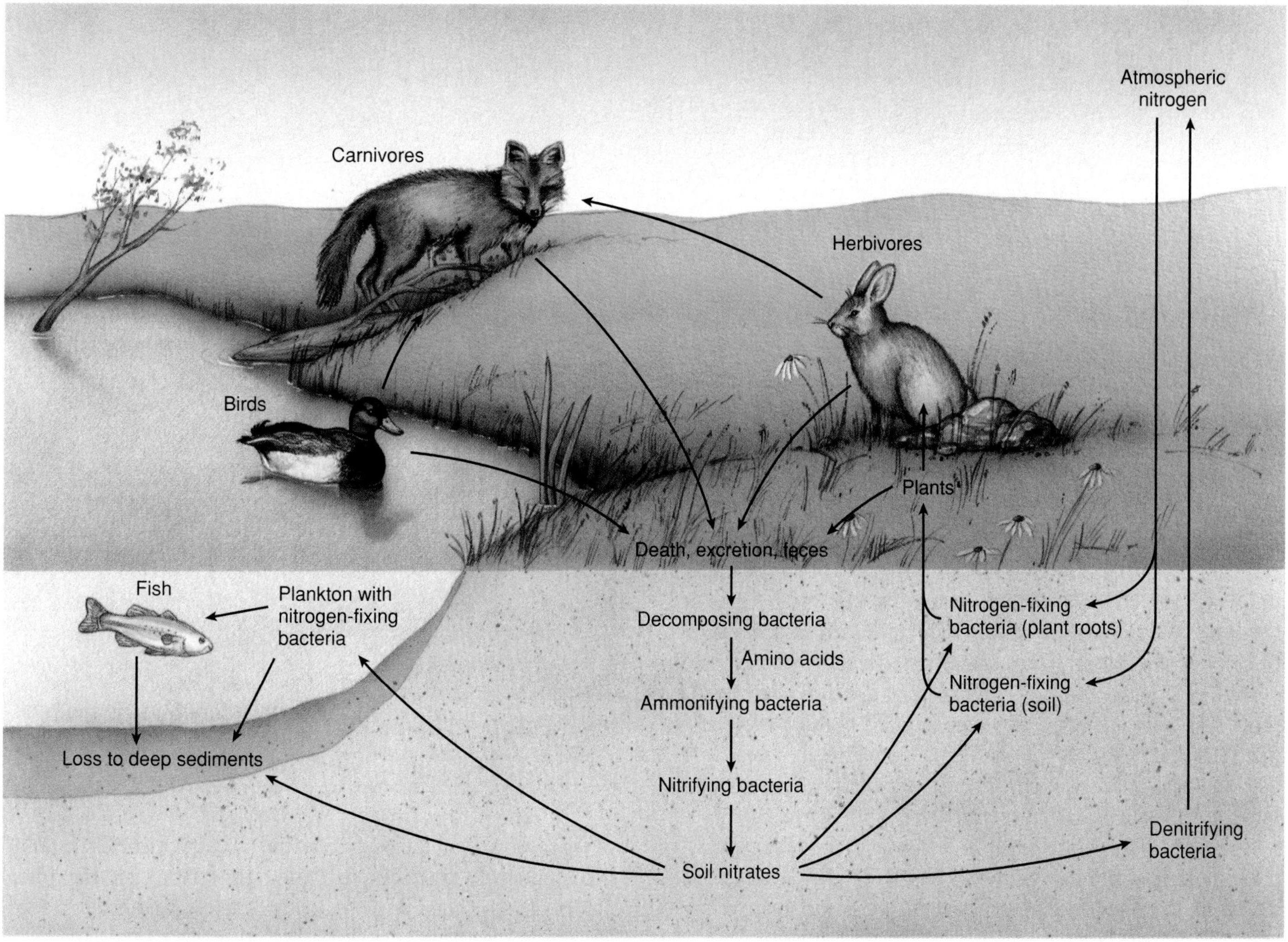

Figure 26.9 The nitrogen cycle.
Relatively few kinds of organisms—all of them bacteria—can convert atmospheric nitrogen into forms that can be used for biological processes.

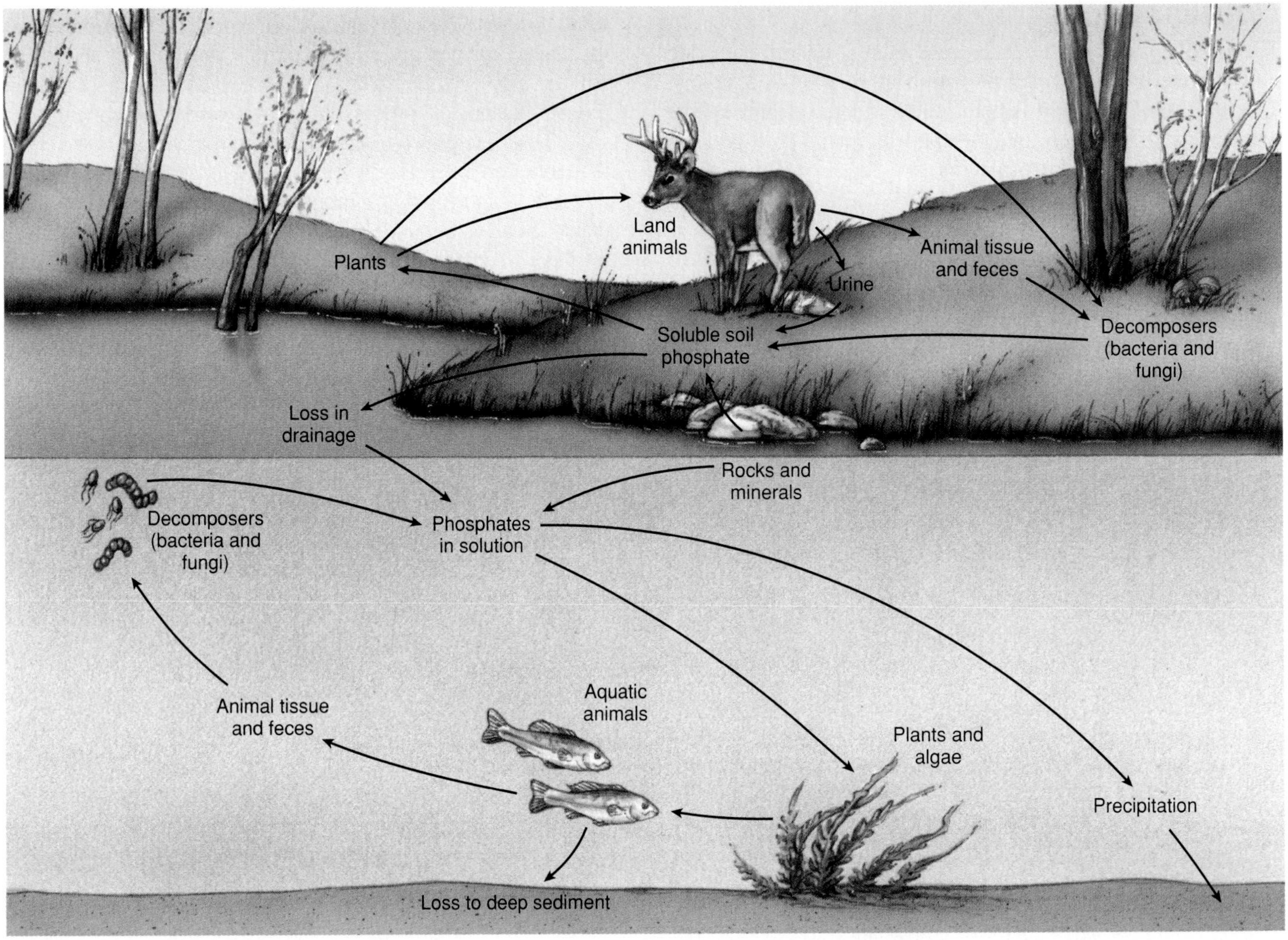

Figure 26.10 The phosphorus cycle.
Phosphorus plays a critical role in plant nutrition; next to nitrogen, phosphorus is the element most likely to be so scarce that it limits plant growth.

The Phosphorus Cycle

Phosphorus is an essential element in all living organisms, a key part of both ATP and DNA. Phosphorus is often in very limited supply in the soil of particular ecosystems, and because phosphorus does not form a gas, none is available in the atmosphere. Most phosphorus exists in soil and rock as the mineral calcium phosphate, which dissolves in water to form phosphate ions (Coca-Cola is a sweetened solution of phosphate ions). These phosphate ions are absorbed by the roots of plants and used by them to build organic molecules like ATP and DNA. When the plants and animals die and decay, bacteria in the soil convert the organic phosphorus back into phosphorus ions, completing the cycle (figure 26.10).

The phosphorus level in freshwater lake ecosystems is often quite low, preventing much growth of photosynthetic algae in these systems. Such ecosystems are particularly vulnerable to the inadvertent addition of phosphorus by human activity. For example, agricultural fertilizers and many commercial detergents are rich in phosphorus. Pollution of a lake by the addition of phosphorus to its waters first produces a green scum of algal growth on the surface of the lake and then, if the pollution continues, proceeds to "kill" the lake. After the initial bloom of rapid algal growth, aging algae die, and bacteria feeding on the dead algae cells use up so much of the lake's dissolved oxygen that fish and invertebrate animals suffocate. Such rapid, uncontrolled growth caused by excessive nutrients in an aquatic ecosystem is called **eutrophication** (figure 26.11).

Figure 26.11 Eutrophication.
These fish suffocated after bacteria, feeding on algal blooms caused by phosphorus pollution, used up oxygen in the lake.

26.4 Major Kinds of Ecosystems

Shaped by evolution, the kinds of animals and plants that live in particular ecosystems are well suited to conditions there—for instance, polar bears are found in the polar Arctic and not in steamy tropical rain forests. The types of animals and plants that live in a particular ecosystem depend in large measure on the physical nature of the habitat—the soils, the terrain, and most particularly, the climate.

How Weather Shapes Ecosystems

The world contains a great diversity of ecosystems because its climate varies a great deal from place to place. On a given day, Miami and Boston often have very different weather. There is no mystery about this. The tropics are warmer than the temperate regions because the sun's rays arrive almost perpendicular (that is, dead on) at regions near the equator, while near the poles they hit the earth at an angle, which spreads them out over a much greater area, thus providing less energy per unit of area (figure 26.12). This simple fact—that because the earth is a sphere some parts of it receive more energy from the sun than others—is responsible for much of the earth's different climates and thus, indirectly, for much of the diversity of its ecosystems.

The earth's annual orbit around the sun and its daily rotation on its own axis are also both important in determining world climate. Because of the daily cycle, the climate at a given latitude is relatively constant. Because of the annual cycle and the inclination of the earth's axis, all parts away from the equator experience a progression of seasons.

Latitude and Rainfall

The major atmospheric circulation patterns result from the interactions between six large air masses. These great air masses occur in pairs, with one air mass of the pair occurring in the northern latitudes and the other occurring in the southern latitudes. These air masses affect climate because the rising and falling of an air mass influence its temperature, which, in turn, influences its moisture-holding capacity.

Near the equator, warm air rises and flows toward the poles (figure 26.13). As it rises and cools, this air loses most of its moisture because cool air holds less water vapor than warm air. (This explains why it rains so much in the tropics.) When this air has traveled to about 30 degrees north and south latitudes, the cool, dry air sinks and becomes reheated, sucking up water like a sponge as it warms and producing a broad zone of low rainfall. It is no accident that all of the great deserts of the world lie near 30 degrees north or 30 degrees south latitude. Air at these latitudes is still warmer than it is in the polar regions, and thus it continues to flow toward the poles. At about 60 degrees north and south latitudes, air rises and cools and sheds its moisture, and such are the locations of the great temperate forests of the world. Finally, another air mass rises at 60 degrees north and south latitudes, producing zones of very low temperatures and precipitation known as the polar regions.

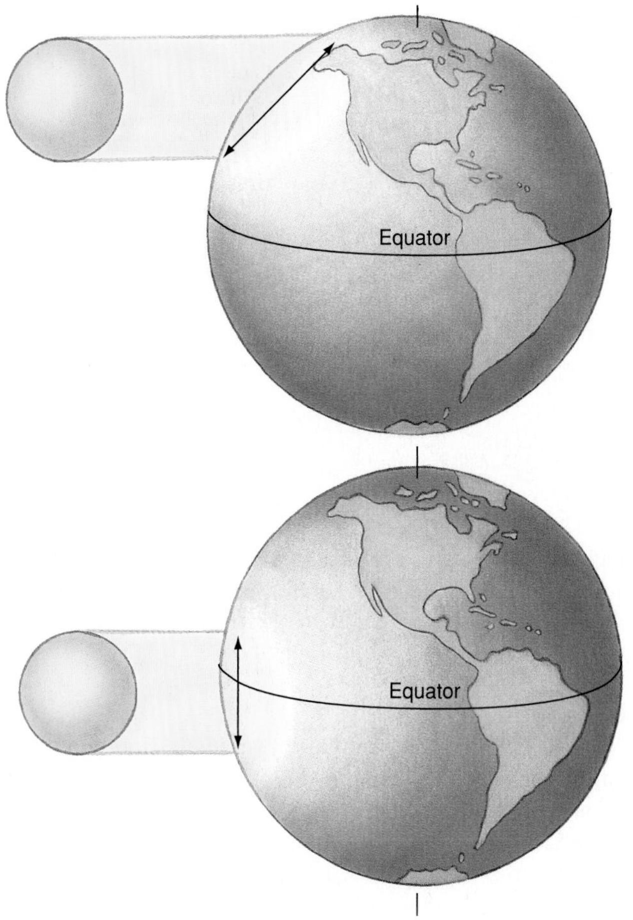

Figure 26.12 Latitude affects climate.
The relationship between earth and sun is critical in determining the nature and distribution of life on earth. The tropics are warmer than the temperate regions because the sun's rays strike at a direct angle, providing more energy per unit of area.

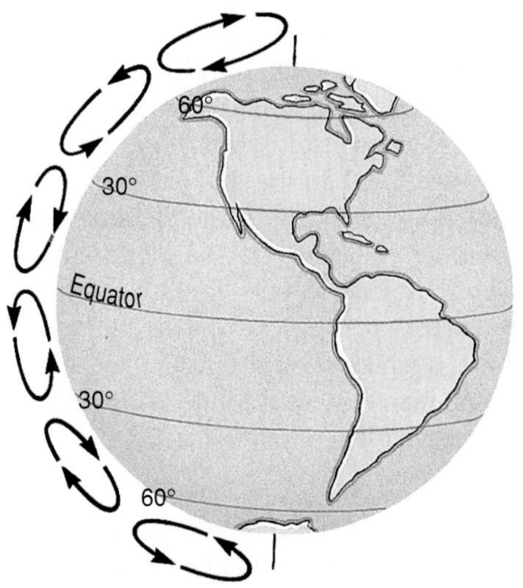

Figure 26.13 Air rises at the equator and then falls.
The pattern of air movement out from and back to the earth's surface forms three pairs of great cycles. At 30 degrees, at 60 degrees, and at the poles, zones of very dry climate are created by descending dry air.

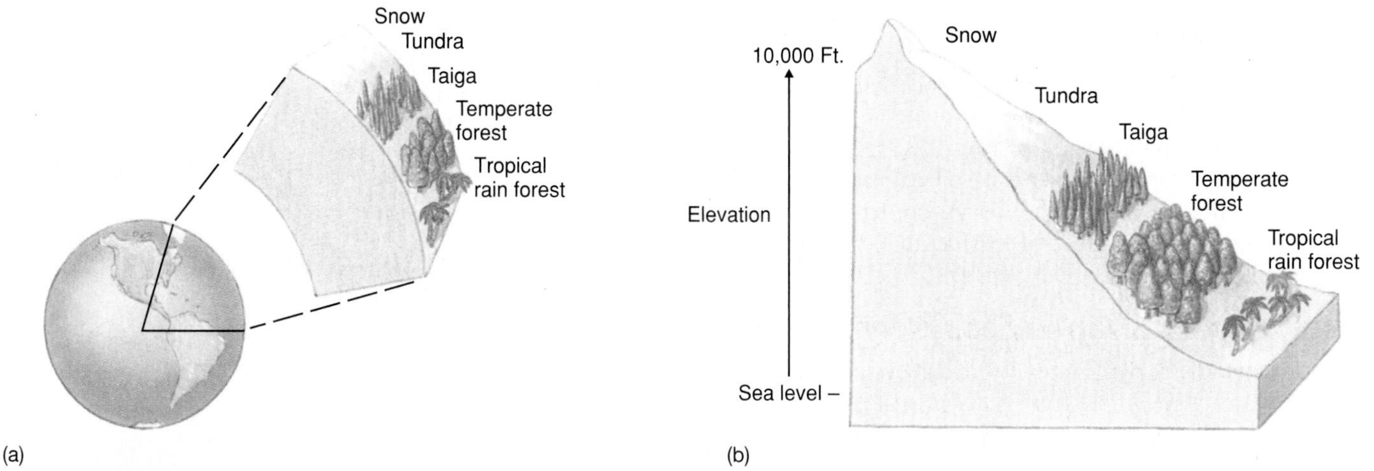

Figure 26.14 How elevation affects ecosystems.
The same land ecosystems that normally occur north and south of the equator at sea level (*a*) can occur in the tropics as elevation increases (*b*). Thus, on a tall mountain in southern Mexico or Guatemala, you might see a sequence of ecosystems such as is illustrated here.

Rain Shadows

If you were to climb a mountain, you would find the weather changing in much the same fashion as it does traveling north or south from the equator, and for much the same reason (figure 26.14). When a moving body of air encounters a mountain, it is forced upward, and as it is cooled at higher elevations the air's moisture-holding capacity decreases, producing rain on the windward side of the mountains—the side from which the wind is blowing. The effect on the other side of the mountain—the leeward side—is quite different. As the air passes the peak and descends on the far side of the mountains, it is warmed, so its moisture-holding capacity increases. Sucking up all available moisture, the air dries up the surrounding landscape, often producing a desert. This effect, called a **rain shadow,** is responsible for deserts such as Death Valley, which is in the rain shadow of Mount Whitney, the tallest mountain in the Sierra Nevada (figure 26.15). Mediterranean climates result when winds blow from a cool ocean onto a warm land during the summer. As a result, the air's moisture-holding capacity is increased and precipitation is blocked, similar to what occurs on the leeward side of mountains. This effect accounts for dry, hot summers and cool, moist winters in areas with a mediterranean climate. Such a climate is unusual on a world scale; in the regions where it occurs, many unusual kinds of plants and animals, often local in distribution, have evolved.

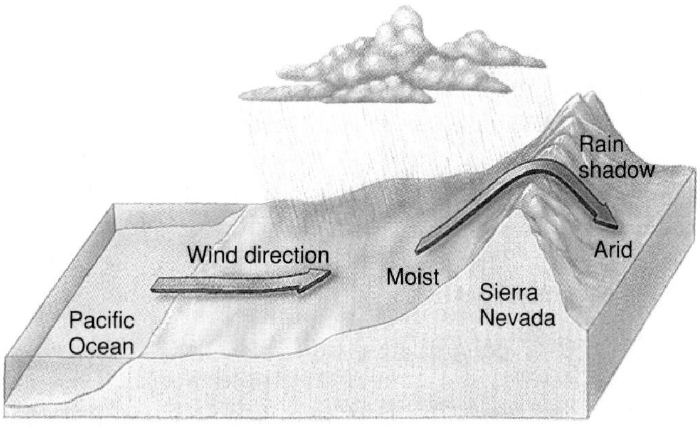

Figure 26.15 The rain shadow effect.
Moisture-laden winds from the Pacific Ocean rise and are cooled when they encounter the Sierra Nevada. As they cool, their moisture-holding capacity decreases and precipitation occurs. As the air descends on the east side of the range, it warms, its moisture-holding capacity increases, and the air picks up moisture from its surroundings. As a result, arid conditions prevail on the east side of these mountains.

Ocean Currents

In the ocean, patterns of water circulation are determined by the patterns of atmospheric circulation, but they are modified by the land masses around which and against which the ocean currents must flow. Oceanic circulation is dominated by huge surface spirals (figure 26.16). These spirals move clockwise in the Northern Hemisphere and counterclockwise in the Southern Hemisphere. They profoundly affect life not only in the oceans but also on coastal lands by the way they redistribute heat. For example, the Gulf Stream, in the North Atlantic, swings away from North America and reaches Europe near the southern British Isles. Because of the Gulf Stream, western

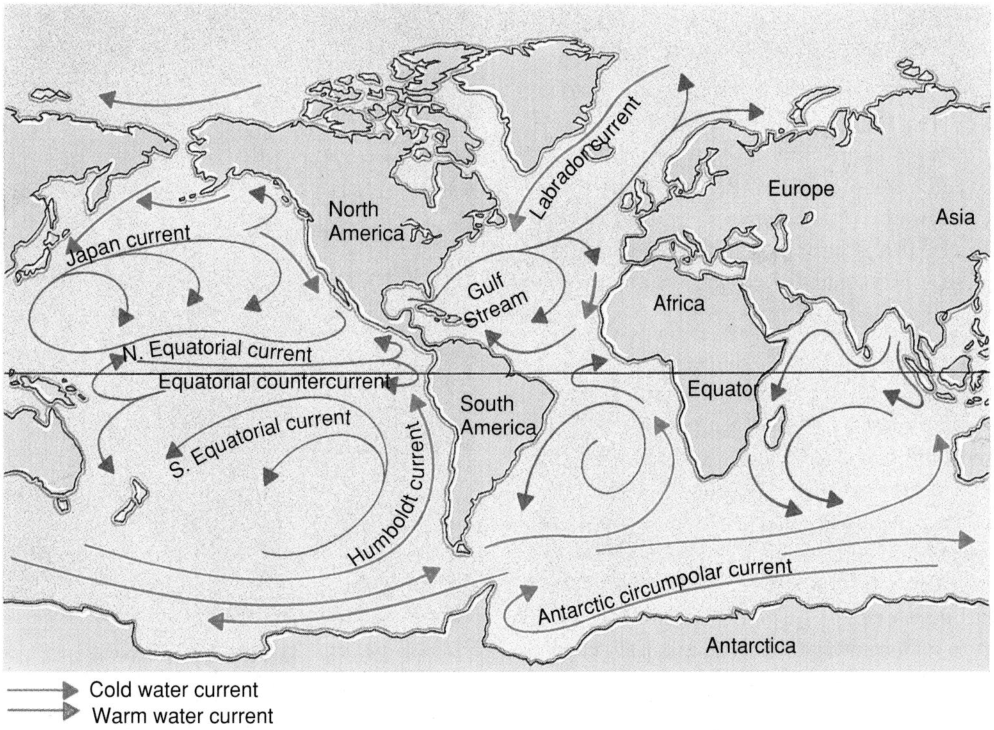

Figure 26.16 Oceanic circulation.
The circulation in the oceans moves in great surface spiral patterns called gyres; it profoundly affects the climate on adjacent lands.

Europe is much warmer and thus more temperate than is eastern North America at the same latitudes. In South America, the Humboldt current carries cold water northward up the west coast and helps to make possible an abundance of marine life that supports the fisheries of Peru and northern Chile. Marine birds, which feed on these organisms, are responsible for the phosphorus-rich guano deposits of these countries (figure 26.17). Mining these deposits for their use as a fertilizer has become a commercial industry. When the coastal waters are not as cold as usual, the devastating phenomenon called El Niño affects the vitality of both marine and terrestrial populations of animals and plants worldwide. For example, El Niño's moist, rainy conditions can cause increased growth of vegetation, reducing the food available to nesting marine birds.

Figure 26.17 These pelicans depend on the Humboldt current.
Seabirds like these Mexican pelicans bring up phosphorus from the deeper layers of the sea by eating fishes and other marine animals and depositing their remains as guano on the rookeries.

Ocean Ecosystems

Most of the earth's surface—nearly three-quarters—is covered by water. The seas have an average depth of more than 3 kilometers, and they are, for the most part, cold and dark. Photosynthetic organisms are confined to the upper few hundred meters, because light does not penetrate any deeper. All organisms that live below this level feed on organic debris that rains downward. The three main kinds of ecosystems are shallow waters, open sea surface, and deep sea waters.

Shallow Waters

Very little of the earth's ocean surface is shallow, mostly that along the shoreline, but this small area contains many more species than other parts of the ocean (figure 26.18). The world's great commercial fisheries occur on banks in the coastal zones, where nutrients derived from the land are more abundant than in the open ocean. Part of this zone consists of the **intertidal region,** which is exposed to the air whenever the tides recede (figure 26.19). Partly enclosed bodies of water, such as those that often form at river mouths and in coastal bays, where the salinity is intermediate between that of seawater and freshwater, are called **estuaries.** Estuaries are among the most naturally fertile areas in the world, often containing rich stands of submerged and emergent plants, algae, and microscopic organisms. They provide the breeding grounds for most of the coastal fish and shellfish that are harvested both in the estuaries and in open water (figure 26.20).

Figure 26.18 Diversity is great in coastal regions.
Fishes and many other kinds of animals find food and shelter among the kelp beds that occur in the coastal waters of temperate regions.

Open Sea Surface

Drifting freely in the upper, better-illuminated waters of the ocean is a diverse biological community of microscopic organisms called the **plankton.** Most of the plankton occurs in the top 100 meters of the sea. Some of the plankton are algae and photosynthetic bacteria, which collectively account for about 40% of all the photosynthesis that takes place on earth. Many fishes swim in these waters as well, feeding on the plankton.

Deep Sea Waters

In the deep waters of the sea, below the top 300 meters, little light penetrates. Very few organisms live there, compared to the rest of the ocean, but those that do include some of the most bizarre organisms found anywhere on earth. Many deep sea inhabitants have bioluminescent (light-producing) body parts that they use to communicate or to attract prey.

Figure 26.19 The intertidal region.
Diverse communities occur in intertidal regions. Many different habitats are created by the pounding of the waves and the periodic drying and flooding as the tides move out and in. Organisms that live in tide pools have adaptations that protect them from the dry air.

(a)

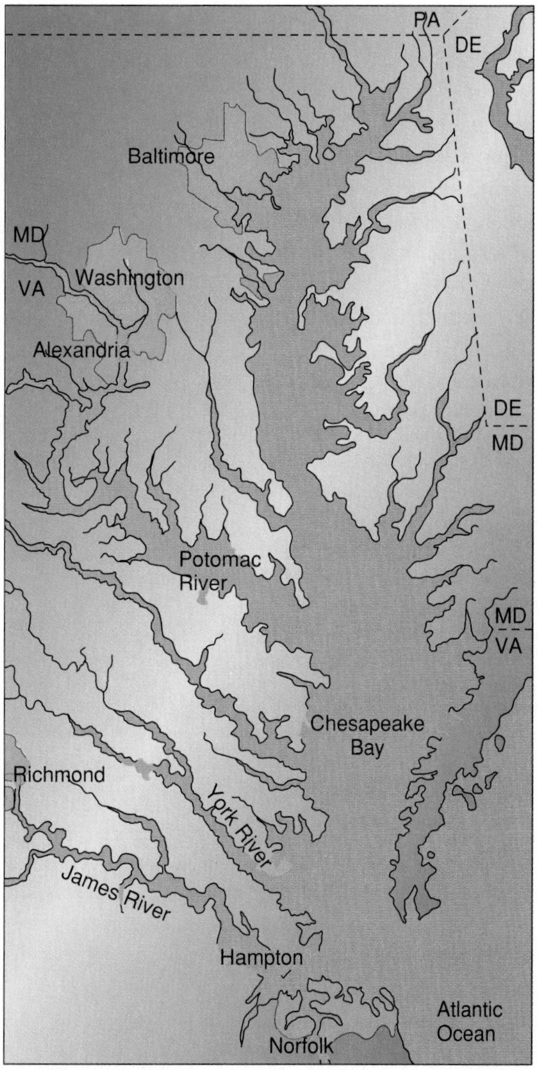

(b)

(c)

(d)

(e)

Figure 26.20 An endangered estuary: Chesapeake Bay.
Chesapeake Bay, which has more than 11,300 kilometers of shoreline, is an estuary and drains more than 166,000 square kilometers in one of the most densely populated and heavily industrialized areas in North America. (*a*) The body of open water is about 320 kilometers long and, at some points, nearly 50 kilometers wide. (*b*) Large metropolitan areas and shipping facilities make Chesapeake Bay one of the busiest natural harbors anywhere. (*c*) The bay is one of the most biologically productive bodies of water in the world. It yielded an annual average of about 275,000 kilograms of fish in the 1960s but only a tenth as much in the 1980s. The human population of the area grew 50% during the same period. (*d*) This grebe is coated from an oil spill off the mouth of the Potomac River. More than 290 oil spills were reported in the bay in one year (1983), with oil transport and commercial shipping expected to double by the year 2020. (*e*) Pesticides, increases in nutrients, and uncontrolled erosion from certain agricultural practices cloud the water and thereby block the light needed for photosynthesis and upset the delicate ecological balance on which the bay's productivity depends. The states that border the bay are cooperating, with the assistance of the Environmental Protection Agency, to try to restore Chesapeake Bay's former productivity.

Freshwater Ecosystems

Freshwater ecosystems (lakes, ponds, and rivers) are distinct from both ocean and land ecosystems, and they are very limited in area. Inland lakes cover about 1.8% of the earth's surface and rivers and streams about 0.3%. All freshwater habitats are strongly connected to land ones, with marshes and swamps constituting intermediate habitats. In addition, a large amount of organic and inorganic material continually enters bodies of freshwater from communities growing on the land nearby (figure 26.21). Many kinds of organisms are restricted to freshwater habitats. When they occur in rivers and streams, they must be able to attach themselves in such a way as to resist or avoid the effects of current or risk being swept away.

Like the ocean, ponds and lakes have three zones in which organisms live: a shallow "edge" zone, an open-water surface zone, and a deep-water zone where light does not penetrate (figure 26.22). **Thermal stratification,** characteristic of the larger lakes in temperate regions, is the process whereby water at a temperature of 4°C (which is when water is most dense) sinks beneath water that is either warmer or cooler. In winter, water at 4°C sinks beneath cooler water that freezes at the surface at 0°C. Below the ice, the water remains between 0° and 4°C, and plants and animals survive there. In spring, as the ice melts, the surface water is warmed to 4°C and sinks below the cooler water, bringing the cooler water to the top with nutrients from the lake's lower regions. This process is known as the **spring overturn** (figure 26.23).

In summer, warmer water forms a layer over the cooler water (about 4°C) that lies below. In the area between these two layers, called the *thermocline,* temperature changes abruptly. Depending on the climate of the particular area, the warm upper layer may become as much as 20 meters thick during the summer. In autumn, its temperature drops until it reaches that of the cooler layer underneath—4°C. When this occurs, the upper and lower layers mix—a process called the **fall overturn.** Therefore, colder waters reach the surfaces of lakes in the spring and fall, bringing up fresh supplies of dissolved nutrients.

Lakes can be divided into two categories, based on their production of organic material. **Eutrophic lakes** have an abundant supply of minerals and organic matter (figure 26.24*a*). Oxygen is depleted below the thermocline in the summer because of the abundant organic material and high rate at which aerobic decomposers in the lower layer use oxygen. These stagnant waters again reach the surface after the fall overturn. In **oligotrophic lakes,** on the other hand, organic matter and nutrients are relatively scarce. Such lakes are often deeper than eutrophic ones, and their deep waters are always rich in oxygen (figure 26.24*b*). Oligotrophic lakes are highly susceptible to pollution from excess phosphorus from such sources as fertilizer runoff, sewage, and detergents.

Figure 26.21 A nutrient-rich stream.
In this stream in the northern coastal mountains of California, as in all streams, much organic material falls or seeps into the water from communities along the edges. This input is responsible for much of the stream's biological productivity.

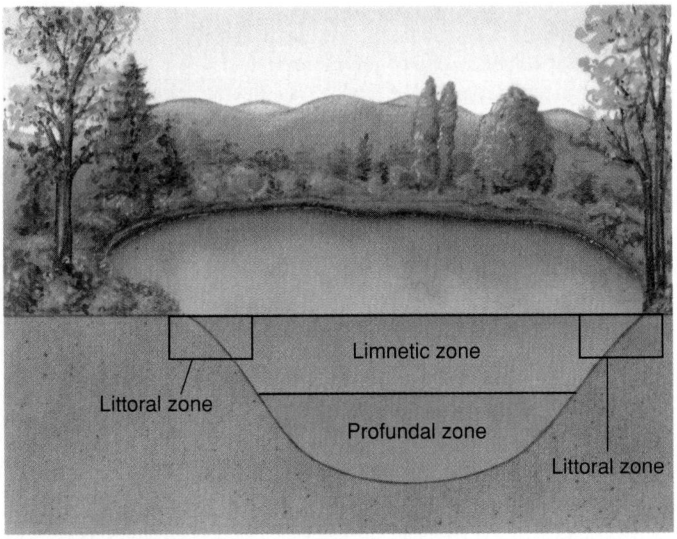

Figure 26.22 The three major freshwater ecosystems.
A shallow "edge" (littoral) zone lines the periphery of the lake where attached algae and their insect herbivores live. An open-water surface (limnetic) zone lies across the entire lake and is inhabited by floating algae, zooplankton, and fish. A dark, deep-water (profundal) zone overlies the sediments at the bottom of the lake. The profundal zone contains numerous bacteria and wormlike organisms that consume dead debris settling at the bottom of the lake.

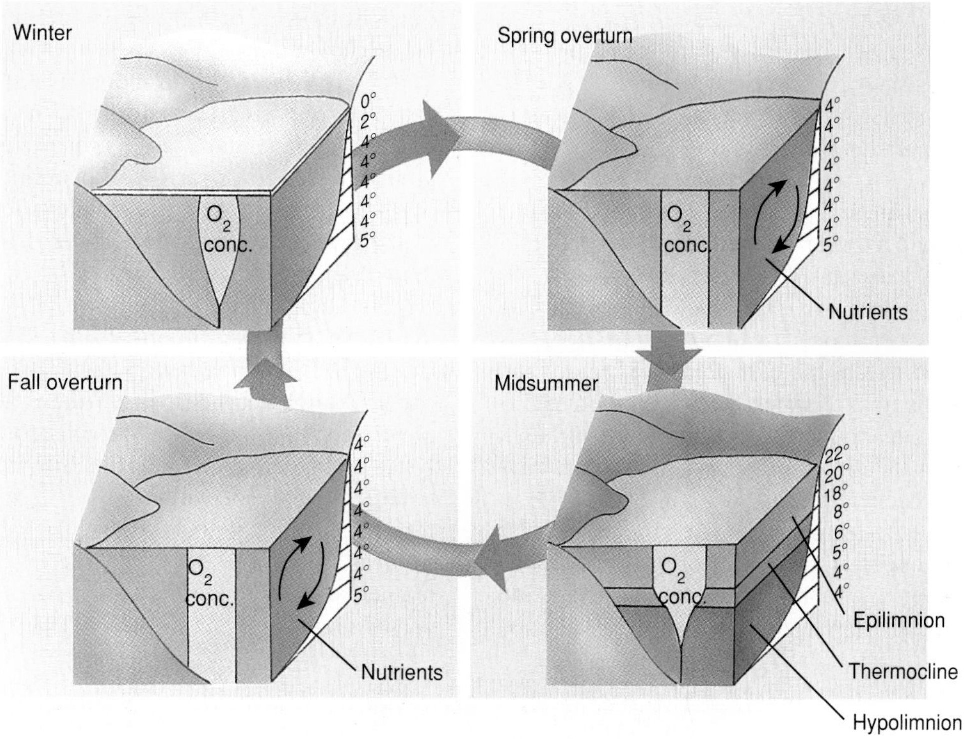

Figure 26.23 Spring and fall overturns in freshwater ponds or lakes.
The pattern of stratification in a large pond or lake in temperate regions is upset in the spring and fall overturns. Of the three layers of water shown in midsummer (*lower right*), the densest water occurs at 4°C (the hypolimnion). The warmer water at the surface is less dense (the epilimnion). The thermocline is the zone of abrupt change in temperature that lies between them. If you have dived into a pond in temperate regions in the summer, you have experienced the existence of these layers directly.

(a)

(b)

Figure 26.24 Eutrophic and oligotrophic bodies of water.
(a) In this eutrophic farm pond, the surface bloom of green algae reflects the plentiful supply of nutrients in the water. (*b*) Lake Tahoe, an oligotrophic lake, lies high in the Sierra Nevada on the border between California and Nevada. The drainage of fertilizers applied to the plantings around residences, businesses, and recreational facilities bordering the lake poses an ever-present threat to the maintenance of the water's deep blue color.

Land Ecosystems

Living on land ourselves, we humans tend to focus much of our attention on terrestrial ecosystems. A **biome** is a terrestrial ecosystem that occurs over a broad area. Each biome is characterized by a particular climate and a defined group of organisms.

While biomes can be classified in a number of ways, the seven most widely occurring biomes are (1) tropical rain forest, (2) savanna, (3) desert, (4) temperate grassland, (5) temperate deciduous forest, (6) taiga, and (7) tundra. The reason that there are seven biomes, and not one or eighty, is that they have evolved to suit the climate of the region, and the earth has seven principal climates. The seven biomes differ remarkably from one another but are consistent within; a particular biome looks the same, with the same creatures living there, wherever it occurs on earth.

There are seven other less widespread biomes: polar ice; mountain zone; temperate evergreen forest; warm, moist evergreen forest; tropical monsoon forest; chaparral; and semidesert. Figure 26.25 shows the geographic distribution of all 14 biomes, both the seven principal ones and the seven less common ones.

If there were no mountains and no climatic effects caused by the irregular outlines of the continents and by different sea temperatures, each biome would form an even belt around the globe. In fact, their distribution is greatly affected by these factors, especially by elevation. Thus, the summits of the Rocky Mountains are covered with a vegetation type that resembles tundra, whereas other forest types that resemble taiga occur farther down. It is for reasons such as these that the distributions of the biomes are so irregular. One trend that is apparent is that those biomes that normally occur at high latitudes also follow an altitudinal gradient along mountains. That is, biomes found far north and far south of the equator at sea level also occur in the tropics but at high mountain elevations.

Distinctive features of the seven major biomes—tropical rain forest, savanna, desert, temperate grassland, temperate deciduous forest, taiga, and tundra—are now discussed in more detail.

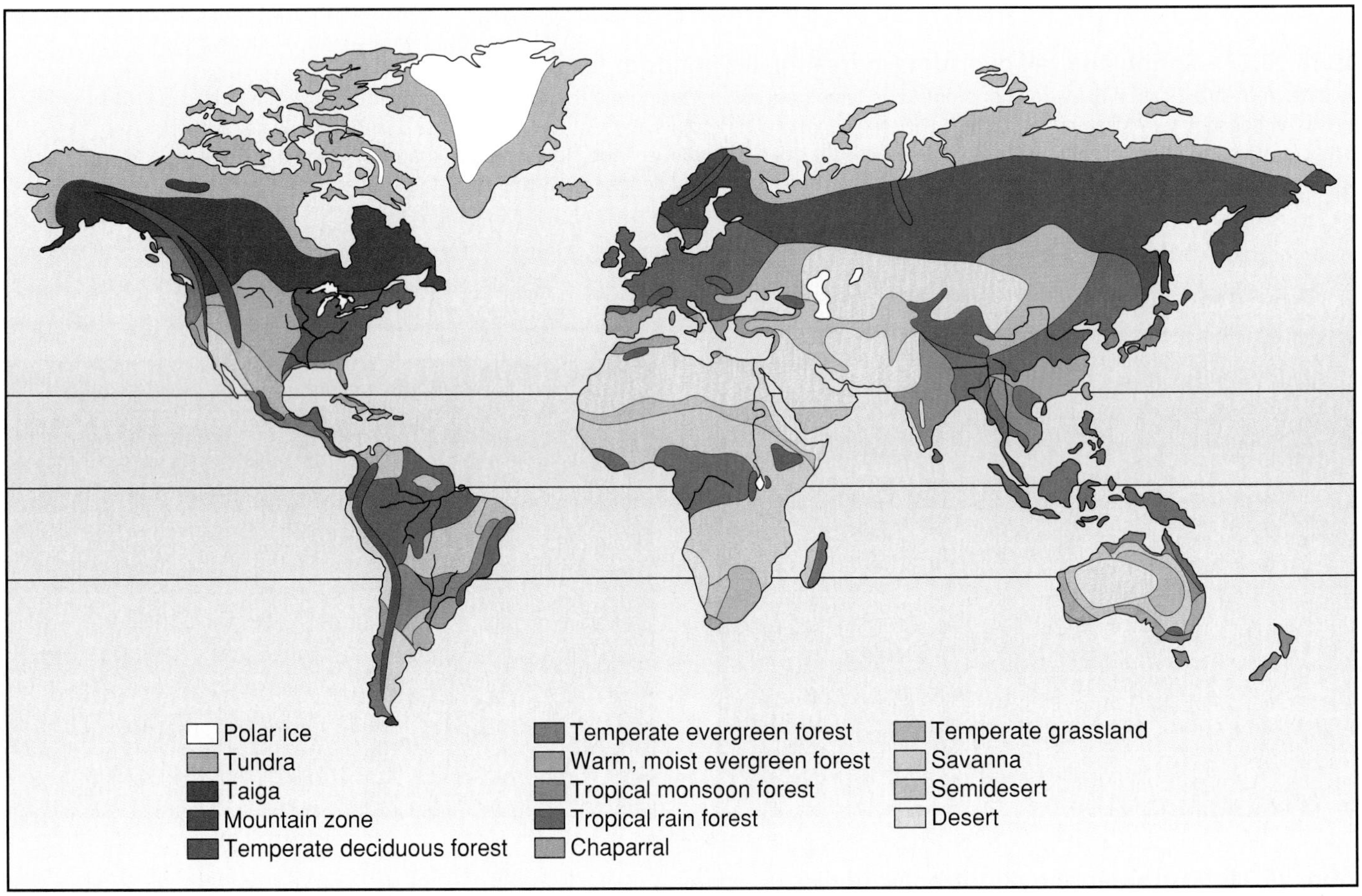

Figure 26.25 Distribution of the earth's biomes.
The seven primary types of biomes are tropical rain forest, savanna, desert, temperate grassland, temperate deciduous forest, taiga, and tundra. In addition, seven less widespread biomes are shown.

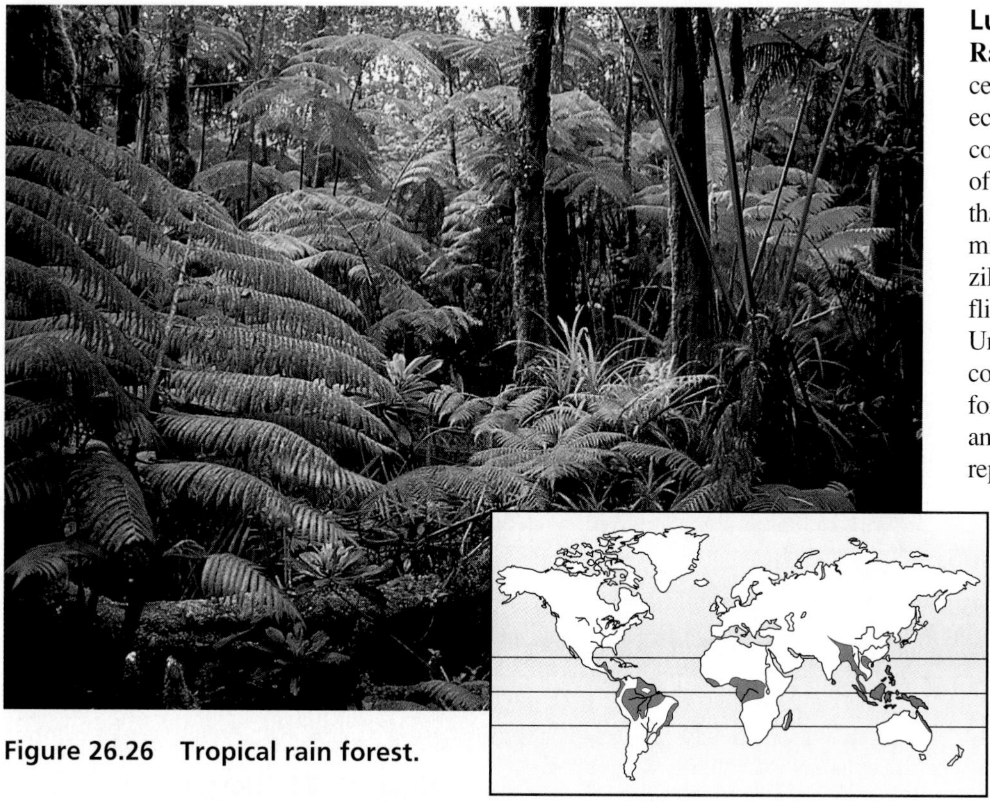

Figure 26.26 Tropical rain forest.

Lush Tropical Rain Forests

Rain forests, which experience over 250 centimeters of rain a year, are the richest ecosystems on earth (figure 26.26). They contain at least half of the earth's species of terrestrial plants and animals—more than 2 million species! In a single square mile of tropical forest in Rondonia, Brazil, there are 1,200 species of butterflies—twice the total number found in the United States and Canada combined. The communities that make up tropical rain forests are diverse in that each kind of animal, plant, or microorganism is often represented in a given area by very few individuals. There are extensive tropical rain forests in South America, Africa, and Southeast Asia. But the world's tropical rain forests are being destroyed, and with them, countless species, many of them never seen by humans. Perhaps a quarter of the world's species will disappear with the rain forests during the lifetime of many of us.

Figure 26.27 Savanna.

Savannas: Dry Tropical Grasslands

In the dry climates that border the tropics are found the world's great grasslands, called **savannas.** Landscapes are open, often with widely spaced trees, and rainfall (75 to 125 centimeters annually) is seasonal. Many of the animals and plants are active only during the rainy season. The huge herds of grazing animals that inhabit the African savanna are familiar to all of us (figure 26.27). Such animal communities occurred in North America during the Pleistocene epoch but have persisted mainly in Africa. On a global scale, the savanna biome is transitional between tropical rain forest and desert. As these savannas are increasingly converted to agricultural use to feed rapidly expanding human populations in subtropical areas, their inhabitants are finding it difficult to survive. The elephant and rhino are now endangered species; lion, giraffe, and cheetah will soon follow them.

Deserts: Burning Hot Sands

In the interior of continents are found the world's great deserts, especially in Africa (the Sahara), Asia (the Gobi) and Australia (the Great Sandy Desert). **Deserts** are dry places where fewer than 25 centimeters of rain falls in a year—an amount so low that vegetation is sparse and survival depends on water conservation (figure 26.28). Plants and animals may restrict their activity to favorable times of the year, when water is present. To avoid high temperatures, most desert vertebrates live in deep, cool, and sometimes even somewhat moist burrows. Those that are active over a greater portion of the year emerge only at night, when temperatures are relatively cool. Some, such as camels, can drink large quantities of water when it is available and then survive long, dry periods. Many animals simply migrate to or through the desert, where they exploit food that may be abundant seasonally.

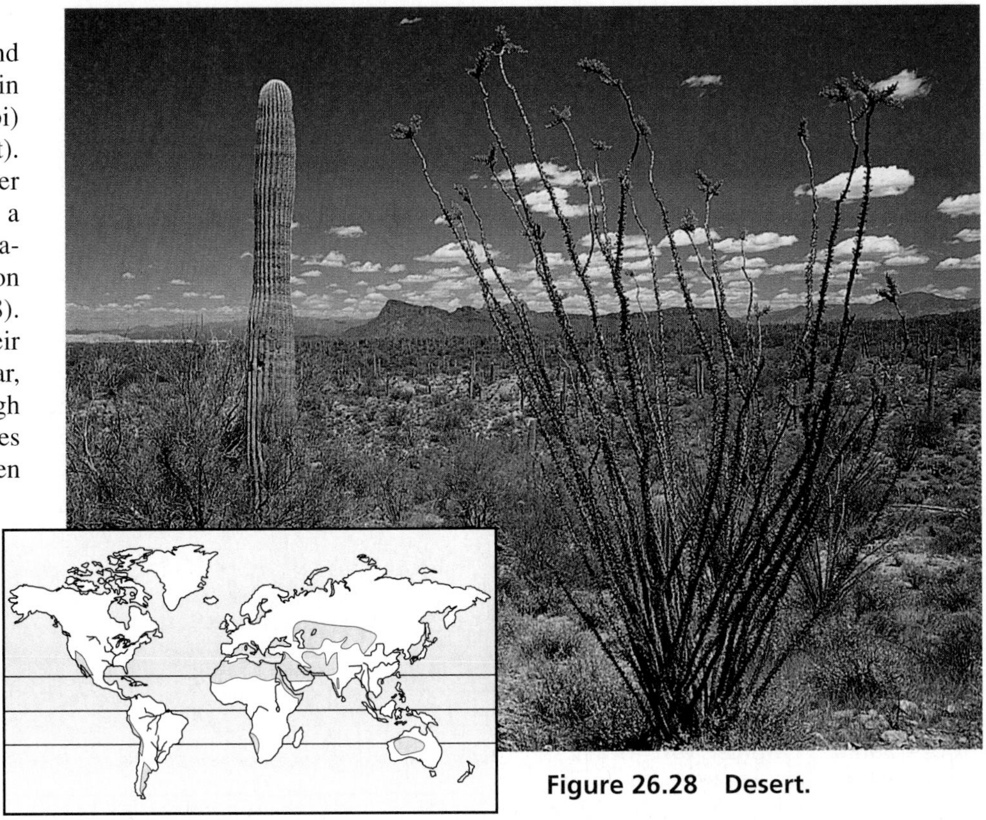

Figure 26.28 Desert.

Grasslands: Seas of Grass

Halfway between the equator and the poles are temperate regions where rich **grasslands** grow. These grasslands once covered much of the interior of North America, and they were widespread in Eurasia and South America as well. Such grasslands are often highly productive when converted to agriculture. Many of the rich agricultural lands in the United States and southern Canada were originally occupied by **prairies,** another name for temperate grasslands. The roots of perennial grasses characteristically penetrate far into the soil, and grassland soils tend to be deep and fertile. Temperate grasslands are often populated by herds of grazing mammals. In North America, the prairies were once inhabited by huge herds of bison and pronghorns (figure 26.29). The herds are almost all gone now, with most of the prairies having been converted to the richest agricultural region on earth.

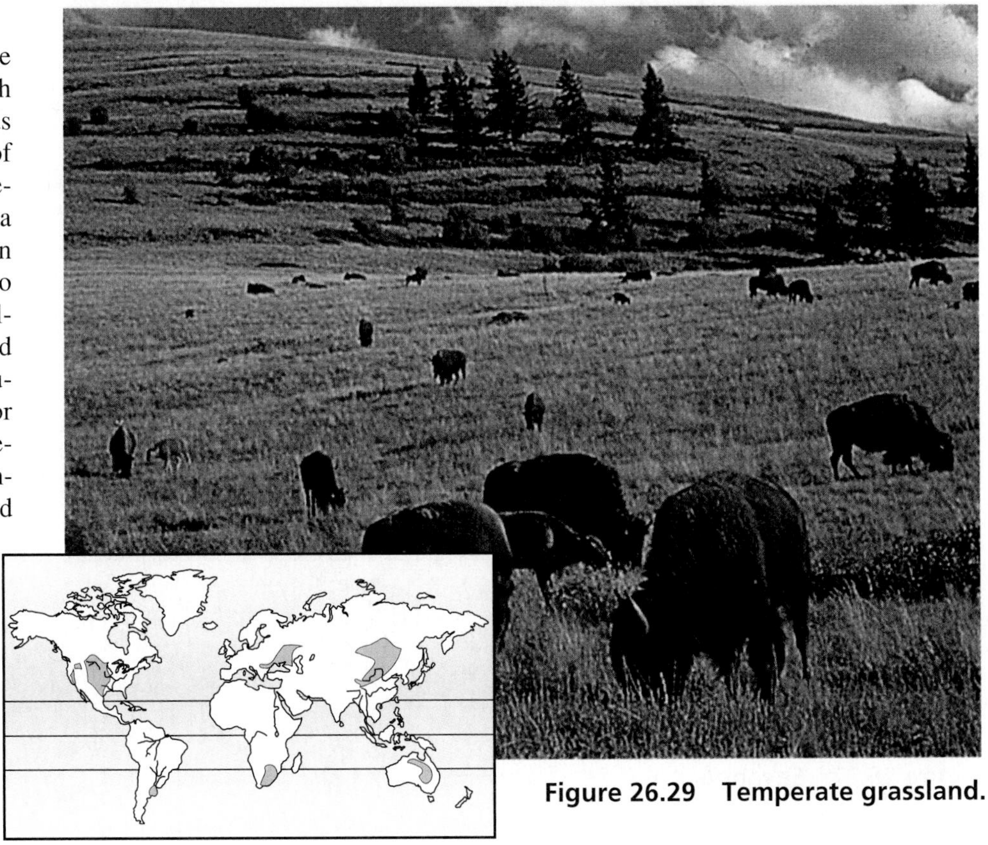

Figure 26.29 Temperate grassland.

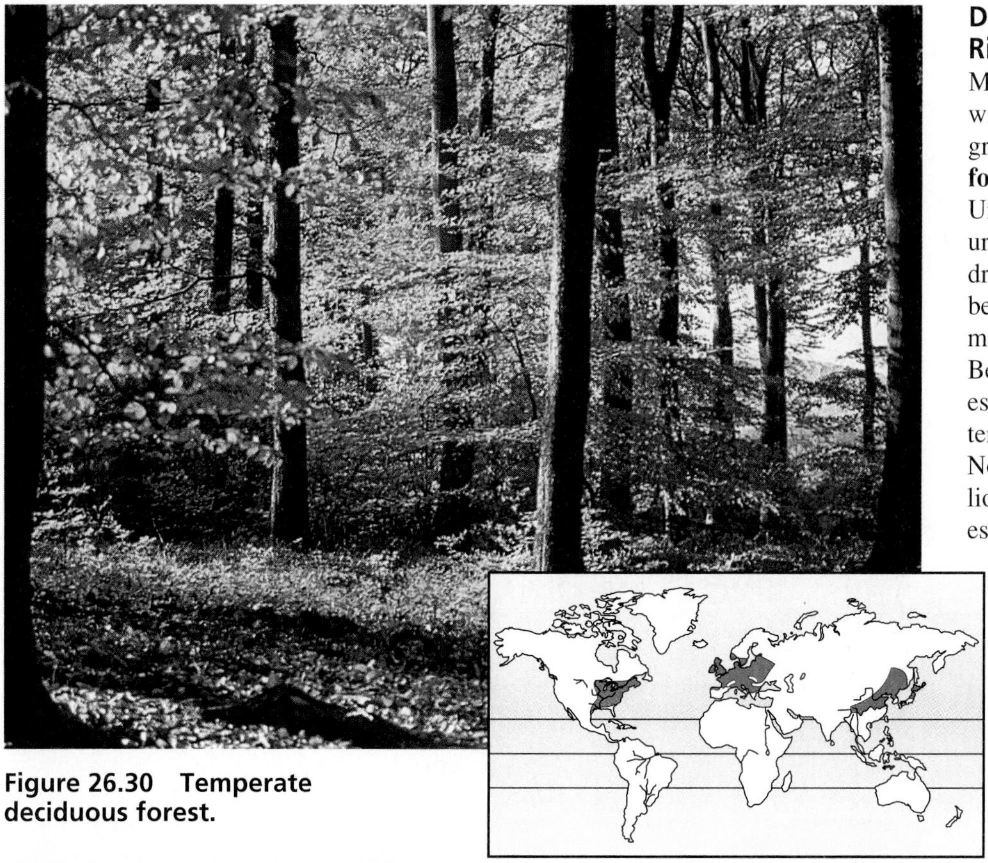

Figure 26.30 Temperate deciduous forest.

Deciduous Forests: Rich Hardwood Forests

Mild climates (warm summers and cool winters) and plentiful rains promote the growth of **deciduous** ("hardwood") **forests** in Eurasia, the northeastern United States, and eastern Canada (figure 26.30). A deciduous tree is one that drops its leaves in the winter. Deer, bears, beavers, and raccoons are the familiar animals of the temperate regions. Because the temperate deciduous forests represent the remnants of more extensive forests that stretched across North America and Eurasia several million years ago, these remaining areas—especially those in eastern Asia and eastern North America—share animals and plants that were once more widespread. Alligators, for example, are found only in China and in the southeastern United States. The deciduous forest in eastern Asia is rich in species because climatic conditions have remained constant.

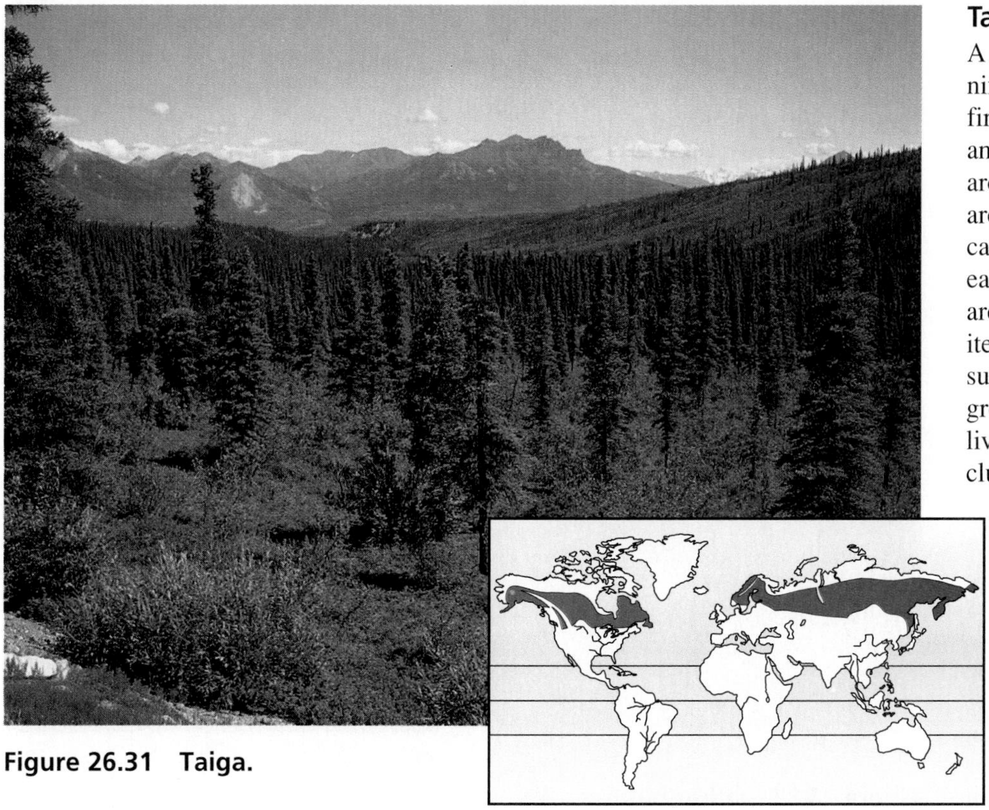

Figure 26.31 Taiga.

Taiga: Trackless Conifer Forests

A great ring of northern forests of coniferous trees (spruce, hemlock, and fir) extends across vast areas of Asia and North America. Coniferous trees are ones with leaves like needles that are kept all year long. This ecosystem, called **taiga,** is one of the largest on earth (figure 26.31). Here, the winters are long and cold, and most of the limited amount of precipitation falls in the summer. Because it has too short a growing season for farming, few people live there. Many large mammals, including elk, moose, deer, and such carnivores as wolves, bears, lynx, and wolverines, live in the taiga. Traditionally, fur trapping has been extensive in this region, which is also important in lumber production. Marshes, lakes, and ponds are common and are often fringed by willows or birches. Most of the trees occur in dense stands of one or a few species.

Tundra: Cold Boggy Plains

In the far north, above the great coniferous forests and below the polar ice, there are few trees. There the grassland, called **tundra,** is open, windswept, and often boggy (figure 26.32). Enormous in extent, this ecosystem covers one-fifth of the earth's land surface. Very little rain or snow falls. When rain does fall during the brief arctic summer, it sits on frozen ground, creating a sea of boggy ground. **Permafrost,** or permanent ice, usually exists within a meter of the surface. Trees are small and are mostly confined to the margins of streams and lakes. Large grazing mammals, including musk-oxen, caribou, reindeer, and carnivores such as wolves, foxes, and lynx, live in the tundra. Lemming populations rise and fall on a long-term cycle, with important effects on the animals that prey on them.

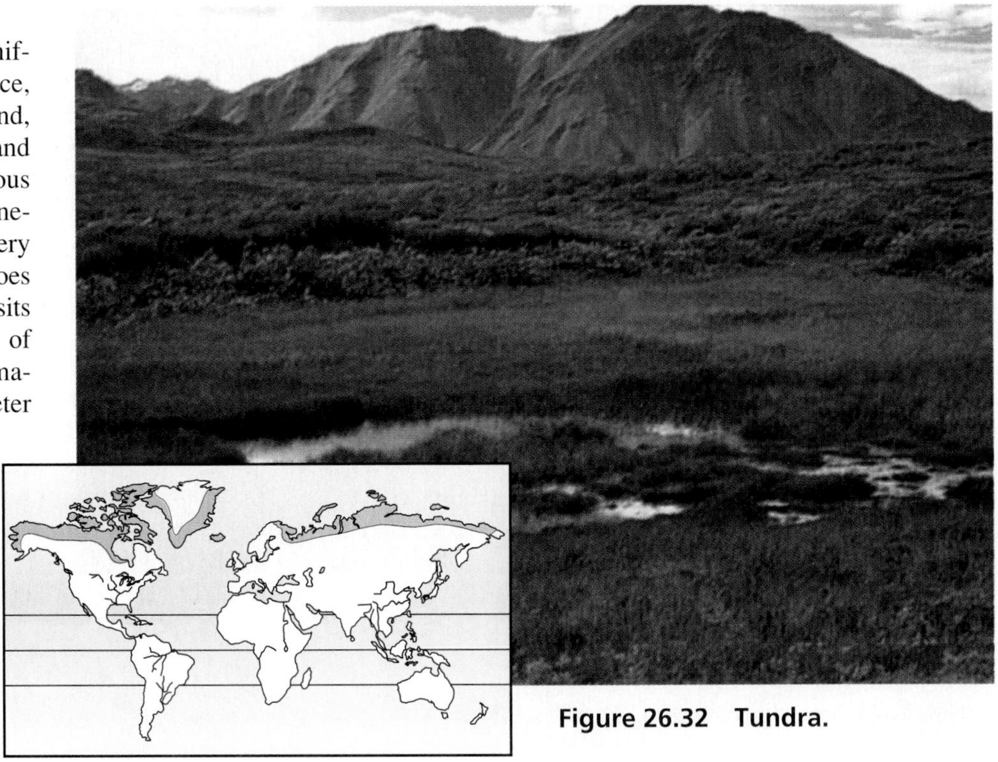

Figure 26.32 Tundra.

Other Biomes

Other biomes include chaparral; polar ice; mountain zone (alpine); temperate evergreen forest; warm, moist evergreen forest; tropical monsoon forest; and semidesert. **Chaparral** consists of evergreen, often spiny shrubs and low trees that form communities in regions with a mediterranean, dry summer climate (figure 26.33*a*). Because of its relatively dry conditions, chaparral is frequently subjected to periodic fires, and interestingly, many plant species found in chaparral can germinate only when they have been exposed to the hot temperatures generated during a fire. **Polar ice** caps lie over the Arctic Ocean in the north and Antarctica in the south (figure 26.33*b*). The poles receive almost no precipitation, so although ice is abundant, freshwater is scarce. In **tropical monsoon forests,** rainfall is very seasonal, heavy during the monsoon season and approaching drought conditions in the dry season (figure 26.33*c*). **Semidesert** areas occur in tropical regions with less rain than monsoon forests but more rain than savannas (figure 26.33*d*).

Figure 26.33 Other biomes.
Other less widespread biomes include (*a*) chaparral, (*b*) polar ice, (*c*) tropical monsoon forest, and (*d*) semidesert.

CHAPTER 26

HIGHLIGHTS

Key Terms | Key Concepts

26.1 What Is an Ecosystem?

Key Terms

ecology 558
habitat 558
ecosystem 558

Key Concepts

- Ecology is the study of how organisms fit into and interact with their environment.
- Within a particular environment, many ecosystems may exist.
- An ecosystem is a self-sustaining group of organisms and the minerals, water, and weather that make up the habitat.

26.2 Energy Flows Through Ecosystems

Key Terms

producer 559
consumer 559
food web 561

Key Concepts

- The flow of energy drives the growth and interaction of organisms within ecosystems.
- Producers are able to capture energy from sunlight by conducting photosynthesis, while consumers must eat plants or animals to obtain their energy.
- Ecologists categorize organisms into trophic or feeding levels based on how they obtain their energy.
- The complex flow of energy among trophic levels is called a food web.

26.3 Materials Cycle Within Ecosystems

Key Terms

evaporation 562
transpiration 562
fossil fuel 564
nitrogen fixation 565

Key Concepts

- Water cycles either by evaporation and condensation or by absorption and transpiration.
- The breaking up of molecules, the burning of wood and fossil fuels, and erosion are ways in which carbon cycles.
- Although nitrogen is abundant in organisms and the atmosphere, much of life relies on the ability of some bacteria to fix nitrogen.

26.4 Major Kinds of Ecosystems

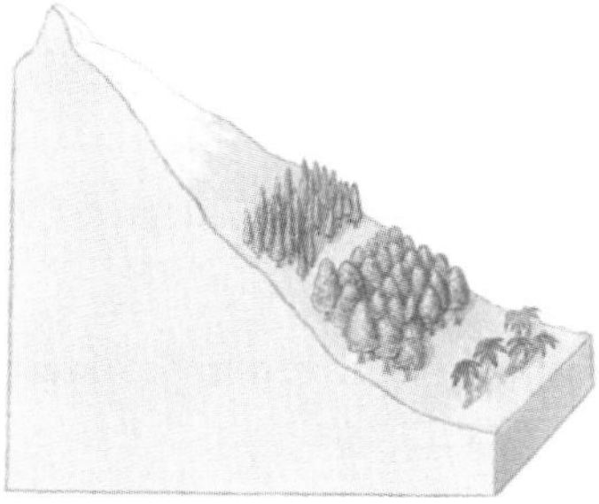

Key Terms

rain shadow 568
plankton 570
biome 574

Key Concepts

- The climate of a particular area will be affected by the intensity of the sun's rays, air currents, elevation, and ocean currents.
- A rain shadow effect is caused by mountains, which force winds upwards, causing them to cool and release their moisture on the windward side of mountains.
- Ocean ecosystems consist of highly diverse coastal areas, open waters that contain plankton, and deep waters where little light penetrates.
- A biome is a climatically defined assemblage of organisms that occurs over a wide area.

26

CONCEPT REVIEW

1. Organisms that eat plant-eaters are
   a. producers.
   b. in trophic level 3.
   c. in trophic level 2.
   d. detritivores.

2. Herbivores are
   a. primary consumers.
   b. secondary consumers.
   c. decomposers.
   d. photosynthetic.

3. The path of energy in complex ecosystems with animals often feeding at various trophic levels is called a
   a. food chain.
   b. mineral cycle.
   c. community.
   d. food web.

4. When a green plant is consumed by another organism, how much of the energy of the plant is converted into the body of the animal that consumes it?
   a. none of it
   b. all of it
   c. 1%
   d. 10%

5. In which of the following cycles does the reservoir of the nutrient exist in mineral form?
   a. carbon cycle
   b. phosphorus cycle
   c. nitrogen cycle
   d. two of the above

6. In the environmental water cycle, water is returned to the atmosphere by
   a. evaporation.
   b. condensation.
   c. transpiration.
   d. precipitation.

7. Which of the following does not account for the variation in climates in different places on earth?
   a. the interaction between large air masses
   b. spring overturn
   c. the rain shadow effect
   d. the earth's annual orbit around the sun

8. In the rain shadow effect, rain falls on the windward side of mountains because
   a. air is cooled as it is forced upwards.
   b. air is warmed as it is forced upwards.
   c. warm air collides with cooler air.
   d. the moisture-holding capacity of air is increased.

9. Most of an ocean's diversity occurs in
   a. the open sea surface.
   b. the shallow waters.
   c. plankton.
   d. the deep sea waters.

10. In the _________ of lakes, cooler water is brought to the top with nutrients from the lake's lower regions.
    a. thermocline
    b. thermal stratification
    c. spring overturn
    d. fall overturn

11. One of the largest ecosystems on earth is
    a. tropical rain forests.
    b. polar ice.
    c. taiga.
    d. chaparral.

12. The difference between an ecosystem and a community is that an ecosystem possesses a _________.

13. The total amount of energy fixed by photosynthesis per unit of time minus the metabolic expense is called the ecosystem's _________.

14. The total weight of all the organisms living in an ecosystem is called the _________ of that ecosystem.

15. Carbon that is tied up in the wood of plants and is gradually transformed by pressure into coal or oil is called _________.

16. Through a process called _________, bacteria convert the plentifully available nitrogen gas into a form that other organisms are able to utilize.

17. Partly enclosed bodies of water where the salinity is intermediate between that of seawater and freshwater are called _________.

18. In _________ lakes, nutrients and organic matter are scarce.

19. The earth's great _________ lie near 30 degrees north or south latitude.

20. The biome that is transitional between tropical rain forest and desert is _________.

Answers to the Concept Review questions appear in Appendix B.

CHALLENGE YOURSELF

1. The net carbon dioxide in the atmosphere is increasing. Where is this carbon dioxide coming from? How could you stop this increase?

2. Why does the net productivity of an ecosystem decrease as it becomes mature?

3. What kinds of biological communities would you expect to find on the windward and leeward sides of a mountain range in an area where the annual precipitation ranged between 20 and 100 centimeters per year and was distributed mainly in one rainy season? How would the height of the mountain range affect the situation?

FOR FURTHER READING

Attenborough, D. *The Living Planet: A Portrait of the Earth.* London: William Collins Sons and BBC, 1984. A beautifully written and illustrated account of the biomes.

Doherty, J. "Alaska's Arctic Refuge." *Smithsonian,* March 1996, 32–43. A visit to a largely untouched ecosystem of great beauty and fragility.

Dybas, C. "The Deep-Sea Floor Rivals Rain Forests in Diversity of Life." *Smithsonian,* January 1996, 96–106. The deep-sea floor is an ecosystem where millions of species of worms live their lives.

"Ecology of Large Rivers." *BioScience* 45 (March 1995). An entire issue devoted to the problems of understanding and managing our nation's large river ecosystems. Timely and informative.

"Frontiers in Biology: Ecology." *Science* 269 (July 1995). An entire issue devoted to recent progress in ecology.

Holloway, M. "Nurturing Nature." *Scientific American,* April 1994, 98–104. The considerable undertaking of the rescue of the Florida Everglades is discussed.

Kusler, J. A., W. J. Mitsch, and J. S. Larson. "Wetlands." *Scientific American,* January 1994, 64–70. The complexity of and vital role played by wetlands are discussed, along with the environmental policy that may or may not protect them.

Norse, E. A. *Ancient Forests of the Pacific Northwest.* Washington, D.C.: The Island Press. 1990. A thorough and interesting account of the controversy surrounding the harvest of ancient forests in the Pacific Northwest of the United States.

Paul, W. M., et al. "The Global Carbon Cycle." *American Scientist* 78 (1990): 310–26. The dynamic responses of natural systems to carbon dioxide may determine the future of the earth's climate.

Rützler, K., and I. Feller. "Caribbean Mangrove Swamps." *Scientific American,* March 1996, 94–99. These complex tropical ecosystems occur worldwide.

Spencer, C. N., B. R. McClelland, and J. A. Stanford. "Shrimp Stocking, Salmon Collapse, and Eagle Displacement." *Bioscience* 41 (1991): 14–21. Cascading interactions in the food web of a large aquatic ecosystem.

Suchanek, T. H. "Temperate Coastal Marine Communities: Biodiversity and Threats." *American Zoologist* 34, 1 (1994): 100–114. A revealing discussion about how the considerable wealth from marine ecosystems is endangered due to mismanagement, pollution, habitat loss, introduced species, climate change, and other factors.

TECHNOLOGY LINKS

The Living World Home Page
http://www.wcbp.com/biology/tlw

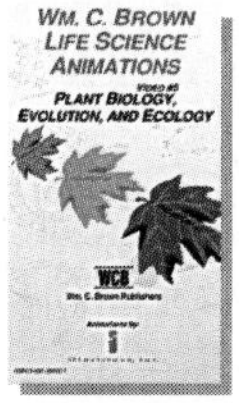

Life Science Animations
Videotape 5
#51 Carbon and Nitrogen Cycles
#52 Energy Flow Through an Ecosystem

27

Living in Ecosystems

CHAPTER OUTLINE

Figure 27.1 Pollination by a bat.
Many flowers have coevolved with other species to facilitate transfer of pollen. Insects are widely known as pollinators, but they're not the only ones. Notice the cargo of pollen on the bat's snout.

The picture of the biological world painted in the previous chapter is a limited one, in which ecosystems are described as if they were chemical factories fueled by a flow of energy from the sun. In fact, this is only part of the story. The organisms themselves also play a major role in shaping the nature of their ecosystem. All of us—evergreens, insects, and humans—share a common history of adaptations and change. Because organisms are able to evolve, the interaction of species with one another creates a complex web of interactions within ecosystems, delicately adjusted to solve the problems of living (figure 27.1).

27.1 Population Dynamics

Ecologists consider groups of organisms at four progressively more inclusive levels of organization. First, individuals of a species that live together are called a **population.** The place in which these individuals live is called their habitat. Populations of different species that live together are called **communities.** A community, together with the nonliving factors with which it interacts, is called an **ecosystem.** An ecosystem regulates the flow of energy, ultimately derived from the sun, and the cycling of the essential elements on which the lives of its plants, animals, and other organisms depend. Finally, the part of the earth where life exists or can be supported is called the **biosphere.** The biosphere is, in effect, a collection of complex ecosystems.

Characteristic Properties of Populations

One of the critical properties of any population is its **population size**—the number of individuals in the population. For example, if an entire species consists of only one or a few small populations, that species is likely to become extinct, especially if it occurs in areas that have been or are being radically changed (figure 27.2). In addition to population size, **population density**—the number of individuals that occur in a unit area, such as per square kilometer—is often an important characteristic. A third significant property is **population dispersion,** the scatter of individual organisms within the population's range. Individuals may be spaced randomly, in clumps, or uniformly (figure 27.3).

(a) (b)

Figure 27.2 Highly endangered populations.
These animals exist in very small populations and are in danger of extinction unless the activities threatening them are brought under control. (*a*) The Sumatran rhinoceros, *Didermoceros sumatrensis,* has been reduced to a few hundred individuals on the island of Sumatra. (*b*) The Mediterranean monk seal, *Monachus monachus.* Fewer than 500 individuals of this species are believed to exist, and they are confined to remote, cliffbound coasts and islands in the Mediterranean.

(a) Random (b) Clumped (c) Uniform

Figure 27.3 Dispersion patterns in populations.
Random dispersion (*a*) is rare, while clumped dispersion (*b*) is more common. Clumped patterns are indicative of an environment in which resources are unevenly distributed. Uniform dispersion (*c*) reflects a population that is evenly spaced throughout the range.

Population Growth

In addition to size, density, and dispersion, another key characteristic of any population is its capacity to grow. Most populations tend to remain relatively constant in number, regardless of how many offspring the individuals produce. Under certain circumstances, population size can increase rapidly for a time. The rate at which a population increases when there are no limits on its rate of growth is called the **innate capacity for increase,** or **biotic potential.** This theoretical rate is almost impossible to calculate, however, because there are usually limits to growth. What biologists in fact calculate is the **realized rate of population increase** (abbreviated ***r***). This parameter is defined as the number of individuals added to the population minus the number lost from it. The number added to it equals the birth rate plus the number of **immigrants** (new individuals entering and residing with the population), while the number lost from it equals the death rate plus the number of **emigrants** (individuals leaving the population). *r* thus equals:

$$r = (\text{birth} + \text{immigration}) - (\text{death} + \text{emigration})$$

Exponential Growth

A population's innate capacity for growth is constant, determined largely by the organisms' physiology. Its actual growth, on the other hand, is not a constant, because *r* depends on both the birth rate and the death rate, and both of these factors change as the population increases in size. Thus to get the population growth rate, *r* must be corrected for population size:

$$\text{population growth rate} = rN,$$

where *r* is the realized rate of population increase and *N* is the number of individuals in the population. In general, as a population increases and begins to exhaust its resources, its death rate rises. The number of individuals grows rapidly at first, and this type of growth is called **exponential growth** (figure 27.4). Soon, however, the rate of increase slows as the death rate begins to rise. Eventually, just as many individuals are dying as are being born. The early rapid phase of population growth lasts only for a short period, usually when an organism reaches a new habitat where resources are abundant. Examples of this phenomenon include algae colonizing a newly formed pond and the first terrestrial organisms that arrive on an island recently thrust up from the sea.

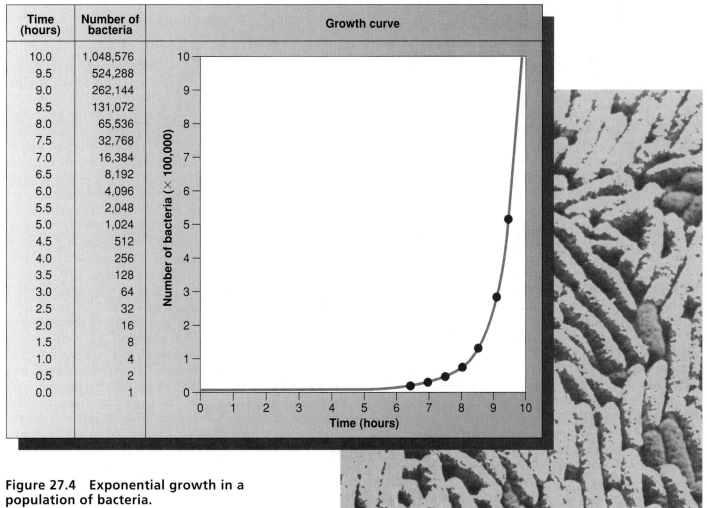

Time (hours)	Number of bacteria
10.0	1,048,576
9.5	524,288
9.0	262,144
8.5	131,072
8.0	65,536
7.5	32,768
7.0	16,384
6.5	8,192
6.0	4,096
5.5	2,048
5.0	1,024
4.5	512
4.0	256
3.5	128
3.0	64
2.5	32
2.0	16
1.5	8
1.0	4
0.5	2
0.0	1

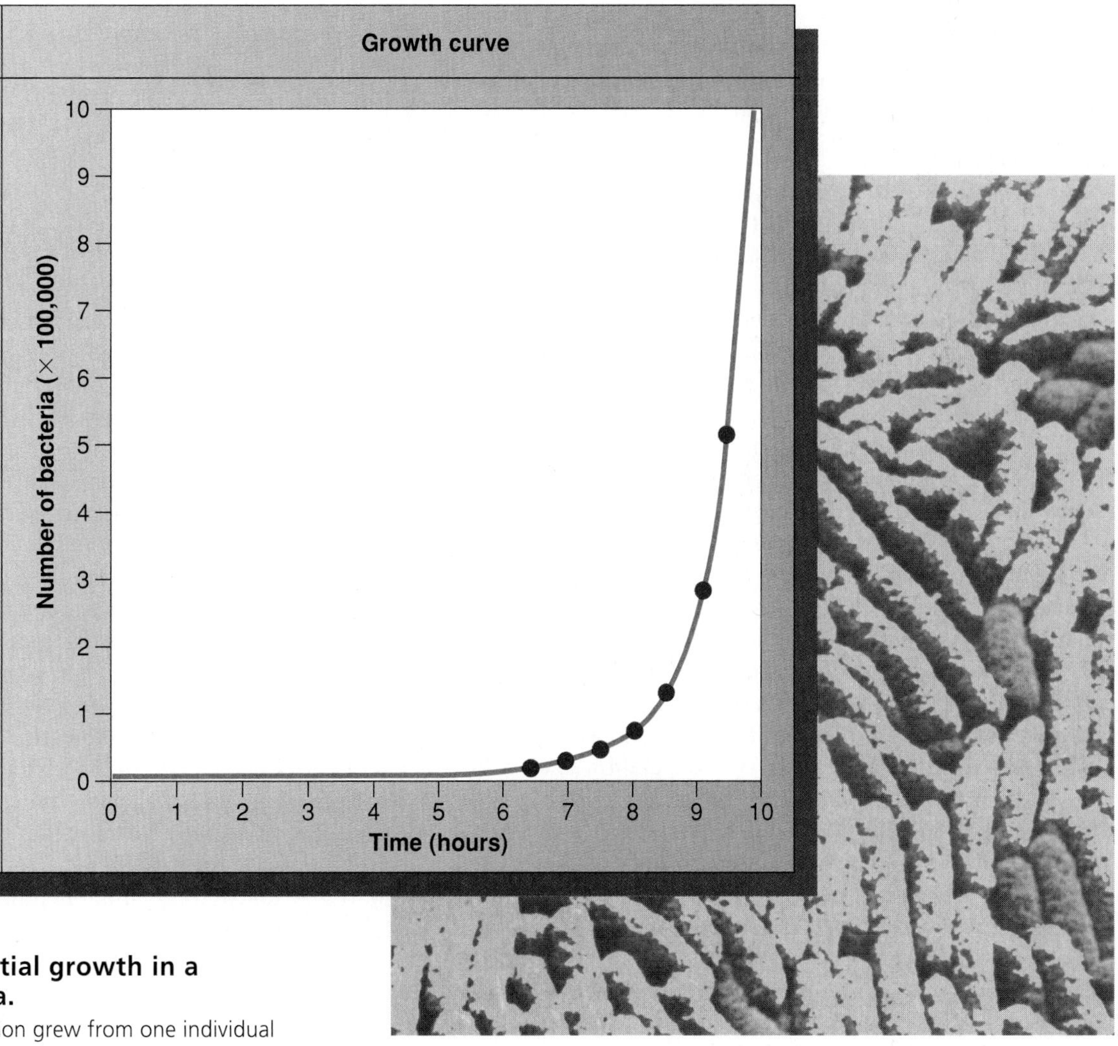

Figure 27.4 Exponential growth in a population of bacteria.
In just 10 hours, this population grew from one individual to over 1 million!

Carrying Capacity

No matter how rapidly new populations grow, they eventually reach an environmental limit imposed by shortages of some important factor, such as space, light, water, or nutrients. A population ultimately stabilizes at a certain size, called the **carrying capacity** of the particular place where the population lives. The carrying capacity, symbolized by K, is the number of individuals that can be supported at that place indefinitely. As carrying capacity is approached, the population's rate of growth slows greatly, because in effect, there is less room for each individual. The growth of a specific population, which is always limited by one or more factors in the environment, can be approximated by the following *logistic growth equation*:

$$\text{population growth rate} = rN\left(\frac{K-N}{K}\right)$$

In other words, the growth of the population under consideration equals the ideal rate of increase (r multiplied by N, the number of individuals present at any one time), adjusted for the amount of resources still available. The adjustment is made by multiplying rN by the fraction of K still unused (K minus N, divided by K). As N increases, the fraction by which r is multiplied becomes smaller and smaller, and the rate of increase of the population declines. Graphically, this relationship is the S-shaped **sigmoid growth curve,** characteristic of biological populations (figure 27.5).

Density-Dependent and Density-Independent Effects

As a population approaches its carrying capacity for a particular habitat, competition for resources, emigration, and accumulation of toxic waste products all increase. The resources for which population members must increasingly compete may include food, shelter, light, mating sites, or any other factor necessary for them to carry out their life cycle and reproduce.

Changes in the intensity of competition that occur as a result of population density are called **density-dependent effects.** Among animals, density-dependent effects are often accompanied by hormonal changes that cause changes in behavior that directly affect ultimate population size. One striking example occurs among migratory locusts (a kind of grasshopper), which when crowded produce hormones that cause them to enter a new, migratory phase. The locusts then take off as a swarm and fly long distances to new habitats. In contrast, **density-independent effects** are caused by such factors as the weather and physical disruption of the habitat—factors that operate regardless of population size.

Agriculture depends to some extent on the characteristics of the sigmoid growth curve. Early in a population's history, resources are not yet limiting the growth of individuals. For best yields, this is the best time to "harvest" a population. Commercial fisheries also attempt to operate so that fish populations are always harvested in the steep, rapidly growing parts of the growth curve, rather than when the population is near the carrying capacity of its habitat or when it is very small. Overharvesting a population that is too small can destroy the population's productivity for many years. In forestry, the harvesting of mature trees tends to exploit the upper part of the sigmoid curve; proper management requires an understanding of the characteristics of the populations being harvested.

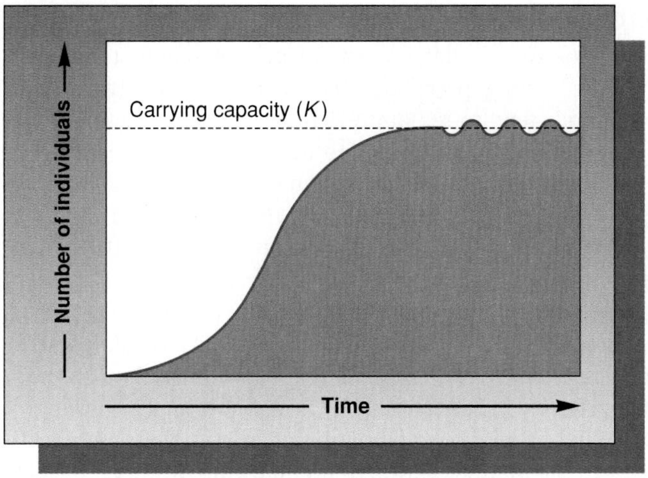

Figure 27.5 The sigmoid growth curve.
The sigmoid growth curve begins with a period of exponential growth like that shown in figure 27.4. When the population approaches its environmental limits (K), growth slows and finally stabilizes, fluctuating around the maximum number of individuals that the environment will hold.

Mortality and Survivorship

A population's intrinsic rate of increase depends on the ages of the organisms in it and the reproductive performance of the individuals in the various age groups. Very young and very old individuals obviously are not as reproductively active as the rest of the population. The speed with which a population can grow depends on how many of its members are reproducing, and thus, on the population's **age distribution.** A population with many reproducing individuals will grow much faster than one with few reproducing individuals (figure 27.6). Age distributions differ greatly from species to species and even, to some extent, from place to place within a given species.

One way to express a population's distribution is the **survivorship curve. Survivorship** is defined as the percentage of an original population that survives at a given age. Survivorship curves are based on data collected in a summary called a **life table.** Life tables originated when insurance companies began studying the human population to find out when individuals in the population were expected to die. Ecologists discovered that this method of recording mortality rates could be applied to other populations. A life table shows how many individuals are alive at any one time and expresses this information in a variety of ways: the number of individuals dying during an interval of time, the

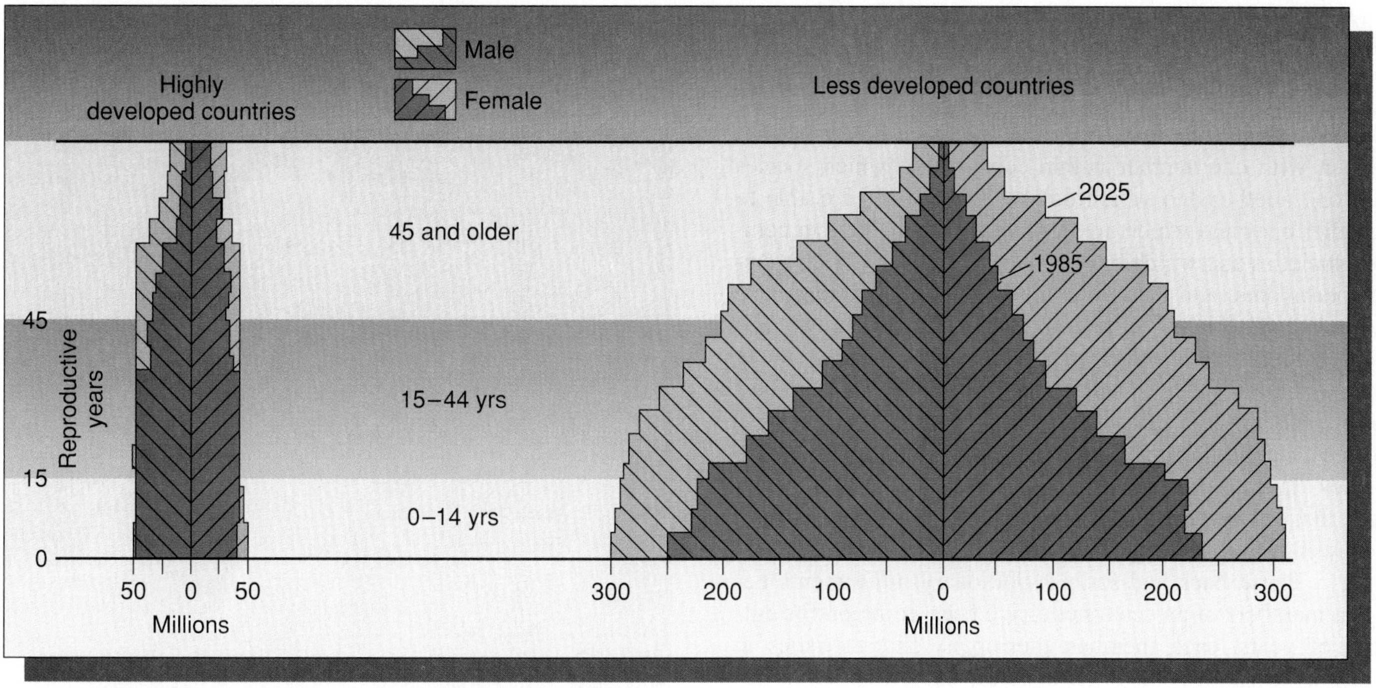

Figure 27.6 The population explosion is hitting less developed countries hardest.
Less developed countries have triangular age distribution profiles, with much of their population yet to enter childbearing age. When all of the young people begin to bear children, the population will experience rapid growth.

mortality rate (the percentage of an original population that are dead at any given age), and an estimation of the remaining life span of an individual at any one time. Life tables can be used to examine mortality patterns in a given population, and when graphed in a survivorship curve, these individual patterns of populations can be grouped into types.

Figure 27.7 shows the three main types of survivorship curves. A type I survivorship curve represents the kind of life cycle found in humans, where even though infants are susceptible to death at relatively high rates, humans' highest mortality rates occur later in life, in their postreproductive years. In a type II survivorship curve, as represented by hydra, individuals are likely to die at any age, producing a straight line. A type III survivorship curve represents the kind of life cycle found in oysters, which produce vast numbers of offspring, only a few of which live to reproduce. Once oyster offspring grow into reproductive individuals, however, their mortality is extremely low.

Many animal and protist populations in nature probably have survivorship curves that lie somewhere between those characteristic of type II and type III, and many plant populations, with high mortality at the seed and seedling stages, are probably closer to type III. Humans have probably approached type I more and more closely through the years, with the birth rate remaining relatively constant or declining somewhat but with the death rate dropping markedly.

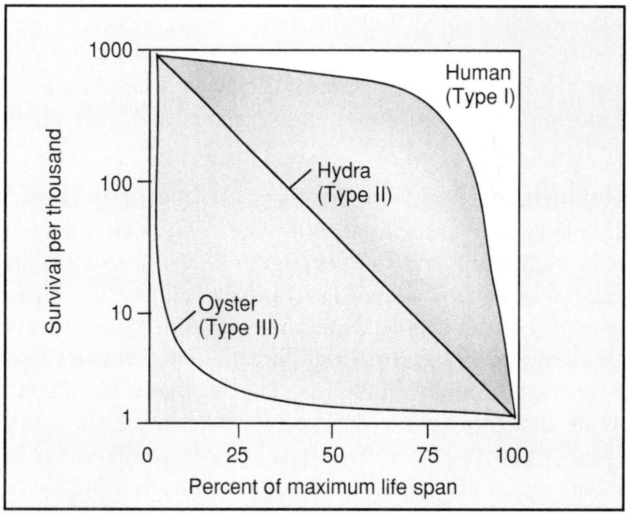

Figure 27.7 Survivorship curves.
The shapes of the curves are determined by the percentages of individuals who die at specific ages. Type I curves represent populations in which most individuals reach their physiologically determined maximum age, as with humans. Type II curves represent populations in which the mortality rate is fairly constant at all ages, as with the freshwater animal called the hydra. Type III curves indicate populations in which mortality is high in the early stages but then declines, as with the oyster.

27.2 How Coevolution Shapes Ecosystems

Evolution shapes the many ways in which animals and plants interact with one another within ecosystems: which species help one another and which do not; which ones are able to eat other ones and which are not; which species live in complex societies and which have solitary members. The producers, herbivores, and carnivores in a mature ecosystem have changed and adjusted to one another continually over millions of years. For example, flowers have evolved many ways to encourage a particular kind of insect to feed on their nectar, rather than on that of any other kind of flower, so that the insect will spread its pollen to other plants of the same species. Insects in turn have evolved a number of special traits that let particular species specialize in particular kinds of flowers when seeking nectar.

These back-and-forth evolutionary adjustments between members of an ecosystem have been going on for millions of years, each member altering its characteristics in response to changes in other species, only to have its response elicit further changes in that species. Biologists call this pattern of challenge-and-response **coevolution.**

Coevolving in Competition: Plants and Their Herbivores

About 250,000 species of plants and algae capture about 1% of the energy that falls on their leaves and convert it into the materials that fuel all life on earth. At least 2 million species of herbivores feed on these plants. The rich tapestry of life on earth has been dictated in large measure by how herbivores go about eating plants and how plants avoid being eaten.

How Plants Defend Themselves from Herbivores

While plants defend themselves in many ways—you may be familiar with thorns and stickers (figure 27.8)—their most important defenses are chemicals. Virtually all plants contain toxic or distasteful chemicals that make eating them unpleasant or dangerous. Over 15,000 different plant chemicals have been characterized by biologists, and as many as 100,000 kinds are thought to exist. Have you ever broken out in a rash from contact with poison ivy? If so, you were the victim of chemical attack by a plant—the ivy produces a sticky oil called urushiol that can cause severe blistering and can persist on clothing for years unless removed with soap. Most herbivores tend to avoid plants that possess defensive chemicals—which is undoubtedly why so many plants possess them.

Related plants often produce the same chemical defenses. Poison ivy, poison oak, and poison sumac all produce urushiol, for example. Plants of the mustard family produce "mustard oils," the substances that give the sharp, distinctive tastes to mustard, cabbage, radish, and horseradish. We humans enjoy the taste, but it is toxic to many kinds of insects.

Figure 27.8 Plant defenses.
The most obvious ways for plants to limit herbivore activities are physical defenses. Thorns, spines, and prickles discourage browsers, although some animals, such as this gerenuk, have learned to overcome these obstacles.

How Herbivores Slip Past Plant Defenses

Over the millions of years that herbivores and plants have shared the earth's ecosystems, herbivores have evolved many ways to overcome the chemical defenses of plants. Because each plant group's chemical defense is different, this coevolution results in herbivores that specialize in feeding on particular kinds of plants that no other herbivore eats.

How have animals managed to overcome the chemical defenses of plants? Cabbage butterflies provide a good example. Plants of the mustard family are avoided by almost all insects because of their toxic chemicals, but the caterpillars of cabbage butterflies eat these plants avidly! This particular species of butterfly has evolved the ability to break down the mustard oils, rendering them harmless. As a result of this evolutionary breakthrough, cabbage

Figure 27.9 Commensalism: sea anemones and clownfish.
These clownfish often form commensal relationships with sea anemones, gaining protection by remaining among the anemones' tentacles and gleaning scraps from the anemones' food.

butterflies have been able to use a new food resource (plants of the mustard family) without any competition from other herbivores, all of which still avoid mustards.

Animals also manufacture defensive chemicals to defend themselves from being eaten by carnivores, as well as offensive chemical weapons to aid in the capture of prey. Bees, wasps, scorpions, and spiders are but a few examples. Have vertebrates played this evolutionary game? Sure. The brightly colored poison dart frogs of South America, for example, produce a very toxic chemical in the mucus that covers their skin. A few micrograms, the weight of a tiny sliver of eyelash, will kill a person if injected into the bloodstream.

Coevolving in Cooperation: Symbiosis

Symbiotic relationships are those in which two kinds of organisms live together. Symbiotic relationships reflect a particularly intimate form of coevolution. The major kinds of symbiotic relationships are (1) **commensalism,** in which one species benefits while the other neither benefits nor is harmed, and (2) **mutualism,** in which both participating species benefit. **Parasitism,** in which one species lives by consuming another, used to be considered another form of symbiosis, but biologists now prefer to consider it a special form of predation (in which the "eater" is much smaller than the "eatee").

Commensalism: Taking Without Hurting

In natural ecosystems, the individuals of one species often become physically attached to those of another. Birds nest in trees, for example, and tropical orchids grow on the branches of other plants. In these cases, the host plant is unharmed, while the organism that grows or nests on it benefits. A well-known example of commensalism involves the relationship between small tropical fishes, called clownfish, and sea anemones, marine animals with stinging tentacles. These fish have evolved the ability to live among the tentacles of the sea anemones, even though the tentacles would quickly paralyze other fish that touched them. The fish gain protection from predators by remaining among the anemones' tentacles, and they also feed on the leftovers from the meals of the host anemone (figure 27.9).

Another striking example of this sort of symbiosis is seen in the deep sea, where barnacles grow in profusion on the backs of whales. While these "passengers" seem to be no trouble to the whale, they gain two advantages: (1) they are protected on the whale's back from predators, and (2) they are carried by the whale to new sources of food (which they filter from water). Of course it is difficult to be certain that the whale receives no benefit from having barnacles on its back—perhaps other whales find it more attractive to be so decorated. Because biologists cannot know everything, any case of commensalism may be a case of mutualism hidden by our ignorance of what is going on. There is no clear-cut boundary between commensalism and mutualism.

Mutualism: Sharing Benefits

Mutualism, in which animals evolve ways of cooperating, has been of fundamental importance in determining the structure of biological communities. In some of the most important examples, photosynthetic organisms provide food in exchange for other favors. Certain fungi and green algae form associations called lichens, in which the fungi garner nutrients for both from rock, while the algae carry out photosynthesis to provide food for both. Mycorrhizae are associations of fungi and the roots of plants in which the fungi obtain nutrients from the soil, while the roots provide food obtained from photosynthesizing leaves. Legumes and certain other plants possess nitrogen-fixing nodules in their roots, which are associations of bacteria and root cells; the bacteria fix atmospheric nitrogen, while the roots provide food obtained from the photosynthesizing leaves.

In diverse ecosystems such as the tropical rain forest, many forms of cooperation have evolved. A striking but by no means unusual example of mutualism is provided by leaf cutter ants. These ants live in enormous colonies that daily send out scavenging parties to bring back fragments cut out of leaves to the underground nest. They don't eat these leaves—instead, special "munchers" chew the leaves up and spread the leaf paste out on the floor of underground chambers. They then inoculate the paste with the spore of a particular species of fungus. A special caste of "farmers" then grow and tend the fungus farm, weeding out any other fungal species that might contaminate the bed. When harvested, the fungi constitute the primary food of the ants and their larvae.

Another relationship of this kind involves ants and aphids. Aphids are small insects that suck fluids from living plants with their piercing mouthparts. They extract a certain amount of the sucrose and other nutrients from this fluid but excrete the rest. Certain ants have taken advantage of this habit by protecting the aphids from predators, carrying the aphids to new plants, and using the honeydew that the aphids excrete as food (figure 27.10).

Figure 27.10 Mutualism: ants and aphids.
These ants are tending to willow aphids, feeding on the "honeydew" that the aphids excrete continuously, moving the aphids from place to place, and protecting them from potential predators.

27.3 How Ecosystems Develop Over Time

An ecosystem can be thought of as a machine with living parts that captures a tiny fraction of the energy bombarded upon it by the sun and uses that energy to modify the environment. As we have seen, the form of an ecosystem is structured by a complex web of biological and environmental interactions, shaped by climate and evolution. Each organism in an ecosystem confronts the challenge of survival in a different way. The sum of all its interactions, how it copes with the physical environment and competes with other organisms, is called that organism's **niche.** A niche is a picture of how an organism lives in an ecosystem.

To understand the niches of an ecosystem we must focus our attention not on worldwide cycles or biological interactions molded by long centuries of coevolution but rather on day-to-day events, the mundane happenings that occur every hour as organisms muddle along trying to get through another day. We will see that much of what happens depends on competition, on what the other guy is doing.

Competing in Ecosystems: The Niche

A niche is an organism's "job" in the ecosystem. A niche is often described in terms of how it affects energy flow through the ecosystem. For example, the niche of a leaf-covered tree is that of a primary producer, while the niche of a caterpillar that eats the leaves is that of a herbivore (a primary consumer), and a blue jay that eats the caterpillar is a carnivore (a secondary consumer). Niches are also often described in terms of space utilization. Some small carnivores, like spiders, hunt insect prey within leaf litter on the forest floor; others, like ladybird beetles, hunt on the stems of plants; and still others, like dragonflies, hunt on the wing in the air. Within every ecosystem, each organism has its own peculiar niche, its particular way of making a living and influencing its surroundings.

Filling the Niche

The niche of an organism includes everything about how it chooses to live—the temperature and humidity it prefers; whether it feeds in the morning, the evening, or all the time; the season it reproduces; what it chooses to eat; and where it searches for its food. The total niche that an organism is potentially able to use within an ecosystem, the entire range of each factor that it is able to exploit, is called its **fundamental niche.**

Dividing the Niche

Sometimes organisms are not able to occupy their entire fundamental niche, because somebody else is using it. We call such situations when two organisms attempt to utilize the same resource **competition.** Competition is the struggle by two organisms to utilize the same resource when there is not enough of the resource to go around. The part of a fundamental niche that an organism actually occupies in an ecosystem is called its **realized niche.** Figure 27.11 shows fundamental and realized niches in two species of barnacle.

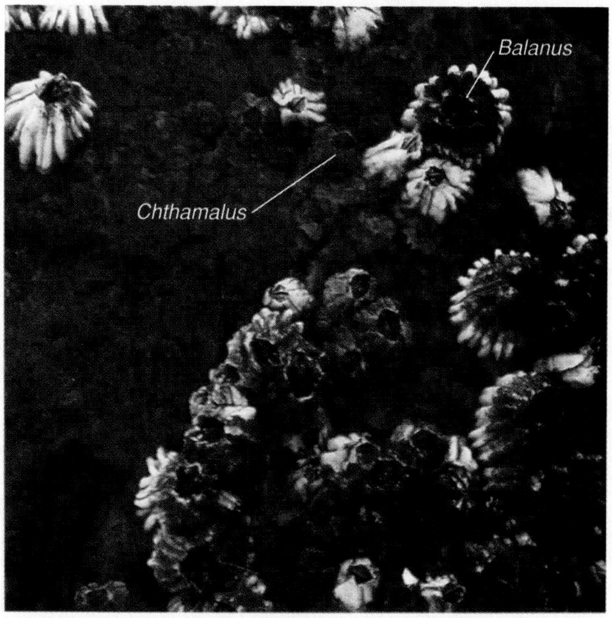

(a)

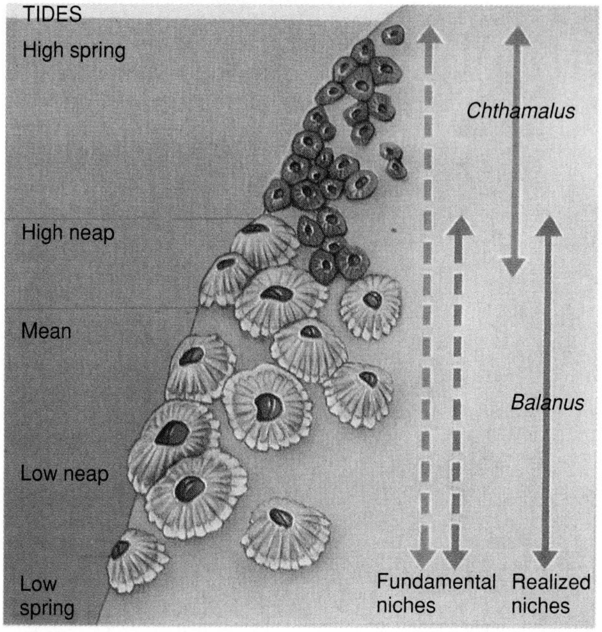

(b)

Figure 27.11 Competition that limits niche use.
(*a*) *Chthamalus* (the smaller, smoother barnacle) and *Balanus* (the larger, ridged barnacle) grow together on a rock along the coast of Scotland. (*b*) Arrows show the distribution of the two species with respect to different water levels. *Chthamalus* can live in both deep and shallow zones (its fundamental niche), but *Balanus* forces *Chthamalus* out of that part of its fundamental niche that overlapped the realized niche of *Balanus*.

When two species compete fiercely for the same resource, one usually wins, driving the other species to extinction within the ecosystem. This process, called **competitive exclusion,** is rare in mature ecosystems, however, because evolution tends to minimize such wasteful competition by favoring the subdivision of niches. Competing species undergo evolutionary changes that make them tend to use different parts of the resource. This sort of competition-avoiding change is called **character displacement.** For example, in one set of field observations, the late Princeton ecologist Robert MacArthur studied five species of warblers—small, insect-eating birds that coexist during part of the year in the forests of the northeastern United States and adjacent Canada. Although they all appeared to be competing for the same resources, MacArthur found that each species actually spent most of its time feeding in different parts of the trees, so each ate different subsets of insects in those trees (figure 27.12). Some of the warbler species fed on insects near the ends of the branches, while others regularly penetrated well into the foliage, and some stayed high on the trees, while others fed on the lower branches. These patterns were recombined in different ways characteristic of each warbler species. As a result of these different feeding habits, each species of warbler actually occupied a different niche; in other words they had different ways of utilizing the environment's resources and thus did not directly compete with one another for limited resources.

In mature ecosystems a species rarely occupies its full fundamental niche, almost always instead being limited to a smaller realized niche. In effect, the occupants of the ecosystem share the wealth. The result is an ecosystem that fosters many more kinds of lifestyles, a more complex, specialized, and far more interdependent ecological system.

During the past decade, there has been a lively debate about the real importance of competition in nature in relation to community structure. Environmental changes greatly affect the outcome of competitive situations. Furthermore, in the kinds of highly diverse situations that occur in nature, there are opportunities for many species to coexist in the same environment, so evaluating the competitive interactions that may be occurring is often very difficult.

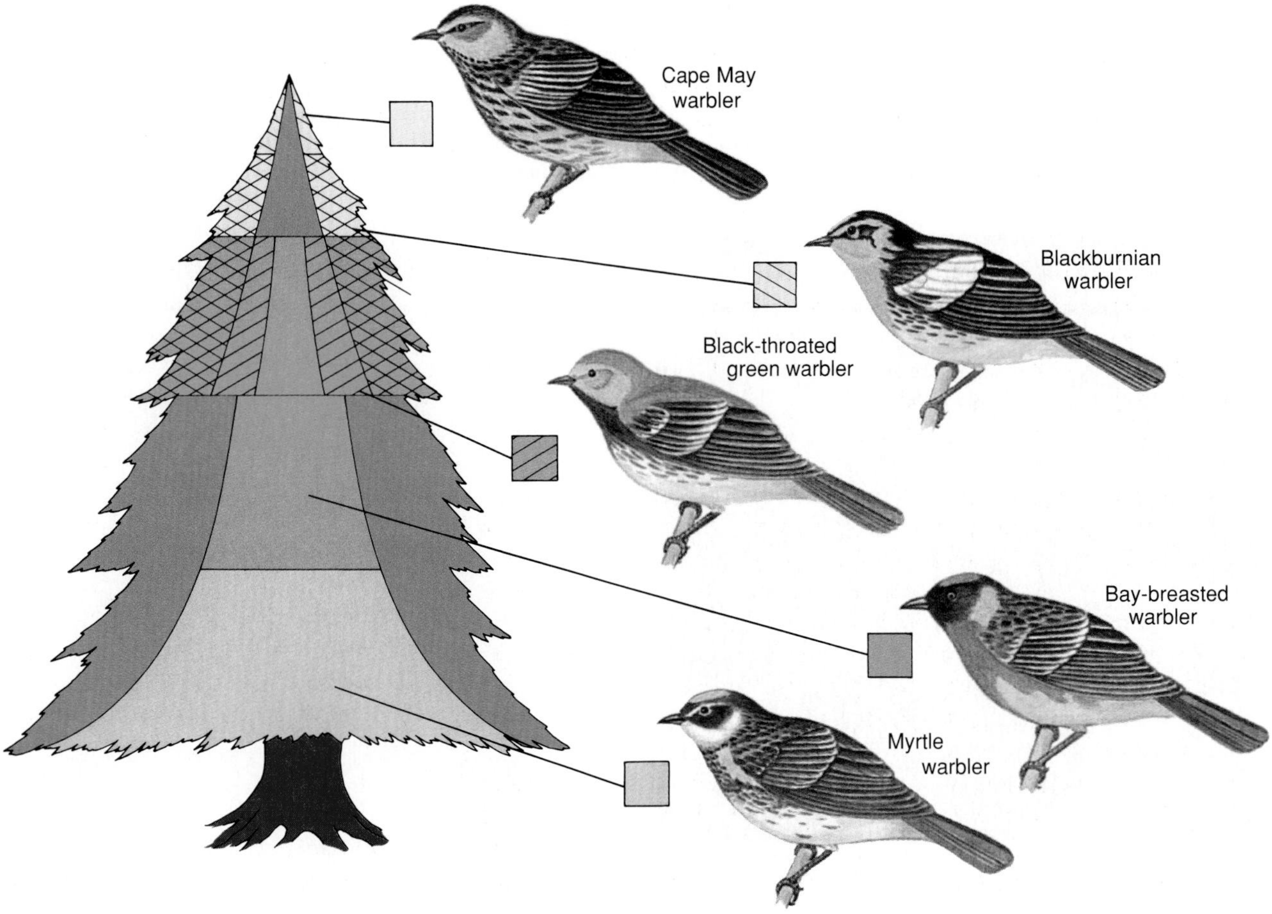

Figure 27.12 Using environmental resources differently avoids competition.
Although all five species of warbler shown here feed on insects in the same spruce trees at the same time, they mainly feed in different parts of the tree and in different ways. Thus, competition is avoided.

Figure 27.13 Wolves chasing a moose—what will be the outcome?
On Isle Royale, in Michigan, a large pack of hungry wolves chased a moose for 2 kilometers. The moose finally turned and faced the wolves, who by that time were exhausted from running through chest-deep snow. The wolves laid down, and the moose walked away. Thus, predator-prey relationships are not always clear-cut. Isle Royale wolves only prey on moose who are old and infirm; rarely are they able to run down a healthy moose.

Predator-Prey Interactions

Predation, like competition, is a factor that may limit population size. In this sense, predation includes everything from one animal capturing and eating another to parasitism. Predation and parasitism are two ends of a biological spectrum between which there is no clearly marked distinction. They are governed by similar principles.

When experimental populations are set up in the laboratory, the predator often exterminates the prey and then becomes extinct itself, having nothing to eat. If refuges are provided for the prey, prey populations will be driven to low levels but can recover. The populations of predators and prey then tend to follow a cyclical pattern: the low population levels of the prey species provide scant food for the predators, which, in turn, become scarce, which allows the prey to recover and again become abundant. Population cycles are characteristic of some species of small mammals, such as lemmings, and may be stimulated, at least in some situations, by their predators.

The predator-prey relationships between large carnivores and grazing mammals are sometimes deceiving (figure 27.13). For example, on Isle Royale in Lake Superior, moose once multiplied freely in isolation. When wolves later reached the island by crossing over the ice in winter, as the moose had done earlier, biologists assumed that the wolves were playing the determining role in controlling the moose population. More careful studies, however, have demonstrated that this is not the case. The moose that the wolves eat are mostly old and diseased and would not have survived long anyway. In general, the moose are controlled by the amount of available food, their diseases, and many factors other than the wolves.

Wolves have been returning slowly to remnants of their former habitat in the United States outside of Alaska, where fewer than 1,300 wolves live. Because wolves kill a small amount of livestock, their presence in farming areas is controversial. Their reintroduction into their former range in Yellowstone National Park has likewise met with opposition from neighboring ranchers. On the other hand, there is no known case in North America of injury to a human being by a wild, nonrabid wolf.

Both plants and animals have devised complex mechanisms with which to ward off predators. As with plants, some animals employ chemical defenses to ward off predators and herald their poisonous nature with dramatic coloration. Still other animals mimic poisonous species by adopting their coloration. The sections that follow examine these defenses in more detail.

Animal Defenses

Some animals are able to feed on plants rich in defensive chemicals and receive an extra benefit. For example, plants of the milkweed family contain cardiac glycosides that protect these plants from most herbivores. Cardiac glycosides have a drastic effect in vertebrate heart function. However, some insects are able to feed on the plants and concentrate the poison in their bodies. Insects that feed regularly on plants of the milkweed family are generally brightly colored. Among them are brightly colored cerambycid beetles, whose larvae feed on the roots of the milkweed plants; bright blue or green chrysomelid beetles; and bright red bugs (order Hemiptera). In some parts of the world, there are also bright red grasshoppers and other very vividly colored insects. These herbivores advertise their poisonous nature by their bright colors, using a warning strategy known as **aposematic coloration.** Aposematic coloration is characteristic of animals with effective defense systems, including not only poisons but also stings, bites, and other means of repelling predators. Such organisms benefit by clearly advertising their defenses—for example, by exhibiting colors not normally found in that particular habitat (figure 27.14).

When caterpillars of monarch butterflies feed on plants of the milkweed family, they concentrate, store, and pass the cardiac glycosides through the chrysalis stage to the adult butterfly and even to the eggs; all stages are then themselves protected from predators! A bird that eats a monarch butterfly quickly regurgitates it and thenceforth avoids the conspicuous orange and black pattern that characterizes the adult monarch (figure 27.15). Locally, however, some birds have acquired the ability to tolerate the protective chemicals and are able to eat monarchs.

Animals that display warning coloration must remain together if the system is to be effective. A lone individual that exhibits warning coloration but is eaten will not deliver a message useful for the survival of other animals with a similar appearance. But if genetically related individuals are similarly colored and live in the same vicinity, the selective advantage is obvious. Such animals tend to live together in family groups. In contrast, some animals use **cryptic coloration**—coloration that blends in with the surroundings—to avoid predation (figure 27.16). Animals that are cryptically colored, or camouflaged, tend not to live in groups because the discovery of one by a predator would offer a valuable clue to the presence of others.

Figure 27.14 Warning: poisonous frog!
Frogs of the family Dendrobatidae, such as this individual, are abundant in the forests of Latin America and are extremely poisonous to vertebrates. More than 200 different chemicals called alkaloids have been isolated and identified in these frogs, and some of them are playing important roles in neuromuscular research. Two groups of Indians in western Colombia obtain a potent poison for blowgun darts from extremely toxic species of these frogs that live in their region.

Figure 27.15 Monarch butterflies are poisonous.
All stages of the monarch butterfly's life cycle are protected from birds and other predators by the poisonous chemicals in the milkweeds and dogbanes on which the butterflies feed as larvae. (*a*) A cage-reared bluejay, which has never seen a monarch butterfly before, eats one. (*b*) The same bird a few minutes later, regurgitating the butterfly. This bird is not likely to eat an orange and black insect again! Both caterpillars (*c*) and adult butterflies (*d*) "advertise" their poisonous nature with warning coloration.

Figure 27.16 A striking example of cryptic coloration.
A young jackrabbit in the desert near Tucson, Arizona, displays both stillness and camouflage.

Mimicry

Many unprotected species have come, during the course of their evolution, to resemble distasteful ones that exhibit warning coloration. Provided that the unprotected animals are present in low numbers relative to those of the species they resemble, the unprotected animals will also be avoided by predators. Such a pattern of resemblance is called **Batesian mimicry.** Many of the best-known examples of Batesian mimicry occur among butterflies and moths. Predators in systems of this kind apparently use visual cues to hunt for their prey. Otherwise, similar patterns of coloration would not offer any protection to species without chemical defenses.

The groups of butterflies that are the models in Batesian mimicry are, not surprisingly, members of groups whose larvae feed on only one or a few closely related plant families that are strongly protected chemically. One well-known mimic is the viceroy butterfly, which resembles the poisonous monarch. The larvae feed on willows and cottonwoods, and neither they nor the adults are distasteful to birds.

In another kind of mimicry, **Müllerian mimicry,** several unrelated but protected animal species come to resemble one another. Thus, a number of different kinds of stinging wasps have black-and-yellow striped abdomens, but they may not all be descended from a common ancestor with similar coloration. In general, yellow-and-black and bright red tend to be common color patterns that warn predators relying on vision to avoid such animals.

In both Batesian and Müllerian mimicry, mimic and model must not only look alike but also act alike to deceive predators (figure 27.17). For example, the members of several families of beetles that resemble wasps behave surprisingly like the wasps they mimic, flying often and actively from place to place. Mimics must also spend most of their time in the same habitats as the their models. If they did not, predators would discover that all of those conspicuous animals are not only easily seen but also quite tasty!

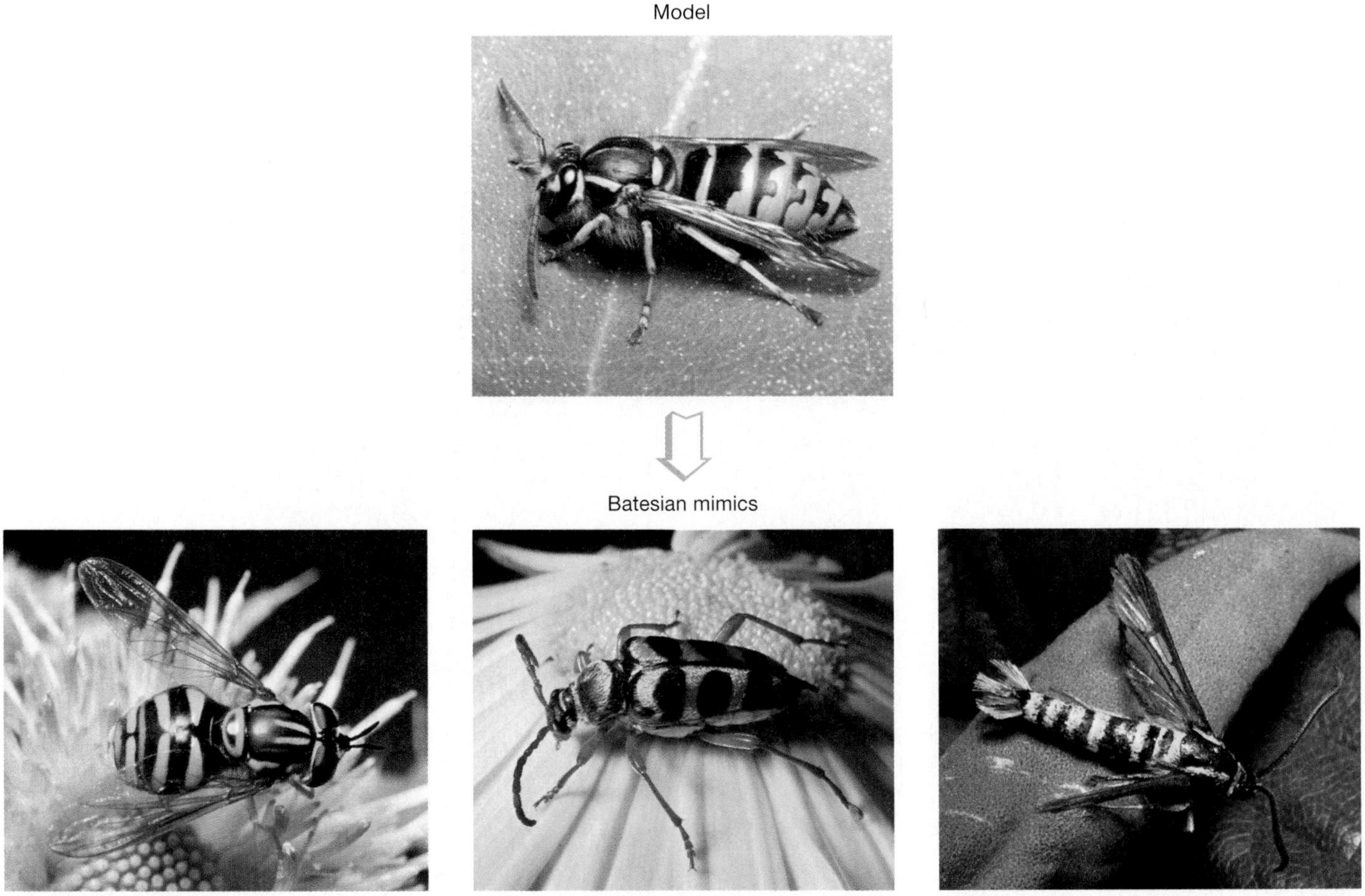

Figure 27.17 Batesian and Müllerian mimicry.
The familiar yellow-and-black stripes of wasps (the model, *top*) form the basis for large Batesian and Müllerian mimicry complexes. Shown are Batesian mimics representing three separate orders of insects, all rarer than wasps and with behavior patterns similar to those of the dangerous wasps they resemble. None of these mimics sting, yet all are conspicuous members of the communities where they occur, flying about actively and remaining in full view at all times. Because these insects all resemble each other, they are Müllerian mimics of each other.

Competition Develops Ecosystems

You can think of the community of organisms that inhabit an ecosystem as a web of competitive interactions. The niche of each species within the ecosystem is linked to many other niches by how it competes for limited resources—that is, it competes with other species for space and materials, it may prey on some species, and it is the prey of others. Because of these relationships, any change that alters the niche of one species can affect many others within the web. Such changes are a natural part of the history of any ecosystem.

The beginnings of ecosystems almost always involve serious disruptions. When an undersea volcano erupts to form a new island, or a receding glacier exposes bare soil, the empty ecosystem is quickly occupied. First, plants invade the area. These first settlers are specialized for life under harsh conditions such as bare rock, and they are able to eke out a living where few others could. They do not remain in the ecosystem very long, however, because their pioneering efforts to make the ground more hospitable encourage invasion by a second wave of plant immigrants. The new arrivals soon outcompete and replace the original inhabitants, only to be replaced in turn by a third wave of species better able to compete in the new environment that the second wave created. And so the process continues. As the ecosystem matures, niches become more and more finely subdivided, species more and more interdependent.

This regular progression of species replacement is called **succession** (figure 27.18). When succession occurs on land without soil where nothing has grown before, it is called **primary succession.** By contrast, succession that occurs in areas where there has been previous growth, as in abandoned fields or forest clearings, is called **secondary succession.** It used to be thought that the stages of a succession were always the same and always led to the same final community of organisms within any particular ecosystem, one called a **climax community.** Biologists now realize that chance plays a major role in the outcome of the many competitive interactions that determine each successional change, and that for this reason, no two successions are exactly alike. Nevertheless, the progression of changes will tend toward similar communities in similar physical conditions. That is why a biome like rain forest or tundra is so similar worldwide.

(a)

(b)

Figure 27.18 Succession.
Mount St. Helens in the state of Washington erupted violently on May 18, 1980. The lateral blast devastated more than 600 square kilometers of forest and recreation lands within 15 minutes. (*a*) This is how an area near Clearwater Creek looked four months after the eruption. (*b*) Five years later, succession was underway at the same spot, with shrubs, blueberries, and dogwoods following the first plants that became established immediately after the blast.

27.4 What Makes an Ecosystem Stable?

Mature ecosystems in nature persist because they are able to cope with changes in climate and with the many conflicting demands of their inhabitants. Climate, the coevolution of species, competition—these driving forces of evolutionary change have molded ecosystems over long periods of time to create the world we see today. However, the most important single influence on natural ecosystems today—human activity—is not acting slowly. Human influence now extends to all parts of the globe, to every blade of grass growing anywhere on our world. The fate of every ecosystem on earth over the next century will be influenced more by its stability in the face of human disruption than by any other factor. It is thus essential that we consider how ecosystems respond to disruption and try to gain an understanding of why some ecosystems are more stable than others.

Biodiversity Promotes Stability

Ecologists have long sought to understand why some ecosystems are more stable than others—better able to avoid permanent change and return to normal after disturbances like land clearing, fire, invasion by plagues of insects, or severe storm damage. Most ecologists now agree that biologically diverse ecosystems are in general more stable than simple ones. Ecosystems with more kinds of different organisms support a more complex web of interactions, and as a result, an alternative niche is more likely to be able to compensate for the effect of a disruption. The number of species in an ecosystem, called **species richness,** is the quantity usually measured by biologists in attempting to characterize an ecosystem's **biodiversity** (figure 27.19).

However, the very complexity of highly diverse ecosystems, while buffering them from everyday insult, also makes them more likely to have particular points of vulnerability that, if damaged, can have far-reaching consequences. A species that interacts with many other elements of an ecosystem in critical ways is called a **keystone species.** Because the loss of a keystone species may affect many other organisms in a diverse community, such species represent points of particular sensitivity in the ecosystem—places where the complex ecological machine can be easily broken. Think of how you would design a computer to protect it from "crashing"—like a complex ecosystem, you would make it with a diverse array of redundant (repeated) circuits that could take over the function of others if the need arose. However, this added complexity has a price: The advanced chip is uniquely vulnerable to damage at those points where functional crossover takes place—a single flaw can destroy the operation of the entire computer, while a simple chip would have lost only one circuit.

Thus, if we wish to preserve natural ecosystems, we should do all we can to preserve their biological diversity, while at the same time realizing that diverse ecosystems can be damaged too.

Figure 27.19 How many species are there?
Scientists are attempting to determine the species richness in the canopy of moist tropical forests, the most diverse biological community on earth. In these experiments, the insects and other animals living in the treetops are sprayed with insecticide and then sampled as they drop onto the sheets below. By such methods, Terry L. Erwin, Jr., of the Smithsonian Institution, has estimated that there may be as many as 30 million kinds of organisms in the world. Taxonomists have thus far recognized only about 1.4 million of these.

No ecosystem, however complex or simple, can be stable in the face of massive disruption. Evolution has provided no mechanism to buffer ecosystems from total destruction by human activity. Unhappily, much of today's disruption of the world's ecosystems is of this drastic sort—replacing forests with grazing land, with "pure stands" of a single kind of tree, with plowed fields for farming, or with asphalt.

Figure 27.20 Jaguars in Central America.
The jaguar is a rare species in Central America. Hunting and habitat destruction by humans have eliminated the jaguar from much of its former range. Jaguars are especially sensitive to habitat fragmentation because they require a large area for their daily activities.

What Promotes Biodiversity?

If diversity is such a good thing, why doesn't evolution drive all ecosystems toward high species richness? Why isn't tundra as diverse as tropical rain forest? If we examine all the species-rich ecosystems on earth, and ask what they have in common that species-poor ecosystems lack, two general trends emerge: large size and tropical latitude.

Size

Larger ecosystems, because they contain a more varied array of physical habitats, are able to support a greater number of different species. Because they offer a more diverse array of potential niches, more species can be packed into them.

This dependence of diversity upon the physical size of an ecosystem has a very important consequence: If you reduce the size of the ecosystem, you reduce the number of species it can support (figure 27.20). In today's world, this happens with increasing frequency. A new road cut across a forest divides the ecosystem in two for the many animals that cannot or do not cross it. In extreme cases, reduction in ecosystem area can produce *faunal collapse,* a situation in which many of the animal species living in the ecosystem become extinct there because the smaller ecosystem is simply unable to divide its resources into that many pieces.

For exactly this reason, in all but the largest of our national parks, a high proportion of the mammals once there have become extinct. Different animals have been lost from different parks, so nothing is entirely lost, but the pattern of species loss is clear. When the parks were formed at the turn of the century, their animals were part of much larger communities, but, walled off from their surroundings by society's growing development around the parks, the animals have been restricted to much smaller areas. The loss of species richness has been the result. If we want to maintain animal diversity for future generations, the parks will have to be managed carefully, and lost species reintroduced. In today's world, there is no longer any true wilderness. Our national parks, like gardens, have to be tended.

Latitude

Compared to the arctic region, tropical areas have many more species, more than 6 million of the estimated 10 million species of organisms that exist on earth. Why should this be so? There are two principal reasons:

1. **Length of growing season.** As a general rule of thumb, ecosystems with more resources are able to subdivide them into more niches, and so maintain a greater number of species. In the tropics, with ample sunlight, warm temperatures, and generous rainfall throughout the year, the growing season never stops. Progressing toward the poles, however, growing seasons shorten, so that polar ecosystems have much less energy to work with in weaving their food webs.
2. **Climatic stability.** The climate in the tropics has not been subjected to glaciers and other major disruptions during the evolutionary past. Thus the unchanging physical conditions have provided a longer evolutionary window in the tropics for specialized relationships to develop.

27.5 How Humans Disrupt Ecosystems

To get some sense of how ecosystems are responding to the massive impact of the twentieth century, it is first important to realize that disruption and disturbance are a natural part of the life of any ecosystem. What is different today is the scale of the disturbance. Ecosystems are being harmed in new ways for which evolution has not prepared any safeguards. Having already converted large stretches of many biomes to agriculture, urban areas, or other uses, human populations are attacking those biomes—such as the tropical rain forests—that are less suitable for exploitation and about which we know much less. There are so many of us that we must manage the whole earth as a single system if those who follow us are to find a stable ecological situation in which they can live.

What We Are Doing Wrong

Human activities harm ecosystems in three principal ways: disrupting physical habitats, lowering species diversity, and reducing competition.

Disrupting Physical Habitats

Land clearing for agriculture and the many chemicals produced by industrial society can alter the physical habitat and have disastrous effects on natural ecosystems. Burning high-sulfur coal produces acid precipitation that harms forests and lakes by lowering the pH of the habitat. Throughout the world, fish and amphibian populations (whose larvae develop in water and are very sensitive to pH) are becoming extinct as a result (figure 27.21). Many other cases will be discussed in the next chapter, but the general point is clear—human industrial society disrupts the physical habitat in many ways, sometimes with far-reaching consequences that are difficult to predict.

Figure 27.21 Acid precipitation kills.
Twin Pond in the Adirondack Mountains of upstate New York is one of the many lakes in the region in which the levels of acidity in the lake have killed the fishes, amphibians, and most of the other kinds of animals and plants that once lived there.

Figure 27.22 Conversion of forest to pasture.
This fire in Brazil, and countless others like it, was set deliberately to clear the forest for livestock grazing. Soon much of Brazil's natural forest will be gone.

Lowering Species Diversity

The rapid intensification of agriculture that has accompanied this century's explosion in the human population has often acted to lower species richness. In the most obvious (and most important) instance, conversion of forest to farmland or pasture (figure 27.22) greatly reduces the number of species from hundreds to one or a few. Similarly, massive **clear-cutting** of the virgin forests such as that now occurring in the Pacific Northwest of the United States, followed by planting with one species of tree as a future lumber "crop," greatly reduces the number of species living in the ecosystem. So does the intensive commercial fishing carried out today in many parts of the earth's oceans. Our high-consumption society is lowering the species diversity of many ecosystems.

Reducing Competition

We often unintentionally eliminate the sort of competition that promotes species richness. Removal of predators, whether wolves or insects, has this effect, as does the intentional or accidental introduction of **exotic species** from other parts of the world. Intensely competitive themselves, these exotic species invade the niches parceled out among many species, displacing all of them and creating a simplified ecosystem in which far less competition occurs. By removing or introducing organisms without regard to niche relationships, our society often wreaks havoc on the complex web of biological relationships within ecosystems (figure 27.23).

(a)

(b)

Figure 27.23 An introduced organism.
(*a*) European purple loosestrife was introduced to North America sometime before 1860. (*b*) It is a rapidly growing plant that tends to take over an area and drive out native species of plants and animals. Purple loosestrife has spread over thousands of square kilometers of marshes and other wetlands in North America.

How to Modify the Environment Successfully

The world's human population is growing explosively, and the pressing future needs of some 6 billion people ensure that the world's environment will continue to be disrupted. Realistically, we cannot expect to eliminate environmental disruption altogether. What we can do is try to minimize the damage, to promote policies that modify the environment in such a way that we get what we need while disturbing the fabric of the earth's ecosystems as little as possible. From what we have considered in this chapter, three general principles emerge:

1. **Minimize physical disruption.** It is important to disrupt the physical habitat as little as possible. Society could obtain most of what it requires without major damage to habitats if proper, informed care were taken.
2. **Maintain biodiversity.** It is essential to avoid lowering species diversity as a consequence of exploitation (figure 27.24). Replacement of forests need not be with pure stands; they are simply an economic convenience for more profitable lumbering.
3. **Avoid reducing competition.** It is important not to intrude into natural ecosystems unless compelled to do so by social necessity or, in some cases, to preserve them. Introductions of exotic species, even to achieve what seem desirable consequences, should be avoided because their long-term consequences cannot be predicted.

Figure 27.24 Escape from extinction.
(*a*) This palm is the last individual of its species, *Hyophorbe verschaffeltii,* known in nature. It is shown here on the ecologically devastated island of Rodrigues, in the Indian Ocean, the same island where that famous flightless pigeon, the dodo, became extinct more than three centuries ago. (*b*) *Hyophorbe verschaffeltii* is not likely to become extinct because it is preserved in cultivation on a university campus on the island of Mauritius.

CHAPTER 27

HIGHLIGHTS

27.1 Population Dynamics

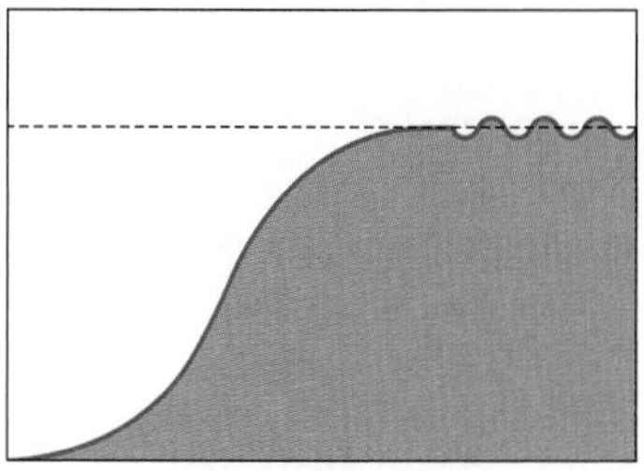

Key Terms

- population 584
- innate capacity for increase 585
- realized rate of population increase 585
- carrying capacity 586

Key Concepts

- A population is a group of individuals of the same species that live together.
- The innate rate at which a population grows is a function of the organism's physiology.
- The realized growth rate of a population is affected by birth and death, immigration and emigration, the carrying capacity, and population density.

27.2 How Coevolution Shapes Ecosystems

Key Terms

- coevolution 588
- symbiotic relationship 589
- commensalism 589
- mutualism 589

Key Concepts

- Different species within ecosystems have coevolved to compete with, interact with, and benefit from each other.
- In symbiotic relationships, two species interact such that only one benefits or both benefit from the interaction.

27.3 How Ecosystems Develop Over Time

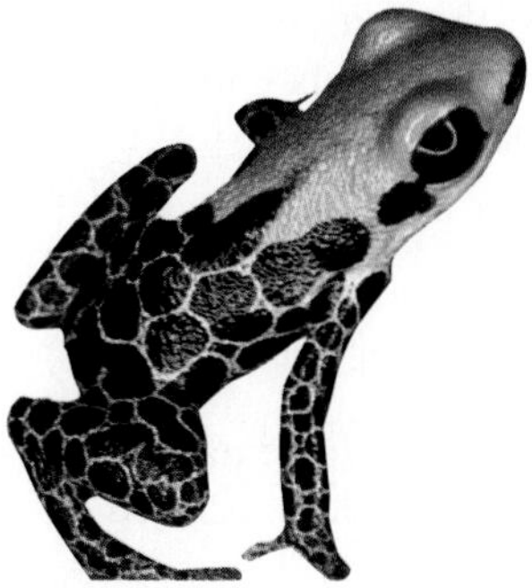

Key Terms

- fundamental niche 591
- aposematic coloration 594
- cryptic coloration 594
- Batesian mimicry 596

Key Concepts

- Within an ecosystem, organisms can potentially occupy a range of habitats and behave in many different ways.
- Organisms usually only occupy part of their fundamental niche due to competition.
- In predator-prey interactions, organisms may advertise their toxicity using aposematic coloration or hide their palatability using cryptic coloration.
- Often palatable species will mimic the coloration of toxic ones to fool predators.

27.4 What Makes an Ecosystem Stable?

Key Terms

- species richness 598
- biodiversity 598

Key Concepts

- The number of different species, or species richness, in an ecosystem determines its biological diversity.
- Ecosystems with biodiversity are more stable, while, at the same time, they may also have points of vulnerability.
- A larger size and equatorial latitude will promote biodiversity in an ecosystem.

27.5 How Humans Disrupt Ecosystems

Key Terms

- clear-cutting 601
- exotic species 601

Key Concepts

- The scale at which humans cause disturbance is often too fast and too large for ecosystems to accommodate.
- Minimizing pollution, the clear-cutting of forests, and the introduction of exotic species will help to preserve our ecosystems.

27

CONCEPT REVIEW

1. _________ patterns of population dispersion are indicative of an environment in which resources are unevenly distributed.
   a. Random
   b. Clumped
   c. Sporadic
   d. Uniform

2. The hormonal change of locusts to their migratory phase, which is brought on by crowding when populations reach large numbers, is an example of
   a. carrying capacity.
   b. exponential growth.
   c. density-dependent effects.
   d. density-independent effects.

3. What is the difference between a type II and a type III survivorship curve?
   a. In type II, mortality is higher in the early stages than it is in type III.
   b. In type III, mortality is the same in early stages as it is in type II.
   c. In type III, mortality is higher in the early stages than it is in type II.
   d. In type II, a large proportion of individuals survive to the maximum age, whereas in type III they do not.

4. Which of the following is *not* an example of mutualism?
   a. barnacles growing on the backs of whales
   b. mycorrhizae
   c. leaf cutter ants and the fungi they grow
   d. ants protecting aphids while using their honeydew as food

5. Adult monarch butterflies are primarily protected from birds by
   a. Müllerian mimicry.
   b. resembling the poisonous viceroy butterfly.
   c. manufacturing poisons from molecules that are toxic to birds.
   d. obtaining poisons from the plants that their larvae eat and retaining them.

6. Animals that are cryptically colored
   a. are distasteful.
   b. tend not to live in groups.
   c. exhibit colors not normally found in the animal's habitat.
   d. live together in family groups.

7. A _________ species is one that interacts with many components of an ecosystem in critical ways.
   a. cryptic
   b. diverse
   c. keystone
   d. competitive

8. Which of the following does *not* promote biodiversity?
   a. introduction of exotic species
   b. larger size of an ecosystem
   c. longer growing season
   d. climatic stability

9. Populations with higher proportions of children are expected to grow more (rapidly or slowly) _________ than populations with lower proportions of children and a greater proportion of individuals of childbearing age.

10. If two species live together with one benefiting, and the other neither benefiting nor being harmed, their relationship is described as _________.

11. Mutualism, commensalism, and parasitism are all _________ relationships, in which two kinds of organisms live together.

12. The part of an organism's fundamental niche that it actually occupies in an ecosystem is called its _________ niche.

13. Competition between species that results in each of them using different parts of a resource is called _________.

14. Poisonous animals advertise their toxicity using warning or _________ coloration.

15. _________ succession occurs in areas where there has been previous growth.

16. Biologists measure the _________ of an ecosystem to characterize its overall biodiversity.

17. The introduction of exotic species acts mainly to reduce _________ in an ecosystem.

Answers to the Concept Review questions appear in Appendix B.

CHALLENGE YOURSELF

1. How can the principle of competitive exclusion be applied to the origin of large numbers of species of certain genera in the Hawaiian Islands? What would you expect to happen in the future to the hundreds of species of *Drosophila* that occur there, leaving aside the possibility of their extinction through the destruction of their habitats by humans?

2. Most chemically protected prey species produce toxins that make them taste bad or make the predator sick. If the prey is "tasted" by an unsuspecting predator, the prey usually dies. If they are dead, how have the chemicals protected them? Would it be an advantage if the prey could produce a toxin that killed its predator? Why or why not?

FOR FURTHER READING

Caro, T. M., and M. K. Laurenson. "Ecological and Genetic Factors in Conservation: A Cautionary Tale." *Science* 263 (January 28, 1994): 485–86. An interesting discussion of how data gained in captivity may not reflect results observed in the wild.

Daily, G. C., and P. R. Ehrlich. "Population, Sustainability, and the Earth's Carrying Capacity." *BioScience,* September 1992, 761–70. An argument for finding ways to sustain the growing human population without undermining the planet's potential for supporting future generations.

Davies, N., and M. Brooke. "Coevolution of the Cuckoo and Its Hosts." *Scientific American,* January 1991, 92–98. The classic story of an evolutionary "arms race" between a parasite and its host.

Devries, P. "Stinging Caterpillars, Ants, and Symbiosis." *Scientific American,* October 1992, 76–82. An article about the intricate symbiosis shared by ants and caterpillars, a symbiosis that says a lot about interdependence within ecosystems and how it arises.

Earle, S. *Sea Change: A Message of the Oceans.* New York: Putnam, 1995. Conservation of the oceans is a critical challenge we all face in the coming century. This book, by a renowned marine biologist, is a call to arms.

Nadkarni, N. M. "Diversity of Species and Interactions in the Upper Canopy of Forest Ecosystems." *American Zoologist* 34, no. 1 (1994): 70–78. How important is diversity to nutrient cycling? Diversity of the forest canopy is explored in this paper.

Nap, J., and T. Bisseling. "Developmental Biology of a Plant-Prokaryote Symbiosis—the Legume Root Nodule." *Science* 250 (November 1990): 948–54. The story of one of earth's most critical symbiotic relationships, nitrogen fixation in plants.

Raven, P., L. Berg, and G. Johnson. *Environment.* Philadelphia: Saunders, 1995. General text on environmental science by the author.

Rennie, J. "Living Together." *Scientific American,* January 1992, 120–33. An excellent overview of how coevolution and symbiosis have shaped life today.

Walker, T. "Butterflies and Bad Taste." *Science News,* June 1991, 348–49. It turns out that viceroy butterflies, the classic example of tasty creatures mimicking an unpalatable one, don't taste good after all.

TECHNOLOGY LINKS

The Living World Home Page
http://www.wcbp.com/biology/tlw

28

Planet Under Stress

CHAPTER OUTLINE

Figure 28.1 A smokestack pollutes the air.
Smokestacks, like this one in River Rouge, Michigan, disrupt the worldwide ecosystem by polluting the air and causing acid rain.

The following words, written by John Donne nearly 400 years ago as part of a meditation on dying, are a prophetic description of life in our world today:

> *No man is an island, entire of itself; every man is a piece of the continent, a part of the main; if a clod be washed away by the sea, Europe is the less; any man's death diminishes me, because I am involved in mankind; and therefore never send to know for whom the bell tolls; it tolls for thee.*

Try reading the quotation again, but for the word "man" substitute the word "ecosystem," and you will see the point. Our world is one ecological continent, one highly interactive biosphere, and damage done to any one ecosystem can have ill effects on many others (figure 28.1).

28.1 Global Change

Burning high-sulfur coal in Illinois kills trees in Vermont, while dumping refrigerator coolants in New York destroys atmospheric ozone over Antarctica and leads to increased skin cancer in Madrid. Biologists call such widespread effects on the worldwide ecosystem **global change.** The pattern of global change that has become evident within the last decade, including chemical pollution, acid rain, the ozone hole, and the greenhouse effect, is one of the most serious problems facing humanity's future.

Pollution

The problem posed by chemical pollution has grown very serious in recent years, both because of the growth of heavy industry and because of an overly casual attitude in industrialized countries. Air pollution is a major problem in the world's cities. In Mexico City, oxygen is sold routinely on corners for patrons to inhale. Cities such as New York, Boston, and Philadelphia are known as brown air cities because the pollutants in the air are usually sulfur oxides emitted by industry. Cities such as Los Angeles, however, are called gray air cities because the pollutants in the air undergo chemical reactions in the sunlight (figure 28.2). A particularly clear example of the casual attitude occurred in Basel, Switzerland, in 1986, when firefighters putting out a fire in a warehouse washed 30 tons of mercury and pesticides into the Rhine River. This poison flowed down the river, all across Germany to Holland and the sea, killing fish, plants, and everything alive in the river. Why were such chemicals stored so close to a river that flows through the heart of Europe?

In another example closer to home, a poorly piloted oil tanker named the *Exxon Valdez* ran aground in Alaska in 1989 and heavily polluted many kilometers of North American coastline and the organisms that live there with oil (figure 28.3). If the tanker had been loaded no higher than the waterline, little oil would have been lost, but it was loaded far higher than that, and the weight of the above-waterline oil forced thousands of tons of oil out the hole in the ship's hull. Why do policies permit overloading like this? In 1971, industries disposed of chemical wastes containing highly toxic dioxin by paying a local businessman in the small town of Times Beach, Missouri, to take it away. The man mixed 18,000 gallons of dioxin with waste oil and then sprayed the oily mix on the dirt roadsides to keep the dust down in summer for two years in succession. Now the entire town is uninhabitable (even its soil will have to be incinerated). What policy permitted industry to dispose of dangerous chemicals so casually?

Figure 28.2 Air pollution.
In this picture of the Los Angeles basin, the air is thick with smog, a mixture of nitrous oxides, ozone, and other pollutants produced by automobile exhausts and industrial emissions.

Plastics

Incidents of pollution such as that of the Rhine, of Alaska's coast, and of Times Beach are stories that can be told countless times throughout the industrial world, from Bhopal in India to Love Canal in New York to the James River in Virginia. Not all of the pollutants that threaten these ecosystems are poisons, immediately toxic to life. Many nontoxic forms of pollution arise as by-products of industry. For example, the polymers known as plastics, which we produce in abundance, break down slowly if at all in nature, because no bacteria can decompose plastics. Thus, virtually all of the plastic items that have ever been produced are still with us. Imagine for a moment every plastic wrapper, Styrofoam cup, and plastic container you have ever used, piled up beside you—where do you think it all went? Nowhere. Every bit of it is still in the ecosystem. Even burning doesn't remove it—burning simply converts plastic to a gas that contaminates the atmosphere.

Figure 28.3 Sea otter, an endangered species.
Oceanic oil spills kill thousands of animals, such as this endangered sea otter, each year and can persist in the ocean for decades.

Agricultural Chemicals

The spread of "modern" agriculture, and particularly the Green Revolution which brought high-intensity farming to the Third World, has caused very large amounts of many kinds of new chemicals to be introduced into the global ecosystem, particularly pesticides, herbicides, and fertilizers. It is no accident that one of the highest rates of cancer per thousand people in the United States is seen in the Mississippi Delta region, where the river draining our country's agricultural heartland empties into the sea.

Obviously, we cannot take a "back to nature" approach—one that ignores the important contributions made to our lives by the intelligent use of chemicals. We must still care for the needs of a very crowded world and feed the additional billions of people who will join us during the next few decades. But it is essential that we use our technology as intelligently as possible and protect the productive capacity of all parts of the earth. Ecosystems cannot absorb chemical punishment indefinitely.

Water Pollution

Water pollution is a very serious consequence of our casual attitude about pollution. "Flushing it down the sink" doesn't work in today's crowded world. There is simply not enough water available to dilute the many substances that the enormous human population produces continuously. Despite improved methods of sewage treatment, lakes and rivers throughout the world are becoming increasingly polluted (figure 28.4).

Figure 28.4 Polluting a freshwater lake.
Varying the number of water-treatment plants that are in operation will affect how long it takes to kill off a freshwater lake, as seen in this screen capture from an interactive CD-ROM. When only one or two plants are operating, the lake can repair the damage and sustain itself. (*Explorations in Human Biology*, Module 16, "Pollution of a Freshwater Lake")

Acid Rain

The smokestacks you see in figure 28.5 are those of the Four Corners power plant in New Mexico. This facility burns coal, sending the smoke high up into the atmosphere through these stacks, each of which is over 65 meters tall. The smoke the stacks belch out contains high concentrations of sulfur, because the coal that the plant burns is rich in sulfur. High-sulfur coal was not a popular fuel in the first half of the nineteenth century, because the sulfur-rich smoke smells awful, like rotten eggs. The sulfur in the smoke also combined chemically with limestone in many public buildings, producing unsightly black surfaces. Tall smokestacks were invented to get around these problems. The high stacks release the sulfur-rich smoke high up in the atmosphere, where winds and air currents disperse and dilute it. The first tall stacks were introduced in Britain in the mid-1950s. Such tall stacks rapidly became popular in the United States and Europe—there are now over 800 of them in the United States alone.

Twenty years later, in the 1970s, ecologists began to report evidence that the tall stacks were not eliminating the problem, just exporting the ill effects elsewhere. Throughout northern Europe, lakes were reported to have suffered drastic drops in biodiversity, some even becoming devoid of life. The trees of the great Black Forest of Germany seemed to be dying—and the damage was not limited to Europe. It turned out that the sulfur introduced into the upper atmosphere by high smokestacks combines up there with water vapor to produce sulfuric acid. When this water later falls back to earth as rain or snow, it carries the sulfuric acid with it. Drifting to the northeast, the acid is taken far from its source, but the result is now clear. When schoolchildren in a nationwide project measured the pH of natural rainwater in 1989, locations around the United States rarely had a pH lower than 5.6, except in the northeastern United States, where rain and snow now have a pH of about 3.8, about 100 times the typical value for the rest of the country. This pollution-acidified precipitation is called **acid rain.**

Acid rain destroys life. Many of the forests of the northeastern United States and Canada have been seriously damaged. In fact, it is now estimated that at least 1.4 million acres of forests in the Northern Hemisphere have been adversely affected (figure 28.6). Not just trees are affected. Thousands of lakes in Sweden and Norway no longer support fish—these lakes are now eerily clear. In the northeastern United States and Canada, too, tens of thousands of lakes are dying biologically as their pH levels fall to below 5.0. Throughout the United States and Europe, local populations of frogs and other amphibians are becoming extinct, their larvae unable to develop properly in acid water.

The solution at first seems obvious: Capture and remove the emissions instead of releasing them into the atmosphere. But there have been serious problems with implementing this solution. First, it is expensive. Estimates of the cost of installing and maintaining the necessary "scrubbers" in the United States are on the order of $5 billion a year. An additional difficulty is that the polluter and the recipient of the pollution are far from one another, and neither wants to pay so much for what they view as someone else's problem. The Clean Air Act revisions of 1990 have begun to address this problem by mandating some cleaning of emissions in the United States, although much still remains to be done worldwide.

Figure 28.5 Tall stacks export pollution.
Tall stacks like those of the Four Corners coal-burning power plant in New Mexico send pollution far up into the atmosphere.

Figure 28.6 Acid rain.
Acid precipitation is killing many of the trees in North American and European forests. Much of the damage is done to the mycorrhizae, fungi growing within the cells of the tree roots. Trees need mycorrhizae in order to extract nutrients from the soil.

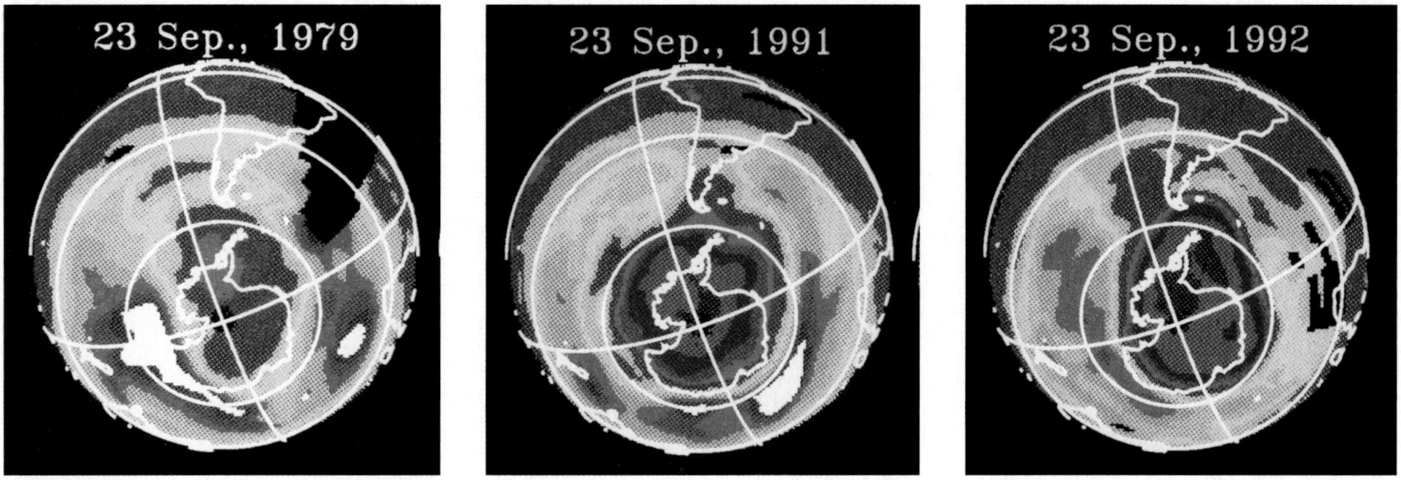

Figure 28.7 Satellite photos of the ozone hole over Antarctica.
In these views of the South Pole from a satellite, the swirling colors represent levels of ozone in the atmosphere. The pink and purple colors indicate the areas with the least amount of ozone. As you can clearly see, there is an ozone hole over Antarctica that grew between 1979 and 1992 to about the size of the United States.

The Ozone Hole

Living things were able to leave the oceans and colonize the surface of the earth only after a protective shield of ozone had been added to the atmosphere by photosynthesis. For 3 billion years before that, life was trapped in the oceans because radiation from the sun seared the earth's surface unchecked. Nothing could survive that bath of destructive energy. Imagine if that shield were taken away. Alarmingly, it appears that we are destroying it ourselves. Starting in 1975, the earth's ozone shield began to disintegrate. Over the South Pole in September of that year, satellite photos revealed that the ozone concentration was unexpectedly less than elsewhere in the earth's atmosphere, as if some "ozone-eater" were chewing it up in the Antarctic sky, leaving a mysterious zone of lower-than-normal ozone concentration, an **ozone hole.** Every year after that, more of the ozone has been depleted, and the hole grows bigger and deeper (figure 28.7).

What is eating the ozone? Scientists soon discovered that the culprit was a class of chemicals that everyone had thought to be harmless: **chlorofluorocarbons (CFCs).** CFCs were invented in the 1920s, a miracle chemical that was stable, harmless, and a near-ideal heat exchanger. Throughout the world, CFCs are used in large amounts as coolants in refrigerators and air conditioners, as the gas in aerosol dispensers, and as the foaming agent in Styrofoam containers. All of these CFCs eventually escape into the atmosphere, but no one worried about this until recently, both because CFCs were thought to be chemically inert and because everyone tends to think of the atmosphere as limitless. But CFCs are very stable chemicals, so over many years they have accumulated in the atmosphere.

It turned out that the CFCs were causing mischief the chemists had not imagined. High over the South and North Poles, nearly 50 kilometers up, where it was very, very cold, the CFCs stuck to frozen water vapor and began to act as catalysts of a chemical reaction. Just as an enzyme carries out a reaction in your cells without being changed itself, so the CFCs began to catalyze the conversion of ozone (O_3) into oxygen (O_2) without being used up themselves. Very stable, the CFCs in the atmosphere just kept at it—little machines that never stop. They are still there, still doing it, today. The drop in ozone worldwide is now over 3%.

Ultraviolet radiation is a serious human health concern. Every 1% drop in the atmospheric ozone content is estimated to lead to a 6% increase in the incidence of skin cancers. At middle latitudes the drop of approximately 3% that has occurred worldwide is estimated to have led to an increase of perhaps as much as 20% in skin cancers, which are one of the more lethal diseases afflicting humans. Humans are relatively resistant to increased ultraviolet radiation, but other organisms, such as the highly susceptible photosynthetic plankton species that are so important to global productivity, are apparently much more highly susceptible.

International agreements to lower the level of CFC production worldwide over the next several decades have been signed, but no one knows if this gradual phase-out will be adequate, or too little too late. The vast majority of the CFCs that have already been manufactured have not yet reached the upper atmosphere, and an enormous additional amount is scheduled to be produced in the coming decades. We can only guess how much of the earth's ozone will be destroyed.

The Greenhouse Effect

For over 150 years, the growth of our industrial society has been fueled by cheap energy, much of it obtained by burning fossil fuels—coal, oil, and gas. Coal, oil, and gas are the remains of ancient plants, transformed by pressure and time into carbon-rich "fossil fuels." When such fossil fuels are burned, this carbon is combined with oxygen atoms, producing carbon dioxide (CO_2). Industrial society's burning of fossil fuels has released huge amounts of carbon dioxide into the atmosphere. As with CFCs, no one paid any attention to this because the carbon dioxide was thought to be harmless and because the atmosphere was thought to be a limitless reservoir, able to absorb and disperse any amount. It turns out neither assumption was true, and in this century, the levels of carbon dioxide in the atmosphere have risen sharply and continue to rise.

What is alarming is that the carbon dioxide doesn't just sit in the air doing nothing. The chemical bonds in carbon dioxide molecules transmit radiant energy from the sun but trap the longer wavelengths of infrared light, or heat, and prevent them from radiating into space. This creates what is known as the **greenhouse effect.** Planets that lack this type of "trapping" atmosphere are much colder than those that possess one. If the earth did not have a "trapping" atmosphere, the average earth temperature would be about −20°C, instead of the actual +15°C.

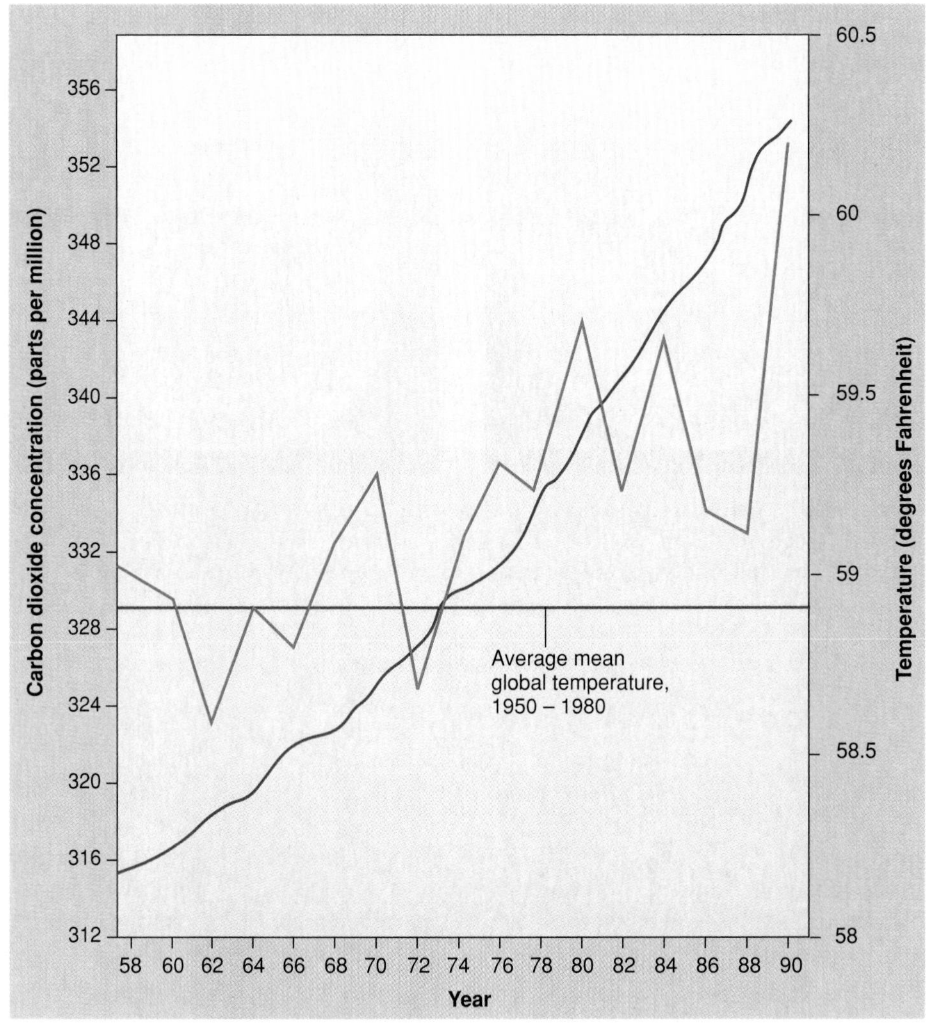

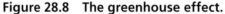

Figure 28.8 The greenhouse effect.
The concentration of carbon dioxide in the atmosphere has shown a steady increase for many years (*blue line*). The *red line* shows the average global temperature for the same period of time. Note the general increase in temperature since the 1950s and, specifically, the sharp rise during the 1980s, the warmest decade in the last 110 years.
Data from Geophysical Monograph, American Geophysical Union, National Academy of Sciences, and National Center for Atmospheric Research.

The earth's greenhouse effect is intensifying with increased fossil fuel combustion and certain types of waste disposal. These activities are increasing the amounts of carbon dioxide, CFCs, nitrogen oxides, and methane—all "greenhouse gases"—in the atmosphere. The rise in average global temperatures during the past century is consistent with increased carbon dioxide concentrations in the atmosphere (figure 28.8); seven of the eight warmest years on record in the United States were in the 1980s. However, scientists have not yet been able to demonstrate with certainty that the temperature increase was caused by the atmospheric changes.

Increases in the amounts of greenhouse gases could increase average global temperatures from 1° to 4°C, which could have serious associated effects in prime agricultural lands, changes in sea levels, and alterations in rain patterns. The projected changes concern average temperatures and cannot be connected with particular hot or cold days. In 1992 and 1993, below-average temperatures were recorded worldwide, but scientists believe that these cooler temperatures were a result of recent volcanic eruptions. Despite these temperature readings, experts feel that the earth is warming now more rapidly than any time in the past. There is considerable disagreement among governments about what ought to be done about global warming. Some governments are encouraging the replacement of fossil fuels by nuclear power, while others are arguing that more study is needed before a proper plan of action can be formed.

28.2 Saving Our Environment

The pattern of global change that is overtaking our world is very disturbing. Human activities are placing a severe stress on the global ecosystem, and we must quickly find ways to reduce the harmful impact. There are four key areas in which it will be particularly important to meet the challenge successfully: reducing pollution, finding other sources of energy, preserving nonreplaceable resources, and curbing population growth.

Reducing Pollution

One of the most encouraging developments of the early 1990s has been the worldwide increase in efforts to reduce pollution. International agreements to limit CFC production is but one example. In most cases, however, pollution problems are more localized—pollution of a local lake by industrial runoff, a particular factory that emits noxious fumes into the atmosphere, a dump in a small town that accumulates hazardous chemicals.

To solve the problem of industrial pollution it is first necessary to understand the cause of the problem. In essence, it is a failure of our economy to set a proper price on environmental health. To understand how this happens, we must think for a moment about money. The economy of the United States (and much of the rest of the industrial world) is based on a simple feedback system of supply and demand. As a commodity gets scarce, the price goes up, and this added profit acts as an incentive for more of the item to be produced; if too much is produced, the price falls and less of it is made because it is no longer so profitable to produce the item.

This system works very well and is responsible for the economic strength of our nation, but it has one great weakness. If demand is set by price, then it is very important that all the costs be included in the price. Imagine that the person selling the item were able to pass off part of the production cost to a third person. The seller would then be able to set a lower price and sell more of the item! Driven by the lower price, the buyer would purchase more than if all the costs had been added into the price.

Unfortunately, that sort of pricing error is what has driven the pollution of the environment by industry over the last century. The true costs of energy and of the many products of industry are composed of direct production costs, such as materials and wages, and of indirect costs, such as pollution of the ecosystem. Economists have identified an "optimum" amount of pollution based on how much it costs to reduce pollution versus the social and environmental cost of allowing pollution (figure 28.9).

The indirect costs of pollution are usually not taken into account. However, the indirect costs do not disappear because we ignore them. They are simply passed on to future generations, which must pay the bill in terms of damage to the ecosystems on which we all depend. Increasingly, the future is now. Our world, unable to support more damage, is demanding that something be done—that we finally pay up.

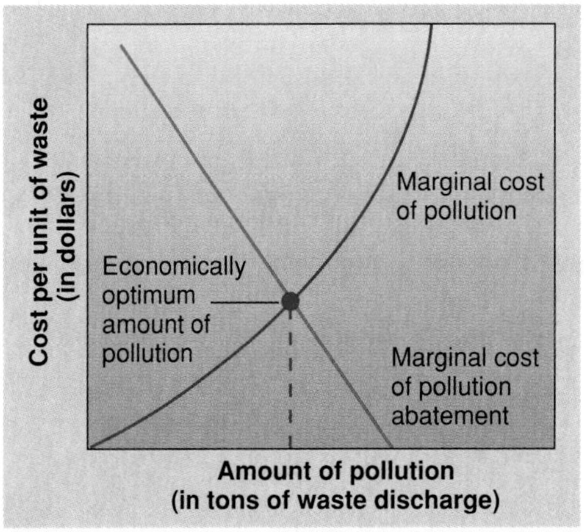

Figure 28.9 Is there an optimum amount of pollution?
Economists identify the "optimum" amount of pollution as the amount whose marginal cost equals the marginal cost of pollution abatement (the point where the two curves intersect). If more pollution than the optimum is allowed, the social cost will be unacceptably high. If less than the optimum amount of pollution is allowed, the economic cost will be unacceptably high.

Antipollution Laws

Two effective approaches have been devised to curb pollution in this country. The first is to pass laws forbidding it. In the last 20 years, laws have begun to significantly curb the spread of pollution by setting stiff standards for what can be released into the environment. For example, all cars are required to have effective catalytic converters to eliminate automobile smog. Similarly, the Clean Air Act of 1990 requires that power plants install scrubbers on their smokestacks to eliminate sulfur emissions. The effect is that the consumer pays to avoid polluting the environment. The costs of the converters make cars more expensive, and the costs of the scrubbers increase the price of the energy. The new, higher costs are closer to the true costs, lowering consumption to more appropriate levels.

Pollution Taxes

A second approach to curbing pollution has been to increase the consumer costs directly by placing a tax on the pollution. The idea here is that the true costs might be so high as to inhibit industrial growth and development, eliminating jobs and depressing economic growth. Besides, true environmental costs are very hard to calculate. How do you put a price on the possibility that sulfur emissions might cause a faraway frog population to become extinct? Instead, some economists advocate an artificial price hike imposed by the government as a tax added to the price of production. This added cost lowers consumption too, but by adjusting the tax, the government can attempt to balance the conflicting demands of environmental safety and economic growth. Such taxes, often imposed as "pollution permits," are becoming an increasingly important part of antipollution laws.

Finding Other Sources of Energy

Global warming from burning coal and oil, the increasing scarcity of oil, and the political problems generated by obtaining it all make it desirable to find alternative energy sources. Many countries are turning to nuclear power for their growing energy needs. In less than 50 years, nuclear power has become a leading source of energy. In 1995, more than 500 nuclear reactors were producing power worldwide. Over 70% of all France's electricity is now produced by nuclear power plants.

Nuclear power plants have not been as popular in this country as in the rest of the world, because we have ample access to cheap coal and because the public fears the consequences of an accident. An explosion at the Three Mile Island nuclear plant in Pennsylvania in 1979 released little radiation into the environment but galvanized these fears. There has been little nuclear power development in this country since then (figure 28.10). However, it seems likely that as global warming becomes a more serious problem, as costs of burning coal are increased to include environmental damage, and as oil becomes more scarce, nuclear power will again become an important energy source in this country.

In theory nuclear power can provide plentiful, cheap energy, but the reality is less encouraging. Nuclear power presents several problems that must be overcome if it is to provide a significant portion of the energy that will fuel our future world.

1. **Safe operation.** Because of the potential for vast radioactive contamination such as occurred when a reactor exploded at Chernobyl in the Ukraine in 1986 (figure 28.11), it is crucial that nuclear power plants be designed "fail-safe" and be sited away from densely populated areas. While new power plant designs are far safer than those of Chernobyl, there is still much room for improvement.
2. **Waste disposal.** Spent nuclear fuel remains very radioactive for thousands of years, and the plant itself quickly wears out—after about 25 years the intense radioactivity makes its metal pipes brittle. As yet, no safe way has been devised to dispose of this growing mountain of highly radioactive material. Plans to bury radioactive wastes in remote locations have not yet proven practical. To date, not a single nuclear power plant in France or the United States has been successfully dismantled and its dangerously radioactive components disposed of safely. Waste disposal remains the single most serious obstacle to the widespread use of nuclear power. Until the problem is solved, nuclear power will remain a promise unfulfilled.
3. **Security.** Some of the uranium in nuclear fuel is converted to plutonium during its decay, and this plutonium could be recovered from spent fuel and used to make nuclear weapons. It will be very important to find ways to effectively guard nuclear power plants against terrorism.

Figure 28.10 Three Mile Island nuclear power plant.
Since a nuclear accident here in 1979, the building of nuclear power stations in the United States has slowed dramatically.

Figure 28.11 A nuclear disaster.
The Russian scientists in this helicopter measure radiation levels over the Chernobyl reactor days after a 1986 explosion blew it apart.

Preserving Nonreplaceable Resources

Among the many ways ecosystems are being damaged, one class of problem stands out as more serious than all the rest: consuming or destroying resources that we cannot replace in the future. While a polluted stream can be cleaned up, no one can restore an extinct species. In the United States, three sorts of nonreplaceable resources are being consumed at alarming rates: topsoil, groundwater, and biodiversity.

Topsoil

The United States is one of the most productive agricultural countries on earth, largely because much of it is covered with particularly fertile soils. Our midwestern farm belt sits astride what was once a great prairie. The soil of that ecosystem accumulated bit by bit from countless generations of animals and plants until, by the time humans came to plow, the rich soil extended down several feet.

We can never replace this rich **topsoil,** the capital upon which our country's greatness is built, yet we are allowing it to be lost at a rate of centimeters every decade. By repeatedly tilling (turning the soil over) to eliminate weeds, we permit rain to wash more and more of the topsoil away, into rivers and eventually out to sea. Our country has lost one-quarter of its topsoil since 1950! New approaches are desperately needed to lessen the reliance on intensive cultivation. Some possible solutions include using genetic engineering to make crops resistant to weed-killing herbicides and terracing to recapture lost topsoil (figure 28.12).

Groundwater

A second resource that we cannot replace is **groundwater,** water trapped beneath the soil within porous rock reservoirs called aquifers. This water seeped into its underground reservoir very slowly during the last ice age over 12,000 years ago. We should not waste this treasure, for we cannot replace it.

In most areas of the United States, local governments exert relatively little control over the use of groundwater. As a result, a very large portion is wasted watering lawns, washing cars, and running fountains. A great deal more is inadvertently being polluted by poor disposal of chemical wastes—and once pollution enters the groundwater, there is no effective means of removing it.

Figure 28.12 Preserving topsoil is the key to successful agriculture.
Loss of topsoil is one of the world's most pressing environmental problems. Contour farming such as seen here is one way to minimize soil runoff.

Biodiversity

The number of species in danger of extinction during your lifetime is far greater than the number that became extinct with the dinosaurs. This disastrous loss of **biodiversity** is important to every one of us, because as these species disappear, so does our chance to learn about them and their possible benefits for ourselves (figure 28.13). The fact that our entire supply of food is based on 20 kinds of plants, out of the 250,000 available, should give us pause. Like burning a library without reading the books, we don't know what it is we waste. All we can be sure of is that we cannot retrieve it. Extinct is forever.

Over the last 20 years, about half of the world's tropical rain forests have been either burned to make pasture land or cut for timber (figure 28.14). Over 6 million square kilometers have been destroyed. Every year the rate of loss increases as the human population of the tropics grows. About 160,000 square kilometers were cut in 1991, a rate greater than .6 hectares (1.5 acres) per second! At this rate, all the rain forests of the world will be gone in your lifetime. In the process, it is estimated that one-fifth or more of the world's species of animals and plants will become extinct—more than a million species. This would be an extinction event unparalleled for at least 65 million years, since the age of the dinosaurs.

You should not be lulled into thinking that loss of biodiversity is a problem limited to the tropics. The ancient forests of the Pacific Northwest are being cut at a ferocious rate today, largely to supply jobs (the lumber is exported), with much of the cost of cutting it down subsidized by our government (the Forest Service builds the necessary access roads, for example). At the current rate, very little will remain in a decade. Nor is the problem restricted to one area. Throughout our country, natural forests are being "clear-cut," replaced by pure stands of lumber trees planted in rows like so many lines of corn. It is difficult to scold those living in the tropics when we ourselves do such a poor job of preserving our own country's biodiversity.

Figure 28.13 Who cares if a plant becomes extinct?
You should. Most corn (*Zea mays*) is genetically uniform and thus difficult to improve by selection (*a*). The field in Mexico shown in (*b*) is dominated by a wild perennial relative, *Zea diploperennis*, discovered in 1977. The two species can be crossed to produce fertile hybrids, and *Z. diploperennis* is resistant to five of the seven main types of viral diseases that damage corn. The production of fertile hybrids of *Z. mays* and *Z. diploperennis* will revolutionize the farming industry, but this great improvement was almost lost. *Z. diploperennis* might have never been found at all, because it originally occurred in only one field. Its single population was nearly destroyed by the clearing of land by humans in the area (c).

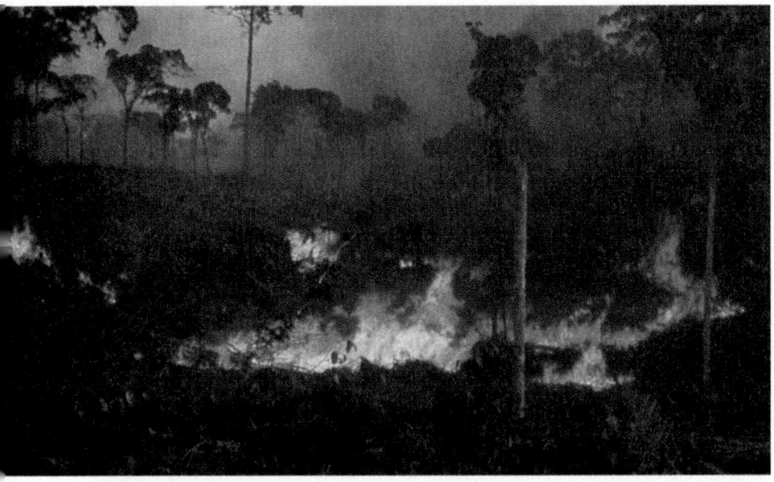

(a)

(c)

(b)

(d)

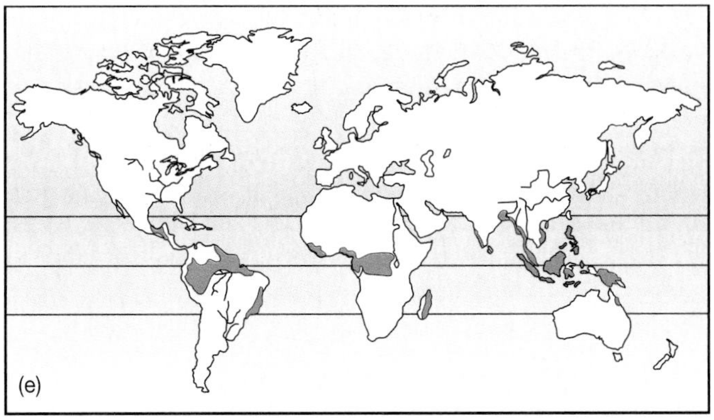

(e)

Figure 28.14 Destroying the tropical rain forests.
(*a*) These fires are destroying the rain forest in Brazil, which is being cleared for cattle pasture. (*b*) The flames are so widespread and so high that their smoke can be viewed from space. This satellite photo shows a plume of smoke generated from the burning of the rain forest. (*c*) The consequences of deforestation can be seen on these elevation slopes in Ecuador, which now support only low-grade pastures where highly productive forest grew in the 1970s. (*d*) As time passes, the consequences of tropical deforestation may become even more severe, as shown here by the extensive erosion in this area of Tanzania from which forest has been removed. (*e*) Map showing world locations of rain forests.

Curbing Population Growth

If we were to solve all the problems mentioned in this chapter, we would merely buy time to address the fundamental problem: There are getting to be too many of us.

Human beings first reached North America at least 12,000 to 13,000 years ago, crossing the narrow straits between Siberia and Alaska and moving swiftly to the southern tip of South America. By 10,000 years ago, when the continental ice sheets withdrew and agriculture first developed, about 5 million people lived on earth, distributed over all the continents except Antarctica. With the new and much more dependable sources of food that became available through agriculture, the human population began to grow more rapidly. By the time of Christ, an estimated 130 million people lived on earth. By 1650, the world's population had doubled, and doubled again, reaching 500 million.

Since 1650, and probably for much longer, the average human birth rate has remained nearly constant, at about 30 births per year per 1,000 people worldwide. However, with the spread of better sanitation and improved medical techniques, the death rate has fallen steadily, to an estimated 10 deaths per 1,000 people per year in 1995. The difference between these two figures amounts to an annual worldwide increase in human population of approximately 1.8%. This number may seem small, but don't be deceived—it leads to a doubling of the world's population in only 39 years!

Figure 28.15 The world's population is centered in mega-cities.
Mexico City, the world's largest city, has over 20 million inhabitants.

One of the most alarming trends taking place in developing countries is the massive movement to urban centers. For example, Mexico City, the largest city in the world, is plagued by smog, traffic, inadequate waste disposal, and other problems—and it will have a population of more than 30 million by the end of the century (figure 28.15). The prospects of supplying adequate food, water, and sanitation to these people are almost unimaginable. The lot of the rural poor, mainly farmers, in Mexico is even worse, and it is no wonder that many Mexicans attempt to move to the United States.

The world population passed five billion people in early 1987, and the annual increase now amounts to about 95 million people. Put another way, more than 260,000 people are added to the world population each day, or more than 180 every minute (figure 28.16). At this rate, the world's population will be well over 6 billion by the year 2000. Such growth cannot continue, because our world cannot support it. Just as a cancer cannot grow unabated in your body without eventually killing you, so humanity cannot continue to grow unchecked in the biosphere without killing it.

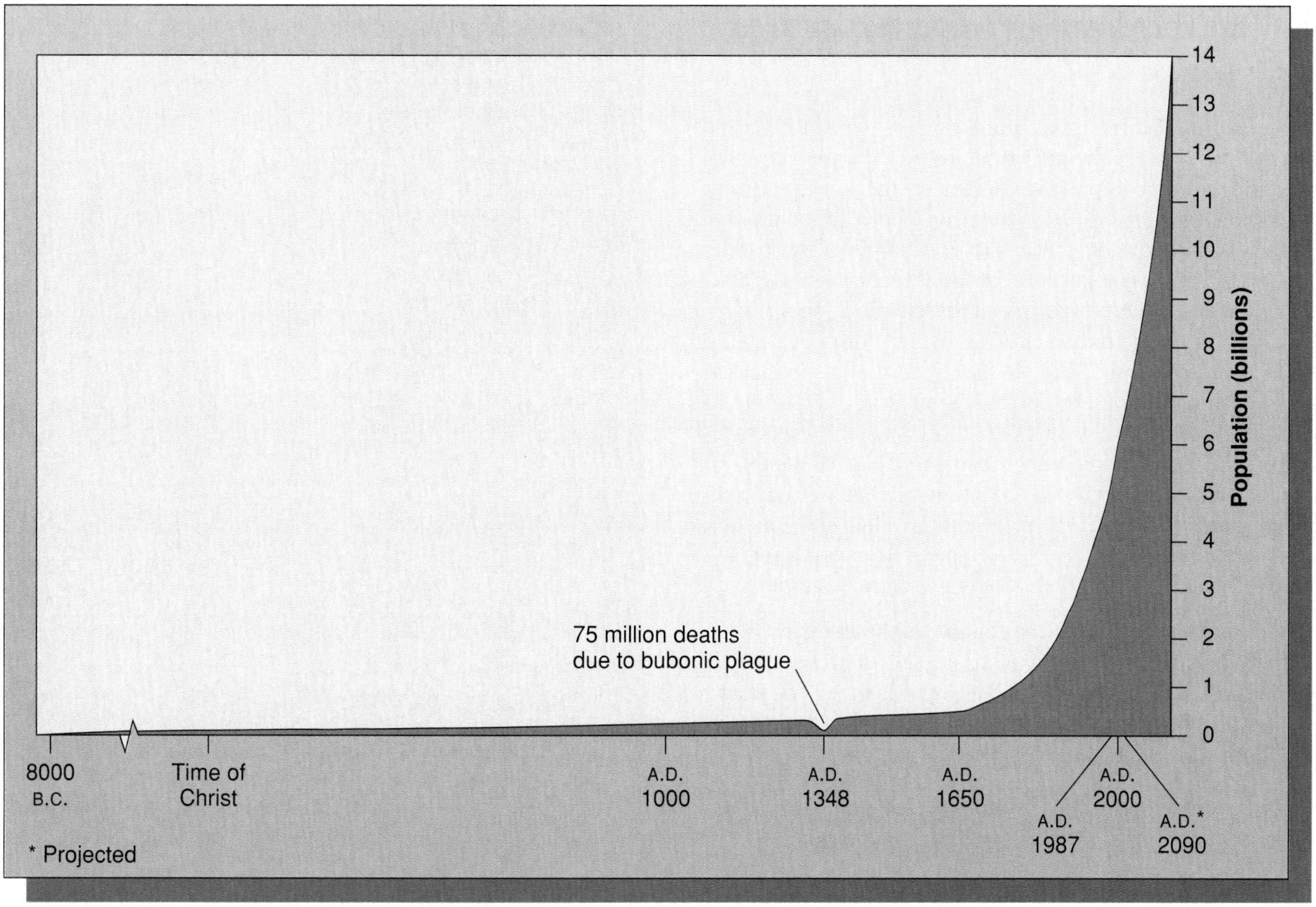

Figure 28.16 Growth curve of the human population.
Over the past 300 years, the world population has been growing steadily. By the year 2000, the earth's human population will exceed 6 billion.

In view of the limited resources available to the human population and the need to learn how to manage those resources well, the first and most necessary step toward global prosperity is to stabilize the human population. One of the surest signs of the pressure being placed on the environment is human utilization of about 40% of the total net global photosynthetic productivity on land. Given that statistic, a doubling of the human population in 43 years poses extraordinarily severe problems. The facts virtually demand restraint in population growth. If and when technology is developed that would allow greater numbers of people to inhabit the earth in a stable condition, the human population can be increased to whatever level might be appropriate.

By the year 2000, about 60% of the people in the world will be living in countries that are at least partly tropical or subtropical. An additional 20% will be living in China, and the remaining 20% in the so-called developed, or industrialized, countries: Europe, Russia, Japan, the United States, Canada, Australia, and New Zealand. Whereas the populations of the developed countries are growing at an annual rate of only about 0.3%, those of the less developed, mostly tropical countries (excluding China) are growing at an annual rate estimated in 1995 to be about 2.2%.

Most countries are devoting considerable attention to slowing the growth rate of their populations, and there are genuine signs of progress. If progress continues, the United Nations estimates that the world's population may stabilize by the close of the next century at a level of 13 – 15 billion people, nearly three times the number living today. No one knows whether the world can support so many people indefinitely. Finding a way to do so is the greatest task facing humanity. The quality of life that will be available for your children in the next century depends to a large extent on our success.

28.3 Solving Environmental Problems

It is easy to become discouraged when considering the world's many environmental problems, but do not lose track of the single most important conclusion that emerges from our examination of these environmental problems—the fact that each of the problems is solvable. A polluted lake can be cleaned up; a dirty smokestack can be altered to remove noxious gas; waste of key resources can be stopped. What are required are a clear understanding of the problem and a commitment to doing something about it. The extent to which American families **recycle** aluminum cans and newspapers is evidence of the degree to which people want to become part of the solution, rather than part of the problem.

If we look at how success was achieved in those instances where environmental problems have been overcome, a simple pattern emerges. Viewed simply, five components are involved in successfully solving any environmental problem:

1. **Assessment.** The first stage of addressing any environmental problem is scientific analysis, gathering information about what is going on. Data must be collected and experiments performed in order to construct a "model" of the ecosystem that describes how it is responding to the situation. Such a model can then be used to make predictions about the future course of events in the ecosystem.
2. **Risk analysis.** Using the results of the scientific investigation as a tool, it is possible to predict the consequences of environmental intervention. It is necessary to evaluate not only the potential for solving the environmental problem but also any adverse effects the action plan might create.
3. **Public education.** When a clear choice can be made among alternative courses of action, the public must be informed. This involves explaining the problem in terms people can understand, presenting the alternative actions available, and describing the probable costs and results of the different choices.
4. **Political education.** The public, through its elected officials, selects a course of action and implements it. Individuals can have a major impact at this stage by exercising their right to vote and be heard. Many voters do not understand the magnitude of what they can achieve by writing letters and supporting special-interest groups.
5. **Follow-through.** The results of any action should be carefully monitored to see if the environmental problem is being solved and, more basically, to evaluate and improve the initial assessment and modeling of the problem. We learn by doing.

Figure 28.17 The future is now.
The girl gazing from this page faces an uncertain future. She is a refugee, an Afghani. The whims of war have destroyed her home, her family, and all that is familiar to her. Her expression carries a message about our own future: The problems humanity faces on an increasingly unstable, overcrowded, and polluted earth are no longer hypothetical. They are with us today and demand solutions.

What You Have to Contribute

You cannot hope to preserve what you do not understand. Solving the world's environmental problems will take the efforts of many kinds of people—politicians, economists, engineers. However, the issues are largely biological, and when all is said and done a knowledge of biology is the essential tool you will need to contribute to the effort. If economists had understood as much ecology as you have learned in these chapters, the world might not be as polluted today. Biological literacy is no longer a luxury for scientists—it has become a necessity for all of us.

We and all living humans obtain our food and the other materials upon which our civilization depends from ecosystems that we share with earth's other creatures. It has been said that we do not inherit the earth from our parents—we borrow it from our children. We must preserve for them a world in which they can live. That is our challenge for the future, and it is a challenge that must be met soon. In many parts of the world, the future is happening right now (figure 28.17).

CHAPTER 28

HIGHLIGHTS

Key Terms | Key Concepts

28.1 Global Change

Key Terms

acid rain 610
ozone hole 611
CFCs 611
greenhouse effect 612

Key Concepts

- Air and water pollution result mainly from increased industrialization, oil spills, the use of agricultural chemicals, and unwise disposal of wastes.
- Acid rain, which is precipitation polluted by sulfuric acid originating from factories, kills trees and lakes by lowering the pH level.
- Chlorofluorocarbons (CFCs) used in coolants, aerosols, and foaming agents are eating the earth's ozone, which exposes life on earth to harmful ultraviolet radiation.
- An increase in carbon dioxide in the atmosphere is responsible for increased global warming, or the greenhouse effect.

28.2 Saving Our Environment

Key Terms

topsoil 615
groundwater 615
biodiversity 616

Key Concepts

- Antipollution laws, pollution taxes, and economic evaluations are underway to reduce pollution.
- Nuclear power provides an alternative source of energy to the burning of coal and oil, although the challenges of operation must be overcome.
- We must slow down our use of topsoil and groundwater, resources that took thousands of years to accumulate.
- With the destruction of tropical and temperate rain forests, organisms with potentially vital roles in the ecosystem are being lost.
- The alarmingly high human population growth rate is at the core of many environmental problems.

28.3 Solving Environmental Problems

Key Terms

recycle 620
public education 620

Key Concepts

- Environmental problems have been overcome through assessment, analysis, education, and follow-through.
- You can do your part by recycling, educating others about the environment, voting, and writing letters.

28

CONCEPT REVIEW

1. Which of the following statements does *not* describe acid rain?
   a. It is sulfuric acid formed from the sulfur-rich smoke emitted by factories and water vapor in the atmosphere.
   b. It has caused dramatic drops in biodiversity throughout the forests and lakes of Europe, Canada, and the northeastern United States.
   c. Wind and air currents dilute acid rain because factories are designed to release it high into the air.
   d. It causes the pH levels of lakes to fall below 5.0, an acidity in which amphibians cannot develop properly.

2. Why is the concentration of ozone less over the North and South Poles?
   a. CFCs accumulate only in areas where the air is cold.
   b. CFCs stick to frozen water vapor and are able to act as catalysts.
   c. CFC use is highest in these areas.
   d. Ultraviolet rays are stronger in these areas.

3. _________ is caused by an accumulation of large amounts of CO_2 and other gases in the atmosphere.
   a. Acid rain
   b. The ozone hole
   c. Skin cancer
   d. The greenhouse effect

4. Some atmospheric CO_2 is needed to keep our planet warm.
   a. true b. false

5. Which of the following is *not* a problem that has prevented nuclear power from becoming more popular?
   a. securing nuclear power plants against terrorism
   b. decreasing the production of acid rain by nuclear power plants
   c. disposal of nuclear wastes
   d. safely operating nuclear power plants

6. Topsoil must be conserved because
   a. it is fertile soil that has accumulated over thousands of years and prevents erosion.
   b. its underground reservoir is being used up at a fast rate.
   c. the agricultural practice of terracing has caused the loss of centimeters per decade.
   d. none of the above.

7. The loss of biodiversity is mainly a problem limited to the tropics.
   a. true b. false

8. What is the proper order for the following steps in solving environmental problems?
   a. public education
   b. follow-through
   c. assessment
   d. political education
   e. risk analysis

9. _________ catalyze the conversion of ozone into oxygen without being used up.

10. Industrial burning of fossil fuels has released huge amounts of _________ into the atmosphere.

11. Two measures that have been taken to reduce pollution are passing antipollution laws and imposing pollution _________.

12. The three sorts of nonreplaceable resources that are being consumed at an alarming rate in the United States are _________, _________, and _________.

13. During the next 30 to 40 years, _________ % of the world's species are expected to become extinct.

14. Tropical forests are being destroyed for three major reasons: firewood gathering, lumbering, and _________.

15. When agriculture first developed about 10,000 years ago, there were about _________ people on earth.

16. The present population of the world is about _________.

17. The world population is growing at an annual rate of _________ %.

Answers to the Concept Review questions appear in Appendix B.

CHALLENGE YOURSELF

1. Can we ever produce enough food and other materials so that population growth will not be a matter of concern? How?
2. Decreased birth rates have ultimately followed decreased death rates in many countries, for example, Germany and Great Britain. Do you think this will eventually occur worldwide and solve our population problems?
3. Some have argued that attempts by the United States to promote lowering of the birth rate in the underdeveloped countries of the tropics is nothing more than economic imperialism, and that it is in the best interest of these countries that their populations grow as rapidly as possible. Give reasons why you agree or disagree with this assessment.

FOR FURTHER READING

Bongaarts, J. "Can the Growing Human Population Feed Itself?" *Scientific American,* March 1994, 36–42. As human numbers surge upward, some experts are alarmed, others optimistic. Who is right?

Brown, L., ed. *State of the World 1996.* New York: W. W. Norton, 1996. A highly recommended, easily read summary of the ecological problems faced by an overcrowded and hungry world, with an excellent chapter on suggested solutions. A new edition appears every year.

Dasgupta, P. S. "Population, Poverty, and the Local Environment." *Scientific American,* February 1995, 40–45. As forests and rivers recede, a child's labor can become more valuable to parents, spurring a vicious cycle that traps families in poverty.

"The Endless Cycle." *Natural History,* May 1990. An entire issue devoted to recycling. Unless we follow nature's example and return used goods to the production cycle, we face a bleak future.

Gore, A. *Earth in the Balance.* Boston: Houghton Mifflin, 1992. A fascinating personal account of the environmental dilemma facing us all.

Myers, N. *Ultimate Security: The Environmental Basis of Political Stability.* New York: W. W. Norton, 1995. A penetrating and thoughtful look at the link between political stability and environmental sustainability.

Reganold, J. P., R. I. Papendick, and J. F. Parr. "Sustainable Agriculture." *Scientific American,* June 1990, 112–20. Nontraditional approaches to agriculture offer both financial and environmental rewards.

Shcherbak, Y. "Ten Years of the Chernobyl Era." *Scientific American,* April 1996, 44–49. The environmental and health effects of the nuclear disaster that blew apart this power plant will last for generations. This report summarizes the effects 10 years after the event.

Toon, O., and R. Turco. "Polar Stratospheric Clouds and Ozone Depletion." *Scientific American,* June 1991, 68–74. The story of how the "ozone hole" forms every spring in the high skies over Antarctica.

White, R. M. "The Great Climate Debate." *Scientific American,* July 1990, 36–43. Outstanding analysis of appropriate actions that we might take now in the face of impending climatic change.

TECHNOLOGY LINKS

The Living World Home Page
http://www.wcbp.com/biology/tlw

Explorations in Human Biology CD
#16 Pollution of a Freshwater Lake

Appendix A
Classification of Organisms

The classification used in this book is explained in chapter 10. Responding to a wealth of recent molecular data, it recognizes two separate kingdoms for the bacteria (prokaryotes): archaebacteria and eubacteria. It divides the eukaryotes into four kingdoms: the diverse and predominantly unicellular Protista and three large, characteristic multicellular groups derived from them: Fungi, Plantae, and Animalia. Viruses, which are considered nonliving, are not included in this appendix but are treated in chapter 11.

Kingdom Archaebacteria

Prokaryotic bacteria; single-celled, cell walls lack muramic acid. Like all bacteria, they lack a membrane-bounded nucleus, sexual recombination, and internal cell compartments. Archaebacterial cells have distinctive membranes and unique rRNA and metabolic cofactors. Many are capable of living in an anaerobic environment rich in CO_2 and H_2.

Archaebacteria: a halobacterium.

Kingdom Eubacteria

Single-celled prokaryotic bacteria; cell walls contain muramic acid. Like archaebacteria, they lack a true nucleus and true internal cell compartments. Their flagella are simple, composed of a single fiber of protein. Their reproduction is predominantly asexual. Eubacteria often form filaments or other forms of colonies. A very diverse group metabolically; over 2,500 species have been named and far more undoubtedly exist.

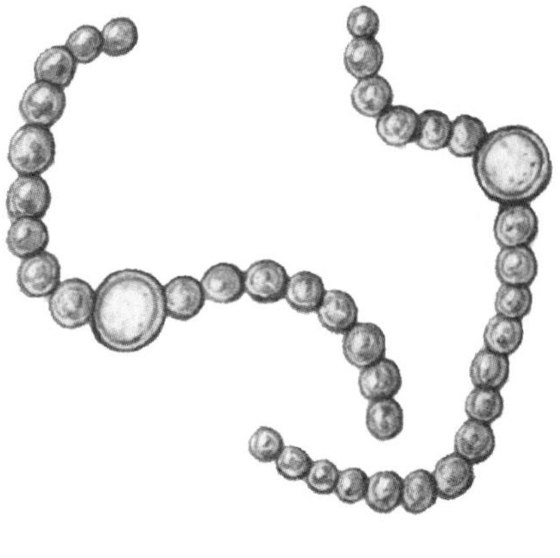

Eubacteria: a cyanobacterium.

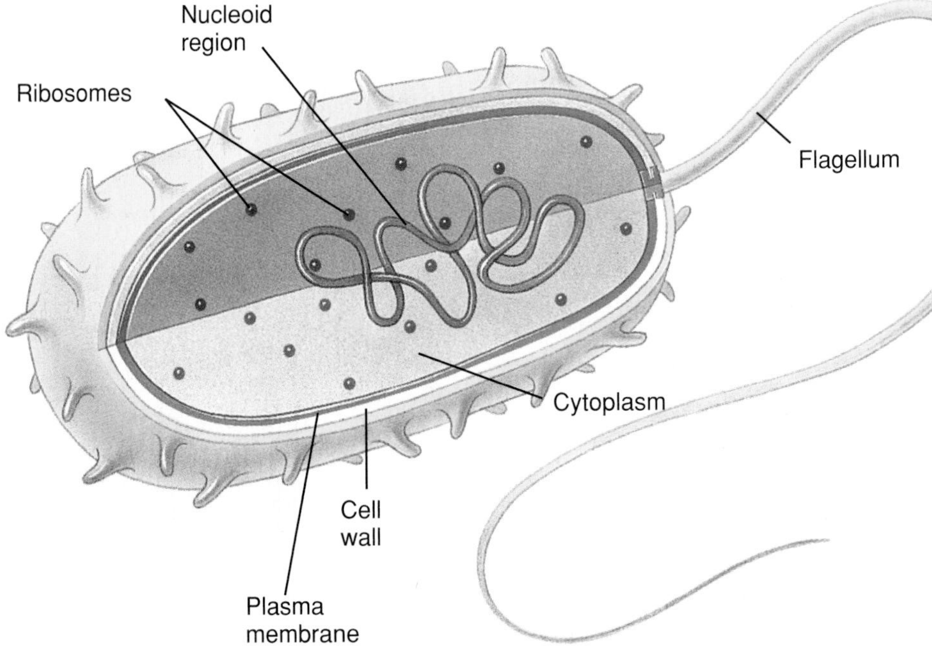

Eubacteria: diagram of a typical cell.

Kingdom Protista

Eukaryotic organisms, including many evolutionary lines of primarily single-celled organisms. Eukaryotes have a membrane-bounded nucleus and chromosomes, sexual recombination, and extensive internal compartmentalization of the cells; their flagella are complex, with 9 + 2 internal organization. They are diverse metabolically but much less so than are bacteria; protists are heterotrophic or autotrophic and may capture prey, absorb their food, or photosynthesize. Reproduction in protists is either sexual, involving meiosis and syngamy, or asexual.

Phylum Caryoblastea

One species of primitive amoeba-like organism, *Pelomyxa palustris,* which lacks mitosis, mitochondria, and chloroplasts.

Phylum Dinoflagellata

Dinoflagellates; unicellular, photosynthetic organisms, most of which are clad in stiff, cellulose plates and have two unequal flagella that beat in grooves encircling the body at right angles. About 1,000 species.

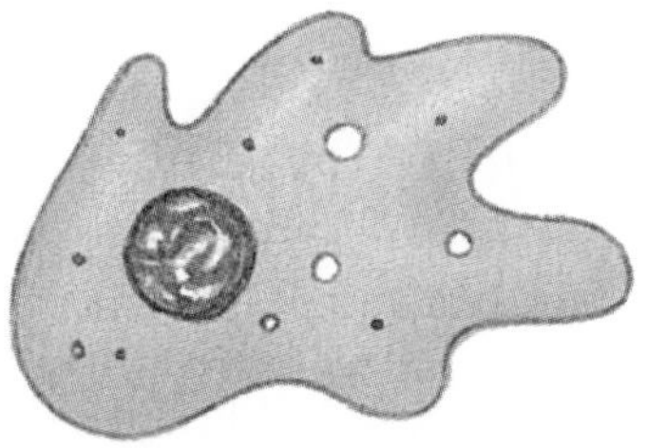

Phylum Rhizopoda: ***Amoeba.***

Phylum Rhizopoda

Amoebas; heterotrophic, unicellular organisms that move from place to place by cellular extensions called pseudopods and reproduce only asexually, by fission. Hundreds of species.

Phylum Sporozoa

Sporozoans; unicellular, heterotrophic, nonmotile, spore-forming parasites of animals. About 3,900 species.

Phylum Acrasiomycota

Cellular slime molds; unicellular, amoeba-like, heterotrophic organisms that aggregate in masses at certain stages of their life cycle and form compound sporangia. About 70 species.

Phylum Myxomycota

Plasmodial slime molds; heterotrophic organisms that move from place to place as a multicellular, gelatinous mass, forming sporangia at times. About 450 species.

Phylum Zoomastigina

Zoomastigotes; a highly diverse phylum of mostly unicellular, heterotrophic, flagellated, free-living or parasitic protists (flagella one to thousands). Thousands of species.

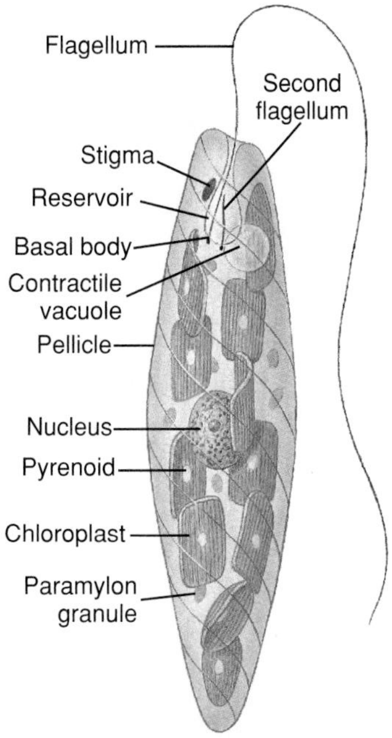

Phylum Zoomastigina: ***Euglena.***

Phylum Euglenophyta

Euglenoids; mostly unicellular, photosynthetic, or heterotrophic protists with two unequal flagella. About 1,000 species.

Phylum Phaeophyta

Brown algae; multicellular, photosynthetic, mostly marine protists with chlorophylls *a* and *c* and a carotenoid that colors the organisms brownish. About 1,500 species.

Phylum Bacillariophyta

Diatoms and related groups; mostly unicellular, photosynthetic organisms with chlorophylls *a* and *c* and fucoxanthin. Diatoms have a unique double shell made of opaline silica. About 11,500 living species.

Phylum Foraminifera

Heterotrophic marine organisms; characteristic feature is a pore-studded shell, or test, some of them brilliantly colored.

Phylum Chlorophyta

Green algae; a large and diverse phylum of unicellular or multicellular, mostly aquatic organisms with chlorophylls *a* and *b,* carotenoids, and starch, accumulated within the plastids (as it also is in plants) as the food storage product. About 7,000 species.

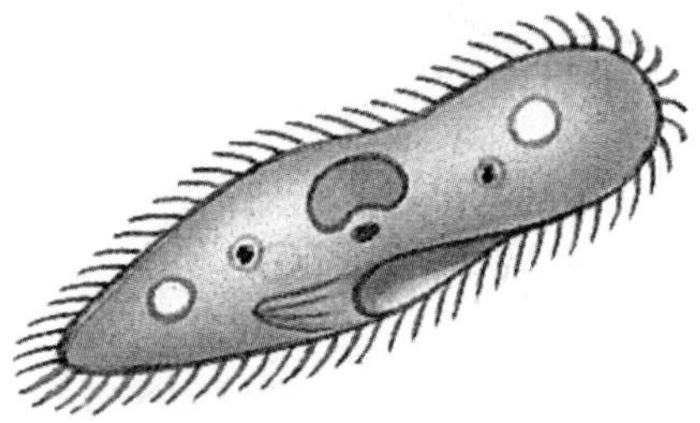

Phylum Ciliophora: ***Paramecium.***

Phylum Ciliophora

Ciliates; diverse, mostly unicellular, heterotrophic protists, characteristically with large numbers of cilia. About 8,000 species.

Phylum Oomycota

Oomycetes; water molds, white rusts, and downy mildews. Aquatic or terrestrial unicellular or multicellular parasites or saprobes that feed on dead organic matter. About 580 species.

Phylum Rhodophyta

Red algae; mostly marine, mostly multicellular protists with chloroplasts containing chlorophyll *a* and phycobilins; no flagellated cells present. About 4,000 species.

Kingdom Fungi

Filamentous, multinucleate, heterotrophic eukaryotes with cell walls rich in chitin; no flagellated cells present. Mitosis in fungi takes place within the nuclei; the nuclear envelope never breaks down. The filaments of fungi grow through the substrate, secreting enzymes and digesting the products of their activity. Septa between the nuclei in the hyphae normally complete only when sexual or asexual reproductive structures are being cut off. Asexual reproduction is frequent in some groups. The nuclei of fungi are haploid, with the zygote the only diploid stage in life cycle. About 77,000 named species. The major groups of fungi used to be called divisions, but currently taxonomists call these groups phyla.

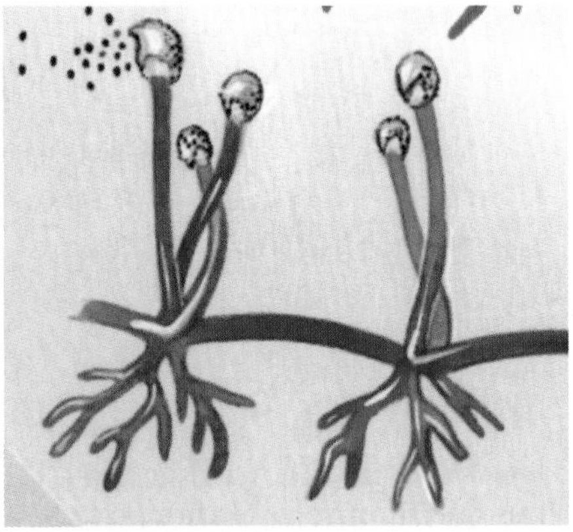

Phylum Zygomycota: haploid spore-bearing stalks.

Phylum Zygomycota

Zygomycetes; bread molds and other microscopic fungi that occur on decaying organic matter. Zygomycetes are the fungal partners in endomycorrhizae. Hyphae aseptate except when forming sporangia or gametangia. Sexual reproduction by equal gametangia containing numerous nuclei; after fusion, zygotes are formed within a zygospore, a structure that often forms a characteristic thick wall. Meiosis occurs during the germination of the zygospore. About 600 species.

Phylum Ascomycota

Ascomycetes; yeasts, molds, many important plant pathogens, morels, cup fungi, and truffles. Hyphae divided by incomplete septa except when asci, the structures characteristic of sexual reproduction, are formed. Dikaryotic hyphae form after appropriate fusions of monokaryotic ones, and eventually differentiate asci, which are often club-shaped, within an ascocarp. Meiosis takes place within asci. About 30,000 named species.

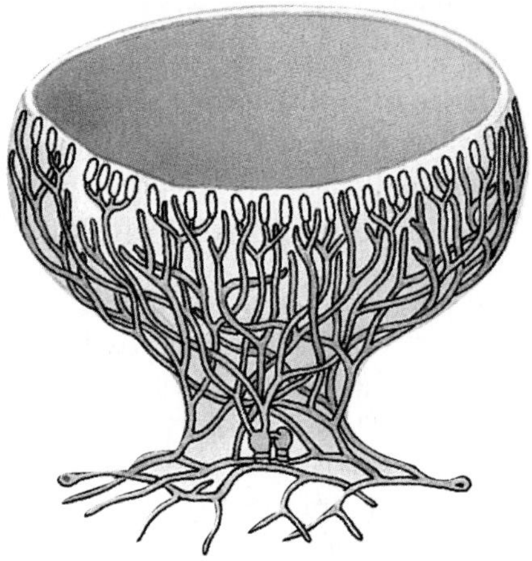

Phylum Ascomycota: ascocarp.

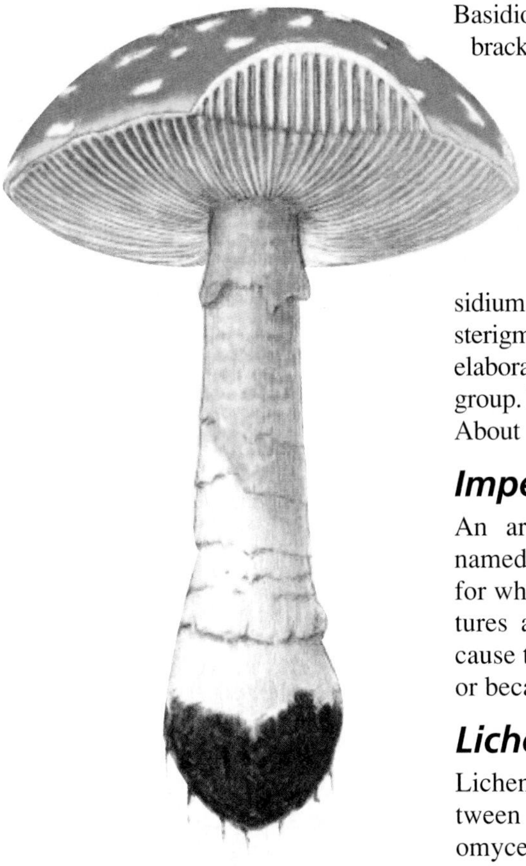

Phylum Basidiomycota: basidiocarp.

Phylum Basidiomycota

Basidiomycetes; mushrooms, toadstools, bracket and shelf fungi, rusts, and smuts. Most ectomycorrhizae involve basidiomycetes; a few involve ascomycetes. Hyphae and life cycle similar to that of ascomycetes but differing in important details. Basidiospores elevated from the basidium on threadlike projections called sterigmata; basidiocarps may be large and elaborate or absent, depending on the group. Meiosis takes place within basidia. About 16,000 named species.

Imperfect Fungi

An artificial group of about 17,000 named species. Most are ascomycetes for which the sexual reproductive structures are not known; this may be because the stages are rare or do not occur or because the fungus is poorly known.

Lichens

Lichens are symbiotic associations between an ascomycete (a few basidiomycetes are also involved) and either a green alga or a cyanobacterium; the fungus provides protection and structure. At least 13,500 species.

Kingdom Plantae

Multicellular, photosynthetic, primarily terrestrial eukaryotes derived from the green algae (phylum Chlorophyta) and, like them, containing chlorophylls *a* and *b,* together with carotenoids, in chloroplasts and storing starch in chloroplasts. The cell walls of plants have a cellulose matrix and sometimes become lignified; cell division is by means of a cell plate that forms across the mitotic spindle. The vascular plants have an elaborate system of conducting cells consisting of xylem (in which water and minerals are transported) and phloem (in which carbohydrates are transported); the mosses have a reduced vascular system, which the liverworts and hornworts (which may not be directly related to the mosses) lack. Plants have a waxy cuticle that helps them to retain water. Most have stomata, flanked by specialized guard cells, which allow water to escape and carbon dioxide to reach the chloroplast-containing cells within their leaves and stems. All plants have an alternation of generations with reduced gametophytes and multicellular gametangia. About 270,000 species. The major groups of plants used to be called divisions, but currently taxonomists call these groups phyla.

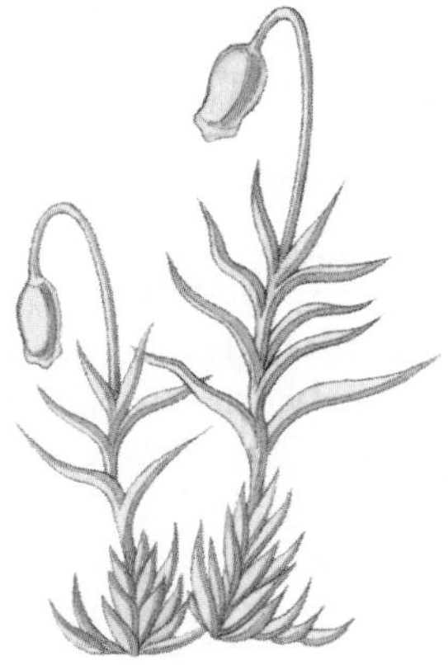

Phylum Bryophyta: a moss.

Phylum Bryophyta

Mosses. Bryophytes have green photosynthetic gametophytes and usually brownish or yellowish sporophytes with little or no chlorophyll. About 16,600 species. Hornworts and liverworts are members of two related phyla.

Phylum Psilophyta

Whisk ferns; a group of vascular plants. Two genera and several species.

Phylum Lycophyta

Lycopods (including clubmosses and quillworts); vascular plants. Five genera and about 1,000 species.

Phylum Sphenophyta

Horsetails; vascular plants. One genus (*Equisetum*) and 15 species.

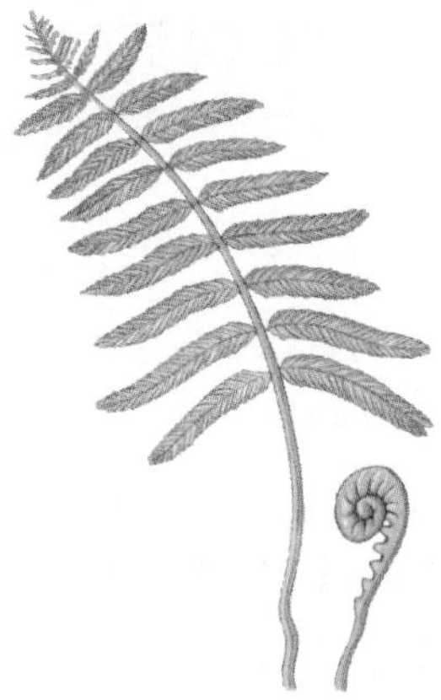

Phylum Pterophyta: a fern.

Phylum Pterophyta

Ferns; vascular plants, often with characteristically divided, feathery leaves (fronds). About 12,000 species.

Phylum Coniferophyta: a conifer.

Phylum Coniferophyta

Conifers; seed-forming vascular plants; mainly trees and shrubs. Archegonia multicellular, antheridia lacking; sperm immotile, carried to the vicinity of the egg by the pollen tube. Gametophytes reduced, held within the ovule or pollen grain. About 50 genera, with about 550 species.

Phylum Cycadophyta

Cycads; tropical and subtropical palmlike gymnosperms. Ten genera and about 100 species.

Phylum Gnetophyta

Gnetophytes, a very diverse group of three genera of gymnosperms. About 70 species.

Phylum Ginkgophyta: a ginkgo.

Phylum Ginkgophyta

One species, the ginkgo or maidenhair tree; a tall deciduous tree with fan-shaped leaves.

Phylum Anthophyta: a fruit tree.

Phylum Anthophyta

Flowering plants, or angiosperms, the dominant group of plants; characterized by a specialized reproductive system involving flowers and fruits. Sperm are immotile, carried to the vicinity of the ovule by the pollen tube. Fertilization is indirect, and the ovules are enclosed at the time of pollination. About 235,000 species.

Kingdom Animalia

Animals are multicellular eukaryotes that characteristically ingest their food. Their cells are usually flexible. In all of the approximately 35 phyla except sponges, these cells are organized into structural and functional units called tissues, which in turn make up organs in most animals. In animals, the cells move extensively during the development of the embryos; the blastula, a hollow ball of cells, forms early in this process and is characteristic of the group. Most animals reproduce sexually; their nonmotile eggs are much larger than their small, flagellated sperm. The gametes fuse directly to produce a zygote and do not divide by mitosis as in plants. More than a million species of animals have been described, and at least several times that many await discovery.

Phylum Porifera

Sponges; animals that mostly lack definite symmetry and possess neither tissues nor organs. About 10,000 species, mostly marine.

Phylum Cnidaria

Corals, jellyfish, hydras; mostly marine, radially symmetrical animals that usually have distinct tissues; two forms, polyps and medusae. About 10,100 species.

Phylum Platyhelminthes: a tapeworm.

Phylum Platyhelminthes

Flatworms; bilaterally symmetrical acoelomates; the simplest animals that have organs. About 13,000 species.

Phylum Nematoda

Nematodes, eelworms, and roundworms; ubiquitous, bilaterally symmetrical, cylindrical, unsegmented, pseudocoelomate worms, including many important parasites of plants and animals. More than 12,000 described species, but the actual number is probably 500,000 or more species.

Phylum Mollusca: a snail.

Phylum Mollusca

Mollusks; bilaterally symmetrical, protostome coelomate animals that occur in marine, freshwater, and terrestrial habitats. Many mollusks possess a shell. At least 110,000 species.

Phylum Annelida

Annelids; segmented, bilaterally symmetrical, protostome coelomates; the segments are divided internally by septa. About 12,000 species.

Phylum Arthropoda: a wasp.

Phylum Arthropoda

Arthropods; bilaterally symmetrical protostome coelomates with a segmented body, chitinous exoskeleton, complete digestive tract, dorsal brain and paired nerve cord, and jointed appendages. Arthropods are the largest phylum of animals, with nearly a million species described and many more to be found.

Phylum Echinodermata: a sea star.

Phylum Echinodermata

Echinoderms; sea stars, brittle stars, sand dollars, sea cucumbers, and sea urchins. Complex deuterostome, coelomate, marine animals that are more or less radially symmetrical as adults. About 6,000 living species.

Phylum Chordata: dinosaurs.

Phylum Chordata

Chordates; bilaterally symmetrical, deuterostome, coelomate animals that have at some stage of their development a notochord, pharyngeal slits, a hollow nerve cord on their dorsal side, and a tail. The best-known group of animals; about 45,000 species.

Appendix B
Answers to Concept Reviews

Chapter 1
1. d. all of the above **2. b.** filter out ultraviolet rays **3. c.** 20% **4. a.** hypothesis **5. d.** theory **6. d.** variables **7. b.** use energy **8. c.** internal environment, stable **9. a.** cell **10. b.** evolution **11.** cold, ozone **12.** observation **13.** control **14.** cellular organization, metabolism, homeostasis, reproduction, and heredity **15.** DNA (deoxyribonucleic acid) **16.** cell **17.** Archaebacteria, Eubacteria, Fungi, Protista, Plantae, Animalia **18.** structure

Chapter 2
1. a. atom **2. a.** radioactive isotope **3. b.** oxidation—loss of an electron **4. b.** three **5. c.** It is absent in water **6. a.** adhesion **7. b.** water-fearing **8. a.** lowers **9. d.** glucose **10. d.** proteins **11. d.** oxygen **12.** creationism **13.** reduction **14.** ionic **15.** three-fourths **16.** two-thirds **17.** cohesion **18.** base **19.** 20 **20.** photosynthesis

Chapter 3
1. c. Life evolved 2 billion years ago **2. d.** DNA **3. a.** eukaryotic **4. b.** energy transfer **5. c.** chromosomes **6. a.** ribosome **7. b.** transport **8. d.** synthesis **9. c, d, a,** then **b** **10. b.** mitochondria **11. a, b,** and **d** **12.** theory **13.** bacteria (prokaryotes) **14.** cytoskeleton **15.** flagella **16.** mitochondria

Chapter 4
1. d. osmosis **2. d** and **e** **3. c.** water enters a hypotonic solution **4. a** **5. b.** exocytosis **6. d** **7. a.** true **8. a.** spindle **9. b.** DNA is replicated **10.** pinocytosis **11.** binary fission **12.** protein (histones) **13.** G_1 **14.** M **15.** cytokinesis

Chapter 5
1. b. lower **2. a.** active **3. b.** ATP **4. c** **5. d.** electrons **6. b** and **d** **7. b.** glycolysis **8. b.** pyruvate **9. d.** oxygen **10. a.** carbon dioxide **11.** endergonic **12.** binding **13.** NADH **14.** pigments **15.** photosynthesis **16.** thylakoid **17.** Krebs **18.** ethyl alcohol

Chapter 6
1. c **2. b.** false **3. b.** one-third **4. a.** phenotype **5. c.** $Ww \times ww$ **6. d.** multiple genes **7. e** then **c** **8. b** **9. c.** 21 **10. a.** hemophilia **11.** homozygous **12.** 25% **13.** independent assortment **14.** epistasis **15.** recombinants **16.** Barr body **17.** Klinefelter's syndrome **18.** amniocentesis

Chapter 7
1. a **2. d.** TTAAGC **3. a.** gene **4. c.** transcription **5. a.** translation **6. b.** false **7. d.** leucine **8. c.** operator site **9. a.** true **10.** DNA **11.** double helix **12.** thymine, cytosine **13.** ribosome **14.** transfer RNA **15.** 64 **16.** enhancers **17.** repressor protein **18.** introns **19.** mutagen

Chapter 8
1. d. a and b **2. b.** shorter **3. c.** plasmids **4. a** **5. d.** amplify a region of DNA **6. b.** cDNA **7. b.** false **8. a** **9. c.** human gene therapy **10.** screening **11.** reverse transcriptase **12.** T_i **13.** growth hormone **14.** enzymes **15.** tumor necrosis factor **16.** Human Genome Project

Chapter 9
1. b **2. d.** Thomas Malthus **3. a.** natural selection **4. a.** true **5. d.** random mating **6. d** **7. b.** false **8. b.** stabilizing **9. d.** hybrid sterility **10.** homologous **11.** fossil **12.** carbon **13.** molecular **14.** appendix **15.** microevolution **16.** races

Chapter 10
1. b, d, g, a, e, f, c **2. d, e, a** **3. c.** sterile **4. d.** ecotypes **5. a.** 1.5 million **6. b.** 10 million **7. c.** common ancestor **8. c.** no tail **9. a.** weight **10.** Carolus Linnaeus **11.** biological species concept **12.** two-thirds **13.** systematics **14.** phylogeny **15.** clade **16.** seeds **17.** crocodiles

Chapter 11
1. a, d **2. c.** an organized nucleus **3. a.** binary fission **4. b.** archaebacteria **5. a.** introns **6. a, c,** and **d** **7. d.** crystallized in solution **8. c.** either RNA or DNA, but not both **9. b.** latency **10.** gram-positive **11.** eubacteria **12.** methanogens **13.** heterocysts **14.** reproduce **15.** tobacco mosaic virus **16.** bacteriophages **17.** respiratory

Chapter 12
1. d. photosynthetic **2. c.** centrioles **3. b.** animals, fungi, or plants **4. a.** an unfertilized egg **5. c.** haploid **6. c.** paramecia **7. d.** green algae **8. a.** sleeping sickness **9. b.** red algae **10.** endosymbiosis **11.** asexual **12.** DNA **13.** dinoflagellates **14.** food **15.** choanoflagellates **16.** red algae (phylum Rhodophyta)

Chapter 13

1. b **2. b.** algae **3. d.** Fungi are multicellular **4. b.** mycelium **5. c.** both asexually and sexually **6. a** **7. d.** *Rhizopus* **8. a.** basidiocarp **9. c.** green algae **10.** multicellularity **11.** hyphae **12.** cellulose **13.** sexual, stress **14.** Ascomycota **15.** *Penicillium* **16.** mycorrhizae, mineral

Chapter 14

1. c **2. b.** larger **3. a.** liverworts **4. c.** sieve elements **5. b.** secondary **6. d.** club moss **7. a.** gymnosperms **8. a.** stamens **9. a.** double fertilization **10.** alternation **11.** sporophyte **12.** *Cooksonia* **13.** apical meristem **14.** carpel **15.** triploid **16.** monocots **17.** fruits

Chapter 15

1. c. stem and leaves **2. a.** are alive at maturity **3. d.** stomata **4. a** **5. b.** gas exchange **6. c.** secondary xylem **7. b.** endodermis **8. a** **9. b and d** **10. b.** mitosis **11. c.** pollen tube **12. d.** birds **13. c.** apical, lateral **14.** sclerenchyma **15.** vascular bundles **16.** transpiration **17.** guard **18.** ethylene **19.** gravitropism **20.** bees **21.** imbibes water

Chapter 16

1. d. Arthropoda **2. b.** choanoflagellates **3. c.** symmetry **4. a.** nematocysts **5. d.** body cavity **6. c.** in the mesoderm **7. b.** Mollusca **8. a.** earthworms **9. b and c** **10. a, b, and c** **11. d.** backbone **12.** Parazoa, Porifera **13.** ectoderm, endoderm, and mesoderm **14.** cephalization **15.** nematodes **16.** mollusks **17.** segmentation **18.** thorax and abdomen **19.** echinodermata **20.** protostomes **21.** Chordata

Chapter 17

1. b, e, a, c, d **2. c.** 470 million **3. c.** plants **4. a(3), b(1), c(2), d(4)** **5. d.** smaller plankton **6. a.** amphibians **7. b.** swim bladders **8. c.** jaws **9. b.** have hair **10. c.** birds **11. d.** drink **12.** trilobites **13.** Permian **14.** lamprey **15.** bony fishes **16.** therapsids **17.** *Archaeopteryx* **18.** countercurrent flow **19.** crocodiles **20.** amniotic

Chapter 18

1. d. tree shrews **2. b.** 40 million **3. c.** prehensile tails **4. b.** false **5. c, d, a, e, b** **6. a.** the use of tools **7. b.** *Ardipithecus ramidus* **8. c.** *Homo erectus* **9. d.** 500,000 **10. a.** Europe **11. a, d, c, e, b** **12.** vision **13.** gibbons **14.** *Australopithecus* **15.** *erectus* **16.** Java **17.** *sapiens* **18.** 13,000 **19.** cultural

Chapter 19

1. c. tissues **2. e.** secretory system **3. a.** movement **4. c.** columnar cells **5. b.** carries oxygen **6. d.** glial **7. b.** dendrites to cell body to axon **8. b.** axial **9. d.** The interior of bones is hollow **10. a.** It is multinucleated **11. d.** troponin **12.** trillion **13.** immune (lymphatic) **14.** epithelial **15.** axon and dendrites **16.** 206 **17.** actin **18.** acetylcholine **19.** ATP

Chapter 20

1. b. capillary **2. c.** They contain many valves **3. a.** delivering oxygen to cells **4. d.** 90% **5. c.** hemoglobin **6. b.** mitral **7. a.** aorta **8. c, e, a, b, d** **9. c.** It is located inside the lungs **10. b.** diffusion **11. b.** increases **12. d** **13.** smooth **14.** 100 micrometers **15.** lymphatic **16.** serum albumin **17.** systolic **18.** bronchioles **19.** brain **20.** carcinogens

Chapter 21

1. a. arginine **2. d.** starch **3. a.** mouth **4. a and e** **5. d.** pancreas **6. c.** compaction **7. c** **8. b.** urea **9. b.** network of capillaries **10. c.** water **11. d.** 1,000,000 **12.** eight **13.** pyloric sphincter **14.** peristalsis **15.** villi **16.** glycogen **17.** bicarbonate **18.** filtration **19.** Henle

Chapter 22

1. b. 1.5 million **2. c** **3. b.** macrophages **4. a.** complement system **5. c.** increased blood flow **6. a.** helper T cell **7. d.** interleukin-1 **8. b.** memory B cells **9. c.** type II diabetes **10. a.** helper T cells **11. c** **12.** dermis **13.** lysozyme **14.** 37 **15.** antibodies **16.** vaccination **17.** myelin sheath **18.** CD4

Chapter 23

1. a. afferent **2. b.** cerebrum **3. d.** coordinating motor reflexes **4. a.** respiration **5. a.** decreases **6. c.** cerebrum **7. b.** axon—synapse—dendrite **8. c.** electrical insulators **9. d.** gravity **10. c.** **11. b.** rods **12.** sodium **13.** action potential **14.** neurotransmitters **15.** stroke **16.** four **17.** semicircular canals **18.** cornea

Chapter 24

1. e. b and c **2. c.** insulin **3. a.** thyroid hormone **4. b.** just above the kidneys **5. a.** aldosterone **6. d** **7. j** **8. i** **9. e** **10. h** **11. g** **12. d** **13. c** **14. f** **15. b** **16. a** **17.** nervous **18.** hypothalamus **19.** steroids **20.** posterior **21.** cortex **22.** calcium **23.** pancreas

Chapter 25

1. a **2. c.** erection **3. c.** fallopian tube **4. b.** decrease **5. d.** LH **6. b.** corpus luteum **7. a.** morula **8. d.** invaginate **9. b.** somites **10. c** **11. b.** false **12. b.** use of a condom **13.** vas deferens **14.** 2 **15.** vagina **16.** prolactin, oxytocin **17.** primitive streak **18.** ectoderm **19.** endoderm **20.** mesoderm **21.** ejaculation **22.** vasectomy

Chapter 26

1. b. in trophic level 3 **2. a.** primary consumers **3. d.** food web **4. d.** 10% **5. b.** phosphorus cycle **6. a.** evaporation **7. b.** spring overturn **8. a** **9. b** **10. c.** spring overturn **11. c.** taiga **12.** physical environment **13.** net primary productivity **14.** biomass **15.** fossil fuel **16.** nitrogen fixation **17.** estuaries **18.** oligotrophic **19.** deserts **20.** savanna

Chapter 27

1. b. clumped **2. c.** density-dependent effects **3. c** **4. a** **5. d** **6. b** **7. c.** keystone **8. a.** introduction of exotic species **9.** rapidly **10.** commensalism **11.** symbiotic **12.** realized **13.** character displacement **14.** aposematic **15.** secondary **16.** species richness **17.** competition

Chapter 28

1. c **2. b** **3. d.** the greenhouse effect **4. a.** true **5. d** **6. a** **7. b.** false **8. c, e, a, d, b** **9.** chlorofluorocarbons (CFCs) **10.** carbon dioxide (CO_2) **11.** taxes **12.** topsoil, groundwater, and biodiversity **13.** 20% **14.** pastureland (grazing) **15.** 5 million **16.** 5 billion **17.** 1.8%

Glossary
Key Terms and Concepts

A

absorption (L. *absorbere,* to swallow down)
The movement of water and of substances dissolved in water into a cell, tissue, or organism.

acid
Any substance that dissociates to form H^+ ions when dissolved in water.

acoelomate (Gr. *a,* not + *koiloma,* cavity)
A bilaterally symmetrical animal not possessing a body cavity, such as a flatworm.

actin (Gr. *actis,* ray)
One of the two major proteins that make up myofilaments (the other is myosin). It provides the cell with mechanical support and plays major roles in determining cell shape and cell movement.

action potential
A single nerve impulse. A transient all-or-none reversal of the electrical potential across a neuron membrane. Because it can activate nearby voltage-sensitive channels, an action potential propagates along a nerve cell.

activation energy
The energy a molecule must acquire to undergo a specific chemical reaction.

active transport
The transport of a solute across a membrane by protein carrier molecules to a region of higher concentration by the expenditure of chemical energy. One of the most important functions of any cell.

adaptation (L. *adaptare,* to fit)
Any peculiarity of structure, physiology, or behavior that promotes the likelihood of an organism's survival and reproduction in a particular environment.

adenosine triphosphate (ATP)
A molecule composed of ribose, adenine, and a triphosphate group. ATP is the chief energy currency of all cells. Cells focus all of their energy resources on the manufacture of ATP from ADP and phosphate, which requires the cell to supply 7 kilocalories of energy obtained from photosynthesis or from electrons stripped from foodstuffs to form 1 mole of ATP. Cells then use this ATP to drive endergonic reactions.

adhesion (L. *adhaerere,* to stick to)
The molecular attraction exerted between the surfaces of unlike bodies in contact, as water molecules to the walls of the narrow tubes that occur in plants.

aerobic (Gr. *aer,* air + *bios,* life)
Oxygen-requiring.

allele (Gr. *allelon,* of one another)
One of two or more alternative forms of a gene.

allele frequency
The relative proportion of a particular allele among individuals of a population. Not equivalent to gene frequency, although the two terms are sometimes confused.

allosteric interaction (Gr. *allos,* other + *stereos,* shape)
The change in shape that occurs when an activator or inhibitor binds to an enzyme. These changes result when specific, small molecules bind to the enzyme, molecules that are not substrates of that enzyme.

alternation of generations
A reproductive life cycle in which the diploid phase produces spores that give rise to the haploid phase and the haploid phase produces gametes that fuse to give rise to the zygote. The zygote is the first cell of the multicellular diploid phase.

alveolus, *pl.* alveoli (L. *alveus,* a small cavity)
One of the many small, thin-walled air sacs within the lungs in which the bronchioles terminate.

amniotic egg
An egg that is isolated and protected from the environment by a more or less impervious shell. The shell protects the embryo from drying out, nourishes it, and enables it to develop outside of water.

anaerobic (Gr. *an,* without + *aer,* air + *bios,* life)
Any process that can occur without oxygen. Includes glycolysis and fermentation. Anaerobic organisms can live without free oxygen.

anterior (L. *ante,* before)
Located before or toward the front. In animals, the head end of an organism.

anther (Gr. *anthos,* flower)
The part of the stamen of a flower that bears the pollen.

antibody (Gr. *anti,* against)
A protein substance produced in the blood by a B cell lymphocyte in response to a foreign substance (antigen) and released into the bloodstream. Binding to the antigen, antibodies mark them for destruction by other elements of the immune system.

anticodon
The three-nucleotide sequence at the end of a tRNA molecule that is complementary to, and base pairs with, an amino acid–specifying codon in mRNA.

antigen (Gr. *anti,* against + *genos,* origin)
A foreign substance, usually a protein, that stimulates lymphocytes to proliferate and secrete specific antibodies that bind to the foreign substance, labeling it as foreign and destined for destruction.

aposematic coloration
An ecological strategy of some organisms that "advertise" their poisonous nature by the use of bright colors.

appendicular skeleton (L. *appendicula,* a small appendage)
The skeleton of the limbs of the human body containing 126 bones.

asexual
Reproducing without forming gametes. Asexual reproduction does not involve sex. Its outstanding characteristic is that an individual offspring is genetically identical to its parent.

atom (Gr. *atomos,* indivisible)
A core (nucleus) of protons and neutrons surrounded by an orbiting cloud of electrons. The chemical behavior of an atom is largely determined by the distribution of its electrons, particularly the number of electrons in its outermost level.

atomic mass
The atomic mass of an atom consists of the combined weight of all of its protons and neutrons.

atomic number
The number of protons in the nucleus of an atom. In an atom that does not bear an electric charge (that is, one that is not an ion), the atomic number is also equal to the number of electrons.

autonomic nervous system (Gr. *autos,* self + *nomos,* law)
The motor pathways that carry commands from the central nervous system to regulate the glands and nonskeletal muscles of the body. Also called the involuntary nervous system.

autosome (Gr. *autos,* self + *soma,* body)
Any of the 22 pairs of human chromosomes that are similar in size and morphology in both males and females.

autotroph (Gr. *autos,* self + *trophos,* feeder)
Self-feeder. An organism that can harvest light energy from the sun or from the oxidation of inorganic compounds to make organic molecules.

axial skeleton
The skeleton of the head and trunk of the human body containing 80 bones.

B

bacterium, *pl.* bacteria (Gr. *bakterion,* dim. of *baktron,* a staff)
The simplest cellular organism. Its cells are smaller and prokaryotic in structure, and they lack internal organization.

Barr body
After Murray L. Barr, Canadian anatomist. The inactivated X chromosome in female mammals that can be seen as a deeply staining body, which remains attached to the nuclear membrane.

basal body
In cells that contain flagella or cilia, a form of centriole that anchors each flagellum.

base
Any substance that combines with H^+ ions. Having a pH value above 7.

Batesian mimicry
After Henry W. Bates, English naturalist. A situation in which a palatable or nontoxic organism resembles another kind of organism that is distasteful or toxic. Both species exhibit warning coloration.

B cell
A lymphocyte that recognizes invading pathogens much as T cells do, but instead of attacking the pathogens directly, it marks them for destruction by the nonspecific body defenses.

bilateral symmetry (L. *bi,* two + *lateris,* side; Gr. *symmetria,* symmetry)
A body form in which the right and left halves of an organism are approximate mirror images of each other.

binary fission (L. *binarius,* consisting of two things or parts + *fissus,* split)
Asexual reproduction of a cell by division into two equal, or nearly equal, parts. Bacteria divide by binary fission.

binomial system (L. *bi,* twice, two + Gr. *nomos,* usage, law)
A system of nomenclature that uses two words. The first names the genus, and the second designates the species.

biomass (Gr. *bios,* life + *maza,* lump or mass)
The total weight of all of the organisms living in an ecosystem.

biome (Gr. *bios,* life + *-oma,* mass, group)
A major terrestrial assemblage of plants, animals, and microorganisms that occur over wide geographical areas and have distinct characteristics. The largest ecological unit.

C

calorie (L. *calor,* heat)
The amount of energy in the form of heat required to raise the temperature of 1 gram of water 1 degree Celsius.

calyx (Gr. *kalyx,* a husk, cup)
The sepals collectively. The outermost flower whorl.

cancer
Unrestrained invasive cell growth. A tumor or cell mass resulting from uncontrollable cell division.

capillary (L. *capillaris,* hairlike)
A blood vessel with a very slender, hairlike opening. Blood exchanges gases and metabolites within capillaries. Capillaries join the end of an artery to the beginning of a vein.

carbohydrate (L. *carbo,* charcoal + *hydro,* water)
An organic compound consisting of a chain or ring of carbon atoms to which hydrogen and oxygen atoms are attached in a ratio of approximately 1:2:1. A compound of carbon, hydrogen, and oxygen having the generalized formula $(CH_2O)_n$, where n is the number of carbon atoms.

carcinogen (Gr. *karkinos,* cancer + -gen)
Any cancer-causing agent.

cardiovascular system (Gr. *kardia,* heart + L. *vasculum,* vessel)
The blood circulatory system and the heart that pumps it. Collectively, the blood, heart, and blood vessels.

carpel (Gr. *karpos,* fruit)
A leaflike organ in angiosperms that encloses one or more ovules. One of the members of the gynoecium.

carrying capacity
The maximum population size that a habitat can support.

catabolism (Gr. *katabole,* throwing down)
A process in which complex molecules are broken down into simpler ones.

catalysis (Gr. *katalysis,* dissolution + *lyein,* to loosen)
The enzyme-mediated process in which the subunits of polymers are held together and their bonds are stressed.

catalyst (Gr. *kata,* down + *lysis,* a loosening)
A general term for a substance that speeds up a specific chemical reaction by lowering the energy required to activate or start the reaction. An enzyme is a biological catalyst.

cell (L. *cella,* a chamber or small room)
The smallest unit of life. The basic organizational unit of all organisms. Composed of a nuclear region containing the hereditary apparatus within a larger volume called the cytoplasm bounded by a lipid membrane.

cellular respiration
The process in which the energy stored in a glucose molecule is released by oxidation. Hydrogen atoms are lost by glucose and gained by oxygen.

central nervous system
The brain and spinal cord, the site of information processing and control within the nervous system.

centromere (Gr. *kentron,* center + *meros,* a part)
A constricted region of the chromosome joining two sister chromatids, to which the kinetochore is attached.

chemical bond
The force holding two atoms together. The force can result from the attraction of opposite charges (ionic bond) or from the sharing of one or more pairs of electrons (a covalent bond).

chemiosmosis
The cellular process responsible for almost all of the adenosine triphosphate (ATP) harvested from eaten food and for all the ATP produced by photosynthesis.

chemoautotroph
An autotrophic bacterium that uses chemical energy released by specific inorganic reactions to power its life processes, including the synthesis of organic molecules.

chiasma, *pl.* chiasmata (Gr. a cross)
In meiosis, the points of crossing-over where portions of chromosomes have been exchanged during synapsis. A chiasma appears as an X-shaped structure under a light microscope.

chromatid (Gr. *chroma,* color + L. *-id,* daughters of)
One of two daughter strands of a duplicated chromosome that is joined by a single centromere.

chromatin (Gr. *chroma,* color)
The complex of DNA and proteins of which eukaryotic chromosomes are composed.

chromosome (Gr. *chroma,* color + *soma,* body)
The vehicle by which hereditary information is physically transmitted from one generation to the next. In a eukaryotic cell, long threads of DNA that are associated with protein and that contain hereditary information.

cilium, *pl.* cilia (L. eyelash)
Refers to flagella, which are numerous and organized in dense rows. Cilia propel cells through water. In human tissue, they move water over the tissue surface.

cladistics
A taxonomic technique used for creating hierarchies of organisms that represent true phylogenetic relationship and descent.

class
A taxonomic category ranking below a phylum (division) and above an order.

clone (Gr. *klon,* twig)
A line of cells, all of which have arisen from the same single cell by mitotic division. One of a population of individuals derived by asexual reproduction from a single ancestor. One of a population of genetically identical individuals.

codominance
In genetics, a situation in which the effects of both alleles at a particular locus are apparent in the phenotype of the heterozygote.

codon (L. code)
The basic unit of the genetic code. A sequence of three adjacent nucleotides in DNA or mRNA that code for one amino acid or for polypeptide termination.

coelom (Gr. *koilos,* a hollow)
A body cavity formed between layers of mesoderm and in which the digestive tract and other internal organs are suspended.

coenzyme
A cofactor that is a nonprotein organic molecule.

coevolution (L. *co-,* together + *e-,* out + *volvere,* to fill)
A term that describes the long-term evolutionary adjustment of one group of organisms to another.

commensalism (L. *cum,* together with + *mensa,* table)
A symbiotic relationship in which one species benefits while the other neither benefits nor is harmed.

community (L. *communitas,* community, fellowship)
The population of different species that live together and interact in a particular place.

competition
Interaction between individuals of two or more species for the same scarce resources. Intraspecific competition is interaction for the same scarce resources between individuals of a single species.

competitive exclusion
The hypothesis that if two species are competing with one another for the same limited resource in the same place, one will be able to use that resource more efficiently than the other and eventually will drive that second species to extinction locally.

complement system
The chemical defense of a vertebrate body that consists of a battery of proteins that become activated by the walls of bacteria and fungi. Complements the cellular defenses.

concentration gradient
The concentration difference of a substance as a function of distance. In a cell, a greater concentration of its molecules in one region than in another.

condensation
The coiling of the chromosomes into more and more tightly compacted bodies begun during the G_2 phase of the cell cycle.

conjugation (L. *conjugare,* to yoke together)
An unusual mode of reproduction that characterizes the ciliates, in which nuclei are exchanged between individuals through tubes connecting them during conjugation.

consumer
In ecology, a heterotroph that derives its energy from living or freshly killed organisms or parts thereof. Primary consumers are herbivores; secondary consumers are carnivores or parasites.

corolla (L. *cornea,* crown)
The petals of a flower, collectively. Usually, the conspicuously colored flower whorl.

cortex (L. bark)
In vascular plants, the primary ground tissue of a stem or root, bounded externally by the epidermis and internally by the central cylinder of vascular tissue. In animals, the outer, as opposed to the inner, part of an organ, as in the adrenal, kidney, and cerebral cortexes.

cotyledon (Gr. *kotyledon,* a cup-shaped hollow)
Seed leaf. Monocot embryos have one cotyledon, and dicots have two.

countercurrent flow
In organisms, the passage of heat or of molecules (such as oxygen, water, or sodium ions) from one circulation path to another moving in the opposite direction. Because the flow of the two paths is in opposite directions, a concentration difference always exists between the two channels, facilitating transfer.

covalent bond (L. *co-,* together + *valare,* to be strong)
A chemical bond formed by the sharing of one or more pairs of electrons.

crossing over
An essential element of meiosis occurring during prophase when nonsister chromatids exchange portions of DNA strands.

cuticle (L. *cutis,* skin)
A very thin film covering the outer skin of many plants.

cytokinesis (Gr. *kytos,* hollow vessel + *kinesis,* movement)
The C phase of cell division in which the cell itself divides, creating two daughter cells.

cytoplasm (Gr. *kytos,* hollow vessel + *plasma,* anything molded)
A semifluid matrix that occupies the volume between the nuclear region and the cell membrane. It contains the sugars, amino acids, proteins, and organelles (in eukaryotes) with which the cell carries out its everyday activities of growth and reproduction.

cytoskeleton (Gr. *kytos,* hollow vessel + *skeleton,* a dried body)
In the cytoplasm of all eukaryotic cells, a network of protein fibers that supports the shape of the cell and anchors organelles, such as the nucleus, to fixed locations.

D

deciduous (L. *decidere,* to fall off)
In vascular plants, shedding all the leaves at a certain season.

dehydration reaction
Water-losing. The process in which a hydroxyl (OH) group is removed from one subunit of a polymer and a hydrogen (H) group is removed from the other subunit.

demography (Gr. *demos,* people + *graphein,* to draw)
The statistical study of population. The measurement of people or, by extension, of the characteristics of people.

density
The number of individuals in a population in a given area.

deoxyribonucleic acid (DNA)
The basic storage vehicle or central plan of heredity information. It is stored as a sequence of nucleotides in a linear nucleotide polymer. Two of the polymers wind around each other like the outside and inside rails of a circular staircase.

depolarization
The movement of ions across a cell membrane that wipes out locally an electrical potential difference.

determinate
Having flowers that arise from terminal buds and thus terminate a stem or branch.

deuterostome (Gr. *deuteros,* second + *stoma,* mouth)
An animal in whose embryonic development the anus forms from or near the blastopore, and the mouth forms later on another part of the blastula. Also characterized by radial cleavage.

dicot
Short for dicotyledon; a class of flowering plants generally characterized by having two cotyledons, netlike veins, and flower parts in fours or fives.

diffusion (L. *diffundere,* to pour out)
The net movement of molecules to regions of lower concentration as a result of random, spontaneous molecular motions. The process tends to distribute molecules uniformly.

dihybrid (Gr. *dis,* twice + L. *hibrida,* mixed offspring)
An individual heterozygous for two genes.

dioecious (Gr. *di,* two + *eikos,* house)
Having male and female flowers on separate plants of the same species.

diploid (Gr. *diploos,* double + *eidos,* form)
A cell, tissue, or individual with a double set of chromosomes.

directional selection
A form of selection in which selection acts to eliminate one extreme from an array of phenotypes. Thus, the genes promoting this extreme become less frequent in the population.

disaccharide (Gr. *dis,* twice + *sakcharon,* sugar)
A sugar formed by linking two monosaccharide molecules together. Sucrose (table sugar) is a disaccharide formed by linking a molecule of glucose to a molecule of fructose.

diurnal (L. *diurnalis,* day)
Active during the day.

division
Traditionally, a major taxonomic group of the plant kingdom comparable to a phylum of the animal kingdom. Today divisions are called phyla.

dominant allele
An allele that dictates the appearance of heterozygotes. One allele is said to be dominant over another if an individual heterozygous for that allele has the same appearance as an individual homozygous for it.

dorsal (L. *dorsum,* the back)
Toward the back, or upper surface. Opposite of ventral.

double fertilization
A process unique to the angiosperms, in which one sperm nucleus fertilizes the egg and the second one fuses with the polar nuclei. These two events result in the formation of the zygote and the primary endosperm nucleus, respectively.

E

ecdysis (Gr. *ekdysis,* stripping off)
The shedding of the outer covering or skin of certain animals. Especially the shedding of the exoskeleton by arthropods.

ecology (Gr. *oikos,* house + *logos,* word)
The study of the relationships of organisms with one another and with their environment.

ecosystem (Gr. *oikos,* house + *systema,* that which is put together)
A community, together with the nonliving factors with which it interacts.

ecotype (Gr. *oikos,* house + L. *typus,* image)
A locally adapted variant of an organism, differing genetically from other ecotypes.

electron
A subatomic particle with a negative electrical charge. The negative charge of one electron exactly balances the positive charge of one proton. Electrons orbit the atom's positively charged nucleus and determine its chemical properties.

electron transport system
A collective term describing the series of membrane-associated electron carriers generated by the citric acid cycle. It puts the electrons harvested from the oxidation of glucose to work driving proton-pumping channels.

element
A substance that cannot be separated into different substances by ordinary chemical methods.

endergonic (Gr. *endon,* within + *ergon,* work)
Describing reactions in which the products contain more energy than the reactants and require an input of usable energy from an outside source before they can proceed. These reactions are not spontaneous.

endocrine gland (Gr. *endon,* within + *krinein,* to separate)
A ductless gland producing hormonal secretions that pass directly into the bloodstream or lymph.

endocrine system
The dozen or so major endocrine glands of a vertebrate.

endocytosis (Gr. *endon,* within + *kytos,* cell)
The process by which the edges of plasma membranes fuse together and form an enclosed chamber called a vesicle. It involves the incorporation of a portion of an exterior medium into the cytoplasm of the cell by capturing it within the vesicle.

endoskeleton (Gr. *endon,* within + *skeletos,* hard)
In vertebrates, an internal scaffold of bone to which muscles are attached.

endosperm (Gr. *endon,* within + *sperma,* seed)
A nutritive tissue characteristic of the seeds of angiosperms that develops from the union of a male nucleus and the polar nuclei of the embryo sac. The endosperm is either digested by the growing embryo or retained in the mature seed to nourish the germinating seedling.

endosymbiotic (Gr. *endon,* within + *bios,* life) theory
Proposes that eukaryotic cells arose from large prokaryotic cells that engulfed smaller ones of a different species, which were not consumed but continued to live and function within the larger host cell. Organelles that are believed to have entered larger cells in this way are mitochondria, chloroplasts, and centrioles.

endothermic
Referring to the ability of animals to maintain a constant body temperature.

energy
The capacity to bring about change, to do work.

enhancer
A site of regulatory protein binding on the DNA molecule distant from the promoter and start site for a gene's transcription.

entropy (Gr. *en,* in + *tropos,* change in manner)
A measure of the disorder of a system. A measure of energy that has become so randomized and uniform in a system that the energy is no longer available to do work.

enzyme (Gr. *enzymos,* leavened; from *en,* in + *zyme,* leaven)
A protein capable of speeding up specific chemical reactions by lowering the energy required to activate or start the reaction but that remains unaltered in the process.

epidermis (Gr. *epi,* on or over + *derma,* skin)
The outermost layer of cells. In vertebrates, the nonvascular external layer of skin of ectodermal origin; in invertebrates, a single layer of ectodermal epithelium; in plants, the flattened, skinlike outer layer of cells.

epistasis (Gr. *epistasis,* a standing still)
An interaction between the products of two genes in which one modifies the phenotypic expression produced by the other.

epithelium (Gr. *epi,* on + *thele,* nipple)
A thin layer of cells forming a tissue that covers the internal and external surfaces of the body. Simple epithelium consists of the membranes that line the lungs and major body cavities and that are a single cell layer thick. Stratified epithelium (the skin or epidermis) is composed of more complex epithelial cells that are several cell layers thick.

erythrocyte (Gr. *erythros,* red + *kytos,* hollow vessel)
A red blood cell, the carrier of hemoglobin. Erythrocytes act as the transporters of oxygen in the vertebrate body. During the process of their maturation in mammals, they lose their nucleus and mitochondria, and their endoplasmic reticulum is reabsorbed.

estrus (L. *oestrus,* frenzy)
The period of maximum female sexual receptivity. Associated with ovulation of the egg. Being "in heat."

estuary (L. *aestus,* tide)
A partly enclosed body of water, such as those that often form at river mouths and in coastal bays, where the salinity is intermediate between that of saltwater and freshwater.

ethology (Gr. *ethos,* habit or custom + *logos,* discourse)
The study of patterns of animal behavior in nature.

euchromatin (Gr. *eu,* good + *chroma,* color)
Chromatin that is extended except during cell division, from which RNA is transcribed.

eukaryote (Gr. *eu,* good + *karyon,* kernel)
A small, membrane-bounded structure that possesses an internal chamber called the cell nucleus. The appearance of eukaryotes marks a major event in the evolution of life, since all organisms on earth other than bacteria are eukaryotes.

eumetazoan (Gr. *eu,* good + *meta,* with + *zoion,* animal)
A "true animal." An animal with a definite shape and symmetry and nearly always distinct tissues.

eutrophic (Gr. *eutrophos,* thriving)
Refers to a lake in which an abundant supply of minerals and organic matter exists.

evaporation
The escape of water molecules from the liquid to the gas phase at the surface of a body of water.

evolution (L. *evolvere,* to unfold)
Genetic change in a population of organisms over time (generations). Darwin proposed that natural selection was the mechanism of evolution.

exergonic (L. *ex,* out + Gr. *ergon,* work)
Describes any reaction that produces products that contain less free energy than that possessed by the original reactants and that tends to proceed spontaneously.

exocytosis (Gr. *ex,* out of + *kytos,* cell)
The extrusion of material from a cell by discharging it from vesicles at the cell surface. The reverse of endocytosis.

exoskeleton (Gr. *exo,* outside + *skeletos,* hard)
An external hard shell that encases a body. In arthropods, comprised mainly of chitin; in vertebrates, comprised of bone.

experiment
The test of a hypothesis. A successful experiment is one in which one or more alternative hypotheses are demonstrated to be inconsistent with experimental observation and are thus rejected.

F

facilitated diffusion
The transport of molecules across a membrane by a carrier protein in the direction of lowest concentration.

family
A taxonomic group ranking below an order and above a genus.

feedback inhibition
A regulatory mechanism in which a biochemical pathway is regulated by the amount of the product that the pathway produces.

fermentation (L. *fermentum,* ferment)
A catabolic process in which the final electron acceptor is an organic molecule.

fertilization (L. *ferre,* to bear)
The union of male and female gametes to form a zygote.

fitness
The genetic contribution of an individual to succeeding generations, relative to the contributions of other individuals in the population.

flagellum, *pl.* flagella (L. *flagellum,* whip)
A fine, long, threadlike organelle protruding from the surface of a cell. In bacteria, a single protein fiber capable of rotary motion that propels the cell through the water. In eukaryotes, an array of microtubules with a characteristic internal 9 + 2 microtubule structure that is capable of vibratory but not rotary motion. Used in locomotion and feeding. Common in protists and motile gametes. A cilium is a small flagellum. The inward movement of certain cell groups from the surface of the blastula.

food web
The food relationships within a community. A diagram of who eats whom.

founder principle
The effect by which rare alleles and combinations of alleles may be enhanced in new populations.

frequency
In statistics, defined as the proportion of individuals in a certain category, relative to the total number of individuals being considered.

fruit
In angiosperms, a mature, ripened ovary (or group of ovaries) containing the seeds. Also applied informally to the reproductive structures of some other kinds of organisms.

G

gamete (Gr. wife)
A haploid reproductive cell. Upon fertilization, its nucleus fuses with that of another gamete of the opposite sex. The resulting diploid cell (zygote) may develop into a new diploid individual, or in some protists and fungi, may undergo meiosis to form haploid somatic cells.

gametophyte (Gr. *gamete,* wife + *phyton,* plant)
In plants, the haploid (n), gamete-producing generation, which alternates with the diploid ($2n$) sporophyte.

ganglion, *pl.* ganglia (Gr. a swelling)
A group of nerve cells forming a nerve center in the peripheral nervous system.

gastrulation
The inward movement of certain cell groups from the surface of the blastula.

gene (Gr. *genos,* birth, race)
The basic unit of heredity. A sequence of DNA nucleotides on a chromosome that encodes a polypeptide or RNA molecule and so determines the nature of an individual's inherited traits.

gene expression
The process in which an RNA copy of each active gene is made, and the RNA copy directs the sequential assembly of a chain of amino acids at a ribosome.

gene frequency
The frequency with which individuals in a population possess a particular gene. Often confused with allele frequency.

genetic code
The "language" of the genes. The mRNA codons specific for the 20 common amino acids constitute the genetic code.

genetic drift
Random fluctuations in allele frequencies in a small population over time.

genetic map
A diagram showing the relative positions of genes.

genetics (Gr. *genos,* birth, race)
The study of the way in which an individual's traits are transmitted from one generation to the next.

genome (Gr. *genos,* offspring + L. *oma,* abstract group)
The genetic information of an organism.

genotype (Gr. *genos,* offspring + *typos,* form)
The total set of genes present in the cells of an organism. Also used to refer to the set of alleles at a single gene locus.

genus, *pl.* genera (L. race)
A taxonomic group that ranks below a family and above a species.

germination (L. *germinare,* to sprout)
The resumption of growth and development by a spore or seed.

gland (L. *glandis,* acorn)
Any of several organs in the body, such as exocrine or endocrine, that secrete substances for use in the body. Glands are composed of epithelial tissue.

glomerulus (L. a little ball)
A network of capillaries in a vertebrate kidney, whose walls act as a filtration device.

gravitropism (L. *gravis,* heavy + *tropes,* turning)
The response of a plant to gravity, which generally causes shoots to grow up and roots to grow down.

greenhouse effect
The process in which carbon dioxide and certain other gases, such as methane, that occur in the earth's atmosphere transmit radiant energy from the sun but trap the longer wavelengths of infrared light, or heat, and prevent them from radiating into space.

guard cells
Pairs of specialized epidermal cells that surround a stoma. When the guard cells are turgid, the stoma is open; when they are flaccid, it is closed.

gymnosperm (Gr. *gymnos,* naked + *sperma,* seed)
A seed plant with seeds not enclosed in an ovary. The conifers are the most familiar group.

H

habitat (L. *habitare,* to inhabit)
The place where individuals of a species live.

half-life
The length of time it takes for half of the carbon-14 present in a sample to be converted to carbon-12.

haploid (Gr. *haploos,* single + *eidos,* form)
The gametes of a cell, tissue, or individual with only one set of chromosomes.

Hardy-Weinberg equilibrium
After G. H. Hardy, English mathematician, and G. Weinberg, German physician. A mathematical description of the fact that the relative frequencies of two or more alleles in a population do not change because of Mendelian segregation. Allele and genotype frequencies remain constant in a random-mating population in the absence of inbreeding, selection, or other evolutionary forces. Usually stated as: If the frequency of allele A is p and the frequency of allele a is q, then the genotype frequencies after one generation of random mating will always be $(p + q)^2 = p^2 + 2pq + q^2$.

Haversian canal
After Clopton Havers, English anatomist. Narrow channels that run parallel to the length of a bone and contain blood vessels and nerve cells.

helper T cell
A class of white blood cells that initiates both the cell-mediated immune response and the humoral immune response; helper T cells are the targets of the AIDS virus (HIV).

herbivore (L. *herba,* grass + *vorare,* to devour)
Any organism that eats plants.

heredity (L. *heredis,* heir)
The transmission of characteristics from parent to offspring.

heterochromatin (Gr. *heteros,* different + *chroma,* color)
That portion of a eukaryotic chromosome that remains permanently condensed and therefore is not transcribed into RNA. Most centromere regions are heterochromatic.

heterokaryon (Gr. *heteros,* other + *karyon,* kernel)
A fungal hypha that has two or more genetically distinct types of nuclei.

heterotroph (Gr. *heteros,* other + *trophos,* feeder)
An organism that does not have the ability to produce its own food. *See also* autotroph.

heterozygote (Gr. *heteros,* other + *zygotos,* a pair)
A diploid individual carrying two different alleles of a gene on its two homologous chromosomes.

hierarchical (Gr. *hieros,* sacred + *archos,* leader)
Refers to a system of classification in which successively smaller units of classification are included within one another.

histone (Gr. *histos,* tissue)
A complex of small, very basic polypeptides rich in the amino acids arginine and lysine. A basic part of chromosomes, histones form the core around which DNA is wrapped.

homeostasis (Gr. *homeos,* similar + *stasis,* standing)
The maintaining of a relatively stable internal physiological environment in an organism or steady-state equilibrium in a population or ecosystem.

homeotherm (Gr. *homeo,* similar + *therme,* heat)
An organism, such as a bird or mammal, capable of maintaining a stable body temperature independent of the environmental temperature. "Warm-blooded."

hominid (L. *homo,* man)
Human beings and their direct ancestors. A member of the family Hominidae. *Homo sapiens* is the only living member.

homologous chromosome (Gr. *homologia,* agreement)
One of the two nearly identical versions of each chromosome. Chromosomes that associate in pairs in the first stage of meiosis. In diploid cells, one chromosome of a pair that carries equivalent genes.

homology (Gr. *homologia,* agreement)
A condition in which the similarity between two structures or functions is indicative of a common evolutionary origin.

homozygote (Gr. *homos,* same or similar + *zygotos,* a pair)
A diploid individual whose two copies of a gene are the same. An individual carrying identical alleles on both homologous chromosomes is said to be homozygous for that gene.

hormone (Gr. *hormaein,* to excite)
A chemical messenger, often a steroid or peptide, produced in a small quantity in one part of an organism and then transported to another part of the organism, where it brings about a physiological response.

hybrid (L. *hybrida,* the offspring of a tame sow and a wild boar)
A plant or animal that results from the crossing of dissimilar parents.

hybridization
The mating of unlike parents of different taxa.

hydrogen bond
A molecular force formed by the attraction of the partial positive charge of one hydrogen atom of a water molecule with the partial negative charge of the oxygen atom of another.

hydrolysis reaction (Gr. *hydro,* water + *lyse,* break)
The process of tearing down a polymer by adding a molecule of water. A hydrogen is attached to one subunit and a hydroxyl to the other, which breaks the covalent bond. Essentially the reverse of a dehydration reaction.

hydrophobic (Gr. *hydro,* water + *phobos,* hating)
Refers to nonpolar molecules, which do not form hydrogen bonds with water and therefore are not soluble in water.

hydroskeleton (Gr. *hydro,* water + *skeletos,* hard)
The skeleton of most soft-bodied invertebrates that have neither an internal nor an external skeleton. They use the relative incompressibility of the water within their bodies as a kind of skeleton.

hypertonic (Gr. *hyper,* above + *tonos,* tension)
Refers to a cell that contains a higher concentration of solutes than its surrounding solution.

hypha, *pl.* hyphae (Gr. *hyphe,* web)
A filament of a fungus. A mass of hyphae comprises a mycelium.

hypothalamus (Gr. *hypo,* under + *thalamos,* inner room)
The region of the brain under the thalamus that controls temperature, hunger, and thirst and that produces hormones that influence the pituitary gland.

hypothesis (Gr. *hypo,* under + *tithenai,* to put)
A proposal that might be true. No hypothesis is ever proven correct. All hypotheses are provisional—proposals that are retained for the time being as useful but that may be rejected in the future if found to be inconsistent with new information. A hypothesis that stands the test of time—often tested and never rejected—is called a theory.

hypotonic (Gr. *hypo,* under + *tonos,* tension)
Refers to the solution surrounding a cell that has a lower concentration of solutes than does the cell.

I

inbreeding
The breeding of genetically related plants or animals. In plants, inbreeding results from self-pollination. In animals, inbreeding results from matings between relatives. Inbreeding tends to increase homozygosity.

incomplete dominance
The ability of two alleles to produce a heterozygous phenotype that is different from either homozygous phenotype.

independent assortment
Mendel's second law: the principle that segregation of alternative alleles at one locus into gametes is independent of the segregation of alleles at other loci. Only true for gene loci located on different chromosomes or those so far apart on one chromosome that crossing over is very frequent between the loci.

industrial melanism (Gr. *melas,* black)
Phrase used to describe the evolutionary process in which initially light-colored organisms become dark as a result of natural selection.

inflammatory response (L. *inflammare,* to flame)
A generalized nonspecific response to infection that acts to clear an infected area of infecting microbes and dead tissue cells so that tissue repair can begin.

integument (L. *integumentum,* covering)
The natural outer covering layers of an animal. Develops from the ectoderm.

interneuron
A nerve cell found only in the middle of the spinal cord that acts as a functional link between sensory neurons and motor neurons.

internode
The region of a plant stem between nodes where stems and leaves attach.

interoception (L. *interus,* inner + Eng. *[re]ceptive*)
The sensing of information that relates to the body itself, its internal condition, and its position.

interphase
That portion of the cell cycle preceding mitosis. It includes the G_1 phase, when cells grow, the S phase, when a replica of the genome is synthesized, and a G_2 phase, when preparations are made for genomic separation.

intron (L. *intra,* within)
A segment of DNA transcribed into mRNA but removed before translation. These untranslated regions make up the bulk of most eukaryotic genes.

ion
An atom in which the number of electrons does not equal the number of protons. An ion does carry an electrical charge.

ionic bond
A chemical bond formed between ions as a result of the attraction of opposite electrical charges.

ionizing radiation
High-energy radiation, such as X rays and gamma rays.

isolating mechanisms
Mechanisms that prevent genetic exchange between individuals of different populations or species. May be behavioral, morphological, or physiological.

isotonic (Gr. *isos,* equal + *tonos,* tension)
Refers to a cell with the same concentration of solutes as its environment.

isotope (Gr. *isos,* equal + *topos,* place)
An atom that has the same number of protons but different numbers of neutrons.

J

joint
The part of a vertebrate where one bone meets and moves on another.

K

karyotype (Gr. *karyon,* kernel + *typos,* stamp or print)
The particular array of chromosomes that an individual possesses.

kinetic energy
The energy of motion.

kinetochore (Gr. *kinetikos,* putting in motion + *choros,* chorus)
A disk of protein bound to the centromere to which microtubules attach during mitosis, linking each chromatid to the spindle.

kingdom
The chief taxonomic category. This book recognizes six kingdoms: Archaebacteria, Eubacteria, Protista, Fungi, Animalia, and Plantae.

L

lamella, *pl.* lamellae (L. a little plate)
A thin, platelike structure. In chloroplasts, a layer of chlorophyll-containing membranes. In bivalve mollusks, one of the two plates forming a gill. In vertebrates, one of the thin layers of bone laid concentrically around the Haversian canals.

learning
The creation of changes in behavior that arise as a result of experience, rather than as a result of maturation.

ligament (L. *ligare,* to bind)
A band or sheet of connective tissue that links bone to bone.

linkage
The patterns of assortment of genes that are located on the same chromosome. Important because if the genes are located relatively far apart, crossing over is more likely to occur between them than if they are close together.

lipid (Gr. *lipos,* fat)
A loosely defined group of molecules that are insoluble in water but soluble in oil. Oils such as olive, corn, and coconut are lipids, as well as waxes, such as beeswax and earwax.

lipid bilayer
The basic foundation of all biological membranes. In such a layer, the nonpolar tails of phospholipid molecules point inward, forming a nonpolar zone in the interior of the bilayers. Lipid bilayers are selectively permeable and do not permit the diffusion of water-soluble molecules into the cell.

littoral (L. *litus,* shore)
Referring to the shoreline zone of a lake or pond or the ocean that is exposed to the air whenever water recedes.

locus, *pl.* loci (L. place)
The position on a chromosome where a gene is located.

loop of Henle
After F. G. J. Henle, German anatomist. A hairpin loop formed by a urine-conveying tubule when it enters the inner layer of the kidney and then turns around to pass up again into the outer layer of the kidney.

lymph (L. *lympha,* clear water)
In animals, a colorless fluid derived from blood by filtration through capillary walls in the tissues.

lymphatic system
An open circulatory system composed of a network of vessels that function to collect the water within blood plasma forced out during passage through the capillaries and to return it to the bloodstream. The lymphatic system also returns proteins to the circulation, transports fats absorbed from the intestine, and carries bacteria and dead blood cells to the lymph nodes and spleen for destruction.

lymphocyte (Gr. *lympha,* water + Gr. *kytos,* hollow vessel)
A white blood cell. A cell of the immune system that either synthesizes antibodies (B cells) or attacks virus-infected cells (T cells).

lyse (Gr. *lysis,* loosening)
To disintegrate a cell by rupturing its cell membrane.

M

macromolecule (Gr. *makros,* large + L. *moliculus,* a little mass)
An extremely large molecule. Refers specifically to carbohydrates, lipids, proteins, and nucleic acids.

macrophage (Gr. *makros,* large + *-phage,* eat)
A phagocytic cell of the immune system able to engulf and digest invading bacteria, fungi, and other microorganisms, as well as cellular debris.

marrow
The soft tissue that fills the cavities of most bones and is the source of red blood cells.

mass
In chemistry, the total number of protons and neutrons in the nucleus of an atom. Approximately equal to the atomic weight.

mass flow
The overall process by which materials move in the phloem of plants.

meiosis (Gr. *meioun,* to make smaller)
A special form of nuclear division that precedes gamete formation in sexually reproducing eukaryotes.

Mendelian ratio
After Gregor Mendel, Austrian monk. Refers to the characteristic 3:1 segregation ratio that Mendel observed, in which pairs of alternative traits are expressed in the F_2 generation in the ratio of three-fourths dominant to one-fourth recessive.

menstruation (L. *mens,* month)
Periodic sloughing off of the blood-enriched lining of the uterus when pregnancy does not occur.

meristem (Gr. *merizein,* to divide)
In plants, a zone of unspecialized cells whose only function is to divide.

mesoderm (Gr. *mesos,* middle + *derma,* skin)
One of the three embryonic germ layers that form in the gastrula. Gives rise to muscle, bone, and other connective tissue; the peritoneum; the circulatory system; and most of the excretory and reproductive systems.

mesophyll (Gr. *mesos,* middle + *phyllon,* leaf)
The photosynthetic parenchyma of a leaf, located within the epidermis. The vascular strands (veins) run through the mesophyll.

metabolism (Gr. *metabole,* change)
The process by which all living things assimilate energy and use it to grow.

metamorphosis (Gr. *meta,* after + *morphe,* form + *osis,* state of)
Process in which form changes markedly during postembryonic development—for example, tadpole to frog or larval insect to adult.

metaphase (Gr. *meta,* middle + *phasis,* form)
The stage of mitosis characterized by the alignment of the chromosomes on a plane in the center of the cell.

metastasis, *pl.* metastases (Gr. to place in another way)
The spread of cancerous cells to other parts of the body, forming new tumors at distant sites.

microevolution (Gr. *mikros,* small + L. *evolvere,* to unfold)
Refers to the evolutionary process itself. Evolution within a species. Also called adaptation.

microfilament (Gr. *mikros,* small + L. *filum,* a thread)
In cells, a protein thread composed of parallel fibers of actin cross-connected by myosin. Their movement results from an ATP-driven shape change in myosin. The contraction of vertebrate muscles and many other kinds of cell movement in eukaryotes result from the movements of microfilaments within cells.

microtubule (Gr. *mikros,* small + L. *tubulus,* little pipe)
In eukaryotic cells, a long, hollow cylinder about 25 nanometers in diameter and composed of the protein tubulin. Microtubules influence cell shape, move the chromosomes in cell division, and provide the functional internal structure of cilia and flagella.

mimicry (Gr. *mimos,* mime)
The resemblance in form, color, or behavior of certain organisms (mimics) to other more powerful or more protected ones (models), which results in the mimics being protected in some way.

mitochondrion, *pl.* mitochondria (Gr. *mitos,* thread + *chondrion,* small grain)
A tubular or sausage-shaped organelle 1 to 3 micrometers long. Bounded by two membranes, mitochondria closely resemble the aerobic bacteria from which they were originally derived. As chemical furnaces of the cell, they carry out its oxidative metabolism.

mitosis (Gr. *mitos,* thread)
The M phase of cell division in which the microtubular apparatus is assembled, binds to the chromosomes, and moves them apart. This phase is the essential step in the separation of the two daughter cell genomes.

mole (L. *moles,* mass)
The atomic weight of a substance, expressed in grams. One mole is defined as the mass of $6.0222 ¥ 10^{23}$ atoms.

molecule (L. *moliculus,* a small mass)
The smallest unit of a compound that displays the properties of that compound.

monocot
Short for monocotyledon; flowering plant in which the embryos have only one cotyledon, the flower parts are often in threes, and the leaves typically are parallel-veined.

monosaccharide (Gr. *monos,* one + *sakcharon,* sugar)
A simple sugar.

morphogenesis (Gr. *morphe,* form + *genesis,* origin)
The formation of shape. The growth and differentiation of cells and tissues during development.

motor endplate
The point where a neuron attaches to a muscle. A neuromuscular synapse.

multicellularity
A condition in which the activities of the individual cells are coordinated and the cells themselves are in contact. A property of eukaryotes alone and one of their major characteristics.

muscle (L. *musculus,* mouse)
The tissue in the body of humans and animals that can be contracted and relaxed to make the body move.

muscle cell
A long, cylindrical, multinucleated cell that contains numerous myofibrils and is capable of contraction when stimulated.

muscle spindle
A sensory end organ that is attached to a muscle and sensitive to stretching.

mutagen (L. *mutare,* to change)
A chemical capable of damaging DNA.

mutation (L. *mutare,* to change)
A change in a cell's genetic message.

mutualism (L. *mutuus,* lent, borrowed)
A symbiotic relationship in which both participating species benefit.

mycelium, *pl.* mycelia (Gr. *mykes,* fungus)
In fungi, a mass of hyphae.

mycology (Gr. *mykes,* fungus)
The study of fungi. A person who studies fungi is called a mycologist.

mycorrhiza, *pl.* mycorrhizae (Gr. *mykes,* fungus + *rhiza,* root)
A symbiotic association between fungi and plant roots.

myofibril (Gr. *myos,* muscle + L. *fibrilla,* little fiber)
A contractile microfilament, composed of myosin and actin, within muscle.

myosin (Gr. *myos,* muscle + *in,* belonging to)
One of two protein components of myofilaments. (The other is actin.)

N

natural selection
The differential reproduction of genotypes caused by factors in the environment. Leads to evolutionary change.

nematocyst (Gr. *nema,* thread + *kystos,* bladder)
A coiled, threadlike stinging process of cnidarians that is discharged to capture prey and for defense.

nephron (Gr. *nephros,* kidney)
The functional unit of the vertebrate kidney. A human kidney has more than 1 million nephrons that filter waste matter from the blood. Each nephron consists of a Bowman's capsule, glomerulus, and tubule.

nerve
A bundle of axons with accompanying supportive cells, held together by connective tissue.

nerve impulse
A rapid, transient, self-propagating reversal in electrical potential that travels along the membrane of a neuron.

neuromodulator
A chemical transmitter that mediates effects that are slow and longer lasting and that typically involve second messengers within the cell.

neuromuscular junction
The structure formed when the tips of axons contact (innervate) a muscle fiber.

neuron (Gr. nerve)
A nerve cell specialized for signal transmission.

neurotransmitter (Gr. *neuron,* nerve + L. *trans,* across + *mitere,* to send)
A chemical released at an axon tip that travels across the synapse and binds a specific receptor protein in the membrane on the far side.

neurulation (Gr. *neuron,* nerve)
The elaboration of a notochord and a dorsal nerve cord that marks the evolution of the chordates.

neutron (L. *neuter,* neither)
A subatomic particle located within the nucleus of an atom. Similar to a proton in mass, but as its name implies, a neutron is neutral and possesses no charge.

neutrophil
An abundant type of white blood cell capable of engulfing microorganisms and other foreign particles.

niche (L. *nidus,* nest)
The role an organism plays in the environment; actual niche is the niche that an organism occupies under natural circumstances; theoretical niche is the niche an organism would occupy if competitors were not present.

nitrogen fixation
The incorporation of atmospheric nitrogen into nitrogen compounds, a process that can be carried out only by certain microorganisms.

nocturnal (L. *nocturnus,* night)
Active primarily at night.

node (L. *nodus,* knot)
The place on the stem where a leaf is formed.

node of Ranvier
After L. A. Ranvier, French histologist. A gap formed at the point where two Schwann cells meet and where the axon is in direct contact with the surrounding intercellular fluid.

nonrandom mating
A phenomenon in which individuals with certain genotypes sometimes mate with one another more commonly than would be expected on a random basis.

notochord (Gr. *noto,* back + L. *chorda,* cord)
In chordates, a dorsal rod of cartilage that forms between the nerve cord and the developing gut in the early embryo.

nucleic acid
A nucleotide polymer. A long chain of nucleotides. Chief types are deoxyribonucleic acid (DNA), which is double-stranded, and ribonucleic acid (RNA), which is typically single-stranded.

nucleosome (L. *nucleus,* kernel + *soma,* body)
The basic packaging unit of eukaryotic chromosomes, in which the DNA molecule is wound around a ball of histone proteins. Chromatin is composed of long strings of nucleosomes, like beads on a string.

nucleotide
A single unit of nucleic acid, composed of a phosphate, a five-carbon sugar (either ribose or deoxyribose), and a purine or a pyrimidine.

nucleus (L. a kernel, dim. Fr. *nux,* nut)
A spherical organelle (structure) characteristic of eukaryotic cells. The repository of the genetic information that directs all activities of a living cell. In atoms, the central core, containing positively charged protons and (in all but hydrogen) electrically neutral neutrons.

O

oocyte (Gr. *oion,* egg + *kytos,* vessel)
A cell in the outer layer of the ovary that gives rise to an ovum. A primary oocyte is any of the 2 million oocytes a female is born with, all of which have begun the first meiotic division.

operon (L. *operis,* work)
A cluster of functionally related genes transcribed onto a single mRNA molecule. A common mode of gene regulation in prokaryotes; it is rare in eukaryotes other than fungi.

order
A taxonomic category ranking below a class and above a family.

organ (L. *organon,* tool)
A complex body structure composed of several different kinds of tissue grouped together in a structural and functional unit.

organelle (Gr. *organella,* little tool)
A specialized compartment of a cell. Mitochondria are organelles.

organism
Any individual living creature, either unicellular or multicellular.

organ system
A group of organs that function together to carry out the principal activities of the body.

osmoconformer
An animal that maintains the osmotic concentration of its body fluids at about the same level as that of the medium in which it is living.

osmoregulation
The maintenance of a constant internal solute concentration by an organism, regardless of the environment in which it lives.

osmosis (Gr. *osmos,* act of pushing, thrust)
The diffusion of water across a membrane that permits the free passage of water but not that of one or more solutes.

osmotic pressure
The increase of hydrostatic water pressure within a cell as a result of water molecules that continue to diffuse inward toward the area of lower water concentration (the water concentration is lower inside than outside the cell because of the dissolved solutes in the cell).

osteoblast (Gr. *osteon,* bone + *blastos,* bud)
A bone-forming cell.

osteocyte (Gr. *osteon,* bone + *kytos,* hollow vessel)
A mature osteoblast.

outcross
A term used to describe species that interbreed with individuals other than those like themselves.

oviparous (L. *ovum,* egg + *parere,* to bring forth)
Refers to reproduction in which the eggs are developed after leaving the body of the mother, as in reptiles.

ovulation
The successful development and release of an egg by the ovary.

ovule (L. *ovulum,* a little egg)
A structure in a seed plant that becomes a seed when mature.

ovum, *pl.* ova (L. egg)
A mature egg cell. A female gamete.

oxidation (Fr. *oxider,* to oxidize)
The loss of an electron during a chemical reaction from one atom to another. Occurs simultaneously with reduction. Is the second stage of the 10 reactions of glycolysis.

oxidative metabolism
A collective term for metabolic reactions requiring oxygen.

oxidative respiration
Respiration in which the final electron acceptor is molecular oxygen.

P

parasitism (Gr. *para,* beside + *sitos,* food)
A symbiotic relationship in which one organism benefits and the other is harmed.

parthenogenesis (Gr. *parthenos,* virgin + Eng. *genesis,* beginning)
The development of an adult from an unfertilized egg. A common form of reproduction in insects.

partial pressures (P)
The components of each individual gas—nitrogen, oxygen, and carbon—that together constitute the total air pressure.

pathogen (Gr. *pathos,* suffering + Eng. *genesis,* beginning)
A disease-causing organism.

pedigree (L. *pes,* foot + *grus,* crane)
A family tree. The patterns of inheritance observed in family histories. Used to determine the mode of inheritance of a particular trait.

peptide (Gr. *peptein,* to soften, digest)
Two or more amino acids linked by peptide bonds.

peptide bond
A covalent bond linking two amino acids. Formed when the positive (amino, or NH_3) group at one end and a negative (carboxyl, or COO) group at the other end undergo a chemical reaction and lose a molecule of water.

peristalsis (Gr. *peri,* around + *stellein,* to wrap)
The rhythmic sequences of waves of muscular contraction in the walls of a tube.

pH
Refers to the concentration of H^+ ions in a solution. The numerical value of the pH is the negative of the exponent of the molar concentration. Low pH values indicate high concentrations of H^+ ions (acids), and high pH values indicate low concentrations.

phagocyte (Gr. *phagein,* to eat + *kytos,* hollow vessel)
A cell that kills invading cells by engulfing them. Includes neutrophils and macrophages.

phagocytosis (Gr. *phagein,* to eat + *kytos,* hollow vessel)
A form of endocytosis in which cells engulf organisms or fragments of organisms.

phenotype (Gr. *phainein,* to show + *typos,* stamp or print)
The realized expression of the genotype. The observable expression of a trait (affecting an individual's structure, physiology, or behavior) that results from the biological activity of proteins or RNA molecules transcribed from the DNA.

pheromone (Gr. *pherein,* to carry + [hor]mone)
A chemical signal emitted by certain animals that signals their reproductive readiness.

phloem (Gr. *phloos,* bark)
In vascular plants, a food-conducting tissue basically composed of sieve elements, various kinds of parenchyma cells, fibers, and sclereids.

phosphodiester bond
The bond that results from the formation of a nucleic acid chain in which individual sugars are linked together in a line by the phosphate groups. The phosphate group of one sugar binds to the hydroxyl group of another, forming an —O—P—O bond.

photon (Gr. *photos,* light)
The unit of light energy.

photoperiodism (Gr. *photos,* light + *periodos,* a period)
A mechanism that organisms use to measure seasonal changes in relative day and night length.

photorespiration
A process in which carbon dioxide is released without the production of ATP or NADPH. Because it produces neither ATP nor NADPH, photorespiration acts to undo the work of photosynthesis.

photosynthesis (Gr. *photos,* light + *-syn,* together + *tithenai,* to place)
The process by which plants, algae, and some bacteria use the energy of sunlight to create from carbon dioxide (CO_2) and water (H_2O) the more complicated molecules that make up living organisms.

phototropism (Gr. *photos,* light + *trope,* turning to light)
A plant's growth response to a unidirectional light source.

phylogeny (Gr. *phylon,* race, tribe)
The evolutionary relationships among any group of organisms.

phylum, *pl.* phyla (Gr. *phylon,* race, tribe)
A major taxonomic category, ranking above a class.

physiology (Gr. *physis,* nature + *logos,* a discourse)
The study of the function of cells, tissues, and organs.

pigment (L. *pigmentum,* paint)
A molecule that absorbs light.

pinocytosis (Gr. *pinein,* to drink + *kytos,* cell)
A form of endocytosis in which the material brought into the cell is a liquid containing dissolved molecules.

plankton (Gr. *planktos,* wandering)
The small organisms that float or drift in water, especially at or near the surface.

plasma (Gr. form)
The fluid of vertebrate blood. Contains dissolved salts, metabolic wastes, hormones, and a variety of proteins, including antibodies and albumin. Blood minus the blood cells.

plasma membrane
A lipid bilayer with embedded proteins that control the cell's permeability to water and dissolved substances.

plasmid (Gr. *plasma,* a form or something molded)
A small fragment of DNA that replicates independently of the bacterial chromosome.

platelet (Gr. dim of *plattus,* flat)
In mammals, a fragment of a white blood cell that circulates in the blood and functions in the formation of blood clots at sites of injury.

pleiotropy (Gr. *pleros,* more + *trope,* a turning)
Describing a gene that produces more than one phenotypic effect.

polarization
The charge difference of a neuron so that the interior of the cell is negative with respect to the exterior.

polar molecule
A molecule with positively and negatively charged ends. One portion of a polar molecule attracts electrons more strongly than another portion, with the result that the molecule has electron-rich (–) and electron-poor (+) regions, giving it magnetlike positive and negative poles. Water is one of the most polar molecules known.

pollen (L. fine dust)
A fine, yellowish powder consisting of grains or microspores, each of which contains a mature or immature male gametophyte. In flowering plants, pollen is released from the anthers of flowers and fertilizes the pistils.

pollen tube
A tube that grows from a pollen grain. Male reproductive cells move through the pollen tube into the ovule.

pollination
The transfer of pollen from the anthers to the stigmas of flowers for fertilization, as by insects or the wind.

polygyny (Gr. *poly,* many + *gyne,* woman, wife)
A mating choice in which a male mates with more than one female.

polymer (Gr. *polus,* many + *meris,* part)
A large molecule formed of long chains of similar molecules.

polymerase chain reaction (PCR)
A process by which DNA polymerase is used to copy a sequence of interest repeatedly, making millions of copies of the same DNA.

polymorphism (Gr. *polys,* many + *morphe,* form)
The presence in a population of more than one allele of a gene at a frequency greater than that of newly arising mutations.

polynomial system (Gr. *polys,* many + [bi]nomial)
Before Linnaeus, naming a genus by use of a cumbersome string of Latin words and phrases.

polyp
A cylindrical, pipe-shaped cnidarian usually attached to a rock with the mouth facing away from the rock on which it is growing. Coral is made up of polyps.

polypeptide (Gr. *polys,* many + *peptein,* to digest)
A general term for a long chain of amino acids linked end to end by peptide bonds. A protein is a long, complex polypeptide.

polysaccharide (Gr. *polys,* many + *sakcharon,* sugar)
A sugar polymer. A carbohydrate composed of many monosaccharide sugar subunits linked together in a long chain.

population (L. *populus,* the people)
Any group of individuals, usually of a single species, occupying a given area at the same time.

posterior (L. *post,* after)
Situated behind or farther back.

potential difference
A difference in electrical charge on two sides of a membrane caused by an unequal distribution of ions.

potential energy
Energy with the potential to do work. Stored energy.

predation (L. *praeda,* prey)
The eating of other organisms. The one doing the eating is called a predator, and the one being consumed is called the prey.

primary growth
In vascular plants, growth originating in the apical meristems of shoots and roots, as contrasted with secondary growth; results in an increase in length.

primary nondisjunction
The failure of homologous chromosomes to separate in meiosis I. The cause of Down syndrome.

primary plant body
The part of a plant that arises from the apical meristems.

primary producers
Photosynthetic organisms, including plants, algae, and photosynthetic bacteria.

primary structure of a protein
The sequence of amino acids that makes up a particular polypeptide chain.

primordium, *pl.* primordia (L. *primus,* first + *ordiri,* begin)
The first cells in the earliest stages of the development of an organ or structure.

productivity
The total amount of energy of an ecosystem fixed by photosynthesis per unit of time. Net productivity is productivity minus that which is expended by the metabolic activity of the organisms in the community.

prokaryote (Gr. *pro,* before + *karyon,* kernel)
A simple bacterial organism that is small and single-celled, lacks external appendages, and has little evidence of internal structure.

promoter
An RNA polymerase binding site. The nucleotide sequence at the end of a gene to which RNA polymerase attaches to initiate transcription of mRNA.

prophase (Gr. *pro,* before + *phasis,* form)
The first stage of mitosis during which the chromosomes become more condensed, the nuclear envelope is reabsorbed, and a network of microtubules (called the spindle) forms between opposite poles of the cell.

protein (Gr. *proteios,* primary)
A long chain of amino acids linked end to end by peptide bonds. Because the 20 amino acids that occur in proteins have side groups with very different chemical properties, the function and shape of a protein is critically affected by its particular sequence of amino acids.

protist (Gr. *protos,* first)
A member of the kingdom Protista, which includes unicellular eukaryotic organisms and some multicellular lines derived from them.

proton
A subatomic particle in the nucleus of an atom that carries a positive charge. The number of protons determines the chemical character of the atom because it dictates the number of electrons orbiting the nucleus and available for chemical activity.

protostome (Gr. *protos,* first + *stoma,* mouth)
An animal in whose embryonic development the mouth forms at or near the blastopore. Also characterized by spiral cleavage.

protozoa (Gr. *protos,* first + *zoion,* animal)
The traditional name given to heterotrophic protists.

pseudocoel (Gr. *pseudos,* false + *koiloma,* cavity)
A body cavity similar to the coelom except that it is unlined.

punctuated equilibrium
A hypothesis of the mechanism of evolutionary change that proposes that long periods of little or no change are punctuated by periods of rapid evolution.

Q

quaternary structure of a protein
A term to describe the way multiple protein subunits are assembled into a whole.

R

radial symmetry (L. *radius,* a spoke of a wheel + Gr. *summetros,* symmetry)
The regular arrangement of parts around a central axis so that any plane passing through the central axis divides the organism into halves that are approximate mirror images.

radioactivity
The emission of nuclear particles and rays by unstable atoms as they decay into more stable forms. Measured in curies, with 1 curie equal to 37 billion disintegrations a second.

radula (L. scraper)
A rasping, tonguelike organ characteristic of most mollusks.

recessive allele
An allele whose phenotype effects are masked in heterozygotes by the presence of a dominant allele.

recombination
The formation of new gene combinations. In bacteria, it is accomplished by the transfer of genes into cells, often in association with viruses. In eukaryotes, it is accomplished by reassortment of chromosomes during meiosis and by crossing over.

reducing power
The use of light energy to extract hydrogen atoms from water.

reduction (L. *reductio,* a bringing back; originally, "bringing back" a metal from its oxide)
The gain of an electron during a chemical reaction from one atom to another. Occurs simultaneously with oxidation.

reflex (L. *reflectere,* to bend back)
An automatic consequence of a nerve stimulation. The motion that results from a nerve impulse passing through the system of neurons, eventually reaching the body muscles and causing them to contract.

refractory period
The recovery period after membrane depolarization during which the membrane is unable to respond to additional stimulation. The period after ejaculation, lasting 20 minutes or longer, during which males lose their erection, arousal is difficult, and ejaculation is almost impossible.

renal (L. *renes,* kidneys)
Pertaining to the kidney.

repression (L. *reprimere,* to press back, keep back)
The process of blocking transcription by the placement of the regulatory protein between the polymerase and the gene, thus blocking movement of the polymerase to the gene.

repressor (L. *reprimere,* to press back, keep back)
A protein that regulates transcription of mRNA from DNA by binding to the operator and so preventing RNA polymerase from attaching to the promoter.

resolving power
The ability of a microscope to distinguish two lines as separate.

respiration (L. *respirare,* to breathe)
The utilization of oxygen. In terrestrial vertebrates, the inhalation of oxygen and the exhalation of carbon dioxide.

resting membrane potential
The charge difference that exists across a neuron's membrane at rest (about 70 millivolts).

restriction endonuclease
A special kind of enzyme that can recognize and cleave DNA molecules into fragments. One of the basic tools of genetic engineering.

restriction fragment-length polymorphism (RFLP)
An associated genetic mutation marker detected because the mutation alters the length of DNA segments.

retrovirus (L. *retro,* turning back)
A virus whose genetic material is RNA rather than DNA. When a retrovirus infects a cell, it makes a DNA copy of itself, which it can then insert into the cellular DNA as if it were a cellular gene.

ribose
A five-carbon sugar.

ribosome
An organelle composed of protein and RNA that translates RNA copies of genes into protein.

RNA polymerase
The enzyme that transcribes RNA from DNA.

S

saltatory conduction
A very fast form of nerve impulse conduction in which the impulses leap from node to node over insulation portions.

sarcoma (Gr. *sarx,* flesh)
A cancerous tumor that involves connective or hard tissue, such as muscle.

sarcomere (Gr. *sarx,* flesh + *meris,* part of)
The fundamental unit of contraction in skeletal muscle. The repeating bands of actin and myosin that appear between two Z lines.

sarcoplasmic reticulum (Gr. *sarx,* flesh + *plassein,* to form, mold; L. *reticulum,* network)
The endoplasmic reticulum of a muscle cell. A sleeve of membrane that wraps around each myofilament.

scientific creationism
A view that the biblical account of the origin of the earth is literally true, that the earth is much younger than most scientists believe, and that all species of organisms were individually created just as they are today.

secondary growth
In vascular plants, growth that results from the division of a cylinder of cells around the plant's periphery. Secondary growth causes a plant to grow in diameter.

second messenger
An intermediary compound that couples extracellular signals to intracellular processes and also amplifies a hormonal signal.

seed
A structure that develops from the mature ovule of a seed plant. Contains an embryo surrounded by a protective coat.

selection
The process by which some organisms leave more offspring than competing ones and their genetic traits tend to appear in greater proportions among members of succeeding generations than the traits of those individuals that leave fewer offspring.

self-fertilization
The transfer of pollen from an anther to a stigma in the same flower or to another flower of the same plant, leading to self-fertilization.

sepal (L. *sepalum,* a covering)
A member of the outermost whorl of a flowering plant. Collectively, the sepals constitute the calyx.

septum, *pl.* septa (L. *saeptum,* a fence)
A partition or cross-wall, such as those that divide fungal hyphae into cells.

sex chromosomes
The X and Y chromosomes, which are different in the two sexes and are involved in sex determination.

sex-linked characteristic
A genetic characteristic that is determined by genes located on the sex chromosomes.

sexual reproduction
Reproduction that involves the regular alternation between syngamy and meiosis. Its outstanding characteristic is that an individual offspring inherits genes from two parent individuals.

shoot
In vascular plants, the aboveground parts, such as the stem and leaves.

sieve cell
In the phloem (food-conducting tissue) of vascular plants, a long, slender sieve element with relatively unspecialized sieve areas and with tapering end walls that lack sieve plates. Found in all vascular plants except angiosperms, which have sieve-tube members.

soluble
Refers to polar molecules that dissolve in water and are surrounded by a hydration shell.

solute
The molecules dissolved in a solution. *See also* solution, solvent.

solution
A mixture of molecules, such as sugars, amino acids, and ions, dissolved in water.

solvent
The most common of the molecules dissolved in a solution. Usually a liquid, commonly water.

somatic cells (Gr. *soma,* body)
All the diploid body cells of an animal that are not involved in gamete formation.

somite
A segmented block of tissue on either side of a developing notochord.

species, *pl.* species (L. kind, sort)
A level of taxonomic hierarchy; a species ranks next below a genus.

sperm (Gr. *sperma,* sperm, seed)
A sperm cell. The male gamete.

spindle
The mitotic assembly that carries out the separation of chromosomes during cell division. Composed of microtubules and assembled during prophase at the equator of the dividing cell.

spore (Gr. *spora,* seed)
A haploid reproductive cell, usually unicellular, that is capable of developing into an adult without fusion with another cell. Spores result from meiosis, as do gametes, but gametes fuse immediately to produce a new diploid cell.

sporophyte (Gr. *spora,* seed + *phyton,* plant)
The spore-producing, diploid ($2n$) phase in the life cycle of a plant having alternation of generations.

stamen (L. thread)
The part of the flower that contains the pollen. Consists of a slender filament that supports the anther. A flower that produces only pollen is called staminate and is functionally male.

steroid (Gr. *stereos,* solid + L. *ol,* from oleum, oil)
A kind of lipid. Many of the molecules that function as messengers and pass across cell membranes are steroids, such as the male and female sex hormones and cholesterol.

steroid hormone
A hormone derived from cholesterol. Those that promote the development of the secondary sexual characteristics are steroids.

stigma (Gr. mark)
A specialized area of the carpel of a flowering plant that receives the pollen.

stoma, *pl.* stomata (Gr. mouth)
A specialized opening in the leaves of some plants that allows carbon dioxide to pass into the plant body and allows water and oxygen to pass out of them.

stratum corneum
The outer layer of the epidermis of the skin of the vertebrate body.

substrate (L. *substratus,* strewn under)
A molecule on which an enzyme acts.

substrate-level phosphorylation
The generation of ATP by coupling its synthesis to a strongly exergonic (energy-yielding) reaction.

succession
In ecology, the slow, orderly progression of changes in community composition that takes place through time. Primary succession occurs in nature over long periods of time. Secondary succession occurs when a climax community has been disturbed.

sugar
Any monosaccharide or disaccharide.

surface tension
A tautness of the surface of a liquid, caused by the cohesion of the liquid molecules. Water has an extremely high surface tension.

surface-to-volume ratio
Describes cell size increases. Cell volume grows much more rapidly than surface area.

symbiosis (Gr. *syn,* together with + *bios,* life)
The condition in which two or more dissimilar organisms live together in close association; includes parasitism, commensalism, and mutualism.

synapse (Gr. *synapsis,* a union)
A junction between a neuron and another neuron or muscle cell. The two cells do not touch. Instead, neurotransmitters cross the narrow space between them.

synapsis (Gr. *synapsis,* contact, union)
The close pairing of homologous chromosomes that occurs early in prophase I of meiosis. With the genes of the chromosomes thus aligned, a DNA strand of one homologue can pair with the complementary DNA strand of the other.

syngamy (Gr. *syn,* together with + *gamos,* marriage)
Fertilization. The union of male and female gametes.

T

taxonomy (Gr. *taxis,* arrangement + *nomos,* law)
The science of the classification of organisms.

T cell
A type of lymphocyte involved in cell-mediated immune responses and interactions with B cells. Also called a T lymphocyte.

tendon (Gr. *tenon,* stretch)
A strap of cartilage that attaches muscle to bone.

tertiary structure of a protein
The three-dimensional shape of a protein. Primarily the result of hydrophobic interactions of amino acid side groups and, to a lesser extent, of hydrogen bonds between them. Forms spontaneously.

test cross
A cross between a heterozygote and a recessive homozygote. A procedure Mendel used to further test his hypotheses.

theory (Gr. *theorein,* to look at)
A well-tested hypothesis supported by a great deal of evidence.

thigmotropism (Gr. *thigma,* touch + *trope,* a turning)
The growth response of a plant to touch.

thorax (Gr. a breastplate)
The part of the body between the neck and the abdomen.

thylakoid (Gr. *thylakos,* sac + *-oides,* like)
A flattened, saclike membrane in the chloroplast of a eukaryote. Thylakoids are stacked on top of one another in arrangements called grana and are the sites of photosystem reactions.

tissue (L. *texere,* to weave)
A group of similar cells organized into a structural and functional unit.

trachea, *pl.* **tracheae** (L. windpipe)
In vertebrates, the windpipe.

tracheid (Gr. *tracheia,* rough)
An elongated cell with thick, perforated walls that carries water and dissolved minerals through a plant and provides support. Tracheids form an essential element of the xylem of vascular plants.

transcription (L. *trans,* across + *scribere,* to write)
The first stage of gene expression in which the RNA polymerase enzyme synthesizes an mRNA molecule whose sequence is complementary to the DNA.

translation (L. *trans,* across + *latus,* that which is carried)
The second stage of gene expression in which a ribosome assembles a polypeptide, using the mRNA to specify the amino acids.

translocation (L. *trans,* across + *locare,* to put or place)
In plants, the process in which most of the carbohydrates manufactured in the leaves and other green parts of the plant are moved through the phloem to other parts of the plant.

transpiration (L. *trans,* across + *spirare,* to breathe)
The loss of water vapor by plant parts, primarily through the stomata.

transposon (L. *transponere,* to change the position of)
A DNA sequence carrying one or more genes and flanked by insertion sequences that confer the ability to move from one DNA molecule to another. An element capable of transposition (the changing of chromosomal location).

tropism (Gr. *trop,* turning)
A plant's response to external stimuli. A positive tropism is one in which the movement or reaction is in the direction of the source of the stimulus. A negative tropism is one in which the movement or growth is in the opposite direction.

turgor pressure (L. *turgor,* a swelling)
The pressure within a cell that results from the movement of water into the cell. A cell with high turgor pressure is said to be turgid.

U

unicellular
Composed of a single cell.

urea (Gr. *ouron,* urine)
An organic molecule formed in the vertebrate liver. The principal form of disposal of nitrogenous wastes by mammals.

urine (Gr. *ouron,* urine)
The liquid waste filtered from the blood by the kidneys.

V

vaccination
The injection of a harmless microbe into a person or animal to confer resistance to a dangerous microbe.

vacuole (L. *vacuus,* empty)
A cavity in the cytoplasm of a cell that is bound by a single membrane and contains water and waste products of cell metabolism. Typically found in plant cells.

variable
Any factor that influences a process. In evaluating alternative hypotheses about one variable, all other variables are held constant so that the investigator is not misled or confused by other influences.

vascular bundle
In vascular plants, a strand of tissue containing primary xylem and primary phloem. These bundles of elongated cells conduct water with dissolved minerals and carbohydrates throughout the plant body.

vascular cambium
In vascular plants, the meristematic layer of cells that gives rise to secondary phloem and secondary xylem. The activity of the vascular cambium increases stem or root diameter.

ventral (L. *venter,* belly)
Refers to the bottom portion of an animal.

vertebrate
An animal having a backbone made of bony segments called vertebrae.

vesicle (L. *vesicula,* a little bladder)
Membrane-enclosed sacs within eukaryotic organisms created by weaving sheets of endoplasmic reticulum through the cell's interior.

vessel element
In vascular plants, a typically elongated cell, dead at maturity, that conducts water and solutes in the xylem.

villus, *pl.* **villi** (L. a tuft of hair)
In vertebrates, fine, microscopic, fingerlike projections lining the small intestine that serve to increase the absorptive surface area of the intestine.

vitamin (L. *vita,* life + *amine,* of chemical origin)
An organic substance required in minute quantities by an organism for growth and activity but that the organism cannot synthesize.

viviparous (L. *vivus,* alive + *parere,* to bring forth)
Refers to reproduction in which eggs develop within the mother's body and young are born free-living.

voltage-gated channel
A transmembrane pathway for an ion that is opened or closed by a change in the voltage, or charge difference, across the cell membrane.

W

water vascular system
The system of water-filled canals connecting the tube feet of echinoderms.

whorl
A circle of leaves or of flower parts present at a single level along an axis.

wood
Accumulated secondary xylem. Heartwood is the central, nonliving wood in the trunk of a tree. Hardwood is the wood of dicots, regardless of how hard or soft it actually is. Softwood is the wood of conifers.

X

xylem (Gr. *xylon,* wood)
In vascular plants, a specialized tissue, composed primarily of elongate, thick-walled conducting cells, that transports water and solutes through the plant body.

Y

yolk (O.E. *geolu,* yellow)
The stored substance in egg cells that provides the embryo's primary food supply.

Z

zygote (Gr. *zygotos,* paired together)
The diploid ($2n$) cell resulting from the fusion of male and female gametes (fertilization).

Credits

Photographs

Table of Contents

1,2: NASA; 3: © Veronika Burmeister/Visuals Unlimited; 4: © Andrew S. Bajer; 5: © Kjell Sandved/Butterfly Alphabet; 6: © Science VU-NIH/Visuals Unlimited; 7: Courtesy Richard Feldmann, NIH; 8: © Stanley Cohen/SPL/Photo Researchers, Inc.; 9: The Library, University College London and Gene Kritsky; 10: © Universitetsbiblioteket, Uppsala; 11: © David M. Phillips/Visuals Unlimited; 12, 13: © John D. Cunningham/Visuals Unlimited; 14: © Rod Planck/Tom Stack and Associates; 15: © Rod Planck/Photo Researchers, Inc.; 16: © Judd Cooney/Phototake; 17: © Bertram G. Murray, Jr./Animals Animals/Earth Scenes; 18: © Michael Nichols/Magnum Photos, Inc.; 19: © T. McCarthy/Unicorn Stock Photos, Inc.; 20: © P. Motta & S. Correr/SPL/Photo Researchers, Inc.; 21: © Peter Arnold/Peter Arnold, Inc.; 22: © Lennart Nilsson, *The Body Victorious,* Bonnier Fakta; 23: © Lennart Nilsson, from *Behold Man;* 24: © Janeart Ltd./The Image Bank; 25: © Tony Stone Images, Chicago; 26: © National Snow & Ice Data Center/SPL/Photo Researchers, Inc. 27: © Merlin D. Tuttle, Bat Conservation International; 28: © John Mielcarek/Dembinsky Photo Associates

Chapter 1

1.1: NASA; 1.2: © Jeanne A. Mortimer, Ph.D.; 1.5: © Tom J. Ulrich/Visuals Unlimited; 1.6a: © T. J. Beveridge/Visuals Unlimited; 1.6b: © J. J. Cardamone, Jr. and B. K. Pugashetti/BPS/Tom Stack & Associates; 1.6c: © Robert Simpson/Tom Stack & Associates; 1.6d: © Marty Snyderman; 1.6e: © James L. Castner; 1.6f: © Michael Fogden/Animals Animals/Earth Scenes; 1.7c: © Don W. Fawcett/Visuals Unlimited; 1.7d: © K. G. Murti/Visuals Unlimited; 1.7e: © Ed Reschke; 1.7h: © Kirtley Perkins/Visuals Unlimited; 1.7i: © Zig Leszczynski/Animals Animals/Earth Scenes; 1.7j left: © Edward S. Ross; 1.7j right: © Alan Nelson/Animals Animals/Earth Scenes; 1.7k: © John D. Cunningham/Visuals Unlimited; 1.7l: © Robert and Jean Pollock; 1.8a: © Joe McDonald/Animals Animals/Earth Scenes; 1.8b: © Kenneth Fink/Photo Researchers, Inc.; 1.8c: © Tom McHugh/Photo Researchers, Inc.; p. 15 icon: NOAA; p. 15 top: © Jeanne A. Mortimer, Ph.D.; p. 15 bottom: © Tom J. Ulrich/Visuals Unlimited

Chapter 2

2.1: NASA; 2.2: Mohr/Darwin Collection; Herbert Rose Barrand, photographer. Reproduced by permission of The Huntington Library, San Marino, California; 2.11a: © Joseph Devenney/The Image Bank; 2.11b: © John Eastcott/Yva Momatiuk/The Image Works; 2.11c: © George I. Bernard/Animals Animals/Earth Scenes; 2.20: © Scott Johnson/Animals Animals/Earth Scenes; 2.30: © Bob McKeever/Tom Stack & Associates; 2.32: NASA; p. 41 icon: NASA; p. 41 middle: © John Eastcott/Yva Momatiuk/The Image Works; p. 41 bottom: NASA

Chapter 3

3.1: © Veronika Burmeister/Visuals Unlimited; 3.9: © SIU/Visuals Unlimited; 3.11a: © J. J. Cardamore/BPS/Tom Stack & Associates; 3.11b: © David M. Phillips/Visuals Unlimited; 3.11c: © Ed Reschke; 3.12c: © Martha J. Powell/Visuals Unlimited; 3.13b: © David Phillips/Visuals Unlimited; 3.15: Courtesy Drs. J. V. Small and G. Rinnerthaler; 3.17a: © Don W. Fawcett/Visuals Unlimited; 3.18a top, 3.18a bottom: Courtesy of Dr. Bessie Huang, Dept. of Cell Biology, The Scripps Research Institute; 3.18b: © Stanley Flegler/Visuals Unlimited; 3.20b: © Alfred Pasiexa/SPL/Photo Researchers, Inc.; 3.21b: Courtesy of Dr. Charles Flickinger, *Medical Cellular Biology,* W. B. Saunders, 1979; 3.22b: © Don W. Fawcett/Visuals Unlimited; 3.23b: Courtesy of Kenneth Miller, Brown University; p. 63 icon: © Veronika Burmeister/Visuals Unlimited

Chapter 4

4.1: © Andrew S. Bajer; 4.5b, 4.5c: © BioPhoto Assoc./Photo Researchers, Inc.; 4.6b: Courtesy Dr. Birgit H. Satir; 4.14: © Lee D. Simon/Photo Researchers, Inc.; 4.16a: © K. G. Murti/Visuals Unlimited; 4.17: © SPL/Photo Researchers, Inc.; 4.19a, 4.19b, 4.19c, 4.19d, 4.19e: © Andrew S. Bajer; 4.20a: © David M. Phillips/Visuals Unlimited; 4.21a, 4.21b: © Cabisco/Visuals Unlimited; p. 81 icon: © Andrew S. Bajer

Chapter 5

5.1: © Kjell Sandved/Butterfly Alphabet; 5.2: © Cleveland P. Hickman; 5.8: © Manfred Kage/Peter Arnold, Inc.; 5.12c: © Manfred Kage/Peter Arnold, Inc.; 5.12d: © Dr. Lewis K. Shumway; 5.16a: © Eric Soder/Tom Stack & Associates; 5.16c: © Erick Soder/Tom Stack & Associates; p.107 icon: © Kjell Sandved/Butterfly Alphabet

Chapter 6

6.1: © Science VU-NIH/Visuals Unlimited; 6.2: Courtesy American Museum of Natural History; 6.3: © Richard Gross/Biological Photography; 6.4a: © John D. Cunningham/Visuals Unlimited; 6.11: Courtesy R. W. Van Norman; 6.15a: From Albert & Blakeslee "Corn and Man," *Journal of Heredity* Vol. 5, pg. 511, 1914, Oxford University Press; 6.19b 1–9, © C. A. Hasenkampf/Biological Photo Service; 6.20: © Dr. Andrew S. Bajer; 6.22a: Courtesy Loris McGavaran, Denver Childrens Hospital; 6.22b: © Richard Hutchings/Photo Researchers, Inc.; 6.24a, 6.24b: © CBS/Phototake; 6.25: © SPL/Photo Researchers, Inc.; 6.27a: The Bettmann Archive; p. 133 icon: © Science VU-NIH/Visuals Unlimited; p. 133 top: Courtesy American Museum of Natural History; p. 133 bottom: © Richard Hutchings/Photo Researchers, Inc.

Chapter 7

7.1: Courtesy Richard Feldmann, NIH; 7.3b: © Lee D. Simon/Photo Researchers, Inc.; 7.3c: Courtesy Prof. A. K. Kleinschmidt; 7.6: © A. C. Barrington Brown/Photo Researchers, Inc.; 7.11: © Sarah Elgin/Washington University; 7.14: Courtesy Oscar L. Miller, Jr.; 7.21: © David Scharp; p. 151 icon: Courtesy Richard Feldmann, NIH

Chapter 8

8.1: © Stanley Cohen/SPL/Photo Researchers, Inc.; 8.2b: Courtesy Bio-Rad Laboratories; 8.3: Courtesy Dr. Ken Culver, photo by John Crawford, National Institutes of Health; 8.5: Courtesy Ralph L. Brinster, University of Pennsylvania, School of Vet. Med.; 8.10: Courtesy Dr. John Sanford, Cornell University; 8.11, 8.12: Monsanto Company; 8.13: Courtesy of Dr. Vernon Pursel; p. 167 icon: © Stanley Cohen/SPL/Photo Researchers, Inc.; p. 167 top: Courtesy Bio-Rad Laboratories; p. 167 bottom: Courtesy Dr. John Sanford, Cornell University

Chapter 9

9.1: The Library, University College London and Gene Kritsky; 9.2: © Christopher Ralling; 9.4: © Cleveland Hickman; 9.5a: © Frank B. Gill/Vireo; 9.5b: © John S. Dunning/Vireo; 9.6: © Mary Evans Picture Library/Photo Researchers, Inc.; 9.18: © Edward S. Ross; 9.19: Courtesy Dr. Victor A. McKusick, Johns Hopkins University; 9.20: © Edward S. Ross; 9.24: Courtesy of the University of Chicago Library/Dept. of Special Collections and Todd L. Savitt; 9.27, 9.28: © Breck P. Kent/Animals Animals/Earth Scenes; 9.30a: © Dan Coffey/The Image Bank; 9.30b: © G. Casperen/Image Bank; 9.30c: © Hank De Lespinasse/Image Bank; 9.30d: © Steve Krongard/The Image Bank; 9.30e: © Chuck Kuhn/The Image Bank; 9.33a, 9.33b: © Edward S. Ross; 9.34a, 9.34b, 9.34c: Courtesy Dr. Kenneth Kaneshiro; 9.34d: © Raymond Mendez/Animals Animals/Earth Scenes; p. 197 icon: The Library, University College London and Gene Kritsky; p. 197 top: © Christopher Ralling; p. 197 bottom: © Breck P. Kent/Animals Animals/Earth Scenes

Chapter 10

10.1: © Universitetsbiblioteket, Uppsala; 10.2a left: © Gerard Fritz/Transparencies, Inc.; 10.2a right: © Times Mirror Higher Education Group, Inc./Bob Coyle, photographer. Specimens courtesy of Dr. John Freeman, Auburn University; 10.2b left: © Werner H. Muller/Peter Arnold, Inc.; 10.2b right: © Richard Parker/Photo Researchers, Inc.; 10.3a: © Dwight R. Kuhn; 10.3b: Heather Angel; 10.3c: © S. Maslowski/Visuals Unlimited; 10.3d: © Henry Ausloos/Animals Animals/Earth Scenes; 10.3e: © John Cancalosi/Peter Arnold, Inc.; 10.3f: © Manfred Danegger/Peter Arnold, Inc.; 10.5a: © Rod Planck/Tom Stack & Associates; 10.5b: © Edward S. Ross; 10.5c: © Brian Parker/Tom Stack & Associates; 10.7a: © Gerard Lacz/Peter Arnold, Inc.; 10.7b: © Ralph Reinhold/Animals Animals/Earth Scenes; 10.7c: © Grant Heilman Photograph, Inc.; 10.8a, 10.8b: © Edward S. Ross; p. 211 icon: © Universitetsbiblioteket, Uppsala

Chapter 11

11.1, 11.2: © David M. Phillips/Visuals Unlimited; 11.4a: © Charles Brinton; 11.5: © Leonard Lessin/Peter Arnold, Inc.; 11.6: © Dwight R. Kuhn; 11.7: © CNRI/SPL/Photo Researchers, Inc.; 11.9a: Reprinted with permission from Donald L. D. Caspar, *Science* 227:773–776 Feb. 1985 Copyright American Association for the Advancement of

Science; 11.10b: © Scott Camazine/Photo Researchers, Inc.; 11.12: © Larry Mulvehill/Photo Researchers, Inc.; p. 227 icon: © David M. Phillips/Visuals Unlimited; p. 227 middle: © Dwight R. Kuhn

Chapter 12

12.1: © John D. Cunningham/Visuals Unlimited; 12.2: Courtesy Dr. Edward W. Daniels, Argonne National Lab, The University of Illinois College of Medicine at Chicago; 12.4: © Manfred Kage/Peter Arnold, Inc.; 12.5: © Bob Evans/Peter Arnold, Inc.; 12.8: © Phil A. Harrington/Peter Arnold, Inc.; 12.9: © Manfred Kage/ Peter Arnold, Inc.; 12.10: © John D. Cunningham/Visuals Unlimited; 12.11: © Rick Harbo; 12.12: © D. P. Wilson/ Eric & David Hosking/Photo Researchers, Inc.; 12.13: © D. P. Wilson/Photo Researchers, Inc.; 12.15: © Kevin Schafer/Peter Arnold, Inc.; 12.16a: © Ed Reschke; 12.16b: © Edward S. Ross; 12.18a: © Brian Parker/Tom Stack & Associates; 12.18b: © Manfred Kage/Peter Arnold, Inc.; 12.19a–f: © Genichiro Higuchi, Higuchi Science Laboratory; p. 245 icon: © John D. Cunningham/ Visuals Unlimited; p. 245 top: © Manfred Kage/Peter Arnold, Inc.; p. 245 middle: © D. P. Wilson/Photo Researchers, Inc.; p. 245 bottom: © Bob Evans/Peter Arnold, Inc.

Chapter 13

13.1: © John D. Cunningham/Visuals Unlimited; 13.2: © Bob Evans/Peter Arnold, Inc.; 13.3: © Jeff Rotman; 13.4: © Kjell Sandved/Butterfly Alphabet; 13.5: © L. West/ Photo Researchers, Inc.; 13.6: © Cabisco/Phototake; 13.8a: © Ed Pembleton; 13.8b: © Kjell Sandved/Butterfly Alphabet; 13.10a: © Michael and Patricia Fogden; 13.10b: © Walt Anderson/Tom Stack & Associates; 13.12: © Barry L. Runk/Grant Heilman; 13.13a: © David M. Phillips/Visuals Unlimited; 13.13b: Courtesy of Roland Wenger, Ph.D., Sandoz Pharma Ltd., Switzerland; 13.14a: © Edward S. Ross; 13.14b: © James Castner; 13.14c: © Edward S. Ross; 13.15a–c, Mark Brundrett; 13.16: © R. Roncadori/Visuals Unlimited; p. 261 icon: © John D. Cunningham/Visuals Unlimited; p. 261 top: © John D. Cunningham/Visuals Unlimited; p. 261 middle: © Jeff Rotman; p. 261 bottom: © Edward S. Ross

Chapter 14

14.1: © Rod Planck/Tom Stack & Associates; 14.2a: © Edward S. Ross; 14.2b: © Robert L. Dunne/Photo Researchers, Inc.; 14.2c: © Hermann Eisenbeiss/Photo Researchers, Inc.; 14.2d: © Rod Planck/Tom Stack & Associates; 14.3: © Terry Ashley/Tom Stack & Associates; 14.5a: © George E. Watson; 14.5b: © Richard Gross/Biological Photography; 14.6: © Kirtley Perkins/Visuals Unlimited; 14.7: Thomas Elias; 14.8: © Edward S. Ross; 14.10: © E. J. Cable/Tom Stack & Associates; 14.12: © Charlie Ott/Photo Researchers, Inc.; 14.13: © Edward S. Ross; 14.14: Courtesy of Harlan P. Banks; 14.15a: © Rod Planck/Tom Stack & Associates; 14.15b: © Edward S. Ross; 14.18a: T. Walker; 14.18b: Courtesy of R. J. Delroit/Agronomy Publications; 14.20: © Brian Parker/Tom Stack & Associates; 14.21a: © Edward S. Ross; 14.21b: © Michael and Patricia Fogden; 14.21c: © Runk Schoenberger/Grant Heilman; 14.21d: © Kjell Sandved/Butterfly Alphabet; 14.23: © Ed Pembleton; 14.25: © Michael and Patricia Fogden; 14.28: © James L. Castner; p. 285 icon: © Rod Planck/Tom Stack & Associates

Chapter 15

15.1: © Rod Planck/Photo Researchers, Inc.; 15.3: © Biophoto Associates/Photo Researchers, Inc.; 15.4: © George Wilder/Visuals Unlimited; 15.5: © Richard H. Gross/Biological Photography; 15.6: © E. J. Cable/Tom Stack & Associates; 15.7d: Courtesy Wilfred Cote, SUNY College of Environmental Forestry; 15.8a: © Randy Moore/Visuals Unlimited; 15.11a: © E. J. Cable/Tom Stack & Associates; 15.12: © Science VU/ R. A. Gregory/Visuals Unlimited; 15.13a: © Dwight R. Kuhn; 15.13b: © John D. Cunningham/Visuals Unlimited; 15.14: Courtesy Ray F. Evert; 15.15: © Carolina Biological Supply Company/Phototake; 15.16b: © Terry Ashley/Tom Stack & Associates; 15.17: © Jim Solliday/Biological Photo Service; 15.18: © E. J. Cable/Tom Stack & Associates; 15.21: © Doug Wechsler/Animals Animals/Earth Scenes; 15.24a: © J. A. L. Cooke/Animals Animals/Earth Scenes; 15.24b: © David M. Dennis/Tom Stack & Associates; 15.24c: © Kjell Sandved/Butterfly Alphabet; 15.28: Sylvan H. Wittwer; 15.29: © John D. Cunningham/Visuals Unlimited; 15.32a: Courtesy of William F. Chissoe, Nobel Microscopy Lab, University of Oklahoma; 15.32b: Courtesy of Dr. Joan Nowicke; 15.33: © Gary Milburn/ Tom Stack & Associates; 15.36a top left: © Edward S. Ross; 15.36a bottom left: © Kirtley-Perkins/Visuals Unlimited; 15.36b top right: © James L. Castner; 15.36b bottom right: © Dwight R. Kuhn; 15.37a top: © Adam Hart-Davis/SPL/Photo Researchers, Inc.; 15.37b bottom: © Helmut Gritscher/Peter Arnold, Inc.; p. 317 icon: © Rod Planck/Photo Researchers, Inc.; p. 317 top: © George Wilder/Visuals Unlimited

Chapter 16

16.1: © Judd Cooney/Phototake; 16.3a: © Jim and Cathy Church; 16.3b: © David J. Wrobel/Biological Photo Service; 16.5a: © Gwen Fidler/Tom Stack & Associates; 16.5b: © William C. Ober, M.D.; 16.5c: © A. Bannister/ Natural History Photographic Agency; 16.10a: © T. E. Adams/Visuals Unlimited; 16.10b: © Stan Elems/Visuals Unlimited; 16.14: © Tom E. Adams/Peter Arnold, Inc.; 16.15: © T. E. Adams/Visuals Unlimited; 16.17a: © Kjell Sandved/Butterfly Alphabet; 16.17b: © Marty Snyderman; 16.19a: © David M. Dennis/Tom Stack & Associates; 16.19b: © Kjell Sandved/Butterfly Alphabet; 16.21: © S. Giordano, III; 16.23: © Franz Lanting/ Minden Pictures; 16.24a: © George & Kathy Dodge; 16.24b: © Anne Moreton/Tom Stack & Associates; 16.25a: © Cleveland P. Hickman; 16.25b: © G. Lacz/ Natural History Photographic Agency; 16.25c: © T. E. Adams/Visuals Unlimited; 16.25d, 16.27a: © Kjell Sandved/Butterfly Alphabet; 16.27b: © Edward S. Ross; 16.27c: © Don Valntini/Tom Stack & Associates; 16.27d: © John Gerlach/Visuals Unlimited; 16.27e: © Edward S. Ross; 16.27f: © J. A. Alcock/Visuals Unlimited; 16.27g: © Cleveland P. Hickman; 16.29a: © Alex Kerstitch/ Visuals Unlimited; 16.29b: © Edward S. Ross; 16.30a: © Alex Kerstitch/Visuals Unlimited; 16.30b: © Carl Roessler/Tom Stack & Associates; 16.30c: © Bill Ober; 16.32a: © Jim Church; 16.32b: © Heather Angel; 16.32c: © Gunter Ziesler/Peter Arnold, Inc.; 16.34: © Eric N. Olson, Ph.D./The University of Texas MD Anderson Cancer Center; p. 351 icon: © Judd Cooney/Phototake, NYC; p. 351 top: © George & Kathy Dodge; p. 351 bottom: © Gunter Ziesler/Peter Arnold, Inc.

Chapter 17

17.1: © Bertram G. Murray, Jr./Animals Animals/Earth Scenes; 17.3a: © John Gurche; 17.3b: © John D. Cunningham/Visual Unlimited; 17.5a–b: *The Age of Reptiles,* a murel by Rudolph F. Zallinger, © 1966, 1975, 1985, 1989, Peabody Museum of Natural History, Yale University, New Haven, Connecticut, USA; 17.6: © E. R. Degginger/Animals Animals/Earth Scenes; 17.8: George C. Page Museum; 17.9: © Michael and Patricia Fogden; 17.11: © Tom McHugh/Photo Researchers, Inc.; 17.13: © Marty Snyderman; 17.15: © John D. Cunningham/ Visuals Unlimited; 17.16: © Peter Scoones/Contact Stock Images; 17.18: © John Shaw/Tom Stack & Associates; 17.19: © Brian Parker/Tom Stack & Associates; 17.21: © Alan Nelson/Animals Animals/Earth Scenes; 17.22: © OSF/Animals Animals/Earth Scenes; 17.28: © Ed Reschke; p. 385 icon: © Bertram G. Murray/Animals Animals/Earth Scenes

Chapter 18

18.1: © Michael Nichols/Magnum Photos, Inc.; 18.2: © Mary Evans Picture Library/Photo Researchers, Inc.; 18.4: © Alan Nelson/Animals Animals/Earth Scenes; 18.6a–d: © Russell A. Mittermeier; 18.8: © John Reader/ SPL/Photo Researchers, Inc.; 18.9a–d: © David L. Brill 1985 Artifact credit: National Museum of Tanzania, Dar es Salaam; 18.10a: © Dan Dreyfus & Associates Photography; 18.10b: © John Reader; 18.11b: © 1994 Tim D. White/Brill Atlanta Housed in National Museum of Ethiopia; 18.12: © John Reader/SPL/Photo Researchers, Inc.; 18.13: © Gordan Gahan/National Geographic Society; 18.14: The Bettman Archive; 18.16a–d: © David L. Brill 1985 Artifact Credit: Musee De L'Homme, Paris; 18.18: © Douglas Waugh/Peter Arnold, Inc.; 18.19: © Alan Carey/Image Works, Inc.; p. 405 icon: © Michael Nichols/Magunum Photos, Inc.; p. 405 top: © Dan Dreyfus & Associates Photography; p. 405 middle: © Gordon Gahan/National Geographic Society; p. 405 bottom: © Alan Carey/The Image Works

Chapter 19

19.1: © T. McCarthy/Unicorn Stock Photos, Inc.; 19.4: Cynthia Turner Alexander/Terry Cockerhan, Synapse Media Production, from C. P. Anthony and G. A. Thibodeau: *Textbook of Anatomy and Physiology,* 10/e, 1979, C. V. Mosby; 19.5a–c: © Ed Reschke; 19.6: © Mandred Kage/Peter Arnold, Inc.; 19.7: © Chuck Brown/Photo Researchers, Inc.; 19.8: © Lennart Nilsson, from *Behold Man;* 19.9: © Edwin A. Reschke; 19.10: © David M. Phillips/Visuals Unlimited; 19.11a–c: © Ed Reschke; 19.13: © Gerard and Agnice Vndys/Photo Researchers, Inc.; 19.16, 19.22a: © Ed Reschke; p. 429 icon: © T. McCarthy/Unicorn Stock Photos, Inc.; p. 429 top: Cynthia Turner Alexander/Terry Cockerhan, Synapse Media Production, from C. P. Anthony and G. A. Thibodeau: *Textbook of Anatomy and Physiology,* 10/e, 1979, C. V. Mosby; p. 429 middle: © David M. Phillips/ Visuals Unlimited; p. 429 bottom: © Lennart Nilsson, from *Behold Man*

Chapter 20

20.1: © P. Motta & S. Correr/SPL/Photo Researchers, Inc.; 20.5, 20.6: © Ed Reschke; 20.8: © Thomas Sims; 20.10: © M. Murayama/BPS/Tom Stack & Associates; 20.16: © Ellen Dirkson/Visuals Unlimited; 20.22: Courtesy American Cancer Society; 20.27: © Laurence Migdale/Photo Researchers, Inc.; p. 451 icon: © P. Motta & S. Correr/SPL/Photo Researchers, Inc.; p. 451 top: © M. Murayama/BPS/Tom Stack & Associates; p. 451 bottom: © Laurence Migdale/Photo Researchers, Inc.

Chapter 21

21.1: © Peter Arnold/Peter Arnold, Inc.; 21.11b: © David Phillips/Visuals Unlimited; 21.11c: Courtesy Susumu Ito; 21.17: © Times Mirror Higher Education Group, Inc./Bob Coyle, photographer; 21.18: © Kerrick James; p. 473 icon: © Peter Arnold/Peter Arnold, Inc.; p. 473 bottom: © Kerrick James

Chapter 22

22.1: © Lennart Nilsson, *The Body Victorious,* Bonnier Fakta; 22.2a–b: George Johnson; 22.4: © Manfred Kage/ Peter Arnold, Inc.; 22.5b: Courtesy Dr. Gilla Kaplan; 22.14: Courtesy National Library of Medicine; 22.16: Courtesy Larry Arlian; 22.18: © Lennart Nilsson; p. 491 icon: © Lennart Nilsson, *The Body Victorious,* Bonnier Fakta; p. 491 middle: Courtesy Larry Arlian; p. 491 bottom: © Lennart Nilsson, *The Body Victorious,* Bonnier Fakta

Chapter 23

23.1, 23.3: © Lennart Nilsson; 23.7: Dr. Marcus E. Rachle, Washington University, McDonnell Center for High Brain Function/Peter Arnold, Inc.; 23.9: © Lennart Nilsson, from *Behold Man;* 23.14: © Springer-Verlag New York, Inc.; 23.15b: © Don W. Fawcett/Photo Researchers, Inc.; 23.19: © Jack Vartoogian; 23.22c: Courtesy Dean E. Hillman; 23.23: © Stephen Dalton/ Animals Animals/Earth Scenes; 23.28b: Courtesy of Beckman Vision Center at UCSF School of Medicine /D. Copenhagen, S. Mittman and M. Maglio; p.517 icon: © Lennart Nilsson, from *Behold Man;* p. 517 top: © Lennart Nilsson; p. 517 middle: Dr. Marcus E. Rachle, Washington University, McDonnell Center for High Brain Function/Peter Arnold, Inc.; p. 517 bottom: © Springer-Verlag New York, Inc.

Chapter 24

24.1: © Janeart Ltd./The Image Bank; 24.2: © Lennart Nilsson, from *Behold Man;* 24.11: © John Paul Kay/Peter Arnold, Inc.; 24.13: © Ed Reschke; p. 533 icon: © Janeart Ltd./The Image Bank; p. 533 top: © Lennart Nilsson, from *Behold Man*

Chapter 25

25.1: © Tony Stone Images/Chicago Inc.; 25.3b, 25.7b: © Lennart Nilsson; 25.9: © Ed Reschke; 25.13a: From Lennart Nilsson, *A Child Is Born*, 1976, Dell Publishing, Bonnierforlagen; 25.13b–d: © Lennart Nilsson; 25.15: © Photo Researchers, Inc.; 25.16a–d: © Times Mirror Higher Education Group, Inc./Bob Coyle, photographer; p. 553 icon: © Tony Stone Images/Chicago Inc.; p. 553 top: © Lennart Nilsson; p. 553 top middle: © Ed Reschke; p. 553 bottom middle: © Lennart Nilsson; p. 553 bottom: © Times Mirror Higher Education Group, Inc./Bob Coyle, photographer

Chapter 26

26.1: National Snow & Ice Data Center/SPL/Photo Researchers, Inc.; 26.2: Photograph by David Swanlund, Courtesy Save-The-Redwoods League; 26.7: © Doug Sokell/Tom Stack & Associates; 26.11: © C. C. Lockwood/Animals Animals/Earth Scenes; 26.17: © Edward S. Ross; 26.18: © Marty Snyderman; 26.19: © Anne Wertheim/Animals Animals/Earth Scenes; 26.20a: © James L. Amos/Peter Arnold, Inc.; 26.20c–e: Courtesy Chesapeake Foundation; 26.21: © Edward S. Ross; 26.24a: © John D. Cunningham/Visuals Unlimited; 26.24b: © Daniel Gotshall; 26.26a: © Michael Graybill and Jan Hodder/Biological Photo Service; 26.27a: © E. R. Degginger/Photo Researchers, Inc.; 26.28a: © Kerrick James; 26.29a: © J. Weber/Visuals Unlimited; 26.30a: © IFA/Peter Arnold, Inc.; 26.31a: © Charlie Ott/ National Audubon Society/Photo Researchers, Inc.; 26.32b: © John Shaw/Tom Stack & Associates; 26.33a: © Tom McHugh/Photo Researchers, Inc.; 26.33b: © Dave Watts/Tom Stack & Associates; 26.33c: © E. R. Degginger/Animals Animals/Earth Scenes; 26.33d: © Gunter Ziesler/Peter Arnold, Inc.; p. 579 icon: National Snow & Ice Data Center/SPL/Photo Researchers, Inc.; p. 579 top: Photograph by David Swanlund, Courtesy Save-The-Redwoods League

Chapter 27

27.1: © Merlin D. Tuttle, Bat Conservation International; 27.2a: Photo by Helmut Diller, World Wide Fund For Nature, Switzerland; 27.2b: Photo by Helmut Diller, World Wide Fund For Nature, Switzerland; 27.8: © Gerald and Buff Corsi/Tom Stack & Associates; 27.9: © Kjell Sandved/Butterfly Alphabet; 27.11a: © Anne Wertheim/Animals Animals/Earth Scenes; 27.13: Courtesy Rolf O. Peterson; 27.14: © James L. Castner; 27.15a–b: © Lincoln P. Brower; 27.15c–d, 27.16, 27.17a–d: © Edward S. Ross; 27.18a–b: © Peter Frenzen; 27.19: © Edward S. Ross; 27.20: © A. Schmidecker/FPG International Corp.; 27.21: © John D. Cunningham/ Visuals Unlimited; 27.22: © Peter May/Peter Arnold, Inc.; 27.23a–b: © John Shaw/Tom Stack & Associates; 27.24a–b: Harold More; p. 603 icon: © Merlin D. Tuttle, Bat Conservation International; p. 603 top: © Kjell Sandved/Butterfly Alphabet; p. 603 top middle: © James L. Castner; p. 603 bottom middle: © Edward S. Ross; p. 603 bottom: © John Shaw/Tom Stack & Associates

Chapter 28

28.1: © John Mielcarek/Dembinsky Photo Associates; 28.2: © Greg Vaughn/Tom Stack & Associates; 28.3: © Norbert Wu 1992; 28.5: © Grant Heilman; 28.6: Courtesy Richard Klein; 28.7: NASA; 28.10: © Byron Augustine/Tom Stack & Associates; 28.11: SOVFOTO/ EASTFOTO; 28.12: © David M. Dennis/Tom Stack & Associates; 28.13a–c: Hugh Iltis; 28.14a: © James Blair, Staff Photographer/National Geographic Society; 28.14b: NASA; 28.14c: Courtesy Olov Hedberg; 28.14d: © Frans Lanting/Minden Pictures; 28.15: © Stephanie Maze/ Woodfin Camp, Inc.; 28.17: © Steve McCurry/National Geographic; p. 621 icon: © John Mielcarek/Dembinsky Photo Associates; p. 621 top: © Norbert Wu 1992; p. 621 middle: Hugh Iltis; p. 621 bottom: © David M. Dennis/ Tom Stack & Associates

Line Art and Text

2.3: Stuart Ira Fox, *Human Physiology*, 4th edition. Copyright © 1993 Times Mirror Higher Education Group, Inc., Dubuque, Iowa. All Rights Reserved. Reprinted by permission.

2.24: Copyright © by Irving Geis.

3.12, 3.23: Ruth Bernstein and Stephen Bernstein, *Biology*. Copyright © 1996 Times Mirror Higher Education Group, Inc., Dubuque, Iowa. All Rights Reserved. Reprinted by permission.

5.4: E. Peter Volpe, *Biology of Human Concerns*, 4th edition. Copyright © 1993 Times Mirror Higher Education Group, Inc., Dubuque, Iowa. All Rights Reserved. Reprinted by permission.

5.12, 5.19: Ruth Bernstein and Stephen Bernstein, *Biology*. Copyright © 1996 Times Mirror Higher Education Group, Inc., Dubuque, Iowa. All Rights Reserved. Reprinted by permission.

6.28: Kent M. Van De Graaff and Stuart Ira Fox, *Concepts of Human Anatomy and Physiology*, 4th ed. Copyright © 1995 Times Mirror Higher Education Group, Inc., Dubuque, Iowa. All Rights Reserved. Reprinted by permission.

7.8: Joan G. Creager, *Human Anatomy and Physiology*, 2d ed. Copyright © 1992 Joan G. Creager. Reprinted by permission of Times Mirror Higher Education Group, Inc., Dubuque, Iowa. All Rights Reserved.

8.17: Copyright © 1995 by *The New York Times Co.* Reprinted by permission.

9.25: Illustration copyright © by Irving Geis.

19.12: John W. Hole, Jr., *Human Anatomy and Physiology*, 5th ed. Copyright © 1990 Times Mirror Higher Education Group, Inc., Dubuque, Iowa. All Rights Reserved. Reprinted by permission.

21.9: Kent M. Van De Graaff, *Human Anatomy*, 3d ed. Copyright © 1992 Times Mirror Higher Education Group, Inc., Dubuque, Iowa. All Rights Reserved. Reprinted by permission.

22.3: John W. Hole, Jr., *Human Anatomy and Physiology*, 5th ed. Copyright © 1990 Times Mirror Higher Education Group, Inc., Dubuque, Iowa. All Rights Reserved. Reprinted by permission.

23.12: Stuart Ira Fox, *Human Physiology*, 4th ed. Copyright © 1993 Times Mirror Higher Education Group, Inc., Dubuque, Iowa. All Rights Reserved. Reprinted by permission.

23.15: Eldon D. Enger, et al., *Concepts in Biology*, 6th ed. Copyright © 1991 Times Mirror Higher Education Group, Inc., Dubuque, Iowa. All Rights Reserved. Reprinted by permission.

23.16: Stuart Ira Fox, *Human Physiology*, 4th ed. Copyright © 1993 Times Mirror Higher Education Group, Inc., Dubuque, Iowa. All Rights Reserved. Reprinted by permission.

23.20: John W. Hole, Jr., *Human Anatomy and Physiology*, 6th ed. Copyright © 1993 Times Mirror Higher Education Group, Inc., Dubuque, Iowa. All Rights Reserved. Reprinted by permission.

23.22: Stuart Ira Fox, *Human Physiology*, 4th ed. Copyright © 1993 Times Mirror Higher Education Group, Inc., Dubuque, Iowa. All Rights Reserved. Reprinted by permission.

Ta 23.5: Eldon D. Enger, et al., *Concepts in Biology*, 6th ed. Copyright © 1991 Times Mirror Higher Education Group, Inc., Dubuque, Iowa. All Rights Reserved. Reprinted by permission.

25.14: John W. Hole, Jr., *Human Anatomy and Physiology*, 6th ed. Copyright © 1993 Times Mirror Higher Education Group, Inc., Dubuque, Iowa. All Rights Reserved. Reprinted by permission.

28.9: *Environment* by Peter H. Raven, Linda R. Berg and George Johnson, copyright © 1993 by Saunders College Publishing, reprinted by permission of the publisher.

Illustrators

Art & Science, Inc.: 1.3, 1.7, 2.8B, 2.9, 2.10, 2.16, 3.4, 3.5, 3.19, 3.24, 4.7, 4.8, 4.9, 4.18, 5.9, 5.11, 5.16, 5.18, 5.24, 5.26, 5.27B–C, 5.29, 6.13, 6.14, 6.15, 7.3, 7.5, 7.10, 8.7, 8.8, 8.9, 9.25, 10.4, 10.9, 10.10, 11.3, 11.8, 12.3, 15.11, 15.23, 16.2, 16.4, 16.6, 16.8, 16.12, 16.16, 16.18, 16.20, 16.22, 16.31, 16.33, 17.10, 18.11, 18.15. 19.14, 20.12, 21.7, 21.10, 21.11, 21.19 left, 23.2, 23.8, 23.17, 23.24, 23.26, 24.3, 24.8, 25.4, 25.13, 26.6, 26.8, 26.9, 26.10, 26.22 and Table and Text Art on pages 15, 63, 107, 205, 211, 218–19, 227, 254, 268–69, 285, 317, 323–25, 351, 365, 367, 369, 372, 373, 385, 451, 473 and 579.

Scott Bodell: 25.7A.

Todd Buck: 25.14.

Margie Caldwell-Gill: 3.10, 8.6, 14.19, 15.34, 18.7, 19.11, and Text Art on page 63.

Barbara Cousins: 1.4, 7.20, 19.24, 20.18, 21.2, 21.20, 23.6, 23.10, 24.12 and Text Art on pages 151, 473, and 517.

Barbara Cousins/Art & Science: 5.10.

Barbara DeGraves: 9.31.

Diphrent Strokes: 2.19, 2.27, 5.13, 5.15, 8.17, 9.10, 9.11, 9.26, 10.6, 10.11, 14.17, 14.27, 15.19, 15.20, 15.22, 15.27, 15.30, 15.31, 15.37, 15.38, 17.2, 17.25, 17.30, 17.32, 18.5, 20.7, 20.21, 20.24, 20.25, 20.26, 21.16, 22.17, 23.29, 23.30, 24.6, 24.7, 25.8, 26.20, 26.25, 26.26, 26.27, 26.28, 26.29, 26.30, 26.31, 26.32, 27.4, 27.5, 27.6, 28.14E, 28.16 and Text Art on pages 211, 317, and 473.

FEH Illustrations: 19.3

Ethan Geehr: 3.12 top, 3.23, 5.19 and Text Art on page 107.

Rob Gordon: 23.15A.

Floyd Hosmer: 3.13, 3.14, 3.20, 3.21, 5.27A, 20.17 and Text Art on pages 63 and 451.

Illustrious, Inc.: 2.3.

Carlyn Iverson/Ethan Geehr/Art & Science: 5.12.

Medical Art Company: 5.23, 8.15, 9.12, 19.2, 24.4, 24.5, 25.5, 25.12 and Text Art on page 533.

William Ober: 2.25, 3.16, 15.2, 20.3, 20.20 and Text Art on page 451.

William Ober/Claire Garrison: 4.2, 7.4, 9.9, 13.9, 15.16, 15.35.

William Ober/Claire Garrison/Diphrent Strokes: 9.32.

Felicia Paras: 23.12, 23.16, 23.22.

Nadine Sokol: 2.4, 2.12, 2.22, 3.6, 4.12, 7.8, 7.17, 19.19, 21.4, 21.13, 22.8, 22.9, 22.11, 23.4, 23.5, 23.13, 23.21, 25.2, 28.8, 28.9 and Text Art on pages 41, 81, 429.

Kevin Somerville: 19.23, 21.6, 21.15, 21.19B, 23.31, 24.9.

Tom Waldrop: 23.30.

Index

A

E

H

I

Q

R

S

T